LINEAR CIRCUITS

To

My Father

DR. CHARLES V. SCOTT

"Adventure on, for, from the littlest clue
Has come whatever worth man ever knew;
The next to lighten all men may be you."

JOHN MASEFIELD

LINEAR CIRCUITS

by

RONALD E. SCOTT

Professor of Electrical Engineering
Northeastern University
Director of Research
Contronics, Inc.

with the editorial assistance of

MARTIN W. ESSIGMANN

Chairman of the Department of Electrical Engineering
Northeastern University

ADDISON-WESLEY PUBLISHING COMPANY

READING, MASSACHUSETTS · MENLO PARK, CALIFORNIA
LONDON · AMSTERDAM · DON MILLS, ONTARIO · SYDNEY

This book is in the

ADDISON-WESLEY SERIES IN THE ENGINEERING SCIENCES

Electrical Engineering

Printed in the United States of America

Library of Congress Catalog Card No. 60–11853

ISBN 0-201-06820-6
JKLMNOPQRS-AL-7987

PREFACE

Linear Circuits is intended to provide a first course in Electrical Engineering for electrical engineering students. The book is designed to achieve a maximum of depth in the circuit-theory area in the minimum possible time. To this end, circuits are presented as linear models and circuit theory as a branch of applied mathematics. The purpose of the book is to provide a foundation in linear circuits, early in the engineering program, upon which the whole undergraduate curriculum can be based. With this background, electronics leads to the circuit theory of active and nonlinear elements; machinery and transducers lead to the circuit theory of time-varying elements; and electromagnetic fields lead to the theory of distributed-parameter circuits. The signal analysis of circuit theory finds an extension in information theory and communications theory, and the stimulus-response relationships of circuit theory are carried to their logical conclusion in systems analysis.

The material covered in the book falls into four main sections:

1. Resistive circuit theory.
2. Transients in the time domain.
3. Steady-state a-c circuit theory.
4. Transients in the frequency domain.

The first two sections are in Part 1 (Time-Domain Analysis), and the second two sections are in Part 2 (Frequency-Domain Analysis). The first section deals with the fundamental laws of circuit theory, the elements of network topology, mesh currents and node voltages, and the methods used for solving one-terminal-pair and two-terminal-pair problems. By using a framework of resistive circuits we are able to present a simple but complete picture of the methods of circuit theory. The second section covers transients in simple circuits and concludes with signal analysis in the time domain and the elements of convolution theory.

The third section considers the response of networks to single-frequency sinusoids. Time-domain and frequency-domain circuits are clearly separated, and all the methods of the first section are applied to the frequency-domain circuits. Magnitude and phase response curves are obtained from both the s-plane and log-db plots. The fourth section treats transients by frequency-domain methods. Both Fourier methods and Laplace methods are presented, with the emphasis being placed on the response of circuits to complex waves of excitation.

There is sufficient material in the two parts for about a year and a half of circuit theory (about 150 class hours). At Northeastern University both parts are covered in four separate terms. The preliminary notes for the book have been used at other schools for both two-semester and three-semester courses. If the students have a strong background of mathematics and physics, some of the material in Part 1 may be omitted. A suggested syllabus for a one-semester course from Part 1 is: Chapters 2, 3, 4, 5, 10, 11, and 12.

The debt which I owe to Professor E. A. Guillemin of the Massachusetts Institute of Technology, one of the truly great teachers of our time, will be obvious to many of you. It is also a pleasure to acknowledge the enthusiastic support of Northeastern University for this writing venture, the help of Arthur Glazer, Kenneth Bourque, and Marcello Carrabes in checking the problem solutions, and the continuous encouragement from my long-suffering wife and family. I am deeply grateful to the staff of the Addison-Wesley Publishing Company for the professional help which they have given me.

R. E. S.

CONTENTS

PART 1: TIME-DOMAIN ANALYSIS

CHAPTER 1

INTRODUCTION

"The goal of science is order. Science is constructed from facts just as a house is constructed from stones, but an accumulation of facts is no more a science than a pile of stones is a house."

Henri Poincaré

In the year 600 B.C. Thales of Miletus, one of the Seven Sages of Greece, rubbed amber with a cloth and observed that it would then attract light objects. The phenomenon was called "electricity," from the Greek word amber. The first recorded use of electricity was by Anteus, a freed-man of the Emperor Tiberius of Rome, who stepped upon an electric eel and was immediately cured of his gout!

Since those early years the knowledge of electricity has multiplied endlessly and so have its uses. Virtually all the conveniences of modern life depend in some way upon electricity. The subject is so comprehensive that a student cannot discover all the facts for himself. In the interest of efficiency he must learn the subject as an ordered science.

Circuit theory is the basic discipline of electrical science. It concerns the transfer of energy from one device to another. It has no regard for the internal structure of the devices, nor for the location of device A with respect to device B. It considers only the amount of energy which passes from one device to the other. Since the principal function of electricity is the transfer and control of energy, it is not surprising that circuit theory plays a key role in electrical engineering.

1–1 Model theory. Every branch of science has its own collection of "models." When the physicist talks of an atom or an electron he does not pretend to be dealing with physical reality, but only with a model whose mathematical equations he understands and whose behavior he can predict with precision. The economist who analyzes the fluctuations of an economy also deals with models. His models are less precise and the conclusions which he draws from them are correspondingly less reliable. Indeed, the "exactness" of a science is a measure of the degree of correspondence between its models and physical reality.

In electrical engineering we are seeking to predict the behavior of real electrical devices interconnected in various ways. To avoid the unbelievable complexity and endless divergences of real physical devices we replace them by idealized mathematical models. There are two points of view regarding models. The first is that a model is a purely mathematical approximation to a real physical device. For example, a real wheel may be described mathematically as a circle of radius R. The second point of view is that the model is an idealized physical device. An "ideal" wheel would be perfectly round and might have zero mass, just like its mathematical counterpart, the circle. Both points of view are useful. The main reason for having models is so that we can describe devices mathematically. However, our physical intuition concerning ideal physical models will often guide and direct our mathematical manipulations. A block diagram of the relationship between physical reality and models is shown in Fig. 1–1. Electric circuits are the idealized models which are used to represent electrical devices.

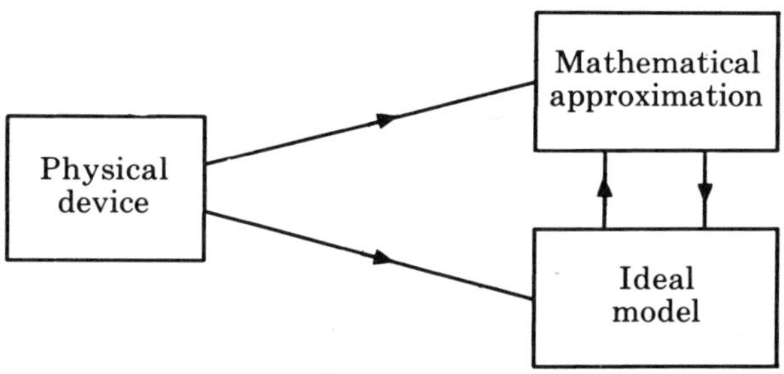

FIG. 1–1. Physical reality and models.

We would not, as engineers, be interested in circuits unless they did correspond more or less exactly to some physical reality. The nature of the correspondence is not easy to predict. The use of idealized circuits is not an exact science, but a subject which must be learned in the laboratory and by experience. However, the accumulated experience of generations of electrical engineers tells us that the correspondence is exceedingly good between such circuits and physical electrical devices. Initially the student must accept this result and learn circuit theory as he once learned the multiplication table. If the student is disappointed to learn that circuit theory is not "real" engineering, he will also be overjoyed to learn that neither does it become obsolete. As he becomes more proficient as an engineer he will use more complex models to represent a given device, but the rules for manipulating them will remain the same. As new physical devices are invented, they too will be reduced to models and treated by the same circuit theory.

1–2 Stimulus and response in physical systems. The ideas of stimulus and response are basic to our concept of a physical world. Intuitively we see examples of cause and effect in every aspect of the world around us.

Some examples are predictable. When we rub a finger on the wall the finger becomes warm, and when we shut our eyes we no longer see. Other examples are less predictable. The weather depends on many factors and cannot be foreseen with certainty. The behavior of human beings varies from time to time and even depends on who conducts the experiment!

In electrical engineering it is possible to isolate the stimuli and to measure the responses. Furthermore, the results can be duplicated from time to time and from person to person. A block diagram of a typical situation is shown in Fig. 1–2. The input stimulus represents energy supplied to the system by an external source, and the output response represents the utilization of the energy at some other point. The idealized system itself is called a *circuit*, or a *network*.

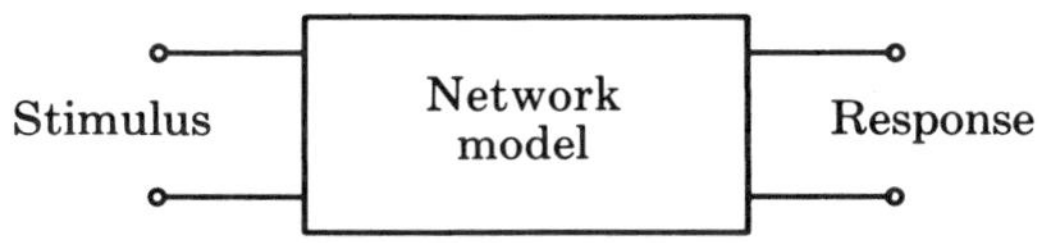

FIG. 1–2. The stimulus-response relationship.

In many networks the output response is directly proportional to the input stimulus. For example, if the input is doubled, the output also doubles. Networks which possess this property are called *linear*. The theory of linear networks, which will be discussed in this book, forms a basic core of network theory and indeed represents the majority of the electrical engineering problems which can be solved by simple analytical methods. The implications of linearity are far-reaching. Consider, for example, a network with an input stimulus and an output response. A second identical stimulus is added at the input. Since the total stimulus is now doubled, the output response also doubles. In effect, the output response is the sum of the two separate responses which would have been obtained if the two sources acted separately. This result can be extended to arbitrary sources at arbitrary points in a network.* It is called the *principle of superposition*. Since the response for a large number of sources acting at once can be obtained as the sum of the responses produced by each of the sources acting alone, it is never necessary to consider more than one source at a time. The single stimulus-response relationship shown in Fig. 1–2 thus represents the most general situation for a linear network. It is the central theme of this book.

The first third of the book discusses a particularly simple type of network in which the response is equal to the stimulus multiplied by a con-

---

* For a mathematical proof of linearity and superposition, see Chapter 3.

stant.* This constant, which summarizes the effect of the network in producing a response from the input stimulus, is called the *response function* of the network. In the second third of the book the response function is no longer a constant, but becomes a function of time. The third section of the book discusses the particularly important case of the stimulus-response relationship when the stimulus is sinusoidal in character.

1–3 Circuit variables: current and voltage. If we were interested in the flow of automobile traffic in a country, we would undoubtedly concentrate on the movements along the trunk highways and ignore the exact distribution within any single city. Similarly, in the realm of electric circuits we are often interested in the flow of electricity from one device to another, and not in the distribution of the energy within any one device. Under these circumstances we are justified in taking the basic building blocks of our model theory as "lumped" elements in which the electrical behavior is summarized by the effect produced at the *terminals* or trunk lines entering the device. The simplest devices have a single pair of terminals. Electricity enters at one terminal and leaves from the other. A typical lumped element and its associated terminal pair are shown in Fig. 1–3. The actual physical device is pictured in Fig. 1–3(a) as occupying some region of space. The network model shown in Fig. 1–3(b) has one pair of terminals with an idealized mathematical connection between them. It has no physical dimensions and no preferred orientation in space; it can be moved around and rotated at will.

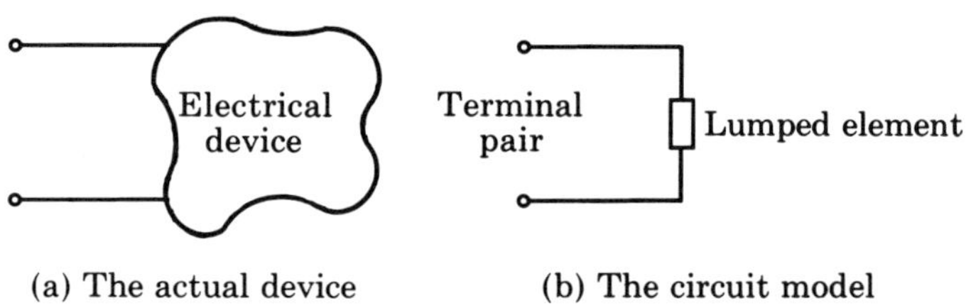

FIG. 1–3. A lumped element and its terminal pair.

Networks are composed of these ideal elements linked together by ideal connectors which absorb no energy at all. In many applications the network models predict the behavior of actual physical devices with acceptable precision. A limitation on the accuracy is imposed by the finite speed of propagation of electrical energy. If the energy is applied to one device,

* These networks are sometimes loosely called direct-current, or d-c, networks. The correct requirement is that there should be no energy-storage elements in the network. There is no restriction on the type of stimulus.

there will be a slight delay before it appears in the next one. If the source is turned off before the energy reaches the next device, a pulse of energy will be effectively trapped on the connecting wire. Since the connector is now storing energy it can no longer be considered ideal, and in fact it must now be treated as an energy-storage element in the circuit. A similar effect occurs within any given device. The energy propagates into the interior at a finite speed, and to account for its location at any given time a large number of internal energy-storage elements must be postulated, some of which do and some of which do not contain energy at any given instant. In many situations electrical energy propagates at essentially the speed of light, 3×10^{10} cm/sec. Usually* we can neglect the time taken for the energy to propagate around our circuits and, when we can, the simple circuit models are accurate enough.

In all of the preceding discussions we have considered only the energy in a circuit. Energy, however, is not a measurable quantity. It is, rather, a derived concept which relates such diverse phenomena as mechanical motion, chemical combination, atomic disintegration, hydraulic flow, and heat. Energy is "measured" by association with some measurable quantity in the system of interest. For example, in mechanical systems energy is measured in terms of mass and velocity, or force and distance.

To measure energy in an electrical system we define a new measurable quantity called *electric charge*. Historically, charges were first observed by rubbing dry substances together. For example, when cat's fur is stroked with rubber charges appear which cause the fur to stand on end. By contact with the cat's fur charge can be transferred to other light bodies, which then repel each other. In 1747 Benjamin Franklin called the charge on cat's fur positive, and the different kind of charge that appeared on the rubber negative. In modern times atomic physicists have pictured electric charge as being one of the building blocks of the universe. The ultimate unit of electric charge is the electron, with a mass of 9.107×10^{-31} kgm and a small negative charge.

Electrons move in orbits about the nucleus of each atom and many of them break loose and move at random in the interatomic spaces. An excess of electrons in a body represents a negative charge, and a deficit of electrons a positive charge. The practical unit of charge is *the coulomb*, which represents 6.24×10^{18} electrons.

In the year 1819, Hans Christian Oersted observed that a flow of electric charge produced a force on a magnetic compass needle, and that the force was proportional to the rate of flow of charge. The measurement of this magnetic effect is a great deal easier than the measurement of the forces

* Problems involving long transmission lines and propagation by radiowaves must be treated by the more general methods of "field theory."

between charges, and for this purpose a new variable, current, equal to the rate of flow of charge, is defined in terms of a unit called the *ampere*: 1 ampere = 1 coulomb/second. Current, rather than charge, is the basic variable of circuit theory.* It is represented by the symbol i or I. In summary:

$$i = \frac{dQ}{dt}, \tag{1–1}$$

where i = current in amperes, Q = charge in coulombs, t = time in seconds.

Two additional comments need to be made about the properties of currents. First, currents have direction. In the traffic analogy, charge was represented by the total number of automobiles in a city. In the same analogy the current would be the number of cars entering or leaving the city per second. Clearly, the number of cars per second has direction. By the same token, the number of coulombs per second, or the current in a circuit, also has direction. *Because current has direction it is customary to consider it as "flowing" in a given direction around a circuit.*†

The second comment concerns the nature of current. In most good conductors of electricity the current consists of a motion of free electrons. In other materials, however, it may be a motion of positive charge, or of positive charge in one direction and negative charge in the other direction. By convention, a positive current is taken to be a flow of positive charge in the direction of a reference arrow used to mark the current direction. The magnetic effect produced by a flow of positive charge is equal to that produced by a flow of negative charge in the opposite direction, that is, the amount of current is the same in either case. Since we can make no distinction between the two cases, it avoids confusion if *current is always considered to be an equivalent flow of positive charge.*

In order to keep track of the direction of a current, it is convenient to assign algebraic reference directions. In the traffic analogy we might assign a plus sign to all traffic entering the city and a minus sign to all traffic leaving the city. A net plus count for the traffic would then mean a movement of cars into the city, while a negative count would mean a movement out of the city. In electric circuits the reference direction for the current is marked by an arrow on a figure. A current of positive charge moving in the direction of the arrow is considered to be positive and if moving in the opposite direction to be negative. Note particularly that *the reference direction may be chosen arbitrarily, and the actual currents can be either positive or negative.*

---

* The total charge passing is not current × time, but the time integral of the current.

† Not only charge, but all the derivatives of charge have reference directions and "flow."

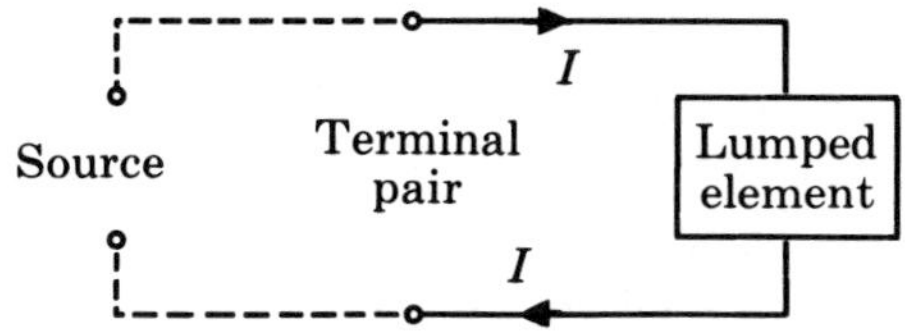

FIG. 1–4. An exact definition of "current."

The given specifications for current are only approximate, for the same reason that the circuits themselves are only approximations to physical devices. Owing to the finite speed of propagation of the electric energy around a circuit the current is not the same at any two points in the circuit. If we wish to define current precisely we must say where we are going to measure it. For our idealized circuit models, however, there can be no confusion, since we can measure the current only at the terminals of the lumped elements. In our model theory we will ignore the finite speed of propagation of the electrical effects and assume that exactly the same amount of current enters one terminal and leaves the other. Figure 1–4 thus gives a precise definition of current for our circuit models. In terms of the traffic analogy, an automobile leaves at the second terminal whenever one enters the first terminal. Actually, the movement is more like that of a merry-go-round. The source does not create charge.* It merely separates the charges which are already present and moves them around the circuit. It should also be clear that the charges themselves do not have to move with the velocity of light. When the "merry-go-round" starts up, the effect of the starting propagates around the circuit at the velocity of light, but the actual motion of the charges can be very slow indeed. We can summarize this discussion by saying that *in our models currents are measured and defined only at the terminals of the lumped elements.*

In our quest for a method of measuring energy in electric circuits we have introduced the concept of electric charge and have defined an electric current as a flow of charge. We still cannot keep track of the energy. An additional variable is required which will measure the amount of energy gained or lost by each element of charge as it passes through a device.† This variable is called *voltage*. It is represented by the symbols e, E or v, V. The unit of voltage is the *volt*: 1 volt = 1 joule/coulomb. In summary:

$$v = \frac{dW}{dQ}, \qquad (1\text{–}2)$$

where v = voltage in volts, W = energy in joules, Q = charge in coulombs.

* The case of atomic disintegration is specifically excluded.

† The total energy is not voltage × charge, but the integral of voltage with respect to charge.

If it is to be useful, voltage must be a readily measurable quantity. A direct mechanical measurement can be made in terms of the force between two charged bodies, or in terms of the deflection of a moving charge as it passes between two charged bodies. More often, however, voltage is measured in terms of the related current it produces in a known circuit.

From the definitions of current and voltage given in Eqs. (1–1) and (1–2), it is apparent that the product of current and voltage is the power associated with an electrical device. Power is measured in joules per second, or watts, and has the symbol p. Thus

$$p = e \cdot i, \tag{1–3}$$

where p = power in watts, e = voltage in volts, i = current in amperes. The dimensions associated with Eq. (1–3) are

$$\text{watts} = \text{volts} \times \text{amperes} = \frac{\text{joules}}{\text{coulomb}} \times \frac{\text{coulombs}}{\text{second}} = \frac{\text{joules}}{\text{second}} \tag{1–4}$$

Voltage does not possess direction in the same sense as does the current. It does, however, have polarity, since energy can either be supplied to the electrical device or received from it. A plus sign for the voltage at the terminal where the current enters indicates that energy is absorbed in the device and, conversely, a negative sign at that terminal indicates that the electrical device is a source of energy. In terms of our traffic analogy the positive voltage represents a hill and the negative voltage a valley. If cars enter the city from a hill and leave by way of a valley they supply a net amount of energy to the city. On the other hand, if they enter through a valley and leave from a hill the city must supply energy to them! *It is convenient to regard voltage as potential energy and voltage levels as elevations or heights.**

A pair of signs, plus and minus, are used at a terminal pair to give a reference direction for the voltage. The positive sign represents an energy level which is higher than that at the negative sign. In terms of potential energy, the positive sign represents an elevation above the negative sign. However, it should be carefully noted *that the reference marks for voltage,* as for current, *are for algebraic reference only.* If an actual voltage has the same polarity as the reference signs it is a positive voltage, and if it has the opposite polarity it is a negative voltage.

Consistent signs must be assigned to current and to voltage if the algebraic sign of the power is to have meaning. Usually current directions

---

* Voltage *is* potential if terminal-pair definitions are used. In circuit theory there is no need to distinguish between the potential (the static, conservative field) and electromotive force (the nonconservative field component).

are assigned at random and the voltage is made to agree. *A plus sign for the voltage is assigned at the terminal where the current enters the device.* If this convention is followed a positive sign for the power will represent energy lost in the device. The voltages are then *voltage drops* produced by the flow of current. As the coulombs "roll downhill" they lose energy to the device.

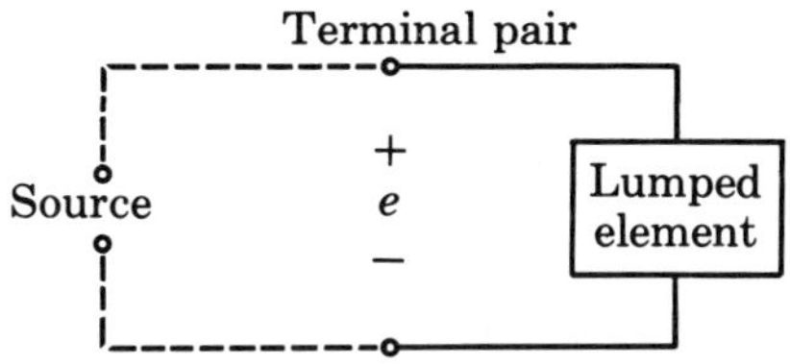

FIG. 1–5. An exact definition of "voltage."

A precise definition of voltage is possible only for idealized circuit models. It is associated with the terminal pair of an ideal lumped element, as shown in Fig. 1–5, where all the electric and magnetic effects are contained within the device and the terminals project into a force-free region. A voltmeter could thus be placed across the terminals and it would read the correct voltage no matter where the leads were placed. In a real physical device it is impossible to confine all the electric and magnetic forces within the lumped element, and the stray forces affect the reading. Furthermore, this effect varies with the location of the voltmeter leads. The placement of voltmeter leads is one of the arts which must be learned in the laboratory.*

We shall not discuss the use of actual meters in this text, but it will be convenient to define ideal meters for the measurement of current voltage and power. An ideal *ammeter* placed in the path of the current gives a reading proportional to the current passing through it. It must absorb no energy from the circuit and thus has no voltage across it. An ideal *voltmeter* is connected across a terminal pair in parallel with a lumped element. The reading is proportional to the difference in the energy levels of the two terminals. Since no energy can be absorbed by the meter, no current must flow through it. An ideal *wattmeter* measures the power entering a terminal pair. It is a combination of a voltmeter and an ammeter, and gives a reading that is proportional to the product of the voltage and the current. These three types of meters, with their symbols and connections, are shown in Fig. 1–6. All three have reference directions marked on them, so that a positive reading will have algebraic significance. Although

* The electromagnetic field is nonconservative, and if voltage is defined as a line integral the path must be specified.

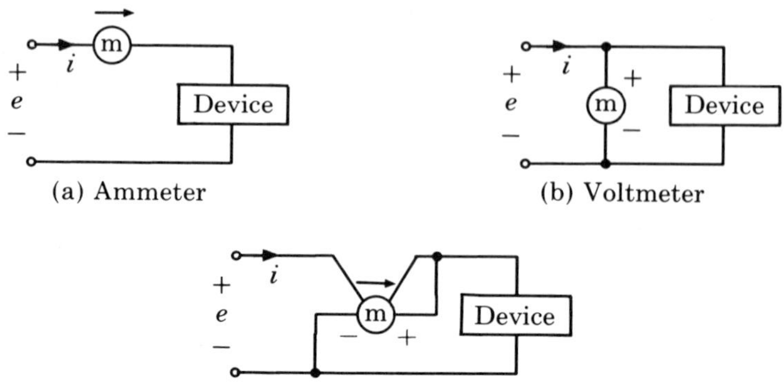

(a) Ammeter (b) Voltmeter (c) Wattmeter

FIG. 1–6. Ideal meters.

practical meters can be constructed to approach the characteristics of ideal meters, no real meter can operate without absorbing some energy from the circuit.

1–4 Circuit parameters: resistance, inductance, and capacitance. Sources. In the preceding sections we have seen that the energy associated with an electrical device is specified in terms of the current through it and the voltage across it. The relationship between current and voltage is determined by the nature of the device. If we force a given current through a device the energy which we must supply will be determined by the device. Alternatively, if we supply charge with a given amount of energy, the number of coulombs which will pass through will be determined by the device.

There are three basic types of linear circuit elements. A *resistor* is an element which dissipates energy (as heat or in some other way). An *inductor* stores energy by virtue of a current through it. A *capacitor* stores energy by virtue of a voltage existing across it. Sources that supply a given number of coulombs per second, regardless of the energy required from them, are called *current sources.* Sources that supply coulombs with a given energy, regardless of the number of coulombs required, are called *voltage sources.* We shall discuss the parameters and the sources in turn.

Resistance. Electrical elements which dissipate energy are called *resistors.* The idealized circuit models are called *resistances.* Typical examples of resistors are the heating elements in stoves, the filaments in electric light bulbs, and the copper wires used to bring electric power into our homes.

The theory of electrical resistors was given by Georg Simon Ohm in the year 1826. He clarified the concepts of current and voltage and showed that for a resistor *the current is proportional to the voltage.* His contemporaries thought so little of his work that he was not made a professor until twenty-two years later. The verdict of history is kinder. The volt-ampere relationship of the resistance element bears the name of Ohm's law, and the unit of resistance is the ohm. For the ideal resistance element

$$e = Ri, \tag{1–5}$$

where e = voltage across the element in volts,
i = current through the element in amperes,
R = resistance in ohms.

An alternative form of Eq. (1–5) is

$$i = e/R = Ge, \tag{1–6}$$

where $G = 1/R$ = conductance in mhos.

The sign convention associated with Eqs. (1–5) and (1–6) requires that in order for the proportionality constant between the current and the voltage to be a positive number, the voltage *must* be taken to be positive at the terminal where the current enters. The symbol for resistance and the sign convention associated with it are shown in Fig. 1–7. If this sign convention is adopted the power dissipated in the resistance will also be positive, and we have

$$p = e \cdot i = i^2R = e^2/R. \tag{1–7}$$

The first form of Eq. (1–7) is convenient when the current and the voltage are both known, the second form when only the current is known, and the third form when only the voltage is known.

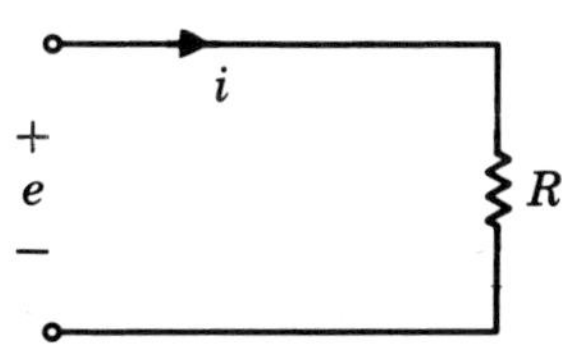

FIG. 1–7. The resistance parameter and its sign convention.

Inductance. The electrical element which stores energy in association with a flow of current is called an *inductor.* The idealized circuit model for the inductor is called an *inductance.* Inductance was first discovered by Michael Faraday in 1831. (Faraday began as a janitor in the Royal Institution of London, but he later rose to become its director.) He observed a magnetic effect in the air surrounding a wire carrying a current, and noted that a compass needle always pointed in the direction of a circle around the wire. Faraday explained the effect in terms of magnetic "lines of force" in the "aether" surrounding the wire. He postulated that the lines of force were stretched into place by the current in the wire, somewhat as though they were elastic bands, and that the energy thus stored in the aether returned to the wire in the form of a generated voltage when the current decayed to zero.*

Practical inductors are usually made of many turns of fine wire wound in a coil or a helix to enhance the magnetic effect without greatly increasing the size of the device. Often the wire is wound on magnetic material, which is capable of storing more energy per ampere than is air. Because of their shapes, inductors are called *coils,* and because of the energy required to establish a current in them, *chokes.*

For the ideal circuit model of an inductor, the voltage is proportional to the rate of change of the current rather than, as in the resistance, to the current iself. The proportionality constant is called the *inductance.* It has the symbol L, and the unit is called a *henry,* after Joseph Henry, an American investigator who obtained many of Faraday's results independently. The mathematical volt-ampere relationship for an inductor is

$$e = L\frac{di}{dt}, \tag{1–8}$$

where

L = inductance in henrys,

e = voltage across the element in volts,

di/dt = rate of change of the current in amperes per second.

An alternative form of Eq. (1–8) is obtained by integrating both sides with respect to the time. To include all possible magnetic effects we must integrate from minus infinity to the present time. Thus

$$i = \frac{1}{L}\int_{-\infty}^{t} e\,dt. \tag{1–9}$$

Both Eq. (1–8) and Eq. (1–9) are linear, since doubling the current in Eq. (1–8) doubles the voltage, and doubling the voltage in Eq. (1–9) doubles the current. The power entering the inductor at any instant is given by

$$p = e \cdot i = Li\frac{di}{dt}\cdot \tag{1–10}$$

When the current is constant, the derivative is zero and no additional energy is stored in the inductor. The "elastic bands" maintain their form

* The lines of force were called *flux linkages* by Maxwell, who justified Faraday's concepts by more rigorous mathematics.

and the energy associated with them is fixed. When the current increases, the derivative of the current has a positive value and the power, as given by Eq. (1–10), is positive. Additional energy is stored in the inductor. The total energy in the inductor at any given time is the integral of the power from minus infinity to that time. It is expressed by

$$W_L = \int_{-\infty}^{t} ei\,dt = \int_{-\infty}^{t} Li\,\frac{di}{dt}\,dt = \int_{0}^{i} Li\,di = \tfrac{1}{2}\,Li^2, \qquad (1\text{–}11)$$

where

W_L = energy stored in the inductor at time t, in joules,

i = current in the inductor at time t in amperes.

Equation (1–11) shows that the *total energy in the inductor depends only on the instantaneous value of the current* and not on the manner in which it was established. It shows further that the energy is a function of the current alone, and could exist even if the voltage were zero. In order for the energy stored in the inductor, as given by Eq. (1–11), to be positive, the current and the voltage must have consistent signs, as shown in Fig. 1–8. The sign convention is easy to remember, since it is the same as the convention for a resistor. The symbol for an inductor is a series of loops reminiscent of a coiled wire.

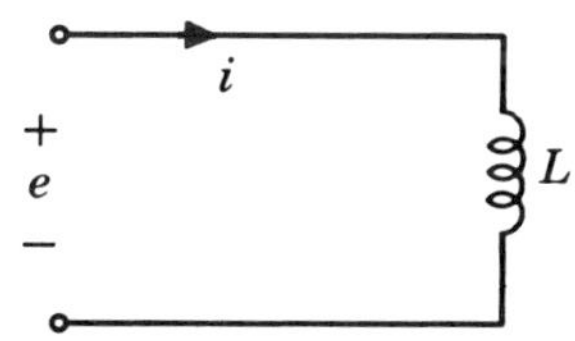

FIG. 1–8. The inductance parameter and its sign convention.

Capacitance. An inductor is a device that stores energy by virtue of a current through it. A *capacitor* also stores energy, but here the energy storage is due to the voltage across the device, and is independent of the current through it. The idealized circuit model of a capacitor is called *capacitance.*

Historically, capacitance is the oldest of the three circuit parameters. In 1745 van Mussenbroek of Leyden performed an experiment for the purpose of preserving static electricity from the decay which it experienced in air. He placed an insulator between two metal sheets, charged it by rubbing, and handed it to a friend, Cunaeus, who received a violent shock! For the first time a method of accumulating electric charge and intensifying its electric effect was available.

A mathematical discussion of the energy storage on a capacitor was given by Simeon Poisson to the French Academy in 1812. He likened the forces between charges to the gravitational force between a mass and the earth, and postulated that the energy of the stored charges was like the potential energy of a mass at rest above the earth. As the conducting plates of a capacitor are made larger, they contain more charge and hence more energy at any given voltage. The closer the plates the greater the force between the charges, and hence the greater the energy stored per unit charge. Thus capacitance is a parameter which is proportional to the size of the plates and inversely proportional to the spacing between them.

The modern form of the "Leyden jar" (capacitor) is similar to the original. Sometimes the material between the conducting plates is air, and sometimes it is an insulator (nonconductor of electricity) such as glass or mica which will store more energy in a smaller volume. In "electrolytic" capacitors the oxidized surface of a metal such as aluminum or tantalum is used for the insulator.

In the ideal circuit model of a capacitor, the voltage is proportional to the charge, or to the integral of the current. The constant of proportionality is the reciprocal of the capacitance, in farads. The total charge on the capacitor is the sum of all the charges which have entered the capacitor since the beginning of time, and therefore the integral of the current must be taken from minus infinity to the present time. The voltage across the capacitance is given by

$$e = \frac{1}{C} Q = \frac{1}{C} \int_{-\infty}^{t} i \, dt, \tag{1-12}$$

where

e = voltage across the capacitance in volts,

Q = charge on the capacitance in coulombs $= \int_{-\infty}^{t} i \, dt$,

i = current in amperes (coulombs per second),

C = capacitance in farads.

The current i is said to "flow" through the capacitor, since a positive charge entering one terminal repels a positive charge at the other terminal.* The derivative of Eq. (1–12) gives an alternative form for the volt-ampere relationship of the capacitance:

$$i = C \frac{de}{dt} \cdot \tag{1-13}$$

---

* Maxwell postulated that a real physical current, which he called the *displacement current,* existed in the region between the plates. The existence of a real displacement current makes possible the propagation of radio waves.

The power entering the capacitance at any time is given by

$$p = e \cdot i = Ce\frac{de}{dt}. \tag{1-14}$$

When the voltage is constant the derivative is zero and the power expressed by Eq. (1-14) is zero. Only by increasing the voltage can additional energy be fed into the capacitance. The total energy in the capacitance is

$$W_C = \int_{-\infty}^{t} e \cdot i\, dt = \int_{-\infty}^{t} Ce\frac{de}{dt}\, dt = \int_{0}^{e} Ce\, de = \tfrac{1}{2}Ce^2, \tag{1-15}$$

where

W_C = energy stored in the capacitance at any time t, in joules,

e = voltage across the capacitance at any time t in volts.

From Eq. (1-15), the energy in the capacitance is seen to be a function of the voltage across it, and not a function of the current at that instant. A sign convention is necessary for the current and voltage in order to make the energy stored in the capacitance, as given by Eq. (1-15), positive. The sign convention and the symbol for capacitance are shown in Fig. 1-9.

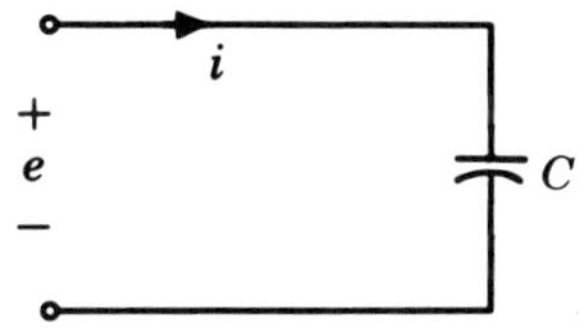

FIG. 1-9. The capacitance parameter and its sign convention.

In electricity, as in mechanics, there are two types of energy storage. The energy stored in an inductor is associated with the motion of electric charge and in many respects is similar to kinetic energy in mechanics. The energy stored in a capacitor is associated with a force between stationary charges and is like potential energy in mechanics. Many of the interesting properties of electrical devices stem from the existence of these two types of energy storage. Wave motion, for example, is possible only when energy can alternate between one form and the other. Resonance phenomena, used in tuning radio and television sets, also depend on the existence of the two types of energy and an alternation between them.

Sources. In addition to the passive circuit elements which store and dissipate energy, we must define ideal or model source elements which can supply energy to our circuits. Idealized sources fall into two categories,

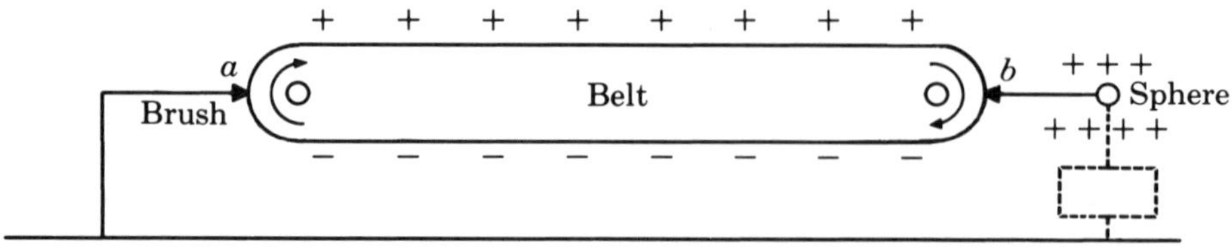

FIG. 1–10. Electrostatic (van de Graaff) generator.

current sources, which supply a given number of coulombs per second regardless of the energy which is required of them, and *voltage sources*, which supply an arbitrary number of coulombs, each possessing a fixed amount of energy.

Historically, *current sources* were discovered first. The earliest electrical effects were obtained by rubbing two materials together. The amount of charge transferred from one material to the other depended only on the materials and the frictional force. A modern version of the frictional generator, called a van de Graaff generator, is shown in Fig. 1–10. A moving belt transports the static charge produced by friction at point a to point b, where it is transferred to the inside of a sphere. The amount of charge deposited on the sphere per second depends on the speed of the belt and the amount of charge produced at a. It is almost independent of the external circuit connected to the sphere. Such devices are used to produce currents at high voltages for nuclear research. The ideal circuit model of a current source produces a fixed current no matter what voltage is across it. The symbol for a current source and its volt-ampere relationship are given in Fig. 1–11. It is necessary to suppose that the current source is connected to some external circuit to complete the path for the current. If the external circuit presents infinite resistance the voltage across the source rises to infinity and the power supplied must also become infinite. An ideal current source must be capable of supplying infinite power, and for this reason alone it is not physically realizable. Many physical sources do, however, behave like current sources when the power drawn from them is small. Later we shall see that ideal sources in conjunction with ideal resistors can represent a wide range of physical sources.

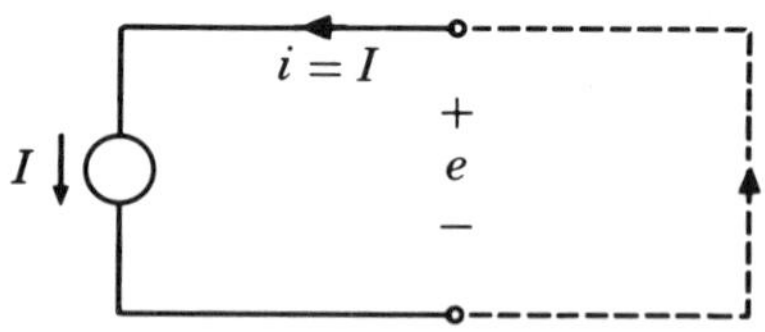

FIG. 1–11. A current source.

The earliest *voltage sources* were discovered by Luigi Galvani in 1791. In the course of dissecting frogs he noticed a peculiar twitching and an electric spark when he touched a nerve with a scalpel. He rightly attributed the effect to electricity, but it remained for Alessandro Volta to explain the origin of the electricity. Volta showed that the electricity was produced whenever two dissimilar metals were immersed in a conducting fluid. (Galvani's frog merely acted as the "fluid" connecting the scalpel and the metal table.) By 1800 Volta had constructed cells of copper and zinc and observed the first continuous electric currents ever produced by man. The electric variable *voltage* was named for him, in honor of this achievement. Modern batteries and storage cells are similar in principle to Volta's original devices. The energy is released by chemical action in the cells, and the amount obtained per coulomb is fixed by the nature of the chemical reaction. As more coulombs are drawn from the device the total activity increases but the energy acquired by each coulomb remains the same.

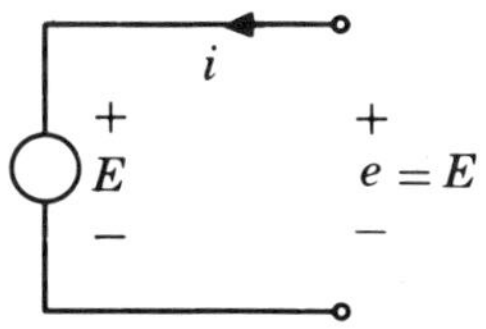

FIG. 1–12. A voltage source.

In the ideal circuit model of a voltage source the voltage across the terminals is constant and independent of the current drawn from it. The symbol for a voltage source and the volt-ampere relationship are shown in Fig. 1–12. When an open circuit is placed across a voltage source the current is zero and no power is drawn from it. The strain on a voltage source occurs when zero resistance is placed across it. The current then rises to infinity and the power delivered by the source must likewise be infinite. Many physical sources behave like voltage sources provided that only small currents are drawn from them. Later we shall see that the combination of ideal voltage sources and ideal resistors can give a better representation of real physical sources.

1–5 The circuit laws $\sum e = 0$ and $\sum i = 0$. All of the basic circuit elements discussed in the previous section have two external terminals. Electric circuits are constructed by connecting these external terminals in various ways. The connections are assumed to have zero resistance and to pass currents without absorbing any power; they are called *short circuits.* Conversely, if there is no connection between two terminals an *open*

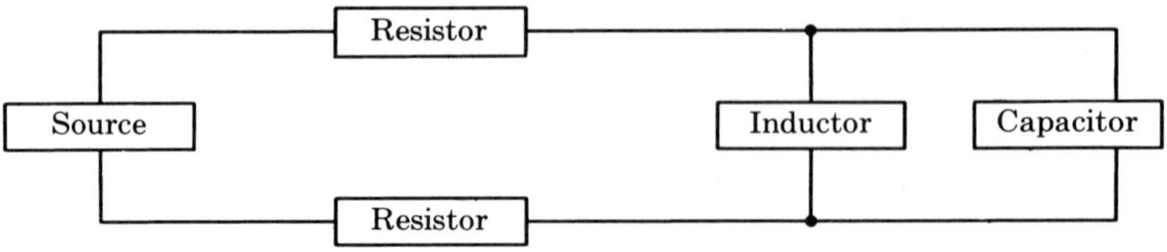

FIG. 1–13. An example of a "circuit."

circuit is said to exist between them. An example of a circuit is shown in Fig. 1–13. Circuits are related to the mathematical science of junctions and elements, called *topology*. The exact location of the elements is of no consequence. The shapes and lengths of the short circuits joining the elements have no significance. In our mathematical model the elements can be turned around and placed in any desired position. They can even be inverted without the wires falling off! This inherent abstraction, which lends flexibility and generality to electric circuits, is by no means trivial. Mechanical engineers, when faced with similar problems, tend to draw actual physical devices. But their dashpots cannot be inverted without spilling the fluid! Because of its simplicity and complete generality, circuit theory has been highly exploited in electrical engineering.

One of the facilities which an electrical engineer must acquire is the ability to recognize the essential nature of a circuit no matter how it is drawn. He must be able to redraw circuits in their simplest form. An example of four equivalent circuits is given in Fig. 1–14.

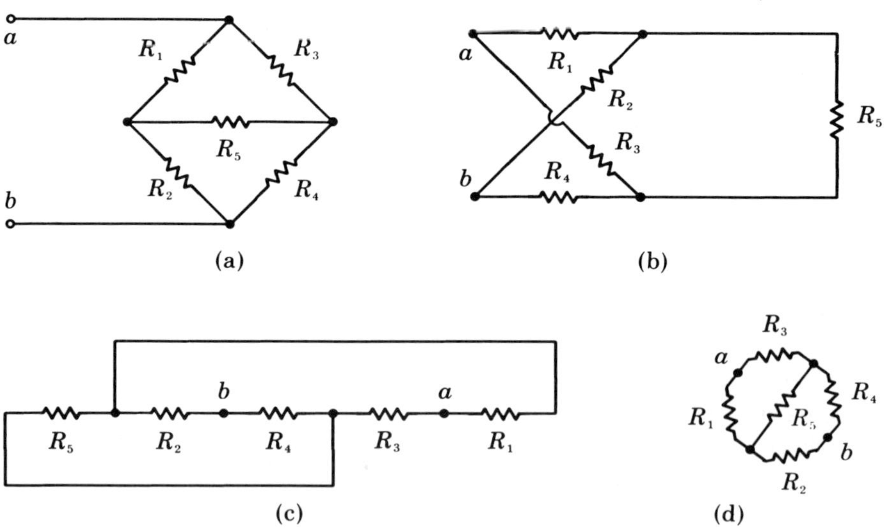

FIG. 1–14. Four equivalent forms of the same circuit.

The basic laws which govern the interconnection of electrical elements were first stated by Gustav Kirchhoff in a footnote to a paper written in 1848. Kirchhoff's laws are a consequence of the laws of conservation of energy and conservation of charge. Since conservation of charge postulates that charge is neither created nor destroyed,* the charge which enters a junction point in a network must either leave instantaneously or be stored there. It cannot be stored, however, because the junction is an infinitesimal mathematical point, and the charge possesses a finite mass and size. Hence the charge which arrives at a junction point at any instant must leave immediately. In terms of currents, the total current entering the junction must be equal to the total current leaving the junction. If currents leaving the junction are taken to be positive and currents entering are taken to be negative, the algebraic sum of the currents at the junction will be zero. Thus

$$\sum i = 0. \tag{1-16}$$

Equation (1-16) is Kirchhoff's current law. An example of its application to a junction point is given in Fig. 1-15. For this junction Kirchhoff's current law is

$$i_3 + i_4 - i_1 - i_2 = 0 \tag{1-17}$$

or, written in terms of currents entering and leaving, it is

$$i_1 + i_2 = i_3 + i_4.$$

The reader should be able to write the equation in either form with no hesitation. Note that the current directions shown in the figure and used in writing the equations are the *assumed* directions. If the actual current is in the reference direction it is positive, and if it is in the opposite direction, it is negative. In the statement of Kirchhoff's law, we must differentiate between the algebraic sign used for current references and the algebraic sign used to differentiate currents entering and leaving a node.

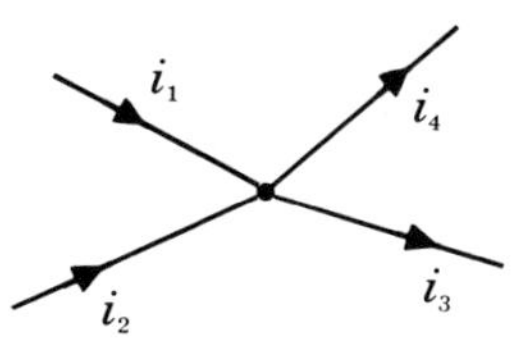

FIG. 1-15. An example of Kirchhoff's current law.

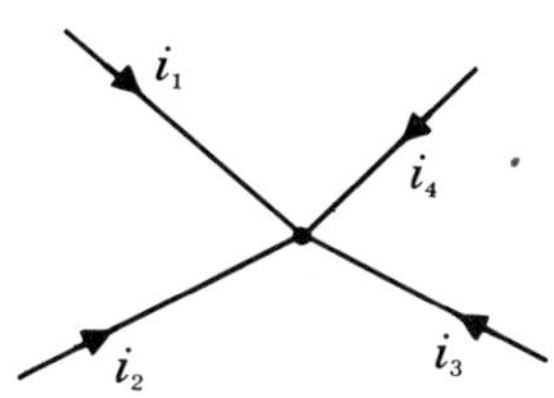

FIG. 1-16. Currents entering a junction.

* Again, atomic disintegrations are excluded from circuit theory.

For example, all the reference arrows might point toward the junction, as in Fig. 1–16. The current law equation for this node is

$$i_1 + i_2 + i_3 + i_4 = 0. \tag{1–18}$$

Equation (1–18) may seem to be impossible, if it is not recalled that the directions shown are reference directions only and that the values of some of the currents can be negative, indicating that their direction is away from the node.

The second Kirchhoff's law springs from the law of conservation of energy. In a gravitational field the energy level of a point A with respect to a point B can be found by moving a unit mass from A to B and noting the energy given up in the process. Similarly, in electric circuits the energy level of point A with respect to point B can be obtained by moving a charge from A to B and noting the energy given up in the process. The energy one coulomb gives up as it moves from point A to point B is the same along any path from A to B. (If it were not, we could make a perpetual motion machine by letting the coulombs slide down the path with the large energy difference and then moving them back along the path with the small energy difference.) Since the energy lost by a coulomb in moving from point A to point B is independent of the path, the energy for any round trip must be zero. This statement is Kirchhoff's voltage law:

$$\sum e = 0. \tag{1–19}$$

As an example of the Kirchhoff voltage law, consider the circuit of Fig. 1–17.

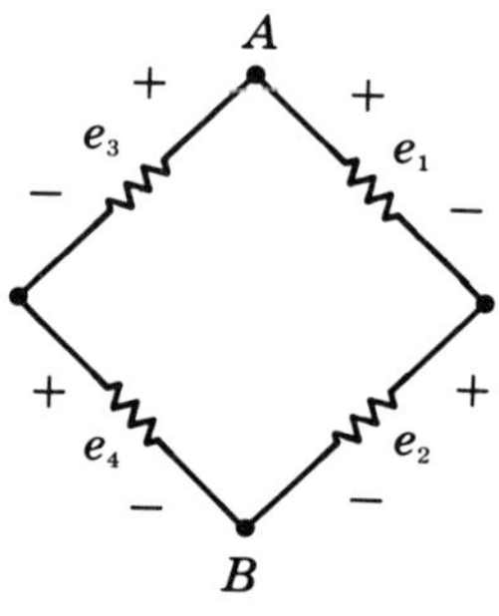

FIG. 1–17. An example of Kirchhoff's voltage law.

Adding up the voltages around the loop, we obtain

$$e_1 + e_2 - e_4 - e_3 = 0. \tag{1–20}$$

As an alternative we might equate the total voltage drop from A to B by the two paths, to obtain

$$e_1 + e_2 = e_3 + e_4.$$

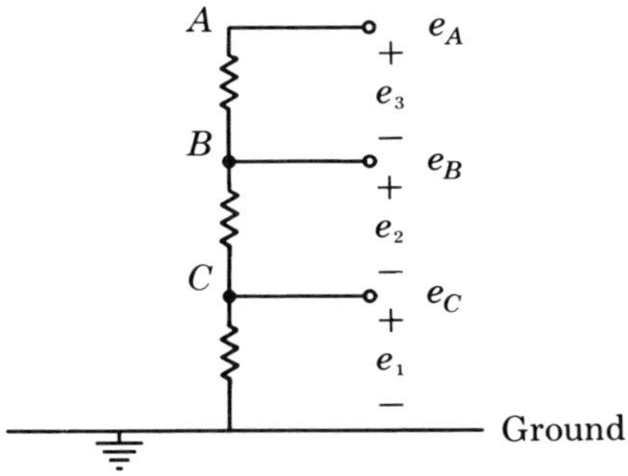

FIG. 1–18. Voltages with respect to ground.

It is essential to be able to write both forms of the equation by inspection of the circuit. As in the case of currents, Kirchhoff's voltage law is written in terms of assumed or reference polarities, and we must distinguish carefully between the algebraic sign used with the voltage reference marks and the algebraic sign used to distinguish voltage rises and voltage drops.

The idea that voltage is analogous to gravitational elevation suggests the possibility of a reference point analogous to sea level. In electrical networks, this potential is called *ground.* We have previously defined voltages across separate elements only, but by means of Kirchhoff's voltage law we can compute the voltage at any point in a network with respect to the ground or reference voltage. The land surveyor has the same problem. He measures the incremental elevations from one point to the next, and to obtain the elevation above sea level he must add a series of incremental results. Figure 1–18 illustrates the procedure and shows the symbol for the ground point. In terms of the voltages across the elements, the voltages of the junctions A, B, and C with respect to ground are

$$e_C = e_1, \qquad e_B = e_1 + e_2, \qquad e_A = e_1 + e_2 + e_3.$$

By analogy with the concept of the ground point, we can define more general currents by means of the equation for Kirchhoff's current law. A particular example of the process is afforded by the circulating currents i_a and i_b in Fig. 1–19. The loop current i_a takes the closed path *abeda*,

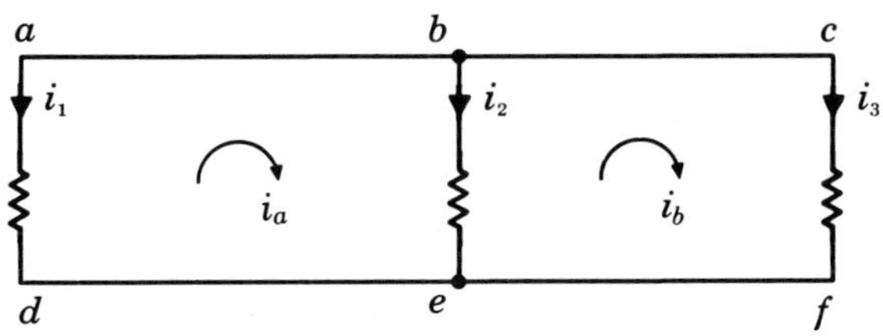

FIG. 1–19. Currents flowing around loops.

and the loop current i_b takes the closed path *bcfeb*. Kirchhoff's current law enables us to define these circulating currents in terms of the currents in the elements, according to the following equations:

$$i_1 = -i_a, \qquad i_2 = i_a - i_b, \qquad i_3 = i_b.$$

One final comment about Kirchhoff's laws concerns the notation for the voltages and currents in the separate elements. If the junctions between the elements are assigned symbols, both the currents and the voltages can be expressed in terms of these symbols by means of a double-subscript notation. Thus the current i_1 in Fig. 1–19 could be called i_{ad}, the current i_2 could be called i_{be}, and the current i_3 called i_{cf}. A similar notation could be used for the voltage drops produced by these currents. Thus v_{ad} or e_{ad} is the voltage drop produced by the current from a to d, and v_{be} or e_{be} is the voltage drop produced by the current i_{be}, and v_{cf} or e_{cf} is the voltage drop produced by the current i_{cf}.*

This chapter has served to introduce the subject of circuit theory. Circuits have been presented as models. The elements of resistance, inductance, and capacitance have been defined accurately and completely by their volt-ampere relationships. These definitions and Kirchhoff's two laws provide the complete foundation of circuit theory. The whole subject will unfold as a natural consequence of these definitions and these laws. No additional *ad hoc* results will be brought in at a later date. The basic rules have been given. We must now learn to play the game!

1–6 Summary.

Model theory

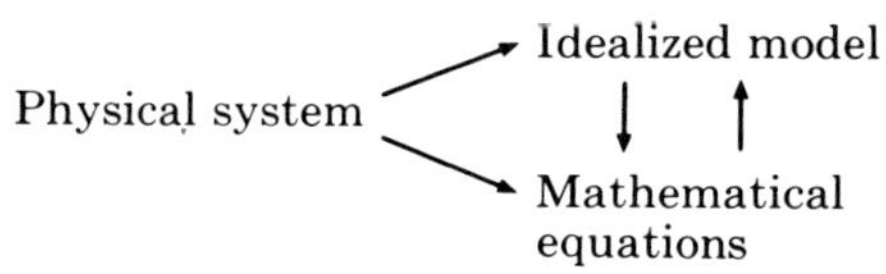

Stimulus-response relations

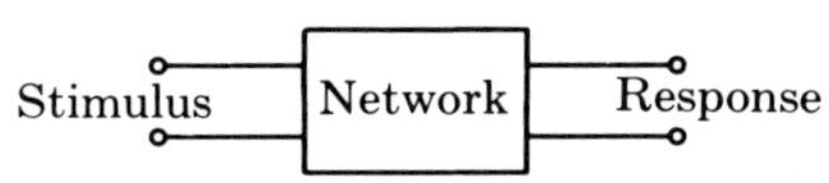

Linearity: response is proportional to stimulus

Response function: relates response and stimulus

* Some authors use v for voltage drops and e for voltage rises. This is nonsense! It is on a par with using different symbols for the assumed currents entering and leaving a node!

Circuit variables

Voltage: e or v, joules per coulomb
Current: i, coulombs per second
Power: $p = e \cdot i$, joules per second, or watts.

Circuit parameters

Resistance 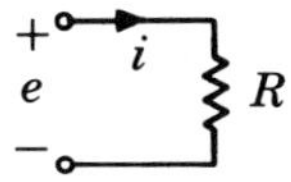

$e = iR,\qquad i = Ge$

Inductance 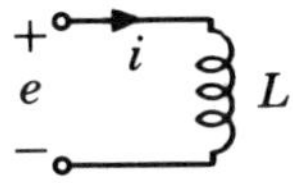

$e = L\dfrac{di}{dt},\qquad i = \dfrac{1}{L}\displaystyle\int_{-\infty}^{t} e\,dt$

Capacitance 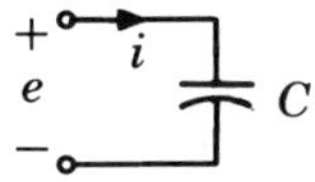

$e = \dfrac{1}{C}\displaystyle\int_{-\infty}^{t} i\,dt,\qquad i = C\dfrac{de}{dt}$

Current source 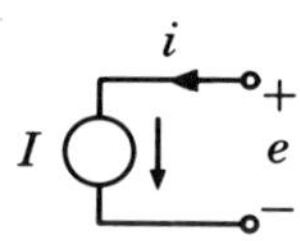

$i = I$

Voltage source 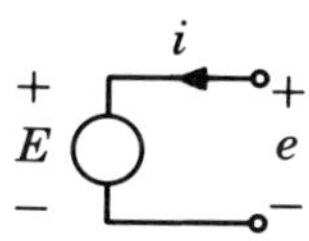

$e = E$

Meters

Voltmeter 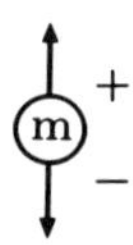

Ammeter 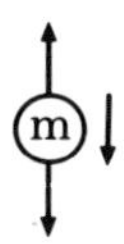

Wattmeter 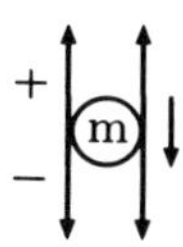

The circuit laws

Current law: $\sum i = 0$

Voltage law: $\sum e = 0$

REFERENCES

Circuit theory has been introduced in this chapter in terms of model theory and the defined characteristics of the ideal elements. It will prove interesting and instructive to explore some alternative introductions to the subject.

1. E. A. Guillemin, *Introductory Circuit Theory*, John Wiley & Sons, New York, 1953. Guillemin introduces the subject through the mathematical theory of topology. The introduction is logical and clear, but it is not for the faint-hearted!

2. MIT EE Staff, *Electric Circuits*, John Wiley & Sons, New York, 1943. This text is a classic attempt to introduce circuit theory from the more general field theory. It has never been popular with students because field theory is immeasurably more difficult than circuit theory.

3. Kerchner and Corcoran, *Alternating Current Circuits*, John Wiley & Sons, New York, 3rd ed., 1951. In this text circuit theory begins with steady-state sine waves and the response of circuits to excitation of this sort.

4. H. H. Skilling, *Electrical Engineering Circuits*, John Wiley & Sons, New York, 1957. Skilling introduces the subject by means of six plausible "experiments." Circuit laws are presented as experimental results.

5. M. E. Van Valkenburg, *Network Analysis*, Prentice-Hall, New Jersey, 1955. While this text is not designed for a first course in circuit theory, it does give a rapid review of the foundations of network theory in Chapter 1 and a few elements of topology in Chapter 2.

6. W. LaPierre, *Introduction to Electrical Engineering*, Harper & Brothers, New York, 1956. LaPierre treats direct-current circuits as his first topic and proceeds from there to steady-state sine waves.

7. Galbraith and Spence, *Fundamentals of Electrical Engineering*, The Ronald Press, New York, 1955. The introduction is essentially a review of the physics underlying electrical engineering.

8. W. R. LePage, *Alternating-Current Circuits*, McGraw-Hill, New York, 1952. As the title implies, this text treats circuit theory from the point of view of steady-state sine waves. It is clearly written.

9. Skroder and Helm, *Circuit Analysis by Laboratory Methods*, Prentice-Hall, New Jersey, 1946. This text is a laboratory manual for circuit theory. The subject is introduced via the laboratory equipment available for building circuits and for testing them.

10. M. B. Reed, *Alternating-Current Circuit Theory*, Harper & Brothers, New York, 1948. A somewhat more advanced and more challenging presentation of the steady-state sinusoid.

EXERCISES*

Section 1–1

1. Trace the energy conversions in

   (a) a mechanical phonograph,
   (b) an electrical phonograph,
   (c) a hydroelectric power station,
   (d) a steam-electric power station,
   (e) a portable gasoline-driven power plant for electricity,
   (f) a diesel-electric locomotive,
   (g) a solar battery,
   (h) a thermostatically controlled electric oven,
   (i) an automobile lighting system.

2. From the following items, pick 10 that are clearly idealized models and not actual physical entities:

an atom	the economic system	a circle
oxygen	a dollar bill	a point
an electron	inertia	the wind
a light beam	a house	love
an electric circuit	a particle	energy
a radiowave	a road	the equator
a map	a line	gasoline

3. Boolean algebra provides a mathematical model for problems involving two choices. For example, an open switch can be represented by a zero, and a closed switch by a one. Consider the two switches in series and give the meaning of the symbolic expressions $(1 + 1)$, $(1 + 0)$, and $(0 + 0)$, both symbolically and physically.

Section 1–2

4. Consider the economy of the United States as a physical system and pair up the stimulus-response relationships below:

Stimulus	*Response*
war	recession
elections	higher wage rates
excess government spending	less graft in government
increased birth rate	inflation
labor-controlled government	higher profit rates
management-controlled government	expanding economy
fear and uncertainty	a stock market boom

5. Consider a physical system made up of a steam-electric generating station supplying lights for a small town. Identify the stimulus and the response, and suggest a method for operating the system.

---

* Answers to Exercises are given at the back of the book.

6. "Viewed as stimulus-response transactions, purchases in a store are linear operations, but large-scale purchases on the stock market are nonlinear." Explain.

Section 1–3

7. Thirty coulombs of positive charge per second are passing through a wire in a direction from a to b. (a) What is the current if the assumed reference direction is from a to b? (b) What is the current if the assumed reference direction is from b to a? (c) How would these answers change if the moving coulombs were negative instead of positive?

8. In moving from a to b, a coulomb of charge changes its energy by 10 joules. Give the voltage of point a with respect to point b if (a) the charge is positive and the energy is lost, (b) the charge is positive and the energy is gained, (c) the charge is negative and the energy is lost, and (d) the charge is negative and the energy is gained.

9. In a certain two-terminal device a positive current of 10 amp enters at terminal a and leaves from terminal b. What is the power absorbed in the device when (a) the voltage at a is 10 volts positive with respect to b? (b) the voltage at b is 10 volts positive with respect to a? (c) the voltage at a is -10 volts positive with respect to b?

10. Which of the devices of Fig. 1–20 are sources (values are in volts and amperes)?

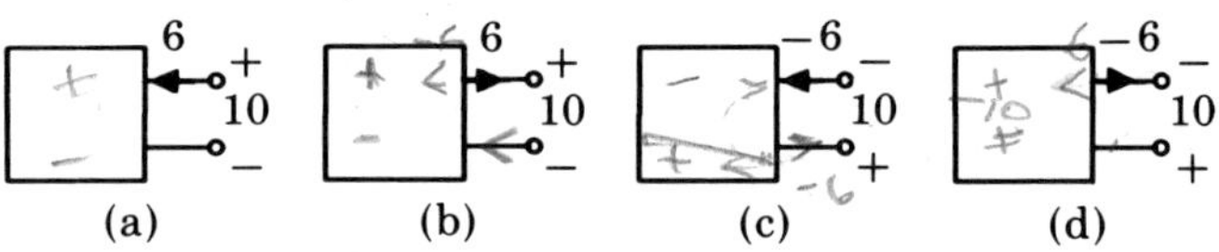

FIGURE 1–20

Section 1–4

11. A certain physical resistor is composed of 22 ft of copper wire with a resistance of 1 ohm per foot, in series with 6 in. of carbon rod with a resistance of 2 ohms per inch. Find the lumped-parameter resistance value, and show the resistance symbol with the proper value marked on it.

12. A certain physical capacitor is composed of two layers of tinfoil separated by a layer of paper. The capacitance per square inch is known to be 10×10^{-12} farad. Find the capacitance of 10 square feet of the combination, and draw the lumped-parameter capacitance symbol with the value marked on it.

13. A certain physical inductor is constructed by winding copper wire with negligible resistance in the form of a solenoidal coil. The inductance increases in proportion to the square of the number of turns. If a 10-turn coil has an inductance of 1×10^{-4} henry, find the inductance of a 1000-turn coil, draw the lumped-parameter symbol, and label it.

14. A voltage source of value 6 volts is connected in series with a resistance of 2 ohms. Find the current in the circuit and the power absorbed by the resistor.

15. A voltage source of 6 volts is connected across a capacitor with a capacitance of 10^{-6} farad. Give the energy stored in the capacitor and describe what happens to the energy when the battery is removed.

16. A current of 10 amp is flowing through an inductor of inductance 1 henry. What is the energy stored in the inductor, and what happens to the energy if the current source is replaced by a short circuit?

17. How much power is delivered by a 10-volt voltage source (a) if it is open-circuited? (b) if a resistance of 1 ohm is placed across it? (c) if it is short-circuited? (d) if it is connected in parallel with a current source of value 1 amp and direction such that the current enters the positive terminal of the voltage source?

18. How much power is delivered by a 10-amp current source (a) if it is open-circuited? (b) if it is short-circuited? (c) if a resistance of 1 ohm is placed in series with it? (d) if a voltage source of 1 volt is placed across it so that the current enters at the negative terminal of the voltage source?

Section 1–5

19. Redraw the circuit of Fig. 1–21(a) so that it will have the form of part (b).

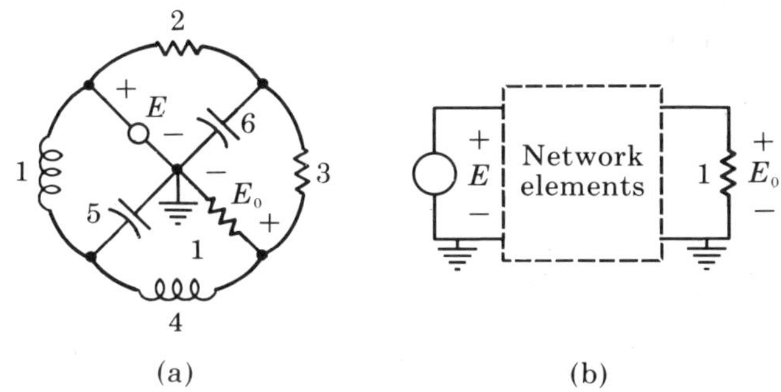

FIGURE 1–21

20. Points *abcd* are junctions in an electric network. The following voltages are known: $V_{ab} = 10$, $V_{bc} = -2$, $V_{ad} = 5$. Find V_{cd}.

21. Points *abcdo* are junctions in an electric network. The following currents are known: $I_{ao} = 4$, $I_{bo} = -3$, $I_{co} = 10$. Find I_{do}.

22. Points *abcd* are junctions in an electric circuit. Point *a* is the ground or reference point. Point *b* is 6 volts above point *a*. Point *c* is 3 volts below point *d*, and point *d* is 10 volts below *b*. Draw the nodes and find the voltages of *b*, *c*, and *d* relative to ground.

23. For Fig. 1–22, express the currents in the elements i_{10}, i_{20}, i_{30}, and i_{40} in terms of the mesh currents i_a, i_b, i_c, and i_d.

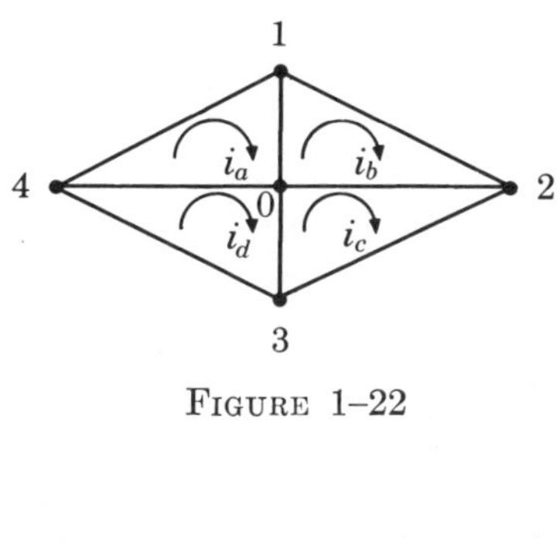

FIGURE 1–22

CHAPTER 2

THE 2*b* EQUATIONS OF ELECTRICAL NETWORKS

"It is idle to expect any great advancement in science from the superinducing and indrafting of new things upon old. We must begin anew from the very foundations, unless we would revolve forever in a circle with mean and contemptible progress."

Book 1, Novum Organum, Francis Bacon, 1620

2–1 Introduction. Some of you may have been exposed to electrical networks before, and you may be anxious to proceed at once to the exciting recent developments in electric circuits. If so, I can only recommend to you the wisdom of Francis Bacon. In this chapter we will inquire into the foundations of electric circuits, the assignation of unknown variables, and the formulation of independent equations which can be used to solve for the unknowns. The electrical networks which we will consider are composed of resistors, inductors, and capacitors connected in some geometrical pattern and excited by current or voltage sources. The problem of *network analysis* is one of finding the currents and voltages in the network elements when the network and the sources are known. Engineers are more often concerned with problems of network *synthesis,* in which the input stimulus and the desired output response are known and a network must be found. Before an engineer can attempt network synthesis, however, he must be very adept at network analysis. Accordingly, most of this book will be devoted to network analysis.

If a person were infinitely clever, he could solve all of the problems of network analysis by the methods of arithmetic. Most of us, however, find it necessary to resort to algebra. The algebraic procedure is always the same. A set of unknown variables is defined, equations are written in terms of these variables, and the equations are then solved by the rules of algebra. This procedure is routine, automatic, and, once learned, may be applied with a minimum of thought. In electrical circuits the unknown variables will be currents and voltages. The laws which are available to write equations for these unknowns are the volt-ampere equations of the elements and Kirchhoff's two laws. All of the methods which we will develop for solving network problems will spring from these laws.

In the first part of this book we will restrict ourselves to networks containing only resistors and sources. In terms of these networks we will be able to formulate general methods for approaching problems, to derive most of the useful network theorems, and to develop a number of specialized artifices for particular networks. The results which we will obtain will be automatically extended to more general networks in later sections.

In this chapter we shall develop the foundations of algebraic circuit theory. We will define the network branches, nodes, and loops, and develop the relationships among them. A network *branch* will be defined as an element, or an element in conjunction with sources of energy.* *Nodes* are the junction points between the branches, and *loops* are independent closed paths within the network. Since each branch contains an element, it requires two unknown variables, the current in the element and the voltage across it. Thus the total number of unknowns in a circuit with b branches must be $2b$. In order to solve for these $2b$ unknowns we must write $2b$ independent equations. For this purpose we have available the volt-ampere equations of the branches, Kirchhoff's current law, and Kirchhoff's voltage law. The volt-ampere equations are relatively simple. There is one for each branch, and the equations are independent of each other because they refer to different branches. The Kirchhoff law equations are not so simple. Both laws can be applied to a network in many different ways, and only some of the resulting equations are independent. In order to resolve the question of independent equations we will find it necessary to resort to a branch of mathematics called *topology*. Topology deals with junctions, branches, and independent paths, and is essential to an understanding of the fundamental properties of electrical networks.

2–2 Series and parallel resistors. A *branch* in an electrical network is an element, or any collection of elements and sources, from which a single pair of terminals protrudes. In the "game" of circuit theory the only variables are voltage and current. Accordingly, the only way a branch can interact with the remainder of a network is through the voltage and current at its terminals. So far as its effect in the network is concerned, the branch is completely and accurately described by the volt-ampere equation relating the current and voltage at its terminals. The simplest form of branch is a single resistor, and for this branch the volt-ampere relation is

$$e = iR.$$

It is often convenient to view combinations of resistors as single compound

---

* Sources are specifically excluded from consideration as branches. We are interested in the branches as an indication of the number of unknown currents and voltages in the system, and the sources are known quantities. The sources represent the excitation of the network and may be considered separately from the network itself.

branches. In fact, whenever the elements within a branch are of the same kind a single equivalent element can be found which has the same volt-ampere equation as the original compound branch, and which therefore can replace the compound branch for computations which involve the remainder of the network.* As examples of compound branches we will consider resistors in series and resistors in parallel.

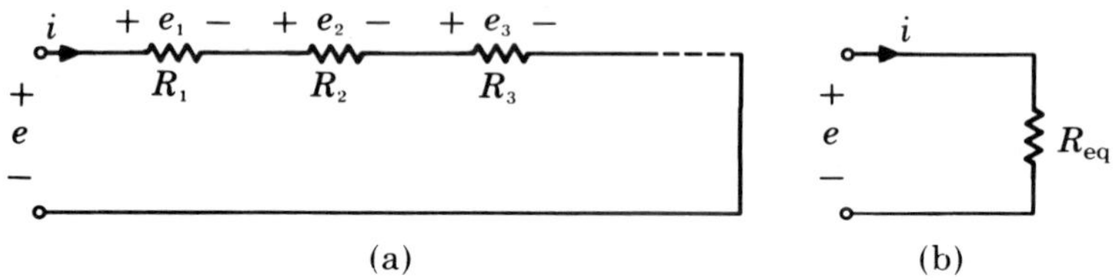

FIG. 2–1. Resistors in series.

Resistors in series. Groups of resistors are in series if they form a single chain between the two terminals of a compound branch. An example is shown in Fig. 2–1(a). We will find a single equivalent resistor to represent this branch, as shown in Fig. 2–1(b). The two branches are equivalent if they have the same volt-ampere equation at their terminals. The equation for the compound branch is best obtained by considering the input current to be known, and computing the resulting voltage. The voltage drops produced in the separate resistors by the current, i, are

$$e_1 = iR_1, \qquad e_2 = iR_2, \qquad e_3 = iR_3, \qquad \ldots$$

The input voltage is the sum of the voltage drops across the separate elements:

$$e = iR_1 + iR_2 + iR_3 + \cdots = i(R_1 + R_2 + R_3 + \cdots). \tag{2–1}$$

The volt-ampere equation of the equivalent resistance is

$$e = iR_{\text{eq}}. \tag{2–2}$$

A comparison of Eqs. (2–1) and (2–2) shows that the equivalent resistance for a number of resistances in series is just the sum of the resistances, or

$$R_{\text{eq}} = R_1 + R_2 + R_3 + \cdots \tag{2–3}$$

It should be noted in passing that we have used all of the circuit laws in the process of finding R_{eq}. Kirchhoff's current law assures us that the

* Clearly, if we replace a compound branch by a single element we set aside any possibility of computing the distribution of currents within the branch.

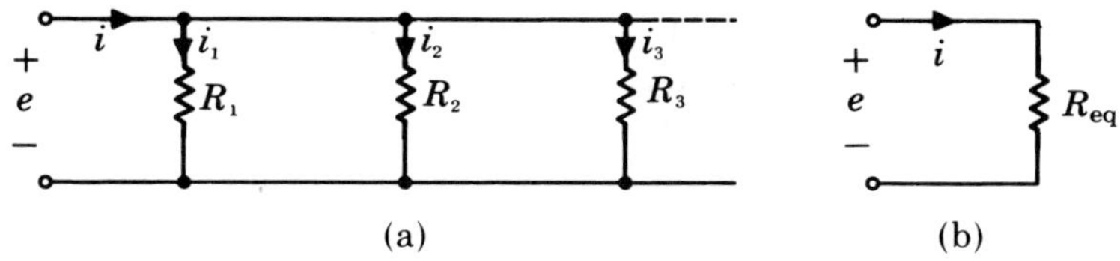

FIG. 2–2. Resistors in parallel.

current in each resistance is the same. The volt-ampere equations of the elements are used to obtain the voltages across the elements, and the final voltage, e, is obtained by means of Kirchhoff's voltage law. It is a characteristic of network analysis that explicitly or implicitly we must use all of these laws to solve any given problem.

Resistors in parallel. A group of resistors are in parallel if each of them is connected directly between two terminals of the compound branch. An example of resistors in parallel is shown in Fig. 2–2(a). The single resistance equivalent to the compound branch is shown in Fig. 2–2(b). The volt-ampere relationship of the compound branch is best obtained by considering the input voltage to be known, and finding the resulting input current. The input voltage appears across each element and the currents in them are

$$i_1 = e/R_1, \qquad i_2 = e/R_2, \qquad i_3 = e/R_3, \qquad \ldots$$

The input current is the sum of the currents in the branches, and is therefore

$$i = \frac{e}{R_1} + \frac{e}{R_2} + \frac{e}{R_3} + \cdots = e\left(\frac{1}{R_1} + \frac{1}{R_2} + \frac{1}{R_3} + \cdots\right). \tag{2–4}$$

The current-voltage relationship for the equivalent resistance is

$$i = e\frac{1}{R_{\text{eq}}}. \tag{2–5}$$

A comparison of Eqs. (2–4) and (2–5) shows that the equivalent resistance is

$$R_{\text{eq}} = \frac{1}{1/R_1 + 1/R_2 + 1/R_3 + \cdots} \tag{2–6}$$

In terms of the conductances $G_{\text{eq}} = 1/R_{\text{eq}}$, $G_1 = 1/R_1$, $G_2 = 1/R_2, \ldots$ Equation (2–6) has a somewhat simpler form:

$$G_{\text{eq}} = G_1 + G_2 + G_3 + \cdots \tag{2–7}$$

Equation (2–7) looks invitingly simple. The total conductance is simply the sum of the conductances of the branches in parallel. Usually, however, the resistance values are recorded in ohms, and the final value is desired in ohms. Under these circumstances Eq. (2–7) represents no net saving in effort, and most people will prefer to use Eq. (2–6) directly.

The case of two resistors in parallel occurs frequently and is worth special consideration. For two resistors,

$$R_{\text{eq}} = \frac{1}{1/R_1 + 1/R_2} = \frac{R_1 R_2}{R_1 + R_2} = \frac{\text{product}}{\text{sum}}. \tag{2–8}$$

As an example of Eq. (2–8) we will find the resistance of a 2-ohm and a 6-ohm resistance in parallel. The equivalent resistance is

$$R_{\text{eq}} = \frac{6 \times 2}{6 + 2} = 1.5 \text{ ohms.}$$

Series-parallel circuits. By successive applications of the rules for combining resistors in series and in parallel it is possible to reduce many complicated branches to single resistors.* As an example we will find the equivalent resistance for the network of Fig. 2–3. The network is reduced from the right-hand side. The first step is to combine the two resistors in series. The equivalent resistance is

$$R_1 = 1 + 1 = 2 \text{ ohms.}$$

The next step is to combine R_1 with the first shunt resistance, with which it is now in parallel. This combination is given by

$$R_2 = \frac{R_1 \times 1}{R_1 + 1} = \tfrac{2}{3} \text{ ohm.}$$

The resultant R_2 is now in series with the second series resistance and can be added to it, to give

$$R_3 = 1 + R_2 = \tfrac{5}{3} \text{ ohm.}$$

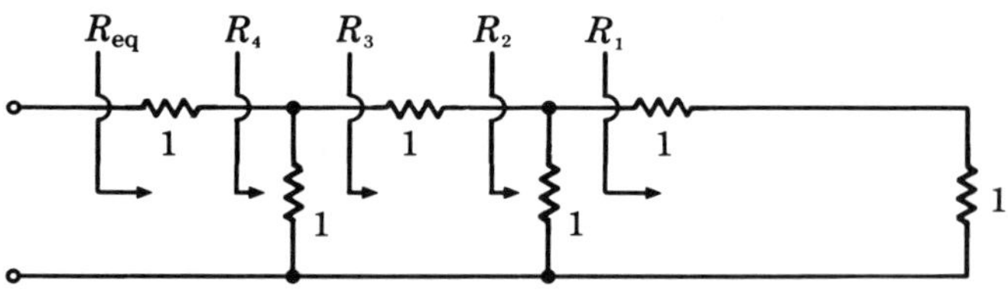

FIG. 2–3. A series-parallel circuit.

* To reduce a general compound branch containing only resistors it is necessary to use the Y-Δ transformation as well as the formulas for series and parallel resistors (see Chapter 4).

The resistance R_3 is then in parallel with a 1-ohm resistance and can be combined with it to give a new resultant,

$$R_4 = \frac{R_3 \times 1}{R_3 + 1} = \tfrac{5}{8} \text{ ohm}.$$

Finally, the resistance R_4 can be added to the remaining series resistance to give the input resistance R_{eq}:

$$R_{eq} = 1 + R_4 = \tfrac{13}{8} \text{ ohms}.$$

The combination of resistors in series and in parallel is easy and will often save a good deal of work. In the next section we shall proceed in a slightly different direction and consider compound branches in which sources as well as resistors occur.

2-3 Source branches. Our definition of a network branch as a section of a network from which a single pair of terminals protrudes includes branches with sources as well as purely resistive branches. These source branches, as well as the resistive branches, are completely defined by the volt-ampere equations at their terminals. We can find this equation for a given branch by applying a current and computing the resultant voltage, or by applying a voltage and computing the resulting current. In order to be consistent with the volt-ampere relations of resistive branches we will take the reference direction for the voltage across the branch to be positive at the terminal where the current enters.

A real physical source can be represented by an ideal voltage source in series with a resistance, or by an ideal current source in parallel with a resistance. Occasionally we may wish to consider voltage sources with no series resistance, or current sources with no parallel resistance. These are "pathological" cases which require special treatment. Initially we shall consider the more usual sources, and later we will develop transformations which will enable us to handle the more unusual cases.

Voltage source and resistance. One of the simplest representations of a real physical source is the ideal voltage source in series with a resistance, as shown in Fig. 2-4(a). Only the terminals *a*–*b* of this device are avail-

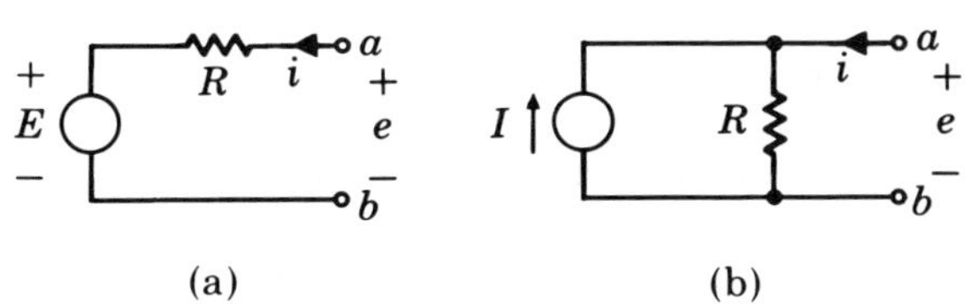

FIG. 2-4. The basic source branches.

able to us for calculations or measurements. The device must be defined, so far as its effect in an electric circuit is concerned, by the volt-ampere equation at these terminals. For convenience we shall obtain the volt-ampere equation by applying a current at the terminals and computing the resulting voltage.* For the directions shown the resulting voltage is

$$e = E + iR. \tag{2-9}$$

An important conclusion that can be drawn from Eq. (2–9) is that when the voltage source is zero, the equation is that of a resistor alone. So far as the circuit is concerned, therefore, *a zero voltage source has been replaced by a short circuit.*

Current source and resistance. A current source in parallel with a resistance is shown in Fig. 2–4(b). The volt-ampere equation at its terminals is again obtained by applying a current and computing the resulting voltage. In this case the voltage is easily seen to be

$$e = IR + iR. \tag{2-10}$$

If the source current I in Eq. (2–10) is zero the volt-ampere equation is that of a resistor alone. In terms of the circuit the source has completely disappeared and *the zero current source is equivalent to an open circuit.* In summary,

$$\text{a dead voltage source} = \text{a short circuit,}$$
$$\text{a dead current source} = \text{an open circuit.}$$

Equivalent sources. An interesting equivalence can be derived from Eqs. (2–10) and (2–9). If $E = IR$ the two equations are identical. Thus, a current source in parallel with a resistance is equivalent to a voltage source in series with the same resistance, provided only that the value of the voltage source is that of the source current multiplied by the resistance. The transformation can be used in either direction, and is so important that it is drawn out explicitly in Fig. 2–5. The numerical values of the equivalent sources are easy to obtain. If the voltage source and the resistance are known, the current source, from dimensional considerations, is the product of the two values. Similarly, the only way an equivalent current source can be obtained from a voltage and a resistance is as the quotient of voltage divided by resistance. The polarities of the equivalent sources are best obtained by inspection. The two sources are equivalent *so far as the external circuit is concerned.* They should be arranged,

* Students often object that the current must come out of the source. Actually it may go either in or out, depending on the remainder of the circuit. We take it to be entering, to be consistent with the directions assumed for the resistance.

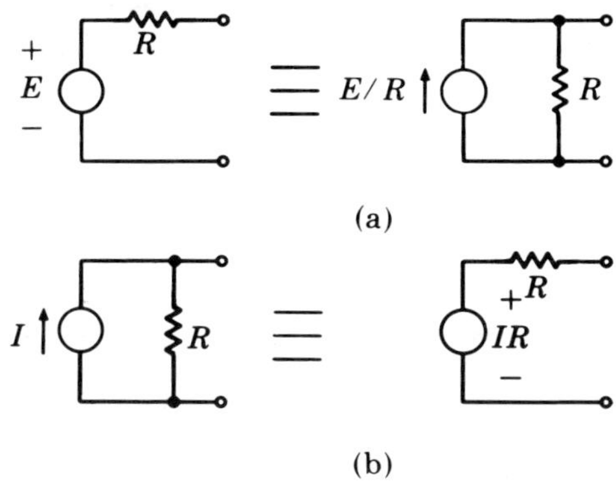

FIG. 2-5. Voltage-to-current and current-to-voltage source transformations.

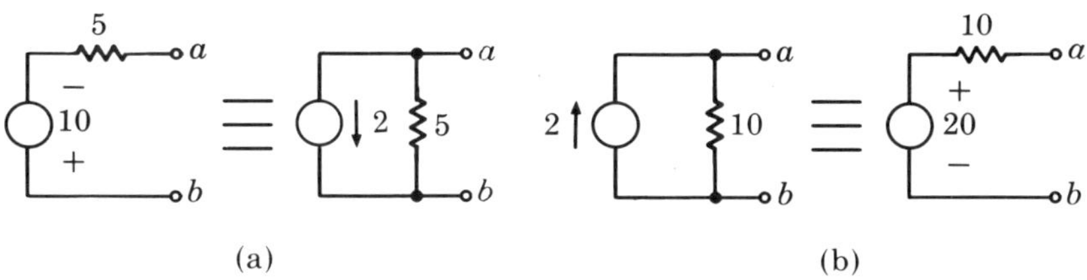

FIG. 2-6. Two examples of source transformations.

therefore, to produce *the same direction of current in the external circuit* when a short circuit is placed across it.

Two numerical examples are shown in Fig. 2-6. In the first example the voltage source of 10 volts in series with the resistance of 5 ohms becomes a current source of 2 amp in parallel with the 5 ohms. The polarity is such that the current would flow from b to a in both cases if the external circuit is passive.* In the second example the current source of 2 amp in parallel with 10 ohms is replaced by a voltage source of 20 volts in series with the 10 ohms. The polarities of the sources in both cases are such that they produce a current flow from a to b in the external circuit.

Series-parallel source branches. It is possible to reduce many series-parallel connections of source branches to a single source and a single resistor. As a first step in this direction we must consider the combination of ideal sources in series and in parallel. Two ideal voltage sources in series can be added algebraically to give one equivalent voltage source. Similarly, two ideal current sources in parallel can be added algebraically to give one resultant current source. These two equivalences are shown in Fig. 2-7(a) and (b). The proof of the equivalence rests on Kirchhoff's

* A passive circuit contains no sources.

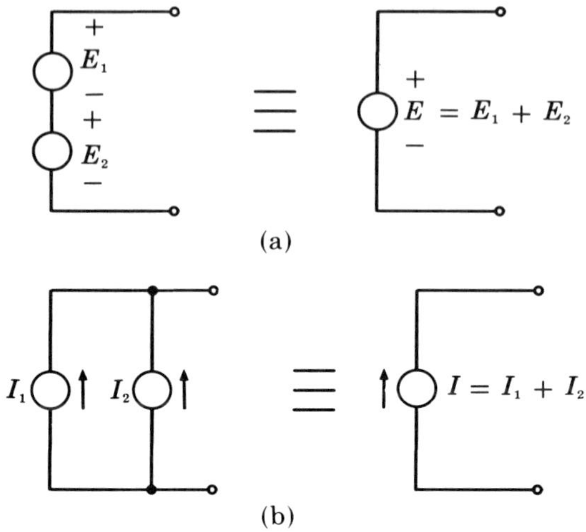

FIG. 2–7. Voltage sources in series and current sources in parallel.

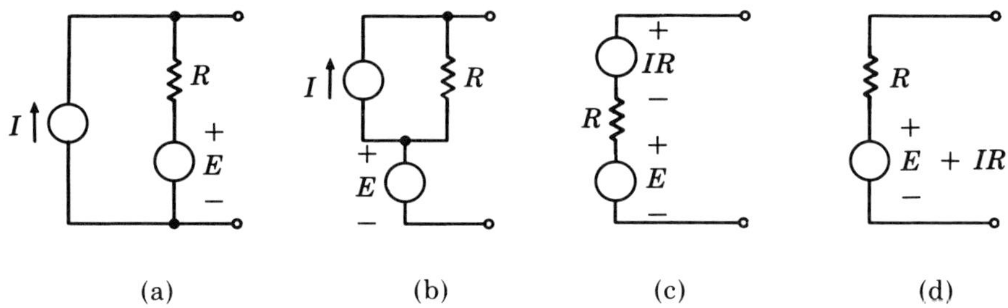

FIG. 2–8. The reduction of a compound source branch.

voltage law for the voltage sources and on Kirchhoff's current law for the current sources.*

As an example of the reduction of a series-parallel connection of source branches to a simpler form, consider the circuit of Fig. 2–8(a). It makes no difference whether the current I is drawn through the voltage source or not, and therefore the circuit can be changed to the form shown in Fig. 2–8(b). In this form a current to voltage source transformation can be applied as shown in Fig. 2–8(c). The two voltage sources in Fig. 2–8(c) can be added to give a resultant voltage of $E + IR$ in series with the resistance R.

* Students sometimes wonder about the combination of two differing voltage sources in parallel, or of two differing current sources in series. By definition, neither of these two conditions is possible for ideal sources. If an attempt is made to reproduce them with real sources, serious overloads will occur.

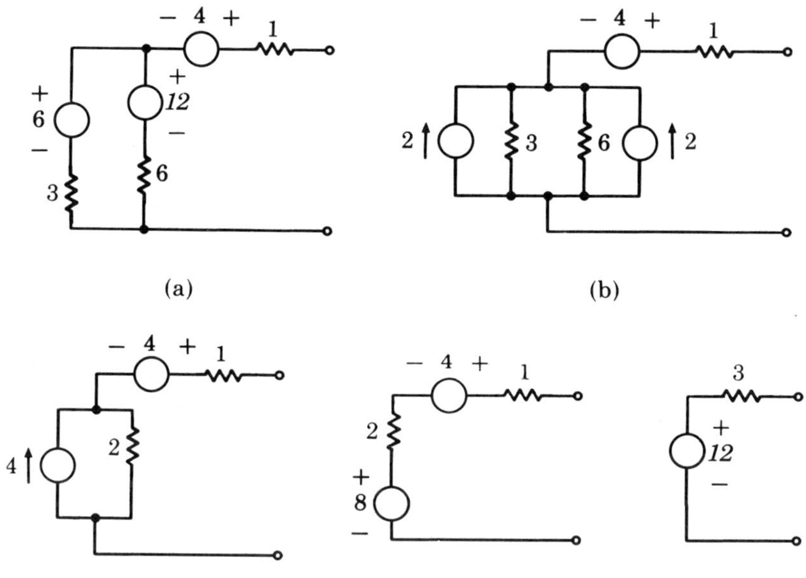

(a) (b) (c) (d) (e)

FIG. 2-9. An example of the reduction of a compound source branch.

A more complicated example is shown in Fig. 2-9(a). As a first step in the reduction process the two voltage source branches in parallel are converted to equivalent current sources, as shown in Fig. 2-9(b). The current sources are added and the two resistors are combined in parallel, as shown in Fig. 2-9(c). To be able to add the two remaining branches it is necessary to convert the current source branch to an equivalent voltage source branch. The final two steps in the process are illustrated in Fig. 2-9(d) and (e). The original compound branch has been reduced to a single resistor of 3 ohms in series with a single voltage source of 12 volts.

The process which has been illustrated here is not completely general. It is, however, very convenient for the series-parallel type of network. The idea of reducing a branch to a single source and a single resistor is more general and will be treated in detail in Chapter 4.*

Voltage source with no series resistor. Real voltage sources always have series resistance along with an ideal voltage source. Sometimes, however, the series resistance is so small that we wish to neglect it in comparison with the other resistances in the problem. An ideal voltage source with no

* The source reductions we have developed here are special cases of two useful and famous theorems called Thévenin's theorem and Norton's theorem.

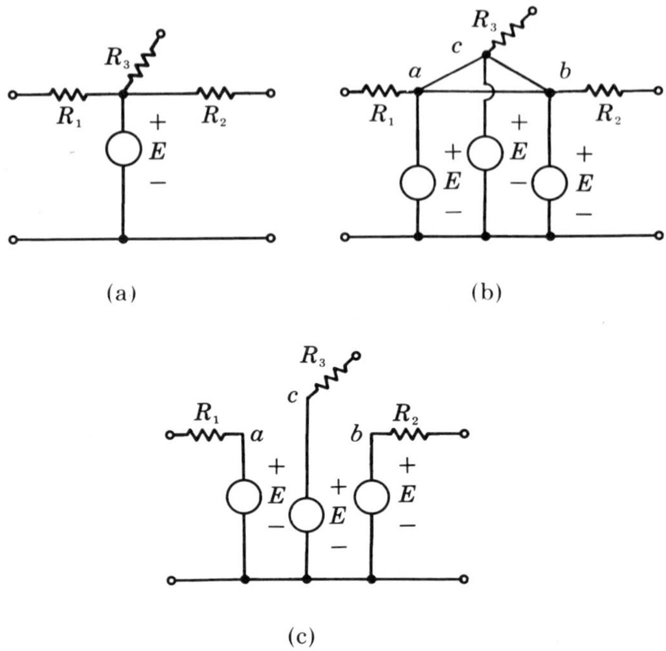

FIG. 2–10. A transformation for voltage sources without resistance.

series resistance poses some problems. For one thing, it is not possible to obtain an equivalent current source, and it also complicates the selection of algebraic unknowns. In order to avoid these problems we will develop in this section a transformation which removes the "ideal" voltage source and replaces it by "real" voltage sources in series with proper resistances. For this purpose consider the diagram in Fig. 2–10(a). The voltage source E has no series resistance. It is "ideal." We can, however, replace it with three identical voltage sources in parallel, as shown in Fig. 2–10(b). We now remove the short circuits joining the positive terminals of the three voltage sources. When they are removed the potentials of points a, b, and c remain identical since the voltage sources are identical. Hence, it does not matter whether the connections are there or not! So we remove them. The final result is that we have replaced the one voltage source with no series resistance with three voltage sources with proper series resistances.

The net effect of the transformation of Fig. 2–10 is to "push the voltage source through the node." The node itself disappears, and instead the voltage source appears in each of the branches radiating from the node. The more general transformation is shown in Fig. 2–11.

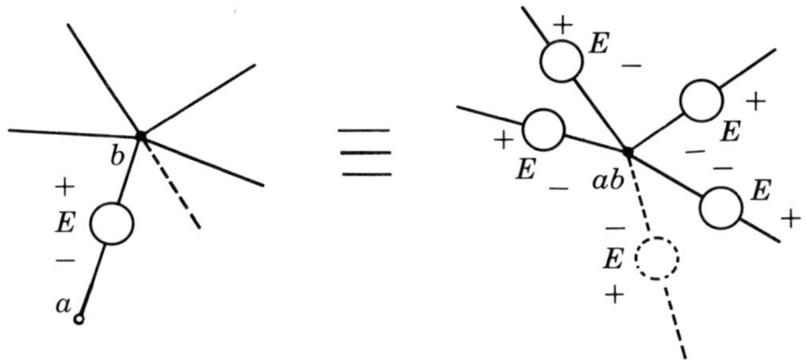

FIG. 2–11. A general form of resistanceless voltage source transformation.

Current source with no parallel resistance. A real physical source can be represented by a current source in parallel with a resistance. Occasionally we may wish to neglect the resistance in parallel because it is so large in comparison with the other resistances in the problem. A current source without a parallel resistance is also an "ideal" case which cannot be converted to a voltage source and which poses certain problems in determining the algebraic unknowns in a general circuit. In order to avoid these problems we will develop a transformation which removes the "ideal" current source and replaces it by a number of "real" ones in parallel with resistances. A typical case is shown in Fig. 2–12(a). The current source appears directly across the nodes a–b, and it has no parallel resistance. Note, however, that the net effect of the current source is to pump I amperes out of the node a and into node b. Any other combination of sources which accomplishes this same result and nothing more is equally acceptable. Such a combination is shown in Fig. 2–12(b). The first source pumps the current from a and deposits it at c. The second source, however, immediately pumps the same current away from c and deposits it at d. The net current delivered to c is thus zero. Similarly, the third source pumps the current out of d and no net current arrives at point d. Finally, the third source deposits the current I at node b and that is the end of the

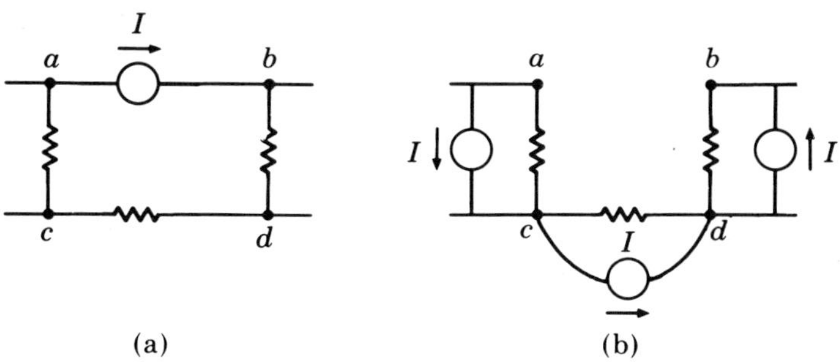

FIG. 2–12. A current source with no parallel resistance.

chain. The final result is that the current is taken from the circuit at a and delivered back at b. Although it calls at a number of other stations on the way, it doesn't stay at any of them!

A transformation of the type shown in Fig. 2–12 replaces a current source bridged across two nodes with a chain of current sources looped around some closed path which also bridges the two nodes. The original current source has no resistance across it, but each source in the chain is associated with the resistance of a branch. A general form of the transformation is shown in Fig. 2–13.

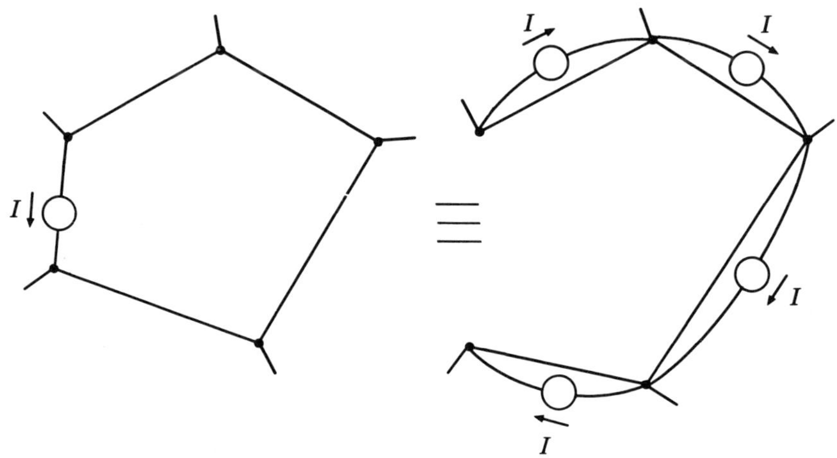

FIG. 2–13. A general current source transformation shown in topological form.

2–4 Problems solvable by arithmetic. Some network problems are so simple that they are always solved by arithmetic. No unknown currents and voltages are assigned, and no equations are written for a solution. The desired response is obtained by inspection! Additional problems can be reduced to trivial arithmetic solutions by the reduction of compound branches and of source branches, as we have discussed in the previous sections. It is desirable to solve as many problems as we can by arithmetic rather than by resorting to the more general algebraic formulations of the next sections.

Voltage source and parallel resistors. An example of a problem which can be solved by inspection is a voltage source in parallel with a number of resistors. A typical example is shown in Fig. 2–14(a). The voltage across each of the resistors is 10 volts, and the currents through them can be written by inspection as

$$i_1 = \tfrac{10}{1} = 10 \text{ amp}, \qquad i_2 = \tfrac{10}{2} = 5 \text{ amp}, \qquad i_3 = \tfrac{10}{3} = 3.3 \text{ amp}.$$

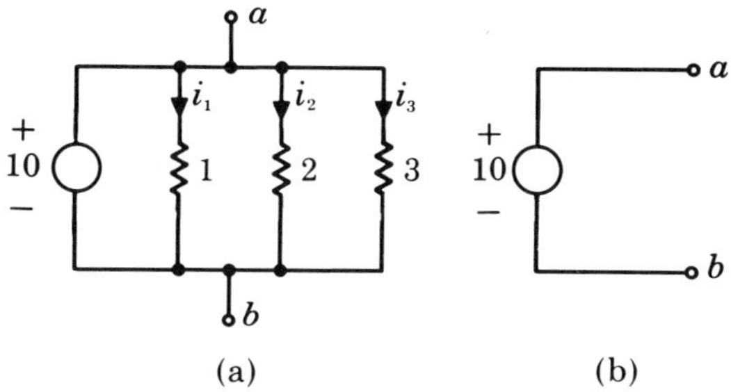

FIG. 2–14. A voltage source in parallel with resistors.

It is interesting to note that these currents would not change if additional resistances were added in parallel with them, nor if this branch were immersed in some more complicated network. In fact, the equivalent source branch between the terminals a and b is simply the 10-volt source shown in Fig. 2–14(b). So far as the distribution of currents and voltages in the remainder of the network is concerned *the resistors in parallel with the voltage source have no effect and can be forgotten entirely.** If the reader is surprised at this result he should reflect that the lights in his house change very little as the neighbors connect and disconnect their lights. The electric power company tries very hard to supply a "voltage source" to its customers, and if it succeeds the electric circuits of all its customers are completely independent.

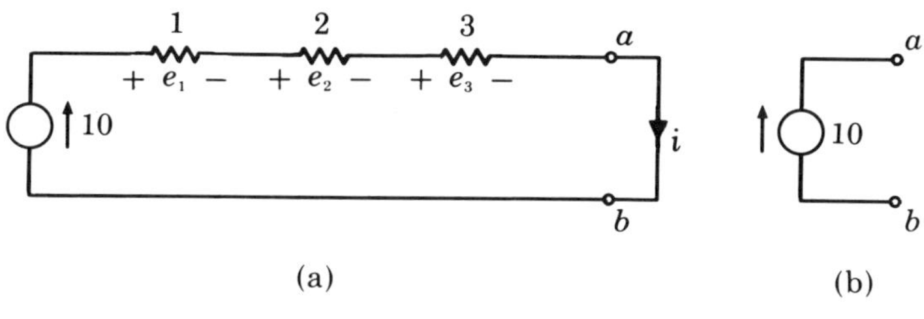

FIG. 2–15. A current source in series with resistors.

Current source and series resistance. A second problem which can be solved by inspection is the current source in series with a number of resistances. A typical example is shown in Fig. 2–15(a). The 10-amp current from the source flows through each of the resistances in turn and the voltage drop across them can be obtained by inspection:

$$e_1 = 10 \times 1 = 10 \text{ volts}, \qquad e_2 = 10 \times 2 = 20 \text{ volts}, \qquad e_3 = 10 \times 3 = 30 \text{ volts}.$$

The total voltage across the source is the sum of e_1, e_2, and e_3, or 60 volts.

---

* From the standpoint of algebra these branches are "dummy" branches.

Instead of a short circuit between the terminals a and b we could imagine some complicated network. So far as this network is concerned a current of 10 amp will flow in at terminal a, and a current of 10 amp will flow out at terminal b. The same effect would be produced by a current source with no series resistance. Thus *the series resistors have no effect on the remainder of the network and can be neglected entirely.** The equivalent source branch is the current source alone, as shown in Fig. 2–15(b). So far as the remainder of the network is concerned the series resistance has simply been erased!

It should be noted, however, that we adopt two completely different points of view concerning resistors in parallel with a voltage source and resistors in series with a current source. When we are interested in the energy distribution in the resistors themselves we treat the problem as a trivial arithmetic one and obtain the answers by inspection. When we are interested in the effect which this branch will produce in the remainder of the network we simply erase the resistors and forget about them altogether! We shall now consider the voltage divider and the current divider as examples of the first point of view and then we will give two examples where the second point of view is more useful.

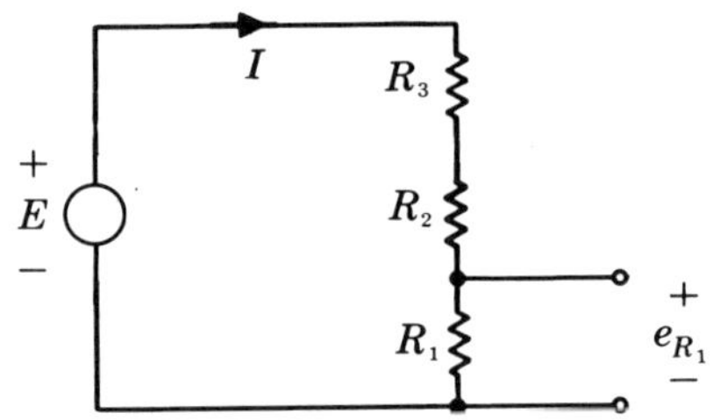

FIG. 2–16. A voltage divider.

The voltage divider. A voltage divider is a voltage source in series with a number of resistances. An example is shown in Fig. 2–16. A very easy solution is obtained by reducing the resistance branch to a single equivalent resistance,

$$R_{\text{eq}} = R_1 + R_2 + R_3.$$

The current from the source can then be found by inspection:

$$I = \frac{E}{R_1 + R_2 + R_3}.$$

Once the current I has been found it is a *known current* and we can find the distribution of voltages as we would for a current source. The

* From the standpoint of algebra these branches are "dummy" branches.

voltage across R_1, for example, is

$$e_{R_1} = R_1 I = \frac{R_1}{R_1 + R_2 + R_3} E. \tag{2–11}$$

Equation 2–11 exhibits the "voltage divider" effect. The voltage across the resistance R_1 is obtained as the same fraction of the total voltage that R_1 is of the total resistance. For a series chain, voltage divides in proportion to resistance. A great many practical devices make use of this result. For example, the volume control in a radio is a voltage divider.*

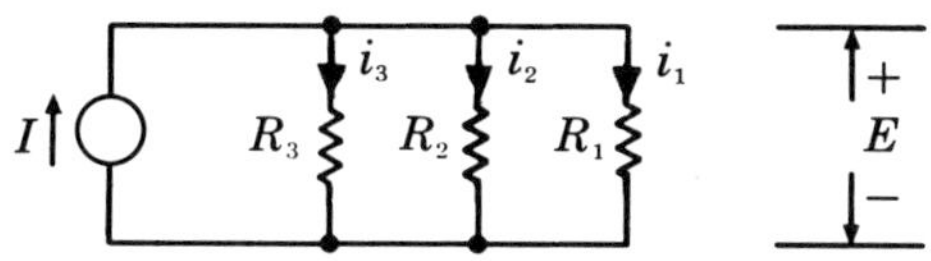

FIG. 2–17. A current divider.

The current divider. A second example which can be reduced to an arithmetic problem by a combination of resistors is the current divider shown in Fig. 2–17. The resistances in parallel are easily reduced to an equivalent branch, with a resistance

$$R_{\text{eq}} = \frac{1}{1/R_1 + 1/R_2 + 1/R_3} \cdot$$

The voltage can then be obtained by inspection, for

$$E = IR_{\text{eq}} = I \frac{1}{1/R_1 + 1/R_2 + 1/R_3} \cdot$$

Once the voltage is known *it can be considered to be a source* and the currents through the various resistors can be obtained by inspection. For example,

$$i_1 = \frac{E}{R_1} = \frac{1/R_1}{1/R_1 + 1/R_2 + 1/R_3} I. \tag{2–12}$$

Equation (2–12) shows the way in which currents divide among resistances in parallel. The current is proportional to the inverse of the resistances, or to the conductances.

The particular case of two resistances occurs frequently and is worth considering separately. If R_3 is removed in the above problem, the current

* Resistances with a variable wiper arm which can make contact at an arbitrary point along the resistance are used as voltage dividers. They are called potentiometers, or "pots" for short.

in the R_1 resistor is

$$i_1 = \frac{1/R_1}{1/R_1 + 1/R_2} I = \frac{R_2}{R_1 + R_2} I. \tag{2-13}$$

Equation (2–13) shows that the current divides as the opposite resistance divided by the sum of the resistances. Voltage dividers and current dividers occur so frequently that Eqs. (2–11) and (2–13) are well worth remembering.

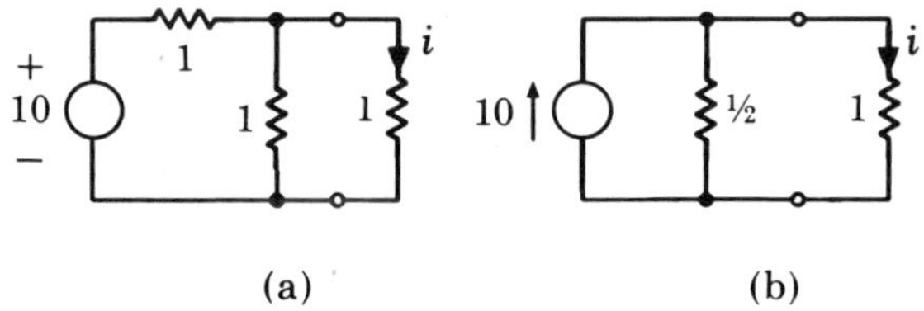

FIG. 2–18. An arithmetic problem solved by branch simplification.

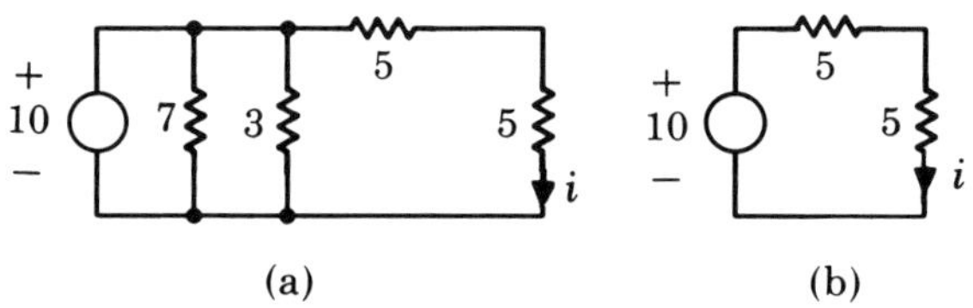

FIG. 2–19. An arithmetic problem solved by branch elimination.

Arithmetic problems by branch simplification. Many problems which appear complicated on the surface can be reduced to arithmetic inspection by the application of the reduction methods for resistive branches and source branches. An example is shown in Fig. 2–18. The original circuit is shown in Fig. 2–18(a). As a first step the voltage source is transformed to a current source and the two parallel resistors combined, as in Fig. 2–18(b). The current through the 1-ohm resistor is then obtained by means of Eq. (2–13). It is

$$i = \frac{\frac{1}{2}}{1 + \frac{1}{2}} \times 10 = 3.3 \text{ amp.}$$

A second example is shown in Fig. 2–19(a). The current i in the 5-ohm resistor is the desired response. So far as this branch is concerned the 7- and the 3-ohm resistances have no effect. The equivalent source branch with these resistors removed is shown in Fig. 2–19(b). From this circuit the desired current is obtained by inspection:

$$i = \frac{10}{5 + 5} = 1 \text{ amp.}$$

As problems get more complicated the idea of reducing them to arithmetic solutions becomes less attractive. At some point it is desirable to resort to the more routine methods of algebra. In order to apply algebra we must answer the following questions:

1. How many unknowns must we specify?
2. Which unknowns should we specify?
3. What equations should we write for these unknowns?

The answers to these questions are not simple, nor are they unique. There are many different ways of setting up the algebra of a problem. As a result of a great deal of work done in the past, a number of very convenient and general algebraic procedures have been derived. To be able to understand them, however, requires the ability to think about networks in a rather general and abstract way. The concepts of algebraic topology discussed in the next section will prove very helpful.

2–5 Algebraic topology. The branch of mathematics called *topology* was originally invented to deal with the problem of islands connected by bridges and the number of different ways by which one could get from one island to another. The bridges have been generalized to lines and the islands to junctions, and topology has become a very abstract field. Electrical networks are composed of branches with junctions between them. They have "topological" properties. Specifically, we are interested in topology as an aid in picking the algebraic unknowns we should use in solving a problem, and as a guide in writing independent equations to solve for these unknowns.

Nodes, branches, loops, and graphs. We have already had some occasion to use the concepts of branch and node. For use with topology, however, we must make our definitions somewhat more precise. Previously we used the term "branch" to represent any section of a network with a pair of terminals protruding from it. We wish now, however, to associate a branch with an algebraic unknown in the circuit, and for this reason we would like to consider each resistance as a separate branch. If it weren't for the presence of sources, we would simply define a branch for each resistance in the circuit. We would like, however, not to count resistances in parallel with voltage sources, or resistances in series with current sources. They do not represent unknowns in the algebraic sense. Sources themselves also represent known values in the circuit and are not counted as algebraic unknowns. Thus we define a *topological branch as a resistance which is left after the sources in the network have been reduced to zero.* Zero voltage sources become short circuits and short out any resistances in parallel with them, and zero current sources become open circuits and remove any resistances in series with them from the circuit. The resistance elements which are thus removed from the circuit are called "dummy" branches.* They do not affect any of the other currents or voltages in the circuit and can be ignored so far as the remainder of the network is concerned. Furthermore, the determination of the current or voltage in a dummy branch is a trivial problem which can be carried out at any time, independently of the rest of the network. In this text, unless specifically stated to the contrary, dummy branches will be excluded from consideration as topological branches.

We have previously thought of a node as any junction point in a network. We would now like to be more specific and to define a *topological node as a junction point between two resistors in a network in which all the sources have been reduced to zero.* It is often convenient to keep track of the branches and the nodes in an electrical network by drawing a *topological graph* in which the branches are represented by lines, and nodes by the junction points between the lines. Two examples of electrical networks and their topological graphs are shown in Fig. 2–20. In Fig. 2–20(a) the voltage source and the current source are "properly" applied with a series resistance and a shunt resistance respectively. In these circumstances the topological graph has very much the same form as the original network. In Fig. 2–20(b) the voltage source and the current source do not have the desired series and shunt resistances. Consequently, when they are removed from the circuit they take some of the other resistances

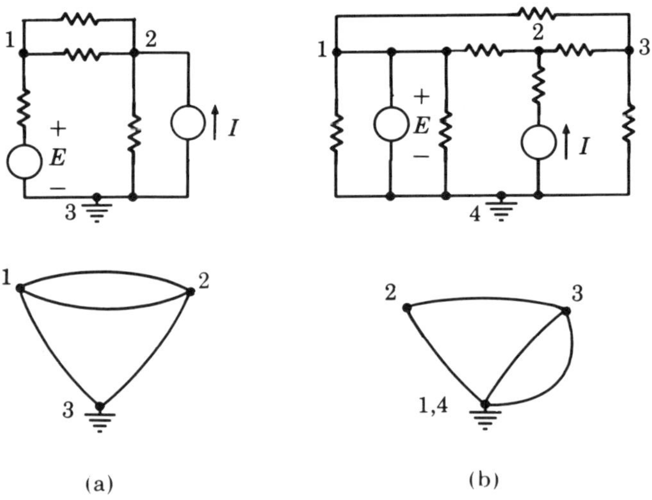

FIG. 2–20. Networks and their topological graphs.

* The sources are also dummy branches and are likewise excluded from topological considerations.

with them, and the topological graph is very different from the original network.

In addition to the branches and the nodes, which we have now defined precisely in a topological sense, we must introduce one more concept, that of the *independent path*, or *loop*. Our interest in closed paths arises from the fact that Kirchhoff's voltage law must be applied around closed paths. We are interested in having the resulting set of equations independent, and hence we are interested in independent paths. The equations will be independent if each one contains some information which is not contained in any other equation. A sufficient, but not necessary, condition for the independence of the loops is that they should each contain a branch which is not in any other loop. The equations will then be clearly independent.*

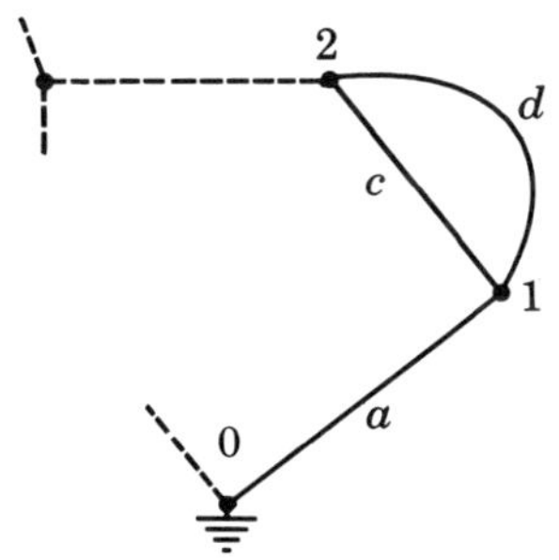

FIG. 2–21. The construction of a topological graph.

In order to find the number of independent loops in a network, we will construct it branch by branch, as shown in Fig. 2–21. We start with the ground or reference node at 0. The first branch which is added contributes one node, but no loops. This branch is marked a and the node 1. Branch c is then added. It contributes an additional node, 2, but no loops. Branch d connects nodes 2 and 1. It produces no new nodes, but it does produce a closed path, or loop, which is independent because it contains the new branch d.

A little reflection will show that if we continue this process we will obtain either a new node or a new independent loop every time we add a branch. If we remember to count the additional ground node with which we started, we have shown that the number of branches must be equal to

* A set of independent equations can be added and subtracted in various ways until a new set is obtained in which each equation contains each of the elements in the network. For such a set of equations it would be necessary to check the determinant to tell if they were independent.

the sum of the number of nodes plus the number of loops, less the ground node.* This result is the basic theorem of topology:

$$b = n + l - 1, \tag{2–14}$$

where b = number of branches, n = number of nodes, and l = number of independent loops.

Meshes. The number of branches and the number of nodes are obvious by inspection even without drawing a topological graph. The number of loops is a more difficult quantity to obtain. It can always be found from Eq. (2–14), but for a certain class of networks, called planar networks, it too can be obtained by inspection from the original network. A *planar network* is one which can be drawn flat on a sheet of paper without requiring that any lines leave the plane of the paper and cross over other lines. Examples of planar and nonplanar networks are shown in Fig. 2–22; (a) and (b) are planar and (c) is nonplanar. It should be observed in passing that the mere presence of a crossed wire is not indicative of a nonplanar network. It must be impossible to draw the network without it.

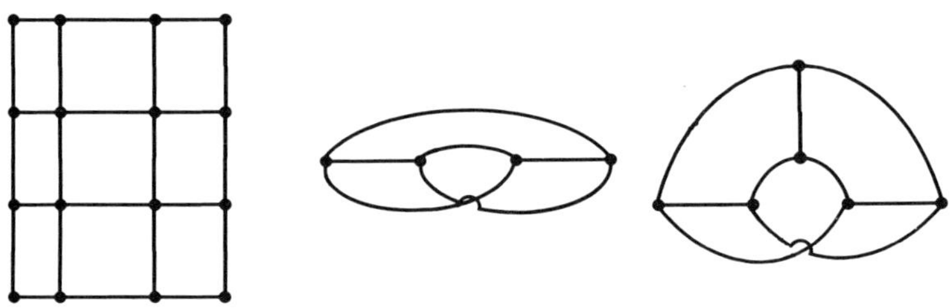

(a) (b) (c)

FIG. 2–22. Planar and nonplanar graphs.

Whenever a graph is planar a suitable set of independent loops can be obtained as the *meshes* in the graph. If the graph is viewed as a fishnet the meshes are the holes, or if the graph is viewed as a large window they are the individual panes of glass. To prove that the meshes are an independent set of loops we will construct a planar graph in a particular way, as shown

* A graph which contains only the node-producing branches and no loop-producing branches is called a "tree." The construction of a tree first would provide a possible but a less general proof of Eq. (2–14) than the one used here. The path of the independent loop is defined exactly by the tree and the added branch. A number of paths are usually possible in the alternative method used here.

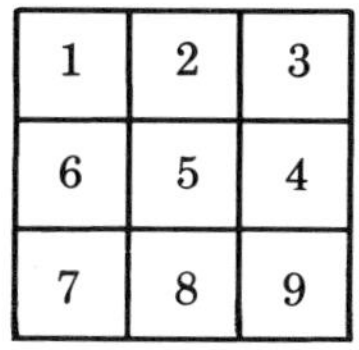

FIG. 2–23. Meshes as independent loops.

in Fig. 2–23. The meshes are added in the order 1, 2, 3, 4, 5, 6, 7, 8, 9. In this order it is obvious that the meshes are independent loops, because the completion of each successive mesh requires the addition of a new element which is not used in any of the previous meshes. It is further obvious that there are just the right number of meshes. Each time a branch is added in this process either a new node or one of the meshes is created. Since our general formula shows that each new branch must be associated with either a node or an independent loop, we are assured that the meshes are a possible selection of independent loops.

As an example of the use of the topological formula we will find the number of branches, meshes, and nodes for the graphs of Fig. 2–20. For the graph of Fig. 2–20(a), by inspection we find

$$b = 4, \qquad n = 3, \qquad l = 2,$$

and

$$b = n + l - 1 = 3 + 2 - 1 = 4.$$

For the graph of Fig. 2–20(b), the same values are obtained by inspection:

$$b = 4, \qquad n = 3, \qquad l = 2.$$

2–6 The $2b$ method of solving network problems. We shall now consider the algebraic solution of a network problem in which there are b branches. These topological branches are counted with the sources in the network dead, and hence each one represents a resistance with an unknown current and an unknown voltage. The total number of unknown algebraic variables in the problem is thus $2b$. To solve for these unknowns we have available the volt-ampere equations of the branches, Kirchhoff's current law, and Kirchhoff's voltage law. In this section we will show that these laws provide the correct number of independent equations to solve for the unknowns. Initially we will consider networks with real physical sources, but later we will extend the method to include resistanceless sources.

The unknown variables. We have defined topological branches as single resistances which remain when all of the sources become zero. In particular, resistances in parallel with voltage sources are shorted out, and resistances in series with current sources are removed from the network. There are thus two unknown quantities in every branch, a voltage and a current. For a circuit with b branches, there are $2b$ unknown variables. We could pick the unknowns as the currents and voltages in the resistors, but it is more convenient, as we shall see below, to pick them as the voltages and the currents associated with each of the branches. A source can then be included by virtue of its effect on the volt-ampere relationship of the branch. In summary,

$$\text{Number of variables} = 2b.$$

The volt-ampere equations. Let us assume that the voltage across each branch and the current entering each branch are the unknowns in our problem. The nature of the branch in each case imposes a single constraint on these variables. In the earlier sections of this chapter we examined the nature of this constraint equation for a number of resistive and source branches. For the sake of completeness, however, the volt-ampere equations are repeated here. For a single resistor,

$$e = iR.$$

For a voltage source in series with a resistor,

$$e = E + iR.$$

For a current source in parallel with a resistor,

$$e = IR + iR,$$

and for a combination of a voltage source E in series with a resistance R and a current source I in parallel with these,

$$e = E + IR + iR. \tag{2–15}$$

So long as we restrict ourselves to branches which contain single resistors the volt-ampere equation cannot be any more complicated than Eq. (2–15).*

Since there are b branches there must be b volt-ampere equations, and since each branch is different the equations are obviously independent. In summary,

$$\text{Number of independent volt-ampere equations} = b.$$

* The case of a resistanceless source can be handled by one of the source transformations discussed above or by methods discussed below.

The $\sum i$ *equations.* The Kirchhoff current law can be written at any single node in a network, but it can be written as well for any subsection of the network. The sum of the currents entering the subsection must be equal to the sum of the currents leaving the subsection. Such a subsection of the network represents a collection of nodes, and is termed a *supernode.* Not all of the current law equations written for the nodes and supernodes in a network will be independent. Our principal problem in this section will be to determine how many of these equations are independent, and to find an easy way to write a set of independent current law equations. Any one set of equations will be satisfactory because the others can be obtained from this set by addition and subtraction.

Let us imagine that we are constructing the network one branch at a time. The reference node and the node at the end of the first branch are not independent, since they involve the same current and no other currents. Each additional node which is added, however, provides a new opportunity for writing the current law, and each of these equations must be independent of the preceding ones, since it involves the branch which was added to create the new node. The addition of branches between existing nodes merely alters the current laws written at these nodes but does not provide an opportunity to write any new laws. Thus the number of independent current laws will be equal to the number of nodes in the network minus the original ground node. Not only do these nodes provide the correct number of current laws but they also provide a suitable set of locations for the application of the law. The Kirchhoff current law equations written at all of the nodes except the ground node will be an independent set. In summary,

$$\text{Number of independent current law equations} = n - 1.$$

The $\sum e$ *equations.* The Kirchhoff voltage law equation can be written around any closed path in a network, including even open-circuit paths through nonexistent branches. Not all of the equations, however, will be independent. In this section we will specify the number which are independent and at the same time indicate a method for writing them.

In the section on topology we defined the concept of the independent loop. In particular, for planar networks, a suitable set of independent loops was the set of meshes formed in the plane. The independent loops or meshes were characterized by the fact that each of them contained some information which was not in the others. One possible set was formed in which each mesh contained a branch which was not in any of the preceding meshes of the set. If the Kirchhoff voltage law equations are written around a set of independent loops or meshes they must be independent, because they will then contain new information in each equation. In a planar network the simplest set of closed paths which will yield an independent set of voltage law equations is the set of meshes. In a nonplanar network any set of independent loops will be satisfactory. In summary,

$$\text{Number of independent voltage law equations} = l.$$

The number of unknowns and the number of equations. The total number of unknowns in a network problem is $2b$, that is, b currents and b voltages. The number of volt-ampere equations which can be written for these unknowns is b. The number of $\sum i$ equations is $n - 1$, and the number of $\sum e$ equations is l. We have shown in the previous sections that these equations are independent. The total number of equations is

$$b + n - 1 + l.$$

By the basic theorem of topology we have shown that $b = n - 1 + l$ and hence the total number of equations is $2b$. We have thus been able to find $2b$ independent equations which we can write for the $2b$ unknowns in a network problem. The unknowns can therefore be found by the simple rules of algebra. It is interesting to observe that any method of solving network problems must explicitly or implicitly satisfy all the $2b$ equations. Initially we will consider a basic method, called the $2b$ method, in which all of the equations are explicitly written down. Later we will develop clever notations which will automatically satisfy some of the equations and require the explicit statement of only a very few of them.

The 2b method for normal sources. When the sources in a network are "normal" they can be included in the volt-ampere equations of the branches. By normal sources we mean voltage sources in series with a resistance, or current sources in parallel with a resistance. When the sources are abnormal, or resistanceless, we must resort to source transformations or to the use of dummy variables. In this section we shall consider the $2b$ method for a network in which the sources are normal, as shown in Fig. 2–24(a). The topological graph is shown in Fig. 2–24(b). In it there are three branches, two nodes, and two meshes. The number of unknown

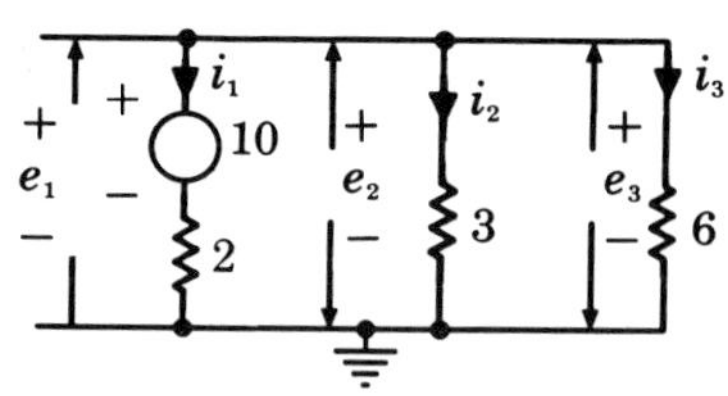

(a)

(b)

FIG. 2–24. An example solved by the $2b$ method.

variables is thus six, and the equations to be written for them include three volt-ampere equations, one $\sum i$ equation, and two $\sum e$ equations.

The variables (2b of them). The six variables are the three branch currents i_1, i_2, and i_3, and the three branch voltages e_1, e_2, and e_3. The current directions are assumed at random, but the voltage polarities are taken to agree with them.

The v-a equations (b, or 3, of them):

$$e_1 = 2i_1 + 10, \qquad e_2 = 3i_2, \qquad e_3 = 6i_3.$$

The $\sum i$ *laws* ($n - 1$, or 1, of them):

$$i_1 + i_2 + i_3 = 0.$$

The $\sum e$ *laws* (l, or 2, of them):

$$e_1 = e_2, \qquad e_2 = e_3.$$

Solution. The above six equations are independent and can be used to solve for the six unknowns, as follows:

From the $\sum i$ law:

$$i_1 + i_2 + i_3 = 0.$$

From the v-a equations:

$$\frac{e_1 - 10}{2} + \frac{e_2}{3} + \frac{e_3}{6} = 0.$$

From the $\sum e$ equations:

$$\frac{e_1}{2} + \frac{e_1}{3} + \frac{e_1}{6} = 5 \qquad \text{and} \qquad e_1 = 5.$$

The other values can be obtained from this one by the use of the original equations,

$$e_1 = e_2 = e_3 = 5,$$

$$i_1 = \frac{e_1 - 10}{2} = -\tfrac{5}{2}, \qquad i_2 = \tfrac{5}{3}, \qquad i_3 = \tfrac{5}{6}.$$

The principal problem associated with the 2b method is the algebraic solution of the simultaneous equations. The shear number of equations makes it difficult to see the most direct route to the answers. Another apparent problem arises when the sources are resistanceless, and we next consider this possibility.

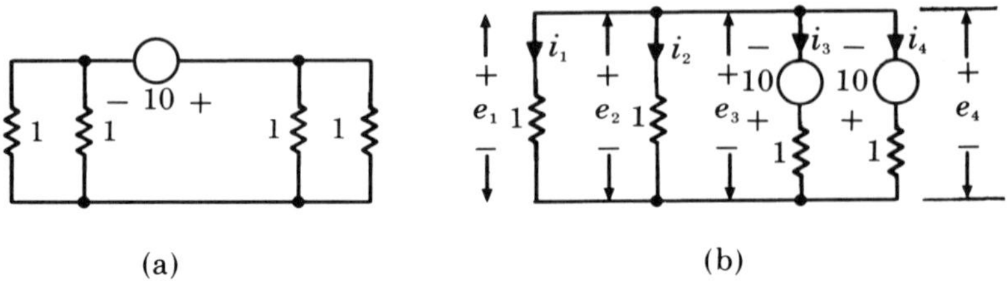

FIG. 2–25. The 2b method for a resistanceless voltage source.

The 2b method for resistanceless sources. An example in which a voltage source appears with no series resistance is shown in Fig. 2–25(a). The same network with a source transformation to move the voltage source through a node is shown in Fig. 2–25(b). The currents in all of the resistances in Fig. 2–25(b) and the voltages across them will have the same values as in (a), but a node has been removed and the current in the original source cannot be found directly. For the network of Fig. 2–25(b),

$$b = 4, \qquad n = 2, \qquad l = 3.$$

The variables are

$$i_1, i_2, i_3, i_4, \qquad \text{and} \qquad e_1, e_2, e_3, e_4.$$

The equations are

$$\begin{aligned} &\text{v-a:}\quad e_1 = i_1, && \Sigma e:\quad e_1 = e_2, && \Sigma i:\quad i_1 + i_2 + i_3 + i_4 = 0.\\ &\qquad e_2 = i_2, && \qquad e_2 = e_3, &&\\ &\qquad e_3 = i_3 - 10, && \qquad e_3 = e_4. &&\\ &\qquad e_4 = i_4 - 10. && && \end{aligned}$$

The solution is

$$\begin{aligned} i_1 + i_2 + i_3 + i_4 &= 0,\\ e_1 + e_2 + e_3 + 10 + e_4 + 10 &= 0,\\ 4e_1 &= -20,\\ e_1 &= -5, \end{aligned}$$

from which

$$e_1 = e_2 = e_3 = e_4 = -5,$$
$$i_1 = -5, \qquad i_2 = -5, \qquad i_3 = 5, \qquad i_4 = 5.$$

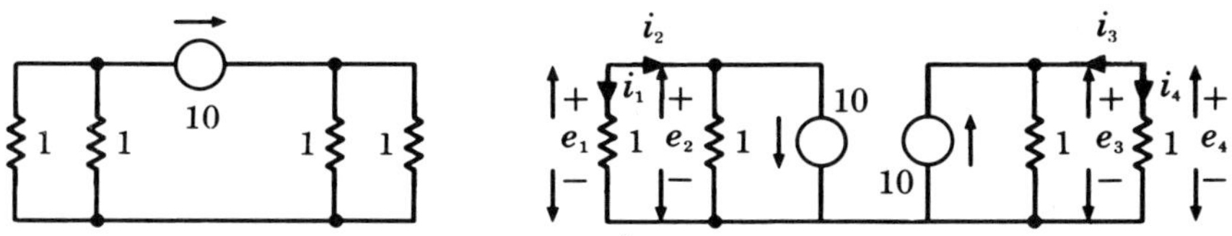

(a) (b)

FIG. 2–26. The 2*b* method for a resistanceless current source.

A second example containing a resistanceless source is shown in Fig. 2–26(a). The same network with a current source transformation that bridges two current sources around a loop is shown in Fig. 2–26(b). The currents and the voltages in all of the resistances will be the same in the two networks, but of course the mesh formed by the 10-amp source in Fig. 2–26(a) is not present in Fig. 2–26(b). Each of the sources in part (b) can be associated with a resistance branch, and the problem can be solved by the usual 2*b* method. In this case,

$$b = 4, \qquad n = 3, \qquad l = 2.$$

The equations are

$$\begin{aligned} \text{v-a:}\quad e_1 &= i_1, & \textstyle\sum e:\quad e_1 &= e_2, & \textstyle\sum i:\quad i_1 + i_2 &= 0,\\ e_2 &= i_2 - 10, & e_3 &= e_4. & i_2 + i_3 &= 0.\\ e_3 &= i_3 + 10, & & & &\\ e_4 &= i_4. & & & & \end{aligned}$$

The solution is

$$i_1 + i_2 = 0, \qquad \text{and} \qquad i_3 + i_4 = 0,$$

$$e_1 + e_2 + 10 = 0, \qquad e_3 + e_3 - 10 = 0,$$

from which from which

$$e_1 = -5 = e_2. \qquad e_3 = e_4 = 5.$$

There is little point in finding the branch currents for Fig. 2–26(b), since they will be different from the currents in the resistances of the original problem. The original currents can be obtained directly from the branch voltages.

The dummy-variable method for resistanceless sources. An alternative method to the source transformations described above can be used to solve problems with resistanceless sources. It is called *the method of dummy variables* because it uses branches, nodes, and meshes which do not exist in the true topological sense. A voltage source without any resistance in series with it forms a bridge across two of the nodes in a circuit. One of these two nodes is a *dummy node*, which ceases to exist when the voltage source is reduced to zero. On the other hand, with the source active the dummy node is present and an additional Kirchhoff current law equation can be written for it which will allow us to solve for the unknown current in the source. The source itself is treated as a dummy branch. It is a separate branch because it no longer appears in the volt-ampere equations of one of the other branches, but it is a dummy branch because the voltage across it is a known quantity and no volt-ampere equation can be written for it. As an example of the dummy variable method for a voltage source with no resistance, consider the circuit of Fig. 2–25(a). In it, geometrically, there are four topological branches and one dummy branch, two topological nodes and one dummy node, and three independent meshes. There are six unknown currents and voltages in the topological branches, and one unknown current in the source branch. The regular six $2b$ equations suffice to solve for the topological unknowns, and the added Kirchhoff current law at the dummy node completes the solution for the current in the source branch. The equations are:

$$
\begin{array}{llll}
\text{v-a:} & e_1 = i_1, \quad & \sum i: \; i_1 + i_2 = i_s, \quad & \sum e: \; e_1 = e_2, \\
& e_2 = i_2, & \qquad i_3 + i_4 = -i_s. & \qquad e_3 = e_4, \\
& e_3 = i_3, & & \qquad e_3 - e_2 = 10. \\
& e_4 = i_4. & &
\end{array}
$$

The solution is

$$i_1 + i_2 + i_3 + i_4 = 0,$$

from which

$$e_1 + e_2 + e_3 + e_4 = 0$$

and

$$2e_2 + 2e_3 = 0,$$

but

$$-e_2 + e_3 = 10,$$

from which

$$e_3 = 5 = e_4$$

and

$$e_2 = e_1 = -5, \text{ as found previously.}$$

A similar situation exists for a current source bridged directly across two nodes in a circuit with no resistance in parallel. When the current source is made zero for the topological diagram it simply disappears and the two nodes remain. The apparent mesh created by the current source, also disappears. It is a *dummy mesh* and not a true topological mesh. We can use the dummy mesh to obtain an additional Kirchhoff voltage law equation which will allow us to introduce an additional variable, namely the voltage across the current source. An example is shown in Fig. 2–26(a). In this problem there are four real branches and one dummy source branch, two topological meshes and one dummy mesh, and three nodes. There are eight current and voltage variables in the branches and an additional voltage across the current source. To solve for these variables we have the regular $2b$ equations plus an additional voltage law equation around the dummy mesh. The equations are:

$$\begin{array}{llll} \text{v-a:} & e_1 = i_1, \quad & \sum i: \; i_1 + i_2 = -10, \quad & \sum e: \; e_1 = e_2, \\ & e_2 = i_2, & \qquad\; i_3 + i_4 = 10. & \qquad\;\; e_3 = e_4, \\ & e_3 = i_3, & & \qquad\;\; e_3 - e_4 = e_s. \\ & e_4 = i_4. & & \end{array}$$

The solution is

$$\begin{array}{lcl} i_1 + i_2 = -10, & \text{and} & i_3 + i_4 = 10, \\ \text{from which} & & \text{from which} \\ e_1 + e_2 = -10, & & e_3 + e_4 = 10, \\ e_1 = e_2 = -5. & & e_3 = e_4 = 5. \end{array}$$

From these values the source voltage e_s can be found. It is

$$e_s = e_3 - e_2 = 5 + 5 = 10 \text{ volts.}$$

The advantage of the dummy-variable method as compared with the source transformation method is that the former leaves the network unaltered while the latter changes it in some respect. One is likely to forget that the network has been changed by the source transformation and to make errors on that account. Both methods are perfectly general, however, and either one can be applied to any problem in network analysis.

2–7 Summary.

Resistors in series $\quad R = R_1 + R_2 + R_3 + \cdots$

Resistors in parallel $\quad R = \dfrac{1}{1/R_1 + 1/R_2 + 1/R_3 + \cdots}\cdot$

Source branches

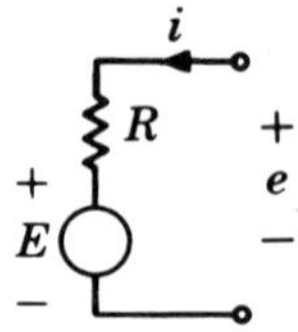

$$e = E + iR$$
("dead" voltage = short)

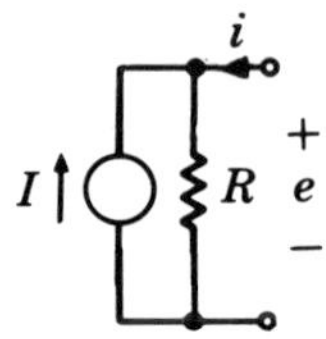

$$e = IR + iR$$
("dead" current = open)

Source transformations

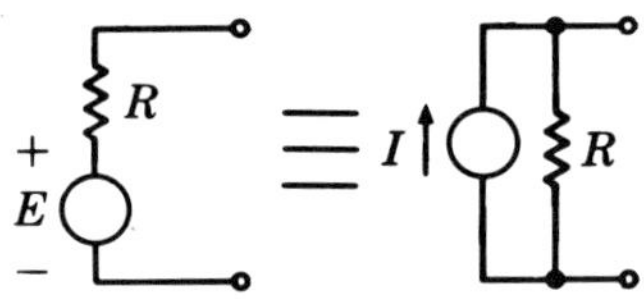

where $E = IR$ or $I = E/R$

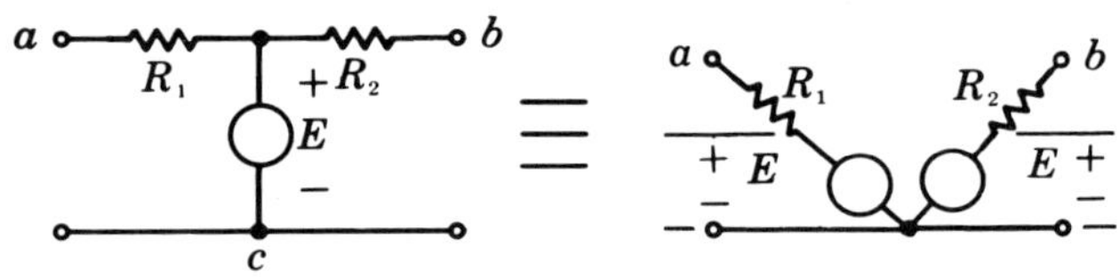

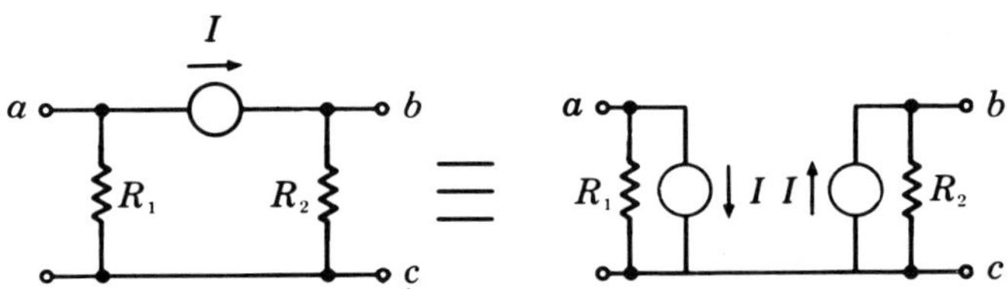

The voltage divider

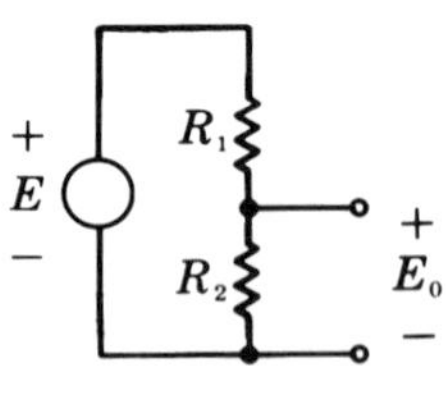

$$E_0 = \frac{R_2}{R_1 + R_2} E$$

The current divider

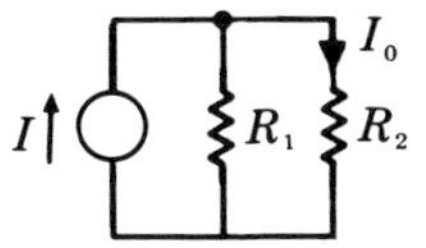

$$I_0 = \frac{R_1}{R_1 + R_2} I$$

Algebraic topology

No. of branches $= b$ (all sources dead)
No. of nodes $= n$
No. of independent nodes $= n - 1$
No. of independent loops $= l$

Basic equation $b = n + l - 1$

The 2b method

No. of unknowns $= 2b$ (b currents and b voltages)

No. of volt-ampere equations $= b$
No. of independent i laws $= n - 1$
No. of independent e laws $= l$
Total number of equations $= b + n - 1 + l = 2b$

REFERENCES

In this book a clear distinction has been made between those problems which are solved by arithmetic and those problems which are solved by algebra. The arithmetic methods of solving circuit problems are covered in almost all elementary circuit theory books. The algebraic methods of solving circuit problems were known to Kirchhoff in 1847, but are covered adequately in very few elementary circuit theory books. References for network algebra and network topology are listed below.

1. G. Kirchhoff, "Über die Auflösung der Gleichungen auf welche man bei der Untersuchung der linearen Vertheilung galvanischer Ströme geführt wird," *Ann. Phys. u. Chem.*, **2,** 72 (1847). Kirchhoff applied mathematical topology to the solution of network problems in 1847. Modern writers usually just rediscover his results.

2. G. Kron, *Tensor Analysis of Networks,* John Wiley & Sons, New York, 1939. Kron was the first of the modern writers to recognize the importance of topology in network theory. He presents the 2b method as the basic or primitive method for networks. The tensor notation which he uses has prevented a wider understanding of his work.

3. P. Le Corbeiller, *Matrix Analysis of Electric Networks,* John Wiley & Sons, New York, 1950. Le Corbeiller translates Kron's work into more usual electrical notation. The level is a little above most undergraduates.

4. E. A. Guillemin, *Introductory Circuit Theory,* John Wiley & Sons, New York, 1953. Chapters 1 and 2 of Guillemin's text present topology in very much the same notation as that of Kron. The level is such that good sophomore students can read the material.

5. Gardner and Barnes, *Transients in Linear Systems,* John Wiley & Sons, New York, 1942. This standard graduate-level text presents a clear introduction to circuit topology. It is singled out because *it treats sources as separate branches* and modifies the topological formulas accordingly.

6. J. L. Synge, "The Fundamental Theorem of Electrical Networks," *Quarterly of Applied Math,* Vol. 9, pp. 113–127 (1951). The basic theorem of topology is presented clearly by a master mathematician.

EXERCISES*

Section 2–2

1. Two resistors, R_1 and R_2, are connected in series. Give an equivalent single resistor when R_1 and R_2 have the following values:

(a) $R_1 = R_2 = 1$ (b) $R_1 = R_2 = 10^6$

(c) $R_1 = 1, \quad R_2 = 0$ (d) $R_1 = 1, \quad R_2 = \infty$

(e) $R_1 = 3, \quad R_2 = 6$

2. Two resistors, R_1 and R_2, are connected in parallel. Give an equivalent single resistor when R_1 and R_2 have the following values:

(a) $R_1 = R_2 = 1$ (b) $R_1 = R_2 = 10^6$

(c) $R_1 = 1, \quad R_2 = 0$ (d) $R_1 = 1, \quad R_2 = \infty$

(e) $R_1 = 3, \quad R_2 = 6$

3. Find an expression for the resistance of three resistors in parallel in the form of a single quotient. Give the equivalent parallel resistance when R_1, R_2, and R_3 have the following values:

(a) $R_1 = R_2 = R_3 = 3$ (b) $R_1 = R_2 = R_3 = 1$

(c) $R_1 = R_2 = 2, \quad R_3 = 1$ (d) $R_1 = 3, \quad R_2 = 6, \quad R_3 = 2$

(e) $R_1 = 1, \quad R_2 = 2, \quad R_3 = 3$

---

* Numerical values are in ohms, volts, and amperes.

4. By using series-parallel combinations, reduce the branch in Fig. 2–27 to a single resistance at terminals a–b.

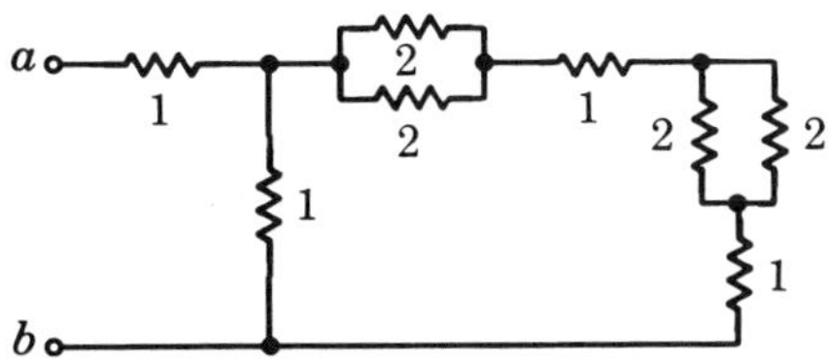

FIGURE 2–27

5. Reduce the combination of resistances in Fig. 2–28 to a single equivalent value at terminals a–b.

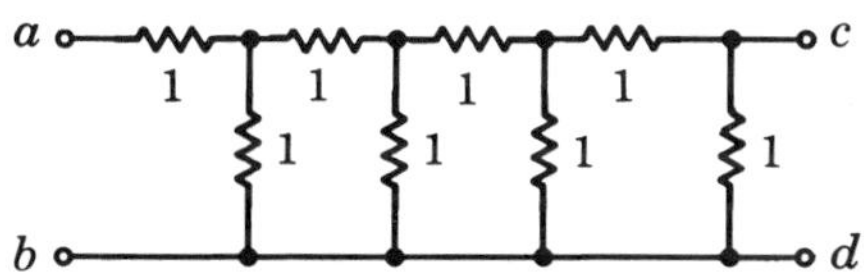

FIGURE 2–28

6. Reduce the combination of resistances in Fig. 2–28 to a single resistance at terminals c–d.

7. Find the input resistance of the circuit of Fig. 2–29 at terminals a–b.

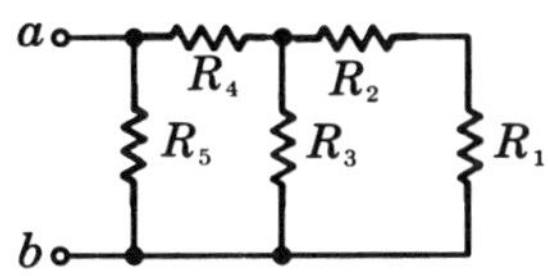

FIGURE 2–29

Section 2–3

8. Write the volt-ampere equations of the source branches in Fig. 2–30.

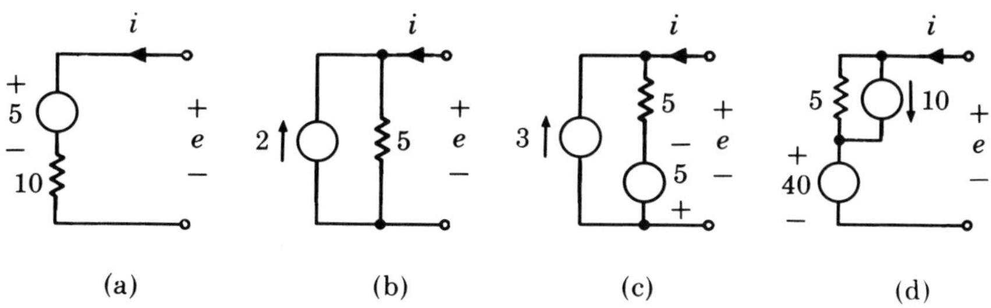

FIGURE 2–30

9. Reduce the branches in Fig. 2–31 to single sources.

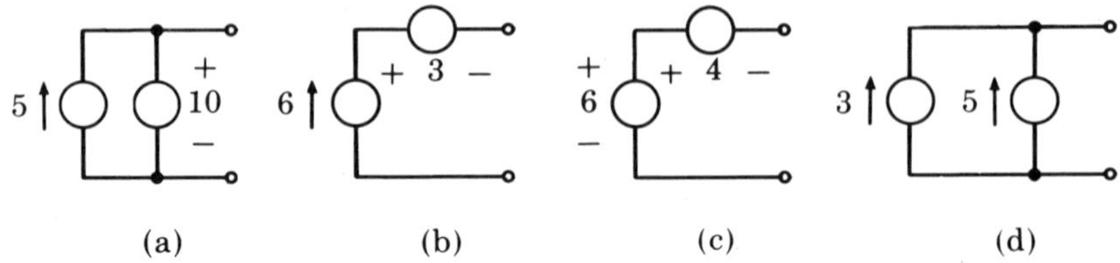

FIGURE 2–31

10. Convert the voltage source branches in Fig. 2–32 to equivalent current source branches.

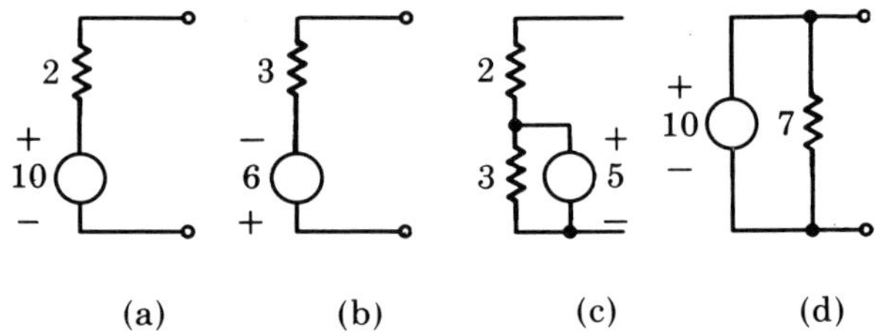

FIGURE 2–32

11. Convert the current source branches in Fig. 2–33 to equivalent voltage source branches.

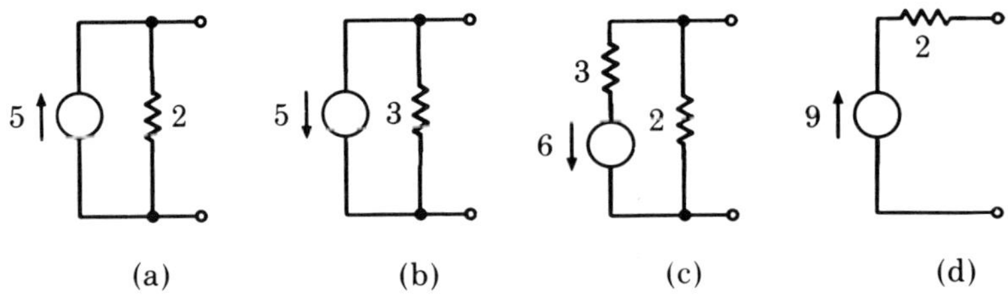

FIGURE 2–33

12. Convert the branches in Fig. 2–34 to single voltage sources in series with single resistors.

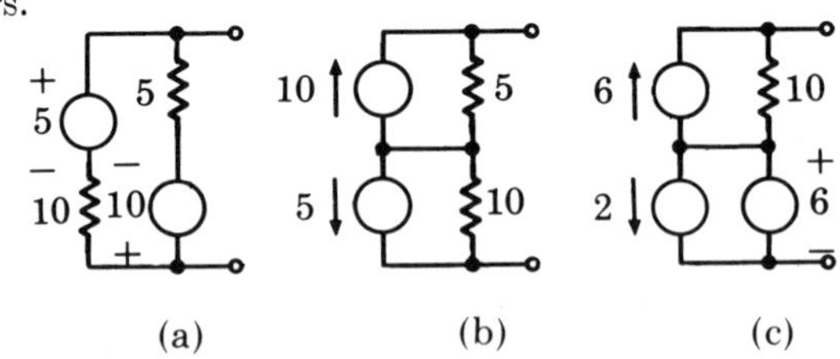

FIGURE 2–34

13. Convert the branches in Fig. 2–35 to single current sources in parallel with single resistors.

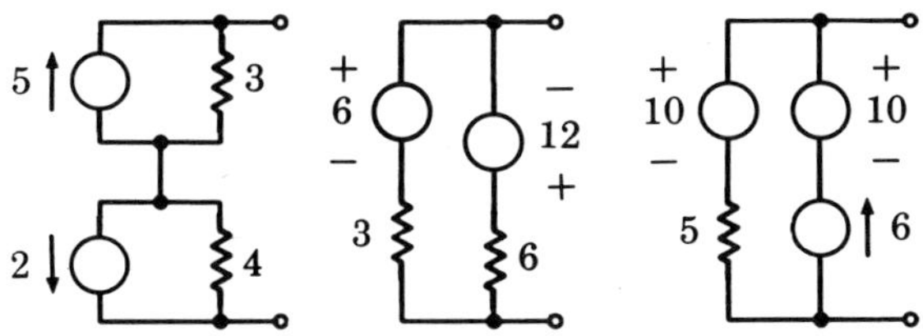

(a) (b) (c)

FIGURE 2–35

14. Without changing the topology of the network (i.e., without combining resistances) reduce the excitation to a single source in the network of Fig. 2–36.

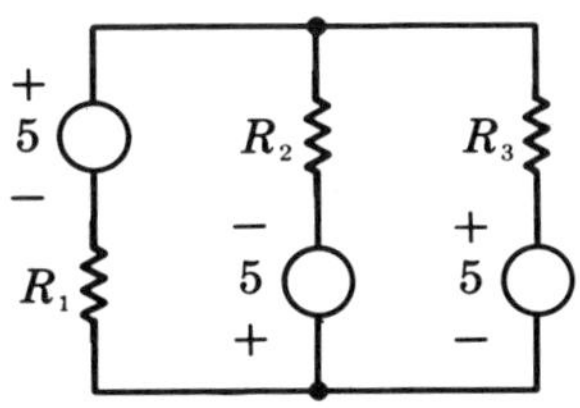

FIGURE 2–36

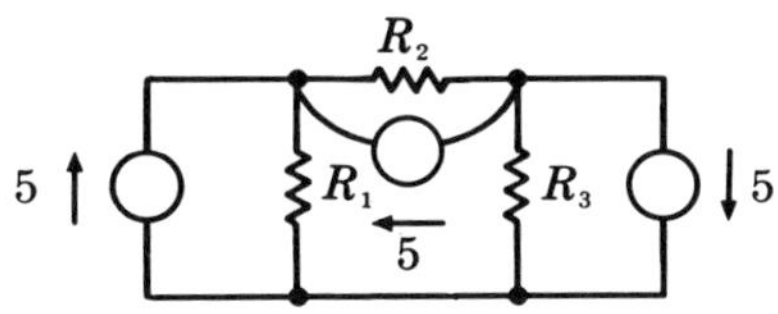

FIGURE 2–37

15. Without combining any of the resistance branches, reduce the excitation in the circuit of Fig. 2–37 to a single source.

16. Without combining any of the resistance branches, reduce the networks of Fig. 2–38 to forms with only one source.

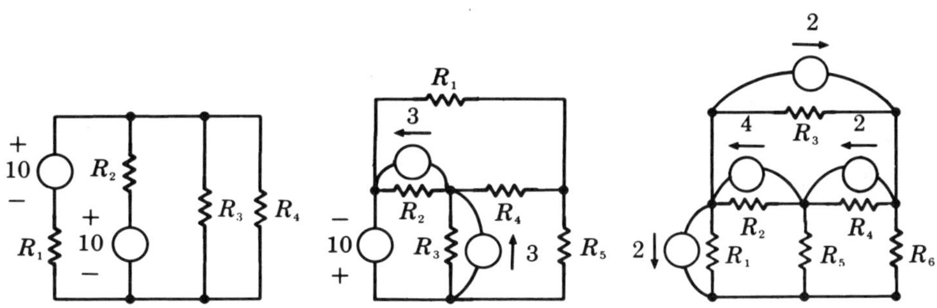

(a) (b) (c)

FIGURE 2–38

Section 2–4

17. Three resistors, $R_1 = 10$, $R_2 = 5$, and $R_3 = 5$ ohms, are connected in series across a 100-volt source. What is the voltage across R_3?

18. Three resistors, $R_1 = 10$, $R_2 = 10$, and $R_3 = 5$ ohms, are connected in parallel across a 10-amp source. What is the current through the 5-ohm resistor?

19. Without writing any algebraic equations, find the voltage e_0 in the circuit of Fig. 2–39.

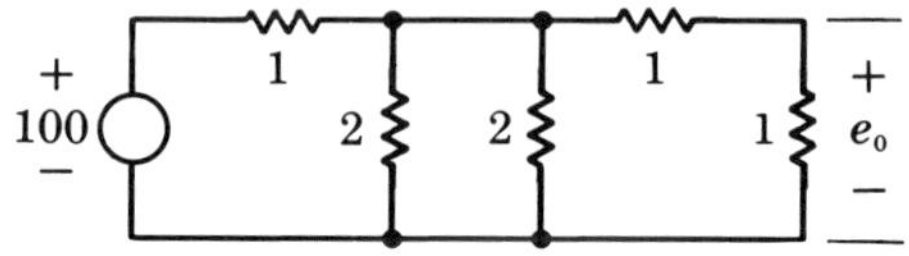

FIGURE 2–39

20. Without writing any algebraic equations, find the current i_0 in the circuit of Fig. 2–40.

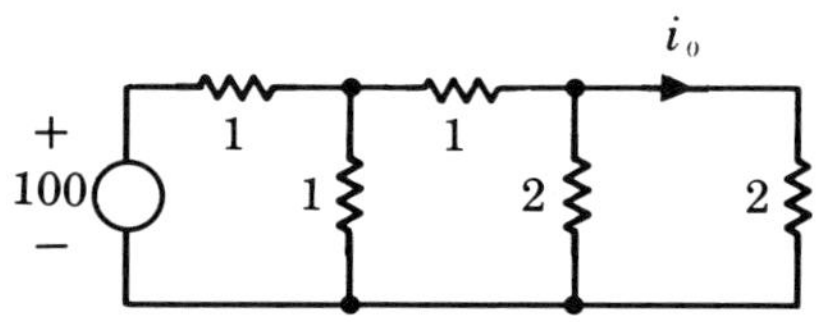

FIGURE 2–40

21. By inspection, find the voltage e_0 in the circuit of Fig. 2–41.

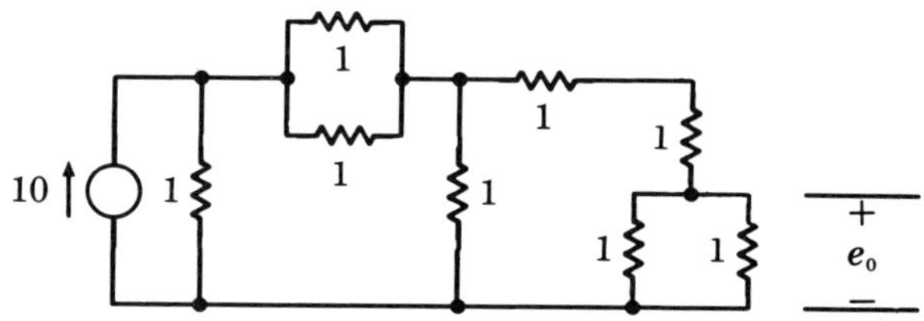

FIGURE 2–41

22. Find e_0 in the circuit of Fig. 2–42.

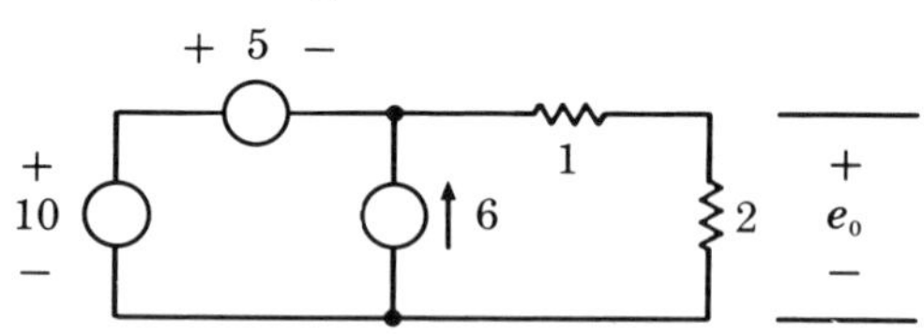

FIGURE 2–42

23. Find e_0 in the circuit of Fig. 2–43.

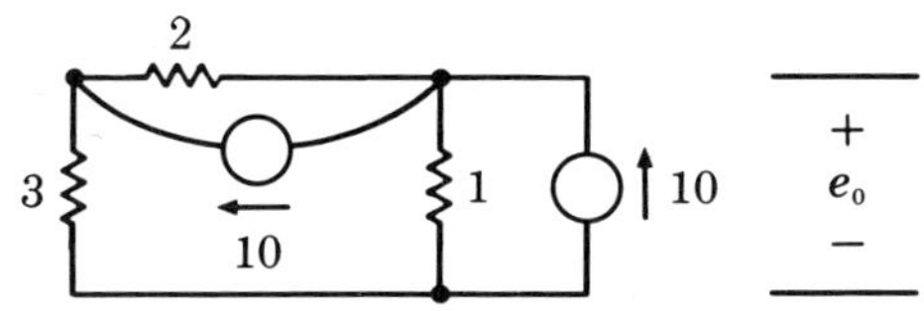

FIGURE 2–43

24. Find e_0 in the circuit of Fig. 2–44.

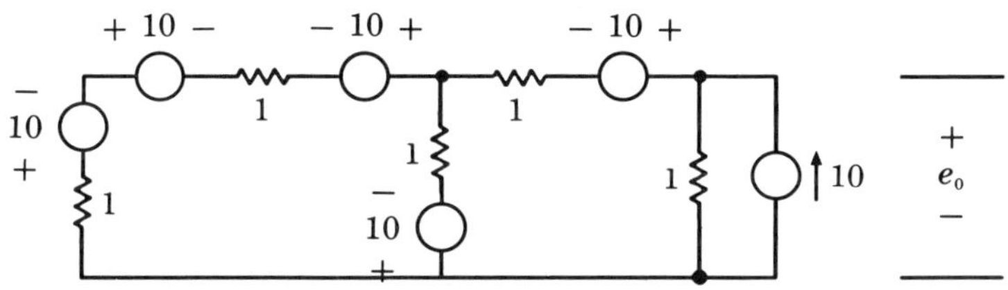

FIGURE 2–44

Section 2–5

25. For the circuit of Fig. 2–45, give the number of branches, the number of nodes, the number of loops, the number of current equations, the number of voltage equations, and the number of volt-ampere equations.

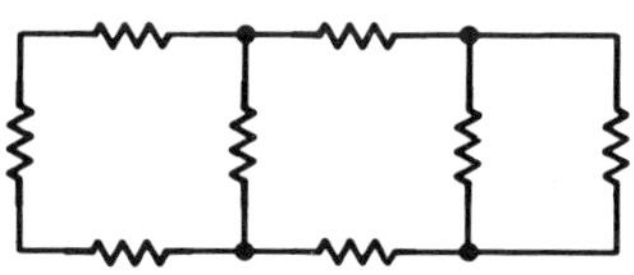

FIGURE 2–45

26. For the circuit of Fig. 2–46, give the number of branches, loops, nodes, current equations, voltage equations, and volt-ampere equations.

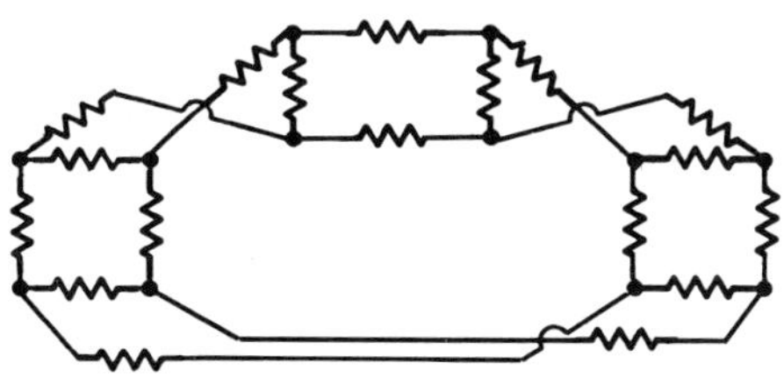

FIGURE 2–46

27. For the circuit of Fig. 2–47, give the number of branches, loops, nodes, current equations, voltage equations, and volt-ampere equations.

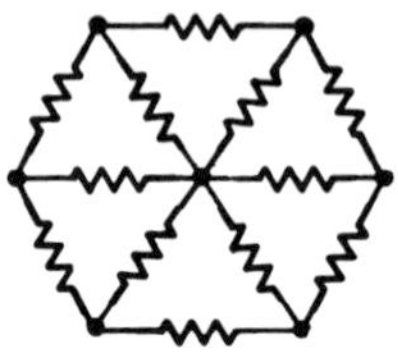

FIGURE 2–47

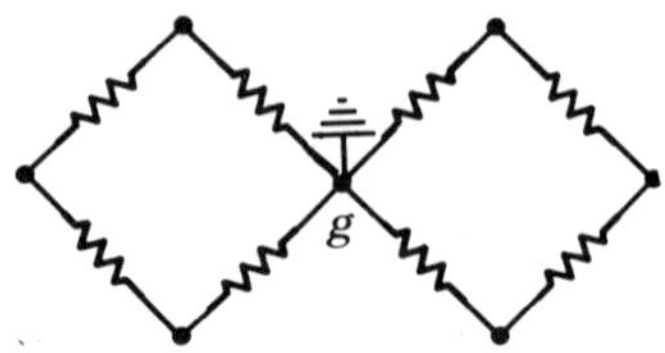

FIGURE 2–48

28. The circuit in Fig. 2–48 has two parts joined at the single node g, but otherwise separate. Find b, n, and l for this network, and show that the basic formula of topology holds. How many times is the node g counted?

29. Which of the four networks of Fig. 2–49 are planar? Redraw those that are planar in a form in which no wires are crossed.

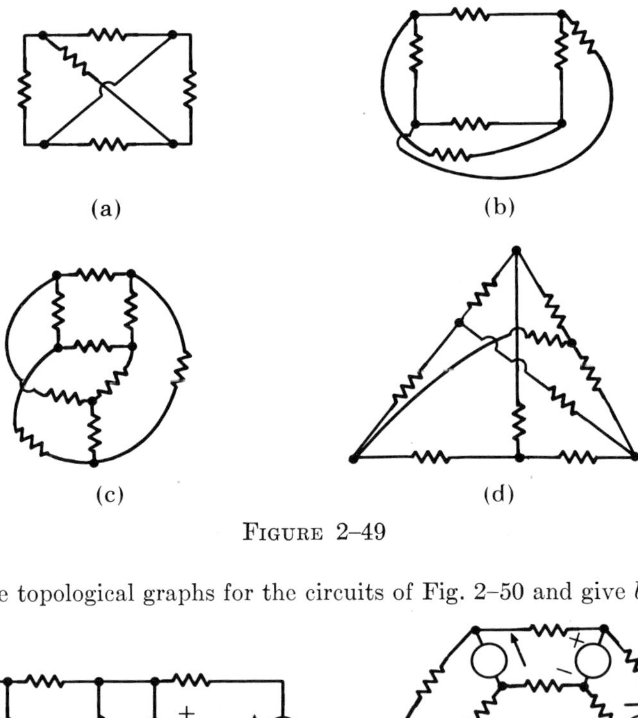

FIGURE 2–49

30. Make topological graphs for the circuits of Fig. 2–50 and give b, n, and l.

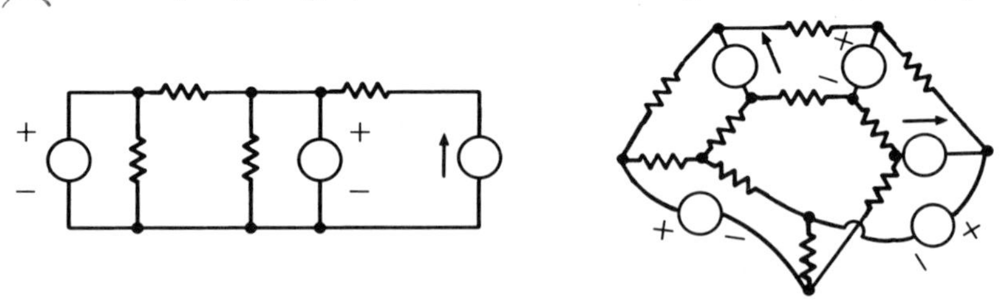

(a) (b)

FIGURE 2–50

31. Make topological graphs of the circuits of Fig. 2–51 and give b, n, and l.

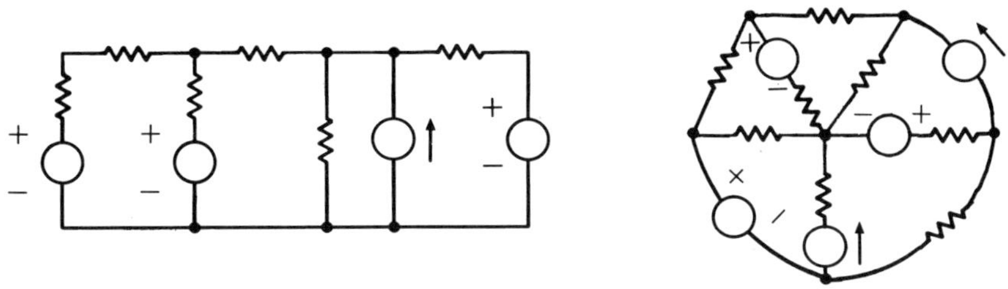

(a) (b)

FIGURE 2–51

32. Make a topological graph for the circuit of Fig. 2–52 and give the values of b, n, and l. Some of the resistances and all of the sources can be considered to be dummy branches. How many dummy branches are there, and how many dummy meshes and dummy nodes are created by these branches? Show that the basic law of topology is applicable to the totality of branches, meshes, and nodes, including the dummy ones.

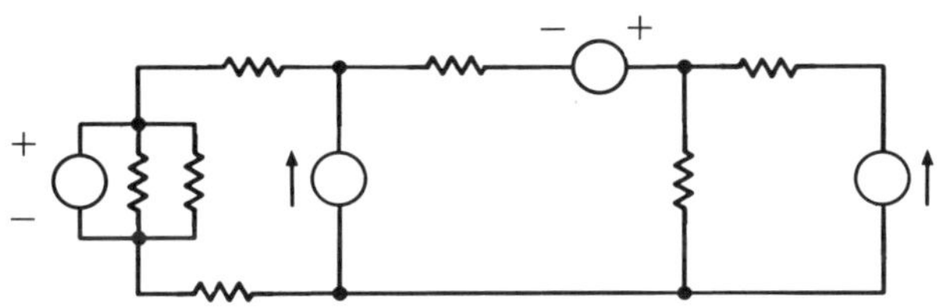

FIGURE 2–52

33. How many topologically independent branches, loops and nodes exist in the circuit of Fig. 2–53? How many dummy branches, loops, and nodes? Verify the basic formula of topology for the totality of branches, meshes, and nodes, including the dummy ones. Find the current in each resistance.

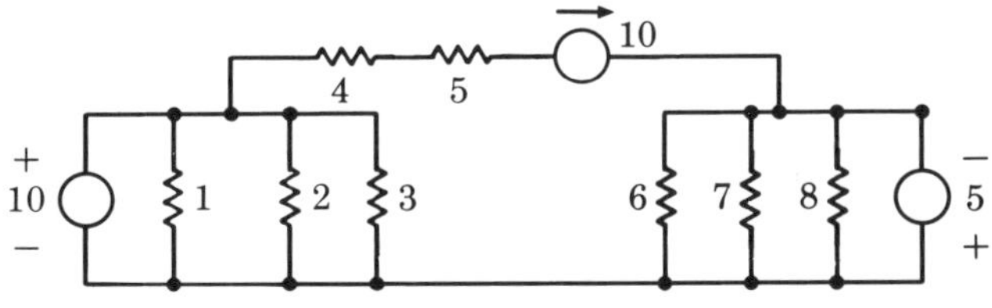

FIGURE 2–53

Section 2–6

34. For the network of Fig. 2–54, find the number of branches, nodes, and loops. Write the $2b$ current equations, voltage equations, and volt-ampere equations, and solve for all the unknown voltages and currents.

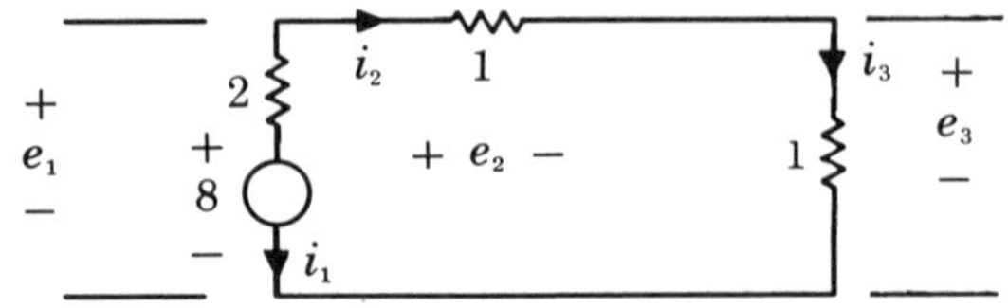

FIGURE 2–54

35. By the $2b$ method, find all the currents and all the voltages in the network of Fig. 2–55.

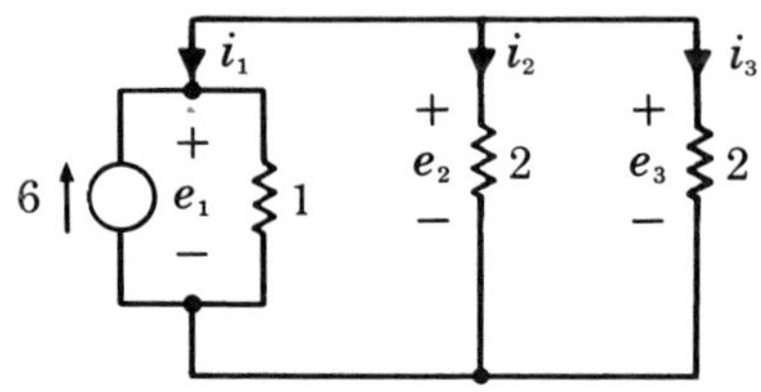

FIGURE 2–55

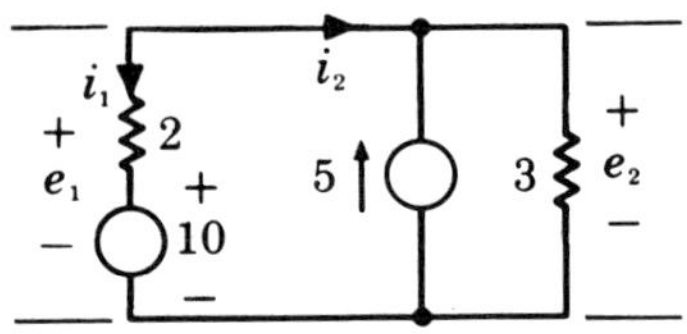

FIGURE 2–56

36. Solve for the currents i_1 and i_2 and the voltages e_1 and e_2 in the network of Fig. 2–56.

37. Make a source transformation which associates the voltage source in the circuit of Fig. 2–57 with two of the resistors. Use the $2b$ method on the resultant circuit to solve for the current in each of the resistances. Also find the current through the source in the original network.

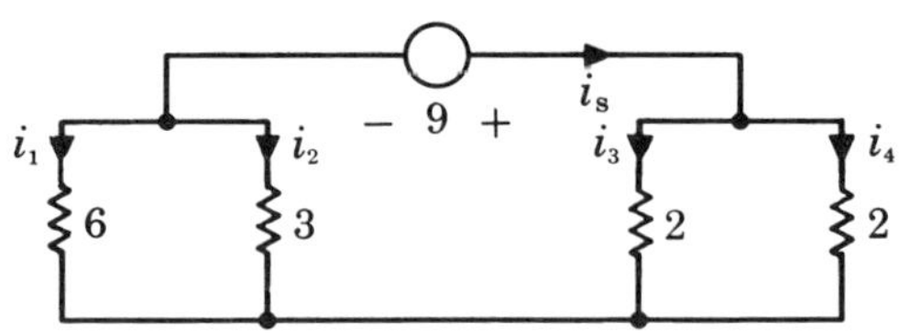

FIGURE 2–57

38. Solve Exercise 37 by treating the voltage source as a dummy branch and writing an extra current law equation at the dummy node which is then created.

39. Make a source transformation which associates the current source with each of a pair of the resistances in the circuit of Fig. 2–58. Use the $2b$ method

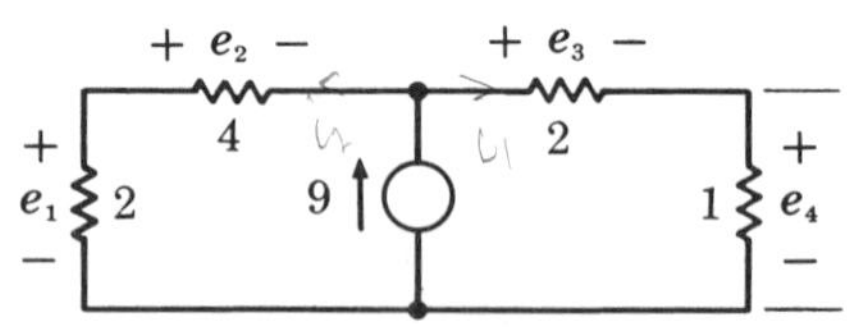

FIGURE 2–58

to solve the resultant network for the voltages across each of the resistances. Also find the voltage across the current source in the original circuit.

40. Solve Exercise 39 by treating the current source as a dummy branch and writing an extra voltage law equation around the dummy mesh thereby created.

41. After making the appropriate source transformations, solve the circuit of Fig. 2–59 for all the currents and voltages by the $2b$ method.

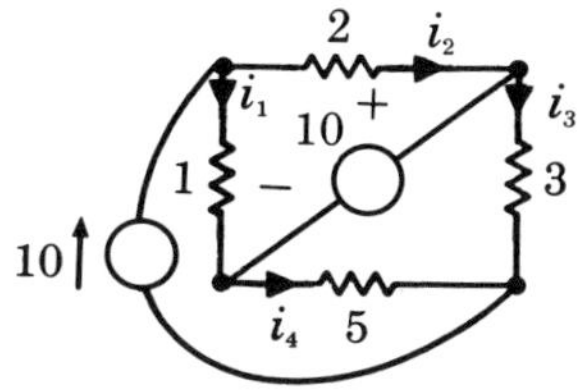

FIGURE 2–59

42. Solve Exercise 41 by treating the sources as dummy branches.

CHAPTER 3

MESH CURRENTS AND NODE VOLTAGES

"Brain succeeds better than brawn."

Ovid, Metamorphoses 111–54

3–1 Introduction. For many centuries it has been recognized that an intellectual approach to a problem is superior to a brute-force approach. The $2b$ method falls in the class of brute-force approaches. It can be used to solve any network problem, but it is laborious because of the great number of equations involved. We can improve on the $2b$ method in two ways. The first is by means of an *efficient notation.* Anyone who has tried to multiply Roman numerals will appreciate the value of an efficient notation! In circuit theory we can choose our variables so cleverly that we can satisfy many of the $2b$ equations simply by virtue of the notation. The second method of reducing labor is by means of an *orderly procedure.* By writing the equations in a systematic fashion we can satisfy many of the $2b$ equations by inspection. As a result of an efficient choice of variables and an orderly procedure for writing down the equations, we can reduce to a bare minimum the number of equations that are written explicitly.

Two methods which meet the requirements of the preceding paragraph are in general use. They are called the *mesh-current method* and the *node-voltage method.* Mesh currents were introduced by Maxwell and have been in use since his time. A *mesh current* is a component of current which circulates in a closed path (like a snake swallowing its tail). It automatically satisfies Kirchhoff's current law, since it both enters and leaves each junction in its path. A *node voltage* is the voltage of a point with respect to some reference point or datum. It is analogous to the elevation of a point above sea level. Any closed path must start and end at the same elevation above sea level, or at the same voltage above datum. Kirchhoff's voltage law is thus automatically satisfied by the node-voltage notation.

The equations which result from the use of mesh-current or node-voltage variables are often used for numerical calculations. In addition, they are convenient for theoretical analysis. As an example of this latter use we will discuss the principle of *duality,* the *linearity theorem,* and the *superposition theorem.* Two networks are said to be *dual* when the mesh-current equations for the first network are numerically identical to the node-voltage equations for the second network. Networks are *linear* when the output response is directly proportional to the input stimulus. When networks are linear, the *superposition* theorem can be used and the response for a number of simultaneous stimuli can be obtained as the sum of the responses which each would produce acting separately.

The importance of linearity and superposition lies in the fact that the most general network analysis for linear networks need only consider a single source and a single response. The relationship between the stimulus and the response characterizes the network for all types of stimuli.

3–2 Mesh currents. The $2b$ equations are basic to any formulation of the algebraic or equilibrium equations for a network. All of the $2b$ equations must be satisfied but not all of them must be stated explicitly. It is possible, for example, to use only the currents as variables and to satisfy the volt-ampere equations by inspection. The number of equations explicitly stated is then reduced to b. If, in addition, we were able to combine the branch currents into a new set of variables which would automatically satisfy the Kirchhoff current law at each of the $n - 1$ independent nodes in the network, we would need to state only $b - n + 1 = l$ of the original $2b$ equations. Since we need an unknown variable for each of these equations, we need l current variables, or one for each independent loop or mesh in the problem. An obvious possibility for the unknowns is the set of l circulating currents which traverse the independent paths.* These components of current close on themselves and in doing so automatically satisfy Kirchhoff's current law, since they enter and leave each node in their paths. Furthermore, since these component currents circulate around independent paths they must be themselves independent. From the above arguments we can conclude that the l circulating currents which flow around the contours of a set of independent loops are a suitable set of independent variables to use in a network problem, and that the l Kirchhoff voltage-law equations provide the means for solving for them.

If the graph is planar, the most obvious set of independent paths is the set of meshes. The mesh currents may be supposed to circulate around the "panes of glass" which fit into the "window panes" of the network. It is so convenient to use the meshes that we will restrict ourselves to mesh currents in most of the following discussion. The method is more general, however, and could be applied to any set of independent loops.

Most readers will be familiar with the resolution of forces and velocities into components. The mesh currents should be viewed as similar resolu-

* An infinite number of other possibilities also exist, but very little use has ever been made of them in network theory. Any combination of branch currents could be a variable. The only restrictions are that the variables used should be independent and equal in number to the equations available for their solution.

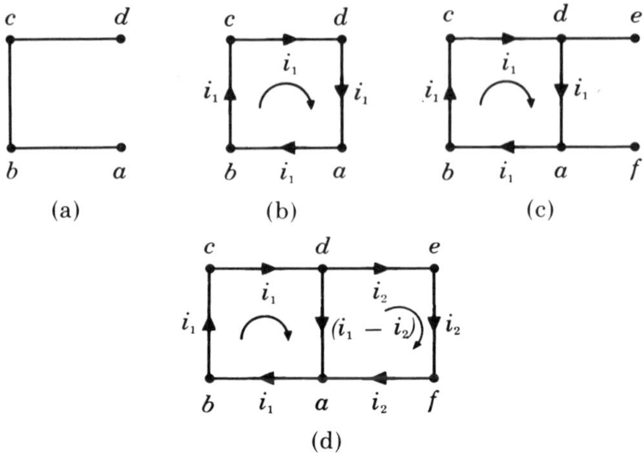

FIG. 3–1. The concept of a circulating current.

tions. They do not represent actual physical currents in a network, but only their mathematical counterparts. To emphasize this component concept it is instructive to build up a set of mesh currents from the branch currents, by means of Kirchhoff's current laws.

In the network of Fig. 3–1(a) a branch current could be defined in each of the three branches. The application of the current law at nodes b and c shows that the three branch currents are equal, and the application of the current law at node d shows that they are all zero. When the loop is completed as in Fig. 3–1(b) a current i_1 can flow around it. One point of view would be to say that the application of the current law at nodes b, c, and d showed that all the branch currents were equal. An alternative point of view would be to state that the current i_1 was a circulating current flowing around the mesh and automatically satisfying the current law at each node on the way. The addition of branches de and af in Fig. 3–1(c) does not change the situation, since no current can flow in either of these branches. The completion of a second loop, however, as in Fig. 3–1(d), allows a second current i_2 to flow in branches de, ef, and fa. The application of Kirchhoff's current law at either node a or node d shows that the current in the middle branch is $i_1 - i_2$ in the downward direction. It is convenient to view this current as the superposition of a current i_1 in the downward direction and i_2 in the upward direction. Then the current i_1 continues to flow around the complete path $abcda$, and the new current i_2 flows around the path $defad$. The two currents i_1 and i_2 circulate in a clockwise direction around their respective meshes and flow in each branch which borders the mesh. The actual current in any branch is the sum of the mesh currents which flow through it. Normally, all the mesh currents are assumed to flow in the *clockwise direction*, and therefore the branch currents appear as differences of mesh currents. As a conclusion of the preceding argument we can say that the incorporation of Kirchhoff's current-law equations into the branch current variables leads to a new set of variables which we can interpret as currents circulating around the meshes.

The following is a summary of the mesh-current method for solving a circuit problem:

1. Sets of clockwise mesh currents are assigned as the unknowns. These currents automatically satisfy Kirchhoff's current laws.

2. Kirchhoff's voltage-law equations are written for each mesh. The volt-ampere equations are used to obtain the equations in terms of the current variables.

3. The mesh currents are found by solving these equations. The branch currents are found from the mesh currents, and the branch voltages from the branch currents.

As a first example of the mesh-current method we will solve the circuit of Fig. 3–2. There is a single mesh in this circuit and we therefore assume a single mesh current flowing around it in a clockwise direction. It is the only current in each branch and therefore the branch currents are equal to the mesh current. The application of Kirchhoff's voltage law around the mesh gives

$$i(R_1 + R_2 + R_3) = E. \tag{3–1}$$

The left side of Eq. (3–1) represents the voltage drop produced in the three resistors by the mesh current i, and the right side represents the voltage source which tends to make current flow in the assumed clockwise direction. A solution of Eq. (3–1) is

$$i = \frac{E}{R_1 + R_2 + R_3}\cdot \tag{3–2}$$

The branch currents are equal to i, and the branch voltages can be obtained by multiplying i by the appropriate value of resistance.

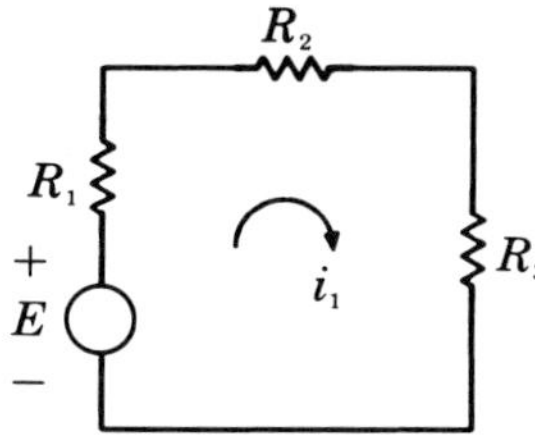

FIG. 3–2. A single-mesh circuit.

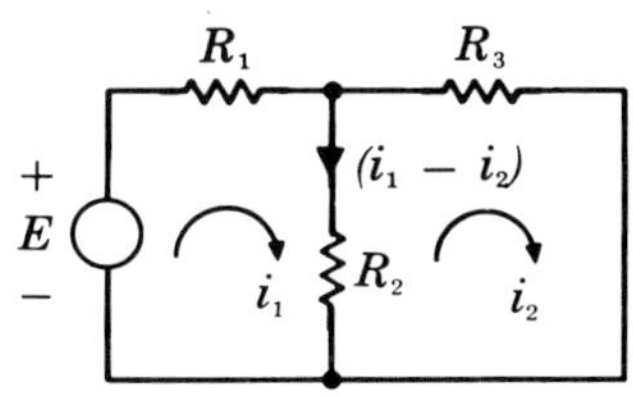

FIG. 3–3. A two-mesh circuit.

It is relatively easy to include the volt-ampere equations by inspection so long as the source is a voltage source. Current sources may be transformed to voltage sources before the equations are written, or they may be handled by the dummy-variable procedure discussed in the next section.

The circuit shown in Fig. 3–3 has two meshes and requires two mesh currents for its solution. If they are taken to be i_1 and i_2 in the clockwise direction, the current in the center branch is $i_1 - i_2$ in the downward direction. A voltage-law equation must be written around each of the meshes, as follows:

$$\text{Mesh 1:}\quad i_1R_1 + (i_1 - i_2)R_2 = E. \tag{3–3}$$

$$\text{Mesh 2:}\quad i_2R_3 - R_2(i_1 - i_2) = 0. \tag{3–4}$$

If we collect the coefficients of i_1 and i_2 in Eqs. (3–3) and (3–4), we obtain*

$$i_1(R_1 + R_2) - i_2(R_2) = E, \tag{3–5}$$

$$-i_1(R_2) + i_2(R_2 + R_3) = 0. \tag{3–6}$$

The first term in Eq. (3–5) is the voltage drop produced in mesh 1 by the mesh current i_1, and the second term is the voltage drop produced in mesh 1 by the current in mesh 2. This latter term is negative relative to the first one because i_2 flows through R_2 in a direction opposite to that of i_1. The term on the right side of the equation is the net source voltage, which tends to make current flow in the assumed clockwise direction. For a source of the opposite polarity, this term would appear with a negative sign.

Equation (3–6) represents the application of the voltage law to mesh 2. The first term is the voltage drop produced in mesh 2 by mesh current i_1. The second term is the voltage drop produced in mesh 2 by its own mesh current i_2. Usually this second term is taken to be positive, in which case the first term is negative. There are no sources in this mesh and therefore the right side of the equation is zero.

It is important to learn to write the mesh equations directly in the form of Eqs. (3–5) and (3–6). The symmetry is useful and will help in avoiding minor errors. Note that all the terms produced by a current in its own mesh are positive, while all the terms produced by a current in any other mesh are negative. Furthermore, the mutual terms are similar, since the branch between two meshes is common to both meshes. (Essentially the process is one of superposition.)

---

* Equations (3–3) and (3–4) are more general in form than are Eqs. (3–5) and (3–6). They would apply if the resistors were nonlinear, but they are less useful in the linear case.

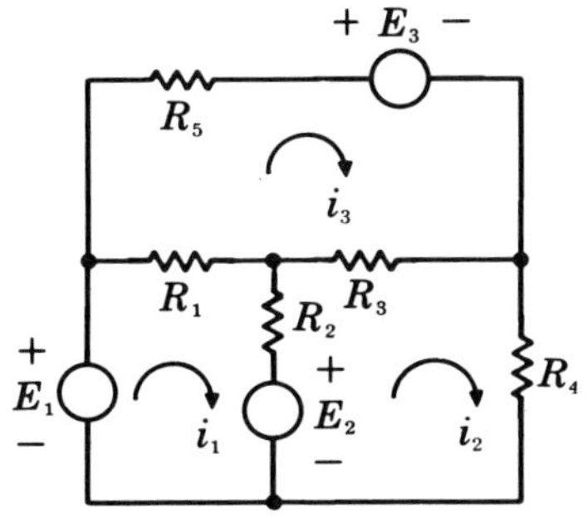

FIG. 3–4. A three-mesh circuit.

The same procedure can be extended to more complicated circuits. A 3-mesh example is shown in Fig. 3–4. The equations for this circuit are

$$\begin{aligned} i_1(R_1 + R_2) - i_2R_2 - i_3R_1 &= E_1 - E_2, \\ -i_1(R_2) + i_2(R_2 + R_3 + R_4) - i_3(R_3) &= E_2, \\ -i_1(R_1) - i_2(R_3) + i_3(R_1 + R_3 + R_5) &= -E_3. \end{aligned} \tag{3–7}$$

The network itself is characterized by the left sides of these equations and the sources by the right sides. The convention regarding mesh currents is so well established that we can drop the variables and describe the network by the array of coefficients alone, as shown in Eq. (3–8).

$$\begin{bmatrix} R_1 + R_2 & -R_2 & -R_1 \\ -R_2 & R_2 + R_3 + R_4 & -R_3 \\ -R_1 & -R_3 & R_1 + R_3 + R_5 \end{bmatrix}. \tag{3–8}$$

Such an array is called a *matrix*. The symmetry is more apparent in the matrix than in the equations themselves. It is a consequence of the clockwise choice of mesh currents and of the orderly way in which the equations were recorded. All the terms on the principal diagonal, running from the upper left to the lower right, are positive. All of the other terms are negative, and the whole matrix is symmetrical about the principal diagonal. Algebraic errors which might spoil the symmetry of the matrix are detected at once.

The mesh-current method can be generalized to the case of an arbitrary number of meshes, k. Clockwise currents are assumed in each mesh and the voltage-law equation is written around each mesh in turn. The resulting equations will have the form

$$\begin{bmatrix} i_1(r_{11}) - i_2(r_{12}) - \cdots - i_k(r_{1k}) = E_1 \\ -i_1(r_{12}) + i_2(r_{22}) - \cdots - i_k(r_{2k}) = E_2 \\ \vdots \\ -i_1(r_{1k}) - i_2(r_{2k}) - \cdots + i_k(r_{kk}) = E_k \end{bmatrix}, \tag{3-9}$$

where

i_1 is the mesh current in mesh 1,

r_{11} is the sum of the resistances through which mesh current 1 flows,

r_{12} is the resistance common to meshes 1 and 2,

E_1 is the net source voltage tending to produce clockwise current in mesh 1,

and so on.

Numerical example. As a numerical example we consider the circuit of Fig. 3-3 with $R_1 = 1$, $R_2 = 2$, and $R_3 = 3$ ohms, and $E = 10$ volts. The numerical mesh equations are

$$i_1(3) - i_2(2) = 10,$$
$$-i_1(2) + i_2(5) = 0,$$

from which

$$i_1 = \frac{\begin{vmatrix} 10 & -2 \\ 0 & 5 \end{vmatrix}}{\begin{vmatrix} 3 & -2 \\ -2 & 5 \end{vmatrix}} = \tfrac{50}{11} \quad \text{and} \quad i_2 = \frac{\begin{vmatrix} 3 & 10 \\ -2 & 0 \end{vmatrix}}{\begin{vmatrix} 3 & -2 \\ -2 & 5 \end{vmatrix}} = \tfrac{20}{11}.$$

The branch currents and the branch voltages can be found from the mesh currents. The current in R_2 is

$$i_1 - i_2 = \tfrac{30}{11}$$

and the voltages across the resistors are

$$e_{R_1} = \tfrac{50}{11},$$
$$e_{R_2} = \tfrac{30}{11} \times 2 = \tfrac{60}{11},$$
$$e_{R_3} = \tfrac{20}{11} \times 3 = \tfrac{60}{11}.$$

3-3 Dummy mesh currents. The mesh equations are written for the sum of voltages around a closed path. A voltage source can therefore be easily included in the equations. A current source, on the other hand, requires special attention. There are two possible approaches:

1. Source transformations which convert the current sources to voltage sources.

2. The use of dummy mesh-current variables.

Source transformations have been discussed in the previous chapter. They can always be used, but they have the disadvantage that they change the form of the network. The dummy-variable method leaves the network unchanged and is therefore usually preferable to the source-transformation method. The concept of dummy mesh currents comes from topology. The topological meshes in a network are those which are present when all the sources are "dead," that is, when the current sources are replaced by open circuits and the voltage sources by short circuits. Usually mesh currents are defined only around the topological meshes. The principal reason for this limitation is that the number of voltage law equations available to solve for the mesh currents is exactly equal to the number of topological meshes. A *dummy mesh current* is a mesh current defined around the dummy mesh created by the presence of a current source in the network.* No voltage-law equation can be written for the dummy mesh current. Instead it is defined in terms of the current source which created it. The defining equation is called a *constraint* equation. The following is a summary of the dummy mesh-current method:

1. Mesh currents are defined around all the apparent meshes in a circuit without regard for whether or not true topological meshes exist.

2. Voltage-law equations are written around each topological mesh.

3. A constraint equation is written for each of the dummy meshes created by each current source.

An example of the solution of a mesh-current problem in which there are current sources will now be given. The circuit is shown in Fig. 3–5(a). The procedure first used will be a source transformation. The circuit of Fig. 3–5(b) is identical to the circuit of Fig. 3–5(a) except for the transformation of the current source in parallel with R_1 to a voltage source in series with R_1. Although the form of the circuit is changed by the transformation, the current through the external resistance R_2 will not be changed. The mesh equation for the circuit of Fig. 3–5(b) is

$$i_1(R_1 + R_2) = E_1 = IR_1,$$

from which the current in R_2 is

$$i_1 = \frac{IR_1}{R_1 + R_2}. \tag{3–10}$$

---

* Recall that current sources create dummy meshes and voltage sources create dummy nodes.

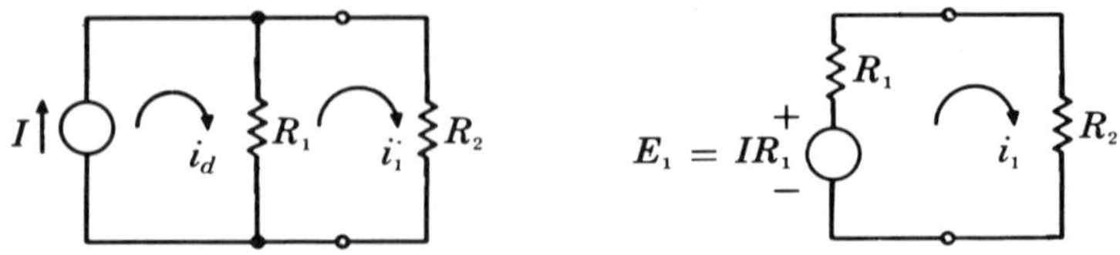

(a) Circuit with dummy mesh (b) Transformed circuit

FIG. 3–5. An example of a dummy mesh.

The disadvantage of the source transformation is that the form of the network is changed. The currents in R_1 in the two circuits are quite different!

The alternative approach, using dummy mesh currents, is shown in Fig. 3–5(a). The mesh current i_1 is assumed in the genuine topological mesh and a dummy mesh current i_d is assumed around the dummy mesh created by the current source. The dummy current i_d is treated as an ordinary mesh current so far as the equation for the i_1 mesh is concerned, but no equation is written around the dummy mesh. Instead, it is merely observed that the source I constrains the dummy mesh current i_d to be equal to I. The two equations are

$$i_1(R_1 + R_2) - i_d(R_1) = 0$$

and

$$i_d = I, \tag{3–11}$$

from which

$$i_1 = \frac{R_1}{R_1 + R_2} I$$

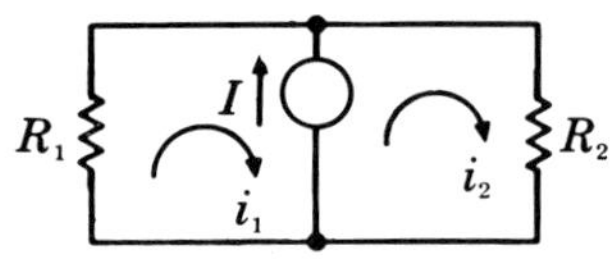

FIG. 3–6. A dummy mesh produced by an internal source.

as before.

The dummy-variable method is perfectly general. A somewhat more complicated example is shown in Fig. 3–6. There are two apparent meshes and currents are defined around them. The topological mesh is the path containing the resistors R_1 and R_2, and it does not agree with either of the mesh currents which have been defined. They are both dummy mesh currents! Two equations can be written for them, however. The first is a voltage-law equation around the topological mesh:

$$i_1R_1 + i_2R_2 = 0. \tag{3–12}$$

The second is a constraint equation for the current which flows downward through the source I. The mesh current i_1 flows downward through the source and the mesh current i_2 flows upward through the source. The net current must be I in the upward direction. Hence

$$i_1 - i_2 = -I. \tag{3–13}$$

A simultaneous solution of Eqs. (3–12) and (3–13) gives the two currents i_1 and i_2 as

$$i_1 = \frac{-R_2}{R_1 + R_2} I \qquad \text{and} \qquad i_2 = \frac{R_1}{R_1 + R_2} I.$$

The two examples which have been solved by the dummy-variable method show that the procedure is easier if the current source occurs in an outside branch of the network. Often a circuit can be redrawn to achieve this form. If so, this method is easier than the source-transformation method, and is preferable because it leaves the network geometry unchanged.* A minor disadvantage of the dummy-variable method is that the equations lack some of the symmetry of the pure mesh equations. Usually, however, this is a small price to pay for the convenience of leaving the network form unchanged.

3–4 The node-voltage method. The voltage variables which have been used in the previous chapters have been branch voltages. A branch voltage is analogous to the elevation of one point on a map with respect to another point. An alternative method of specifying elevations is with respect to a reference level, usually sea level. A similar alternative exists in electrical networks. We can describe the voltage level of each node with respect to some datum, or reference point, which is usually called the "ground." The civil engineer uses elevations above sea level in order to suppress any errors in his measurements. If he recorded the elevations between points he would have to check all possible routes to a given point to make sure that all the branch elevations added up to the same value! Node-to-datum voltages are used in electrical engineering for much the same reason. They automatically satisfy Kirchhoff's voltage law around any path. When we use node-to-datum voltages as our algebraic unknowns we do not have to write the voltage-law equations explicitly. If we consider that the volt-ampere equations are trivial and can be written by inspection we are left with only

$$b - l = n - 1 \tag{3–14}$$

equations. Since we must have one variable for each equation, we need $n - 1$ variables. The voltages of the $n - 1$ nodes with respect to the

* The difference between merely rearranging the network and changing it by means of transformations and equivalent circuits should be clearly understood.

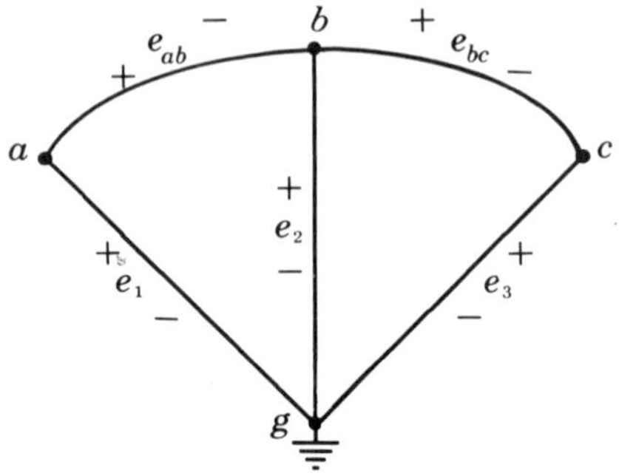

FIG. 3–7. Node-to-datum voltages and Kirchhoff's voltage law.

ground node are a suitable set. There are $n - 1$ of them and they are independent, since the nodes are independent.*

As an example of the way in which the node-to-datum voltages satisfy Kirchhoff's voltage laws, consider the circuit of Fig. 3–7. The voltages at nodes a, b, and c with respect to the ground node g are e_1, e_2, and e_3 respectively. The voltage between a and b is the difference between the voltage at a and the voltage at b, and similarly the voltage between b and c is the difference between the voltages at b and at c. Thus

$$e_{ab} = e_1 - e_2 \qquad \text{and} \qquad e_{bc} = e_2 - e_3. \tag{3–15}$$

Equations (3–15) are merely statements of Kirchhoff's voltage law around loops a and b respectively. The laws are automatically satisfied by the node-to-datum definition of the variables. In a more general circuit the branch voltages will always be determined by the node-to-datum voltages. Some of the branch voltages will be directly equal to node voltages. Others will be obtained as differences between two node voltages. Kirchhoff's voltage law around any closed path will be automatically satisfied.

The following is a procedure for using the node-to-datum (hereafter called just node) voltages:

1. A datum node is selected and a set of node voltages with respect to this datum is assigned as unknowns. The Kirchhoff voltage laws are automatically satisfied by these variables.

2. A Kirchhoff current-law equation is written at each of these nodes. The volt-ampere equations are included by inspection and the equations are written in terms of the node voltages.

3. The equations are solved for the node voltages and the branch currents, and branch voltages are found from the node voltages.

---

* Other sets of voltages between supernodes can be used, but they are less common than the node-to-datum set.

As a first example of the method we consider the circuit of Fig. 3–8. The lower node will be assumed as ground or datum, and the upper node will be assigned the unknown value e_a. The current-law equation at the upper node yields

$$e_a\left(\frac{1}{R_1} + \frac{1}{R_2}\right) = I \tag{3–16}$$

or

$$e_a(G_1 + G_2) = I, \tag{3–17}$$

where

$$G_1 = \frac{1}{R_1} \qquad \text{and} \qquad G_2 = \frac{1}{R_2}.$$

The left side of Eq. (3–16) represents the currents flowing away from the node voltage e_a, and the right side represents the current source flowing into the node and tending to make the node positive. Current sources can be included directly in the equations. Voltage sources pose a problem which must be solved by source transformations or by the dummy-voltage method discussed below.

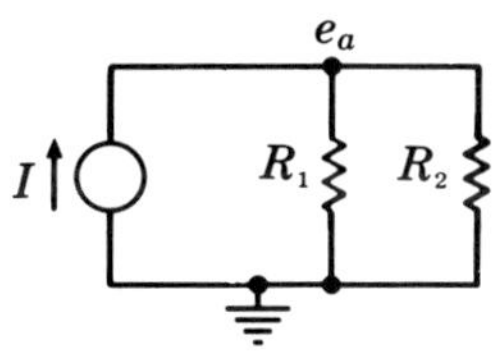

FIG. 3–8. A one-node-voltage problem.

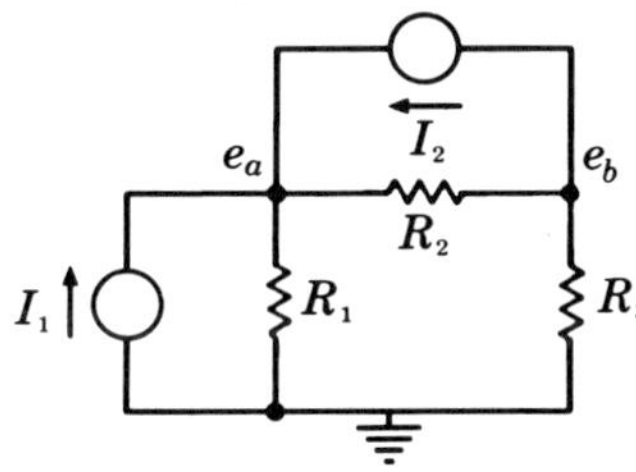

FIG. 3–9. A two-node-voltage problem.

A second example of the node-voltage method will be the solution of the circuit shown in Fig. 3–9. In this circuit there are two unknown nodes e_a and e_b, in addition to the ground node. The current-law equations are

$$e_a\,\frac{1}{R_1} + (e_a - e_b)\,\frac{1}{R_2} - I_1 - I_2 = 0 \tag{3–18}$$

and

$$(e_b - e_a)\,\frac{1}{R_2} + e_b\,\frac{1}{R_3} + I_2 = 0. \tag{3–19}$$

Equation (3–18) represents the sum of the currents flowing away from the node voltage e_a. The first term is the current through R_1 and the second is the current through R_2. The remaining terms are the source currents. Equation (3–19) represents the currents flowing away from node e_b. The first term is the current through R_2, the second term is the current through R_3, and the third term is the source current. (Note that the voltage across R_2 is either $e_a - e_b$, or $e_b - e_a$, depending on which polarity is assumed for the reference voltage across the branch.) It is usually more convenient to rewrite Eqs. (3–18) and (3–19) in the form*

$$e_a\left(\frac{1}{R_1} + \frac{1}{R_2}\right) - e_b\,\frac{1}{R_2} = I_1 + I_2 \tag{3–20}$$

and

$$-e_a\,\frac{1}{R_2} + e_b\left(\frac{1}{R_2} + \frac{1}{R_3}\right) = -I_2. \tag{3–21}$$

An alternative interpretation of Eqs. (3–20) and (3–21) is very useful. The first term in Eq. (3–20) represents the sum of the currents flowing away from the node voltage e_a when all the other node voltages are equal to zero, or connected to the ground point. The coefficient which multiplies e_a is thus the sum of the conductances which radiate from the node. In fact, if a circle is drawn around the node each resistor which it intersects will be included in the coefficient. The second term of Eq. (3–20) is the current which flows away from the node voltage e_a when e_b is given a positive value and when e_a is grounded or zero. It is a negative term because the current obviously flows into the node and not away from it. The coefficient of e_b is the mutual conductance which joins the node e_a to the node e_b. The terms on the right side of the equation are the source currents which enter the node and tend to give the voltage its assumed positive value.

Equation (3–21) can be interpreted similarly. The first term is the current which the voltage e_a would produce at node b. The second term is the total current which would flow from e_b if it were given a positive value and the other nodes were all grounded. The term on the right side is the net source current tending to make e_b positive. Since the source current is directed away from the node it has a negative sign in this equation.

The symmetry of Eqs. (3–20) and (3–21) is useful. All of the elements on the principal diagonal, from the upper left to the lower right, are positive. These elements are "self" elements, having to do with the effect of a voltage at its own node. All of the terms off the principal diagonal are negative, and the coefficients are equal in pairs, since the conductance between node a and node b is the same as the conductance between node b and node a. The student should learn to write the equations directly in the form of Eqs. (3–20) and (3–21). They are easier to use than the

---

* Equations (3–18) and (3–19) are more general, but Eqs. (3–20) and (3–21) are more useful in a linear problem.

other equations and the symmetry provides an automatic check on minor algebraic errors.

The symmetry is even more obvious with respect to the array of the coefficients in the equations. This array is called a *conductance parameter matrix* because the elements have the dimensions of conductance. For Eqs. (3–20) and (3–21) the matrix is

$$\begin{bmatrix} 1/R_1 + 1/R_2 & -1/R_2 \\ -1/R_2 & \dfrac{1}{R_2} + \dfrac{1}{R_3} \end{bmatrix}. \tag{3–22}$$

In terms of conductances the matrix is written as

$$\begin{bmatrix} G_1 + G_2 & -G_2 \\ -G_2 & G_2 + G_3 \end{bmatrix}. \tag{3–23}$$

The node-voltage method can be generalized to the case of k independent nodes. The node voltages are all assumed to be positive with respect to the datum and the current-law equations at each node in turn are written:

$$\begin{aligned} e_1(g_{11}) - e_2(g_{12}) - \cdots & -e_k(g_{1k}) = I_1, \\ -e_1(g_{12}) + e_2(g_{22}) - \cdots & -e_k(g_{2k}) = I_2, \\ \vdots \qquad & \\ -e_1(g_{1k}) - e_2(g_{2k}) - \cdots & +e_k(g_{kk}) = I_k, \end{aligned} \tag{3–24}$$

where

e_1 is the node voltage at node 1,

g_{11} is the total conductance radiating from node 1,

g_{12} is the conductance between nodes 1 and 2,

I_1 is the net source current tending to make a positive voltage at node 1,

and so on.

Numerical example. The complete solution of a problem by the node-voltage method will be illustrated by means of a numerical example. Consider the circuit of Fig. 3–9 with $I_1 = 2$ amp, $I_2 = 1$ amp, $R_1 = 1$ ohm, $R_2 = 2$ ohms, and $R_3 = 3$ ohms. The node-voltage equations are

$$\begin{aligned} e_a(1 + \tfrac{1}{2}) - e_b(\tfrac{1}{2}) &= 1 + 2 = 3, \\ -e_a(\tfrac{1}{2}) + e_b(\tfrac{1}{2} + \tfrac{1}{3}) &= -1, \end{aligned}$$

from which

$$e_a = \frac{\begin{vmatrix} 3 & -\frac{1}{2} \\ -1 & \frac{5}{6} \end{vmatrix}}{\begin{vmatrix} \frac{3}{2} & -\frac{1}{2} \\ -\frac{1}{2} & \frac{5}{6} \end{vmatrix}} = 2 \quad \text{and} \quad e_b = \frac{\begin{vmatrix} \frac{3}{2} & 3 \\ -\frac{1}{2} & -1 \end{vmatrix}}{\begin{vmatrix} \frac{3}{2} & -\frac{1}{2} \\ -\frac{1}{2} & \frac{5}{6} \end{vmatrix}} = 0.$$

The voltage across R_2 is $e_a - e_b = 2$, and the currents in the three resistors are

$$i_{R_1} = \frac{e_a}{R_1} = 2 \text{ amp}, \qquad i_{R_2} = \frac{e_a - e_b}{R_2} = 1 \text{ amp},$$

$$i_{R_3} = \frac{e_b}{R_3} = 0 \text{ amp}.$$

3–5 Dummy node variables. Node-voltage equations are written for the sum of the currents leaving nodes. Current sources can therefore be included directly. Voltage sources, however, require special attention. There are two alternative approaches:

1. Source transformations which convert all voltage sources to current sources.

2. The use of dummy node-voltage variables.

The subject of source transformations has been discussed thoroughly in the preceding chapter. Source transformations are always applicable, but they have the disadvantage that they change the form of the network. The dummy node-voltage method, on the other hand, leaves the form of the network unchanged. Dummy node voltages are based on topological considerations. Each voltage source can be considered to introduce a dummy node into the circuit. Each dummy node can be included in the equations by means of a dummy node voltage. The current-law equations are written only at genuine topological nodes, and the dummy node voltage is defined by the voltage constraint of the source. The following is a summary of the dummy node-voltage method:

1. Node voltages are defined at each node in the network, including the dummy nodes created by voltage sources.

2. Current-law equations are written at each genuine topological node or supernode. (Topologically, the two terminals of each voltage source are shorted together. They belong to the same supernode.)

3. A constraint equation is written for each of the dummy nodes created by the voltage sources.

A simple circuit which will be used to illustrate the solution of problems involving voltage sources by the node-voltage method is shown in Fig. 3–10. The original circuit is shown in part (a), and the circuit with a

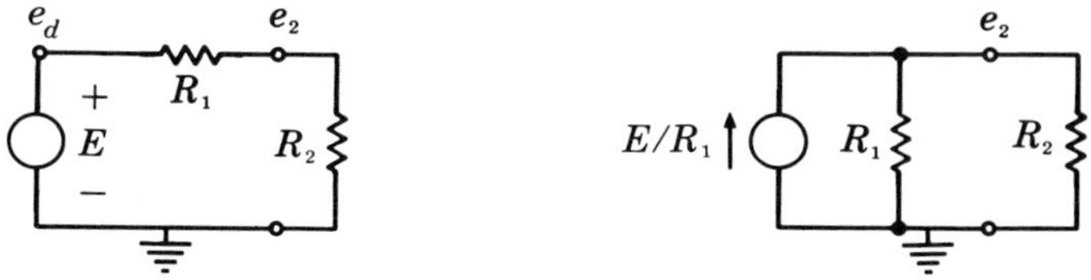

(a) Circuit with dummy node (b) Transformed circuit

FIG. 3–10. An example of a dummy node.

transformed source is shown in (b). The node-variable method can be applied directly to the circuit of (b). The node equation is

$$e_2\left(\frac{1}{R_1}+\frac{1}{R_2}\right)=\frac{E_1}{R_1}, \tag{3–25}$$

from which

$$e_2 = E_1\,\frac{R_2}{R_1+R_2}. \tag{3–26}$$

The voltage e_2, as given by Eq. (3–26), is correct for the circuits of both Fig. 3–10(a) and (b). The voltage across the resistor R_1 in the two circuits is not the same, however. The dummy-variable method is applied directly to the original circuit. A dummy node voltage e_d is assumed at the terminal of the voltage source. The constraint equation is simply

$$e_d = E. \tag{3–27}$$

The node equation at the topological node e_2 is written just as though e_d were a real node voltage. It is

$$e_2\left(\frac{1}{R_1}+\frac{1}{R_2}\right)-e_d\,\frac{1}{R_1}=0. \tag{3–28}$$

A combination of Eqs. (3–27) and (3–28) gives the same value of e_2 as before:

$$e_2 = E_1\,\frac{R_2}{R_1+R_2}. \tag{3–29}$$

The dummy-variable method as used in Eq. (3–28) automatically makes a source transformation in the equations, without requiring any change in the network itself. It is usually preferable to the source transformation method. Whenever the voltage source is connected to ground, as in this example, the dummy variable is defined directly by the source and the

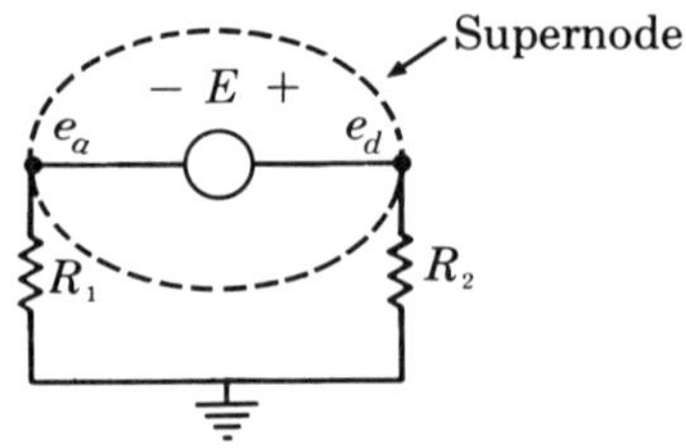

FIG. 3–11. A dummy node produced by a floating source.

method is particularly simple. When the voltage source is not connected to ground the method can still be applied, but the constraint equation is more complicated. As an example, consider the circuit of Fig. 3–11. Topologically, it has just one unknown node and a ground node. One of the apparent nodes is a dummy. It doesn't matter which of the nodes is taken to be the unknown and which the dummy. For the arrangement shown the dummy voltage is defined by the constraint equation

$$e_d = e_a + E. \tag{3–30}$$

The node-voltage equation must be written at a genuine topological node. The topological node is actually the combination of e_a and e_d which is obtained by collapsing the voltage source to zero. The same effect is obtained by combining e_a and e_d into the dotted supernode, as shown in Fig. 3–11. The Kirchhoff current-law equation for this supernode is obtained by equating to zero the sum of the currents leaving it. Thus it is

$$\frac{e_a}{R_1} + \frac{e_d}{R_2} = 0. \tag{3–31}$$

From Eqs. (3–30) and (3–31) we obtain

$$e_a = -\frac{R_1}{R_1 + R_2} E \qquad \text{and} \qquad e_d = \frac{R_2}{R_1 + R_2} E.$$

It is more difficult to write the dummy node equations when the voltage source is not connected to ground. Whenever possible it is desirable to shift the ground point to make the analysis easier.

3–6 Duality. In this section we shall explore the similarity between the general mesh equations (Eqs. 3–9) and the general node equations (Eqs. 3–24). This similarity is based on an important network principle called *duality*. Two networks are said to be dual when the mesh equations for one are numerically equal to the node equations for the other.* We discuss duality at this point because of the light which it sheds on the mesh and node equations. In later chapters it will reappear in many forms. Every problem which is solved and every theorem which is proved has a dual which may well be as useful as the original.

Duality has the nature of a transformation from the original network to the dual network. The transformation is accomplished in the following fashion:

1. A set of mesh equations with numerical coefficients is written for the original network.
2. The variables are changed from current to voltage.
3. The resulting node equations are interpreted as the dual network.

The first two steps are straightforward. The third step is really network synthesis, and we must show that it is always possible to find the dual network. First of all, let us consider the mutual elements between two meshes in the original network. In the transformed equations these terms will occur in the position of mutual conductances between the corresponding pairs of node voltages. They may be interpreted as physical resistances between the nodes with values which are numerically equal in mhos to the original resistance values in ohms. Each resistance which appears only in a self-resistance term in the mesh equations will transform to a conductance which appears only in the self-conductance term in the dual equations. Such an element can be interpreted as a conductance to ground in the dual network. Voltage sources in the mesh equations transform to current sources in the node equations, with values which are numerically the same. Current sources in the mesh equations result in dummy mesh equations which transform to dummy node equations in the dual set. The dummy node equations describe voltage sources which are numerically equal to the current sources in the original network.†

A number of consequences of duality can be deduced from the foregoing discussion. Some of them are included in Table 3–1. Since duality is a unique one-to-one transformation the property of duality is a mutual one between two networks, and the transformation can be carried out in either direction. Alternatively, two applications of the duality transformation bring us back to the original network.

---

* The property of duality is more general than this. Any set of equilibrium equations written for a planar network can be changed into a numerically identical set for a dual network by changing all the current variables to voltages, and all the voltage variables to currents.

† Only planar networks have duals. Nonplanar networks have three or more loop currents passing through a single resistance. The dual would require a single resistance between three or more nodes.

TABLE 3–1

Original network		*Dual network*
current i	transforms to	voltage e
voltage e	transforms to	current i
source current I	transforms to	source voltage E
source voltage E	transforms to	source current I
resistance R	transforms to	conductance G
inductance L	transforms to	capacitance C
capacitance C	transforms to	inductance L
series branches	transform to	parallel branches
parallel branches	transform to	series branches
open circuits	transform to	short circuits
short circuits	transform to	open circuits
mesh currents	transform to	node voltages
node voltages	transform to	mesh currents

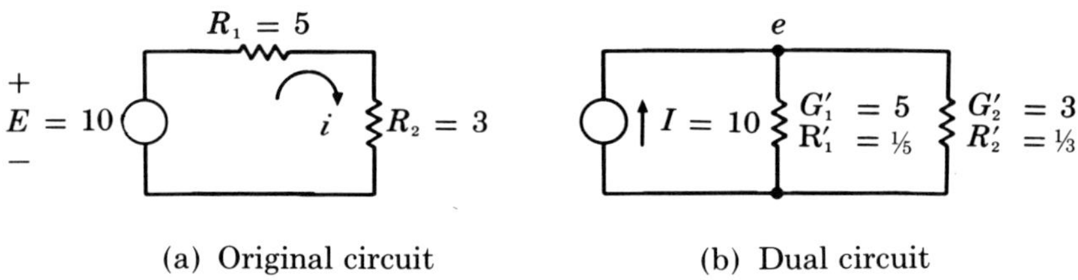

FIG. 3–12. A circuit and its dual.

As an example of duality consider the circuit of Fig. 3–12(a). The mesh equation for this circuit is

$$i(R_1 + R_2) = E \tag{3–32}$$

or, numerically,

$$i(5 + 3) = 10. \tag{3–33}$$

The duality transformation changes Eq. (3–33) to

$$e(5 + 3) = 10. \tag{3–34}$$

Equation (3–34) can be interpreted as the node-voltage equation for two parallel resistances driven by a current source, for which the following equation holds:

$$e(G'_1 + G'_2) = I. \tag{3–35}$$

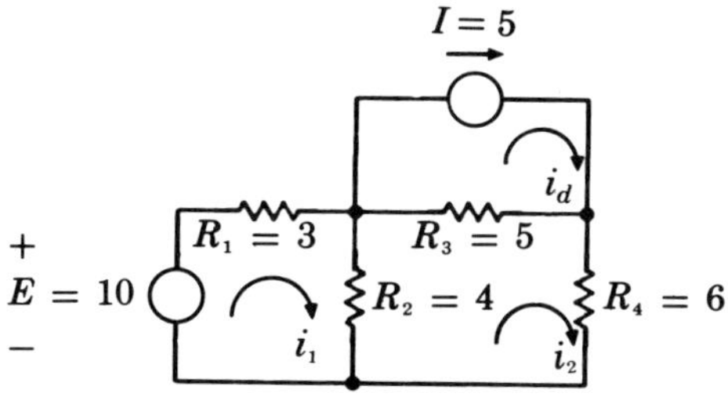

(a) Original circuit

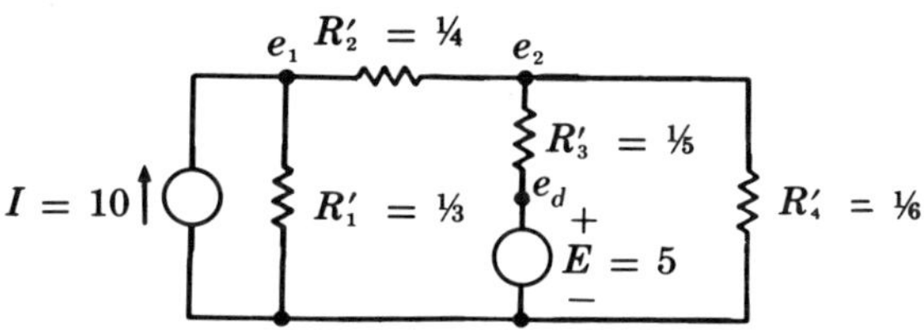

(b) Dual circuit

FIG. 3–13. The dual of a circuit with a dummy mesh.

The values of the elements in the dual circuit are

$$G_1' = 5 \quad \text{or} \quad R_1' = \tfrac{1}{5}\text{ ohm},$$

$$G_2' = 3 \quad \text{or} \quad R_2' = \tfrac{1}{3}\text{ ohm},$$

and

$$I = 10\text{ amp}.$$

The dual circuit is shown in Fig. 3–12(b).

A second example of the duality principle will be given for the circuit of Fig. 3–13(a). In this circuit there are two genuine meshes and one dummy mesh. The corresponding dual circuit will have two real node voltages and one dummy node voltage. The mesh equations for the original circuit are

$$\begin{aligned} i_1(3+4) - i_2(4) &= 10, \\ -i_1(4) + i_2(4+5+6) - i_d(5) &= 0, \end{aligned} \tag{3–36}$$

and the constraint equation is

$$i_d = 5.$$

The dual equations are obtained by changing the current variables into voltages, as follows.

$$\begin{aligned} &e_1(3+4) - e_2(4) = 10, \\ &-e_1(4) + e_2(4+5+6) - e_d(5) = 0, \\ &e_d = 5. \end{aligned} \tag{3–37}$$

The dual network is obtained from Eq. (3–37) by a process of synthesis. (Note that it is better to leave each element as a separate numerical value, since each element in the original circuit will appear as a dual element in the dual circuit.) The first step in the synthesis procedure is the location of a ground node, the two real nodes, and the dummy node. These points may be drawn more or less at random. The next step is the determination of the mutual elements. The mutual conductance between nodes 1 and 2 is evidently 4 mhos. A $\frac{1}{4}$-ohm resistance can therefore be installed between the nodes. Now the self-conductances can be determined. The total conductance radiating from node 1 is 7 mhos. Four mhos go to node 2. The remaining 3 mhos must go to ground. We can therefore add a $\frac{1}{3}$-ohm resistance to ground from node 1. The first equation is completed by the addition of a current source of 10 amp feeding into the node from the ground. In the second equation there are two mutual elements. The 4-mho resistance between nodes 1 and 2 has already been installed. An additional mutual element of 5 mhos occurs between the dummy node and node 2. The corresponding resistance value is $\frac{1}{5}$ ohm. The remaining 6 mhos represent a $\frac{1}{6}$-ohm resistance to ground. There is no current source in the second equation of (3–37). There remains only the dummy node, and the third equation tells us that it is a voltage source of 5 volts. Since the dummy mesh current is clockwise, or positive, the dummy node voltage is positive with respect to the ground node. The complete network is shown in Fig. 3–13(b).

The technique we have just discussed can always be used to find the dual of a planar network. We will now examine a geometrical adaptation of this method which arrives at the final dual network without the necessity of writing down any of the equations. Each mesh current in the original network transforms to a node voltage in the dual network. We can therefore locate the correct nodes by placing a dot in the center of each mesh. The process is valid for both real and dummy meshes. Each element that occurs between two meshes must appear in dual form between the corresponding nodes in the dual circuit. These elements can be located geometrically by drawing lines between the two nodes in such a way that they cut each element that exists between the corresponding meshes. Each line is then replaced by the dual of the element it cuts.

Only the elements which occur in the mutual branches between meshes will be included in the above construction. The remaining elements are on the outside of the original network. In the dual network they represent elements which go to ground from the nodes. It is apparent that the geometrical equivalent of the ground node is a ring around the outside of the diagram. Each line that cuts an element between a node and this ground ring can be replaced by the dual of the element.

An additional problem occurs relative to the polarity of the source elements. The conventional positive direction for mesh currents is clockwise, and the conventional polarity for the node voltages is positive with respect to ground. *Thus a voltage or a current source which tends to produce a clockwise current around a mesh in the original circuit should be replaced in the dual circuit by the dual source element with a polarity which tends to make the node at the center of the mesh positive.*

As an example of the geometrical process, the duals of the two previous circuits are obtained geometrically in Fig. 3–14. In part (a) the node is assumed in the center of the mesh and a ground node which completely encircles the figure is drawn. Lines from the node to the circle cut the two resistances and the source. Each line is replaced by the dual of the element it cuts. The original voltage source tended to produce clockwise current in the mesh and therefore the current source in the dual circuit must enter the node and tend to make it positive.

In Fig. 3–14(b) two real nodes are installed in the real meshes and a dummy node in the dummy mesh. The lines joining the nodes are replaced by the duals of the elements which they cut. The ground node encircles the diagram and has connections to each of the three nodes.

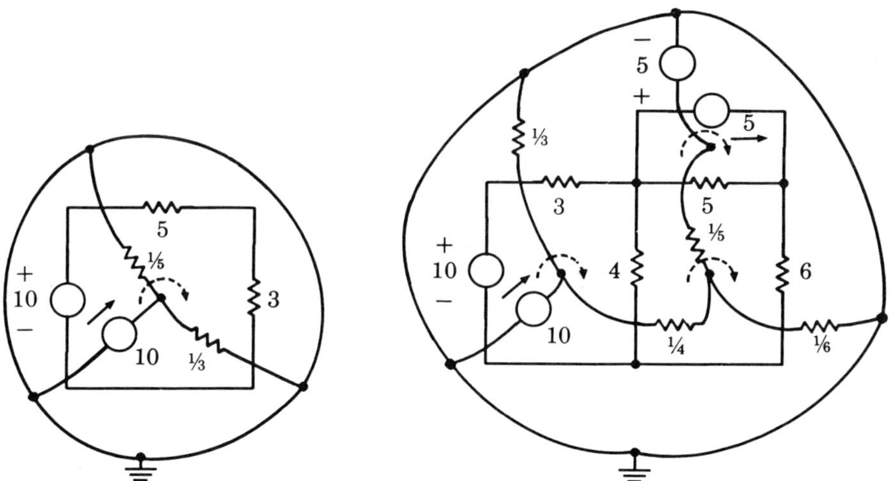

(a) The dual of Fig. 3-12(a) (b) The dual of Fig. 3-13(a)

Fig. 3–14. The geometrical determination of dual circuits.

The voltage source in mesh 1 of the original circuit tends to produce clockwise current, and hence the current source in the dual network must enter the corresponding node. The current source in the original network flows in the clockwise direction in the dummy node, and hence the corresponding dummy voltage must be positive with respect to ground.

The geometrical method of obtaining the dual circuit is surprisingly easy after a little practice, and it is usually preferable to writing the equations explicitly.

3–7 Linearity and superposition. Mesh and node equations are useful for solving specific network problems. They also provide a method for discussing networks in a more generalized way, and for deriving important theoretical relationships. Of these the most important and far reaching are the theorems of *linearity* and *superposition*. Linearity states that the response at any point in a network is proportional to the stimulus which is causing it. Superposition is an extension of the linearity principle to the case of multiple excitation. When the superposition principle applies, the output response from a number of sources acting simultaneously is simply the sum of the responses which would be produced by each of the sources acting alone with all of the other sources dead. The superposition theorem includes the linearity theorem as a special case since increasing the value of a stimulus is equivalent to adding a second stimulus to it.*

Proof. To prove the linearity and superposition principles we will assume a general set of mesh equations. (Node equations could have been used equally well.) We will suppose that all the current sources have been converted to equivalent voltage sources. In the section on source transformations we have already shown that a source transformation does not change the response in the network external to the source itself. If the response within the source branch is desired one can switch to the general node equations, for which the current sources remain unchanged and the voltage sources are transformed instead. The general mesh equations, from Eq. (3–9), are

$$\begin{aligned} i_1(r_{11}) - i_2(r_{12}) - \cdots - i_k(r_{1k}) &= E_1, \\ -i_1(r_{12}) + i_2(r_{22}) - \cdots \qquad\qquad &= E_2, \\ \vdots \qquad\qquad\qquad\qquad & \\ -i_1(r_{1k}) - i_2(r_{2k}) - \cdots \quad i_k(r_{kk}) &= E_k. \end{aligned} \tag{3–38}$$

---

* The mathematical definition of *linearity* covers both the linearity and superposition theorems. Thus, for a linear function,

$$f(ax) = af(x) \tag{1}$$

and

$$f(x+y) = f(x) + f(y) \tag{2}$$

The solution for a response current is of the form

$$i_1 = \frac{\begin{vmatrix} E_1 & -r_{12} \cdots & -r_{1k} \\ E_2 & r_{22} & \\ \vdots & & \\ E_k & -r_{2k} & -r_{kk} \end{vmatrix}}{\begin{vmatrix} r_{11} & -r_{12} \cdots & -r_{1k} \\ -r_{12} & r_{22} \cdots & \\ \vdots & & \\ -r_{1k} & -r_{2k} \cdots & r_{kk} \end{vmatrix}} \tag{3–39}$$

or

$$i_1 = \frac{E_1}{\Delta}\begin{vmatrix} r_{22} \cdots r_{2k} \\ \vdots \\ -r_{2k} \cdots r_{kk} \end{vmatrix} + \frac{(-E_2)}{\Delta}\begin{vmatrix} -r_{12} \cdots & -r_{1k} \\ -r_{23} & \\ \vdots & \\ -r_{2k} \cdots & r_{kk} \end{vmatrix} + \cdots, \tag{3–40}$$

where Δ is the denominator determinant. (For a resistive network Δ will never be zero.) Equation (3–40) is obtained by a Laplace expansion of the numerator determinant of Eq. (3–39). Each term in Eq. (3–40) is identical to the term which would have been written if only the source involved in that term were acting and all the other sources were zero. The total response is the sum of the responses which would be produced by each of the sources acting separately, with the other sources equal to zero. Equation (3–40) is thus a verification of the superposition theorem.

The linearity theorem is proved by noting that the response produced by any one of the sources acting alone is directly proportional to the value of that source in Eq. (3–40). Both the denominator determinant and the numerator determinant are constants depending only on the values of the resistances in the network.

There would not be much advantage in using the superposition theorem if one solved a problem by means of the general equations and determinants. The advantage lies in the flexibility which superposition lends to other methods of solving the problem. We must therefore know exactly what we mean by the statement that the response for any one source is computed with the "other sources dead." *A dead voltage source is a zero voltage constraint between two points or, in other words, a short circuit. A dead current source means that there is zero current flowing between two points or, in other words, an open circuit between the two points.* In applications of superposition it is important to know whether to replace sources by open or short circuits.

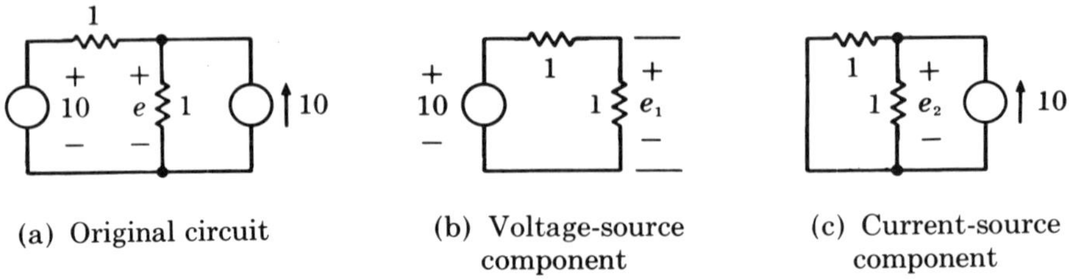

FIG. 3–15. A circuit solved by superposition.

As an example of the use of the superposition theorem we will compute the output voltage in the circuit of Fig. 3–15(a). We can consider this voltage to be composed of two parts, e_1 and e_2, contributed by the voltage source and the current source respectively. The component e_1 is computed from the circuit of Fig. 3–15(b). It is

$$e_1 = \tfrac{1}{2} \times 10 = 5 \text{ volts.}$$

The component e_2 is computed from the circuit of Fig. 3–15(c). It is

$$e_2 = 10 \times \tfrac{1}{2} = 5 \text{ volts.}$$

The total output voltage e is the sum of these two components:

$$e = e_1 + e_2 = 10 \text{ volts.}$$

Superposition is particularly useful when the components produced by the separate sources can be obtained by inspection, as in the problem just solved. If each response must be obtained by mesh or node equations it it is better to apply them directly to the original circuit.

A second type of superposition problem is illustrated in Fig. 3–16. It is known that the voltage E_3 is zero when I_1 is 1 amp and E_2 is 1 volt. Also, when I_1 is zero and E_2 is 10 volts, E_3 becomes 1 volt. The problem

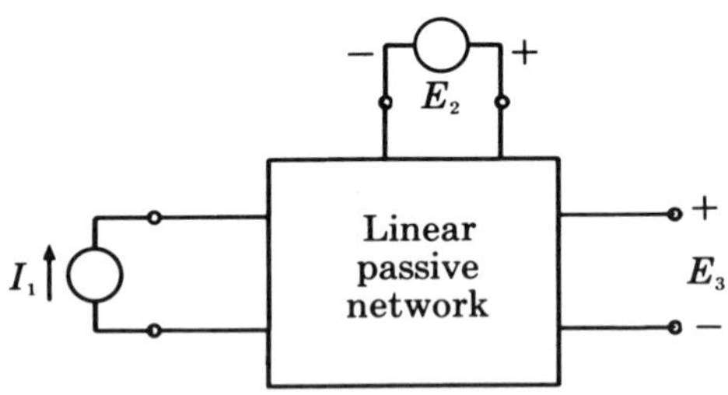

FIG. 3–16. An example of superposition.

is to find E_3 when I_1 is 10 amp and E_2 is zero. The mathematical statement of the superposition principle for this problem is

$$E_3 = K_1 I_1 + K_2 E_2,$$

where K_1 is the proportionality constant between the current I_1 and the voltage E_3 (it has the dimensions of resistance) and K_2 is the proportionality constant between the voltage E_2 and the voltage E_3 (it is dimensionless). The data given in the problem allow us to find the two constants K_1 and K_2 and thus to find the output for any combination of inputs. The equations are

$$0 = K_1 \times 1 + K_2 \times 1,$$

$$1 = K_1 \times 0 + K_2 \times 10,$$

from which $K_2 = -K_1 = \frac{1}{10}$. The general expression for E_3 is thus

$$E_3 = -\tfrac{1}{10} I_1 + \tfrac{1}{10} E_2.$$

When $I_1 = 10$ amp and E_2 is zero,

$$E_3 = -1 \text{ volt}.$$

The above solution is typical of a type of reasoning which will be used frequently in subsequent chapters. Four possible types of proportionality constants will occur, resistances and dimensionless voltage ratios as in this problem, and also conductances and dimensionless current ratios.

A direct application of the principle of linearity is *normalization* in circuit problems. For example, since the output response in a network is directly proportional to the value of the source, it is convenient to use source values of 1 volt and 1 amp, and to scale up the value of the response to agree with any other source value. The concept of normalization can be extended to the resistance values as well. The resulting circuit is a "linear model" of the original problem. The normalization can be carried out in two ways. If the voltages are left alone and all the resistances are multiplied by a constant, all of the currents will be divided by the same constant. This result can be verified in Eq. (3–40), but it is intuitively obvious from Ohm's law. Alternatively, the currents can be left unchanged, and then, if the resistances are multiplied by a constant, the voltages will be multiplied by the same constant. In many problems both the source and resistance values are scaled and a single normalization constant is used to convert the normalized solution to the solution for the original problem. The steps in this process are:

1. Resistances are scaled to values near 1 ohm.

2. The source is scaled to 1 volt or 1 amp. Only one source can be considered at a time.

3. The solution to the normalized problem is obtained.

4. A scale factor is computed which converts the normalized results to the answers for the original problem.

Normalization is useful in analysis and essential for any sort of rational network synthesis.

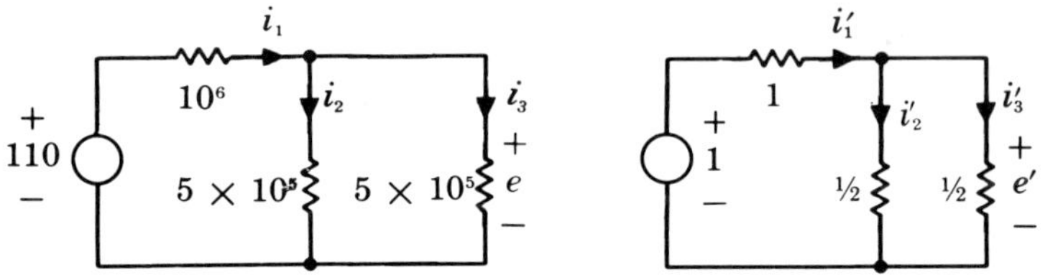

(a) The original circuit (b) The normalized circuit

FIG. 3–17. A problem solved by normalization.

As an example of this process, consider the circuit of Fig. 3–17(a). The scaled circuit or linear model is shown in (b). The source has been reduced by a factor of 110, and the resistance values by a factor of 10^6. The values of the current in the normalized circuit are

$$i_1' = 0.8 \text{ amp}, \quad i_2' = 0.4 \text{ amp}, \quad i_3' = 0.4 \text{ amp}, \quad e' = 0.2 \text{ volts}. \tag{3–41}$$

The currents in the original problem are 110 times larger than the currents in Eq. (3–11) because of the source normalization, and 10^6 times smaller because of the resistance normalization. They are therefore

$$i_1 = 0.8 \times 110 \times 10^{-6} = 88 \times 10^{-6} \text{ amp},$$

$$i_2 = 0.4 \times 110 \times 10^{-6} = 44 \times 10^{-6} \text{ amp},$$

$$i_3 = 0.4 \times 110 \times 10^{-6} = 44 \times 10^{-6} \text{ amp}.$$

The voltage in the original problem is 110 times as large as that given in Eq. (3–41) because of the source increase, and it is unchanged by the change in resistance. It is therefore

$$e = 0.2 \times 110 = 22 \text{ volts}.$$

If the circuit had been driven by a current source the resistance normalization would have changed the voltages and not the currents. A change in units can accomplish much the same results as normalization and is essentially equivalent to it. Common unit multipliers used in electrical work are the following:

M, Mega: multiplies the size of the unit by 10^6,

K, Kilo: multiplies the size of the unit by 10^3,

m, milli: multiplies the size of the unit by 10^{-3},

μ, micro: multiplies the size of the unit by 10^{-6}.

As an example, 1 megohm (written 1 M) is 10^6 ohms, and 1 microvolt (written 1 μ-volt) is 10^{-6} of a volt. Certain combinations of units are self-consistent. For example, a useful set in electronics is volts, kilohms, and milliamps.

We have considered a number of special applications of linearity and superposition. The real importance of the theorems, however, transcends these examples. It lies in the flexibility which these theorems lend to linear circuit analysis. The most general problem which need be considered is one with a single source and a single response. Multiple sources can be handled by superposition. The single "stimulus-response" relationship thus becomes the central theme of linear circuit theory, and the remainder of the book will be devoted to it. The stimulus-response relationship characterizes the network. By making use of this concept we will be able to solve rather complex networks involving energy storage devices and arbitrary forms of stimulating functions.

3–8 Summary.

Mesh currents. The equilibrium equations for the mesh currents are

$$\begin{aligned} i_1r_{11} - i_2r_{12} - \cdots - i_kr_{1k} &= E_1, \\ -i_1r_{12} + i_2r_{22} - \cdots - i_kr_{2k} &= E_2, \\ &\vdots \\ -i_1r_{1k} - i_2r_{2k} - \cdots + i_kr_{kk} &= E_k, \end{aligned}$$

where

r_{11} = total resistance around mesh 1,

r_{12} = resistance between mesh 1 and mesh 2,

E_1 = net source voltage producing clockwise current in mesh 1,

and so on.

Dummy mesh currents. Every current source creates a dummy mesh in a circuit.

Dummy mesh | *Constraint equation*

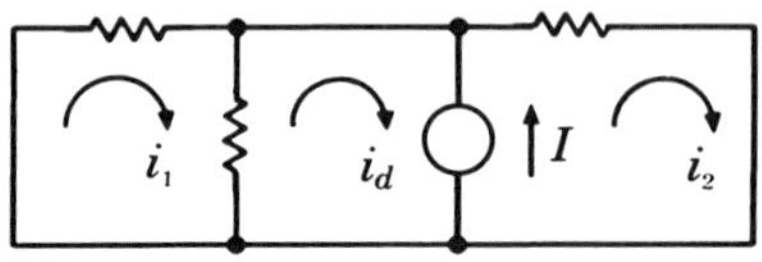

$$i_2 - i_d = I$$

Node voltages. The equilibrium equations for the node voltages are

$$\begin{aligned} e_1 g_{11} - e_2 g_{12} - \cdots - e_k g_{1k} &= I_1, \\ -e_1 g_{12} + e_2 g_{22} - \cdots - e_k g_{2k} &= I_2, \\ &\vdots \\ -e_1 g_{1k} - e_2 g_{2k} - \cdots + e_k g_{kk} &= I_k, \end{aligned}$$

where

g_{11} = total conductance radiating from node 1,

g_{12} = conductance between node 1 and node 2,

I_1 = net source current tending to make the voltage at node 1 positive,

and so on.

Dummy node voltages. Every voltage source creates a dummy node in a circuit.

Dummy node | *Constraint equation*

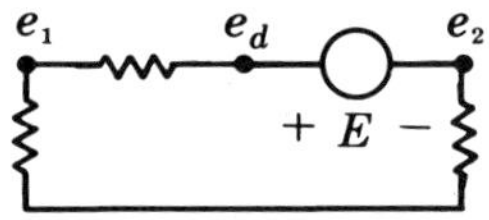

$$e_d = e_2 + E$$

Linearity. In a linear network the response is directly proportional to the stimulus.

For voltage source: $E_0 = K_1 E_s$ or $I_0 = K_2 E_s$,

where

E_s = the source stimulus,

E_0 = output voltage response,

I_0 = output current response,

K_1 = a dimensionless constant,

K_2 = a constant with dimensions of conductance.

For current source: $E_0 = K_3 I_s$ or $I_0 = K_4 I_s$,

where

I_s = the source stimulus,

E_0 = output voltage response,

I_0 = output current response,

K_3 = a constant with the dimensions of resistance,

K_4 = a dimensionless constant.

Superposition. In a linear network the response for two or more sources acting simultaneously is the sum of the responses for each source acting alone with the other voltage sources short circuits and the other current sources open circuits.

$$E_0 = K_1 I_1 + K_2 E_1 + \cdots,$$

$$I_0 = K_3 I_1 + K_4 E_1 + \cdots,$$

where

E_0 = the output voltage response,

I_0 = the output current response,

I_1 = a current source stimulus,

E_1 = a voltage source stimulus,

K_1 = a constant with dimensions of resistance,

K_2 = a dimensionless constant,

K_3 = a dimensionless constant,

K_4 = a constant with dimensions of conductance.

Scaling of units

M, Mega: multiplies size of unit by 10^6

K, Kilo: multiplies size of unit by 10^3

m, milli: multiplies size of unit by 10^{-3}

μ, micro: multiplies size of unit by 10^{-6}

REFERENCES

Mesh currents are traditional in electrical engineering texts. Node voltages, promoted by Professor Guillemin of M.I.T., are now found in most modern texts. The treatment of dummy variables given here is original, although occasional references to special cases will be found in the literature. Additional references on these basic methods are the following.

1. E. A. Guillemin, *Introductory Circuit Theory*, John Wiley & Sons, New York, 1953. Mesh and node methods are presented in the framework of topology and are spread through the first three chapters. Duality is also presented as a general topological concept.

2. M. B. Reed, *Alternating-Current Circuit Theory*, Harper Brothers, New York, 1956. A more advanced treatment relying on matrix theory. One of the few books which gives the transformations to be used with resistanceless sources.

3. Seshu and Balbanian, *Linear Network Analysis*, John Wiley & Sons, New York, 1959. A graduate-level treatment which follows Reed's approach. An example is given which involves a dummy variable.

4. LePage and Seely, *General Network Analysis*, McGraw-Hill, New York, 1952. A treatment of mesh and node equations in terms of steady-state sinusoidal impedances and elementary matrices. Only simple examples are considered.

5. Kerchner and Corcoran, *Alternating Current Circuits*, John Wiley & Sons, New York, 3rd ed, 1951. In this standard text, mesh equations are presented as a special technique for solving three-phase circuits. Node equations are briefly mentioned and little used.

6. H. H. Skilling, *Electrical Engineering Circuits*, John Wiley & Sons, New York, 1957. Mesh and node equations are presented in terms of steady-state impedances. An example contains a dummy variable.

7. H. G. Booker, *An Approach to Electrical Science*, McGraw-Hill, New York, 1959. Network theory is developed initially for capacitive networks on the theory that lossless elements are more basic than "lossy" ones. This approach gives perspective to network theory.

8. MIT EE Staff, *Electric Circuits*, John Wiley & Sons, New York, 1943. This text contains one of the earliest discussions of node equations. It is very straightforward. Source transformations are used to obtain voltage sources in mesh analysis, and current sources in node analysis.

EXERCISES*

Section 3–2

1. Write and solve the equations for the mesh currents in the circuit of Fig. 3–18.

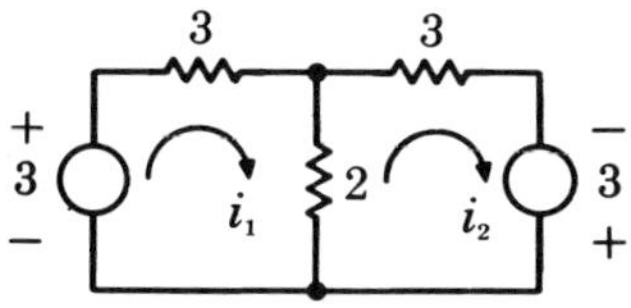

FIGURE 3–18

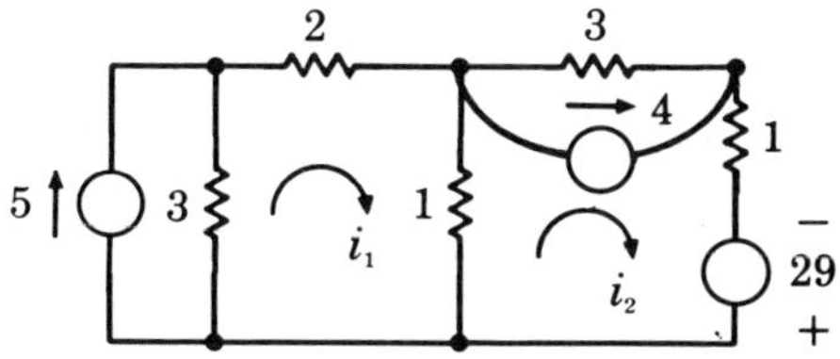

FIGURE 3–19

2. After making the necessary source transformations, write and solve the equations for the mesh currents in the circuit of Fig. 3–19.

3. Write and solve the mesh current equations for the circuit of Fig. 3–20, using source transformations if necessary.

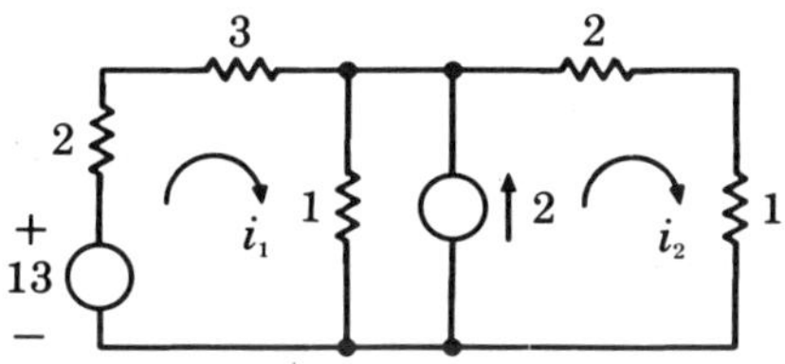

FIGURE 3–20

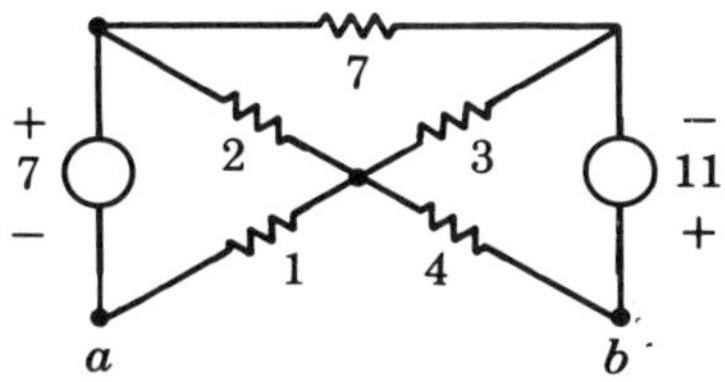

FIGURE 3–21

4. Set up the mesh equations for the circuit of Fig. 3–21 and use them to find the voltage drop from point a to point b.

5. Write the mesh equations for the circuit of Fig. 3–22 and solve for the current in the 12-ohm resistor. Redraw the circuit with the 12-ohm resistor below the diagram and find the current in it by inspection.

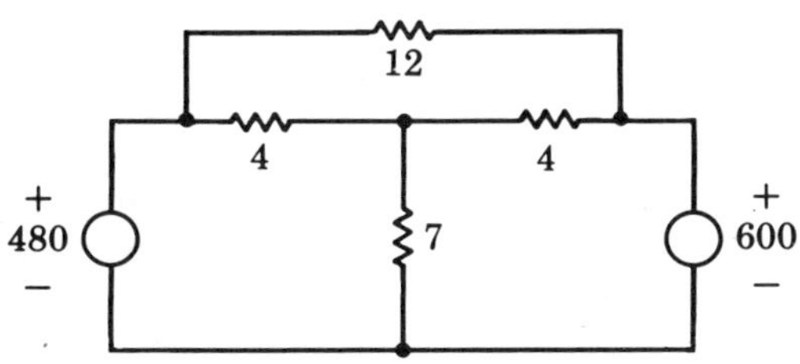

FIGURE 3–22

---

* Numerical values are in ohms, henrys, farads, volts, amperes.

6. Write the mesh equations for the circuit of Fig. 3–23 and use them to find the current and the power in each resistor. Be sure to give answers for the original circuit and not a transformed version of it.

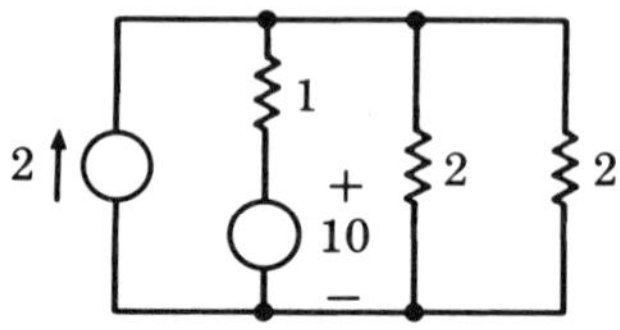

FIGURE 3–23

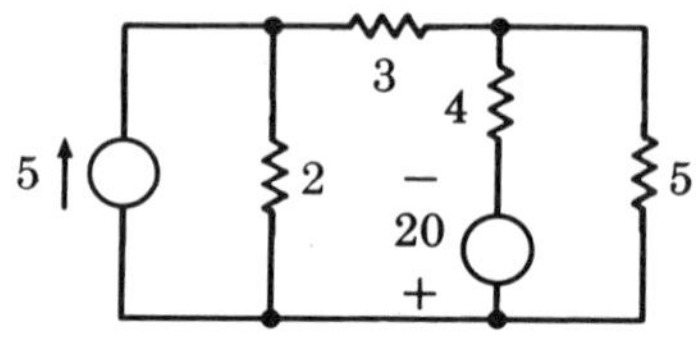

FIGURE 3–24

7. By means of the mesh equations, find the power supplied by the 5-amp source in the circuit of Fig. 3–24.

8. Find the values of the mesh currents in the circuit of Fig. 3–25.

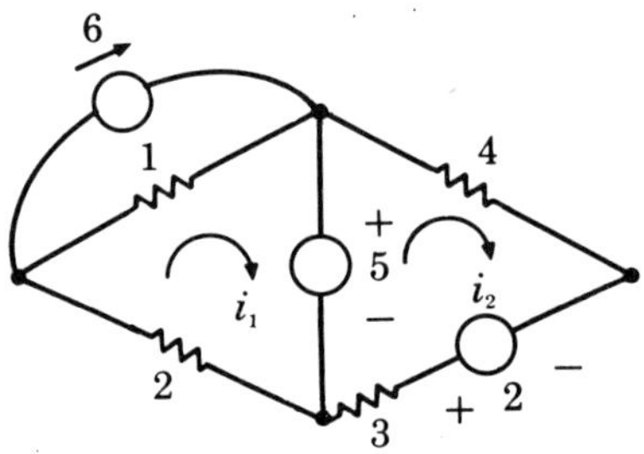

FIGURE 3–25

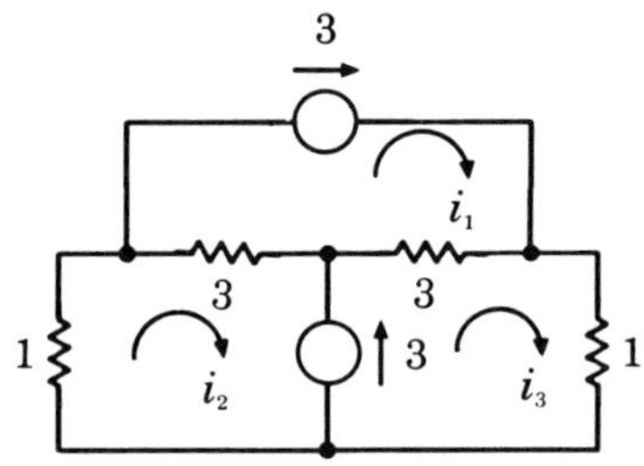

FIGURE 3–26

Section 3–3

9. Solve Exercise 2 by means of dummy mesh currents instead of source transformations.

10. Solve Exercise 3 by means of dummy mesh currents instead of source transformations.

11. Solve Exercise 6 by means of dummy mesh currents instead of source transformations.

12. Solve Exercise 7 by means of dummy mesh currents instead of source transformations.

13. The circuit of Fig. 3–26 involves two resistanceless sources. If it is to be solved by source transformations the sources must be replaced with equivalent source chains bridged around meshes. It is easier to solve the problem by means of dummy mesh currents. Find i_1, i_2, and i_3.

14. If a current source occurs in an outside loop the dummy mesh method is particularly simple. In the circuit of Fig. 3–27 this effect can be accomplished by bridging the current source in the third dimension. A dummy mesh current can be assumed in the dummy mesh, which is at right angles to the plane of the paper. Find i_1 and i_2, using this technique.

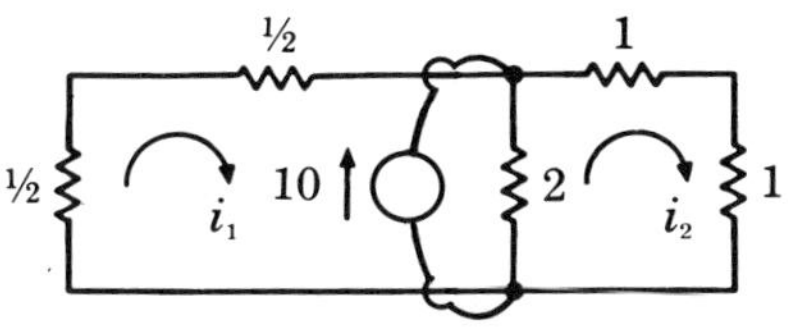

FIGURE 3–27

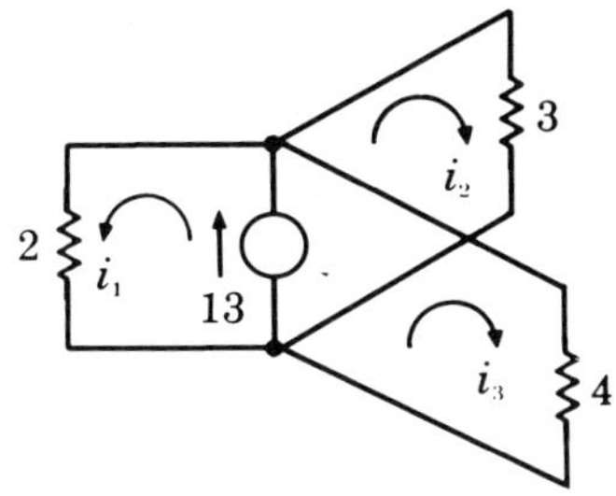

FIGURE 3–28

15. Solve for the currents i_1, i_2, and i_3 in the circuit of Fig. 3–28, using dummy mesh currents.

16. Find the currents i_1, i_2, i_3, and i_4 in the circuit of Fig. 3–29, using dummy mesh currents.

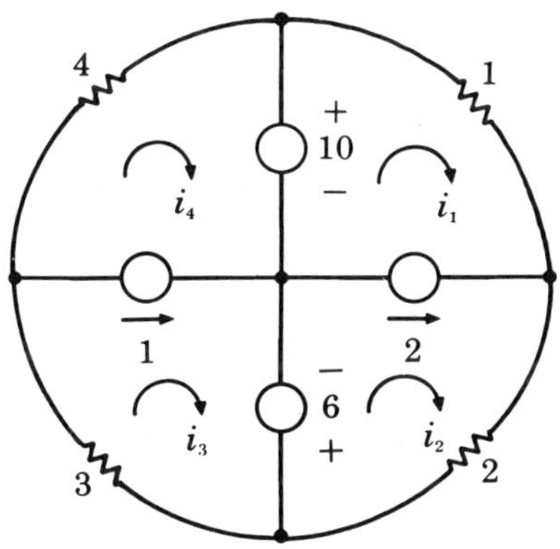

FIGURE 3–29

Section 3–4

17. Write and solve the node-voltage equations for the circuit of Fig. 3–30.

18. After making the necessary source transformations, write and solve the node-voltage equations for the circuit of Fig. 3–31.

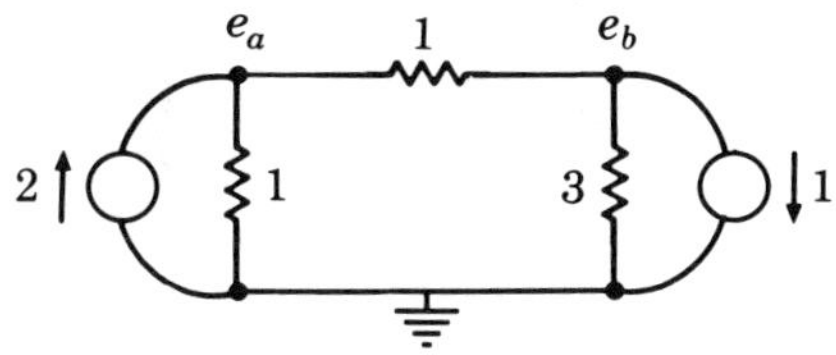

FIGURE 3–30

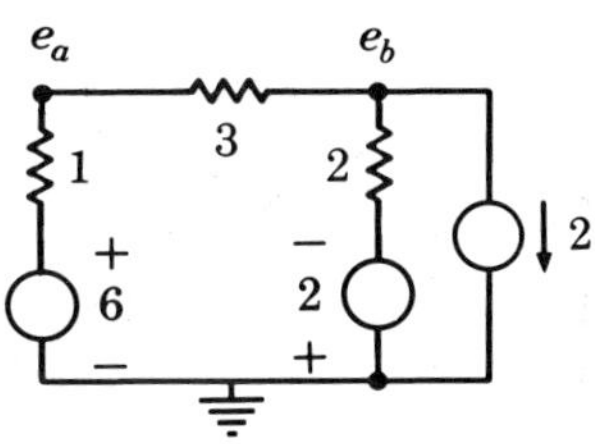

FIGURE 3–31

19. Solve for e_a and e_b in the circuit of Fig. 3–32, using any source transformations which are necessary.

20. Use the node-voltage method to set up the equations for the circuit of Fig. 3–33, and solve for the three voltages e_a, e_b, and e_c.

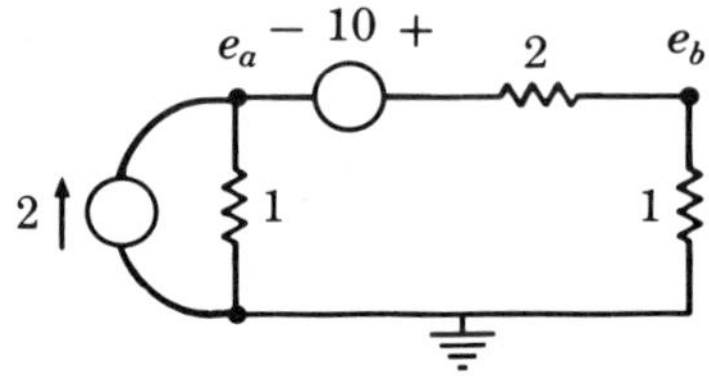

FIGURE 3–32

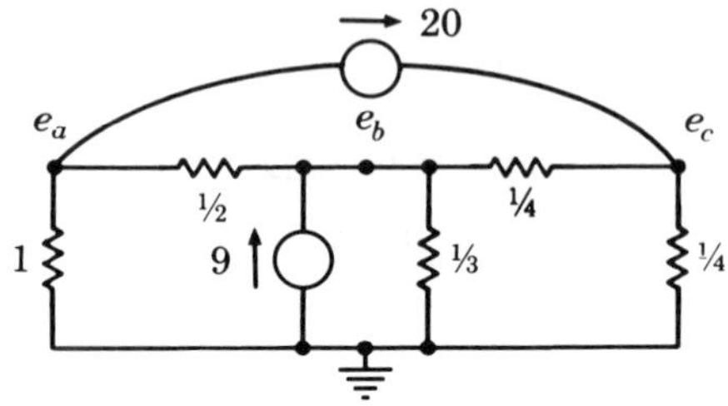

FIGURE 3–33

21. Write the node-voltage equations for the circuit of Fig. 3–34 and from them find e_a, e_b, and the power delivered by the 10-volt source.

22. Write the node equations and solve for the voltage e_2 in the circuit of Fig. 3–35. Redraw the circuit with the 5-ohm resistor and its two connected resistors below the remaining resistor. Now find the voltage across the 5-ohm resistor by inspection.

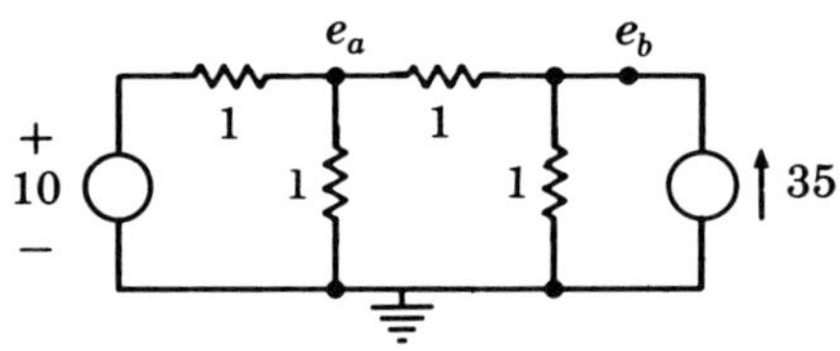

FIGURE 3–34

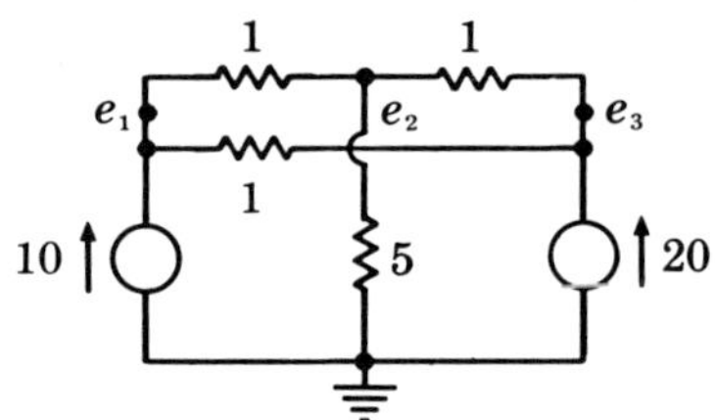

FIGURE 3–35

23. Solve for the node voltages e_1 and e_2 in the circuit of Fig. 3–36.

24. Find the values of the node voltages e_a and e_b in the circuit of Fig. 3–37.

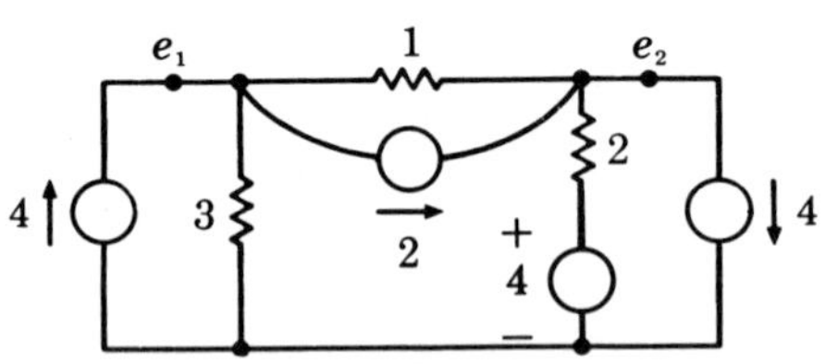

FIGURE 3–36

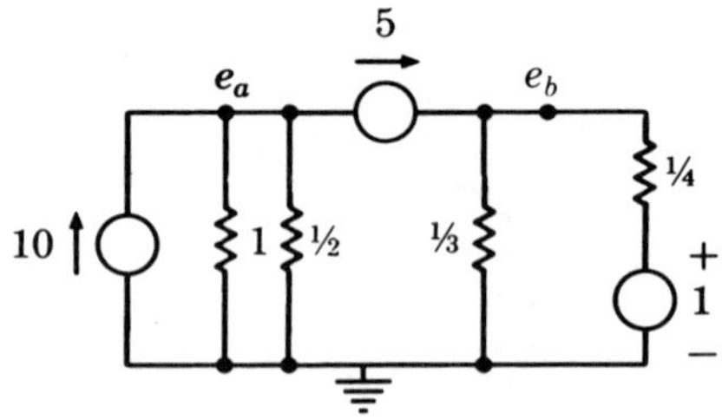

FIGURE 3–37

Section 3–5

25. Solve Exercise 18 by means of a dummy node voltage instead of a source transformation.

26. Solve Exercise 19 by means of a dummy node voltage instead of a source transformation.

27. Solve Exercise 21 by means of a dummy node voltage instead of a source transformation.

28. Solve Exercise 23 by means of a dummy node voltage and without any source transformations.

29. Solve for the voltages e_a, e_b, and e_c in the circuit of Fig. 3–38, using the dummy voltage technique and writing an equation at the supernode containing e_a and e_c.

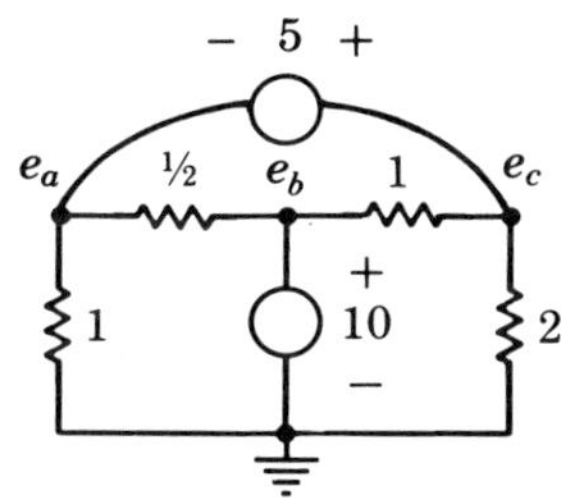

FIGURE 3–38

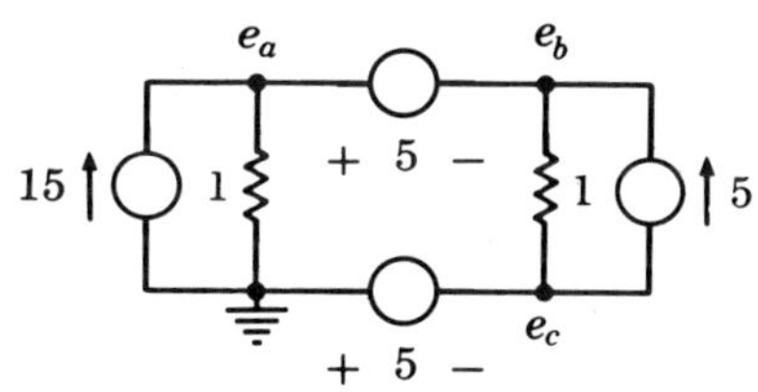

FIGURE 3–39

30. Solve for the node voltages e_a, e_b, and e_c in the circuit of Fig. 3–39, using the dummy-voltage method.

31. Find the voltages e_a, e_b, and e_c in the circuit of Fig. 3–40 by means of node-voltage equations and a constraint equation for the dummy variable.

32. Solve for the voltages e_a, e_b, e_c, and e_d in the circuit of Fig. 3–41.

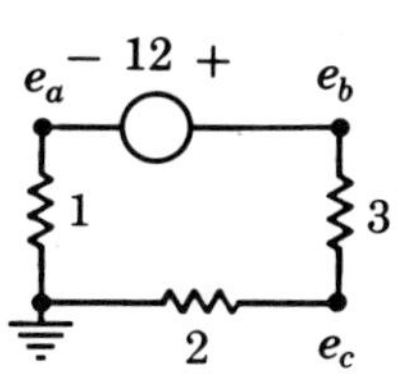

FIGURE 3–40

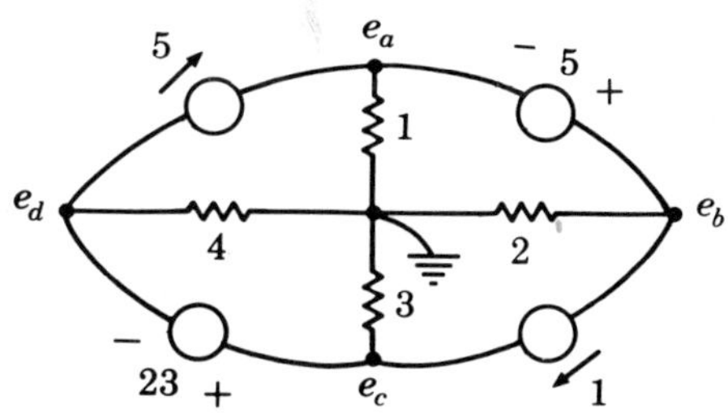

FIGURE 3–41

Section 3–6

33. Write the mesh equations for the circuit of Fig. 3–42. Transform these equations into the equations of the dual circuit. Construct the dual circuit (be sure the values are in ohms, volts, amperes).

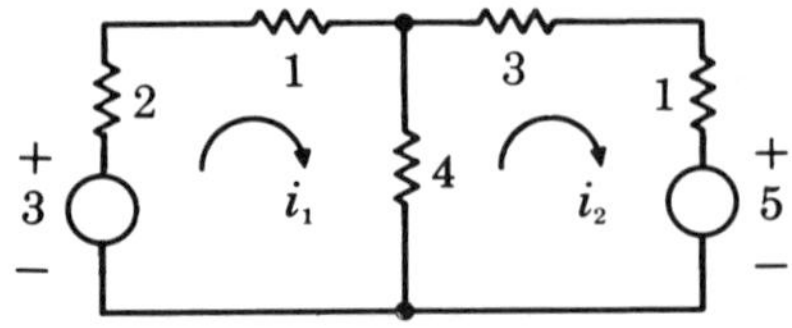

FIGURE 3–42

34. Obtain the dual of the circuit in Exercise 33 by the geometrical method.

35. Write the mesh equations and the constraint equations for the dummy variables in the circuit of Fig. 3–43. Transform these equations to the dual form. Construct the dual circuit.

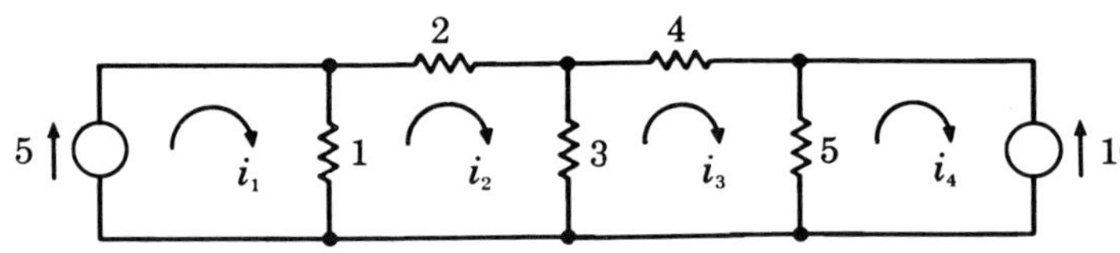

FIGURE 3–43

36. Obtain the dual of the circuit in Exercise 35 by the geometrical method.

37. Find the dual of the network of Fig. 3–44 and then obtain the dual of the dual network.

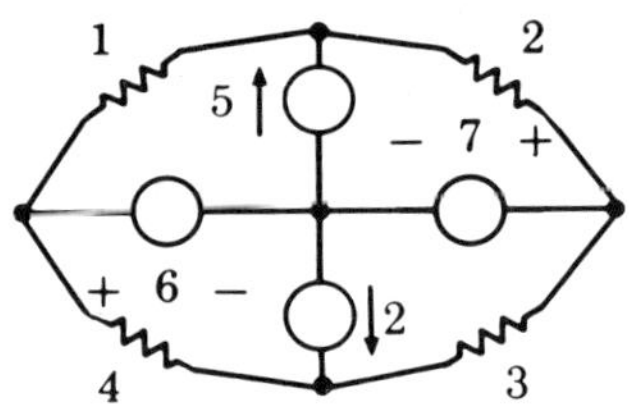

FIGURE 3–44

38. For the circuit of Fig. 3–45 write the three topological mesh equations and the constraint equation for the current source. Transform these equations to dual form and construct the dual network from the equations.

39. Obtain the dual of the circuit in Exercise 38 by the geometrical method.

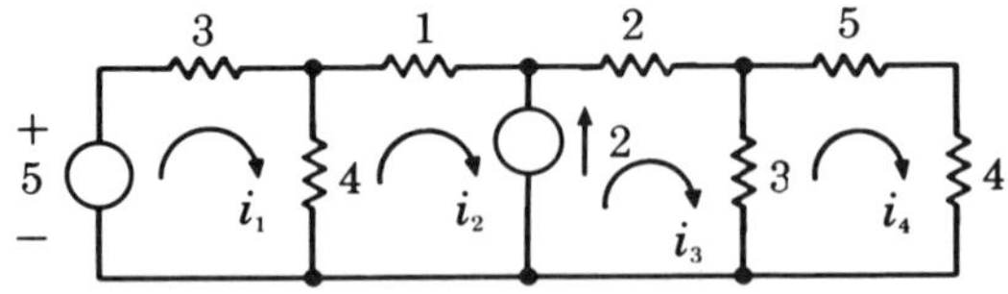

FIGURE 3–45

40. Obtain the dual of the network of Fig. 3–46. How many independent nodes are there in the original network? How many independent meshes are there in the dual network?

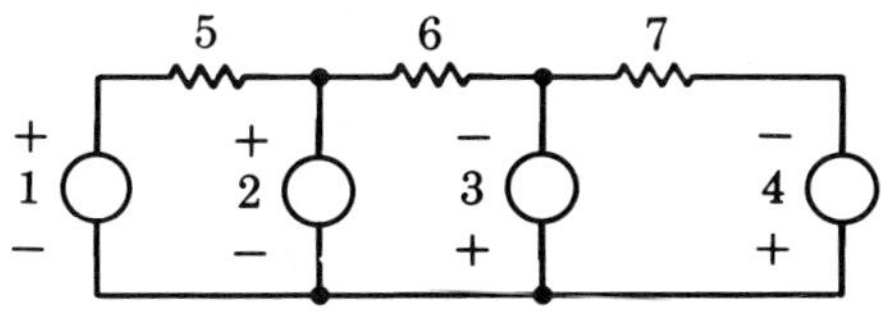

FIGURE 3–46

Section 3–7

41. (a) Find the value of the current in the 3-ohm resistance in the circuit of Fig. 3–47. (b) What does this current become when the input voltage is raised to 100 volts? (c) What does this current become if the input voltage is raised to 100 volts and the resistance values are simultaneously increased by 10^6?

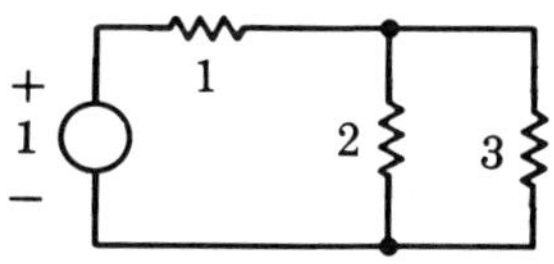

FIGURE 3–47

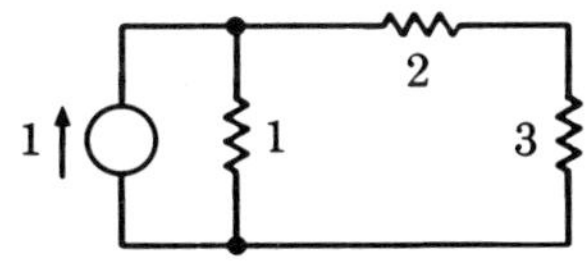

FIGURE 3–48

42. (a) Compute the voltage across the 3-ohm resistor in the circuit of Fig. 3–48. (b) What does this voltage become if the driving current is reduced to 1 ma? (c) What does this voltage become if the driving current is reduced to 1 ma and simultaneously the resistance values are increased by a factor of 10^6?

43. (a) Normalize the circuit of Fig. 3–49 to give a maximum resistance of 6 ohms. Compute the normalized current i. (b) Plot a curve of the current i in the original circuit as the driving voltage is changed from zero to 100 volts.

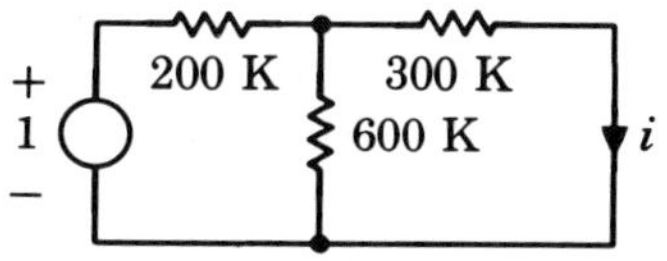

FIGURE 3–49

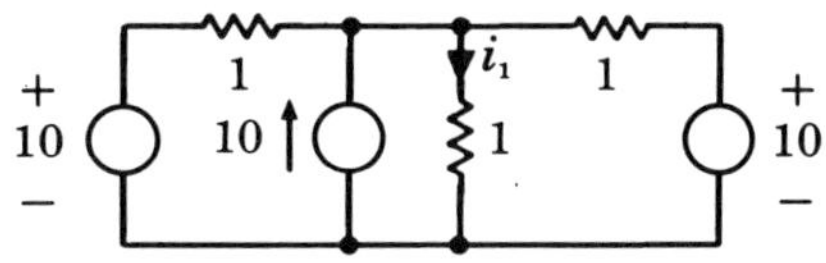

FIGURE 3–50

44. Discuss the following pairs of units from the standpoint of normalization, and give the third unit in each case. (a) milliamperes and kilohms, (b) milliamperes and kilovolts, (c) microamperes and volts, (d) kilovolts and megohms.

45. Find the current i_1 in the circuit of Fig. 3–50, using the principle of superposition. What currents are drawn from the voltage sources when all three sources are acting?

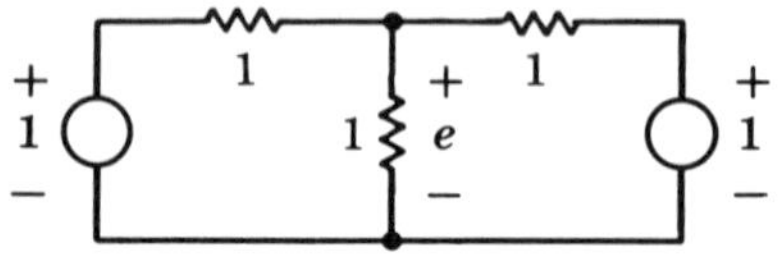

FIGURE 3–51

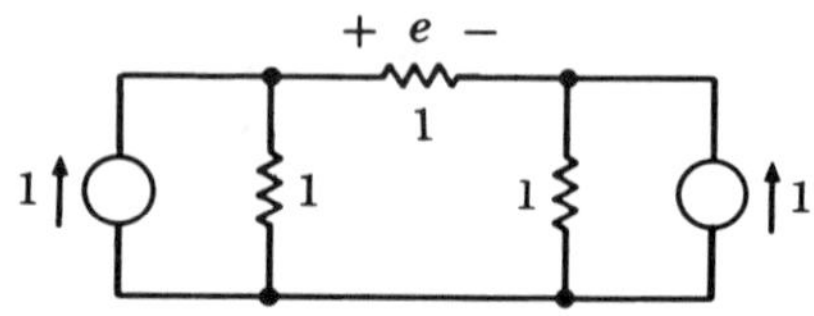

FIGURE 3–52

46. Find the voltage e in the circuit of Fig. 3–51 by using the superposition theorem. If the source at the right is reversed, what does this voltage become?

47. Find the voltage e in the circuit of Fig. 3–52, using the superposition theorem. If the source at the right is reversed, what does this voltage become?

48. (a) By means of superposition, find the current in the resistor in the circuit of Fig. 3–53. (b) With both sources acting, give the power from the voltage source, the power from the current source, and the power used in the resistor. (c) Formulate and solve the dual-circuit problem.

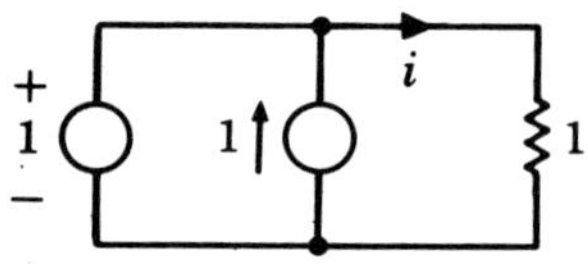

FIGURE 3–53

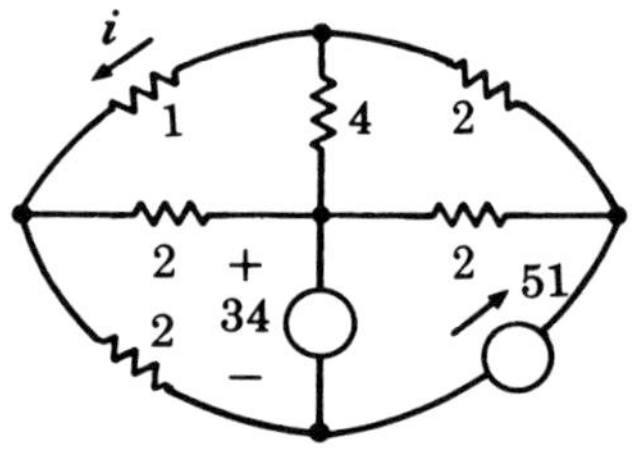

FIGURE 3–54

49. Find the current i through the 1-ohm resistor in the circuit of Fig. 3–54 by means of the superposition theorem.

50. A linear passive circuit has two sets of terminal-pairs. When 10-volt sources are applied to both of them the input currents are each 1 amp. When one of the voltages is dropped to 5 volts the current from the other source drops to zero. What value of voltage at the source with zero current would cause the current to drop to zero if the voltage at the other terminal were left at 10 volts?

51. Two current sources I_1 and I_2 are applied to a network and a voltage e is read at a third terminal-pair. When $I_1 = 1$ and $I_2 = 0$, the voltage e is 1 volt. When $I_1 = 0$ and $I_2 = 1$ the voltage e is -2 volts. What is the voltage e when $I_1 = -1$ amp and $I_2 = 2$ amp?

52. It is possible that the superposition method will result in an answer which is the difference between two large quantities. If so, computational care must be used in order to obtain an answer which is not grossly in error. As an example, find the current i in the circuit of Fig. 3–55 by means of superposition.

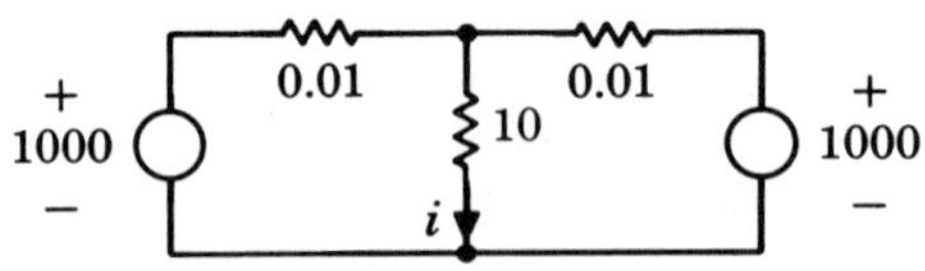

FIGURE 3–55

CHAPTER 4

METHODS OF CHARACTERIZING ONE-TERMINAL-PAIR RESISTIVE NETWORKS

"I sometimes think that never blows so red
The Rose, as where some buried Caesar bled;
That every Hyacinthe the Garden wears
Dropt in her lap from some once lovely head."

Rubáiyát of Omar Khayyám

4–1 Introduction. In this chapter we shall begin to investigate the special properties of linear circuits. Almost no area of applied mathematics has been developed to the level of sophistication of this one, and we can properly join old Omar the tentmaker in appreciating the sufferings of those who prepared the way for us! As a consequence of linearity and superposition in a network we need to consider no more than a single excitation function at a time. Two classes of problem arise, however, depending on where the response is measured. In a *one-terminal-pair network* the stimulus and the response occur at the same pair of terminals. One of them is a current and the other a voltage. In a *two-terminal-pair network* the stimulus is applied at one pair of terminals and the response is measured at another.

In this chapter we shall consider one-terminal-pair networks. They take the form of branches, which may be simple or compound and which are described mathematically, or characterized, by the equivalent resistance of the branch. In earlier chapters we dealt with some special cases of equivalent resistance. Now we shall consider two general approaches:

1. The application of a current or a voltage source and the computation of the resulting voltage or current.

2. The reduction of the branch by successive applications of series-parallel and wye-delta formulas.

The only essentially new element in these methods is the *wye-delta transformation*, which will be developed in this chapter along with special methods for handling symmetrical networks and ladder-type networks.

When a one-terminal-pair network contains sources as well as resistances, it can still be reduced to an equivalent simple branch which now contains a source as well as a resistance. If the branch is taken to be a voltage source in series with a resistance we have *Thévenin's theorem*, and if it is taken as a current source in parallel with a resistance we have

Norton's theorem. Occasionally it is desirable to substitute a voltage source for a known value of voltage at some point in a circuit, or a current source for a known value of current. The equivalence seems obvious, but it is dignified by the name *substitution theorem.*

4–2 Equivalent resistance by source application. An equivalent resistance is defined to be one which draws the same current as the original network when it is excited by the same voltage.* The ratio of the voltage across a terminal-pair to the current flowing into it gives the value of the equivalent resistance. In the laboratory resistance is measured by applying a current source and measuring the resulting voltage, or by applying a voltage source and measuring the resulting current.† Analytically, the same procedure is followed. A current source is applied to the terminal-pair and the resulting voltage is computed, or a voltage source is applied and the resulting current is computed. We have used this technique in a limited way in the previous chapters. Now we will combine it with the powerful methods of mesh and node equations in order to achieve greater generality.

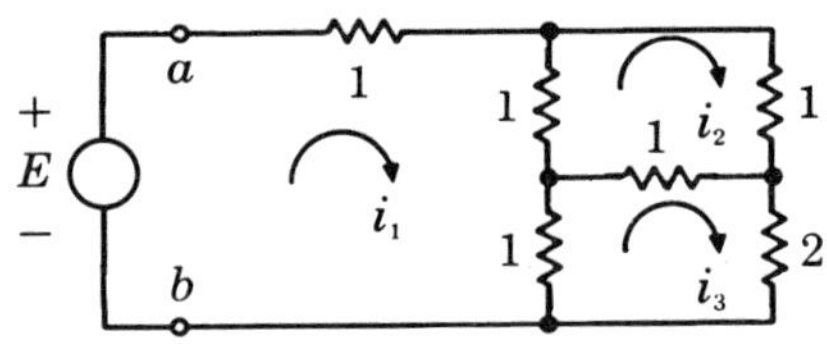

FIG. 4–1. Resistance computation by means of mesh currents.

Mesh-equation method. The natural source for the mesh equations is a voltage source. Accordingly, if we wish to use the mesh-equation method to find the resistance at some pair of terminals we usually apply a voltage source and solve for the mesh current which flows through it. Since resistance is the ratio of voltage to current, the exact value of the voltage source is of no consequence. It can be taken as one volt or it can be left in literal form. To illustrate the method, we will find the resistance between the terminals a and b in the circuit of Fig. 4–1.‡ This circuit cannot be reduced to a single resistance by application of series-parallel methods. However, mesh equations provide a straightforward solution. These

---

* See Section 2–2.

† Actually, we simply measure the current and the voltage at the terminals. A measured current is equivalent to a current source.

‡ The resistance between the terminals is often colloquially referred to as the resistance "looking in at the terminals."

equations are

$$
\begin{aligned}
i_1(3) - i_2(1) - i_3(1) &= E,\\
-i_1(1) + i_2(3) - i_3(1) &= 0, \qquad (4\text{–}1)\\
-i_1(1) - i_2(1) + i_3(4) &= 0.
\end{aligned}
$$

The desired mesh current through the source is

$$
i_1 = \frac{\begin{vmatrix} E & -1 & -1 \\ 0 & 3 & -1 \\ 0 & -1 & 4 \end{vmatrix}}{\begin{vmatrix} 3 & -1 & -1 \\ -1 & 3 & -1 \\ -1 & -1 & 4 \end{vmatrix}} = \tfrac{11}{24}E,
$$

and the equivalent resistance is

$$
R = \frac{E}{i_1} = \tfrac{24}{11}\text{ ohms.}
$$

The mesh-current method is perfectly general and can always be used to find the equivalent resistance between a pair of terminals. Sometimes the node-voltage method will be easier, however, and sometimes the best method is a step-by-step reduction of the network to a single resistance.

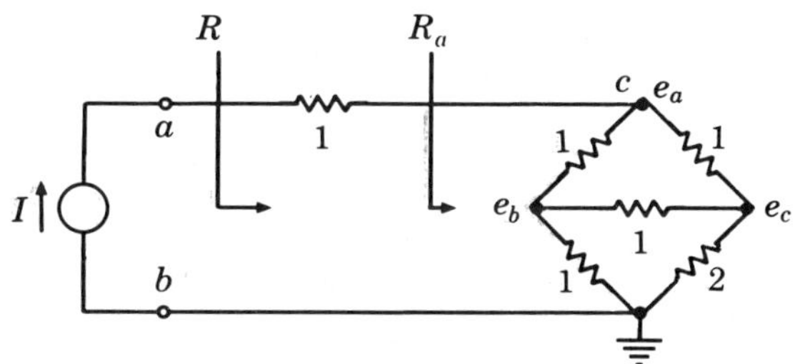

FIG. 4–2. Resistance computation by means of node voltages.

Node-equation method. If we wish to use node equations to find the equivalent resistance at a pair of terminals we apply a current source and compute the voltage across it. As before, the exact value of the source is immaterial, since the resistance is obtained as a ratio. The method will be illustrated by means of the circuit of Fig. 4–2, which is identical to the one solved previously by the mesh method. The resistance in series with the current source has no effect, and we will find only the resistance R_a between the terminals c and b. There are three nodes in the circuit, whose equations are

$$\begin{aligned} e_a(2) - e_b(1) - e_c(1) &= I, \\ -e_a(1) + e_b(3) - e_c(1) &= 0, \\ -e_a(1) - e_b(1) + e_c(\tfrac{5}{2}) &= 0. \end{aligned} \tag{4–2}$$

The voltage is

$$e_a = \frac{\begin{vmatrix} I & -1 & -1 \\ 0 & 3 & -1 \\ 0 & -1 & \frac{5}{2} \end{vmatrix}}{\begin{vmatrix} 2 & -1 & -1 \\ -1 & 3 & -1 \\ -1 & -1 & \frac{5}{2} \end{vmatrix}} = \tfrac{13}{11} I,$$

and the resistance is

$$R_a = \frac{e_a}{I} = \tfrac{13}{11} \text{ ohms.}$$

The input resistance, as before, is

$$R = 1 + R_a = \tfrac{24}{11} \text{ ohms.}$$

The node-voltage method of finding a resistance is equally as general as the mesh-current method. The choice between them should be based on the topology of the network, with a view to requiring the smallest number of equations. Sometimes the use of a dummy mesh or a dummy node will reduce the number of equations. For example, the resistance R_a in Fig. 4–3 can be found by applying a voltage source and writing node equations.

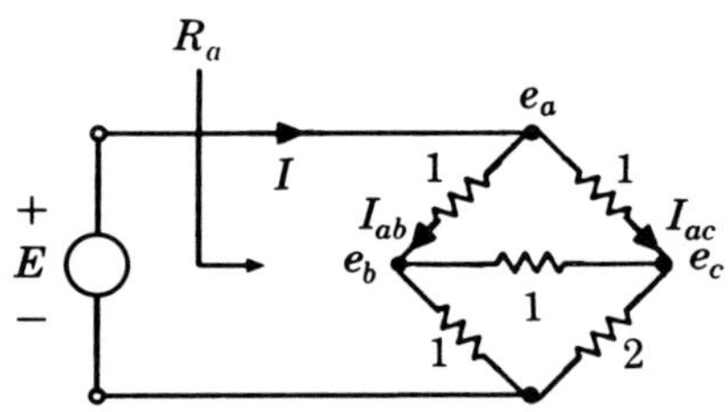

FIG. 4–3. Resistance computation by means of a dummy node voltage.

The circuit is the same as the right-hand portion of the circuit of Fig. 4–2. The node e_a is a dummy node, and only two node equations need be written. They are

$$\begin{aligned} e_b(3) - e_c(1) &= e_a/1 = E, \\ -e_b(1) + e_c(\tfrac{5}{2}) &= e_a/1 = E, \end{aligned} \tag{4–3}$$

from which

$$e_b = \frac{\begin{vmatrix} E & -1 \\ E & \frac{5}{2} \end{vmatrix}}{\begin{vmatrix} 3 & -1 \\ -1 & \frac{5}{2} \end{vmatrix}} = \tfrac{7}{13}E \quad \text{and} \quad e_c = \frac{\begin{vmatrix} 3 & E \\ -1 & E \end{vmatrix}}{\begin{vmatrix} 3 & -1 \\ -1 & \frac{5}{2} \end{vmatrix}} = \tfrac{8}{13}E. \tag{4-4}$$

The branch currents and the input current can be found from the node voltages in Eq. (4–4). Thus

$$I_{ab} = \frac{e_a - e_b}{1} = E(1 - \tfrac{7}{13}) = \tfrac{6}{13}E,$$

$$I_{ac} = \frac{e_a - e_c}{1} = E(1 - \tfrac{8}{13}) = \tfrac{5}{13}E,$$

and the total current is

$$I = I_{ab} + I_{ac} = \tfrac{11}{13}E,$$

from which the resistance, as before, is found to be

$$R_{ab} = \frac{E}{I} = \tfrac{13}{11} \text{ ohms.}$$

The hybrid method used in this problem may be less work in some instances than the mesh-current method or the node-voltage method, but it is a good deal less straightforward. It appeals most to those who are skilled in circuit theory.

4–3 Equivalent resistance by network reduction. Resistances in series and resistances in parallel can be combined into equivalent resistances. We have already used this technique in simplifying a number of series-parallel networks.* Not all networks, however, can be reduced by the application of series and parallel formulas. The network shown in Figs. 4–1, 4–2, and 4–3 falls into this catagory. A more general formula, called a Y–Δ transformation, is required.

The Y–Δ *transformation.* The Y–Δ transformation is more general than the series-parallel formula because it relates to the behavior of a three-terminal network instead of a two-terminal network. A Y network is composed of three resistances connected to a common point O, as shown in Fig. 4–4(a). A Δ network consists of three resistances connected in a triangular form, as shown in Fig. 4–4(b). When we are concerned only with the effect of these networks upon some external network we can find values

---

* See Sections 2–2, 2–3, and 2–4.

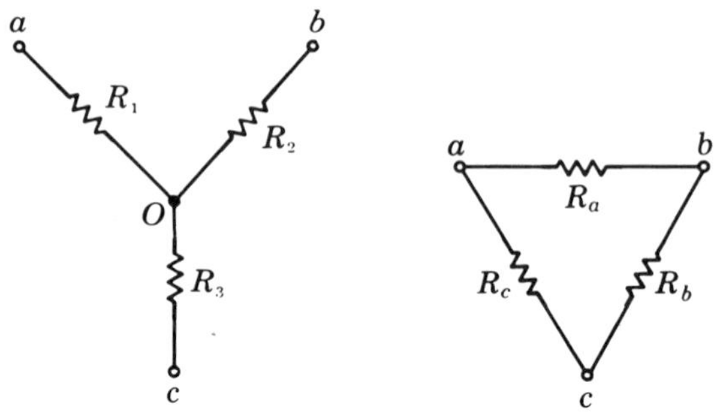

(a) The Y network (b) The Δ network

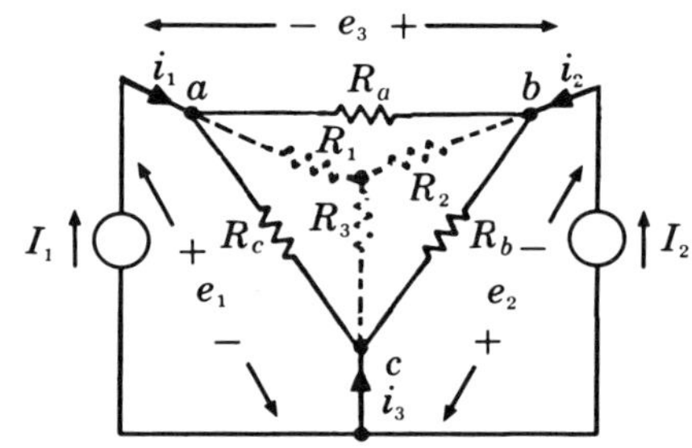

(c) The excitation of a Y or a Δ

FIG. 4–4. The basic Y and Δ networks.

of the resistances which will make the Y network and the Δ network equivalent. The nature of the equivalence is illustrated in Fig. 4–4(c). For either the Y or the Δ network connected between the points a, b, and c, we must obtain the same current values i_1, i_2, and i_3, and the same voltage values e_1, e_2, and e_3.

In order to establish the equivalence we can assume that we know the currents i_1, i_2, and i_3, and can therefore replace them with current sources. From the given currents each of the networks must generate the same voltages e_1, e_2, and e_3. Kirchhoff's laws simplify the problem.* The whole network in either case is a supernode and

$$i_1 + i_2 + i_3 = 0 \qquad \text{or} \qquad i_3 = -(i_1 + i_2). \tag{4–5}$$

Equation (4–5) shows that we can excite the network with only two current sources, at i_1 and i_2, and that the current i_3 must provide a return

* For a more complete discussion, see the reciprocity theorem, Section 5–2.

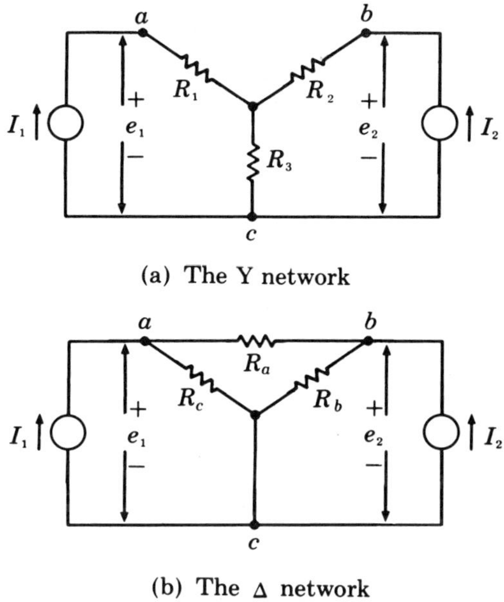

FIG. 4–5. The Y–Δ transformation.

path for these two currents. Similarly, the sum of the voltages must be equal to zero:

$$e_1 + e_2 + e_3 = 0 \qquad \text{or} \qquad e_3 = -(e_1 + e_2) \tag{4–6}$$

Equation (4–6) indicates that if the two responses e_1 and e_2 are known, the response e_3 is simply their sum. We can therefore safely ignore e_3.

In summary, we will establish the equivalence of the Y and the Δ by driving terminal-pairs *a–c* and *b–c* with current sources and requiring that the voltages across the terminals be the same in the Y and the Δ. The two equivalent networks with their excitations are shown in Fig. 4–5. The responses can be written down by inspection for each network by means of the superposition theorem. For the Y network the voltages are

$$\begin{aligned} e_1 &= (R_1 + R_3)I_1 + R_3 I_2, \\ e_2 &= R_3 I_1 + (R_2 + R_3)I_2, \end{aligned} \tag{4–7}$$

and for the Δ network they are

$$\begin{aligned} e_1 &= \frac{R_c(R_a + R_b)}{R_a + R_b + R_c} I_1 + \frac{R_c R_b}{R_a + R_b + R_c} I_2, \\ e_2 &= \frac{R_c R_b}{R_a + R_b + R_c} I_1 + \frac{R_b(R_a + R_c)}{R_a + R_b + R_c} I_2. \end{aligned} \tag{4–8}$$

The coefficients in Eqs. (4–7) and (4–8) must be identical, and thus

$$
\begin{aligned}
R_1 + R_3 &= \frac{R_c(R_a + R_b)}{R_a + R_b + R_c}, \\
R_3 &= \frac{R_b R_c}{R_a + R_b + R_c}, \\
R_2 + R_3 &= \frac{R_b(R_a + R_c)}{R_a + R_b + R_c}.
\end{aligned}
\tag{4–9}
$$

From Eq. (4–9) we can obtain

$$
\begin{aligned}
R_1 &= \frac{R_a R_c}{R_a + R_b + R_c}, \\
R_2 &= \frac{R_a R_b}{R_a + R_b + R_c}, \\
R_3 &= \frac{R_b R_c}{R_a + R_b + R_c}.
\end{aligned}
\tag{4–10}
$$

Inspection of Fig. 4–4(c) makes it apparent that a general expression for the Y resistance is

$$
\text{Y resistance} = \frac{\text{Product of adjacent } \Delta \text{ resistances}}{\text{Sum of } \Delta \text{ resistances}}. \tag{4–11}
$$

Alternatively, Eq. (4–9) could have been solved for the Δ resistors and the following equations would have been obtained:

$$
\begin{aligned}
R_a &= \frac{R_1 R_2 + R_2 R_3 + R_3 R_1}{R_3}, \\
R_b &= \frac{R_1 R_2 + R_2 R_3 + R_3 R_1}{R_1}, \\
R_c &= \frac{R_1 R_2 + R_2 R_3 + R_3 R_1}{R_2}.
\end{aligned}
\tag{4–12}
$$

Again an inspection of Fig. 4–4(c) reveals that the Δ resistances are given by

$$
\Delta \text{ resistance} = \frac{\text{Sum of the products of the Y resistances taken in pairs}}{\text{The opposite Y resistance}} \tag{4–13}
$$

It is also possible to convert Eq. (4–12) to conductances, and obtain

$$
\begin{aligned}
G_a &= \frac{G_1 G_2}{G_1 + G_2 + G_3}, \\
G_b &= \frac{G_2 G_3}{G_1 + G_2 + G_3}, \qquad (4\text{–}14) \\
G_c &= \frac{G_3 G_1}{G_1 + G_2 + G_3}.
\end{aligned}
$$

Equation (4–14) is in the form

$$
\Delta \text{ conductance} = \frac{\text{Product of adjacent Y conductances}}{\text{Sum of Y conductances}}. \qquad (4\text{–}15)
$$

Equations (4–13) and (4–15) are equivalent and the choice between them is dictated only by personal preference. Some people like to work always in ohms, and some people like the similarity of Eqs. (4–11) and (4–15).

When all the resistances in the Y are equal, the equivalent Δ resistances are also equal, and are equal to three times the Y resistances. Similarly, if the Δ resistances are equal the Y resistances, which are equivalent, are also equal, and are equal to one-third of the Δ resistances. This symmetrical transformation, shown in Fig. 4–6, is a useful one to remember as a check on one's memory of the general transformation.

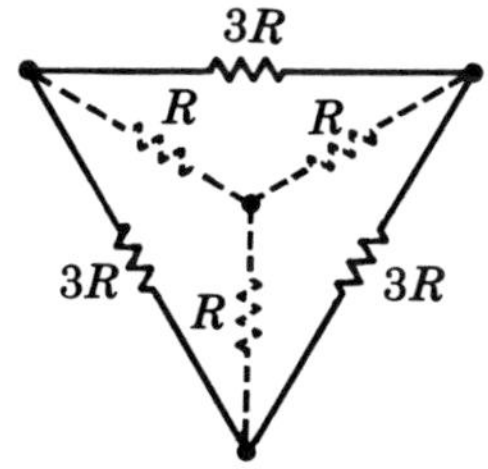

FIG. 4–6. The symmetrical Y–Δ transformation.

As an example of the reduction of a complex branch to a single resistance by means of the Y–Δ transformation and the series-parallel formulas, we consider the circuit of Fig. 4–7(a). We have already found the resistance of this circuit by several other methods. The first step in the reduction procedure is a Y–Δ transformation of the section of the network enclosed by *bde*. The resultant circuit is shown in Fig. 4–7(b). The node *c* has been removed and the circuit has a series-parallel form. The node *d* can be removed by combining resistances in parallel and then

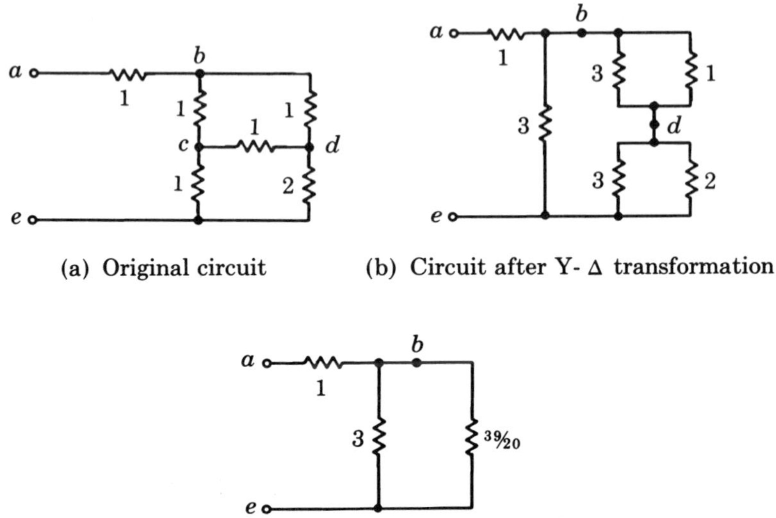

(a) Original circuit (b) Circuit after Y- Δ transformation

(c) Circuit after series- parallel transformation

FIG. 4–7. The reduction of a one-terminal-pair network to a single resistance by Y–Δ and series-parallel transformations.

in series. The circuit then appears as in Fig. 4–7(c), and the input resistance can be found by inspection to be

$$R_{ae} = \tfrac{24}{11} \text{ ohms,}$$

as in the previous examples.

A combination of Y–Δ or Δ–Y transformations and the series-parallel formulas can be used to reduce any resistance network to a single equivalent resistance. This method is very easy when the Y–Δ transformation is symmetrical, as in this example. When the Y–Δ transformations are unsymmetrical this method is seldom any less work than the mesh- and node-equation methods. Nevertheless, an example of an unsymmetrical Δ–Y transformation is shown in Fig. 4–8. In order to reduce Fig. 4–8(a) we have picked the Δ *acda* and have changed it to a Y with three resistances R_1, R_2, and R_3, and a new node 0. The resistance values (Eq. 4–11) are

$$R_1 = \frac{3 \times 1}{3 + 2 + 1} = \tfrac{1}{2} \text{ ohm,}$$

$$R_2 = \frac{3 \times 2}{3 + 2 + 1} = 1 \text{ ohm,}$$

$$R_3 = \frac{2 \times 1}{3 + 2 + 1} = \tfrac{1}{3} \text{ ohm.}$$

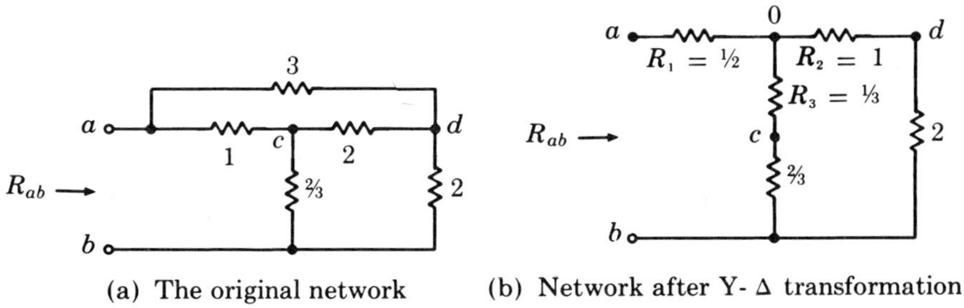

FIG. 4–8. The reduction of an unsymmetrical network by Y–Δ methods.

From the new equivalent circuit shown in Fig. 4–8(b) the input resistance can be obtained by inspection:

$$R_{ab} = \tfrac{1}{2} + \frac{1 \times 3}{1 + 3} = \tfrac{5}{4} \text{ ohms.}$$

4–4 The resistance of symmetrical networks. Problems which possess natural symmetry occur in all branches of science, and network theory is no exception. An exploitation of symmetry properties can reduce the effort required to solve a complicated problem such as may occur in three-phase power systems, motor and generator windings, electrical filters, and the electrical analogs of field-theory problems. The recognition of symmetry is part of the "art" of network theory. There are a few general rules which can be applied, but beyond that the network theorist must use his own ingenuity. Some things to look for in a network are the following:

1. *A branch with no voltage across it.* Such a branch can be removed or replaced by a short circuit without altering the rest of the network.
2. *A branch with no current in it.* Such a branch can also be removed or replaced by a short circuit without altering the rest of the network.
3. *Two nodes at the same potential.* A short circuit can be placed between these nodes without altering the rest of the circuit.
4. *Two adjacent mesh currents which are clockwise and equal.* The branch between the two meshes can be opened up because it has no current in it.

In general, a network should be redrawn and altered to make it look as symmetrical as possible in a physical sense, although electrical symmetry and physical symmetry are not coincident. Resistances can be split into two resistances in series, or into two resistances in parallel. Sources similarly can be replaced by series or parallel combinations. The following theorem will help to uncover electrical symmetry:

If a source is reversed in polarity all the currents and voltages produced by it are reversed in polarity but unchanged in magnitude.

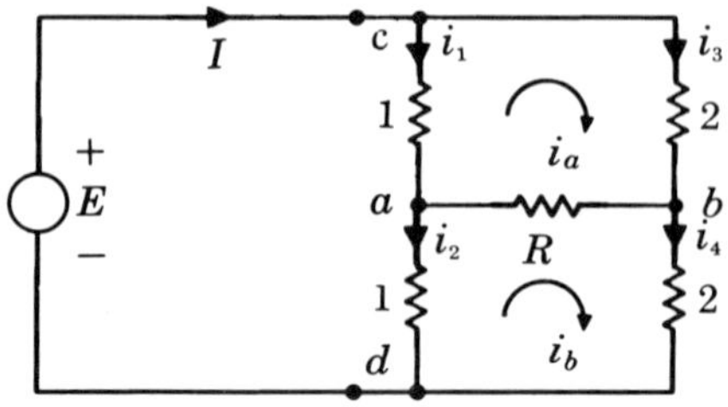

FIG. 4–9. A symmetrical bridge circuit.

The balanced bridge shown in Fig. 4–9 is an example of a symmetrical circuit. We can uncover the symmetry by any of the four ways listed above.

Method 1. Branch voltage. The branch R has no voltage across it and can therefore be removed from the circuit or replaced by a short circuit. This conclusion is easiest to reach by noting that if R is removed from the circuit the voltages at points a and b are equal. If R is replaced by an open circuit the resistance between c and d is

$$R_{cd} = \frac{2 \times 4}{2 + 4} = \tfrac{4}{3} \text{ ohms.}$$

If R is replaced by a short circuit the resistance between c and d is

$$R_{cd} = \frac{2 \times 1}{2 + 1} + \frac{2 \times 1}{2 + 1} = \tfrac{4}{3} \text{ ohms.}$$

Method 2. Branch current. The currents i_1 and i_2 are equal and hence the current through the branch R is zero and the branch can be removed or shorted as before. To see that $i_1 = i_2$, we reverse the source. Because of the symmetry, the current i_{da} which then flows must have the same magnitude as the original current i_1. Thus

$$(i_{da})_{\text{source reversed}} = (i_1)_{\text{original source}}.$$

Changing the source to its original polarity would reverse i_{da} and give

$$(i_{da})_{\text{original source}} = -(i_1)_{\text{original source}}.$$

However, by definition i_{da} is the negative of i_2 and hence

$$i_2 = i_1.$$

Method 3. Node voltage. The node voltages e_a and e_b are equal and hence a short circuit can be placed between them. To prove that e_a and e_b are

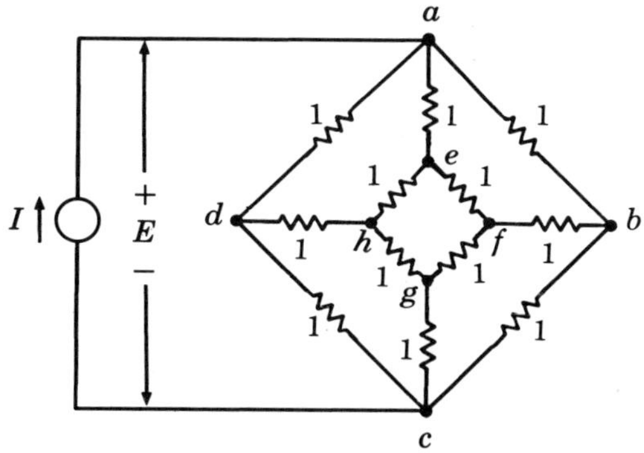

(a) Original network

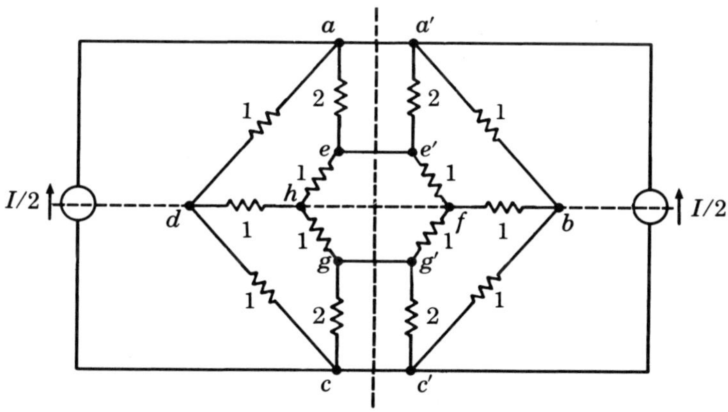

(b) A more symmetrical form of the original network

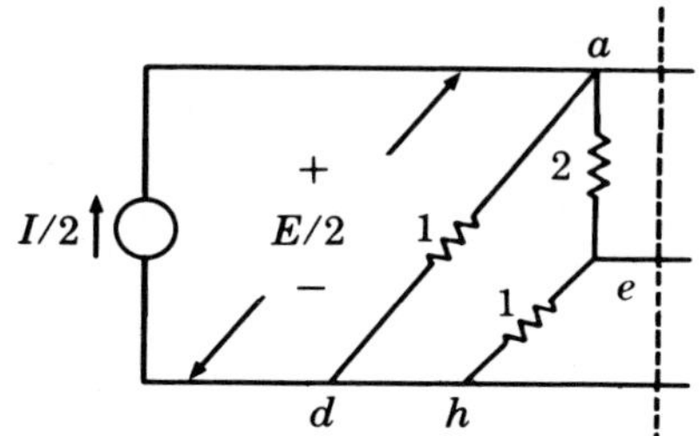

(c) A symmetrical section of the network

Fig. 4–10. An example with short-circuit and open-circuit symmetry.

equal, we must first show that $i_1 = i_2$, as in the previous section. Then, since the resistances are equal, $e_a = E/2$. Similarly, $i_3 = i_4$, and since the resistances are again equal, $e_b = E/2$, which is the same value as e_a.

Method 4. Mesh current. The mesh currents are equal and hence the current in R is zero and it may be replaced by an open circuit. To show that the mesh currents i_a and i_b are equal, we need only write the mesh equations:

$$i_a(4) - i_b(1) = I, \qquad -i_a(1) + i_b(4) = I.$$

The two equations are perfectly symmetrical and hence the two currents must be equal. (Symmetrical equations are a sufficient, but not a necessary, condition for the equality of the variables.)

A second example of symmetry, which will illustrate the creation of symmetrical forms by breaking up resistances and sources, is shown in Fig. 4–10(a). The circuit is redrawn in a more symmetrical form in (b). It is now symmetrical about a horizontal line drawn through its center, and about a vertical line drawn through its center. Because the right half and the left half of the network are identical it is easy to see that no current flows in the lines a–a', e–e', g–g' or c–c'. They can therefore be opened and the network is split in two. By means of the theorem concerning reversed source polarities and the symmetry of the upper and lower halves of the network it is easy to show that

$$i_{ad} = i_{ab} = i_{dc} = i_{bc}, \qquad i_{ae} = i_{gc}, \qquad i_{eh} = i_{ef} = i_{hg} = i_{fg}$$

and hence that the potentials of e_d, e_h, e_f, and e_b are all equal to $E/2$. A shorting line can be passed through these points and only one-quarter of the network need be considered in finding the input resistance. This section of the network is shown in Fig. 4–10(c). The input resistance is

$$R = \frac{E/2}{I/2} = \frac{1 \times 3}{1 + 3} = \tfrac{3}{4}\text{ ohm}.$$

The examples of symmetry which have been considered here are representative only. In particular, it should be observed that since we have been finding input resistance we have been exciting the networks with a single source. Special types of symmetry can occur when there are several sources present. Some of these will be considered in the chapter on response functions of two-terminal-pair networks and some of them must be reserved until we have considered more general types of exciting functions.

4–5 The resistance of ladder networks. We are singling out ladder networks for special attention for two reasons. The first is their practical importance. Many circuits, particularly in radio and television, have a

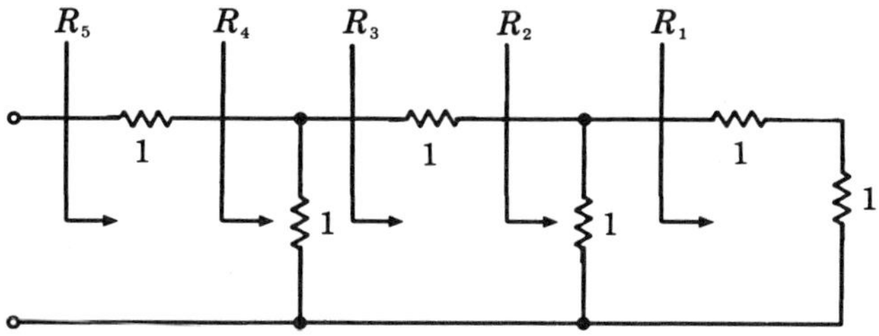

FIG. 4-11. The resistance of a ladder network by series-parallel transformations.

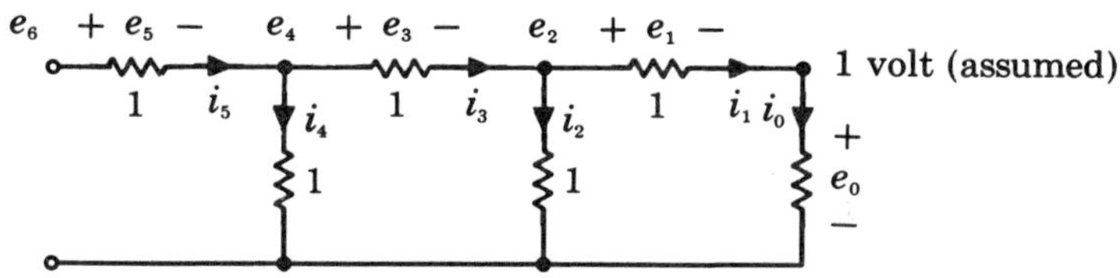

FIG. 4-12. Input resistance of a ladder by the linearity principle.

ladder form. A ladder has the advantage that no element is more than one node away from the ground, and therefore the effect of unwanted stray capacitance is minimized. A second reason is that ladders are particularly amenable to special techniques.

A ladder network is shown in Fig. 4-11. The input resistance R_5 can be obtained by successive applications of the series and parallel resistance formulas. For example,

$$R_1 = 1 + 1 = 2 \text{ ohms},$$

$$R_2 = \frac{1 \times 2}{1 + 2} = \tfrac{2}{3} \text{ ohm},$$

$$R_3 = 1 + R_2 = 1\tfrac{2}{3} = \tfrac{5}{3} \text{ ohms},$$

$$R_4 = \frac{1 \times R_3}{1 + R_3} = \frac{\frac{5}{3}}{1 + \frac{5}{3}} = \tfrac{5}{8} \text{ ohm},$$

$$R_5 = 1 + R_4 = 1 + \tfrac{5}{8} = \tfrac{13}{8} \text{ ohms}.$$

A second technique for solving ladder networks is illustrated in Fig. 4-12. The voltage e_0 is assumed to be 1 volt and the input voltage and current required to produce this condition are computed. The computations follow a step-by-step procedure in which it is not necessary to solve any equations. The input resistance is the ratio of e_6 to i_5. The computations are as follows:

$$
\begin{aligned}
e_0 &= 1 \text{ volt (assumed)},\\
i_0 &= e_0/1 &&= 1 \text{ amp},\\
i_1 &= i_0 &&= 1 \text{ amp},\\
e_1 &= i_1 \times 1 &&= 1 \text{ volt},\\
e_2 &= e_0 + e_1 &&= 1 + 1 = 2 \text{ volts},\\
i_2 &= e_2/1 &&= 2/1 \quad = 2 \text{ amp},\\
i_3 &= i_1 + i_2 &&= 1 + 2 = 3 \text{ amp},\\
e_3 &= i_3 \times 1 &&= 3 \times 1 = 3 \text{ volts},\\
e_4 &= e_3 + e_2 &&= 3 + 2 = 5 \text{ volts},\\
i_4 &= e_4/1 &&= 5/1 \quad = 5 \text{ amp},\\
i_5 &= i_4 + i_3 &&= 5 + 3 = 8 \text{ amp},\\
e_5 &= i_5 \times 1 &&= 8 \times 1 = 8 \text{ volts},\\
e_6 &= e_5 + e_4 &&= 8 + 5 = 13 \text{ volts}.
\end{aligned}
$$

The input resistance, as before, is

$$R = \frac{e_6}{i_5} = \tfrac{13}{8} \text{ ohms}.$$

The exact value assumed for e_0 is of no consequence. Because of the principle of linearity a change in e_0 would be accompanied by the same percentage change in all the currents and voltages in the circuit and the ratio of the input voltage to the input current would be unchanged. It is usually possible to perform the calculations directly on the circuit diagram, and the method is quick and easy to use.

The same technique can be applied to double ladder networks of which the twin T and the bridged T are examples. We assume the output voltage to be 1 volt, as before, and in addition we must assign one of the

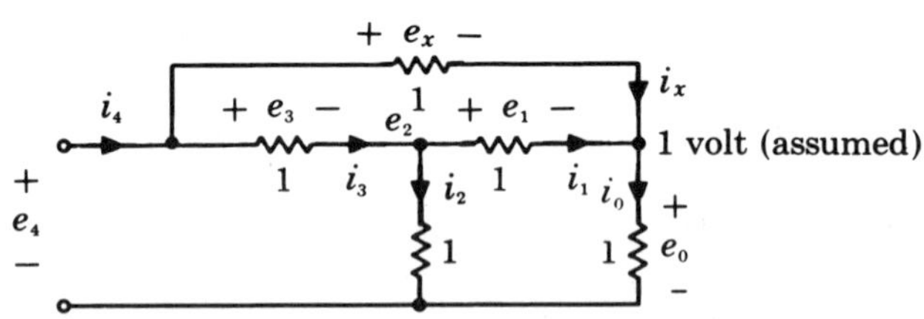

FIG. 4–13. A bridged T network treated by the "ladder method."

currents feeding the output node to be an unknown x amperes. If we proceed as before we will obtain an expression for the input voltage from each of the two paths. The equality of these expressions gives us an equation to determine the unknown current x, and then the problem is solved. The circuit of Fig. 4–13 will serve to illustrate the details. An output voltage of 1 volt and a current of x amperes in the top resistance are assumed. Along the bottom ladder,

$$\begin{aligned}
i_0 &= 1 \text{ amp}, \\
i_1 &= (1 - x) \text{ amp}, \\
e_1 &= 1(1 - x) \text{ volts}, \\
e_2 &= e_1 + e_0 = 1 + (1 - x) = (2 - x) \text{ volts}, \\
i_2 &= e_2/1 = \frac{(2 - x)}{1} = 2 - x \text{ amp}, \\
i_3 &= i_1 + i_2 = (1 - x) + (2 - x) = 3 - 2x \text{ amp}, \\
e_3 &= i_3 \times 1 = (3 - 2x) \times 1 = 3 - 2x \text{ volts}, \\
e_4 &= e_3 + e_2 = (3 - 2x) + (2 - x) = 5 - 3x \text{ volts}.
\end{aligned}$$

Along the upper path,

$$\begin{aligned}
e_x &= i_x \times 1 = (x) \times 1 = x \text{ volts}, \\
e_4 &= e_0 + e_x = 1 + x \text{ volts}.
\end{aligned}$$

The two expressions for e_4 must be equal, and therefore

$$5 - 3x = 1 + x \qquad \text{or} \qquad x = 1.$$

(This circuit is actually a balanced bridge, and this value could have been seen by inspection.)

The input voltage is

$$e_4 = 1 + x = 2 \text{ volts}$$

and the input current is

$$i_4 = i_3 + i_x = 3 - 2x + x = 2 \text{ amp},$$

so that the input resistance is

$$R = \frac{e_4}{i_4} = \tfrac{2}{2} = 1 \text{ ohm}.$$

It is not certain that one saves much effort in this process over a straightforward solution by mesh currents or node voltages, but it is more fun!

4–6 Branches with sources. Thévenin's theorem and Norton's theorem. In Chapter 2 we observed that series-parallel branches containing sources could be reduced to a single equivalent resistance and a single equivalent source.* In this section we will derive a more general theorem, called Thévenin's theorem, which will reduce any linear network branch to a single source in series with a passive branch. A general statement of Thévenin's theorem is as follows:

Any linear two-terminal network branch with sources can be replaced by an equivalent voltage source in series with an equivalent passive branch. The voltage source is the voltage measured at the terminals when nothing is connected to the external circuit, and the passive branch is the "dead" network, in which all of the sources have been reduced to zero. In a resistive network the "dead" network is a single equivalent resistance.

Thévenin's theorem is the volt-ampere relationship of a general network branch containing sources. The theorem is nothing more than a simple application of the superposition principle. Consider the general branch shown in Fig. 4–14(a). The volt-ampere relationship can be obtained by exciting the circuit with a current source and measuring the resultant voltage. By the superposition theorem this voltage is composed of two parts:

$$e = \begin{matrix}\text{voltage from external}\\ \text{current source}\end{matrix} + \begin{matrix}\text{voltage from sources}\\ \text{within the network}\end{matrix}. \tag{4–16}$$

The voltage produced by the external current source is measured with the sources inside the network placed equal to zero. In other words, it is the voltage which the current I would produce in the "dead" network. The voltage from the sources within the network is obtained with the external current source "dead," which is to say that it is the voltage which exists across the terminals when they are open. We have thus interpreted Eq. (4–16) in the form of Fig. 4–14(b). The voltage e produced at the terminals of the network is the sum of two components, one of which is the voltage which the current would produce in the "dead" network and the other of which is the open-circuit voltage across the terminals. For a network of sources and resistances the "dead" network reduces to a single equivalent resistance and the Thévenin equivalent circuit takes the form shown in Fig. 4–14(c).

Norton's theorem is the dual of Thévenin's theorem. It can be stated as follows:

Any linear two-terminal network branch with sources can be replaced by an equivalent current source in parallel with an equivalent passive branch. The current source is the current which would flow from the terminals if they were short-circuited, and the passive branch is the "dead" net-

* Section 2–3, especially Figs. 2–8 and 2–9.

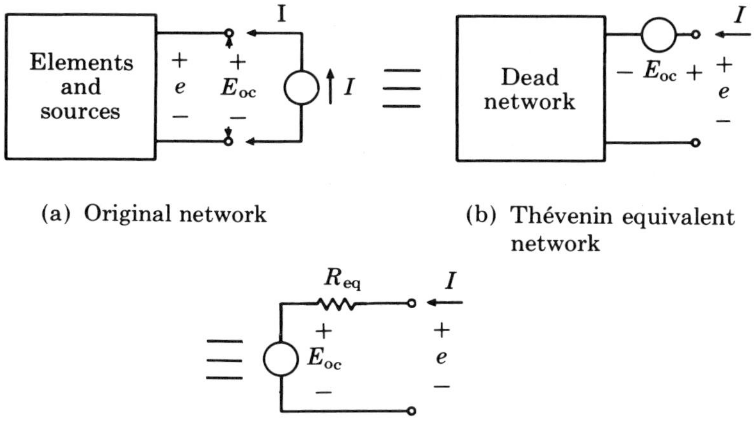

(a) Original network

(b) Thévenin equivalent network

(c) Thévenin equivalent for a resistive network

FIG. 4–14. Thévenin's theorem.

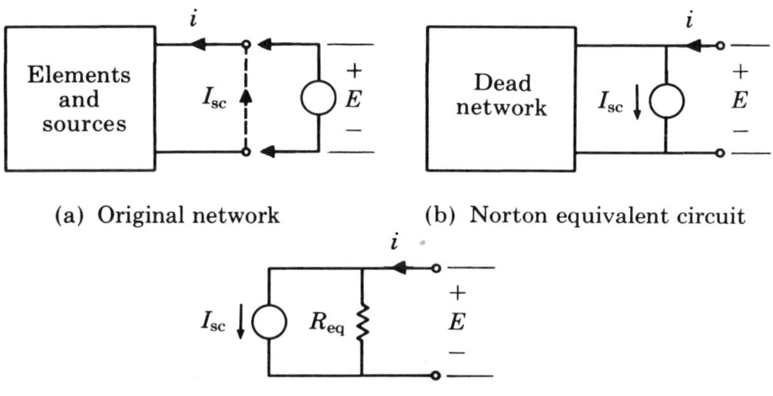

(a) Original network

(b) Norton equivalent circuit

(c) Norton circuit for a resistive network

FIG. 4–15. Norton's theorem.

work, with all of the sources reduced to zero. In a resistive network the "dead" network is a single equivalent resistance.

To prove Norton's theorem we obtain the volt-ampere relationship of the general branch when it is excited by a voltage source, as shown in Fig. 4–15(a). By the principle of superposition the current i which flows from the source is the sum of two components:

$$i = \begin{array}{c}\text{current from the external}\\ \text{voltage source}\end{array} + \begin{array}{c}\text{current from sources}\\ \text{within the network}\end{array}. \tag{4–17}$$

The current produced by the external voltage source is measured with all of the sources within the network placed equal to zero. It is the current which the voltage source E would produce in the "dead" network. The current from sources within the network is obtained with the external voltage source "dead," and a "dead" voltage source is a short-circuit across the terminals. We can thus interpret Eq. (4–17) in the form of Fig. 4–15(b). The input current i is the sum of two components, one of which is the current produced in the dead network by the voltage E, and the other of which is the short-circuit current available at the terminals. If the network is resistive, the "dead" network can be replaced by a single equivalent resistance, and the Norton equivalent circuit takes the form shown in Fig. 4–15(c).*

The source polarity for Thévenin's and Norton's theorems is usually best obtained by inspection. The equivalent circuit must produce current flow *in the external circuit* in the same direction as in the original problem.

To prove Thévenin's and Norton's theorems we used only the superposition theorem. Thus Thévenin's and Norton's theorems may be expected to hold for any linear circuit for which superposition holds.† They hold also regardless of the linearity or nonlinearity of any circuits to which the linear branch may be connected. Because of their generality these theorems are the most useful of all the network theorems.

Thévenin's and Norton's theorems provide, for example, the most direct way of making the current-source to voltage-source transformations.‡ Consider the circuit of Fig. 4–16(a). The short-circuit current produced by the voltage source is

$$I_{sc} = \frac{E}{R},$$

and the resistance at the terminals with the voltage source dead is simply

$$R_{eq} = R.$$

The Norton equivalent circuit is thus a current source of value E/R in parallel with the resistance R, as shown in Fig. 4–16(b). Similarly, Thévenin's theorem can be applied to the current source in parallel with a resistance, as shown in Fig. 4–16(c). The open-circuit voltage is

$$E_{oc} = IR$$

* The Norton circuit is also just a source transformation of Thévenin's circuit.

† The theorems are simply applications of the mathematical principle of linearity.

‡ See Section 2–3 for an earlier derivation of these transformations.

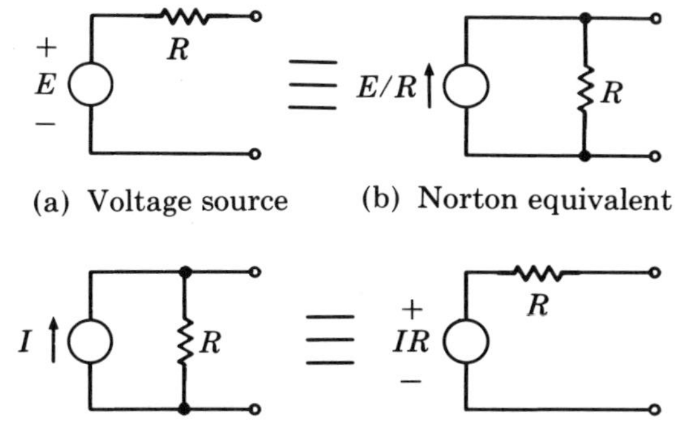

(a) Voltage source (b) Norton equivalent

(c) Current source (d) Thévenin equivalent

FIG. 4–16. Source transformations by Thévenin's and Norton's theorems.

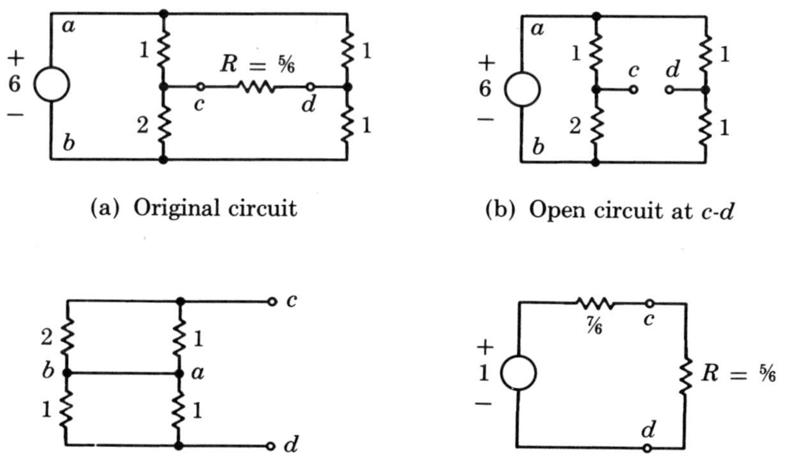

(a) Original circuit (b) Open circuit at c-d

(c) Resistance at c-d (d) Equivalent Thévenin circuit

FIG. 4–17. The complete solution of a problem by Thévenin's theorem.

and the resistance at the terminals with the current source open is

$$R_{\text{eq}} = R.$$

The Thévenin equivalent circuit is thus a voltage source of value IR in series with a resistance R, as shown in Fig. 4–16(d).

Thévenin's and Norton's theorems are more general, however, than the current-source to voltage-source transformations. As a further example we will use Thévenin's theorem to find the current in the resistance R in Fig. 4–17(a). The procedure will be to find a Thévenin equivalent circuit at the terminals c–d and then to apply the resistance R to this circuit. The open-circuit voltage can be found from the circuit of Fig. 4–17(b). The node voltages are

$$e_c = \tfrac{2}{3} \times 6 = 4 \text{ volts} \qquad \text{and} \qquad e_d = \tfrac{1}{2} \times 6 = 3 \text{ volts.}$$

The open-circuit voltage across the terminals c–d is

$$e_c - e_d = 1 \text{ volt.}$$

The resistance at the terminals c–d can be found from the circuit of Fig. 4–17(c). It is

$$R_{\text{eq}} = \frac{2 \times 1}{2 + 1} + \frac{1 \times 1}{1 + 1} = \tfrac{7}{6} \text{ ohms.}$$

The Thévenin equivalent circuit is shown in Fig. 4–17(d) to the left of the terminals c–d, and the resistance $R = \frac{5}{6}$ ohm is applied to the right of these terminals. Since the Thévenin equivalent circuit accurately represents the circuit so far as its effect at the terminals c–d is concerned, the current in the $\frac{5}{6}$-ohm resistance will be the same in Fig. 4–17(a) and (d). It is

$$i = \frac{1}{(\frac{7}{6}) + (\frac{5}{6})} = \tfrac{1}{2} \text{ amp.}$$

Any other method of solution would involve a great deal more work.

A second useful application of Thévenin's and Norton's theorems is to the measurement problem in electrical circuits. When we insert physical voltmeters and ammeters into the circuit, we disturb the circuit, and the readings we obtain are not those which existed previously. By taking two sets of readings on different scales, we can compute the correct values by means of Thévenin's and Norton's theorems. Consider the example shown in Fig. 4–18(a). The voltage across the terminals is measured with two nonideal voltmeters. The first meter has a resistance of 100,000 ohms and gives a reading of 45 volts. The second meter has a resistance of 50,000 ohms and gives a reading of 30 volts. (The meters are calibrated to read correctly the voltage across their terminals regardless of how it is obtained.)

The Thévenin equivalent circuit for the device is shown in Fig. 4–18(b). The problem is to find the open-circuit voltage which existed before the meters were placed across the terminals. From the first meter reading,

$$E = \frac{100{,}000}{100{,}000 + R} \times E_{\text{oc}} = 45 \text{ volts,} \tag{4–18}$$

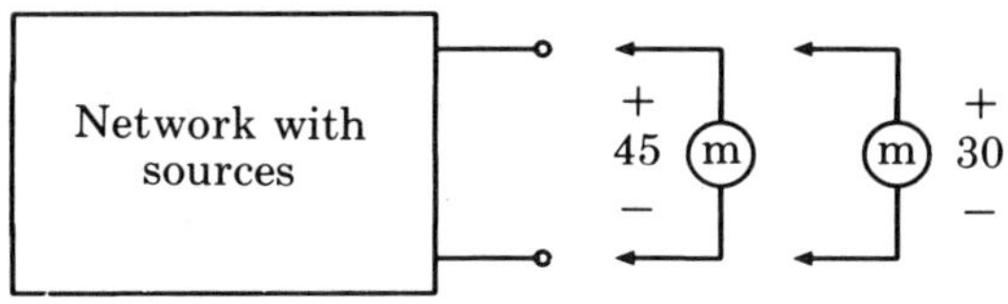

(a) Voltage measured with finite resistance voltmeters

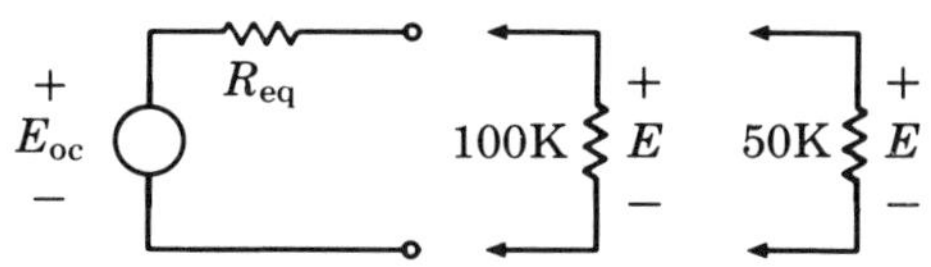

(b) Equivalent circuit of network and meters

FIG. 4–18. The application of Thévenin's theorem to the measurement of voltage with finite-resistance meters.

and from the second,

$$E = \frac{50{,}000}{50{,}000 + R} \times E_{oc} = 30 \text{ volts.} \tag{4–19}$$

A simultaneous solution of Eqs. (4–18) and (4–19) gives the required value:

$$E_{oc} = 90 \text{ volts.}$$

A similar application of Norton's theorem can be made to the case of a nonideal ammeter. A number of other applications of Thévenin's and Norton's theorems will be given in later chapters. In general, they are useful when the desired solution for a problem involves only a small fraction of the total possible number of currents and voltages. Many practical problems fall in this category.

4–7 The substitution theorem. We observed previously that a perfect voltage source can be obtained in the laboratory by measuring the voltage and adjusting it to the desired value. A similar remark holds for a perfect current source. These statements are particular examples of the *substitution theorem:*

> *A* known *voltage in a circuit can be replaced by an ideal voltage source, and a* known *current can be replaced by an ideal current source.*

The substitution theorem itself is a special case of the symmetry principles which were discussed in Section 4–4. Consider a resistance R_{ab}

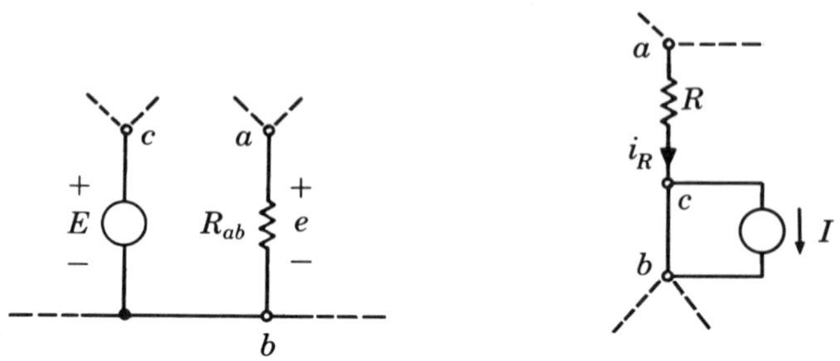

(a) Voltage source substitution (b) Current source substitution

FIG. 4–19. The substitution theorem.

connected between points a and b in a network, as shown in Fig. 4–19(a). A voltage source is connected to point b. If the voltage rise from b to c is the same as the rise from b to a, the points c and a will be at the same potential. They can be shorted together. The resistance R_{ab} is then in parallel with a voltage source and can be removed, so far as the rest of the circuit is concerned. The net result is the replacement of the resistance by a voltage source with the same value of voltage across it.

A similar argument holds for the current source in Fig. 4–19(b). Suppose that a current i_R is flowing in the resistance from point a to point b. A current source can be inserted in parallel with the short-circuit between points c and b without changing anything. If the current source I is the same as the current i_R, there will be no current in the connection between b and c, and it can be removed. The resistance R is then in series with the current source and it can be removed so far as the rest of the circuit is concerned. The net result is that the resistance R has been replaced by an ideal current source with the same value of current through it.

The proof of the substitution theorem is easy. Our problem will be to find a use for it! *It cannot be applied unless the solution to the problem is already known.* Its applications in linear, passive circuits are apt to be artificial for this reason, but surprisingly enough it turns out to be the cornerstone of linear, active circuit theory involving transistors and vacuum tubes.* The reason for its importance is simple. It allows a complicated circuit to be broken up into smaller sections which can be handled individually and then recombined into the over-all circuit. The procedure is illustrated in Fig. 4–20. Two networks are connected in cascade and excited at the input. The output voltage from the second network is the required response. A solution by means of the substitution theorem has the following steps:

* See, for example, Angelo, *Electronic Circuits*, McGraw-Hill (1958).

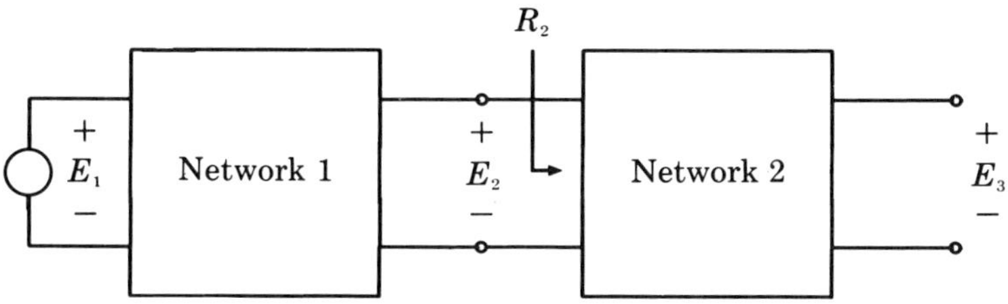

FIG. 4-20. An illustration of the use of the substitution theorem.

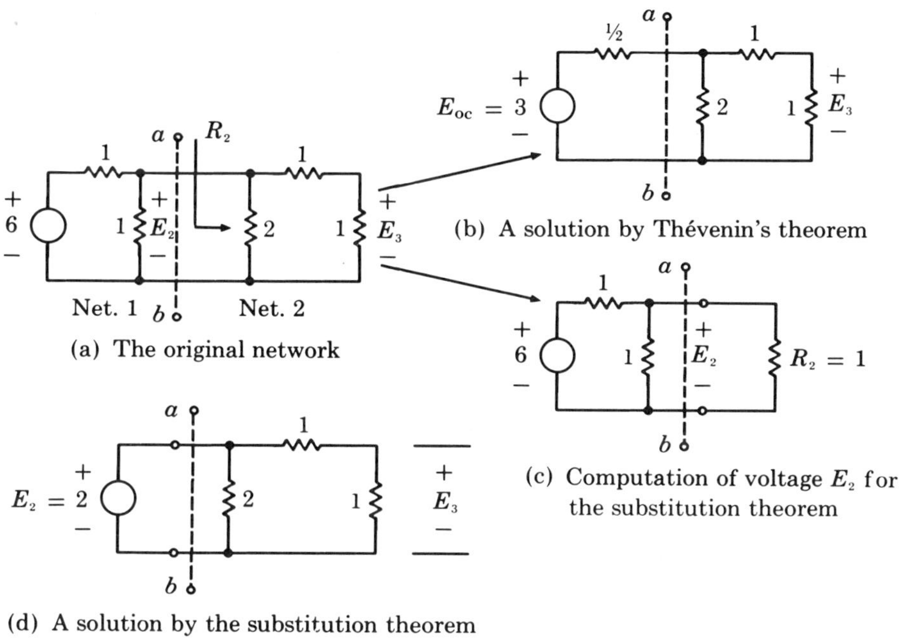

FIG. 4-21. Solutions by Thévenin's theorem and the substitution theorem.

1. The input resistance of network 2 is computed.

2. This resistance is connected to network 1 and the voltage E_2 is obtained. (It is important to note that this voltage is the same as E_2 in the original network.)

3. Since the voltage E_2 is known, it is replaced by an ideal voltage source, in accordance with the substitution theorem.

4. The output E_3 is computed from network 2, driven by the source E_2. The key element in this process is *that one never considers more than one of the networks at a time.* A single difficult problem is traded for two easier ones. A numerical example is provided by Fig. 4-21(a). The solution by

the substitution theorem and by Thévenin's theorem are given side by side. The reasoning processes used in the two methods are very different, and it is important to fully appreciate these differences. The equivalent Thévenin circuit for network 1 is shown in Fig. 4–21(b). The output voltage E_3 which occurs when network 2 is connected to this equivalent circuit can be obtained by inspection:

$$E_3 = 1 \text{ volt.}$$

Figure 4–21(c) shows the same circuit treated by the substitution theorem. The input resistance of network 2 is 1 ohm. This resistance is connected to the output of network 1 and the voltage across it is obtained. It is

$$E_2 = 2 \text{ volts.}$$

By the substitution theorem this voltage can be replaced by a voltage source, as shown in Fig. 4–21(d), and from this circuit the output voltage E_3 is found to be, as before,

$$E_3 = 1 \text{ volt.}$$

The importance of the method of reasoning given here cannot be overestimated. It leads directly to the partitioning of complicated circuits into smaller subsections which can be described in terms of an input resistance, an output resistance, and a transfer function. This approach is always used in vacuum-tube and transistor problems and in complicated "system" problems where some of the blocks may not even be electrical in nature.

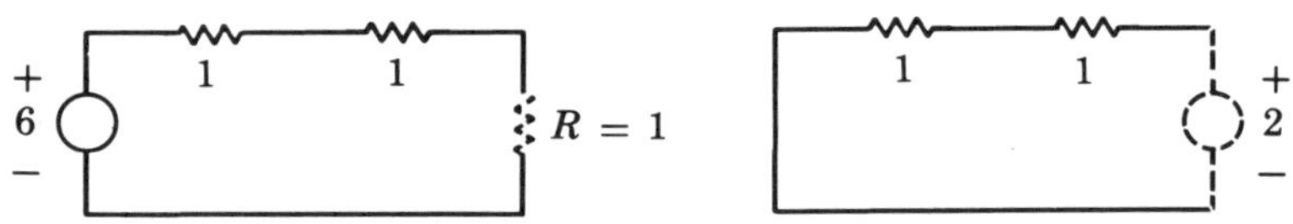

(a) A circuit to which the resistance $R = 1$ has been added

(b) The change in current computed by the substitution theorem

FIG. 4–22. The effect of adding a resistance computed by means of the substitution theorem.

A second application of the substitution theorem is shown in Fig. 4–22. The problem is to find the effect on the rest of the circuit of adding an extra resistance, R, in series with the others. The straightforward solution would be to find the currents before and after the resistance is added and to subtract them to find the effect of the resistance R. A solution which uses the substitution theorem instead is the following:

1. The voltage across R is found from the circuit of Fig. 4–22(a).

2. The resistance R is replaced by a voltage source, by the substitution theorem.

3. The effect of R is computed from this voltage source by the superposition theorem. (In order that the source should go to zero with R a voltage source must be used for the addition of a series resistance and a current source for the addition of a parallel resistance.)

For the circuit of Fig. 4–22(a) the current which flows is 2 amp and the voltage across R is

$$E_R = 2 \text{ volts.}$$

In accordance with the substitution theorem, this voltage is replaced by a voltage source in Fig. 4–22(b). Since the voltage source is directly proportional to R, it would become zero if R were removed. Hence that component of the current in the circuit which is contributed by the addition of the resistance must be given by the current produced by this source. It is

$$i_s = 1 \text{ amp}$$

in the opposite direction from the original current. Thus the addition of the resistance R has caused the current to decrease by 1 amp.

The application of this method in a more complicated problem has the advantage of focusing attention on the effects of adding the element. In fact, the changes produced by adding the element are obtained at every point in the network. Additional applications of the substitution theorem will be considered in later chapters.

4–8 Summary.

Equivalent resistance by source application

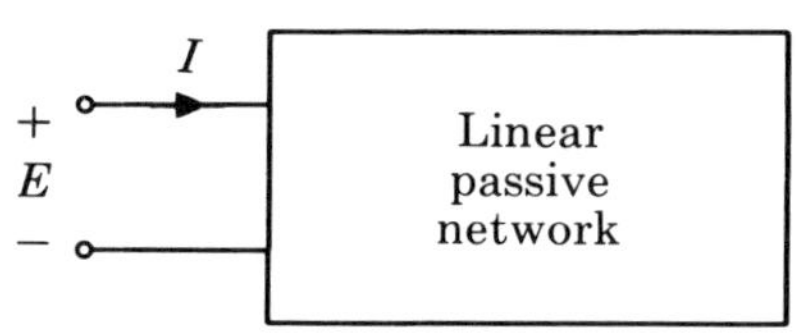

(a) Apply E and compute I
(b) Apply I and compute E

$$R = \frac{E}{I}$$

Equivalent resistance by network reduction

(a) Series

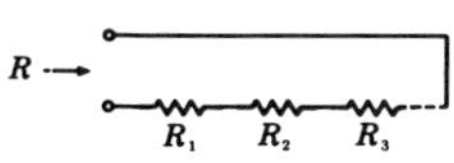

$$R = R_1 + R_2 + R_3 + \cdots$$

(b) Parallel

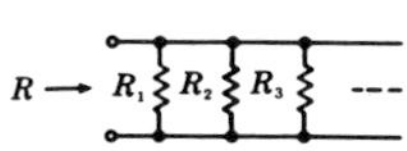

$$R = \frac{1}{1/R_1 + 1/R_2 + 1/R_3} \cdots$$

(c) Y–Δ

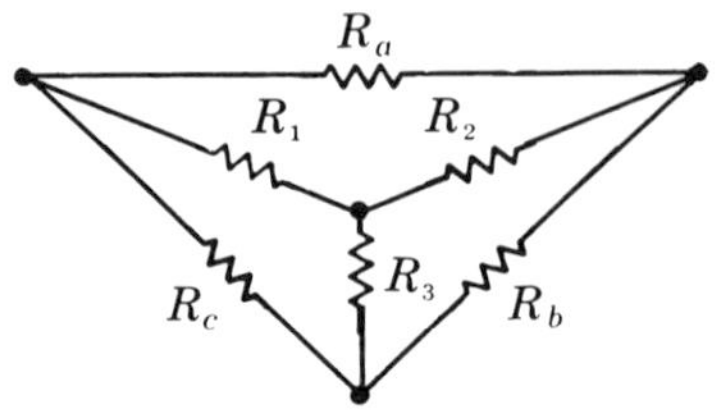

$$R_1 = \frac{R_a R_c}{R_a + R_b + R_c}$$

$$R_2 = \frac{R_a R_b}{R_a + R_b + R_c}$$

$$R_3 = \frac{R_b R_c}{R_a + R_b + R_c}$$

$$R_a = \frac{R_1 R_2 + R_2 R_3 + R_3 R_1}{R_3}$$

$$R_b = \frac{R_1 R_2 + R_2 R_3 + R_3 R_1}{R_1}$$

$$R_c = \frac{R_1 R_2 + R_2 R_3 + R_3 R_1}{R_2}$$

Resistance of symmetrical networks (symmetry principles)

(a) Short or open branches with no voltage across them.
(b) Short or open branches with no current in them.
(c) Short two nodes with the same potentials.
(d) Open two adjacent meshes with the same mesh currents.
(e) Reversing a source reverses all currents and voltages.

The resistance of ladder networks

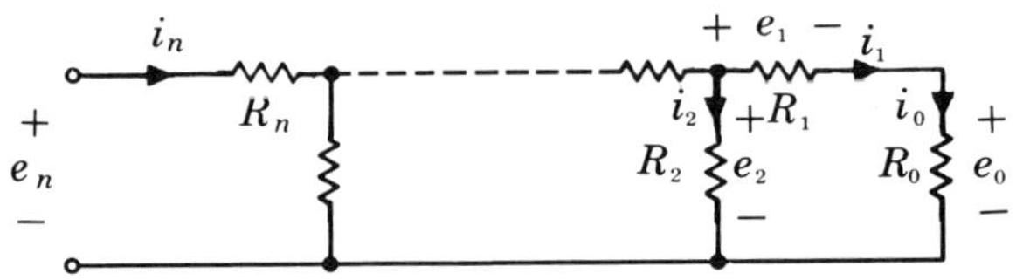

$$R = \frac{e_n}{i_n}$$

$$\begin{aligned} e_0 &= 1 \text{ volt (assumed)} \\ i_0 &= e_0/R_0 \\ i_1 &= i_0 \\ e_1 &= i_1 R_1 \\ e_2 &= e_0 + e_1 \\ i_2 &= e_2/R_2 \\ &\vdots \\ e_n &\ldots \\ i_n &\ldots \end{aligned}$$

Thévenin's theorem

"A two-terminal network with sources can be replaced by the open-circuit voltage in series with the input resistance of the dead network."

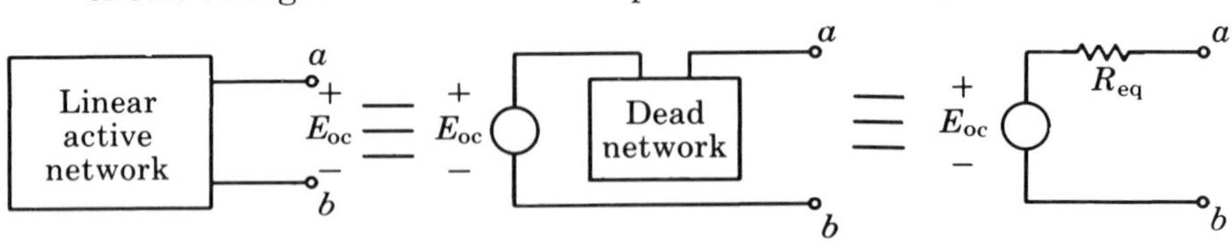

Norton's theorem

"A two-terminal network with sources can be replaced by the short-circuit current in parallel with the input resistance of the dead network."

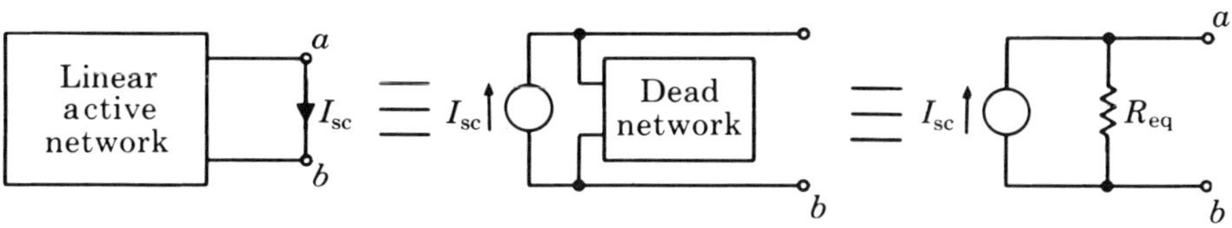

The substitution theorem

"A *known* voltage can be replaced by a voltage source, and a *known* current by a current source."

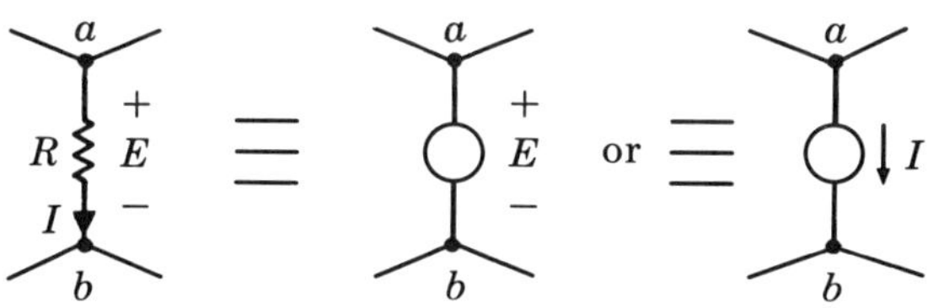

REFERENCES

Most elementary textbooks on circuit theory discuss the equivalent resistance of two-terminal networks. The treatment in this chapter is distinguished by the generality of the proofs of the theorems and by a more thorough discussion of symmetry than is usually found. References for special topics are listed below.

1. M. B. Reed, *Alternating-Current Circuit Theory,* Harper & Brothers, New York, 1948. Reed gives a particularly complete discussion of the theorems of electric circuits, including several which we have not included. All of the proofs are by matrices and determinants and in steady-state a-c notation.

2. E. A. Guillemin, *Introductory Circuit Theory,* John Wiley & Sons, New York, 1953. Guillemin gives a very general discussion of Y–Δ transformations in terms of mesh and node equations. There is no formal discussion of the substitution theorem, but it is used in a number of examples. Some discussion of symmetry is given.

3. H. H. Skilling, *Electrical Engineering Circuits,* John Wiley & Sons, New York, 1957. Skilling discusses most of the theorems given by Reed and the proofs are more elementary. The Thévenin proof is made to depend on the linearity of the external circuit. Several of the examples we have given under the substitution theorem are considered as separate theorems.

4. MIT EE Staff, *Electric Circuits*, John Wiley & Sons, New York, 1943. In this standard text it is noted that Helmholtz really discovered Thévenin's theorem! Norton's theorem is not explicitly stated.

5. W. B. Boast, *Principles of Electric and Magnetic Circuits*, Harper & Brothers, New York, 1950. Boast gives a particularly good discussion of Y–Δ transformations and is one of the few authors to include the general star-to-mesh transformation.

6. LePage and Seeley, *General Network Analysis*, McGraw-Hill, New York, 1952. Thévenin's theorem and Norton's theorem are discussed in a general way by means of determinants. The Y–Δ transformation is mentioned only in the section on two-terminal-pair networks.

EXERCISES*

Section 4–2

1. Find the equivalent resistance of the network of Fig. 4–23 at the terminals a–b by applying a voltage source and writing mesh equations.

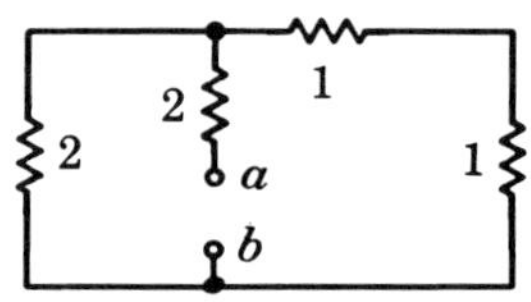

FIGURE 4–23

2. Repeat Exercise 1, using a current source as the excitation and writing mesh equations to solve for the unknowns.

3. Find the resistance of the network in Exercise 1 by applying a current source and writing node equations to solve for the unknowns.

4. Find the resistance of the network of Exercise 1 by applying a voltage source and writing node equations to find the unknowns.

5. Find the equivalent input resistance of the bridge circuit of Fig. 4–24 by applying a voltage source and writing mesh equations to solve for the unknowns.

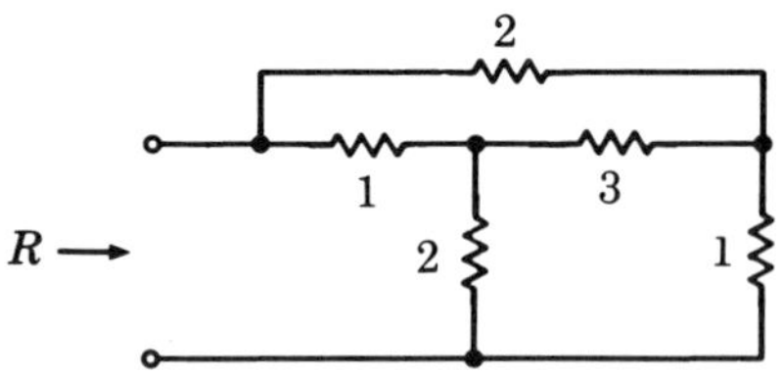

FIGURE 4–24

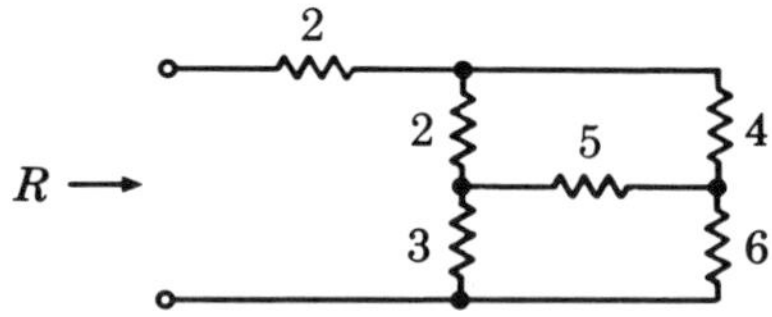

FIGURE 4–25

* All numerical values are in ohms, volts, amperes.

6. Repeat Exercise 5, but change the exciting source to a current source. Mesh equations should be written again for the unknowns.

7. Find the input resistance of the network of Exercise 5 by applying a current source and writing the node equations.

8. Find the input resistance of the network in Exercise 5 by applying a voltage source and writing the node equations.

9. Find the resistance of the network of Fig. 4–25 by means of mesh currents. Pick your own type of excitation.

10. Repeat Exercise 9, but use node equations. Pick either type of excitation.

11. Find the input resistance of the network of Fig. 4–26.

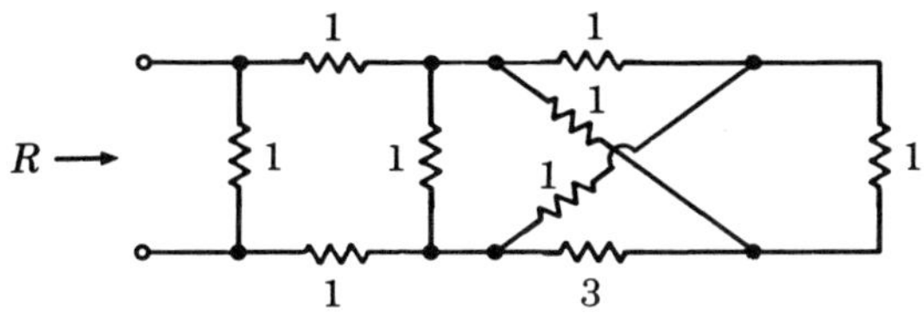

FIGURE 4–26

Section 4–3

12. Change the networks of Fig. 4–27 from Y to Δ.

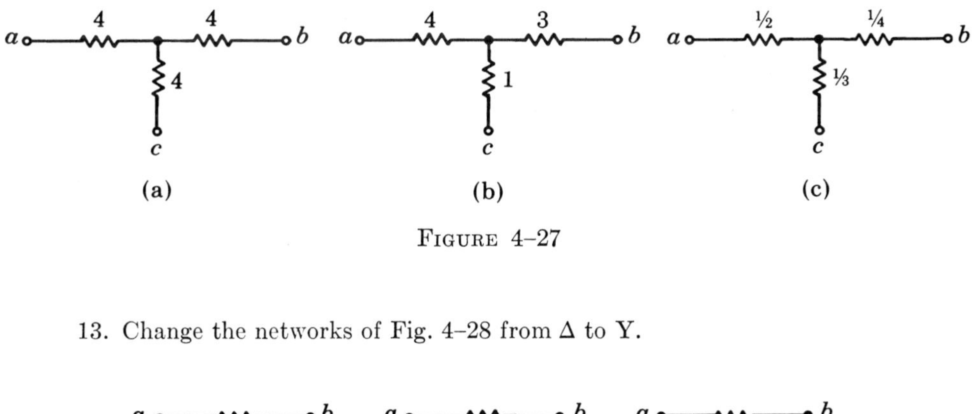

FIGURE 4–27

13. Change the networks of Fig. 4–28 from Δ to Y.

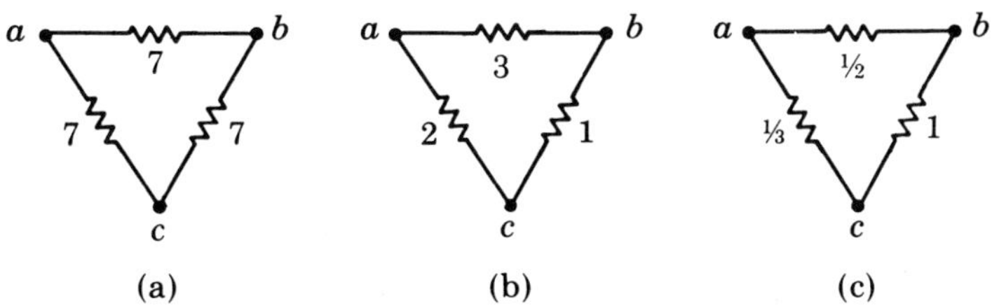

FIGURE 4–28

14. Reduce the network of Fig. 4–29 to a three-resistance Y between the points a, b, and c.

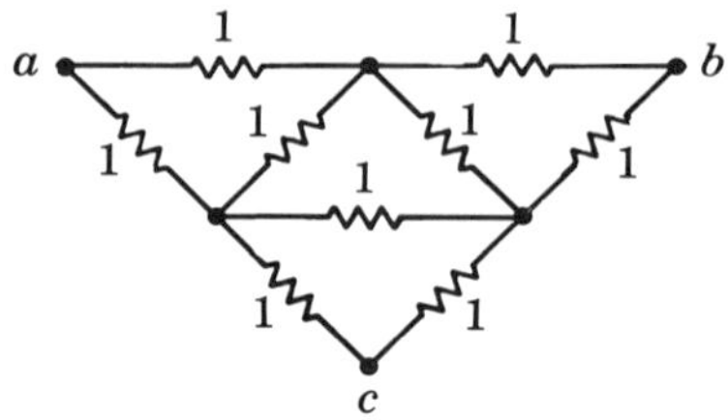

FIGURE 4–29

15. Reduce the network of Fig. 4–23 to a single resistance by means of series-parallel transformations.

16. Reduce the network of Fig. 4–24 to a single resistance by means of series-parallel and Y–Δ transformations.

17. Reduce the network of Fig. 4–25 to a single resistance by means of series-parallel and Y–Δ transformations.

18. Reduce the network of Fig. 4–26 to a single resistance by means of series-parallel and Y–Δ transformations.

19. Find the input resistance of the circuit of Fig. 4–30 by using several Δ–Y transformations.

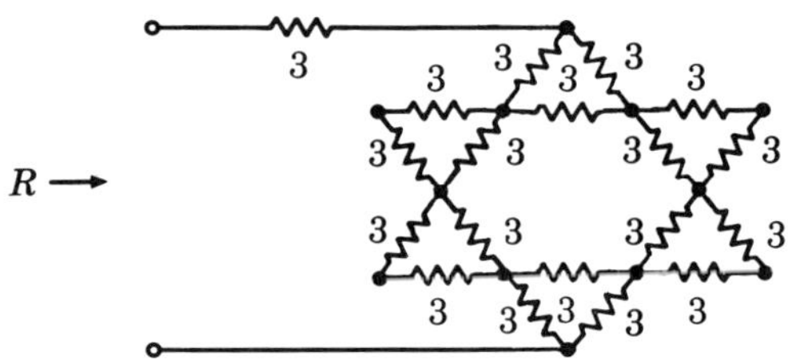

FIGURE 4–30

20. Find the input resistance of the circuit of Fig. 4–31 by means of series-parallel and Y–Δ transformations which reduce the network to a single resistance.

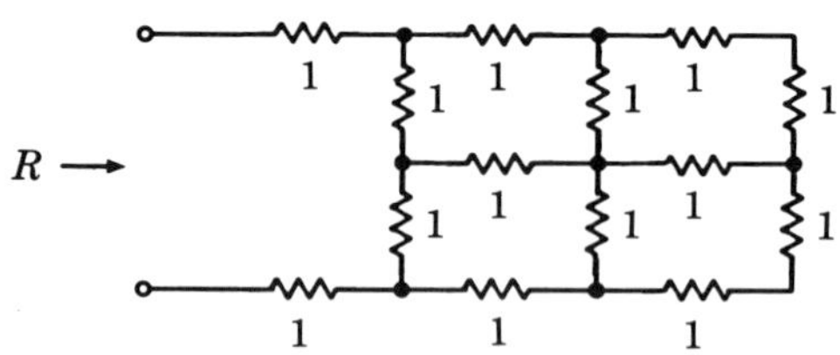

FIGURE 4–31

Section 4–4

21. Solve Exercise 9 by means of symmetry principles.
22. Solve Exercise 19 by means of symmetry principles.
23. Solve Exercise 20 by means of symmetry principles.
24. Find the resistance of the circuit of Fig. 4–32.

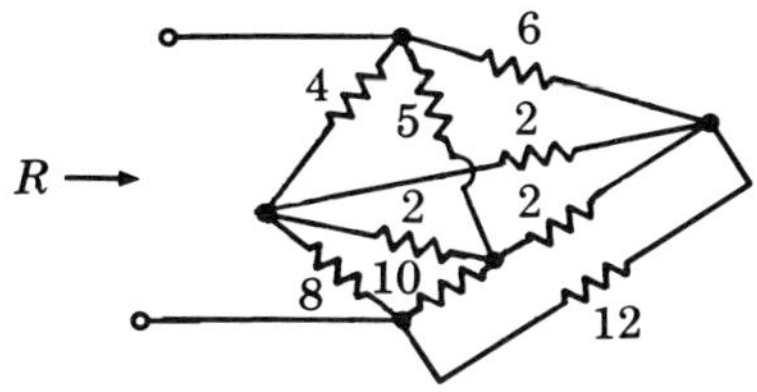

FIGURE 4–32

25. A cube is constructed by soldering together twelve 1-ohm resistors to represent the twelve edges. Find the resistance between any two adjacent corners.

26. For the cube of Exercise 25, find the resistance between any pair of diagonally opposite corners.

27. For the cube of Exercise 25, find the resistance between any two diagonally opposite corners of any one face of the cube.

28. When a network is symmetrical but the excitation is not, it is usually possible to resolve the excitation into symmetrical and antisymmetrical components. The network of Fig. 4–33(a) is an example. Show that the voltage E in Fig. 4–33(a) is the sum of the voltage E_1 in Fig. 4–33(b) and E_2 in Fig. 4–33(c). By finding the voltages E_1 and E_2, obtain the input resistance to the original circuit.

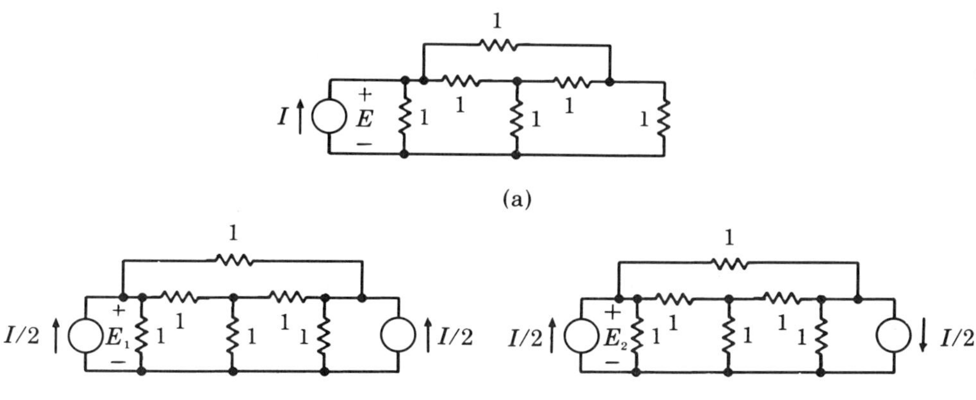

FIGURE 4–33

Section 4–5

29. Find the input resistance of the ladder network of Fig. 4–34 by successive applications of the series-parallel formulas.

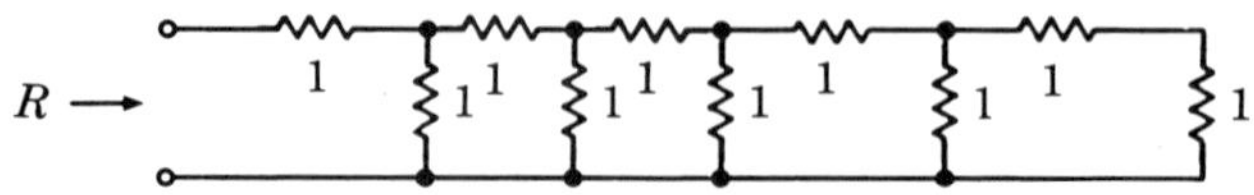

FIGURE 4–34

30. Repeat Exercise 29, using the "ladder method" of an assumed voltage at the end of the ladder from which a consistent set of currents and voltages are obtained.

31. Use the series-parallel and Y–Δ transformations to reduce the "double ladder" of Fig. 4–35 to a single resistance at the input terminals.

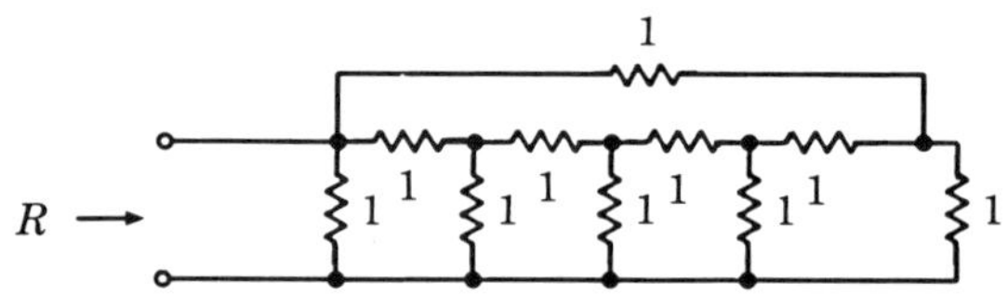

FIGURE 4–35

32. Find the equivalent input resistance of the circuit in Exercise 31 by means of the "ladder method" of an assumed voltage.

33. Find the input resistance of the ladder network of Fig. 4–36 by the assumed voltage method. Check the answer by combining resistances in series and in parallel.

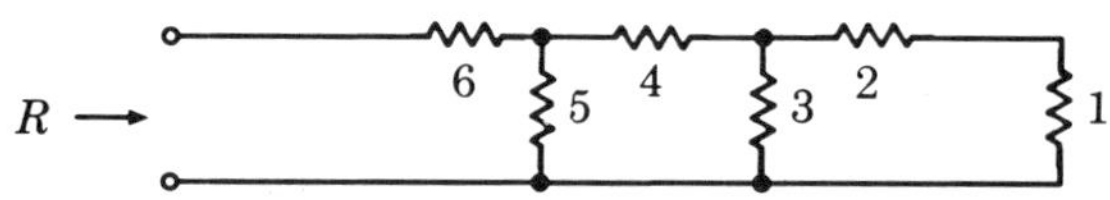

FIGURE 4–36

34. Find the input resistance of the circuit of Fig. 4–37 by the ladder method. (*Hint*: It is easier to start with an assumed voltage of two.)

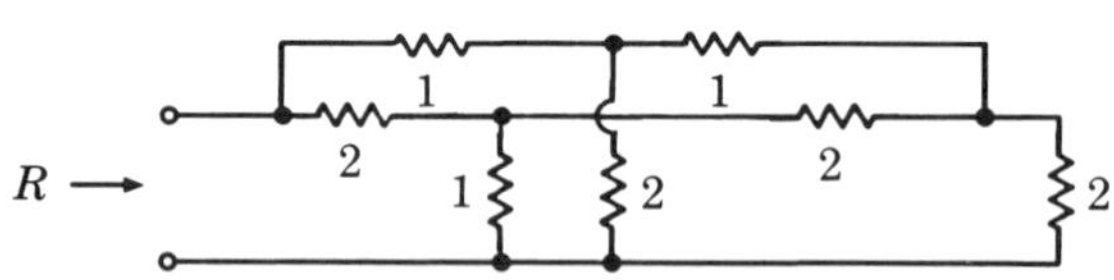

FIGURE 4–37

35. Find the input resistance of the circuit of Fig. 4–38 by assuming a voltage of 1 volt across the right-hand resistance and working backward to the input terminals. Verify your answer by solving the same problem as a symmetrical network.

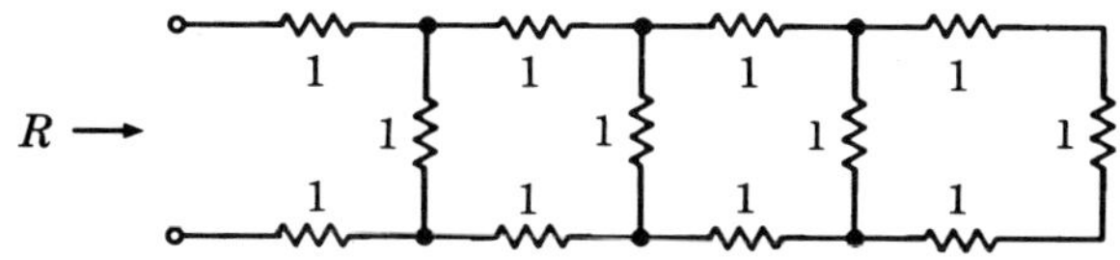

FIGURE 4–38

36. Starting with an assumed voltage of 4 volts across the right-hand resistance, find the input resistance of the circuit of Fig. 4–39.

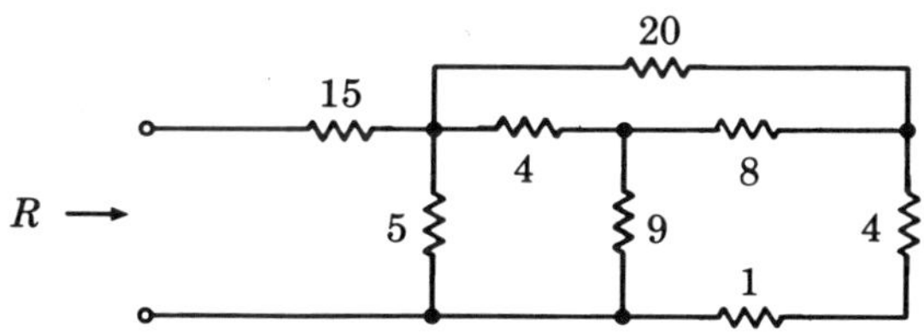

FIGURE 4–39

Section 4–6

37. (a) Find the Thévenin equivalent for the network of Fig. 4–40 at the terminal-pair a–b. (b) Convert the Thévenin equivalent to a current source in parallel with a resistance. (c) Obtain the Norton equivalent circuit by applying a short circuit directly to the original network.

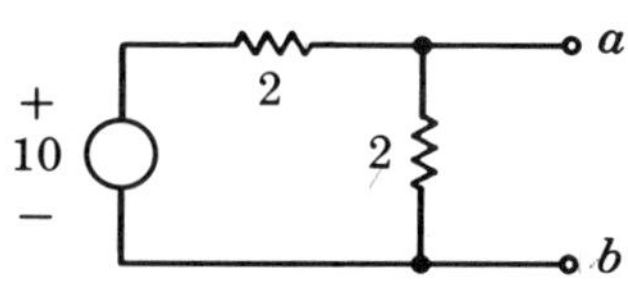

FIGURE 4–40

38. Give the Thévenin and the Norton equivalent circuits for the network of Fig. 4–41 at the terminals a–b.

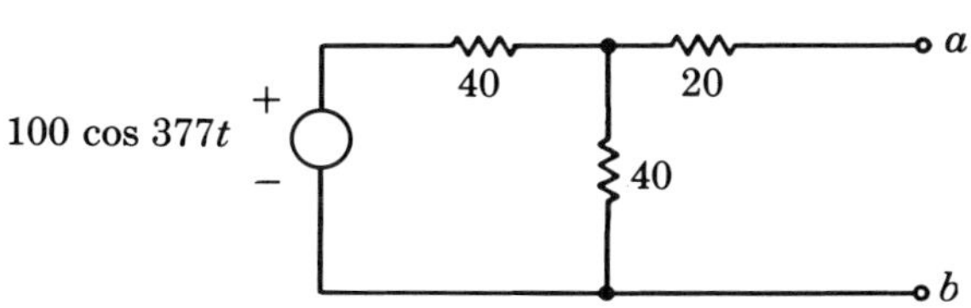

FIGURE 4–41

39. Find both the Thévenin and Norton equivalent circuits for the network of Fig. 4–42 at the terminals a–b.

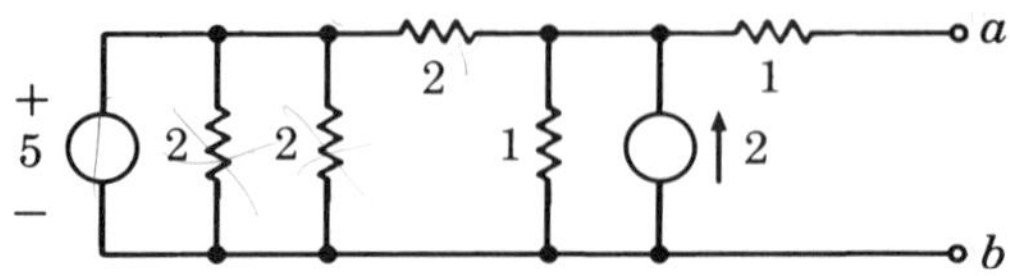

FIGURE 4–42

40. Give the Thévenin and the Norton equivalent circuits which hold for the terminal-pair a–b in the circuit of Fig. 4–43.

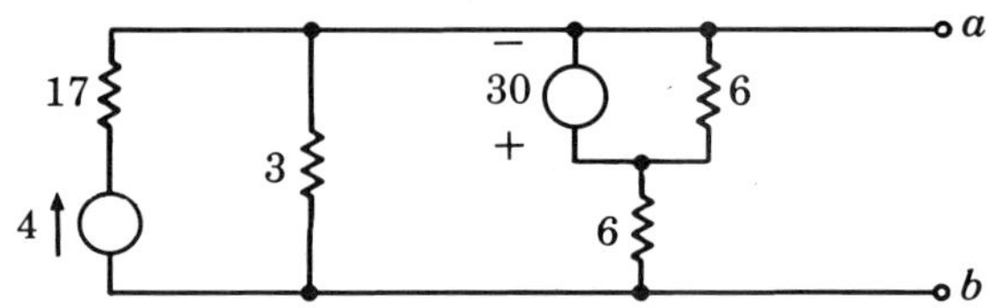

FIGURE 4–43

41. Directly from the circuit of Fig. 4–44 compute the open-circuit voltage, the short-circuit current, and the equivalent resistance at the terminal-pair a–b. From these give the Thévenin and Norton equivalent circuits.

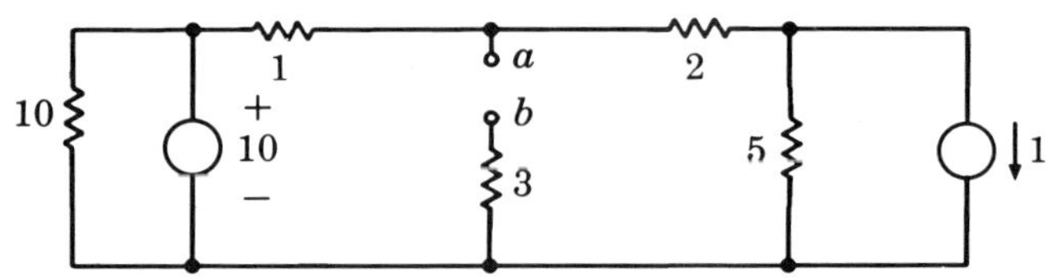

FIGURE 4–44

42. Obtain the Thévenin equivalent circuit at the terminals a–b in the network of Fig. 4–45 by (a) a direct computation of the open-circuit voltage and the equivalent resistance, (b) a reduction of the network to a single branch by means of series-parallel formulas and voltage-to-current source transformations.

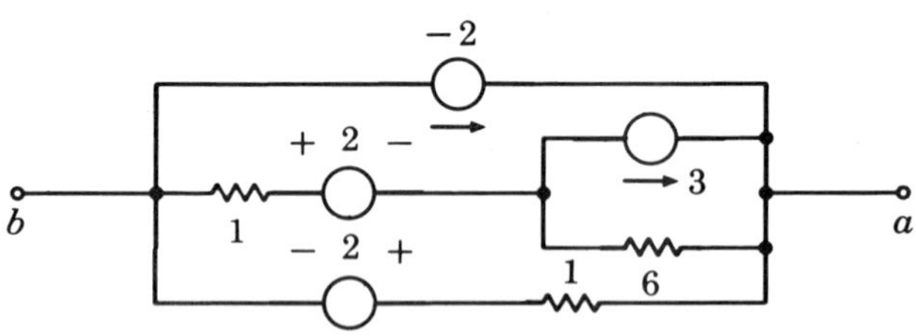

FIGURE 4–45

43. Obtain the Norton equivalent circuit at the terminals a–b in the network of Fig. 4–46 by (a) a direct computation of the short-circuit current and the equivalent resistance, (b) a reduction of the network to a single branch by means of series-parallel and current-to-voltage source transformations.

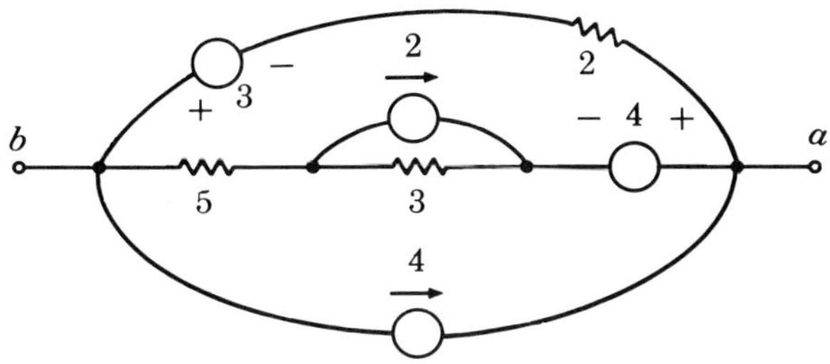

FIGURE 4–46

44. Obtain both the Norton and Thévenin equivalent circuits for the network of Fig. 4–47 at the terminals a–b by direct calculations on the original network.

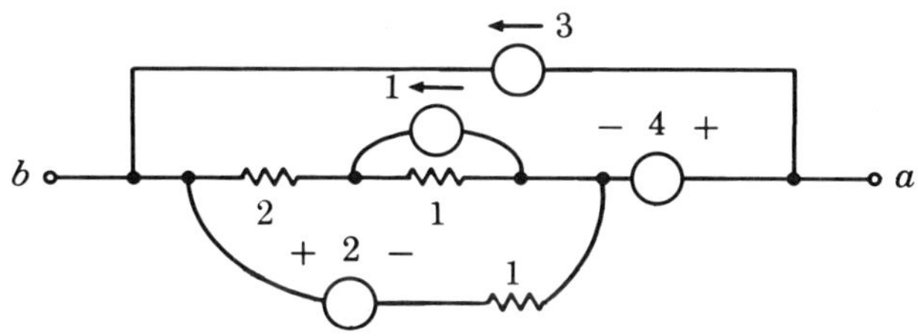

FIGURE 4–47

45. (a) Find the Thévenin equivalent circuit for the bridge circuit of Fig. 4–48 at the terminals a–b. (b) If a zero-resistance ammeter were placed across a–b, what current would flow through it? (c) If a 1-ohm ammeter were placed across a–b, what current would flow through it?

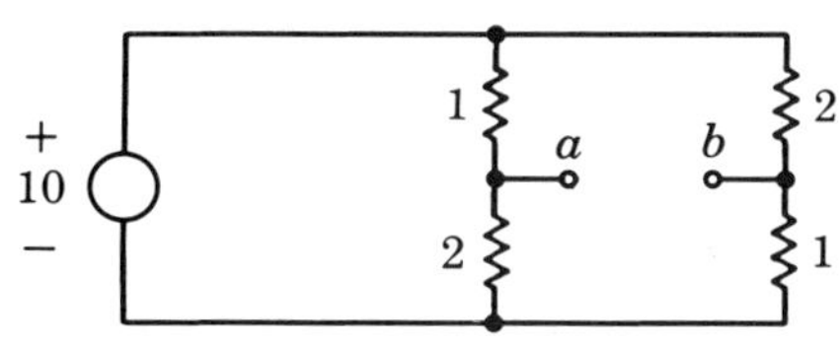

FIGURE 4–48

46. Find both the Thévenin and the Norton equivalent circuits for the network of Fig. 4–49 at the terminals a–b.

47. Find both the Thévenin and the Norton equivalent circuits for the network of Fig. 4–50 at the terminals a–b.

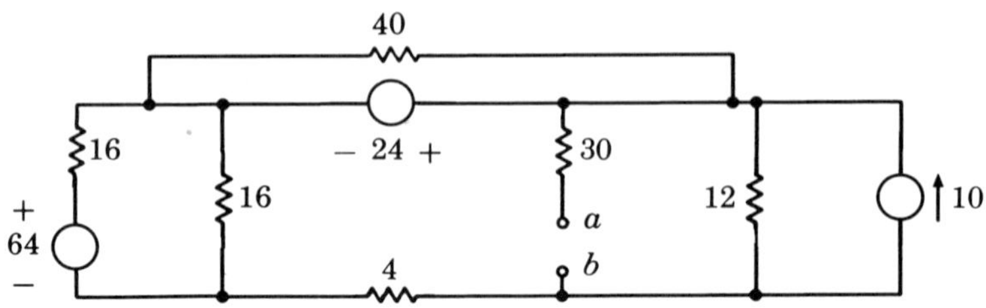

FIGURE 4–49

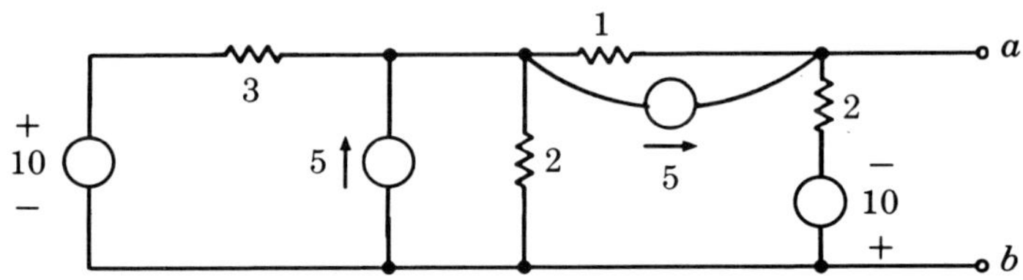

FIGURE 4–50

48. Find the Thévenin equivalent circuit for the network of Fig. 4–51 at the terminals a–b. What is the Norton equivalent circuit?

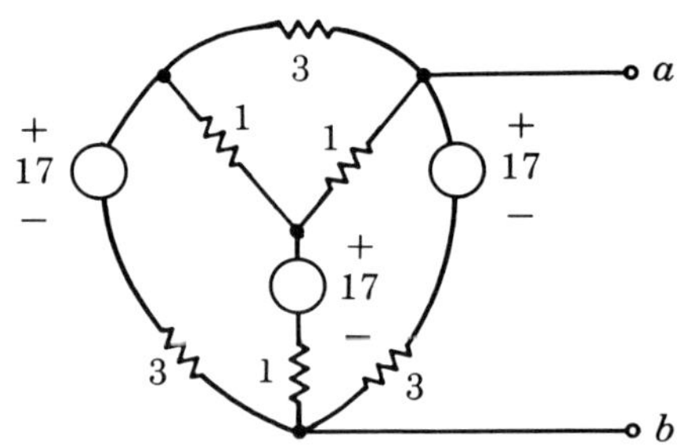

FIGURE 4–51

Section 4–7

49. (a) Solve for the input resistance of network B in Fig. 4–52. (b) Find the voltage E_2 when network B is connected to network A. (c) Find the voltage E_0 by means of the substitution theorem.

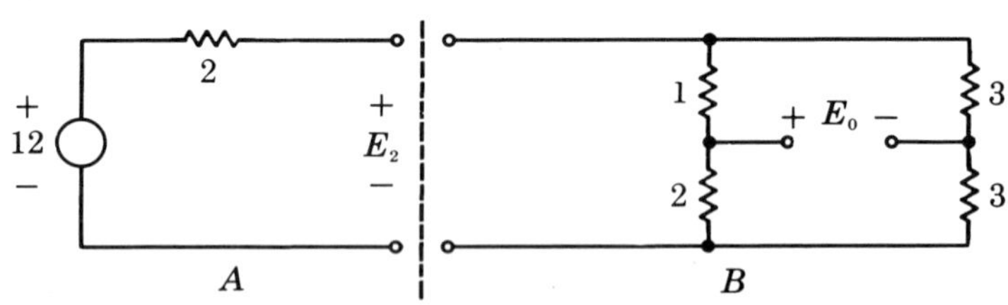

FIGURE 4–52

50. (a) In Fig. 4–53, find the input resistance of network B, (b) find the voltage E_2 when network B is connected to network A, (c) find the current I_0 by means of the substitution theorem.

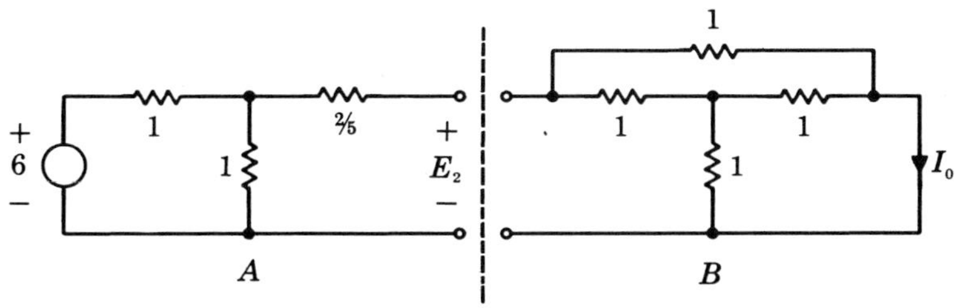

FIGURE 4–53

51. In Fig. 4–54, find the voltage e_a by finding the equivalent resistances of the two sections of the network, and then solve for the voltage e_0 by means of the substitution theorem.

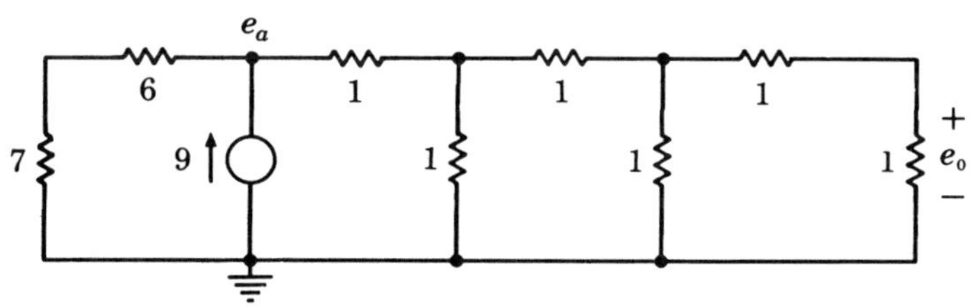

FIGURE 4–54

52. In Fig. 4–55, find the current i_a by finding the equivalent resistances of the two sections of the network, and then solve for the voltage e_0 by means of the substitution theorem.

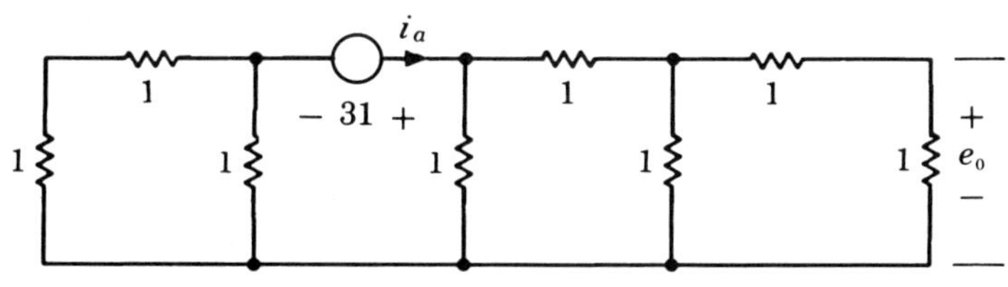

FIGURE 4–55

53. In the laboratory, certain measurements are made on the "black box" of Fig. 4–56. The resistance at the input terminals is found to be 5 ohms. The resistance at the output terminals is found to be 10 ohms. (In both cases the other terminals are left open-circuited.) When a voltage source of 5 volts is applied to E_1, the output voltage E_2 reads 1 volt. If a 6-volt source with an internal resistance of 1 ohm were applied to the input terminals, what output voltage E_2 would be obtained?

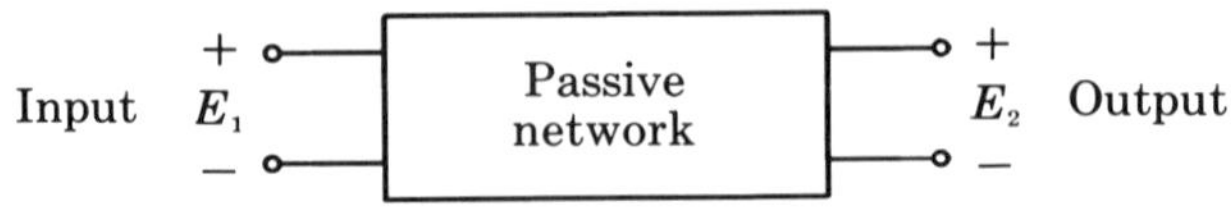

FIGURE 4–56

54. In the circuit of Fig. 4–57 the resistance R is changed from 1 ohm to 2 ohms. Use the substitution theorem to find the change in source current caused by this change in resistance.

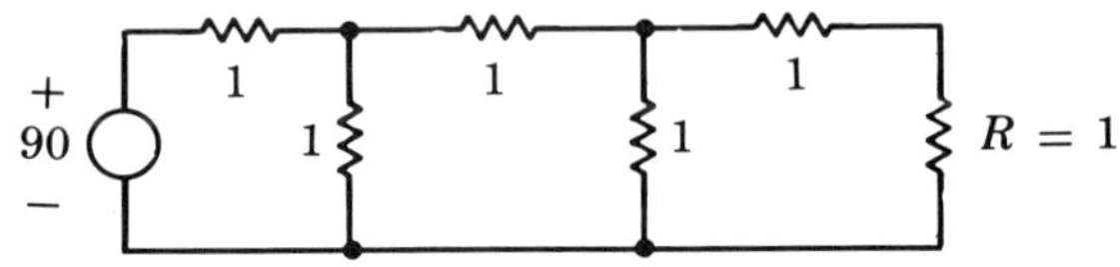

FIGURE 4–57

55. In the circuit of Fig. 4–58 a resistance of 1 ohm is added in parallel with the output resistance R. Use the substitution theorem to find the corresponding change in the source current.

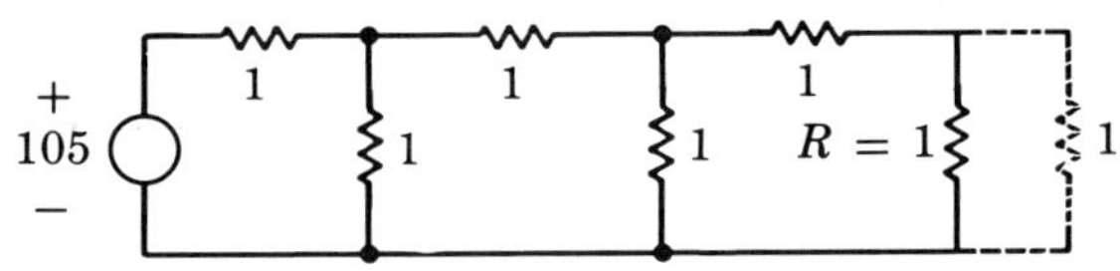

FIGURE 4–58

56. An ammeter with resistance R_m is used to measure the current in a circuit. The resistance looking back into the network at the terminals of the ammeter is R. Use the substitution theorem to show that the correct current reading (i.e., the reading of a zero-resistance ammeter) is given by

$$I \text{ (correct)} = I \text{ (meter)} \left(1 + \frac{R_m}{R}\right).$$

57. A voltmeter with resistance R_m is used to measure the voltage in a circuit. The resistance looking back into the network at the terminals of the voltmeter is R. Use the substitution theorem to show that the correct voltage reading (i.e., the reading of an infinite-resistance voltmeter) is given by

$$E \text{ (correct)} = E \text{ (meter)} \left(1 + \frac{R}{R_m}\right).$$

CHAPTER 5

METHODS OF CHARACTERIZING TWO-TERMINAL-PAIR RESISTIVE NETWORKS

"Love of an idea is love of God. Understanding is love."

Frank Lloyd Wright

5–1 Introduction. In this chapter we shall extend the idea of linear circuits to the case of the two-terminal-pair network. We have already observed that linearity and superposition make it unnecessary for us to analyze anything more complicated than a two-terminal-pair network. This chapter therefore represents the culmination of our study of resistance networks. The rest of the book merely extends these ideas to circuits which involve energy storage.

The complete description of a two-terminal-pair network involves more than one stimulus-response relationship. If we apply a stimulus at one of the terminal-pairs we produce a response at both that terminal-pair and the second one. Excitation of the second terminal-pair involves two more relationships. Apparently a total of four stimulus-response relationships are involved. For most networks two of the four are identical, however, by the *reciprocity theorem*.

When the network is resistive the stimulus-response relationships are simply equivalent resistances, and therefore a two-terminal-pair resistive network can be reduced to three equivalent resistances.* The Ohm's law relationships between the voltages and currents at the two terminal-pairs also involve three constants. They can be written as two equations for the voltages in terms of the currents, or two equations for the currents in terms of the voltages.

In this chapter we shall find the equivalent resistances of a number of two-terminal-pair networks by a process of series-parallel and Y–Δ transformations and also by application of the Ohm's law equations. The equivalent networks can be presented in the form of a Y (often called a T), or a "Δ" (often called a π). The usual Y–Δ transformation relates them. Neither of these networks is the most general form of two-terminal-pair equivalent circuits. Negative elements can occur in either of them.† The

* Three elements are sufficient if the input and output terminal-pairs can have a common ground. Otherwise six are required.

† These circuits are not realizable with positive elements, but they are perfectly satisfactory for computational purposes.

lattice is a more general form, and will be considered in the symmetrical case. The lattice elements are always positive.

The special techniques for symmetrical networks and ladder networks apply to the two-terminal-pair network as well as to the one-terminal-pair network.

5–2 Equivalent circuits of two-terminal-pair networks. The reciprocity theorem. In the previous chapter we discussed how a one-terminal-pair resistive network can be reduced to a single equivalent resistance and can be described completely by a single volt-ampere equation at the terminals. It would be helpful if a two-terminal-pair network could be reduced to two resistances, but such is not the case. Three is a minimum number, and even this number cannot be achieved unless the input terminal-pair and the output terminal-pair have a common terminal. Fortunately, most practical networks fall in this category or can be treated as if they did. If the input and the output do not possess a common ground, six resistances are required in the equivalent circuit.

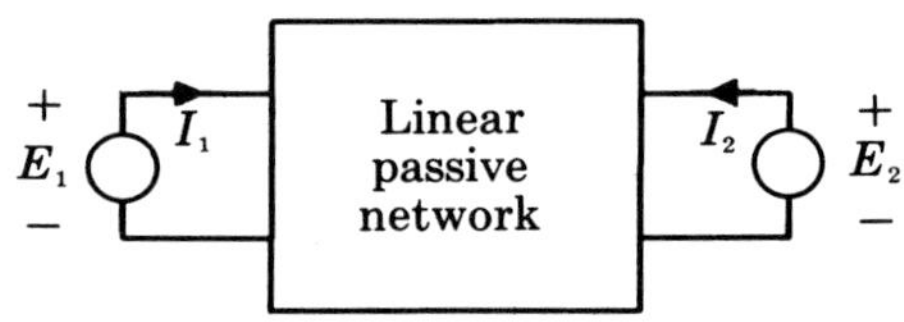

FIG. 5–1. Two-terminal-pair network for which reciprocity holds.

The reciprocity theorem. The reciprocity theorem concerns the stimulus-response relationships of a two-terminal-pair network, such as is shown in Fig. 5–1. Let us suppose that we have excited the network with a voltage source at E_1 and a voltage source at E_2. We can then write a set of mesh equations to describe the complete behavior of the system. They will have the following form:

$$
\begin{array}{rllll}
i_1r_{11} & -\ i_2r_{12} & -\ i_3r_{13} & - \cdots & -i_nr_{1n} = E_1, \\
-i_1r_{12} & +\ i_2r_{22} & -\ i_3r_{23} & - \cdots & -i_nr_{2n} = E_2, \\
-i_1r_{13} & -\ i_2r_{23} & +\ i_3r_{33} & - \cdots & -i_nr_{3n} = 0, \\
\vdots & & & & \\
-i_1r_{1n} & -\ i_2r_{2n} & -\ r_3r_{3n} & - \cdots & \quad i_nr_{nn} = 0.
\end{array}
\tag{5–1}
$$

The only two sources in the whole system are E_1 and E_2. First of all, we will let $E_2 = 0$ and compute the current i_2 which then flows in the short circuit.

It is

$$i_2 = \frac{-E_1}{\Delta}\begin{vmatrix} -r_{12} & -r_{23} \cdots & -r_{2n} \\ -r_{13} & r_{33} \cdots & -r_{3n} \\ \vdots & & \\ -r_{1n} & -r_{3n} \cdots & r_{nn} \end{vmatrix}, \tag{5-2}$$

where Δ is the determinant of the coefficients on the left side of the equation. Now we will let $E_1 = 0$ and compute the current i_1 which flows in it in response to the voltage E_2:

$$i_1 = \frac{-E_2}{\Delta}\begin{vmatrix} -r_{12} & -r_{13} \cdots & -r_{1n} \\ -r_{23} & r_{33} \cdots & -r_{3n} \\ \vdots & & \\ -r_{2n} & -r_{3n} \cdots & r_{nn} \end{vmatrix}. \tag{5-3}$$

The determinants of Eqs. (5-2) and (5-3) are the same except for an interchange of the first row and first column. It is easy to verify from the Laplace expansion for a determinant that this interchange does not change the value of the determinant, and hence that the two determinants are equal. A necessary condition for this equality, of course, is the symmetry of the original determinant of the coefficients. The element between mesh 1 and mesh 2 must look the same when viewed from either mesh. All resistances, of course, meet this requirement, but not all network elements do. Those that do are called *bilateral* elements and those that do not are called unilateral elements.* When the two determinants are equal we have shown that the current produced at i_2 per volt at E_1 is the same as the current produced at i_1 per volt at E_2. Since any two branches in a network can be selected as the input and the output, a general statement of the reciprocity theorem is:

An ideal voltage source and an ideal ammeter in any two branches of a passive network which is linear and bilateral can be interchanged without changing the numbers read on either of them.

An analogous argument can be carried out with current sources as the excitation for the network and voltages as the responses. Node equations are written for the network in a general way and the following conclusions can be reached:

A current source and a voltmeter across any two node-pairs in a passive network which is linear and bilateral can be interchanged without changing the numbers read on either of them.

---

* Many nonlinear elements are not bilateral, but the only linear element which is not bilateral is the gyrator.

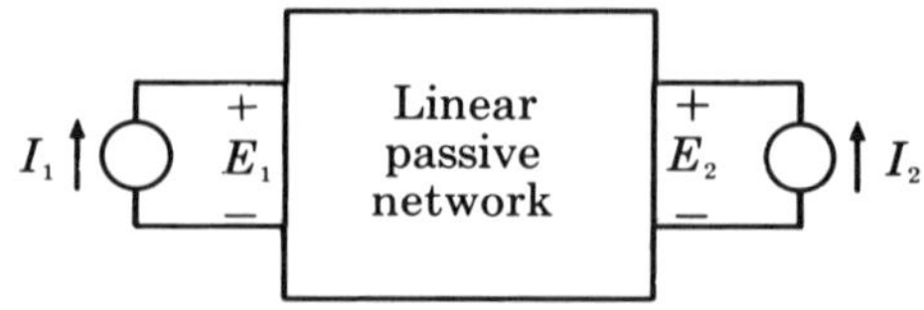

FIG. 5–2. Two-terminal-pair network with current excitation.

These two statements constitute the *reciprocity theorem.* It is easy to remember, since the elements which are interchanged must have the same internal impedance. In the first statement both the voltage source and the ammeter have zero resistance. In the second both the current source and the voltmeter have infinite resistance.

Three-resistance Y or T networks. Now let us return to our central problem of an equivalent circuit for the general two-terminal-pair network, as shown in Fig. 5–2. Let us consider the two currents as the excitations and the two voltages as the responses. By superposition, we can write

$$\begin{aligned} E_1 &= r_{11}I_1 + r_{12}I_2, \\ E_2 &= r_{21}I_1 + r_{22}I_2, \end{aligned} \tag{5–4}$$

where the r's are simply constants of proportionality [in general not the same as those in Eq. (5–1)] obtained by determining the response due to a unit value of each source, with the other source given zero value. Thus

$$\begin{aligned} r_{11} &= \frac{E_1}{I_1} \quad \text{with} \quad I_2 = 0 \\ r_{12} &= \frac{E_1}{I_2} \quad \text{with} \quad I_1 = 0, \\ r_{21} &= \frac{E_2}{I_1} \quad \text{with} \quad I_2 = 0, \\ r_{22} &= \frac{E_2}{I_2} \quad \text{with} \quad I_1 = 0. \end{aligned} \tag{5–5}$$

Since a zero-current source is simply an open circuit, the resistance r_{11} is the input resistance of the network with the output open-circuited. The resistance r_{22} is the output resistance of the network with the input open-circuited. The constant r_{12} is the voltage E_1 produced at the open-circuited input terminals per ampere of output current I_2, and the constant r_{21} is the voltage E_2 produced at the open-circuited output terminals per ampere of input current I_1. By the reciprocity theorem these two constants are equal, or

$$r_{12} = r_{21}, \tag{5–6}$$

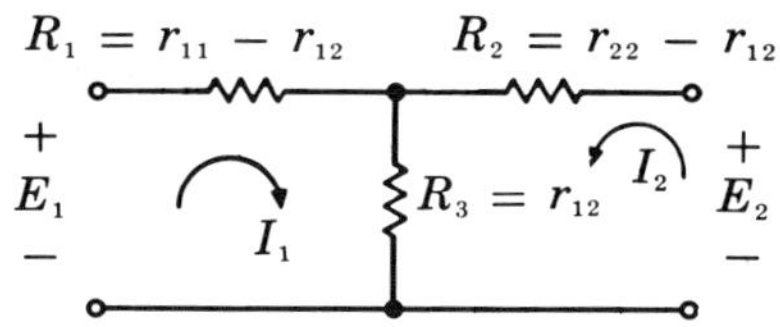

FIG. 5–3. Two-terminal-pair equivalent T circuit.

and we can write the volt-ampere relationships (or Ohm's law equations) for the two-terminal-pair network as

$$\begin{aligned} E_1 &= r_{11}I_1 + r_{12}I_2, \\ E_2 &= r_{12}I_1 + r_{22}I_2. \end{aligned} \tag{5–7}$$

Usually the r's are obtained from Eq. (5–5) by direct calculation. Then Eq. (5–7) represents a complete algebraic characterization of the device, so far as its effect on the rest of the circuit is concerned.* As written, it describes the voltages produced by currents into the two terminal-pairs. Just as Ohm's law is valid for both types of excitation, however, so is Eq. (5–7). If the voltages are the input excitations, the currents can be obtained by solving the equations. The three constants r_{11}, r_{12}, and r_{22} completely describe the network behavior for any type of excitation.

When the network is resistive, Eq. (5–7) represents a three-resistance network, as shown in Fig. 5–3. From the equations, we find that the mutual resistance R_3 between the two meshes in the network is equal to r_{12}:

$$R_3 = r_{12}. \tag{5–8}$$

The total self-resistance in mesh 1 is r_{11}, and therefore

$$R_1 = r_{11} - r_{12}. \tag{5–9}$$

Similarly,

$$R_2 = r_{22} - r_{12}. \tag{5–10}$$

The equivalent Y or T circuit, as it is usually called, is perfectly general in a mathematical sense. However, in a physical sense it is not, since any of the resistances given by Eqs. (5–8), (5–9), and (5–10) can turn out to be negative. A negative value of R_3 results if the output voltage is opposite in polarity from the input voltage, and a negative value of R_1 or R_2 occurs when r_{12} is larger than r_{11} or r_{22}. We need not be disturbed by the negative r's, since we do not have to build the real circuit.† For

* Provided that the input and the output terminals have a common ground or can be treated as if they did.

† Negative resistances are not physically realizable in passive circuits; they represent sources of energy and hence require "active" circuits.

purposes of computation these r's are just as good as positive ones; they mean only that the T equivalent circuit is not the most general form for a two-terminal-pair passive network. Later we will show that the lattice is the most general form.

The T circuit of Fig. 5–3 is, of course, identical to a Y circuit, and the Y–Δ transformation we developed in the preceding chapter applies to it. The transformed circuit in the Δ form is usually called a π when it relates to two-terminal-pair networks. We could derive the π by means of a Y–Δ transformation, but it is more instructive to develop it directly from the general two-terminal-pair network of Fig. 5–2. In the process we will obtain an alternate form of the Ohm's law relationships for two-terminal-pair networks.

Three-resistance Δ or π networks. In order to obtain the T network representation of a two-terminal-pair network we considered the excitation to be the currents and the responses to be the voltages. It is equally appropriate to interchange these. If the voltages are the excitations in Fig. 5–2 and the currents are the responses, by means of the superposition theorem we can write

$$\begin{aligned} I_1 &= g_{11}E_1 + g_{12}E_2, \\ I_2 &= g_{21}E_1 + g_{22}E_2, \end{aligned} \tag{5–11}$$

where the g's are coefficients of proportionality between the individual sources and the individual responses, given by

$$\begin{aligned} g_{11} &= \frac{I_1}{E_1} \quad \text{with} \quad E_2 = 0, \\ g_{12} &= \frac{I_1}{E_2} \quad \text{with} \quad E_1 = 0, \\ g_{21} &= \frac{I_2}{E_1} \quad \text{with} \quad E_2 = 0, \\ g_{22} &= \frac{I_2}{E_2} \quad \text{with} \quad E_1 = 0. \end{aligned} \tag{5–12}$$

The constant g_{11} is the input admittance of the network with a short circuit on the output terminal-pair. It should be observed particularly that g_{11} is not the reciprocal of r_{11} of Eq. (5–5), since the output terminals are open-circuited for the measurement of r_{11}. Similarly, g_{22} is the admittance at the output terminal-pair, and it is not the reciprocal of r_{22}, since there is a short circuit across the input terminals for the measurement of g_{22} and an open circuit for the measurement of r_{22}. The constant g_{12} is current produced at the input per volt applied to the output, and the constant g_{21} is the current produced at the output per volt at the input. These two quantities are equal by the reciprocity theorem:

$$g_{12} = g_{21}. \tag{5-13}$$

As a result of Eq. (5-13) the general Ohm's law relationships of Eq. (5-11) can be simplified to

$$\begin{aligned} I_1 &= g_{11}E_1 + g_{12}E_2, \\ I_2 &= g_{12}E_1 + g_{22}E_2. \end{aligned} \tag{5-14}$$

Equation (5-14) represents an alternate form of the volt-ampere equations of a two-terminal-pair network given in Eq. (5-7). It is equally valid and equally general. Although we derived it by assuming the voltages as excitations, it is equally true if the currents are the excitations. It merely expresses the constraints the network itself imposes on the currents and voltages at its terminals.* Since Eq. (5-14) and Eq. (5-7) are equivalent, it is possible to derive one from the other, and we shall do so in a later section. We content ourselves here with deriving the equivalent π circuit of Fig. 5-4 from Eq. (5-14). From the equation, the mutual conductance is g_{12}, and hence the resistance R_b between the two nodes is $-1/g_{12}$, or

$$R_b = \frac{-1}{g_{12}}. \tag{5-15}$$

The negative sign occurs because of the direction assumed for I_2. The total conductance at node 1 is g_{11}. The mutual conductance is $-g_{12}$. The conductance to ground at node 1 is therefore $g_{11} + g_{12}$, and thus

$$R_a = \frac{1}{g_{11} + g_{12}}. \tag{5-16}$$

Similarly, the total conductance at node 2 is g_{22} and the net conductance

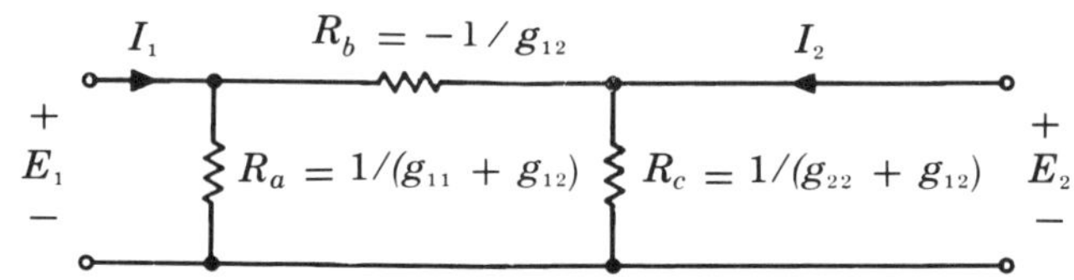

FIG. 5-4. Two-terminal-pair equivalent π circuit.

---

* Note that it is not possible to supply both an arbitrary current and an arbitrary voltage to the *same* terminal-pair of a network. Only one can be supplied externally; the network determines the other.

to ground is $g_{22} + g_{12}$. Thus

$$R_c = \frac{1}{g_{22} + g_{12}}. \tag{5-17}$$

The π circuit is not any more general than the T circuit. Any of the resistances given by Eqs. (5–15), (5–16), and (5–17) can turn out to be negative.

We will have occasion in a later section to compute the r's and the g's of some basic networks and to find the relationship between them. Before we do this, however, we shall consider the direct reduction of some two-terminal-pair networks to three equivalent resistances, and some of the difficulties encountered in this process. For example, three resistances are not sufficient to represent the two-terminal-pair network if we care about the voltages between the input and the output terminal-pairs.

5–3 The direct reduction of three- and four-terminal networks. A two-terminal-pair network can be either a three-terminal network or a four-terminal network. Three-terminal networks have a common terminal, usually the ground, between the input and the output terminal pairs. A four-terminal network does not have this connection; each of the four terminals is independent. Three-terminal networks are more common than

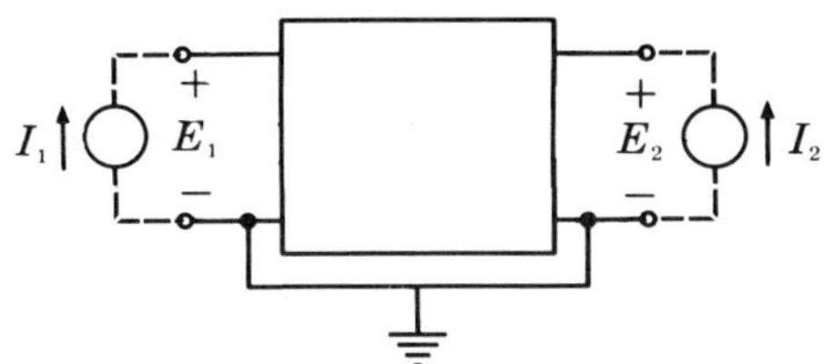

(a) Three-terminal network

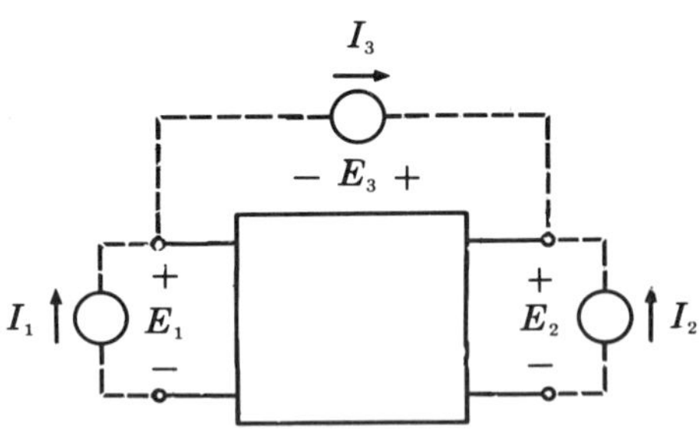

(b) Four-terminal network

FIG. 5–5. Three- and four-terminal networks.

four-terminal networks. In addition, four-terminal networks can be reduced to equivalent three-terminal networks if we are interested only in the input and output voltages and do not care about the voltages between the input terminal pair and the output terminal pair. Examples of three- and four-terminal networks are shown in Fig. 5–5. There are only two independent excitations and two independent responses in the three-terminal network. If the excitations are current sources the voltages are given by

$$\begin{aligned} E_1 &= r_{11}I_1 + r_{12}I_2, \\ E_2 &= r_{12}I_1 + r_{22}I_2, \end{aligned} \tag{5–18}$$

as in Eq. (5–7). Three constants are required to express the effect of the network, and the network can be reduced to three resistances. A four-terminal network, however, can be given independent excitations at three terminal-pairs and there are three independent responses. For current excitations, as in Fig. 5–5(b), the voltage responses are

$$\begin{aligned} E_1 &= r_{11}I_1 + r_{12}I_2 + r_{13}I_3, \\ E_2 &= r_{12}I_1 + r_{22}I_2 + r_{23}I_3, \\ E_3 &= r_{13}I_1 + r_{23}I_2 + r_{33}I_3. \end{aligned} \tag{5–19}$$

Nine constants are necessary to express the effect of the network in Eq. (5–19), but because of the reciprocity theorem three pairs of them are equal and only six of them are different. We can conclude from this result that six resistances are necessary to represent an equivalent circuit for a general four-terminal network. In summary:

A three-terminal network has three equivalent resistances.

A four-terminal network has six equivalent resistances.

A moment's reflection shows that these statements are reasonable. The most general equivalent network would contain a resistance between each pair of terminals. For three terminals there are just three resistances, but for four terminals there are six. Three-terminal networks can always be reduced to three equivalent resistances by series-parallel and Y–Δ transformations. Four-terminal networks require a more general form of the Y–Δ transformation, called the *star-mesh* transformation.

Star-mesh transformation. A general star, such as the one shown in Fig. 5–6(a), can be reduced to a general mesh which eliminates the node E_0, as shown in Fig. 5–6(b). The general mesh has a resistance between each pair of its nodes. In terms of the node voltages, the currents in

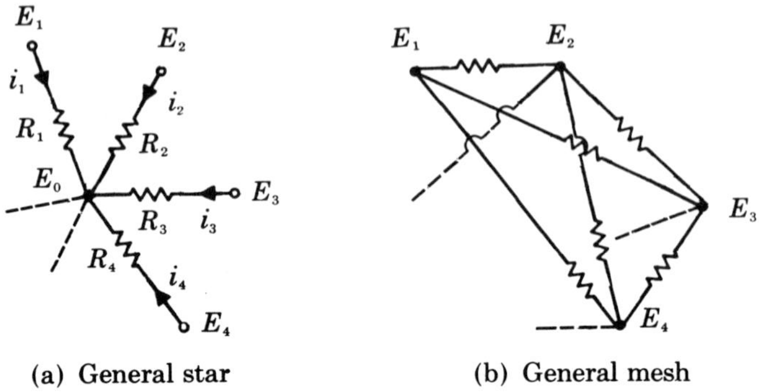

FIG. 5–6. The general star-mesh transformation.

the star are

$$
\begin{aligned}
i_1 &= G_1(E_1 - E_0),\\
i_2 &= G_2(E_2 - E_0),\\
i_3 &= G_3(E_3 - E_0),\\
&\vdots\\
i_n &= G_n(E_n - E_0),
\end{aligned}
\tag{5–20}
$$

from which

$$
\begin{aligned}
i_1 + i_2 + \cdots + i_n = -(G_1 + G_2 + G_n + \cdots + G_n)E_0 \\
+ G_1E_1 + G_2E_2 + G_3E_3 + \cdots + G_nE_n.
\end{aligned}
$$

By Kirchhoff's law, the sum of the currents is equal to zero, and hence

$$
E_0 = \frac{G_1}{\sum G} E_1 + \frac{G_2}{\sum G} E_2 + \cdots + \frac{G_n}{\sum G} E_n. \tag{5–21}
$$

From Eqs. (5–20) and (5–21) the current i_1 can be expressed in terms of the mesh node voltages:

$$
\begin{aligned}
i_1 &= G_1E_1 - G_1E_0\\
&= G_1E_1 - \frac{G_1^2}{\sum G} E_1 - \frac{G_1G_2}{\sum G} E_2 - \frac{G_1G_3}{\sum G} E_3 - \cdots - \frac{G_1G_n}{\sum G} E_n\\
&= \frac{G_1G_2}{\sum G}(E_1 - E_2) + \frac{G_1G_2}{\sum G}(E_1 - E_2) + \cdots + \frac{G_1G_n}{\sum G}(E_1 - E_n).
\end{aligned}
\tag{5–22}
$$

Equation (5–22) represents a resistance between node E_1 and each of the other nodes, whose value is given by

$$G\text{ (mesh)} = \frac{\text{Product of adjacent } G\text{'s (star)}}{\text{Sum of the } G\text{'s (star)}}. \tag{5-23}$$

The formula in Eq. (5–23) can be extended to the other nodes of the system by symmetry. It is identical in form to the Y–Δ formula given in Eq. (4–15). Observe, however, that there is no corresponding Δ–Y equation in the general case. The reason is obvious. If the general star has n resistances, the equivalent mesh of Eq. (5–23) will have $n(n-1)/2$ resistances. Thus, four Y resistances become six Δ resistances. The mesh-star transformation cannot exist, in general, because six resistances cannot be compressed into four.

Examples of the use of the general star-mesh transformation will be given in the following sections on the reduction of three- and four-terminal networks.

Three-terminal network reduction. Three-terminal networks can always be reduced to three equivalent resistances by successive applications of series-parallel and Y–Δ transformations. An example is shown in Fig. 5–7(a). As a first step, the node d is removed by a Y–Δ transformation; the network of Fig. 5–7(b) results. The parallel resistances are combined to give three $\frac{3}{4}$-ohm equivalents, and a Δ–Y transformation is applied to these to give the network of Fig. 5–7(c). The final three-resistance equivalent is shown in Fig. 5–7(d).

Other three-terminal networks can be reduced to three resistances in a similar fashion. The exact order of the Y–Δ, the series, and the parallel transformations will vary from case to case. The only general rule is to leave the input and output terminals alone!

Four-terminal network reduction. The reduction of four-terminal networks is more difficult than the reduction of three-terminal networks.

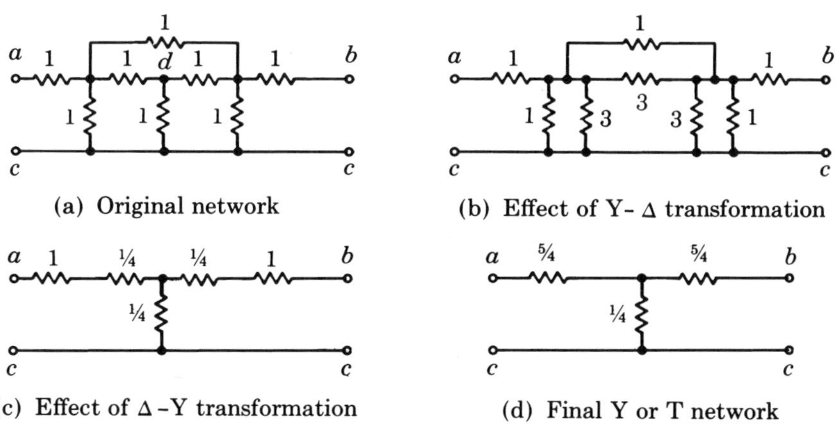

FIG. 5–7. The reduction of a three-terminal network to three resistances.

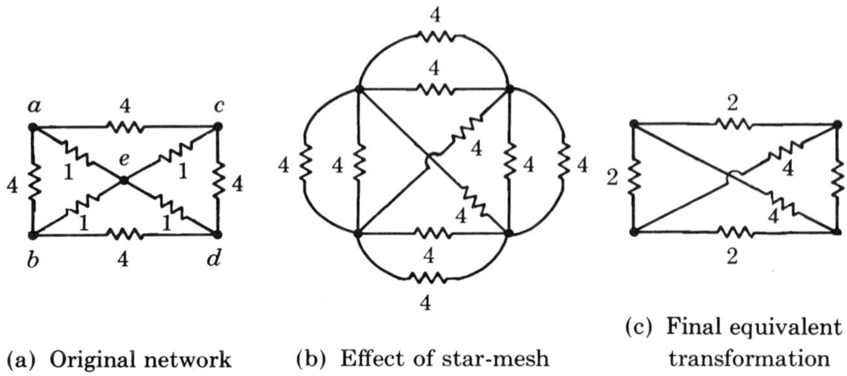

FIG. 5–8. The reduction of a four-terminal network to six resistances.

Often the general star-mesh transformation is required. An example is shown in Fig. 5–8(a). The general star-mesh transformation is applied to remove the node e. This result cannot be accomplished by Y–Δ transformations without removing one of the nodes a, b, c, d at the same time. The four resistances emanating from e are replaced by six that join the four nodes a, b, c, and d in pairs. Because of the symmetry the resistances are equal and are given by

$$G \text{ (mesh)} = \frac{\text{Product of adjacent } G\text{'s (star)}}{\text{Sum of the } G\text{'s (star)}} = \frac{1 \times 1}{4},$$

or

$$R \text{ (mesh)} = 4 \text{ ohms.}$$

The transformed network is shown in Fig. 5–8(b), and a six-resistance equivalent in Fig. 5–8(c). (Because of the symmetry this network can actually be reduced to four resistances; see Bartlett's bisection theorem, Section 5–6.)

Other four-terminal networks can be reduced in a similar fashion. In general, the resulting network will contain six resistances, although in special cases it may contain less, as in the example here.* In some circumstances the four-terminal networks can be reduced to three-terminal networks, as we shall now see.

The reduction of four-terminal networks to three-terminal networks. In Section 5–2 we showed that two-terminal-pair networks could be represented by three resistances if we were interested only in the input and output voltages and not in the voltage between the input and output

* The six-resistance form is an unbalanced lattice, or bridge circuit.

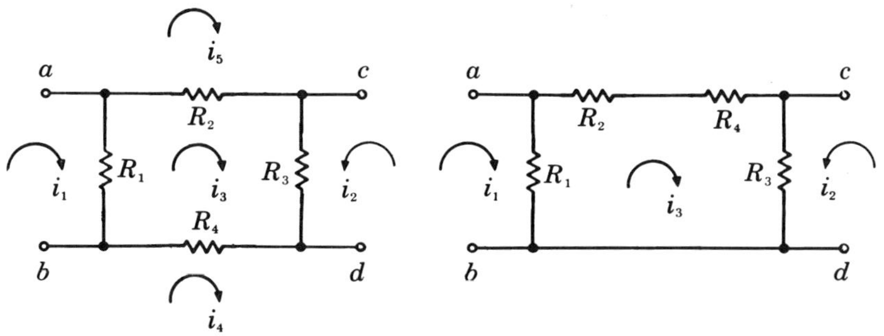

(a) Four-terminal network (b) Equivalent three-terminal network

FIG. 5–9. A four-terminal network which can be reduced to three.

terminal-pairs. The three-terminal equivalent can always be found from the general equations developed in Section 5–2, but sometimes it is easily obtained from the network itself, as in the examples which will be discussed now. Consider the circuit of Fig. 5–9(a). The input terminals are a and b; the output terminals are c and d. The meaning of the statement that only the input and the output terminals are of interest is that the external mesh currents shown as i_4 and i_5 are zero. *External to the network, there can be no connection between the output terminals and the input terminals.* If this condition is satisfied the system can be described by the three mesh currents i_1, i_2, and i_3. Moving the resistance R_3 from the bottom line to the top line does not affect the mesh equations, and therefore does not affect the relationship between the input terminals and the output terminals. The equivalent three-terminal circuit is given in Fig. 5–9(b).

A similar situation holds for the circuit of Fig. 5–10(a). The statement that only the input voltage and the output voltage are of interest is equivalent to stating that there are no currents in R_1, R_2, R_3, and R_4 except the input and output mesh currents. The mesh equations written for these currents therefore are not altered by moving R_3 and R_4 from

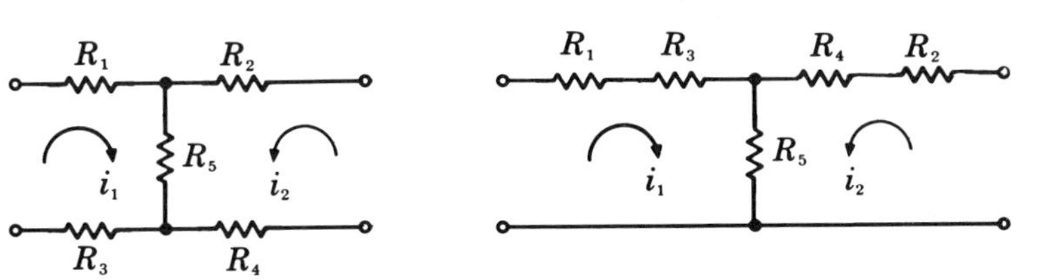

(a) Four-terminal network (b) Equivalent three-terminal network

FIG. 5–10. A second four-terminal network which can be reduced to three.

the bottom line to the top line, and a three-resistance equivalent circuit is obtained, as in Fig. 5–10(b).

The analysis of the lattice or bridge network is more difficult and is perhaps best treated by the general equations. The particular case of a symmetrical lattice is treated in Section 5–6.

The methods discussed in connection with the circuits of Figs. 5–9 and 5–10 can be extended to any number of internal meshes. A typical example is shown in Fig. 5–11(a). None of the mesh-current equations would be changed if the resistances in the bottom row were moved into the top row. The resulting three-terminal network is shown in Fig. 5–11(b).* The three-terminal network is equivalent to the four-terminal network only with respect to the input and the output terminal pairs. If it is to be used to replace the four-terminal network there should be no external connections between the two terminal pairs.

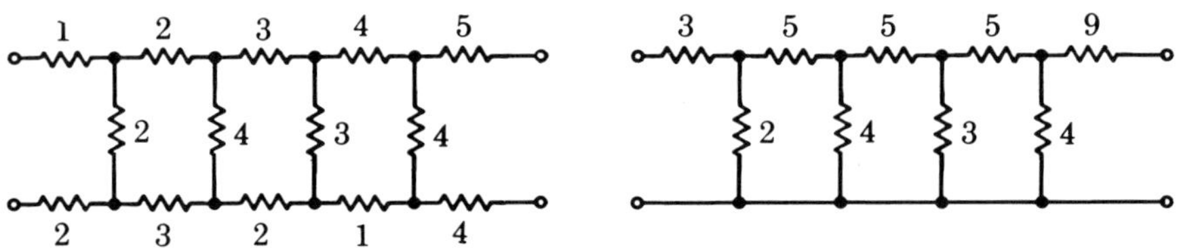

(a) Four-terminal network (b) Equivalent three-terminal network

FIG. 5–11. A more general four-terminal network which can be reduced to three.

A network which cannot be easily transformed from four terminals to three is shown in Fig. 5–12. Networks of this type are best transformed by means of the r parameters or the g parameters, as discussed in the next section.

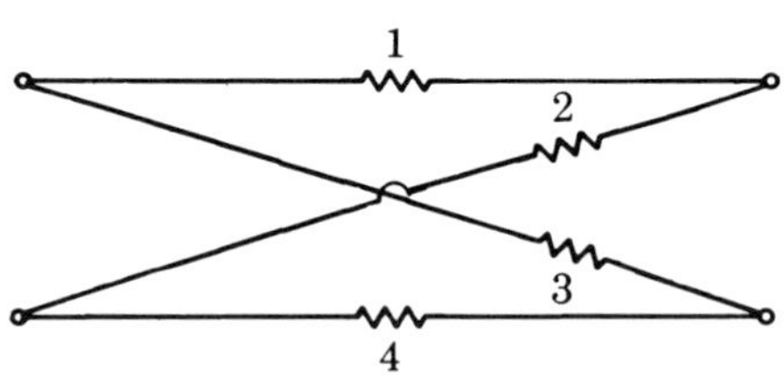

FIG. 5–12. A four-terminal network not easily reduced to three.

* For the transformation of "balanced" or symmetrical four-terminal networks to three-terminal networks, see Section 5–5 on symmetry.

5-4 The reduction of two-terminal-pair networks by r and g parameters.* For many networks the most direct way of obtaining an equivalent three-terminal network from an original four-terminal network is by means of the r or the g parameters. Here we shall compute the r and g parameters for simple networks, and use them to find equivalent T and π networks in three-terminal form. Since there is a relationship between the T and the π networks, there is obviously a relationship between the r and the g parameters, and this is what we will investigate in this section.

The r parameters. In Section 5-2 we used the superposition theorem to show that a two-terminal-pair network could always be represented by the equations

$$E_1 = r_{11}I_1 + r_{12}I_2, \qquad E_2 = r_{12}I_1 + r_{22}I_2. \tag{5-24}$$

The r's are parameters which completely specify the behavior of the network at its terminals.† They describe the relation between the terminal currents and voltages of the network in a manner that may be regarded as a generalization of Ohm's law for a single resistance. The r parameters can be computed for any two-terminal-pair network. The T equivalent circuit obtained from them may or may not have positive elements. An example of each possibility is given below.

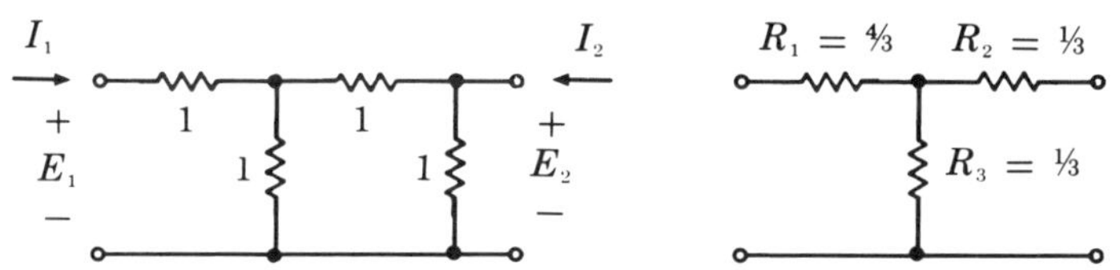

(a) Original network (b) Equivalent network

FIG. 5-13. Network conversion by r parameters.

Example 1. A two-terminal-pair network which can be reduced to three equivalent resistances is shown in Fig. 5-13(a). The r parameters will be obtained and from them the equivalent T. The constant r_{11} is the input resistance of the network with the output terminals open-circuited. It can be obtained by applying an input current of 1 amp and measuring the input voltage which it produces. Alternatively, it can be obtained by combining the resistances in the network. Here

$$r_{11} = \frac{E_1}{I_1} = \tfrac{5}{3} \text{ ohms.}$$

* Many other sets of parameters are possible, but these are the basic ones.
† A matrix of the r's represents the coefficients of Eq. (5-24).

The constant r_{12} is obtained by applying an input current of 1 amp and computing the output voltage with the second pair of terminals open-circuited. By the reciprocity theorem, the transfer resistances in both directions are the same. Here

$$r_{12} = \frac{E_2}{I_1} = \tfrac{1}{3}\text{ ohm.}$$

The constant r_{22} is the input resistance at the second terminal-pair of the network with an open circuit at the first terminal pair:

$$r_{22} = \frac{E_2}{I_2} = \tfrac{2}{3}\text{ ohm.}$$

The equivalent T network can be synthesized directly from Eq. (5–24). The result is shown in Fig. 5–13(b). The resistance values are given by

$$\begin{aligned} R_1 &= r_{11} - r_{12} = \tfrac{5}{3} - \tfrac{1}{3} = \tfrac{4}{3}\text{ ohm,} \\ R_2 &= r_{22} - r_{12} = \tfrac{2}{3} - \tfrac{1}{3} = \tfrac{1}{3}\text{ ohm,} \\ R_3 &= r_{12} = \tfrac{1}{3}\text{ ohm.} \end{aligned}$$

Example 2. A second example of reduction to a T circuit is shown in Fig. 5–14(a). For ease in computation the lattice form of Fig. 5–14(a) is redrawn as a bridge in Fig. 5–14(b). In this circuit each branch has a resistance of 4 ohms and the input resistance with an open circuit at the output is therefore

$$r_{11} = 2\text{ ohms.}$$

The output resistance with an open circuit at the input is the parallel combination of 5 and 3 ohms, or

$$r_{22} = \tfrac{15}{8}\text{ ohms.}$$

To obtain the transfer resistance a current of 1 amp is applied at the input. It divides equally between the two paths, and the voltage at a is therefore $\frac{1}{2}$ volt and at b, 1 volt. The output voltage is $-\frac{1}{2}$ volt and

$$r_{12} = \frac{E_2}{I_1} = -\tfrac{1}{2}\text{ ohm.}$$

The equivalent circuit is shown in Fig. 5–14(c). The resistance values are

$$\begin{aligned} R_1 &= r_{11} - r_{12} = 2 + \tfrac{1}{2} = \tfrac{5}{2}\text{ ohms,} \\ R_2 &= r_{22} - r_{12} = \tfrac{15}{8} + \tfrac{1}{2} = \tfrac{19}{8}\text{ ohms,} \\ R_3 &= r_{12} = -\tfrac{1}{2}\text{ ohm.} \end{aligned}$$

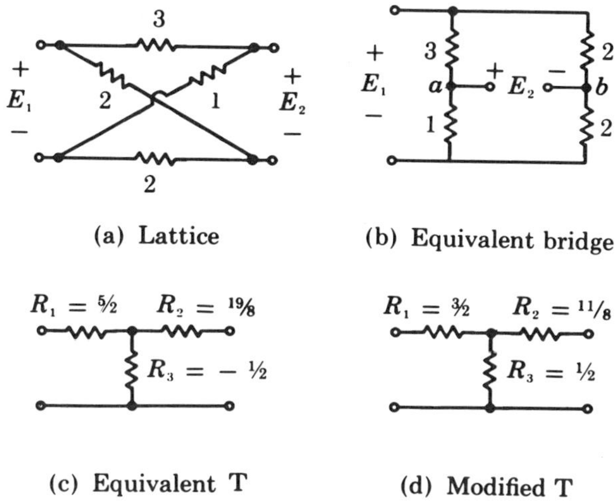

FIG. 5–14. The conversion of a lattice to a T.

The circuit of Fig. 5–14(c) is a perfect representation of the original circuit. The negative sign in the mutual element is a consequence of the polarity chosen for the output. If the assumed direction of E_2 is reversed, the sign of r_{12} is reversed but the other parameters stay the same. Then

$$R_1 = r_{11} - r_{12} = 2 - \tfrac{1}{2} = \tfrac{3}{2} \text{ ohms},$$
$$R_2 = r_{22} - r_{12} = \tfrac{15}{8} - \tfrac{1}{2} = \tfrac{11}{8} \text{ ohms},$$
$$R_3 = r_{12} = \tfrac{1}{2} \text{ ohm}.$$

The new equivalent circuit is shown in Fig. 5–14(d). The only difference between the circuits of Fig. 5–14(c) and (d) is the polarity of the output voltage in response to a positive stimulus at the input. Bridge, or lattice, structures have this flexibility in the polarity of the output voltage because they are four-terminal networks. A T circuit is a three-terminal network, and can achieve this generality only by the use of a negative resistance.

The g parameters. The g parameters are equally as general as the r parameters and can always be used to represent a two-terminal-pair network. The circuit which is obtained naturally from the g parameters is a π circuit. The basic equations for the g's are*

$$\begin{aligned} I_1 &= g_{11}E_1 + g_{12}E_2, \\ I_2 &= g_{12}E_1 + g_{22}E_2. \end{aligned} \tag{5–25}$$

* A matrix of the g's represents the coefficients of Eq. (5–25).

The π equivalent circuit is synthesized from Eq. (5–25) by considering these to be a pair of node-voltage equations. The elements are most naturally obtained as conductances, which may or may not be all positive. In Example 3 we will find the g parameters and the π equivalent circuit for the same network used in Example 1.

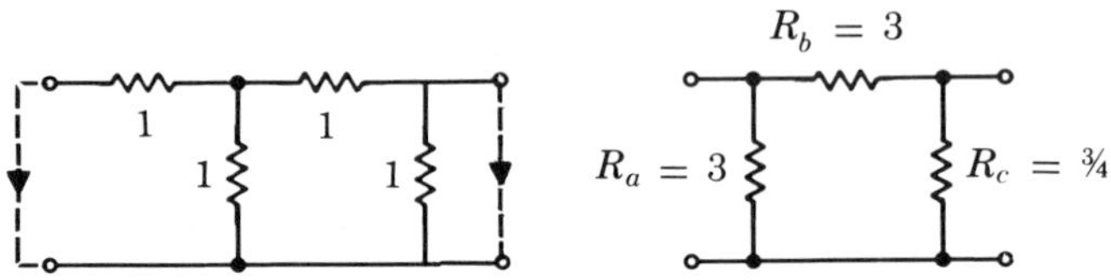

(a) Original network (b) Equivalent π network

FIG. 5–15. Network conversion by g parameters.

Example 3. The network of Fig. 5–15(a) is to be reduced to an equivalent π circuit. The constant g_{11} is the input conductance of the network with a short circuit at the output terminal-pair. It can be obtained by applying a voltage of 1 volt at the input and computing the resulting input current, or it can be obtained by a direct reduction of the network to a single resistance whose reciprocal is g_{11}. Here

$$g_{11} = \frac{I_1}{E_1} = \tfrac{2}{3} \text{ mho.}$$

The constant g_{12} is the current in the short circuit at terminal-pair 2 when 1 volt is applied to terminal-pair 1. It is called the short-circuit transfer conductance, and is, of course, the same for both network directions, by the reciprocity theorem. Here

$$g_{12} = \frac{I_2}{E_1} = -\tfrac{1}{3} \text{ mho.}$$

The negative sign is a consequence of the assumed polarity of the current I_2 which is into the network at the output terminal-pair. For three-terminal networks the current I_2 which flows in response to a positive value of 1 volt at the input will always be negative, and hence g_{12} will always be negative. For four-terminal networks it can have either polarity. The constant g_{22} is similar to g_{11}. It is the output short-circuit conductance of the network. It can be computed by applying 1 volt to the output terminal-pair and a short circuit to the input terminal-pair and finding the current which flows into the network. Alternatively, it can be found by reducing the network to a single equivalent resistance or conductance. Here

$$g_{22} = \tfrac{5}{3} \text{ mhos.}$$

The node equations for the equivalent network are thus

$$\begin{aligned} \tfrac{2}{3}E_1 - \tfrac{1}{3}E_2 &= I_1, \\ -\tfrac{1}{3}E_1 + \tfrac{5}{3}E_2 &= I_2. \end{aligned} \tag{5-26}$$

The network synthesized from these equations is shown in Fig. 5–15(b).* The resistance values are

$$R_a = \frac{1}{g_{11} + g_{12}} = \frac{1}{\frac{2}{3} - \frac{1}{3}} = 3 \text{ ohms},$$

$$R_b = \frac{-1}{g_{12}} = \frac{-1}{-\frac{1}{3}} = 3 \text{ ohms},$$

$$R_c = \frac{1}{g_{22} + g_{12}} = \frac{1}{\frac{5}{3} - \frac{1}{3}} = \tfrac{3}{4} \text{ ohm}.$$

The networks of Figs. 5–15(b) and 5–13(b) are two equivalent forms for the same network. They are, of course, related by the Y–Δ transformation, and we could have obtained the π from the T in this fashion. We have preferred to use the g-parameter method only in order to gain familiarity with it. In some problems the g parameters are a good deal simpler than the r parameters and it is convenient to be able to use either of them. Obviously, there is a direct relationship between the g and the r parameters as well as between the resulting networks. We will now obtain the equations relating the g's and the r's.

Relationship of the r and g parameters. The relationship between the r and the g parameters can be obtained by starting with a network which is defined by one set of parameters, and solving the network equations for the other set. Suppose that the network is defined by

$$\begin{aligned} E_1 &= r_{11}I_1 + r_{12}I_2, \\ E_2 &= r_{12}I_1 + r_{22}I_2. \end{aligned} \tag{5-27}$$

---

* The g-parameter method is not completely similar to the r-parameter method because the two mesh currents are not both clockwise. The dual of two node voltages positive with respect to ground is two mesh currents in a clockwise direction.

We can solve these equations for the currents in terms of the voltages:

$$I_1 = \frac{\begin{vmatrix} E_1 & r_{12} \\ E_2 & r_{22} \end{vmatrix}}{\begin{vmatrix} r_{11} & r_{12} \\ r_{12} & r_{22} \end{vmatrix}} = \frac{r_{22}}{r_{11}r_{22} - r_{12}^2} E_1 - \frac{r_{12}}{r_{11}r_{22} - r_{12}^2} E_2,$$

$$I_2 = \frac{\begin{vmatrix} r_{12} & E_1 \\ r_{12} & E_2 \end{vmatrix}}{\begin{vmatrix} r_{11} & r_{12} \\ r_{12} & r_{22} \end{vmatrix}} = \frac{-r_{12}}{r_{11}r_{22} - r_{12}^2} E_1 + \frac{r_{11}}{r_{11}r_{22} - r_{12}^2} E_2. \tag{5–28}$$

Equations (5–28) are in the general form of

$$\begin{aligned} I_1 &= g_{11}E_1 + g_{12}E_2, \\ I_2 &= g_{12}E_1 + g_{22}E_2. \end{aligned} \tag{5–29}$$

We can match the corresponding coefficients of Eqs. (5–28) and (5–29) to obtain*

$$\begin{aligned} g_{11} &= \frac{r_{22}}{r_{11}r_{22} - r_{12}^2}, \\ g_{12} &= \frac{-r_{12}}{r_{11}r_{22} - r_{12}^2}, \\ g_{22} &= \frac{r_{11}}{r_{11}r_{22} - r_{12}^2}. \end{aligned} \tag{5–30}$$

Alternatively, we might reverse the process and start with the network defined in terms of the g's. We would then obtain the following equations for the r's in terms of the g's:

$$\begin{aligned} r_{11} &= \frac{g_{22}}{g_{11}g_{22} - g_{12}^2}, \\ r_{12} &= \frac{-g_{12}}{g_{11}g_{22} - g_{12}^2}, \\ r_{22} &= \frac{g_{11}}{g_{11}g_{22} - g_{12}^2}. \end{aligned} \tag{5–31}$$

As a verification of Eq. (5–30) we will compute the g's of Example 3 above from the r's of Example 1. The r values are

$$r_{11} = \tfrac{5}{3}, \qquad r_{12} = \tfrac{1}{3}, \qquad r_{22} = \tfrac{2}{3}.$$

---

* Provided that $r_{11}r_{22} - r_{12}^2 \neq 0$.

The g values computed by means of Eq. (5–30) are

$$g_{11} = \frac{\frac{2}{3}}{\frac{5}{3} \times \frac{2}{3} - (\frac{1}{3})^2} = \frac{\frac{2}{3}}{1} = \frac{2}{3},$$

$$g_{12} = \frac{-\frac{1}{3}}{1} = -\frac{1}{3},$$

$$g_{22} = \frac{\frac{5}{3}}{1} = \frac{5}{3}.$$

These values of the g's agree exactly with those computed directly from the circuit in Example 3.

5–5 Symmetry in two-terminal-pair networks. A general two-terminal-pair network is shown in Fig. 5–16. The terminal-pair a–b is the input, and the terminal-pair c–d is the output. If the network is symmetrical about the center line x–y through the two terminal-pairs it is said to possess "balanced" symmetry. Balanced networks can be converted directly to three-terminal networks. The network may also be symmetrical about the center line m–n. In this case the input and the output terminal-pairs are indistinguishable and the network is said to possess "transfer" symmetry. Networks which possess transfer symmetry can be solved by means of the symmetry principles stated in the previous chapter. The presence of symmetry reduces the number of parameters required to describe the network.

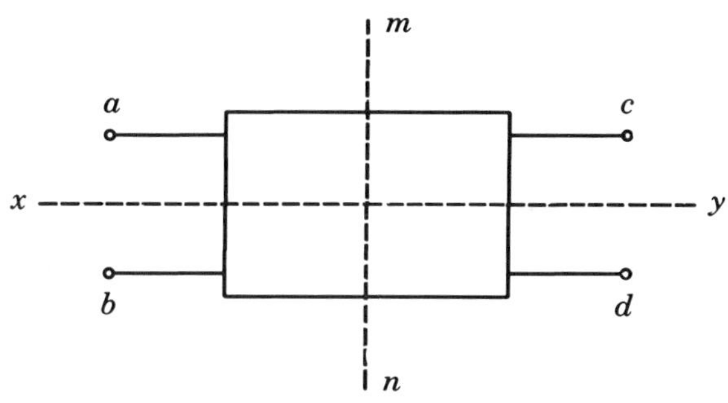

FIG. 5–16. Symmetry in two-terminal-pair networks.

Balanced symmetry. An example of a four-terminal network with balanced symmetry is given in Fig. 5–17(a). Because of the symmetry the center points of the shunt branches, marked x, y, and z, will be at the same potential as the midpoint of the exciting sources at either end. By the principles of symmetry, a shorting bar can therefore be connected through the points g, h, x, y, and z, and the circuit of Fig. 5–17(b) is obtained. It is then convenient to use the linearity theorem and to double

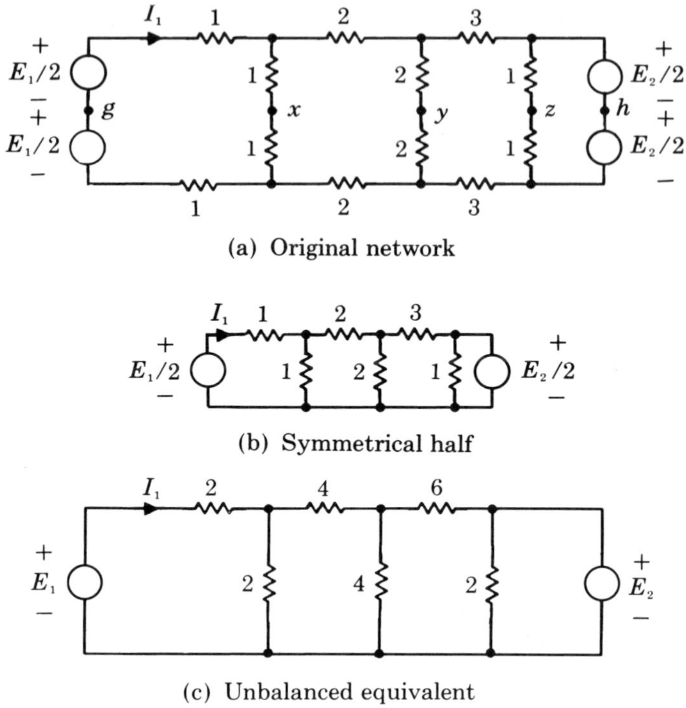

FIG. 5–17. A network with balanced symmetry converted to unbalanced form.

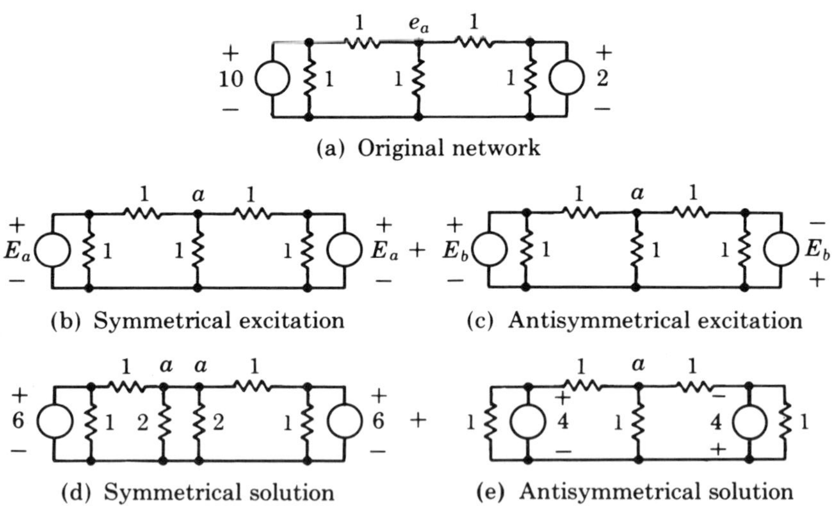

FIG. 5–18. A symmetrical network with unsymmetrical excitation.

the voltages and the resistances, leaving the currents unchanged, as shown in Fig. 5–17(c).

The net result of the method of balanced symmetry is to convert the original four-terminal network to an equivalent three-terminal network. Note, however, that this conversion is *different* from that discussed in Section 5–3. There the three-terminal equivalent networks were only mathematically similar to the four-terminal networks. Here they are also physically similar. If a three-resistance equivalent circuit is obtained for the circuit of Fig. 5–17(b), the addition of a symmetrical lower half will convert it into the exact equivalent circuit of the original four-terminal network.

Transfer symmetry. A three-terminal network with transfer symmetry is shown in Fig. 5–18(a). We wish to find the voltage e_a. The problem would be symmetrical if it were not for the excitation. The trick is to replace the original excitation by the sum of the two symmetrical cases shown in Fig. 5–18(b) and (c). In Fig. 5–18(b) the excitation is symmetrical, but in (c) it is antisymmetrical. If the sum of these two excitations is to equal the original excitation, then

$$E_a + E_b = 10, \qquad E_a - E_b = 2,$$

from which

$$E_a = 6 \qquad \text{and} \qquad E_b = 4.$$

The solutions for the two symmetrical problems must then be obtained. The first one is shown in Fig. 5–18(d). There is no current in the line a–a and therefore it can be opened up. The voltage at a then is

$$e_a \text{ (symmetrical)} = \tfrac{2}{3} \text{ of } 6 = 4 \text{ volts}.$$

The solution of the antisymmetrical case is shown in Fig. 5–18(e). Because of the symmetry the two mesh currents are equal and there is no current in the middle branch. It can therefore be shorted, and the voltage at a is

$$e_a \text{ (antisymmetrical)} = 0 \text{ volts}.$$

The voltage e_a in the original problem is the sum of the two components which have been obtained. Thus,

$$\begin{aligned} e_a &= e_a \text{ (symmetrical)} + e_a \text{ (antisymmetrical)} \\ &= 4 + 0 \\ &= 4 \text{ volts}. \end{aligned}$$

The method used here for resolving an unsymmetrical excitation into the sum of a symmetrical and an antisymmetrical excitation is perfectly general and usually is less work than a more straightforward solution.

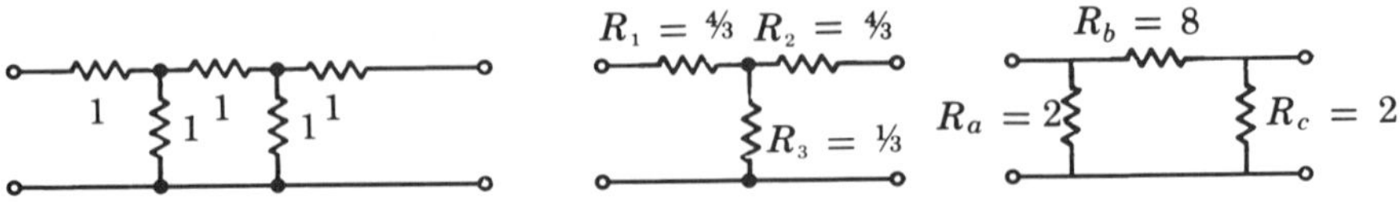

(a) Symmetrical circuit (b) Equivalent T (c) Equivalent π

FIG. 5–19. The equivalent T and π for a symmetrical circuit.

Symmetry and the r parameters. The r parameters relate to three-terminal networks and therefore are concerned only with transfer symmetry. If the network possesses transfer symmetry, the open-circuit resistances measured at each end of the network will be the same. The network is therefore described by two response functions, r_{11} and r_{12}. An example which illustrates this statement is shown in Fig. 5–19(a). For this circuit,

$$r_{11} = r_{22} = \tfrac{5}{3} \text{ ohms},$$

$$r_{12} = \tfrac{1}{3} \text{ ohm}.$$

The equivalent T circuit of Fig. 5–19(b) has the resistance values

$$R_1 = R_2 = r_{11} - r_{12} = \tfrac{5}{3} - \tfrac{1}{3} = \tfrac{4}{3} \text{ ohms},$$

$$R_3 = r_{12} = \tfrac{1}{3} \text{ ohm}.$$

There are still three resistances in the T circuit, but two of them are identical in value. The symmetry of the network is apparent from the T circuit.

Symmetry and the g parameters. Transfer symmetry reduces the number of independent g parameters to two, just as it did for the r parameters. Since the input and output terminal-pairs are indistinguishable, the input conductance measured with a short on the output must be identical to the output conductance measured with a short on the input. Since $g_{11} = g_{22}$, it is possible to describe the system with only two response functions, g_{11} and g_{12}. For the network of Fig. 5–19(a),

$$g_{11} = g_{22} = \tfrac{5}{8} \text{ mho},$$

$$g_{12} = -\tfrac{1}{8} \text{ mho}.$$

The equivalent π network is shown in Fig. 5–19(c). The values of the resistances are

$$R_a = R_c = \frac{1}{g_{11} + g_{12}} = \frac{1}{\frac{5}{8} - \frac{1}{8}} = 2 \text{ ohms},$$

$$R_b = \frac{-1}{g_{12}} = 8 \text{ ohms}.$$

The π circuit of Fig. 5–19(c) could also have been derived from the T circuit of Fig. 5–19(b) by means of a Y–Δ transformation.

5–6 The symmetrical lattice. The symmetrical lattice is a four-terminal circuit which possesses both balanced and transfer symmetry. As a result of these two types of symmetry the six resistances which are necessary to represent a general four-terminal network are reduced to four and even these four consist of two pairs of equal resistances. Thus only two parameters are necessary to specify a balanced lattice, a typical example of which is shown in Fig. 5–20(a). An alternative form of the lattice is drawn in Fig. 5–20(b), which displays its character as a bridge circuit.

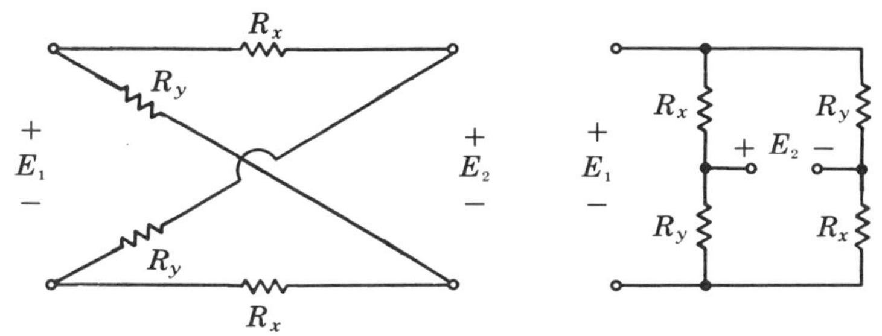

(a) A symmetrical lattice (b) The equivalent bridge form

FIG. 5–20. The symmetrical lattice.

The equivalent T circuit. The r parameters can be obtained by inspection from the bridge circuit of Fig. 5–20(b). They are

$$r_{11} = r_{22} = \tfrac{1}{2}(R_x + R_y),$$
$$r_{12} = \tfrac{1}{2}(R_y - R_x). \tag{5–32}$$

An equivalent T circuit can be obtained directly from the r parameters. The circuit is shown in Fig. 5–21(a). The resistance values are

$$R_1 = R_2 = r_{11} - r_{12} = R_x,$$
$$R_3 = r_{12} = \tfrac{1}{2}(R_y - R_x). \tag{5–33}$$

If R_x is greater than R_y the output voltage in Fig. 5–20(b) is negative. The corresponding effect can be achieved in the T circuit of Fig. 5–21(a) only by means of a negative shunt resistance. The lattice is more general than the T, since it has the option of either polarity of output voltage. If the polarity of the output voltage is not important, the alternative

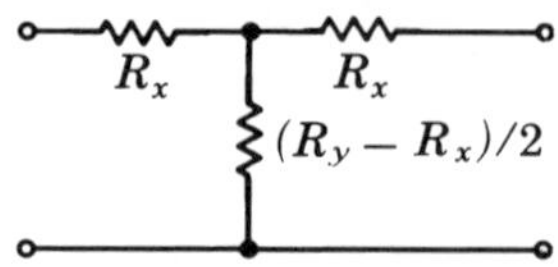

(a) T circuit for positive output $(R_y > R_x)$

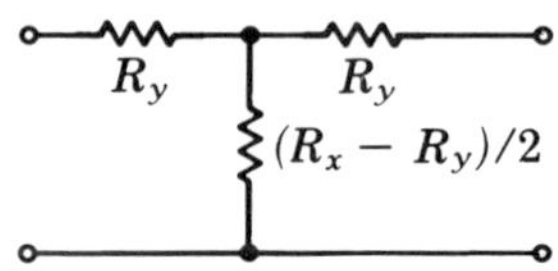

(b) T circuit for negative output $(R_x > R_y)$

FIG. 5–21. The T circuit equivalents for the symmetrical lattice.

equivalent circuit of Fig. 5–21(b) can be used. The negative resistance is removed at the expense of changing the polarity of the output voltage.

The lattice equations in (5–32) can also be solved for the lattice resistances in terms of the r parameters in the form

$$R_y = r_{11} + r_{12}, \qquad R_x = r_{11} - r_{12}. \tag{5–34}$$

For a symmetrical circuit it is easy to show that r_{11} must be equal to or greater than r_{12}. Consider the general symmetrical circuit of Fig. 5–22(a), driven by a current source I_1. The input circuit is shown in Fig. 5–22(b). The input resistance is

$$r_{11} - r_{12} + \frac{r_{12}(r_{11} - r_{12})}{r_{12} + r_{11} - r_{12}} = r_{11} - \frac{r_{12}^2}{r_{11}}.$$

Since the input resistance must be positive for a passive circuit, we have

$$r_{11} \geq \frac{r_{12}^2}{r_{11}}$$

and thus

$$r_{11} \geq r_{12}. \tag{5–35)*}$$

By virtue of Eq. (5–35) we can conclude that the lattice resistances R_x and R_y can always be found in a positive form (r_{11} is a driving point resistance, of course, and must be positive). For this reason the lattice

---

* In the unsymmetrical case $r_{11}r_{22} \geq r_{12}^2$ and r_{12} can be greater than either r_{11} or r_{22}, but not both.

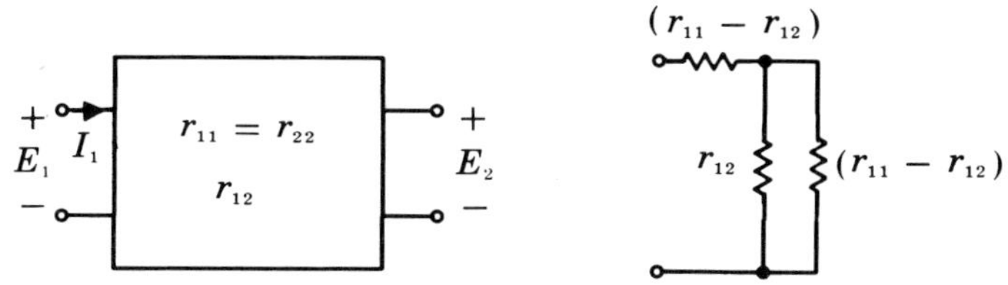

(a) General symmetrical circuit (b) Input circuit

FIG. 5–22. The general symmetrical circuit.

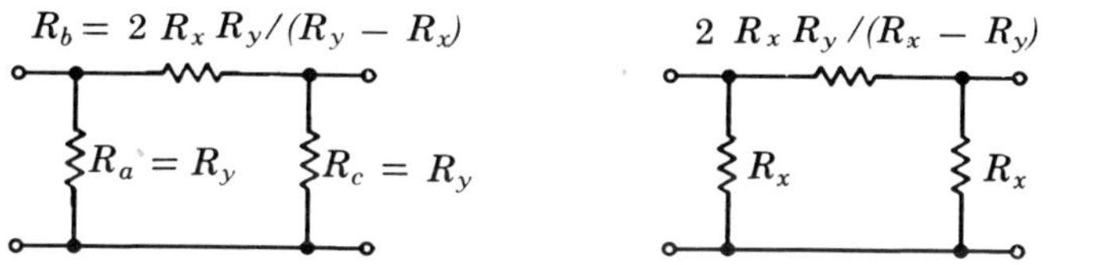

(a) π circuit when $R_y > R_x$ (b) π circuit when $R_x > R_y$

FIG. 5–23. The equivalent π circuits for the symmetrical lattice.

is sometimes referred to as the most general symmetrical two-terminal-pair network.

The equivalent π circuit. The g parameters for the lattice circuit can be obtained from Fig. 5–20(b). They are

$$g_{11} = g_{22} = \frac{R_x + R_y}{2R_xR_y} = \tfrac{1}{2}(G_x + G_y),$$
$$g_{12} = \frac{R_x - R_y}{2R_xR_y} = \tfrac{1}{2}(G_y - G_x). \tag{5–36}$$

The equivalent π circuit is determined directly from the g values in Eq. (5–36). The circuit is shown in Fig. 5–23(a). The values are

$$R_a = R_c = \frac{1}{g_{11} + g_{12}} = R_y,$$
$$R_b = \frac{-1}{g_{12}} = \frac{2R_xR_y}{R_y - R_x}. \tag{5–37}$$

An alternative form of the circuit, obtained by reversing the polarity of the output voltage from the lattice, is shown in Fig. 5–23(b). As with the T circuit, a negative resistance can be removed at the expense of reversing the polarity of the output voltage.

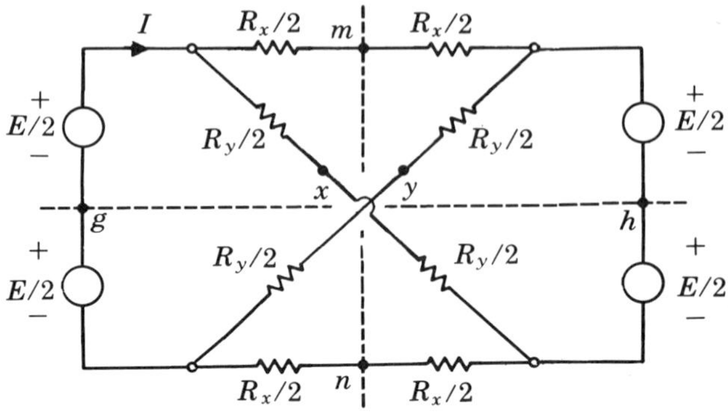

(a) Symmetrical excitation of symmetrical lattice

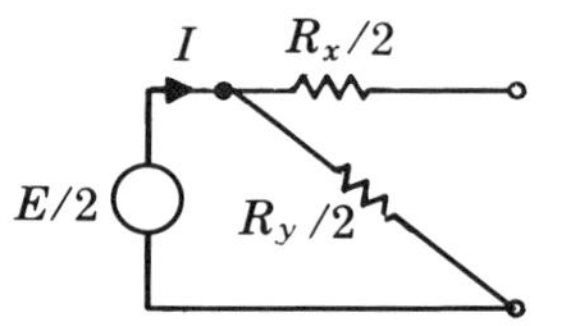

(b) Equivalent input circuit

FIG. 5–24. Bartlett's bisection theorem for R_y.

The lattice equations of (5–36) can also be solved for the lattice elements in terms of the g parameters, with the result that

$$R_x = \frac{1}{g_{11} - g_{12}}, \qquad R_y = \frac{1}{g_{11} + g_{12}}. \tag{5–38}$$

It is easy to show that $g_{11} > g_{12}$ and that R_x and R_y will therefore always be positive resistances if a lattice is formed from the g parameters.

Bartlett's bisection theorem. Balanced lattice networks find a great deal of use in electrical filters, partly for the simplicity and flexibility which they exhibit. Bartlett's bisection theorem is a method of reducing any symmetrical two-terminal-pair network to an equivalent symmetrical lattice. It is essentially a specific application of the symmetry principles discussed in Section 5–5. First of all, let us consider the symmetry of the lattice itself, as illustrated in Fig. 5–24(a).* We will apply symmetrical voltage excitation at the two ends. Because of symmetry a ground plane can be passed through points g, x, y, and h. Furthermore, there is no

* Because of the symmetry the lower arms are usually indicated by broken lines.

current in R_x and therefore it can be opened up. The input current can thus be computed from the circuit of Fig. 5–24(b). This current is

$$I \text{ (symmetrical)} = \frac{E/2}{R_y/2} = \frac{E}{R_y},$$

from which (5–39)

$$R_y = \frac{E}{I \text{ (symmetrical)}}.$$

Equation (5–39) gives the lattice resistance R_y in terms of the current which flows into the network when it is symmetrically excited by a 1-volt source. Since this ratio is also the input resistance of the left half of the network with the straight-through connections opened, it is also called R_{och}, where the initials stand for open-circuit-half.

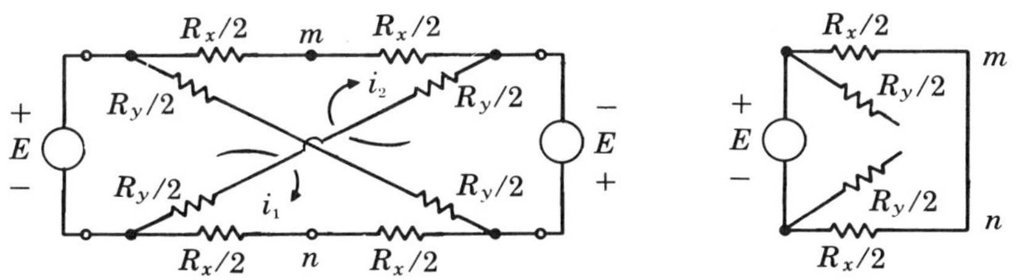

(a) Antisymmetrical excitation of symmetrical lattice (b) Equivalent input circuit

FIG. 5–25. Bartlett's bisection theorem for R_x.

Now let us apply antisymmetrical excitation to the lattice, as shown in Fig. 5–25(a). The two loop currents i_1 and i_2 are clearly equal because of the symmetry and therefore no current flows in the diagonal resistance. A similar argument holds for the other diagonal resistance. The diagonals can therefore be opened up. Furthermore, the points m and n are at the same potential and can be joined by a short circuit. The result is the circuit of Fig. 5–25(b). The input current is

$$I \text{ (antisymmetrical)} = \frac{E}{R_x},$$

from which (5–40)

$$R_x = \frac{E}{I \text{ (antisymmetrical)}}.$$

Equation (5–40) gives the lattice resistance R_x in terms of the current which flows into the network when it is antisymmetrically excited by a 1-volt source. The ratio can also be interpreted as the resistance of the

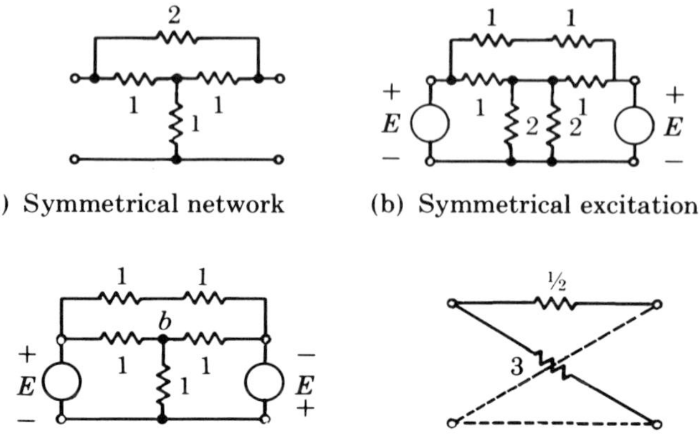

(a) Symmetrical network (b) Symmetrical excitation

(c) Antisymmetrical excitation (d) Equivalent lattice

FIG. 5–26. An equivalent lattice found by Bartlett's bisection theorem. [By convention, the symmetrical lower arms in (d) are indicated by broken lines.]

left half of the network with the straight-through connections shorted. It is therefore called the short-circuit-half resistance, R_{sch}.*

An example of the use of Bartlett's bisection theorem to find the lattice arms which are equivalent to a symmetrical two-terminal-pair network is given in Fig. 5–26(a). The arm R_y is found from the symmetrically excited network of Fig. 5–26(b). There is no current in the straight-through connections and they can be opened. The resistance of the remaining half-network is R_y. The value for the network of Fig. 5–26(b) is

$$R_y = 3 \text{ ohms} = R_{\text{och}}.$$

The arm R_x is found from the antisymmetrically excited network of Fig. 5–26(c). In this network there is no current in the shunt resistance and it can be replaced by a short circuit. Furthermore, the voltage at point b is midway between $+E$ and $-E$ and therefore can be shorted to ground. The input resistance of the remaining half-network is R_x. In this case it has the value

$$R_x = \tfrac{1}{2} \text{ ohm} = R_{\text{sch}}.$$

The equivalent lattice is shown in Fig. 5–26(d). The procedure used here can always be used to find the arms of an equivalent lattice for any symmetrical two-terminal-pair network. If all the connections between the

* The names short-circuit-half or open-circuit-half do not describe what happens to the crossed connection, which may require opening or shorting independently of the straight-through connections.

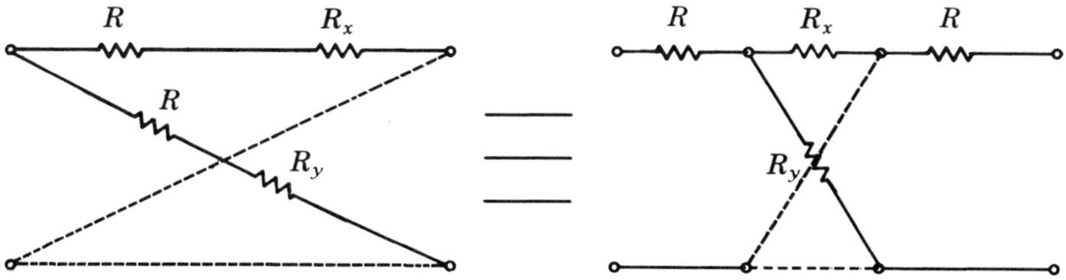

(a) Lattice with common resistance (b) Equivalent circuit with common resistance removed

FIG. 5–27. The removal of common resistance from the lattice arms.

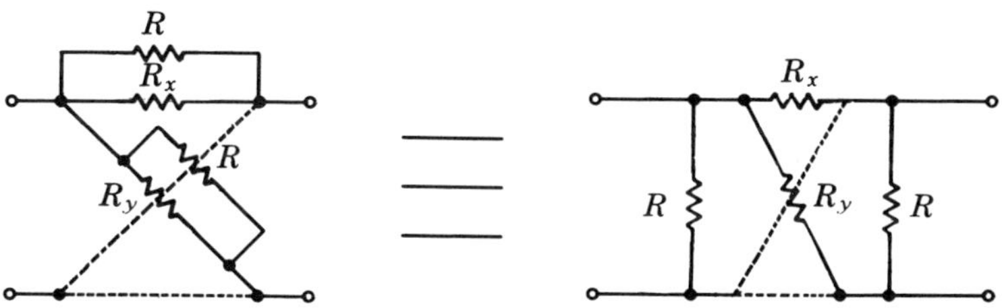

(a) Lattice with common conductance (b) Equivalent circuit with common conductance removed

FIG. 5–28. The removal of common conductance from the lattice arms.

two halves of the network are straight through, as for a T or a π network, the procedure is very simple. Crossed connections, e.g., in the lattice itself, can introduce difficulties. Each case should be resolved on its own merits by means of symmetrical and antisymmetrical excitation and the general principles of symmetry.

The inverse of Bartlett's bisection theorem. Bartlett's bisection theorem allows us to find an equivalent lattice to represent some symmetrical two-terminal-pair network. It is often convenient to be able to reverse this process and to proceed from the lattice to an equivalent unbalanced or three-terminal form. Two basic theorems which facilitate this process are illustrated in Figs. 5–27 and 5–28. The proof of the equivalence shown in Fig. 5–27 rests on Bartlett's bisection theorem. Both networks have the same values for R_{och} and R_{sch} and therefore they must be equivalent. A similar argument holds for the two networks in Fig. 5–28.

The application of the two theorems of Figs. 5–27 and 5–28 will be illustrated for the network of Fig. 5–29(a). The procedure is to remove common series or shunt elements until one of the branches has been reduced to an open circuit or a short circuit. The lattice then degenerates

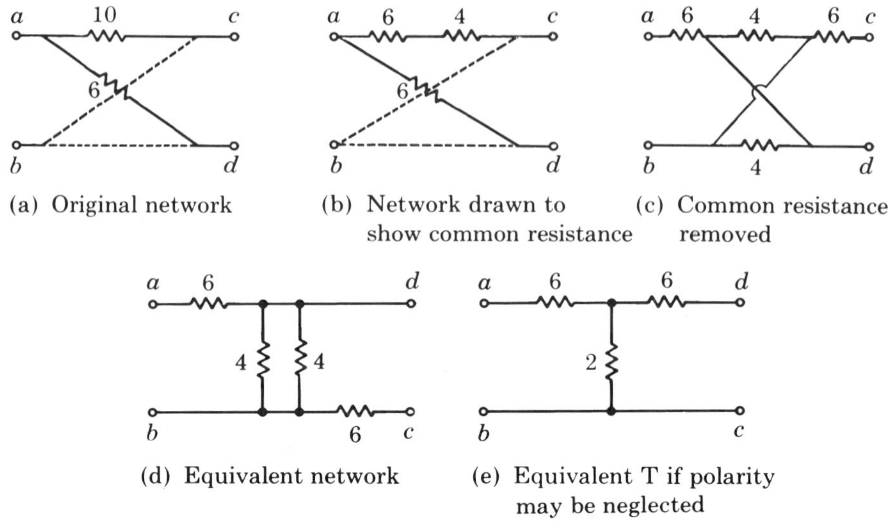

FIG. 5–29. The reduction of a symmetrical lattice to a T by means of the inverse bisection theorem.

to a T or a π, which can be presented in the form of an unbalanced circuit. Since the lattice is more general than the T or the π, it may be necessary to reverse the polarity of the output voltage, i.e., to interchange the series and the shunt arms in the lattice. If the network of Fig. 5–29(a) is redrawn as in Fig. 5–29(b), the common 6-ohm resistance in the series and shunt arms is apparent. By the theorem of Fig. 5–27 this common element can be removed from the lattice, and the circuit of Fig. 5–29(c) results. In this circuit the shunt arms are short circuits. It is really the T of Fig. 5–29(d), in which however, the polarity of the output terminals has been reversed. The final unbalanced circuit is shown in Fig. 5–29(e). It is equivalent to the original lattice if only the input and the output voltages are of interest and if the polarity of the output voltage can be interchanged.

So long as the lattice elements are all of the same kind (i.e., they are all resistances here), the lattice can be reduced to a T or a π in a single step. For a general *R–L–C* network the reduction may not always be possible, since the series and shunt arms may be different types of element.

Our discussion of two-terminal-pair networks has placed a good deal of emphasis upon symmetry. It might be objected that we have placed undue emphasis on symmetrical networks. However, the practical network problem is one of design or synthesis, and it can usually be accomplished with a symmetrical network. Indeed, most of the advanced theory of network design is based upon the use of symmetrical network sections.

5-7 Summary.

Reciprocity theorem. A current source and a voltmeter, or a voltage source and an ammeter can be interchanged in a circuit without any change in the readings.

The r parameters

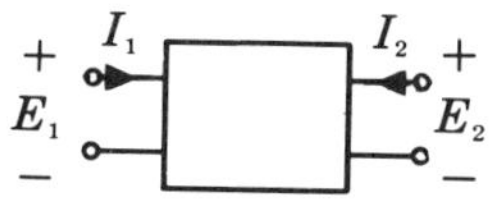

$$E_1 = r_{11}I_1 + r_{12}I_2$$

$$E_2 = r_{21}I_1 + r_{22}I_2$$

$$r_{11} = \frac{E_1}{I_1}\bigg|\, I_2 = 0 \qquad r_{22} = \frac{E_2}{I_2}\bigg|\, I_1 = 0$$

$$r_{12} = \frac{E_1}{I_2}\bigg|\, I_1 = 0 \qquad r_{21} = \frac{E_2}{I_1}\bigg|\, I_2 = 0$$

$$r_{12} = r_{21} \text{ (reciprocity)}$$

The g parameters

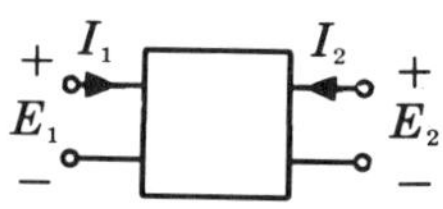

$$I_1 = g_{11}E_1 + g_{12}E_2$$

$$I_2 = g_{21}E_1 + g_{22}E_2$$

$$g_{11} = \frac{I_1}{E_1}\bigg|\, E_2 = 0 \qquad g_{22} = \frac{I_2}{E_2}\bigg|\, E_1 = 0$$

$$g_{12} = \frac{I_1}{E_2}\bigg|\, E_1 = 0 \qquad g_{21} = \frac{I_2}{E_1}\bigg|\, E_2 = 0$$

$$g_{12} = g_{21} \text{ (reciprocity)}$$

The T circuit

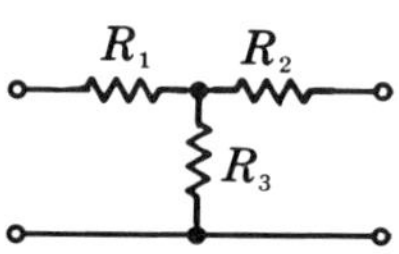

$$R_1 = r_{11} - r_{12}$$

$$R_2 = r_{22} - r_{12}$$

$$R_3 = r_{12}$$

The π circuit

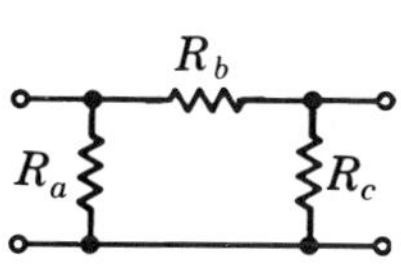

$$R_a = \frac{1}{g_{11} + g_{12}}$$

$$R_b = \frac{-1}{g_{12}}$$

$$R_c = \frac{1}{g_{22} + g_{12}}$$

Star-mesh transformation

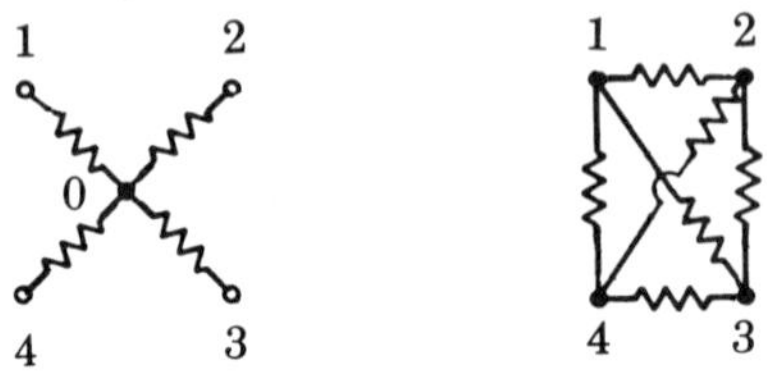

(a) Star (b) Mesh

$$G\text{ (mesh)} = \frac{\text{Prod. of adjacent } G\text{'s}}{\text{Sum of } G\text{'s (star)}}$$

Four-terminal to three-terminal transformation

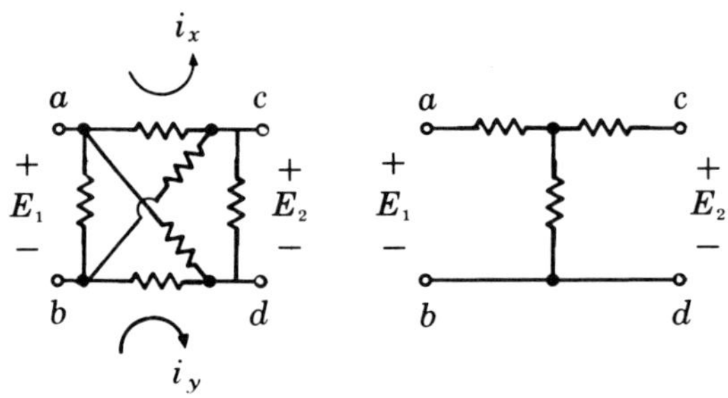

(a) Four terminal (b) Three terminal

Conditions for transformation

1. $i_x = i_y = 0$.
2. Polarity of E_2 not important.

Method

1. Find r's or g's for network a.
2. Derive T or π from r's or g's.

Relationship of r and g parameters

$$g_{11} = \frac{r_{22}}{r_{11}r_{22} - r_{12}^2} \qquad g_{22} = \frac{r_{11}}{r_{11}r_{22} - r_{12}^2} \qquad g_{12} = \frac{-r_{12}}{r_{11}r_{22} - r_{12}^2}$$

$$r_{11} = \frac{g_{22}}{g_{11}g_{22} - g_{12}^2} \qquad r_{22} = \frac{g_{11}}{g_{11}g_{22} - g_{12}^2} \qquad r_{12} = \frac{-g_{12}}{g_{11}g_{22} - g_{12}^2}$$

Symmetry

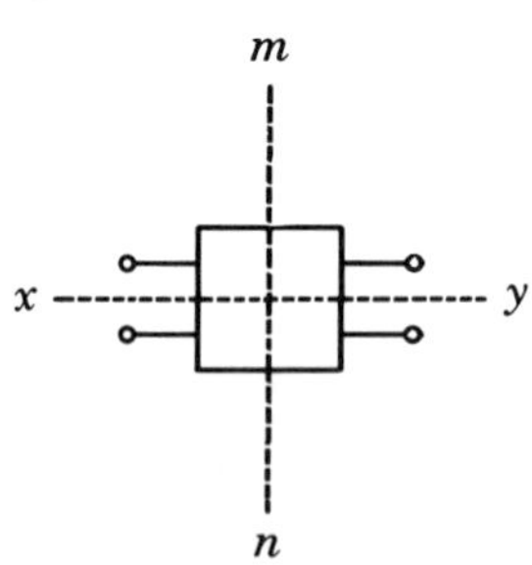

Balanced symmetry

1. Network symmetrical about x–y.
2. Network reduces to three terminals.

Transfer symmetry

1. Network symmetrical about m–n.
2. $r_{11} = r_{22}$.
3. $g_{11} = g_{22}$.

Symmetrical excitation

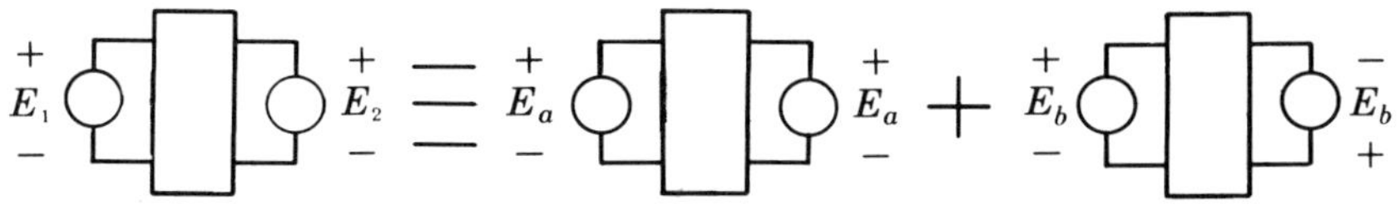

$$E_a = \tfrac{1}{2}(E_1 + E_2) \qquad E_b = \tfrac{1}{2}(E_1 - E_2)$$

Symmetrical lattice

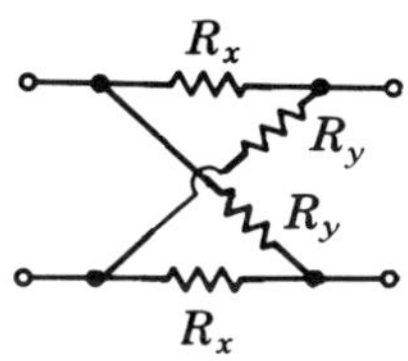

$$R_x = r_{11} - r_{12} = \frac{1}{g_{11} - g_{12}}$$

$$R_y = r_{11} + r_{12} = \frac{1}{g_{11} + g_{12}}$$

Bartlett's bisection theorem (to find lattice arms)

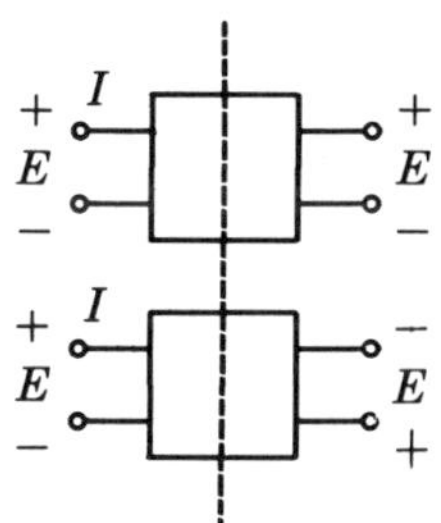

$$R_x = \frac{E}{I\ \text{(antisymmetrical)}} = R_{sch}$$

$$R_y = \frac{E}{I\ \text{(symmetrical)}} = R_{och}$$

Inverse bisection theorem

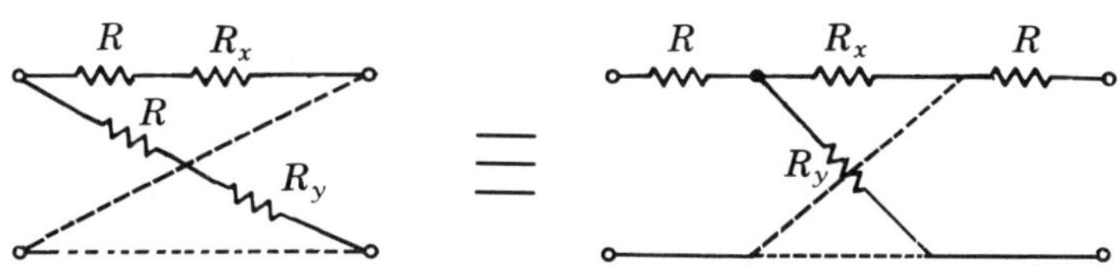

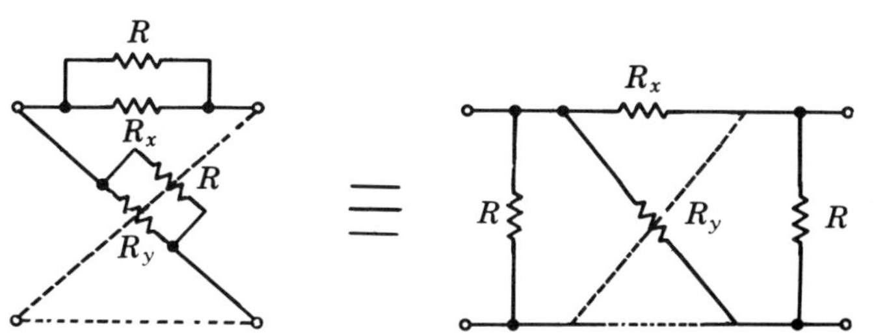

REFERENCES

The subject of two-terminal-pair networks is very well covered in the literature. In addition to the r parameters and the g parameters discussed in this chapter, there are the hybrid or h parameters, the general-circuit or $ABCD$ parameters, the iterative parameters, the image parameters, and the scattering parameters. Many of them are developed for particular applications, such as lossless filters, transmission lines or waveguides, and are best discussed in connection with these applications. The r parameters and the g parameters are basic, the others can be derived from them. In addition to its basic nature, the treatment in this chapter pays more attention to symmetry than is found in most other texts. Specific references for further reading are listed below.

1. E. A. Guillemin, *Communications Networks,* Vol. 2, John Wiley & Sons, New York, 1935. This text is the standard reference in English on two-terminal-pair networks. The only set of parameters not covered in detail is the scattering set. Chapter 4 gives the equations relating most of the sets of parameters.

2. E. A. Guillemin, *Introductory Circuit Theory,* John Wiley & Sons, New York, 1953. This text does not treat the two-terminal-pair network in detail, but Chapter 3 does include some interesting manipulations of the various network forms.

3. Kerchner and Corcoran, *Alternating Current Circuits,* John Wiley & Sons, New York, 3rd ed., 1951. Kerchner and Corcoran emphasize the image parameters and give an introduction to filter theory in terms of them.

4. M. B. Reed, *Electric Network Synthesis,* Prentice-Hall, New Jersey, 1955. In this text Reed gives a complete account of the image parameter method of filter theory. It is, of course, a specialized work for the designer of filters.

5. Seshu and Balbanian, *Linear Network Analysis,* John Wiley & Sons, New York, 1959. This text gives one of the few general discussions of the scattering parameters. They are confined usually to texts on microwave circuits and to original literature, such as the *IRE-PCGT,* Vol. 3, June 1956.

6. W. B. Boast, *Principles of Electric and Magnetic Circuits,* Harper & Brothers, New York, 1950. Boast gives an interesting method for checking the accuracy of a complex network reduction by applying arbitrary voltages as tests at selected points in the network and its equivalent reduction.

7. A. C. Bartlett, *The Theory of Artificial Lines and Filters,* John Wiley & Sons, New York, 1930. Bartlett's bisection theorem is stated here.

EXERCISES*

Section 5–2

1. When the voltage E is 10 volts, the current i is 1 amp in the circuit of Fig. 5–30. If a voltage of 100 volts is applied to the terminals c–d, and a short circuit is placed across terminals a–b, what current will flow in the short circuit?

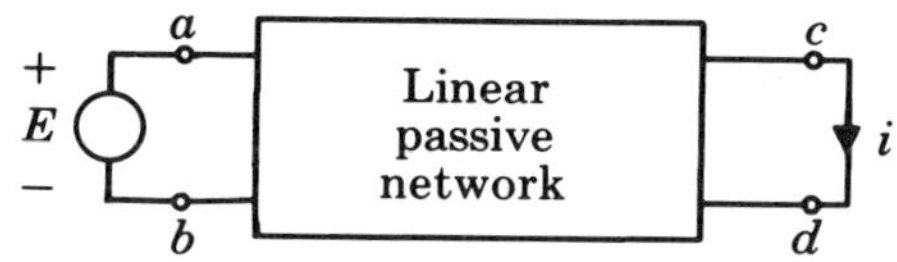

FIGURE 5–30

2. When the current I is 10 amp in the circuit of Fig. 5–31 a voltage of −2 volts appears across the terminals c–d. If a current of 20 amp is applied at terminals c–d and if the current source at terminals a–b is removed, what voltage will appear at terminals a–b? Give the polarity of the voltage, assuming that the current enters at c and that terminal a is positive for the voltage measurement.

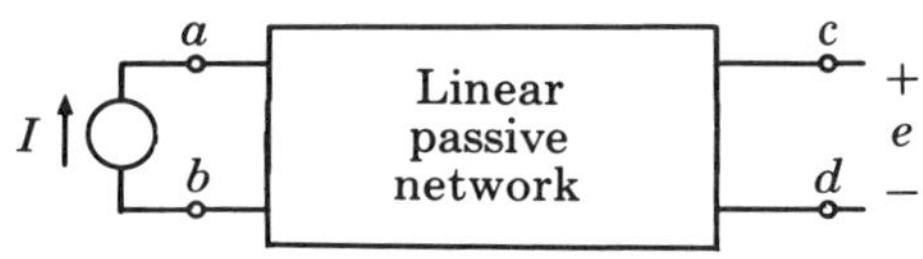

FIGURE 5–31

3. Verify the reciprocity theorem for the circuit of Fig. 5–32 by computing the current which is flowing in the ammeter m, and then interchanging the voltage source and the ammeter and recomputing the current in the ammeter.

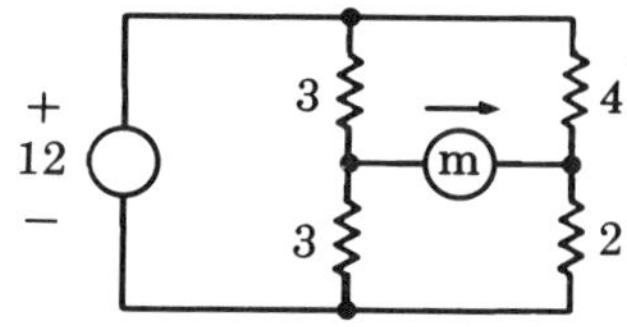

FIGURE 5–32

4. Verify the reciprocity theorem for the circuit of Fig. 5–33 by computing the voltage that would be measured by the voltmeter m, and then interchanging the current source and the voltmeter and recomputing the voltage read by the voltmeter.

5. A linear passive network has a pair of input terminals and a pair of output terminals. When the input current is 2 amp, the input voltage is 10 volts and the output voltage is 5 volts. If the current source were moved to the output

* Numerical values are in ohms.

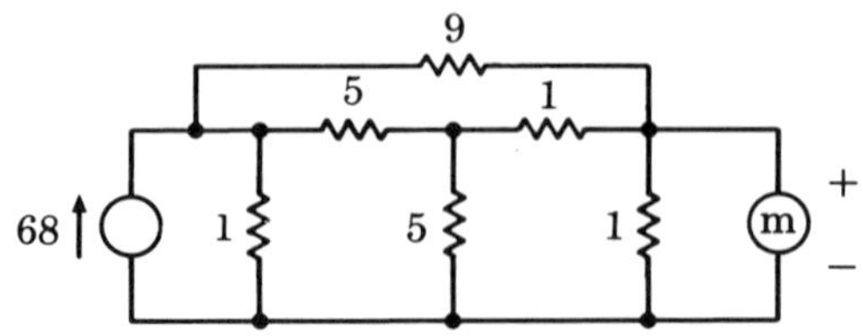

FIGURE 5–33

terminals and if a 5-ohm resistance were placed across the input terminals, what current would flow in the 5-ohm resistance?

6. A linear passive network has a pair of input terminals and a pair of output terminals. When the input voltage is 10 volts the input current is 5 amp, and the current in a short circuit across the output terminals is simultaneously 1 amp. If the voltage source is moved to the output terminals and a 2-ohm resistance is placed across the input terminals, what voltage will occur across the 2-ohm resistance?

7. When the voltage E_1 in the circuit of Fig. 5–34 is 10 volts, the current i_1 is 5 amp and the voltage e_2 is 2 volts. A 1-ohm resistance added across terminals c–d reduces the voltage there to 1 volt. Obtain the r parameters which describe the circuit, and then find what current would flow in the 1-ohm resistance if it were placed across terminals a–b and the 10-volt source were placed across terminals c–d.

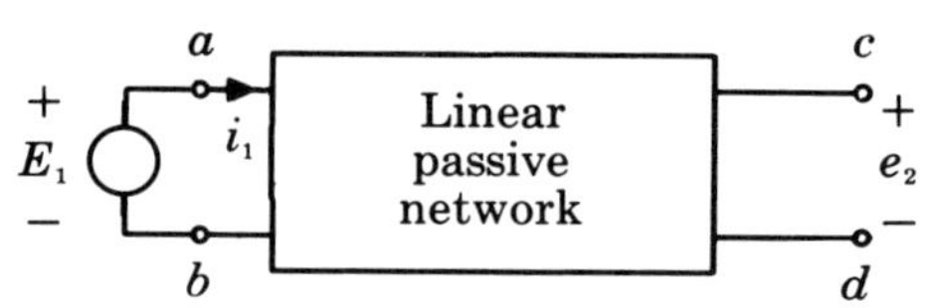

FIGURE 5–34

8. A current source of value 10 amp is applied to the input terminals of a two-terminal-pair network. The corresponding input voltage is 20 volts, and the current flowing in a short circuit at the output terminals is 4 amp at the same time. If the resistance across the output terminals is changed to 1 ohm, the current changes to 2 amp. Find the current which would flow in a 1-ohm resistor placed across the input terminals if the output terminals are driven by a 5-amp current source.

Section 5–3

9. Reduce the two-terminal-pair network of Fig. 5–35 to three equivalent resistances by means of network transformations.

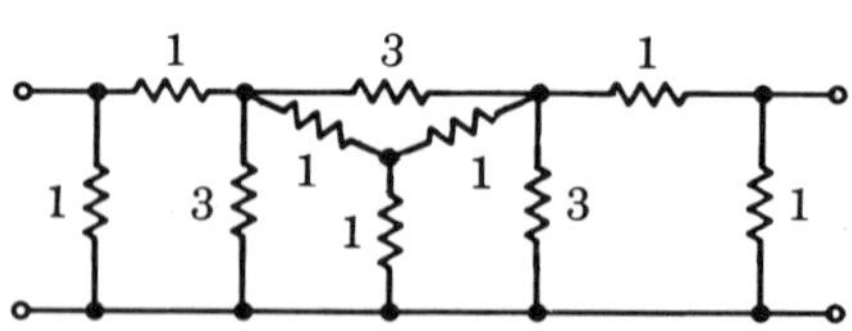

FIGURE 5–35

10. Reduce the two-terminal-pair network of Fig. 5–36 to three equivalent resistances by Y–Δ and series-parallel transformations.

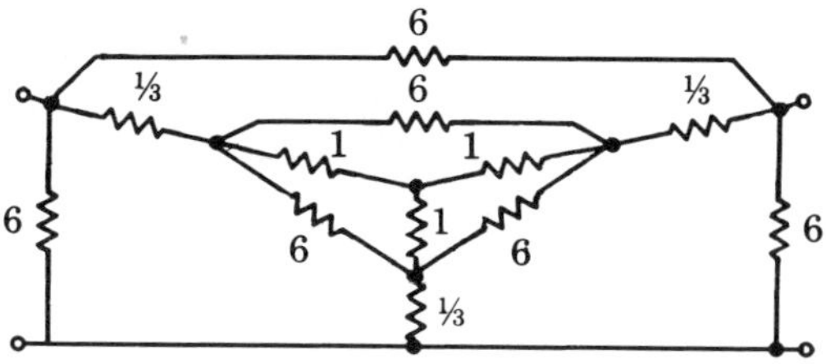

FIGURE 5–36

11. Find a T circuit equivalent for the bridged T circuit of Fig. 5–37.

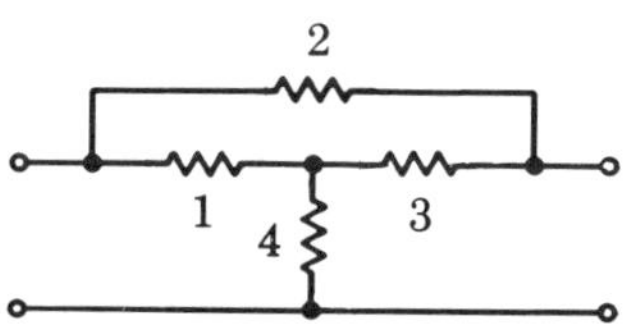

FIGURE 5–37

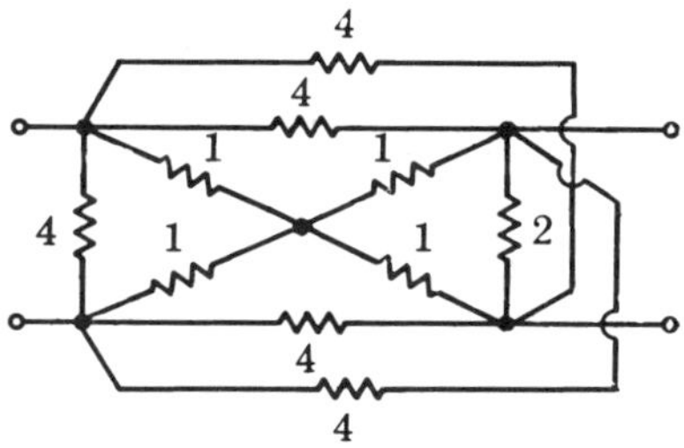

FIGURE 5–38

12. Reduce the four-terminal network of Fig. 5–38 to a minimum number of resistances while preserving all four terminals.

13. Convert the four-element star network of Fig. 5–39 to a six-element lattice form.

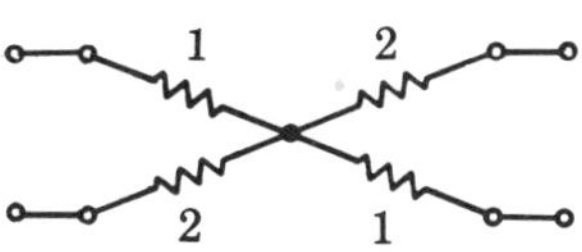

FIGURE 5–39

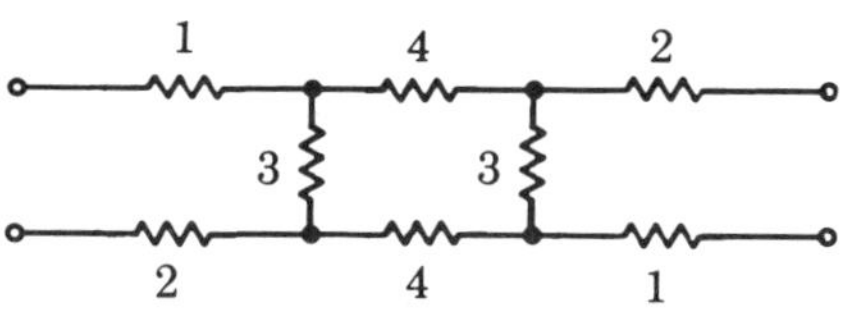

FIGURE 5–40

14. Reduce the four-terminal network of Fig. 5–40 to an equivalent T circuit.

15. Convert the network of Fig. 5–41 from four-terminal to three-terminal form and find an equivalent T by network transformations.

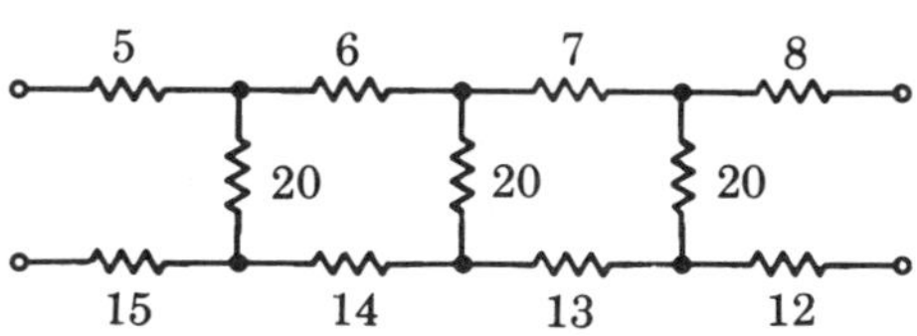

FIGURE 5–41

16. Convert a T circuit with each arm equal to 2 ohms, to a six-element lattice by the following procedure: (a) Convert the T to an equivalent symmetrical four-terminal network. (b) Apply two Y–Δ transformations. (c) Finally, use a star-to-mesh transformation.

17. A π circuit has a series arm of 8 ohms and two shunt arms of 2 ohms each. Convert this circuit to a six-element lattice by the following procedure: (a) Convert the π to an equivalent symmetrical four-terminal network. (b) Apply a general star-to-mesh transformation.

Section 5–4

18. Find the r parameters for the circuit of Fig. 5–42 and from them derive an equivalent three-resistance T circuit.

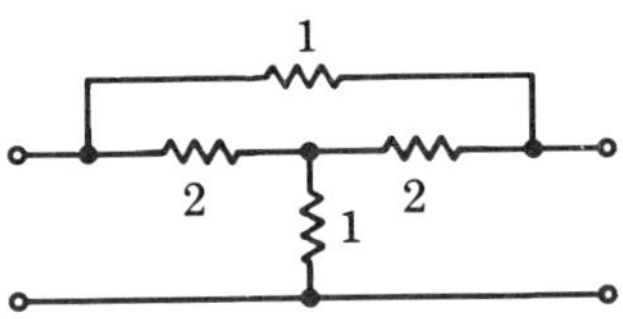

FIGURE 5–42

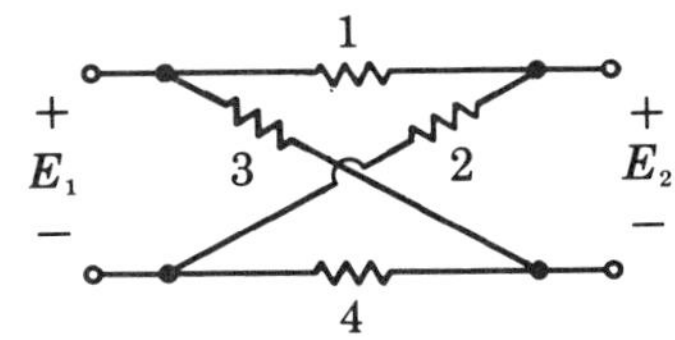

FIGURE 5–43

19. Find the g parameters for the network of Exercise 18, and from them derive the equivalent π network.

20. Find the r parameters for the network of Fig. 5–43 and give an equivalent three-resistance T. What happens to the r parameters and to the T network if the polarity of the output voltage E_2 is assumed in the opposite direction?

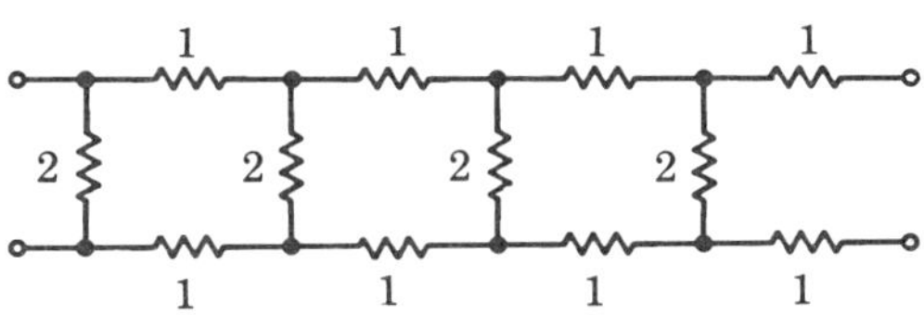

FIGURE 5–44

21. Find the g parameters for the network of Fig. 5–43 and give the equivalent π network.

22. Find the r parameters and the equivalent T network for the network of Fig. 5–44.

23. Find the g parameters and the equivalent π network for the one shown in Fig. 5–44.

24. The two-terminal-pair network of Fig. 5–45 is defined by the r parameters, r_{11}, r_{12}, r_{22}. The load resistance R is connected to the output terminals. Show that the ratios E_1/I_1 and E_2/I_1 with the load connected are

$$R_1 = r_{11} - \frac{r_{12}^2}{r_{22} + R} = \frac{E_1}{I_1}$$

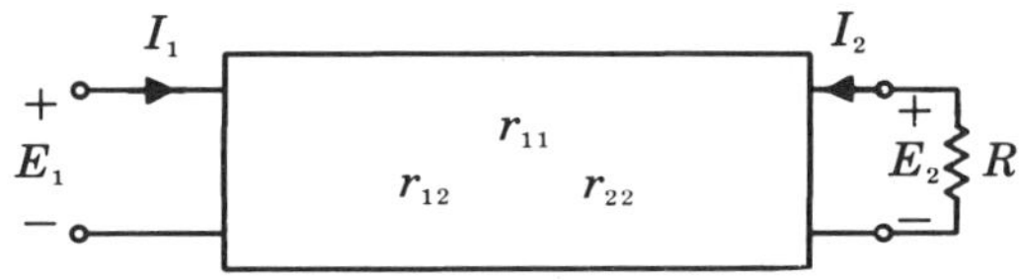

FIGURE 5–45

and

$$R_{12} = \frac{r_{12}R}{r_{22} + R} = \frac{E_2}{I_1}.$$

25. A two-terminal-pair network is defined by the g parameters g_{11}, g_{12}, and g_{22}. A load with conductance G is connected to the output terminals. Show that the ratios I_1/E_1 and I_2/E_1 with the load connected are

$$G_1 = g_{11} - \frac{g_{12}^2}{g_{22} + G} = \frac{I_1}{E_1}$$

and

$$G_{12} = \frac{g_{12}G}{g_{22} + G} = \frac{I_2}{E_1}.$$

Section 5–5

26. Change the four-terminal network of Fig. 5–46 to a three-terminal network by means of symmetry principles, and then obtain an equivalent T network.

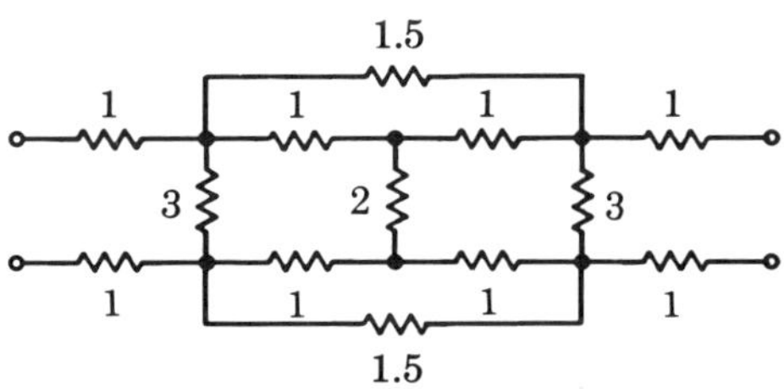

FIGURE 5–46

27. Find an equivalent T for the network of Fig. 5–47.

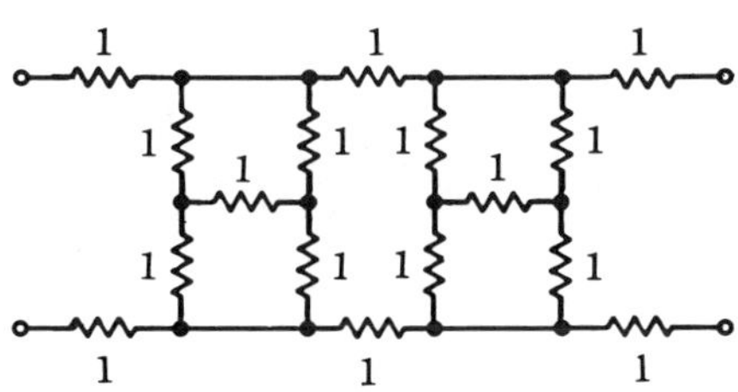

FIGURE 5–47

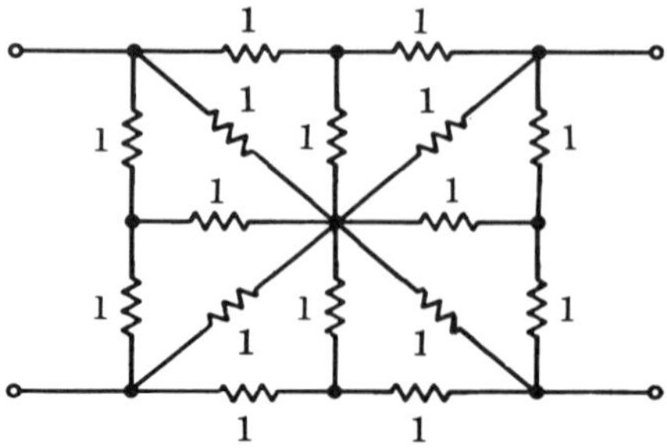

FIGURE 5–48

28. By means of symmetry reduce the network of Fig. 5–48 to a π network.
29. What is the voltage e_a in the circuit of Fig. 5–49?

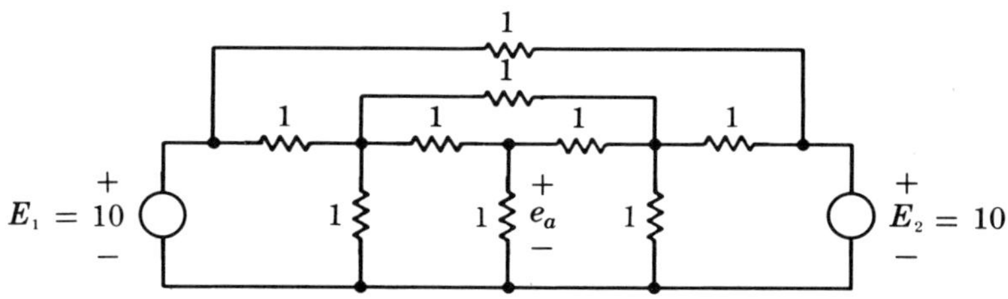

FIGURE 5–49

30. If the polarity of the output voltage in the circuit of Exercise 29 is reversed, what does the voltage e_a become?
31. Find the current from the 10-volt source in the circuit of Fig. 5–50 by utilizing the symmetry and resolving the excitation into symmetrical components.

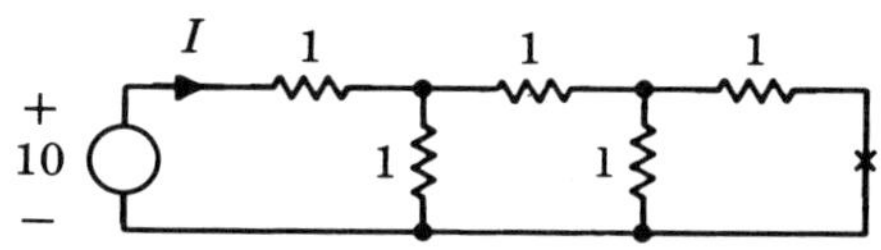

FIGURE 5–50

32. Find the voltage across the 10-amp source in the circuit of Fig. 5–51 by utilizing the symmetry and resolving the excitation into symmetrical components.

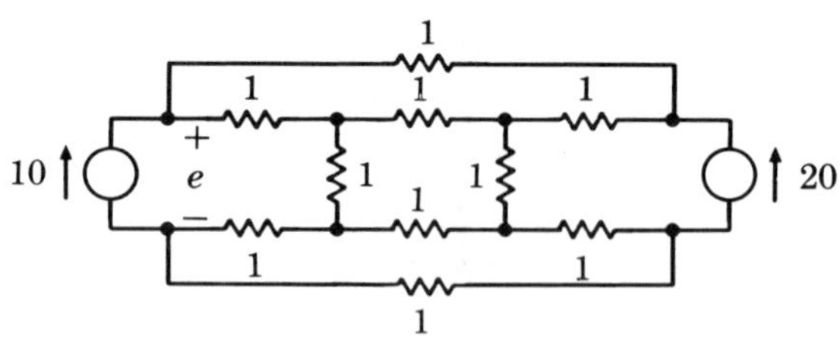

FIGURE 5–51

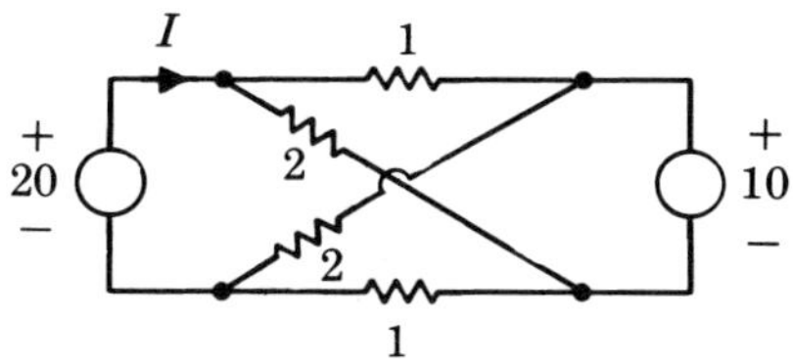

FIGURE 5–52

33. Find the current from the 20-volt source in the circuit of Fig. 5–52, using symmetrical components of the excitation.

34. Make use of symmetry and symmetrical excitation to find r_{11}, r_{22}, and r_{12} for the circuit of Fig. 5–53.

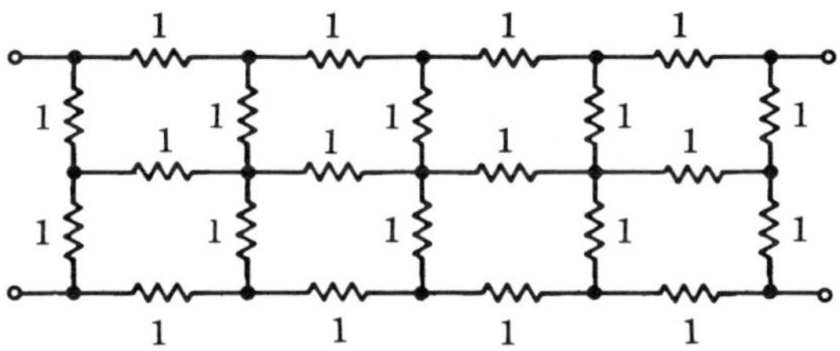

FIGURE 5–53

35. By means of symmetry find g_{11}, g_{22}, and g_{12} directly from the circuit of Fig. 5–53.

Section 5–6

36. Find the equivalent r and g parameters for the circuit of Fig. 5–54 and give the equivalent T and π circuits.

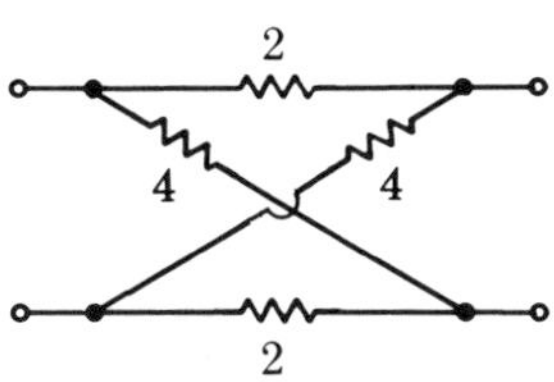

FIGURE 5–54

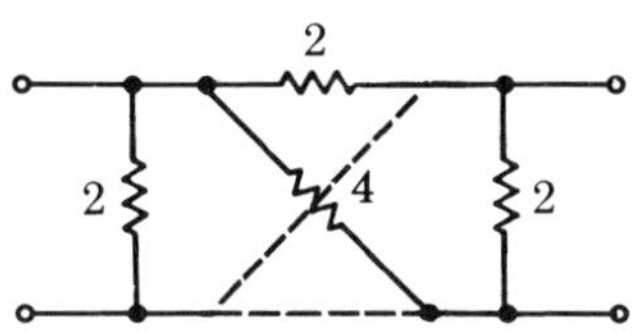

FIGURE 5–55

37. Repeat Exercise 36 with the series and shunt arms in the lattice interchanged. Use negative resistances in the equivalent circuits.

38. Show that the Y–Δ transformation holds even for the negative elements of Exercise 37.

39. Find the equivalent T circuit for the lattice of Fig. 5–55.

40. Find an equivalent π circuit directly from the circuit of Exercise 39.

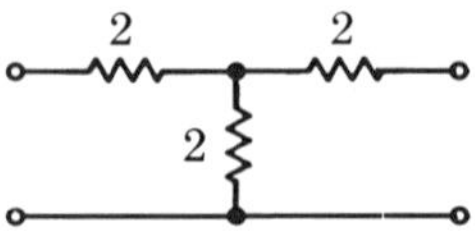

Figure 5–56

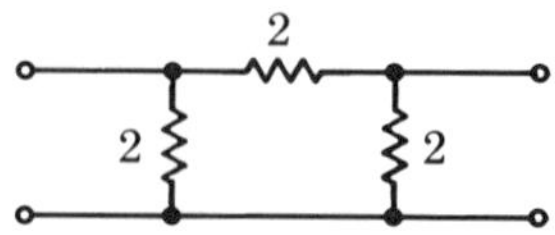

Figure 5–57

41. Use Bartlett's bisection theorem to find the equivalent lattice for the network of Fig. 5–56.

42. Use Bartlett's bisection theorem to find the equivalent lattice for the network of Fig. 5–57.

43. Use Bartlett's bisection theorem to find the equivalent lattice for the network of Fig. 5–58.

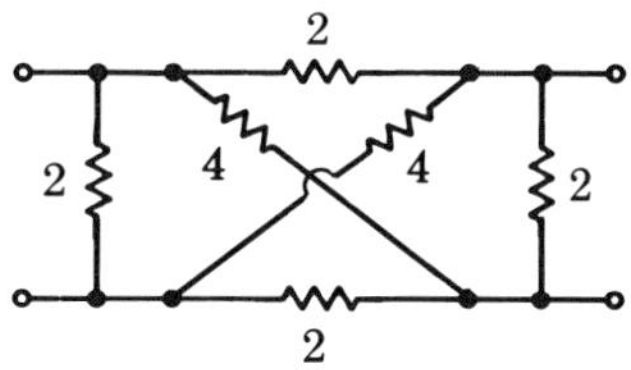

Figure 5–58

44. Find the equivalent lattice for the network of Fig. 5–59 by means of Bartlett's bisection theorem. (It is not easy to break this network into halves. It is easy enough to solve for the symmetrical and antisymmetrical currents.)

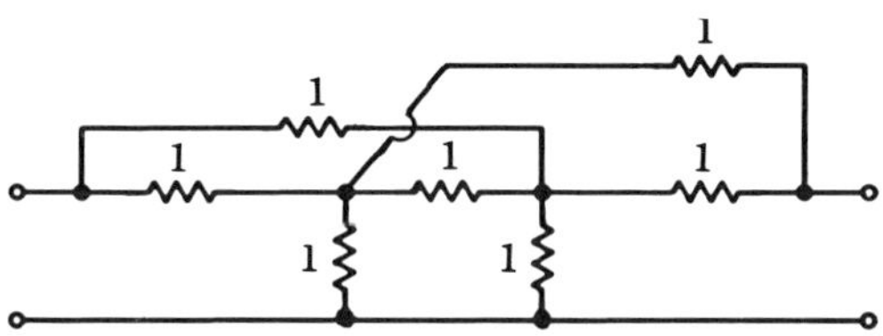

Figure 5–59

45. Change the lattice of Fig. 5–60 to a T by means of the inverse bisection theorem.

46. Change the lattice of Fig. 5–61 to a T by means of the inverse bisection theorem. Change the output polarity to obtain positive elements if necessary.

47. Find equivalent π networks for the lattices of Exercises 45 and 46 by applying the inverse bisection theorem directly to the networks.

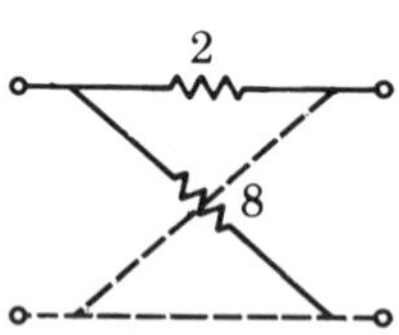

Figure 5–60

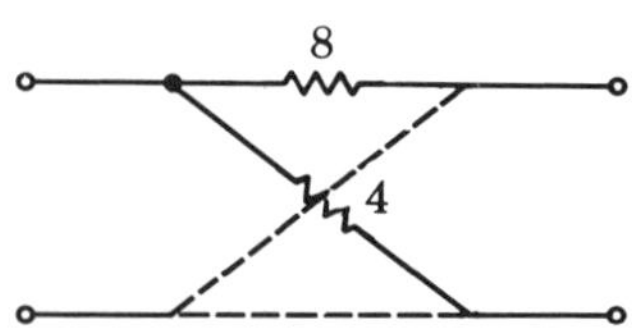

Figure 5–61

CHAPTER 6

POWER AND ENERGY IN RESISTIVE CIRCUITS

"The inevitable comes to pass by effort."
Oliver Wendell Holmes

Effort, force, and energy lie behind all accomplishments in the real physical world. Electricity is important to us because of the energy which it carries. Indeed, we know very little about electricity except that it will carry energy and do things for us. In previous chapters we have stayed away from concepts of energy, and instead considered the concepts of current and voltage. We did so because we were interested in the "quality" of the energy as well as its quantity. Voltage and current became the primary circuit variables and energy and power were secondary. The underlying concept is still energy, however, and in this chapter we return to it. Since resistive circuits do not store energy, we may deal with *power*, or rate of flow of energy.

6–1 Power in terms of branch currents and voltages. Current is a measure of the rate at which electric charge flows through a device. Voltage is a measure of the amount of energy associated with each unit of charge. Thus the product of voltage and current gives the rate at which energy is being absorbed or given off by the device. Current is measured in coulombs per second and voltage in joules per coulomb, and the product of the two is the power in joules per second, or watts. Thus the power, p, is

$$p = e \times i. \tag{6–1}$$

A second equation which is often useful expresses the algebraic requirement that the total power generated by the sources in a network must be equal to the total power absorbed by the passive elements. Since most of the elements are passive, we will take the power lost in a passive element to be positive and the power supplied by a source to be negative. Then

$$\sum p = 0. \tag{6–2}$$

Obviously, there is a sign convention for Eqs. (6–1) and (6–2). We used it in Chapter 1 to define the polarity of the voltage, but for convenience it is repeated in Fig. 6–1(a). Positive power is dissipated in the element if the voltage and the current have the signs shown. The application of the

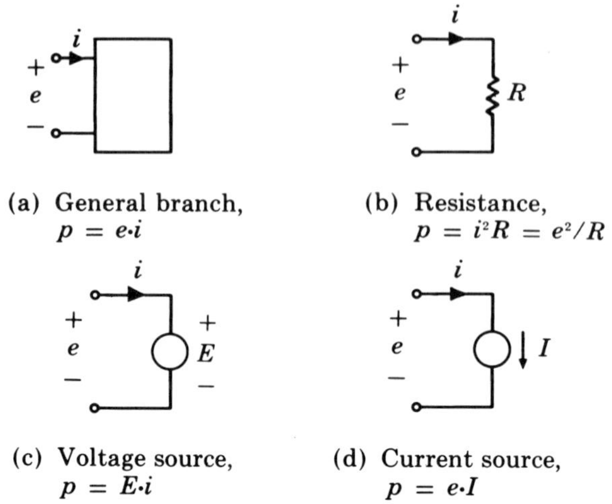

(a) General branch, $p = e \cdot i$

(b) Resistance, $p = i^2R = e^2/R$

(c) Voltage source, $p = E \cdot i$

(d) Current source, $p = e \cdot I$

FIG. 6–1. The sign conventions for power absorbed.

formula to a resistive branch is shown in Fig. 6–1(b). The power is positive, and has a value given by

$$p_R = e \times i = i^2R = \frac{e^2}{R} \cdot \tag{6–3}$$

The alternative forms in Eq. (6–3) are obtained by substituting the volt-ampere equation of the resistance in the original formula. The power associated with the voltage source of Fig. 6–1(c) is

$$p_V = e \times i = E \times i. \tag{6–4}$$

Equation (6–4) has only this one form. If the source is supplying energy the current will be negative and the power absorbed will also be negative. The power associated with the current source of Fig. 6–1(d) is

$$p_I = e \times i = e \times I. \tag{6–5}$$

Equation (6–5) also has only one form. If the source is supplying energy the voltage will be negative and the power absorbed by the source will also be negative.

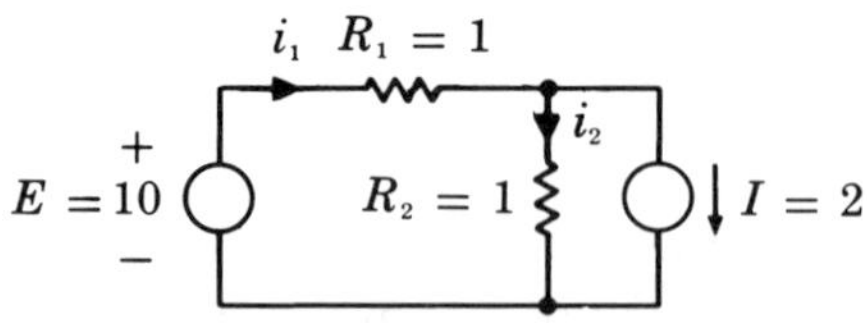

FIG. 6–2. Power computations by means of branch currents.

The basic method for obtaining the power in a circuit is to solve for the currents and the voltages in each of the branches and to add algebraically the power absorbed by and the power supplied to the network. We shall use the circuit of Fig. 6–2 to illustrate this method. The currents and the voltages in each branch are computed by means of the superposition theorem. They are

$$\begin{aligned} i_1 &= 5 + 1 = 6 \text{ amp}, \\ i_2 &= 5 - 1 = 4 \text{ amp}, \\ e_1 &= i_1 R_1 \quad = 6 \text{ volts}, \\ e_2 &= i_2 R_2 \quad = 4 \text{ volts}. \end{aligned}$$

The power dissipated in the two resistances can now be found:

$$\begin{aligned} p_{R_1} &= i_1^2 R_1 = 6^2 \times 1 = 36 \text{ watts}, \\ p_{R_2} &= i_2^2 R_2 = 4^2 \times 1 = 16 \text{ watts}. \end{aligned}$$

The total power lost in the two resistances is 52 watts. The power dissipated in the two sources is $p_V + p_I$, where

$$\begin{aligned} p_V &= E \times i_1 = 10 \times -6 = -60 \text{ watts}, \\ p_I &= e_2 \times I = 2 \times 4 = 8 \text{ watts}. \end{aligned}$$

The net power absorbed by the sources is thus -52 watts, or the power supplied by the sources is $+52$ watts. The total algebraic power is

$$\sum p = 52 - 52 = 0.$$

In this example the two resistances and the current source are absorbing power, and the voltage source is supplying all the power. *The sign conventions must be observed carefully, because a source does not always supply power!* One advantage of the method of finding the power absorbed or delivered by each branch is that the individual relations as well as the gross over-all relation are then known. A disadvantage of this method is that more effort is required to find all the branch powers than to find the single resultant circuit power by the methods discussed below.

6–2 Power in terms of mesh currents and node voltages. Quadratic forms. If a circuit problem has been solved by means of mesh currents or node voltages, the total power absorbed by the network can be obtained directly from these values, without recourse to the branch variables. In practical problems it is usually better to obtain the branch currents and voltages from the mesh currents or node voltages, because the power in each element can then be computed. The "quadratic-form" method discussed here is principally useful in theoretical work.

Quadratic form for mesh currents. The general mesh equations are

$$\begin{array}{rl} i_1 r_{11} - i_2 r_{12} - i_3 r_{13} - \cdots = e_1 & \times\, i_1, \\ -i_1 r_{12} + i_2 r_{22} - i_3 r_{23} - \cdots = e_2 & \times\, i_2, \\ \vdots & \\ -i_1 r_{1n} - i_2 r_{2n} - i_3 r_{3n} - \cdots = e_n & \times\, i_n. \end{array} \tag{6–6}$$

The first of these equations is multiplied by i_1, the second by i_2, the third by i_3, and so on. The sum of the terms on the right side is the power supplied by the voltage sources in the circuit. Accordingly, the expression on the left side must be the power absorbed by the resistances in the circuit. It has the form

$$\begin{aligned} p = i_1^2 r_{11} &- 2 i_1 i_2 r_{12} - 2 i_1 i_3 r_{13} - \cdots \\ &+ i_2^2 r_{22} - 2 i_2 i_3 r_{23} - \cdots \\ &\qquad\quad + i_3^2 r_{33} - \cdots \\ &\qquad\qquad \vdots \end{aligned} \tag{6–7}$$

Equation (6–7) gives the total power absorbed by the circuit, in terms of the mesh currents. Since it represents the power absorbed by a linear, passive circuit it must be positive for all values of the currents, both positive and negative. Indeed, it must be positive for a random selection of the currents, since Eq. (6–6) can then be satisfied by an appropriate choice of voltages in each mesh. Such an expression is called a *positive definite quadratic form.* It is of theoretical interest because it expresses a condition that must be satisfied by a set of self- and mutual resistances if they are to represent a physically realizable circuit. It sets some boundaries on network synthesis procedures. At the moment we will use it only to compute the power absorbed by a typical circuit, as shown in Fig. 6–3.

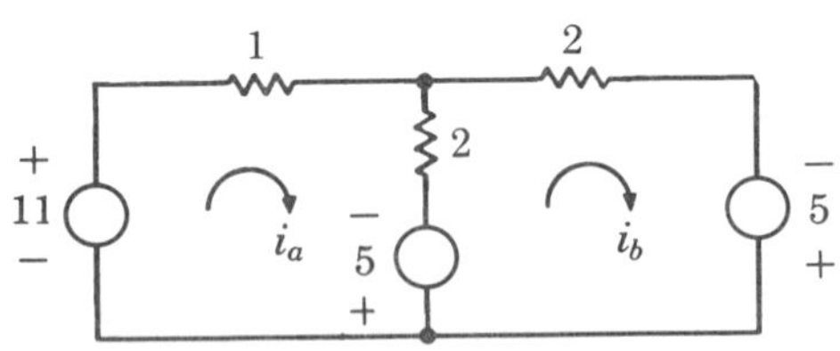

FIG. 6–3. Power computations by means of mesh currents; a quadratic form.

The mesh equations for this circuit are

$$i_a(3) - i_b(2) = 16, \qquad -i_a(2) + i_b(4) = 0,$$

from which $i_a = 8$ amp and $i_b = 4$ amp. The quadratic form, Eq. (6–7), gives the power directly in terms of these mesh currents. It is

$$\begin{aligned} p &= i_a^2(3) - 2i_a i_b(2) + i_b^2(4) \\ &= 8^2 \times 3 - 2 \times 8 \times 4 \times 2 + 4^2 \times 4 \\ &= 128 \text{ watts.} \end{aligned}$$

As a check on this result, we note that the i^2R losses in the three resistances are

$$p = 8^2 \times 1 + 4^2 \times 2 + 4^2 \times 2 = 128 \text{ watts,}$$

and the power supplied by the sources is

$$p = 11 \times 8 + 4 \times 5 + 4 \times 5 = 128 \text{ watts.}$$

Quadratic form for node voltages. A similar quadratic form that gives the power absorbed in a network in terms of the node voltages can be derived from the general node equations

$$\begin{array}{rl|l} e_1g_{11} - e_2g_{12} - e_3g_{13} - \cdots & = i_1 & \times e_1, \\ -e_1g_{12} + e_2g_{22} - e_3g_{23} - \cdots & = i_2 & \times e_2, \\ \vdots & & \\ -e_ng_{1n} - e_2g_{2n} - e_3g_{3n} - \cdots & = i_n & \times e_n. \end{array} \tag{6–8}$$

The first of these equations is multiplied by e_1, the second by e_2, and so on. The sum of the terms on the right side then gives the total power supplied to the network from all the current sources. The expression on the left side must be the power absorbed by the resistances in the network:

$$\begin{aligned} p = e_1^2g_{11} - 2e_1e_2g_{12} - 2e_1e_3g_{13} - \cdots \\ + e_2^2g_{22} - 2e_2e_3g_{23} - \cdots \\ + e_3^2g_{33} - \cdots \end{aligned} \tag{6–9}$$

Equation (6–9) is a second positive-definite quadratic form. For any arbitrary choice of the e's it must give a positive value of p. We will use it to compute the power absorbed in the circuit of Fig. 6–4. The node equations for this circuit are

$$e_a(2) - e_b(1) = -1, \qquad -e_a(1) + e_b(2) = 0,$$

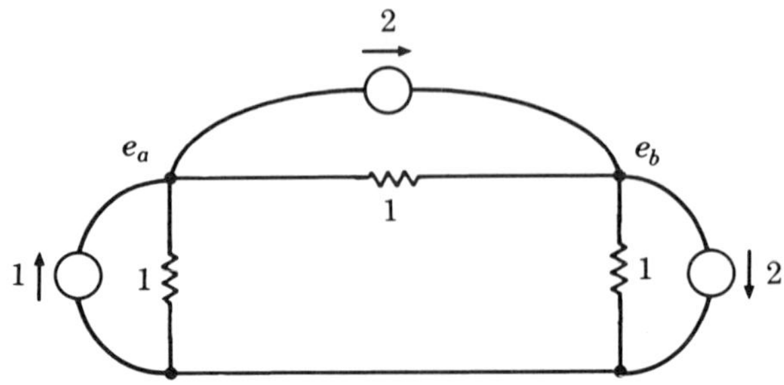

FIG. 6–4. Power computations by means of node voltages; a quadratic form.

from which $e_a = -\frac{2}{3}$ volt and $e_b = -\frac{1}{3}$ volt. The power lost in the resistances, according to Eq. (6–9), is

$$\begin{aligned} p &= e_a^2 g_{11} - 2e_a e_b g_{12} + e_b^2 g_{22} \\ &= \tfrac{4}{9} \times 2 - 2 \times \tfrac{2}{3} \times \tfrac{1}{3} \times 1 + \tfrac{1}{9} \times 2 \\ &= \tfrac{2}{3} \text{ watt.} \end{aligned}$$

As a check, the power computed from the e^2/R losses is found to be

$$p = \tfrac{4}{9} \times 1 + \tfrac{1}{9} \times 1 + \tfrac{1}{9} \times 1 = \tfrac{2}{3} \text{ watt,}$$

and the power drawn from the current sources is

$$p = -1 \times \tfrac{2}{3} + 2 \times \tfrac{1}{3} + 2 \times \tfrac{1}{3} = \tfrac{2}{3} \text{ watt.}$$

The positive-definite quadratic form gives a direct method of computing the power absorbed by a circuit when the mesh currents or the node voltages are known. It does not give the power in the separate elements and therefore is not often used in practical problems. It is of interest mainly for the insight which it gives into the network, and as a background for network synthesis.

6–3 Power in equivalent one-terminal-pair circuits. In Chapter 4 we investigated equivalent circuits for one-terminal-pair networks in some detail. We found that a passive circuit* could always be reduced to a single resistance, and that an active one could be reduced to a single current or voltage source in conjunction with a resistance. It is appropriate at this point to investigate the power relations for equivalent one-terminal-pair circuits. Obviously, the equivalence would be of little use unless it held for power as well as for voltage or current.

---

* A passive circuit is one with no sources present.

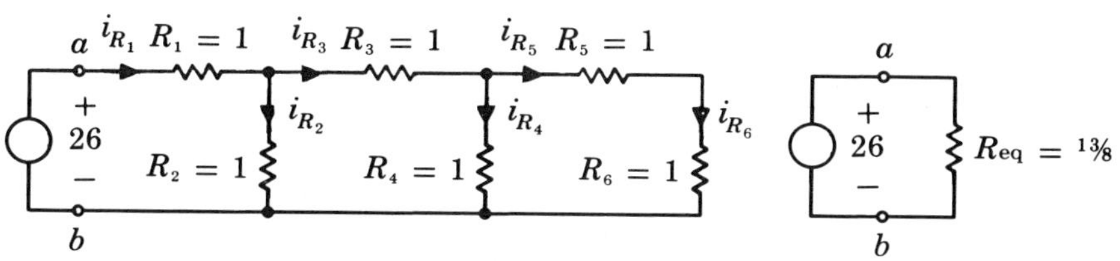

(a) Original circuit (b) Equivalent resistance

FIG. 6–5. Power computations from an equivalent resistance.

Power in an equivalent resistance. If a one-terminal-pair network is reduced to a single equivalent resistance, the power which will be absorbed by this resistance is the same as the power which would have been absorbed by the original network. The argument for this conclusion comes from Eq. (6–2), rewritten in the form

$$\text{Net source power} = \text{net power absorbed.} \tag{6–10}$$

The equivalent one-terminal-pair network has the same volt-ampere equation as the original network. Therefore it will draw the same current from a voltage source and the same power from the source. Since the source power must be equal to the network power, the two networks must draw equal power. An example which illustrates this relationship is shown in Fig. 6–5(a). The branch currents in this network are

$$i_{R_1} = 16 \text{ amp}, \qquad i_{R_2} = 10 \text{ amp}, \qquad i_{R_3} = 6 \text{ amp},$$
$$i_{R_4} = 4 \text{ amp}, \qquad i_{R_5} = i_{R_6} = 2 \text{ amp}.$$

The corresponding power absorbed in each branch is

$$\begin{aligned} p_{R_1} &= 16^2 \times 1 = 256 \text{ watts} \\ p_{R_2} &= 10^2 \times 1 = 100 \text{ watts} \\ p_{R_3} &= 6^2 \times 1 = 36 \text{ watts} \\ p_{R_4} &= 4^2 \times 1 = 16 \text{ watts} \\ p_{R_5} &= 2^2 \times 1 = 4 \text{ watts} \\ p_{R_6} &= 2^2 \times 1 = \underline{4 \text{ watts}} \end{aligned}$$

and the total power is $p_t = 416$ watts.

The equivalent resistance for the network at the terminals a–b is $\frac{13}{8}$ ohms, as may be verified by combining resistances in series and in parallel. This resistance is shown directly across the voltage source in Fig. 6–5(b). The power it draws is

$$p_{eq} = \frac{26^2}{13/8} = 416 \text{ watts.}$$

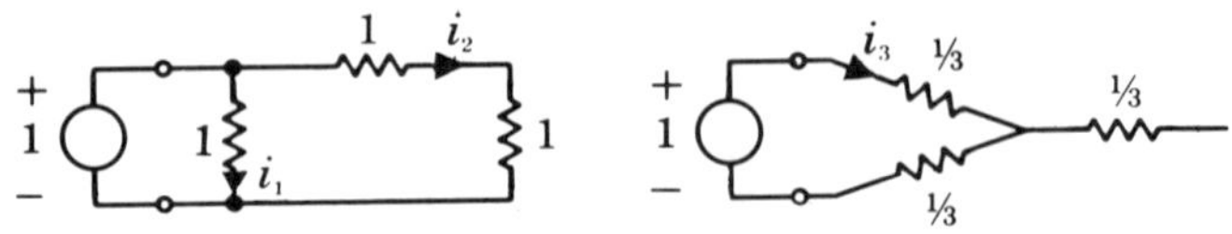

(a) Original network (b) Transformed network

FIG. 6–6. Power computations after a Y–Δ transformation.

As a second example, consider the Y–Δ transformation shown in Fig. 6–6. The circuit of Fig. 6–6(b) is obtained from the circuit of (a) by means of a Y–Δ transformation. The currents in the original circuit are

$$i_1 = 1 \text{ amp}, \qquad i_2 = \tfrac{1}{2} \text{ amp},$$

and the power absorbed by the resistances is

$$p = i_1^2 \times 1 + i_2^2 \times 2 = 1^2 \times 1 + \frac{1}{2^2} \times 2 = 1.5 \text{ watts}.$$

The current in the equivalent circuit of Fig. 6–6(b) is

$$i_3 = 1.5 \text{ amp},$$

and the power absorbed by the resistances in this circuit is

$$p = i_3^2 \times \tfrac{2}{3} = (1.5)^2 \times \tfrac{2}{3} = 1.5 \text{ watts}.$$

The power absorbed by the equivalent circuit is equal to the power absorbed by the original circuit. The power in the individual resistances, however, is very different.

Power from equivalent sources. It is often convenient to use an equivalent source branch in a network problem. Thévenin's and Norton's theorems are examples of equivalent source branches. Since the equivalent sources have the same volt-ampere equations at their terminals as the original networks do, they must supply the same power to an external circuit. Hence, according to Eq. (6–10), they must deliver the same power. *The power absorbed by the equivalent resistance in the source branch will in general, however, be very different.* When an equivalent source is used the equivalence is good only for the network external to the source. An example is shown in Fig. 6–7. The original network is given in Fig. 6–7(a) and the Thévenin equivalent source branch in (b). The currents in the original circuit are

$$i_1 = 1 \text{ amp}, \qquad i_2 = 10 \text{ amp}.$$

The power in the load resistance R_2, external to the source branch, is

$$p_{R_2} = i_2^2 R_2 = 100 \text{ watts},$$

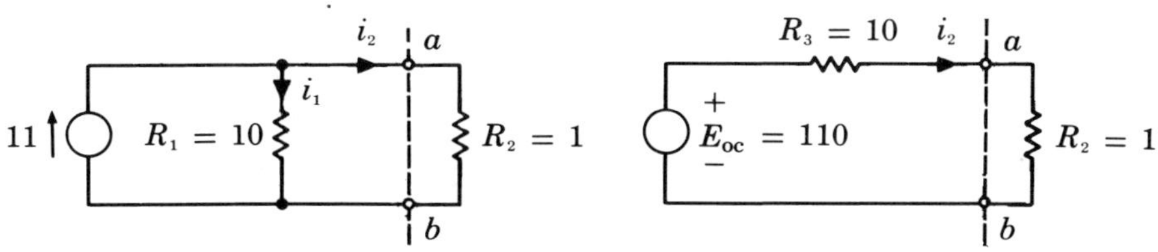

(a) Original circuit (b) Thévenin equivalent circuit

FIG. 6–7. Power from a Thévenin equivalent circuit.

and the power lost in the internal resistance of the source is

$$p_{R_1} = i_1^2 R_1 = 10 \text{ watts.}$$

The power generated by the ideal current source is

$$p_I = 11 \times 10 = 110 \text{ watts.}$$

In the Thévenin equivalent circuit of Fig. 6–7(b) the current is

$$i_2 = 10 \text{ amp.}$$

The power in the load resistance external to the source transformation is, as before,

$$p_{R_2} = i_2^2 \times R_2 = 100 \text{ watts.}$$

The power in the internal resistance of the equivalent source is

$$p_{R_3} = i_2^2 \times R_3 = 1000 \text{ watts,}$$

and the power generated by the ideal voltage source is

$$p_V = 110 \times 10 = 1100 \text{ watts.}$$

As predicted, the power in the external load is the same whether the original source or the equivalent source is used. However, the power lost in the equivalent source branch and the power generated by the ideal source in the equivalent source branch are very different from that lost in and generated by the original source. This is an important point to remember. It is not surprising, after all, since the whole form of the equivalent source is different from that of the original one. The equivalent source always has a single resistance. The original source might have had any number of them. *In general, equivalent sources give the correct power and energy relationships only at their terminals.*

A second example of a source transformation is shown in Fig. 6–8. The circuit of Fig. 6–8(a) is transformed to the circuit of (b) by means

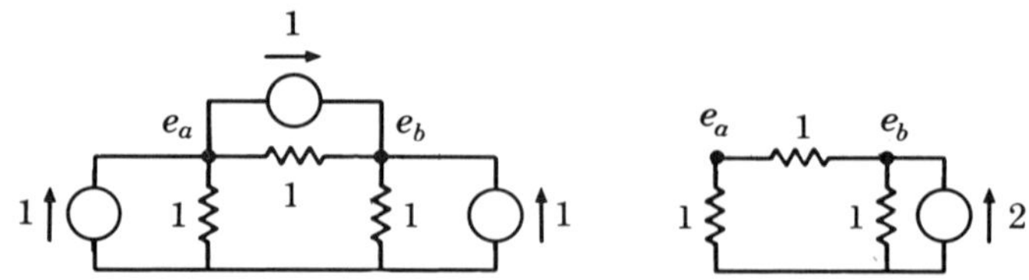

(a) Original circuit (b) Equivalent circuit

FIG. 6–8. Power from a circuit after a source transformation.

of a general current source transformation. The voltages in the original circuit can be obtained by superposition. They are

$$e_a = \tfrac{2}{3} + \tfrac{1}{3} - \tfrac{1}{3} = \tfrac{2}{3} \text{ volt,}$$

$$e_b = \tfrac{2}{3} + \tfrac{1}{3} + \tfrac{1}{3} = \tfrac{4}{3} \text{ volts.}$$

The powers from the sources are

$$p_{I_1} = I_1 \times e_a = 1 \times \tfrac{2}{3} = \tfrac{2}{3} \text{ watt,}$$

$$p_{I_2} = I_2 \times (e_b - e_a) = 1 \times \tfrac{2}{3} = \tfrac{2}{3} \text{ watt,}$$

$$p_{I_3} = I_3 \times e_b = 1 \times \tfrac{4}{3} = \tfrac{4}{3} \text{ watts,}$$

from which the total power is found to be $p_t = \frac{8}{3}$ watts.
The node voltages in the circuit of Fig. 6–8(b) have the same values as the node voltages in the circuit of Fig. 6–8(a), and the power from the source is

$$p_I = 2 \times \tfrac{4}{3} = \tfrac{8}{3} \text{ watts.}$$

The total source power is the same in the two circuits. The powers from the individual sources in the two cases, however, are very different. The general rule for both equivalent resistances and equivalent sources is that the power and energy relationships are correct *only at the terminals.*

6–4 Power in equivalent two-terminal-pair circuits. A three-resistance network is equivalent to a two-terminal-pair network in the sense that it has the same volt-ampere relationships at the two terminal-pairs. Thus the original network and the equivalent will draw the same current from the same external voltage sources, and the same power. The equivalence does not extend into the interior of the networks, however, and there is no relationship between the power in one of the equivalent resistances and the power in a resistance of the original network. In this section we shall consider an example of the power in a two-terminal-pair network

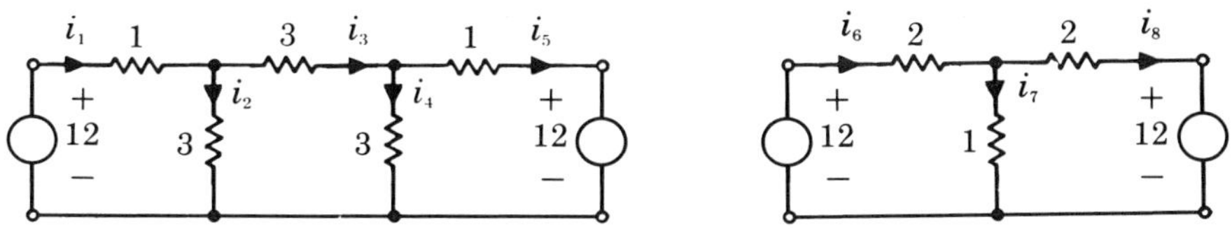

(a) Original two-terminal-pair network

(b) Equivalent network

FIG. 6–9. Power from an equivalent two-terminal-pair network.

and its equivalent, and discuss power in terms of the r and g parameters.

Power in a two-terminal-pair network. A two-terminal-pair network with voltage excitation at both terminal-pairs is shown in Fig. 6–9(a). The currents in this network can be obtained easily by the application of symmetry principles. They are

$$i_1 = 3 \text{ amp}, \qquad i_2 = 3 \text{ amp}, \qquad i_3 = 0, \qquad i_4 = 3 \text{ amp},$$
$$i_5 = -3 \text{ amp}.$$

The powers in the corresponding resistances are

$$\begin{aligned} p_1 &= i_1^2 \times 1 = 9 \text{ watts} \\ p_2 &= i_2^2 \times 1 = 27 \text{ watts} \\ p_3 &= i_3^2 \times 1 = 0 \text{ watts} \\ p_4 &= i_4^2 \times 1 = 27 \text{ watts} \\ p_5 &= i_5^2 \times 1 = 9 \text{ watts} \end{aligned}$$

and the total power is $\qquad p_t = 72$ watts.

The equivalent three-resistance circuit is shown in Fig. 6–9(b), where the exciting sources are the same as in Fig. 6–9(a). The currents in this circuit are

$$i_6 = 3 \text{ amp}, \qquad i_7 = 6 \text{ amp}, \qquad i_8 = -3 \text{ amp},$$

and the corresponding powers are

$$\begin{aligned} p_6 &= i_6^2 \times 2 = 18 \text{ watts} \\ p_7 &= i_7^2 \times 1 = 36 \text{ watts} \\ p_8 &= i_8^2 \times 2 = 18 \text{ watts} \end{aligned}$$

and the total power is $\qquad p_t = 72$ watts.

The total power absorbed by the equivalent circuit in this example is the same as the total power absorbed by the original network. There is no connection between the powers in the individual resistances. Although the example considered here is a three-terminal network, the arguments

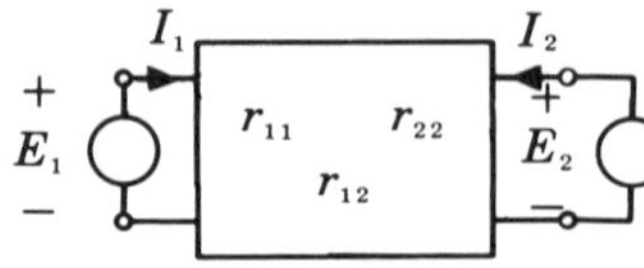

FIG. 6–10. Power from the general two-terminal-pair parameters.

apply equally well to a four-terminal network excited at only two terminal-pairs.

*Power in terms of the r parameters.** A general two-terminal-pair network can be described in terms of three r parameters. For the network of Fig. 6–10, the r-parameter formulation is

$$E_1 = r_{11}I_1 + r_{12}I_2, \qquad E_2 = r_{12}I_1 + r_{22}I_2. \tag{6–11}$$

If the first of these equations is multiplied by I_1 and the second by I_2 and the results are added, the expression on the left side will be the power delivered to the network by external sources. Therefore the expression on the right side must be the power absorbed in the network. This expression is

$$p = r_{11}I_1^2 + 2r_{12}I_1I_2 + r_{22}I_2^2. \tag{6–12}$$

The two squared terms in Eq. (6–12) are always positive. The middle term can be either positive or negative, depending on the signs of I_1, I_2, and r_{12}. For the particular case $I_1 = \pm I_2$, the power absorbed by the network is

$$p = (r_{11} + r_{22} \pm 2r_{12})I_1^2. \tag{6–13}$$

Since this power must be positive,

$$r_{11} + r_{22} \pm 2r_{12} \geq 0. \tag{6–14}$$

A second inequality involving the r parameters can be obtained from Eq. (6–11) if the voltage E_2 is made equal to zero. Then, from the second of equations (6–11),

$$r_{12}I_1 = -r_{22}I_2,$$

and from the first,

$$\frac{E_1}{I_1} = r_{11} + \frac{r_{12}I_2}{I_1} = r_{11} - \frac{r_{12}^2}{r_{22}}. \tag{6–15}$$

Equation (6–15) gives the input resistance of the network with a short circuit on the output terminal-pair. If the input current is I_1, the input power is

$$p = I_1^2\left(r_{11} - \frac{r_{12}^2}{r_{22}}\right). \tag{6–16}$$

---

* Do not confuse with mesh-current resistances of Section 6–2.

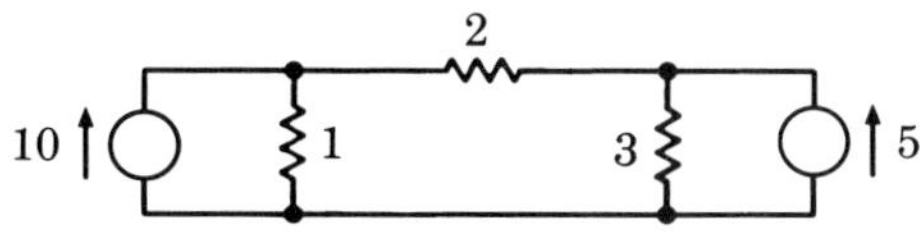

FIG. 6–11. Power from the r parameters.

Since this expression must be positive,

$$r_{11}r_{22} - r_{12}^2 \geq 0. \tag{6–17}$$

We now give an example of power computations from the r parameters with respect to the circuit of Fig. 6–11. For this circuit,

$$r_{11} = \tfrac{5}{6}, \qquad r_{12} = \tfrac{1}{2}, \qquad r_{22} = \tfrac{3}{2},$$

and the power absorbed, according to Eq. (6–12), is

$$\begin{aligned} p &= \tfrac{5}{6} \times 10^2 + 2 \times \tfrac{1}{2} \times 10 \times 5 + \tfrac{3}{2} \times 5^2 \\ &= \frac{1025}{6} \text{ watts.} \end{aligned}$$

It is interesting also to verify Eqs. (6–14) and (6–17). For Eq. (6–14),

$$r_{11} + r_{22} - 2r_{12} = \tfrac{5}{6} + \tfrac{3}{2} - 2 \times \tfrac{1}{2} = \tfrac{4}{3},$$

and for Eq. (6–17),

$$r_{11}r_{22} - r_{12}^2 = \tfrac{5}{6} \times \tfrac{3}{2} - (\tfrac{1}{2})^2 = 1.$$

The only condition under which these two equations would be fulfilled with the equals sign would be if $r_{11} = r_{22} = r_{12}$. In this case the network reduces to a single shunt resistance.

*Power in terms of the g parameters.** The g parameters satisfy much the same power relationships as do the r parameters. A general two-terminal-pair network can be represented by the equations

$$I_1 = g_{11}E_1 + g_{12}E_2, \qquad I_2 = g_{12}E_1 + g_{22}E_2. \tag{6–18}$$

If the first of these equations is multiplied by E_1 and the second one by E_2 and the results are added, the expression on the left side will be the power delivered to the network by the external sources, and therefore the expression on the right side must be the power absorbed in the network. It is

$$p = g_{11}E_1^2 + 2g_{12}E_1E_2 + g_{22}E_2^2. \tag{6–19}$$

* Do not confuse with node-voltage conductances of Section 6–2.

For the particular case of $E_1 = \pm E_2$, Eq. (6–19) becomes

$$p = E_1^2(g_{11} + g_{22} \pm 2g_{12}).$$

Since the power must be positive,

$$g_{11} + g_{22} \pm 2g_{12} \geq 0. \tag{6–20}$$

For all three-terminal networks and for most four-terminal networks, g_{12} is negative and the plus sign in Eq. (6–20) is the significant sign.

A second relationship which must be satisfied by the g's can be obtained by letting $I_2 = 0$ in Eq. (6–18). From the second equation,

$$g_{12}E_1 = -g_{22}E_2,$$

and from the first equation

$$\frac{I_1}{E_1} = g_{11} + g_{12}\frac{E_2}{E_1} = g_{11} - \frac{g_{12}^2}{g_{22}}. \tag{6–21}$$

Equation (6–21) represents the input conductance of the network with an open circuit at the output. If a voltage E_1 is applied, the input power is

$$p = E_1^2\left(g_{11} - \frac{g_{12}^2}{g_{22}}\right).$$

Since the input power must be positive,

$$g_{11}g_{22} - g_{12}^2 \geq 0. \tag{6–22}$$

The use of the g parameters to compute the power absorbed by a network will be illustrated for the network of Fig. 6–12. The g values are

$$g_{11} = \tfrac{5}{11}, \qquad g_{22} = \tfrac{3}{11}, \qquad g_{12} = -\tfrac{2}{11}.$$

The power absorbed in the network, by Eq. (6–19), is

$$\begin{aligned} p &= \tfrac{5}{11} \times 10^2 - 2 \times \tfrac{2}{11} \times 10 \times 5 + \tfrac{3}{11} \times 5^2 \\ &= \tfrac{375}{11} \text{ watts.} \end{aligned}$$

The numerical value of Eq. (6–20), for the positive sign in the equation, is

$$\tfrac{5}{11} + \tfrac{3}{11} - \tfrac{2}{11} = \tfrac{6}{11},$$

and for Eq. (6–22) is

$$\frac{5}{11} \times \frac{3}{11} - \frac{2^2}{11^2} = \frac{1}{11}.$$

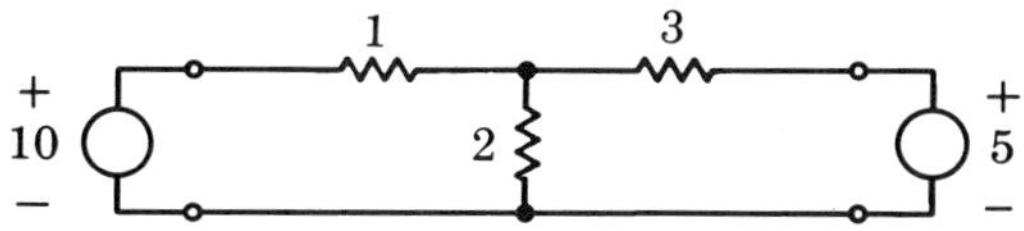

FIG. 6–12. Power from the g parameters.

In order that these equations should be satisfied with the equals sign, g_{11} must equal g_{22} and both must equal g_{12}. The network reduces to a single series resistance for this condition.

6–5 The maximum-power theorem. An ideal voltage source is capable of supplying infinite power if a short circuit is placed across it, and an ideal current source is capable of supplying infinite power if an open circuit is placed across it. A real physical source cannot supply infinite power and therefore must have some "internal" resistance associated with it. A voltage source has a series resistance, and a current source has a parallel resistance. We shall now investigate the conditions under which these real sources will deliver maximum power to an external load resistance.

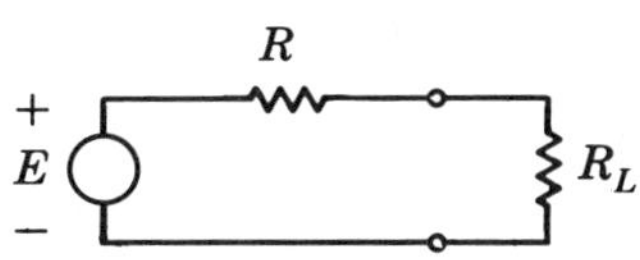

FIG. 6–13. Voltage source with internal resistance and external load resistance.

Voltage source and resistance. A voltage source, with its internal resistance R and an external load resistance R_L, is shown in Fig. 6–13. The source and its internal resistance can be an actual voltage source or they can represent a Thévenin equivalent circuit for some more complicated situation. For the discussion here, both the voltage and the internal resistance will be considered to be fixed and the load resistance R_L will be the variable which is to be altered to obtain maximum power.* The current in the circuit is

$$i = \frac{E}{R + R_L},$$

and the power delivered to the resistance R_L is

$$p_L = i^2 R_L = \frac{E^2 R_L}{(R + R_L)^2}. \tag{6–23}$$

* Other constraints are possible and are important in some problems.

To find the maximum value of p_L we differentiate it with respect to R_L:

$$\frac{dp_L}{dR_L} = E^2 \frac{(R + R_L)^2 - 2R_L(R + R_L)}{(R + R_L)^4}.$$

The maximum occurs when the derivative is zero, or at

$$R_L = R. \tag{6–24}$$

The condition for maximum power is therefore that the load resistance should be equal to the internal resistance of the source. It is then said to "match" the source. The value of the maximum power when $R_L = R$ is

$$p_{\max} = \frac{E^2}{4R}. \tag{6–25}$$

The value given in Eq. (6–25) is also called the *available power*, since it is the maximum that is available from a given source and its internal resistance.

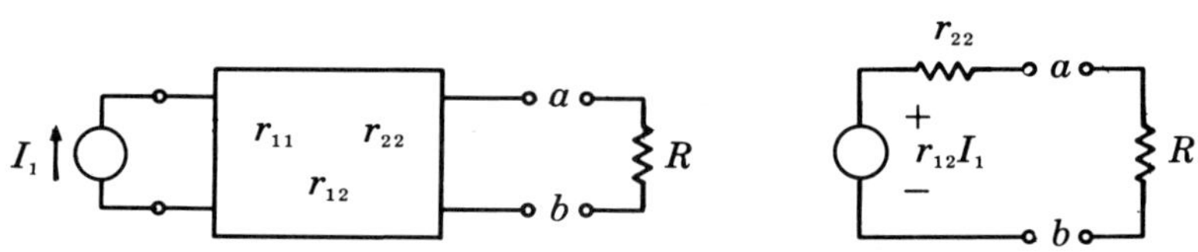

(a) General two-terminal-pair network (b) Equivalent output circuit

FIG. 6–14. Maximum power from a two-terminal-pair network and r parameters.

As an example, we compute the maximum available power from the general two-terminal-pair network shown in Fig. 6–14(a). The network is driven by a current source and is described by the r parameters. The equivalent Thévenin output circuit is a voltage source of value $r_{12}I_1$ in series with a resistance of r_{22}, as shown in Fig. 6–14(b). The maximum power occurs when $R = r_{22}$, and its value is

$$p_{\max} = \frac{r_{12}^2 I_1^2}{4r_{22}}.$$

If the circuit had been driven by a voltage source instead of a current source, it would have been necessary to use the short-circuit, or g parameters.

Current source and resistance. A similar maximum power theorem holds for a current source I, in parallel with a resistance R which is considered

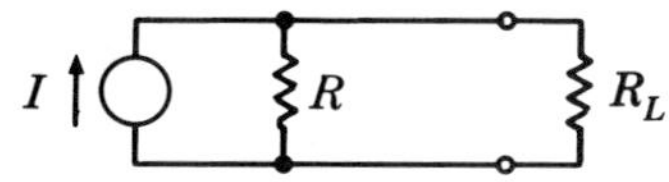

FIG. 6-15. Current source with internal resistance and external load resistance.

to be internal to the source, and a load resistance R_L in parallel with the source, as shown in Fig. 6-15. The voltage across the load resistance is

$$e = \frac{IRR_L}{R + R_L},$$

and the power delivered to the external load is

$$p_L = \frac{e^2}{R_L} = \frac{I^2R^2R_L}{(R + R_L)^2}. \tag{6-26}$$

The derivative of this expression is

$$\frac{dp_L}{dR_L} = \frac{IR^2[(R + R_L)^2 - 2R_L(R + R_L)]}{(R + R_L)^4},$$

and the maximum power occurs when the derivative is zero, at

$$R_L = R. \tag{6-27}$$

According to Eq. (6-27), the maximum power occurs when the load resistance is equal to the internal resistance of the source, just as for the voltage source. The value of the maximum power is

$$p_{\max} = \frac{I^2R}{4} \quad \text{or} \quad \frac{I^2}{4G}. \tag{6-28}$$

As an example of the use of the maximum-power theorem for a current source, we consider the general two-terminal-pair network shown in Fig. 6-16(a). The equivalent Norton circuit for the output terminal-pair is shown in Fig. 6-16(b). The network is driven by a voltage source

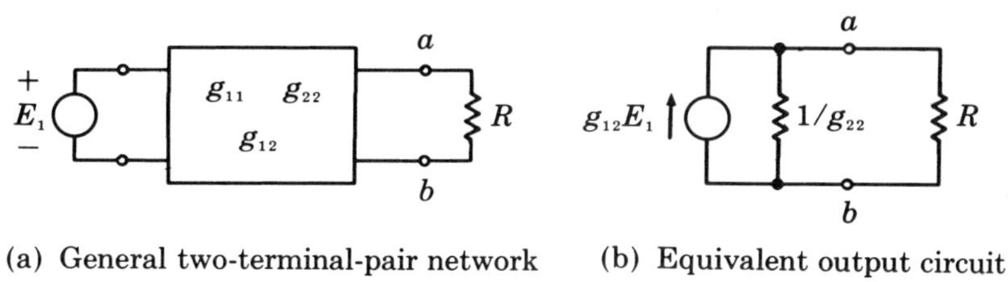

FIG. 6-16. Maximum power from a two-terminal-pair network and the g parameters.

which is assumed to remain constant, and is described by a set of g parameters. The maximum power occurs at

$$R_L = \frac{1}{g_{22}},$$

and its value is

$$p_{\max} = \frac{g_{12}^2 E_1^2}{4g_{22}}.$$

The concept of matching for maximum power is important in almost all communications circuits. The power levels are usually low, and the important thing is to obtain the maximum signal power regardless of the efficiency of the system. Actually, under matched conditions half the power generated is delivered to the load and half is dissipated in the internal resistance of the generator. The efficiency is only 50 percent! In high-power circuits, where the energy itself is of major importance, one cannot afford the luxury of "matching." Then the important thing is not that a maximum power be delivered, but that it be delivered at maximum efficiency.

6–6 Power and the superposition theorem. *The superposition theorem does not apply to power.* Superposition is a property which depends on the linearity of the quantity in question, and power is a quadratic rather than a linear quantity. An example to illustrate this fact is shown in Fig. 6–17.

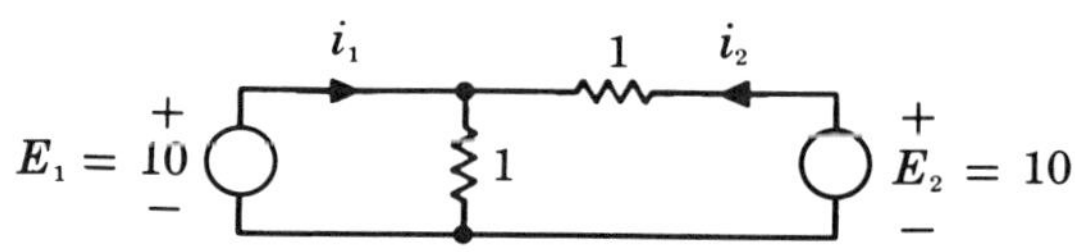

FIG. 6–17. Power from two sources.

The power from the source E_1 acting alone is

$$p_1 = \frac{E_1^2}{1/2} = 200 \text{ watts}.$$

The power from source E_2 acting alone is

$$p_2 = \frac{E_2^2}{1} = 100 \text{ watts}.$$

A superposition of these two powers gives

$$p_1 + p_2 = 300 \text{ watts}.$$

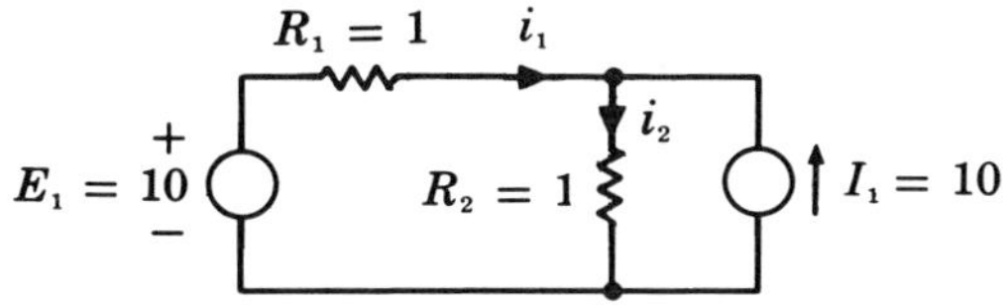

FIG. 6-18. Power from a single current source and a single voltage source.

The actual currents in the circuit when the two sources are acting simultaneously are

$$i_1 = 10 \text{ amp}, \qquad i_2 = 0.$$

The correct value of the power from the two sources is therefore

$$p = E_1 i_1 + E_2 i_2 = 10 \times 10 + 0 = 100 \text{ watts}.$$

The total power delivered to the circuit by the two sources acting together is very different from the sum of the powers which the two sources supply when they are acting separately. *Superposition does not apply.*

There is one peculiar case where the superposition of power does apply, and that is for a single current source combined with a single voltage source. A general proof of this fact is left as an exercise, but an example is shown in Fig. 6-18. The power from the voltage source acting alone is

$$p_V = \frac{e^2}{2} = 50 \text{ watts}.$$

The power from the current source acting alone is

$$p_I = i^2 \times \tfrac{1}{2} = 50 \text{ watts}.$$

The sum of the powers from these two sources acting separately is

$$p_V + p_I = 50 + 50 = 100 \text{ watts}.$$

With both sources acting simultaneously the currents in the circuit are

$$i_1 = 0, \qquad i_2 = 10 \text{ amp}.$$

The power absorbed by the resistances is therefore

$$p = i_1^2 R_1 + i_2^2 R_2 = 0 + 100 = 100 \text{ watts}.$$

The power from the voltage source is

$$p_V = 10 \times 0 = 0,$$

and the power from the current source is

$$p_I = 10 \times 10 = 100 \text{ watts.}$$

The total power computed by superposition for the single current and voltage source agrees with the correct value computed from the actual currents and voltages in the circuit. The values from the individual sources, however, are not correct.

In general, superposition should not be used to compute power. Instead, the actual branch currents in the circuit should be computed and the power obtained from them. *Superposition can be used to compute currents because they are linear. It cannot be used for power because power is not linear.*

It has not been possible in this chapter for us to exhaust the subject of power and energy in electric circuits because we have been dealing with resistive circuits in which there is no energy stored. In subsequent chapters we will introduce the energy storage elements, inductance and capacitance, and we will add the energy storage effects to the energy dissipation effects we have considered here.

6–7 Summary.

Basic formulas

$$p = e \times i$$

$$\Sigma p = 0$$

Power in terms of branch variables

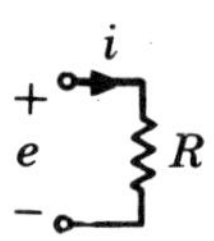

$$p_R = e \times i = i^2 R = \frac{e^2}{R}$$

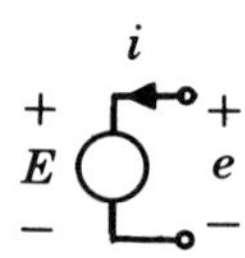

$$p_E = e \times i = E \times i \quad \text{(negative for source)}$$

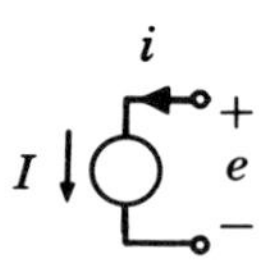

$$p_I = e \times i = e \times I \quad \text{(negative for source)}$$

Power from mesh and node equations (positive definite quadratic forms)

$$\begin{aligned} p = i_1^2 r_{11} - 2i_1 i_2 r_{12} &- 2i_1 i_3 r_{13} - \cdots \\ + i_2^2 r_{22} &- 2i_2 i_3 r_{23} - \cdots \\ &+ i_3^2 r_{33} - \cdots \end{aligned}$$

$$p = e_1^2 g_{11} - 2e_1e_2g_{12} - 2e_1e_3g_{13} - \cdots$$
$$+ e_2^2 g_{22} - 2e_2e_3g_{23} - \cdots$$
$$+ e_3^2 g_{33} - \cdots$$

Power in equivalent two-terminal circuits

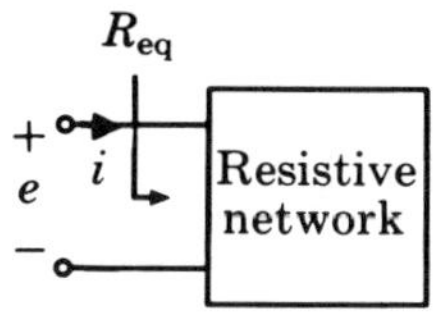

$$p = i^2 R_{\text{eq}} = \frac{e^2}{R_{\text{eq}}} \qquad \text{(power absorbed)}$$

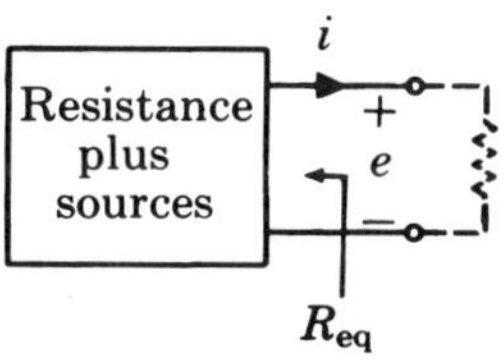

$$p = e \times i = E_{\text{oc}} \times i - i^2 R_{\text{eq}} \qquad \text{(power delivered)}$$
$$= e \times I_{\text{sc}} - \frac{e^2}{R_{\text{eq}}}$$

Power in equivalent two-terminal-pair circuits (r parameters and g parameters)

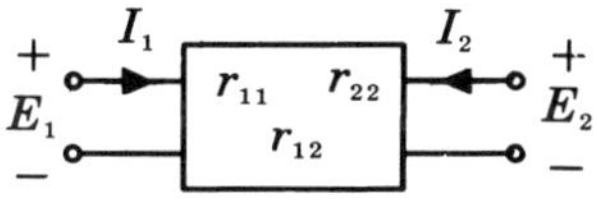

$$p = r_{11}I_1^2 + 2r_{12}I_1I_2 + r_{22}I_2^2$$

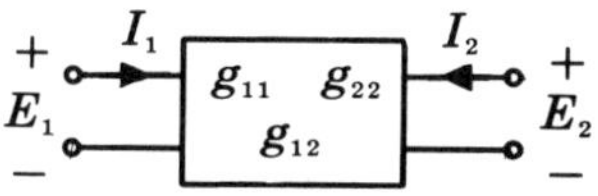

$$p = g_{11}E_1^2 + 2g_{12}E_1E_2 + g_{22}E_2^2$$

Marimum-power theorem

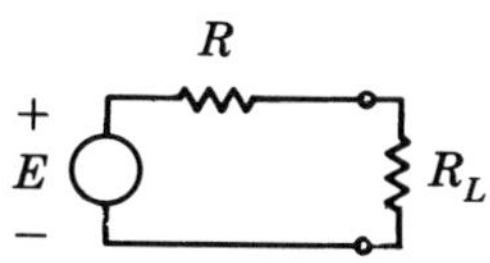

$$R_L = R, \qquad p_{\max} = \frac{E^2}{4R}$$

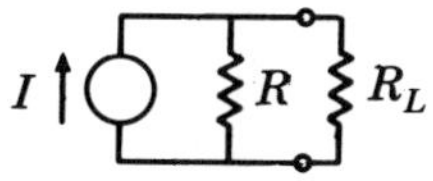

$$R_L = R, \qquad p_{\max} = \frac{I^2 R}{4}$$

Superposition of power

The superposition theorem does not apply to power except for the case of a single current source combined with a single voltage source.

REFERENCES

Most texts do not discuss power and energy until they have covered alternating-current circuits. The treatment in this chapter has been able to look at several interesting aspects of power and energy, untrammelled by the added complexities of alternating currents and energy-storage elements. References for specific topics are listed below.

1. E. A. GUILLEMIN, *Introductory Circuit Theory*, John Wiley & Sons, New York, 1953. Guillemin discusses quadratic forms and shows the equivalence of Kirchhoff's laws and the more general Lagrange equations.

2. BIRKHOFF and MACLANE, *A Survey of Modern Algebra*, Macmillan and Co., New York, 1955. A good reference for the mathematical aspects of positive definite quadratic forms.

3. GALBRAITH and SPENCE, *Fundamentals of Electrical Engineering*, The Ronald Press, New York, 1955. One of the few texts which gives a separate section on power and energy in resistive circuits.

4. H. G. BOOKER, *An Approach to Electrical Science*, McGraw-Hill, New York 1959. A completely different approach from the one used in this text. Booker views energy storage and lossless systems as fundamental and does not cover power in resistance circuits until fairly late in the book.

5. KERCHNER and CORCORAN, *Alternating Current Circuits*, John Wiley & Sons, New York, 3rd ed., 1951. This text begins with a discussion of alternating currents, average power, and root-mean-square values.

6. W. B. BOAST, *Principles of Electric and Magnetic Circuits*, Harper & Brothers, New York, 1950. Boast includes a chapter on practical sources of electric energy, including most of the common types of generators.

EXERCISES*

Section 6–1

1. Compute the current in each resistance in the circuit of Fig. 6–19, the power lost in each resistance, and the power supplied by each source.

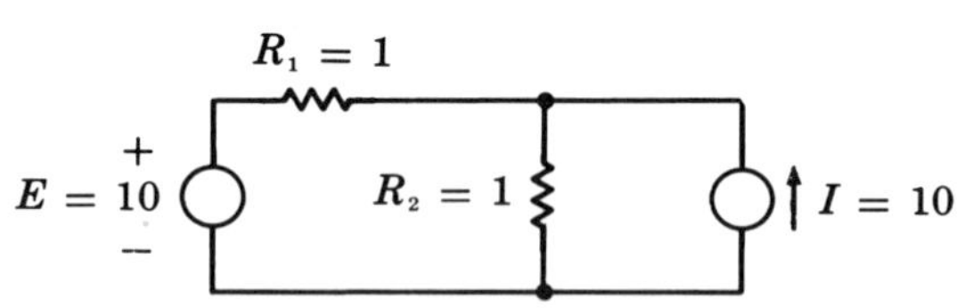

FIGURE 6–19

* Numerical values are in ohms.

2. Compute the current in each resistance in the circuit of Fig. 6–20 and, from these values, the total power dissipated in the circuit.

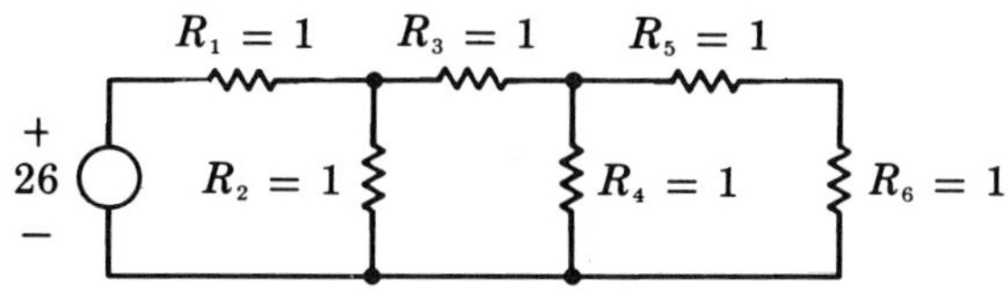

FIGURE 6–20

3. Find the current in each resistance in the circuit of Fig. 6–21 and the total power dissipated. Give also the power from the sources.

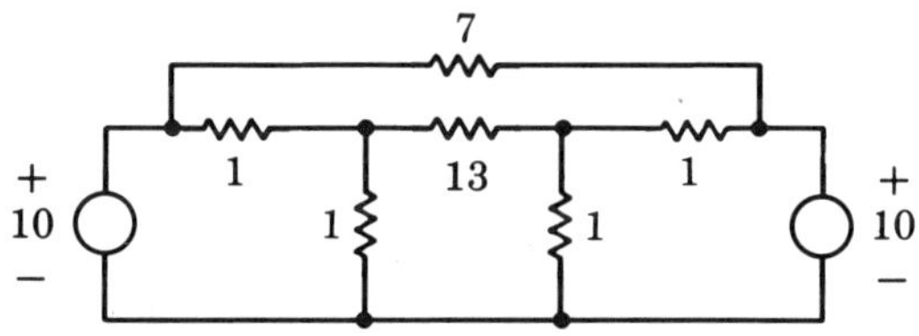

FIGURE 6–21

4. Find the power absorbed in the resistances in the circuit of Fig. 6–22. Also give that delivered by each source.

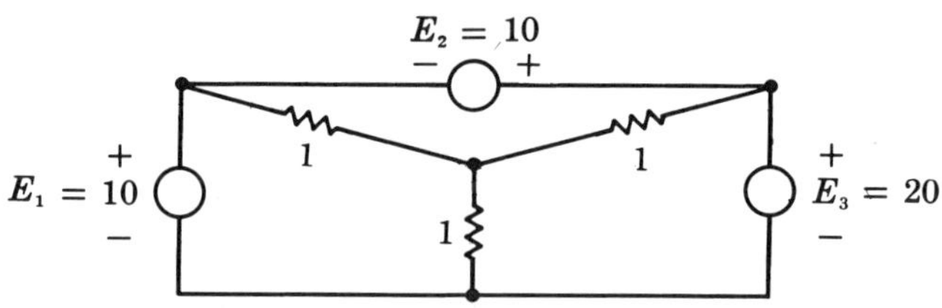

FIGURE 6–22

5. Find the current in each resistance in the circuit of Fig. 6–23 and the total power dissipated. Check the result by finding the power delivered by each source.

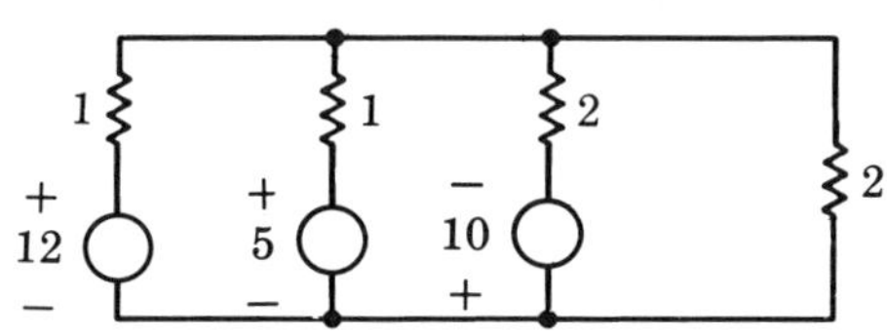

FIGURE 6–23

6. Find the current in each resistance in the circuit of Fig. 6–24 and the total power dissipated. Check the result by finding the power delivered by each source.

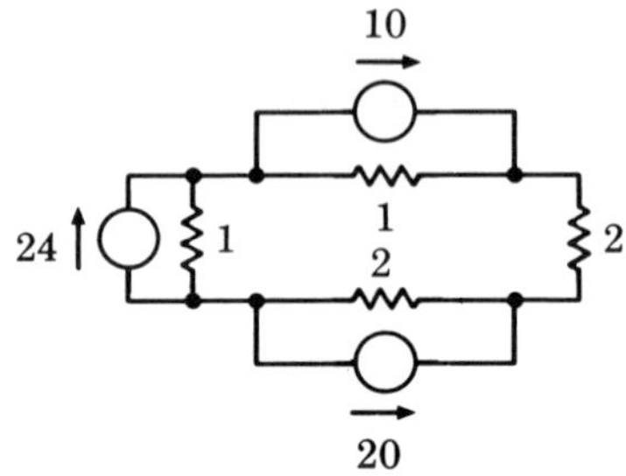

FIGURE 6–24

7. Which of the sources in the network of Fig. 6–25 are delivering power and which are absorbing power? Give the amount in each case.

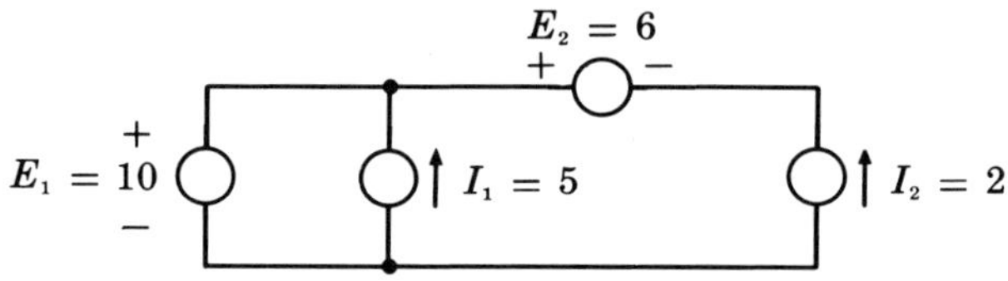

FIGURE 6–25

8. Which of the sources in the network of Fig. 6–26 are delivering power and which are absorbing power? Give the amount in each case.

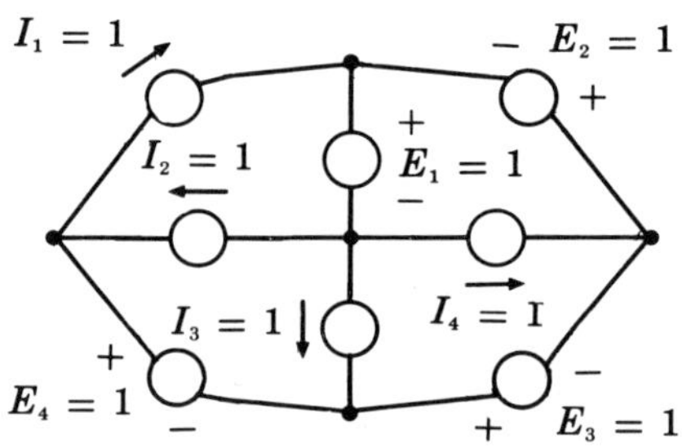

FIGURE 6–26

Section 6–2

9. A certain circuit has three meshes. The current in a clockwise direction in each mesh is 1 amp. The circuit constants are $r_{11} = r_{22} = r_{33} = 3$ ohms, and $r_{12} = r_{23} = r_{31} = 1$ ohm. Find the power which is absorbed, by means of the quadratic form.

10. Add the voltage sources which would be required in the meshes of Exercise 9, and find the power they are delivering.

11. A circuit has three nodes and a reference (ground) node. The voltages at the first two nodes have values of plus one, and the voltage at the third node is zero. The circuit constants are $g_{11} = g_{22} = g_{33} = 2$ mhos, and $g_{12} = g_{23} = g_{31} = 1$ mho. Find the power being dissipated in the circuit.

12. Add the current sources which would be required at the three nodes to give the values indicated in Exercise 11, and find the power they deliver to the circuit.

13. Solve for the mesh currents in the circuit of Fig. 6–27 and find the power dissipated, by means of the quadratic form. Verify your answer by means of the source power and the i^2R losses in the branches.

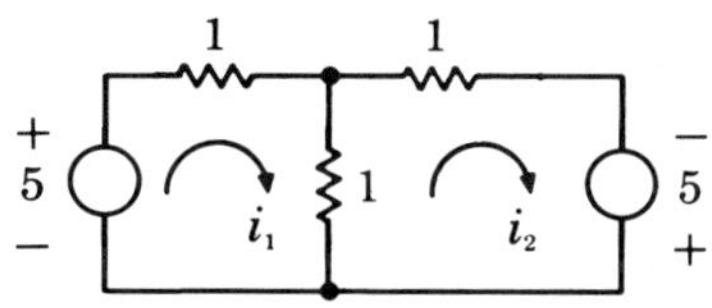

FIGURE 6–27

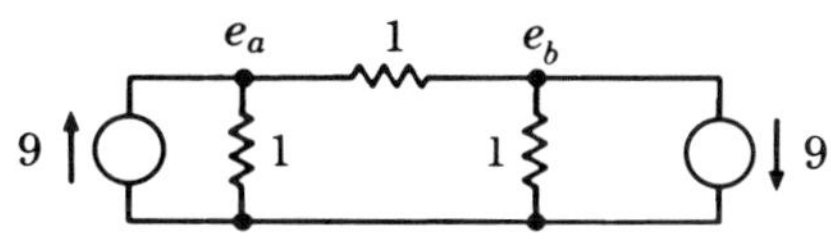

FIGURE 6–28

14. Solve for the node voltages in the circuit of Fig. 6–28 and find the power dissipated, by means of the quadratic form. Verify your answer by means of the source power and the branch losses.

15. Find the mesh currents in the circuit of Fig. 6–29 and the power dissipated by means of the quadratic form.

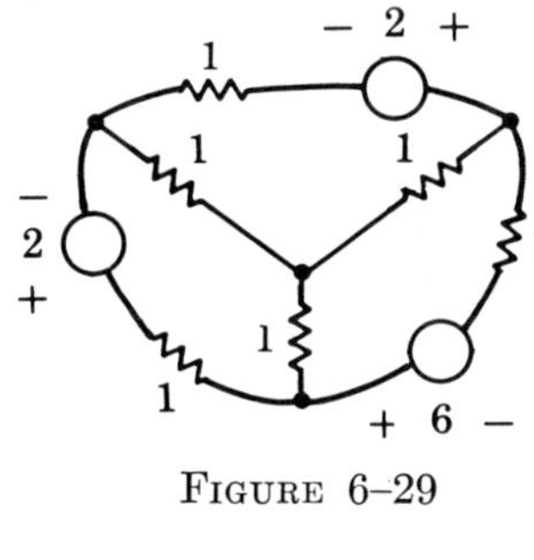

FIGURE 6–29

16. Find the node voltages in the circuit of Fig. 6–30 and the power dissipated by means of the quadratic form.

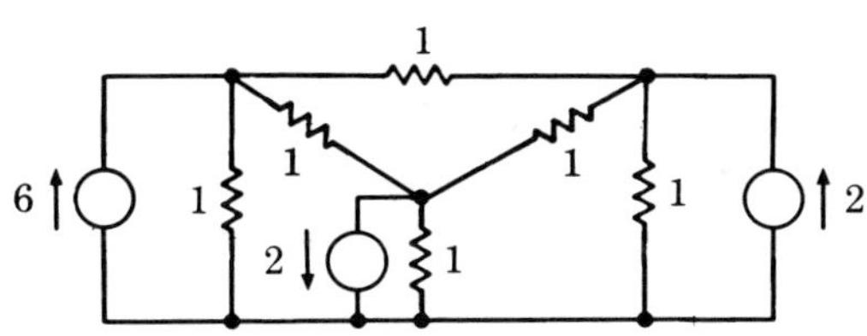

FIGURE 6–30

Section 6–3

17. Find the equivalent resistance of two resistors in series. Show that the equivalent resistance will draw exactly the same power as the series combination, when connected to either a current or a voltage source.

18. Find the equivalent resistance of two resistors in parallel. Show that the equivalent resistance will draw exactly the same power as the parallel combination, when connected to either a current or a voltage source.

19. Transform the three-resistance Δ of Fig. 6–31 to a Y. Show that the Δ and the Y draw exactly the same power when connected to the same current or voltage sources at the input and the output.

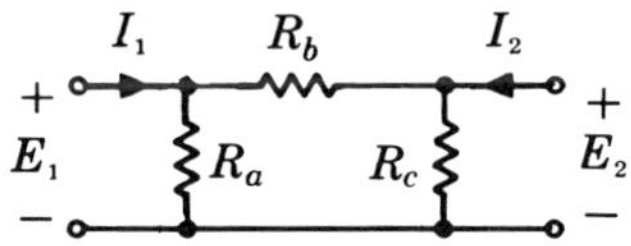

FIGURE 6–31

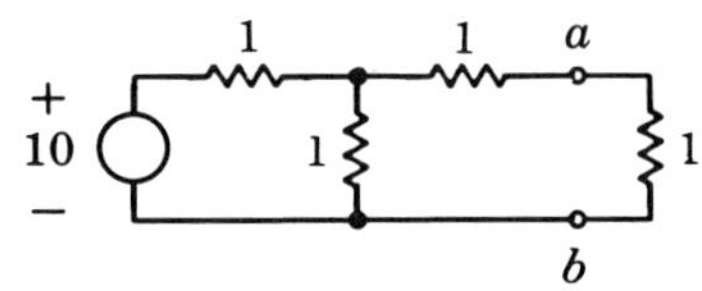

FIGURE 6–32

20. Find the Thévenin equivalent for the circuit of Fig. 6–32 at the terminals a–b. Show that the power in the load resistance external to the equivalent circuit is the same in the two cases, but that the power supplied and the power lost to the left of the terminals a–b are different.

21. Find the Norton equivalent circuit for the section of the network of Fig. 6–33 to the left of the terminals a–b. Show that the power in the resistance external to the equivalent circuit is the same in the two cases, but that the power generated and the power lost in the network to the left of the terminals a–b are different.

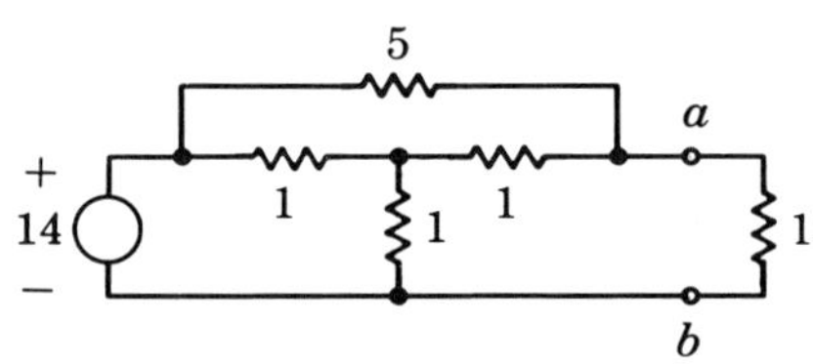

FIGURE 6–33

22. For the voltage-to-current source transformation of Fig. 6–34 find the power lost in the internal resistance R for both cases. For what value of R_L are the internal losses equal in the two cases? When are the voltage-source losses greater, and when are the current-source losses greater?

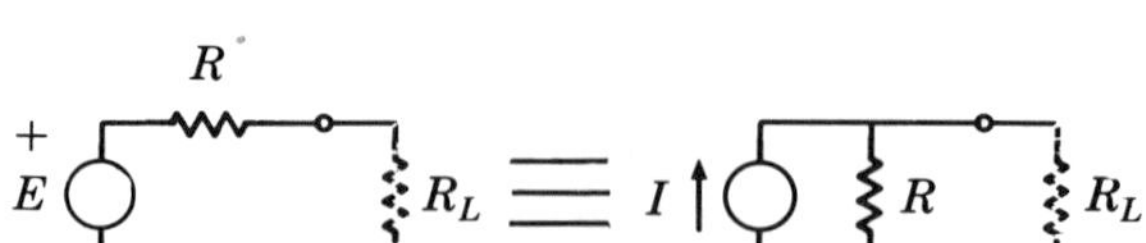

FIGURE 6–34

23. Show that the total power delivered to the network remains the same when a voltage source is "pushed through a node" in the general source transformation of Fig. 6–35.

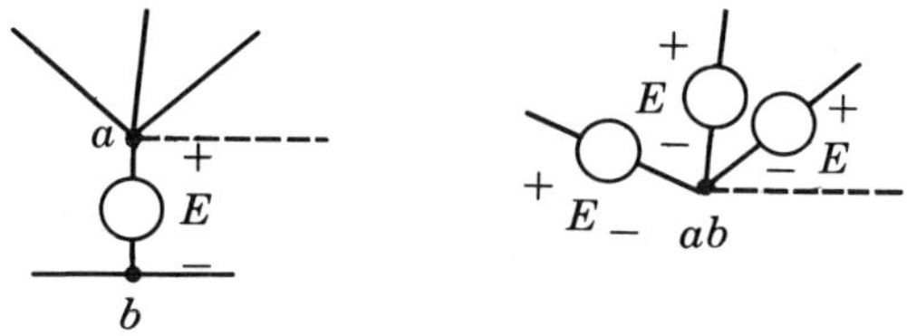

FIGURE 6–35

24. Show that the total power delivered to the network remains the same when a current source is "bridged around a loop" in the general source transformation of Fig. 6–36.

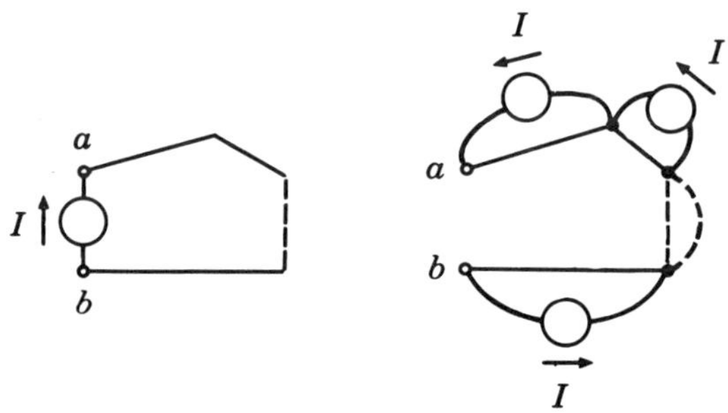

FIGURE 6–36

Section 6–4

25. Reduce the circuit of Fig. 6–37 to an equivalent π. Calculate the power dissipated in the original circuit, and in the equivalent π.

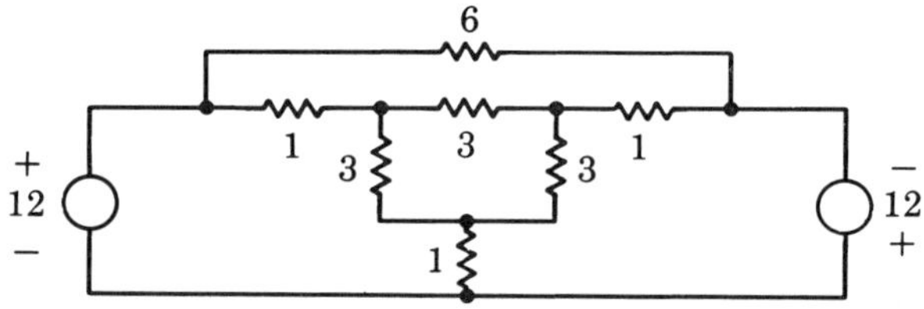

FIGURE 6–37

26. Repeat Exercise 25 with the right-hand source reversed in polarity.

27. For the circuit of Fig. 6–38 find the r parameters and the power absorbed in terms of them.

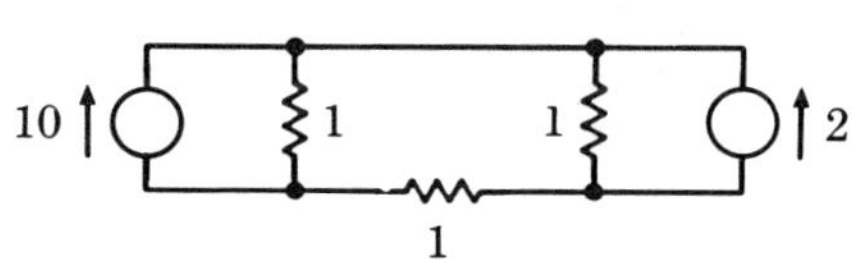

FIGURE 6–38

28. Repeat Exercise 27 with the right-hand source reversed in polarity.

29. Find the g parameters for the circuit of Fig. 6–39 and from them find the power absorbed by the network.

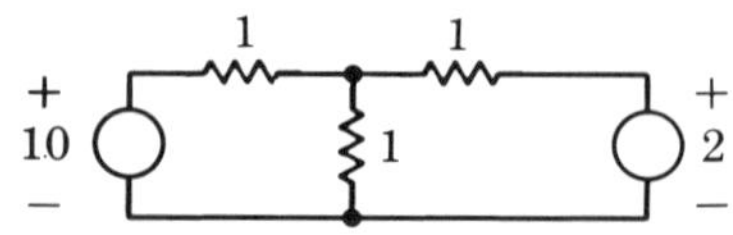

FIGURE 6–39

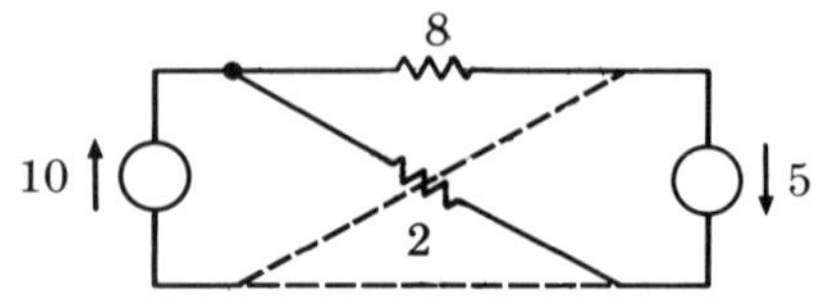

FIGURE 6–40

30. Repeat Exercise 29 with the polarity of the right-hand source reversed.

31. Find the r parameters for the circuit of Fig. 6–40 and from them obtain the power dissipated in the network.

32. Make a T circuit for the network of Fig. 6–40 and from it obtain the power dissipated. Also obtain a T circuit with positive element values, which will give the correct power dissipated.

33. Find the g parameters for the circuit of Fig. 6–41 and make an equivalent π circuit. Find the power dissipated, from the g parameters and from the π circuit.

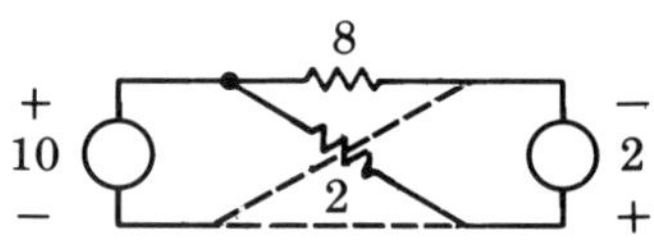

FIGURE 6–41

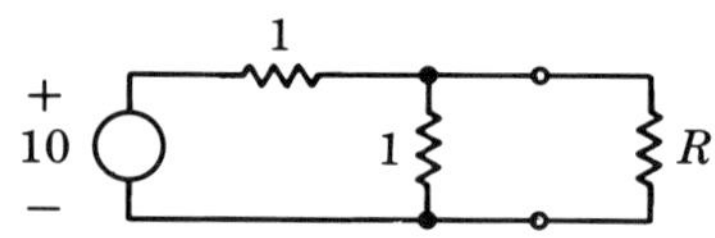

FIGURE 6–42

Section 6–5

34. For what value of the resistance R will the circuit of Fig. 6–42 absorb maximum power, and what is the value of the maximum power?

35. What value of R will receive the maximum power in the circuit of Fig. 6–43? What is the value of the maximum power?

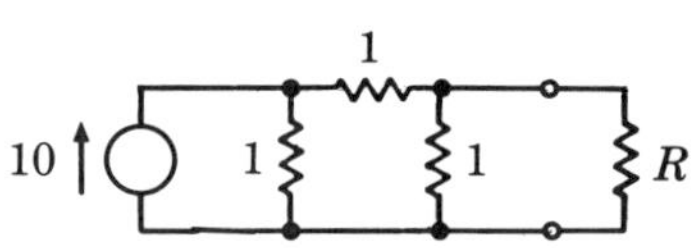

FIGURE 6–43

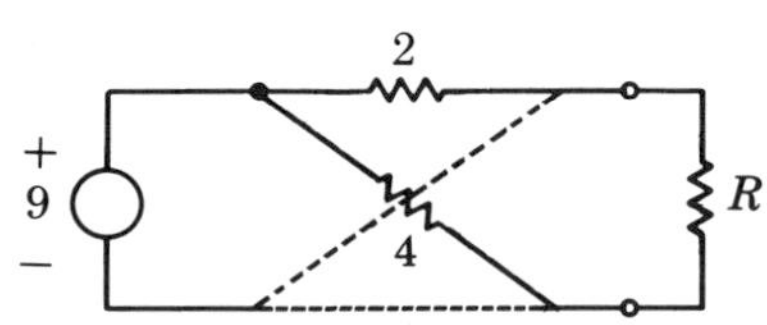

FIGURE 6–44

36. Adjust the value of R in the circuit of Fig. 6–44 to obtain maximum power. What is the value of the maximum power?

37. If the 9-volt voltage source in Fig. 6–44 is changed to a 9-amp current source, how will the answer to Exercise 36 change?

38. A two-terminal-pair network is described by the r parameters $r_{11} = r_{22} = 2$, and $r_{12} = 1$. It is driven by a 10-amp current source at the input. What resistance across the output terminals will result in a "match," and what will be the value of the maximum power obtained in the resistance?

39. If the drive at the input terminals in Exercise 38 is changed from a current source to a voltage source of 10 volts, how will the answers change?

40. A two-terminal-pair network is described by the g parameters $g_{11} = g_{22} = 2$, and $g_{12} = 1$. It is driven by a 10-volt voltage source at the input terminals. What resistance placed across the output terminals will receive maximum power, and what is the value of the maximum power?

41. If the drive at the input terminals in Exercise 40 is changed from a 10-volt voltage source to a 10-amp current source, how will the answers change?

Section 6–6

42. Compute the power delivered by source E_1 in the circuit of Fig. 6–45 with E_2 dead, and the power delivered by E_2 with E_1 dead. By computing the correct power, show that the sum of these two does not give the power absorbed in the circuit.

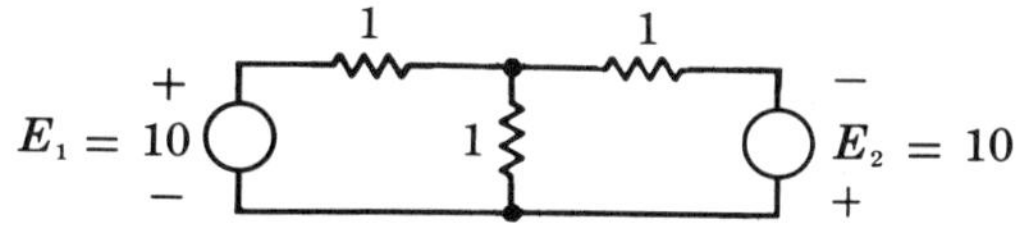

FIGURE 6–45

43. Compute the power delivered by the source I_1 with the source I_2 dead in the circuit of Fig. 6–46, and the power delivered by I_2 with the source I_1 dead. By computing the correct power absorbed in the circuit with both sources acting, show that superposition of power does not apply to this circuit.

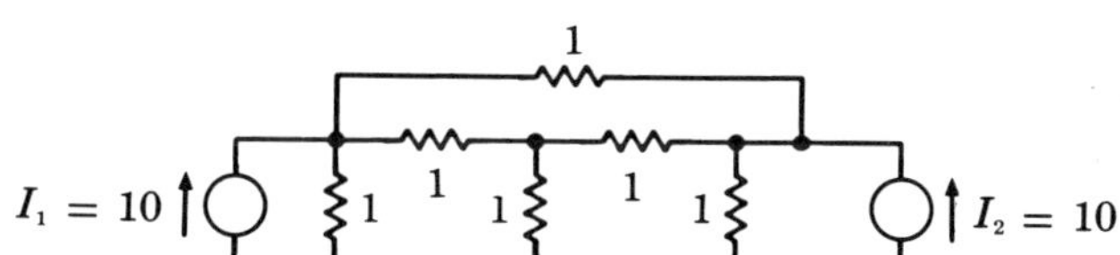

FIGURE 6–46

44. Compute the power from the voltage source in the circuit of Fig. 6–47 with the current source dead, and the power from the current source with the voltage source dead. By computing the correct power when both sources are acting, show that superposition of power does hold for this circuit.

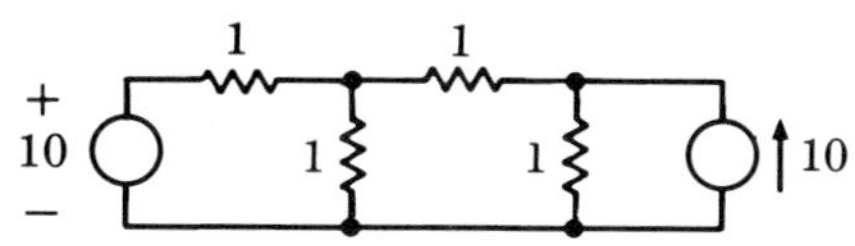

FIGURE 6–47

45. For the circuit of Fig. 6–48, plot a graph of the power delivered by the voltage source as a function of the current from the current source.

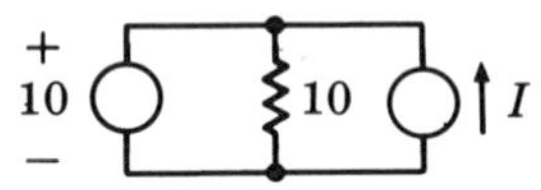

FIGURE 6–48

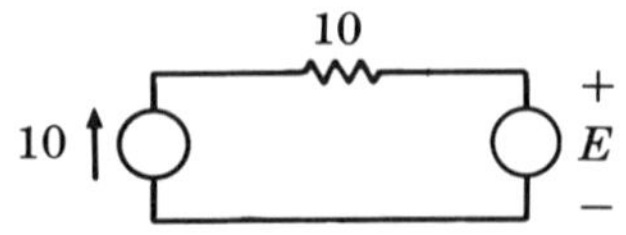

FIGURE 6–49

46. For the circuit of Fig. 6–49, plot a graph of the power delivered by the current source as a function of the voltage of the voltage source.

47. Show that the superposition of power holds for the circuit of Fig. 6–50. Examine the situation when the resistance is zero.

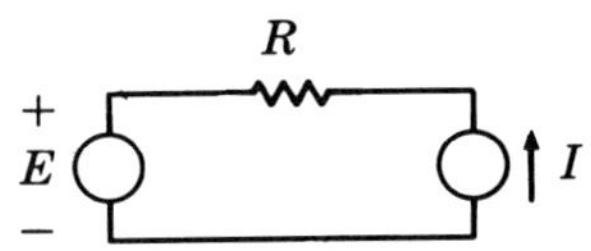

FIGURE 6–50

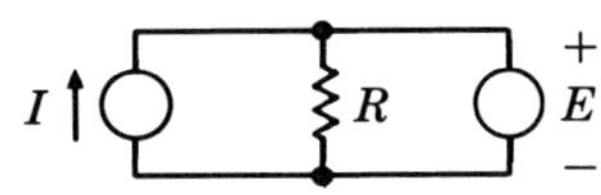

FIGURE 6–51

48. Show that the superposition of power holds for the circuit of Fig. 6–51. Examine the situation when the resistance is infinite.

CHAPTER 7

AN INTRODUCTION TO ENERGY STORAGE IN ELECTRIC CIRCUITS

"A new idea is first attacked as absurd; then it is admitted to be true, but obvious; finally it is seen to be so important that its adversaries claim that they themselves discovered it."

William James: Pragmatism—a new name for some old ways of thinking.

7–1 Introduction. The traditional approach to circuit theory has been to discuss alternating current circuits before considering the more basic transient behavior. The reversal of this order has been bitterly attacked in some quarters. Nevertheless, this reversal gives a better appreciation of the whole field of circuit theory, with its diverse applications to modern technology, and it is the approach we shall follow. Transient circuits are a natural result of the introduction of inductance and capacitance, the energy-storage elements, into our circuits. These elements differ from resistance by virtue of their volt-ampere relationships, the equations for which contain integrals and derivatives of the current and voltage. The equations for circuits containing these elements thus must also contain integrals and derivatives. They are called *differential* equations (or integro-differential equations). The solutions are not difficult to obtain but, unlike the solutions of resistive circuits, they are functions of time and not merely constants.

Most of what we have learned in the previous chapters about resistive circuits can be applied directly to the more general circuits containing energy-storage elements. The volt-ampere equations are, of course, different, but Kirchhoff's laws remain unchanged. The $2b$ equations, the mesh equations, and the node equations are written in the same way as they were for resistive circuits. A two-terminal circuit is still described by a single stimulus-response relationship and a two-terminal-pair circuit by three of them. *The stimulus-response relationships, however, are no longer constants (resistances). They are functions of time.* Any known time function can be used for the stimulus, but the most common stimuli are *step-function, impulse,* and *sinusoidal stimuli.* A *step-function* stimulus is a suddenly applied constant current or voltage.* An *impulse* is a very large pulse of

* The step function represents a single jump. Mathematicians use the term for any piecewise constant function.

current or voltage applied for a very short time. A *sinusoidal* stimulus is a current or voltage which varies periodically, like a sine function. These three functions are used because they can be generated easily in the laboratory, and because more arbitrary inputs can easily be resolved into summations of steps, impulses, or sine waves. Linearity and superposition hold for circuits with inductance and capacitance, and hence the response to an arbitrary input stimulus can be obtained as the summation of a number of the test responses. If the resolution of the input is accomplished with steps or impulses, the problem is said to be solved in the *time domain*. If it is accomplished by means of sine waves of varying frequencies, the problem is said to be solved in the *frequency domain*. For the next few chapters we will concentrate on problems solved in the time domain. They are also called *transient* problems because, in general, the time functions eventually decay to zero.

The condition of a network which is characterized by certain amounts of energy stored in various of its elements is called an *energy state of the network*. Both the total amount of energy and the distribution of it are of interest. When we change a network from one energy state to another, the change is not instantaneous. The network reacts in a characteristic way, and the change is accomplished only after a transient disturbance which may affect all the currents and voltages in the circuit. The form of this disturbance is determined by the nature of the network involved, and is called its *natural behavior* or sometimes its *transient behavior*. In essence, the network has a certain type of response which it prefers above all others, and whenever it is excited in any way it tends to produce a response of this sort. A common way of observing the natural behavior of a network is to store some energy on a capacitance or in an inductance and then to allow the network to perform its own maneuvers. The system will decay from the initial energy state to one of zero energy as the initial energy is used up, and the currents and voltages which occur will represent the natural (transient) behavior of the network.

A mechanical analogy may help to make the concept of natural behavior clearer. We can store initial energy in a pendulum, for example, by lifting it up or by striking it. In either case, the pendulum will continue to oscillate with an amplitude which gradually decays to zero as the initial energy is used up in wind friction. These damped oscillations are the natural behavior of the pendulum system.

The concept of the "natural behavior of a system" is one of the most far-reaching in all of physics, and we shall make use of it in many ways. We will study the natural behavior of a number of simple circuits, and we will be able to solve many simple problems by inspection. Later we will relate the natural behavior to the fascinating phenomena of resonance.

7-2 The volt-ampere relationships and the energy stored in electric circuits. The volt-ampere relationships of the elements were discussed in Section 1-4. The basic definitions are repeated here for convenience.

Element	*Symbol*	*Direct equation*	*Inverse equation*	
Resistance	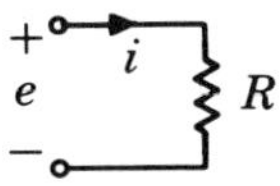	$e = iR$	$i = Ge$	(7-1)
Inductance	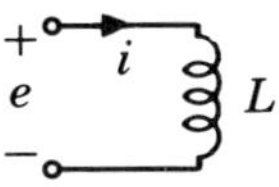	$e = L\dfrac{di}{dt}$	$i = \dfrac{1}{L}\displaystyle\int_{-\infty}^{t} e\,dt$	(7-2)
Capacitance	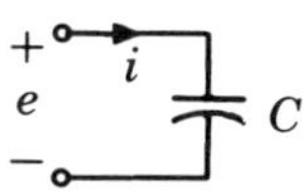	$e = \dfrac{1}{C}\displaystyle\int_{-\infty}^{t} i\,dt$	$i = C\dfrac{de}{dt}$	(7-3)

In the first six chapters we have been concerned with only the first of these three elements. In the present chapter we must extend our methods of analysis to include the energy-storage elements, inductance and capacitance. First of all, we should observe that the volt-ampere relationships of Eqs. (7-1), (7-2), and (7-3) represent constraints which the elements impose on the currents and voltages. Either the current or the voltage can be considered to be the stimulus, and the other the response. By analogy with the resistance case, the equation for the voltage in terms of the current has been called the direct equation, and the equation for the current in terms of the voltage has been called the inverse equation. Neither should be considered more basic than the other, however, and *the volt-ampere equations must be so thoroughly learned that they can be used with equal facility in either form.* The volt-ampere equations are the abc's of circuit theory; we cannot proceed further unless we are completely familiar with them.

Flux linkage and charge. The volt-ampere equations above are written in terms of the current and the voltage. These variables are basic in circuit theory and are the ones most often used. Occasionally, however, it is convenient to have a special name and a special symbol for the integral of the voltage, and another one for the integral of the current. The integral of the voltage is called the *flux linkage,* and it is represented by the symbol λ. Thus

$$\lambda = \int_{-\infty}^{t} e\,dt. \tag{7-4}$$

In order to explain the origin of the name flux linkage we note from Eq. (7–4) and Eq. (7–2) that

$$i = \frac{1}{L} \times \lambda. \tag{7–5}$$

If we take the derivative of this equation and multiply it by the inductance L, we obtain

$$L \frac{di}{dt} = \frac{d\lambda}{dt}.$$

However, from the first form of Eq. (7–2), the inductance times the derivative of the current is the voltage, and hence

$$e = L \frac{di}{dt} = \frac{d\lambda}{dt}. \tag{7–6}$$

Stated in words, Eq. (7–6) gives the voltage across an inductance as the inductance times the rate of change of the current, or as the rate of change of the flux linkage. The latter form was the one preferred by Faraday, who viewed flux linkages as elastic bands circling the wires of the inductance. The current stretches the elastic bands and stores energy in them. As the current changes, the number of elastic bands which are stretched also changes, and the voltage is proportional to the rate at which these elastic bands appear. When the current decays to zero the elastic bands disappear into the wire, and the rate at which they do so gives the voltage produced. According to this view, the voltage across an inductance is proportional to the rate of change of the flux linkages. In circuit theory we do not have much occasion to use this point of view. However, it will be encountered again in the study of magnetic fields.

The integral of the current is perhaps a more familiar concept than the integral of the voltage. It is the electric charge Q. By definition,

$$Q = \int_{-\infty}^{t} i \, dt. \tag{7–7}$$

In terms of the charge Q the voltage across a capacitance can be written as

$$e = \frac{1}{C} Q. \tag{7–8}$$

In some physics texts Eq. (7–8) is taken as the definition of a capacitance, rather than Eq. (7–3), which we have used. The derivative of Eq. (7–8) is

$$C \frac{de}{dt} = \frac{dQ}{dt}. \tag{7–9}$$

Since the rate of change of the charge with respect to time is the current in coulombs per second, Eq. (7–9) is just the inverse form of Eq. (7–3).

Duality. Duality for resistive circuits was discussed in Section 3–6. Duality is a transformation in which currents and voltages are interchanged. Thus the mesh-current equations for a circuit transform into the node-voltage equations for the dual circuit.

Because of the symmetry of Eqs. (7–2) and (7–3) it is apparent that an interchange of current and voltage merely interchanges inductance and capacitance. Thus inductance and capacitance are dual quantities. The dual of a 5-henry inductance is a 5-farad capacitance, and the dual of a $\frac{1}{3}$-farad capacitance is a $\frac{1}{3}$-henry inductance. The duality transformation can be effected by writing the mesh equations of the original network (assumed to be planar), changing the currents to voltages, and interpreting the resulting equations as the node-voltage equations of the dual network. Alternatively, the dual network can be obtained by the geometrical method discussed in Chapter 3. An example of a duality transformation will now be given. The original network is shown in Fig. 7–1(a). The mesh equation for this circuit is obtained by equating the voltage drops around the circuit to the value of the exciting source. Thus

$$e_L + e_R + e_C = e(t)$$

or

$$3\frac{di}{dt} + 5i + \tfrac{1}{4}\int_{-\infty}^{t} i\,dt = e(t). \tag{7–10}$$

If voltages and currents are interchanged, Eq. (7–10) becomes

$$3\frac{de}{dt} + 5e + \tfrac{1}{4}\int_{-\infty}^{t} e\,dt = i(t). \tag{7–11}$$

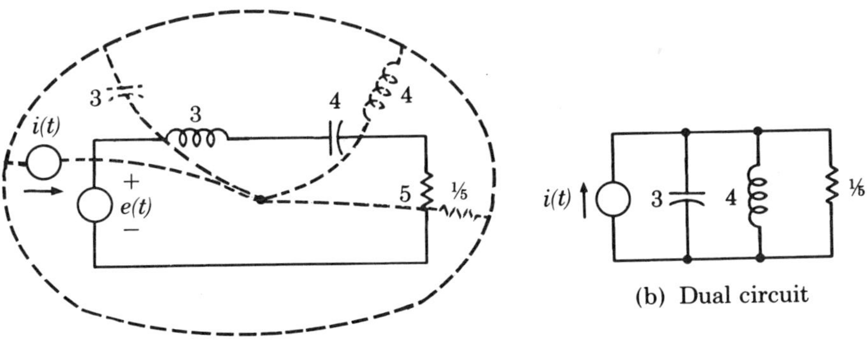

(b) Dual circuit

(a) Original circuit

FIG. 7–1. Duality in energy-storage circuits.

This equation can be interpreted as the circuit of Fig. 7–1(b), consisting of a parallel combination of $C = 3$ farads, $L = 4$ henries, and $R = \frac{1}{5}$ ohm. The broken lines in Fig. 7–1(a) represent the geometrical dual. If each broken line is replaced by the dual of the element through which it passes, the circuit of Fig. 7–1(b) is obtained. The current source is directed into the node because the voltage source tended to produce clockwise mesh current.

Energy in L and in C. Instantaneous power is given by the product of voltage and current. The total energy absorbed in a given circuit depends on the power, or the rate at which energy is absorbed, and the time for which it is absorbed. If the power is constant the energy is simply the product of the power and the time. Thus

$$W = p \times T,$$

where p = power in watts, T = time in seconds, W = energy in watt-seconds, or joules. A more usual unit of energy is the kilowatt-hour, which is the energy absorbed if a thousand watts are applied for one hour. When the power is not constant, it is necessary to sum the contributions obtained in each interval of time. If the intervals of time are made vanishingly small, the result is the integral

$$W = \int_{t_1}^{t_2} e \times i\, dt, \tag{7–12}$$

where t_1 and t_2 are the beginning and the ending of the interval over which the energy is to be measured. Usually we must account for the energy which has been stored in the inductance or the capacitance at any time in the past, and therefore t_1 must be taken as $-\infty$. Thus the energy stored in an inductance is

$$W_L = \int_{-\infty}^{t} e \times i\, dt = \int_{-\infty}^{t} L \frac{di}{dt}\, i\, dt. \tag{7–13}$$

It is possible to change the variable in Eq. (7–13) from time to current. If the current is zero at $-\infty$, and at time t has a value of I, the integral becomes

$$W_L = \int_0^I Li\, di = \tfrac{1}{2}LI^2. \tag{7–14}$$

The significant thing about Eq. (7–14) is that the energy stored in the inductance depends only on the instantaneous value of the current and not on the past history of the device, except as that past history has contributed to the current. It should be noted also that the energy in the inductance is in no way affected by the voltage across it, except as the voltage affects the current. A useful conclusion which can be drawn from Eq. (7–14) is that an inductance can be analyzed without a full knowledge of its past history. Indeed, all we need to know is the current which it is carrying at the start of our problem.

A similar situation holds for the capacitance. The energy stored in a capacitance is

$$W_C = \int_{-\infty}^{t} e \times i\,dt = \int_{-\infty}^{t} e \times C\,\frac{de}{dt}\,dt. \tag{7–15}$$

For all well-behaved functions of time, the variable in Eq. (7–15) can be changed from time to voltage. The voltage at $t = -\infty$ is assumed to be zero, and the voltage at $t = t$ is E. The integral then becomes

$$W_C = \int_0^E Ce\,de = \tfrac{1}{2}CE^2. \tag{7–16}$$

The energy in the capacitance depends only on the voltage across it at the instant in question. A principal significance of Eq. (7–16) is that it is not necessary to know the whole past behavior of the capacitance in order to be able to predict its future behavior in a given circuit. The past behavior is summed up in the energy stored in the capacitance at the beginning of the new problem. We now examine the question of initial energy storage more fully.

Initial energy theorem. The integral for the flux linkage in a circuit, as given in Eq. (7–4), has limits which run from minus infinity to the present time. The integral can be likened to the balance in your bank account. The present balance is obtained as the summation (or integration) of all the deposits and withdrawals which you have made over the history of the account. Minus infinity is the time at which the account was started. The money in the account is analogous to the flux linkages, and the present balance of the account is analogous to the net number of flux linkages present at time t. It is obviously inconvenient to deal with the complete history of the account for each new deposit or withdrawal. Instead, a new balance is obtained at some time t, usually the beginning of each month, and this balance sums up the complete past history of the account. A similar summary can be used for the inductance at some time t, usually taken to be $t = 0$. We could deal with the flux linkages, but it is more convenient in circuit problems to deal directly with the voltage and with the current. From Eq. (7–2),

$$i = \frac{1}{L}\int_{-\infty}^{t} e\,dt = \frac{1}{L}\int_{-\infty}^{0} e\,dt + \frac{1}{L}\int_{0}^{t} e\,dt. \tag{7–17}$$

The integral from minus infinity to zero in Eq. (7–17) represents the flux linkages which have been stored in the inductance from the time it was created until the reference time $t = 0$. The net number of flux linkages at this time, divided by the inductance, is the existing current. Thus the first term in Eq. (7–17) represents the initial current I_0 in the inductance at $t = 0$, and we can write

$$i = I_0 + \frac{1}{L}\int_0^t e\,dt. \tag{7–18}$$

The whole past history of the inductance is summed up in the current which it is carrying at $t = 0$, and the second term is the current which would flow in an "unfluxed" or dead inductance. In effect, *an inductance with initial current can be replaced by an inductance without initial current in parallel with a current source, switched in at the initial instant, which has a value equal to the initial current in the inductance.*

Initial energy stored in a capacitance can be treated similarly. From Eq. (7–3), the voltage across a capacitance is

$$e = \frac{1}{C}\int_{-\infty}^t i\,dt = \frac{1}{C}\int_{-\infty}^0 i\,dt + \frac{1}{C}\int_0^t i\,dt. \tag{7–19}$$

The integral from minus infinity to zero in Eq. (7–19) represents the charge which has been stored in the capacitance from the time it was created until the reference time, $t = 0$. The charge added can be either plus or minus, and hence the net charge at some given time can be either plus or minus. The net charge, divided by the capacitance, is the voltage E_0 which exists across the capacitance at $t = 0$, and hence we can write

$$e = E_0 + \frac{1}{C}\int_0^t i\,dt. \tag{7–20}$$

The whole past history of the capacitance is summed up in the value of the voltage which exists across it at the beginning of a problem. The second term represents the change in the voltage across the capacitance produced by the currents which flow during the given problem. The change is identical to the voltage which the currents would produce in an uncharged capacitance. In effect, *a capacitance with initial stored energy can be replaced by an uncharged (dead) capacitance in series with a voltage source, switched in at the initial instant, which has a value equal to the initial value of the voltage across the capacitance.*

From a philosophical point of view, integration represents "memory" in a circuit. The integral sums the contributions which it has received and "remembers" the sum total of them. An inductance "remembers" the flux linkages which have been stored in it and a capacitance "remembers" the charges which have been stored in it. A derivative, on the other hand, represents a prediction of the future. If the present value of a quantity and the rate at which it is changing are known, its value in the immediate future can be predicted. Thus, if we know the voltage across an inductance and therefore the derivative of the current through it, we can predict the value of the current at a future instant. Similarly, if we know the current which is flowing into a capacitance and hence the rate of change of the voltage across it, we can predict the voltage at a future instant. In order to predict the future values we must, of course, know the values at the beginning of the problem, and in the next section we will show that they can be easily obtained.

7–3 Computations at $t = 0$ and at $t = \infty$. Perhaps you have wondered about the effect of stray inductance and capacitance in the resistive circuits we have studied in earlier chapters. Actually, they seldom matter because after a period of time the inductance has a constant current in it and a fixed energy storage, and the capacitance has a constant voltage across it and a fixed energy storage. In fact, after a long time in any circuit the transient variations attendant on the redistribution of energy die out and the voltage across each inductance and the current through each capacitance becomes zero.* An element which has current through it but no voltage across it is indistinguishable from a short circuit, and an element which has a voltage across it but no current through it is indistinguishable from an open circuit. Thus, a long time after a switching operation (usually specified as $t = \infty$), the inductance behaves like a short circuit, and the capacitance behaves like an open circuit.

A similar argument will show that at the instant just following a switching operation (usually called $t = 0$) an R–L–C circuit reduces to a pure resistance circuit. For this argument it will be assumed that any initial energy storages have been converted to equivalent constant sources applied at the switching instant.† Thus all the inductances and capacitances can be treated as "dead" elements at the switching instant. The current in an inductance is given by

$$i_L = \frac{1}{L}\int_{-\infty}^{t} e\,dt.$$

The value from minus infinity to zero is taken care of by the constant current source in parallel with the inductance. We are interested in the voltage just after the switching operation, which we can designate as

* Idealized circuits with no energy loss must be considered as having a finite but arbitrarily small resistance.

† It is assumed here that no infinite values of voltage and current are present. Impulses will be treated in a later chapter.

time $t = 0^+$. Thus the current in the inductance is

$$i_L = \frac{1}{L}\int_0^{0^+} e\,dt. \tag{7–21}$$

The integration in Eq. (7–21) occurs over zero base line, and hence the total area under the curve is zero. (The only exception would be if the voltage were infinite in the interval.*) Thus the current through the inductance is zero. Since the voltage across it can have any finite value, the inductance is behaving like an open circuit.

Similarly, if the initial energy stored in a capacitance is replaced by a constant voltage source in series with the capacitance, the voltage across the "dead" capacitance must have the form

$$e_C = \frac{1}{C}\int_0^{0^+} i\,dt. \tag{7–22}$$

The value of the integral in Eq. (7–22) is zero because the range of the integration time is zero, and the value of the current during this time is finite. Since the voltage is zero, while the current can have any finite value, the capacitance is like a short circuit. A summary of the behavior of inductance and capacitance at $t = 0$ and at $t = \infty$ is given in Fig. 7–2. At $t = 0$ the inductance is a current source with a value equal to the initial current in the inductance; at $t = \infty$ it is a short circuit. At $t = 0$ the capacitance is a voltage source with a value equal to the initial voltage in the capacitance; at $t = \infty$ it is an open circuit.

The computation of "final" values (values at $t = \infty$) for the circuit of Fig. 7–3(a) proceeds as follows. After all the transients have died out the currents and voltages can be found from the purely resistive circuit of Fig. 7–3(b). The values are

$$i_L = 5 \text{ amp}, \qquad e_C = 5 \text{ volts}.$$

The energy stored in the two elements at $t = \infty$ can be computed from these values, but the values themselves are determined strictly by the resistive circuit.

We now compute the "initial" values (values at $t = 0$) just after a switching operation, for the circuit of Fig. 7–4(a). The switch is closed upon the direct voltage source at $t = 0$. The currents and voltages are to be found at the instant just after the switch is closed, $t = 0^+$. At this instant the inductance is an open circuit and the capacitance is a short circuit. The equivalent resistive circuit for computations at $t = 0^+$ is shown in Fig. 7–4(b). The currents and voltages computed from this circuit are

$$i_1 = 5 \text{ amp}, \qquad i_2 = 2.5 \text{ amp}, \qquad i_3 = 2.5 \text{ amp}, \qquad e_L = 5 \text{ volts}.$$

---

* See last footnote on previous page.

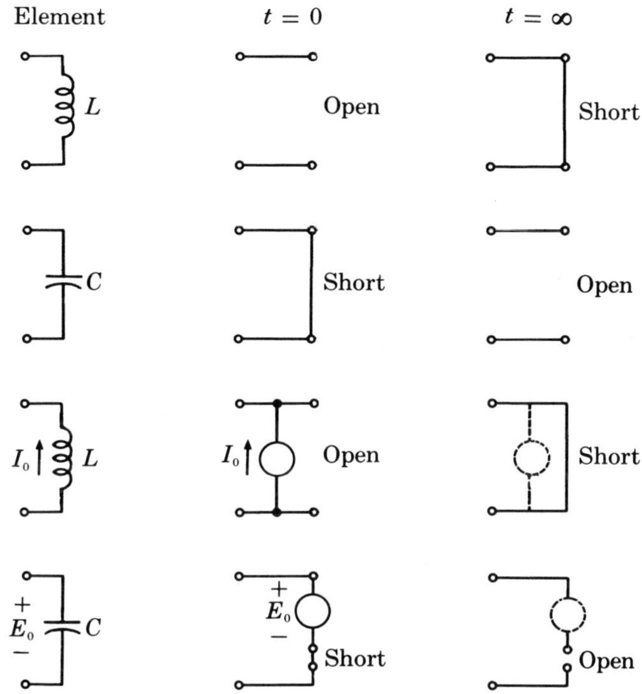

FIG. 7-2. The behavior of inductance and capacitance at $t = 0$ and $t = \infty$.

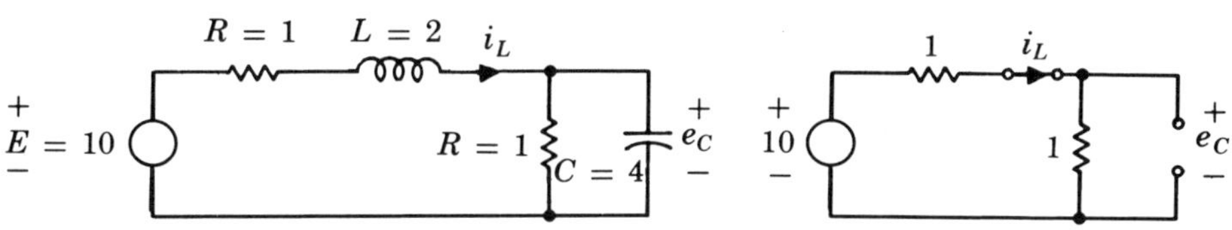

(a) The complete circuit (b) The circuit at $t = \infty$

FIG. 7-3. An example of computations at $t = \infty$

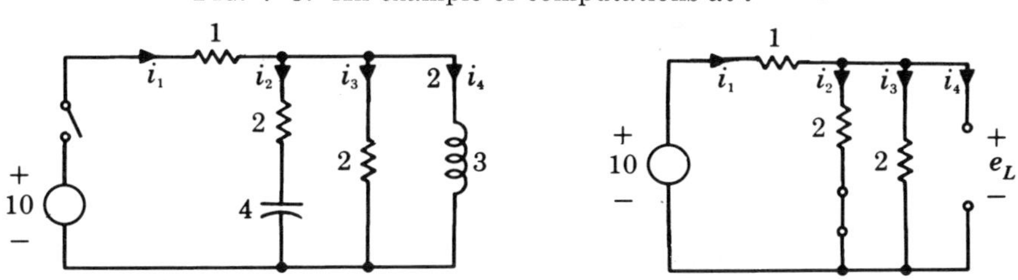

(a) Complete circuit, switch closed at $t = 0$ (b) Equivalent resistive circuit at $t = 0^+$

FIG. 7-4. An example of computations at $t = 0^+$.

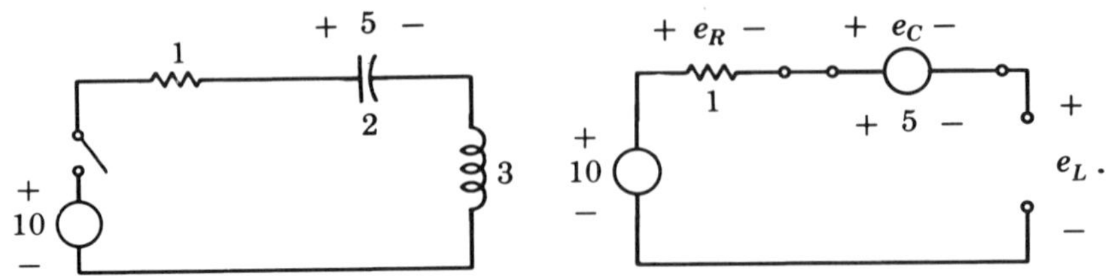

(a) Circuit with initial energy storage, switch closed at $t = 0$

(b) Equivalent circuit for $t = 0^+$

FIG. 7–5. Computations at $t = 0^+$ for a circuit with initial energy storage.

A second example of computations at $t = 0^+$ is shown in Fig. 7–5(a). The initial energy stored in the capacitance is replaced by a constant voltage source. The equivalent resistive circuit at $t = 0^+$ is shown in Fig. 7–5(b). Because the inductance is an open circuit, the current is

$$i(0^+) = 0.$$

Although the current is zero, its derivative is not zero. By Kirchhoff's law, the voltage across the inductance is

$$e_L = 5 \text{ volts},$$

and hence

$$\frac{di}{dt} = \frac{1}{L} e_L = \frac{5}{3} \text{ volts}.$$

(It is essential to know the volt-ampere equations well enough to use them with facility in the reverse direction, as we have done here.) We do not have to stop with the first derivative. By using Kirchhoff's laws and the volt-ampere equations of the elements, and by differentiating them as many times as necessary, we can find as many derivatives in succession as we wish.* Here we shall pursue the matter only as far as the second derivative of the current. In practice it is seldom necessary to go any further than this.

By Kirchhoff's law, for the circuit of Fig. 7–5(a)

$$e_L + e_R + e_C = 10$$

or

$$L \frac{di}{dt} + Ri + \frac{1}{C} \int_{-\infty}^{t} i \, dt = 10.$$

---

* An expression for the unknown current or voltage can then be written, of course, in terms of Maclaurin's series.

The integral from $-\infty$ to 0 is the initial voltage of 5 volts on the capacitance, and hence we can write

$$L\frac{di}{dt} + Ri + \frac{1}{C}\int_0^t i\,dt = 5 \text{ volts.}$$

This result could perhaps have been obtained more directly from the equivalent circuit of Fig. 7–5(b). To obtain a second derivative we must differentiate both sides of this equation. In order to avoid any questions which might arise right at the switching instant, we will perform the differentiation at $t = 0^+$. Since the upper limit of the integral is a variable, the derivative of the integral is the integrand. After $t = 0$ the applied voltages are constant and the derivative of the right side of the equation is zero. Thus we obtain

$$L\frac{d^2i}{dt^2} + R\frac{di}{dt} + \frac{i}{C} = 0,$$

and the value of the second derivative at $t = 0$ is

$$\begin{aligned}\frac{d^2i}{dt^2} &= \frac{1}{L}\left(-R\frac{di}{dt} - i\right)\\ &= \tfrac{1}{3}(-1 \times \tfrac{5}{3} - 0)\\ &= -\tfrac{5}{9}\text{ amp/sec/sec.}\end{aligned}$$

At $t = 0^+$ and at $t = \infty$ the most complicated circuits involving resistance, inductance, and capacitance reduce to pure resistance circuits and can be solved with relative ease by means of the methods we developed for direct-current circuits. Most of the methods for solving general circuit problems make use of this fact. For the direct solution of differential equations the initial values of the variable and its derivatives must be known. For the method of matching the known characteristic behavior of the network to the boundary conditions at $t = 0$ and at $t = \infty$, the initial and final values are, of course, essential. Even for the transform methods considered in the latter part of this book, the initial conditions must be specified.

7–4 The heritage from resistive networks. Most, but not all, of the material on resistive networks we have covered in the preceding chapters can be applied directly to general R–L–C networks. It is important, however, to know what is applicable and what is not. In this section we shall consider the concepts which can be applied in general, whether the networks have energy storage or not, and in the next section we will extend some of the less general concepts to the energy storage case.

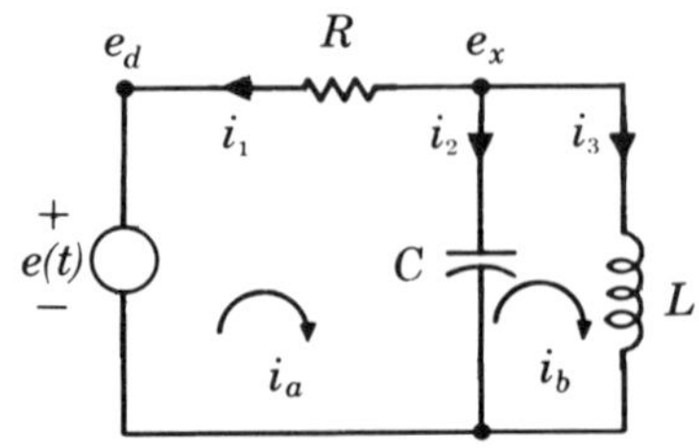

FIG. 7–6. A circuit with resistance, inductance, and capacitance.

The 2b equations. All of the topological considerations of branches, nodes, and loops are applicable to R–L–C circuits. Kirchhoff's laws are unchanged. Only the volt-ampere equations change. Thus the $2b$ equations can be written in the same fashion as they were for resistive networks. The result, however, is a set of simultaneous differential equations, instead of the algebraic equations obtained previously. We will not be concerned for the moment with the solution of these equations, but only with their formulation. The general solution of linear differential equations will be discussed in the next chapter, and particular cases will be elaborated in subsequent chapters. As an example of the formulation of the $2b$ equations, consider the network of Fig. 7–6. In this network there are three branches, two meshes, and two nodes. There are thus six $2b$ equations:

Volt-ampere equations:

$$e_1 = i_1 R + e(t), \qquad e_2 = \frac{1}{C}\int_{-\infty}^{t} i_2\,dt, \qquad e_3 = L\,\frac{di_3}{dt}.$$

Current law equation:

$$i_1 + i_2 + i_3 = 0.$$

Voltage law equations:

$$e_1 = e_2 = e_3.$$

We will make no attempt to solve the equations at this point, nor to simplify them by substitution. It is well to remember, however, that any method of writing network equations must explicitly or implicitly satisfy all of these equations.

The mesh equations. Each mesh equation represents the sum of the voltages around a mesh, produced by all the mesh currents in the circuit and by the sources in the mesh. As an example, the mesh equations for the circuit of Fig. 7–6 will be given. Two clockwise mesh currents, i_a and i_b, are assumed in the two meshes. The currents must satisfy the following equations:

$$Ri_a + \frac{1}{C}\int_{-\infty}^{t} i_a\,dt - \frac{1}{C}\int_{-\infty}^{t} i_b\,dt = e(t), \tag{7-23}$$

$$-\frac{1}{C}\int_{-\infty}^{t} i_b\,dt + L\,\frac{di_b}{dt} + \frac{1}{C}\int_{-\infty}^{t} i_b\,dt = 0. \tag{7-24}$$

The first two terms in Eq. (7–23) represent the voltage drops produced in mesh 1 by the current i_a flowing around it. The third term represents the voltage drop produced in mesh 1 by the current i_b flowing in mesh 2. This term is negative, since the voltage has opposite polarity from the voltage drops produced by the current i_a. The voltage on the right side of the equation is positive, since it tends to make current flow in the assumed direction. The first term in Eq. (7–24) is the voltage produced in mesh 2 by the current i_a in mesh 1. It is negative because it is in the opposite direction from the voltages produced by i_b in its own mesh. The second and third terms of the equation are the voltage drops produced by the mesh current i_b in its own mesh. They are always taken to be positive. There is no voltage on the right side of this equation, since there is no source in this mesh. It is possible, of course to write these equations in terms of the branch current $i_a - i_b$ which flows downward through the capacitance, but this method is not recommended. It is better to learn to write them directly in terms of the mesh currents.

The mesh equations lack some of the symmetry of those for the purely resistive case, but not all of it. The self terms are still positive and the mutual terms are negative. If current sources exist in the mesh equations, it is recommended that they be handled by means of dummy variables, rather than by source transformations. Source transformations in the differential equations are possible, but they are rather awkward.

The node equations. Node equations represent the application of Kirchhoff's current law to each node. As an example, the node equations will be written for the circuit of Fig. 7–6. There is one real node, e_x, in this circuit, and one dummy node, e_d. A summation of the currents flowing away from the node e_x gives

$$C\,\frac{de_x}{dt} + \frac{1}{L}\int_{-\infty}^{t} e_x\,dt + \frac{1}{R}\,e_x - \frac{1}{R}\,e_d = 0, \tag{7-25}$$

and the dummy node equation is

$$e_d = e(t). \tag{7-26}$$

The first three terms in Eq. (7–25) represent the currents flowing away from the node e_x when it has a voltage and all the other nodes are at zero voltage. The last term represents the current which flows into the node

e_x when a voltage is applied to e_d, with all other nodes at ground potential. The term on the right side of the equation is zero because there is no current source at the node.

In this particular case a source transformation could have been used, since the element associated with the source was a resistance. Actually, source transformations can be developed for the other elements, but the dummy-variable method is usually more convenient. If the volt-ampere equations of the elements have been thoroughly learned, there will be no difficulty in writing the node or the mesh equations for the most complicated R–L–C circuit. The solution of the equations is another story, to which a good deal of the remainder of this book will be devoted.

Linearity theorem. The inductance and the capacitance are "linear," just as the resistance was linear. The response at any point in a network is directly proportional to the stimulus which caused it. The principal difference between the resistive network and the R–L–C network is that in the latter the response is a function of time, whereas in the resistive network it was a constant. If the stimulus is increased, the response is increased by a proportionate amount at each instant of time after the application of the stimulus. Note that two separate experiments are implied. (1) A stimulus is applied and a response is measured as a function of time. (2) The network is allowed to return to its initial state and a larger value of the stimulus is applied. A second response is measured as a function of time. If the time origins of the two response curves are made to coincide, the second curve will appear to be an enlarged replica of the first. At each instant the value of the second response will be greater than the value of the first response by exactly the factor by which the second stimulus was greater than the first stimulus.

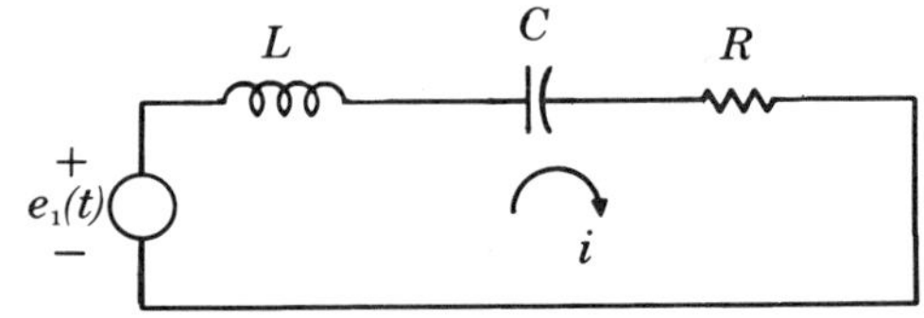

FIG. 7–7. A one-mesh linear circuit.

We shall now discuss linearity with respect to the circuit of Fig. 7–7 and extend the argument to the general case by induction. The mesh equation for the circuit of Fig. 7–7 is

$$L\frac{di}{dt} + Ri + \frac{1}{C}\int_{-\infty}^{t} i\,dt = e_1(t). \tag{7–27}$$

We will suppose that we have found the solution to this equation and that it is the function $i_1(t)$. If $i_1(t)$ is substituted into the left side of Eq. (7–27) it must produce identically the function $e_1(t)$. Thus

$$L\frac{di_1}{dt} + Ri_1 + \frac{1}{C}\int_{-\infty}^{t} i_1\,dt \equiv e_1(t). \tag{7–28}$$

Equation (7–28) is not really an equation, but an *identity*. If $i_1(t)$ is multiplied by a constant K in this expression, each term is multiplied by K, and the whole expression is equal to K times $e_1(t)$. Thus

$$L\frac{dKi_1}{dt} + RKi_1 + \frac{1}{C}\int_{-\infty}^{t} Ki_1\,dt = K\left(L\frac{di_1}{dt} + Ri_1 + \frac{1}{C}\int_{-\infty}^{t} i_1\,dt\right)$$

$$\equiv Ke_1(t). \tag{7–29}$$

Equation (7–29) can be interpreted in reverse. If a voltage $Ke_1(t)$ is applied the solution will be $Ki_1(t)$. This statement is the linearity principle for this problem. *If the stimulus is increased by a factor oj K, the response will be increased by the same factor.*

The proof of the linearity theorem can be extended to the general case by induction. For a given problem with one source applied, the left side of the mesh equation will consist of a number of terms, like the left side of Eq. (7–27). There will be a number of mesh currents involved and a large number of R's, L's, and C's. If the solutions are known, a number of terms like Eq. (7–28) can be formed. Each current can then be multiplied by K, as in Eq. (7–29). By removing the constant K from each term an expression which is K times the original response is obtained. *If the stimulus is multiplied by K the response is also multiplied by K.** This expression is essentially the linearity theorem.

Superposition theorem.† Superposition is a slightly more general form of the linearity theorem. It applies to several sources. We will discuss superposition with respect to the circuit of Fig. 7–7 and extend the argument to the general case by means of induction. The two sources, $e_1(t)$ and $e_2(t)$, are applied separately to the circuit of Fig. 7–7. We will suppose that we have found the currents produced in the two cases, and that they are $i_1(t)$ and $i_2(t)$. The following two equations must be satisfied identically:

---

* We have assumed that the R's, L's, and C's are constants. The proof can be extended to the case where they are functions of time, but not to the case where they are functions of current or voltage, where linearity does not apply.

† Both the linearity theorem and the superposition theorem are a result of the mathematical property of linearity.

$$L\frac{di_1}{dt} + Ri_1 + \frac{1}{C}\int_{-\infty}^{t} i_1\,dt \equiv e_1(t), \tag{7–30}$$

and

$$L\frac{di_2}{dt} + Ri_2 + \frac{1}{C}\int_{-\infty}^{t} i_2\,dt \equiv e_2(t). \tag{7–31}$$

The sum of these two identities is

$$L\frac{d}{dt}(i_1 + i_2) + R(i_1 + i_2) + \frac{1}{C}\int_{-\infty}^{t}(i_1 + i_2)\,dt \equiv e_1(t) + e_2(t). \tag{7–32}$$

Interpreted in reverse, Eq. (7–32) shows that when the two sources $e_1(t)$ and $e_2(t)$ are applied simultaneously, the equation is satisfied by a solution which is the sum of the two currents obtained when the sources are applied separately, with the other one equal to zero. This statement is the superposition theorem.

The general case will involve additional currents and additional terms on the left side of the equation, but the same sort of arguments apply. Furthermore, once superposition has been proved for two sources, it can be extended to any finite number by induction.

*Thévenin's theorem and Norton's theorem.** Thévenin's and Norton's theorems represent the application of superposition to the volt-ampere equation describing a general two-terminal network with sources of the form shown in Fig. 7–8(a). Since superposition is applicable to R–L–C circuits, these two theorems must also apply. The general form of Thévenin's theorem is shown in Fig. 7–8(b). The open-circuit voltage is now a function of time, and the dead network must be left in its general

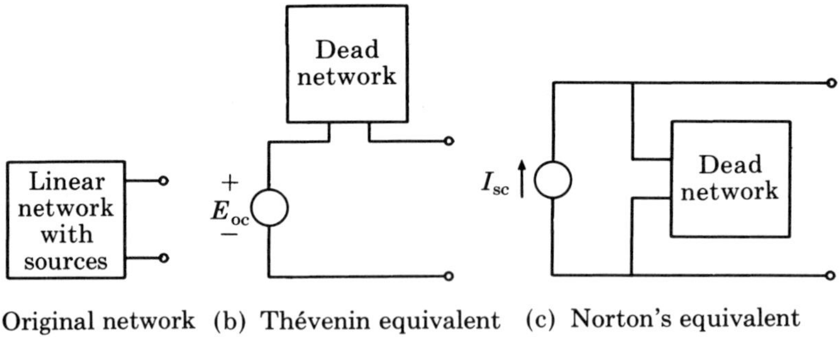

FIG. 7–8. Thévenin and Norton equivalents for general energy-storage circuits.

* Thévenin's and Norton's theorems are applications of the mathematical property of linearity.

form. It can no longer be reduced to a single element. Norton's theorem is shown in its general form in Fig. 7–8(c). The short-circuit current is now a function of time, and the dead network in parallel with it cannot be reduced to a single element.

Thévenin's and Norton's theorems do not offer much simplification for the general network problem. In fact, they may complicate it. The reason is that the network itself is not simplified by the application of the theorems, and it may be quite difficult to find the network response to a complex function of time such as the open-circuit voltage or the short-circuit current. In special cases the two theorems are useful. If all of the elements within the Thévenin or Norton branch are of the same kind, a great simplification can be effected.*

7–5 Stimulus-response relationships in energy storage circuits. We have seen that a one-terminal-pair resistive network can be reduced to a single element which we have described as the stimulus-response relationship for the network. Two-terminal-pair resistive networks can be described by three resistances, or three stimulus-response relationships. We must separate the process of finding these relationships, which is called *characterizing* the network, and the process of using them, which is called the *utilization* of the network. We have characterized resistive networks by equivalent resistances to obtain simple mathematical expressions to represent the network. This basic philosophy can be extended to energy-storage networks, but the stimulus-response relationships *become functions of time,* and have no direct representation as equivalent elements.

One-terminal-pair networks. A general one-terminal-pair network is shown in Fig. 7–9(a). If the network is resistive it can be represented or *characterized* by a single equivalent resistance which is obtained by applying a one-ampere current and measuring the resulting voltage, according to the equation

$$R_{\text{eq}} = \frac{E_1}{I_1}. \tag{7–33}$$

Another way of viewing Eq. (7–33) is to call R_{eq} the response voltage produced by a one-ampere current at the input. It is this point of view which is most easily extended to the energy storage case. As an alternative, of course, one can apply a voltage and measure the resulting current, according to the equation

$$\frac{1}{R_{\text{eq}}} = \frac{I_1}{E_1}. \tag{7–34}$$

* The two theorems are also useful in the transform methods we will discuss in later chapters.

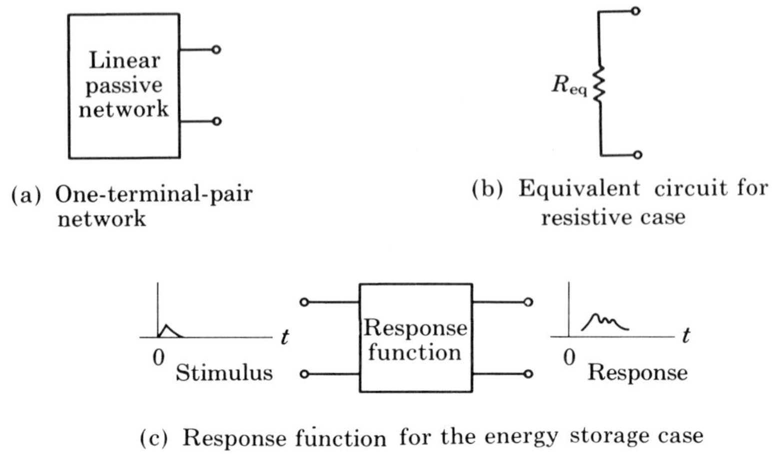

FIG. 7–9. Response functions for one-terminal-pair networks.

For the resistive case it makes no difference which of Eqs. (7–33) and (7–34) is used, but philosophically there is a difference, that shows up when there are energy storages in the circuit. In either case, however, the resistive network is *characterized* by a single equivalent resistance, as shown in Fig. 7–9(b).

The *utilization* of the equivalent resistance is so direct and simple for the resistive circuit that we have paid very little attention to it. Once the equivalent resistance has been obtained, it can be used to replace the original network and to compute the response for any arbitrary stimulus which is applied.

To find a response function that characterizes an energy-storage network, we apply a known function of time as the stimulus,* and we measure the response as a function of time. When the stimulus is a voltage the response function will be different from that obtained when the stimulus is a current. There is a relationship between the two time functions, of course, but it is not the simple reciprocal which held for the resistive case. For the moment it is perhaps better to view the two cases as separate stimulus-response relationships of the type shown in Fig. 7–9(c). It is even convenient to separate the terminals, so that both the input and output waveforms can be shown.†

We might speculate about the utility of the response functions described in the preceding paragraph. They certainly cannot be used in the way equivalent resistances are used! And yet they do contain all of the information about the network necessary to find its response for any arbi-

* Usually an impulse function.

† The one-terminal-pair responses are often called *driving-point* responses.

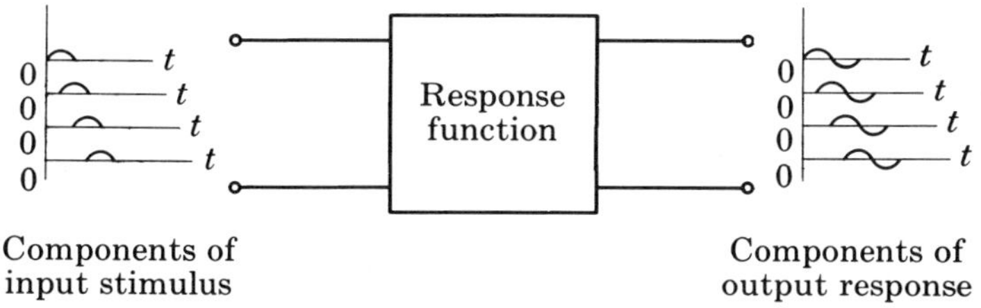

FIG. 7-10. The method of utilizing a response function for a general stimulus.

trary input stimulus. The procedure, illustrated in a general way in Fig. 7-10, depends upon superposition. The input time function is decomposed into a series of time functions that resemble the test stimulus. Each of these components has a response similar to that of the test stimulus, and their sum is the desired output.

In the process described in the preceding paragraph we encounter for the first time the idea of breaking down a complex waveform into simpler components. This process is called *signal analysis.* It will play an important role in some of the later chapters of this book.

Two-terminal-pair networks. A general two-terminal-pair network is shown in Fig. 7-11. In the resistive case it can be represented by three equivalent resistances. The three resistances are the responses obtained when constant currents or constant voltages are applied at the terminal-pairs. For current excitation the representation is

$$\begin{aligned} E_1 &= r_{11}I_1 + r_{12}I_2, \\ E_2 &= r_{12}I_1 + r_{22}I_2. \end{aligned} \tag{7-35}$$

The three stimulus-response relationships are r_{11}, r_{12}, and r_{22}. Four such relationships appeared originally, but two were equal, by the reciprocity theorem. The *characterization* of the resistive network in terms of three constants is complete. It can be *utilized* in the form of Eq. (7-35) for all subsequent computations regarding the network.

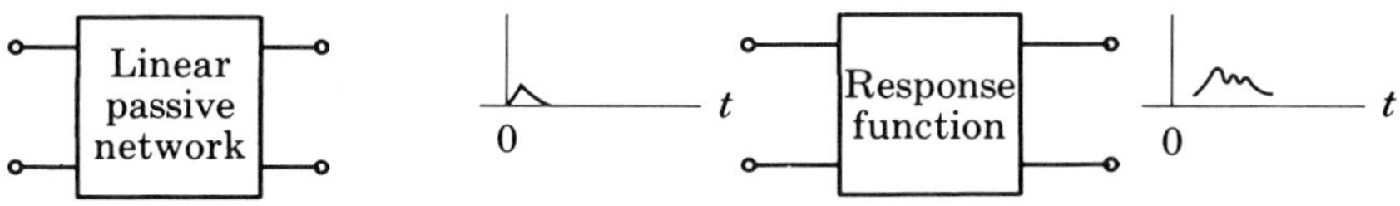

(a) Two-terminal-pair network (b) Three response functions of this form are required to specify a two-terminal-pair network

FIG. 7-11. Response functions for two-terminal-pair networks.

The characterization of an energy-storage network with two terminal-pairs is almost as easy as for a resistive network. A given time function of current is applied at the input, and the resulting time functions at the input and the output are measured. The same time function of current is applied at the output, and the input and output voltage time functions are measured. By the reciprocity theorem, two of these time functions will be the same.* Thus three time functions are sufficient to characterize a general three-terminal network. If voltage excitation is used instead of current excitation, three additional time functions are required, corresponding to the three g parameters instead of to the three r parameters. The time functions for voltage excitation are related to the time functions for current excitation, but the relation is a subtle one and we shall not pursue it at this point. It is better to view each excitation function and each response as a separate problem of the sort shown in Fig. 7–11(b).

The stimulus-response functions for an energy-storage network must be utilized one at a time, in any case. Each applied input must be resolved into components which resemble the test function, and each output must be obtained as the sum of the outputs which would be produced by the separate components. There is little advantage in considering more than one stimulus-response relationship† at a time, and for the next few chapters we shall restrict ourselves to this type of problem. Each stimulus-response relationship applies to a particular network configuration. The nature of the test source, the location of response, and the condition of the terminal-pair remote from the excitation are all part of the specification of the network problem.

7–6 Mutual inductance. Mutual inductance is an important circuit parameter. It has no influence on circuits where the currents and voltages are constant, but it can play a decisive role in circuits where the currents and voltages vary as functions of time. Historically, mutual inductance, along with self-inductance, was discovered by Michael Faraday in 1831. Faraday found that when the current in one wire was varied a voltage would appear in a second wire close to, but not touching, the first wire. He rightly attributed the effect to the magnetic field produced by the current in the first wire. In Faraday's view (Section 1–4) the magnetic field consisted of lines of force surrounding the current-carrying wire. These stretched "elastic bands" store the energy of the magnetic field. When the current is increased the elastic bands spread out from the wire

* We will not pause to prove the reciprocity theorem for energy-storage networks. We will make very little use of it for the next few chapters, and a general proof in the transform notation is trivially simple.

† The stimulus-response relationships for two-terminal-pairs are often called *transfer responses*.

like ripples in a pool of water, and when the current is decreased the elastic bands collapse back into the wire. As they do so they give up their energy to the current in the wire in the form of an induced voltage, whose magnitude is proportional to the number of lines which collapse into the wire, or cut it, in the words of Faraday. Any other wire in the magnetic field is also cut by the changing lines of force, and has a similar voltage induced in it. By showing that a physical motion of a magnetic field would cause the same sort of voltage to be produced, Faraday convinced himself that it was the motion of the lines of force that produced the voltage.

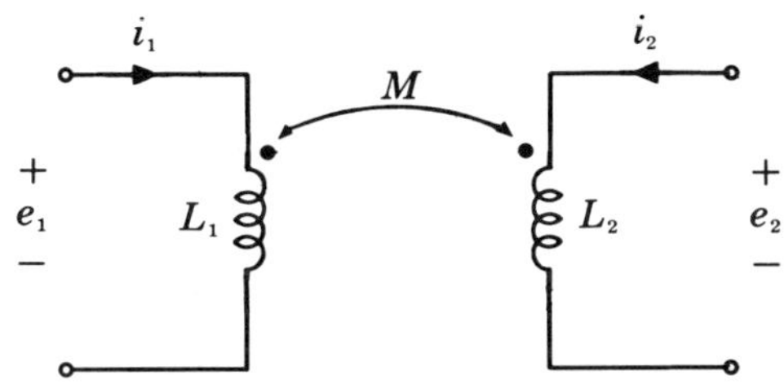

FIG. 7–12. Two coils with mutual inductance between them.

The circuit symbol for two coils with mutual inductance between them is shown in Fig. 7–12. The volt-ampere equations for the coupled coils are defined for current sources i_1 and i_2 as the stimuli, and the voltages e_1 and e_2 as the responses. These volt-ampere equations are the Ohm's law equations for the coils, and once defined they hold for any type of excitation. The current i_1 is applied first, with i_2 equal to zero. The voltage produced in each coil is proportional to the rate of change of i_1. Then the current i_2 is applied, with the current i_1 equal to zero. The voltage produced across each coil is then proportional to the rate of change of the current i_2. By the superposition theorem, the total voltage e_1 and the total voltage e_2 are given by

$$\begin{aligned} e_1 &= L_1 \frac{di_1}{dt} + M \frac{di_2}{dt}, \\ e_2 &= M \frac{di_1}{dt} + L_2 \frac{di_2}{dt}. \end{aligned} \tag{7–36}$$

In Eq. (7–36), L_1 is the constant of proportionality between the rate of change of current in coil 1 and the voltage across it. When the second coil is open-circuited, L_1 represents the normal inductance of the first coil. Similarly, L_2 is the constant of proportionality between the current derivative in the second coil and the voltage produced by it. If the first coil is open-circuited, L_2 represents the inductance of the second coil. The constant M gives the relationship between the voltage produced in the second coil by a rate of change of current in the first coil, or the voltage produced in the first coil by a rate of change of current in the second coil. By the reciprocity theorem, these two expressions are equal. The constants L_1 and L_2 are called the *self-inductances* of the coil, and the constant M is called the *mutual inductance* of the coil.

The polarity of the voltage produced by the mutual inductance poses a problem we have not previously encountered. The actual polarity of the voltage can be reversed by changing the direction of the windings in the second coil, and the assumed or reference polarities can be taken in either direction. In a diagram, coupled coils (transformers) are usually marked with dots to show the relative winding directions. When the current enters at the dot in one winding and is increasing positively, the voltage induced in the other winding is positive at the other dot. To use this dot notation, the assumed directions for the currents and voltages must also be marked on the circuit diagram. The assumed direction for the current is also the assumed direction for the positive derivative of the current. Thus if the current enters at one dot, the voltage induced at the other dot will be positive.

For the circuit shown, the current enters at the dot in the first coil. The voltage induced in the second coil is positive at the other dot, and for the assumed direction of the current it is a voltage drop with the same polarity as the $L_2(di_2/dt)$ term. Similarly, the current in the secondary coil enters at one dot and the voltage induced in the primary is positive at the other. This voltage is in the same direction as the voltage drop produced by the self-inductance, and therefore must have the same sign as the $L_1(di_1/dt)$ term. By the reciprocity theorem, the two mutual-inductance terms must have the same sign as well as the same magnitude, and it is therefore not necessary to find both of them, except as a check on the accuracy of the work.

The polarity of the mutual-inductance term is so important that an equivalent circuit which places the sign of the mutually induced voltages in evidence will be given. The assumed current enters at the dot in the primary coil of Fig. 7–13(a), and the voltage produced by this current is positive at the dot in the second coil. In fact, the mutually induced voltage appears as a generator in series with the inductance of the secondary winding. The assumed current in the secondary enters at the undotted terminal, and the voltage induced by it in the primary coil is positive at the undotted end. The equivalent voltage generator is shown in series with the self-inductance of the primary. The equivalent circuit of Fig. 7–13(b) is useful in most problems involving mutual inductance. In general, the equations should be set up using mesh currents, and not node voltages. The mutual-inductance parameters have been defined on the basis of currents, and they do not fit very well into node-voltage equations.

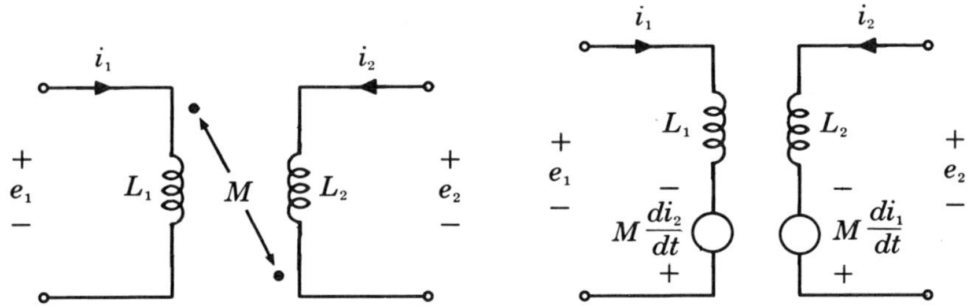

(a) Coupled coils (b) Equivalent circuit with induced-voltage-generators

FIG. 7–13. The induced-voltage-generator equivalent circuit for coupled coils.

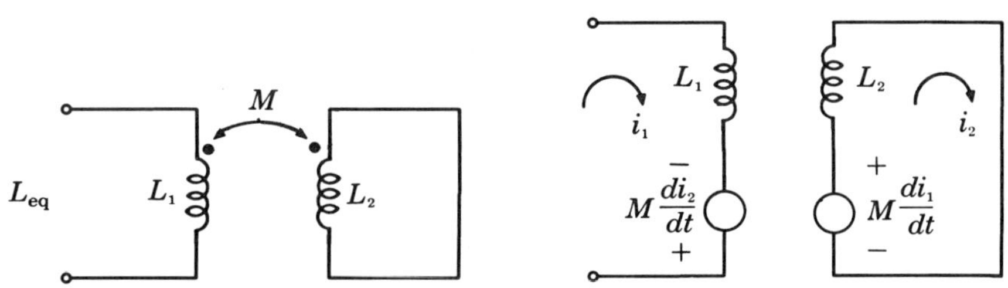

(a) Coils with secondary shorted (b) Equivalent circuit

FIG. 7–14. Two coils with the secondary short-circuited.

As an example of a problem involving mutual inductance, we will find the equivalent inductance of the two coupled coils of Fig. 7–14(a) when the secondary is short-circuited. We assume clockwise mesh currents in each mesh and draw the equivalent circuit of Fig. 7–14(b). The primary current enters at the dotted end of the coil and therefore the induced voltage in the secondary is positive at the dotted end. The secondary current enters at the undotted end and the induced voltage in the primary is positive at the undotted end. The mesh equations are

$$
\begin{aligned}
L_1 \frac{di_1}{dt} - M \frac{di_2}{dt} &= e_1, \\
-M \frac{di_1}{dt} + L_2 \frac{di_2}{dt} &= 0.
\end{aligned}
\tag{7-37}
$$

Although these equations are differential equations, they can be solved easily for the derivatives of the currents. Thus

$$\begin{aligned}\frac{di_1}{dt} &= \frac{L_2}{L_1L_2 - M^2}\,e_1,\\ \frac{di_2}{dt} &= \frac{-M}{L_1L_2 - M^2}\,e_1.\end{aligned} \tag{7–38}$$

The equivalent input inductance is the ratio of the input voltage to the derivative of the input current:

$$L_{\text{eq}} = \frac{e_1}{di_1/dt} = \frac{L_1L_2 - M^2}{L_2}. \tag{7–39}$$

Coefficient of coupling. It has probably already become apparent that there are some limitations on the size of the mutual inductance relative to the self-inductances of the coils which it couples. Clearly, there is a maximum amount of "coupling" which can occur between the two coils when all the flux produced by either of them links the other as well. Two coils wound on an iron core come very close to achieving this result. We can find the limitation on the mutual inductance by computing the energy stored in the circuit of Fig. 7–14.* The input power is

$$p = e \times i = L_{\text{eq}}\frac{di_1}{dt} \times i_1 \tag{7–40}$$

and the energy stored in a time t is

$$W = \int_0^t e_1 \times i_1\,dt = L_{\text{eq}}\int_0^t i_1\frac{di_1}{dt}\,dt = \tfrac{1}{2}L_{\text{eq}}i_1^2. \tag{7–41}$$

The energy given by Eq. (7–41) must be positive, since if it were negative the inductance could act as a source of energy indefinitely. Physically, it is a passive element and cannot do this. Therefore L_{eq} must be positive. Hence, from Eq. (7–39),

$$L_1L_2 - M^2 \geq 0, \qquad \text{or} \qquad M \leq \sqrt{L_1L_2}. \tag{7–42}$$

The maximum value of M occurs when it is equal to $\sqrt{L_1L_2}$. The minimum value is zero when the coils are very far apart. The degree of coupling of the coils can be expressed by the ratio of M to its maximum value, which is called the coefficient of coupling, K, given by

$$K = \frac{M}{\sqrt{L_1L_2}}. \tag{7–43}$$

* A similar restriction holds for a series connection of coils, but it is less restrictive.

The maximum value of K is unity when the coils are coupled so closely that all the flux is common to both of them. The minimum value of K is zero when the coils are coupled so loosely that there is no flux common to both of them.

This chapter has been written to bridge the gap between purely resistive circuits and the more general energy-storage circuits which will be considered in the remainder of the book. The volt-ampere relationships of the energy-storage elements have been examined in detail, and the dual nature of inductance and capacitance has been noted. It has been shown that at $t = 0$ and at $t = \infty$ the general circuits reduce to purely resistive forms which can be solved by the methods of the preceding chapters. So far as formulating the equations for general networks is concerned, the $2b$ method, the mesh equations, and the node equations can be carried over directly from resistive circuits. The differential equations which are obtained are more difficult to solve than the algebraic equations of resistive networks. In Chapter 9 we shall consider the problem of solving these differential equations.

A principal difference between resistive circuits and energy-storage circuits is that the response functions of one-terminal-pair and two-terminal-pair resistive networks are constants, while the response functions for energy-storage networks are functions of time. After we have considered differential equations in general, we will obtain the response functions of a number of common networks.

Another difference between resistive and energy-storage circuits is the presence of mutual inductance in the latter. Mutual inductance is essentially a two-terminal-pair quantity and we have delayed its introduction until this chapter because we did not need it in resistive circuit theory.

7–7 Summary.

The volt-ampere equations

Resistance 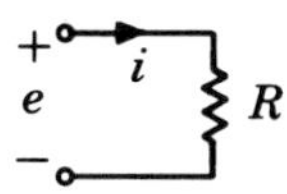

$e = Ri$ $\qquad i = Ge$

Inductance 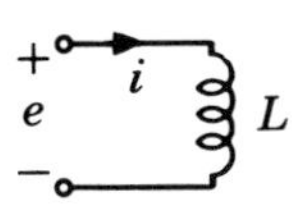

$$e = L\frac{di}{dt} \qquad i = \frac{1}{L}\int_{-\infty}^{t} e\,dt$$

Capacitance 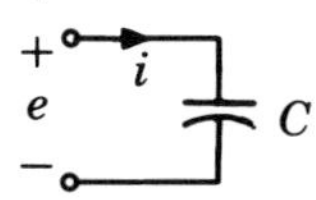

$$e = \frac{1}{C}\int_{-\infty}^{t} i\,dt \qquad i = C\frac{de}{dt}$$

Computations at $t = 0$ and $t = \infty$

Element	$t = 0^+$	$t = \infty$

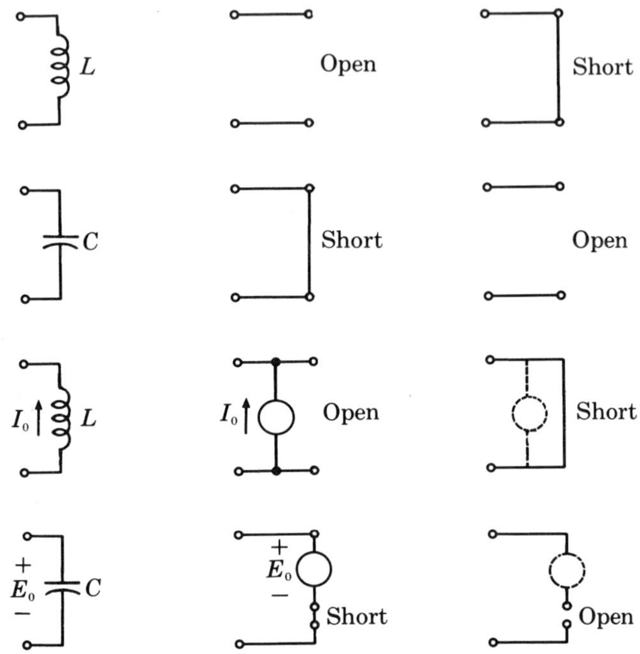

Mesh equation

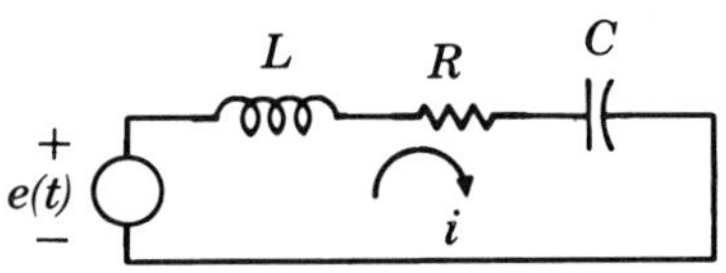

$$L\frac{di}{dt} + Ri + \frac{1}{C}\int_{-\infty}^{t} i\,dt = e(t)$$

Node equation

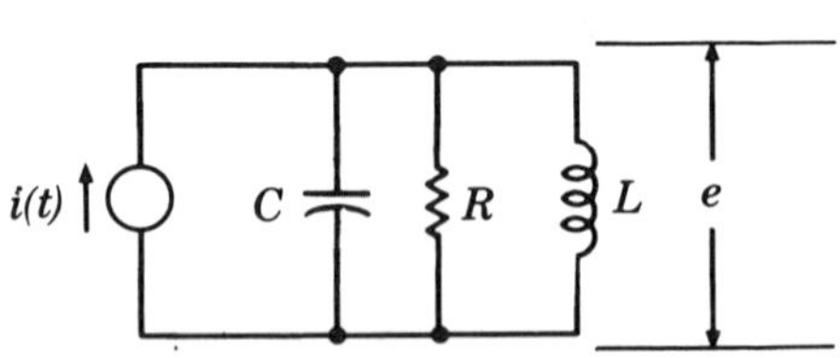

$$C\frac{de}{dt} + \frac{1}{R}e + \frac{1}{L}\int_{-\infty}^{t} e\,dt = i(t)$$

Stimulus-response relationship

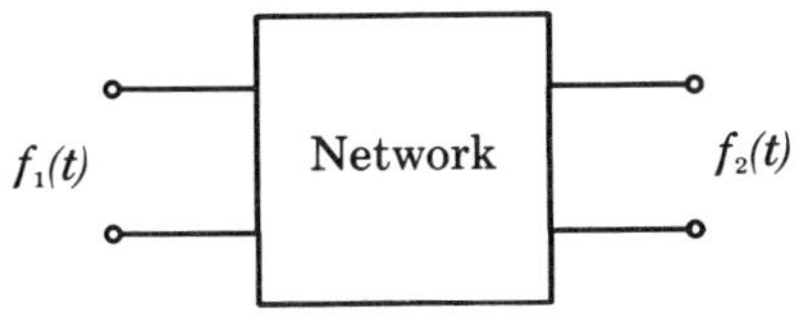

Driving point function

$f_1(t)$ and $f_2(t)$ at same terminal-pair.

Transfer function

$f_1(t)$ and $f_2(t)$ at different terminals.

Mutual inductance

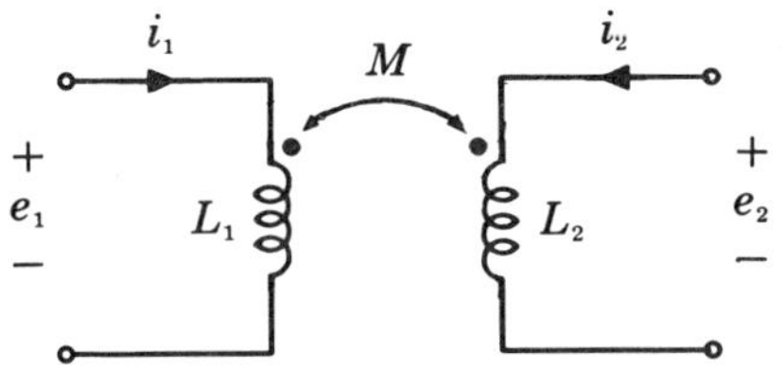

$$e_1 = L_1 \frac{di_1}{dt} + M \frac{di_2}{dt}$$

$$e_2 = M \frac{di_1}{dt} + L_2 \frac{di_2}{dt}$$

Coefficient of coupling

$$K = \frac{M}{\sqrt{L_1 L_2}} \leq 1$$

REFERENCES

The best references for the formulation of the basic differential equations of electric circuits are the books which present the theory of Laplace transforms. Mutual inductance is treated in elementary textbooks, but usually only for the steady-state sinusoidal case. Specific references are listed below.

1. D. K. Cheng, *Analysis of Linear Systems,* Addison-Wesley, Reading, Mass., 1959. Cheng gives a brief but excellent treatment of the formulation of the differential equations for circuits, the evaluation of initial conditions, and the general properties of the differential equations.

2. E. Weber, *Linear Transient Analysis,* John Wiley & Sons, New York, 1954. This text covers essentially the same material as does Cheng's.

3. Gardner and Barnes, *Transients in Linear Systems,* John Wiley & Sons, New York, 1942. The first of the modern texts on Laplace transforms. The treatment of source transformations is especially complete.

4. M. E. Van Valkenburg, *Network Analysis,* Prentice-Hall, New Jersey, 1955. A slightly more elementary version of Laplace transforms. Includes initial conditions and mutual inductance.

5. Seshu and Balabanian, *Linear Network Analysis,* John Wiley & Sons, New York, 1959. This text is one of the few to discuss initial conditions in the presence of impulses, a topic reserved for Chapter 8 here.

6. E. A. Guillemin, *Introductory Circuit Theory,* John Wiley & Sons, New York, 1953. Guillemin gives the general volt-ampere relationships of the ele-

ments and an elementary treatment of circuits based on the singularity functions, a topic reserved for Chapter 8 of this text. Guillemin treats mutual inductance by a matrix method instead of the dot convention.

7. LEWIS and GOODHEART, *Basic Electric Circuit Theory,* The Ronald Press, New York, 1957. This text gives a treatment of mutual inductance in both the differential-equation form and the steady-state form. It also proves the reciprocity theorem for mutual inductance in the transient case.

EXERCISES*

Section 7–2

1. Sketch the currents in the circuits of Fig. 7–15 as functions of time. The switches are closed at $t = 0$.

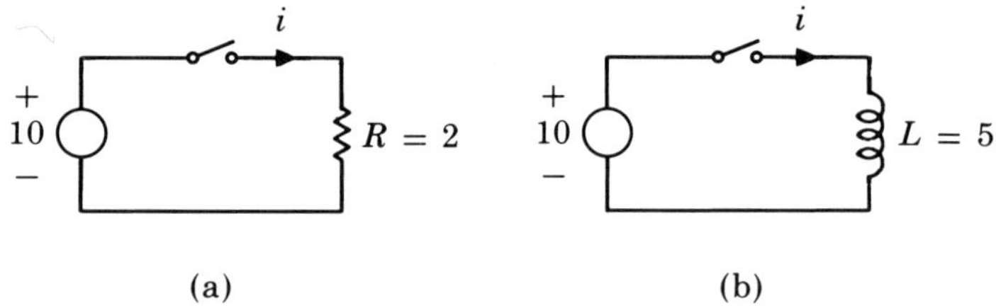

FIGURE 7–15

2. Sketch the voltages which appear across the elements in the circuits of Fig. 7–16 as functions of time. The switches are opened at $t = 0$.

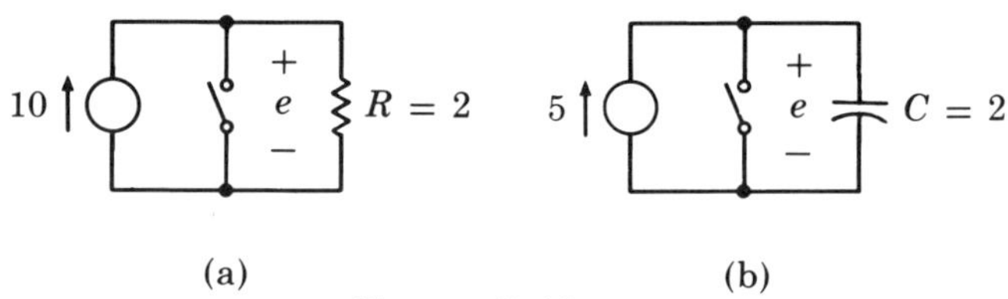

FIGURE 7–16

3. The switches in the circuits of Fig. 7–17 are closed at $t = 0$. Find the analytic expressions for the currents which flow after $t = 0$.

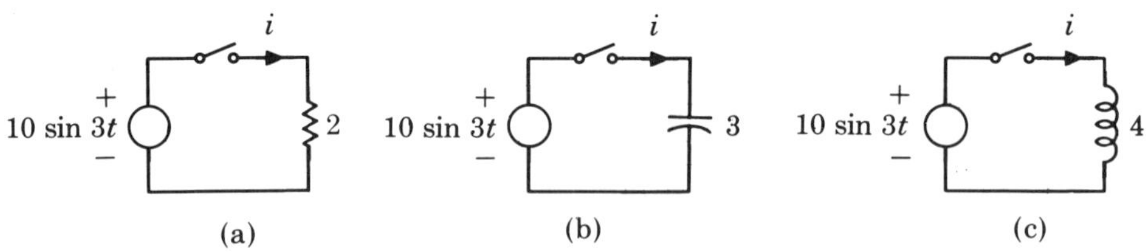

FIGURE 7–17

* Numerical values are in ohms, henrys, and farads.

4. The switches in the circuits of Fig. 7–18 are opened at $t = 0$. Find the analytic expressions for the voltages which appear across the elements for $t > 0$.

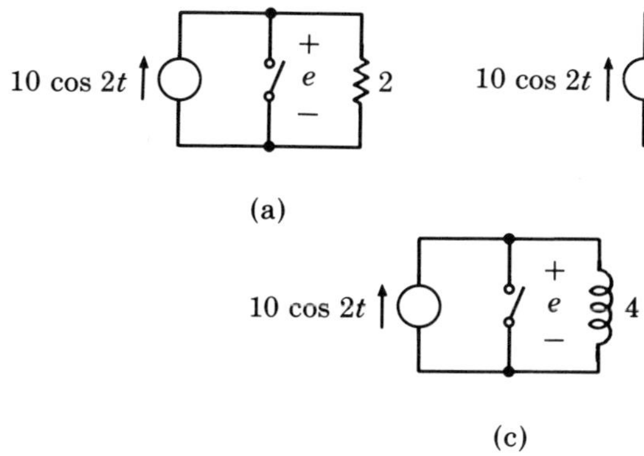

FIGURE 7–18

5. The switch in the circuit of Fig. 7–19 is closed at $t = 0$. How much charge flows from the battery to the capacitor? How long does it take?

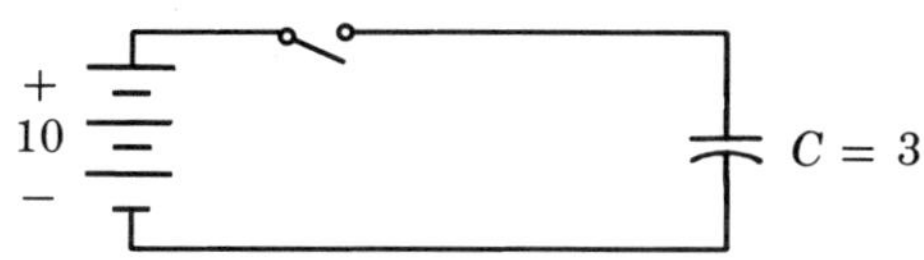

FIGURE 7–19

6. The switch in the circuit of Fig. 7–20 is opened at $t = 0$. How many flux linkages are created in the inductance? How long before they appear?

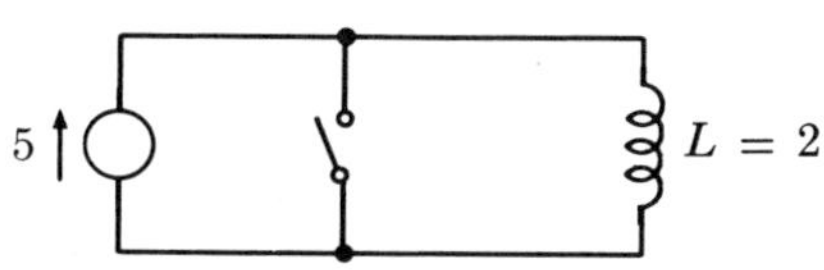

FIGURE 7–20

7. A constant current of 4 amp is fed into a capacitance of 2 farads, starting at $t = 0$. How much energy is stored after 10 sec? How much after 100 sec? (The capacitance has no leakage resistance.)

8. A constant voltage of 6 volts is connected across an inductance of 2 henries. What is the energy stored after 5 sec? What is it after 10 sec? (Assume a perfect inductance with no series resistance.)

9. Find the dual of the circuit of Fig. 7–21.

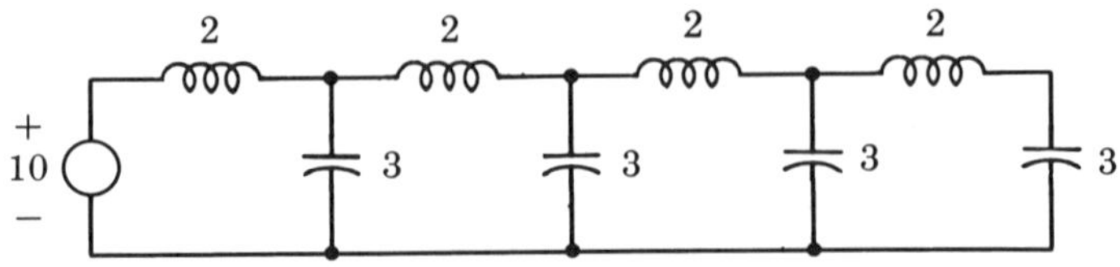

FIGURE 7–21

10. Find the dual of the circuit of Fig. 7–22.

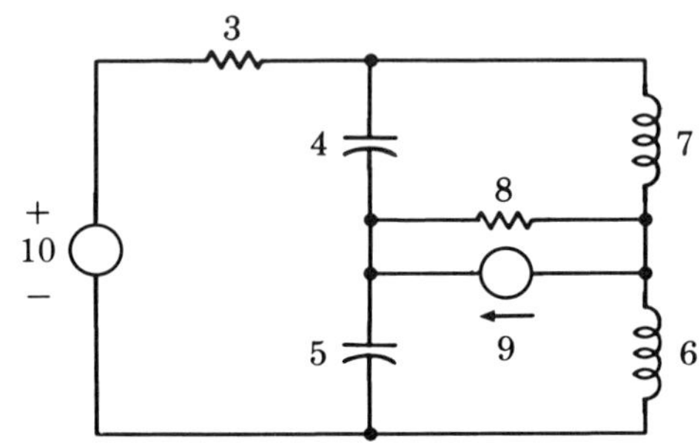

FIGURE 7–22

Section 7–3

11. The switch in the circuit of Fig. 7–23 is closed at $t = 0$. Compute the voltage across each inductance at $t = 0$ and the voltage across each capacitance at $t = \infty$.

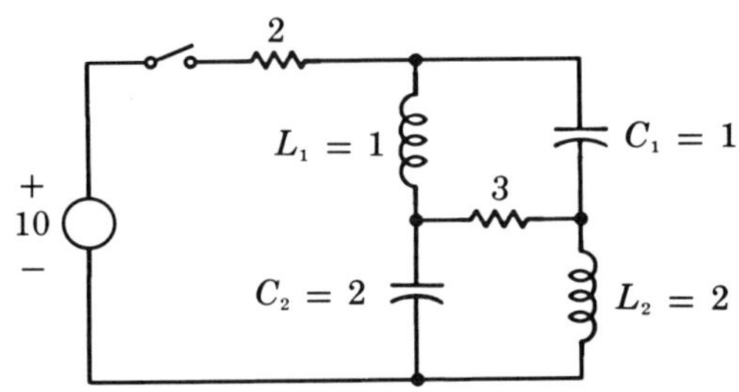

FIGURE 7–23

12. A direct voltage of 10 volts is suddenly applied to a series R–L–C circuit at $t = 0$. There is no energy stored at this time. The circuit values are $R = 1$ ohm, $C = 2$ farads, and $L = 8$ henries. Find the current and its first derivative at the instant after the switch is closed, and find the current at $t = \infty$.

13. A direct current of 10 amp is suddenly switched into a parallel R–L–C circuit in which there is no prior energy storage and in which the circuit values are $R = 1$ ohm, $L = 2$ henries, and $C = 8$ farads. Find the voltage and its first derivative at the instant after the switch is closed, and find the voltage at $t = \infty$.

14. A series R–L circuit has values of $R = 2$ ohms and $L = 3$ henries, and an initial current of 10 amp. Find the value of the first derivative and the second

derivatives of the current at the initial instant, and the value of the current at $t = \infty$.

15. A series R–C circuit has values of $R = 2$ ohms and $C = 3$ farads. There is an initial voltage of 10 volts on the capacitance. Find the initial value of the current and of its first two derivatives. Give also the value of the current at $t = \infty$.

16. The switch in the circuit of Fig. 7–24 is closed at $t = 0$. Find the initial value of the voltage across the resistance R and of its first derivative. Give also the value at $t = \infty$. (A steady value of current has been reached before the switch is closed.)

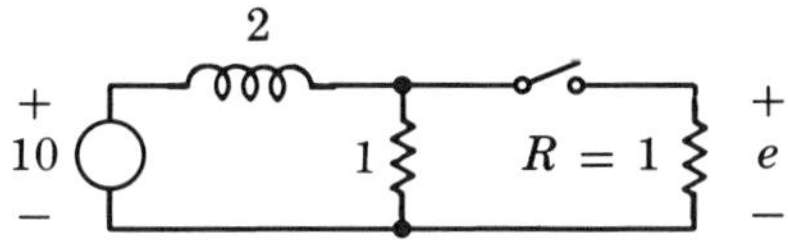

FIGURE 7–24

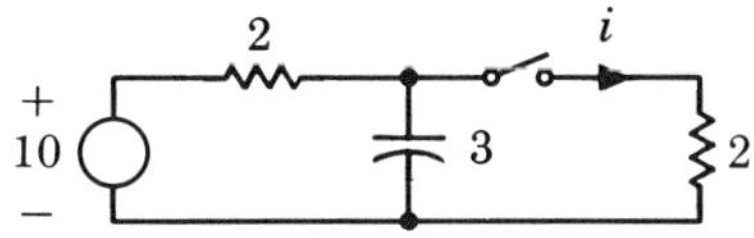

FIGURE 7–25

17. The switch in the circuit of Fig. 7–25 is closed at $t = 0$. Give the initial value of the current i and of its first derivative. What is the value of the current at $t = \infty$? (A steady value of voltage has been reached before the switch is closed.)

18. The switch in the circuit of Fig. 7–26 is opened at $t = 0$. Previous to this time a steady-state condition has been attained. Find the voltage and the first derivative of the voltage across the capacitance just after the switch is opened. What value will it have at $t = \infty$?

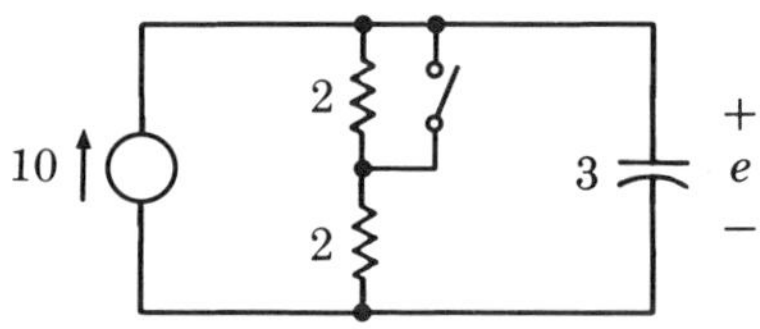

FIGURE 7–26

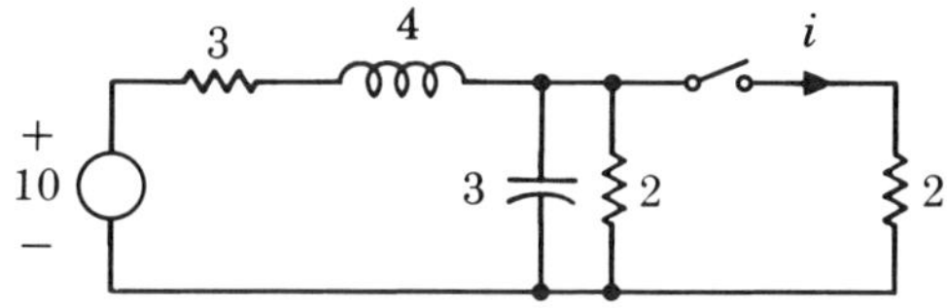

FIGURE 7–27

19. The switch in the circuit of Fig. 7–27 is closed at $t = 0$, at which time steady-state conditions are present in the network to the left of the switch. Find the initial current through the switch, its first derivative, and its value at $t = \infty$.

20. The switch in the circuit of Fig. 7–28 is opened at $t = 0$. Find the initial value of the voltage, $e(t)$, and its first three derivatives.

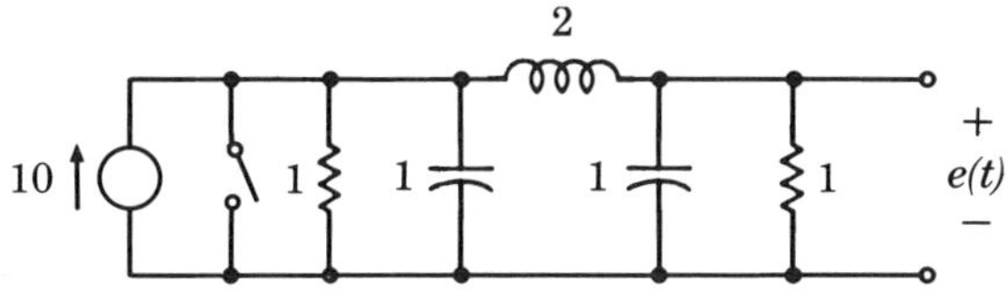

FIGURE 7–28

Section 7–4

21. Write the $2b$ equations for the circuit of Fig. 7–29.

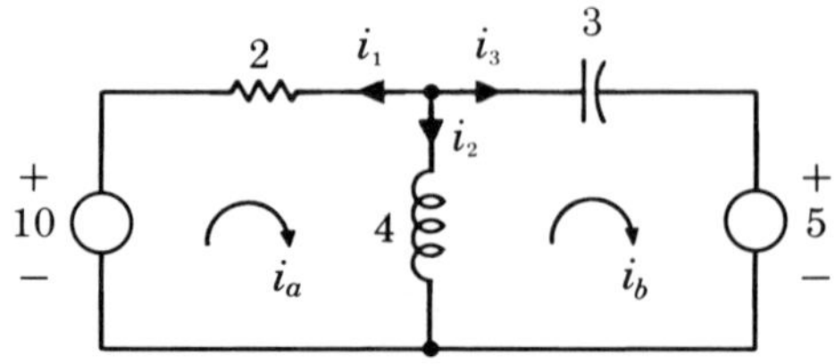

FIGURE 7–29

22. Write the mesh equations for the circuit of Fig. 7–29.
23. Write the node equations for the circuit of Fig. 7–30.

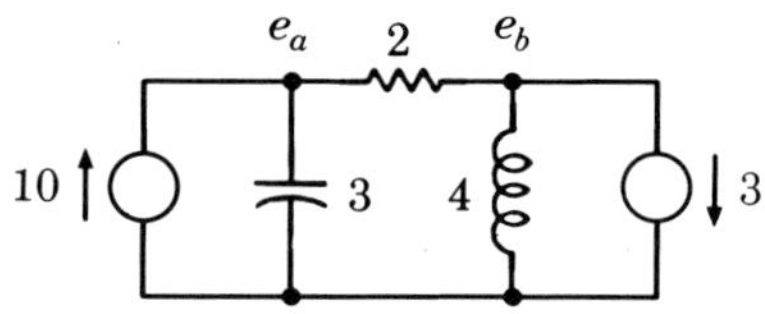

FIGURE 7–30

24. Write the mesh equations for the circuit of Fig. 7–31. Use a dummy mesh current.

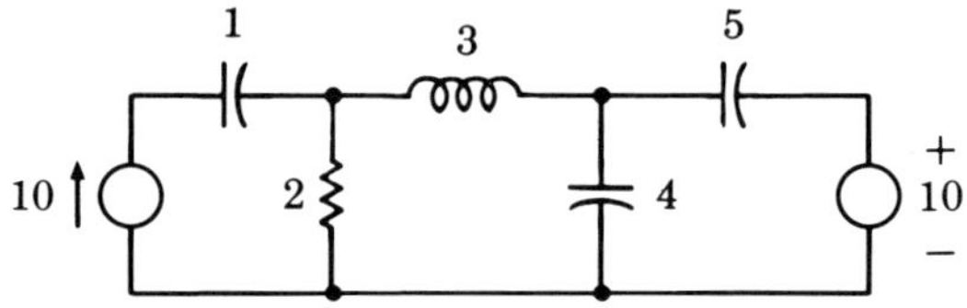

FIGURE 7–31

25. Write the node equations for the circuit of Fig. 7–32. Use a dummy node voltage.

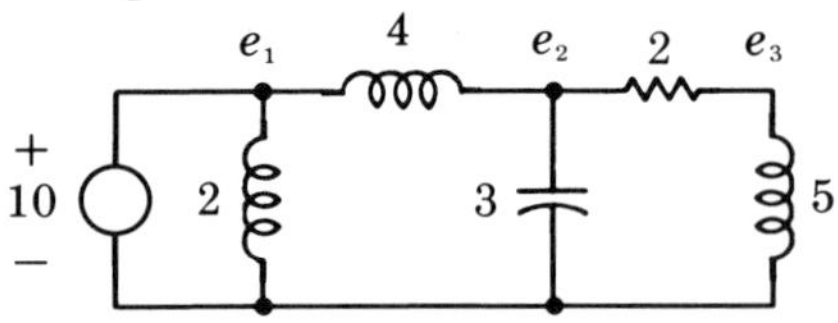

FIGURE 7–32

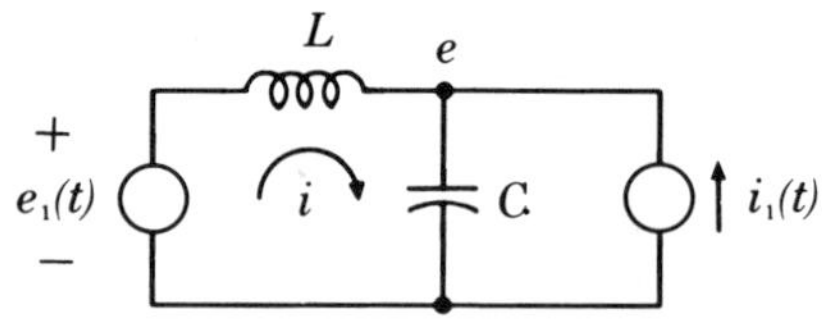

FIGURE 7–33

26. Write the equations necessary for a solution of the circuit of Fig. 7–33 by the superposition principle.

27. Initial energy stored, when it is associated with an integral in the differential equation, can be removed as a constant value or a d-c source. Using this principle, write the mesh equations for the circuit of Fig. 7–34.

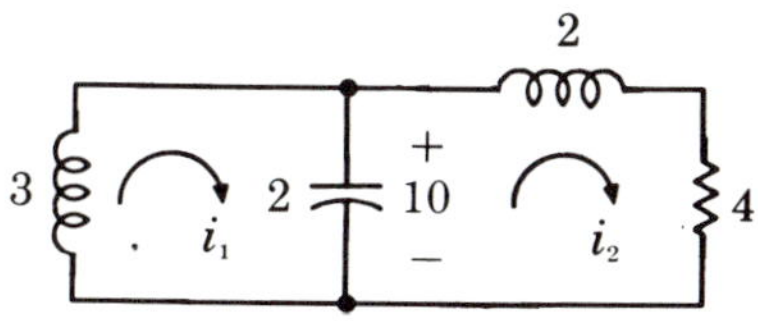

FIGURE 7–34

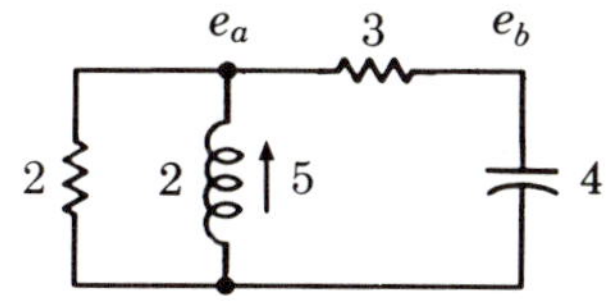

FIGURE 7–35

28. Write the node equations for the circuit of Fig. 7–35, taking out the initial energy storage in the form of a constant source.

29. It is not always convenient to treat initial energy storages as sources. Show that the initial energy stored on the inductance determines the initial value of the current in the circuit of Fig. 7–36, and that the initial value of the energy stored on the capacitance determines the initial value of the derivative of the current in the circuit.

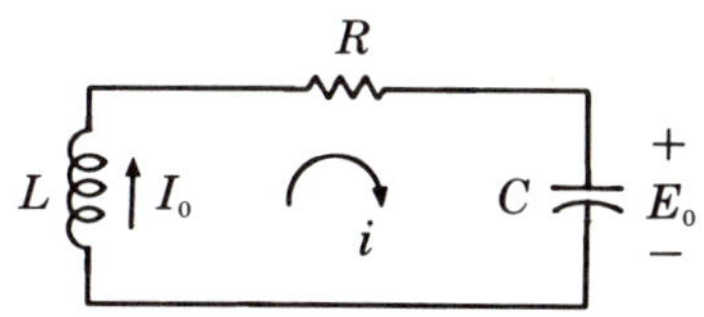

FIGURE 7–36

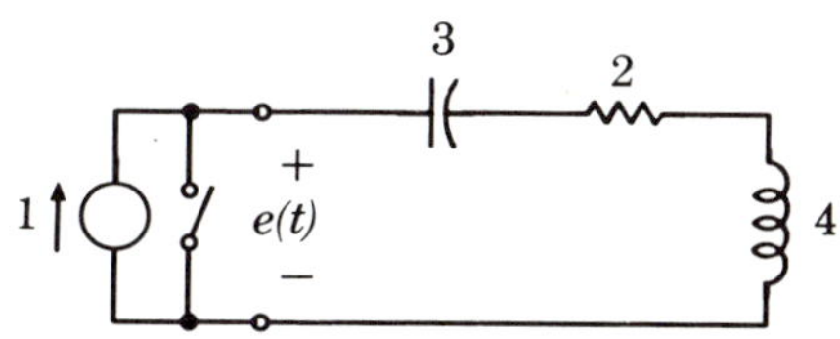

FIGURE 7–37

30. Formulate and solve the problem which is the dual of Exercise 29.

Section 7–5

31. A one-terminal-pair network can be characterized by the response voltage obtained when a direct current of 1 amp is suddenly applied to it. Find the response voltage for the network of Fig. 7–37. Comment on the voltage across the inductance.

32. The network of Fig. 7–37 can be characterized equally well by the response current it draws from a suddenly applied unit voltage. Write, but do not solve, the differential equation for this current.

33. Find the response current of the network of Fig. 7–38 to a suddenly applied voltage of 1 volt. Comment on the current in the capacitance. (The answer should be a function of time good for $t > 0$.)

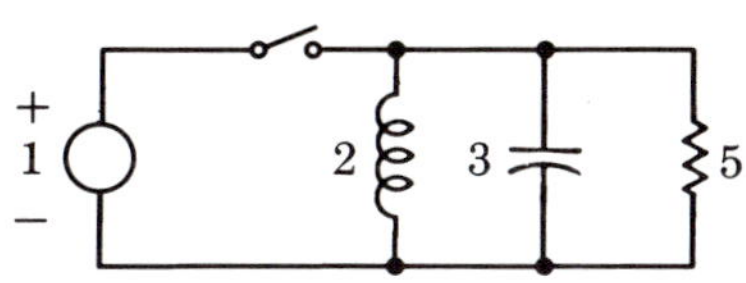

FIGURE 7–38

34. The network of Fig. 7–38 can be characterized equally well by its response to a suddenly applied current of 1 amp. Write the differential equation for this voltage function.

35. A two-terminal-pair network can be characterized by two driving-point response functions and a transfer response function. The test function to be used as the stimulus in the circuit of Fig. 7–39 is a 1-amp current, suddenly applied at $t = 0$. Find the two driving-point voltage response functions and the transfer voltage response function, with the other terminals open-circuited in each case.

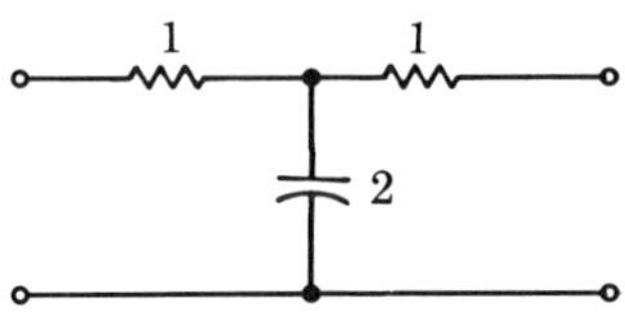

FIGURE 7–39

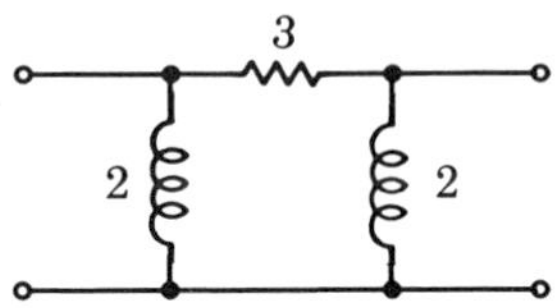

FIGURE 7–40

36. The test function which is used as the stimulus in characterizing the network of Fig. 7–40 is a 1-volt voltage source suddenly applied at $t = 0$. Find the driving-point current responses and the transfer current response, with the other terminals short-circuited in each case.

37. The test function for the circuit of Fig. 7–41 is a 1-volt voltage source. Write the differential equations for the current response functions, with the other terminals short-circuited.

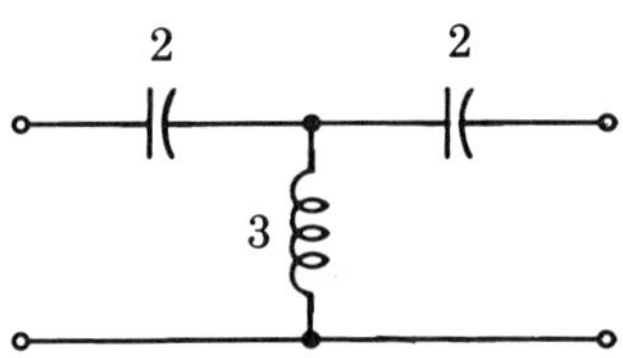

FIGURE 7–41

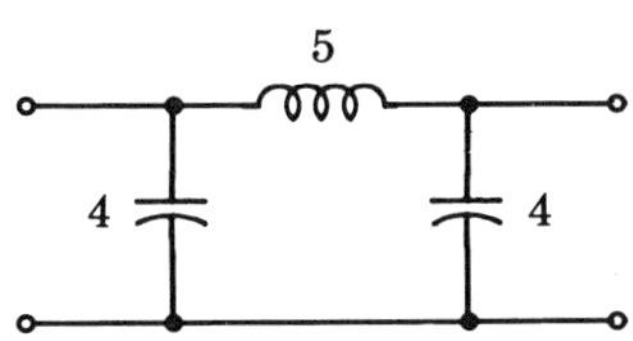

FIGURE 7–42

38. The test stimulus for the circuit of Fig. 7–42 is a unit current applied suddenly at $t = 0$. Write the differential equations from which the driving-point and transfer voltage response functions could be obtained.

Section 7–6

39. Write the differential equations for the mesh currents in the circuit of Fig. 7–43.

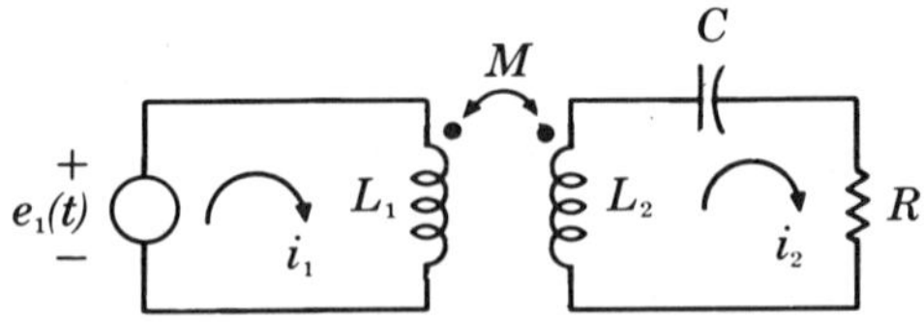

FIGURE 7–43

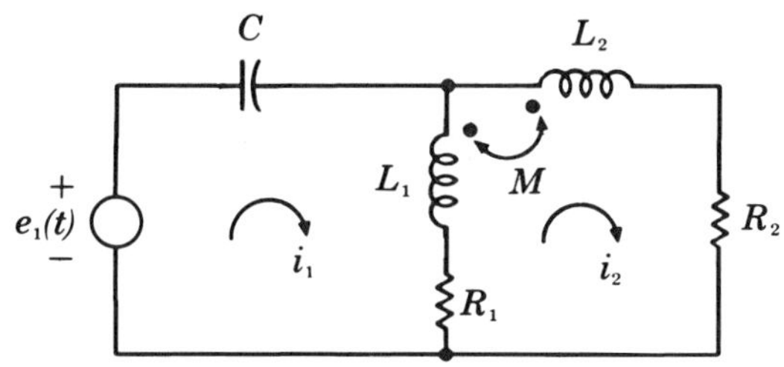

FIGURE 7–44

40. Write the differential equations for the mesh currents in the circuit of Fig. 7–44.

41. Find the equivalent inductance of two inductances in series with mutual inductance between them, as shown in the circuit of Fig. 7–45.

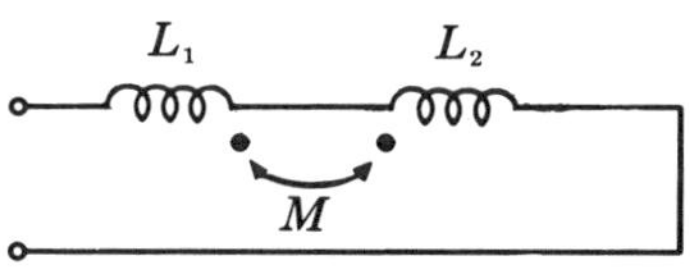

FIGURE 7–45

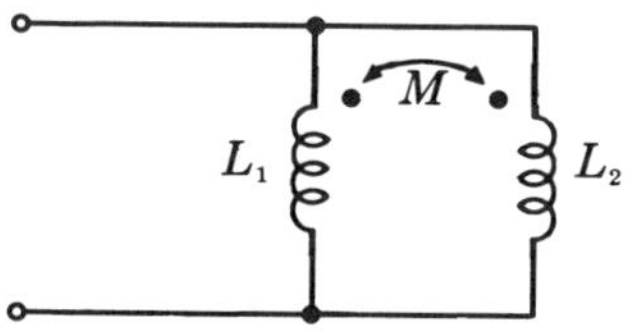

FIGURE 7–46

42. Repeat Exercises 39 through 41 with the polarity dot for the mutual inductance at the other end of the second coil.

43. Find the equivalent inductance of the parallel combination of Fig. 7–46.

44. Repeat Exercise 43 with the polarity dot on the other end of the second coil.

45. Find an equivalent T circuit for the two coupled coils with a common ground shown in the circuit of Fig. 7–47.

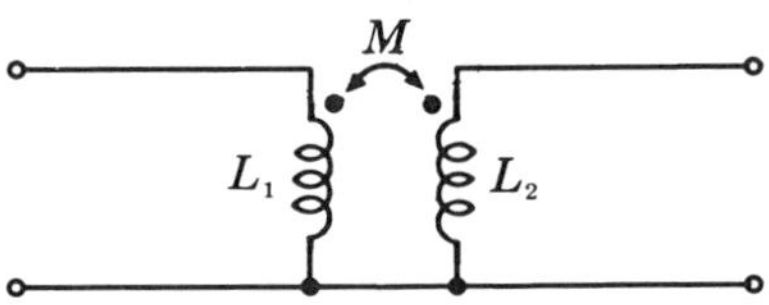

FIGURE 7–47

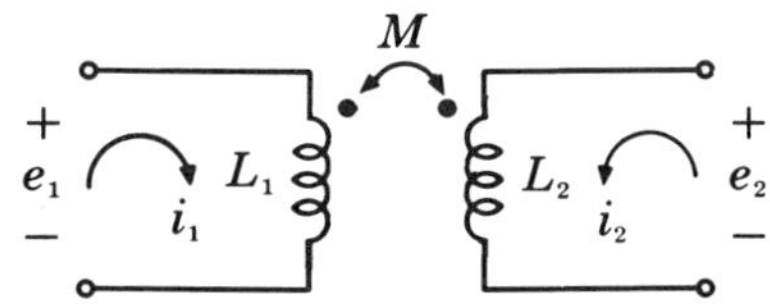

FIGURE 7–48

46. Repeat Exercise 45 with the polarity dot on the other end of the second coil.

47. The two mutually coupled coils of Fig. 7–48 can be represented by three "inverse inductance" parameters which are defined by the following equations:

$$i_1 = \Gamma_1 \int_{-\infty}^{t} e_1\,dt + \Gamma_m \int_{-\infty}^{t} e_2\,dt, \qquad i_2 = \Gamma_m \int_{-\infty}^{t} e_1\,dt + \Gamma_2 \int_{-\infty}^{t} e_2\,dt.$$

Find the values of the Γ's for the circuit. (Γ = gamma.)

48. Repeat Exercise 47 for the opposite polarity of mutual inductance.

CHAPTER 8

DIFFERENTIATION, INTEGRATION, AND SINGULARITY FUNCTIONS

"A mathematician is a man who is willing to assume everything except responsibility."

T. von Kármán

Mathematics is an area of idealization. If a mathematical development is to apply to a physical problem, an element of approximation must enter at some point. It is natural for the physicist and engineer, who are dealing with both real devices and the mathematical descriptions of them, to combine the two areas and to develop an "intuitive mathematics" which relies on physical reasoning for some of its validity. This approach is particularly useful in the areas of integration and differentiation, and is almost essential for the application of the so-called singularity functions.

8–1 Introduction. It is not the purpose of this chapter to present a complete course in calculus. The student should already be familiar with the elements of differentiation and integration. Experience has shown, however, that the concept of differentiation and integration as continuous "operators" is not easily grasped, and although this viewpoint is perhaps not rigorous enough or general enough to be presented in a mathematics course, it is extremely useful in many engineering problems. It is important to learn the meaning of rigor, but we must also understand those concepts that sharpen our intuitive reasoning processes.

Differentiation and integration can be fitted into the stimulus-response relationship by representing them as "black boxes" which perform operations upon an input to produce an output. The input stimulus is usually a single-valued function of time, and the output of the black box is the instantaneous value of the derivative or the integral of the input waveform. Although this point of view is purely mathematical, it is very closely related to electric circuits. For example, electrical energy storage elements can be used in analog computers to perform the mathematical operations of differentiation and integration.

The latter part of this chapter deals with the singularity functions. Here the split between mathematical rigor and useful intuition is even more pronounced. A rigorous treatment of the singularity functions is not afforded by the usual Riemann calculus, but requires the more advanced Lebesgue calculus, which is seldom taught to undergraduates.

However, the singularity functions are very useful in many practical problems, because they represent simple idealizations of practical switching operations. Most of the difficulties of singularity functions are of a mathematical nature and arise from the assumption of an action taking place in zero time. In intuitive reasoning, we never quite reach this limit. A switching operation takes place in a very short interval, but the interval is still finite. Mathematical problems which involve singularity functions can always be resolved by allowing the functions to spread out over a short interval and to occur in a definite sequence. The objective is to obtain the advantages of idealized switching operations, which take place in zero time, without introducing insolvable mathematical difficulties.

8-2 Differentiation of functions. Since differentiation is less difficult than integration, we shall discuss it first. The functional and operational concept of differentiation is illustrated in Fig. 8-1. The input to the black box is the function $f_1(t)$, and the output is the function $f_2(t)$. Differentiation is represented by the black box, which operates on the input so that

$$f_2(t) = \frac{d}{dt}[f_1(t)] = f_1'(t). \tag{8-1}$$

Equation (8-1) states that the output function $f_2(t)$ is at each instant of time equal to the slope, or derivative, of the input function $f_1(t)$. If $f_1(t)$ is a straight line, the slope is a constant equal to the rise divided by the run. If $f_1(t)$ is a more general curve, the slope at any instant is the slope of the tangent drawn to the curve. Note that the derivative *of a point* has no meaning. The nature of the curve on either side of the point determines the derivative; the curve must be known before the derivative *at the point* can be obtained.

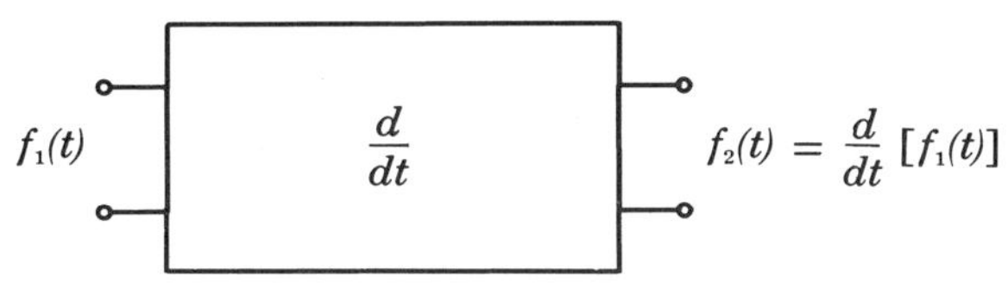

FIG. 8-1. Differentiation as a continuous operation.

Graphical differentiation. A typical curve and its derivative are shown in Fig. 8-2. To the left of the origin the original curve $f(t)$ and its derivative $f'(t)$ are both zero. Between $t = 0$ and $t = 1$ the original curve rises with a constant slope of $+1$. The derivative curve has therefore a constant value of $+1$. Between $t = 1$ and $t = 3$ the original curve drops with a slope of -1, and the corresponding portion of the derivative curve also

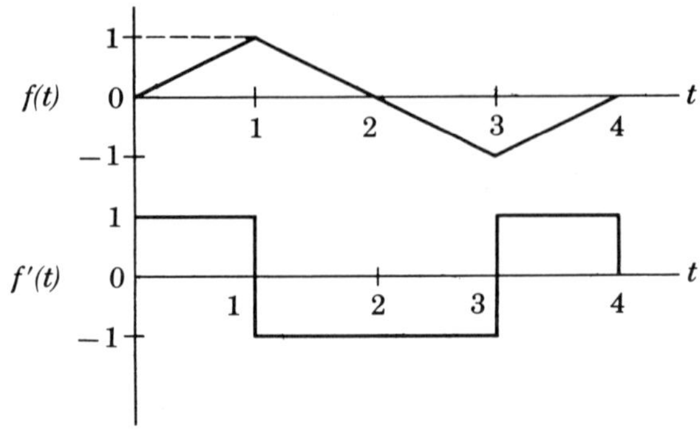

FIG. 8–2. A typical curve and its derivative.

has a value of -1. From $t = 3$ to $t = 4$ the original curve rises again with a slope of $+1$, and thus the derivative curve again has a value of $+1$. To the right of $t = 4$ the original curve and its derivative are both zero. Note that a function can be zero *at a point* and still have a derivative. For example, at the point $t = 2$ in Fig. 8–2, the curve $f(t)$ is zero but its derivative has a value of -1.

The original function shown in Fig. 8–2 looks rather simple, since it consists of straight lines. Nevertheless, to express it mathematically is not done easily without the singularity functions which are discussed in a later section, because the derivatives of the function are discontinuous at the sharp corners.

Analytical differentiation. If an original function can be expressed as a simple mathematical expression, the derivative function can usually be found in standard tables of derivatives. An example is the function

$$f_1(t) = 3 \sin 3t. \tag{8–2}$$

Tables of derivatives show that

$$d\,(\sin x) = \cos x\,dx.$$

Applying this formula to the function of Eq. (8–2) gives the derivative

$$f_2(t) = \frac{d}{dt}\,[f_1(t)] = 9 \cos 3t. \tag{8–3}$$

The functions $f_1(t)$ and $f_2(t)$, or $f'(t)$, are plotted in Fig. 8–3 from $t = 0$ to $t = \pi$. The derivative curve can also be obtained by drawing a tangent at each point on the original curve and then plotting the slopes of the tangents. Thus, at $t = \pi/6$ the slope of the tangent is zero, and the corresponding value of the derivative curve is zero. At $t = \pi/4$ the slope

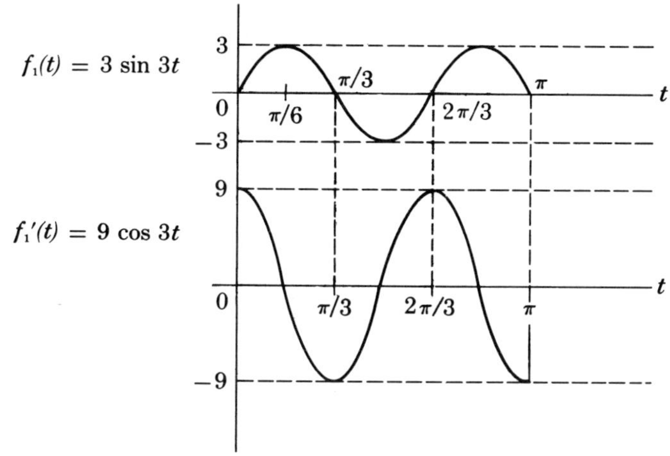

FIG. 8-3. The function $f(t) = 3 \sin 3t$ and its derivative.

of the tangent is negative and has a value of approximately −6.3. Equation (8-3) yields a more accurate value:

$$f_2\left(t = \frac{\pi}{4}\right) = 9 \cos\left(\frac{3\pi}{4}\right) = -9 \times 0.707 = -6.363.$$

In general it can be said that the operation of taking a derivative is a straightforward procedure which can be accomplished either analytically or graphically. It is best to think of the graphical procedure as the basic method and to consider the analytical procedure only a method of accomplishing the graphical procedure without actually drawing the curve.

8-3 Integration of functions. Integration is more difficult than differentiation, because the value of an integral at a point involves the entire past history of the function and not just the nature of the function in the immediate vicinity of the point. Another problem in integration is the use of the *indefinite* integral, which is frequently misunderstood by students. We shall use only the definite integral in this textbook, and it will always have a definite meaning: the area bounded by the curve and the two limits specified by the integral. Usually the lower limit will be minus infinity, which we interpret merely as a time in the past prior to which the function is zero; the upper limit will be a variable equal to time. Thus the basic integral measures the area under a function from the instant the function began in the remote past until the present time. As time changes, the area changes, and thus traces out a function which is the integral of the given function.

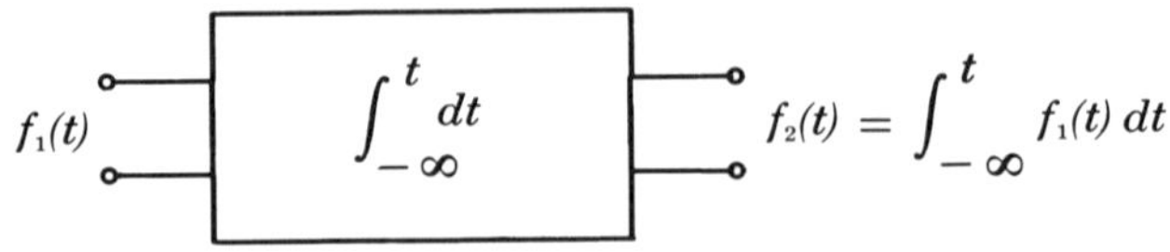

FIG. 8–4. Integration as a continuous operation.

The operational concept of integration is shown in Fig. 8–4. The input function to the black box is $f_1(t)$. The operation of integration consists in finding the area under the function curve from the time it began until the present time, t, which is a variable. As time changes the area changes; at any given instant the value of the area is the output function $f_2(t)$. The operation of integration is identical in concept to the accumulation of cash in a bank account. The input function $f_1(t)$ is analogous to the record of deposits and withdrawals. The output function $f_2(t)$ is analogous to the instantaneous amount, or balance, which is present in the account. The balance is the algebraic sum of all the deposits and withdrawals which have occurred since the instant in the past when the account was opened.

Expressed mathematically, the definite integral of a function is

$$f_2(t) = \int_{-\infty}^{t} f_1(t)\,dt. \tag{8–4}$$

Equation (8–4) states that the value of the function $f_2(t)$ at some time t is the area under the curve $f_1(t)$ from the beginning of time up to the present time t. In any actual problem, it is possible to find a time in the past so remote that the function $f_1(t)$ is zero to the left of it. Very often this point is $t = 0$, but to be more general we shall refer to the beginning as $t = a$. Then, the functional integral is

$$f_2(t) = \int_{a}^{t} f_1(t)\,dt. \tag{8–5}$$

Graphical integration. An example of a function which can be integrated easily is shown in Fig. 8–5. The function $f_1(t)$ to be integrated is zero until $t = 0$, has a value of unity from $t = 0$ to $t = 5$, and is zero to the right of $t = 5$. The operational integral plots a value which accumulates from the time when the function was identically zero up to a variable present time. The area accumulated to the left of $t = 0$ is zero. Therefore, the value of the integral curve at $t = 0$ is zero.* At $t = 1$, an area of unity has accumulated; thus the value of the integral curve at this point is 1. At $t = 2$, the total accumulated area is 2; the value of the integral

* Integrals with respect to time are often designated as $f^{-1}(t)$.

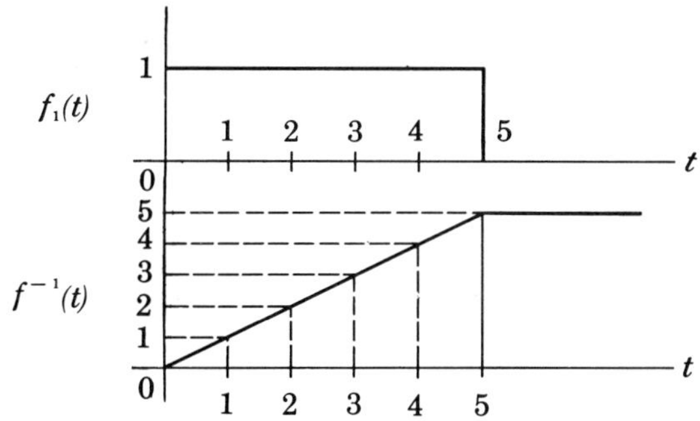

FIG. 8-5. Integration of a function.

curve is therefore 2. At $t = 3$, the integral curve equals 3, at $t = 4$ it equals 4, and so on. Note that the integral curve rises linearly. Thus, at $t = \frac{1}{2}$, the accumulated area is $\frac{1}{2}$, and the value of the integral curve is $\frac{1}{2}$, and so on. To the right of $t = 5$ the integral curve maintains a value of 5. In terms of the banking analog, a deposit of 1 dollar is made each month for five months, and thus the total accumulation is 5 dollars. This amount remains undisturbed unless further deposits or withdrawals are made.

If the curve to be integrated is nonlinear, it can be approximated by a series of rectangles of varying heights, and the integral can be obtained by adding the areas of the rectangles. Machines called *planimeters* are available for performing such graphical integration, but often it is sufficiently accurate merely to count squares under the curve after it is plotted on graph paper.

Analytic integration. If the analytic form of the original function is known, the integral of the function can be obtained by the use of tables of integrals. The entries in the tables are usually indefinite integrals. An indefinite integral is identical to that of Eq. (8-4), except that an arbitrary constant is added to account for the integral which may have accumulated before time $t = -\infty$. Although physical functions are zero at minus infinity, mathematical functions need not be. Thus mathematicians include an arbitrary constant in the indefinite integral to cover all cases. No practical difficulty arises, because most integrations are effected between two limits, the constant at minus infinity canceling out. Consider the integral of Eq. (8-5),

$$f_2(t) = \int_a^t f_1(t)\, dt,$$

which represents the area under the curve $f_1(t)$ from the point a to the point t. This area equals the difference between the area under the curve from minus infinity to t and the area under the curve from minus infinity to a. We can therefore write

$$f_2(t) = \int_{-\infty}^{t} f_1(t)\,dt - \int_{-\infty}^{a} f_1(t)\,dt. \tag{8-6}$$

The integral from minus infinity to t is tabulated as a functional value plus a constant to include the value of the integral at minus infinity. If this functional value is listed as

$$\int f_1(t)\,dt = F_1(t) + C, \tag{8-7}$$

we can write Eq. (8–6) as

$$f_2(t) = [F_1(t) + C] - [F_1(a) + C],$$

or

$$f_2(t) = F_1(t) - F_1(a). \tag{8-8}$$

When evaluating definite integrals, we can ignore the arbitrary constants given in the tables of indefinite integrals. The value under a curve is simply the value of the indefinite integral at the upper limit minus its value at the lower limit. As an example, consider the function

$$f_1(t) = \begin{cases} 3 \sin 3t & \text{for} \quad t > 0, \\ 0 & \text{for} \quad t < 0. \end{cases}$$

The integral of this function is zero for $t < 0$; for $t > 0$ the integral is

$$f_2(t) = \int_0^t 3 \sin 3t\,dt. \tag{8-9}$$

The tabulated indefinite integral is

$$\int \sin x\,dx = -\cos x + C. \tag{8-10}$$

If we substitute $x = 3t$ in the right-hand member of Eq. (8–10), we can evaluate the integral of Eq. (8–9) as follows:

$$f_2(t) = \left(-\frac{\cos 3t}{1}\right) - \left(-\frac{\cos 0}{1}\right),$$

or

$$f_2(t) = 1 - \cos 3t.$$

The function $3 \sin 3t$ and its integral are plotted in Fig. 8–6. From $t = 0$ to $t = \pi/3$ the function is positive, and the area which accumulates in the integral curve increases. Since at the beginning the function has small values, each successive integrating step contributes very little area; thus the integral curve rises slowly. In the neighborhood of $t = \pi/6$ the func-

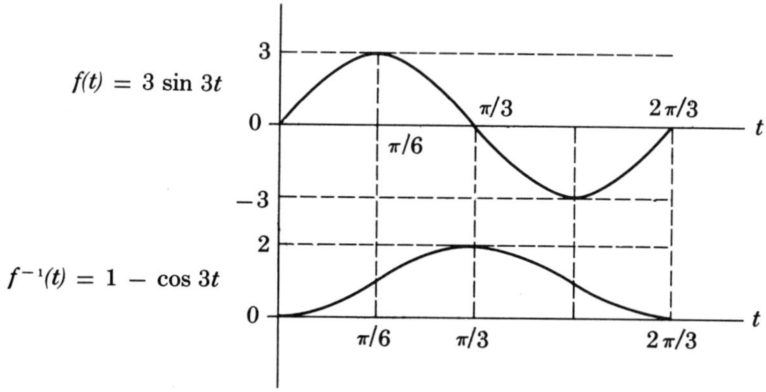

FIG. 8–6. The function $f(t) = 3 \sin 3t$ and its integral.

tion reaches its maximum value, and the integral curve rises at its maximum rate. This rate decreases as the function decreases, until the maximum value of the integral is reached at $t = \pi/3$. Beyond this point the function becomes negative, and in the integration negative area is subtracted from the positive area, until the net accumulated area is zero at $t = 2\pi/3$. This means that the positive and negative areas of the function, to the left of $t = 2\pi/3$, are exactly equal.

An approximate integration of the function in Fig. 8–6 can bc obtained by counting squares under its curve. The student should view as basic the concept of an integral as an area and consider tabulated values of integrals only a means of avoiding the counting of squares.

8–4 The relationship of differentiation and integration. Differentiation and integration as defined in the preceding sections are inverse operations. Thus, if we take a given curve, integrate it, and then differentiate the result, we obtain the original curve. Similarly, if we first differentiate and then integrate, we return to the original curve. These results will be proved by a study of the two curves in Fig. 8–7. In this illustration the lower curve is the integral of the upper curve. In the time interval Δ, the

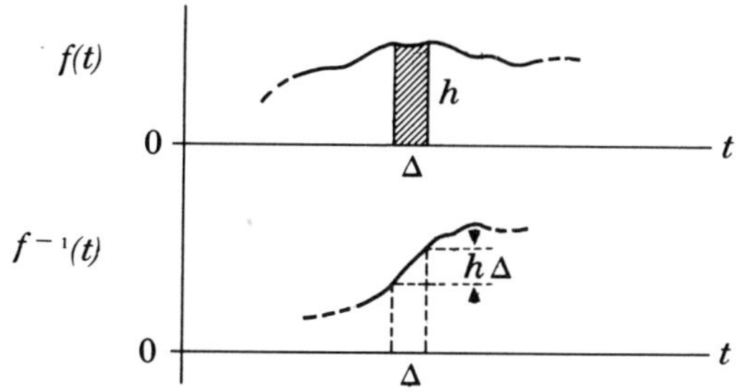

FIG. 8–7. The relationship between the derivative and the integral.

increment of area contributed by the upper curve is $h\Delta$. This value is added to the height of the integral curve at the end of the interval Δ. Now let us reverse the process and obtain the slope of the integral curve. Over the interval Δ the slope is given by the rise in the curve over the corresponding run, that is,

$$\frac{h\Delta}{\Delta} = h,$$

which is the value of the original curve. Since the process works either way, the derivative curve is the inverse of the integral curve and vice versa.

If we consider functions extending from minus infinity to the present and if we consider the integral curve to be zero at minus infinity, then the two curves are exact inverses. (The addition of a constant to an integral produces a curve which has the same derivative as the original curve.)

Figure 8–8 shows an example of the inverse relationship between integral and derivative curves. The center curve is the integral of the top curve, and the lowest curve is the derivative of the center curve. Note that the lowest curve is identical to the top curve. The example in Fig. 8–8 meets the condition that the integral curve shall extend over all time; thus the inverse relationship here is an exact one.

A second example of the inverse relationship is provided by the analytic expression

$$\int_0^x \sin x\, dx = 1 - \cos x, \tag{8–11}$$

and the derivative of the integral curve in Eq. (8–11):

$$\frac{d}{dx}(1 - \cos x) = \sin x. \tag{8–12}$$

Note in Eq. (8–12) that the derivative of the integral curve in Eq. (8–11) equals the original function, $\sin x$. The two functions $1 - \cos x$ and $\sin x$ are exact inverses. The sine and cosine functions by themselves are not.

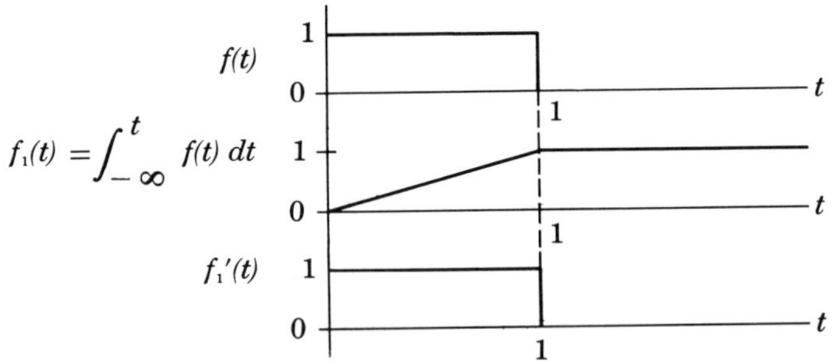

FIG. 8–8. The inverse nature of integral and derivative curves.

To return to general terms, there is a constant in the integration process whose value depends on the point where the integration begins. Mathematical functions which have a value at minus infinity or which are indeterminate at minus infinity, like the sine and cosine, must be sectioned or cut off at some point by means of the lower limit on the integral, as is done in Eq. (8–11). Only then can the mathematical functions represent physically realizable functions, and only then are the operations of differentiation and integration exact inverses of each other.

8–5 The singularity functions. Since the publication of Oliver Heaviside's *Electrical Papers* in 1892, electrical engineers and physicists have made use of unit switching functions, or *singularity functions*, on a more or less empirical basis. Heaviside, who viewed mathematics as an experimental science, reasoned in this fashion: If a certain mathematical procedure produces correct results, then it is justified. The correctness of the results can usually be checked independently. The important thing is to be able to obtain results. Needless to say, this point of view was (and still is) anathema to mathematicians. A mathematician starts with certain premises and then by a chain of unassailable reasoning reaches certain conclusions, which are correct by virtue of the nature of the reasoning. For years mathematicians insisted that no such functions as singularity functions could exist, but engineers continued to use them to obtain useful results. The dilemma was resolved with the publication of Laurent Schwartz's *Theorie des Distributions* in 1951. This French mathematician invented a special class of functions called *distribution functions.* They are not ordinary functions, being defined only in abstract space, but they do give a rigorous mathematical representation of singularity functions. We shall approach singularity functions from an intuitive point of view.

Singularity functions are approximations of actual switching waveforms. In fact, we idealize switching waveforms in much the same fashion as we idealize network elements, and for much the same reason. It is easier to solve a problem in which a switch has only two positions, open and closed, than it is to solve the actual problem which involves a complicated transition between the two states. The mathematical difficulties associated with idealized switching functions arise from the assumption that the switching takes place in zero time. In our intuitive approach we shall avoid these difficulties by never quite reaching the limit of zero time. In fact, if a switch is closed at zero time, we shall find it convenient to break this instant into three parts: 0^-, the instant just before the switch is closed; 0, the instant during which the switch is being closed; and 0^+, the instant just after the switch is closed. These three instants are separated by intervals which are negligibly short, but nevertheless finite.

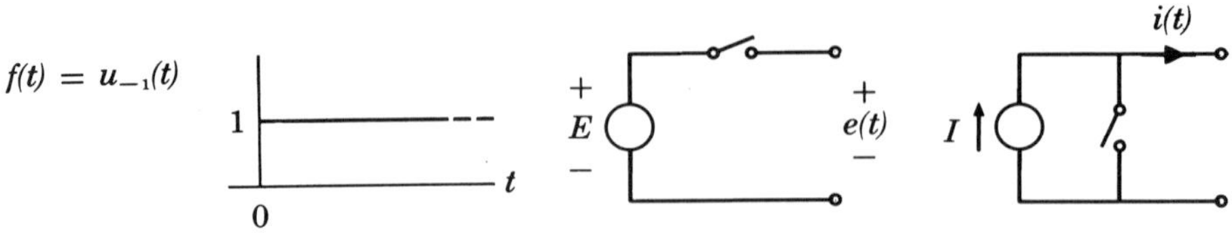

(a) The step function (b) Voltage step (c) Current step

FIG. 8–9. The unit step function.

The step function. A constant current or a constant voltage switched into a circuit at time $t = 0$ can be represented by a *step function.* The unit step function is denoted by the symbol $u_{-1}(t)$ and defined by the equations

$$u_{-1}(t) = \begin{cases} 0 & \text{for} \quad t < 0, \\ 1 & \text{for} \quad t > 0. \end{cases} \tag{8–13}$$

A graph of the function is shown in Fig. 8–9(a). The function is zero for negative values of time and unity for positive values of time. The switching operation occurs in the short interval between 0^-, when the function is zero, and 0^+, when it is unity. The value of the function precisely at zero is indeterminate. A voltage source switched into a circuit at $t = 0$ can be represented by means of the unit step function. Consider the source shown in Fig. 8–9(b). Let us say that the switch is closed at $t = 0$. The voltage $e(t)$, which is zero before $t = 0$ and equal to E for t greater than zero, can be expressed analytically as

$$e(t) = Eu_{-1}(t). \tag{8–14}$$

In Eq. (8–14) the function $u_{-1}(t)$ multiplies E by zero for negative values of t and by unity for positive values of t. The net result is simply to cut off $e(t)$ for negative values of t.

A current source [Fig. 8–9(c)] switched into a circuit at $t = 0$ can also be expressed in terms of the unit step function. Prior to $t = 0$ the switch is closed, and the current from the source passes through the short circuit. At $t = 0$ the switch is opened, and the current passes through the external circuit. The current $i(t)$, which is zero before $t = 0$ and equal to I for t greater than zero, can be expressed analytically as

$$i(t) = Iu_{-1}(t). \tag{8–15}$$

In Eq. (8–15) the function $u_{-1}(t)$ causes the current to be zero for negative values of t.

The ramp function. Since the circuit elements can produce integrals and derivatives, we shall be interested in the integrals and derivatives of the

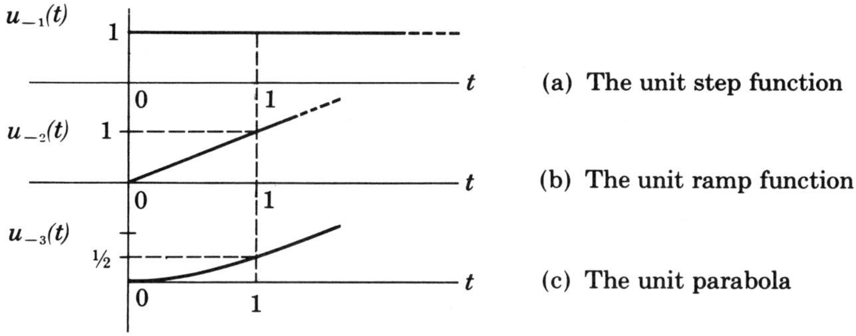

FIG. 8–10. The integrals of the unit step function.

step function. The integral of the step function is called the *ramp function.* The unit ramp function is denoted by the symbol $u_{-2}(t)$ and is obtained as follows:

$$u_{-2}(t) = \int_{-\infty}^{t} u_{-1}(t)\,dt = \int_{0}^{t} dt$$

or (8–16)

$$u_{-2}(t) = \begin{cases} t & \text{for} \quad t > 0, \\ 0 & \text{for} \quad t < 0. \end{cases}$$

A graph of the unit ramp function is shown in Fig. 8–10(b). The slope of the function is unity and its value at $t = 1$ is unity. The integral of a step function with height A is a ramp function having a value of A at $t = 1$, and thus a slope of A.

Other singularity functions can be obtained by taking further integrals. The *unit second-order function,* the integral of the unit ramp, is represented by the symbol $u_{-3}(t)$ and is obtained as follows:

$$u_{-3}(t) = \int_{-\infty}^{t} u_{-2}(t)\,dt = \int_{0}^{t} t\,dt$$

or (8–17)

$$u_{-3}(t) = \begin{cases} t^2/2 & \text{for} \quad t > 0, \\ 0 & \text{for} \quad t < 0. \end{cases}$$

The graph of the unit second-order function [Fig. 8–10(c)] is zero for negative values of t and a parabola for positive values of t. Additional higher-order functions can be obtained, but they occur rather rarely in circuit theory.

The impulse function. The derivative of the unit step function is called the *impulse function.* Figure 8–11(a) shows a unit step function plotted on

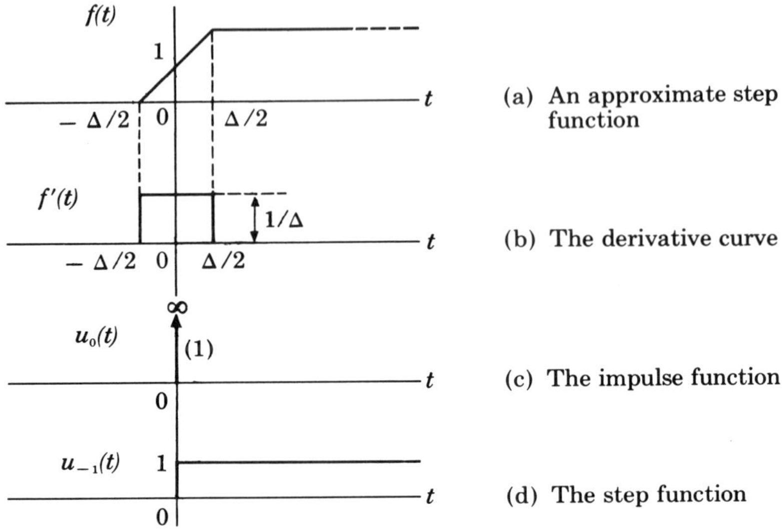

FIG. 8–11. The derivative of the step function: the impulse function.

an expanded scale. The derivative [Fig. 8–11(b)] is zero everywhere except between 0^- and 0^+. If we denote this interval as Δ, the derivative over the interval equals $1/\Delta$. As Δ approaches zero the value of the derivative approaches infinity. Thus the impulse function [Fig. 8–11(c)] is zero everywhere except at $t = 0$; at this point it is infinite. The area under the impulse can be obtained from Fig. 8–11(b). It is

$$\frac{1}{\Delta} \times \Delta = 1. \tag{8–18}$$

The significance of Eq. (8–18) lies in the fact that the area under the derivative of the step function is independent of the rise-time, Δ, over which the step occurs. It is convenient to characterize the impulse by the area under it rather than by its height, which is infinite. The graphical symbol for the impulse is shown in Fig. 8–11(c). The number in parentheses beside the impulse denotes the area under the impulse. The notation $u_0(t)$ is used for the unit impulse.

Since the derivative of the step function is an impulse function, the integral of the impulse function should be a step function. If we integrate the impulse function of Fig. 8–11(c), we obtain no area to the left of the origin. As we go from 0^- to 0^+, however, we pick up the area under the impulse; thus the integral curve jumps from 0 to 1. Since there are no further contributions to the right of the origin, the value of the integral curve remains at unity. This integral curve, shown in Fig. 8–11(d), is obviously the unit step function. The derivative of a step function with height A is an impulse function with area A. When we integrate the impulse, we pick up an area A at the origin, and thus the integral curve jumps up by an amount A.

No trouble will arise in dealing with the impulse function if the sequence of events which occur at 0^-, 0, and 0^+ is kept clearly in mind. It should also be remembered that the impulse is characterized by its area rather than by its height. Thus an impulse of current is represented by an area whose dimensions are *amp·sec,* or *coulombs;* an impulse of voltage is represented by an area whose dimensions are *volt·sec,* or *flux linkages.* It may appear that the impulse is a nonphysical concept, but *it is neither more so nor less so than the idealized step function.* The impulse can be approximated in practice by any large pulse which occurs in a short time. Until, in various problems, we determine more exactly just what constitutes a short time we shall find it convenient to think of the impulse as occurring in essentially zero time. Later we shall be able to use impulses which extend over finite intervals of time.

As shown in Fig. 8–11(c), we use the notation $u_0(t)$ for the unit impulse. By definition,

$$u_0(t) = \frac{d}{dt} u_{-1}(t). \tag{8–19}$$

The symbol $u_0(t)$ makes the unit impulse, rather than the unit step, the basic singularity function, which is the customary attitude in more advanced treatments of singularity functions. Even in this elementary textbook we shall come to find that the response of a network to a unit impulse is a more basic function than its response to a unit step function.

Additional singularity functions can be defined as the derivatives of the unit impulse function. For example, the *unit doublet* is

$$u_1(t) = \frac{d}{dt} u_0(t). \tag{8–20}$$

In order to be able to take a derivative of the unit impulse function, we shall view it as a triangular pulse of height $1/\Delta$ and width 2Δ, as shown in Fig. 8–12(a). This is permissible, since the exact shape of an impulse function does not matter so long as its area does not change. Thus the derivative curve [Fig. 8–12(b)] consists of a positive pulse followed by a negative pulse. The pulses have magnitudes of $1/\Delta^2$ and areas of $1/\Delta$. In the limit as Δ approaches zero, the areas become infinite. Thus the unit doublet [Fig. 8–12(c)] consists of an infinite positive impulse followed by an infinite negative impulse in such a way that the second integral has a value of unity. The doublet is characterized by its second integral and has the dimensions of amp·sec^2 for a current doublet, and volt·sec^2 for a voltage doublet.

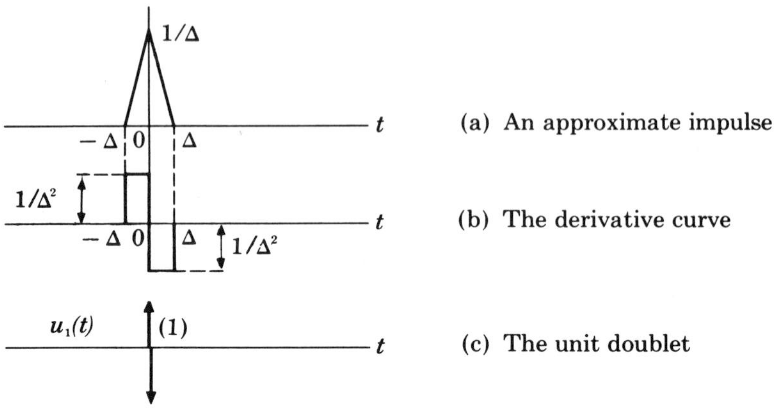

FIG. 8–12. The derivative of the impulse function.

Other singularity functions can be obtained by taking further derivatives of the unit impulse function. In these cases, we would follow a procedure similar to that used in obtaining the unit doublet; that is, we would assume a finite value of Δ and change the shapes of the pulses to make them easily differentiable. In practice, singularity functions of higher order than the doublet seldom occur in circuit theory.

8–6 The use of singularity functions to represent waveforms. The singularity functions we defined in the preceding section are used to represent certain waveforms produced by switching operations which occur at time $t = 0$. In this section we shall extend their applicability to switching operations which occur at any time t. In addition, we shall show how arbitrary waveforms can be approximately represented by trains of step functions or impulse functions.

Singularity functions at arbitrary times. The unit step function $u_{-1}(t)$ is zero for negative t and unity for positive t. Similarly, a unit step function of some other argument, $u_{-1}(x)$ for example, is zero for negative values of the argument and unity for positive values of the argument. In order to change the time t at which a step function occurs, we merely change the variable. Thus if we replace x in the step function $u_{-1}(x)$ by $t - a$, we obtain the unit step function

$$u_{-1}(t - a). \tag{8–21}$$

This function is zero when its argument is negative, that is, for t less than a; the function has a value of unity when its argument is positive, that is, when t is greater than a. The result of this substitution is to produce a step function which begins at $t = a$, as shown in Fig. 8–13.

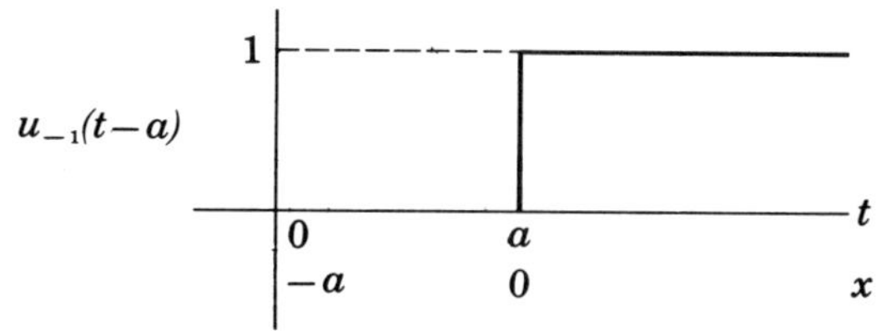

FIG. 8–13. A delayed step function.

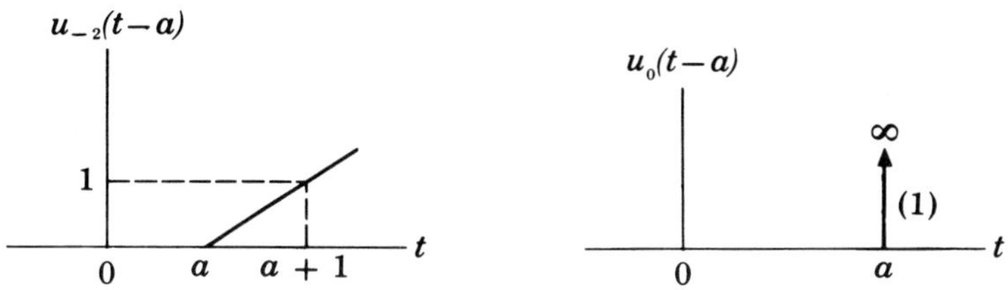

(a) Delayed ramp function **(b) Delayed impulse function**

FIG. 8–14. A delayed unit ramp function and unit impulse function.

Similarly, a unit ramp function which begins at $t = a$ is

$$u_{-2}(t - a), \tag{8–22}$$

and a unit impulse function which begins at $t = a$ is

$$u_0(t - a). \tag{8–23}$$

The graphs of these functions are shown in Figs. 8–14(a) and 8–14(b).

Sectioned waveforms. One of the most useful applications of the unit step function is to section or cut off a function for negative values of time. Most mathematical expressions extend over all time, that is, from minus infinity to plus infinity. Most actual problems, however, begin at a particular time, usually called zero time, and their solutions are valid only for positive time. One way in which we can indicate that a solution is not valid for negative time is to multiply it by the unit step function. The product is zero for negative time, since the unit step function is zero; the solution remains unchanged for positive time, since the unit step function is unity. Figure 8–15 illustrates this multiplication. The waveform shown in Fig. 8–15(a) is a sine wave which extends over all time. Its mathematical expression is

$$f(t) = \sin t. \tag{8–24}$$

The waveform shown in Fig. 8–15(b) is a sine wave which is zero for

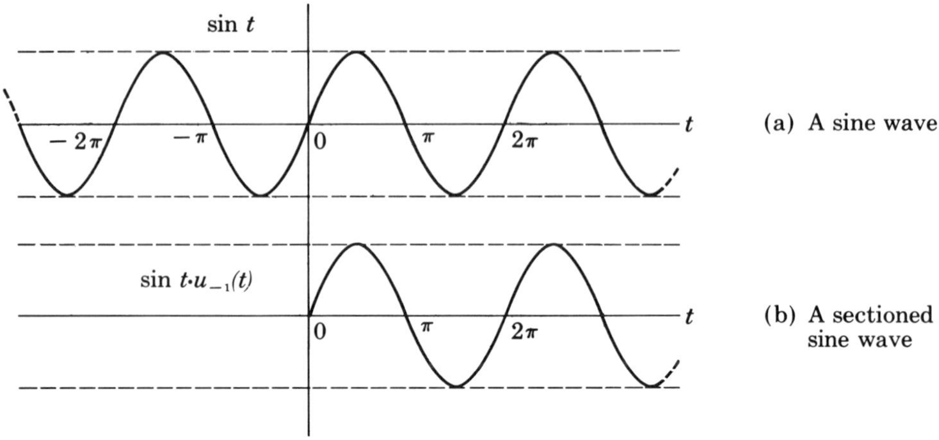

FIG. 8–15. The use of a step function to section a waveform.

negative time. It is said to be *sectioned at the origin.* Its mathematical expression is

$$f(t) = \sin t \times u_{-1}(t). \tag{8–25}$$

Essentially, multiplication by the function $u_{-1}(t)$ avoids the necessity for a separate note stating that the solution applies only for t greater than zero.

The resolution of periodic waveforms. We can express many periodic waveforms in terms of the delayed singularity functions which we have discussed. To achieve this, we view a periodic waveform as the result of a superposition of many successive singularity functions, each function delayed somewhat from the one preceding it. An example is shown in Fig. 8–16. The square wave of Fig. 8–16(a) can be produced by a summation of the unit step functions shown in Fig. 8–16(b). The first step function is positive and occurs at $t = 0$, giving the square-wave function a value of unity from $t = 0$ to $t = 1$; at $t = 1$ a second, negative step function occurs which subtracts from the first to leave a net value of zero for the square wave between $t = 1$ and $t = 2$; at $t = 2$ a third, positive step function occurs which, when added to the first two, produces a value of unity for the square wave from $t = 2$ to $t = 3$. At $t = 3$ a negative step function again occurs, and so on. We can obtain a mathematical expression for the square wave of Fig. 8–16(a) by adding the component step functions shown in Fig. 8–16(b). Thus,

$$f(t) = u_{-1}(t) - u_{-1}(t - 1) + u_{-1}(t - 2) - u_{-1}(t - 3) + \cdots \tag{8–26}$$

We may consider each step function in Eq. (8–26) to be a separate source.

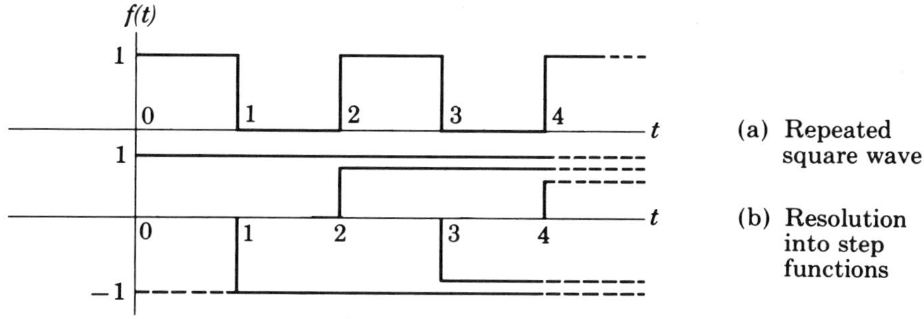

FIG. 8–16. The resolution of a periodic square wave into step functions.

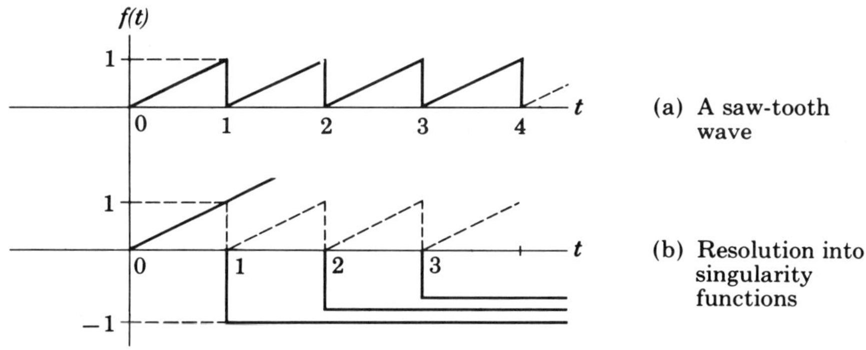

FIG. 8–17. The resolution of a saw-tooth wave into singularity functions.

We can then obtain the response each source produces in a network by considering all the other sources to be zero. The sum of all such individual responses is the response of the network to the original function.

A slightly more complicated example is shown in Fig. 8–17. From $t = 0$ to $t = 1$ the saw-tooth function [Fig. 8–17(a)] is represented by a unit ramp function which rises to a value of 1. At $t = 1$ a negative unit step function occurs which reduces the saw-tooth function to zero. The initial ramp continues to rise, however, and at $t = 2$, the net value is again unity. A second, negative unit step function occurs at $t = 2$, and so on. The mathematical expression for the saw-tooth function, obtained from Fig. 8–17(b), is

$$f(t) = u_{-2}(t) - u_{-1}(t - 1) - u_{-1}(t - 2) - u_{-1}(t - 3) - \cdots . \tag{8–27}$$

The over-all response of a network to the function of Eq. (8–27) can again be obtained by the superposition of the responses produced by the separate singularity functions.

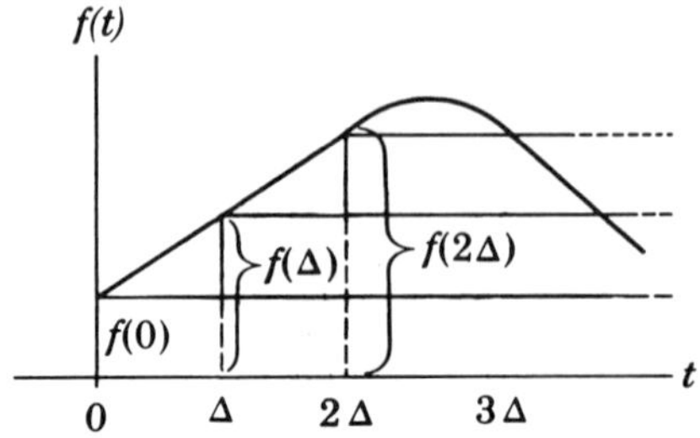

FIG. 8–18. The resolution of an arbitrary function into step functions.

The representation of arbitrary waveforms by step-function trains. A series of step functions with increasing delays is called a *step-function train.* It is possible to obtain an approximate expression for an arbitrary waveform in terms of such a train. Figure 8–18 shows a typical waveform, which we shall represent by a train of step functions with a spacing Δ between them. The initial step function occurs at $t = 0$ and has a height of $f(0)$. A mathematical expression for it is

$$f_1(t) = f(0) \times u_{-1}(t).$$

At $t = \Delta$, a second step function is added to bring the representative waveform into numerical agreement with the given waveform. The height of the second step function is the difference between the value of the given waveform at $t = \Delta$ and the value of the first step function, or $f(0)$. Thus we obtain for the second step function

$$\begin{aligned} f_2(t) &= [f(\Delta) - f(0)] \times u_{-1}(t - \Delta) \\ &= \Delta\left[\frac{f(\Delta) - f(0)}{\Delta}\right] \times u_{-1}(t - \Delta) \\ &= \Delta f'(\Delta) \times u_{-1}(t - \Delta), \end{aligned}$$

where $f'(\Delta)$ is the derivative of the given function, evaluated at $t = \Delta$. It can be shown that a third step function added at $t = 2\Delta$ has the value

$$f_3(t) = \Delta f'(2\Delta) \times u_{-1}(t - \Delta),$$

and that expressions for additional step functions are obtained in a similar manner. Thus a mathematical expression for the given waveform, in terms of the step functions, is

$$\begin{aligned} f(t) &= f(0)u_{-1}(t) \\ &\quad + \Delta[f'(\Delta) \times u_{-1}(t - \Delta) + f'(2\Delta) \times u_{-1}(t - 2\Delta) + \cdots]. \end{aligned} \tag{8–28}$$

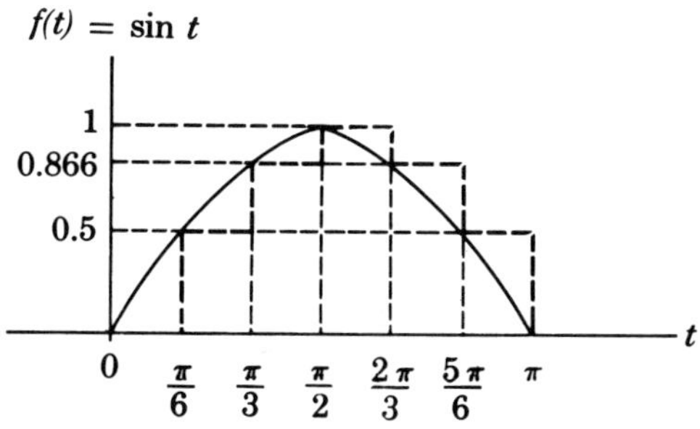

FIG. 8–19. The resolution of a sinusoidal pulse into step functions.

Equation (8–28) can be viewed as an approximate expression for the given waveform. The approximation becomes better as Δ becomes smaller. In the limit, Eq. (8–28) becomes an accurate expression for the given function and has the form of an integral. We shall have occasion to use both Eq. (8–28) and its integral form (in which Δ becomes infinitesimal).

Another example of the step-function approximation is shown in Fig. 8–19. Here, the approximation to the given sinusoidal pulse, indicated by the broken lines, is expressed by

$$\begin{aligned} f(t) &= 0.5u_{-1}(t - \pi/6) + 0.366u_{-1}(t - \pi/3) + 0.134u_{-1}(t - \pi/2) \\ &\quad - 0.134u_{-1}(t - 2\pi/3) - 0.366u_{-1}(t - 5\pi/6) - 0.5u_{-1}(t - \pi). \end{aligned} \tag{8–29}$$

If we conclude that Eq. (8–29) is not a sufficiently accurate representation of the sinusoidal pulse, we can use a larger number of step functions, spaced more closely together.

The representation of arbitrary waveforms by impulse trains. The representation of an arbitrary waveform by an impulse train is actually easier to achieve than the representation by a step-function train. It must be remembered, however, that an impulse is simply a short pulse which is relatively high with respect to its width. The exact shape of the pulse is of no consequence; only its area is significant. Thus an arbitrary waveform can be represented by impulses if it is first divided into a series of sections, as shown in Fig. 8–20. For convenience we take the base line of each section to be Δ in length. The area of the first section is approximately $f(0) \times \Delta$ and can be represented by an impulse occurring at some point within the interval $t = 0$ to $t = \Delta$. For convenience we assume that the impulse occurs at the beginning of the interval. The impulse representing the first area is

$$f_1(t) = \Delta \times f(0) \times u_0(t).$$

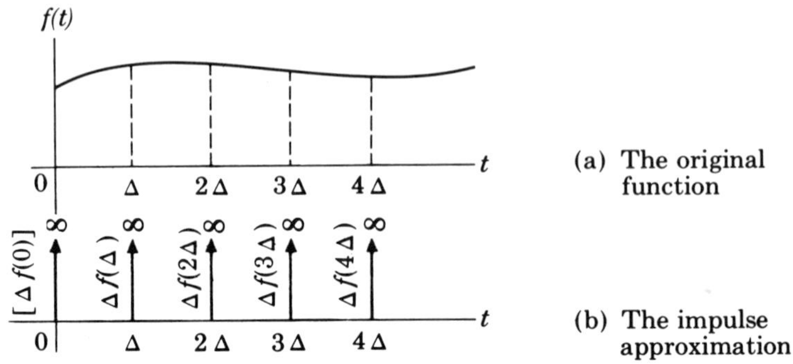

FIG. 8–20. The resolution of an arbitrary function into impulse functions.

The area of the second section is approximately $f(\Delta) \times \Delta$. The impulse representing it, occurring at the beginning of the time interval, is

$$f_2(t) = \Delta \times f(\Delta) \times u_0(t - \Delta).$$

Similarly, the impulse representing the third area is

$$f_3(t) = \Delta \times f(2\Delta) \times u_0(t - 2\Delta).$$

Thus the given waveform is represented by

$$f(t) = \Delta[f(0) \times u_0(t) + f(\Delta) \times u_0(t - \Delta) + f(2\Delta) \times u_0(t - 2\Delta) + \cdots]. \tag{8–30}$$

Equation (8–30) can also be viewed in two ways. As it stands, it is an approximate expression for an original waveform. If Δ is allowed to approach zero, it becomes an accurate expression for the given waveform and has the form of an integral. We shall have occasion to use both points of view in our later work. It may seem strange to replace a smooth curve by a series of infinite impulses, and the equivalence may seem somewhat forced. However, if we consider that the smooth torque produced by an automobile engine is really the result of a series of impulsive explosions, the concept may appear more reasonable. Since most physical systems tend to smooth out the effects of impulses, a relatively crude approximation gives a reasonably accurate answer. Another example of an approximation by an impulse train is shown in Fig. 8–21. Here, the pulse of Fig. 8–21(a) is divided into four intervals, and the area of each is obtained. The approximate waveform [Fig. 8–21(b)] consists of four impulses, having values of $\frac{7}{2}$, $\frac{5}{2}$, $\frac{3}{2}$, and $\frac{1}{2}$. Thus the given pulse is represented by

$$f(t) = \tfrac{7}{2}u_0(t) + \tfrac{5}{2}u_0(t - \Delta) + \tfrac{3}{2}u_0(t - 2\Delta) + \tfrac{1}{2}u_0(t - 3\Delta). \tag{8–31}$$

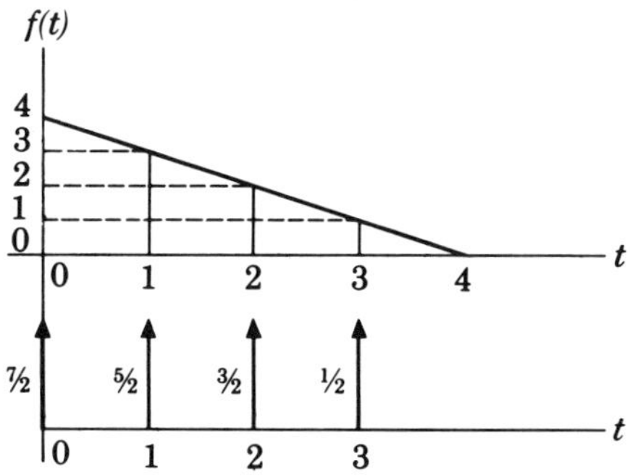

(a) The original pulse

(b) The impulse approximation

FIG. 8–21. An example of the resolution of a waveform into impulses.

Again, if we believe that Eq. (8–31) is not a sufficiently accurate representation for the given pulse, we can use more impulses, spaced closer together. One technique for testing accuracy in these cases is to solve the problem twice, using more impulses in the second calculation, to see by how much the answer changes. If it doesn't change by much, the original spacing of impulses is considered satisfactory.

We shall use singularity functions a great deal in our future work. As a first step, we shall use them to characterize a number of simple networks. This will consist of applying step functions and impulses to the networks and noting the corresponding response functions.* Knowing the response of a network to an impulse or step function, we can then find the response to more arbitrary input functions by resolving the input function into impulse trains or step-function trains, utilizing the methods developed in this chapter. In order to find step and impulse responses of simple networks, we shall need to use the differential equations discussed in Chapter 9.

8–7 Summary.

Operational derivative (slope at time t)

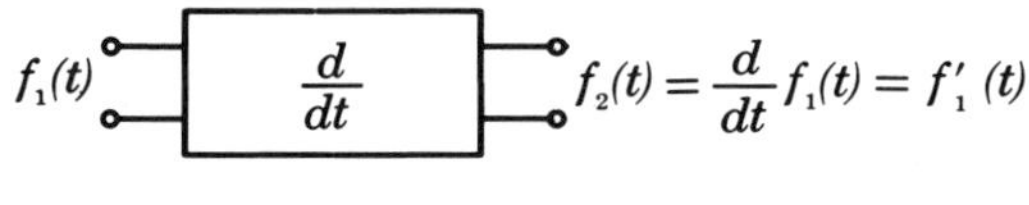

Operational integral (area up to time t)

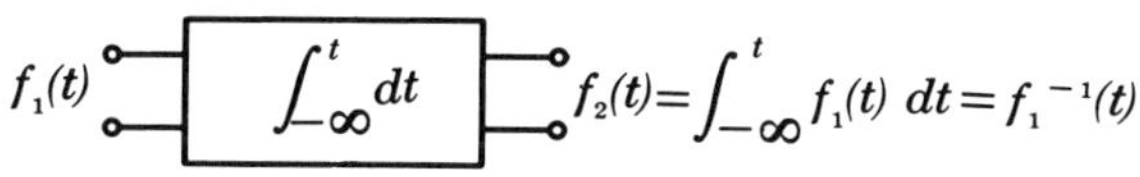

* In physics, impulse responses are usually called *Green's functions*. They are widely used both for functions of time and for functions of space.

Relationship between derivative and integral

$$\frac{d}{dt}[f^{-1}(t)] = f(t)$$

$$\int_{-\infty}^{t} f'(t) = f(t)$$

The singularity functions

$$u_n(t) = \frac{d}{dt}[u_{n-1}(t)] = \int_{-\infty}^{t} u_{n+1}(t)\,dt$$

$\vdots$

Unit ramp $\quad u_{-2}(t) = \begin{cases} 0 \text{ for } t < 0 \\ t \text{ for } t > 0 \end{cases}$

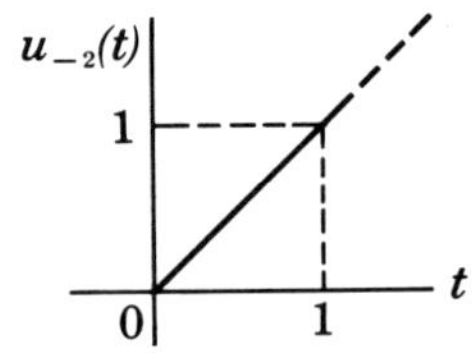

Unit step $\quad u_{-1}(t) = \begin{cases} 0 \text{ for } t < 0 \\ 1 \text{ for } t > 0 \end{cases}$

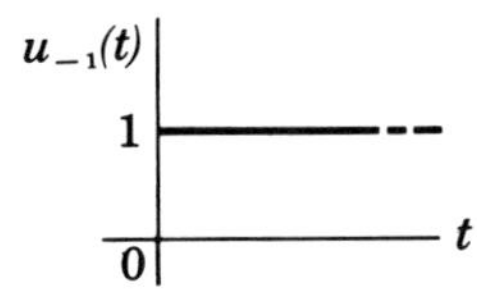

Unit impulse $\quad u_0(t) = \begin{cases} 0 \text{ for } t > 0 \\ 0 \text{ for } t < 0 \\ \infty \text{ for } t = 0 \end{cases}$

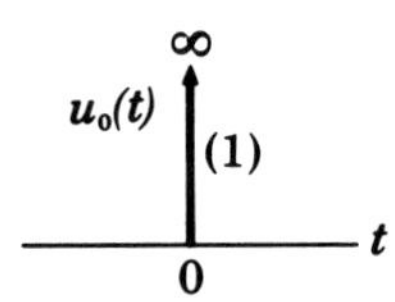

Unit doublet $\quad u_1(t) = \begin{cases} 0 \text{ for } t > 0 \\ 0 \text{ for } t < 0 \\ \infty \text{ for } t = 0^- \\ -\infty \text{ for } t = 0^+ \end{cases}$

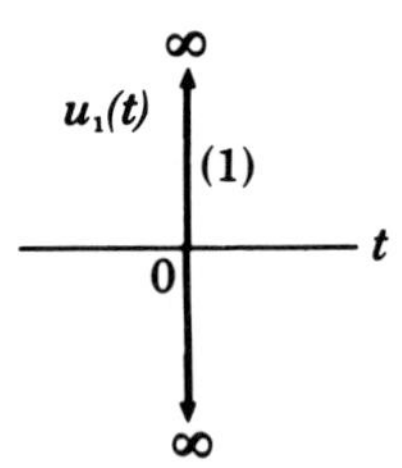

$\vdots$

Delayed singularity functions

$$u_n(t - a) = u_n(t) \text{ delayed to start at } t = a$$

Step approximation

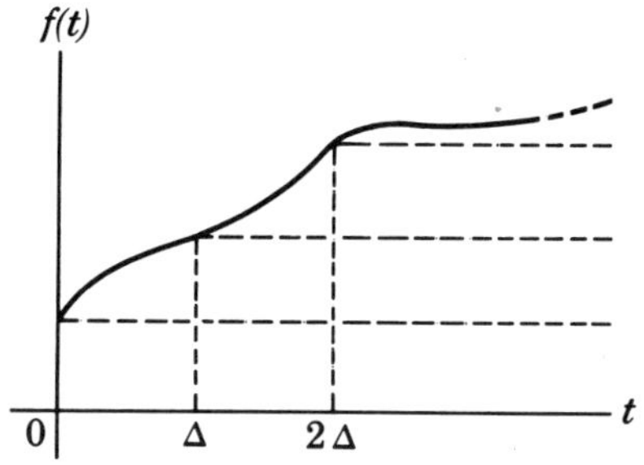

$$f(t) \cong f(0)u_{-1}(t) + [f(\Delta) - f(0)]u_{-1}(t - \Delta) + [f(2\Delta) - f(\Delta)]u_{-1}(t - 2\Delta) \cdots$$
$$= f(0)u_{-1}(t) + \Delta[f'(\Delta)u_{-1}(t - \Delta) + f'(2\Delta)u_{-1}(t - 2\Delta) + \cdots]$$

Impulse approximation

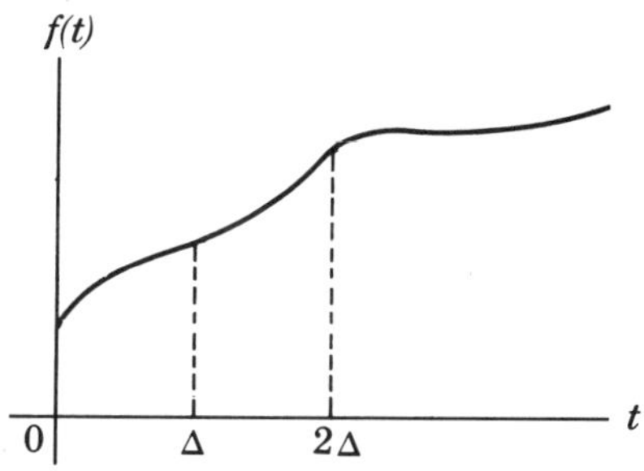

$$f(t) \cong \Delta[f(0)u_0(t) + f(\Delta)u_0(t - \Delta) + f(2\Delta)u_0(t - 2\Delta) + \cdots]$$

REFERENCES

The subjects in this chapter are mathematical. However, the best reference material will be found not in the mathematics field but in the field of electrical engineering. Functional integrals and derivatives are not stressed in elementary calculus because of the importance ascribed to the indefinite integral. The concept of integration and differentiation as "time operators" has been developed largely in the field of analog computers. The idea of impulse modulation and sampling comes from the field of digital computers. The whole field of the singularity functions has been avoided by conventional mathematicians. It is only within the last few years that mathematicians have acknowledged that they could exist. Listed below are a few selected references for this material.

1. G. B. THOMAS, JR., *Calculus and Analytical Geometry*, Addison-Wesley, Reading, Mass., 1954. A good introductory text to the calculus. Integration is discussed from a number of points of view, both mathematical and physical.

2. D. K. Cheng, *Analysis of Linear Systems*, Addison-Wesley, Reading, Mass., 1959. This text gives an excellent discussion of the impulse function and points out why distribution functions must be used in a rigorous discussion. A chapter on Z-transforms which use impulse sampling is also included.

3. E. A. GUILLEMIN, *Introductory Circuit Theory*, John Wiley & Sons, New York, 1953. Guillemin points out clearly the advantages of using the definite integral from minus infinity to t rather than the indefinite integral. He also gives an elementary introduction to the singularity functions.

4. E. A. GUILLEMIN, *The Mathematics of Circuit Analysis*, John Wiley & Sons, New York, 1949. This graduate-level text gives a semi-rigorous discussion of the singularity functions.

5. L. SCHWARTZ, *Theorie des Distributions*, Hermann et Cie, Paris, 1950. The original rigorous treatment of singularity functions.

6. B. FRIEDMAN, *Principles and Techniques of Applied Mathematics*, John Wiley & Sons, New York, 1956. Friedman gives a useful account of distribution functions.

7. KORN and KORN, *Analog Computers*, McGraw-Hill, New York, 1956. This standard text on analog computers stresses the operational meaning of integration. This point of view is very useful, especially in understanding differential equations.

8. W. K. LINVILL, "Sampled-data Control System," *Trans. AIEE*, Vol. 70, Part II, pp. 1779–1788 (1951).

9. RAGAZZINI and ZADEH, "The Analysis of Sampled-Data Systems," *Trans. AIEE*, Vol. 71, Part II, pp 225–232 (1952). This paper and that of Reference 8 show the relationship of impulse approximations to the conventional theory of numerical analysis, and introduce a convenient technique for performing it, called Z-transforms.

10. H. A. HELM, "The Z-transform," *B.S.T.J.*, Vol. XXXVIII (Jan. 1959). A rigorous treatment of singularity functions and Z-transforms by mean of Lebesgue integrals.

EXERCISES

Section 8–2

1. Sketch the derivative of the function shown in Fig. 8–22.

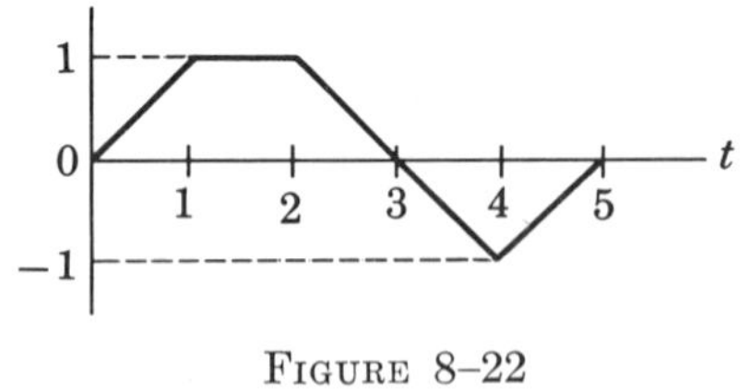

FIGURE 8–22

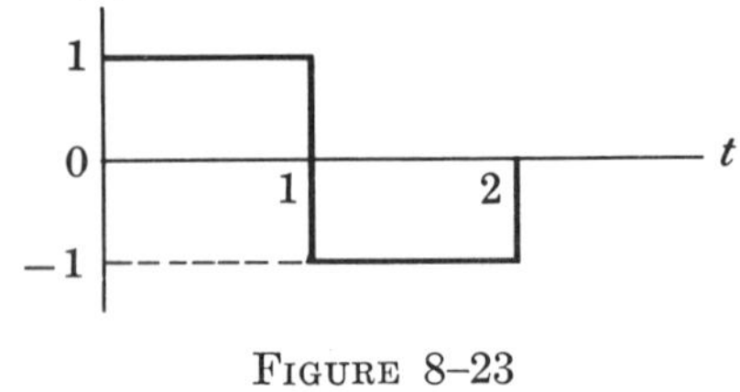

FIGURE 8–23

2. Sketch the derivative of the function shown in Fig. 8–23.
3. Sketch the derivative of the function shown in Fig. 8–24.

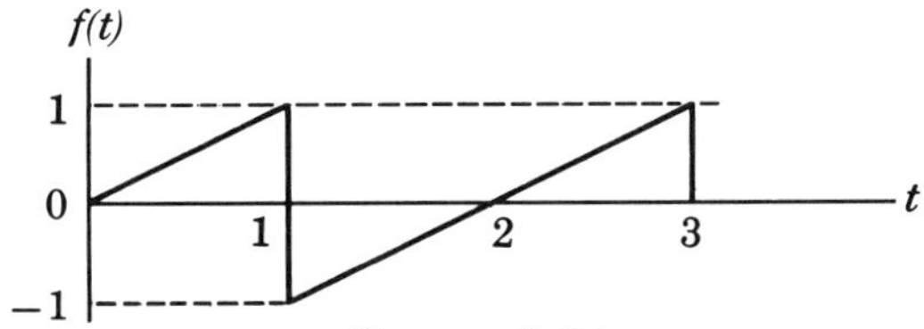

FIGURE 8–24

4. Sketch the function $f(t) = \sin t$ from $t = 0$ to $t = 2\pi$. Also, obtain and sketch its derivative over the same range.

5. Sketch the function $f(t) = (1 - e^{-t})$ and its derivative from $t = 0$ to infinity.

6. Obtain the first, second, and third derivatives of the function $f(t) = e^{2t}$. Sketch each over all values of time.

Section 8–3

7. Sketch the integral of the waveform shown in Fig. 8–22.
8. Sketch the integral of the waveform shown in Fig. 8–23.
9. Sketch the integral of the waveform shown in Fig. 8–24.
10. Sketch the function $f(t) = \sin t$ from $t = 0$ to $t = 2\pi$. Assuming that the function is zero over the rest of time, obtain and sketch the integral.
11. A function is zero for negative values of t and equal to $e^{-t/10}$ for positive values of t. Obtain and sketch the integral.
12. Sketch the first, second, and third integrals of the function shown in Fig. 8–25.

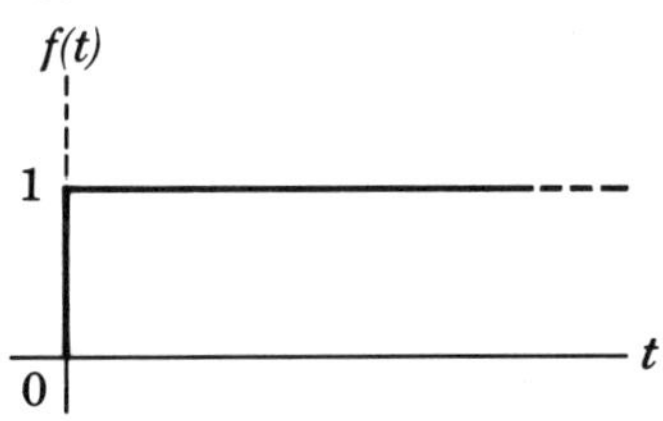

FIGURE 8–25

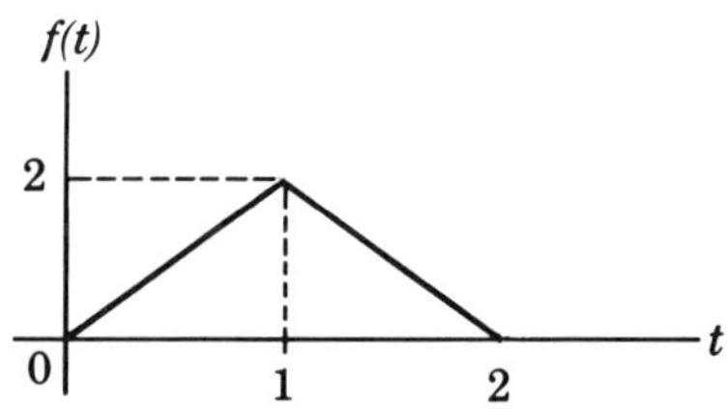

FIGURE 8–26

13. Differentiate the function shown in Fig. 8–26. Integrate the resulting waveform to obtain the original function.

Section 8–4

14. Differentiate the function $f(t) = \sin t$. Integrate the resulting expression from $t = 0$ to $t = t$. Comment on the inverse property of the curves.
15. Integrate the function $f(t) = \sin t$ from $t = 0$ to $t = t$. Differentiate the resulting expression, and comment on the inverse nature of the curves.
16. Differentiate the function $f(t) = \cos t$. Integrate the resulting expression from $t = 0$ to $t = t$. Comment on the inverse nature of the curves.

17. Differentiate the function $f(t) = e^t$. Integrate the resulting curve from minus infinity to t. Do you obtain the original curve?

18. Repeat Exercise 17 but integrate from $t = 0$ to $t = t$. Explain any discrepancy which occurs.

19. A curve $f(t)$ is defined by

$$f(t) = \begin{cases} e^t & \text{for} \quad t > 0, \\ 0 & \text{for} \quad t < 0. \end{cases}$$

Note that the derivative of this curve is infinite at $t = 0$. What area must be enclosed under the infinite derivative at $t = 0$ to make the integral of the derivative of $f(t)$ equal to the original function?

Section 8–5

20. Sketch the derivative of the function shown in Fig. 8–27. Find the area under the derivative curve. What does the area become when a is allowed to approach zero? What does the derivative become?

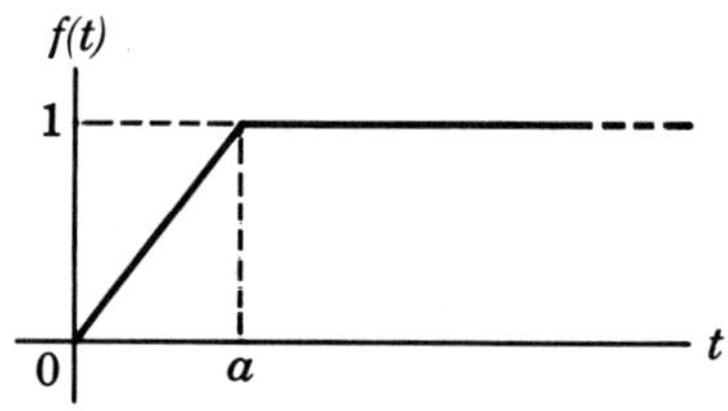

FIGURE 8–27

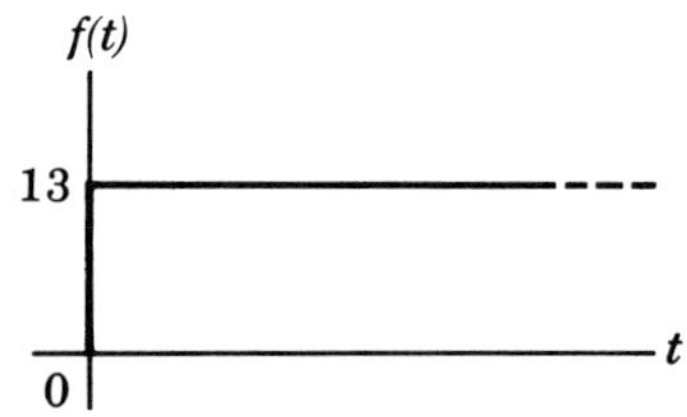

FIGURE 8–28

21. Write an analytic expression for the waveform shown in Fig. 8–28.

22. Find an analytic expression for the derivative of the waveform shown in Fig. 8–28. Sketch the derivative.

23. Find an analytic expression for the integral of the waveform shown in Fig. 8–28. Sketch the integral.

24. Sketch the function $f(t) = \cos t \times u_{-1}(t)$. Obtain the derivative from the sketch. Also, obtain the derivative analytically by the formula for the derivative of the product of two functions.

25. Find the integral of the function $f(t) = \sin t \times u_{-1}(t)$ from minus infinity to t by the formula for integration by parts, and also by absorbing $u_{-1}(t)$ in the limits of the integral.

26. Sketch the following two functions, their derivatives, and their integrals:

$$\text{(a)} \quad u_0(t) \times \cos t, \qquad \text{(b)} \quad u_0(t) \times \sin t.$$

Section 8–6

27. Express the waveform of Fig. 8–23 in terms of singularity functions. Express the integral and derivative of the waveform in the same manner.

28. Express the waveform of Fig. 8–24 in terms of singularity functions. Express the integral and derivative of the waveform in the same manner.

29. Find an expression for the pulse shown in Fig. 8–29 in terms of singularity functions.

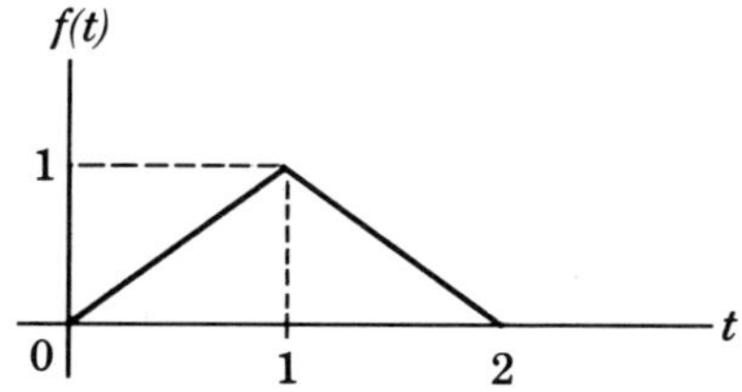

FIGURE 8–29

30. Write an analytical expression for the waveform shown in Fig. 8–30 in terms of singularity functions.

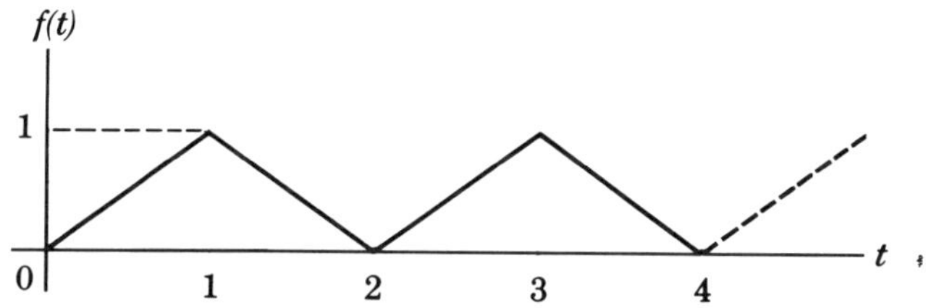

FIGURE 8–30

31. The pulse shown in Fig. 8–31 consists of the first half-cycle of a sine wave. Obtain an expression for it in terms of sinusoids and singularity functions.

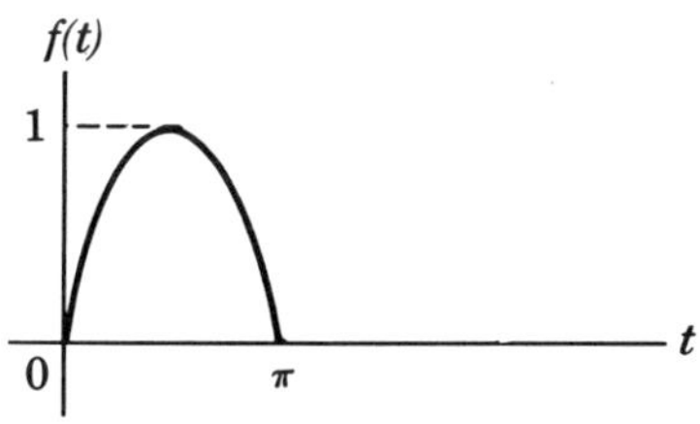

FIGURE 8-31

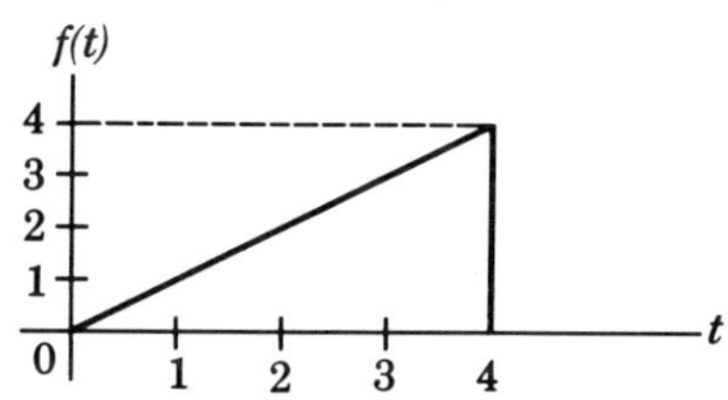

FIGURE 8–32

32. Sketch the waveform of the following expression:

$$f(t) = u_{-1}(t) + u_{-1}(t-1) + u_{-1}(t-2) + \cdots.$$

33. Sketch the waveform of the following expression:

$$f(t) = u_{-2}(t) - 2u_{-2}(t-1) + u_{-2}(t-2).$$

34. Sketch the waveform obtained when the functions in Exercises 32 and 33 are multiplied together.

35. Write an expression for the triangular wave of Fig. 8–32 in terms of singularity functions. Also, obtain an approximate expression in terms of five equally spaced step functions.

36. Obtain the integrals and the derivatives of the two expressions for the triangular wave of Fig. 8–32. What conclusion can you draw?

CHAPTER 9

THE DIFFERENTIAL EQUATIONS OF ELECTRIC CIRCUITS

"The theory of differential equations is the most important branch of mathematics. It occupies a central position from which lines of development extend in many directions."

Sophus Lie

9–1 Introduction. In this chapter we shall digress from circuit theory to study differential equations, which will appear in the discussions of later chapters. Another reason for studying differential equations is that they provide a link between the problems which we shall solve in electric circuit theory and similar problems in a host of other fields. In a very real sense differential equations are central to the study of mathematics, physics, and all the applied sciences. The starting point for a study of almost any branch of physics or applied science is a differential equation. The starting point for a study of electric circuits is the linear differential equation with constant coefficients.

Linear differential equations with constant coefficients are perhaps the simplest of all differential equations.* We have already written some typical equations of this type and have verified that the linearity and superposition theorems are applicable to them. We shall make frequent use of these theorems in solving the equations of this chapter. Since differential equations represent circuits, the concepts of stimulus and response are still applicable. The simplest sort of stimulus for a network is provided by energy initially stored in one of the circuit elements. The stored energy causes currents to flow in the circuit. As the currents flow, the energy is dissipated in the resistances, and eventually the currents decrease to zero. The response of a network excited by an initial energy storage and then left undistributed is a characteristic of the network. The response is known as the network's *natural behavior*, or its *transient response*, because the currents and voltages decrease to zero after an interval of time. The term *force-free behavior* is also applied to this response function, since it is produced by the network itself without an externally applied source. From a mathematical standpoint the natural behavior of a network is the solution of the differential equation with all the sources equated to zero.

* The coefficients may vary as functions of the independent variable and the equations remain linear. We shall not treat this case.

This solution is called the *complementary function.* All four of these names are in general use, and we shall make no distinction among them.

The response of a network to excitation by an impulse source is very similar to the network's natural behavior. The impulse exists only between $t = 0^-$ and $t = 0^+$. After this interval it is zero, just as it would be for obtaining the complementary function of the differential equation. The only way in which the impulse affects the network is to store some energy during the impulse interval. The stored energy then produces the natural behavior of the network, after the impulse is over. Thus, *after* $t = 0^+$ *the impulse response of a network is equal to its natural behavior.* This is one of the reasons for the importance of the impulse response.

The response of a network to excitation by a step function can be found by integration of the impulse response. The linearity theorem extends to integration and differentiation of the stimulus and response, since they are linear operations. Alternatively, the response to a step function can be obtained directly from the appropriate differential equation. In this case a final value, or *steady-state solution,* exists which is proportional to the excitation and does not decrease to zero with time. The steady-state value is simply the solution for the network at $t = \infty$; it is identical to the direct-current value treated in earlier chapters. The *complete solution* of a network differential equation is the sum of the natural behavior and the steady-state solution. Since the steady-state solution by itself satisfies the differential equation, it might be asked why we bother with the transient solution. The answer is: Because the steady-state solution by itself does not satisfy the initial conditions (conditions at $t = 0$) in the circuit. The transient solution provides a smooth transition from the initial energy state of the network, as represented by the initial values of the currents and voltages, to the final energy state, as represented by the final values of the currents and voltages.

We have already learned that it is possible to resolve a more general excitation into impulse trains or step-function trains. We can therefore treat the more general excitation by superposition. Also, it is possible to solve the differential equation directly for the response to the more general excitation. As a brief illustration, consider the application to a network of a sinusoidal type of excitation. In this case the complete solution of the differential equation is the transient solution plus a solution which is of the same nature as the excitation function. The latter is still called a steady-state solution, although it is not a constant. A more accurate term for it is *the forced solution.* Mathematically, the forced solution is called a *particular integral* of the differential equation.

An example which clarifies some of the concepts we have just introduced is provided by the pendulum, or swing. This system and the R–L–C circuit satisfy the same differential equation. Energy can be initially stored in the system in either of two ways. If the pendulum is lifted up from its rest position and held, it is given potential energy; if the pendulum in its rest position is given a velocity, it acquires kinetic energy. In either case if the pendulum is left undisturbed, after the initial energy storage, it swings back and forth with a constant frequency until the stored energy is dissipated by windage and friction. These decaying oscillations constitute the natural behavior of the system. If a step function of force, for example the wind, is suddenly applied to the pendulum, the result is twofold: (1) transient oscillations occur which eventually die out; and (2) the pendulum acquires a steady-state offset in its position. It is also possible to drive the pendulum back and forth at any desired frequency by applying a force at this frequency. In this case the pendulum oscillates at the desired frequency after the transient oscillations at the natural frequency have disappeared.

9–2 A first-order system excited by internal energy storage. An electric circuit with a single energy-storage element is represented by a first-order differential equation. The problem, however, is not completely described by the differential equation. The energy initially stored must also be stated. A knowledge of the stored energy is required to compute the initial value of the variable of interest in the problem. In summary, the specification of a problem concerned with single energy storage requires

1. The statement of the differential equation.
2. The initial value of the variable.

The differential equation is written by using the methods for resistive circuits which we have already discussed. The initial value of the variable is computed (by the methods described in Chapter 7) from the purely resistive circuit which holds at $t = 0$. Our principal concern at this point is the solution of the differential equation.

R–L circuit. An example of a single-energy-storage circuit is the resistance-inductance combination shown in Fig. 9–1. The differential equation for this circuit is

$$L\frac{di}{dt} + Ri = 0. \tag{9–1}$$

To solve the equation we separate the variables:

$$\frac{di}{i} = \frac{-R}{L}\,dt. \tag{9–2}$$

The current function on the left side of Eq. (9–2) is identically equal to the time function on the right side. The integrals are equal provided we integrate each function between corresponding limits. Thus if we inte-

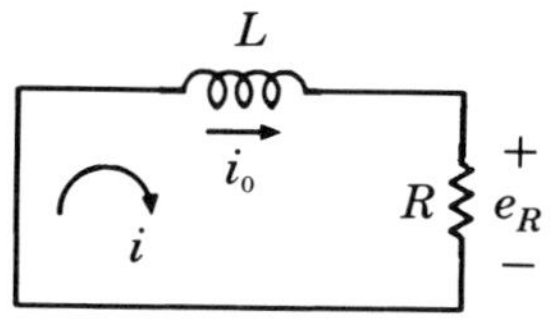

FIG. 9–1. An R–L circuit with initial energy storage.

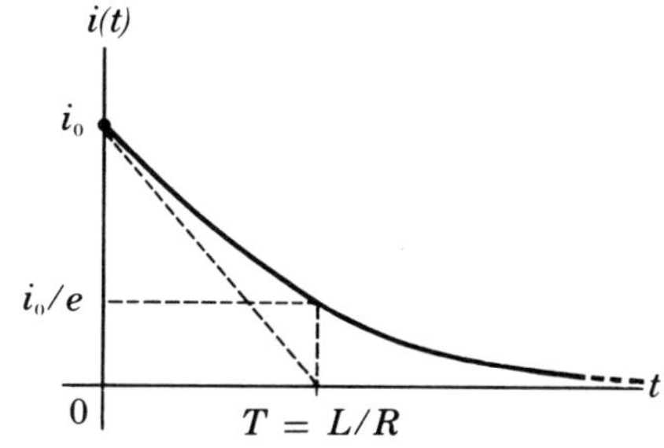

FIG. 9–2. The damped exponential curve $i = i_0 e^{-Rt/L}$

grate from $t = 0$ to $t = t$ on the right side, we must integrate from the value of the current at $t = 0$, namely, i_0, to the value of the current at $t = t$, namely, i. Accordingly

$$\int_{i_0}^{i} \frac{di}{i} = \int_0^t \left(\frac{-R}{L}\right) dt,$$

from which

$$\log i - \log i_0 = \frac{-R}{L} t,$$

or

$$i = i_0 e^{-Rt/L}. \tag{9–3}$$

Equation (9–3) shows that the natural behavior of the current in the R–L circuit is a damped exponential curve which starts at the initial value i_0 and decays to zero, as shown in Fig. 9–2. At $t = L/R$ the value of the current is

$$i = i_0 e^{-1} = \frac{i_0}{e}, \tag{9–4}$$

or about one-third of the initial value. The time L/R is called the *time constant* of the circuit. An alternative interpretation of the time constant can be obtained from the derivative of the current,

$$\frac{di}{dt} = \frac{-Ri_0}{L} e^{-Rt/L}$$

the value of which at $t = 0$ is

$$\frac{di(0)}{di} = \frac{-Ri_0}{L}. \tag{9–5}$$

If the exponential curve continued to fall (after $t = 0$) at the rate given by Eq. (9–5) for one time constant, L/R seconds, it would drop by i_0 amperes. Thus the time constant is the time at which the extension of the initial derivative cuts the t-axis.

In Fig. 9–1, the voltage across the resistance, and hence that across the inductance, is

$$e_R = Ri_0e^{-Rt/L}. \tag{9–6}$$

All the currents and voltages in the circuit have equations in the form of damped exponentials, which are characterized by their initial values and time constants. Thus the natural or transient behavior of every current or voltage in the circuit of Fig. 9–1 can be represented by an equation of the form

$$f(t) = Ae^{-t/T}, \tag{9–7}$$

where A = initial value of the function,
$T = L/R$ (time constant).

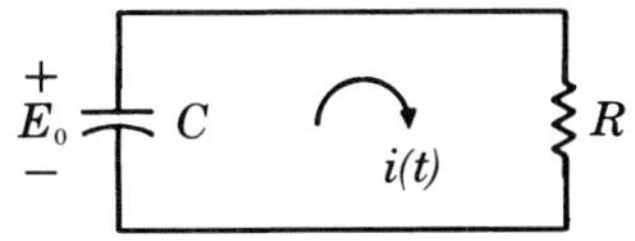

FIG. 9–3. An R–C circuit with initial energy storage.

R–C circuit. The R–C circuit shown in Fig. 9–3 is also represented by a differential equation of the first order:

$$Ri + \frac{1}{C}\int_{-\infty}^{t} i\,dt = 0,$$

or

$$R\frac{di}{dt} + \frac{i}{C} = 0. \tag{9–8}$$

Equation (9–8), being similar to Eq. (9–1), is solved in a similar manner. The solution is

$$i = I_0e^{-t/RC} = I_0e^{-t/T} \tag{9–9}$$

where I_0 = initial value of the current,
$T = RC$ (time constant).

The initial value of the current is computed from the state of the circuit at $t = 0$. At this instant, the voltage on the capacitance is E_0. Since this voltage also appears across the resistance, the initial current is E_0/R.

Thus, as a function of time the current is

$$i = \frac{E_0}{R} e^{-t/T}. \tag{9–10}$$

In the circuit of Fig. 9–3 the voltage across R and the voltage across C are both given by expressions having the form

$$e_R = E_0 e^{-t/RC}. \tag{9–11}$$

The natural behavior of a single-energy-storage circuit involving capacitance is a damped exponential, having the form

$$f(t) = Ae^{-t/T}, \tag{9–12}$$

where A = initial value of the variable,
T = RC (time constant).

All the currents and voltages are given by expressions having the form shown in Eq. (9–12). This equation is identical to Eq. (9–7) for the inductive circuit, except that the time constant is now RC instead of L/R.

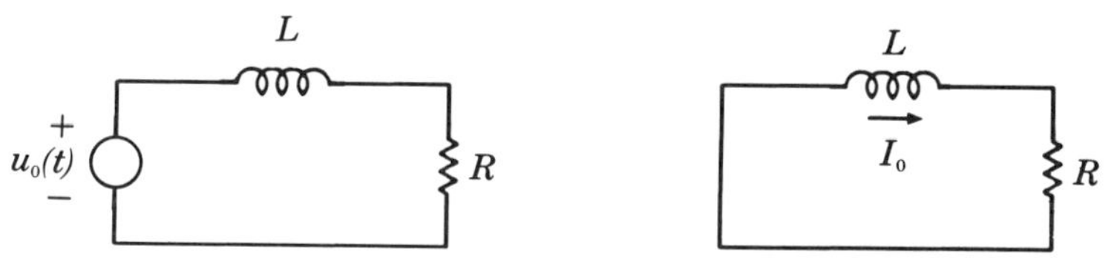

(a) Circuit excited by an impulse (b) Circuit for $t > 0$

FIG. 9–4. An R–L circuit excited by an impulse.

Impulse excitation of the inductive circuit. Since an impulse is zero for t greater than zero it does not appear in the differential equation, which describes the variable only for t greater than zero. The impulse affects the solution only through the initial value of energy storage which it establishes during its short active interval. An inductive circuit excited by an impulse is shown in Fig. 9–4(a). The state of the circuit for t greater than zero is shown in Fig. 9–4(b). The differential equation for the circuit of Fig. 9–4(a) is

$$L\frac{di}{dt} + Ri = u_0(t).$$

For t greater than zero this equation becomes

$$L\frac{di}{dt} + Ri = 0, \tag{9–13}$$

and the solution is

$$i = I_0 e^{-Rt/L}. \tag{9–14}$$

The impulse can affect the solution only through the value of I_0. To make this clear we direct our attention to the three time instants 0^-, 0, and 0^+. At $t = 0^-$, before the impulse strikes, the circuit is at rest and the current is zero. At $t = 0$, when the impulse is acting, the inductance behaves as an open circuit, and all the voltage appears across it. The current established in the inductance between $t = 0^-$ and $t = 0^+$ is

$$i(0^+) = \frac{1}{L}\int_{0^-}^{0^+} u_0(t)\,dt. \tag{9–15}$$

To solve for the current in Eq. (9–15) we must find the area under the impulse curve from 0^- to 0^+. By definition, this area is the value of the impulse, which in Eq. (9–15) is unity. Thus

$$i(0^+) = \frac{1}{L}\cdot \tag{9–16}$$

Substituting the value of the current in Eq. (9–16) for I_0 in Eq. (9–14), the general expression for the natural behavior of the circuit, results in the impulse response

$$i = \frac{1}{L}e^{-tR/L}. \tag{9–17}$$

Thus the current response to an impulse voltage has the same form as the natural behavior of the circuit.

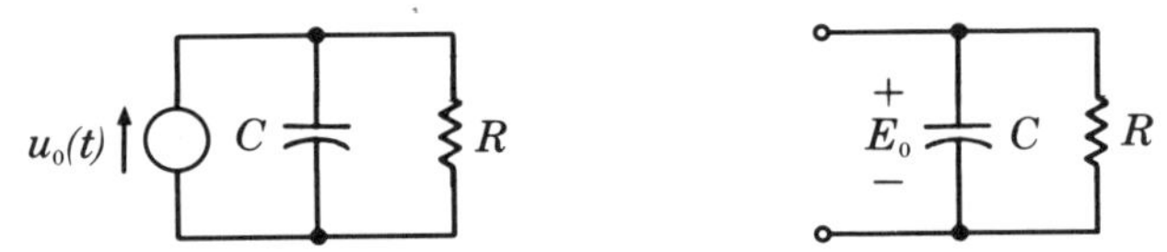

(a) Circuit excited by an impulse (b) Circuit for $t > 0$

FIG. 9–5. An R–C circuit excited by an impulse.

Impulse excitation of the capacitive circuit. An R–C circuit excited by a parallel impulse of current is shown in Fig. 9–5(a). The state of the circuit for t greater than zero is shown in Fig. 9–5(b). The differential equation for the circuit of Fig. 9–5(a) is

$$C\frac{de}{dt} + \frac{e}{R} = u_0(t), \tag{9–18}$$

and for the circuit of Fig. 9–5(b) is

$$C\frac{de}{dt} + \frac{e}{R} = 0. \tag{9–19}$$

The solution of Eq. (9–19) is

$$e = E_0 e^{-t/RC}. \tag{9–20}$$

The impulse can affect Eq. (9–20) only through the initial value of the voltage, E_0. The impulse stores energy in the capacitance between $t = 0^-$ and $t = 0^+$, and this energy then excites the circuit. At $t = 0^-$ the circuit is at rest, and there is no charge on the capacitance. At $t = 0$, when the impulse strikes, the capacitance acts as a short circuit, and all the current flows into it. The voltage produced by the action of the impulse is

$$e_C(0^+) = \frac{1}{C}\int_{0^-}^{0^+} u_0(t)\,dt. \tag{9–21}$$

Since the area under the impulse curve from $t = 0^-$ to $t = 0^+$ is unity

$$e_C(0^+) = \frac{1}{C}\cdot \tag{9–22}$$

Substituting $1/C$ for the initial voltage, E_0, in Eq. (9–20), we obtain the expression for the impulse response of the circuit:

$$e = \frac{1}{C}e^{-t/RC}. \tag{9–23}$$

Again, the impulse response has the same analytical form as the natural behavior of the circuit.

In the classical theory of differential equations emphasis is placed on the natural, or force-free, behavior of a differential equation. In circuit theory the impulse response is used more often than the natural behavior, since a stimulus-response relationship of this sort can characterize a network. Apart from a constant multiplier, however, the two functions are identical. In any case the constant multiplier depends on the exact nature of the excitation and the exact point at which the response is required. The same exponential term appears in all the natural responses. It is truly a characteristic of a network.

9–3 First-order system excited by a step function. If a first-order circuit is suddenly excited by a constant value (step function), it produces a response which is the sum of its natural behavior and a steady-state value proportional to the amplitude of the excitation. The steady-state value, which remains after the natural behavior has disappeared, is the value of

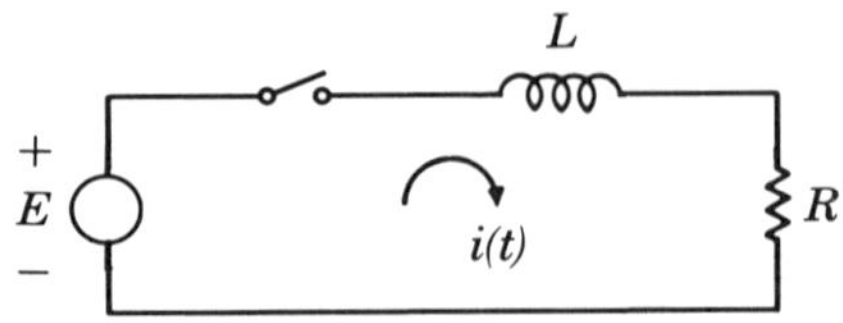

FIG. 9–6. An R–L circuit excited by a step function of voltage.

the circuit response at $t = \infty$. Since the circuit is resistive at $t = \infty$ the steady-state value is obtained by solving a resistive problem of the sort described in the earlier chapters. It is possible, of course, for the steady-state value to be zero. Then, the step-function response consists solely of the transient response. In general, however, a constant is added to the expression to account for the steady-state value.

A series L–R circuit excited by a step function of voltage is shown in Fig. 9–6. Assume the switch is closed at $t = 0$. For t greater than zero the differential equation is

$$L\frac{di}{dt} + Ri = E. \tag{9–24}$$

When the variables are separated

$$\frac{di}{E/R - i} = \frac{R}{L}\,dt. \tag{9–25}$$

Since the two functions in Eq. (9–25) are equal, the right side can be integrated from $t = 0$ to $t = t$, and the left side from the current value at $t = 0$ to the current value at $t = t$. Thus

$$\int_{I_0}^{i} \frac{di}{E/R - i} = \int_0^t \frac{R}{L}\,dt,$$

from which

$$\log\left(\frac{E/R - i}{E/R - I_0}\right) = \frac{-R}{L}\,t$$

and since I_0 is zero for the given excitation,

$$i = \frac{E}{R}\,(1 - e^{-Rt/L}). \tag{9–26}$$

The first term in Eq. (9–26) is the steady-state, or constant, value and the second term is the transient. At $t = \infty$ the inductance is a short circuit, and the current is E/R. The transient term has the same form as Eq. (9–7) which expresses the natural behavior of the R–L circuit. The function in Eq. (9–26) is zero when $t = 0$ and rises to a final value of E/R

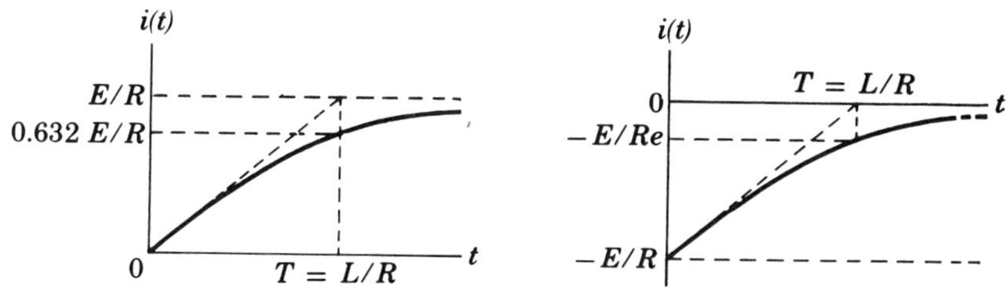

(a) The rising exponential curve (b) Curve (a) minus E/R

FIG. 9–7. The rising exponential curve $i(t) = (E/R)(1 - e^{-Rt/L})$.

when t is very large. The transition between the two states is accomplished smoothly by means of the natural response function $e^{-Rt/L}$, as indicated by Fig. 9–7(a). In one time constant the curve comes within approximately one-third of its final value; to be accurate, it reaches $(1 - 1/e)$ of the final value, E/R. It can be seen that our second interpretation of the time constant is also valid here. If the derivative at $t = 0$ is extended it reaches the final value at $t = L/R$, that is, in one time constant. The curve is, in truth, no more than a damped exponential starting at a negative value and decaying to zero. This interpretation of the curve is shown in Fig. 9–7(b). In Fig. 9–7(a) the axes are shifted downward so that the final value is E/R rather than zero.

The step response can also be obtained as the integral of the impulse response, but this point of view belongs more to circuit theory than to differential equations, and we shall explore it in a later chapter.

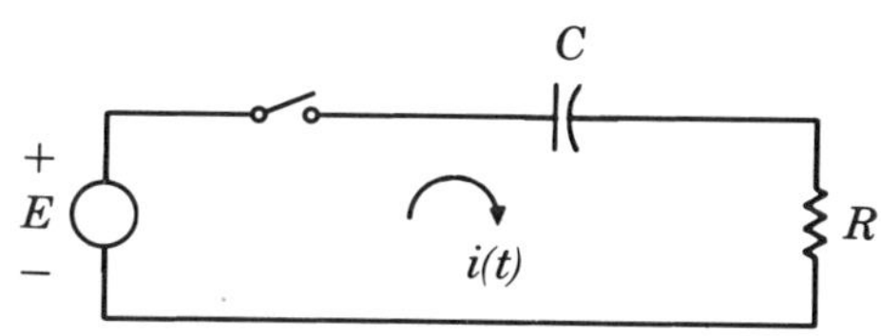

FIG. 9–8. An R–C circuit excited by a step function of voltage.

As a second example of the response of a circuit to a suddenly applied constant excitation we shall consider the circuit of Fig. 9–8. Again, the switch is closed at $t = 0$. For t greater than zero the differential equation is

$$Ri + \frac{1}{C}\int_{-\infty}^{t} i\,dt = E$$

or

$$R\frac{di}{dt} + \frac{i}{C} = 0. \tag{9–27}$$

Since this differential equation is identical to Eq. (9–8) the solution is

$$i = I_0 e^{-t/RC}. \tag{9–28}$$

In this case the steady-state solution is zero. A transient is excited, however, since there is an initial value of current. At $t = 0$ the capacitance acts as a short circuit, the voltage appears across the resistance, and the current is

$$i_0 = \frac{E}{R}. \tag{9–29}$$

Substituting E/R for I_0 in Eq. (9–28) we obtain the step response

$$i = \frac{E}{R} e^{-t/RC}. \tag{9–30}$$

A step response may have a constant term, as in Eq. (9–26), or it may not have a constant term, as in Eq. (9–30), depending on the value of the response at $t = \infty$. A transient term consisting of a damped exponential is always present, its amplitude depending on the value of the response at $t = 0$. Essentially, the transient makes up the difference between the value of the response at $t = 0$ and the steady-state value at $t = \infty$. The importance of the initial response and the final response in determining the complete solution cannot be overstressed.

9–4 First-order system excited by a sinusoidal wave. A general excitation function can be resolved into step functions or impulses and treated by the method of superposition. It can also be treated directly, and the differential equation can be solved for the given driving function. As an example of the direct method we shall discuss sinusoidal excitation. Figure 9–9 shows a series R–L circuit to which a sinusoidal voltage is applied at $t = 0$. The differential equation is

$$L\frac{di}{dt} + Ri = E \sin \omega t. \tag{9–31}$$

Remembering our previous solutions we look for a response function for Eq. (9–31) which is the sum of a transient solution and a steady-state

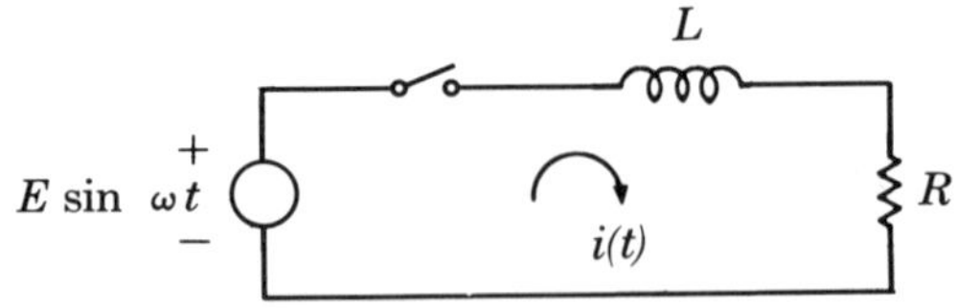

FIG. 9–9. An R–L circuit excited by a sinusoidal voltage.

solution. The transient solution, according to Eq. (9–3), has the form

$$i_t = Ae^{-Rt/L}. \tag{9–32}$$

The steady-state solution is not a constant in this case, but rather is a sinusoid having the same frequency as the forcing function. A general expression for such a sinusoid, which can have an arbitrary amplitude and phase relative to the driving function, is

$$i_{ss} = B_1 \cos \omega t + B_2 \sin \omega t. \tag{9–33}$$

Its derivative is

$$\frac{di_{ss}}{dt} = -B_1\omega \sin \omega t + B_2\omega \cos \omega t. \tag{9–34}$$

The steady-state solution postulated in Eq. (9–33) and its derivative must satisfy Eq. (9–31). Substitution results in

$$L(-B_1\omega \sin \omega t + B_2\omega \cos \omega t) + R(B_1 \cos \omega t + B_2 \sin \omega t) = E \sin \omega t.$$

When we rearrange terms,

$$\sin \omega t(-LB_1\omega + RB_2 - E) + \cos \omega t(LB_2\omega + RB_1) = 0. \tag{9–35}$$

If our assumed steady-state solution is indeed a solution, Eq. (9–35) must be satisfied at all instants of time. For Eq. (9–35) to be zero at all instants the two quantities enclosed by parentheses must be zero. Thus the conditions which we must satisfy are

$$LB_2\omega + RB_1 = 0,$$

$$-LB_1\omega + RB_2 - E = 0,$$

from which

$$B_1 = \frac{-\omega LE}{(\omega L)^2 + R^2},$$

$$B_2 = \frac{RE}{(\omega L)^2 + R^2}. \tag{9–36}$$

In terms of the values of B_1 and B_2 the steady-state solution is

$$i_{ss} = \frac{-\omega LE}{(\omega L)^2 + R^2} \cos \omega t + \frac{RE}{(\omega L)^2 + R^2} \sin \omega t. \tag{9–37}$$

The complete solution is the sum of the steady-state solution, Eq. (9–37), and the transient solution, Eq. (9–32):

$$i = Ae^{-Rt/L} + \frac{E}{(\omega L)^2 + R^2} (-\omega L \cos \omega t + R \sin \omega t). \tag{9–38}$$

The arbitrary constant, A, in Eq. (9–38) is evaluated by setting $t = 0$ in the equation. At this instant the current is zero, since the voltage is zero and the inductance is an open circuit. Thus

$$i(0) = 0 = A + \frac{E}{(\omega L)^2 + R^2}(-\omega L),$$

and

$$A = \frac{\omega LE}{(\omega L)^2 + R^2}. \tag{9–39}$$

The final expression for the current is

$$i = \frac{\omega LEe^{-Rt/L}}{(\omega L)^2 + R^2} + \frac{E}{(\omega L)^2 + R^2}(-\omega L \cos \omega t + R \sin \omega t). \tag{9–40}$$

The first term in Eq. (9–40) is the transient term. With the passage of time it decays to zero, leaving only the second term. The second term is a sinusoid having a frequency equal to that of the applied driving voltage and a magnitude and phase which depend on the values of the circuit constants. The cosine and the sine terms can, of course, be combined into a single term by means of trigonometric manipulation. At the moment we are concerned not with such details, but only with the general method of obtaining the solution. In summary

1. The transient solution is obtained in terms of an arbitrary constant.
2. A steady-state solution is obtained which is of the same nature as the driving function.*
3. The complete solution is the sum of the steady-state solution and the transient solution.
4. The arbitrary constant is evaluated from the value of the variable at $t = 0$.

Resonance and the method of variation of parameters. When the driving function has the same form as the natural behavior of the circuit the phenomenon called resonance appears, creating a minor problem. At resonance the steady-state solution continues to increase with time and must be found by the method of variation of parameters. An example of a differential equation in which the driving function has the same form as the natural behavior of the circuit is

$$\frac{di}{dt} + i = e^{-t} \quad \text{with} \quad i(0) = 0. \tag{9–41}$$

---

* The trial solution which must be assumed is, in general, the sum of the driving function, its first derivative, its second derivative, and so on, so long as the derivatives are independent. Each term requires an arbitrary constant.

There is no difficulty in obtaining the transient solution. It is, as for any driving function,

$$i_t = Ae^{-t}. \tag{9–42}$$

Obtaining the steady-state solution poses a problem, however, since the assumption of a steady-state solution of the same form as the driving function results in an indeterminate equation. To avoid this difficulty we assume a steady-state solution of the form

$$i_{ss} = u(t)e^{-t}, \tag{9–43}$$

where $u(t)$ is a function to be determined and not just a constant. The derivative is

$$\frac{di_{ss}}{dt} = e^{-t}u'(t) - e^{-t}u(t). \tag{9–44}$$

If Eq. (9–43) is to be a solution it must satisfy Eq. (9–41):

$$e^{-t}u'(t) + e^{-t}u(t) - e^{-t}u(t) = e^{-t},$$

from which

$$u'(t) = 1 \qquad \text{and} \qquad u(t) = t + C, \tag{9–45}$$

where C is a constant. The steady-state solution is therefore

$$i_{ss}(t) = e^{-t}(t + C), \tag{9–46}$$

and the complete solution is

$$i(t) = te^{-t} + Ce^{-t} + Ae^{-t}. \tag{9–47}$$

The two constants C and A are really just a single constant, and they can be found by applying to Eq. (9–47) the condition at $t = 0$. Thus

$$i(0) = A + C = 0,$$

and the final expression for the current is

$$i(t) = te^{-t}. \tag{9–48}$$

The method we used to obtain this solution is called *variation of parameters*, since we allowed the constant multiplier in Eq. (9–43) to become a function of time. The method of variation of parameters can always be used to resolve the problem created in a circuit by resonance.

9–5 Second-order system excited by initial energy storage. We cannot integrate a second-order equation as easily as we can a first-order equation. The second-order solution, however, has the same general form as the first-order solution. The response to an arbitrary excitation consists of two parts: a transient, or homogeneous, solution and a steady-state solution, or particular integral. The homogeneous solution is the solution of the differential equation without an external driving force connected to the circuit. In other words it is the natural behavior of the circuit produced by internally stored energy. In this section we shall investigate the natural behavior of a second-order system. The natural behavior of a first-order system contains one arbitrary constant to accommodate the single, arbitrary energy storage. The natural behavior of a second-order system contains two arbitrary constants, one for each of the arbitrary energy storages. In order to evaluate these two constants we need to know the values of the two energy storages in the circuit, or more conveniently the values of the response function and its first derivative at $t = 0$. Thus the specification of a two-energy-storage system by means of a second-order differential equation requires

1. The statement of the differential equation.
2. The initial value of the variable and of its first derivative.

The nature of the solution is more complex than that of the first-order solution. It can vary from a damped-exponential form when circuit losses are large to an oscillatory form when circuit losses are small. The student may well shudder at the increased complexity facing him as he proceeds from the first-order system to one of the second order. What will a third-order system be like? Fortunately, not much worse than the second order. The natural behavior of the higher-order systems can be resolved into exponential, sinusoidal, and damped sinusoidal components which are similar to those encountered in the second-order system.

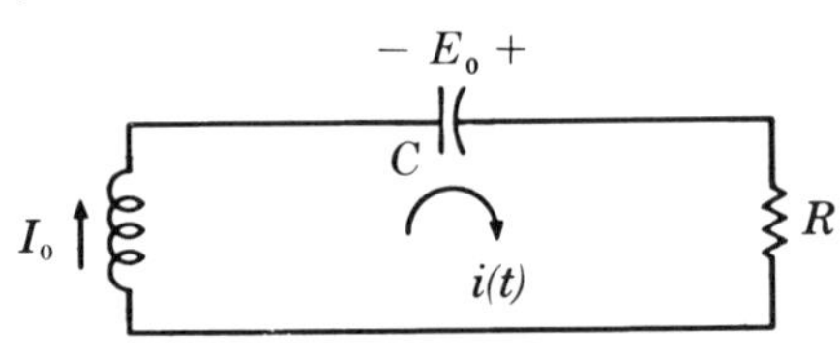

FIG. 9–10. An R–L–C circuit excited by an initial energy storage.

A second-order circuit excited by energy initially stored in the elements is shown in Fig. 9–10. The differential equation for the circuit is

$$L\frac{di}{dt} + Ri + \frac{1}{C}\int_{-\infty}^{t} i\,dt = 0$$

or

$$L\frac{d^2i}{dt^2} + R\frac{di}{dt} + \frac{i}{C} = 0. \tag{9–49}$$

The initial values are determined by the energy stored in the inductance and capacitance. The initial current is

$$i(0) = I_0, \tag{9–50}$$

and the initial voltage on C is

$$\frac{1}{C}\int_{-\infty}^{0} i\,dt = E_0.$$

At $t = 0$ Eq. (9–49) becomes

$$L\frac{di(0)}{dt} + Ri(0) + E_0 = 0$$

or

$$\frac{di}{dt} = -\frac{1}{L}(RI_0 + E_0). \tag{9–51}$$

Since we need two arbitrary constants in our second-order solution we shall try a function consisting of two first-order solutions and having the form

$$i = A_1e^{p_1t} + A_2e^{p_2t}, \tag{9–52}$$

where the A's and p's are constants to be found. The first and second derivatives of Eq. (9–52) are

$$\begin{aligned}\frac{di}{dt} &= A_1p_1e^{p_1t} + A_2p_2e^{p_2t},\\ \frac{d^2i}{dt^2} &= A_1p_1^2e^{p_1t} + A_2p_2^2e^{p_2t}.\end{aligned} \tag{9–53}$$

If Eq. (9–52) is a solution it must satisfy the differential equation. Substituting in Eq. (9–49) we obtain

$$A_1e^{p_1t}\left(Lp_1^2 + Rp_1 + \frac{1}{C}\right) + A_2e^{p_2t}\left(Lp_2^2 + Rp_2 + \frac{1}{C}\right) = 0. \tag{9–54}$$

Since neither $A_1e^{p_1t}$ nor $A_2e^{p_2t}$ can be zero if we are to have a second-order solution, the two terms in parentheses must be zero. Thus each of the p's must satisfy an equation of the form

$$p^2 + \frac{R}{L}p + \frac{1}{LC} = 0. \tag{9–55}$$

The two p values are

$$p_{1,2} = -\frac{R}{2L} \pm \sqrt{\left(\frac{R}{2L}\right)^2 - \frac{1}{LC}}$$

or

$$p_{1,2} = -\alpha \pm \sqrt{\alpha^2 - \omega_0^2}\,, \tag{9-56}$$

where

$$\alpha = \frac{R}{2L} \quad \text{and} \quad \omega_0^2 = \frac{1}{LC}\,. \tag{9-57}$$

Equation (9–56) gives the values of p_1 and p_2 which make Eq. (9–52) a solution of the differential equation. The two constants A_1 and A_2 are obtained from the initial conditions in each problem. There are three general types of solution, depending on whether $\alpha > \omega_0$, $\alpha = \omega_0$, or $\alpha < \omega_0$. These three solutions are called respectively the overdamped case, the critically damped case, and the underdamped case. We shall treat each of them.

Overdamped case ($\alpha > \omega_0$). When $\alpha > \omega_0$ the two values of p are real and negative, and the solution consists of the sum of two damped exponential terms. We shall obtain the transient solution for the initial conditions of

$$i(0) = I_0 \quad \text{and} \quad \frac{di(0)}{dt} = 0 \text{ (i.e., } E_0 = I_0R).$$

From Eq. (9–52) the general solution is

$$i(t) = A_1e^{p_1t} + A_2e^{p_2t}, \tag{9-58}$$

and the derivative is

$$\frac{di}{dt} = A_1p_1e^{p_1t} + A_2p_2e^{p_2t}. \tag{9-59}$$

The values of the A's are determined from the initial conditions and Eqs. (9–58) and (9–59). At $t = 0$

$$i(0) = I_0 = A_1 + A_2,$$

$$i'(0) = 0 = A_1p_1 + A_2p_2,$$

from which

$$A_1 = \frac{p_2I_0}{p_2 - p_1} \quad \text{and} \quad A_2 = \frac{-p_1I_0}{p_2 - p_1}\,.$$

The transient solution for the overdamped case is therefore

$$i(t) = I_0\frac{p_2}{p_2 - p_1}\,e^{p_1t} - I_0\frac{p_1}{p_2 - p_1}\,e^{p_2t}, \tag{9-60}$$

where $p_1 = -\alpha + \sqrt{\alpha^2 - \omega_0^2}\,,$

$p_2 = -\alpha - \sqrt{\alpha^2 - \omega_0^2}\,.$

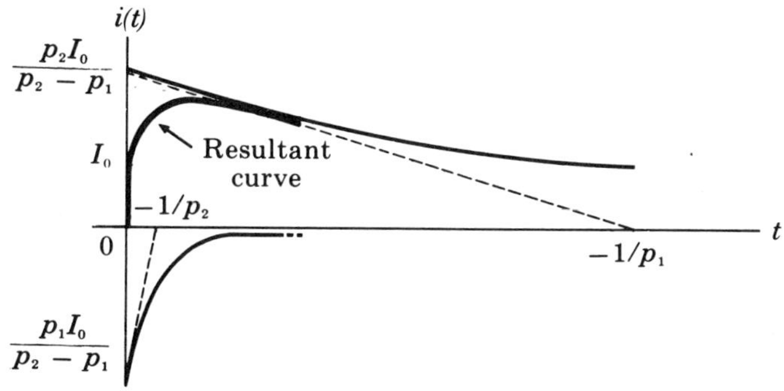

FIG. 9–11. The overdamped current $i(t) = [p_2I_0/(p_2 - p_1)]e^{p_1t} + [p_1I_0/(p_2 - p_1)]e^{p_2t}$.

In Eq. (9–60) both p_1 and p_2 are negative, and the difference $p_2 - p_1$ is negative. The coefficients of the exponentials thus have the signs shown. The magnitude of p_1 is less than the magnitude of p_2, and hence the time constant of the first exponential term in Eq. (9–60) is longer than the time constant of the second, or negative, term. A sketch of the response function is shown in Fig. 9–11. The difference between the negative and positive curves at $t = 0$ is I_0. The beginning of the curve is influenced most strongly by the negative exponential, which has a short time constant. The final disposition of the curve is almost entirely determined by the positive exponential, which has a long time constant. It is characteristic of the response of overdamped systems that a pulse is produced whose rise time is controlled by one time constant and whose fall time is controlled by another.

Critically damped case ($\alpha = \omega_0$). It might seem that the case of $\alpha = \omega_0$ would be the simplest to solve since the p values are equal and only one term occurs in the transient solution. The reverse is true, however, since we need two terms for a second-order solution in order to account for the two arbitrary energy storages. Somehow we must find a second solution! Let us restrict ourselves to the initial conditions considered in the case $\alpha > \omega_0$, that is, the initial value of the current is I_0 and the initial derivative is zero. Then, according to Eq. (9–60), the general solution for the current is

$$\frac{i(t)}{I_0} = \frac{p_2}{p_2 - p_1} e^{p_1 t} - \frac{p_1}{p_2 - p_1} e^{p_2 t}. \tag{9–61}$$

If in Eq. (9–61) we let $p_1 = p_2$, the mathematical difficulty is at once evident. The two exponentials become equal, but their coefficients become infinite, making the difference indeterminate. To evaluate the indeterminacy we let

$$p_1 = -\alpha, \qquad p_2 = p_1 + \delta,$$

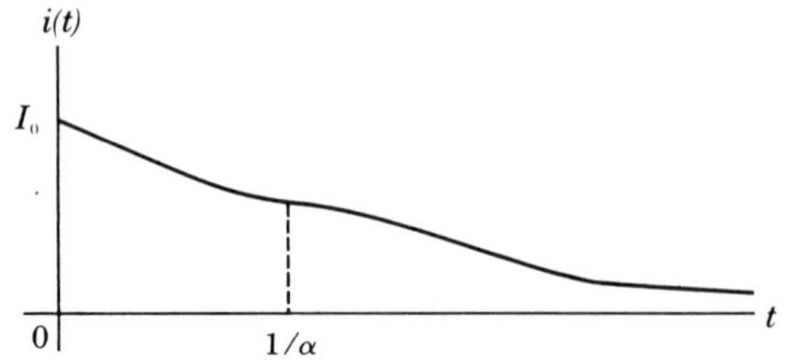

FIG. 9–12. The critically damped current $i(t) = I_0 e^{-\alpha t}(1 + \alpha t)$.

where δ is a small number which later will be allowed to become zero. Substituting $p_1 + \delta$ for p_2 in Eq. (9–61), we obtain

$$
\begin{aligned}
\frac{i(t)}{I_0} &= \frac{p_1 + \delta}{p_1 + \delta - p_1} e^{p_1 t} - \frac{p_1}{p_1 + \delta - p_1} e^{p_1 t + \delta t} \\
&= \frac{e^{p_1 t}}{\delta} [p_1 + \delta - p_1(1 + \delta t + \cdots)] \\
&= e^{p_1 t}\left(\frac{\delta - \delta p_1 t}{\delta} + \cdots\right). \qquad (9\text{–}62)
\end{aligned}
$$

Now, as δ is allowed to approach zero Eq. (9–62) approaches the limit

$$\frac{i(t)}{I_0} = e^{-\alpha t}(1 + \alpha t). \qquad (9\text{–}63)$$

The waveform for Eq. (9–63) is sketched in Fig. 9–12. Note that the maximum value occurs at $t = 0$. Equation 9–63 was developed for a particular set of initial conditions. In the more general case the terms $e^{-\alpha t}$ and $te^{-\alpha t}$ remain, but the multiplying constants are arbitrary, as in the following expression

$$i(t) = e^{-\alpha t}(A_1 + A_2 t). \qquad (9\text{–}64)$$

The principal new result in Eq. (9–64) is the inclusion of a term multiplied by time t. This term could have been obtained more generally by the method of variation of parameters.

Underdamped case ($\alpha < \omega_0$). When $\alpha < \omega_0$ the p-roots become conjugate complex numbers, and the whole character of the solution changes. For this case it is convenient to write the p values in the form

$$
\begin{aligned}
p_1 &= -\alpha + j\sqrt{\omega_0^2 - \alpha^2} = -\alpha + j\omega_d, \\
p_2 &= -\alpha - j\sqrt{\omega_0^2 - \alpha^2} = -\alpha - j\omega_d.
\end{aligned} \qquad (9\text{–}65)
$$

To observe the effect of the p values given in Eq. (9–65) we shall again consider the series R–L–C circuit with the initial conditions $i(0) = I_0$ and $i'(0) = 0$. As given in Eq. (9–61) the general solution is

$$\frac{i(t)}{I_0} = \frac{p_2}{p_2 - p_1} e^{p_1 t} - \frac{p_1}{p_2 - p_1} e^{p_2 t}.$$

Substitution of the p values from Eq. (9–65) gives

$$\frac{i(t)}{I_0} = \frac{-\alpha - j\omega_d}{-2j\omega_d} e^{-\alpha t} e^{j\omega_d t} - \frac{-\alpha + j\omega_d}{-2j\omega_d} e^{-\alpha t} e^{-j\omega_d t}$$

$$= \frac{e^{-\alpha t}}{2j\omega_d} [\alpha(e^{j\omega_d t} - e^{-j\omega_d t}) + j\omega_d(e^{j\omega_d t} + e^{-j\omega_d t})]. \tag{9–66}$$

By Euler's formula

$$\frac{e^{+j\theta} + e^{-j\theta}}{2} = \cos\theta, \qquad \frac{e^{j\theta} - e^{-j\theta}}{2j} = \sin\theta.$$

Therefore Eq. (9–66) can be written in the form

$$\frac{i(t)}{I_0} = e^{-\alpha t}\left(\frac{\alpha}{\omega_d} \sin \omega_d t + \cos \omega_d t\right) \cdot \tag{9–67}$$

An alternative form of Eq. (9–67) can be obtained by trigonometric manipulation. If we let $\alpha = \omega_0 \sin\theta$ and $\omega_d = \omega_0 \cos\theta$, we can write Eq. (9–67) in the form

$$\frac{i(t)}{I_0} = \frac{e^{-\alpha t}\omega_0}{\omega_d} (\cos \omega_d t \cos\theta + \sin \omega_d t \sin\theta)$$

$$= \frac{\omega_0}{\omega_d} e^{-\alpha t} \cos(\omega_d t - \theta). \tag{9–68}$$

A sketch of the waveform of Eq. (9–68) is shown in Fig. 9–13. Notice that it is a damped sine wave. The envelope of the curve is the damped exponential $e^{-\alpha t}$. The initial value of the current is I_0, and the period of the sinusoidal oscillation is $t_d = 2\pi/\omega_d$. The phase angle is $-\theta$, which means that the cosine term has not reached its maximum value at $t = 0$. In the absence of the damping term the cosine function would reach a peak at a positive angle of θ radians. In a later chapter we shall investigate the sinusoidal response function in more detail. At the moment we are interested in it only as a solution of the second-order equation when damping is less than critical. Referring to Eq. (9–68) we now define the quantities which we have used.

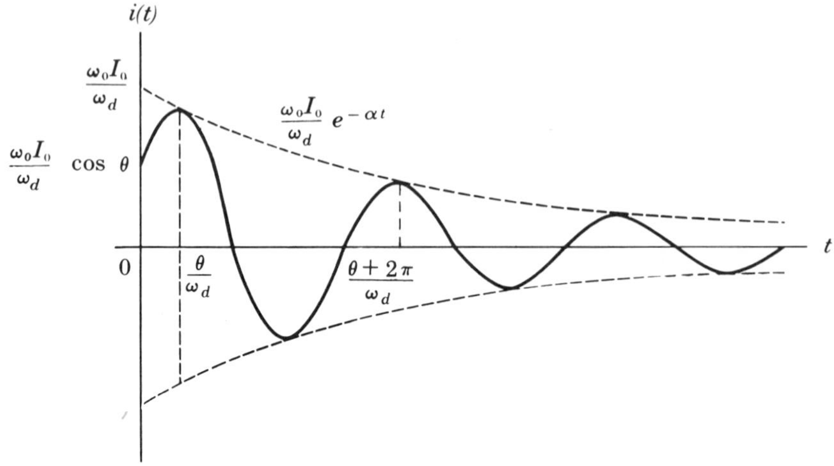

FIG. 9–13. The underdamped current $i(t) = (\omega_0 I_0/\omega_d)\cos(\omega_d t - \theta)e^{-\alpha t}$.

ω_d is the damped natural frequency in radians per second. It is the number of radians through which the oscillation passes each second.

ω_0 is the value of ω_d when the resistance term which produces the damping becomes zero. It is the undamped natural frequency in radians per second.

α is the damping constant, or the reciprocal of the time constant of the envelope curve.

For different initial conditions the nature of the second-order response function remains the same, but the amplitude and the phase of the resulting damped sine wave change.

At this point we shall summarize the types of natural behavior which can occur in a second-order differential equation of the form

$$\frac{d^2x}{dt^2} + 2\alpha\frac{dx}{dt} + \omega_0^2 x = 0. \tag{9–69}$$

Case 1: Overdamped ($\alpha > \omega_0$).

$$x = A_1 e^{p_1 t} + A_2 e^{p_2 t}, \tag{9–70}$$

where

$$p_1 = -\alpha + \sqrt{\alpha^2 - \omega_0^2},$$

$$p_2 = -\alpha - \sqrt{\alpha^2 - \omega_0^2}.$$

Case 2: Critically damped ($\alpha = \omega_0$).

$$x = e^{-\alpha t}(A_1 + A_2 t). \tag{9–71}$$

Case 3: Underdamped ($\alpha < \omega_0$).

$$x = e^{-\alpha t}(A_1 \sin \omega_d t + A_2 \cos \omega_d t) \tag{9–72}$$

or

$$x = Ae^{-\alpha t} \cos (\omega_d t + \theta),$$

where

$$\omega_d = \sqrt{\omega_0^2 - \alpha^2}\,.$$

In all cases A_1 and A_2 are constants determined by the exact form of the initial conditions. In the alternative expression for case 3, A_1 and A_2 are replaced by the amplitude A and the phase angle θ.

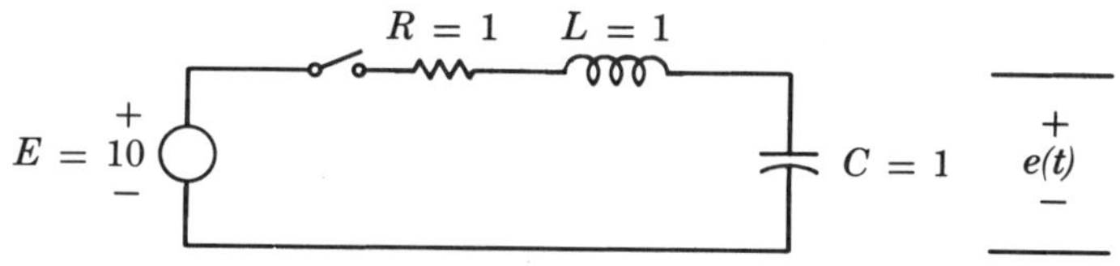

FIG. 9–14. An *R–L–C* circuit excited by a step function of voltage.

9–6 Second-order system excited by a step function. The response of a second-order system excited by a step function consists of two parts. The first part is the steady-state, or direct-current solution, which we have obtained in earlier chapters; the second part is the transient term which is added to match the steady-state values to the given initial conditions. We shall find the voltage produced across the capacitance in the circuit of Fig. 9–14, as an illustration of the general procedure. After the switch is closed at $t = 0$ the differential equation for the current in the circuit is

$$\frac{di}{dt} + i + \int_{-\infty}^{t} i\,dt = E. \tag{9–73}$$

The voltage across the capacitance is the integral of the current, or the current is the derivative of the voltage. Hence we may write

$$\frac{d^2e}{dt^2} + \frac{de}{dt} + e = E. \tag{9–74}$$

For this equation

$$\alpha = \tfrac{1}{2}, \qquad \omega_0 = 1,$$

$$\omega_d = \sqrt{\omega_0^2 - \alpha^2} = \frac{\sqrt{3}}{2} = 0.866.$$

The initial conditions, obtained from the circuit of Fig. 9–14, are

$$e(0) = 0 \qquad \text{and} \qquad \frac{de}{dt}(0) = 0.$$

The solution of Eq. (9–74), subject to the initial conditions, will be obtained in several steps.

Steady-state solution. At $t = \infty$ the inductance is a short circuit and the capacitance is an open circuit. The voltage across the capacitance is therefore

$$e_{ss} = 10 \text{ volts.} \tag{9–75}$$

Effectively, the two derivatives in the differential equation are simply neglected.

Transient solution. The transient solution for Eq. (9–74), with the right-hand side equated to zero, is given by Eq. (9–72), since $\alpha < \omega_0$:

$$e(t) = e^{-1/2t}(A_1 \sin 0.866t + A_2 \cos 0.866t). \tag{9–76}$$

Complete solution. The complete solution is the sum of the transient solution and the steady-state solution. Thus

$$e(t) = e^{-t/2}(A_1 \sin 0.866t + A_2 \cos 0.866t) + 10.$$

Its derivative is

$$\frac{de}{dt} = e^{-t/2}(+0.866A_1 \cos 0.866t - A_2 0.866 \sin 0.866t)$$
$$- \frac{e^{-t/2}}{2}(A_1 \sin 0.866t + A_2 \cos 0.866t).$$

The constants A_1 and A_2 are obtained from the initial conditions. At $t = 0$

$$e(0) = 0 = A_2 + 10,$$
$$e'(0) = 0 = 0.866A_1 - 0.5A_2,$$

from which

$$A_2 = -10 \qquad \text{and} \qquad A_1 = -5.77.$$

Final solution. Once the arbitrary constants have been determined, we can write the final solution. In this case

$$e(t) = 10 - e^{-t/2}(5.77 \sin 0.866t + 10 \cos 0.866t).$$

The first term is the steady-state value. The second term is the transient solution. The exponential multiplying factor causes the second term to become negligibly small after a few seconds, leaving only the steady-state value.

The details of the problem we have just solved are less important than the ideas involved. We cannot proceed very far in circuit theory without the concepts of steady-state solution and transient solution, and these concepts are formulated accurately only in terms of the differential equation. The differential equation also makes clear the importance of computations at $t = 0$ and at $t = \infty$. The arbitrary constants in the general solution are determined by the initial conditions, and the values at $t = \infty$ are the steady-state solutions for step-function excitation.

We could also obtain a complete solution for a sinusoidal driving function, as we did for the first-order differential equation. The method is straightforward, but a considerable amount of algebra and trigonometry is involved. Later we shall study the sinusoidal steady state in some detail and learn to solve such problems by the very powerful methods of complex algebra. These methods are so superior to trigonometric methods that we would be wasting our time if we were to delve deeply into the sinusoidal steady state at this point.

9–7 Summary.

First-order equation—transient solution

Equation

$$\frac{dx}{dt} + \frac{1}{T}x = 0$$

Initial value

$$x(0) = A$$

Solution

$$x = Ae^{-t/T}$$

First-order equation—step function excitation

Equation

$$\frac{dx}{dt} + \frac{x}{T} = C \qquad \text{for} \quad t > 0.$$

Initial value

$$x(0) = X_0$$

Final value

$$x(\infty) = TC$$

Solution

$$x = \underbrace{Ae^{-t/T}}_{\text{Transient}} + \underbrace{B}_{\text{Steady state}},$$

$$x(0) = A + B, \qquad x(\infty) = B.$$

First-order equation—sinusoidal excitation

Equation

$$\frac{dx}{dt} + \frac{x}{T} = C_1 \sin \omega t + C_2 \cos \omega t$$

Initial value

$$x(0) = X_0$$

Solution

$$x = \underbrace{Ae^{-t/T}}_{\text{Transient}} + \underbrace{B_1 \sin \omega t + B_2 \cos \omega t,}_{\text{Steady state}}$$

$$x(0) = A + B_2,$$

where A, B_1, and B_2 are constants.

Second-order equation—transient solution

Equation

$$\frac{d^2x}{dt^2} + 2\alpha \frac{dx}{dt} + \omega_0^2 x = 0$$

Initial values

$$x(0) = X_0 \quad \text{and} \quad \frac{dx(0)}{dt} = X_0'$$

Overdamped solution ($\alpha > \omega_0$)

$$x = A_1 e^{p_1 t} + A_2 e^{p_2 t},$$

where $p_1 = -\alpha + \sqrt{\alpha^2 - \omega_0^2}$,

$p_2 = -\alpha - \sqrt{\alpha^2 - \omega_0^2}$,

A_1 and A_2 are constants.

Critically damped solution ($\alpha = \omega_0$)

$$x = Ae^{-\alpha t}(A_1 + A_2 t), \quad \text{with} \quad \alpha = \omega_0$$

Underdamped solution ($\alpha < \omega_0$)

$$x = e^{-\alpha t}(A_1 \sin \omega_d t + A_2 \cos \omega_d t)$$

$$= Ae^{-\alpha t} \cos (\omega_d t + \theta),$$

where $\omega_d = \sqrt{\omega_0^2 - \alpha^2}$,

A_1, A_2, A, and θ are constants.

Second-order equation—step-function excitation

Equation

$$\frac{d^2x}{dt^2} + 2\alpha\frac{dx}{dt} + \omega_0^2 x = C \qquad \text{for} \quad t > 0$$

Initial values

$$x(0) = X_0 \qquad \text{and} \qquad \frac{dx(0)}{dt} = X_0'$$

Final value

$$x(\infty) = C/\omega_0^2$$

Overdamped solution

$$x = A_1 e^{p_1 t} + A_2 e^{p_2 t} + B$$

Critically damped solution

$$x = e^{-\alpha t}(A_1 + A_2 t) + B$$

Underdamped solution

$$x = e^{-\alpha t}(A_1 \sin \omega_d t + A_2 \cos \omega_d t) + B$$

$$= Ae^{-\alpha t}[\cos(\omega_d t + \theta)] + B,$$

where $B = x(\infty) = C/\omega_0^2$.

Second-order equation—sinusoidal excitation

Equation

$$\frac{d^2x}{dt^2} + 2\alpha\frac{dx}{dt} + \omega_0^2 x = C_1 \sin \omega t + C_2 \cos \omega t$$

Initial values

$$x(t = 0) = X_0 \qquad \text{and} \qquad \frac{dx}{dt}(t = 0) = X_0'$$

Overdamped solution

$$x = A_1 e^{p_1 t} + A_2 e^{p_2 t} + B_1 \sin \omega t + B_2 \cos \omega t$$

Critically damped solution

$$x = e^{-\alpha t}(A_1 + A_2 t) + B_1 \sin \omega t + B_2 \cos \omega t$$

Underdamped solution

$$x = e^{-\alpha t}(A_1 \sin \omega_d t + A_2 \cos \omega_d t) + B_1 \sin \omega t + B_2 \cos \omega t$$

$$= e^{-\alpha t} A \cos(\omega_d t + \theta) + B_1 \sin \omega t + B_2 \cos \omega t$$

REFERENCES

In this chapter we have tried to present differential equations from a point of view somewhere between that of the mathematician and of the electrical engineer. The mathematical point of view has the advantage that it is general and applies to equations which arise from all sorts of physical problems. The electrical engineering point of view has the advantage that it provides concrete illustrations of many of the abstract concepts of differential equations. Thus the natural behavior is the transient response of a circuit, and the particular integral is the steady-state solution. The following references give additional material on both points of view.

1. SALVADORI and SCHWARTZ, *Differential Equations In Engineering Problems*, Prentice-Hall, New York, 1954. This relatively modern text gives a very complete discussion of linear equations and many examples from dynamics, statics, heat transfer, chemical action, and electric circuits.

2. H. B. PHILLIPS, *Differential Equations*, John Wiley & Sons, New York, 1922. A standard text in applied mathematics. It is devoted to linear equations and their applications in physical problems.

3. E. A. GUILLEMIN, *Communications Networks*, Vol. 1, John Wiley & Sons, New York, 1931. This text is the most complete treatment of linear equations, from the point of view of the electrical engineer, which is available in the English language. It is very readable, even in the sections devoted to advanced material.

4. G. B. THOMAS, JR., *Calculus and Analytical Geometry*, Addison-Wesley, Reading, Mass., 1954. The last chapter of this introductory text gives a brief account of the essentials of linear differential equations.

5. D. K. CHENG, *Analysis of Linear Systems*, Addison-Wesley, Reading, Mass., 1959. Chapter 2 gives a good summary of linear differential equations from the classical point of view. The remainder of the text is devoted to solutions by means of Laplace transforms.

EXERCISES

Section 9–2

1. A current is given by the equation

$$i(t) = 10e^{-t/10}.$$

Find the current values at $t = 0$, $t = 10$, $t = 20$, and $t = \infty$, and sketch the curve. What is the time constant?

2. Find the derivative of the curve given in Exercise 1, and sketch it. What is the time constant? If the curve in Exercise 1 were to continue as a straight line with the same slope as that of the actual curve at $t = 0$, how long would it take to reach zero?

3. Solve the following differential equation, subject to the given initial condition:

$$6\frac{di}{dt} + 4i = 0, \qquad i(0) = 5.$$

4. Solve the following differential equation, subject to the given initial condition:

$$2\frac{de}{dt} + e = 0, \qquad e(0) = -10.$$

5. At $t = 0$ the current in the circuit of Fig. 9–15 is 6 amp. If the circuit is left to its own devices what is the current as a function of time?

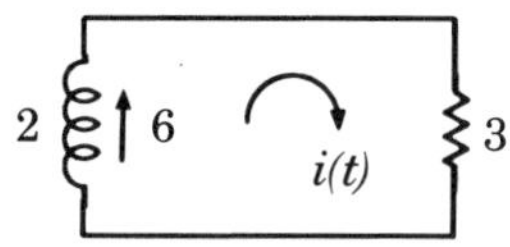

FIGURE 9–15

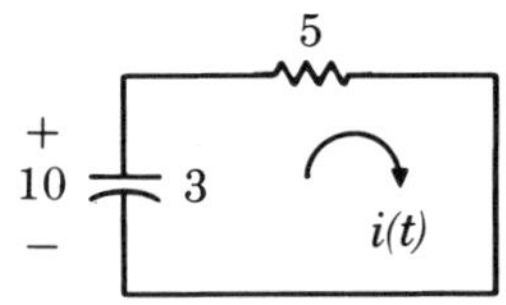

FIGURE 9–16

6. At $t = 0$ the voltage on the capacitance in the circuit of Fig. 9–16 is 10 volts. If the circuit is left to its own devices what is the current as a function of time?

7. Before $t = 0$ a current of 5 amp is flowing in the inductance and short circuit of Fig. 9–17. The switch is opened at $t = 0$. What are the currents i_1 and i_2 as functions of time after this instant?

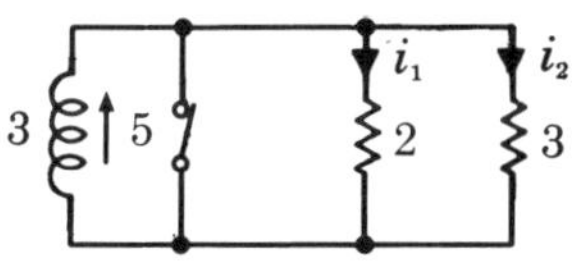

FIGURE 9–17

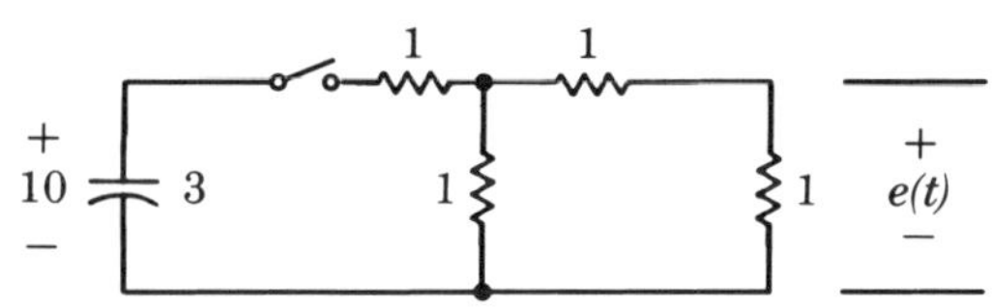

FIGURE 9–18

8. The capacitance in the circuit of Fig. 9–18 has a charge which produces a voltage of 10 volts. The switch is closed at $t = 0$. What is the voltage $e(t)$ as a function of time after this instant?

Section 9–3

9. A voltage is given by the equation

$$e(t) = 10(1 - e^{-t/10}).$$

Find the voltage values at $t = 0$, $t = 10$, $t = 20$, and $t = \infty$, and sketch the curve. What is the time constant?

10. Find the derivative of the curve given in Exercise 9, and sketch it. What is its time constant? If the curve in Exercise 9 were to continue as a straight line with the same slope as that of the actual curve at $t = 0$, how long would it take to reach the final value of 10?

11. Solve the following differential equation, subject to the given initial condition:

$$5\frac{di}{dt} + 3i = 7, \qquad i(0) = 0.$$

12. Solve the following differential equation, subject to the given initial condition:

$$\frac{de}{dt} + 3e = 10, \qquad e(0) = 0.$$

13. The switch in the circuit of Fig. 9–19 is closed at $t = 0$. What is the current as a function of time?

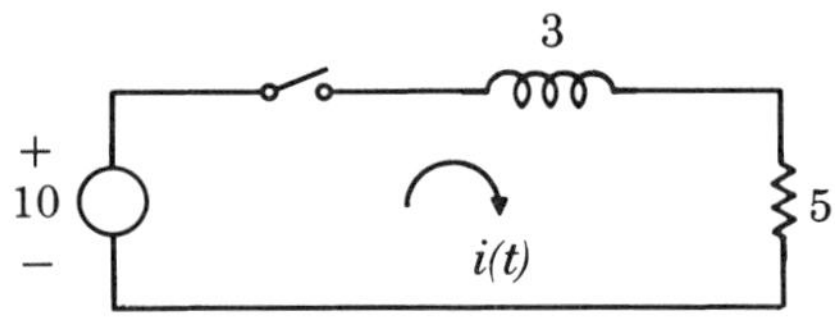

FIGURE 9–19

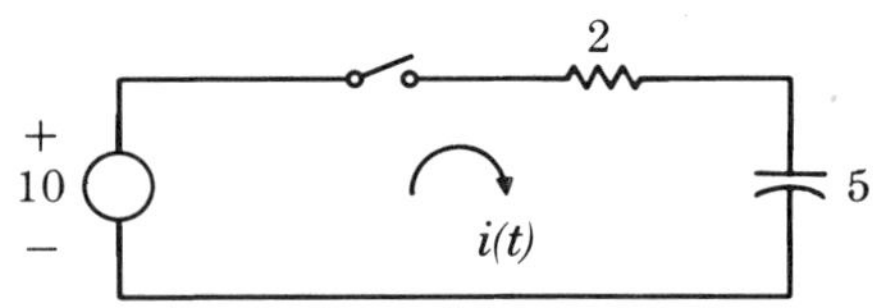

FIGURE 9–20

14. The switch in the circuit of Fig. 9–20 is closed at $t = 0$. What is the current as a function of time?

15. The switch in the circuit of Fig. 9–21 is opened at $t = 0$. What is the voltage $e(t)$?

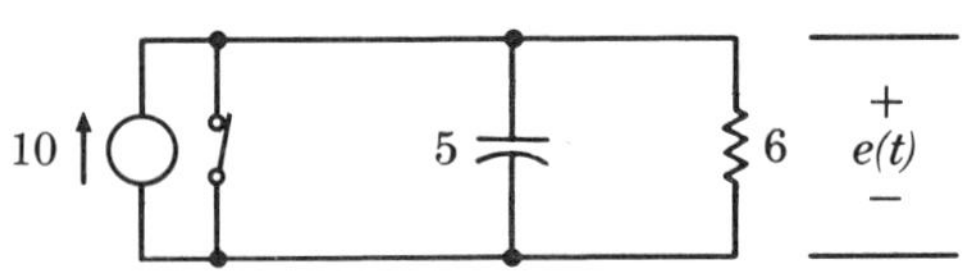

FIGURE 9–21

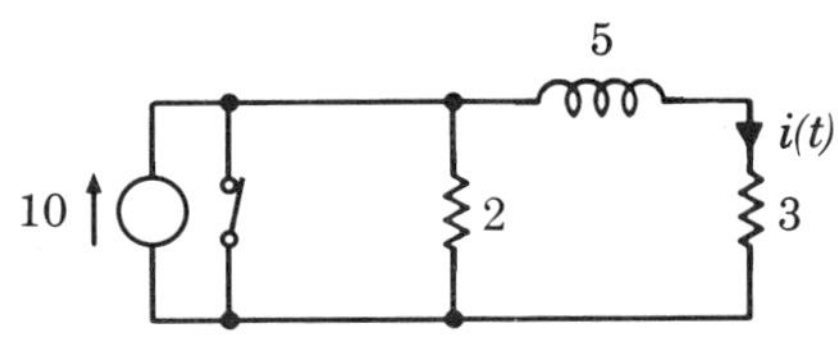

FIGURE 9–22

16. The switch in the circuit of Fig. 9–22 is opened at $t = 0$. Find the current $i(t)$ in the inductance as a function of time.

Section 9–4

17. A current is given by the equation

$$i(t) = 10 \sin (6\pi t + 30°).$$

Find the current value at $t = 0$, the maximum value, the period of the wave, the frequency in cycles per second, and the angular frequency in radians per second. What is the phase angle of this wave with respect to the wave $\cos 6\pi t$? Sketch the waveform over a few cycles.

18. Find the derivative of the waveform in Exercise 17, and sketch it. Of the quantities frequency, angular frequency, period, amplitude, and phase, which are changed by the operation of differentiation? Which remain the same?

19. A voltage is given by the equation

$$e(t) = 10e^{-t/10} + 5 \sin t.$$

Which is the transient term and which is the steady-state term? How long does it take the transient to decay to 1% of the peak value of the steady-state term?

20. Solve the following differential equation, subject to the given initial condition:

$$\frac{di}{dt} + i = \sin t, \qquad i(0) = 0.$$

21. Solve the following differential equation, subject to the given initial condition:

$$\frac{di}{dt} + i = \cos t, \qquad i(0) = 0.$$

22. Solve the following differential equation by means of superposition and the results of Exercises 20 and 21. Also, obtain a direct solution.

$$\frac{di}{dt} + i = \cos (t + 45°), \qquad i(0) = 0.$$

23. The switch in the circuit of Fig. 9–23 is closed at $t = 0$. Find the current as a function of time, and separate the steady-state and transient components.

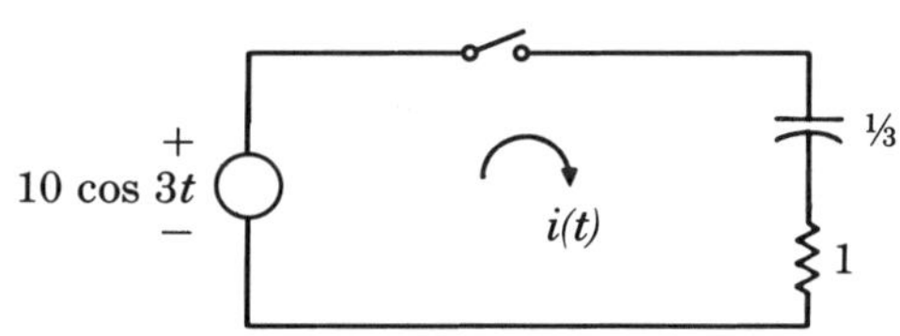

FIGURE 9–23

24. The switch in the circuit of Fig. 9–24 is opened at $t = 0$. Find the voltage $e(t)$ as a function of time, and separate the transient and steady-state components.

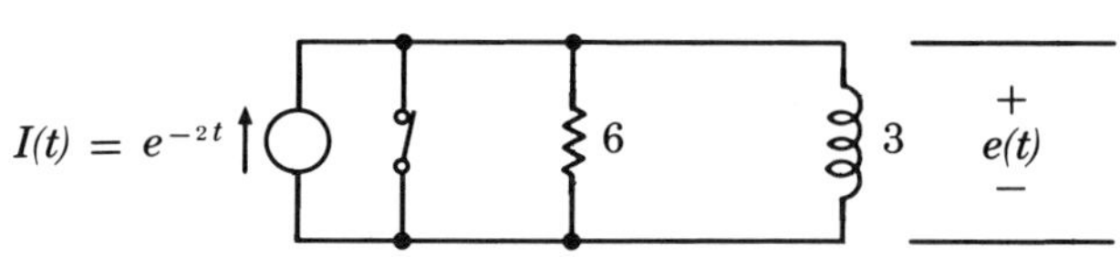

FIGURE 9–24

Section 9–5

25. A current is given by the equation

$$i(t) = 10e^{-t/e} - 10e^{-t}.$$

Find the current values at $t = 0$ and $t = \infty$. What is the maximum value and where does it occur? Sketch the curve.

26. A voltage is given by the equation

$$e(t) = 2te^{-t/2}.$$

Find the voltage values at $t = 0$ and $t = \infty$. What is the maximum value and where does it occur? Sketch the curve.

27. A current is given by the equation

$$i(t) = e^{-t/2} \cos(2\pi t + 45°).$$

Find the current values at $t = 0$, $t = 1$, $t = 2$, and $t = \infty$. What is the time constant of the envelope? What is the period of the wave and what is its frequency? What is the phase at $t = 0$? Sketch the waveform.

28. Solve the following differential equation, subject to the given initial conditions:

$$\frac{d^2i}{dt^2} + 5\frac{di}{dt} + 6i = 0,$$

$$i(0) = 0, \qquad i'(0) = 10.$$

29. Solve the following differential equation, subject to the given initial values:

$$\frac{d^2e}{dt^2} + 4\frac{de}{dt} + 4e = 0,$$

$$e(0) = 0, \qquad e'(0) = 10.$$

30. Solve the following differential equation, subject to the given initial conditions:

$$\frac{d^2i}{dt^2} + 8\frac{di}{dt} + 25i = 0,$$

$$i(0) = 10, \qquad i'(0) = -10.$$

31. The initial value of the current in the circuit of Fig. 9–25 is 10 amp. The initial voltage on the capacitance is zero. If the circuit is left to its own devices what is the current as a function of time?

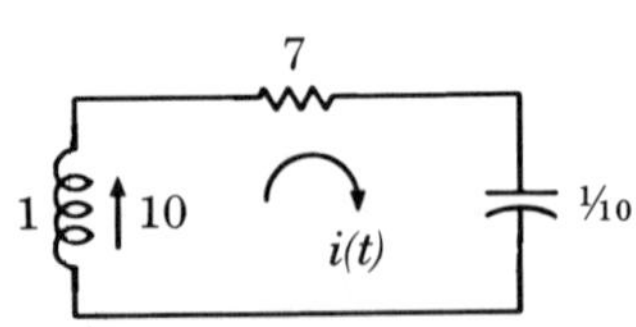

FIGURE 9–25

32. The initial current in the inductance in the circuit of Fig. 9–26 is 10 amp and the voltage on the capacitance is 5 volts. If the circuit is left to its own devices what is the voltage $e(t)$?

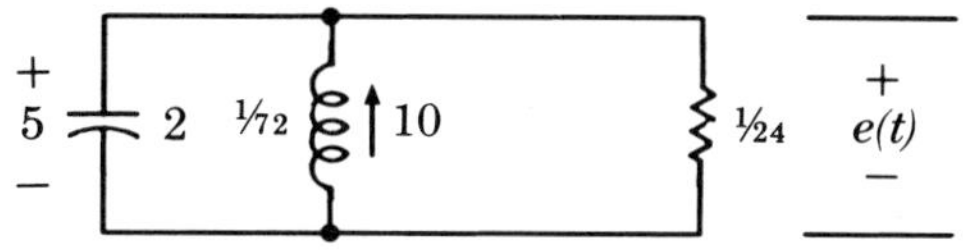

FIGURE 9–26

33. The initial current in the inductance in the circuit of Fig. 9–27 is 10 amp and the voltage on the capacitance is 75 volts, with the polarity shown. The circuit is then left to its own devices. What is the current $i(t)$?

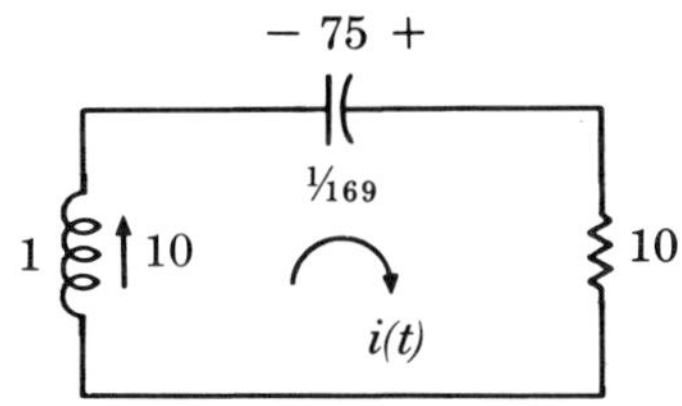

FIGURE 9–27

Section 9–6

34. A voltage is given by the equation

$$e(t) = 10 - 10e^{-t/2}\cos 2\pi t.$$

Find the voltage values at $t = 0$ and $t = \infty$. Which is the steady-state term and which is the transient term? Give the time constant of the envelope of the transient term and the frequency of the oscillation. Sketch the over-all curve.

35. A current is given by the equation

$$i(t) = 10 - 5e^{-t} + 5e^{-t/10}.$$

Find the current values at $t = 0$ and $t = \infty$ and the values of the two time constants. Sketch the over-all curve.

36. A voltage is given by the equation

$$e(t) = 10\cos 6t + 10\sin 6t + 5e^{-t}\cos 2t + 5e^{-t}\sin 2t.$$

Find the following: the transient term, the steady-state term, the initial value of the voltage and of its first derivative, the amplitude of the steady-state term, and the amplitude of the transient term.

37. Solve the following differential equation, subject to the given initial conditions:

$$\frac{d^2i}{dt^2} + 5\frac{di}{dt} + 4i = 8, \qquad i(0) = 0, \qquad i'(0) = 0.$$

38. Solve the following differential equation, subject to the given initial values:

$$\frac{d^2e}{dt^2} + 6\,\frac{de}{dt} + 9e = 9,$$

$$e(0) = e'(0) = 0.$$

39. Solve the following differential equation, subject to the given initial conditions:

$$\frac{d^2i}{dt^2} + 1.414\,\frac{di}{dt} + i = 10,$$

$$i(0) = i'(0) = 0.$$

40. The switch in the circuit of Fig. 9–28 is closed at $t = 0$. No energy is previously stored in the circuit. Find the current as a function of time.

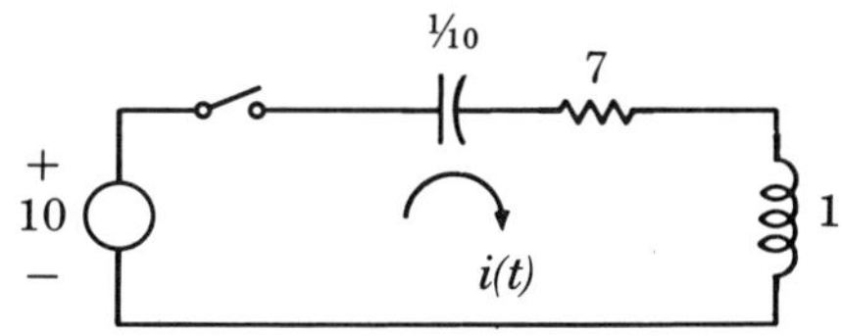

FIGURE 9–28

41. The switch in the circuit of Fig. 9–29 is opened at $t = 0$. No energy is previously stored in the circuit. Find the current in the inductance as a function of time.

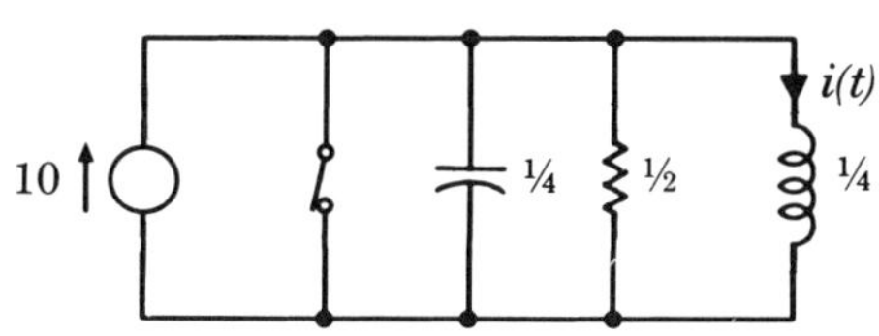

FIGURE 9–29

42. The switch in the circuit of Fig. 9–30 is closed at $t = 0$. No energy is previously stored in the circuit. Find the current as a function of time.

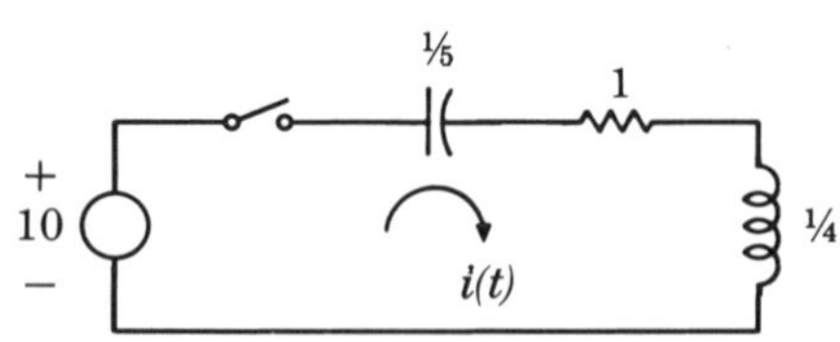

FIGURE 9–30

CHAPTER 10

THE TIME RESPONSE FUNCTIONS OF SINGLE ELEMENTS

"Mathematics is an experimental science. Since the answers can always be verified, it is not important how they are obtained."

Oliver Heaviside

10–1 Introduction. The singularity functions were introduced into electrical engineering by Oliver Heaviside. We shall follow his lead and use them to test and describe our networks. Like Heaviside, we shall be more interested in obtaining useful results than in justifying the rigor of our methods. We can apply our intuition to the singularity functions by assuming that they occur not instantaneously but over a short period of time.

Single-element networks perform the operations of integration, differentiation, and multiplication by a constant. If we excite them with singularity functions the responses are also singularity functions. We shall be concerned with both step and impulse responses of the circuit elements, since step and impulse functions are most commonly used for excitation.

The energy supplied to a network can arise from an externally applied source, or from energy stored in the network at the beginning of the problem. It makes very little difference to the behavior of the network whether the energy is externally or internally supplied. Indeed, we can show that the internal energy storages at the beginning of a problem can be replaced by step or impulse sources without changing the currents and voltages of the network. This equivalence allows us a choice in the method of solving problems. For the solution of differential equations, as in Chapter 9, we usually remove step and impulse sources and account for their effects by computing the initial values of the variables of interest. In the network-theory point of view we are more likely to replace the initial energy-storage values by equivalent sources, since we can treat these sources by the usual stimulus-response techniques.

Initially we shall treat single elements. Then we shall extend our methods to cover multiple-element circuits composed, however, of one type of element. We have already treated in some detail in earlier chapters the case of pure resistance networks. Networks composed of pure capacitance or pure inductance can be treated by exactly the same methods as those for pure resistance networks. In fact there is a resistance analog for each pure capacitance or pure inductance network. Such networks are not very common by themselves, but they form important building blocks in the creation or analysis of more complex networks.

10–2 The step response of resistance, inductance, and capacitance. We shall obtain the step responses of the elements before the impulse responses, because they are inherently simpler and represent many practical switching problems. In all our examples we shall have a dual point of view in mind: The step response may be the solution to an actual physical problem, or it may be merely a method of characterizing the network so that we can later compute its response to a more general input function. The step response is more important than the impulse response from the first point of view but less important from the second.

Resistance. The resistance element has the property of relating the voltage and current associated with it by a constant multiplier. It does not change the waveform. Thus a step-function excitation applied to the element gives rise to a step-function response with a different amplitude. That is, if a current $i(t) = Iu_{-1}(t)$ is applied to a resistance of value R, the voltage response is

$$e(t) = IRu_{-1}(t). \tag{10–1}$$

If a voltage $e(t) = Eu_{-1}(t)$ is applied to a resistance of value R, the resulting current is

$$i(t) = \frac{E}{R}\,u_{-1}(t). \tag{10–2}$$

Inductance. The inductance element has the volt-ampere equations

$$e_L = L\,\frac{di}{dt} \qquad \text{and} \qquad i_L = \frac{1}{L}\int_{-\infty}^{t} e\,dt.$$

If a step function stimulus is applied to an inductance, the response is either the integral or the derivative of the stimulus. When a step function of voltage, $e(t) = Eu_{-1}(t)$, is the stimulus, the response current is the integral, or*

$$i_L = \frac{E}{L}\,u_{-2}(t). \tag{10–3}$$

When the input is a current step, $i(t) = Iu_{-1}(t)$, the response voltage is the derivative, or

$$e_L = LIu_0(t). \tag{10–4}$$

According to Eq. (10–3), if a voltage of value E is suddenly connected to an inductance, the current builds up at a linear rate of E/L amp per sec. It continues to increase so long as the voltage is present and would even-

* The student is cautioned against memorizing these results. There are too many of them! They should be worked out as needed from the volt-ampere equations.

tually become infinite. Equation (10–4) shows that if a step function of current is forced through an inductance, an infinite voltage is produced across the inductance for a very short time. The LI flux linkages equal the area under the curve of voltage versus time.

Both these response functions have practical applications. The starter motor in a stalled car represents practically a pure inductance. When the battery is connected to it, the current rises to a very large value. It is eventually limited by the internal resistance of the battery and the residual resistance of the wires. Equation (10–3) describes the situation until the resistive effects become important. In Eq. (10–4), the very high voltage which exists for a short time is the basis of the ignition system used with internal combustion engines. A step current is applied to a coil, and the very large voltage which occurs across it produces the spark.

Capacitance. The capacitance element has the volt-ampere equations

$$e_C = \frac{1}{C}\int_{-\infty}^{t} i\,dt \qquad \text{and} \qquad i_C = C\,\frac{de}{dt}.$$

If the input is a step current, $i(t) = Iu_{-1}(t)$, the voltage response is

$$e_C = \frac{I}{C}\,u_{-2}(t). \tag{10–5}$$

The voltage in Eq. (10–5) rises linearly at a rate of I/C volts per sec. This result is utilized in producing *sweep* voltages to move the light spot across cathode ray and television tubes. The voltage is allowed to rise linearly to a maximum value, and then the capacitance is shorted to reduce the voltage to zero. During any one sweep the voltage is described by Eq. (10–5).

If the input is a step voltage, $e(t) = Eu_{-1}(t)$, the current response is

$$i_C = ECu_0(t). \tag{10–6}$$

When the capacitance is connected across the constant voltage it becomes charged in a very short time to the value of the voltage source. The amount of charge required to produce the voltage is

$$Q = EC. \tag{10–7}$$

In Eq. (10–7) the charge is the integral of the area under the curve of current versus time given by Eq. (10–6). The important thing is that the capacitance must be charged to the correct voltage in a short time. The amount of charge is important, but the exact shape of the current versus time curve is of very little consequence. For many applications the current can be approximated by an impulse which provides the right amount of charge for the capacitance in zero time. One practical application of this sort of impulse current is in "flash" photography. A capacitance is charged to a high voltage and then discharged through the low resistance of a gas tube. The resultant impulse current causes the gas tube to produce a flash of visible light.

The response of a single-element circuit to a step function is a step, an impulse, or a ramp function. The nature of the response characterizes the element. If the input step is known to be a current or a voltage and if the resulting response is known, the value and the nature of the network element can be computed.

10–3 The impulse response of resistance, inductance, and capacitance. Although the step response can be used to characterize a network, the impulse response is more useful. We have already hinted at the reason in Chapter 9. The impulse response is the natural transient behavior of the system, while the step response may contain a steady-state term as well as the transient term. The transient response truly characterizes the network; the steady-state response depends on the nature of the driving function. Some students find the impulse response more difficult to visualize than the step response. In this connection it is well to remember that an *impulse current is a charge which flows during a short time,* and an *impulse voltage is a flux linkage which attaches itself to a circuit in a brief time.*

Resistance. A resistance element changes an impulse current into an impulse voltage and vice versa, but it does not change the nature of the time function. If the stimulus is the current $i(t) = Qu_0(t)$, the response is the voltage

$$e_R(t) = QRu_0(t). \tag{10–8}$$

If the stimulus is the voltage $e(t) = \lambda u_0(t)$, the response is the current

$$i_R(t) = \frac{\lambda}{R}\, u_0(t). \tag{10–9}$$

In Eq. (10–8) the impulse current of Q coulombs is changed into an impulse voltage of QR flux linkages. In Eq. (10–9) the impulse voltage of λ flux linkages is changed into an impulse current of λ/R coulombs.*

Inductance. When the impulse voltage $e(t) = \lambda u_0(t)$ is applied to an inductance the response current is

$$i_L = \frac{1}{L}\int_{-\infty}^{t} \lambda u_0(t)\, dt = \frac{\lambda}{L}\, u_{-1}(t). \tag{10–10}$$

---

* One problem with the singularity functions is that they are not dimensionless. For example, $u_0(t)$ has the dimensions 1/time and the multiplying constant has the dimensions of volt·seconds or ampere·seconds.

The impulse voltage causes a current to start instantly in the inductance, and the current continues to flow forever. Both these conclusions may seem strange. We have said that the inductance acts as an open circuit when a voltage is suddenly applied to it, and we would not expect to get current in the open circuit. The problem is that of the irresistible force and the immovable object. The applied impulse voltage is infinite, the resistance of the inductance at $t = 0$ is infinite, and thus the current is indeterminate. The indeterminacy is evaluated by the integration in Eq. (10–10). The flux linkages are created in the inductance in zero time, and they establish the finite current of λ/L amperes. It is as though we have a paper bag full of flux linkages which we throw at the coil at time $t = 0$. The bag breaks when it strikes the coil, and the flux linkages attach themselves to the coil in essentially zero time. Since each flux linkage is associated with a current, the current is established in zero time. It should be noted that even in this description the inductance acts as an open circuit at $t = 0$.

The current which flows indefinitely in the circuit is a consequence of the assumption of an ideal inductance, that is, one that has no energy losses associated with it. Most physical inductors are constructed from wire which has loss characteristics; therefore the energy is gradually dissipated and the current decays to zero. Certain "superconductors" have been constructed which at very low temperatures verify the theory. If a current is created in a superconductive coil, which has zero losses below a particular temperature, the current continues for months or even years with no measurable decay. We are interested in the ideal inductance not so much because we hope to use one but because the ideal inductance and other ideal elements, such as resistance, can give us very good approximations for actual physical devices.

When the impulse current $i(t) = Qu_0(t)$ is applied to an inductance the response voltage is

$$e(t) = L\,\frac{d[Qu_0(t)]}{dt} = LQu_1(t). \tag{10–11}$$

The impulse current produces a doublet of voltage. While the infinite current is being established an infinite voltage is produced across the inductance, in such a direction that the voltage opposes the current flow; then, when the current is removed an infinite voltage is produced in the opposite direction, and this voltage tends to keep the current flowing. Since the dimensions of the function $u_1(t)$ are $1/\text{time}^2$, those of the multiplying coefficient (LQ) are therefore volt·seconds^2. The doublet is somewhat less common in practice than the step function and the impulse function.

Capacitance. When an impulse current $i(t) = Qu_0(t)$ is applied to a capacitance the response voltage is

$$e(t) = \frac{1}{C}\int_{-\infty}^{t} Qu_0(t)\,dt = \frac{Q}{C}\,u_{-1}(t). \tag{10–12}$$

The impulse current causes a voltage to appear immediately on the capacitance, and the voltage remains there indefinitely. We have said that the capacitance acts as a short circuit at $t = 0$, and it is surprising to find that a voltage is generated across the short circuit. It is again a matter of an irresistible force and an immovable object. The capacitance is a short circuit but the impulse current is infinite. Thus the voltage is indeterminate and must be evaluated by the integral of Eq. (10–12). The voltage rises on the capacitance, because the coulombs suddenly appear there. It is as though we have a bucket of coulombs which we suddenly dump into the capacitance at $t = 0$. As soon as the coulombs appear on the capacitance the voltage rises to its new value.*

We are much more used to the idea of perfect capacitance than we are to the idea of perfect inductance, and it is probably not surprising to learn that voltage remains on the capacitance indefinitely. A perfect capacitance has zero leakage, or an infinite resistance across its plates, and the coulombs find no closed path by which to leave the capacitance. Practical capacitors can very closely approach the ideal capacitance; they can hold their charge for hours, days, or even months.

When the impulse voltage $e(t) = \lambda u_0(t)$ is applied to a capacitance the response current is

$$i(t) = C\,\frac{d[\lambda u_0(t)]}{dt} = C\lambda u_1(t). \tag{10–13}$$

The impulse voltage produces a doublet of current. While the infinite voltage is being established a large current must flow in the positive direction. In order for the voltage to drop to zero again the current must reverse and carry away the charge which accumulated on the capacitance. The dimensions of the function $u_1(t)$ are $1/\text{time}^2$, and those of the multiplying coefficient $(C\lambda)$ are therefore ampere·seconds^2. The doublet current is somewhat more rare than the step or impulse current.

The impulse responses which we have considered in this section are somewhat more difficult than the step responses discussed in the preceding

---

* A capacitance may be likened to a bathtub, and the coulombs which fill the capacitance may be likened to water which fills the tub. A step current applied to the capacitance is analogous to a tap turned on in the tub. The tub fills at a linear rate. An impulse current applied to the capacitance is analogous to a bucket of water dumped suddenly into the tub. The water level rises in essentially zero time.

section. The mathematical expressions are sufficiently straightforward, but the circuit interpretations involve some indeterminate forms. A voltage impulse applied to an inductance starts up a current in zero time, even though the inductance is an open circuit at $t = 0$. A current impulse applied to a capacitance instantly produces a voltage across it, even though the capacitance is a short circuit at $t = 0$. We shall make a good deal of use of these results in later sections.

10–4 Initial conditions as step and impulse sources. We learned in Chapter 7 that it is not necessary to know the complete past history of an energy-storage element in order to use it in a circuit. In fact, the only thing which must be known is the energy stored at the beginning of the problem. In Chapter 7 we noted that this energy can be represented by a source, at least so far as computations at $t = 0$ are concerned. An initial voltage on a capacitance is equivalent to a direct voltage in series with the capacitance, and an initial current in an inductance is equivalent to a direct current in parallel with the inductance. In this section we shall investigate in more detail this source equivalence for initial energy storage and show that the equivalent sources are actually singularity functions.

Capacitance. The capacitance element is completely described by the volt-ampere equation

$$e_C = \frac{1}{C}\int_{-\infty}^{t} i\,dt = \frac{1}{C}\int_{-\infty}^{0} i\,dt + \frac{1}{C}\int_{0}^{t} i\,dt. \tag{10–14}$$

In Eq. (10–14) the past history of the capacitance is summed up by the integral with limits $-\infty$ and 0. The integral with limits 0 and t is exactly that satisfied by an uncharged capacitance switched into a circuit at $t = 0$. Thus the volt-ampere equation of the capacitance is the sum of a constant equal to the value of the voltage on the capacitance at $t = 0$ and the volt-ampere relationship of an uncharged capacitance. The constant, however, is not the customary numerical constant but a step function which has a constant value for t greater than zero. The physical equivalence is shown in Fig. 10–1(a). When the switch is in position 1 the charged capacitance is in the circuit; when the switch is in position 2 the uncharged capacitance in series with the voltage source E_0 is in the circuit. There is no way that the interchange can be accomplished without the switch. *The switch is an essential part of the equivalence.* In fact, the switch and the voltage source together produce a step function of voltage in series with the capacitance. We can therefore write Eq. (10–14) in the form

$$e_C = E_0 u_{-1}(t) + \frac{1}{C}\int_{0}^{t} i\,dt. \tag{10–15}$$

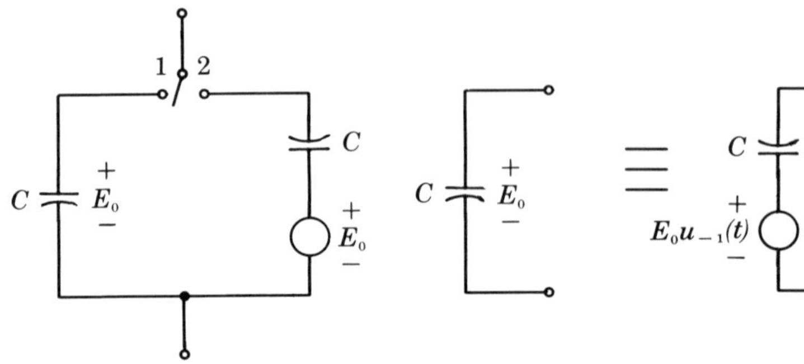

(a) Two equivalent branches at $t = 0$

(b) The voltage-source equivalence

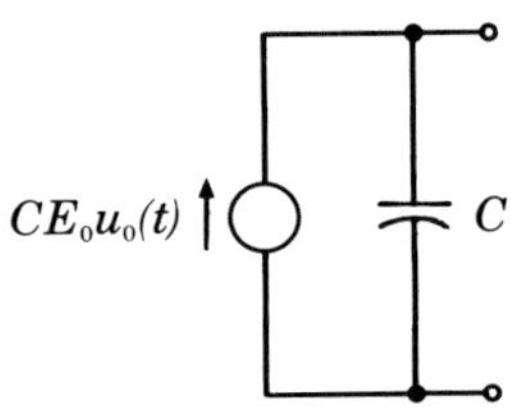

(c) The current-source equivalence

FIG. 10–1. Equivalent sources for the initial energy on a capacitance.

The equivalent circuit represented by Eq. (10–15) is shown up by Fig. 10–1(b). A step function of voltage with an amplitude equal to the value of the voltage on the capacitance at $t = 0$ is placed in series with an uncharged capacitance. The combination is indistinguishable from the original charged capacitance.

One reason for replacing the switch by the step function is so that further operations can be performed on the circuit. For example the Norton equivalent circuit for the step voltage in series with the capacitance is shown in Fig. 10–1(c). The dead network in parallel with the current source is simply the uncharged capacitance. The short-circuit current is computed by means of the volt-ampere equation of the capacitance:

$$I_{sc} = C\,\frac{de}{dt} = CE_0u_0(t). \tag{10–16}$$

Thus the charged capacitance can be represented by an uncharged capacitance in parallel with an impulse current source of CE_0 coulombs. In a sense this equivalence is the inverse of the impulse-establishment of initial energy storages which we considered in Chapter 9. Between $t = 0^-$ and $t = 0^+$ the impulse current source in Fig. 10–1(c) dumps CE_0 coulombs in the capacitance and causes its voltage to rise to the initial value of E_0. This result is independent of the external circuit, since the capacitance acts as a short circuit at $t = 0$, and all the current flows into it.*

In summary, *we can replace a capacitance C charged to a voltage E_0 at $t = 0$ by an uncharged capacitance in series with a step voltage source $E_0 u_{-1}(t)$ or by an uncharged capacitance in parallel with an impulse current source $CE_0 u_0(t)$.*

Inductance. The volt-ampere equation which completely describes the inductance element is

$$i_L = \frac{1}{L}\int_{-\infty}^{t} e\,dt = \frac{1}{L}\int_{-\infty}^{0} e\,dt + \frac{1}{L}\int_{0}^{t} e\,dt. \tag{10–17}$$

In Eq. (10–17) the integral with limits $-\infty$ and 0 is the flux in the inductance at $t = 0$. This integral divided by the inductance value L is the initial current in the inductance, I_0. Thus an inductance with an initial current is equivalent to an unfluxed inductance in parallel with a constant current of value I_0. That this current is really a step function can be seen from Fig. 10–2(a). When the switch is in position 1 the inductance with the initial current is in the circuit; when it is in position 2 the unfluxed inductance in parallel with the current source I_0 is connected into the circuit. If the two branches are to be equivalent the switch must be moved at $t = 0$, which means that *the switch is an integral part of the operation.* The current source I_0 and the switching operation can be combined to produce a step function, and Eq. (10–18) can be rewritten to obtain

$$i_L = I_0 u_{-1}(t) + \frac{1}{L}\int_{0}^{t} e\,dt. \tag{10–18}$$

Equation (10–18) represents an equivalent circuit consisting of a step-function source of current in parallel with an unfluxed inductance, as shown in Fig. 10–2(b). If we apply Thévenin's theorem to the circuit of Fig. 10–2(b) we obtain the circuit of Fig. 10–2(c). The dead network is the unfluxed inductance. The open-circuit voltage is computed from the volt-ampere equation for the inductance:

$$E_{oc} = L\frac{di}{dt} = LI_0 u_0(t). \tag{10–19}$$

---

* The external circuit cannot be an uncharged capacitance, because the two capacitances would be in parallel in the original circuit.

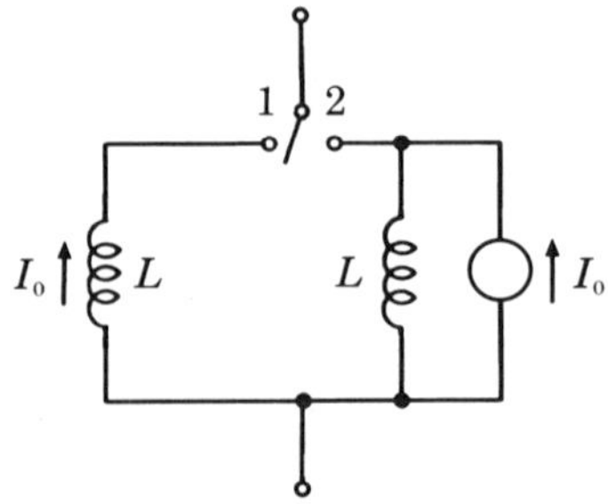

(a) Two equivalent branches at $t = 0$

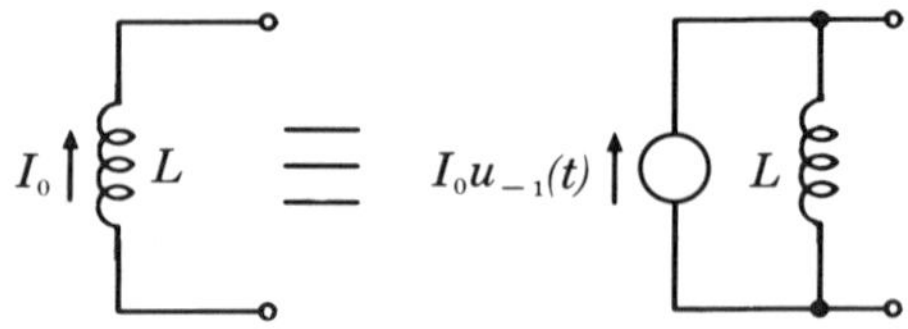

(b) The current-source equivalence

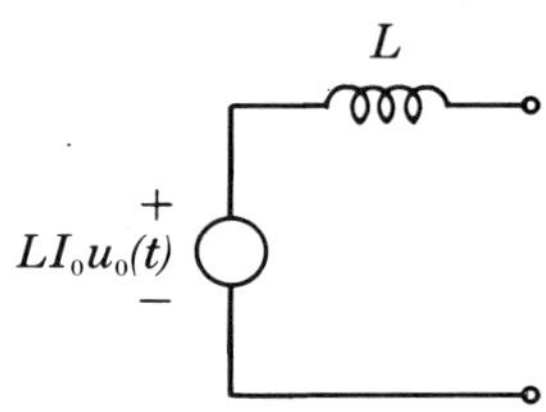

(c) The voltage-source equivalence

FIG. 10–2. Equivalent sources for the initial energy in an inductance.

The inductance with the initial current can thus be represented by an unfluxed inductance in series with an impulse voltage source of value LI_0 flux linkages. Between $t = 0^-$ and $t = 0^+$ the voltage source throws the LI_0 flux linkages into the inductance, and by $t = 0^+$ a current of value I_0 is established in the inductance. This result is independent of the external circuit, since the inductance acts as an open circuit at $t = 0$ and all the voltage appears across it.*

In summary, *an inductance L with an initial current I_0 can be replaced by an unfluxed inductance L in parallel with a current source $I_0 u_{-1}(t)$ or by an unfluxed inductance L in series with a voltage source $LI_0 u_0(t)$.*

The step-function and impulse-function sources equivalent to initial energy storages, which we have discussed in this section, are very useful. Combined with the superposition theorem they make it possible to handle circuits with initial energy stored in them by exactly the same methods which are used in obtaining the stimulus-response relationships for dead networks. We have achieved one step forward in unifying our treatment

---

* The external circuit cannot have a series inductance unless it also has the current I_0 in it.

of electric networks. We shall also find the equivalent sources useful for solving problems which involve the redistribution of initial energy in purely inductive circuits or purely capacitive circuits. In the next section our attention will be turned to such networks.

10–5 One-element-type networks. Networks composed of elements of a single type are fairly common. We have already considered in some detail the case of purely resistive networks. Purely capacitive networks and purely inductive networks also occur in practice. In addition it is possible in some instances to separate a general network into sections which are purely inductive or purely capacitive and to reduce these sections by means of Thévenin's theorem or Norton's theorem. In general it can be said that the solution of problems involving purely inductive or purely capacitive networks is no more difficult than the solution of purely resistive networks. However, the presence of energy initially stored in the purely inductive or purely capacitive circuit can complicate a problem slightly.

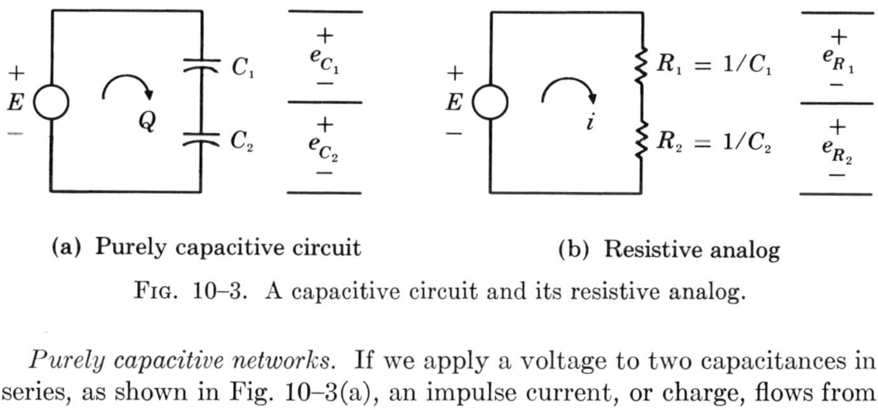

(a) Purely capacitive circuit (b) Resistive analog

FIG. 10–3. A capacitive circuit and its resistive analog.

Purely capacitive networks. If we apply a voltage to two capacitances in series, as shown in Fig. 10–3(a), an impulse current, or charge, flows from the source and produces a voltage on each of the capacitances. The sum of these two voltages must add up to the source voltage, E. We can easily solve for the values of the two voltages by substituting the variable Q for the variable i in the volt-ampere equations for capacitance. In terms of Q these equations are

$$e_C = \frac{Q}{C}, \qquad Q = Ce_C. \tag{10–20}$$

By Kirchhoff's voltage law

$$e_{C_1} + e_{C_2} = E$$

or

$$\frac{Q}{C_1} + \frac{Q}{C_2} = E,$$

from which

$$Q = \frac{C_1C_2E}{C_1 + C_2}.$$

Thus

$$e_{C_1} = \frac{C_2E}{C_1 + C_2}, \qquad e_{C_2} = \frac{C_1E}{C_1 + C_2}. \tag{10–21}$$

According to Eq. (10–21) the voltage across each capacitance varies inversely with the value of the capacitance.

It is possible to solve many problems involving purely capacitive circuits by utilizing Eq. (10–21) and treating charge as the variable. However, we have accumulated a store of theory and techniques for solving purely resistive circuits such that another point of view is more fruitful. That is, we shall make a resistive analog for the capacitive circuit. If we compare Eq. (10–20) with the volt-ampere equation for a resistive element, we see the following similarity:

$$e_C = \frac{1}{C}Q, \qquad e_R = Ri. \tag{10–22}$$

If the value of the resistance in Eq. (10–22) is numerically equal to the reciprocal of the value of the capacitance and if the two voltages are equal, the current in the resistive circuit is analogous to the charge in the capacitive circuit. Kirchhoff's two laws are equally valid for the integrals and derivatives of currents and voltages as for the currents and voltages themselves. We can therefore draw the following conclusions: If we make a resistive circuit in which the geometry is the same as that of a capacitive circuit and if each resistance is equated to the reciprocal of the value of the corresponding capacitance, the equations for the two circuits are numerically identical. The voltages in the two circuits are also identical, and the currents in the resistive circuit are numerically equal to the charges in the capacitive circuit. The two circuits are said to be analogs. In practice we convert the capacitive circuit to its resistive analog, solve for the currents and voltages in the analog, and convert the results back to values for the original circuit.

The resistive analog for the circuit in Fig. 10–3(a) is shown in Fig. 10–3(b). The current in the analog is

$$i = \frac{E}{1/C_1 + 1/C_2},$$

and the two voltages are

$$\begin{aligned} e_{R_1} &= \frac{1/C_1E}{1/C_1 + 1/C_2} = \frac{C_2E}{C_1 + C_2}, \\ e_{R_2} &= \frac{1/C_2E}{1/C_1 + 1/C_2} = \frac{C_1E}{C_1 + C_2}. \end{aligned} \tag{10–23}$$

Note that each voltage in Eq. (10–23) is the same as the voltage across the corresponding capacitance in Eq. (10–21).

Once the resistive analog has been obtained it can be solved in any convenient way. Mesh equations, node equations, Thévenin's theorem, Norton's theorem, Y–Δ transformations, and all the other methods for resistive circuits can be used. It would be impractical to redevelop all these methods for the notation of purely capacitive circuits!

One major difficulty with the resistive analog is that it applies only for voltage sources! There is nothing in the analog circuit which corresponds to current in the capacitive circuit.* The difficulty can be overcome by modifying our analog. Equation (10–22) can be written in the form

$$i_C = C\,\frac{de}{dt}, \qquad i_R = \frac{1}{R}\,e. \tag{10–24}$$

Again, Eq. (10–24) suggests an analog. If the same geometry is maintained in the two circuits and if each capacitance is replaced by a resistance with a value which is the reciprocal of the value of the capacitance, the voltages in the resistive circuit are numerically equal to the derivatives of the voltages in the capacitive circuit. A capacitive circuit to which this analog applies is shown in Fig. 10–4(a). A current source is applied to two parallel capacitances, and it is desired to know how the current divides between them. The resistance analog is shown in Fig. 10–4(b). The currents in the two resistances are

$$\begin{aligned} i_{R_1} &= \frac{1/C_2}{1/C_1 + 1/C_2} \times 10 = \frac{10C_1}{C_1 + C_2}, \\ i_{R_2} &= \frac{1/C_1}{1/C_1 + 1/C_2} \times 10 = \frac{10C_2}{C_1 + C_2}. \end{aligned} \tag{10–25}$$

In this form of the analog the currents in the two circuits are equal and therefore the currents in the capacitances are also given by Eq. (10–25).

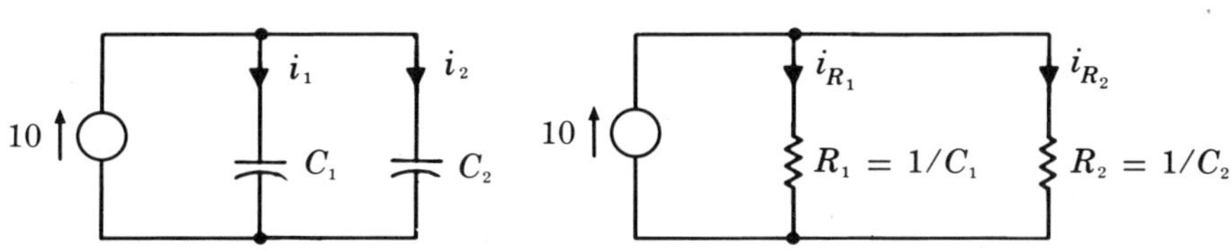

(a) Capacitive circuit excited by a current source

(b) Resistive analog

FIG. 10–4. The resistive analog for a capacitive circuit excited by a current.

---

* The method of treating charge as the variable won't do either, since current is now the given quantity.

The voltage in the resistive circuit is analogous to the derivative of the voltage in the capacitive circuit. The actual voltage in the capacitive circuit is a ramp function rising linearly with time.

In summary, a purely capacitive network can be solved by the use of a resistance analog in which the geometry is unchanged and in which each capacitance is replaced by a resistance with a value which is the reciprocal of the value of the capacitance. If the source is a voltage source, the voltages everywhere in the two circuits are identical. If the source is a current source, the currents everywhere in the two circuits are identical. If both a current and a voltage source are present, the problem can be solved by the method of superposition.

When there is energy initially stored on one of the capacitances, an equivalent source must be used. An example is shown in Fig. 10–5(a). The capacitance C_1 is charged to 10 volts and is then connected in parallel with the capacitance C_2. The voltage across the combination is desired. The equivalent step-function source for the initial energy storage is shown in Fig. 10–5(b). The two capacitances in this circuit are uncharged. The resistive analog is shown in Fig. 10–5(c). The current in this circuit is

$$i = \frac{10}{\frac{1}{3} + \frac{1}{2}} = 12 \text{ amp},$$

and the voltages across the resistances are

$$e_1 = \tfrac{1}{3} \times 12 = 4 \text{ volts},$$

$$e_2 = \tfrac{1}{2} \times 12 = 6 \text{ volts}.$$

Since the voltages are preserved in this analog, the voltages across the corresponding capacitances in the circuit of Fig. 10–5(b) are also 4 volts and 6 volts. The voltage across capacitance C_1 must be combined algebraically with the 10-volt source to obtain the actual voltage across capacitance C_1 in Fig. 10–5(a). The net value is 6 volts. Thus each capaci-

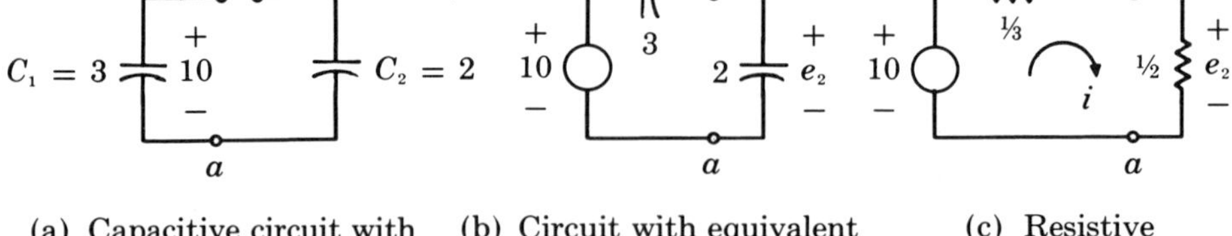

(a) Capacitive circuit with initial energy storage (b) Circuit with equivalent step voltage source (c) Resistive analog

FIG. 10–5. The resistive analog for a capacitive circuit with initial energy storage.

tance in Fig. 10–5(a) has 6 volts across it after the switch is closed, and indeed both voltages must be equal since they are in parallel.

Inductance. A purely inductive network can also be converted to a resistive analog. Again there are two analogs, depending on whether the network includes a current or voltage source. We shall consider first the network with a current source. The volt-ampere equation for the inductance is

$$i_L = \frac{1}{L}\int_{-\infty}^{t} e\,dt = \frac{\lambda}{L}, \tag{10–26}$$

where λ is the flux linkage. This equation is similar to the volt-ampere equation for resistance:

$$i_R = \frac{e}{R}. \tag{10–27}$$

If the resistance in Eq. (10–27) is numerically equal to the inductance of Eq. (10–26) and if the currents are equal in the two cases, the voltage in the resistive circuit is analogous to the flux linkage in the inductive circuit. Thus if each inductance in a purely inductive circuit is changed to a resistance with a value in ohms equal to the value of the inductance in henrys, the equations for the two circuits are identical, providing current and voltage are used as variables in the resistive circuit and current and flux linkage in the inductive circuit. In practical use of the analog the inductive circuit is converted to a resistive analog, the currents and voltages are found in the resistive circuit, and the results are converted back to values for the original circuit. An example of a purely inductive circuit is shown in Fig. 10–6(a), and its resistive analog is shown in Fig. 10–6(b). The currents in the analog are

$$i_1 = \tfrac{2}{5} \times 10 = 4 \text{ amp},$$

$$i_2 = \tfrac{3}{5} \times 10 = 6 \text{ amp}.$$

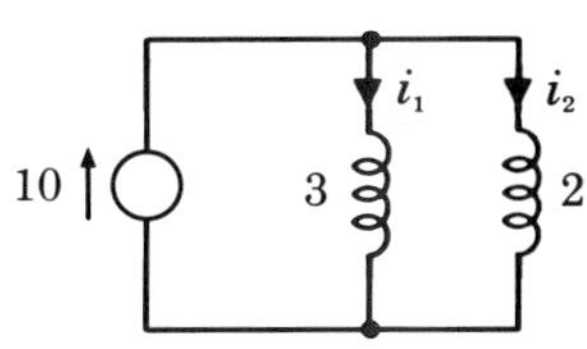

(a) The inductive circuit excited by a current source

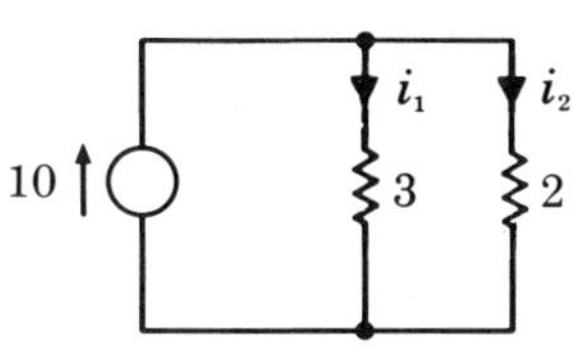

(b) The resistive analog

FIG. 10–6. The resistive analog for an inductive circuit excited by a current source.

Since in this form of the analog the currents are preserved, the currents in the inductive circuit have the same values as those in the resistive circuit.

A second form of the analog must be used if the inductive circuit is excited by a voltage source. In this case we write the volt-ampere equation for inductance in the form

$$e_L = L\,\frac{di}{dt},$$

and compare it with the volt-ampere equation of resistance in the form

$$e_R = Ri.$$

If the resistances in the analog are made numerically equal to the inductances and if the voltage sources are the same, the voltages everywhere in both circuits are equal, and the current in the resistive analog is numerically equal to the derivative of current in the inductive circuit. An example of a purely inductive circuit with a voltage source is shown in Fig. 10–7(a). A direct voltage of 10 volts is connected in series with two inductances, and it is desired to determine the voltage division between the inductances. The resistive analog is shown in Fig. 10–7(b). The voltages across the resistances are

$$e_1 = \tfrac{2}{5} \times 10 = 4 \text{ volts},$$

$$e_2 = \tfrac{3}{5} \times 10 = 6 \text{ volts}.$$

Since the voltages are preserved in the analog, the voltages across the two inductances of Fig. 10–7(a) are also four and six volts. The current through the inductances is a ramp function whose slope is numerically equal to the current in the resistive analog.

Energy initially stored in one of the inductances of a circuit should be transformed into a step current source in parallel with the inductance, and the solution for this arrangement should be obtained. An example

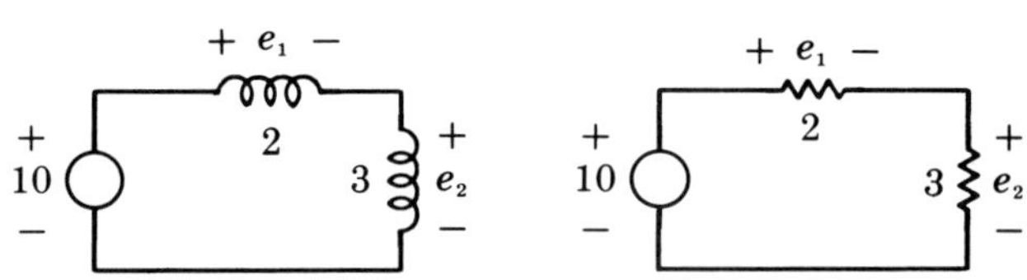

(a) The inductive circuit excited by a voltage source

(b) The resistive analog

FIG. 10–7. The resistive analog for an inductive circuit excited by a voltage source.

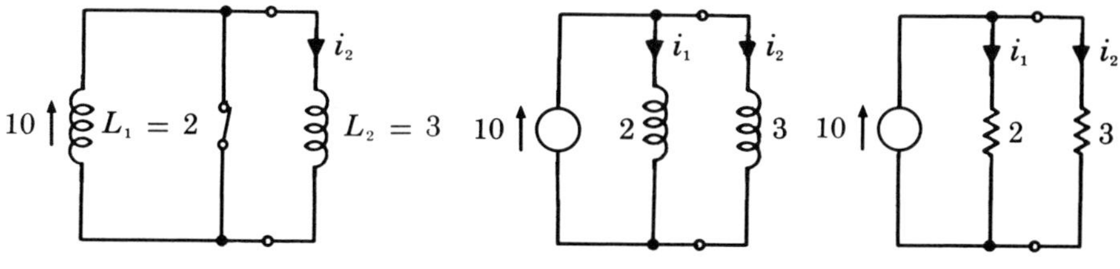

(a) Inductive circuit with initial energy (b) Circuit with equivalent step current source (c) Resistive analog

FIG. 10–8. A resistive analog for an inductive circuit with initial energy.

is shown in Fig. 10–8(a). The 10-ampere current in L_1 is changed to an equivalent current source in parallel with the inductance, as shown in Fig. 10–8(b). The resistive analog of this circuit is shown in Fig. 10–8(c). The currents in the analog are

$$i_1 = \tfrac{3}{5} \times 10 = 6 \text{ amp},$$

$$i_2 = \tfrac{2}{5} \times 10 = 4 \text{ amp}.$$

These currents are also the currents in the inductances of Fig. 10–8(b). The inductance L_1, however, includes the 10-ampere current source. The net current in this inductance is therefore 10 − 6, or 4 amp. Thus both inductances in Fig. 10–8(a) have a current of 4 amp after the switch is opened.

In summary, the purely inductive network can be treated by means of a resistance analog in which every inductance is replaced by an identical resistance counterpart. If the source is a current source, the currents everywhere in the analog circuit are identical to the corresponding currents in the original circuit. If the source is a voltage source, the voltages everywhere in the analog are identical to the corresponding voltages in the original circuit. Initial currents in inductances should be transformed into current sources in parallel with the inductances. If both current and voltage sources are present in a problem, the solution can be obtained by superposition.

The resistance analogs which have been developed in this section are particular cases of a rather general principle: Whenever two physical systems satisfy differential equations which are numerically the same, the systems are said to be analogs. Because of the ease with which electrical quantities can be measured, electrical analogs are often constructed to represent complicated mechanical, thermal, and chemical systems. The behavior of a system is then investigated by means of measurements made upon the analog.

10–6 Summary.

Step function responses of single elements

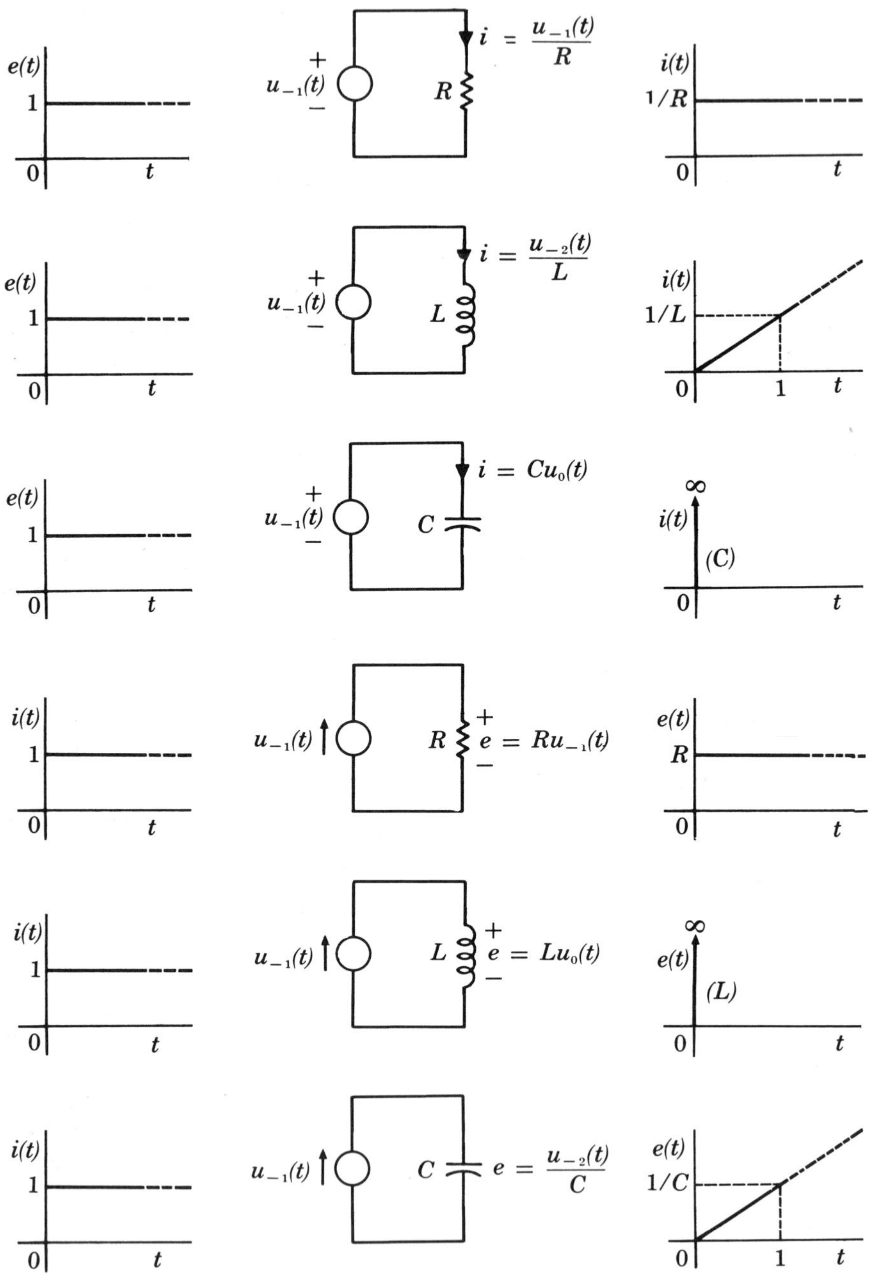

Impulse responses of single elements

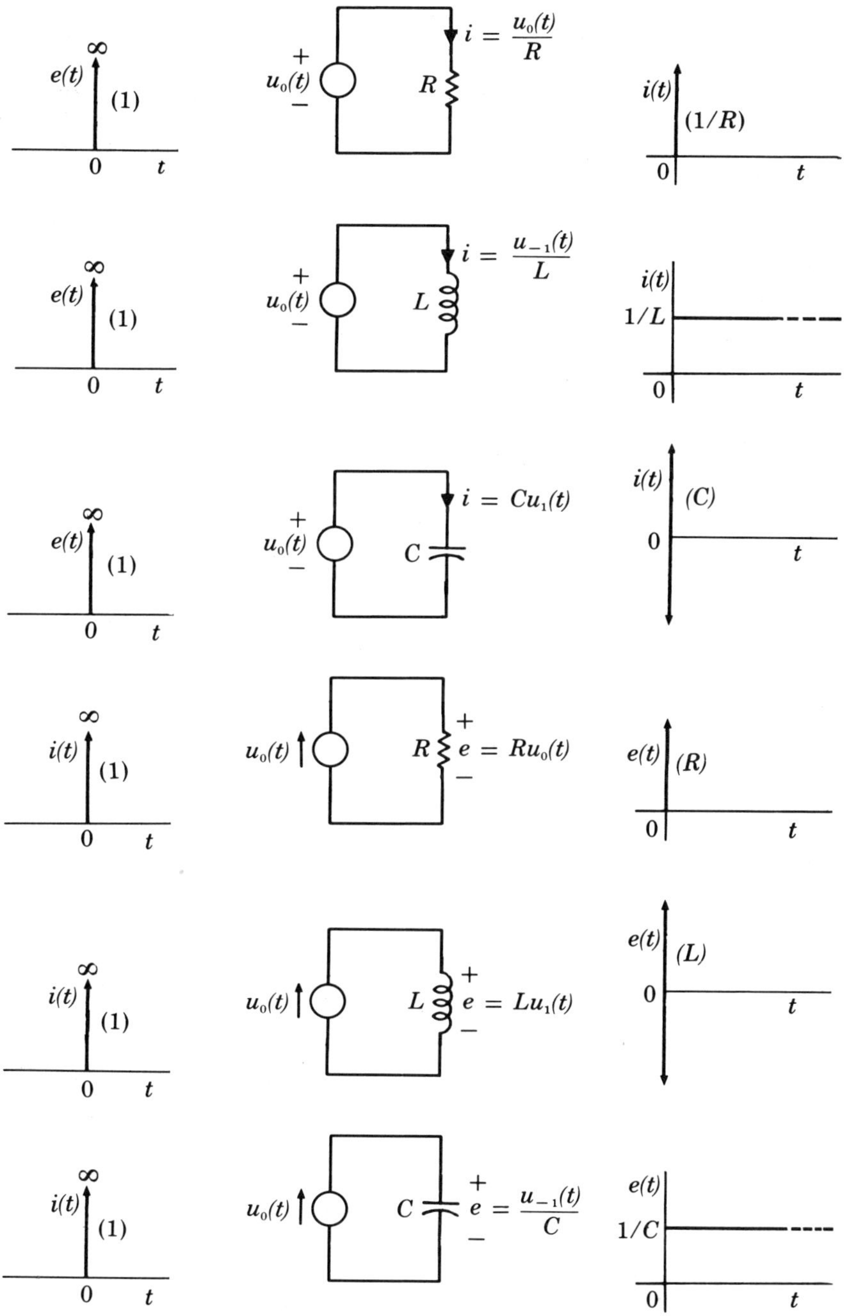

Initial conditions as singularity functions

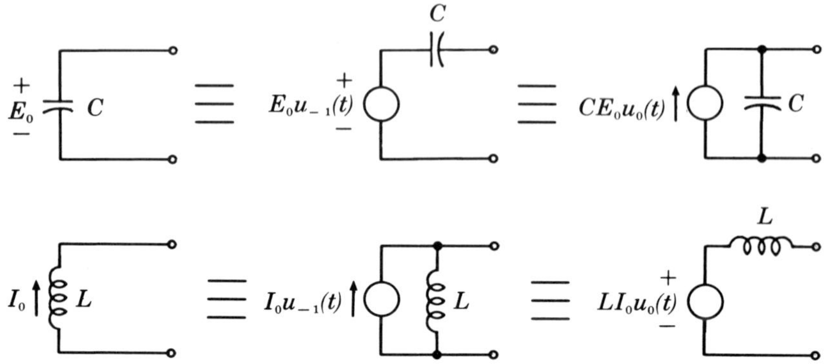

One-element-type circuits

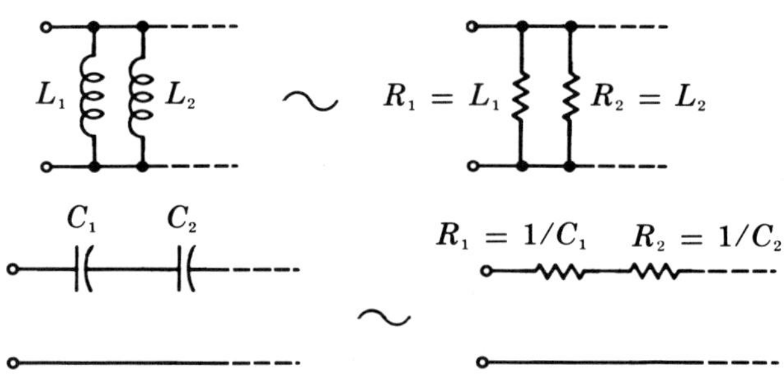

If the source is a voltage source the circuit and the analog have the same voltages. If the source is a current source the analog and the circuit have the same currents.

REFERENCES

The single-element circuits treated in this chapter are mathematical models. The responses from these circuits occur as approximations to real responses, and as component parts in the computation of real responses. Some authors do not treat the single-element circuits at all, preferring to deal only with "practical circuits." Additional references to both points of view are given below.

1. E. A. GUILLEMIN. *Introductory Circuit Theory,* John Wiley & Sons, New York, 1953. Chapter 4 of Guillemin's book gives a treatment of single-element circuits with both step and impulse excitation.

2. M. E. Van Valkenburg, *Network Analysis*, Prentice-Hall, New Jersey, 1955. The R–L and the R–C circuits represent the simplest case which Van Valkenburg considers, whether with differential equations or transforms.

3. H. G. Booker, *An Approach to Electrical Science*, McGraw-Hill, New York, 1959. The first section of this book is concerned with pure capacitance networks and the second with pure resistance networks. The third section contains a chapter on pure inductance networks.

4. H. H. Skilling, *Electrical Engineering Circuits*, John Wiley & Sons, New York, 1957. Skilling's approach is geared closely to physical circuits which could actually be demonstrated to the students.

5. Brenner and Javid, *Analysis of Electric Circuits*, McGraw-Hill, New York, 1959. Brenner and Javid begin their book with a discussion of single-element circuits. They consider impulses only as responses.

6. Lago and Waidelich, *Transients in Electrical Circuits*, The Ronald Press, New York, 1958. This text also begins with single-element circuits.

7. Walsh and Miller, *Introductory Electric Circuits*, McGraw-Hill, New York, 1960. Walsh and Miller begin with d-c circuits and then treat the one-element energy-storage circuits.

EXERCISES*

Section 10–2

1. A step voltage of value $e(t) = 100u_{-1}(t)$ is applied to (a) a resistance of 10 ohms, (b) an inductance of 3 henrys, and (c) a capacitance of $\frac{1}{6}$ farad. In each case give the resulting currents in analytic form and sketch them.

2. A step current of $i(t) = 100u_{-1}(t)$ is applied to (a) a resistance of 3 ohms, (b) an inductance of 7 henrys, and (c) a capacitance of 4 farads. In each case give analytical expression for the resulting voltage and sketch it.

3. In the circuits of Fig. 10–9 the switches are closed at $t = 0$. Give analytical expressions for the resulting currents in terms of the singularity functions.

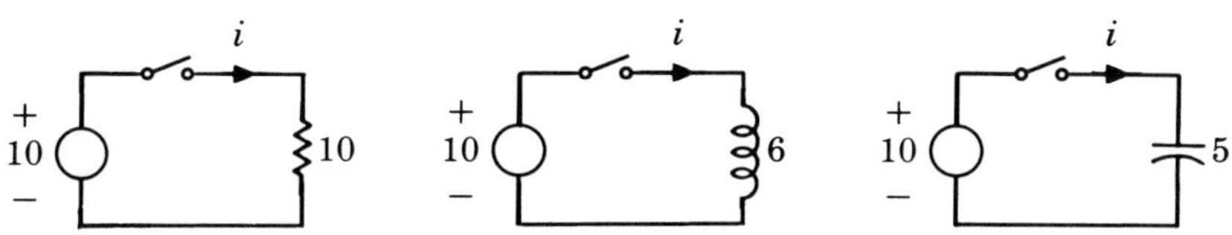

(a) (b) (c)

Figure 10–9

---

* Numerical values are in volts, amperes, ohms, henrys, and farads.

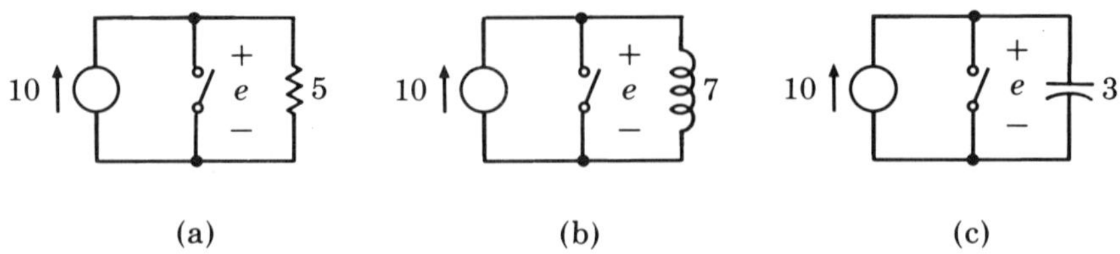

FIGURE 10–10

4. In the circuits of Fig. 10–10 the switches are opened at $t = 0$. Give expressions for the voltages in terms of the singularity functions.

5. The voltage $e(t) = 10u_{-1}(t) - 10u_{-1}(t-1)$ is applied to the parallel combination of $R = 5$, $L = 5$, and $C = \frac{1}{5}$. Give the analytical expression for the input current and sketch it.

6. The current $i(t) = 10u_{-1}(t) - 10u_{-1}(t-1)$ is applied to the series combination of $R = \frac{1}{5}$, $L = \frac{1}{5}$, and $C = 5$. Give the analytical expression for the voltage across the combination and sketch the waveform.

7. A voltage source $e(t) = 10 \cos 2t$ is suddenly switched at $t = 0$ across (a) a resistance of 10 ohms, (b) an inductance of 5 henrys, and (c) a capacitance of $\frac{1}{2}$ farad. In each case give the analytic expression for the current which flows.

8. A current source $i(t) = 5 \sin 3t$ is suddenly switched at $t = 0$ across (a) a resistance of 5 ohms, (b) an inductance of 3 henrys and (c) a capacitance of 2 farads. In each case give the analytical expression for the voltage produced across the element.

9. A certain two-terminal device has the following current response to a step-voltage stimulus:

$$i(t) = 5u_{-1}(t) + 6u_0(t) + 3u_{-2}(t).$$

Give a possible circuit which could represent the device.

10. A certain two-terminal device has the following voltage response to an applied step-current stimulus.

$$e(t) = 3u_{-1}(t) + \tfrac{1}{3}u_0(t) + \tfrac{7}{8}u_{-2}(t).$$

Give a possible circuit which could represent the device.

Section 10–3

11. An impulse voltage $e(t) = 100u_0(t)$ is applied to (a) a resistance of 10 ohms, (b) an inductance of 3 henrys, and (c) a capacitance of 5 farads. Write an analytical expression for the current which flows in each case and sketch it.

12. An impulse current $i(t) = 13u_0(t)$ is applied to (a) a resistance of 4 ohms, (b) an inductance of 5 henrys, and (c) a capacitance of $\frac{1}{3}$ farad. Write an analytical expression for the voltage which occurs in each case and sketch it.

13. The voltage $e(t) = 10u_0(t) - 10u_0(t-1)$ is applied to the parallel combination of $R = 2$, $L = 5$, and $C = \frac{1}{4}$. Give the analytical expression for the current drawn from the source and sketch it.

14. The current $i(t) = 10u_0(t) - 10u_0(t-1)$ is applied to the series combination of $R = \frac{1}{2}$, $L = \frac{1}{4}$, and $C = 5$. Give the analytical expression for the voltage produced across the combination and sketch it.

15. A voltage $e(t) = 10u_0(t) - 10u_1(t)$ is applied to a parallel combination of $R = 2$, $L = 2$, and $C = \frac{1}{2}$. Write an expression for the input current and sketch it.

16. A current $i(t) = 5u_0(t) - 10u_{-1}(t)$ is applied to a series combination of $R = 2$, $C = \frac{1}{4}$, and $L = 1$. Find the resulting voltage as a function of time, and sketch it.

17. The response of a two-terminal network to an input impulse voltage is the current

$$i(t) = 10u_{-1}(t) + 5u_0(t) + 3u_1(t).$$

What is a possible network which would give this response?

18. The response of a two-terminal network to an impulse-current excitation is the voltage

$$e(t) = 7u_1(t) + \tfrac{1}{3}u_0(t) + \tfrac{1}{15}u_{-1}(t).$$

What is a possible network which would have this response?

19. An impulse current $10u_0(t)$ is applied to the parallel combination of $C_1 = 3$ farads and $C_2 = 2$ farads. What current flows in each capacitance, and what voltage appears across the combination?

20. An impulse voltage $18u_0(t)$ is applied to the series combination of $L_1 = 3$ and $L_2 = 6$ henrys. What voltage appears across each, and what current flows from the source?

Section 10–4

21. The switch in the circuit of Fig. 10–11 is closed at $t = 0$. Give the voltage-source equivalent circuit for the network to the left of the terminals a–b. Give also the current-source equivalent circuit.

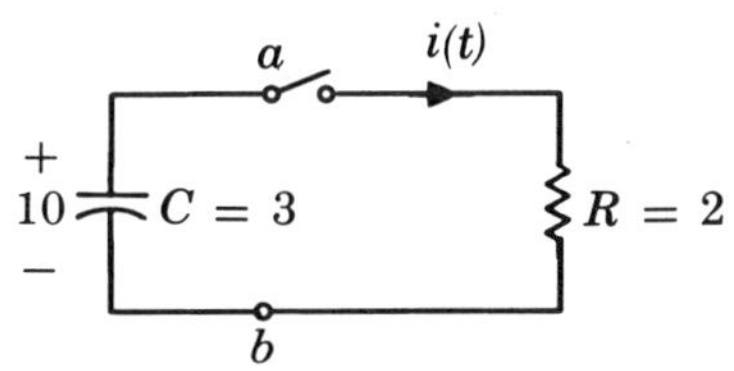

FIGURE 10–11

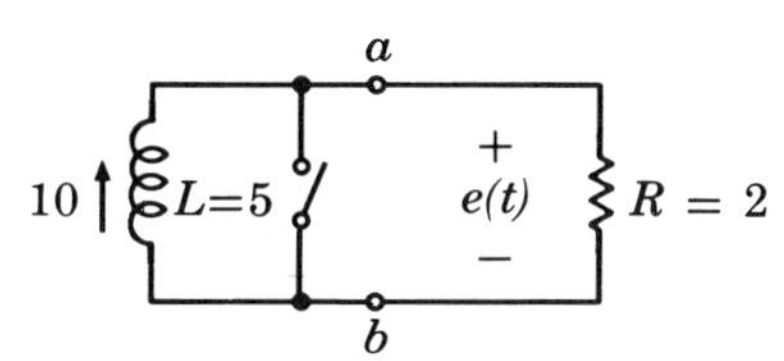

FIGURE 10–12

22. The switch in the circuit of Fig. 10–12 is opened at $t = 0$. Draw the equivalent voltage-source circuit for the network to the left of the terminals a–b. Also draw the equivalent current-source circuit.

23. Show that the circuits used to represent the initial energy stored in the capacitance and inductance in Exercises 21 and 22 can be obtained by applying Thévenin's and Norton's theorems at the terminal-pair a–b in each case.

24. What is the initial current which flows at $t = 0^+$ in the circuit of Fig. 10–11? What is the derivative of this current at $t = 0^+$?

25. What is the initial voltage produced across the resistance at $t = 0^+$ in the circuit of Fig. 10–12? What is the derivative of this voltage at $t = 0^+$?

26. (a) In the circuit of Fig. 10–13 compute the current at $t = 0^+$ by replacing the impulse voltage and the inductance by an inductance with an initial energy storage. (b) As an alternative, treat the inductance as an open circuit from $t = 0^-$ to $t = 0^+$, and compute the current established in it by the infinite impulse voltage by means of the voltage-current equation for the inductance.

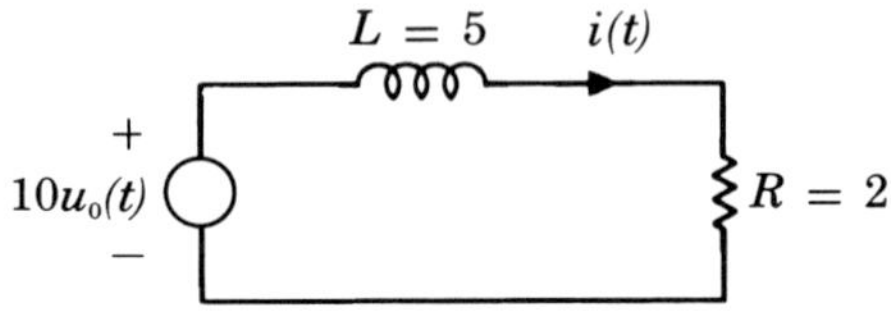

FIGURE 10–13

27. (a) In the circuit of Fig. 10–14 compute the initial voltage at $t = 0^+$ by replacing the impulse current source and the uncharged capacitance by the equivalent charged capacitance. (b) Obtain the same result by treating the capacitance as a short circuit for the interval from $t = 0^-$ to $t = 0^+$ and evaluate the voltage established in it by the infinite current by means of the volt-ampere equation of the capacitance.

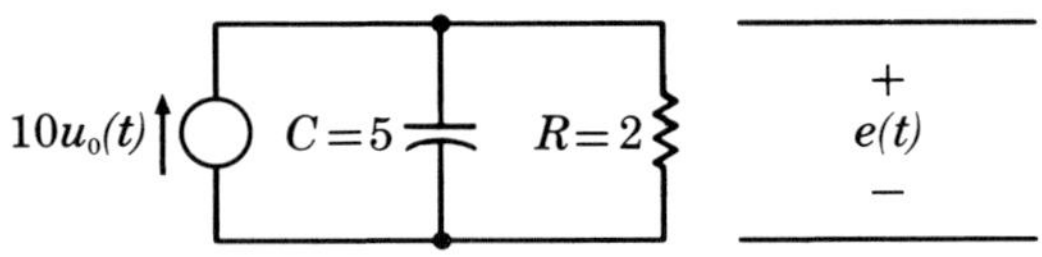

FIGURE 10–14

28. Find the values of the voltage $e(t)$ and its first derivative at $t = 0^+$ in the circuit of Fig. 10–15.

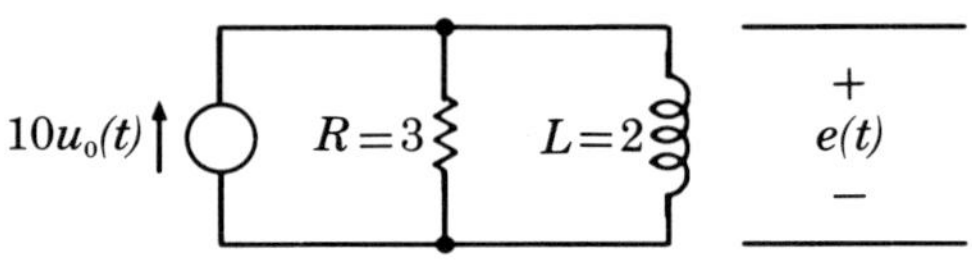

FIGURE 10–15

29. In the circuit of Fig. 10–16 find the values of the current $i(t)$ and its first derivative at $t = 0^+$.

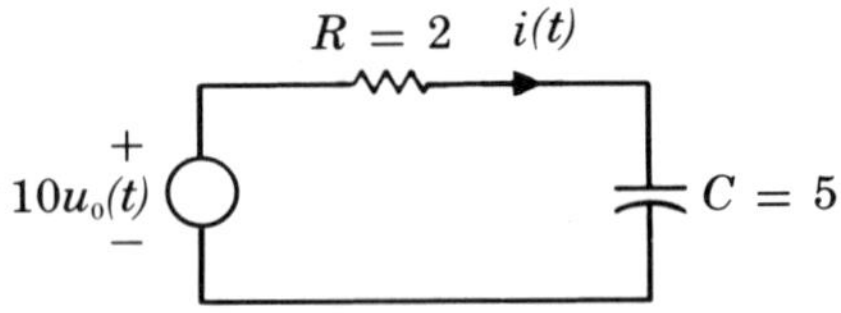

FIGURE 10–16

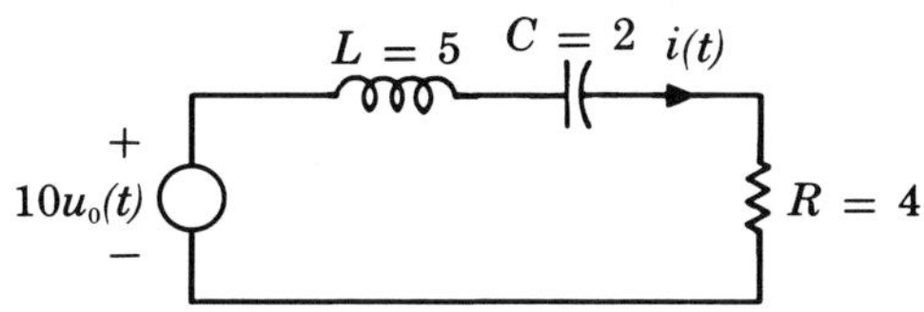

FIGURE 10–17

30. Find $i(0^+)$ and $di(0^+)/dt$ for the circuit of Fig. 10–17.
31. Find $e(0^+)$ and $de(0^+)/dt$ in the circuit of Fig. 10–18.

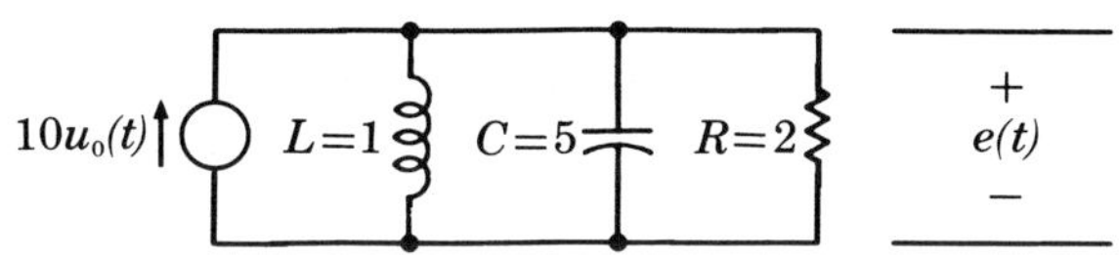

FIGURE 10–18

Section 10–5

32. Find the equivalent capacitance C which exists at terminal-pair a–b in the circuit of Fig. 10–19.
33. Find the equivalent inductance L which exists at terminal-pair a–b in the circuit of Fig. 10–20.

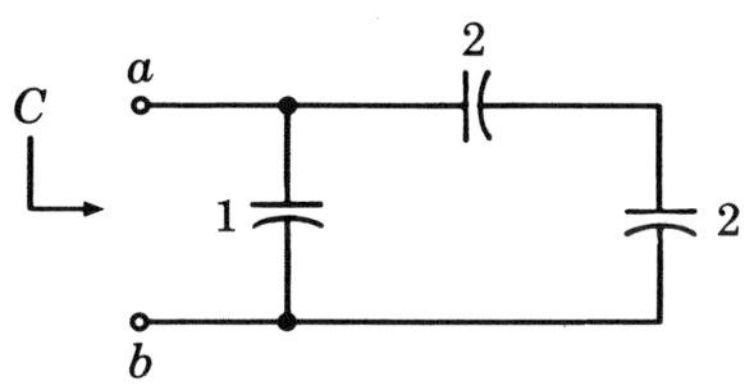

FIGURE 10–19

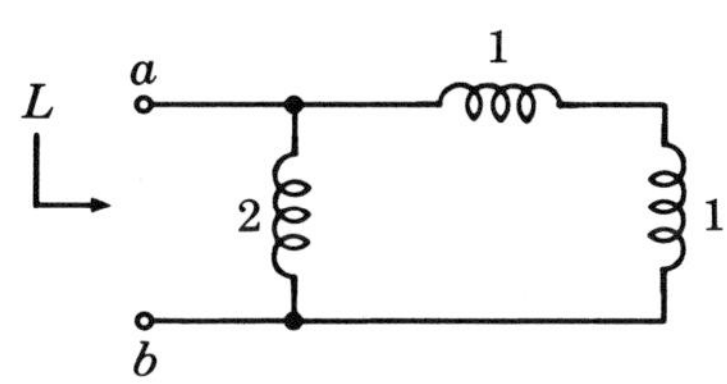

FIGURE 10–20

34. Find the equivalent capacitance C at terminal-pair a–b in the circuit of Fig. 10–21.

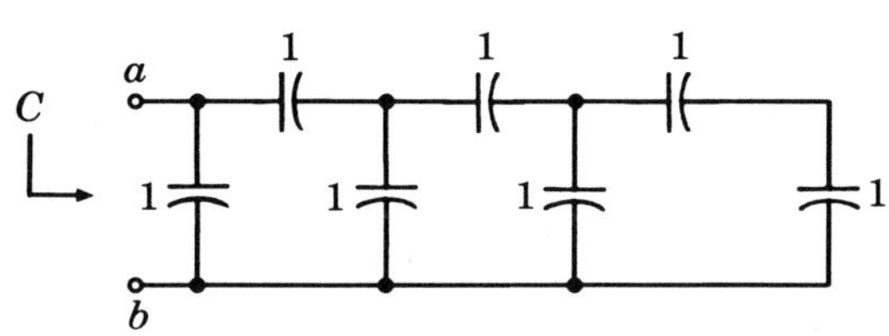

FIGURE 10–21

35. Find the equivalent inductance L at the terminal-pair a–b in the circuit of Fig. 10–22.

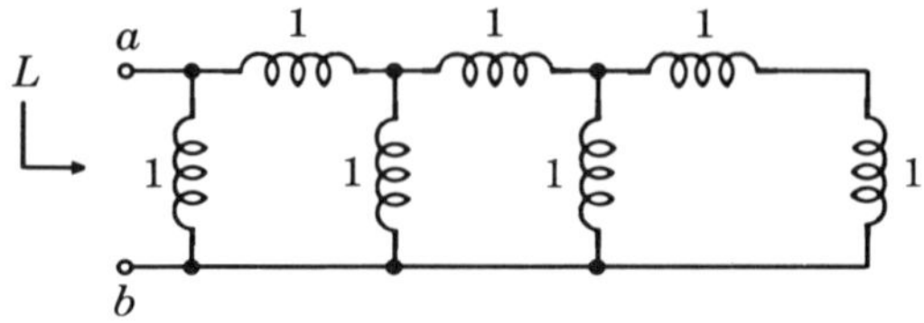

FIGURE 10–22

36. Find the equivalent capacitance of the network of Fig. 10–23 at terminal-pair a–b.

37. Find the equivalent inductance at the terminal-pair a–b in the network of Fig. 10–24.

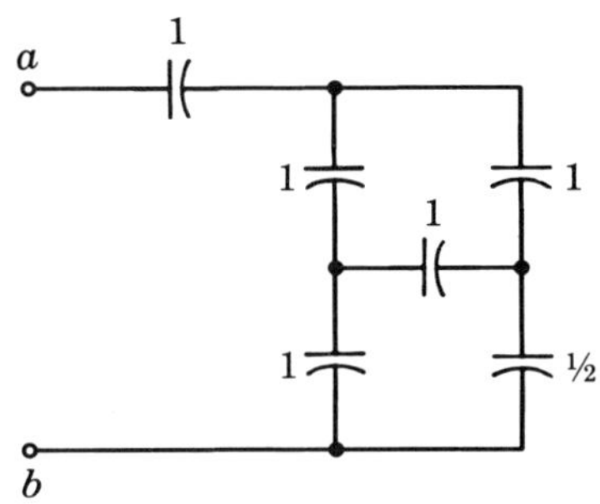

FIGURE 10–23

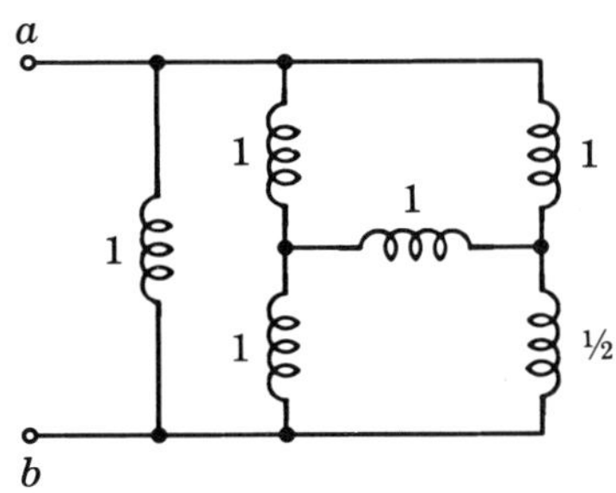

FIGURE 10–24

38. In the circuit of Fig. 10–25 the switch is closed at $t = 0$. The voltage on the capacitance at this time is 10 volts. What is the voltage $e(t)$ after the switch is closed and what is the current which flows?

39. In the circuit of Fig. 10–26 the switch is opened at $t = 0$. Previous to this instant there is a current of 10 amp through L_1 and the switch. What is the current $i(t)$ after the switch is opened, and what is the voltage which occurs across the two inductances?

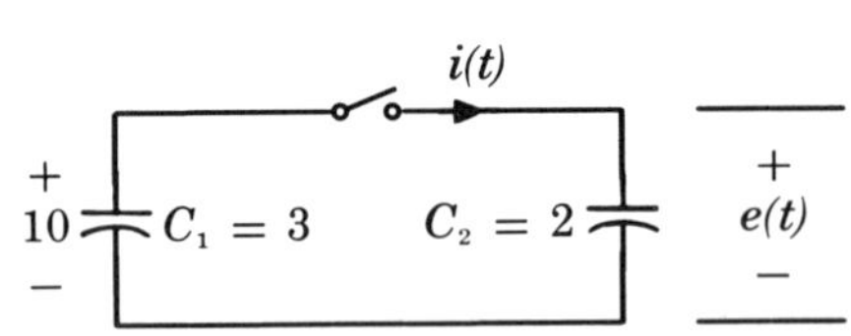

FIGURE 10–25

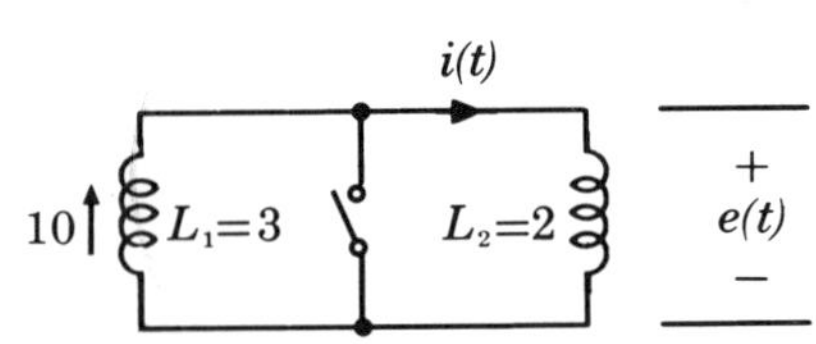

FIGURE 10–26

40. In the circuit of Fig. 10–27 the capacitance C_1 has a voltage of 5 volts, the capacitance C_2 has a voltage of 10 volts, and the capacitance C_3 has a voltage of -6 volts. The three capacitances are connected at $t = 0$. What will the voltage be across the combination for $t > 0$?

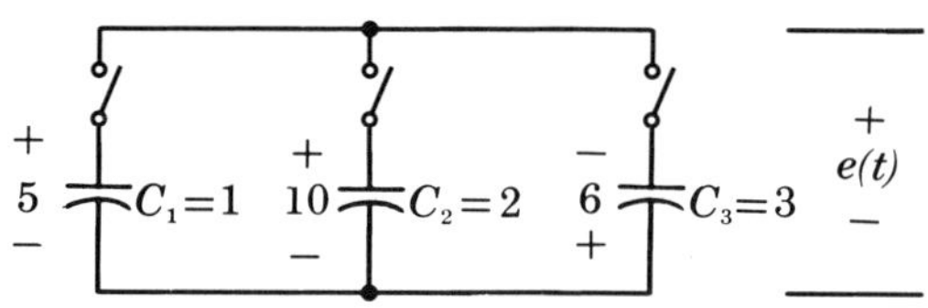

FIGURE 10–27

41. In the circuit of Fig. 10–28 the three switches are opend at $t = 0$. Previous to this time there is a current in L_1 of 5 amp, a current in L_2 of 6 amp, and a current in L_3 of -10 amp. What current will flow through the three series inductances for $t > 0$?

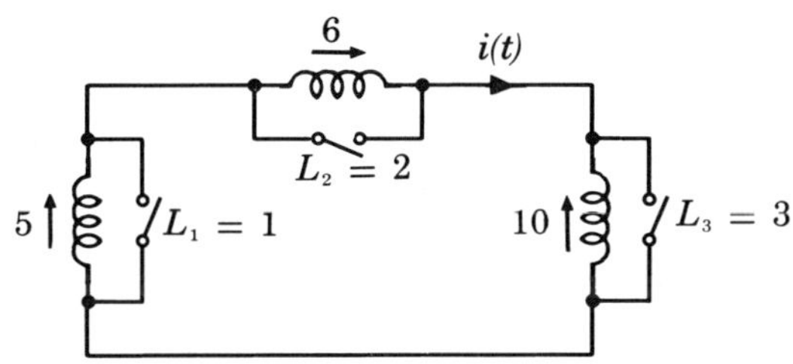

FIGURE 10–28

CHAPTER 11

THE TIME RESPONSE FUNCTIONS OF *R–L* AND *R–C* CIRCUITS

"To him that hath shall be given, and from him that hath not shall be taken away even that which he hath."

Matthew 13: 10–12

11–1 Introduction. Many physical problems are characterized by an exponential type of behavior in which the derivative of the function is proportional to the value of the function. Like the human behavior alluded to in the biblical quotation, two types of exponential behavior are possible: A function may be growing exponentially or it may be decaying exponentially. Compound interest, chain letters, and population growth are examples of positive exponentials. Physical systems with an energy-storage element and an energy-loss element are examples of negative exponentials. Politicians and financiers tend to prefer positive exponentials. Physicists and engineers prefer negative exponentials and often go to great lengths to prevent positive exponentials, or unstable behavior, from occurring!

When we discussed the differential equations of electric circuits in Chapter 9 we learned that the behavior of a system with one energy-storage element is characterized by a single time constant, which is L/R for an inductive circuit and RC for a capacitive circuit. When a system is excited by internal energy storage the currents and voltages decay exponentially to zero from some initial value, at a rate determined by the time constant of the system. The initial value of the variable and the time constant describe the response completely.

When a system is excited by an impulse, energy is stored in the inductance or capacitance in the interval between $t = 0^-$ and $t = 0^+$. After this interval, with the circuit left undisturbed, the behavior produced is identical to the natural behavior of the system. Again, the response can be completely described by the initial value of the variable and the time constant.

When a system is excited by a step function the response contains a steady-state constant value as well as the decaying exponential term. This steady-state value is the value of the function as t approaches infinity, or the final value of the function. The step response starts at an initial value and rises or decays to its final value at an exponential rate. Thus the response is completely described by the initial value of the variable, the final value of the variable, and the time constant.

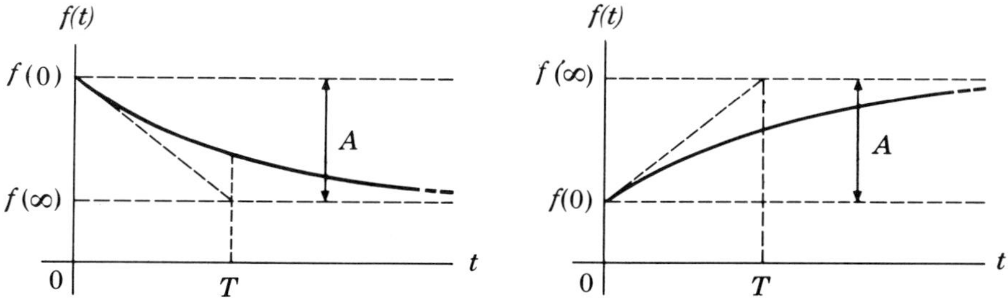

$f(t) = Ae^{-t/T} + f(\infty)$ $\qquad$ $f(t) = A(1-e^{-t/T}) + f(0)$

(a) The decaying exponential curve $\qquad$ **(b) The rising exponential curve**

FIG. 11–1. The fundamental exponential curves.

The two general types of exponentials which occur in stable physical problems are shown in Fig. 11–1. The curve in Fig. 11–1(a) starts at an initial value of $f(0)$ and decays at an exponential rate to $f(\infty)$. The equation for such a curve is

$$f(t) = [f(0) - f(\infty)]e^{-t/T} + f(\infty). \tag{11–1}$$

Equation (11–1) is the ordinary decaying exponential of Eq. (9–12) with its time axis shifted upward by the value of $f(\infty)$. The curve of Fig. 11–1(b) starts at $f(0)$ and rises at an exponential rate to a final value of $f(\infty)$. Its equation is

$$f(t) = [f(\infty) - f(0)](1 - e^{-t/T}) + f(0). \tag{11–2}$$

Equation (11–2) is the rising exponential curve of Eq. (9–26) with its time axis shifted upward by the value of $f(0)$.

Equations (11–1) and (11–2) represent the most general curves which describe the behavior of circuits with a single energy-storage element and a single energy-loss element. In this chapter we shall take these curves as our starting point. We shall not write the differential equations for circuits and solve them. Instead, we will match the known values of $f(0)$, $f(\infty)$, and the time constant, obtained from the circuits, to the known forms of the solutions, as given by Eqs. (11–1) and (11–2). We call this procedure "the method of matching the boundary conditions."

The point of view which we have presented here is slightly more sophisticated than that usually taken in elementary mathematics courses. The values of $f(0)$, $f(\infty)$, and the time constant must be obtained in any case. What we save therefore is the repeated solution of the differential equation. We have solved the first-order linear differential equation once and have found its general solution. We need never solve it again!

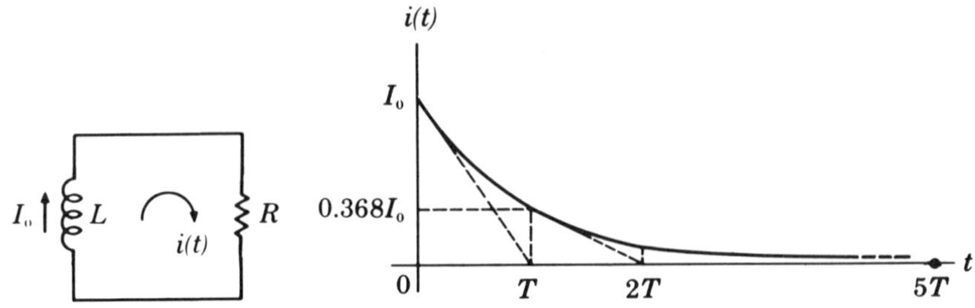

(a) An $R-L$ circuit (b) The response function

FIG. 11–2. The natural response of an R–L circuit.

11–2 The impulse response of R–L circuits. The natural behavior of a circuit is produced when energy is stored within it and it is then left undisturbed. An impulse stores energy in a circuit during the interval between $t = 0^-$ and $t = 0^+$ and therefore excites the natural behavior of the circuit. Before we consider specific impulse responses let us apply the method of matching the boundary conditions to the natural response of the circuit in Fig. 11–2(a). The initial current in the circuit is known to be I_0, and the time constant of the circuit is L/R. The final value of the current at $t = \infty$ must be zero, since all the energy is dissipated. Equation (11–1) with $f(\infty) = 0$ describes this behavior. Thus

$$i(t) = f(0)e^{-t/T}. \tag{11–3}$$

In this case $f(0) = I_0$ and the time constant, T, is L/R. Hence

$$i(t) = I_0 e^{-Rt/L} \qquad \text{for} \quad t > 0. \tag{11–4}$$

Equation (11–4) is sketched in Fig. 10–2(b). Note that the curve can indeed be sketched from the initial value, the final value, and the time constant. After about five time constants the curve has reached its final value of zero, within the accuracy of the plot. At one time constant it is about $\frac{1}{3}$ (more accurately $e^{-1} = 0.367879$) of its initial value, and at time zero it has its initial value. These three points are sufficient to prepare a rough sketch. If additional points are desired they can be obtained easily. At two time constants the value of the curve is about $\frac{1}{3}$ of the value at one time constant, and so on.

An R–L circuit excited by an impulse is shown in Fig. 11–3(a). The circuit for $t > 0$ is shown in Fig. 11–3(b). The only way in which the impulse can affect the circuit for $t > 0$ is through the initial current which it establishes in the inductance between $t = 0^-$ and $t = 0^+$, the interval during which the impulse is acting. During this interval the inductance is an open circuit and the impulse voltage in Fig. 11–3(a)

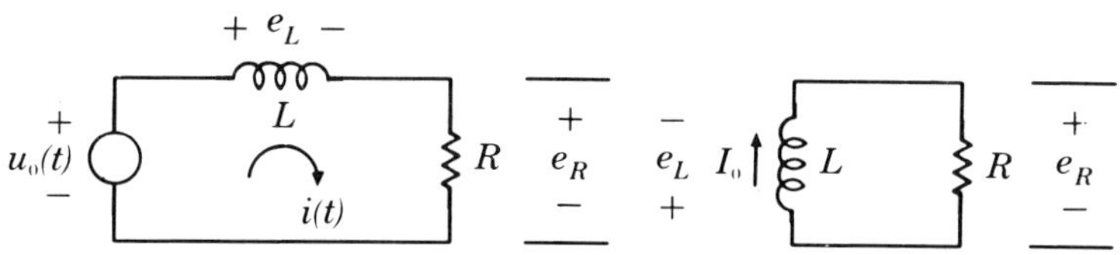

(a) Impulse-excited circuit (b) The circuit for $t > 0$

FIG. 11–3. An R–L series circuit excited by an impulse voltage.

appears across it.* The current established in the inductance is obtained from the integral

$$i_L(0^+) = \frac{1}{L}\int_{0^-}^{0^+} e\,dt = \frac{1}{L}\int_{0^-}^{0^+} u_0(t)\,dt. \tag{11–5}$$

The integral of Eq. (11–5) directs us to find the area under a curve, between the limits 0^- and 0^+, whose area is unity. The answer, of course, is unity. Thus

$$i_L(0^+) = \frac{1}{L}\cdot \tag{11–6}$$

In the circuit of Fig. 11–3(b) the initial current given by Eq. (11–6) decays to zero with a time constant of L/R. Its equation is

$$i_L(t) = \frac{1}{L}e^{-Rt/L} \qquad \text{for} \quad t > 0, \tag{11–7}$$

which is plotted in Fig. 11–4(a).

The voltage across the inductance is interesting. For $t > 0$ the inductance and the resistance are in parallel, and the voltage across the inductance is the negative of the voltage across the resistance, the assumed

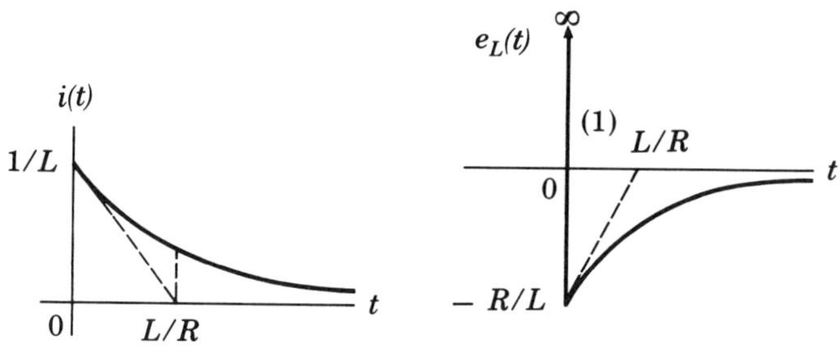

FIG. 11–4. The response functions for the R–L circuit of Fig. 11–3.

* A finite voltage also appears across R, but the voltage across L is still infinite.

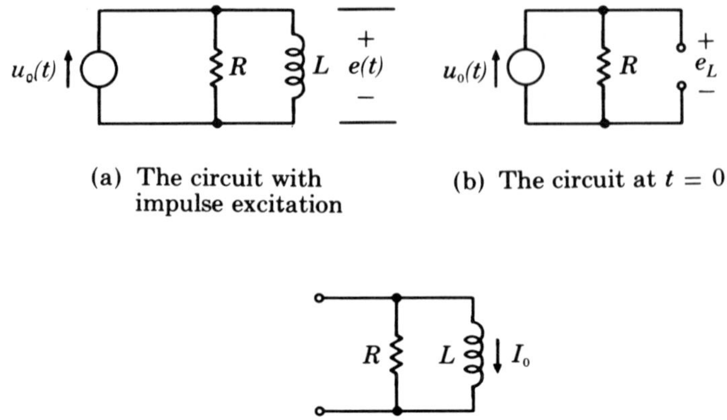

(a) The circuit with impulse excitation

(b) The circuit at $t = 0$

(c) The circuit for $t > 0$

FIG. 11–5. A parallel R–L circuit excited by an impulse current.

polarities being shown in Fig. 11–3(b). This voltage across the inductance can be obtained by multiplying Eq. (11–7) by R. In addition to this voltage there is the original impulse which occurred across the inductance at $t = 0$. Thus the complete expression for the voltage across the inductance is

$$e_L(t) = u_0(t) - \frac{R}{L} e^{-Rt/L} \cdot u_{-1}(t). \tag{11–8}$$

The second term in Eq. (11–8) is multiplied by the function $u_{-1}(t)$ to make it zero for negative values of t and to avoid the necessity of writing a note to this effect. A plot of Eq. (11–8) is shown in Fig. 11–4(b).

The case of an impulse current source in parallel with an inductance and resistance is shown in Fig. 11–5(a). In this circuit it is necessary to distinguish very carefully the instants $t = 0$ and $t = 0^+$. At exactly $t = 0$ the impulse strikes, and the inductance is an open circuit, as shown in Fig. 11–5(b). By $t = 0^+$ the impulse has become inactive, and the dead current source is an open circuit. Energy has been stored in the inductance in the interval from $t = 0^-$ to $t = 0^+$, and the circuit takes the form shown in Fig. 11–5(c). This circuit is identical to that in Fig. 11–3(b). The problem here lies in the calculation of the initial value of the current in the inductance. Referring to Fig. 11–5(b) we note that the impulse current passes through the resistance, producing the voltage

$$e = Ru_0(t).$$

This voltage also appears across the inductance, and thus the current established in the inductance between $t = 0^-$ and $t = 0^+$ is

$$i_L(0^+) = \frac{1}{L}\int_{0^-}^{0^+} Ru_0(t)\,dt = \frac{R}{L}\cdot$$

This current exists at $t = 0^+$ and excites the circuit of Fig. 11–5(c). The initial voltage is negative and has the value

$$e(0^+) = -R^2/L.$$

The final value for the voltage is

$$e(\infty) = 0,$$

and the time constant, T, is L/R. Thus for $t > 0$ the voltage across the parallel circuit is

$$e(t) = \frac{-R^2}{L}\,e^{-Rt/L}.$$

In addition to this voltage there is the impulse $Ru_0(t)$ which occurs at $t = 0$. The complete expression for the voltage is

$$e(t) = Ru_0(t) - \frac{R^2}{L}\,e^{-Rt/L}\cdot u_{-1}(t). \tag{11–9}$$

A sketch of the plot for Eq. (11–9) is shown in Fig. 11–6.

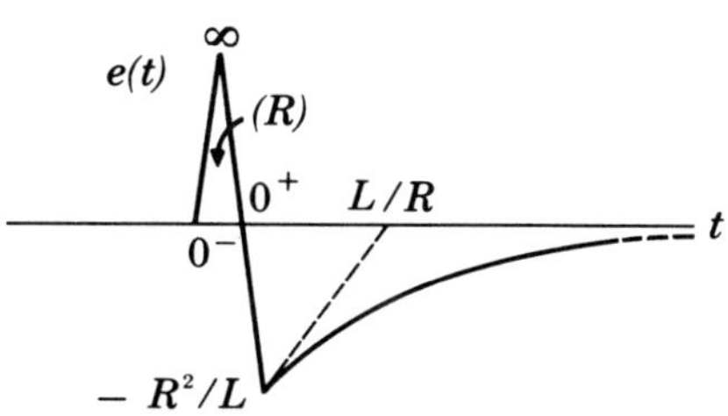

FIG. 11–6. The response function for the circuit of Fig. 11–5.

The impulse responses obtained in this section are perhaps more difficult than the step response which will be discussed in the next section. To avoid impulses some students always deal with step functions and obtain impulse responses by differentiating the step responses. Impulses are really not that bad!

11–3 The step response of *R*–*L* circuits. The step response of a single-energy-storage circuit, like the impulse response, can be obtained by computing the initial value of the variable, the final value of the variable,

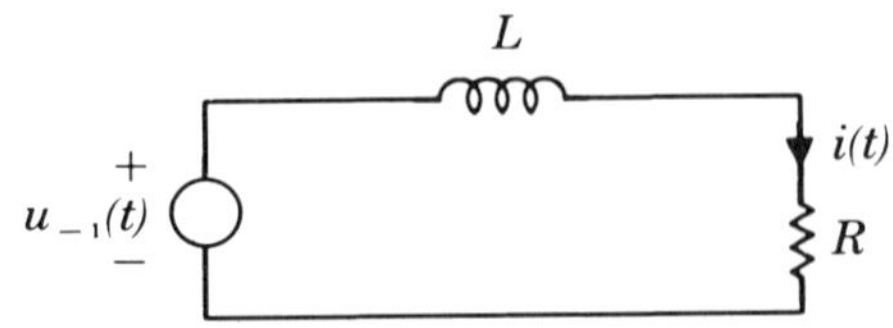

FIG. 11–7. A series R–L circuit excited by a unit step of voltage.

and the time constant. The exponential curve can do nothing but rise or fall from an initial value to a final value and can always be expressed in terms of these values, according to Eq. (11–1) or (11–2). We shall now apply the method of matching the initial and final values, or boundary conditions, to the computation of the step responses of R–L circuits.

A series R–L circuit excited by a step function of voltage is shown in Fig. 11–7. At $t = 0^+$ the inductance is an open circuit, and the current is

$$i(0^+) = 0.$$

At $t = \infty$ the inductance is a short circuit, the current is

$$i(\infty) = 1/R,$$

and the time constant is

$$T = L/R.$$

Since $f(\infty)$ is larger than $f(0)$ the response equation has the form of Eq. (11–2). It is somewhat simpler, however, since $f(0) = 0$. The expression for the current is

$$\begin{aligned} i(t) &= \frac{1}{R}(1 - e^{-Rt/L}) \qquad \text{for} \quad t > 0 \\ &= \frac{1}{R}(1 - e^{-Rt/L})u_{-1}(t). \end{aligned} \tag{11–10}$$

A plot of Eq. (11–10) is shown in Fig. 11–8. The curve can easily be sketched from the initial value, the final value, and the time constant. In about five time constants the curve is essentially equal to its final value of

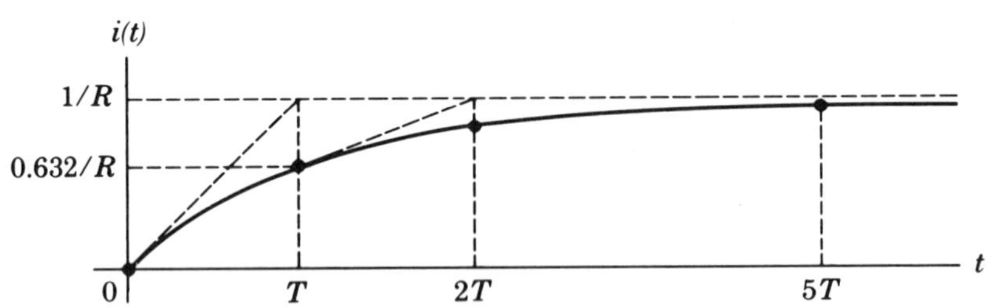

FIG. 11–8. The response function for the circuit of Fig. 11–7.

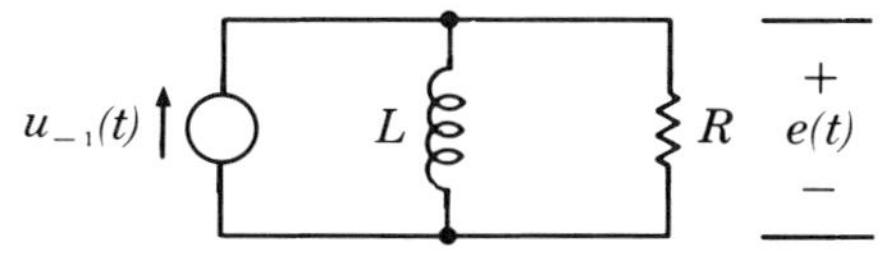

FIG. 11–9. A parallel R–L circuit excited by a step current.

$1/R$. In one time constant it has reached about $\frac{2}{3}$ [more accurately $(1 - e^{-1}) = 0.632121$] of its final value. Additional points can easily be obtained. After two time constants the curve has risen about $\frac{2}{3}$ of the distance from the one-time-constant value to the final value and so on.

The rising exponential curve we have just discussed is characteristic of many step-function responses which possess a final value at $t = \infty$. In other cases the final value is zero and the step response is a decaying exponential. A circuit with a response of this sort is shown in Fig. 11–9. At $t = 0$ the inductance is an open circuit, and the unit step current passes through the resistance. The voltage produced at $t = 0^+$ is

$$e(0^+) = R.$$

At $t = \infty$ the inductance is a short circuit. The step current flows through the short circuit, and the voltage produced is

$$e(\infty) = 0.$$

The time constant is

$$T = L/R.$$

The response is completely determined by the initial value, the final value, and the time constant. Its equation has the form of Eq. (11–1) with $f(\infty) = 0$:

$$e(t) = Re^{-Rt/L} \cdot u_{-1}(t). \tag{11–11}$$

This equation is similar to Eq. (11–7), whose plot is shown in Fig. 11–4(a).

The current in the inductance of Fig. 11–9 can be obtained by integrating the voltage of Eq. (11–11). It is better, however, to compute the initial and final values of the current and to write the current equation independently. For this circuit we have

$$i_L(0^+) = 0, \qquad i_L(\infty) = 1, \qquad T = L/R,$$

and the current equation is

$$i(t) = 1(1 - e^{-Rt/L})u_{-1}(t). \tag{11–12}$$

This equation has the rising exponential form plotted in Fig. 11–8.

The step responses discussed here are fairly typical. It is very seldom that the complete form of Eq. (11–1) or (11–2) is required.

11–4 The ramp response of R–L circuits. The ramp response can be obtained by integrating the step response. Since we have not obtained the general solution of a first-order equation with a ramp function as the driving source it will be convenient to start with the step response. We are interested in the ramp response primarily for the additional light it throws on the meaning of the time constant. If a ramp-function stimulus is applied to a one-time-constant circuit the steady-state response function is also a ramp, and it is delayed with respect to the stimulus by one time constant.

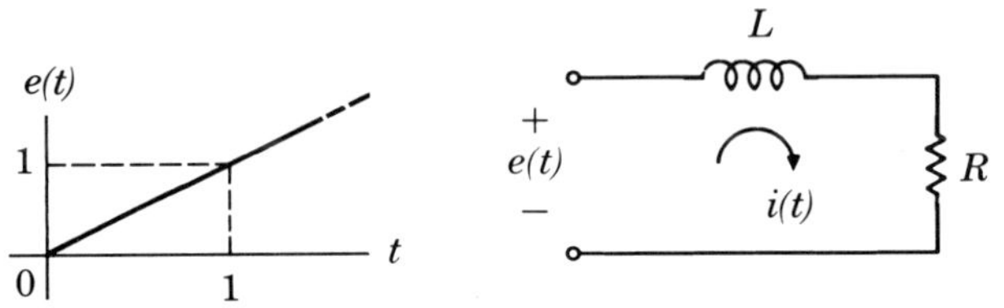

FIG. 11–10. A series R–L circuit excited by a ramp of voltage.

An R–L series circuit excited by a ramp function of voltage is shown in Fig. 11–10. The response of the circuit to a unit step function will be obtained first. According to Eq. (11–10) it is

$$i(t) = \frac{1}{R}(1 - e^{-Rt/L})u_{-1}(t). \tag{11–13}$$

The response of the circuit to a ramp function is the integral of Eq. (11–13) from $-\infty$ to t. Since the function $u_{-1}(t)$ makes the integrand zero for negative values of t, it is necessary to integrate only from zero to t. Hence the ramp response is

$$\begin{aligned} i_{\text{ramp}} &= \int_0^t \frac{1}{R}(1 - e^{-Rt/L})\,dt \\ &= \frac{1}{R}\left[t - \frac{L}{R}(1 - e^{-Rt/L})\right]. \end{aligned} \tag{11–14}$$

After about five time constants the exponential term in Eq. (11–14) becomes negligible, and the steady-state behavior ensues. The steady-state current is

$$i_{ss}(t) = \frac{1}{R}(t - T), \tag{11–15}$$

where $T = L/R$ is the time constant.

A sketch of the curve for Eq. (11–15) is shown in Fig. 11–11. Note that for the ramp-function stimulus the final response is a ramp function which

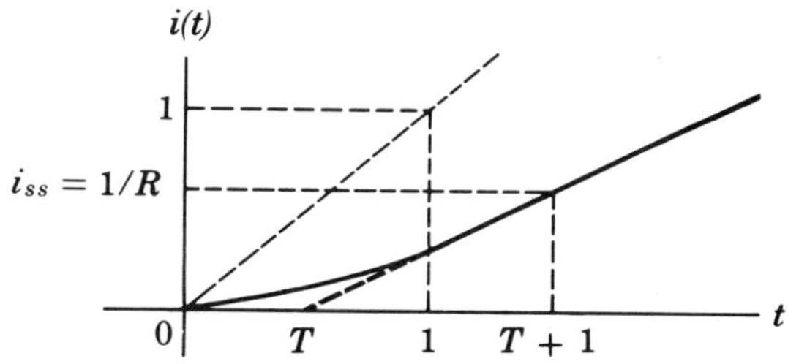

FIG. 11–11. The ramp function response for the circuit of Fig. 11–10.

is delayed with respect to the stimulus. The delay is obtained by projecting the steady-state curve of Eq. (11–15) back to the zero-current or time axis, which the projection cuts at $t = T$.* Thus *the response function is delayed with respect to the stimulus by exactly one time constant.*

The characteristic we have just emphasized provides a rough guide in the design of many measurement and control systems. If a meter has a response time constant of T it will follow a changing input signal, but will lag by approximately T seconds. If the input is changing slowly the T-second delay results in very little error in the reading of the meter. If the input is changing rapidly, however, the T-second delay results in a large error. For a given meter and a given allowable error the maximum permissible rate of change of the input signal is the ratio of the allowable error (absolute and not relative) to the time constant of the meter. A similar consideration is true for control systems. The faster the rate of change of the stimulus, the shorter must be the time constant of the system if the error is to remain constant.

The steady-state response to a ramp stimulus does not necessarily contain a term proportional to time. If a ramp current instead of the step current is applied to the parallel L–R circuit of Fig. 11–9 the voltage response is given by the integral of Eq. (11–11). Thus

$$
\begin{aligned}
e_{\text{ramp}} &= \int_0^t Re^{-Rt/L}\,dt \\
&= L(1 - e^{-Rt/L}). \qquad (11\text{–}16)
\end{aligned}
$$

The response curve of Eq. (11–16) rises at an exponential rate from zero to a final value of L volts. The equation does not have a term proportional to time. The steady-state response is a constant rather than a ramp function.

---

* The concept of delay may become clearer if the resistance R in Eq. (11–15) is assumed to be 1 ohm. Then the input and output are identical except for the time delay.

Because ramp response functions occur relatively infrequently it is hardly worth while to develop a technique for directly finding the steady-state ramp response. Instead, it is usually better to obtain the ramp response as the integral of the step response, as we have done here.

11–5 The impulse response of R–C circuits. An R–C circuit is also a circuit with a single energy-storage element and a single energy-loss element. The response functions which occur in R–C circuits are essentially the same as those which occur in L–R circuits. The principal difference is that the time constant is now $T = RC$ instead of $T = L/R$. We proved this fact in Chapter 9, but we can equally well use the principle of duality to prove it directly.

If R is changed to G and L to C the dual of the R–L circuit is obtained. The voltage response in the dual circuit is the same as the current response in the original circuit. In addition, the same type of functional behavior is obtained. The time constant of the dual circuit changes from L/R to C/G. Since we prefer to deal with ohms rather than mhos C/G becomes $R'C$, where R' is the resistance of the dual circuit.

The expressions for the natural behavior and the impulse response of an R–C circuit have the general form of Eq. (11–1), with $f(\infty)$ equal to zero:

$$f(t) = f(0)e^{-t/T} \quad \text{for} \quad t > 0, \tag{11–17}$$

where $T = RC$ is the time constant.

In order to write an answer by inspection, therefore, it is necessary to compute only the initial value of the variable. For the case of excitation by initial energy storage, as shown in Fig. 11–12, the computation is particularly simple. The initial voltage E_0 also exists across the resistance, and hence the initial current is

$$i(0) = E_0/R.$$

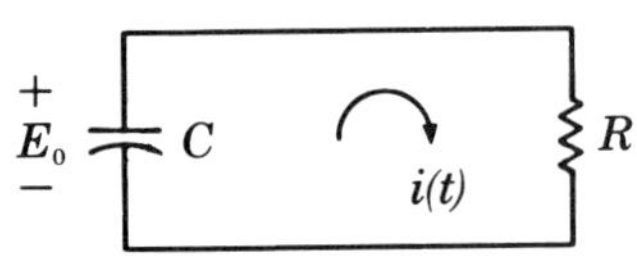

FIG. 11–12. The natural response of an R–C circuit.

The value of the current at infinity is

$$i(\infty) = 0,$$

and the time constant is

$$T = RC.$$

The system fits Eq. (11–17), and the current is

$$i(t) = \frac{E_0}{R} e^{-t/RC} \cdot u_{-1}(t). \tag{11–18}$$

Equation (11–18) represents a current which starts with a value of E_0/R and decays to zero exponentially.

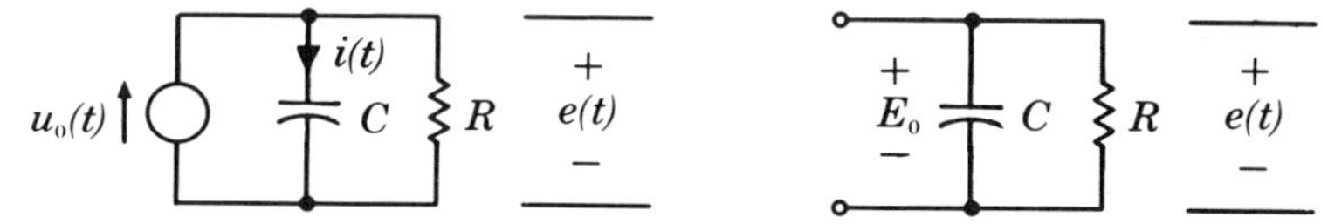

(a) An $R-C$ circuit driven by an impulse (b) The circuit for $t > 0$

FIG. 11–13. The impulse response of a parallel R–C circuit.

Impulse excitation affects the R–C circuit by the energy it stores during the interval $t = 0^-$ to $t = 0^+$. For $t > 0$ the response is identical to the natural behavior. For example, in the circuit of Fig. 11–13(a) a current impulse is applied to a parallel combination of R and C. After $t = 0$ the impulse is zero, and the dead current source is an open circuit. For $t > 0$ the circuit is therefore that shown in Fig. 11–13(b). The principal problem now is the computation of the initial voltage on the capacitance. At $t = 0^-$ the circuit is at rest; at $t = 0$ the impulse strikes, and the capacitance acts as a short circuit. During the interval $t = 0^-$ to $t = 0^+$ current flows into the capacitance and produces the voltage

$$e(0^+) = \frac{1}{C}\int_{0^-}^{0^+} u_0(t)\, dt. \tag{11–19}$$

Equation (11–19) directs us to find the area under a curve between the limits 0^- and 0^+. But the curve is an impulse defined to have an area of unity between the limits of integration. Therefore the area is unity, and the voltage is

$$e(0^+) = \frac{1}{C}\cdot$$

The final value of the voltage is

$$e(\infty) = 0,$$

and the time constant is

$$T = RC.$$

The system fits Eq. (11–17), and therefore the voltage is

$$e(t) = \frac{1}{C}e^{-t/RC}\cdot u_{-1}(t). \tag{11–20}$$

The damped exponential curve of Eq. (11–20) is plotted in Fig. 11–14(a).

The current in the capacitance of Fig. 11–13(a) can be obtained by differentiating Eq. (11–20), but it is simpler to obtain it directly. At $t = 0$ the capacitance receives the impulse current. After $t = 0$, current

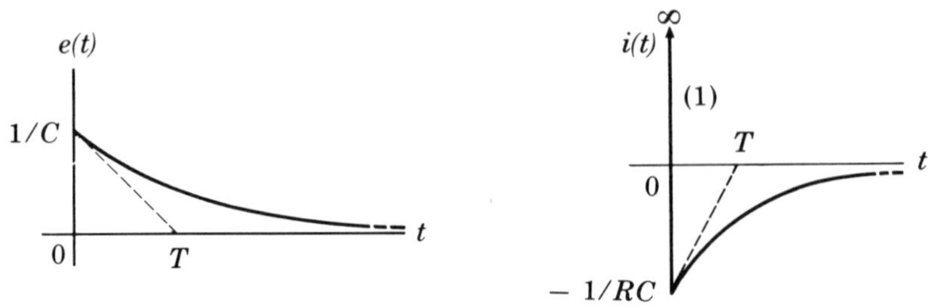

(a) The voltage response (b) The capacitance current response

FIG. 11–14. The response functions for the circuit of Fig. 11–13.

flows out of the capacitance. It starts with a value of $-1/RC$ and decays to zero, with a time constant of RC. The complete expression for the current in the capacitance is therefore

$$i(t) = u_0(t) - \frac{1}{RC}e^{-t/RC} \cdot u_{-1}(t). \tag{11–21}$$

The curve for Eq. (11–21) is sketched in Fig. 11–14(b). The initial impulse in Eq. (11–21) arises from the derivative of the jump which occurs in Eq. (11–20) at $t = 0$. This equation is zero for negative values of time, and suddenly at $t = 0$ its value jumps to $1/C$. The second term in Eq. (11–21) is simply the derivative of Eq. (11–20). Equation (11–21) can also be obtained from Eq. (11–20) by formal differentiation of the two time functions. Thus

$$i(t) = C\frac{de}{dt} = C\frac{d}{dt}\left[\frac{e^{-t/RC}}{C} \cdot u_{-1}(t)\right]$$

$$= e^{-t/RC} \cdot u_0(t) - \frac{1}{RC}e^{-t/RC} \cdot u_{-1}(t). \tag{11–22}$$

In Eq. (11–22) the unit impulse causes the first term to be zero at all time t except at $t = 0$. At this instant the exponential in the first term has a value of unity. Therefore Eq. (11–22) can be written in exactly the same form as Eq. (11–21).

A series R–C circuit excited by an impulse is shown in Fig. 11–15(a). The circuit for $t = 0$, the instant when the impulse strikes, is shown in Fig. 11–15(b). We note that the capacitance acts as a short circuit. After $t = 0^+$ the impulse is inactive, the dead voltage source is replaced by a short circuit, and the capacitance is charged. The circuit for $t > 0$ is shown in Fig. 11–15(c). The problem now is essentially one of computing the initial current in the circuit. Precisely at $t = 0$, when the voltage

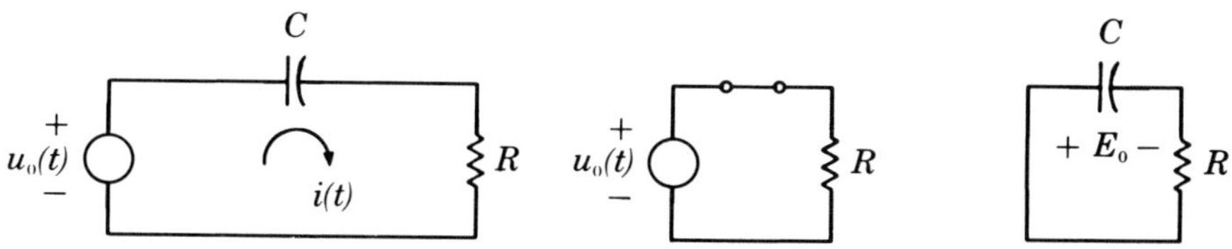

(a) Circuit excited by an impulse (b) Circuit at $t = 0$ (c) Circuit for $t > 0$

FIG. 11–15. A series R–C circuit excited by an impulse voltage.

impulse strikes and the capacitance is a short circuit, the current is

$$i(0) = \frac{1}{R} u_0(t). \tag{11–23}$$

This current passes through the capacitance and in doing so produces a voltage on it given by

$$\begin{aligned} e_C &= \frac{1}{C} \int_{0^-}^{0^+} \frac{1}{R} u_0(t)\, dt \\ &= \frac{1}{RC} \cdot \end{aligned} \tag{11–24}$$

In the circuit of Fig. 11–15(c) this voltage has a negative polarity, and it causes a negative current to flow with an initial value of

$$i(0^+) = -\frac{1}{R^2C} \cdot \tag{11–25}$$

The final value of the current is zero, and the time constant is RC. Thus the complete expression for the current is

$$i(t) = \frac{1}{R} \cdot u_0(t) - \frac{1}{R^2C} e^{-t/RC} \cdot u_{-1}(t). \tag{11–26}$$

Equation (11–26) consists of a positive impulse and a negative exponential which decays to zero with a time constant of RC. A sketch of the plot for Eq. (11–26) is shown in Fig. 11–16.

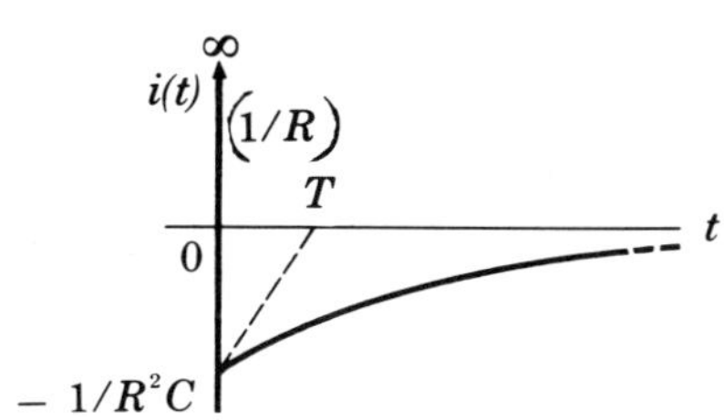

FIG. 11–16. The response current for the circuit of Fig. 11–15.

The impulse responses which we obtained in this section are slightly more difficult than the step responses which we will discuss next. Since they occur frequently in practice, however, the student should learn to deal with them directly.

11–6 The step response of R–C circuits. When a step voltage or current is applied to a circuit a source is present at $t = 0$, and theoretically at least, it will continue to supply energy for all time. Thus there may be a constant value of the variable at $t = \infty$. The steady-state value at $t = \infty$ combines with the natural behavior to produce the step response of the circuit. If we know the time constant and the values of the variable at $t = 0$ and $t = \infty$ we can write the equation for the step response by inspection. The general form of the equation is given by Eq. (11–1) or (11–2), which can often be simplified to

$$f(t) = f(0)e^{-t/T} \qquad \text{for} \quad t > 0, \tag{11–27}$$

and

$$f(t) = f(\infty)(1 - e^{-t/T}) \qquad \text{for} \quad t > 0. \tag{11–28}$$

Either of these equations can be written by inspection when the values of the variable at $t = 0$ and $t = \infty$ and the time constant are known.

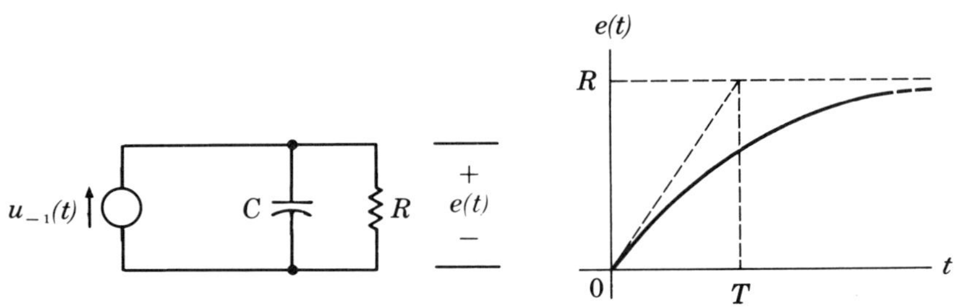

(a) Circuit (b) The voltage response

FIG. 11–17. A parallel R–C circuit excited by a step function of current.

Consider now Fig. 11–17(a). The source is connected at $t = 0$, and a unit step of current is applied to the parallel R–C circuit. At $t = 0$ the capacitance is a short circuit, and the voltage produced across it by the finite current is zero. Thus

$$e(0^+) = 0.$$

After a long interval the capacitance becomes an open circuit, and the current flows through the resistance R. The final voltage is

$$e(\infty) = R,$$

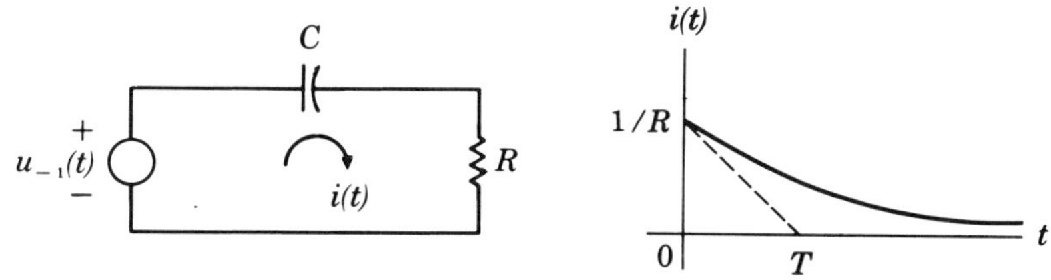

(a) Circuit (b) The current response

FIG. 11–18. A series R–C circuit excited by a step function of voltage.

and the time constant of the circuit is

$$T = RC.$$

Since these data fit Eq. (11–28) the current is

$$i(t) = R(1 - e^{-t/RC})u_{-1}(t). \tag{11–29}$$

This current starts at zero and rises at an exponential rate to a final value of R amperes. A sketch of the curve is shown in Fig. 11–17(b).

A series R–C circuit excited by a step voltage is shown in Fig. 11–18(a). At $t = 0$ the capacitance is a short circuit, and the current is

$$i(0^+) = 1/R.$$

After a long interval the capacitance becomes an open circuit, and the final value of the current is

$$i(\infty) = 0.$$

The time constant of the circuit is

$$T = RC.$$

These data fit Eq. (11–27), and thus the current is

$$i(t) = \frac{1}{R}e^{-t/RC} \cdot u_{-1}(t). \tag{11–30}$$

The step responses of R–C circuits are relatively easy to obtain. The circuits at $t = 0$ or $t = \infty$ reduce to purely resistive circuits, permitting initial or final values to be easily obtained. Circuits with combinations of step-function excitation and initial energy storage can be treated by superposition.

11–7 More general single-time-constant circuits. The circuits which we have considered in the preceding sections consist of one energy-storage element and one resistive element. The same methods can be used for circuits consisting of one energy-storage element and any number of resistances. The resistive part of such a circuit will be a two-terminal network, and it can be reduced to a single resistance by the methods which we developed in earlier chapters.

Occasionally, circuits which have more than one energy-storage element can be treated as single-energy-storage systems. Two capacitances in parallel, for example, can be combined into a single capacitance, or two inductances in series can be combined into a single inductance. If a network can be separated into two sections, one which is purely inductive or purely capacitive and one which is purely resistive, and if the two sections are joined by only two wires, the network can be treated as a single-energy-storage circuit. The resistive branch is reduced to an equivalent resistance with or without equivalent sources, and the energy-storage branch is reduced to a single equivalent branch with or without equivalent sources. Either Thévenin's theorem or Norton's theorem can be used to compute the equivalent source values. The equivalent resistance can be obtained by any of the methods discussed in earlier chapters, and the equivalent energy-storage element can be obtained by means of the resistance analog discussed in Chapter 10. The single time constant of the network is obtained in terms of the equivalent energy-storage element and the equivalent resistance.

The only response which can be obtained by the procedure we have outlined is that of the equivalent circuit. If responses are required at other points in the original network they can be obtained by means of the substitution theorem. Once the voltage or current in the equivalent network has been obtained it can be replaced by an equivalent source. The distribution of the voltages and currents within the resistive network or the energy-storage network can then be obtained by the procedures of a one-element-type problem.

A circuit which illustrates the more general single-time-constant problem is shown in Fig. 11–19(a). If this circuit is broken at terminal-pair a–a the circuit to the left is purely resistive, and the circuit to the right is purely inductive. A Thévenin equivalent circuit for the resistive section of the network consists of the impulse voltage $5u_0(t)$ in series with a resistance of

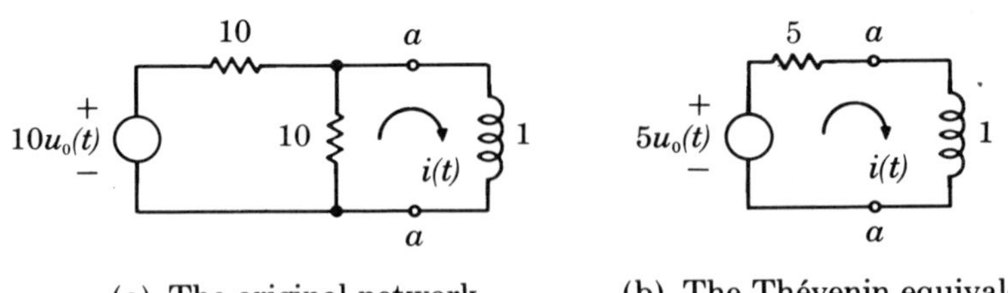

(a) The original network (b) The Thévenin equivalent

FIG. 11–19. A more complicated one-time-constant problem solved by Thévenin's theorem.

5 ohms. This network along with the original inductance is shown in Fig. 11–19(b). The initial current established by the impulse in the inductance of Fig. 11–19(b) is

$$i(0^+) = \frac{1}{1}\int_{0^-}^{0^+} 5u_0(t)\,dt = 5 \text{ amp.}$$

The final current is

$$i(\infty) = 0,$$

and the time constant of the circuit is

$$T = \tfrac{1}{5} \text{ sec.}$$

The current in the inductance is therefore

$$i(t) = 5e^{-5t} \cdot u_{-1}(t). \tag{11–31}$$

The current given by Eq. (11–31) is a decaying exponential. Since the inductive part of the circuit was not changed, this current is also the current in the inductance of the original circuit.

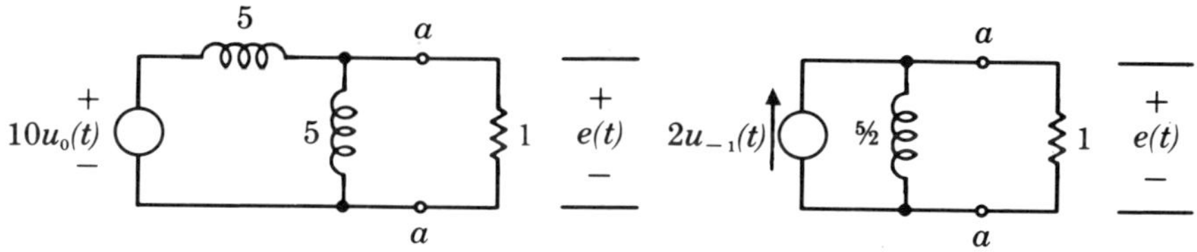

(a) The original network (b) The Norton equivalent

FIG. 11–20. A one-time-constant problem with two inductances.

A circuit which apparently contains two energy-storage elements is shown in Fig. 11–20(a). However, if the circuit is broken at terminal-pair a–a the part to the left is purely inductive and the part to the right is purely resistive. A Thévenin equivalent circuit or a Norton equivalent circuit can be obtained for the inductive section. The dead network in either case consists of the two 5-henry inductances in parallel, or a net inductance of $\frac{5}{2}$ henrys. (In more complicated circuits the net inductance can be found by means of a resistance analog.) The short-circuit current across the terminals a–a is

$$i_{sc} = \frac{1}{L}\int_{-\infty}^{t} e\,dt = \tfrac{1}{5}\int_{0}^{t} 10u_0(t)\,dt$$

$$= 2u_{-1}(t). \tag{11–32}$$

The Norton equivalent circuit for the inductive section, plus the resistance, which remains unchanged, is shown in Fig. 11–20(b). The voltage across the resistance can be obtained from this circuit. At $t = 0$ the inductance is an open circuit and all the current flows through the resistance. Hence the voltage is

$$e(0^+) = 2 \text{ volts.}$$

At $t = \infty$ the inductance is a short circuit, and all the current flows through it. The voltage produced is

$$e(\infty) = 0,$$

and the time constant of the equivalent circuit is

$$T = \tfrac{5}{2} \text{ sec.}$$

In terms of these three values the voltage equation can be written by inspection:

$$e(t) = 2e^{-2t/5} \cdot u_{-1}(t). \tag{11–33}$$

The decaying exponential voltage of Eq. (11–33) is the voltage across terminal-pair a–a in both the equivalent circuit and the original circuit. It is therefore the desired voltage across the resistance in the original circuit.

A circuit in which it is necessary to use the substitution theorem, as well as Thévenin's theorem or Norton's theorem, is shown in Fig. 11–21(a). The problem is to find the voltage at the right end of the ladder when the step voltage $10u_{-1}(t)$ is applied to the left end. First, we must convince ourselves that we are dealing with a single-time-constant problem. To do this we break the circuit at terminal-pair a–a. The circuit to the left is purely capacitive, and the circuit to the right is purely resistive. We can obtain a single capacitance and single source to represent the circuit to the left of a–a, and a single resistance to represent the circuit to the right of a–a. The Thévenin equivalent circuit for the capacitive section is the voltage $5u_{-1}(t)$ in series with a capacitance of 2 farads. (In more complicated examples it may be preferable to solve this problem by means of a resistance analog.) The equivalent resistance for the circuit to the right of a–a is a resistance of value $\tfrac{5}{3}$ ohms. The complete equivalent circuit is shown in Fig. 11–21(b). The voltage e_a can be obtained from this circuit. At $t = 0$ the capacitance acts as a short circuit, and all the voltage appears across the resistance. Thus

$$e_a(0^+) = 5 \text{ volts.}$$

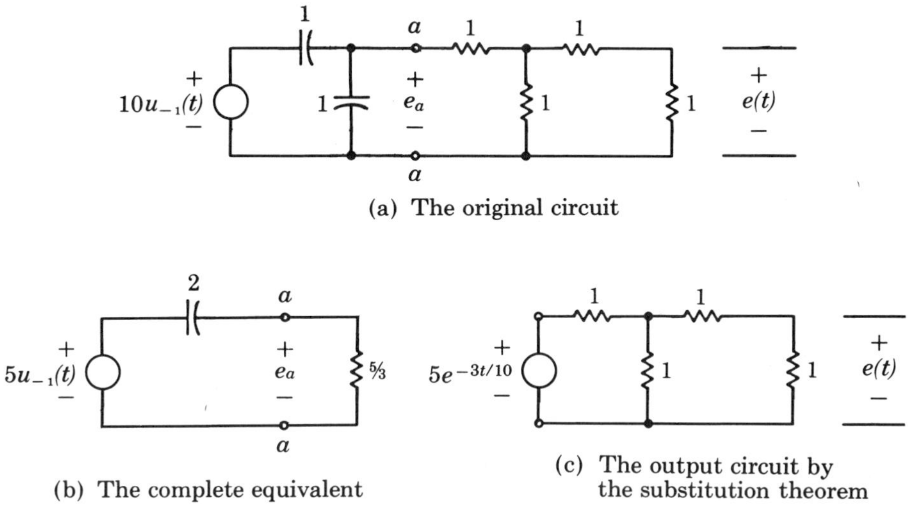

FIG. 11–21. A general form of the single-time-constant circuit.

After a long interval the capacitance acts as an open circuit, and the voltage e_a is

$$e_a(\infty) = 0.$$

The time constant, obtained from the equivalent capacitance and equivalent resistance, is

$$T = 2 \times \tfrac{5}{3} = \tfrac{10}{3} \text{ sec.}$$

In terms of these three values the response equation can be written by inspection:

$$e_a(t) = 5e^{-3t/10} \cdot u_{-1}(t). \tag{11–34}$$

The response voltage e_a of Eq. (11–34) is not the voltage in which we are interested, but here we can turn to the substitution theorem. Now that e_a is known we can replace it by a voltage source having the same value, without changing the circuit. Once we have a voltage source across terminal-pair a–a we can treat each half of the network separately. So far as the resistance section of the network is concerned the voltage e_a completely determines all the internal currents and voltages. The capacitance section is simply another network in parallel with the same voltage source, and it does not alter the source in any way. Therefore in order to find the voltage $e(t)$ we can solve the circuit shown in Fig. 11–21(c). The ladder-network method is perhaps the most convenient. If we assume a voltage of one volt at the output and proceed back through the network we find that an input of five volts is required to provide the one-volt output.

For five volts the output is one volt; for $5e^{-3t/10}$ volts the output is

$$
\begin{aligned}
e(t) &= \tfrac{1}{5} \times 5e^{-3t/10} \\
&= e^{-3t/10} \text{ volts} \qquad \text{for} \quad t > 0. \qquad (11\text{–}35)
\end{aligned}
$$

Equation (11–35) gives the voltage across the output resistance of the ladder when the step voltage of Fig. 11–21(a) is applied at the input. Voltages at other points can be obtained in a similar manner.

The most important aspect of the problem we have just solved is the recognition that it is a single-time-constant problem. When one is in doubt the only real test is whether or not the network can be reduced to a single resistance and a single energy-storage element. If it can be, one is concerned with a single-energy-storage problem, and the response function is of the damped exponential type discussed in this chapter.

The ability to solve simple single-time-constant circuits by inspection is a useful one. A surprising number of the practical circuits which one encounters are simply circuits of this type. For example, as a first approximation, most motors and generators and most coupling circuits for vacuum tubes and transistors are single-time-constant circuits.

11–8 Summary.

General types of exponential behavior

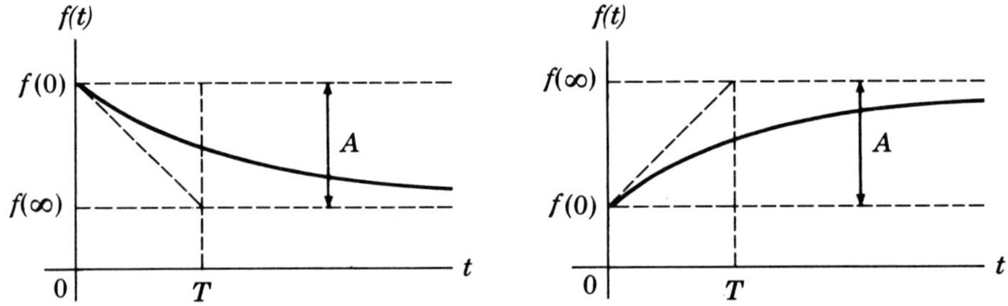

$$f(t) = Ae^{-t/T} + f(\infty) \quad \text{for} \quad t > 0$$

$$f(t) = A(1 - e^{-t/T}) + f(0) \quad \text{for} \quad t > 0$$

The impulse response of R–L circuits

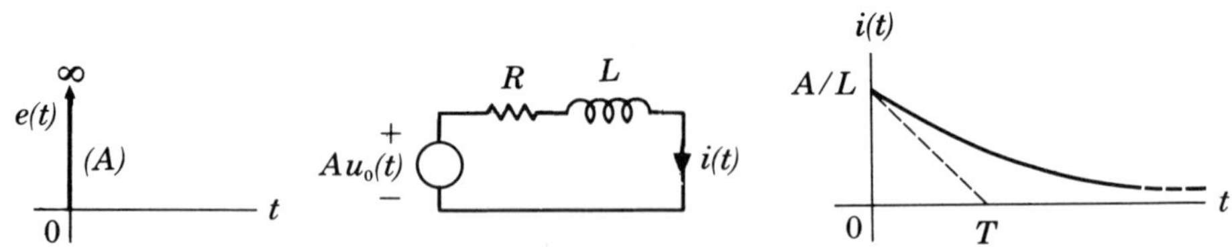

$$i(t) = \frac{A}{L} e^{-Rt/L} \qquad \text{for} \quad t > 0$$

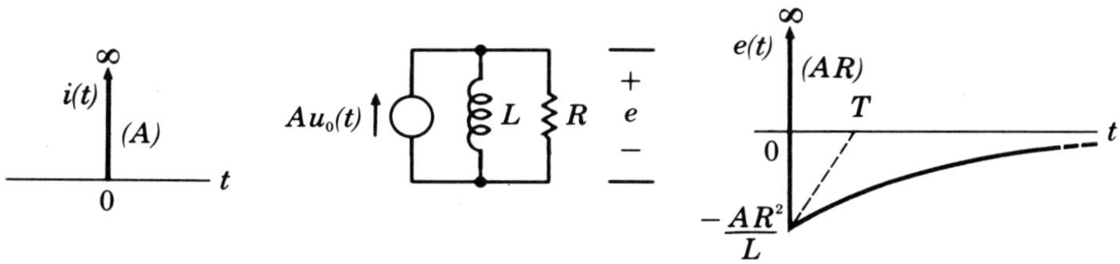

$$e(t) = ARu_0(t) - \frac{AR^2}{L} e^{-Rt/L} \cdot u_{-1}(t)$$

The step response of R–L circuits

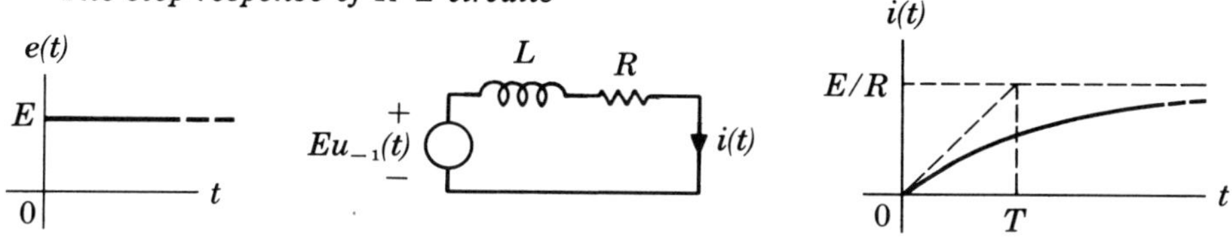

$$i(t) = \frac{E}{R}(1 - e^{-Rt/L})u_{-1}(t)$$

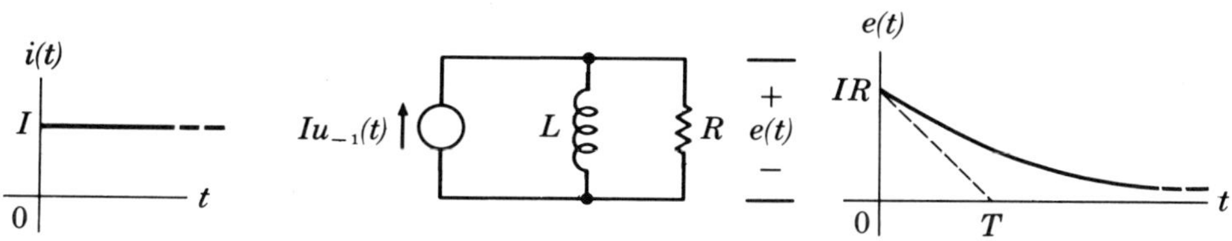

$$e(t) = IRe^{-Rt/L} \cdot u_{-1}(t)$$

The impulse response of R–C circuits

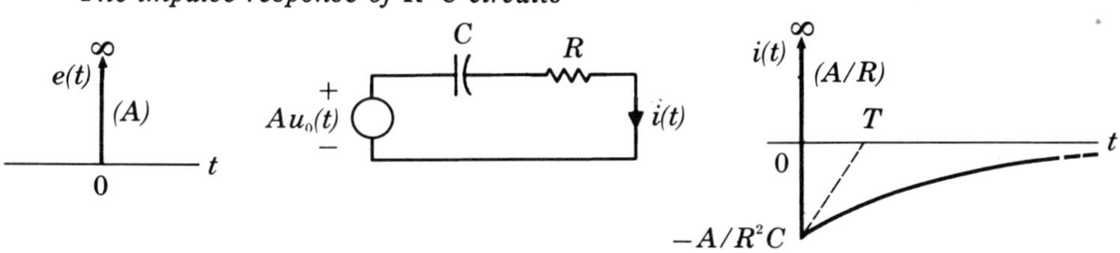

$$i(t) = \frac{A}{R} u_0(t) - \frac{A}{R^2C} e^{-t/RC} \cdot u_{-1}(t)$$

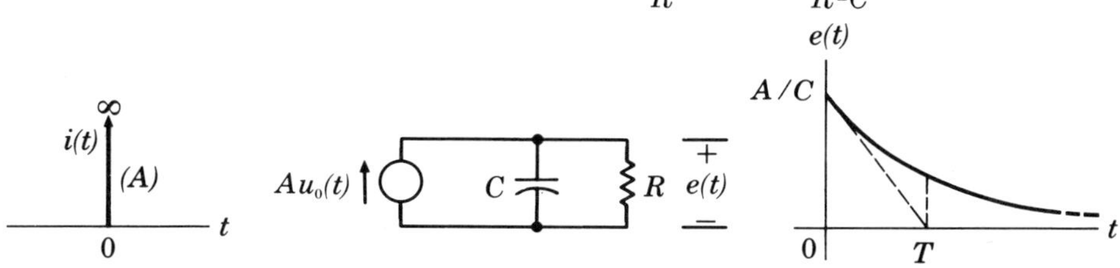

$$e(t) = \frac{A}{C} e^{-t/RC} \cdot u_{-1}(t)$$

The step response of R–C circuits

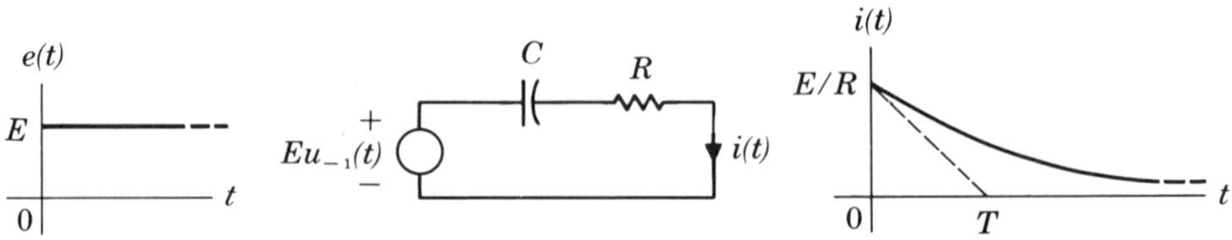

$$i(t) = \frac{E}{R}e^{-t/RC}\cdot u_{-1}(t)$$

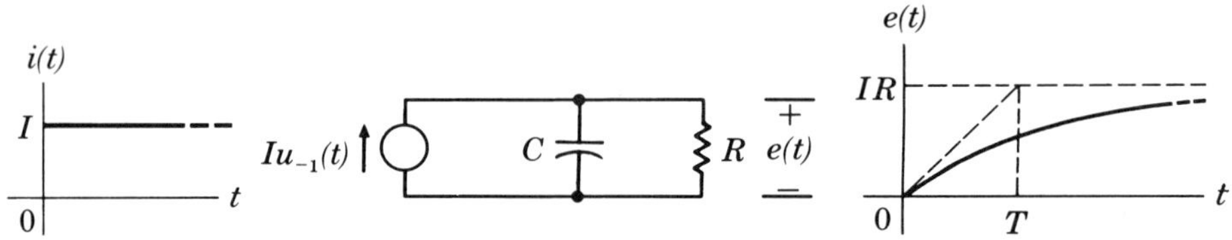

$$e(t) = IR(1 - e^{-t/RC})u_{-1}(t)$$

Responses to ramp excitation

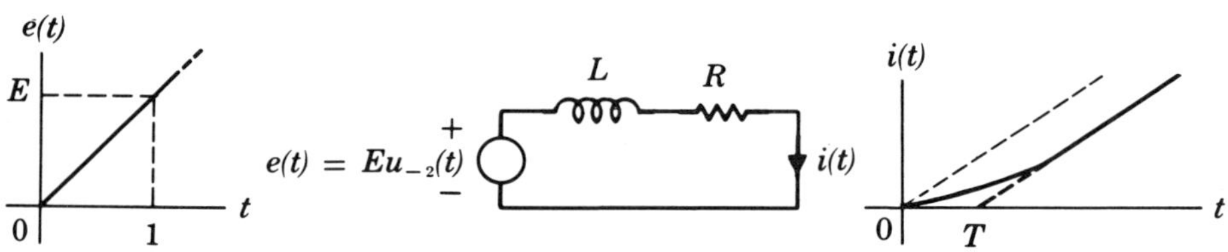

$$i(t) = \frac{E}{R}\left[t - \frac{L}{R}(1 - e^{-Rt/L})\right]u_{-1}(t)$$

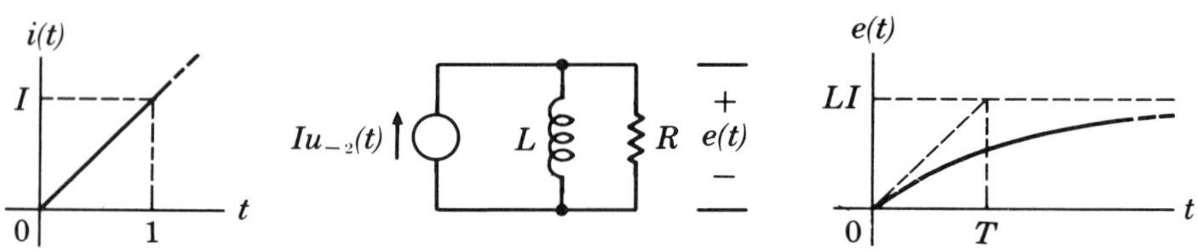

$$e(t) = LI(1 - e^{-Rt/L})u_{-1}(t)$$

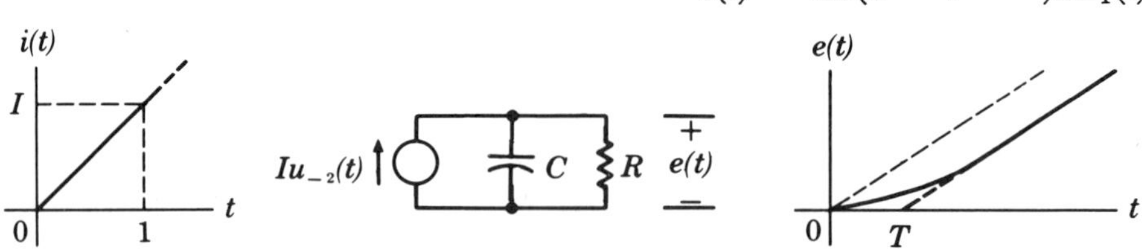

$$e(t) = IR[t - RC(1 - e^{-t/RC})]u_{-1}(t)$$

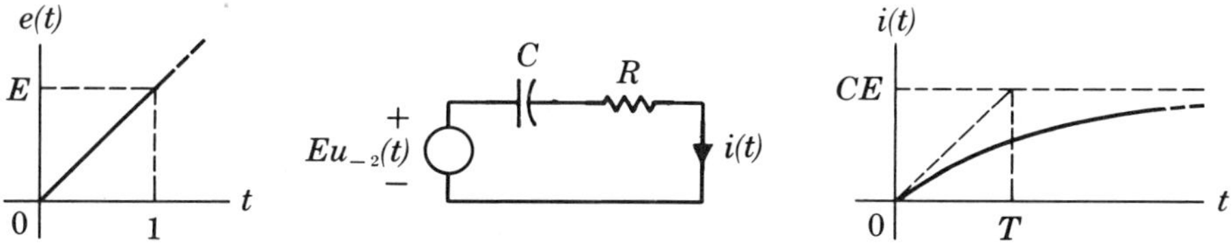

$$i(t) = CE(1 - e^{-t/RC})u_{-1}(t)$$

More general single-time-constant circuits

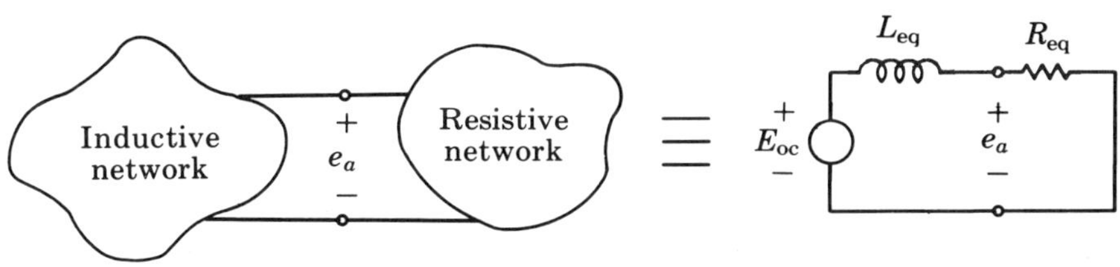

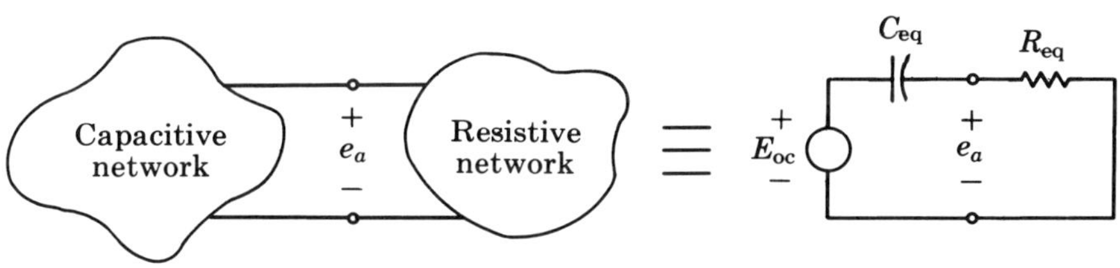

REFERENCES

More attention is paid to one-energy-storage circuits in this chapter than in most other texts. One-energy-storage circuits occur frequently in practice, and the methods developed here are the preferred ones for solving them. Differential equations and Laplace transforms are not suitable for these circuits, and amount to "driving the tack with the sledge hammer."

1. E. A. Guillemin, *Introductory Circuit Theory*, John Wiley & Sons, New York, 1953. The method of writing the transient response by inspection, in terms of the initial value, final value, and time constant, is implied in Guillemin's book.

2. MIT EE Staff, *Electric Circuits*, John Wiley & Sons, New York, 1943. This text gives the classical treatment of R–L and R–C circuits with step excitation, but not with impulse excitation.

3. BRENNER and JAVID, *Analysis of Electric Circuits,* McGraw-Hill, New York, 1959. Brenner and Javid treat the source-free behavior of networks as something apart from the response to step and impulse excitation. This approach is common in the mathematical treatment of linear differential equations.

4. KERCHNER and CORCORAN, *Alternating Current Circuits,* John Wiley & Sons, New York, 3rd ed, 1951. In this text emphasis is placed on steady-state sinusoidal analysis. Transients are included in the last chapter.

5. H. H. SKILLING, *Electrical Engineering Circuits,* John Wiley & Sons, New York, 1957. The R–L and R–C circuits appear in various places throughout this text.

EXERCISES*

Section 11–2

1. The current in a certain circuit is given by the following function of time:

$$i(t) = 10e^{-t/6} \qquad \text{for } t > 0.$$

(a) Compute the values of the current at $t = 0$, 6, 12, and ∞ sec.

(b) Sketch $i(t)$ as a function of time.

(c) Compute $di(t)/dt$ for $t = 0$.

(d) If the current continued to decay at its initial rate, how long would it take to reach zero?

2. An impulse voltage of 6 volt·sec occurs at $t = 0$. This source is connected in series with an R of 3 ohms and an L of 4 henrys. Find the time constant, the value of $i(0^+)$, and the value of $i(\infty)$. Sketch the function, and write its analytic expression.

3. An impulse current source of 4 amp·sec is connected in parallel with $R = 4$ ohms and $L = 5$ henrys. Give the time constant, the initial value of the voltage, and the value of the voltage at $t = \infty$. Sketch the voltage as a function of time and write an analytic expression for it, including the impulse at the origin.

4. In the circuit of Fig. 11–22 the current in the inductance at $t = 0$ is 6 amp. What is the current as a function of time? [*Note:* There is no point in converting the initial energy storage to a source, since the initial value of the current is the desired quantity.]

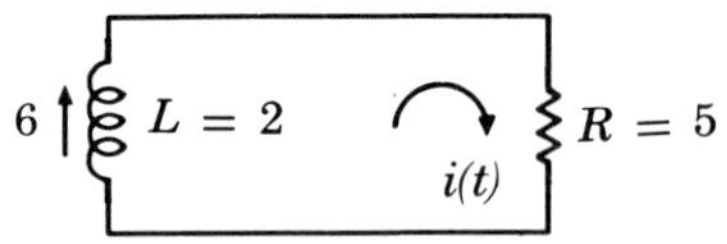

FIGURE 11–22

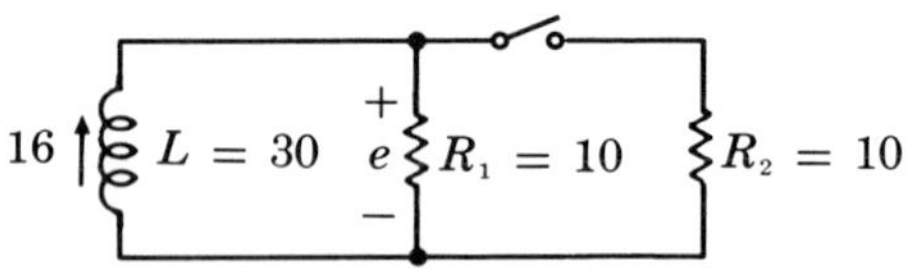

FIGURE 11–23

---

* Numerical values are in volts, amperes, ohms, henrys, and farads.

5. In the circuit of Fig. 11–23 the current in the inductance at $t = 0$ is 16 amp. At the end of 3 seconds the switch is closed. What is the voltage across the resistance R_1 as a function of time?

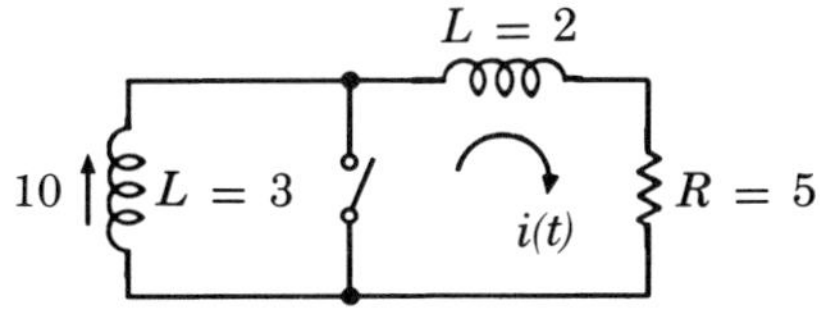

FIGURE 11–24

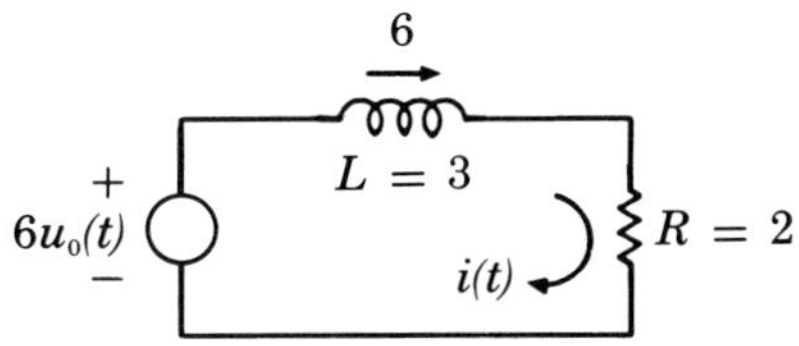

FIGURE 11–25

6. In the circuit of Fig. 11–24 the current in the 3-henry inductance at $t = 0$ is 10 amp. The switch is opened at $t = 0$. What is the current in the circuit for $t > 0$?

7. In the circuit of Fig. 11–25 the current in the inductance just prior to the occurrence of the impulse is 6 amp. Find the current in the circuit as a function of time after the impulse occurs.

8. The current in a series R–L circuit is $i(t) = 100e^{-t/10} \cdot u_{-1}(t)$. The value of R is 2 ohms and the value of L is 20 henrys. An impulse voltage occurs at $t = 10$ sec which instantly reduces the current to zero. What is the value of the impulse? What impulse current in parallel with the elements would accomplish the same result?

9. In the circuit of Fig. 11–26 the switch is closed at $t = 0^+$, just after the impulse is over. What is the voltage $e(t)$ for $t > 0$, and what is the current in the switch?

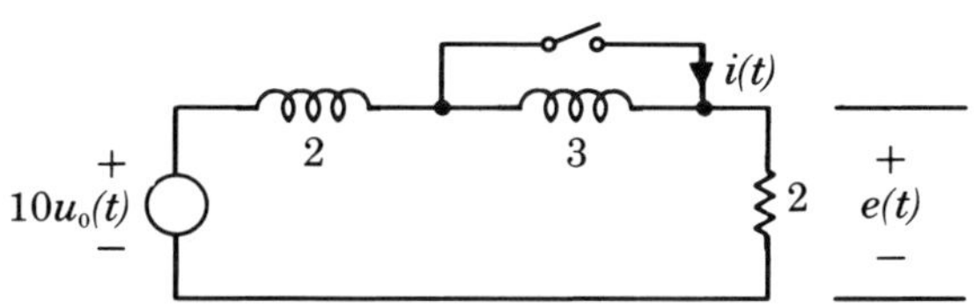

FIGURE 11–26

10. The switch in the circuit of Fig. 11–27 is closed at $t = 0$, when the current in L_1 is 10 amp. Find the voltage $e(t)$ and the current in L_2 for $t > 0$.

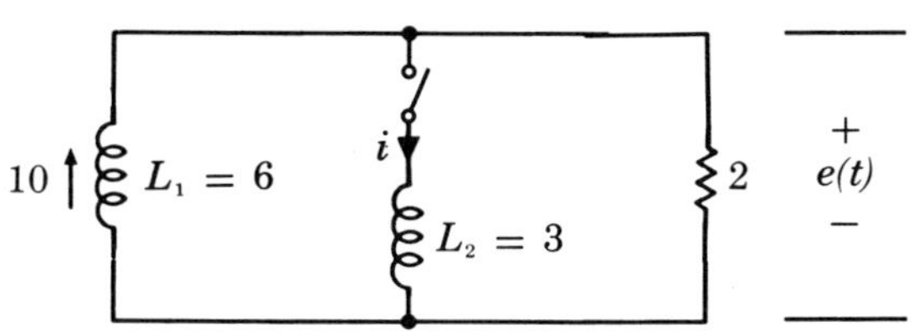

FIGURE 11–27

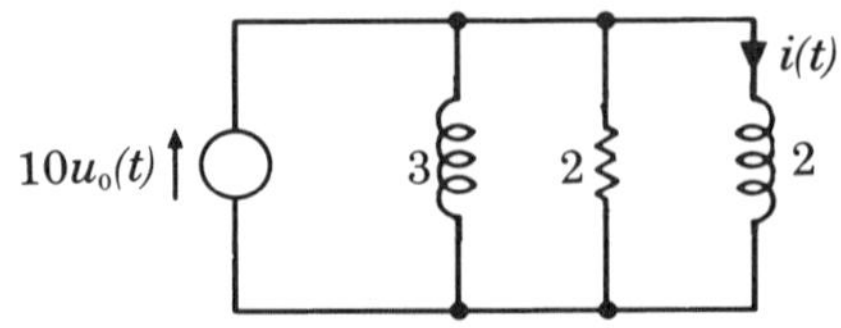

FIGURE 11–28

11. Find the current $i(t)$ in the circuit of Fig. 11–28.

Section 11–3

12. A certain voltage is given by the following function of time:

$$e(t) = 10(1 - e^{-t/6})u_{-1}(t).$$

(a) Compute the values of the voltage at $t = 0$, 6, 12, and ∞ sec.
(b) Sketch $e(t)$ as a function of time.
(c) Compute $de(t)/dt$ for $t = 0$.
(d) If the voltage continued to rise at its initial rate, how long would it take to reach its final value of 10?
(e) Comment on the meaning of the time constant in this circuit.

13. A 10-volt d-c source is switched in series with $R = 2$ ohms and $L = 3$ henrys at $t = 0$. Find the time constant, the current at $t = 0^+$, and the current at $t = \infty$. Sketch the current as a function of time and give the analytic expression for it.

14. A 6-amp d-c source is switched in parallel with an L of 2 henrys and an R of 4 ohms at $t = 0$. Give the time constant and the value of the voltage at $t = 0$, and at $t = \infty$. Sketch the curve and write its analytic expression.

15. A 4-henry inductance and a 3-ohm resistance are connected in parallel. The current in the inductance is 2 amp at $t = 0$.

(a) Replace the initial current in the inductance by a parallel current source and obtain the current in the resistance as a function of time.
(b) Compute the initial value of the current, the final value of the current, and the time constant and thus obtain the current directly.

16. In the circuit of Fig. 11–29 the current in the inductance is -6 amp at $t = 0$.

(a) Find the current as a function of time by computing the initial value, final value, and time constant.
(b) Find the current by superposition.

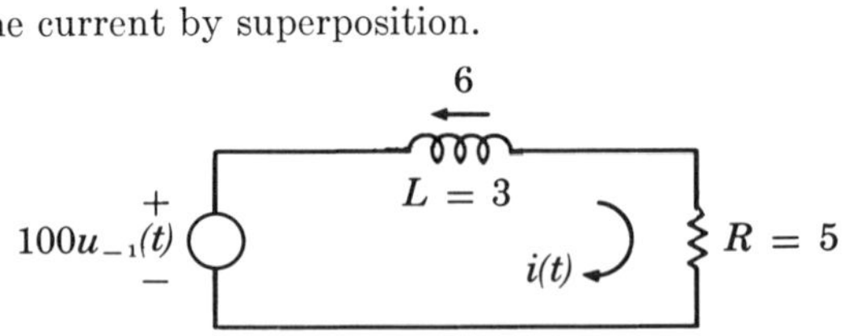

FIGURE 11–29

17. In the circuit of Fig. 11–30, what value of the step-function voltage would result in a constant current of 5 amp for all time greater than zero?

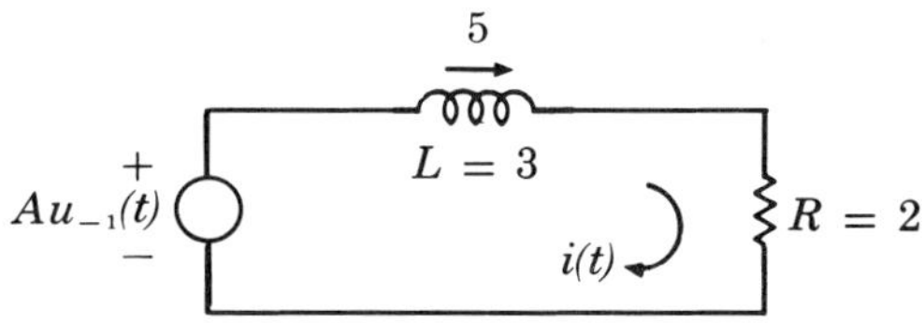

FIGURE 11–30

18. (a) Find the current in the circuit of Fig. 11–31 as a function of time by computing the initial value, the final value, and the time constant. (b) Find the current by superposition.

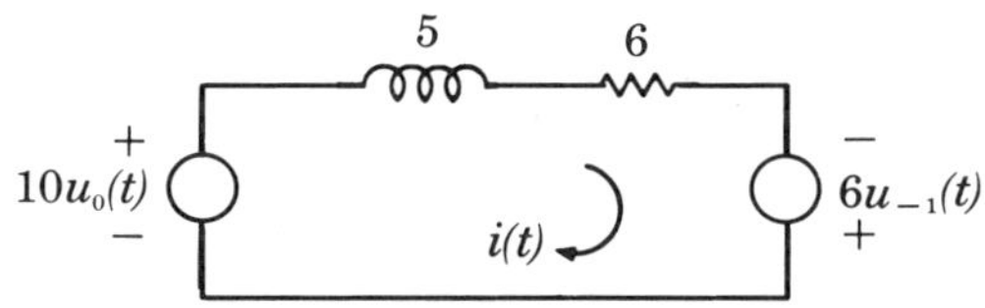

FIGURE 11–31

19. (a) Find the current in the inductance in the circuit of Fig. 11–32 by computing its initial value, its final value, and its time constant. (b) Verify the result of part (a) by superposition.

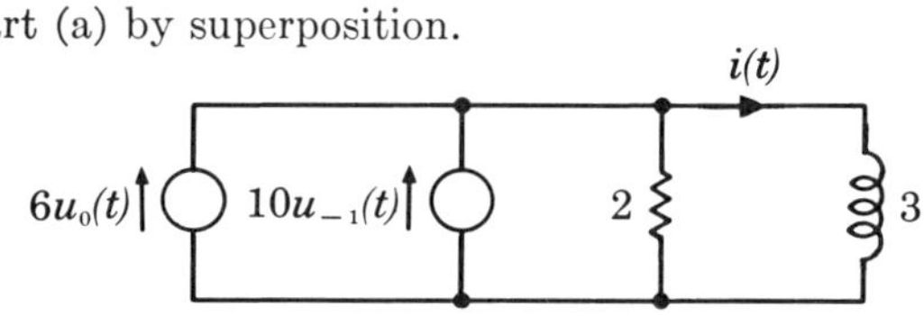

FIGURE 11–32

20. In the circuit of Fig. 11–33 find the current in L_2 as a function of time.

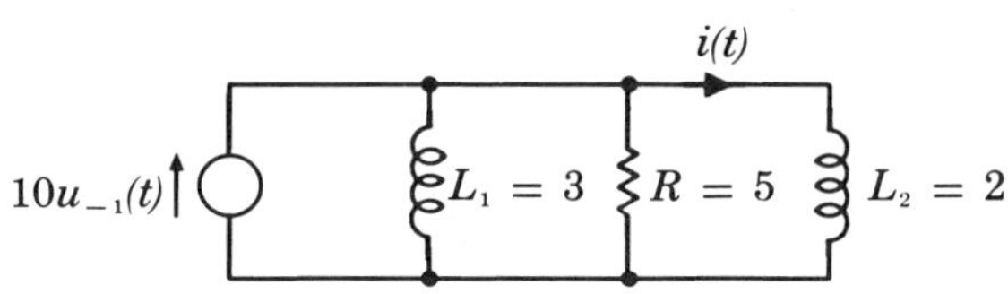

FIGURE 11–33

21. In the circuit of Fig. 11–34 a steady state is reached with the switch open. Then at $t = 0$ the switch is closed. Find $i(t)$ as a function of time from this instant.

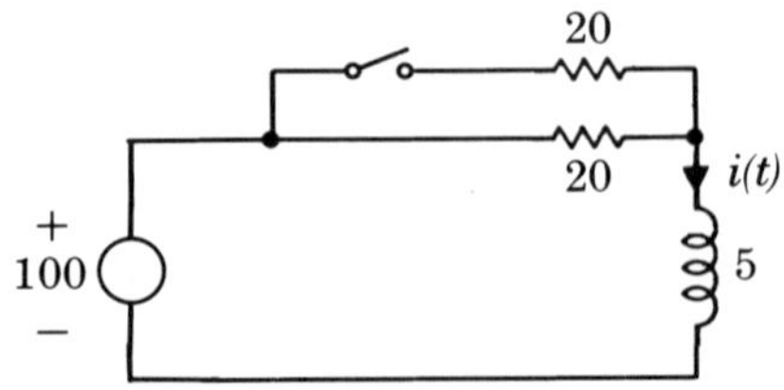

FIGURE 11–34

22. In the circuit of Fig. 11–35 a steady state is reached with the switch open. At $t = 0$ the switch is closed. Write an expression for the current in the inductance as a function of time from this instant.

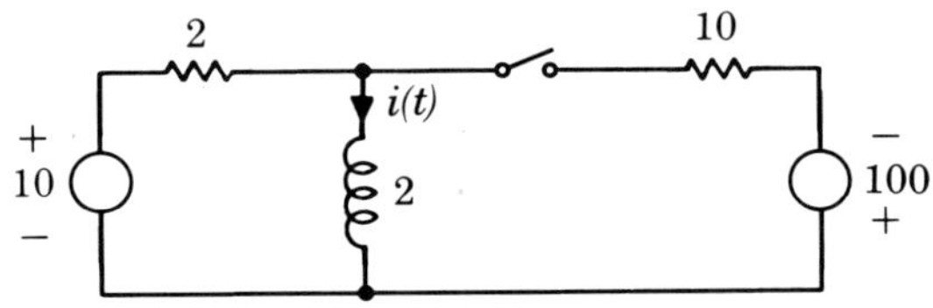

FIGURE 11–35

Section 11–4

23. A ramp voltage is applied to the circuit of Fig. 11–36. Find the voltage across the resistance as a function of time. Find also the steady-state, or ramp-function, component of this voltage.

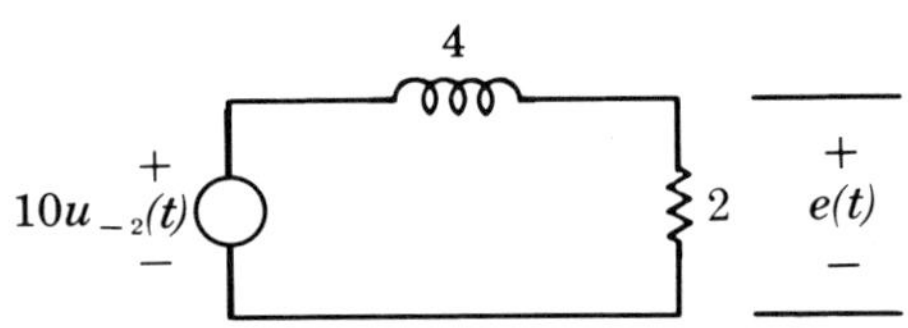

FIGURE 11–36

24. A voltmeter has a time constant of 1 sec. It is to be used to follow a varying d-c input voltage. What is the maximum rate of change of the input voltage that can be allowed if the error due to the meter's lagging time constant is to be less than 1 volt?

25. A ramp voltage $e(t) = 10u_{-2}(t)$ is applied to a series circuit consisting of an inductance of 1 microhenry and a resistance of 10 kilohms. The inductance is the stray inductance of the leads, and it is desired that the voltage across the resistance should follow the input voltage exactly. What is the steady-state error in volts after the initial transient has died down, and how long is it before the transient dies down to $1/e^2$ of its initial value?

26. A ramp current $i(t) = 10u_{-2}(t)$ is applied to the parallel combination of $R = 2$ ohms and $L = 10$ henrys. Find the voltage across the circuit.

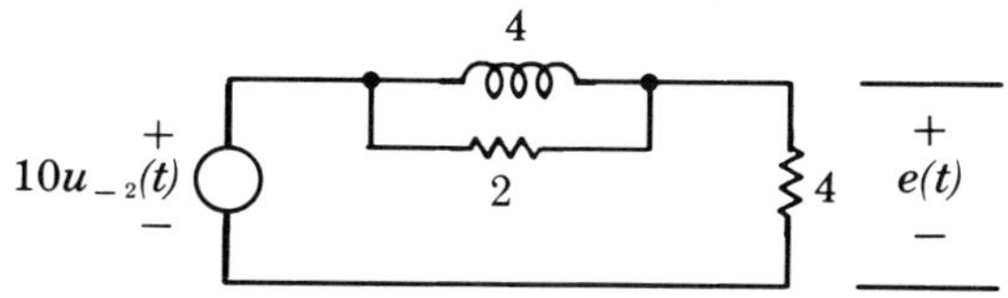

FIGURE 11–37

27. Find the voltage $e(t)$ in the circuit of Fig. 11–37.

28. Find the current $i(t)$ which flows in the circuit of Fig. 11–38 as a result of the combined step-function and ramp-function sources.

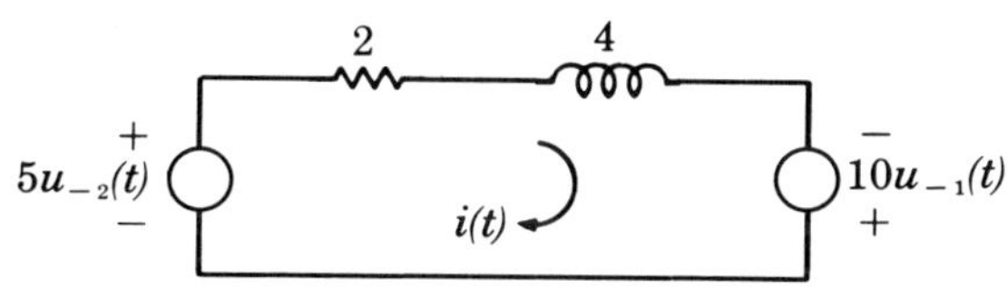

FIGURE 11–38

29. In the circuit of Fig. 11–39 the inductance represents the deflection coil of a television tube. The current in it must be a linear ramp, or sweep, with a value $i(t) = 10t$. What voltage source will produce the required value?

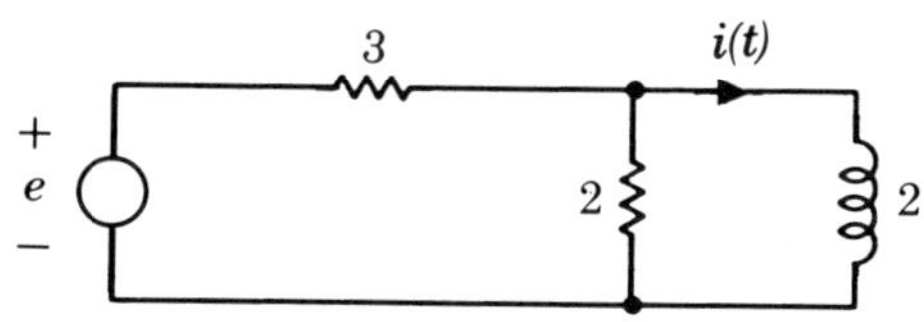

FIGURE 11–39

Section 11–5

30. An impulse current of 5 amp·sec occurs at $t = 0$. This source is connected in parallel with $R = 6$ ohms and $C = 1$ farad. Give the time constant and the values of the voltage produced at $t = 0$ and $t = \infty$. Sketch the voltage $e(t)$, and write an analytic expression for it.

31. A capacitance of 2 farads is charged to an intial voltage of 10 volts at $t = 0$. At this time it is connected in series with a resistance of 2 ohms. Give the time constant, the current at $t = 0$, the current at $t = \infty$, a sketch of the current, and an analytic expression for the current as a function of time.

32. An impulse voltage source $e(t) = 10u_0(t)$ is connected in series with a resistance of 2 ohms and a capacitance of $\frac{1}{2}$ farad. Give the time constant, the current at $t = 0^+$, and the current at $t = \infty$. Sketch the current and give an analytic expression for it. Include the impulse which occurs at $t = 0$.

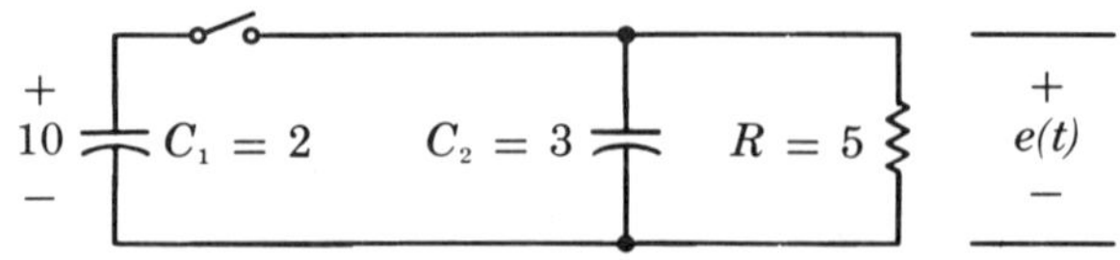

FIGURE 11–40

33. In the circuit of Fig. 11–40 the switch is closed at $t = 0$. At this time there is a voltage on C_1 of 10 volts. Find the voltage $e(t)$ for $t > 0$.

34. In the circuit of Fig. 11–41 the voltage on the capacitance just before the impulse strikes is 10 volts. Find the current in the resistance for $t > 0$.

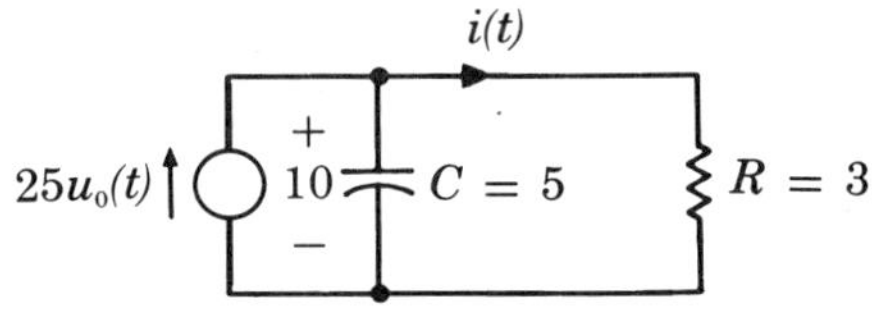

FIGURE 11–41

35. The voltage across a parallel R–C circuit with $R = 2$ and $C = 3$ is

$$e(t) = 50e^{-t/6} \cdot u_{-1}(t).$$

At $t = 6$ sec a current impulse in parallel with the two elements reduces the voltage to a permanent value of zero. What is the value of the impulse? What impulse voltage in series with the two elements would accomplish the same result?

36. In the circuit of Fig. 11–42 the switch is opened just after the impulse is over. What is the voltage $e(t)$ as a function of time measured from the opening of the switch? What is the voltage e_x across the switch?

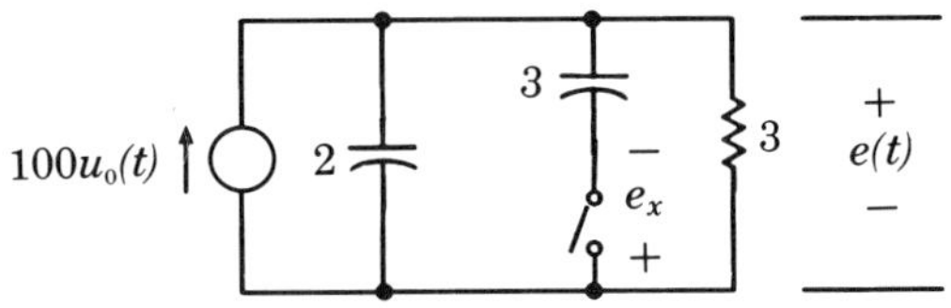

FIGURE 11–42

37. The switch in the circuit of Fig. 11–43 is opened at $t = 0$. At this time there is a voltage of 10 volts on C_1 and no voltage on C_2. Find the current $i(t)$ and the voltage across C_2 as functions of time.

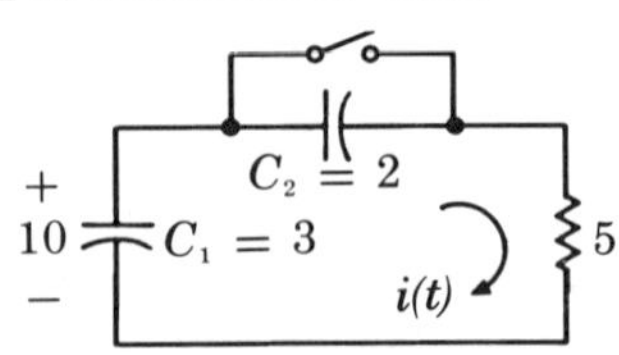

FIGURE 11–43

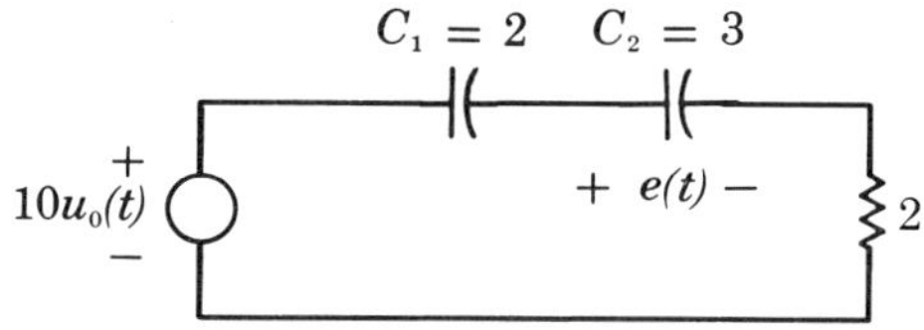

FIGURE 11–44

38. In the circuit of Fig. 11–44 find the voltage $e(t)$ as a function of time.
39. Find the voltage $e(t)$ as a function of time in the circuit of Fig. 11–45.

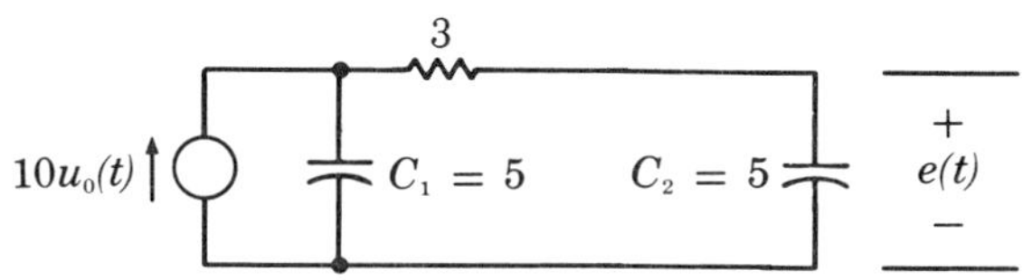

FIGURE 11–45

Section 11–6

40. A 6-volt d-c source is switched in series with a 2-ohm resistance and a 3-farad capacitance at $t = 0$. Give the time constant, the current at $t = 0^+$, the current at $t = \infty$, a sketch of the current as a function of time, and the analytic expression for the current.

41. A 5-amp d-c source is switched in parallel with a 3-ohm resistance and a 2-farad capacitance. Give the time constant, the voltage at $t = 0^+$, the voltage at $t = \infty$, the analytic expression for the current, and a sketch of the current as a function of time.

42. A 3-farad capacitance and a 5-ohm resistance are connected in series. At $t = 0$ there is a voltage on the capacitance of 15 volts.
   (a) Replace the initial energy storage by a step-voltage source, and find the current in the circuit.
   (b) Find the current by a direct computation of its initial value, final value, and time constant.

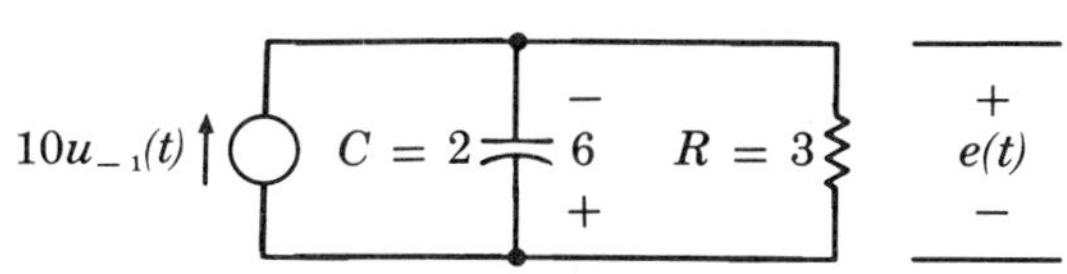

FIGURE 11–46

43. In the circuit of Fig. 11–46 there is a voltage of −6 volts on the capacitance at $t = 0$, when the step-current source is applied. Find the voltage across the resistance as a function of time for $t > 0$.

44. The capacitance in the circuit of Fig. 11–47 has a voltage of 5 volts at $t = 0$, when the step voltage is applied. Find the voltage on the capacitance as a function of time.

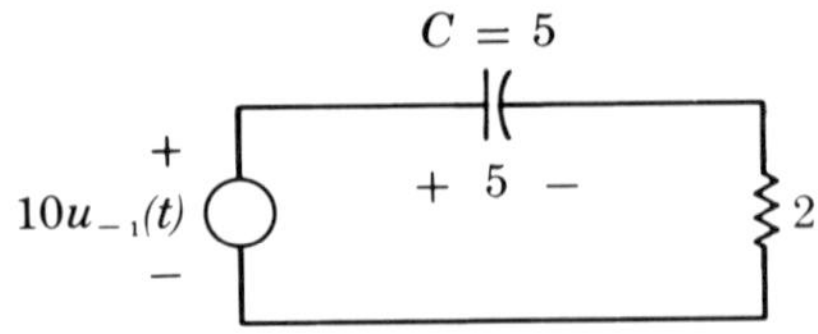

FIGURE 11–47

45. (a) Solve for the voltage $e(t)$ in the circuit of Fig. 11–48 by computing its initial value, final value, and time constant. (b) Verify the result of part (a) by superposition.

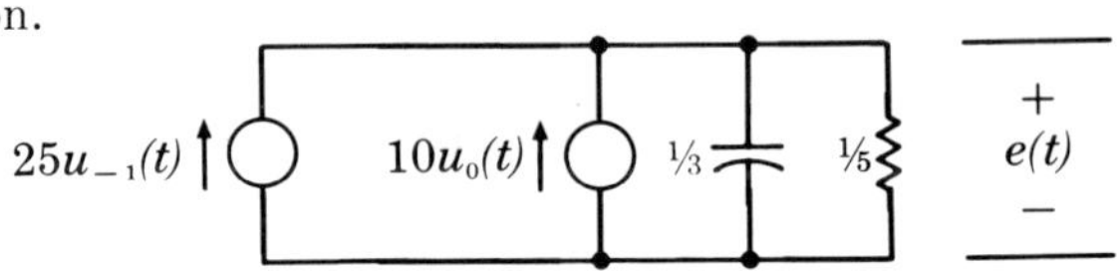

FIGURE 11–48

46. A step voltage $e(t) = 10u_{-1}(t)$ and an impulse voltage $e(t) = 15u_0(t)$ are connected in series in an additive polarity and applied to a series circuit consisting of a capacitance of $C = 3$ farads and a resistance of $R = 5$ ohms. Find the voltage produced across the capacitance.

47. A step voltage $e(t) = u_{-1}(t)$ is applied to a series combination of a capacitance $C_1 = 2$ farads, a capacitance $C_2 = 3$ farads, and a resistance $R = 5$ ohms. Give the voltage produced across the capacitance C_2.

48. The switch in Fig. 11–49 is closed at $t = 0$ after a steady state has been attained in the remainder of the circuit. Find the voltage $e(t)$ across the capacitance as a function of time from the moment of the closing of the switch.

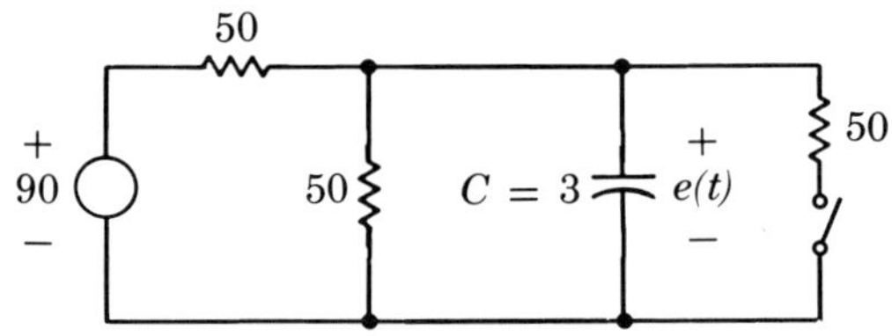

FIGURE 11–49

49. The switch in the circuit of Fig. 11–50 is opened at $t = 0$ after a steady state has been set up by the 100-volt source. Give the voltage $e(t)$ as a function of time, starting from the opening of the switch.

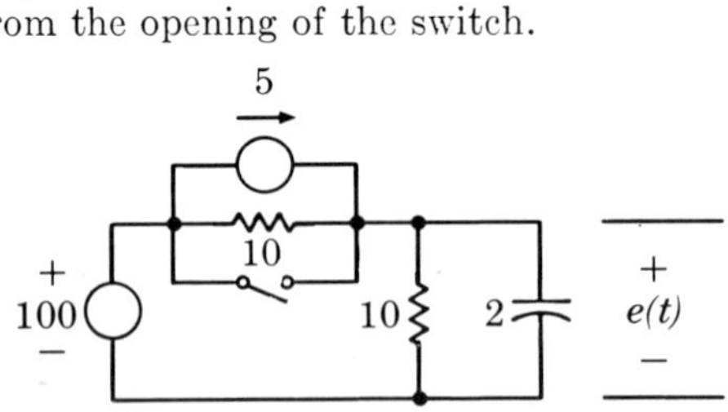

FIGURE 11–50

Section 11–7

50. Find the voltage $e(t)$ in the circuit of Fig. 11–51.

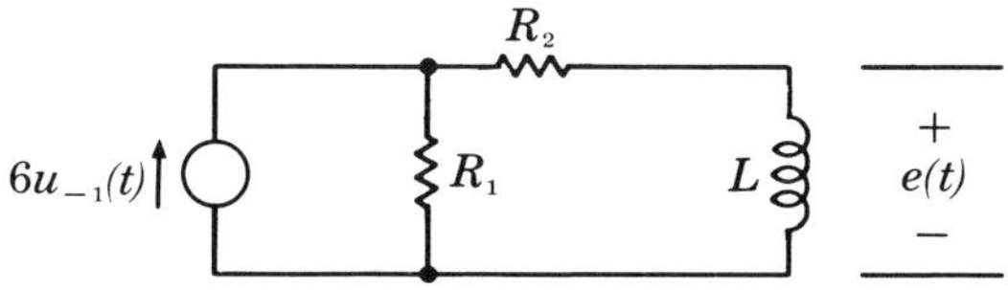

FIGURE 11–51

51. Find the current $i(t)$ in the circuit of Fig. 11–52.

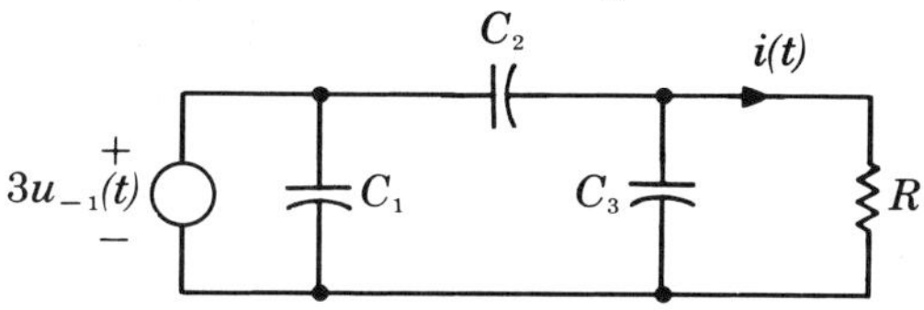

FIGURE 11–52

52. Find the voltage $e(t)$ in the circuit of Fig. 11–53.

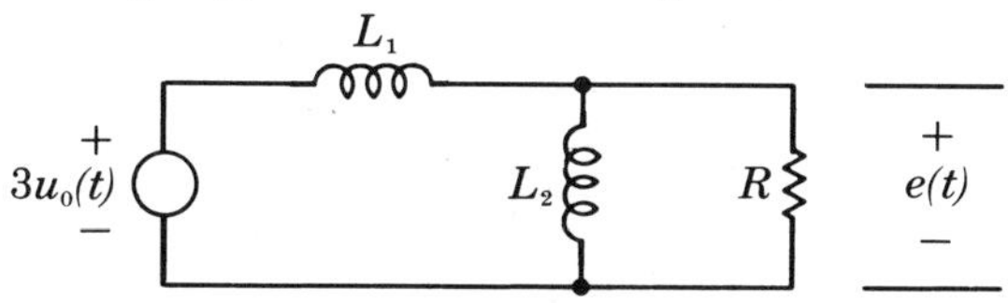

FIGURE 11–53

53. Find the voltage $e(t)$ in the circuit of Fig. 11–54.

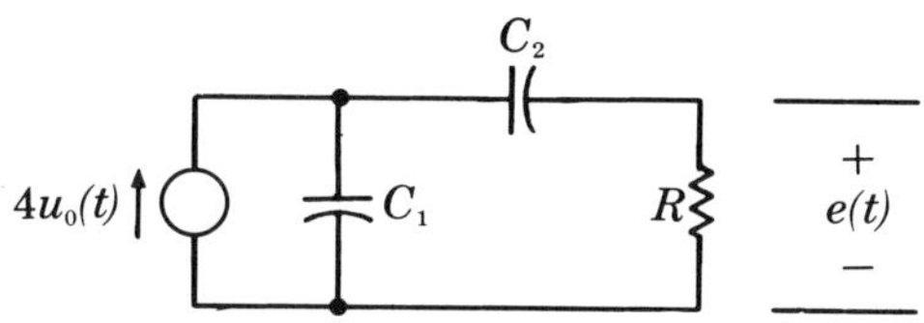

FIGURE 11–54

54. For Fig. 11–55 draw an equivalent circuit at terminal-pair a–b which contains a single C and a single R. Find the voltage e_{ab}, and from this voltage obtain the output voltage $e(t)$.

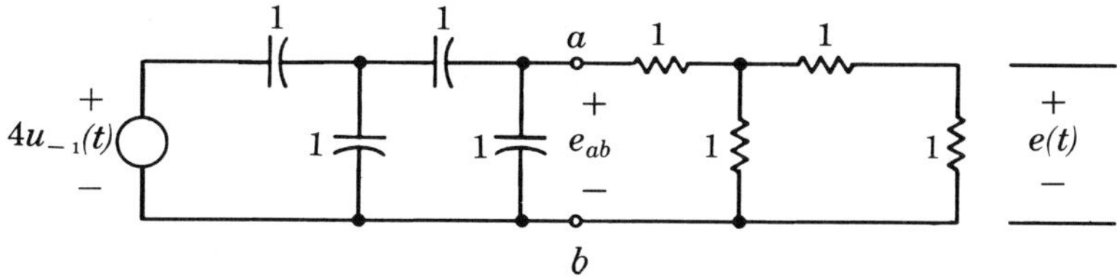

FIGURE 11–55

55. For Fig. 11–56 draw an equivalent circuit at terminal-pair a–b which contains only one L and one R and a voltage source. Obtain the voltage e_{ab}, and from this value obtain the current $i(t)$.

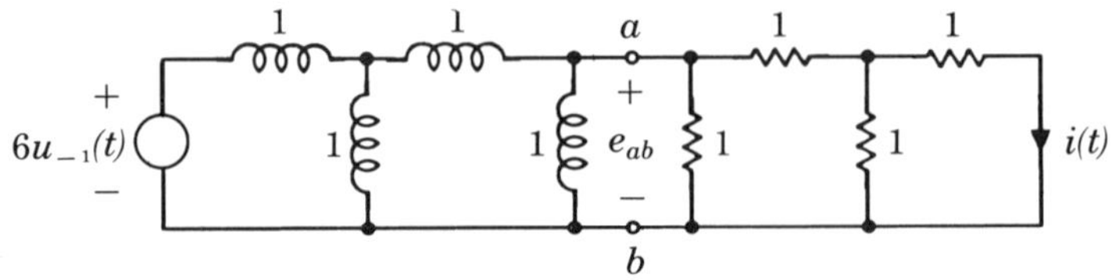

FIGURE 11–56

56. Find the node voltages e_a, e_b, e_c, e_d, and e_e, in the circuit of Fig. 11–57.

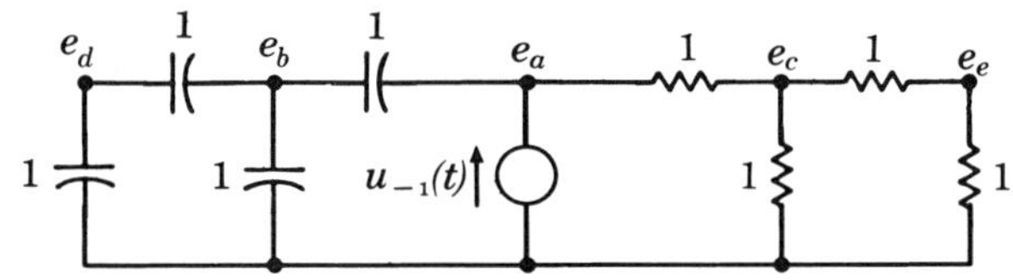

FIGURE 11–57

57. In the circuit of Fig. 11–58 find the node voltages e_a, e_b, e_c, and e_d.

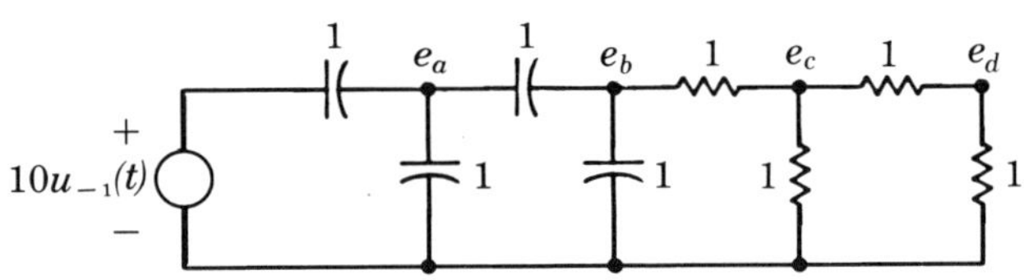

FIGURE 11–58

58. Find the node voltages e_a, e_b, e_c, and e_d in the circuits of Fig. 11–59.

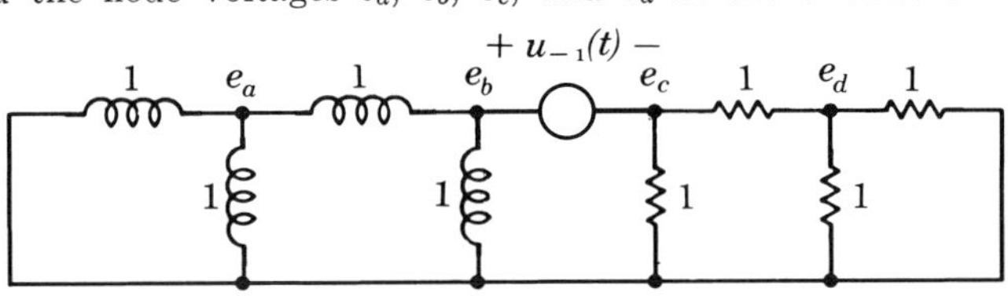

FIGURE 11–59

59. Find the node voltages e_a, e_b, e_c, and e_d in the circuit of Fig. 11–60.

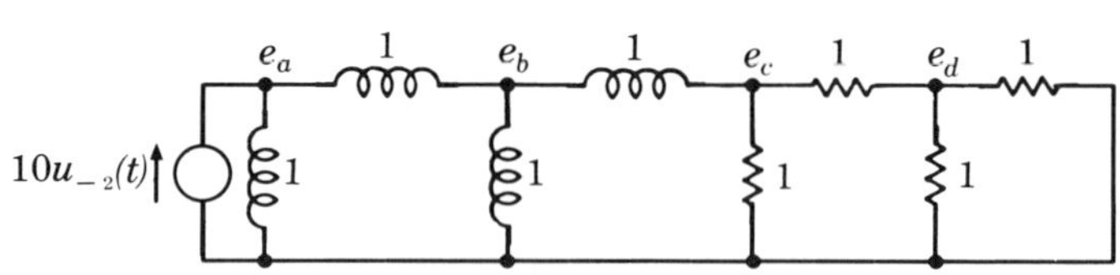

FIGURE 11–60

CHAPTER 12

THE TIME RESPONSE FUNCTIONS OF *L–C* AND *R–L–C* NETWORKS

"Men eat birds, birds eat worms, worms eat men again."

Samuel Butler

12–1 Introduction. The conservation of energy is one of the basic laws of the universe. Many energy changes are cyclical, like that observed by Samuel Butler. Electric circuits which contain two types of energy-storage elements exhibit such cyclical behavior. Energy is first stored in the electric field of a capacitance, is transferred to the magnetic field of an inductance, and is then returned to the electric field of the capacitance. The process continues indefinitely until the energy is dissipated by the resistive losses of the system. Many of the interesting properties of electrical systems, such as resonance and wave propagation, are consequences of the two types of energy which can be stored and the "swapping" of energy from one type to the other and back again.

A circuit with resistance, inductance, and capacitance is described by a second-order differential equation of the form

$$\frac{d^2y}{dt^2} + 2\alpha\frac{dy}{dt} + \omega_0^2 y = f(t). \tag{12–1}$$

In Chapter 9 we obtained the basic solution for this equation, in the form

$$y = A_1e^{p_1t} + A_2e^{p_2t} + y_{ss}(t) \qquad \text{for} \quad t > 0. \tag{12–2}$$

The first two terms in Eq. (12–2) represent the natural behavior of the system, and the last term represents the steady-state response forced on the system by the driving function $f(t)$. The constants A_1 and A_2 are chosen so that the solution satisfies the initial conditions in the system. Essentially, they are determined by the initial energy storages in the system. The transient or natural-behavior terms can take several forms, depending on the values of α and ω_0 in the differential equation. The p-values are given by

$$\begin{aligned} p_1 &= -\alpha + \sqrt{\alpha^2 - \omega_0^2}, \\ p_2 &= -\alpha - \sqrt{\alpha^2 - \omega_0^2}, \end{aligned} \tag{12–3}$$

where $\omega_0 = \dfrac{1}{\sqrt{LC}}$,

$$\alpha = \begin{cases} R/2L & \text{for a series } R\text{–}L\text{–}C \text{ circuit,} \\ 1/2RC & \text{for a parallel } R\text{–}L\text{–}C \text{ circuit.} \end{cases}$$

When $\alpha > \omega_0$ an overdamped or exponential type of solution is produced. When $\alpha = \omega_0$ a critically damped solution is produced, and when $\alpha < \omega_0$ an oscillatory type of solution is produced.

We shall take an attitude toward the second-order differential equation the same as that we took toward the first-ordere quation. We shall consider that we have solved it once and for all time and that the general solution can be adapted to a particular problem by evaluating the constants A_1 and A_2 from the initial conditions of the problem. Since there are two constants to be evaluated, there must be two initial conditions. Usually the two initial conditions are the initial value of the variable and the initial value of the first derivative of the variable. These values can be computed from the initial energy storages in the circuit and the resistive circuit which holds at $t = 0$ by the methods we discussed in Chapter 7. For the response to internal energy storages, as well as for the impulse response, the steady-state solution is zero. For the step response the steady-state solution can be obtained from the value of the variable at $t = \infty$.

We shall begin by considering an ideal lossless circuit consisting of inductance and capacitance. In this case the energy swaps back and forth indefinitely, and what we call the transient solution does not die out. In order to separate the transient and steady-state terms in this case we shall find it convenient to suppose that the circuit resistance is not exactly zero, but only vanishingly small. All real passive circuits actually have at least a small resistance.

12–2 The impulse response of an L–C circuit. A lossless L–C circuit is the electrical analog of a lossless (frictionless) mechanical pendulum. In the electric circuit the energy swaps back and forth from magnetic to electrical energy; in the mechanical system it oscillates between potential energy, in the position of maximum elevation, and kinetic energy, in the position of maximum velocity. An intuitive understanding of the operation of the electric circuit can be obtained from Fig. 12–1. The five illustrations show the state of the same circuit at successive instants of time. In Fig. 12–1(a) the energy is stored in the capacitance. No current is flowing, and no energy is stored in the inductance. The voltage on the capacitance causes a current to start flowing in the positive direction, and at the instant represented by Fig. 12–1(b) the capacitance is discharged and a maximum current is flowing in the inductance. No energy

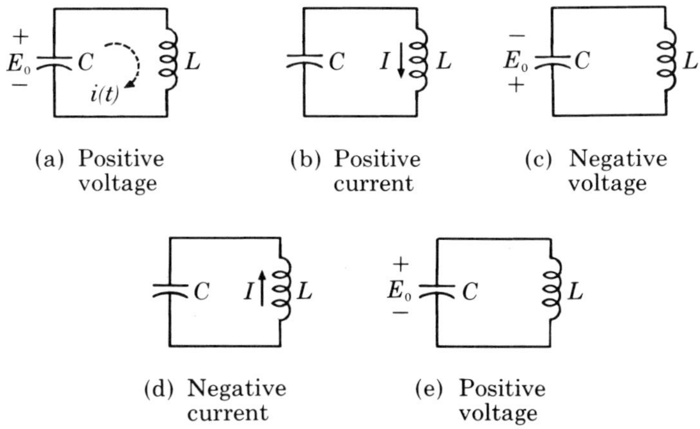

FIG. 12–1. The oscillation of energy in an L–C circuit.

is stored in the capacitance at this instant; it is all stored in the inductance. The operation does not stop here, however, because of the "inertia" of the current in the inductance. The current continues to flow until some force stops it. Hence it continues until the capacitance is charged in the opposite direction, as shown in Fig. 12–1(c). At this instant the current is zero, and all the energy is again stored in the capacitance. The voltage on the capacitance then starts a current flowing in the opposite direction, and eventually all the energy is transferred back to the inductance, as shown in Fig. 12–1(d). Again, the inertial property of the inductance causes the current to continue until it charges the capacitance with the original polarity, as shown in Fig. 12–1(e). From this instant the process repeats again indefinitely.

Since the behavior discussed in the preceding paragraph is excited by energy initially stored in the system, it represents the natural or transient response of the system. Mathematically, the value of α is zero, and the p-values in Eq. (12–2) are pure imaginaries. The two imaginary exponential terms can be combined to produce a sinusoid with an arbitrary amplitude and phase angle, or the sum of a sine term with an arbitrary amplitude and a cosine term with an arbitrary amplitude. The natural response function is therefore*

$$\begin{aligned} f(t) &= A_1 \cos \omega_0 t + A_2 \sin \omega_0 t \\ &= A \cos (\omega_0 t + \theta) \qquad \text{for} \quad t > 0, \end{aligned} \tag{12–4}$$

where $\omega_0 = 1/\sqrt{LC}$,
A_1, A_2, A, and θ are constants determined by the initial conditions.

* This equation is also Eq. (9–72) with $\alpha = 0$.

We will assume a natural response function of the form given by Eq. (12–4) and shall find the arbitrary constants from the initial values of the function and its first derivative. For our first example, consider the circuit of Fig. 12–1(a), which has an initial voltage E_0 on the capacitance. This voltage can be treated as a step function applied in series with the two elements. The inductance at $t = 0$ acts as an open circuit, and therefore the initial current is

$$i(0^+) = 0.$$

The voltage across the inductance is E_0, however, and the rate of change of current is

$$\frac{di(0^+)}{dt} = \frac{E_0}{L}.$$

The assumed form of the current function for $t > 0$ is*

$$i(t) = A_1 \cos \omega_0 t + A_2 \sin \omega_0 t, \tag{12–5}$$

from which

$$\frac{di}{dt} = -A_1\omega_0 \sin \omega_0 t + A_2\omega_0 \cos \omega_0 t. \tag{12–6}$$

When $t = 0^+$ Eqs. (12–5) and (12–6) must yield the initial values of the current and its derivative. Hence

$$i(0^+) = A_1 = 0,$$

$$\frac{di(0^+)}{dt} = A_2\omega_0 = \frac{E_0}{L},$$

from which

$$A_1 = 0, \qquad A_2 = \frac{E_0}{\omega_0 L}.$$

Having evaluated the constants in Eq. (12–5), we obtain

$$i(t) = \frac{E_0}{\omega_0 L} \sin \omega_0 t \qquad \text{for} \quad t > 0. \tag{12–7}$$

A sketch of the response function of Eq. (12–7) is given in Fig. 12–2. Note that the oscillation begins at zero. The function can be reduced to zero for negative values of t by the use of the unit step function. Thus an alternative form of Eq. (12–7) is

$$i(t) = \frac{E_0}{\omega_0 L} \sin \omega_0 t \cdot u_{-1}(t). \tag{12–8}$$

* Since all the matching of initial conditions is done at $t = 0^+$, the function $u_{-1}(t)$ can be dropped.

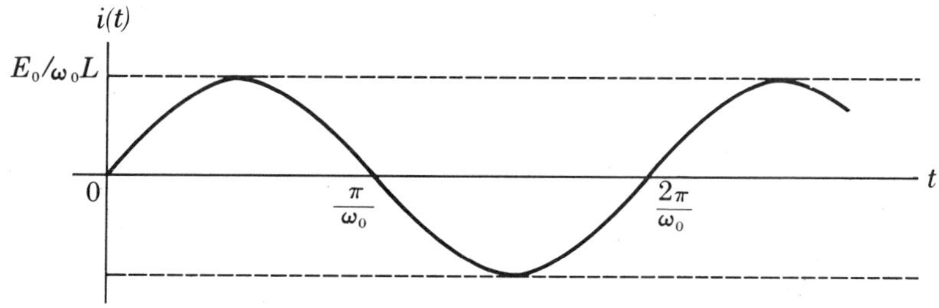

FIG. 12–2. A sketch of the response function $i(t) = (E_0/\omega_0 L) \sin \omega_0 t \cdot u_{-1}(t)$.

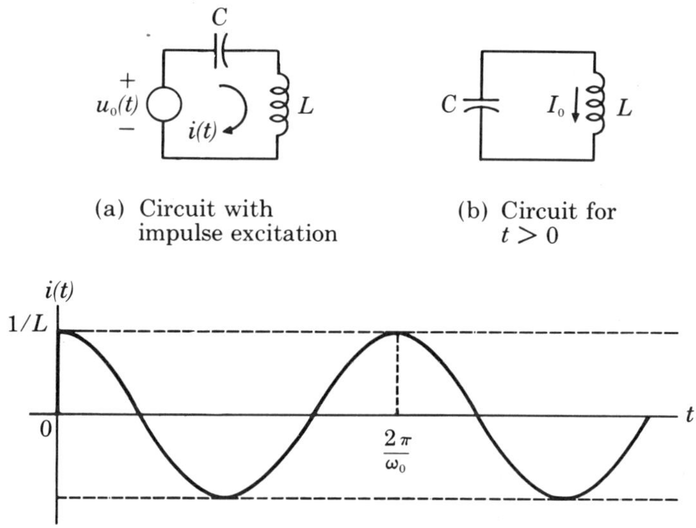

(c) The current response function

FIG. 12–3. A series L–C circuit excited by a voltage impulse.

The maximum amplitude of the oscillation is $E_0/\omega_0 L$ and its period is $2\pi/\omega_0$. If the energy in the circuit is originally stored in the inductance, a cosine term instead of a sine term appears in Eq. (12–8). If both types of energy are initially present, both sine and cosine terms appear in the natural response function.

Another L–C circuit, in which the natural response is excited by an impulse voltage, is shown in Fig. 12–3(a). Again, we shall assume the form of the solution and evaluate the arbitrary constants from the initial conditions in the problem. At $t = 0$, when the impulse strikes, the capacitance is a short circuit, and the inductance is an open circuit. The impulse voltage appears across the inductance. Although the effective resistance is infinite the effective voltage is also infinite. The current produced is indeterminate, and must be evaluated from the integral which gives the current in the inductance:

$$i_L(0^+) = \frac{1}{L}\int_{0^-}^{0^+} e\,dt = \frac{1}{L}\int_{0^-}^{0^+} u_0(t)\,dt = \frac{1}{L}\cdot \qquad (12\text{–}9)$$

The area under the impulse curve from $t = 0^-$ to $t = 0^+$ is unity, as we have had occasion to observe before. By $t = 0^+$ the impulse is over, and the derivative of the current is determined by the energy stored in the circuit by the impulse. The circuit which holds for $t > 0$ is shown in Fig. 12–3(b). At $t = 0^+$ in this circuit the capacitance is a short circuit, and the voltage across it is zero. The inductance is in parallel with the capacitance, and the voltage across it must also be zero. Hence

$$\frac{di(0^+)}{dt} = \frac{e_L}{L} = 0.$$

We will assume a general form for the current and its first derivative, as given by Eqs. (12–5) and (12–6). At $t = 0^+$ the values of the equations must match those which we have computed from the circuit. Therefore

$$i(0^+) = A_1 = 1/L,$$

$$\frac{i(0^+)}{dt} = A_2\omega_0 = 0,$$

from which

$$A_1 = 1/L, \qquad A_2 = 0.$$

The current response function for the circuit of Fig. 12–3(a) is therefore

$$i(t) = \frac{1}{L}\cos\omega_0 t \cdot u_{-1}(t). \qquad (12\text{–}10)$$

This function is plotted in Fig. 11–3(c). It starts at $t = 0$ with a maximum value of $1/L$ and oscillates sinusoidally with a period of $2\pi/\omega_0$.

A parallel L–C circuit excited by an impulse current is the dual of the series L–C circuit we have just discussed. Thus the solution is the same as Eq. (12–10) with i changed to e and L changed to C.

12–3 The step response of an L–C circuit. The step response of an L–C circuit consists of the transient response plus a constant or steady-state value. A general expression for the step response is therefore

$$f(t) = A_1\cos\omega_0 t + A_2\sin\omega_0 t + B \qquad \text{for} \quad t > 0, \qquad (12\text{–}11)$$

where $\omega_0 = 1/LC$,
A_1, A_2, and B are constants.

The first two terms in Eq. (12–11) constitute the transient response, and the third term is the steady-state value. A minor problem arises here, because the transient response does not die out with time. The original oscillations in the circuit, produced when the source is switched in, continue undiminished in amplitude and thus obscure the steady-state value. To avoid this difficulty it is convenient to imagine a very small amount of dissipation in the circuit, which causes the transient response to damp out as time approaches infinity. Then, the steady-state value B is the value of the variable at $t = \infty$. The dissipation added is vanishingly small and does not have any effect on B.

The approach to the solution is similar to that we used for the first-order system. We know the general type of behavior which the circuit produces and can evaluate the arbitrary constants in this general solution by computing the initial and final values of the variable from the resistive circuits at $t = 0$ and $t = \infty$. The principal difference is that in the second-order system the initial value of both the function and its first derivative must be obtained. Solving the circuit of Fig. 12–4 will illustrate the method.

At $t = 0$, when the step voltage is applied, the capacitance is a short circuit, and the inductance is an open circuit. The initial current is therefore

$$i(0^+) = 0.$$

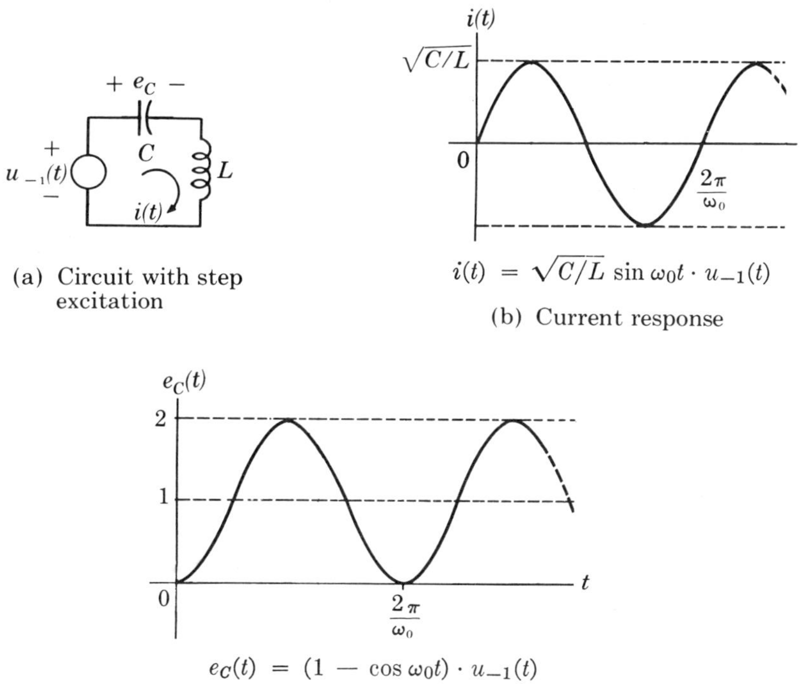

(a) Circuit with step excitation

(b) Current response

(c) Capacitive voltage response

FIG. 12–4. The step response of a series L–C circuit.

The applied voltage appears across the inductance. Hence the rate of change of current is

$$\frac{di(0^+)}{dt} = \frac{e_L}{L} = \frac{1}{L}.$$

At $t = \infty$ the capacitance is an open circuit, and the inductance is a short circuit. If we imagine enough dissipation to damp out the oscillations, the current at $t = \infty$ is

$$i(\infty) = 0.$$

The solution for the current response is assumed to be of the form given by Eq. (12–11):

$$i(t) = A_1 \cos \omega_0 t + A_2 \sin \omega_0 t + B \qquad \text{for} \quad t > 0, \tag{12–12}$$

from which

$$\frac{di}{dt} = -A_1\omega_0 \sin \omega_0 t + A_2\omega_0 \cos \omega_0 t.$$

The three circuit conditions which we have computed are sufficient to determine the three constants A_1, A_2, and B. Since we assumed that the transients have died out

$$i(\infty) = 0 = B.$$

At $t = 0^+$

$$i(0^+) = A_1 = 0,$$

$$\frac{di(0^+)}{dt} = A_2\omega_0 = 1/L,$$

from which

$$A_1 = 0, \qquad A_2 = 1/L\omega_0.$$

The current-response function of the circuit in Fig. 12–4(a) is therefore

$$i(t) = \frac{1}{L\omega_0} \sin \omega_0 t \qquad \text{for} \quad t > 0$$

or

$$i(t) = \sqrt{\frac{C}{L}} \sin \omega_0 t \cdot u_{-1}(t). \tag{12–13}$$

The response function of Eq. (12–13) is sketched in Fig. 12–4(b). It is a sine wave starting at $t = 0$ and having a maximum amplitude of $\sqrt{C/L}$ and a period of $2\pi/\omega_0$.

The voltage across the capacitance in Fig. 12–4(a) can be obtained by integrating Eq. (12–13) from $t = 0$ to $t = t$. It can also be obtained by the same sort of procedure we have used to obtain the current response.

The general form of the function is known, and we can evaluate the constants from the initial and final values. At $t = 0$ the capacitance is a short circuit, and the voltage across it is zero. The current through the capacitance is also zero, and thus the derivative of the voltage is zero:

$$e_C(0^+) = 0, \qquad \frac{de_C(0^+)}{dt} = 0.$$

Since both the function and its derivative are zero at $t = 0$, it might appear that the response is zero. Such is not the case, however, because the response includes the steady-state term B. At $t = \infty$ the capacitance is an open circuit, and the inductance is a short circuit. Since the applied voltage appears across the capacitance

$$e_C(\infty) = 1 = B. \tag{12–14}$$

The general expression for the voltage across the capacitance is

$$e_C(t) = A_1 \cos \omega_0 t + A_2 \sin \omega_0 t + B, \tag{12–15}$$

from which

$$\frac{de_C}{dt} = -A_1\omega_0 \sin \omega_0 t + A_2\omega_0 \cos \omega_0 t.$$

The value of B is given by Eq. (12–14). The two initial conditions, applied to Eq. (12–15), yield

$$e_C(0^+) = A_1 + B = 0,$$

$$\frac{de_C(0^+)}{dt} = A_2\omega_0 = 0,$$

from which

$$A_1 = -B = -1, \qquad A_2 = 0.$$

The complete voltage-response function is

$$e_C(t) = (1 - \cos \omega_0 t) \qquad \text{for} \quad t > 0$$

or

$$e_C(t) = (1 - \cos \omega_0 t)u_{-1}(t). \tag{12–16}$$

The response function of Eq. (12–16) is sketched in Fig. 12–4(c). It is a negative cosine wave with its horizontal axis displaced upwards by unity. The wave starts at zero and reaches a peak value of 2. The period of the cosine oscillation is $2\pi/\omega_0$.

A parallel L–C circuit excited by a unit step of current is the dual of the series L–C circuit we have just solved. The expression for the voltage across the parallel L–C combination is the dual of Eq. (12–13); the expression for the current in the inductance is the dual of Eq. (12–16).

Essentially the L–C response functions which we have obtained in this section are particular cases of the more general R–L–C response functions which we shall next investigate. All real physical circuits have resistance, and the oscillations in them eventually die out. A circuit in which there are a large number of oscillations within one time constant of the exponential envelope curve can be considered to be lossless for the first few cycles of the oscillation. A quantity called the Q of the circuit is a measure of this property. By definition

$$Q = \frac{\omega_0}{2\alpha}. \tag{12–17}$$

The larger the Q, the more nearly a system is lossless, and the more nearly the undamped oscillations approximate the true natural behavior. We shall meet the quantity Q in a number of other applications. It is an important natural parameter of a system.

12–4 The impulse response of the series *R–L–C* circuit. The step from the two-element circuits we have investigated in the preceding sections of this chapter to the three-element circuit involving resistance, inductance, and capacitance leads to increased complexity. There are two reasons for the added complexity. The first is that the nature of the response depends on the relative values of the three elements. It is natural that it should. The R–L–C circuit must be able to represent an R–L circuit (with its damped exponential type of response) to which a small amount of C has been added. Since the amount of C added can be vanishingly small the R–L–C response must be able to approximate very closely a damped exponential. Alternatively, the R–L–C circuit can be viewed as an L–C circuit (with its pure sine-wave response) to which a small amount of resistance has been added. Again, since the resistance added can be very small, the R–L–C response must be able to represent an almost undamped sine wave.

The second reason for the added complexity is that the nature of the response depends on the geometry of the circuit. For the two-element circuit the natural response function is the same with the elements connected in series or in parallel. For the R–L–C circuit it makes a difference whether the elements are connected in series or in parallel. The same differential equation applies to both circuits, but the values of resistance, inductance, and capacitance enter the equation differently in the two cases. In this section we shall consider the series R–L–C circuit. The basic differential equation for the circuit can be written in the standard form:

$$\frac{d^2y}{dt^2} + 2\alpha\frac{dy}{dt} + \omega_0^2 y = 0. \tag{12–18}$$

The parameters in the equation are

$$\alpha = R/2L, \qquad \omega_0^2 = 1/LC. \tag{12–19}$$

The nature of the response function depends on the relative values of α and ω_0. In Chapter 9 we obtained an overdamped response, a critically damped response, and an underdamped response having the forms:

Overdamped response ($\alpha > \omega_0$)

$$y = A_1 e^{p_1 t} + A_2 e^{p_2 t} \qquad \text{for} \quad t > 0, \tag{12–20}$$

where $p_{1,2} = -\alpha \pm \sqrt{\alpha^2 - \omega_0^2}$.

Critically damped response ($\alpha = \omega_0$)

$$y = e^{-\alpha t}(A_1 + A_2 t) \qquad \text{for} \quad t > 0. \tag{12–21}$$

Underdamped response ($\alpha < \omega_0$)

$$\begin{aligned} y &= e^{-\alpha t}(A_1 \cos \omega_d t + A_2 \sin \omega_d t) \\ &= A e^{-\alpha t} \cos (\omega_d t + \theta) \qquad \text{for} \quad t > 0, \end{aligned} \tag{12–22}$$

where $\omega_d = \sqrt{\omega_0^2 - \alpha^2}$.

In this section we shall assume that these general forms of the response functions are known and that the problem is to find the arbitrary constants A_1, A_2, A, and θ from the initial conditions. Since there are two constants to determine in each case, two initial conditions are required. Usually the initial values of the variable and its first derivative are used.

A typical student reaction to the response functions of a second-order system is that they are too complicated to learn by memory and that it is easier to solve the differential equation for each separate problem. Such is not the case. The response functions are relatively simple. They involve only exponentials and sine waves. Furthermore, they are closely tied to the physical behavior of the circuits and can be recalled by physical reasoning. The fundamental reason for learning the response functions of the second-order system is to be able to resolve the responses of higher-order systems into summations of the responses of first-order and second-order systems. It will not be necessary to learn response functions more complicated than those of the second order, but it is necessary to learn them!

For our first example, consider the series R–L–C circuit shown in Fig. 12–5. This circuit is excited by energy initially stored in the capacitance. In order to evaluate the constants in the natural behavior we must determine the initial value of the current and of its first derivative. At

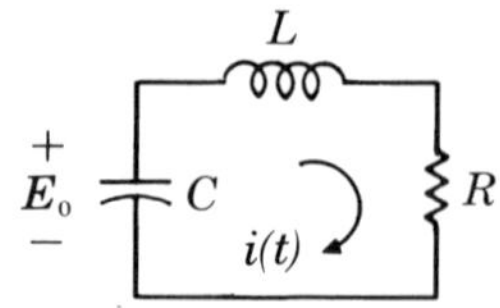

FIG. 12–5. A series R–L–C circuit excited by initial energy stored on the capacitance.

$t = 0$ the inductance is an open circuit, and the charged capacitance is equivalent to a voltage source in series with a short circuit. The current is

$$i(0^+) = 0.$$

Since the voltage across the resistance is zero the voltage across the inductance must be equal to the voltage across the capacitance. The rate of change of current is therefore

$$\frac{di}{dt} = \frac{e_L}{L} = \frac{E_0}{L}.$$

Having the initial conditions we now proceed to evaluate the constants in the three possible response functions.

Overdamped response ($\alpha > \omega_0$). If the circuit is overdamped the response function has the form

$$i(t) = A_1 e^{p_1 t} + A_2 e^{p_2 t}, \tag{12–23}$$

from which

$$\frac{di}{dt} = A_1 p_1 e^{p_1 t} + A_2 p_2 e^{p_2 t}.$$

At $t = 0$

$$i(0^+) = A_1 + A_2 = 0,$$

$$\frac{di(0^+)}{dt} = A_1 p_1 + A_2 p_2 = E_0/L,$$

from which

$$A_1 = \frac{-E_0}{L(p_2 - p_1)}, \qquad A_2 = -A_1.$$

The overdamped response function is therefore

$$i(t) = \frac{-E_0}{L(p_2 - p_1)} (e^{p_1 t} - e^{p_2 t}) \qquad \text{for} \quad t > 0. \tag{12–24}$$

Both p_1 and p_2 in Eq. (12–24) are negative, and the magnitude of p_2 is greater than the magnitude of p_1. The difference $p_2 - p_1$ is therefore negative and the multiplying coefficient in the equation is positive. The

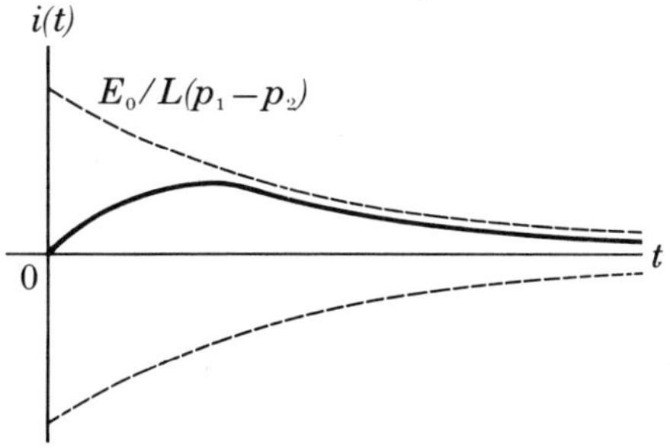

$$i(t) = \frac{E_0}{L(p_1 - p_2)}[e^{p_1 t} - e^{p_2 t}]$$

(a) Overdamped response

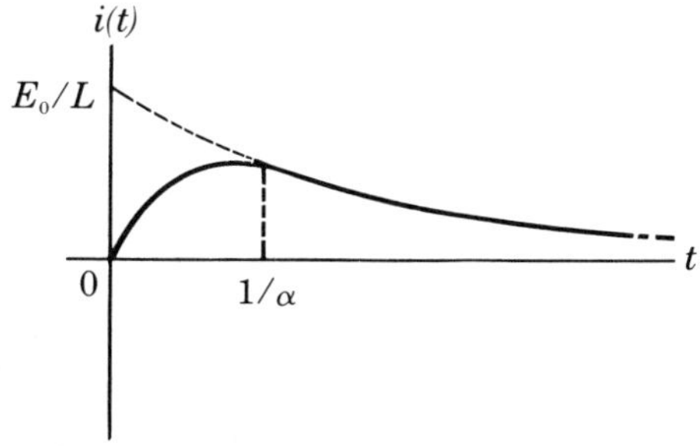

$$i(t) = \frac{E_0}{L}te^{-\alpha t}$$

(b) Critically damped response

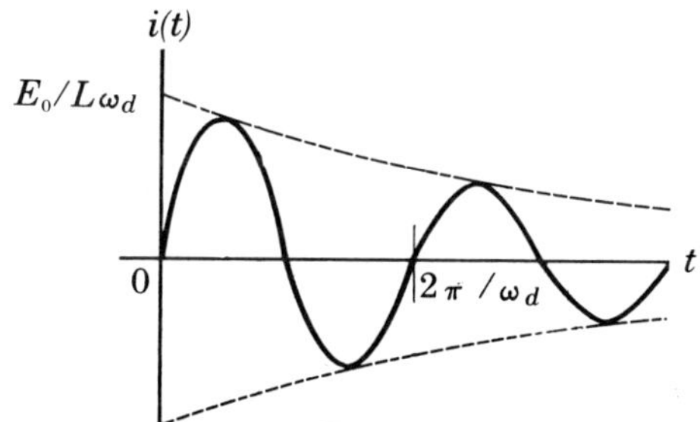

$$i(t) = \frac{E_0}{L\omega_d}e^{-\alpha t}\sin\omega_d t$$

(c) Underdamped response

FIG. 12–6. The natural response functions for the circuit of Fig. 12–5.

time constants of the two exponential terms are the reciprocals of the magnitudes of p_1 and p_2. Thus the time constant of the first term is larger than the time constant of the second term. The over-all response starts at zero, as shown in Fig. 12–6(a). The initial rise time of the curve is determined by the value of p_2. The later values of the curve are determined primarily by the value of p_1.

Critically damped response ($\alpha = \omega_0$). If the circuit is critically damped the response function has the form

$$i(t) = e^{-\alpha t}(A_1 + A_2 t),$$

from which

$$\frac{di}{dt} = e^{-\alpha t}A_2 - \alpha e^{-\alpha t}(A_1 + A_2 t).$$

At $t = 0$

$$i(0^+) = A_1 = 0,$$

$$\frac{di(0^+)}{dt} = A_2 - \alpha A_1 = E_0/L,$$

from which

$$A_1 = 0, \qquad A_2 = E_0/L.$$

The critically damped response function is therefore

$$i(t) = \frac{E_0}{L} te^{-\alpha t} \qquad \text{for} \quad t > 0. \tag{12–25}$$

At $t = 0$ the value of the current in Eq. (12–25) is zero, and at $t = \infty$ it is again zero. The maximum value of the current is $E_0 e^{-1}/L\alpha$, and it occurs at $t = 1/\alpha$, which is one time constant on the exponential envelope curve. A sketch of the response function is shown in Fig. 12–6(b). The general form of the curve is similar to the overdamped case, and a smooth transition occurs between them. Approximately critically damped response functions are often sought in control applications where it is desired to have the system respond as rapidly as possible, but where overshoots and oscillations about the final value are not tolerable.

Underdamped response ($\alpha < \omega_0$). If the circuit is underdamped the response function has the form

$$i(t) = e^{-\alpha t}(A_1 \cos \omega_d t + A_2 \sin \omega_d t),$$

from which

$$\frac{di}{dt} = e^{-\alpha t}(-A_1\omega_d \sin \omega_d t + A_2\omega_d \cos \omega_d t) - \alpha e^{-\alpha t}(A_1 \cos \omega_d t + A_2 \sin \omega_d t).$$

At $t = 0$

$$i(0^+) = A_1 = 0,$$

$$\frac{di(0^+)}{dt} = A_2\omega_d - A_1\alpha = E_0/L,$$

from which

$$A_1 = 0, \qquad A_2 = E_0/L\omega_d.$$

The underdamped response function is therefore

$$i(t) = \frac{E_0}{L\omega_d} e^{-\alpha t} \sin \omega_d t \qquad \text{for} \quad t > 0. \tag{12–26}$$

The function of Eq. 12–26 is sketched in Fig. 12–6(c). The exponential envelope of the oscillations has a maximum value of $E_0/L\omega_d$ at $t = 0$, and it decays to zero with a time constant of $1/\alpha$. The oscillations start at zero (when $t = 0$) and have a constant period of $2\pi/\omega_d$.*

* The discovery that damped oscillations have a constant period was first made by Galileo, by watching the chandeliers in church!

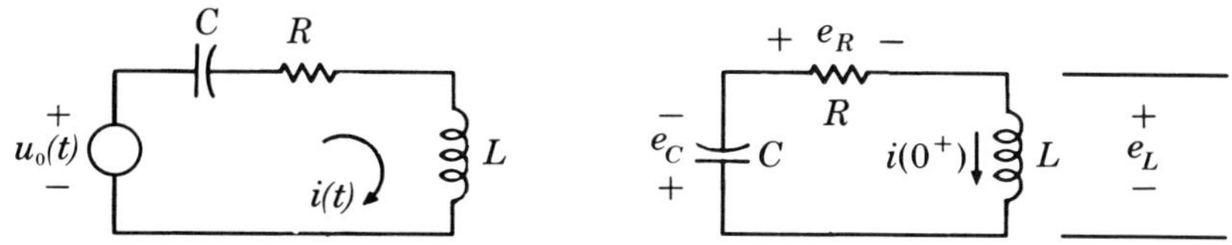

(a) Circuit with impulse voltage (b) Circuit for $t > 0$

FIG. 12–7. An R–L–C circuit excited by an impulse voltage.

The solution for any one set of parameters is fairly simple. The complexity occurs mainly because of the possibility of overdamped, underdamped, and critically damped types of response function.

An R–L–C circuit excited by an impulse voltage is shown in Fig. 12–7(a). The current-response functions in this circuit for $t > 0$ are essentially the same as the functions obtained for the circuit of Fig. 12–5. The principal difference lies in the nature of the initial conditions. For impulse excitation both the current and its first derivative have values at $t = 0^+$. At $t = 0^-$ the circuit is at rest. At $t = 0$, when the impulse strikes, the inductance is an open circuit and the capacitance is a short circuit. All the impulse voltage appears across the inductance. A finite voltage does occur across the resistance, but this voltage is negligible in comparison with the infinite voltage of the impulse. Between $t = 0^-$ and $t = 0^+$ the impulse voltage creates a current in the inductance whose value is

$$i(0^+) = \frac{1}{L}\int_{0^-}^{0^+} u_0(t)\,dt = \frac{1}{L}\cdot \tag{12–27}$$

The derivative of the current at $t = 0^+$ must be computed from the circuit which holds for $t > 0$, after the impulse is over. This circuit is shown in Fig. 12–7(b). To begin, the voltage on the capacitance is zero, because as yet no charge has accumulated. The voltage across the resistance is

$$e_R = R \times \frac{1}{L}\cdot$$

By Kirchhoff's voltage law the voltage across the inductance is the negative of the voltage across the resistance:

$$e_L = -\frac{R}{L}\cdot$$

The derivative of the current is found from the voltage across the inductance:

$$\frac{di(0^+)}{dt} = \frac{e_L(0^+)}{L} = -\frac{R}{L^2}\cdot \tag{12–28}$$

The initial conditions which we have evaluated in Eqs. (12–27) and (12–28) can now be combined with the known form of the natural behavior of the series R–L–C circuit to give the impulse-response function.

The general response functions which are obtained are similar to those of Eqs. (12–24), (12–25), and (12–26). The numerical values of the constants are different, but the procedures for obtaining them are the same as those we have been using. We will not repeat the computations here, for the results are summarized at the end of the chapter. Actually it is the method which is important and not the results. The student should make a practice of adapting the general solution to a particular problem in numerical form, rather than substituting in a particular equation in terms of literal coefficients. At the cost of a very small increase in arithmetic the student can thus make himself independent of long lists of formulas.

12–5 The impulse response of the parallel *R–L–C* circuit. The parallel R–L–C circuit also satisfies a second-order differential equation of the form

$$\frac{d^2y}{dt^2} + 2\alpha\frac{dy}{dt} + \omega_0^2 y = 0.$$

The value of ω_0 is the same as that for the series R–L–C circuit, but the value of α is different:

$$\omega_0 = 1/\sqrt{LC}, \qquad \alpha = 1/2RC. \tag{12–29}$$

The general forms of the solutions for the underdamped, critically damped, and overdamped responses given by Eqs. (12–20), (12–21), and (12–22) apply to the parallel circuit as well as to the series circuit. Actually the parallel circuit driven by a current impulse is the dual of the series circuit driven by a voltage impulse, and the solutions are also duals. Usually, instead of using duality, it is preferable to solve the problem directly by matching the initial conditions to the known form of the transient behavior. Because of the three possible types of behavior, the actual solution in any given problem is best carried out numerically. As an example, for the circuit of Fig. 12–8 we shall determine the initial conditions, and then shall select values for the parameters and solve for the response function.

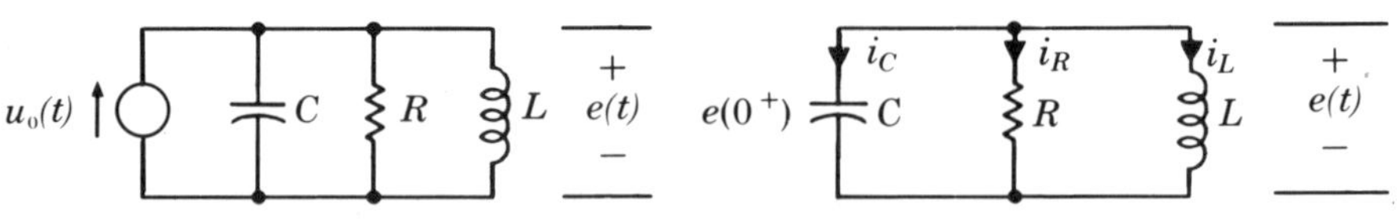

(a) Circuit excited by impulse (b) Circuit for $t > 0$

FIG. 12–8. An R–L–C circuit excited by an impulse current.

At $t = 0^-$ in this circuit everything is at rest. At $t = 0$ the impulse strikes, with the capacitance acting as a short circuit and the inductance acting as an open circuit. All the impulse current enters the capacitance. A finite current flows in the resistance, but since the capacitive current during the impulse is infinite it is not sensibly diminished by the current in the resistance. The voltage acquired by the capacitance, while the impulse current is flowing from $t = 0^-$ to $t = 0^+$, is

$$e_C(0^+) = \frac{1}{C}\int_{0^-}^{0^+} u_0(t)\,dt = \frac{1}{C}\cdot \tag{12-30}$$

The derivative of the voltage at $t = 0^+$ must be obtained from the circuit shown in Fig. 12-8(b), which holds after the impulse is over. No current is flowing in the inductance at $t = 0^+$, but a current

$$i_R(0^+) = \frac{1}{RC}$$

is flowing in the resistance. By Kirchhoff's current law this same current must be flowing out of the capacitance. Hence

$$i_C(0^+) = -1/RC.$$

The rate of change of the voltage depends on the current from the capacitance and is given by

$$\frac{de(0^+)}{dt} = \frac{i_C}{C} = -1/RC^2. \tag{12-31}$$

Suppose that $L = C = R = 1$. Then

$$\omega_0 = \frac{1}{\sqrt{LC}} = 1, \qquad \alpha = \frac{1}{2RC} = \tfrac{1}{2}.$$

Since $\alpha < \omega_0$ the response is of the underdamped type. The damped natural frequency is

$$\omega_d = \sqrt{\omega_0^2 - \alpha^2} = \sqrt{1 - \tfrac{1}{4}} = \frac{\sqrt{3}}{2} = 0.866.$$

For variety we will assume that the solution fits the cosine form of the underdamped response function, in which the unknown constants are the amplitude and the phase angle of the sinusoid. Thus

$$e(t) = Ae^{-\alpha t}\cos(\omega_d t + \theta) \qquad \text{for} \quad t > 0, \tag{12-32}$$

from which

$$\frac{de}{dt} = -A\omega_d e^{-\alpha t} \sin (\omega_d t + \theta) - \alpha A e^{-\alpha t} \cos (\omega_d t + \theta).$$

The initial conditions, from Eqs. (12–30) and (12–31), are

$$\begin{aligned} e(0^+) &= \frac{1}{C} = 1 \text{ volt}, \\ \frac{de(0^+)}{dt} &= -\frac{1}{RC^2} = -1 \text{ volt/sec}. \end{aligned} \tag{12–33}$$

Equation (12–32) and its derivative must reduce to the initial values given by Eqs. (12–33) when $t = 0$. Hence

$$\begin{aligned} e(0^+) &= A \cos \theta = 1, \\ \frac{de(0^+)}{dt} &= -A\omega_d \sin \theta - \alpha A \cos \theta = -1. \end{aligned} \tag{12–34}$$

Equations (12–34) are transcendental. They can be solved by isolating the sine and cosine. When the values of α and ω_d are inserted the equations become

$$A \cos \theta = 1,$$

$$A(0.866) \sin \theta + \frac{A}{2} \cos \theta = 1,$$

from which

$$\begin{aligned} A \cos \theta &= 1, \\ A \sin \theta &= \frac{1}{1.732}. \end{aligned} \tag{12–35}$$

The ratio of these equations gives an expression for the angle θ:

$$\frac{\sin \theta}{\cos \theta} = \tan \theta = \frac{1}{1.732}.$$

The magnitude of the angle is 30°. If A is taken to be positive $\sin \theta$ is positive and $\cos \theta$ is positive. The angle is therefore in the first quadrant:

$$\theta = 30°.$$

The value of A can now be obtained:*

$$A = \frac{1}{\cos \theta} = 1.15.$$

---

* A can also be obtained by squaring and adding Eqs. (12–35).

The final expression for the impulse response of the circuit of Fig. 12–8(a), with all the parameters equal to unity, is

$$e(t) = 1.15e^{-\alpha t} \cos (0.866t + 30°) \qquad \text{for} \quad t > 0. \tag{12–36}$$

The advantage of the form of the impulse response of Eq. (12–36) is that it is easier to plot than the sum of a sine term and cosine term. The procedure for obtaining this form is fairly straightforward, but it does involve a good deal of trigonometry instead of the algebra required by the alternative form. Either form is perfectly correct, and there are trigonometric formulas for changing from one form to the other. We will not dwell on this point now, since we shall study the methods of vector algebra in a later chapter; trigonometric manipulations are easily performed by means of vector algebra. At this point we are concerned with the basic concepts, namely, that there is a response function characteristic of a given network and that the arbitrary constants in this response function can be determined from the initial energy storages in the network, through the initial values of the variable.

12–6 The step response of the series *R–L–C* circuit. The step response of the series R–L–C circuit differs little in complexity from the impulse response or the natural behavior. A steady-state value can occur in the circuit, and this appears as a constant added to the solution. The R–L–C transient decays with time, and therefore the constant can be obtained as the value of the variable at infinity.* Depending on the values of the parameters the response function can be overdamped, critically damped, or underdamped and will have one of the following forms:

Overdamped response ($\alpha > \omega_0$)

$$f(t) = A_1 e^{p_1 t} + A_2 e^{p_2 t} + B \qquad \text{for} \quad t > 0, \tag{12–37}$$

where

$$p_{1,2} = -\alpha \pm \sqrt{\alpha^2 - \omega_0^2},$$
$$\alpha = R/2L,$$
$$\omega_0 = 1/\sqrt{LC}.$$

Critically damped response ($\alpha = \omega_0$)

$$f(t) = e^{-\alpha t}(A_1 + A_2 t) + B \qquad \text{for} \quad t > 0. \tag{12–38}$$

* The lossless case required us to insert some dissipation in the circuit to make the transient die out, but here it does so naturally.

Underdamped response ($\alpha < \omega_0$)

$$f(t) = e^{-\alpha t}(A_1 \cos \omega_d t + A_2 \sin \omega_d t) + B \qquad \text{for} \quad t > 0, \tag{12–39}$$

or

$$f(t) = Ae^{-\alpha t} \cos (\omega_d t + \theta) + B \qquad \text{for} \quad t > 0, \tag{12–40}$$

where

$$\omega_d = \sqrt{\omega_0^2 - \alpha^2}.$$

In Eqs. (12–37), (12–38), (12–39), and (12–40) the constant B represents the value of the variable at $t = \infty$, and the constants A_1 and A_2 are determined by the initial conditions of the problem.

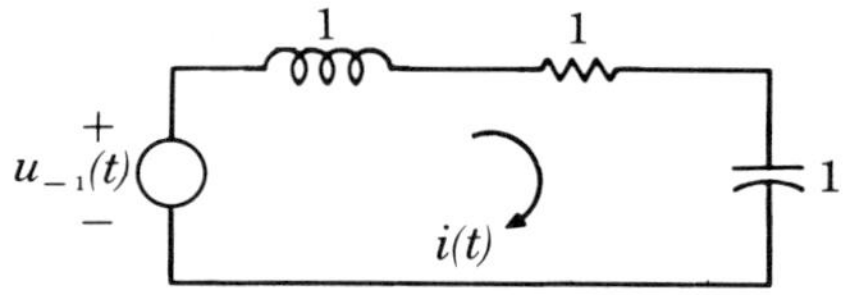

FIG. 12–9. The step response of a series R–L–C circuit.

As an example to illustrate the method of solution we will obtain the step response of the series R–L–C circuit in Fig. 12–9, in which $R = L = C = 1$ and the applied voltage is a unit step. For this circuit

$$\alpha = \frac{R}{2L} = \tfrac{1}{2},$$

$$\omega_0 = \frac{1}{\sqrt{LC}} = 1,$$

$$\omega_d = \sqrt{\omega_0^2 - \alpha^2} = 0.866.$$

Since $\alpha < \omega_0$ the response function is underdamped, and we must use Eq. (12–39) or Eq. (12–40). We shall use Eq. (12–39). The initial conditions in the problem can be obtained from the circuit. At $t = 0$, when the step is applied, the inductance is an open circuit and the capacitance is a short circuit. The current is therefore

$$i(0^+) = 0. \tag{12–41}$$

The applied voltage appears across the inductance and determines the rate of change of the current:

$$\frac{di(0^+)}{dt} = \frac{e_L}{L} = \frac{1}{L} \text{ amp/sec.} \tag{12–42}$$

The response function has the form

$$i(t) = e^{-\alpha t}(A_1 \cos \omega_d t + A_2 \sin \omega_d t) + B, \tag{12–43}$$

from which

$$\frac{di}{dt} = e^{-\alpha t}(-A_1\omega_d \sin \omega_d t + A_2\omega_d \cos \omega_d t) - \alpha e^{-\alpha t}(A_1 \cos \omega_d t + A_2 \sin \omega_d t).$$

When we match the initial conditions of Eqs. (12–41) and (12–42) with the initial values of Eq. (12–43) and its derivative we obtain

$$i(0^+) = A_1 + B = 0,$$
$$\frac{di(0^+)}{dt} = A_2\omega_d - A_1\alpha = \frac{1}{L}. \tag{12–44}$$

The value of B can be obtained from the value of the current in the circuit at $t = \infty$. Since the capacitance is an open circuit the current is zero, and thus

$$B = i(\infty) = 0.$$

From Eq. (12–44) we obtain

$$A_1 = 0, \qquad A_2 = \frac{1}{\omega_d L} = 1.15.$$

The current response to the applied step voltage is therefore

$$i(t) = 1.15e^{-t/2} \sin 0.866t \qquad \text{for} \quad t > 0. \tag{12–45}$$

The function of Eq. (12–45) is a damped sine wave. The time constant of the envelope is 2 seconds and the period of the wave is 7.22 seconds. Very few cycles are produced before the wave becomes vanishingly small. If we had used the alternative form for the response function, given by Eq. (12–40), we would have obtained a cosine wave with a phase angle of $-90°$ and the same envelope function. The two expressions are, of course, equivalent. In the example considered here, the steady-state response happened to be zero. We shall consider an example in the next section in which the steady-state response has a constant non-zero value.

12–7 The step response of a parallel *R–L–C* circuit. The response of a parallel R–L–C circuit to a step-current source is the same as the response of a series R–L–C circuit to a step-voltage source. It is better, however, to solve the parallel circuit problem in detail, using duality only to provide a check for the answer. The parallel R–L–C circuit response functions to a step-current source are similar to the impulse response functions, except for the addition of a constant to represent the steady-state value at $t = \infty$. The possible forms of the functions are given by Eqs. (12–37), (12–38), (12–39), and (12–40) where

$$p_{1,2} = -\alpha \pm \sqrt{\alpha^2 - \omega_0^2}, \tag{12–46}$$

$$\alpha = \frac{1}{2RC}, \tag{12–47}$$

$$\omega_0 = \frac{1}{\sqrt{LC}}, \tag{12–48}$$

$$\omega_d = \sqrt{\omega_0^2 - \alpha^2}. \tag{12–49}$$

As an example of the application of these equations we shall find the current through the inductance in the circuit of Fig. 12–10, which is excited by a unit step current. For the given values of the parameters in the circuit

$$\alpha = \frac{1}{2RC} = \frac{1}{2},$$

$$\omega_0 = \frac{1}{\sqrt{LC}} = 1,$$

$$\omega_d = \sqrt{\omega_0^2 - \alpha^2} = 0.866.$$

Since $\alpha < \omega_0$ the response is underdamped. Therefore

$$i(t) = e^{-\alpha t}(A_1 \cos \omega_d t + A_2 \sin \omega_d t) + B, \tag{12–50}$$

from which

$$\frac{di}{dt} = e^{-\alpha t}(-A_1\omega_d \sin \omega_d t + A_2\omega_d \cos \omega_d t) - \alpha e^{-\alpha t}(A_1 \cos \omega_d t + A_2 \sin \omega_d t).$$

The value of B can be obtained from the circuit response at $t = \infty$. At this time the capacitance is an open circuit, and the inductance is a short circuit. All the current from the source must pass through the inductance.

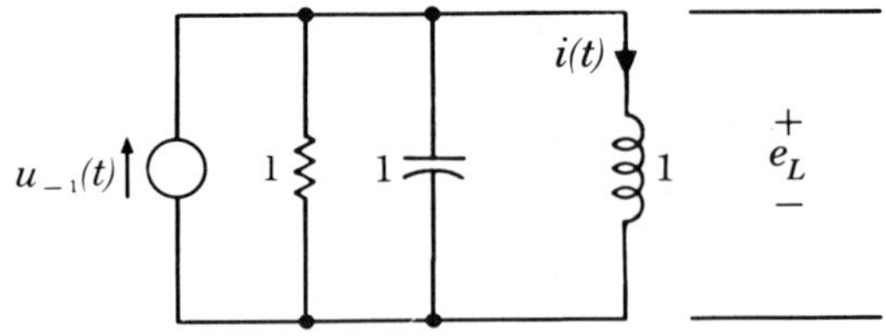

FIG. 12–10. The step response of a parallel R–L–C circuit.

It follows that

$$B = i(\infty) = 1 \text{ amp.} \tag{12–51}$$

The constants A_1 and A_2 are obtained from the initial conditions. At $t = 0$ the capacitance is a short circuit and the inductance is an open circuit. All the current flows into the capacitance, and none flows into the inductance. Hence

$$i(0^+) = 0. \tag{12–52}$$

Furthermore, at $t = 0$ the voltage across the capacitance is zero, since there is as yet no charge in it, and therefore the voltage across the inductance is also zero. The rate of change of the current is obtained from the voltage across the inductance:

$$\frac{di(0^+)}{dt} = \frac{e_L}{L} = 0. \tag{12–53}$$

Because Eq. (12–50) and its derivative must satisfy the boundary conditions at $t = 0$ we find

$$i(0^+) = A_1 + B = 0,$$

$$\frac{di(0^+)}{dt} = A_2\omega_d - A_1\alpha = 0,$$

from which

$$A_1 = -B = -1, \qquad A_2 = -\alpha/\omega_d.$$

The complete expression for the current response in the inductance is

$$i(t) = e^{-t/2}\left(-\cos\omega_d t - \frac{\alpha}{\omega_d}\sin\omega_d t\right) + 1 \qquad \text{for} \quad t > 0$$

$$= 1 - e^{-t/2}\,(\cos 0.866t + 0.577\sin 0.866t). \tag{12–54}$$

In Eq. (12–54) the first term is the steady-state value, and the second term is the transient. The transient is a damped sine wave with an envelope-time-constant of 2 seconds and an oscillatory period of 7.22 seconds.

A great many other response functions are possible in a parallel R–L–C network, depending on the values of R, L, and C. Most of the general types are given in the summary at the end of the chapter. The method of obtaining a response is more important than the results, for it can be applied to any set of numerical data and will yield the correct transient and steady-state response functions. Actually, it is not even the numerical answers which are of interest to us, but the concepts which lie behind them. We have considered at some length the steady-state and the transient response functions. We have observed that the nature of the transient terms is a characteristic of a given type of network. Only the relative values of the various terms are controlled by the particular way in which a network is excited. The nature of its transient response is thus a very basic description of a network. We will investigate this point in more detail in the next section.

12–8 The transient p-plane. Before proceeding with new material we shall summarize briefly what we have learned about the natural behavior of electric circuits. A circuit with a single energy-storage element results in a first-order differential equation, and its natural response function is a damped exponential. A general expression for the damped exponential is

$$f(t) = Ae^{pt}, \tag{12–55}$$

where p is a negative real number.

A system containing two independent energy-storage elements, for example an inductance and a capacitance, is described by a second-order differential equation. The natural response function requires two exponential terms of the sort given by Eq. (12–55) and has the general form

$$f(t) = A_1e^{p_1t} + A_2e^{p_2t}. \tag{12–56}$$

In Eq. (12–56) the actual nature of the response depends on the p-values. If they are both negative real numbers the response is the sum of two damped exponentials; if they are pure imaginary numbers the response is an undamped sinusoid; and if they are complex numbers the response is a damped sinusoid.

Systems with n independent energy-storage elements are described by a differential equation of order n, and the general solution is*

$$f(t) = A_1e^{p_1t} + A_2e^{p_2t} + \cdots + A_ne^{p_nt}. \tag{12–57}$$

Each separate term in Eq. (12–57) is determined by the value of p. If p is a negative real number the response is a damped exponential; if p is a pure imaginary the response of a pair of the terms is an undamped sinusoid; and if p is a complex number two terms can be combined to give a damped sinusoidal response. The p-values determine the nature of the response functions. The arbitrary multiplying constants associated with each term are determined by the values of the variable and its first $n - 1$ derivatives at $t = 0^+$. *The arbitrary constants are characteristic of the excitation and not of the network. It is the p-values which are characteristic of the network and which in fact completely describe it so far as its stimulus-response relationship is concerned.*

The importance of the p-values cannot be overestimated. They represent the natural transient behavior of the network. They tell us how the

---

* See the Appendix, Transients in More Complex Circuits.

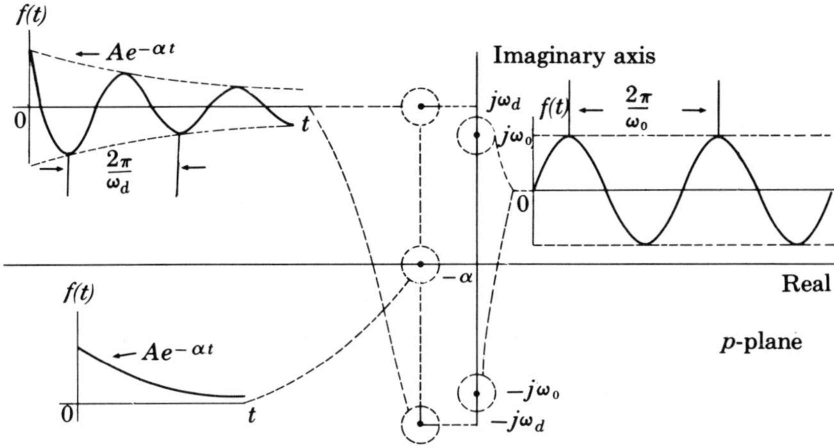

FIG. 12–11. The complex-frequency plane and the various types of transient responses.

network would "like" to behave. Usually a network is driven by some forcing function. The response obtained depends on how closely the forcing function agrees with the natural transient response. Networks, like people, respond with enthusiasm when the forcing function asks them, so to speak, to do that which comes naturally. Indeed, the response of a system to a given stimulus is a direct function of the closeness of the stimulus to the natural behavior of the system.

The dimensions of p must be the reciprocal of time or frequency, in order that the power of the exponential be a dimensionless number. Because of their dimensions the p-values are called the transient natural frequencies of the network. In general the p-values are complex and are therefore complex natural resonant frequencies. It is convenient to record the p-values as points in a complex plane. They must exist as negative real values or as complex conjugate pairs with negative real parts for circuits with resistance. Several representative p-values are shown in Fig. 12–11, along with the sort of transient responses which they represent. A p-value or natural frequency of $-\alpha$ represents a damped exponential with a time constant of $1/\alpha$. A pair of values on the imaginary axis at $\pm j\omega_0$ represents an undamped sine wave of angular frequency ω_0 and period $2\pi/\omega_0$. A pair of complex values at $-\alpha \pm j\omega_d$ represents a damped sinusoid with an envelope-time-constant of $1/\alpha$, an angular frequency of ω_d, and a period of $2\pi/\omega_d$.

For the moment, the complex frequency plane in which we have plotted values of natural frequencies is to be viewed as simply a convenient place to record essential information about the transient behavior of a circuit. After we have studied the sinusoidal steady-state behavior of networks we will return to the complex frequency plane in order to investigate the important phenomenon of resonance. The transient natural frequencies will again return to prominence and take a central role in our treatment of Fourier and Laplace transforms.

12–9 Summary.

Natural behavior of L–C circuit

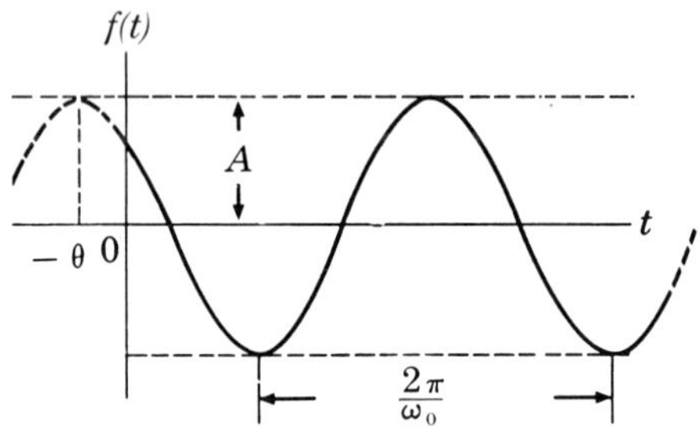

$$\begin{aligned} f(t) &= A \cos(\omega_0 t + \theta) \\ &= A_1 \cos \omega_0 t + A_2 \sin \omega_0 t \end{aligned}$$

$$\begin{aligned} \omega_0 &= 1/\sqrt{LC} \\ A_1 &= A \cos \theta \\ A_2 &= -A \sin \theta \end{aligned}$$

Impulse response of L–C circuit

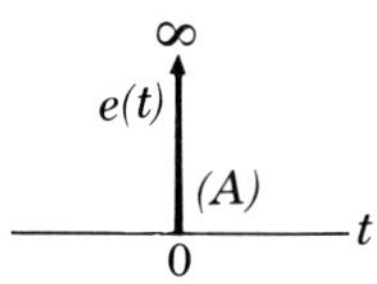

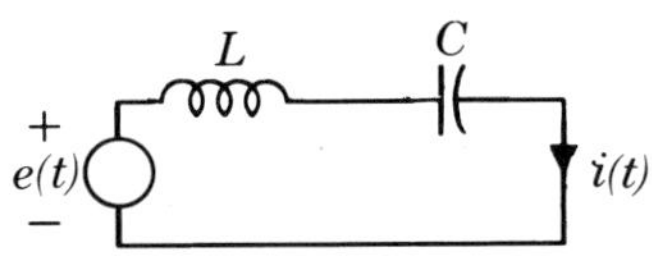

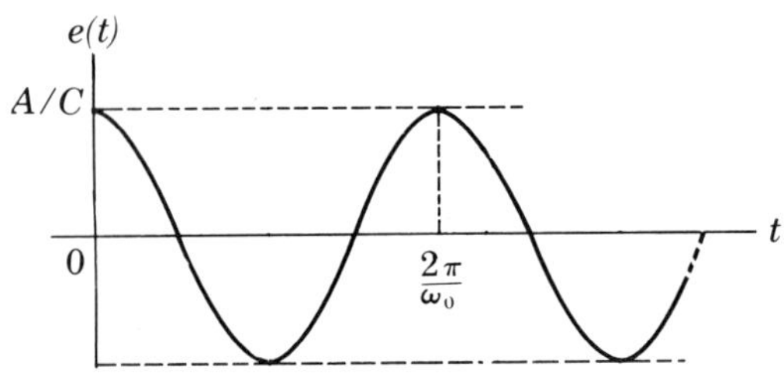

$$i(t) = \frac{A}{L} \cos \omega_0 t$$

$$\text{for} \quad t > 0$$

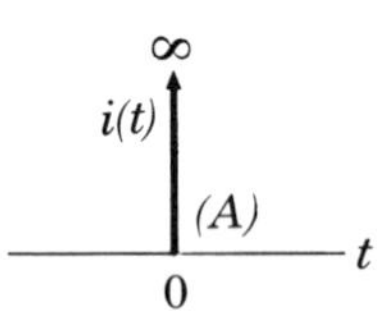

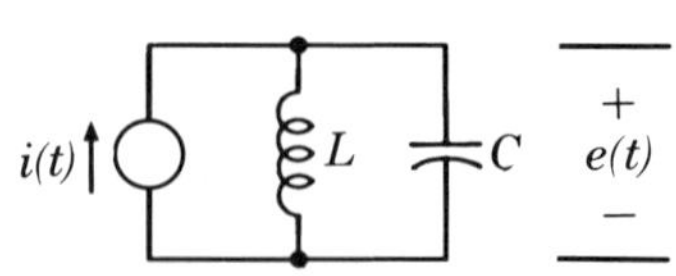

(Response waveform shown on page 409.)

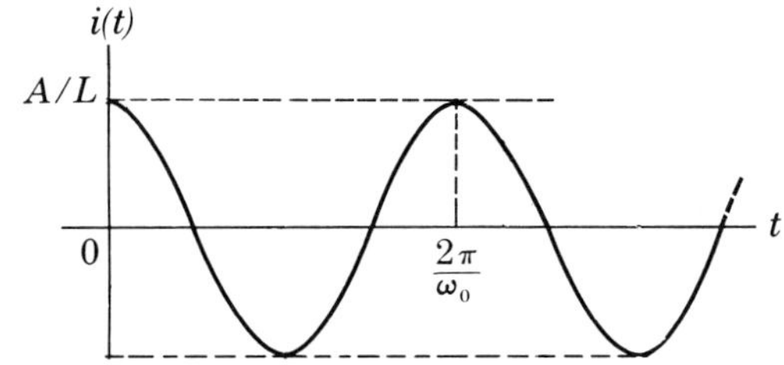

$$e(t) = \frac{A}{C}\cos \omega_0 t$$

$$\text{for} \quad t > 0$$

Step response of L–C circuit

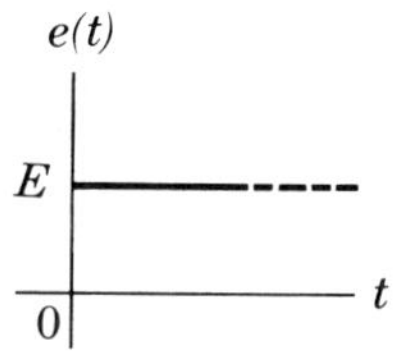

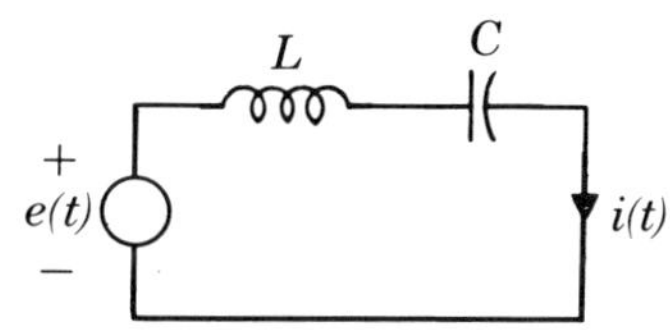

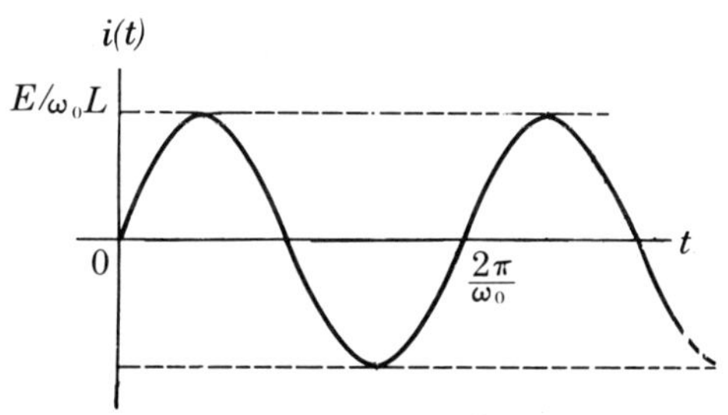

$$i(t) = \frac{E}{\omega_0 L}\sin \omega_0 t$$

$$\text{for} \quad t > 0$$

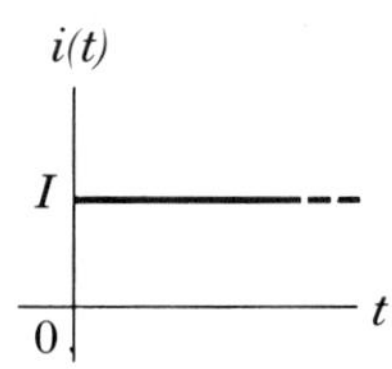

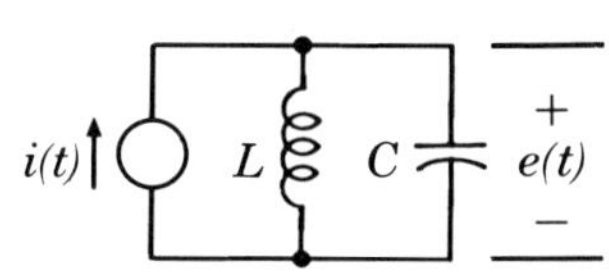

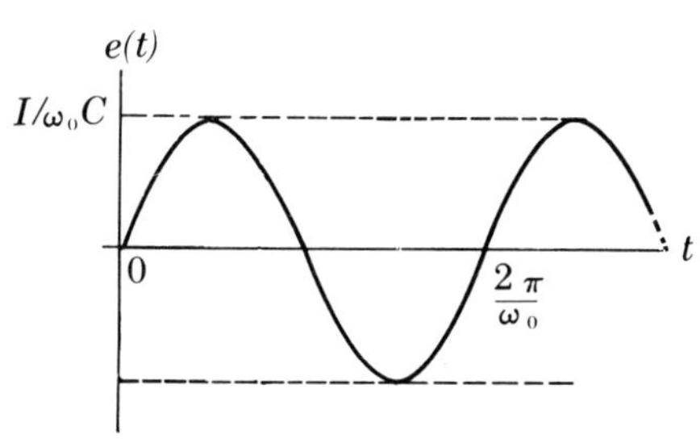

$$e(t) = \frac{I}{\omega_0 C}\sin \omega_0 t$$

$$\text{for} \quad t > 0$$

Natural behavior of R–L–C circuit

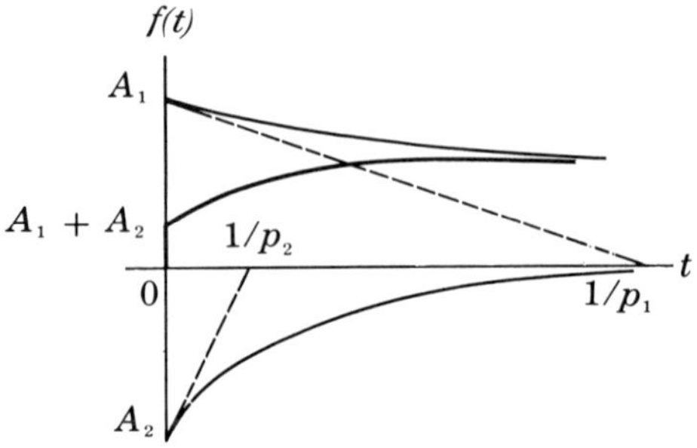

(a) Overdamped

$$f(t) = A_1 e^{-p_1 t} + A_2 e^{-p_2 t} \quad \text{for} \quad t > 0$$

$$p_1 = -\alpha + \sqrt{\alpha^2 - \omega_0^2}$$

$$p_2 = -\alpha - \sqrt{\alpha^2 - \omega_0^2}$$

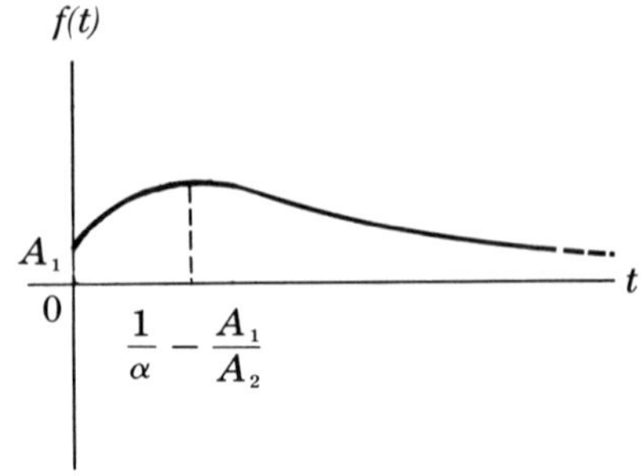

(b) Critically damped

$$f(t) = e^{-\alpha t}(A_1 + A_2 t) \quad \text{for} \quad t > 0$$

$$\omega_0 = 1/\sqrt{LC}$$

$$\alpha = \begin{cases} R/2L \text{ for series circuit} \\ 1/2RC \text{ for parallel circuit} \end{cases}$$

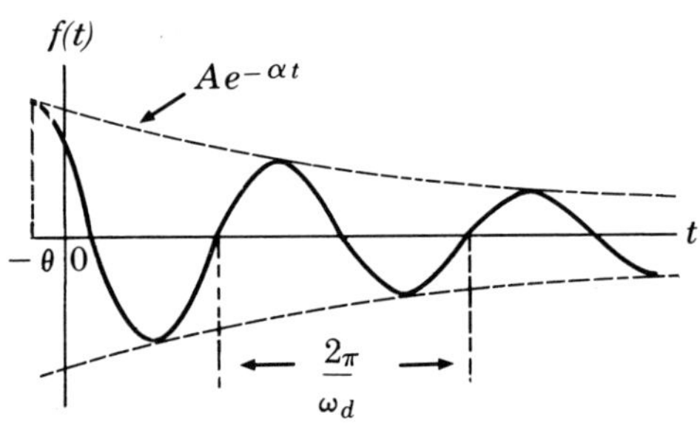

(c) Underdamped

$$f(t) = Ae^{-\alpha t} \cos(\omega_d t + \theta)$$
$$= e^{-\alpha t}(A_1 \cos \omega_d t + A_2 \sin \omega_d t) \quad \text{for} \quad t > 0$$

$$A_1 = A \cos \theta$$

$$A_2 = -A \sin \theta$$

$$\omega_d = \sqrt{\omega_0^2 - \alpha^2}$$

Impulse response of series R–L–C circuit ($\alpha = R/2L$)

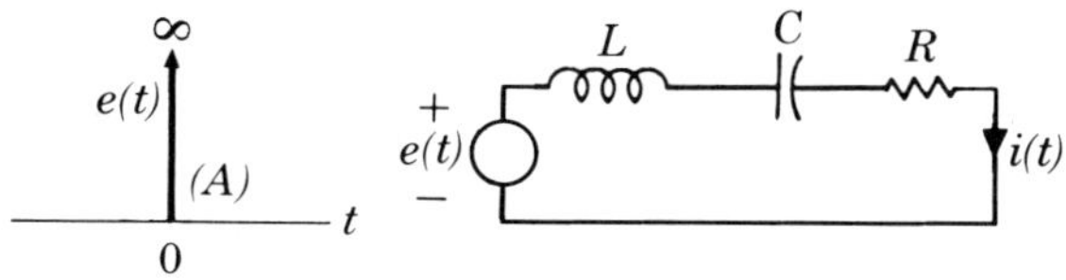

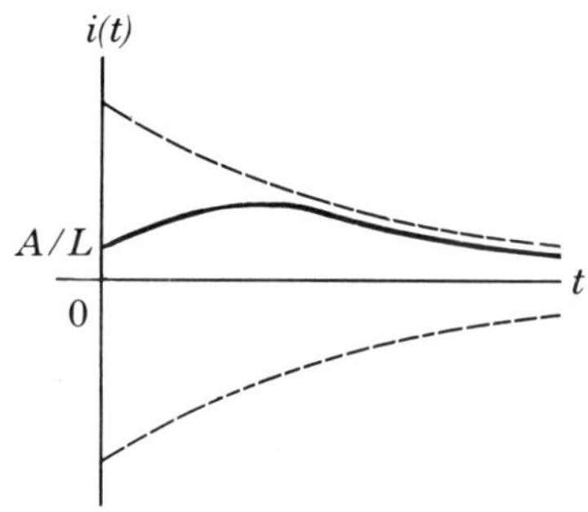

(a) Overdamped

$$i(t) = \frac{A}{2L\sqrt{\alpha^2 - \omega_0^2}}\,(p_1 e^{p_1 t} - p_2 e^{p_2 t}) \quad \text{for} \quad t > 0$$

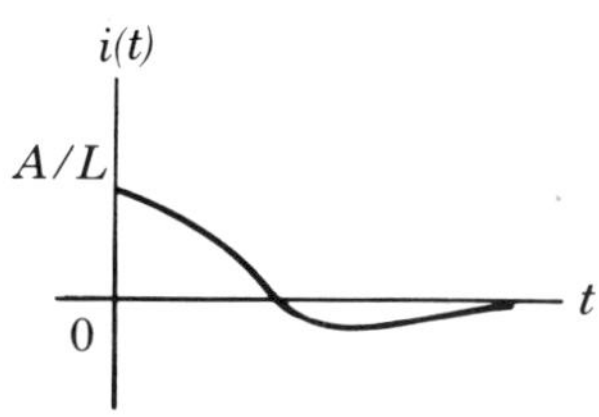

(b) Critically damped

$$i(t) = \frac{A}{L}\,e^{-\alpha t}(1 - \alpha t) \quad \text{for} \quad t > 0$$

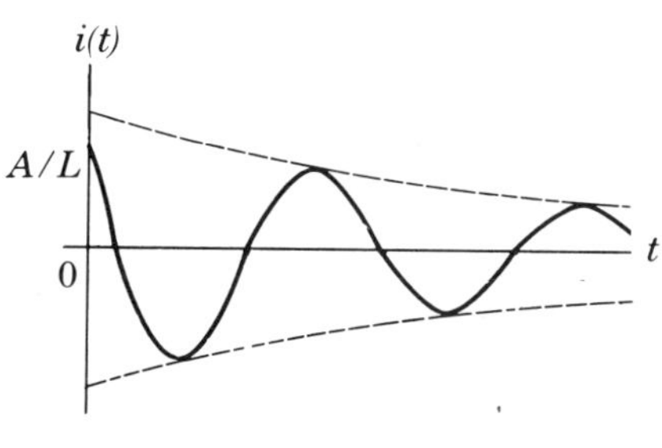

(c) Underdamped

$$i(t) = \frac{A\omega_0}{L\omega_d}\,e^{-\alpha t}\cos(\omega_d t + \theta) \quad \text{for} \quad t > 0$$

$$\theta = \cos^{-1}\frac{\omega_d}{\omega_0}$$

Impulse response of parallel R–L–C circuit ($\alpha = 1/2RC$)

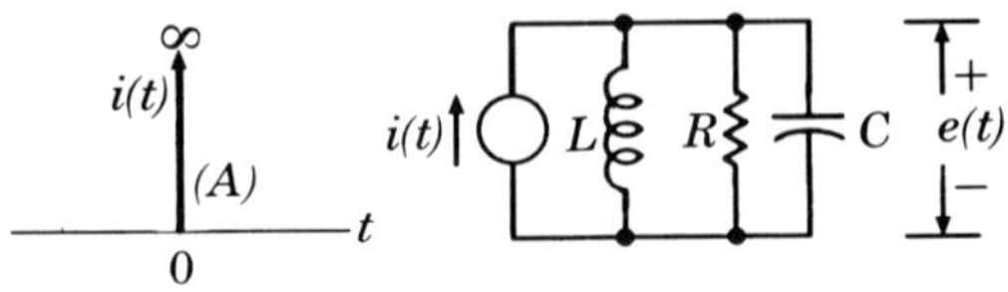

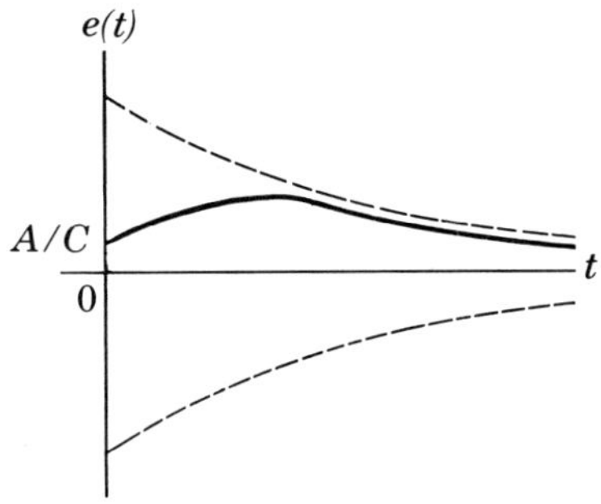

(a) Overdamped

$$e(t) = \frac{A}{2C\sqrt{\alpha^2 - \omega_0^2}}(p_1 e^{p_1 t} - p_2 e^{p_2 t}) \quad \text{for} \quad t > 0$$

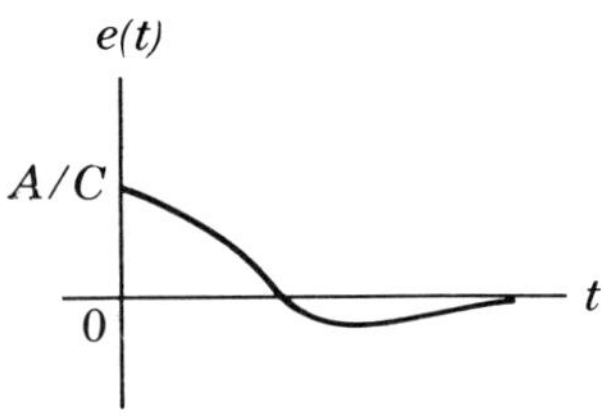

(b) Critically damped

$$e(t) = \frac{A}{C}e^{-\alpha t}(1 - \alpha t) \quad \text{for} \quad t > 0$$

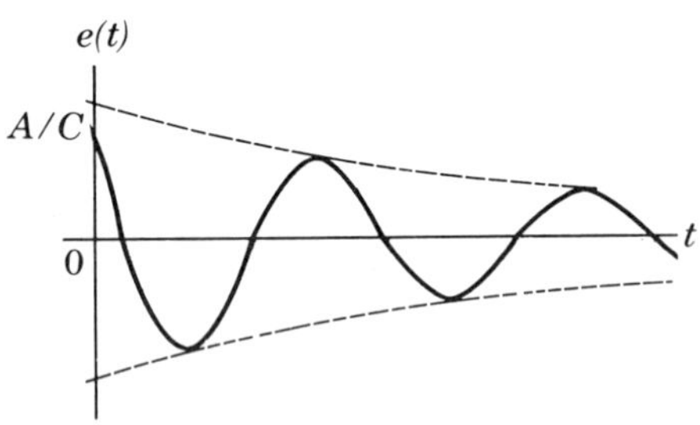

(c) Underdamped

$$e(t) = \frac{A\omega_0}{C\omega_d}e^{-\alpha t}\cos(\omega_d t + \theta) \quad \text{for} \quad t > 0$$

$$\theta = \cos^{-1}\frac{\omega_d}{\omega_0}$$

Step response of series R–L–C circuit ($\alpha = R/2L$)

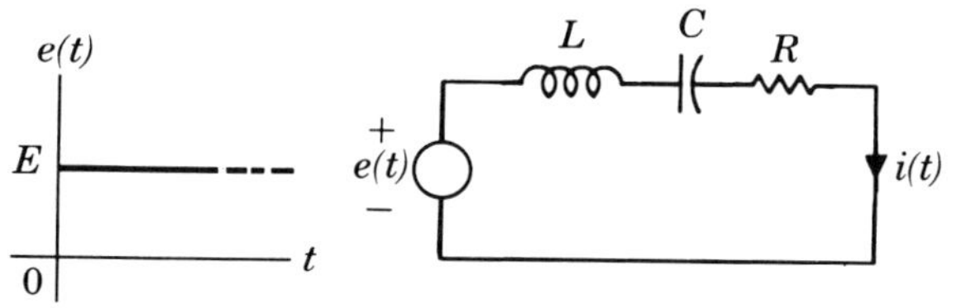

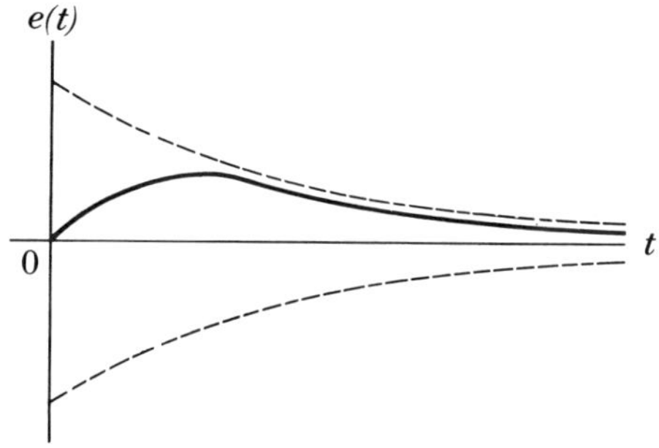

(a) Overdamped

$$i(t) = \frac{E}{2L\sqrt{\alpha^2 - \omega_0^2}}\,(e^{p_1 t} - e^{p_2 t}) \quad \text{for} \quad t > 0$$

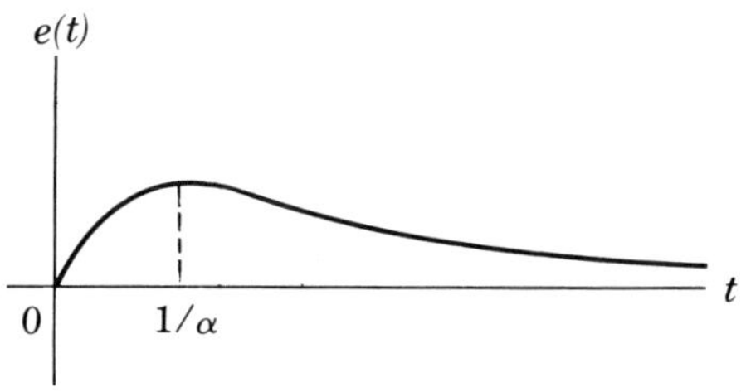

(b) Critically damped

$$i(t) = \frac{E}{L}\,te^{-\alpha t} \qquad \text{for} \quad t > 0$$

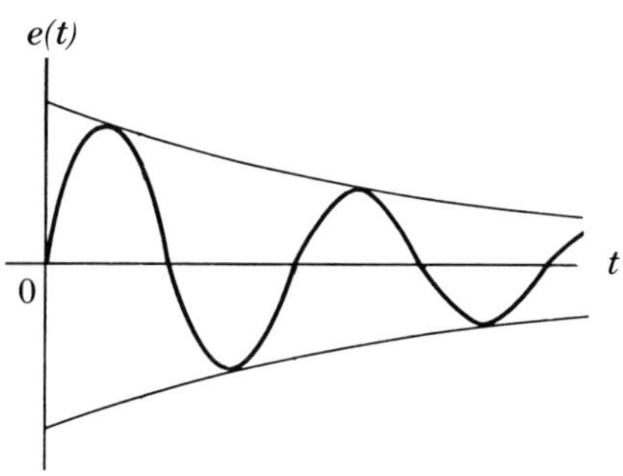

(c) Underdamped

$$i(t) = \frac{E}{L\omega_d}\,e^{-\alpha t}\sin\omega_d t \quad \text{for} \quad t > 0$$

Step response of parallel R–L–C circuit ($\alpha = 1/2RC$)

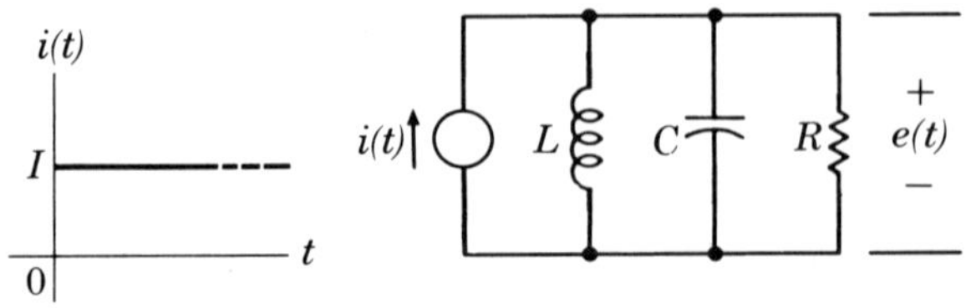

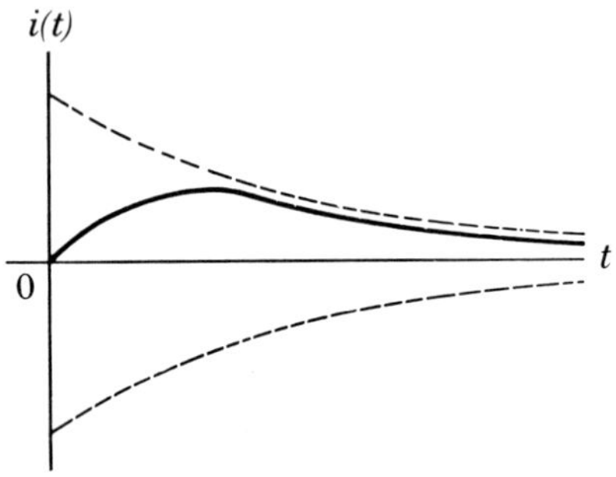

(a) Overdamped

$$e(t) = \frac{I}{2C\sqrt{\alpha^2 - \omega_0^2}}\,(e^{p_1 t} - e^{p_2 t}) \qquad \text{for} \quad t > 0$$

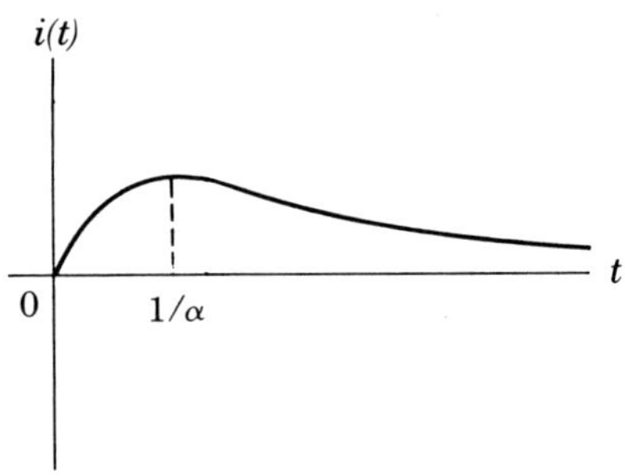

(b) Critically damped

$$e(t) = \frac{I}{C}\,te^{-\alpha t} \qquad \text{for} \quad t > 0$$

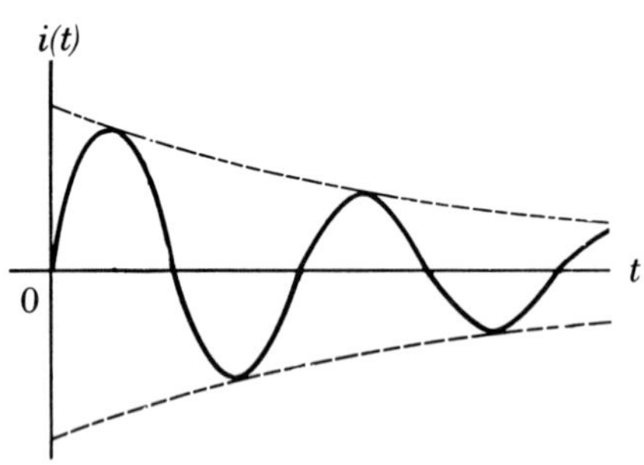

(c) Underdamped

$$e(t) = \frac{I}{C\omega_d}\,e^{-\alpha t}\sin\omega_d t \qquad \text{for} \quad t > 0$$

The complex-frequency plane

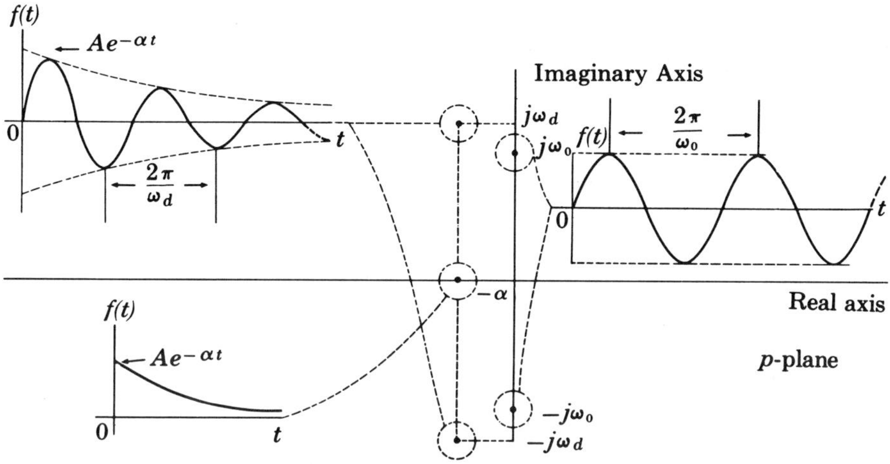

REFERENCES

Chapter 12 continues the presentation of transients in terms of initial values, final values, and natural frequencies. The method of writing the transient expression by inspection is implied in other texts but not explicitly discussed. Alternative points of view will be found in the references listed here.

1. E. A. Guillemin, *Introductory Circuit Theory*, John Wiley & Sons, New York, 1953. Guillemin develops the *L–C* and *R–L–C* transients principally to expose the types of natural behavior which a circuit can exhibit. Both step and impulse responses are treated.

2. MIT EE Staff, *Electric Circuits*, John Wiley & Sons, New York, 1943. This text presents *R–L–C* circuits from the point of view of differential equations. Only the step response is considered.

3. Brenner and Javid, *Analysis of Electric Circuits*, McGraw-Hill, New York, 1959. Brenner and Javid separate the source-free behavior and the step and impulse responses.

4. D. K. Cheng, *Analysis of Linear Systems*, Addison-Welsey, Reading, Mass., 1959. The Laplace transform approach of this book can be justified for two-energy storages. The student should appreciate the physical nature of the natural modes, however, before he becomes enmeshed in the mechanics of the Laplace transform.

5. H. H. Skilling, *Electrical Engineering Circuits*, John Wiley & Sons, New York, 1957. Skilling uses the *R–L–C* circuit to discuss the natural modes of a circuit and to introduce the Laplace transform.

6. Lago and Waidelich, *Transients in Electrical Circuits*, The Ronald Press, New York, 1958. Chapter 3 discusses the basic two-energy-storage circuit.

EXERCISES*

Section 12–2

1. A sinusoidal current is given by $i(t) = 10 \cos (2\pi t + 45°)$.
   (a) Sketch the waveform.
   (b) Give the maximum amplitude.
   (c) Give the frequency.
   (d) Give the period.
   (e) Give the angular frequency ω.
   (f) Give the phase relative to a cosine function.

2. A sinusoidal voltage is given by

$$e(t) = 10^{-6} \cos (2\pi \times 10^6 t) + 10^{-6} \sin (2\pi \times 10^6 t).$$

Sketch the waveform and give its maximum amplitude, its frequency, and its phase relative to a cosine.

3. A series circuit is composed of $L = \frac{1}{4}$ henry, and $C = \frac{1}{16}$ farad. At $t = 0$ there is a current of 1 amp in the inductance and no voltage on the capacitance.
   (a) Find $i(0^+)$, $di(0^+)/dt$, and ω_0.
   (b) Match the values in part (a) to the known form of the transient, $i(t) = A_1 \cos \omega t + A_2 \sin \omega t$.

4. A series circuit with $L = \frac{1}{4}$ henry and $C = \frac{1}{16}$ farad is excited by an initial voltage of 1 volt on the capacitance. There is no current in the inductance at $t = 0$. Find $i(0^+)$, $di/(0^+)/dt$, ω_0, and a solution for the current in the circuit.

5. In the circuit of Fig. 12–12 the current is 1 amp and the voltage on the capacitance is 1 volt at $t = 0$. Find $i(t)$.

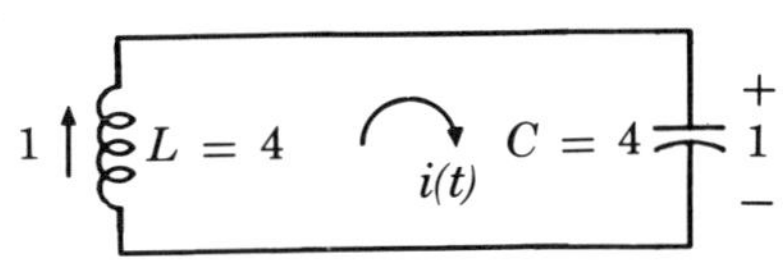

FIGURE 12–12

6. An impulse voltage source of 10 volt·sec is applied in series with an inductance of $\frac{1}{4}$ henry and a capacitance of $\frac{1}{4}$ farad. There is no initial energy stored in the circuit. Find $i(0^+)$, $di(0^+)/dt$, ω_0, and the expression for $i(t)$.

7. An impulse current source of 10 amp·sec is applied in parallel with an inductance of $\frac{1}{4}$ henry and a capacitance of $\frac{1}{4}$ farad. There is no initial energy stored in the circuit. Find $e(0^+)$, $de(0^+)/dt$, and ω_0, and from these values obtain $e(t)$.

8. In the circuit of Fig. 12–13 the switch is closed at $t = 0$, when the current in L_1 is 10 amp and the voltage across C is zero. Find the voltage on C as a function of time from the instant the switch is closed. Also find the current through the switch.

---

* Numerical values are in volts, amperes, ohms, henrys, and farads.

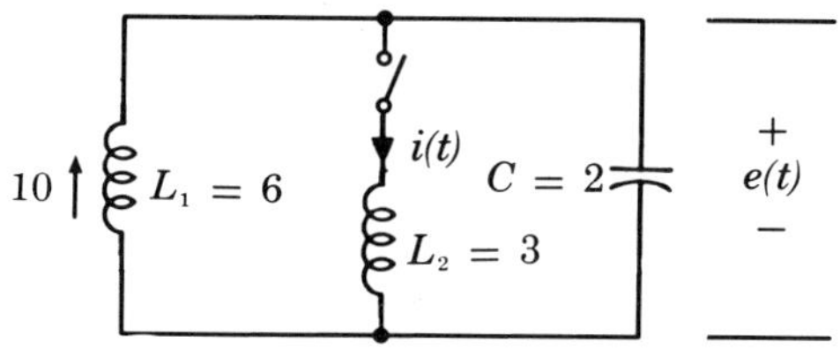

FIGURE 12–13

9. Find the current $i(t)$ in the circuit of Fig. 12–14 for $t > 0$.

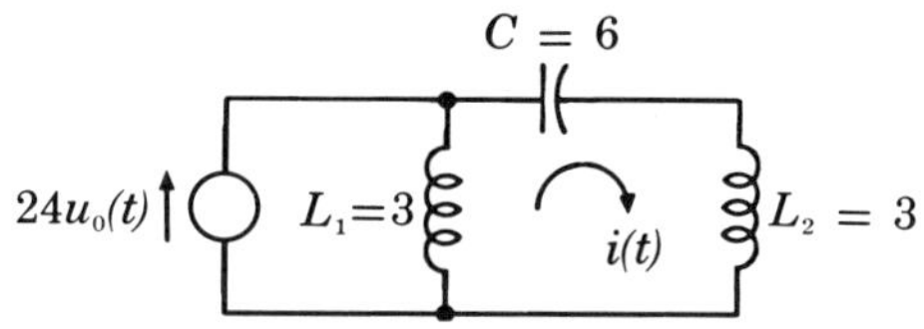

FIGURE 12–14

Section 12–3

10. A 10-volt battery is switched in series with a 3-henry inductance and a 12-farad capacitance at $t = 0$. There is no energy in the circuit at $t = 0$.

(a) Find $i(0^+)$, $di(0^+)/dt$, and ω_0.

(b) Find $i(\infty)$ (assuming a small resistance).

(c) Find $i(t)$.

11. A 10-volt battery is switched in series with a 3-henry inductance and a 12-farad capacitance at $t = 0$. There is no current in the circuit at this instant, but there is a voltage on the capacitance of 5 volts, with a polarity which opposes the applied voltage around the mesh. Find $i(0^+)$, $di(0^+)/dt$, $i(\infty)$, and ω_0. From these values obtain $i(t)$.

12. A voltage source of zero resistance is connected in series with a 3-henry inductance and a 12-farad capacitance. The voltage of the source is zero until $t = 0$, at which time it rises from 0 to 10 volts. At this moment, because of a previous switching operation, there is a current of 5 amp in the circuit, which leaves the positive terminal of the voltage source, and the voltage on the capacitance is zero. Find $i(0^+)$, $di(0^+)/dt$, $i(\infty)$, and ω_0. From these values obtain $i(t)$.

13. The switch in the circuit of Fig. 12–15 is closed at $t = 0$. At this time there is no energy stored in the circuit. (a) Find the voltage $e(t)$ by finding the current and integrating. (b) Find the voltage from $e(0^+)$, $de(0^+)/dt$, $e(\infty)$, and ω_0.

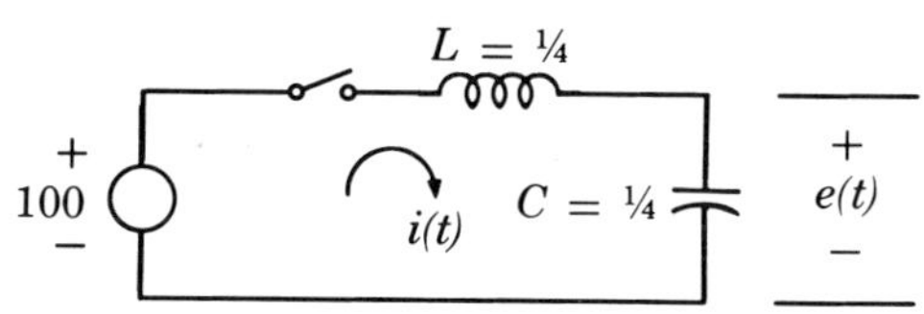

FIGURE 12–15

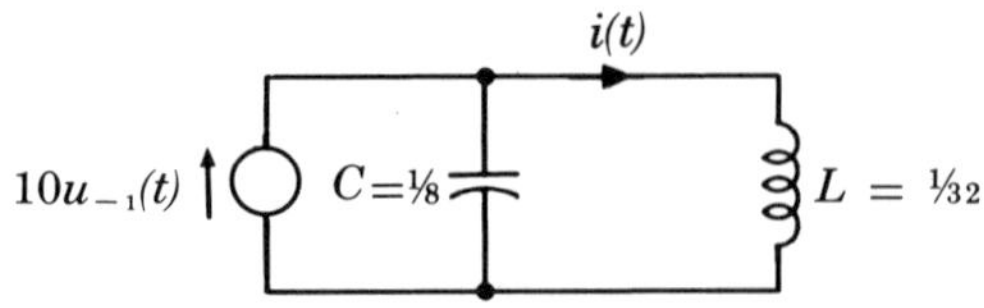

FIGURE 12–16

14. Find the current $i(t)$ in the circuit of Fig. 12–16.

15. When the source is applied at $t = 0$ in the circuit of Fig. 12–17 there is no voltage on the capacitance, but there is a current of 2 amp in the mesh. Find the currents in the two inductances for $t > 0$.

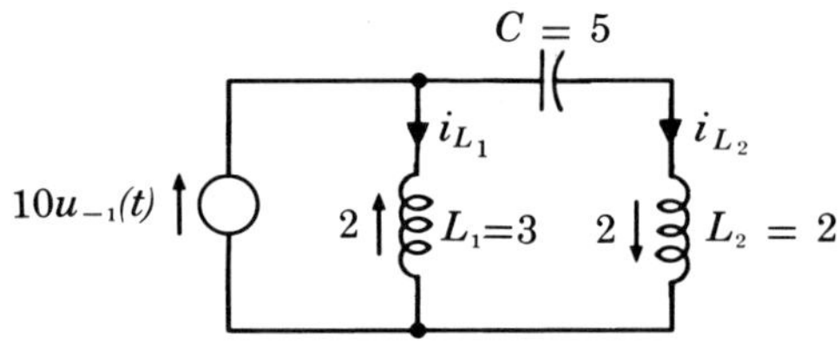

FIGURE 12–17

16. In the circuit of Fig. 12–18 the step voltage source is applied at $t = 0$. At this time there is a voltage of 10 volts on C_2, no voltage on C_1, and no current in the inductance. What is the voltage across C_1 as a function of time?

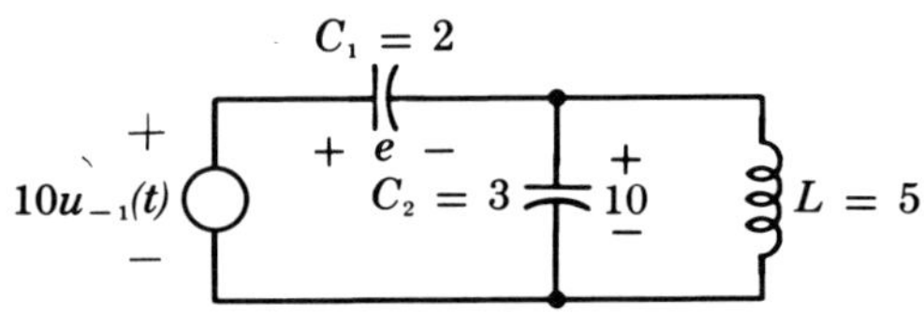

FIGURE 12–18

17. If there were no initial energy storages in the circuit of Exercise 16, what would the voltage $e(t)$ across the capacitance C_1 be?

18. Find the voltage $e(t)$ in the circuit of Fig. 12–19: (a) by superposition and (b) by computing $e(0^+)$, $de(0^+)/dt$, and $e(\infty)$.

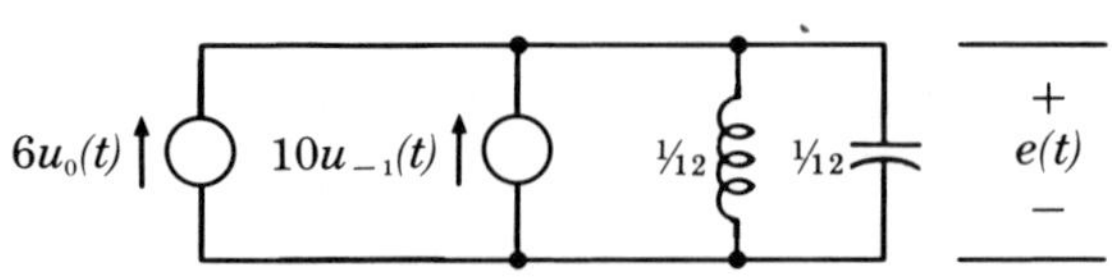

FIGURE 12–19

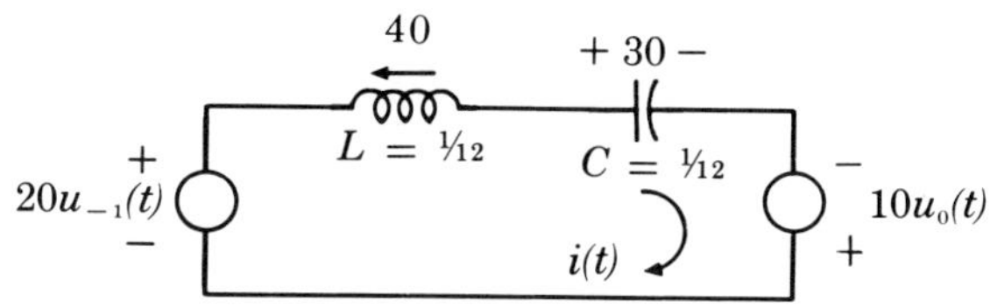

FIGURE 12–20

19. In the circuit of Fig. 12–20 there is an initial current in the inductance of −40 amp and an initial voltage on the capacitance of 30 volts. The two sources act at $t = 0$. What is the current as a function of time?

Section 12–4

20. A damped sinusoidal current is given by $i(t) = 10e^{-t} \cos(2\pi t + 45°)$.
   (a) Sketch the waveform.
   (b) Give the angular frequency, cyclic frequency, and period.
   (c) What is the time constant of the envelope?
   (d) What is the phase relative to a cosine?

21. A series circuit consists of an inductance of 10 henrys, a resistance of 2 ohms, and a capacitance of 5 farads. At $t = 0$ the voltage on the capacitance is 10 volts, and the current is zero.
   (a) Give the values of $i(0^+)$, $di(0^+)/dt$, and $i(\infty)$.
   (b) Assume a response current of the form $i(t) = Ae^{-\alpha t}\cos(\omega_d t + \theta) + B$ and evaluate the constants.
   (c) Sketch the current waveform.

22. A series circuit consisting of an inductance of 10 henrys, a resistance of 2 ohms, and a capacitance of 5 farads has an initial current of 10 amp, with a voltage of 10 volts on the capacitance, at $t = 0$. The direction of the current is positive and is discharging the capacitance. (a) Find the values of $i(0^+)$, $di(0^+)/dt$, and $i(\infty)$. (b) Find $i(t)$.

23. Solve Exercise 22 by means of the superposition theorem.

24. The circuit of Fig. 12–21 is at rest when it is excited by an impulse voltage. Give the current as a function of time from this instant.

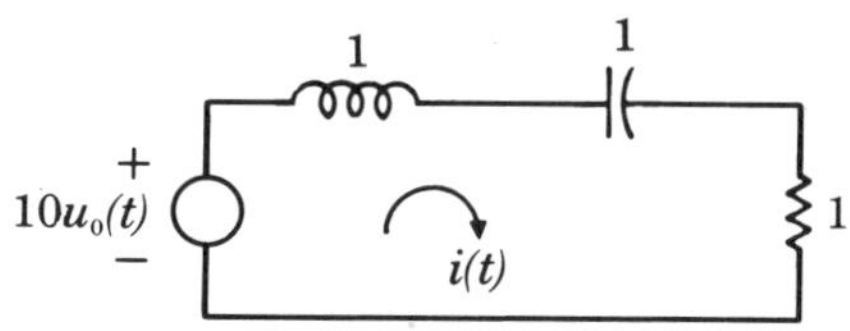

FIGURE 12–21

25. A series circuit consists of $R = \frac{5}{2}$ ohms, $L = \frac{1}{4}$ henry, and $C = \frac{1}{4}$ farad. At $t = 0$ the current is 2 amp and the voltage across the capacitance is 10 volts. The direction of the current is positive and it is discharging the capacitance.
   (a) Find $i(0^+)$, $di(0^+)/dt$, and $i(\infty)$.
   (b) Match the values from part (a) to the general overdamped expression, $i(t) = A_1e^{p_1t} + A_2e^{p_2t} + B$.

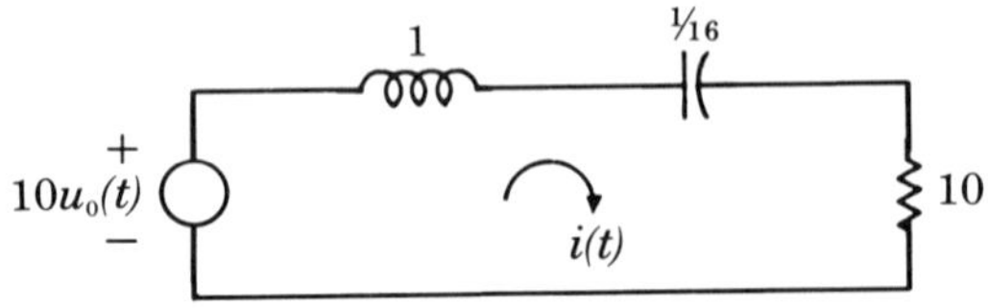

FIGURE 12–22

26. In the circuit of Fig. 12–22 there is no initial energy stored. Find the current $i(t)$.

27. The elements in the circuit of Fig. 12–23 can be changed to produce critical damping.

(a) Find the values of α, ω_0, and ω_d for the given circuit.
(b) Change L in the original circuit to give critical damping.
(c) Change C in the original circuit to give critical damping.
(d) Change R in the original circuit to give critical damping.
(e) Find the current which would flow in the circuit with the resistance adjusted to give critical damping.

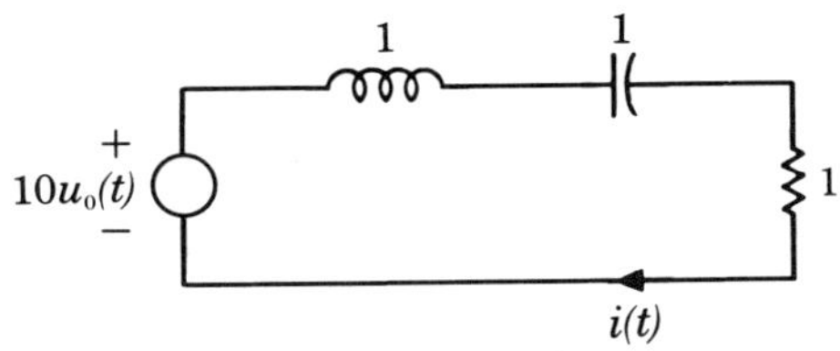

FIGURE 12–23

28. In the circuit of Fig. 12–24 the voltage on the capacitance at $t = 0$ is 9 volts, and there is no current flowing. Find the current for $t > 0$.

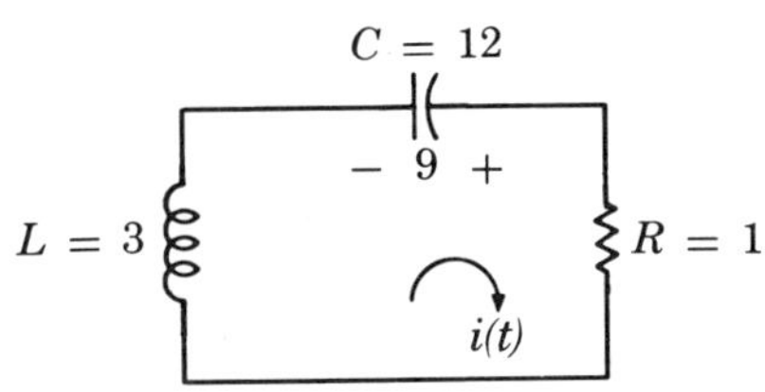

FIGURE 12–24

Section 12–5

29. At $t = 0$ in the circuit of Fig. 12–25 there is a current of 5 amp in the inductance and no voltage on the capacitance. (a) Find $e(0^+)$, $de(0^+)/dt$, and $e(\infty)$. (b) Match the values of part (a) to a general response of the form $e(t) = Ae^{-\alpha t}\cos(\omega_d t + \theta) + B$.

30. In the circuit of Fig. 12–25 at $t = 0$ there is a voltage $e(t)$ of 5 volts across the capacitance and no current in the inductance. Give the voltage $e(t)$ which will ensue.

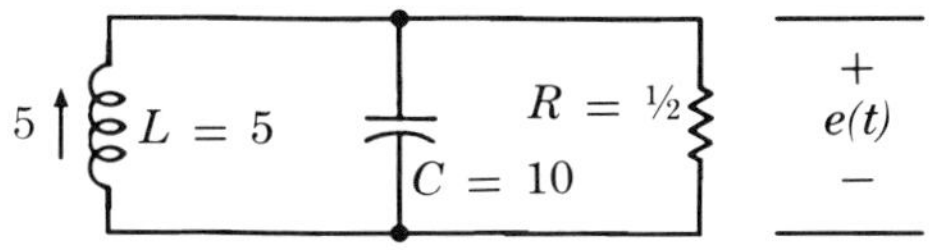

FIGURE 12–25

31. In the circuit of Fig. 12–25 at $t = 0$ the current in the inductance is 5 amp and the voltage across the capacitance is 5 volts. (a) Find $e(t)$ by superposing the results of Exercises 29 and 30. (b) Evaluate the voltage directly from its initial value and the initial value of its derivative.

32. Find $e(t)$ in the circuit of Fig. 12–26.

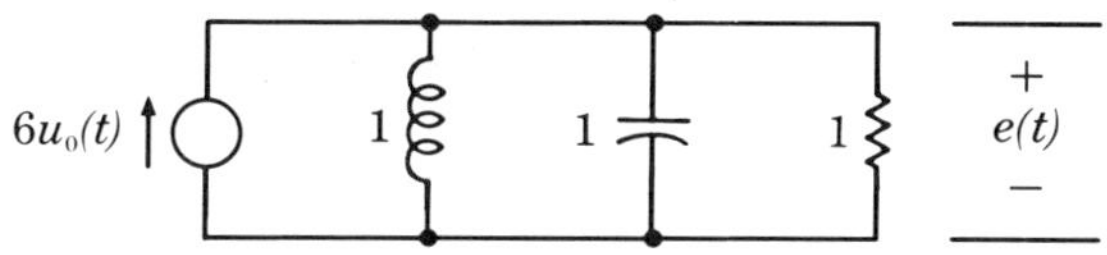

FIGURE 12–26

33. Find the voltage $e(t)$ in the circuit of Fig. 12–27.

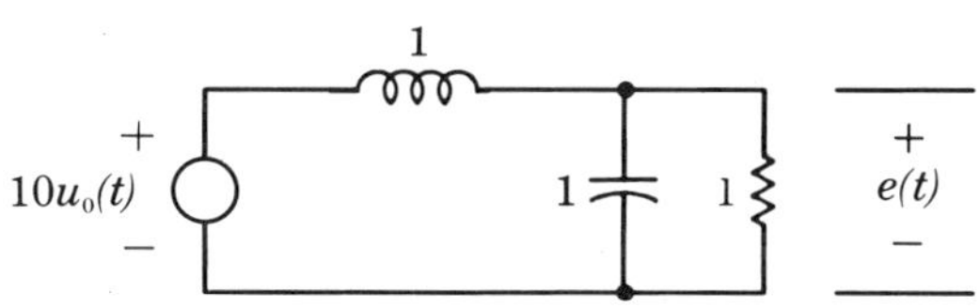

FIGURE 12–27

34. Compute ω_0 and α for the circuit of Fig. 12–28 and find the type of response which it must produce. If the initial voltage on the capacitance is 10 volts, and the initial current in the inductance is 50 amp, what is the voltage $e(t)$ across the resistance?

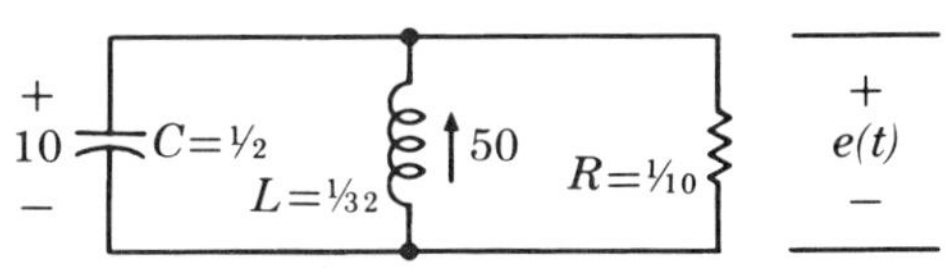

FIGURE 12–28

35. Find $e(t)$ in the circuit of Fig. 12–29.

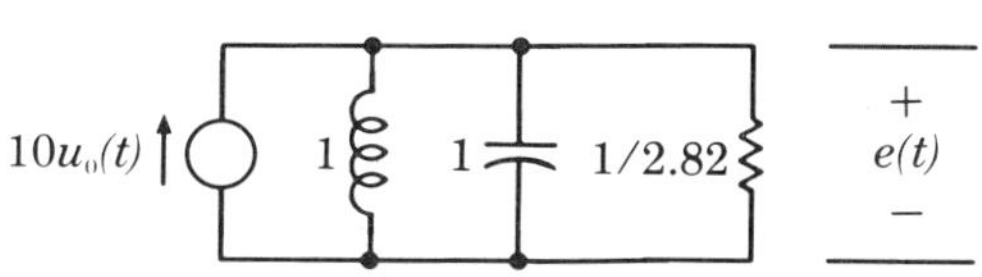

FIGURE 12–29

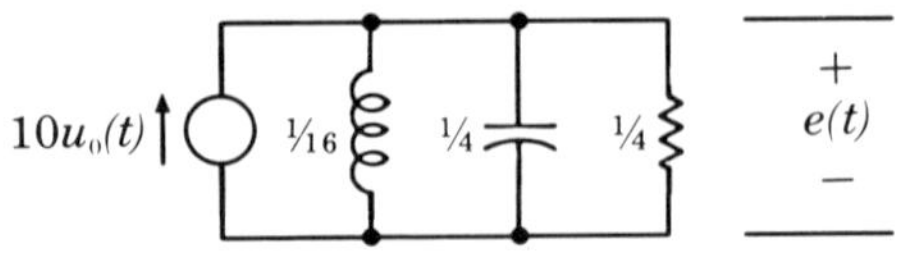

FIGURE 12–30

36. Find $e(t)$ in the circuit of Fig. 12–30.

37. In the circuit of Fig. 12–31 find the value of C which will give critical damping. For this value of C find the current $i(t)$ in the inductance.

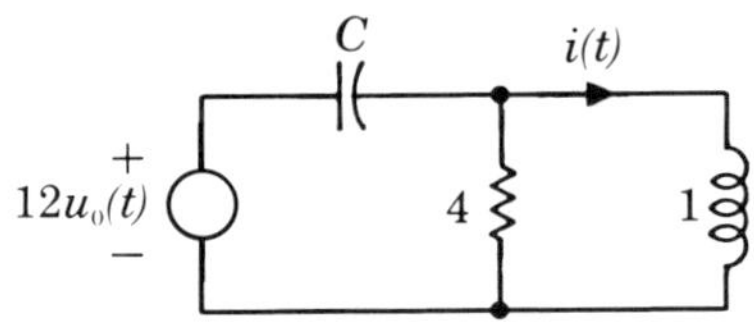

FIGURE 12–31

Section 12–6

38. The switch in the circuit of Fig. 12–32 is closed at $t = 0$. (a) Find $i(0^+)$, $di(0^+)/dt$, and $i(\infty)$. (b) Match the values in part (a) to a general step response of the form $i(t) = Ae^{-\alpha t}\cos(\omega_d t + \theta) + B$.

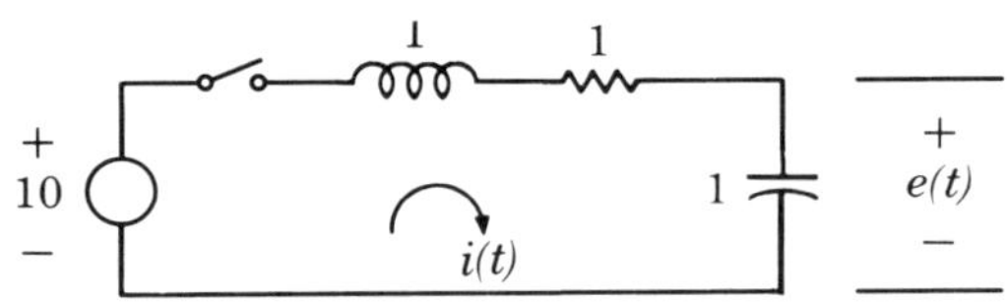

FIGURE 12–32

39. Find the voltage $e(t)$ across the capacitance in Exercise 38: (a) by integrating the current $i(t)$ and (b) by matching the initial and final values.

40. A step voltage $e(t) = 100u_{-1}(t)$ is applied to a series R–L–C circuit with $L = 10$ henrys, $R = 2$ ohms, and $C = 5$ farads. The initial current in the circuit is zero but there is an initial voltage of 50 volts on the capacitance in a direction which opposes the applied source. Find the voltage on this capacitance as a function of time.

41. A step voltage $e(t) = 100u_{-1}(t)$ is applied to a series R–L–C circuit in which $L = 10$ henrys, $R = 2$ ohms, and $C = 5$ farads. At the instant $t = 0$ there is a current in the circuit of 50 amp in a direction which leaves the positive terminal of the source. There is no voltage on the capacitance. Find the voltage on the capacitance as a function of time measured from this instant.

42. The switch in the circuit of Fig. 12–33 is closed at $t = 0$. At this time there is no energy stored in the circuit. (a) Find $i(0^+)$, $di(0^+)/dt$, and $i(\infty)$. (b) Match the response to the general overdamped expression $i(t) = A_1e^{p_1t} + A_2e^{p_2t} + B$.

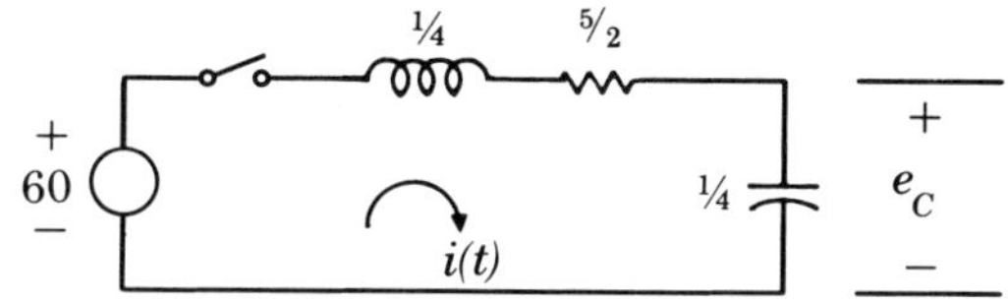

FIGURE 12–33

43. Find the voltage e_C on the capacitance in Exercise 42 (a) by integrating the current and (b) by matching the initial and final values.

44. Consider a series R–L–C circuit excited by a step function of voltage $e(t) = Eu_{-1}(t)$. Show that if $\alpha \gg \omega_0$ the response reduces to that of a series R–L circuit with no C.

45. What value of C in the circuit of Fig. 12–34 will give critical damping? Find the current as a function of time when C has its critical value.

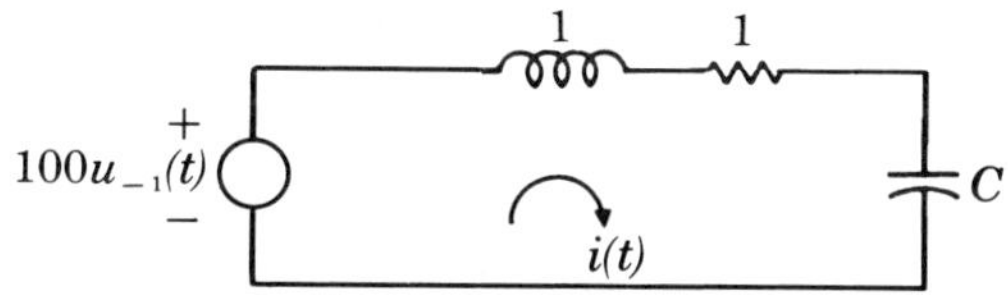

FIGURE 12–34

46. In the circuit of Fig. 12–35 the step voltage is applied at $t = 0$. At this time there is a current of 50 amp in the circuit and a voltage of 50 volts on the capacitance. Find the voltage on C as a function of time, subsequent to this instant.

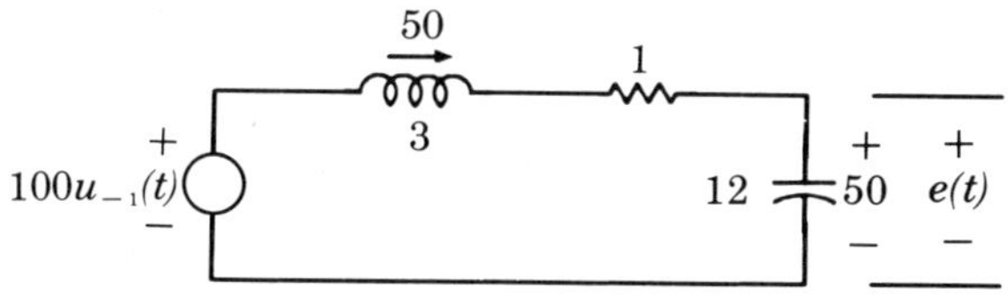

FIGURE 12–35

Section 12–7

47. The switch in the circuit of Fig. 12–36 is opened at $t = 0$. (a) Find the values of $e(0^+)$, $de(0^+)/dt$, and $e(\infty)$. (b) Match these values to a step-function response of the form $e(t) = Ae^{-\alpha t} \cos (\omega_d t + \theta) + B$.

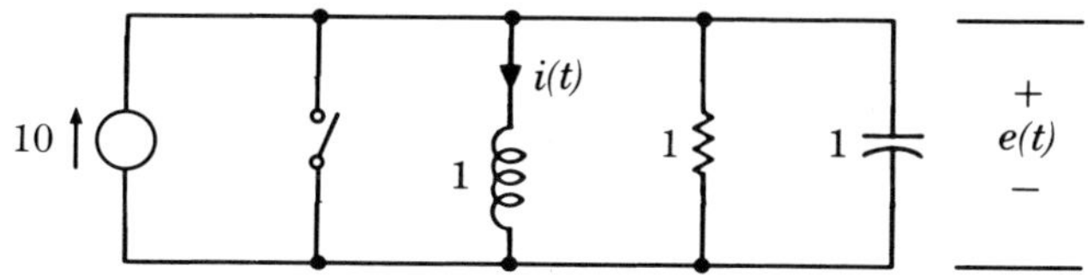

FIGURE 12–36

48. Find the current $i(t)$ in the inductance of Fig. 12–36: (a) by integrating the voltage response and (b) by matching directly to the initial and final values.

49. A step current of $i(t) = 50u_{-1}(t)$ is applied to a parallel R–L–C circuit with $L = 10$ henrys, $R = 1$ ohm, and $C = 5$ farads. The initial voltage on the capacitance is zero but there is a 100-amp current in the inductance entering the same node as the current source. Find the current in the inductance as a function of time from the initial instant.

50. In the circuit of Fig. 12–37 the 100-amp current from the source is flowing through the inductance. The switch is closed at $t = 0$, at which time there is a voltage on the capacitance of 100 volts. Find the current in the inductance as a function of time from the instant when the switch is closed.

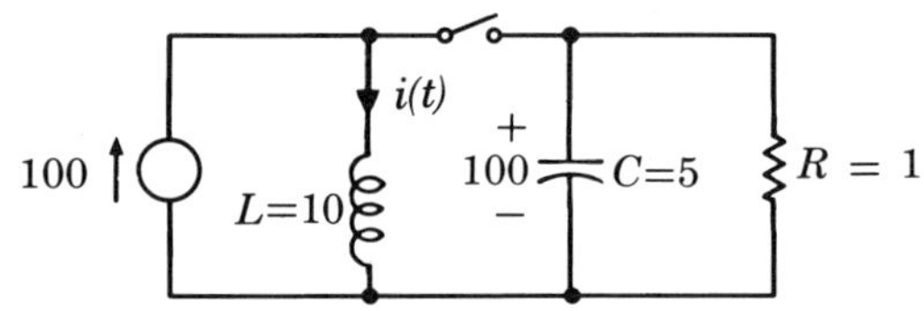

FIGURE 12–37

51. The switch in the circuit of Fig. 12–38 is opened at $t = 0$. There is no initial energy stored in the circuit. (a) Find $e(0^+)$, $de(0^+)/dt$, and $e(\infty)$. (b) Match the solution to the general overdamped expression $e(t) = A_1 e^{p_1 t} + A_2 e^{p_2 t} + B$.

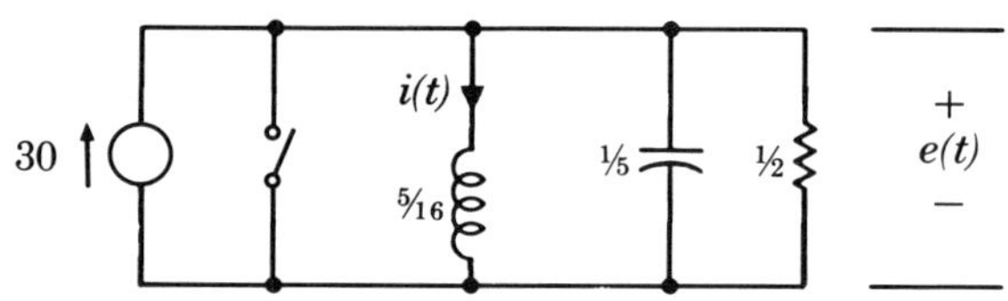

FIGURE 12–38

52. Find the current $i(t)$ in the inductance of Fig. 12–38: (a) by integrating the expression for the voltage and (b) by matching directly to the initial and final values.

53. Adjust the value of R in the circuit of Fig. 12–39 to give critical damping, and find the voltage $e(t)$ which is obtained for this condition.

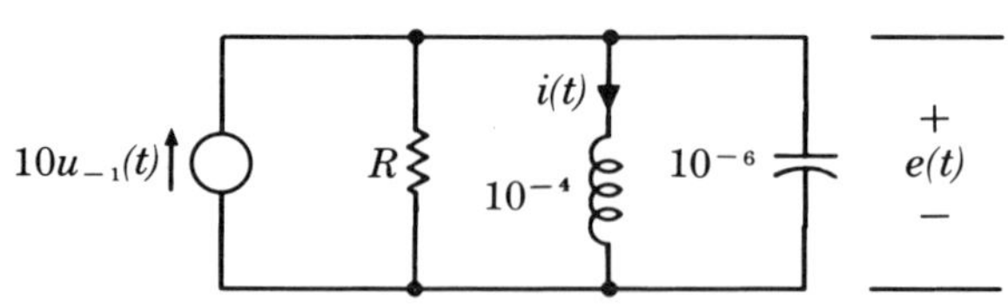

FIGURE 12–39

54. Find the current in the inductance in the critically damped circuit of Exercise 53.

55. A parallel R–L–C circuit is excited by a step-current source $i(t) = Iu_{-1}(t)$. Show that if $\alpha \gg \omega_0$ the voltage response reduces to that of a parallel R–C circuit.

Section 12–8

56. A series R–L circuit with $R = 2$ ohms and $L = 4$ henrys is excited by an impulse voltage source. Sketch the transient natural frequencies in the p-plane, and write a general expression for the transient response.

57. A series R–C circuit with $R = 3$ ohms and $C = 4$ farads is excited by an impulse voltage source. Write a general expression for the transient response for $t > 0$, and sketch the transient natural frequencies in the p-plane.

58. A series L–C circuit with $L = 5$ henrys and $C = 20$ farads is excited by a voltage impulse. Write a general expression for the transient response, and sketch the transient natural frequencies in the p-plane.

59. A series R–L–C circuit with $R = 1$ ohm, $C = 1$ farad, and $L = 1$ henry is excited by a voltage impulse. Write the general response function in exponential form, and sketch the p-values in the p-plane.

60. A parallel R–C circuit with $R = 2$ ohms and $C = 3$ farads is excited by an impulse current source. Write the general transient response in exponential form, and sketch the p-values in the p-plane.

61. A parallel R–L circuit with $R = 6$ ohms and $L = 3$ henrys is excited by an impulse current source. Write the general transient response function in exponential form, and sketch the p-values in the p-plane.

62. A parallel L–C circuit with $L = \frac{1}{4}$ henry and $C = \frac{1}{16}$ farad is excited by a current impulse. Write the general response function in exponential form, and sketch the p-values in the p-plane.

63. A parallel R–L–C circuit with $R = 1$ ohm, $L = 1$ henry, and $C = 1$ farad is excited by an impulse current. Write the general response function in exponential form, and sketch the p-values in the p-plane.

64. The natural response frequencies of a circuit determine its response to initial energy storages and impulse excitation. A particular circuit has one p-value of $-\frac{1}{2}$, and no others. A given capacitance in the circuit has a voltage of 6 volts at $t = 0.^+$ What is its voltage as a function of time?

65. The natural response frequencies of a circuit are known to be $p_1 = -1 + j2$ and $p_2 = -1 - j2$. For a particular excitation the current in a given branch is zero and has a derivative of 10 amp/sec at $t = 0.^+$ What is the complete response as a function of time? (The excitation is initial energy storage.)

66. The natural response frequency of a two-terminal circuit with an open circuit at the terminal-pair is $p = -2$. With a short circuit across the terminal-pair it is $p = -3$.

(a) Give the form of the response current when the circuit is driven by an impulse voltage.

(b) Give the form of the response voltage when the circuit is driven by an impulse current.

(c) How would the answers to parts (a) and (b) change if the sources were step functions instead of impulses?

CHAPTER 13

TIME-DOMAIN SOLUTIONS FOR ARBITRARY EXCITING FUNCTIONS

"He that woo a widow must not dally,
He must make hay while the sun doth shine,
He must not stand with her, shall I, shall I,
But boldly say, widow, thou must be mine."

The Sketch Book, Washington Irving

13–1 Introduction. A small boy soon learns that a hammer is more effective if it is used to strike a sharp blow on a nail than if it is used merely to push against the nail. In substance he has learned that there are some problems in which impulses are more effective than step functions. The quotation from Washington Irving suggests another, and this chapter is devoted to still another. We shall resolve an arbitrary excitation function into a series of impulses, and we shall find the response of a network to the arbitrary excitation by summing the responses to the impulse functions.

We can obtain an approximate answer if we divide the input function into a series of pulses and then approximate each pulse by an impulse. If the pulses have a width which is small relative to the time constants of the network, the accuracy of the approximation will be fairly good. In the limit we can take an infinite number of pulses, with a width which becomes vanishingly small. The summation of the infinite number of impulse responses becomes an integral, and the integral is an exact representation of the network response. This integral is called the *superposition integral* or the *convolution integral.*

In this chapter we shall derive the convolution integral and learn how to use it, both analytically and graphically. The convolution integral has a number of useful properties which lend themselves to rapid and approximate solutions of complicated problems. In addition it leads to a formulation of network problems suitable for computation by large-scale digital computers. The convolution integral is an example of the powerful mathematical method of "Kernels," which is used in electrical engineering, and of the Green's function method, which is very widely applied in physics.

We shall consider only the simplest examples of convolution theory. That is, the functions with which we shall concern ourselves will be impulse responses which decay with time and network input functions of the same general class. It is possible to extend the theory to include periodic waveforms and waveforms defined by statistical parameters, but we will not do so here. It is also possible to extend the theory to include impulse responses of several variables. Such functions are important in three-dimensional problems of physics.

13-2 The equivalence of pulses and impulses. Our whole treatment of impulses has been based on the fact that they are approximations to pulses. Essentially we have defined an impulse as a narrow pulse of width δ and height $1/\delta$ and therefore of unit area. We have said that the exact width and the exact height are not important, but that the area under the impulse is important. The area is important because the amount of energy stored in the circuit while the impulse is acting depends on the area. The energy-storage elements effectively perform the operation of integration, and they evaluate the area under the impulse. So long as we are not concerned with the time interval during which energy is being stored in the circuit, we do not have to concern ourselves with the exact shape of the impulse but only with the area under it. Another way of stating this fact is to say that the energy-storage elements in the circuit smooth the effect of impulse excitation, making it immaterial whether we consider the excitation to be an ideal impulse or an equivalent finite pulse.

Now, however, we must be more quantitative about the approximation involved in replacing a pulse by an impulse. There is an error involved, and it increases with the width of the pulse. As we shall show, the amount of error depends on the ratio of the pulse width to the time constant of the circuit. Figure 13-1(a) shows a series R-L circuit excited by an impulse.

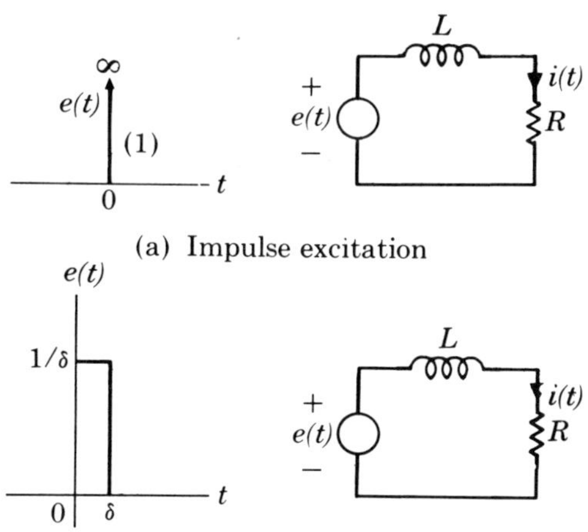

(b) Pulse excitation

FIG. 13-1. A comparison of pulse and impulse excitation for an R-L circuit.

At $t = 0$ the impulse voltage appears across the inductance and by $t = 0^+$ has established a current in it of $1/L$ amp. The current decays exponentially to zero with a time constant of $T = L/R$. The expression for the current is

$$i(t) = \frac{1}{L} e^{-t/T} \qquad \text{for} \quad t > 0. \tag{13–1}$$

A pulse with the same unit area and exciting the same circuit is shown in Fig. 13–1(b). In the interval from $t = 0$ to $t = \delta$ the circuit response is that which would be produced by a step of height $1/\delta$. Mathematically

$$i(t) = \frac{1}{\delta R}(1 - e^{-t/T}) \qquad \text{for} \quad 0 < t < \delta. \tag{13–2}$$

In the interval from $t = 0$ to $t = \delta$ a current is established in the inductance. After this interval the source is removed and the current decays to zero according to the natural behavior of the circuit. From Eq. (13–2) the value of the peak current, which occurs at $t = \delta$, is

$$i(\delta) = \frac{1}{\delta R}(1 - e^{-\delta/T}). \tag{13–3}$$

If δ is small relative to the time constant we can approximate Eq. (13–3) with

$$\begin{aligned} i(\delta) &= \frac{1}{\delta R}\left(1 - 1 + \frac{\delta}{T} - \frac{\delta^2}{2T^2} + \cdots\right) \\ &= \frac{1}{L}\left(1 - \frac{\delta}{2T}\right). \end{aligned} \tag{13–4}$$

From this value the current in the circuit of Fig. 13–1(b) decays exponentially to zero with the same time constant as the current in the circuit of Fig. 13–1(a). At $t = \delta$ the value of the current in the circuit of Fig. 13–1(a) can be obtained from Eq. (13–1):

$$\begin{aligned} i(\delta) &= \frac{1}{L} e^{-\delta/T} \\ &= \frac{1}{L}\left(1 - \frac{\delta}{T}\right). \end{aligned} \tag{13–5}$$

The conclusion we can draw from Eqs. (13–4) and (13–5) is that the impulse response is below the pulse response at $t = \delta$ by an amount

$$\epsilon = \frac{\delta}{2T}. \tag{13–6}$$

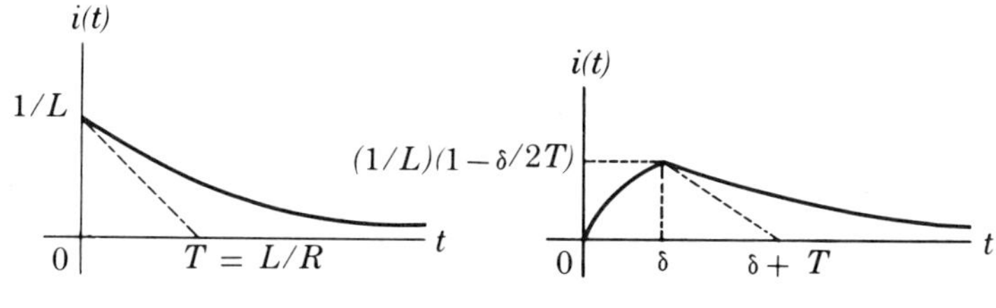

(a) The impulse response (b) The pulse response

FIG. 13-2. The impulse-response current and the pulse-response current for the circuit of Fig. 13-1.

After $t = \delta$ the two curves follow almost identical paths. The impulse response is shown in Fig. 13-2(a) and the pulse response in Fig. 13-2(b). The impulse response rises to its maximum value in zero time. The pulse response does not reach its maximum value until $t = \delta$. After $t = \delta$, however, the two curves will be indistinguishable if

$$\delta \ll T. \tag{13-7}$$

Equation (13-7) provides a quantitative guide for the approximation of a pulse by an impulse. *The pulse length must be short relative to the time constant of the system to which it will be applied.* If there are several time constants involved a safe procedure is to determine the shortest one, and to make the pulse length short relative to it.

A second example of the difference between the impulse response and the pulse response is afforded by the circuit of Fig. 13-1. The voltage across the inductance, which includes an impulse, is a fairly stringent test of the approximation. For the circuit of Fig. 13-1(a) the voltage across the inductance is first the applied impulse and then a negative exponential which starts at a value of $-R/L$ and decays to zero. The complete expression for it is

$$e_L(t) = u_0(t) - \frac{R}{L} e^{-t/T} \cdot u_{-1}(t). \tag{13-8}$$

A sketch of the function of Eq. (13-8) is shown in Fig. 13-3(a). For the circuit of Fig. 13-1(b) the voltage across the inductance starts at a value of $1/\delta$ and decays exponentially to zero for δ seconds. At $t = \delta$ the input pulse drops by $1/\delta$ volts. We can consider the drop to be a new step function suddenly applied at this instant and obtain its effect by superposition. For the sudden change in the pulse the inductance is an open circuit, the resistance parameter is finite, and hence all the change appears across the inductance. At $t = \delta$ therefore the voltage across the inductance drops negatively by $1/\delta$ volts; after this instant it returns exponentially to zero. The net amount by which the voltage becomes negative is equal to the

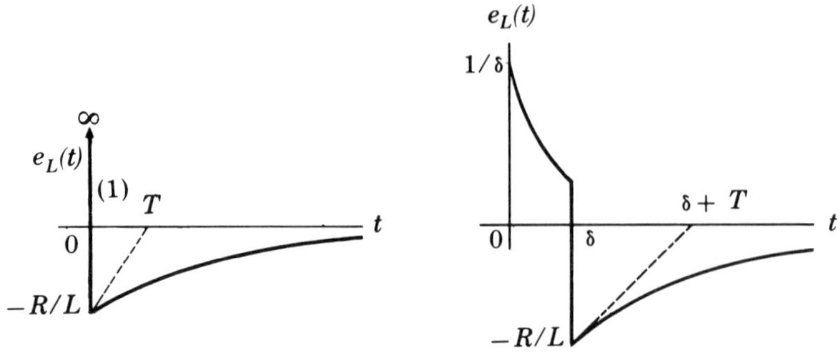

(a) The impulse response (b) The pulse response

FIG. 13–3. The impulse-response voltage and the pulse-response voltage for the circuit of Fig. 13–1.

amount by which it has dropped from its original value in δ seconds. Between $t = 0$ and $t = \delta$ the voltage across the inductance is described by

$$e_L(t) = \frac{1}{\delta} e^{-t/T}. \tag{13–9}$$

At time $t = \delta$ it has the value

$$e_L(\delta) = \frac{1}{\delta} e^{-\delta/T} = \frac{1}{\delta}\left(1 - \frac{\delta}{T} + \cdots\right). \tag{13–10}$$

The voltage drop is therefore approximately $1/T$. For $t > \delta$ the voltage across the inductance is very nearly

$$e_L(t) = -\frac{R}{L} e^{-(t-\delta)/T}. \tag{13–11}$$

The combined curve for Eqs. (13–9) and (13–11) is shown in Fig. 13–3(b). As long as δ is small relative to the time constant, the curve of Fig. 13–3(b) is essentially the same as that for the impulse response. For $t > \delta$ both curves are negative exponentials, starting at $-R/L$ and decaying to zero. At $t = 0$ the impulse response includes an impulse of unit area. The pulse response contains an approximate impulse which is spread over δ seconds. Since its height is approximately $1/\delta$, its area is approximately unity. Again, the criterion for agreement between the impulse response and the pulse response is that the pulse width be short relative to the time constant of the system to which it is applied.

In a pure resistance network, in which there is no averaging by energy-storage elements, the approximation is poor. However, the solution of a pure resistance network is trivial. The output response is just the input stimulus multiplied by a constant. We would not need to use the impulse response to characterize such a circuit!

13–3 The superposition integral or convolution integral. In this section we shall obtain the formula which gives the accurate response function for an arbitrary input stimulus, in terms of the impulse response of the system. This formula is an integral called the superposition integral or the convolution integral. The term superposition is more descriptive of the way in which we shall obtain the integral, but the term convolution is used more often.

We shall suppose that the impulse response of a network has been obtained, as in Fig. 13–4(a), and that it is represented by $h(t)$. The actual input function is divided into a series of pulses of width Δ. The first pulse has an approximate area of $f(0) \cdot \Delta$, the second $f(\Delta) \cdot \Delta$, the third $f(2\Delta) \cdot \Delta$, and so on. Each pulse is now replaced by an impulse having the same area and occurring at the beginning of the pulse. (It might appear more logical to place the impulses at the midpoints of the pulses, but it makes no difference since Δ will approach zero.) An approximate expression for the input function $f_1(t)$ of Fig. 13–4(b), in terms of these impulses, is†

$$f_1^*(t) = \Delta f_1(0)u_0(t) + \Delta f_1(\Delta)u_0(t - \Delta) + \Delta f_1(2\Delta)u_0(t - 2\Delta) + \cdots. \tag{13–12}$$

Equation (13–12) represents a train of impulses whose areas are proportional to the values of the input function at successive instants of time. Each of these impulses, when applied to the network, produces an impulse response modified in amplitude by the area of the related impulse, and shifted in time so that it begins at the instant when the related impulse strikes the network. Thus an approximate expression for the output response function is

$$\begin{aligned} f_2^*(t) &= \Delta f_1(0)h(t) + \Delta f_1(\Delta) \cdot h(t - \Delta) + \Delta f_1(2\Delta) \cdot h(t - 2\Delta) + \cdots \\ &= \sum_{n=1}^{\infty} \Delta f_1(n\Delta)h(t - n\Delta). \end{aligned} \tag{13–13}$$

Equation (13–13) expresses the output response function as the summation of a number of impulse response functions of various amplitudes and instants of occurrence. In order to obtain an exact expression for the output function we let the width of the pulses, Δ, approach zero. If the time constant of the network is not zero, the impulse response and the pulse response will then be identical. At the same time we take n larger

† See Section 8–6 for a discussion of this type of equation.

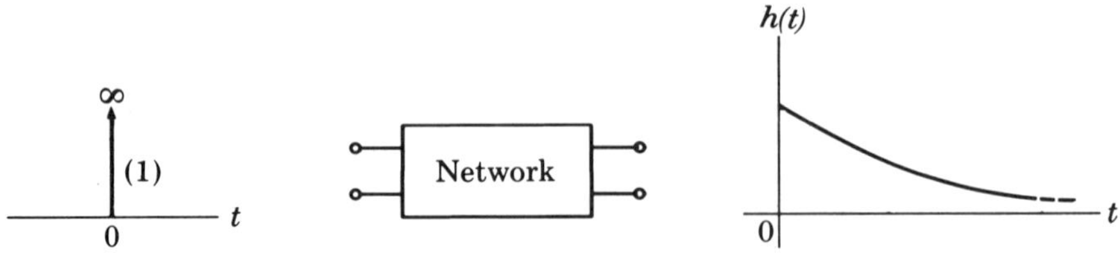

(a) Impulse response

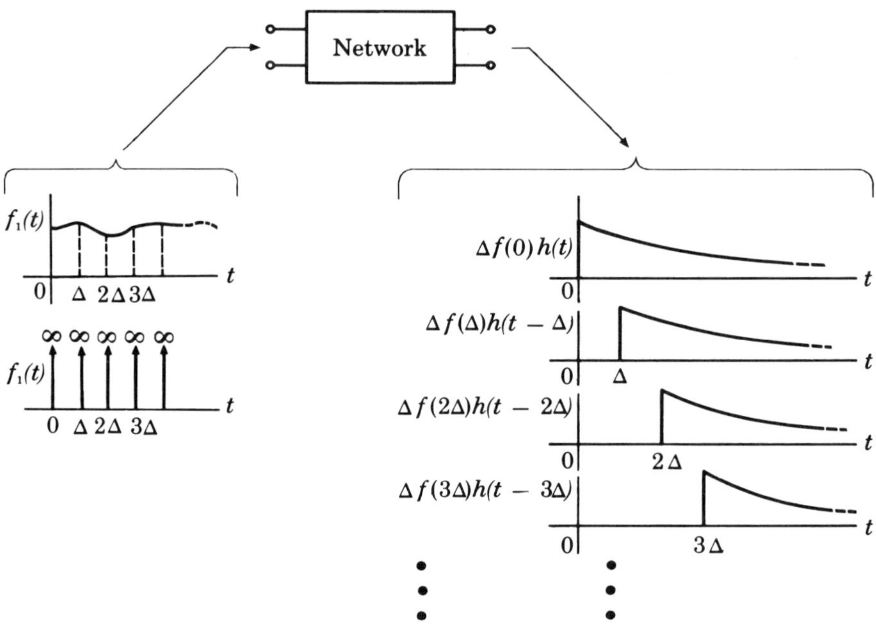

(b) Response to arbitrary input

FIG. 13–4. The network response as the superposition of a number of impulse responses.

and larger in order to sample the input function over its whole range. In the limit $n\Delta$ becomes a continuous variable, and Δ, the spacing between the samples, becomes a differential length. Thus we take

$$n\Delta = T, \qquad \Delta = dT.$$

In the limit the summation of Eq. (13–13) becomes an integral, and the output response function can be written in the form

$$f_2(t) = \int_0^\infty f_1(T)h(t - T)\, dT. \tag{13–14}$$

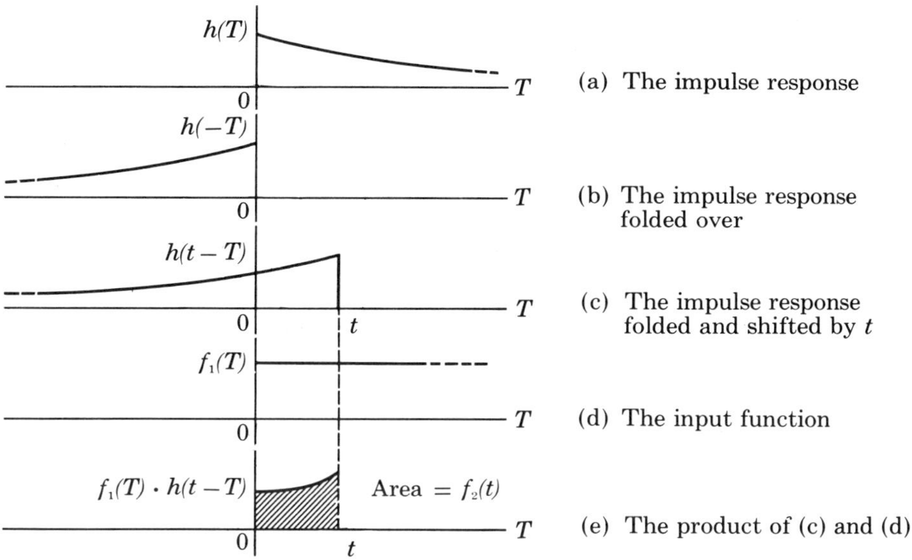

FIG. 13-5. Graphical convolution and the scanning function.

Equation (13-14) is the superposition or convolution integral. It is a parametric integral. To obtain a particular value of $f_2(t)$, for example $f_2(t_1)$, the value of t_1 is inserted in the integral, and a complete integration with respect to the variable T is performed from $T = 0$ to $T = \infty$. Actually the integration does not have to proceed beyond $T = t$, since at this point the argument of the impulse response becomes negative, and the impulse response becomes zero. Thus we can write

$$f_2(t_1) = \int_0^{t_1} f_1(T)h(t_1 - T)\,dT. \tag{13-15}$$

Each point obtained for $f_2(t)$ requires the evaluation of a complete integral of the type given by Eq. (13-15). We shall not actually evaluate the integrals in this fashion, but it is essential to understand the process which would be involved if we did.

A graphical interpretation of the convolution integral of Eq. (13-14) is also of interest. A typical function $h(T)$ is plotted in Fig. 13-5(a). Since the integration is performed with respect to the variable T, we also use T to plot our curves. The curve $h(-T)$ is shown in Fig. 13-5(b). Values previously obtained for $+T$ now occur at $-T$. The net result is a folding of the curve $h(T)$ about the axis of ordinates. Thus the curve $h(-T)$ is a mirror image of the curve $h(T)$, extending to the left and toward minus infinity. In order to produce the function $h(t - T)$, the curve $h(-T)$ must be shifted to the right by t seconds. The value $h(0)$ then occurs when $T = t$. The shifted curve is shown in Fig. 13-5(c), and beneath it, in

Fig. 13–5(d), is plotted the function $f_1(T)$. The convolution integral bids us to multiply the functions of Figs. 13–5(c) and 13–5(d) and to obtain the area under the product function. The product function is shown in Fig. 13–5(e); the hatched area under it is the value of the convolution integral for the selected value of t. As t varies from zero to infinity the function of Fig. 13–5(c) shifts to the right. As it shifts, the product function changes and hence the value of $f_2(t)$ changes. Essentially the function $h(t - T)$ slides over or *scans* the function $f_1(T)$. It is therefore sometimes called a *scanning function*. The name convolution comes from the interpretation which we have just discussed. Convolution means folding, and the essential element in this graphical interpretation is the folding over of the function $h(T)$ to obtain the scanning function $h(t - T)$.

We are more likely to evaluate the convolution integral by the same sort of procedure which we used to derive it rather than by analytical or graphical procedures. We will often be content with an approximate answer which can be obtained by representing the input stimulus by a small number of impulse functions. Let us consider the circuit shown in Fig. 13–6, for which the stimulus is a voltage and the response is a current. The impulse response, shown in Fig. 13–6(a), is

$$i(t) = h(t) = e^{-t} \cdot u_{-1}(t). \tag{13–16}$$

The stimulus function, $f_1(t)$, is the pulse shown in Fig. 13–6(b). For this particular example, an accurate response function can be obtained. The input function at $t = 0$ is a step of unit height. The response current is

$$i(t) = 1 - e^{-t} \qquad \text{for} \quad 0 < t < 1. \tag{13–17}$$

At $t = 1$ the input function is removed, and the circuit is left to itself. The current decays from its value at $t = 1$ to a final value of zero, with a time constant of unity. The maximum value of the current, which occurs at $t = 1$, is

$$i_{\text{max}} = 0.635.$$

A sketch of the response function is shown in Fig. 13–6(b).

An approximation of the response function of Fig. 13–6(b) can be obtained by dividing the input pulse into four pulses having a height of one unit, a width of one-quarter unit, and thus an area of one-quarter unit. These four pulses are then replaced by four impulses of value $\frac{1}{4}$ each, occurring at $t = \frac{1}{4}$, $t = \frac{1}{2}$, $t = \frac{3}{4}$, and $t = 1$. The response for each of these impulses is a damped exponential curve which is zero until the impulse strikes, which achieves a value of one-quarter at this instant, and which then decays to zero with a time constant of unity. The approximate solution is shown in Fig. 13–6(c), that is, the four impulses at the input of the circuit and the four impulse responses at the output. The

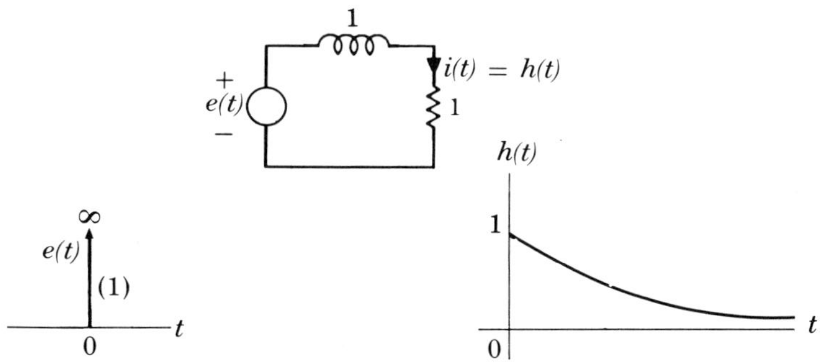

(a) The impulse response

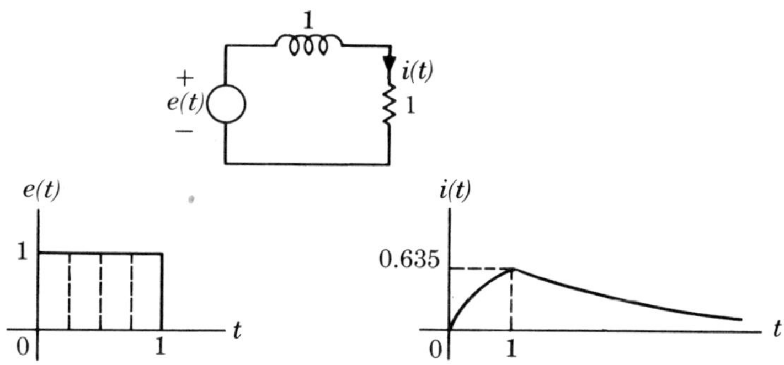

(b) The response to a pulse

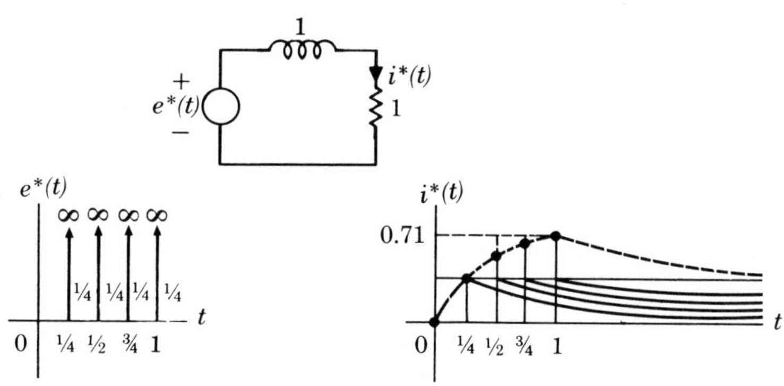

(c) The approximate response by convolution

FIG. 13–6. An approximate method of convolution.

sum of the individual response functions is also shown. It achieves a maximum value of 0.71 at $t = 1$ and then decays with the proper time constant. The approximate response curve is thus within about 10 percent of the correct curve of Fig. 13–6(b). Often this sort of accuracy is all that a problem warrants. The data may not be known with any more accuracy, or it may be important only to obtain a rough idea of the nature of the response function. When the correct response function is not known, an indication of the accuracy of an approximation can be obtained by using a larger number of impulses, in a second computation, to represent the input function. If the answer does not change by much, the approximation is acceptable.

Additional approximate methods of performing the convolution operation will be given later. Before we proceed with these methods, however, we shall consider some examples in which the convolution integral can be evaluated analytically. When both the impulse response and the input stimulus are simple functions of time the convolution integral can be evaluated by direct integration. The accurate answer is to be preferred to an approximate one, whenever it does not require an exhorbitant amount of work.

13–4 Analytical evaluation of the convolution integral. The analytical evaluation of the convolution integral is often difficult without an extensive table of integrals, but for a few simple functions the integral is fairly easy to evaluate. In general the input function does not start until $t = 0$, and hence it is desirable to multiply it by the function $u_{-1}(t)$ in order to section it, mathematically, at the origin. Similarly the response to an impulse which occurs at $t = 0$ is a function which starts at $t = 0$ and which should be sectioned at the origin by multiplying it by the function $u_{-1}(t)$. The $u_{-1}(t)$ term automatically makes the impulse response zero until the impulse is applied. When impulses are applied at instants after $t = 0$, it is important that the response functions be equal to zero until the corresponding impulses are applied. The step function automatically takes care of this problem for us.

A simple network is shown in Fig. 13–7(a), excited by an impulse. At $t = 0$, when the impulse is applied, the inductance acts as an open circuit. All the voltage appears across the inductance and by $t = 0^+$ causes a current to be established in the inductance of

$$i(0^+) = \frac{1}{L}\int_{0^-}^{0^+} u_0(t)\,dt = 1 \text{ amp.}$$

The current then decays to zero exponentially with a time constant, $T = L/R$, of 1 second. A sketch of the impulse response is also shown in

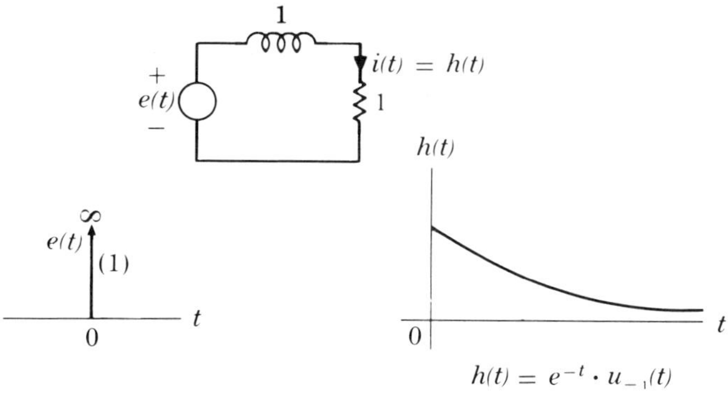

(a) Impulse response

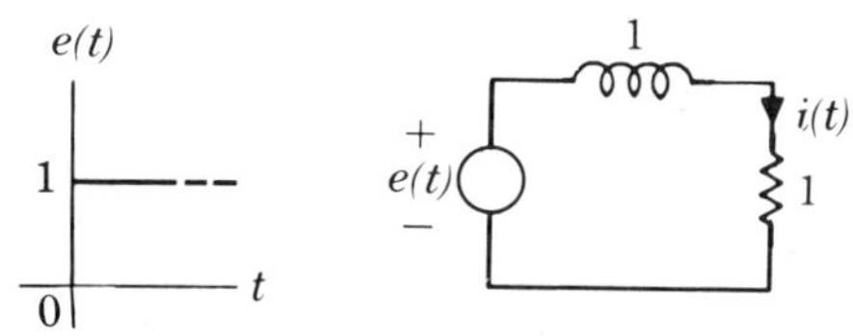

(b) Circuit driven by a step function

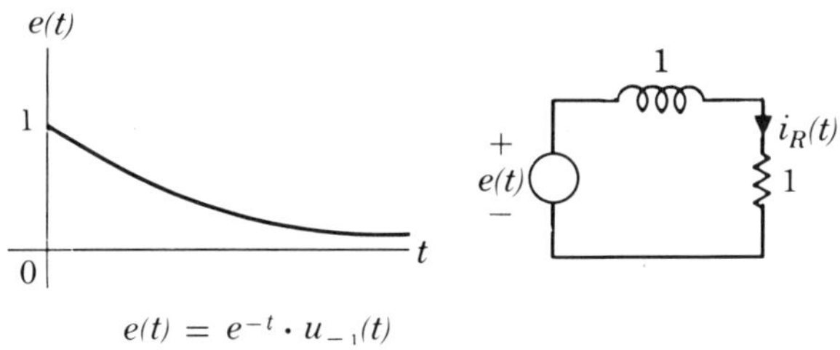

(c) Circuit driven by an exponential function

FIG. 13-7. Analytical convolution.

Fig. 13-7(a). Its mathematical expression, including the step function which makes it zero before $t = 0$, is

$$h(t) = e^{-t} \cdot u_{-1}(t). \tag{13-18}$$

The impulse response of Eq. (13-18) describes, or characterizes, the network. Knowing it, we can compute the response for any other input function. For example, consider finding the response for a unit step of voltage, as shown in Fig. 13–7(b). According to the convolution integral the output current response is given by

$$i(t) = \int_0^\infty u_{-1}(T)h(t - T)\,dT. \tag{13–19}$$

In the h function, given by Eq. (13–18), the variable t must be replaced by the variable $t - T$ in *both* the e^{-t} term and the $u_{-1}(t)$ term. Thus

$$i(t) = \int_0^\infty u_{-1}(T)e^{-(t-T)} \cdot u_{-1}(t - T)\,dT. \tag{13–20}$$

The effect of the two u_{-1} functions in Eq. (13–20) is merely to constrict the limits of the integration. The function $u_{-1}(T)$ is zero for negative values of T and unity for positive values of T. It makes doubly certain that the integration starts at $T = 0$. Such a function, in fact, makes it possible to change the limits of integration to minus infinity or to plus infinity without changing the value of the current. In either case the *actual* integration starts at zero. The function $u_{-1}(t - T)$ is zero when $t - T$ is negative and unity when $t - T$ is positive. The switchover occurs at $t = T$. Thus for $T < t$ the function is unity, and for $T > t$ the function is zero. The effect of the term $u_{-1}(t - T)$ is therefore to make the integrand in Eq. (13–20) equal to zero for $T > t$. In effect the integration is stopped at $T = t$ and does not need to be carried to infinity. Thus

$$\begin{aligned} i(t) &= \int_0^t e^{-(t-T)}\,dT \\ &= [e^{-(t-T)}]_0^t \\ &= 1 - e^{-t}. \end{aligned} \tag{13–21}$$

The result in Eq. (13–21) can easily be checked.* The step function is the integral of the impulse function, and the step function response is the integral of the impulse response from $t = 0$ to $t = t$.

As a second example let us consider the same circuit driven by the input voltage $e(t) = e^{-t} \cdot u_{-1}(t)$, as shown in Fig. 13–7(c). This input function is identical in form to the natural behavior of the system, and when a circuit is excited by its own natural response function the phenomenon of resonance occurs. We are asking the circuit to do what it does naturally, and it responds with enthusiasm. The solution of the differential equation for this case is rather difficult (see Section 9–4), and

---

* The result applies for $t > 0$ and hence should be multiplied by $u_{-1}(t)$. In general, the multiplying factor is $u_{-1}(t - a)$, where a is the beginning of the output function.

special response functions must be invented. The solution comes easily from the convolution integral, which in its general form gives

$$i_R(t) = \int_0^\infty e^{-T} \cdot u_{-1}(T) e^{-(t-T)} \cdot u_{-1}(t - T)\, dT. \tag{13-22}$$

The $u_{-1}(T)$ function has no effect on the integral for $T > 0$, and the $u_{-1}(t - T)$ function causes the integrand to be zero for $T > t$. Thus Eq. (13-22) reduces to

$$\begin{aligned} i_R(t) &= \int_0^t e^{-T} e^{-(t-T)}\, dT \\ &= \int_0^t e^{-t}\, dT \\ &= te^{-t}. \end{aligned} \tag{13-23}$$

The multiplying function, t, in Eq. (13-23) enters as a result of the integration and is not "pulled out of a hat" as it is in the solution of the differential equation.

The method of analytically evaluating the convolution integral which we have used here is particularly simple when the impulse response function is an exponential, or when it can be reduced to exponential form, for example when it is a sine, a cosine, a hyperbolic sine, or a hyperbolic cosine function. It is possible, also, to express more general functions as infinite series of exponential functions. These series are part of the general mathematical subject of orthogonal series and are too complex to be considered here. Instead, we shall consider some general theorems about the convolution integral, and then we shall use them to perform approximate evaluations of the convolution integral.

13-5 Extensions of the convolution theorem. The convolution integral gives the response function of a network when the impulse response of the network and the stimulus function are known. A number of theorems allow us to operate in various ways upon these two functions in order to simplify the evaluation of the integral.

Theorem 1: The interchange of the input function and the impulse response. For the basic relationship portrayed in Fig. (13-8) the input function is $f_1(t)$, the impulse response of the network is $h(t)$, and the response of the

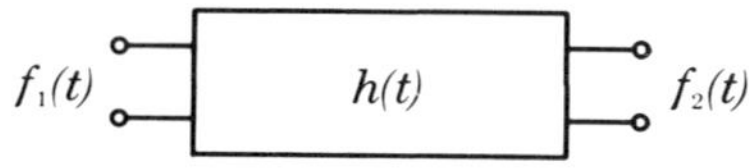

FIG. 13-8. The basic relationship for convolution.

network to the input $f_1(t)$ is $f_2(t)$. The response is expressed by the convolution integral

$$f_2(t) = \int_0^{\infty} f_1(T)h(t - T)\,dT.$$

Since the function $f_1(t)$ is zero until the beginning of the problem, that is, $t = 0$, we can change the lower limit of the integral to minus infinity without changing the value of $f_2(t)$. Hence

$$f_2(t) = \int_{-\infty}^{\infty} f_1(T)h(t - T)\,dT. \tag{13–24}$$

In Eq. (13–24) we make the substitution

$$t - T = z, \qquad dT = -dz. \tag{13–25}$$

When $T = -\infty$, $z = +\infty$ and vice versa. Hence the integral of Eq. (13–24) can be written in the form

$$f_2(t) = \int_{\infty}^{-\infty} f_1(t - z)h(z)(-dz). \tag{13–26}$$

Interchanging the limits changes the sign of the integral, and therefore

$$f_2(t) = \int_{-\infty}^{\infty} f_1(t - z)h(z)\,dz. \tag{13–27}$$

The variable in the integral of Eq. (13–27) is z. When the limits are inserted, however, the z disappears and the integral becomes a function only of t. The variable z is called a dummy variable; it is simply a variable of integration and could be replaced by any other letter without changing the result. In order to make Eq. (13–27) appear more like the original convolution integral we will replace z by T. (It is not the same T as that in the original integral, but in both cases the T's are dummy variables and do not appear in the final answer.) Thus

$$f_2(t) = \int_{-\infty}^{\infty} f_1(t - T)h(T)\,dT. \tag{13–28}$$

A comparison of Eq. (13–28) with the original convolution integral of Eq. (13–24) shows that the functions $f_1(t)$ and $h(t)$ have been interchanged. Convolution is thus seen to be a mutual property of the two relevant functions, and it doesn't matter which is considered to be the input and which is considered to be the impulse response. In summary*

$$f_2(t) = \int_{-\infty}^{\infty} f_1(T)h(t - T)\,dT = \int_{-\infty}^{\infty} h(T)f_1(t - T)\,dT. \tag{13–29}$$

* For functions $f_1(t)$ and $h(t)$ which are zero before $t = 0$, the range of integration in Eq. (13–29) can be from 0 to t.

Equation (13–29) is sometimes written with a special notation indicating convolution:

$$f_2(t) = f_1(t)*h(t) = h(t)*f_1(t). \tag{13–30}$$

In words, Eq. (13–30) states that the output $f_2(t)$ is the convolution of $f_1(t)$ and $h(t)$ or the convolution of $h(t)$ and $f_1(t)$.

Theorem 2: The derivative of the output response. The derivative of the output response function can be expressed as the derivative of the convolution integral. Thus for the situation of Fig. (13–8),

$$f_2(t) = \int_{-\infty}^{\infty} f_1(T)h(t - T)\,dT,$$

$$\frac{df_2}{dt} = \int_{-\infty}^{\infty} f_1(T)h'(t - T)\,dT, \tag{13–31}$$

or

$$f_2(t) = \int_{-\infty}^{\infty} f_1(t - T)h(T)\,dT,$$

$$\frac{df_2}{dt} = \int_{-\infty}^{\infty} f_1'(t - T)h(T)\,dT. \tag{13–32}$$

Stated in words, Eqs. (13–31) and (13–32) tell us that the derivative of the output response can be obtained by convolving: (1) the input function with the derivative of the impulse response; or (2) the derivative of the input function with the impulse response. In the notation of Eq. (13–30)

$$\begin{aligned} f_2'(t) &= f_1(t)*h'(t) = f_1'(t)*h(t) \\ &= h'(t)*f_1(t) = h(t)*f_1'(t). \end{aligned} \tag{13–33}$$

Theorem 3: The integral of the response function. For convenience we will use the symbol $f^{-1}(t)$ to represent the functional integral $\int_{-\infty}^{t} f_1(t)\,dt$. Referring to the relationship of Fig. 13–8 we again write

$$f_2(t) = \int_{-\infty}^{\infty} f_1(T)h(t - T)\,dT.$$

When both sides of this equation are integrated with respect to t the result is

$$\int_{-\infty}^{t} f_2(t)\,dt = \int_{-\infty}^{\infty} f_1(T)\left[\int_{-\infty}^{t} h(t - T)\,dt\right] dT$$

or

$$f_2^{-1}(t) = \int_{-\infty}^{\infty} f_1(T)h^{-1}(t - T)\,dT. \tag{13–34}$$

Using the second form of the convolution integral and performing the same operations we obtain

$$f_2^{-1}(t) = \int_{-\infty}^{\infty} f_1^{-1}(t - T)h(T)\,dT. \tag{13–35}$$

Stated in words, Eqs. (13–34) and (13–35) tell us that the integral of the output response can be obtained by convolving: (1) the input function with the integral of the impulse response; or (2) the impulse response with the integral of the input function. In the notation of Eq. (13–30)

$$\begin{aligned} f_2^{-1}(t) &= f_1(t)*h^{-1}(t) = f_1^{-1}(t)*h(t) \\ &= h^{-1}(t)*f_1(t) = h(t)*f_1^{-1}(t). \end{aligned} \tag{13–36}$$

Theorem 4: Integrations and differentiations of the response. We shall represent n differentiations of a function as $f^n(t)$ and m integrations of a function as $f^{-m}(t)$. Using this notation, when we repeatedly differentiate and integrate in theorems 2 and 3 we obtain

$$f_2^n(t) = f_1(t)*h^n(t) = f_1^n(t)*h(t), \tag{13–37}$$

and

$$f_2^{-m}(t) = f_1(t)*h^{-m}(t) = f_1^{-m}(t)*h(t). \tag{13–38}$$

In essence Eqs. (13–37) and (13–38) show that any number of derivatives or integrals of the output function $f_2(t)$ can be obtained by taking the same number of integrals or derivatives of either of the two time functions which are convolved to give $f_2(t)$.

Integrations and differentiations can be combined in the following manner. First, we find the nth derivative of the output function, in the form

$$f_2^n(t) = \int_{-\infty}^{\infty} f_1(T)h^n(t - T)\,dT. \tag{13–39}$$

Next, we interchange the functions $f_1(t)$ and $h(t)$, according to theorem 1. The result is

$$f_2^n(t) = \int_{-\infty}^{\infty} f_1(t - T)h^n(T)\,dT. \tag{13–40}$$

Finally, on the function of Eq. (13–40) we perform m integrations, which result in

$$f_2^{n-m}(t) = \int_{-\infty}^{\infty} f_1^{-m}(t - T)h^n(T)\,dT. \tag{13–41}$$

According to Eq. (13–41) the integrations and differentiations performed on the input function or the impulse response directly affect the output function. For the particular case in which the number of integrations

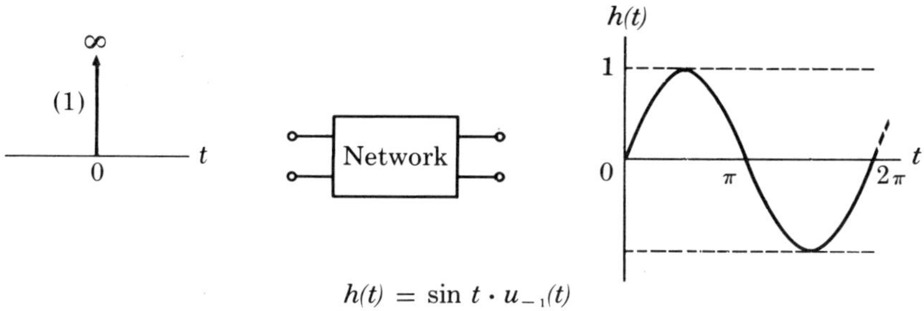

(a) The impulse response

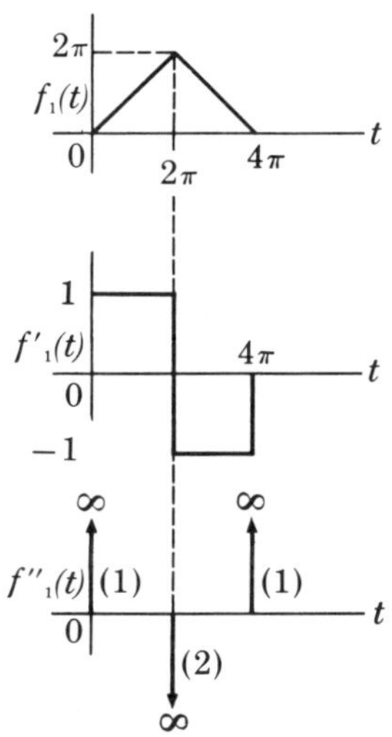

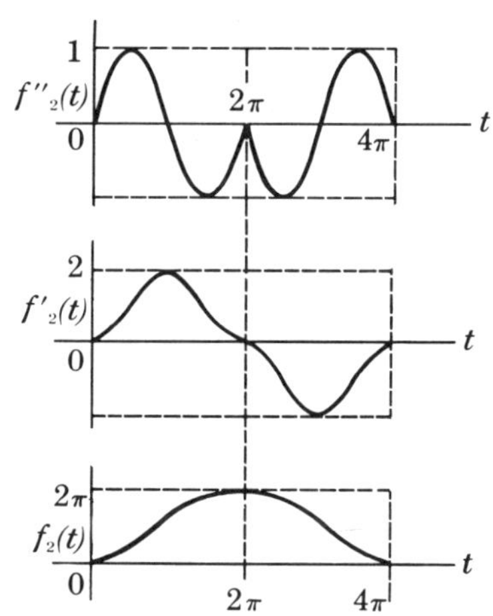

(b) The input function and its derivatives

(c) The second derivative of the output function and its integrals

FIG. 13-9. An approximate method of convolution.

of one of these function equals the number of differentiations of the other function, the output function itself is obtained. This interesting result can be expressed in general terms with the notation of Eq. (13–30):

$$f_2^{n-m}(t) = f_1^n(t)*h^{-m}(t) = f_1^{-m}(t)*h^n(t). \tag{13–42}$$

Equation (13–42) tells us that we can operate on the input function and the impulse response function in various ways before performing the convolution. In particular we are often interested in differentiating one of the functions until it is represented by impulses. *The convolution of a waveform with an impulse is particularly simple. It is equal to the waveform!*

Let us now consider the example shown in Fig. 13–9, which will illustrate the use of Eq. (13–42) to evaluate convolution integrals. The network shown in Fig. 13–9(a) has an impulse response expressed by

$$h(t) = \sin t \cdot u_{-1}(t). \tag{13–43}$$

The input function $f_1(t)$ is shown in Fig. 13–9(b), along with its first two derivatives. The first derivative equals $+1$ between $t = 0$ and $t = 2\pi$ and -1 between $t = 2\pi$ and $t = 4\pi$. The second derivative is a unit impulse at $t = 0$, a negative impulse of value 2 at $t = 2\pi$, and a unit impulse at $t = 4\pi$. Since the second derivative of the input function consists of nothing but impulses the convolution of the second derivative with any other function is very simple. We shall convolve it with the impulse response and obtain the second derivative of the output function, $f_2(t)$, according to the formula

$$f_2''(t) = f_1''(t)*h(t). \tag{13–44}$$

The unit impulse at the origin produces an output $f_2''(t)$ between $t = 0$ and $t = 2\pi$ which is given by Eq. (13–43), namely, $\sin t$. At $t = 2\pi$ a negative impulse of value 2 is applied, and a negative impulse response is produced. This response is a sine wave of the same frequency as that between $t = 0$ and $t = 2\pi$, but with an amplitude twice as large. Thus between $t = 2\pi$ and $t = 4\pi$ the second sine wave cancels the first, producing a net sine wave which is negative and has unit amplitude. At $t = 4\pi$ a second, positive unit impulse occurs, producing another sine wave which exactly cancels the resultant of the first two sine waves for $t > 4\pi$. The complete response function, $f_2''(t)$, is shown in Fig. 12–9(c).

In order to obtain the output function $f_2(t)$ the curve we have just obtained must be integrated twice. The first integral, $f_2'(t)$, is represented by the function $1 - \cos t$ in the interval from $t = 0$ to $t = 2\pi$, and by the mirror image of this function in the interval from $t = 2\pi$ to $t = 4\pi$. The integral of $f_2'(t)$ gives the desired response function $f_2(t)$. It is a positive pulse with a maximum value of 2π at $t = 2\pi$, as shown in Fig. 13–9(c). For the interval from $t = 0$ to $t = 2\pi$, the analytical expression for it is

$$f_2(t) = t - \sin t. \tag{13–45}$$

The function $f_2(t)$ is symmetrical about the ordinate $t = 2\pi$, returning to zero in the interval from $t = 2\pi$ to $t = 4\pi$.

The choice of convolving the second derivative of the input function with the impulse response function is only one of the possibilities in this problem. We might have integrated the impulse response function twice and then convolved the result with the second derivative of the input function. This operation would yield the output function $f_2(t)$ directly.

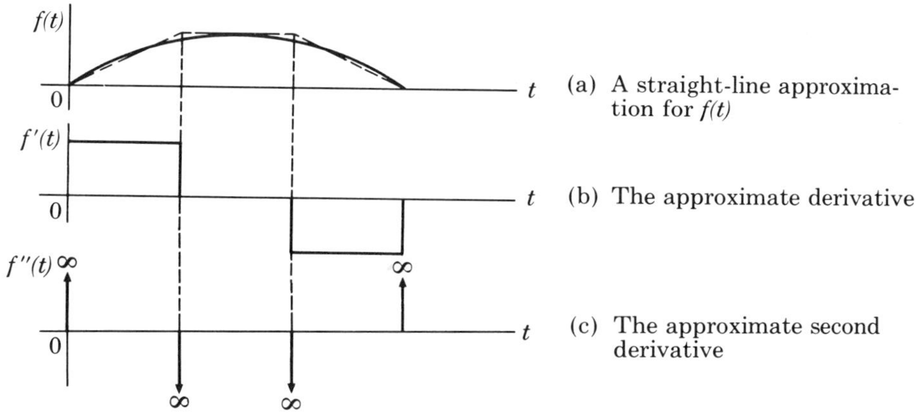

FIG. 13–10. A representation of the second derivative of a function by impulses.

The procedure involved in the example of Fig. 13–9 can be extended to include arbitrary input functions. If we approximate an input function by straight-line segments and then differentiate twice we will obtain a series of impulse functions, as illustrated in Fig. 13–10. Here the original function is shown in Fig. 13–10(a). The broken lines represent an approximation to the original function, which is formed entirely by a few straight segments. The first derivative of the approximate function is shown in Fig. 13–10(b) and the second derivative in Fig. 13–10(c). The convolution of this second derivative with the impulse response of any network is an easy procedure; it is simply a summation of several impulse responses starting at different instants and is readily obtained by graphical or numerical procedures. The response of the network to the original function is the second integral of the function obtained by convolution. In this procedure the differentiation of the input does violence to the function, but the integration of the output smooths it again.

The response function obtained by the convolution method we have just discussed is only an approximation to the correct response function. The more accurately the original function is represented by straight lines, the better is the approximation. A major advantage of the method is that an estimate of the magnitude of the error in the output can easily be obtained. The difference between the original curve and the straight-line approximation is plotted as a separate curve. This curve is called the error function, and the response of the network to the error function is the error in the response. This error can be computed by the same procedure we used to obtain the response for the original function of Fig. 13–10. That is, the error is approximated by straight-line segments, differentiated twice, and convolved with the impulse response of the network. The second integral of the function obtained in this way is a close approximation to the error in the computed response function.

Convolution methods are becoming increasingly popular in network theory. Indeed they are ideally suited to the present state of the art. They allow the network analyst to obtain an approximate idea of the nature of the solution, with very little numerical effort. The same network formulations can be processed by large-scale digital computers whenever more accurate answers are warranted.

13–6 Summary.

Impulse response

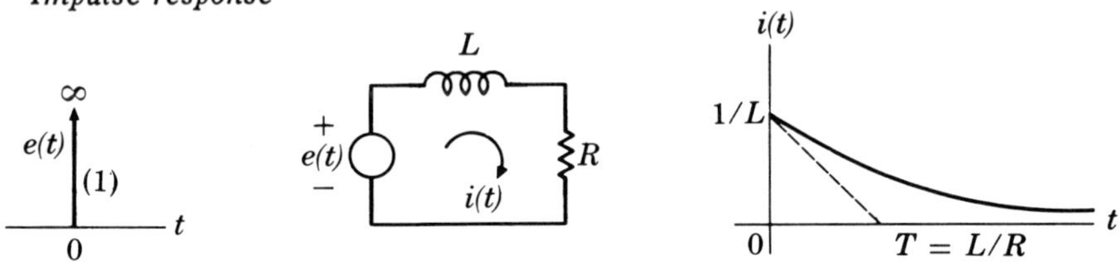

Pulse response

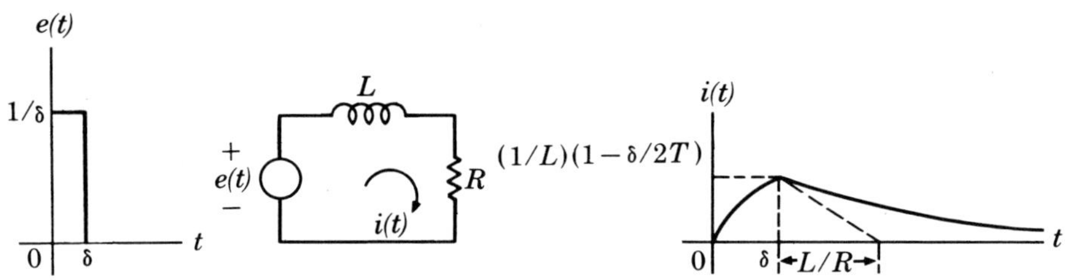

The convolution integral

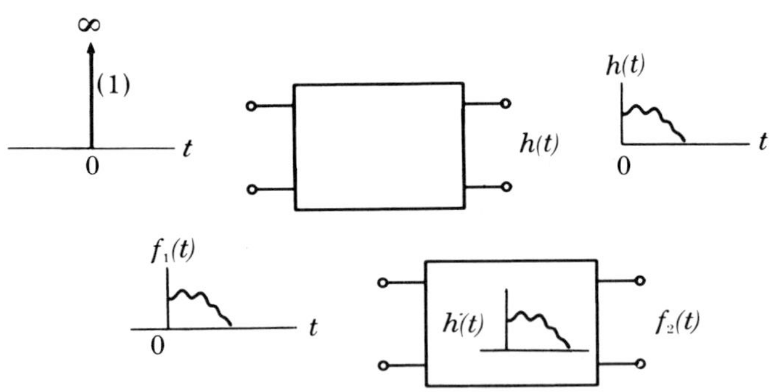

$$f_2(t) = \int_{-\infty}^{\infty} f_1(T)h(t - T)\,dT = f_1(t)*h(t)$$

$$= \int_{-\infty}^{\infty} h(T)f_1(t - T)\,dT = h(t)*f_1(t)$$

Theorems on convolution

1. Interchange of functions

$$f_1(t)*h(t) = h(t)*f_1(t)$$

2. Differentiation

$$\frac{d[f_2(t)]}{dt} = f_1(t)*h'(t) = f_1'(t)*h(t)$$

3. Integration

$$\int_{-\infty}^{t} f_2(t)\,dt = f_1(t)*h^{-1}(t) = f_1^{-1}(t)*h(t)$$

4. Combined integration and differentiation

$$f_2^{m-n}(t) = f_1^m(t)*h^{-n}(t) = f_1^{-n}(t)*h^m(t)$$

The scanning function (graphical convolution)

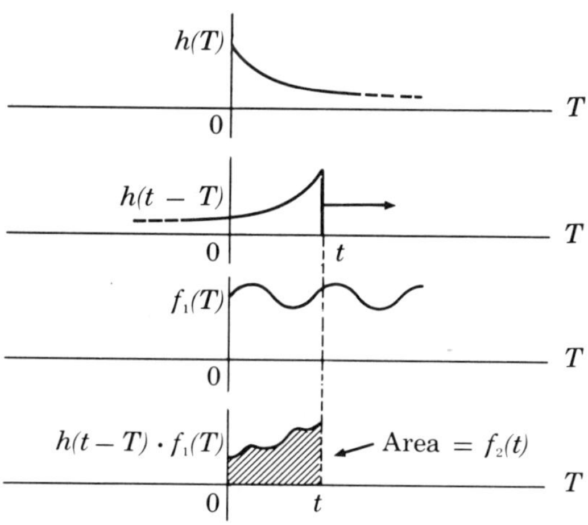

Analytical convolution

$$f_2(t) = \int_{-\infty}^{\infty} f_1(t - T)\cdot u_{-1}(t - T)h(T)\cdot u_{-1}(T)\,dT$$

$$= \int_0^t f_1(t - T)h(T)\,dT$$

Approximate convolution

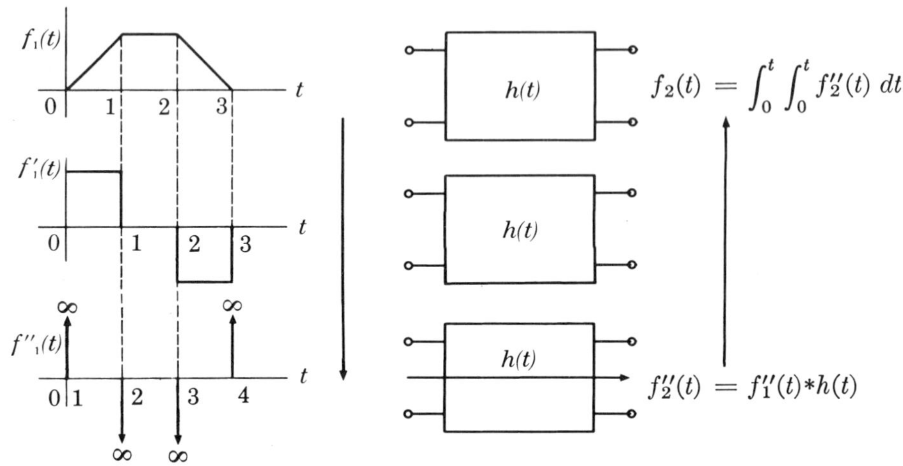

REFERENCES

Chapter 13 deals with the solution of network problems in the time domain, for rather general types of excitation. This topic is omitted in most circuit theory texts, although it is potentially as important as the frequency-domain or transform method. Most of the references given here are graduate-level texts.

1. J. A. Aseltine, *Transform Method in Linear System Analysis,* McGraw-Hill, New York, 1958. Convolution theory is developed in a mathematical fashion. A distinction is made between Fourier and Laplace convolution.

2. D. K. Cheng, *Analysis of Linear Systems,* Addison-Wesley, Reading, Mass., 1959. This text covers approximately the same ground as does Aseltine's. It includes proofs of convolution based on the superposition of impulses and on the superposition of step functions.

3. Seshu and Balabanian, *Linear Network Analysis,* John Wiley & Sons, New York, 1959. These authors devote a chapter to time-domain solutions of network problems, under the heading of integral solutions.

4. M. E. Van Valkenburg, *Network Analysis,* Prentice-Hall, New Jersey, 1955. In this text, convolution is stated without proof and used in a number of examples.

5. J. G. Truxal, *Control System Synthesis,* McGraw-Hill, New York, 1955. Truxal gives an excellent over-all treatment of time-domain methods. In addition to convolution he covers impulse trains and Z-transforms.

6. F. W. Bubb, "A New Linear Operational Calculus," Air Force Technical Rpt 6581, May 1951, Wright Air Development Center, Dayton, Ohio. Bubb introduces the methods of impulse trains for the solution of linear problems.

7. Ragazinni and Zadeh, "The Analysis of Sampled-Data Systems," *Trans. AIEE,* Vol. **71,** Part 2 (1952). The original paper on the Z-transform notation. The method goes back to Laplace, but the notation introduced by Ragazinni and Zadeh has been applied in a wide range of fields.

EXERCISES*

Section 13–2

1. A voltage pulse of height $10/\delta$ and duration δ sec is applied to an inductance of 5 henrys at $t = -\delta$ sec. Find the current produced for $t > 0$ and compare it with the current produced by an impulse voltage with the same area occurring at $t = 0$.

2. A current pulse of height $10/\delta$ and duration δ sec is applied to an inductance of 5 henrys at $t = 0$. Write an expression for the resulting voltage and compare it with the voltage produced by an impulse current of the same area occurring at $t = 0$. [*Note:* A doublet is characterized by its second integral.]

3. A pulse of current of width δ sec and height $10/\delta$ is applied to a capacitance of 5 farads. Find the voltage response as a function of time and compare it with the response produced by an impulse of the same area occurring at $t = 0$.

4. A voltage pulse of height $10/\delta$ and width δ sec is applied to a capacitance of 5 farads. Find the current and compare it with the current produced by an impulse voltage of the same area occurring at $t = 0$.

5. The voltage pulse in the circuit of Fig. 13–11 has a height of $10/\delta$ and a width of δ sec. It begins at $t = -\delta$ and ends at $t = 0$. Find the response voltage $e_2(t)$ produced by this source for $t > 0$ and compare it with the response produced by an impulse voltage of the same area occurring at $t = 0$. For what range of δ are the two responses equal within 1%?

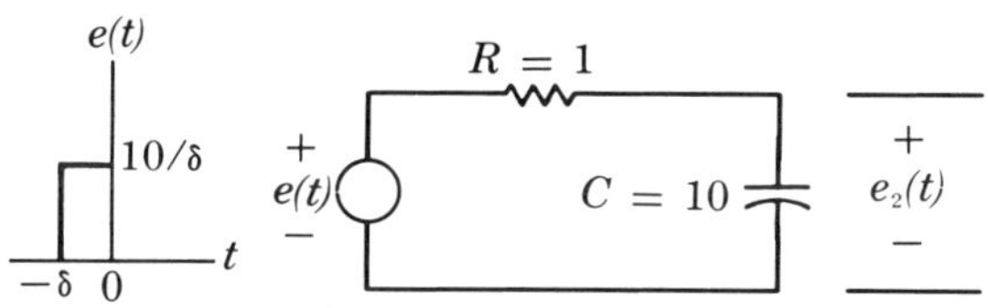

FIGURE 13–11

6. Find the voltage across the resistance in the circuit of Exercise 5. Compare this result with the response to an equivalent impulse occurring at $t = 0$. Find the value of δ which will result in a 1% difference between the two responses (a) so far as the area of the initial pulse is concerned and (b) so far as the exponential decay is concerned.

7. The impulse response and the pulse response in the circuit of Fig. 13–12 are the same if the pulse width δ is short with respect to the period $T = 2\pi\sqrt{LC}$. What are the two responses for $t > 0$? Obtain expressions for the current in each case.

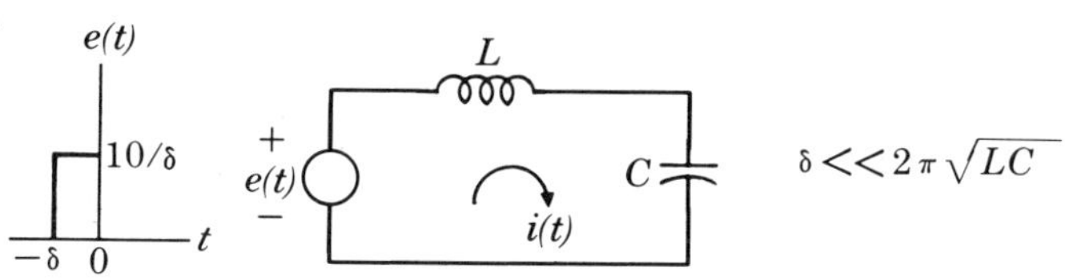

FIGURE 13–12

---

* Numerical values are in volts, amperes, ohms, henrys, and farads.

8. Find the value of δ in the circuit of Fig. 13–12 which will result in a 1% difference between the pulse and impulse responses so far as the maximum amplitudes of the sine waves are concerned. [*Hint:* Use energy.]

Section 13–3

9. In the circuit of Fig. 13–13 find the current response to the voltage pulse in the regions from $t = 0$ to $t = 4$ and from $t = 4$ to $t = \infty$. Also find the response by replacing the pulse by four impulses at $t = 1, 2, 3,$ and 4 sec.

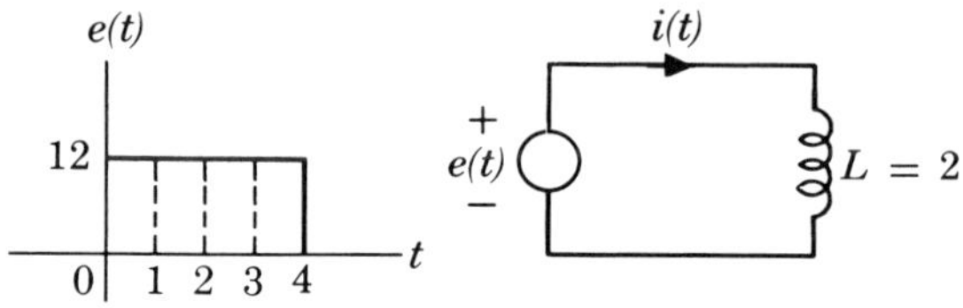

FIGURE 13–13

10. Find the voltage across the capacitance in the circuit of Fig. 13–14 for $t > 0$. Find the approximate solution by replacing the pulse by 4 impulses at $t = 0, -1, -2,$ and -3 sec.

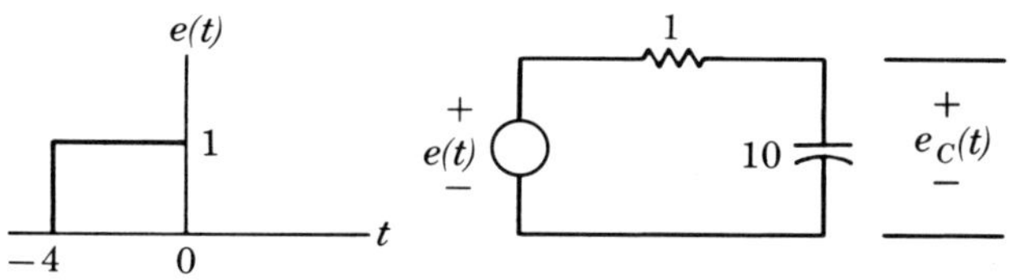

FIGURE 13–14

11. Repeat the approximate solution for the circuit of Exercise 10: (a) by considering the impulses to occur at the beginning of the intervals which they represent and (b) by considering the impulses to occur at the middle of the intervals which they represent.

12. The circuit of Fig. 13–15 is excited by the exponential voltage pulse e^{-t}. The accurate solution for the current is $i(t) = te^{-t} \cdot u_{-1}(t)$. Find the approximate convolution of the input and the impulse response by replacing the input function by impulses at $t = 0.25, 0.5, 0.75, 1.00,$ and 1.25 sec. What is the maximum value of the response pulse as obtained by the approximate method?

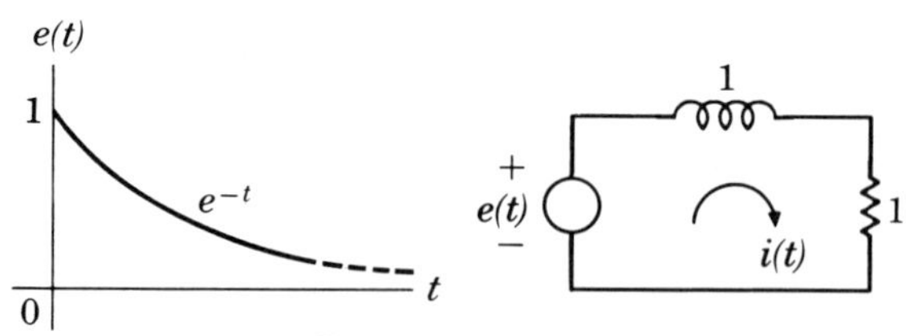

FIGURE 13–15

13. A current pulse of width 2π sec is applied to the L–C circuit of Fig. 13–16. The time $t = 0$ coincides with the end of the pulse. Find the voltage response across the circuit for $t > 0$ by means of an approximate convolution of the impulse response of the circuit with a four-impulse approximation to the pulse.

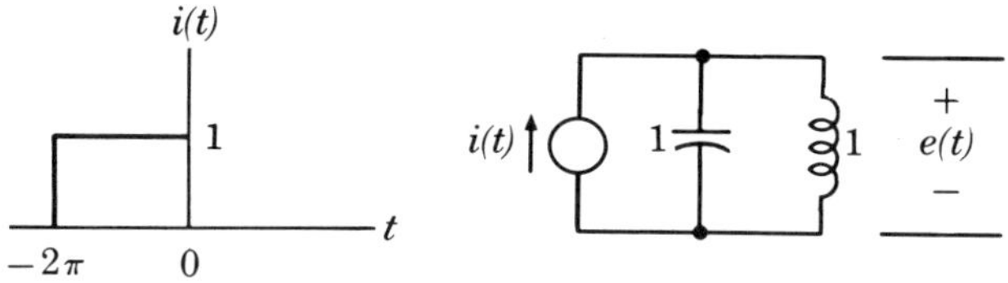

FIGURE 13–16

14. The output voltage of a given circuit is expressed by

$$e_2(t) = \int_0^\infty e_1(T)h(t - T)\,dT,$$

where $e_1(t)$ is the input voltage and $h(t) = e^{-t} \cdot u_{-1}(t)$ is the impulse response.

(a) If $e_1(t)$ is the impulse $u_0(t)$ what is the initial value ($t = 0^+$) of $e_2(t)$ and what is its value at $t = \infty$?

(b) If $e_1(t)$ is the step $u_{-1}(t)$ what is the initial value of $e_2(t)$ and what is its value at $t = \infty$?

15. Repeat Exercise 14 if the impulse response is $h(t) = u_0(t) - e^{-t} \cdot u_1(t)$.

16. The impulse response of a certain network is a pulse of height unity and duration 1 sec. The input function is the step $u_{-1}(t)$. Find the output $f_2(t)$ by graphical convolution. In the process sketch $h(t)$, $h(t - T)$, $f_1(T)$, the product $h(t - T) \cdot f_1(T)$, and $f_2(t)$. Give the analytic expressions for $f_2(t)$ in the regions $0 < t < 1$ and $t > 1$.

17. The impulse response of a network is $h(t) = u_{-1}(t) - u_{-1}(t - 1)$. The input function is $f_1(t) = 100e^{-t/100}$. Find the output in the regions $0 < t < 1$ and $t > 1$ by means of the graphical scanning and integrating procedure.

18. The impulse response of a network is $h(t) = u_{-1}(t) - 2u_{-1}(t - 1) + u_{-1}(t - 2)$. The input is the step function $f_1(t) = u_{-1}(t)$. Find the output function by means of the graphical scanning procedure, and sketch it.

Section 13–4

19. Perform an analytical convolution of the two functions

$$f_1(t) = u_{-1}(t), \qquad f_2(t) = (1 - e^{-t}) \cdot u_{-1}(t),$$

(a) with $f_1(t)$ considered to be the input and (b) with $f_2(t)$ considered to be the input.

20. Convolve the two functions

$$f_1(t) = e^{-5t} \cdot u_{-1}(t), \qquad f_2(t) = e^{-10t} \cdot u_{-1}(t),$$

(a) with $f_1(t)$ as the input and (b) with $f_2(t)$ as the input.

21. Find the convolution of two functions

$$f_1(t) = e^{-t}u_{-1}(t), \qquad f_2(t) = u_{-1}(t - 1).$$

Note that the output will not start at $t = 0$, and therefore add a step function to keep it zero until the appropriate time.

22. Convolve the two functions

$$f_1(t) = e^{-t}u_{-1}(t - 1), \qquad f_2(t) = u_{-1}(t - 1).$$

At what time is the input first applied? What is the delay in the network? When does the output first appear? Apply the appropriate $u_{-1}(t)$ function.

23. Convolve the input pulse $u_{-1}(t) - u_{-1}(t - 1)$ with the impulse response function $h(t) = e^{-t} \cdot u_{-1}(t)$. Be careful to restrict the ranges of validity of the components of the output function.

24. Obtain the convolution of $\cos t$ and $\cos t$ if both functions start at $t = 0$.

25. Convolve $\sin t$ with $\sin t$ if both functions start at $t = 0$.

26. Convolve $\cos t$ with e^{-t} considering both functions to start at $t = 0$.

27. Convolve $\sin t$ with e^{-t} considering both functions to start at $t = 0$.

28. The response of a network to the step function excitation $u_{-1}(t)$ is $e^{-t} \cdot u_{-1}(t)$. What is the response to a suddenly applied function of the form $(\sin t + \cos t)u_{-1}(t)$?

Section 13–5

29. The impulse response of a network is $h(t) = e^{-t} \cdot u_{-1}(t)$. The input function is $f_1(t) = u_{-1}(t) - u_{-1}(t - 1)$.

(a) Convolve the two functions by taking the derivative of the input and the integral of the impulse response.

(b) Convolve the derivative of the input function with the impulse response and integrate the result to obtain the desired output function.

30. The impulse response of a network is $h(t) = \cos t \cdot u_{-1}(t)$. The input function $f_1(t)$ is shown in Fig. 13–17. Obtain the output function by convolving the second derivative of the input with the second integral of the impulse response.

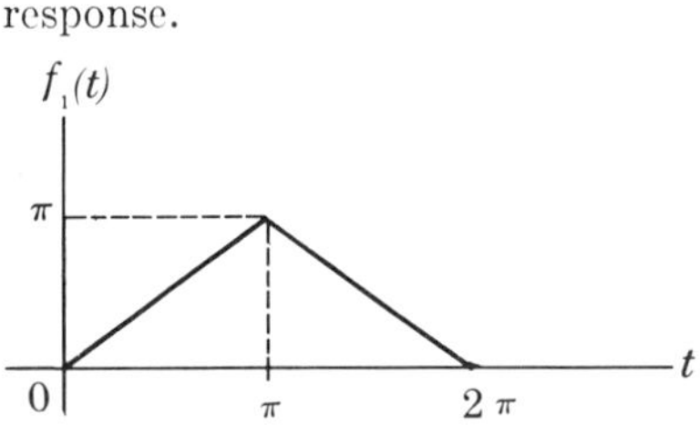

FIGURE 13–17

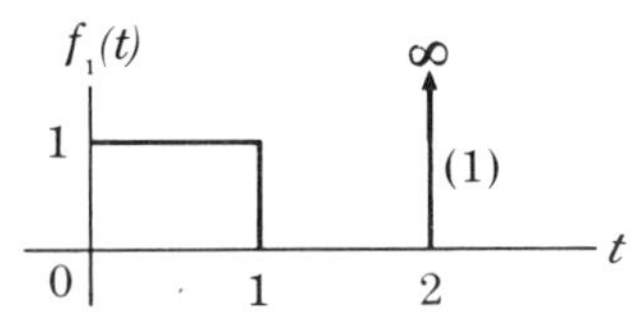

FIGURE 13–18

31. The impulse response of a network is $h(t) = e^{-t} \cdot u_{-1}(t)$. The input function $f_1(t)$ is shown in Fig. 13–18. Obtain the output time function by using superposition plus the technique of convolving derivatives with integrals.

32. Use the technique of convolving integrals and derivatives to convolve the function $f_1(t)$ shown in Fig. 13–19 with an impulse response $h(t) = e^{-t} \cdot u_{-1}(t)$.

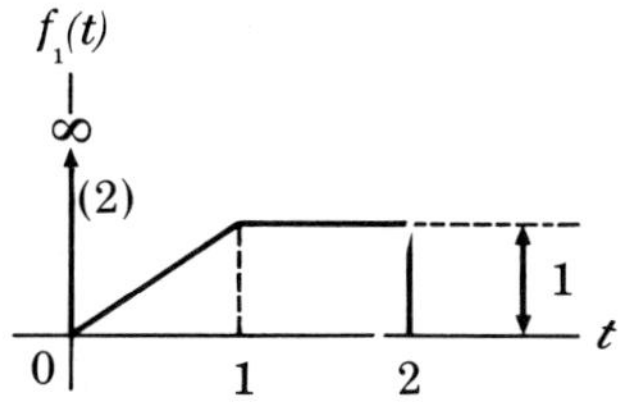

FIGURE 13–19

33. In the design of "sweep" circuits the desired output function is a perfect ramp function. Often the impulse response of the circuit is of the form $h(t) = e^{-t} \cdot u_{-1}(t)$. A perfect ramp output can be obtained by using a combination of an impulse, a step, and a ramp function at the input, as shown in Fig. 13–20. If the step has a height of unity, and the ramp has a slope of unity, what area A should be used in the impulse?

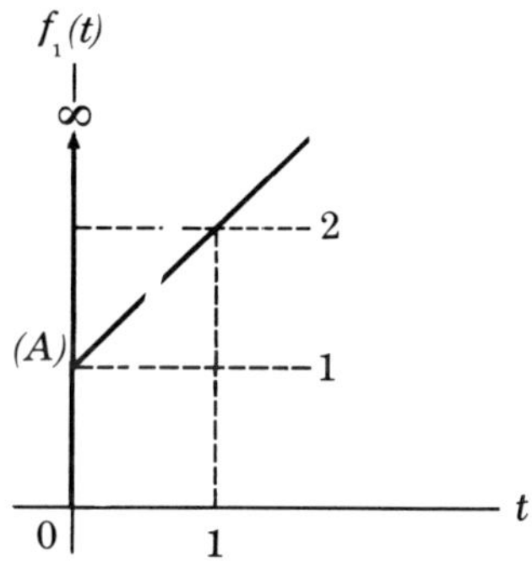

FIGURE 13–20

34. A network has an impulse response which is the doublet pulse $h(t) = u_{-1}(t) - 2u_{-1}(t-1) + u_{-1}(t-2)$. Show that such a network represents an approximate differentiator by applying a ramp function to it and computing the output by convolution theory.

35. The impulse response of a network is $h(t) = e^{-t} \cdot u_{-1}(t)$. The input function is the approximation to a step function, $f_1(t)$, shown in Fig. 13–21. Using convolution methods find the approximate difference between the actual response and the response to a perfect step function.

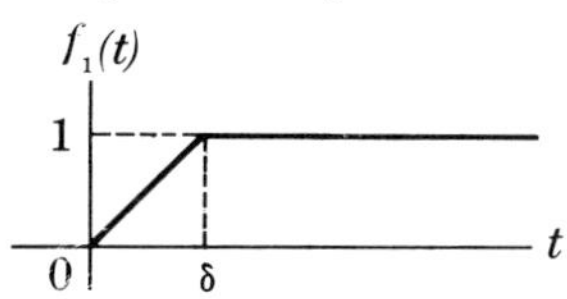

FIGURE 13–21

36. Repeat Exercise 35 by finding the input error function and convolving it with $h(t)$ to obtain the error in the output.

37. An input pulse to a network is represented by the numerical data

t	0	1	2	3	4	5	6
$f_1(t)$	0	1	2	2	4	1	0 .

The second integral of the impulse response is represented by the numerical data

t	0	1	2	3	4
$h^{-2}(t)$	0	1	1	0	0 .

Represent the input data by a straight-line approximation and obtain the output function by numerical convolution.

CHAPTER 14

POWER AND ENERGY IN CIRCUITS WITH ENERGY-STORAGE ELEMENTS

"Do not underestimate the power of poetry. Poets are the unacknowledged legislators of the world."

Percy Bysshe Shelley

14–1 Introduction. The power of poetry lies in its ability to sway the minds of man. Electricity is almost as subtle as poetry can be, but its power is manifested in more physical ways. In most of our work we have been dealing with voltage and current. We must not lose sight of the central significance of power and energy, however. The importance of electricity lies in what it can do for man. In this chapter we shall extend the concepts of Chapter 6 to include circuits which have energy-storage elements as well as energy-dissipation elements.

The pure resistance circuits of Chapter 6 are capable only of dissipating energy. They can transmit energy from one place to another, and they can convert it to some other form (usually heat), but they can not store energy. Circuits with resistance and capacitance or circuits with resistance and inductance can both store and dissipate energy. When energy is initially stored in them, currents, as well as voltages, are produced which gradually dissipate the energy in the resistive elements. Whenever the circuit is changed from one energy state to another, these transient currents flow. In addition to the initial and final energy states, we shall be interested in the energy lost by the transient currents.

Circuits with pure inductance and pure capacitance, and thus no energy losses, produce "transient" currents which do not die out with time. In the course of these transients the energy "swaps" back and forth between the inductance and the capacitance in much the same way that kinetic and potential energy are interchanged in a pendulum system. Lossless, or "conservative" systems, are particularly adapted to treatment from the energy point of view. At the instant all the energy is stored in the inductance the maximum current must occur in the inductance, and at the instant all the energy is stored in the capacitance the maximum voltage must occur across it.

A circuit with resistance, inductance, and capacitance exhibits some of the characteristics of both the single-energy-storage circuits and the lossless circuits. Energy can swap back and forth between the inductance and the capacitance, but as it does so energy is lost in the resistance.

Hence the transient currents and voltages die out with time. For such a circuit we are interested in the initial and final energy states and in the energy lost because of the transient currents.

The impulse function poses certain problems when we wish to consider energy relations. Since the impulse is infinite, the power during the impulse is infinite. The energy delivered by an impulse can be either zero or finite, depending on the circuit element which is involved, and part of the energy may apparently vanish into thin air. This latter phenomenon occurs in lossless circuits and is a consequence of infinite current flowing in zero resistance.* A subtle form of this problem occurs when energy is exchanged between two lossless elements of the same sort. Impulses are produced, and part of the energy is lost.

14–2 Energy and power in single-element circuits. We have had occasion in prior chapters to consider energy and power in single-element circuits. Indeed initial energy storages have been one of the most frequent and important methods of exciting our circuits. It will be convenient to review briefly some of this material and to use it in a slightly different way, in which power and energy are the end results.

Resistance. The resistance element is characterized by the volt-ampere equation

$$e = iR. \tag{14–1}$$

Voltage has dimensions of energy (joules) per coulomb, and current has dimensions of coulombs per second; hence power is the product of current and voltage. The power dissipated in a resistance is

$$p = e \times i = i^2 R = \frac{e^2}{R} \quad \text{joules per second, or watts.} \tag{14–2}$$

The energy dissipated during any interval of time is the integral of the power with respect to time. Thus

$$W_R(t) = \int_0^t (e \times i)\, dt \quad \text{joules.} \tag{14–3}$$

Equation (14–3) gives the energy dissipated in the resistance from time $t = 0$ to time $t = t$. This energy is a function of time. Another form of Eq. (14–3) is sometimes useful:

$$p = \frac{dW_R}{dt} \quad \text{watts.} \tag{14–4}$$

---

* In any lossless system it is impossible to conserve both energy and momentum. We conserve momentum (charge or flux linkage).

In Eq. (14–4) power is expressed as the rate of change of energy. If both power and energy are plotted as functions of time the power curve is the derivative of the energy curve, and the energy curve is the functional integral of the power curve.

Inductance. The inductance element is defined by the volt-ampere equation

$$e = L\frac{di}{dt}. \tag{14–5}$$

The power drawn by an inductance is

$$p_L = e \times i = L\frac{di}{dt} \times i. \tag{14–6}$$

The power of Eq. (14–6) is proportional to the product of the current and the rate of change of the current. Power flows into the inductance only when the current is changing. The energy stored in the inductance is

$$W_L(t) = \int_0^t (e \times i)\, dt = \int_0^t \left(L\frac{di}{dt} \times i\right) dt. \tag{14–7}$$

We can change the variable in Eq. (14–7) from time to current. The integration then occurs from the current at $t = 0$ to the current at $t = t$, namely, i. The two dt's effectively cancel, and if $i(0) = 0$ then

$$W_L(i) = \int_0^i Li\, di = \tfrac{1}{2}Li^2. \tag{14–8}$$

The energy given by Eq. (14–8) is a function of the current through the inductance. The rate of change of the energy is the power entering the inductance:

$$p_L = \frac{d}{dt}\left(\tfrac{1}{2}Li^2\right) = L \times i \times \frac{di}{dt}. \tag{14–9}$$

The power obtained in Eq. (14–9) from the rate of change of the stored energy is equal to the power obtained in Eq. (14–6) from the volt-ampere product.

Capacitance. The capacitance element is defined by the volt-ampere equation

$$i = C\frac{de}{dt}. \tag{14–10}$$

The power which flows into a capacitance is

$$p_C = e \times i = C\frac{de}{dt} \times e. \tag{14–11}$$

The power of Eq. (14–11) is proportional to the product of the voltage and the derivative of the voltage. Power flows into the capacitance only when the voltage across it is changing. The total energy stored at any instant is

$$W_C(t) = \int_0^t (e \times i)\, dt = \int_0^t \left(C\frac{de}{dt} \times e\right) dt. \tag{14–12}$$

If we change the variable in Eq. (14–12) from time to voltage, the lower limit is the voltage at $t = 0$, which we take to be zero, and the upper limit is the voltage at $t = t$, which we simply call e. The total energy then is

$$W_C(e) = \int_0^e (C \times e)\, de = \tfrac{1}{2}Ce^2. \tag{14–13}$$

The energy in the capacitance, as given by Eq. (14–13), depends only on the voltage across it. The rate of change of this energy is the power flowing into the capacitance:

$$p_C = \frac{d}{dt}\left(\tfrac{1}{2}Ce^2\right) = C \times e \times \frac{de}{dt}\cdot \tag{14–14}$$

The power obtained as the rate of change of the stored energy in Eq. (14–14) is identical to the power obtained as the volt-ampere product in Eq. (14–11). The energy stored on the capacitance can increase only if the voltage increases; furthermore, the increase is the result of the positive value of the power obtained when the voltage has a positive derivative.

Examples of power and energy. As an example of the use of the power and energy relationships we have just derived we shall find the power and energy as functions of time for the circuit of Fig. 14–1. The current in this circuit is

$$i(t) = \frac{1}{L}u_{-2}(t) = \frac{t}{L} \qquad \text{for} \quad t > 0. \tag{14–15}$$

The power is the product of the voltage and current:

$$p = \frac{u_{-2}(t) \times u_{-1}(t)}{L} = \frac{t}{L} \qquad \text{for} \quad t > 0. \tag{14–16}$$

The energy is the integral of the power:

$$W = \int_0^t \frac{t}{L}\, dt = \frac{t^2}{2L}\cdot \tag{14–17}$$

Equation (14–16) shows that the power absorbed by the inductance increases linearly with time. At the same time the total energy is increas-

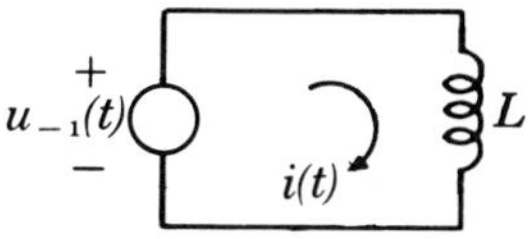

FIG. 14-1. Power and energy relationships in an inductance.

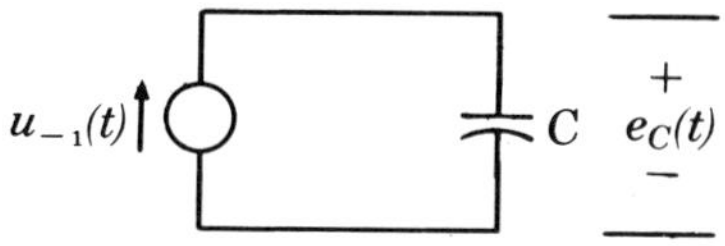

FIG. 14-2. Power and energy relationships in a capacitance.

ing along a square-law curve. Alternatively, Eq. (14-17) can be obtained from the instantaneous current in the inductance, since

$$W = \tfrac{1}{2}Li^2 = \frac{L}{2} \times \frac{t^2}{L^2} = \frac{t^2}{2L}. \tag{14-18}$$

For a second example of the use of the energy equations consider the circuit of Fig. 14-2. In this circuit a step current is applied in parallel with a capacitance. The voltage across the capacitance is

$$e_C = \frac{1}{C}\int_0^t u_{-1}(t)\,dt = \frac{u_{-2}(t)}{C}. \tag{14-19}$$

The power entering the capacitance is

$$p_C = e \times i = \frac{u_{-2}(t) \times u_{-1}(t)}{C} = \frac{t}{C} \qquad \text{for} \quad t > 0, \tag{14-20}$$

and the energy is

$$W_C = \int_0^t \frac{t}{C}\,dt = \frac{t^2}{2C} \tag{14-21}$$

or

$$W_C = \tfrac{1}{2}Ce^2 = \frac{C}{2} \times \frac{t^2}{C^2} = \frac{t^2}{2C}. \tag{14-22}$$

In this example the voltage across the capacitance increases linearly with time, as the current continues to pour coulombs into the capacitance. The power also increases linearly with time, but the energy increases as the square of the time. As the voltage builds up on the capacitance it takes increasingly more energy to pile each additional coulomb on the plates. Since the coulombs arrive at a constant rate, the energy builds up very rapidly.

We shall defer the consideration of impulse sources to a later section, because they involve some additional complex limiting processes in which a very small resistance must be added to an energy-storage element.

14–3 Power and energy in *R–L* and *R–C* circuits. When we have several elements in a circuit we are interested in the distribution of energy among them as a function of the time. A general relationship which must apply is

$$\text{Energy lost} + \text{Energy stored} = \text{Energy supplied.} \tag{14–23}$$

The derivative of this equation must also apply:

$$\text{Power lost} + \frac{d}{dt}(\text{Energy stored}) = \text{Power supplied.} \tag{14–24}$$

It is fairly evident that the energy supplied must be either stored or dissipated in the resistances of the circuit. It is perhaps not quite so obvious that the power supplied is absorbed by heating the resistances or increasing the energy stored. The application of these general relationships is best illustrated by a concrete example.

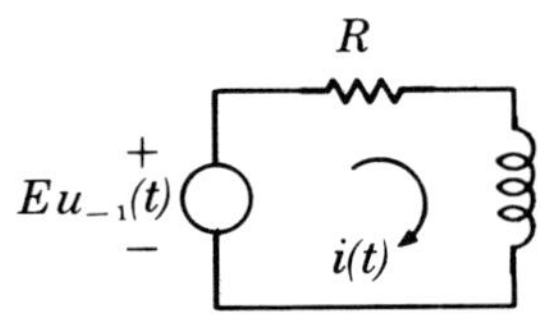

FIG. 14–3. Power and energy relationships in an *R–L* circuit.

In Fig. 14–3 we have a step voltage applied to a resistance and an inductance in series. The current is

$$i(t) = \frac{E}{R}(1 - e^{-t/T}) \qquad \text{for} \quad t > 0, \tag{14–25}$$

where $T = L/R$. We will verify the power relationship first. The power lost in the resistance as a function of time is

$$p_R = i^2R = \frac{E^2}{R}(1 - 2e^{-t/T} + e^{-2t/T}). \tag{14–26}$$

The energy stored in the inductance as a function of time is

$$W_L = \tfrac{1}{2}Li^2 = \frac{LE^2}{2R^2}(1 - 2e^{-t/T} + e^{-2t/T}), \tag{14–27}$$

and the power flowing into it is

$$\begin{aligned} p_L = \frac{dW_L}{dt} &= \frac{LE^2}{2R^2}\left(\frac{2}{T}e^{-t/T} - \frac{2}{T}e^{-2t/T}\right) \\ &= \frac{E^2}{R}e^{-t/T}(1 - e^{-t/T}). \end{aligned} \tag{14–28}$$

The power supplied by the source is

$$p_S = e \times i = E \times \frac{E}{R}(1 - e^{-t/T})$$

$$= \frac{E^2}{R}(1 - e^{-t/T}). \tag{14–29}$$

The sum of Eq. (14–26), or the power lost in the resistance, and Eq. (14–28), or the power entering the inductance, is

$$p_R + p_L = \frac{E^2}{R}(1 - e^{-t/T}). \tag{14–30}$$

Equation (14–30), which gives the total power absorbed by the circuit, agrees with Eq. (14–29), which gives the power supplied by the source. Hence the power-balance equation is satisfied.

To verify the energy-balance equation we need to compute the energy lost in the resistance as a function of time and the energy supplied by the source. In Eq. (14–27) we have an expression for the energy stored in the inductance as a function of time. The energy lost in the resistance is

$$W_R = \int_0^t i^2 R\,dt = \int_0^t \frac{E^2}{R}(1 - 2e^{-t/T} + e^{-2t/T})\,dt$$

$$= \frac{E^2}{R}\left[t - 2T(1 - e^{-t/T}) + \frac{T}{2}(1 - e^{-2t/T})\right]. \tag{14–31}$$

The energy supplied by the source is

$$W_S = \int_0^t (e \times i)\,dt = \int_0^t \frac{E^2}{R}(1 - e^{-t/T})\,dt$$

$$= \frac{E^2}{R}[t - T(1 - e^{-t/T})]. \tag{14–32}$$

The sum of the energy lost in the resistance, Eq. (14–31), and the energy stored in the inductance, Eq. (14–27), is

$$W_R + W_L = \frac{E^2}{R}\left[t - 2T(1 - e^{-t/T}) + \frac{T}{2}(1 - e^{-2t/T})\right]$$

$$+ \frac{E^2}{R}\left(\frac{T}{2} - Te^{-t/T} + \frac{T}{2}e^{-2t/T}\right)$$

$$= \frac{E^2}{R}[t - T(1 - e^{-t/T})]. \tag{14–33}$$

The sum of the energy lost in the resistance up to time t and the energy stored in the inductance at time t, Eq. (14–33), is equal to the energy supplied by the source between $t = 0$ and $t = t$, Eq. (14–32). The energy-balance equation is therefore satisfied. Often we are interested in the energy at a particular instant. In this case the desired value of t has only to be substituted in the above equations. In the circuit of Fig. 14–3 there is a direct current which continues to flow at $t = \infty$, and the total amount of energy supplied by the source therefore can become infinite. At any instant, however, the energy balance is still maintained.

14–4 Power and energy in *L–C* circuits. Energy methods of analysis are particularly useful for L–C circuits, because no energy is lost and it is relatively easy to determine the disposition of energy in the energy-storage elements. For any lossless circuit

$$\text{Energy stored} = \text{Energy supplied.} \tag{14–34}$$

An expanded form of this equation is

$$\text{Capacitive energy} + \text{Inductive energy} = \text{Energy supplied,} \tag{14–35}$$

from which

$$\frac{d}{dt}\,(\text{Capacitive energy} + \text{Inductive energy}) = \text{Power supplied.} \tag{14–36}$$

For the case of excitation by initial energy storage no power is supplied and Eq. (14–36) becomes

$$\frac{d}{dt}\,(\text{Capacitive energy} + \text{Inductive energy}) = 0. \tag{14–37}$$

Equation (14–37) shows that when the capacitive energy increases, the inductive energy must decrease and vice versa. The energy must swap back and forth between the two elements.

We shall illustrate the use of these energy-balance equations by means of the circuit of Fig. 14–4. Here there is no external energy source, and

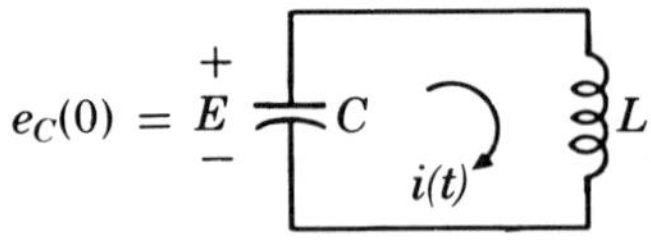

FIG. 14–4. Power and energy relationships in an L–C circuit.

the behavior is determined entirely by the energy initially stored on the capacitance. The current and voltage, as obtained in Chapter 13, are

$$\begin{aligned} e(t) &= E \cos \omega t, \\ i(t) &= \frac{E}{\omega L} \sin \omega t, \end{aligned} \tag{14–38}$$

where $\omega = 1/\sqrt{LC}$. The energy stored in the capacitance is given by

$$W_C = \tfrac{1}{2}Ce^2 = \frac{CE^2}{2} \cos^2 \omega t, \tag{14–39}$$

and the energy stored in the inductance is given by

$$W_L = \tfrac{1}{2}Li^2 = \frac{E^2}{2\omega^2 L} \sin^2 \omega t = \frac{CE^2}{2} \sin^2 \omega t. \tag{14–40}$$

The total energy stored at any instant is

$$W_C + W_L = \frac{CE^2}{2} (\cos^2 \omega t + \sin^2 \omega t) = \tfrac{1}{2}CE^2. \tag{14–41}$$

The derivative of Eq. (14–41) is, of course, equal to zero, as it should be since no power is supplied.

It will be instructive to examine the energy balance for the circuit of Fig. 14–4 in additional detail. At $t = 0$ all the energy is stored in the capacitance, and none is in the inductance. Mathematically

$$\begin{aligned} W_C(0) &= \tfrac{1}{2}CE^2, \\ W_L(0) &= 0. \end{aligned} \tag{14–42}$$

After one-quarter of a cycle of the oscillation, the voltage is zero and the current is maximum. All the energy is now stored in the inductance, and none is in the capacitance. Thus

$$\begin{aligned} W_C\left(\frac{\pi\sqrt{LC}}{2}\right) &= 0, \\ W_L\left(\frac{\pi}{2}\sqrt{LC}\right) &= \tfrac{1}{2}Li_{\max}^2. \end{aligned} \tag{14–43}$$

Since no energy is lost in the circuit, the energy stored in the inductance according to Eq. (14–43) must be equal to the energy stored in the capacitance according to Eq. (14–42). This equality provides a means of determining the maximum value of the current:

$$\tfrac{1}{2}Li_{\max}^2 = \tfrac{1}{2}CE^2,$$

from which

$$i_{\max} = E\sqrt{\frac{C}{L}} = \frac{E}{\omega L}. \tag{14–44}$$

Equations (14–42) and (14–44) make it possible for us to write by inspection the expressions for the current and voltage in the circuit. From Eq. (14–42) the peak value of the sinusoidal wave for the voltage must be E. Since the voltage starts at this peak value and then drops to zero, the phase angle must be that of a cosine. Hence we can write, by inspection,

$$e(t) = E\cos\omega t. \tag{14–45}$$

The current starts at zero at $t = 0$ and rises to a peak value of $E/\omega L$, obtained from Eq. (14–44). The wave for the current must therefore be a sine wave of peak amplitude $E/\omega L$:

$$i(t) = \frac{E}{\omega L}\sin\omega t. \tag{14–46}$$

The voltage and current of Eqs. (14–45) and (14–46) are plotted in Figs. 14–5(a) and 14–5(b), respectively. The energy functions of Eqs. (14–39) and (14–40) are plotted in Fig. 14–5(c). The sum of the two at all times is $CE^2/2$.

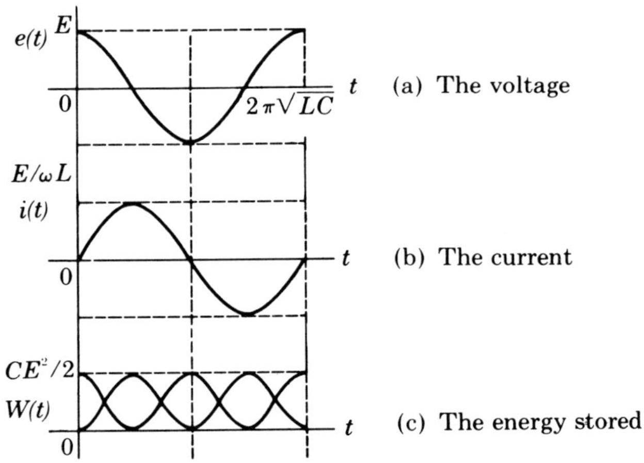

FIG. 14–5. The voltage, current, and energy relationships for the circuit of Fig. 14–4.

A second circuit in which the energy method of analysis is useful is shown in Fig. 14–6. We assume here that the switch K is moved to position 1 for one second, then moved to position 2 for one second, and finally placed in the neutral position. The problem is to find the voltage on the

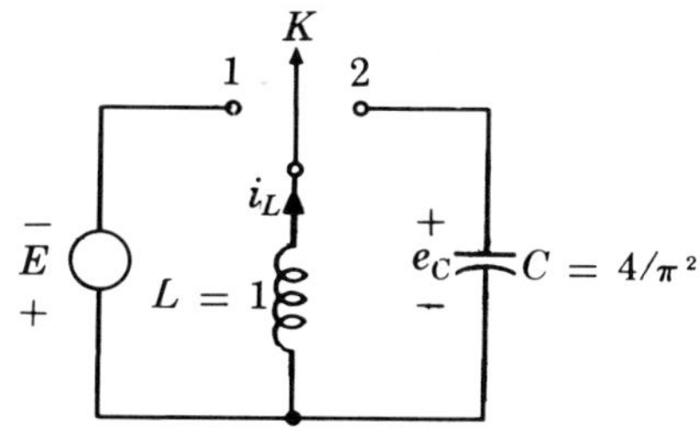

FIG. 14–6. The transference of energy from inductance to capacitance.

capacitance after the switching operation. After the first second the current in the inductance is

$$i_L = \int_0^1 E\,dt = E[t]_0^1 = E \text{ amp.}$$

The energy stored is

$$W_L = \tfrac{1}{2}Li^2 = \tfrac{1}{2}LE^2 \text{ joules.}$$

The frequency of the oscillation, if it were allowed to continue after $t = 1$, would be

$$f = \frac{\omega}{2\pi} = \frac{1}{2\pi\sqrt{LC}} = \tfrac{1}{4} \text{ cycle/sec,}$$

and the corresponding period is

$$T = \frac{1}{f} = 4 \text{ sec.}$$

The switch is left in position 2 for one second, or one-quarter of a period. In one-quarter of a period all the energy which was stored in the inductance moves to the capacitance. The voltage on the capacitance can be obtained from the energy relationship

$$\tfrac{1}{2}Ce_{\text{max}}^2 = \tfrac{1}{2}Li_{\text{max}}^2 = \tfrac{1}{2}LE^2,$$

from which

$$\frac{e_{\text{max}}}{E} = \sqrt{\frac{L}{C}}. \tag{14–47}$$

Equation (14–47) shows that the voltage on the capacitance can be made very large if the ratio of inductance to capacitance is large. This principle is used in vibrator power supplies to obtain high direct voltages from low direct voltages. Essentially the same principle is used to obtain the spark in an automobile ignition system.

14–5 Power and energy relationships in *R–L–C* circuits. The *R–L–C* circuit is more difficult to analyze than the *L–C* circuit because of the energy lost in the resistance. The general energy-balance equation is

$$\text{Energy lost} + \text{Energy stored} = \text{Energy supplied}, \tag{14–48}$$

from which

$$\text{Power lost} + \frac{d}{dt}(\text{Energy stored}) = \text{Power supplied}. \tag{14–49}$$

The application of Eqs. (14–48) and (14–49) will be illustrated with the circuit of Fig. 14–7. In this circuit the step voltage is applied at $t = 0$ and the response current is (see Chapter 13)

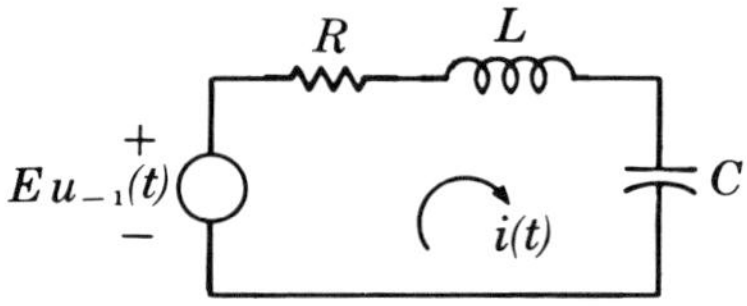

FIG. 14–7. Power and energy relationships in an *R–L–C* circuit.

$$i(t) = \frac{E}{\omega_d L} e^{-\alpha t} \sin \omega_d t \tag{14–50}$$

where

$$\alpha = R/2L,$$
$$\omega_0^2 = 1/LC,$$
$$\omega_d^2 = \omega_0^2 - \alpha^2.$$

Since the transient dies out at infinity and since there is no direct current at infinity, we can compute the total amount of energy supplied by the source:

$$W_S = \int_0^\infty (e \times i)\, dt = \frac{E^2}{\omega_d L} \int_0^\infty e^{-\alpha t} \sin \omega_d t\, dt.$$

The above integral is tabulated and has the value $\omega_d/(\alpha^2 + \omega_d^2)$. Hence

$$W_S = \frac{E^2}{\omega_d L} \frac{\omega_d}{\omega_0^2} = CE^2. \tag{14–51}$$

The total energy dissipated in the resistance from $t = 0$ to $t = \infty$ is

$$W_R = \int_0^\infty i^2 R\, dt = \frac{E^2 R}{\omega_d^2 L^2} \int_0^\infty e^{-2\alpha t} \sin^2 \omega_d t\, dt$$

$$= \frac{E^2 R}{2\omega_d^2 L^2} \left(\int_0^\infty e^{-2\alpha t}\, dt - \int_0^\infty e^{-2\alpha t} \cos 2\omega_d t\, dt \right)$$

$$= \frac{E^2 R}{2\omega_d^2 L^2} \left[\frac{1}{2\alpha} - \frac{\alpha}{2(\alpha^2 + \omega_d^2)} \right]$$

$$= \frac{CE^2}{2}. \tag{14–52}$$

The energy in the inductance at $t = \infty$ is zero, since the current is zero, and the energy in the capacitance is

$$W_C = \frac{CE^2}{2}. \tag{14–53}$$

The sum of the energy stored, Eq. (14–53), and the energy lost in the resistance, Eq. (14–52), is equal to the energy supplied by the source, Eq. (14–51).

It is significant that the energy lost in the resistance is independent of the value of the resistance. If we view the circuit as a method of charging a capacitance, we see that half of the energy is stored in the capacitance and half of it is lost in the resistance. Furthermore, we cannot decrease the amount of energy which is lost by making the resistance smaller. The losses are the same whatever resistance is present. They occur even when the resistance is vanishingly small! We shall have occasion to explore this point further in the next section.

The energy balance we calculated for the circuit of Fig. 14–7 applies only for the total energy supplied by the source to the circuit. An energy balance likewise holds for each instant of time. The only difference is that the integrals are evaluated over the range $t = 0$ to $t = t$ instead of $t = 0$ to $t = \infty$.

14–6 Energy and power in circuits involving impulses. The presence of impulse currents or voltages in a circuit poses special problems for energy and power calculations, since the power is infinite during the interval that the impulse is acting. The total energy supplied or absorbed must be evaluated by a limiting process. Let us begin our discussion by considering the effect of impulses on the separate elements resistance, inductance, and capacitance.

Resistance. A resistance excited by an impulse voltage is shown in Fig. 14–8. The current is

$$i(t) = \frac{A}{R} u_0(t), \tag{14–54}$$

and the power is

$$p_R = \frac{A^2}{R} u_0^2(t). \tag{14–55}$$

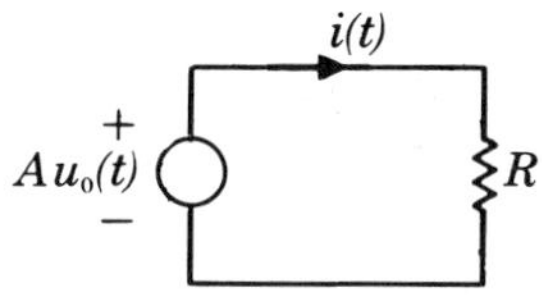

FIG. 14–8. Power and energy for an impulse-excited resistance.

The power in the resistance is infinite during the interval that the impulse is acting; it is zero for all other time.

The energy dissipated in the resistance is the integral of the power versus time curve. Hence

$$W_R = \int_{-\infty}^{t} \frac{A^2}{R} u_0^2(t)\, dt. \tag{14–56}$$

In order to evaluate Eq. (14–56) we must replace the impulse by a pulse of height $1/\delta$ and width δ. The value of the integral is then obtained in the limit as δ approaches zero. In terms of δ

$$W_R = \frac{A^2}{R}\int_0^\delta \frac{1}{\delta^2}\,dt = \frac{A^2}{R\delta}. \tag{14–57}$$

As δ approaches zero, in Eq. (14–57), the energy dissipated in the resistance becomes infinite. The infinite energy required from an impulse acting on resistance is one more indication that the impulse is an idealization. In practice, the value of δ will remain finite and thus the energy of Eq. (14–57) will remain finite, although it can become very large. The energy associated with impulse functions is not always infinite, as we shall learn in the following sections.

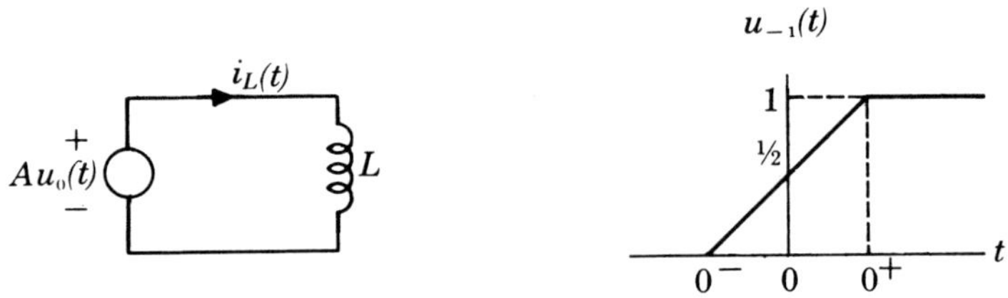

(a) Impulse-excited inductance (b) The step function at $t = 0$

FIG. 14–9. Energy relationships in an impulse-excited inductance.

Inductance. An inductance driven by an impulse voltage is shown in Fig. 14–9(a). The current in the circuit is given by the integral

$$i_L(t) = \frac{1}{L}\int_{-\infty}^{t} Au_0(t)\,dt = \frac{A}{L}\,u_{-1}(t). \tag{14–58}$$

For $t > 0$ the current of Eq. (14–58) is equal to A/L, and the energy stored in the inductance is

$$W_L = \tfrac{1}{2}Li^2 = \frac{A^2}{2L}.$$

The energy supplied by the source is more difficult to compute. It is given by the integral

$$W_S = \int_{-\infty}^{t} (e \times i)\,dt = \frac{A^2}{L}\int_{-\infty}^{t} [u_0(t) \times u_{-1}(t)]\,dt. \tag{14–59}$$

Many students replace the step function $u_{-1}(t)$ in the integral of Eq. (14–59) by unity and thus obtain a value for the integral which is too large by a factor of two. The correct value of $u_{-1}(t)$ is one-half. That this is so can best be seen by referring to Fig. 14–9(b). The step function rises in value from zero to one over the interval from $t = 0^-$ to $t = 0^+$. Right at $t = 0$ it has a value of one-half. Another way of viewing the matter is to note that the impulse has a constant value over the interval from $t = 0^-$ to $t = 0^+$ and that it must be multiplied by the average value of the step function, which again is one-half. With the correct value for $u_{-1}(t)$, the equation for the energy supplied by the source becomes

$$W_S = \frac{A^2}{2L}. \tag{14–60}$$

This value agrees with that for the energy stored in the inductance.

Not everyone, however, is satisfied with this argument. Suppose it is the step function which is the excitation, they say, and the impulse which is the response, as shown in Fig. 14–10(a). Let us analyze the circuit.

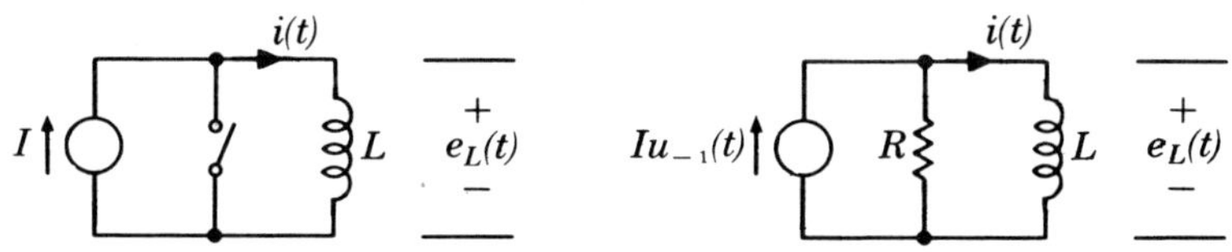

(a) Switch opened at $t = 0$ (b) The "true" equivalent circuit

FIG. 14–10. An example in which the switch dissipates the same amount of energy as is stored in the inductance.

When the switch is opened the current is

$$i(t) = Iu_{-1}(t),$$

and the voltage across the inductance is

$$e_L(t) = L\frac{di}{dt} = LIu_0(t). \tag{14–61}$$

When the current I is finally established the energy stored in the inductance is

$$W_L = \tfrac{1}{2}LI^2, \tag{14–62}$$

and the apparent energy supplied by the source is

$$W_S = \int_{-\infty}^{t} LI^2 u_0(t) u_{-1}(t)\, dt.$$

In this case, or so the argument runs, the step function $u_{-1}(t)$ is surely unity, and the energy from the source is

$$W_S = LI^2. \tag{14–63}$$

The apparent discrepancy between Eqs. (14–62) and (14–63) is explained only by assuming that *the switch has a resistance* which causes the current applied to the inductance to have a finite rise time. The switch absorbs the energy which is apparently lost. The "true" circuit is shown in Fig. 14–10(b). We shall compute the energy relationships in this circuit for a finite value of R, and then observe what happens as R is made infinite. To begin, the voltage is

$$e = IRe^{-Rt/L}u_{-1}(t). \tag{14–64}$$

The energy lost in the resistance is

$$W_R = \int_0^\infty I^2Re^{-2Rt/L}\,dt = \tfrac{1}{2}LI^2. \tag{14–65}$$

When the current I is finally established the energy stored in the inductance is again

$$W_L = \tfrac{1}{2}LI^2. \tag{14–66}$$

The energy supplied by the source is

$$W_S = \int_{-\infty}^\infty (e \times i)\,dt = \int_0^\infty I^2Re^{-Rt/L}\,dt = LI^2. \tag{14–67}$$

The energy balance is now satisfied, since the sum of the energy lost in the resistance, Eq. (14–65), and the energy stored in the inductance, Eq. (14–66), is equal to the energy supplied by the source, Eq. (14–67). What is even more significant is that the energy balance in no way depends on R. The same equations apply even when R is infinite. Thus, *a switch resistance, even if it is infinite, dissipates one-half the energy supplied by the source!* This matter is of some importance in power systems where large currents flowing into inductive loads must be switched into and out of circuits. Arcs occur across the switch contacts, and an oil or air blast must be used to dissipate the energy.

Capacitance. The problem we have just resolved for inductance has a capacitive counterpart. Consider the circuit of Fig. 14–11, in which the capacitance is excited by an impulse current. The current is

$$i(t) = Au_0(t),$$

and the voltage which it produces is given by the integral

$$e_C(t) = \frac{1}{C}\int_{-\infty}^t Au_0(t)\,dt = \frac{A}{C}\,u_{-1}(t). \tag{14–68}$$

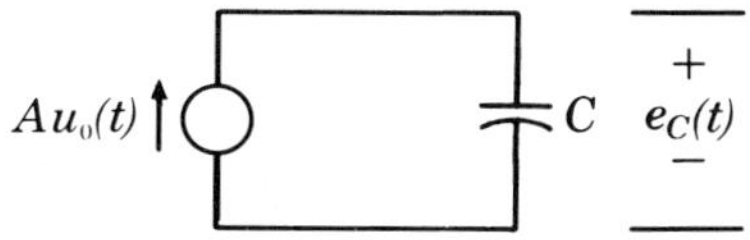

FIG. 14–11. A capacitance excited by an impulse current.

For $t > 0$ the value of the voltage is A/C, and the energy stored in the capacitance is

$$W_C = \tfrac{1}{2}Ce^2 = \frac{A^2}{2C}.$$

The energy supplied by the source is

$$W_S = \int_{-\infty}^{t} (e \times i)\, dt = \frac{A^2}{C}\int_{-\infty}^{t} [u_0(t) \times u_{-1}(t)]\, dt. \tag{14–69}$$

As in the inductive case we must replace the step function $u_{-1}(t)$ in the integral of Eq. (14–69) by one-half, since that is its average value over the range $t = 0^-$ to $t = 0^+$, the interval during which the impulse is acting. Hence we obtain

$$W_S = \frac{A^2}{2C}. \tag{14–70}$$

This value of energy agrees with that for the energy stored in the capacitance.

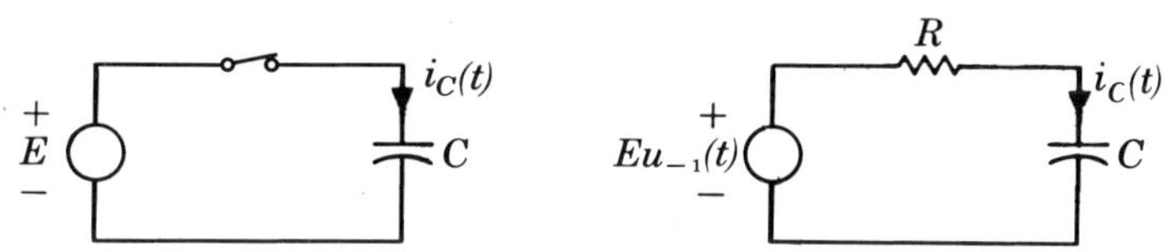

(a) Switch closed at $t = 0$ (b) The "true" equivalent circuit

FIG. 14–12. An example in which the switch dissipates the same amount of energy as is stored in the capacitance.

The inverse situation shown in Fig. 14–12(a) is perhaps more difficult. When the switch is closed an impulse of current flows, which is given by

$$i_C(t) = C\frac{de}{dt} = CEu_0(t). \tag{14–71}$$

This impulse charges the capacitance to the voltage E. The energy stored is

$$W_C = \tfrac{1}{2}CE^2, \tag{14–72}$$

and the apparent energy supplied by the source is

$$W_S = \int_{-\infty}^{t} CE^2 u_0(t) u_{-1}(t)\, dt.$$

In this case the step function is applied first and exists during the entire duration of the impulse. The integral therefore has the value

$$W_S = CE^2. \tag{14–73}$$

The discrepancy between Eqs. (14–72) and (14–73) is explained only by the assumption that there is some resistance in the switch and that the rise time of the voltage is finite. The "true" circuit is shown in Fig. 14–12(b). The energy relationships for this circuit shed light on the energy which apparently disappears in the circuit of Fig. 14–12(a). The current in the circuit of Fig. 14–12(b) is

$$i = \frac{E}{R} e^{-t/RC} u_{-1}(t). \tag{14–74}$$

The total energy lost in the resistance during the charging of the capacitance is

$$W_R = \int_0^{\infty} \frac{E^2}{R} e^{-2t/RC}\, dt = \frac{E^2 C}{2}. \tag{14–75}$$

The significance of Eq. (14–75) is that it is independent of the resistance. The energy lost in the resistance does not depend on the value of the resistance and would have the same value no matter how small the resistance became. *Even if the switch resistance is nominally zero, one-half the energy will be lost in it!* Although the resistance is zero, the current is infinite during the impulse. It is a case again of the irresistible force and the immovable object; the result is indeterminate. We have been able to resolve the indeterminacy only by a limiting process.*

It is true that the energy relationships in circuits with impulse currents or voltages are not as straightforward as those for circuits with finite currents and voltages. It is also true, however, that such problems can always be resolved by replacing the impulses by large pulses of finite duration and the step functions by functions with finite rise times. In the last two examples we have observed the fact that it is not possible to conserve both momentum (charge or flux linkage) and energy in a lossless circuit. Of the two we choose to conserve momentum and to violate the conservation of energy. In a real physical problem energy is lost in the resistance which we neglect in the lossless case.

---

* The classical form of this problem concerns the energy relationships when one capacitance is charged from another in a lossless circuit.

14-7 Summary.

One-element circuits	*Power*	*Energy*
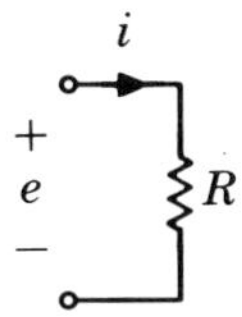	$p_R = e \times i = i^2 R = \dfrac{E^2}{R}$	$W_R = \int_0^t p_R \, dt$
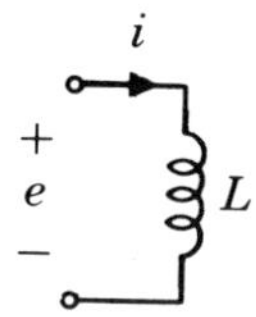	$p_L = e \times i = L \times i \times \dfrac{di}{dt}$	$W_L = \frac{1}{2}Li^2$
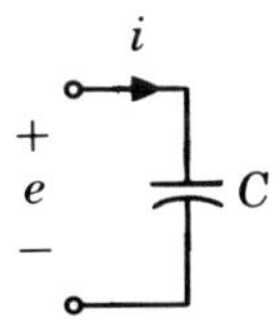	$p_C = e \times i = C \times e \times \dfrac{de}{dt}$	$W_C = \frac{1}{2}Ce^2$

Resistance and inductance

$$i^2 R + \frac{d}{dt}\left(\frac{Li^2}{2}\right) = e_s \times i_s$$

Resistance and capacitance

$$i^2 R + \frac{d}{dt}(Ce^2) = e_s \times i_s$$

Inductance and capacitance

$$\begin{aligned} \frac{d}{dt}\left(\tfrac{1}{2}Li^2 + \tfrac{1}{2}Ce^2\right) &= e_s \times i_s \\ &= 0 \text{ for internal excitation} \end{aligned}$$

Resistance, inductance, and capacitance

$$i^2 R + \frac{d}{dt}\left(\tfrac{1}{2}Li^2 + \tfrac{1}{2}Ce^2\right) = e_s \times i_s$$

Power and energy in impulses

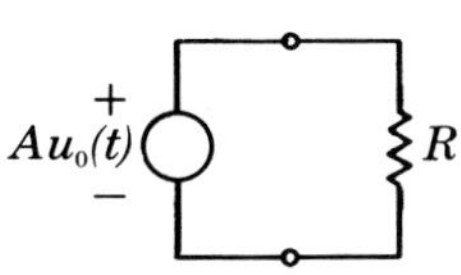

$$p_R = \frac{A^2}{R} u_0^2(t) \qquad\qquad W_R = \infty$$

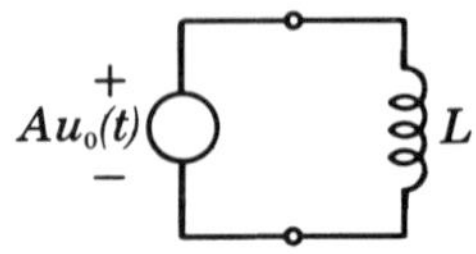

$$p_L = \frac{A^2}{L} u_0(t) \times u_{-1}(t) \qquad W_L = \frac{A^2}{2L}$$

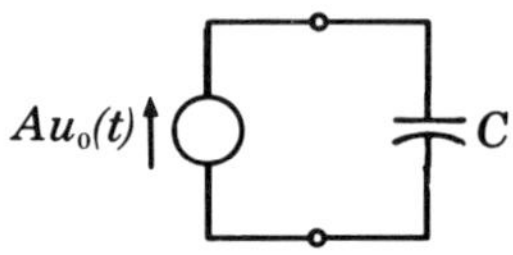

$$p_C = \frac{A^2}{C} u_0(t) \times u_{-1}(t) \qquad W_C = \frac{A^2}{2C}$$

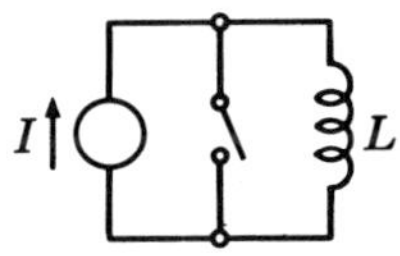

$$p_L = LI^2 u_0(t) \times u_{-1}(t) \qquad W_L = \tfrac{1}{2}LI^2$$

$$W_{\text{switch}} = \tfrac{1}{2}LI^2$$

$$W_{\text{source}} = LI^2$$

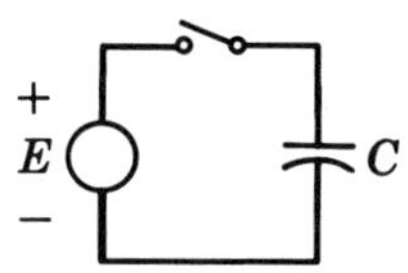

$$p_C = CE^2 u_0(t) \times u_{-1}(t) \qquad W_C = \tfrac{1}{2}CE^2$$

$$W_{\text{switch}} = \tfrac{1}{2}CE^2$$

$$W_{\text{source}} = CE^2$$

REFERENCES

The approach to power and energy relationships in this chapter extends the material of Chapter 6 to circuits involving resistance, inductance, and capacitance. The power and energy computations, however, remain secondary to the computations of the circuit quantities current and voltage. Some physics texts and some field-theory texts in electrical engineering give power and energy a more central role.

1. H. G. Booker, *An Approach to Electrical Science,* McGraw-Hill, New York, 1959. This text emphasizes the physics approach to circuits and treats power and energy as the basic quantities. Thus lossless circuits, in which energy is conserved, are treated before resistive circuits.

2. Clement and Johnson, *Electrical Engineering Science,* McGraw-Hill, New York, 1960. Like Booker in some respects, these authors treat circuits from the point of view of physics.

3. E. A. Guillemin, *Communications Networks,* Vol. 2, John Wiley & Sons, New York, 1935. Guillemin gives an introduction to Lagrange's equations as applied to networks.

4. Seshu and Balabanian, *Linear Network Analysis* John Wiley & Sons, New York, 1959. This text gives a rather complete treatment of the general quadratic forms and the energy functions of networks in the transient state.

5. LEWIS and GOODHEART, *Basic Electric Circuit Theory,* The Ronald Press, New York, 1958. Lewis and Goodheart give a treatment of the energy relationships of coupled coils under transient excitation.

6. BRENNER and JAVID, *Analysis of Electric Circuits,* McGraw-Hill, New York, 1959. Brenner and Javid treat power and energy mainly in steady-state a-c analysis.

7. LAGO and WAIDELICH, *Transients in Electrical Circuits,* The Ronald Press, New York, 1958. This text gives the energy relationships of typical R–L, R–C, and R–L–C circuits.

EXERCISES*

Section 14–2

1. For the circuit of Fig. 14–13 find:
   (a) The currents i_1, i_2, i_3, and i_4.
   (b) The powers P_1, P_2, P_3, and P_4 in the branches.
   (c) The total energy dissipated in the first 10 seconds.
   (d) The energy supplied by the source in the first 10 seconds.
   (e) The equivalent resistance of the circuit, and the energy that would be dissipated in it in the first 10 seconds.

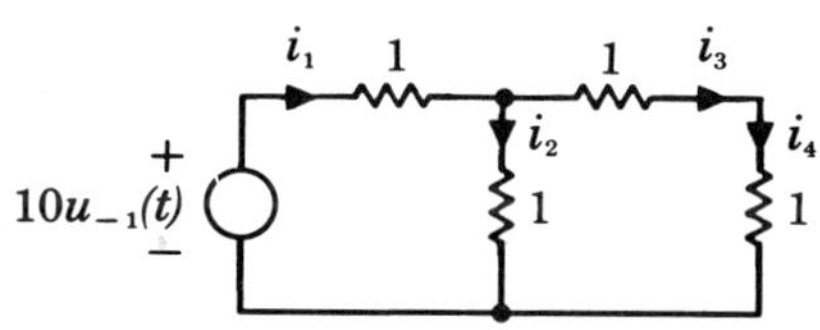

FIGURE 14–13

2. In the circuit of Fig. 14–14 the voltage source and the current source are constant or d-c values.
   (a) For what value of the current source I is the power supplied by the current source zero?
   (b) For what value of I is the power supplied by the voltage source zero?
   (c) For what value of I is the power supplied by the current source equal to the power supplied by the voltage source.

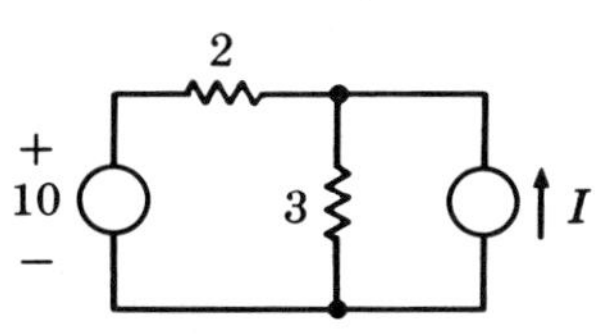

FIGURE 14–14

* Numerical values are in volts, amperes, ohms, henrys, and farads.

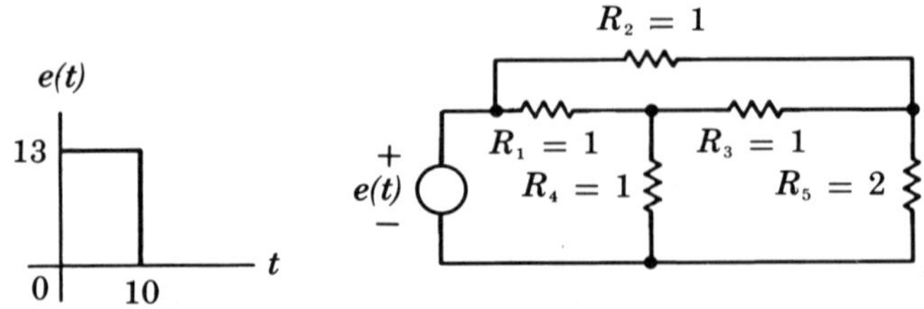

FIGURE 14–15

3. In the circuit of Fig. 14–15 find the total energy absorbed in each resistance from the 13-volt input pulse.

4. Find the energy stored in each of the inductances in the circuit of Fig. 14–16 for $t > 0$.

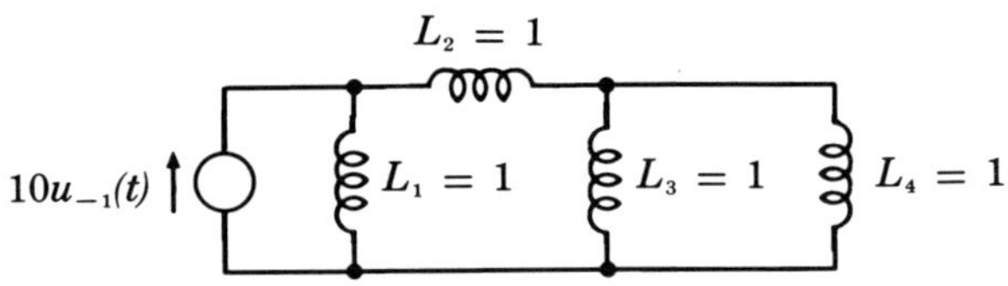

FIGURE 14–16

5. In the circuit of Fig. 14–17 give the energy in each inductance as a function of time. Give also the energy delivered by the source as a function of time.

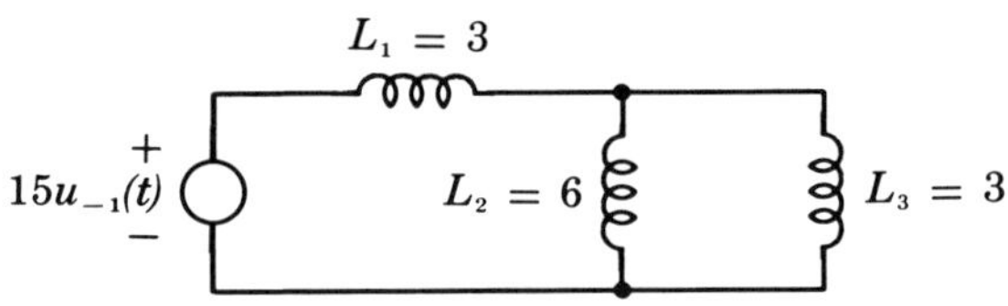

FIGURE 14–17

6. In the circuit of Fig. 14–18 consider the interval from $t = 0$ to $t = 1$ and find the average energy stored in each inductance and the average energy supplied by the source.

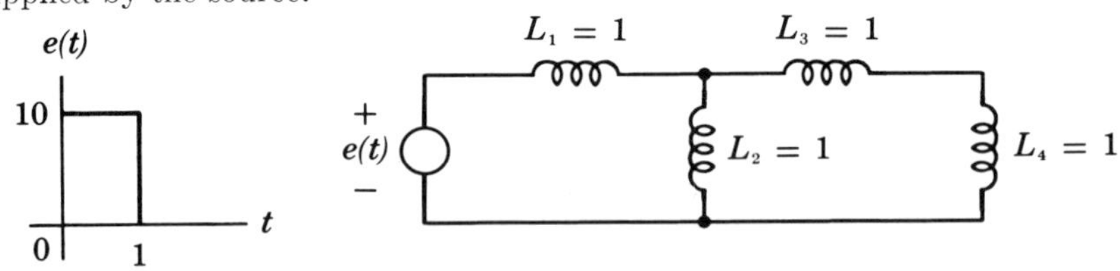

FIGURE 14–18

7. Find the energy stored in each capacitance in the circuit of Fig. 14–19 for $t > 0$.

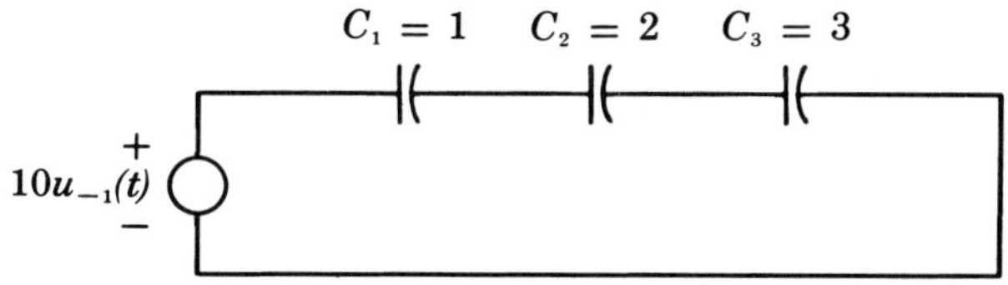

FIGURE 14–19

8. In the circuit of Fig. 14–20 find the energy stored on each capacitance as a function of time, and the energy delivered by the source as a function of time. Also find the power absorbed by each capacitance as a function of time, and the power supplied by the source as a function of time.

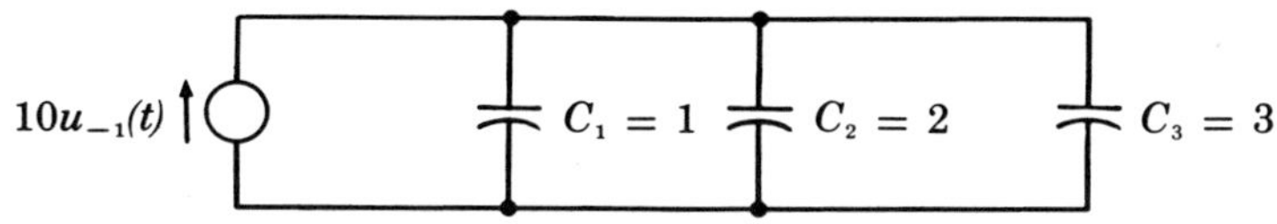

FIGURE 14–20

9. In the circuit of Fig. 14–21 find the power and the energy absorbed by each capacitance as a function of time, and also the power and the energy supplied by the source.

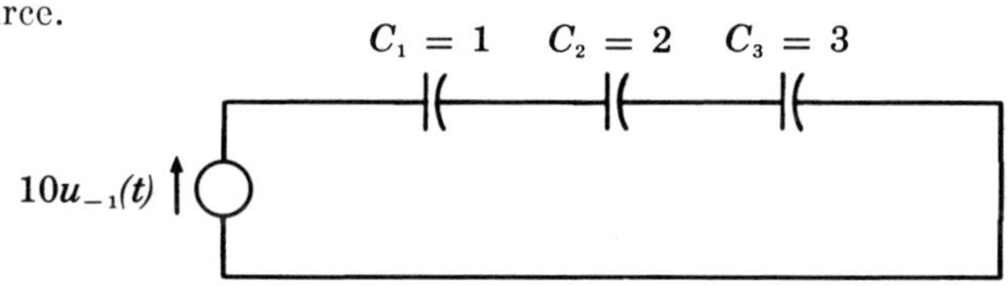

FIGURE 14–21

Section 14–3

10. A 4-henry inductance is connected in series with a 2-ohm resistance. The current in the circuit at $t = 0$ is 10 amp. Find the initial energy stored in the inductance and show that this energy is absorbed in the resistance as the circuit comes to rest.

11. At $t = 0$ a 12-volt battery is suddenly connected in series with a 4-ohm resistance and a 2-henry inductance. Find the energy stored in the inductance as a function of time, the energy lost in the resistance from $t = 0$ to $t = t$, and the energy supplied by the source.

12. In the circuit of Fig. 14–22 find the energy supplied by the source, the energy stored in the inductance, and the energy lost in the resistance. What fraction of the total energy supplied is lost in the resistance?

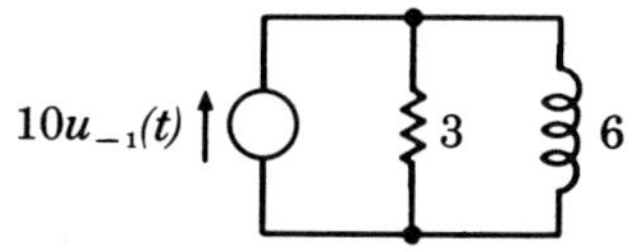

FIGURE 14–22

13. The voltage pulse in the circuit of Fig. 14–23 causes some energy to be stored in the inductance, but the energy is finally dissipated in the resistance. Find the peak value of the energy stored in the inductance and show that the energy lost in the resistance for $t > 2$ is equal to this value. Also show that the energy supplied by the source is equal to the total energy lost in the resistance over all time.

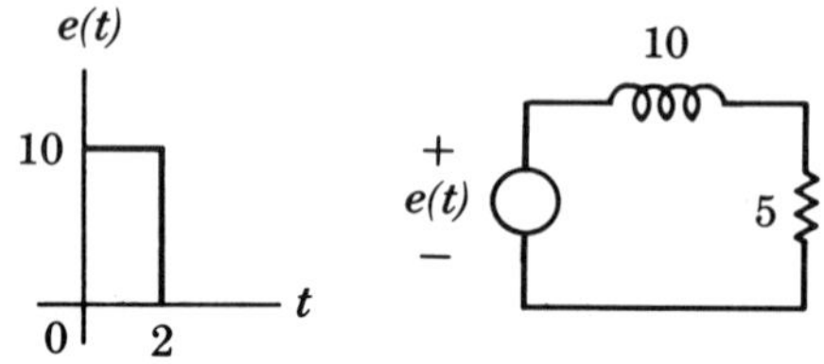

FIGURE 14–23

14. A capacitance of 3 farads is connected in series with a resistance of 2 ohms. At $t = 0$ there is a voltage of 10 volts on the capacitance. What is the initial energy stored on the capacitance and what is the total energy lost in the resistance?

15. At $t = 0$ a 10-volt battery is suddenly connected in series with a capacitance of 2 farads and a resistance of 5 ohms. Compute the final energy stored in the capacitance and the total energy lost in the resistance. What is the relationship between these two quantities?

16. In the circuit of Fig. 14–24 show that the power supplied by the source is balanced at all times by the power absorbed in the capacitance and the resistance.

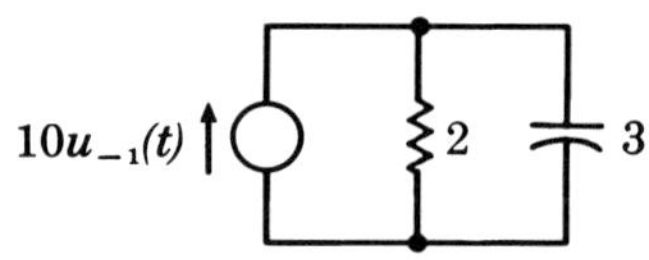

FIGURE 14–24

17. In the circuit of Fig. 14–25 find the final value of the energy stored in each of the capacitances and also give the value of the energy lost in the resistances.

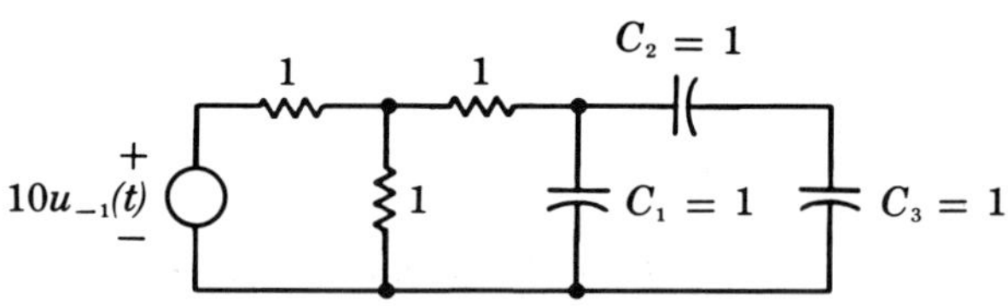

FIGURE 14–25

Section 14–4

18. A $\frac{1}{4}$-farad capacitance and a $\frac{1}{16}$-henry inductance are connected in series. The initial voltage on the capacitance is 12 volts and the initial current is zero. Find the initial energy in the capacitance, the maximum energy in the inductance, and the maximum value of the current. Write the current by inspection.

19. A $\frac{1}{8}$-farad capacitance and a $\frac{1}{2}$-henry inductance are in series. The initial value of the current is 12 amp, and the initial value of the voltage on the capacitance is zero. From the initial energy stored, find the maximum value of the voltage on the capacitance. Write the current, and the voltage on the capacitance, by inspection.

20. A $\frac{1}{3}$-henry inductance and a $\frac{1}{12}$-farad capacitance are connected in series as shown in Fig. 14–26. The initial current is 6 amp and the initial voltage on the capacitance is 12 volts, with the polarities shown. Compute the total initial energy stored, and obtain from this value the maximum value of the current and the maximum value of the voltage across the capacitance. Write time expressions for the current and the voltage, referred to the instant at which the current is a maximum.

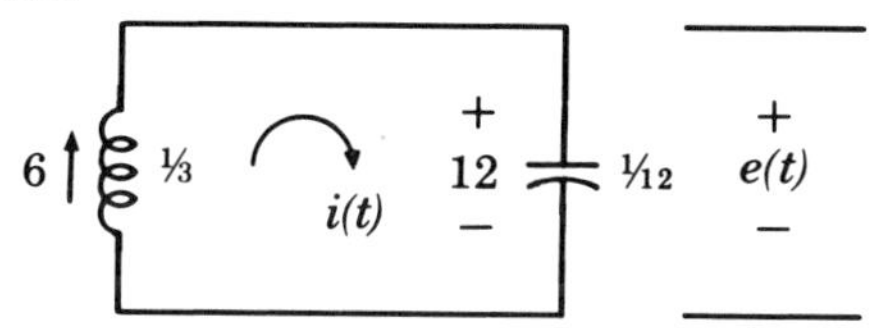

FIGURE 14–26

21. A voltage source $10u_{-1}(t)$ is connected in series with an inductance of 3 henrys and a capacitance of 12 farads. Find the energy stored in the capacitance, the energy stored in the inductance, and the energy supplied by the source as functions of time.

22. A current source $12u_{-1}(t)$ is connected in parallel with an inductance of $\frac{1}{2}$ henry and a capacitance of $\frac{1}{8}$ farad. Compute the power absorbed by the inductance, the power absorbed by the capacitance, and the power supplied by the source as functions of time.

23. In the circuit of Fig. 14–27 find the energy stored in each inductance and the energy stored in the capacitance as functions of time.

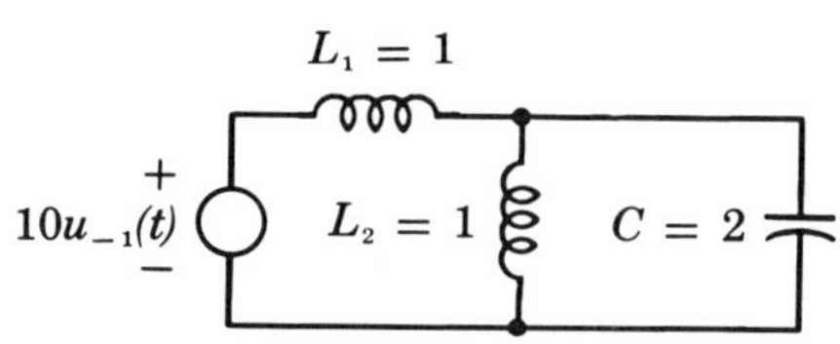

FIGURE 14–27

24. Compute the energy delivered to the L–C circuit of Fig. 14–28 by the voltage pulse.

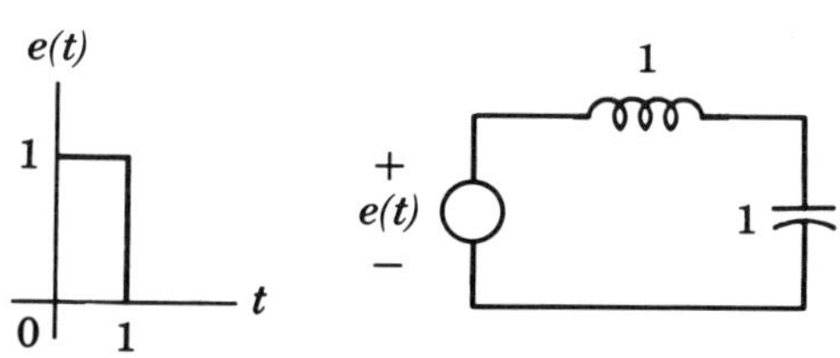

FIGURE 14–28

25. Convert the two sources in the circuit of Fig. 14–29 to equivalent energy storages in the inductance and the capacitance and thus find the total energy stored in the circuit at $t = 0^+$. What is the peak value of the current in the circuit, and what is its frequency?

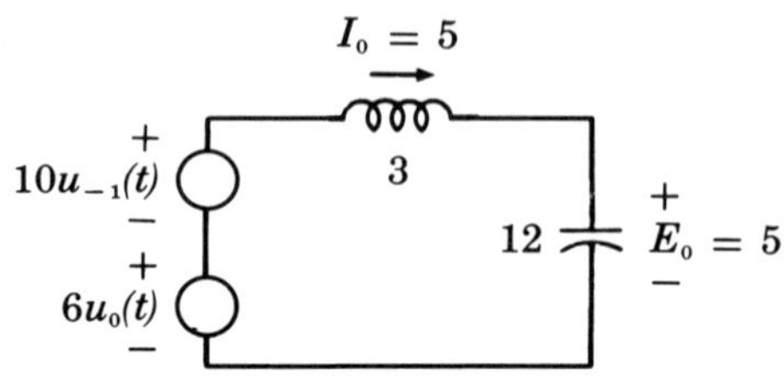

FIGURE 14–29

Section 14–5

26. The switch in the circuit of Fig. 14–30 is closed at $t = 0$. Find the initial energy stored in the capacitance and the total energy lost in the resistance from $t = 0$ to $t = \infty$.

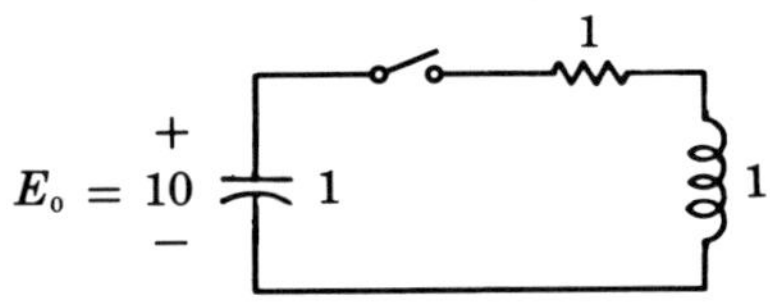

FIGURE 14–30

27. Find the value of resistance in the circuit of Fig. 14–30 which gives critical damping. Show that the initial energy stored in the capacitance and the energy lost in the resistance are balanced for this case.

28. If the resistance in the circuit of Fig. 14–30 is changed to 2.83 ohms the circuit becomes overdamped. Find the current for this case and show that the energy balance is still preserved.

29. In the circuit of Fig. 14–31 find the initial energy stored, the voltage $e(t)$ as a function of time, and the total energy lost in the resistance. The initial energy on the capacitance is zero.

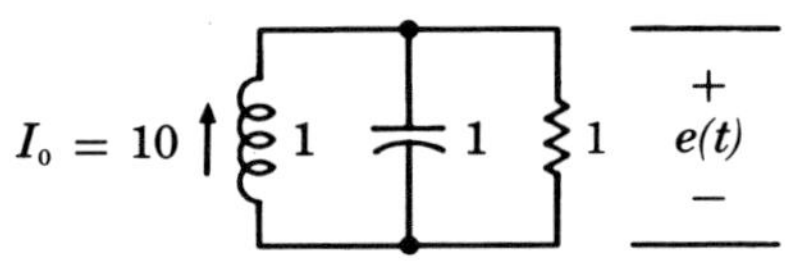

FIGURE 14–31

30. Repeat Exercise 29 with the circuit resistance changed to $\frac{1}{2}$ ohm.

31. Repeat Exercise 29 with the circuit resistance changed to 0.353 ohm.

32. In the circuit of Fig. 14–32 the initial current is 10 amp. Find the current as a function of time, the power lost in the resistance as a function of time, and the total energy lost in the resistance from $t = 0$ to $t = \infty$.

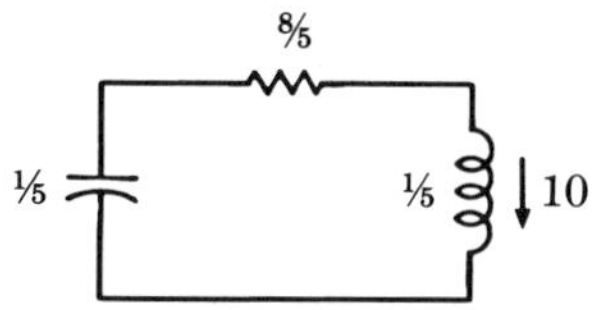

FIGURE 14–32

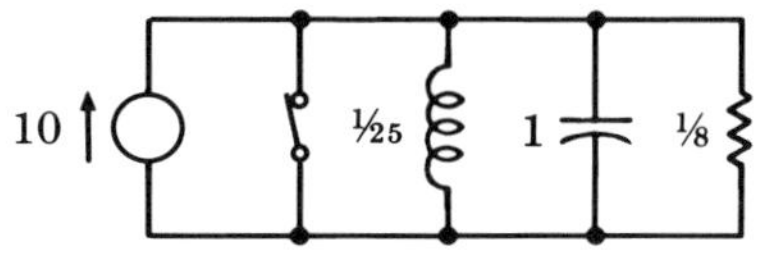

FIGURE 14–33

33. The switch in the circuit of Fig. 14–33 is opened at $t = 0$. Compute the energy supplied from the source to the circuit, the energy lost in the resistance, and the final value of the energy stored in the inductance. Verify the energy balance equation.

34. Repeat Exercise 33 with the same source value, but with an inductance of $\frac{1}{4}$ henry, a capacitance of $\frac{1}{9}$ farad, and a resistance of $\frac{3}{4}$ ohm.

35. The switch in the circuit of Fig. 14–34 is closed at $t = 0$. Compute the energy supplied by the source, the energy lost in the resistance, the final value of the energy stored in the capacitance, and verify the energy balance equation.

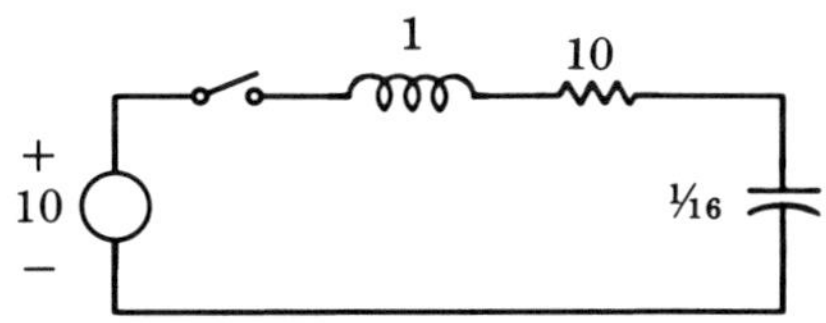

FIGURE 14–34

Section 14–6

36. A short pulse of current is applied to a resistance of 2 ohms. Compute the power during the pulse and the total energy dissipation for:

(a) A pulse of 100 amp for 1 microsec.

(b) A pulse of 200 amp for $\frac{1}{2}$ microsec.

(c) An impulse of the same area as the pulses of parts (a) and (b).

37. A short pulse of current is applied to a capacitance of 2 farads. Compute the power during the pulse and the total energy stored for:

(a) A pulse of 100 amp for 1 microsec.

(b) A pulse of 200 amp for $\frac{1}{2}$ microsec.

(c) An impulse of the same area as the pulses of parts (a) and (b).

38. Repeat Exercise 37 except that the pulse is a voltage and the element is a 2-henry inductance.

39. The switch in the circuit of Fig. 14–35 is closed at $t = 0$. Compute the energy supplied by the source, the energy stored in the capacitance, and the energy lost in the switch.

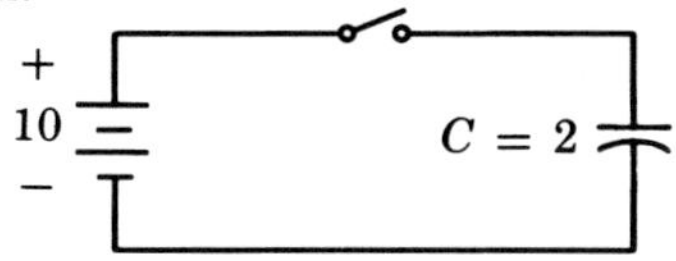

FIGURE 14–35

40. In the circuit of Fig. 14–36 the source is a constant current of 10 amp. The switch is opened at $t = 0$. Compute the energy from the source, the energy stored in the inductance, and the energy lost in the switch.

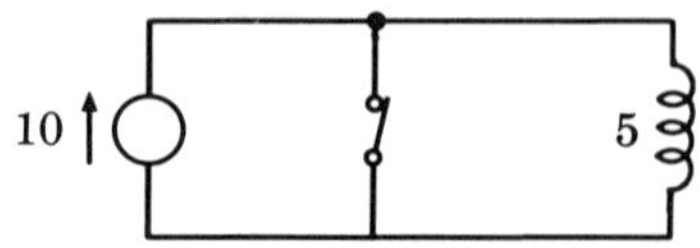

FIGURE 14–36

41. In the circuit of Fig. 14–37 the capacitance C_1 has a voltage of 100 volts at the instant the switch is closed, and the capacitance C_2 has no voltage.

(a) Find the current which flows in the circuit.
(b) Find the total energy lost in the resistance.
(c) Find the energy stored in C_1 and C_2 and check the energy balance equation.
(d) Examine the situation as the resistance R is allowed to approach zero ohms.

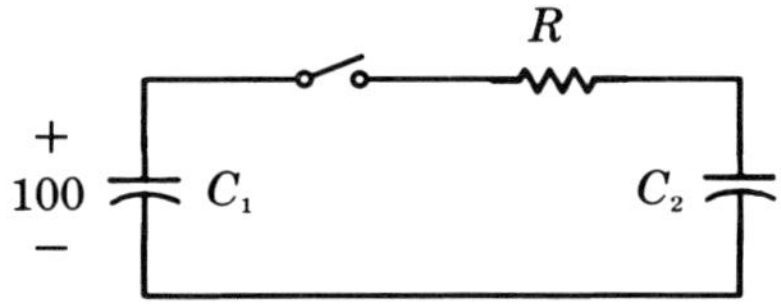

FIGURE 14–37

42. Before $t = 0$ in the circuit of Fig. 14–38 the switch is closed and a current of 100 amp flows through L_1 and the switch. At $t = 0$ the switch is opened.

(a) Find the voltage $e(t)$.
(b) Find the energy lost in the resistance.
(c) Find the energy stored in the two inductances.
(d) Examine the situation as the resistance R is allowed to approach infinite ohms.

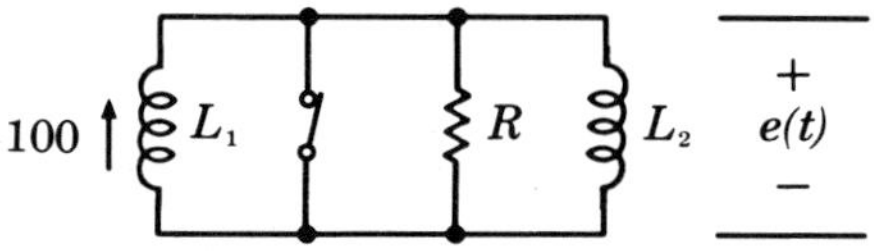

FIGURE 14–38

43. An impulse voltage $10u_0(t)$ is applied in series with a resistance of 1 ohm and an inductance of 5 henrys.

(a) Find the current which flows.
(b) Find the energy from the source.
(c) Find the initial energy stored in the inductance at $t = 0^+$.
(d) Find the total energy lost in the resistance.

CHAPTER 15

INTRODUCTION TO SINE WAVES AND THE FREQUENCY DOMAIN

"Eternity is a pendulum whose balance wheel says unceasingly only the two words, in the silence of a tomb, 'always! never! never! always! . . .'"

Jacques Bridaine

15–1 Introduction. The motion of a pendulum has exerted a fascination for human minds since the first savage watched the swaying of the first tree branch. The smooth sinusoidal motion, back and forth, seems to express some secret of the universe. Indeed nature "loves" the sinusoid. From the motions of the stars and the planets to the tides on earth, and from our heartbeats to our psychological states, we are surrounded by cyclical phenomena. The sinusoidal waveform is one that occurs frequently wherever such phenomena are found. It is the electrical waveform most easily generated by physical devices. Most of the electric power systems in the world produce approximately sinusoidal waveforms, and most of our communications systems also rely on sinusoidal waves. We could justify the development of special techniques for handling sinusoidal excitation of our networks on this basis alone. In addition, however, Fourier's theorem allows us to express arbitrary waveforms in terms of sinusoids and thereby extends the usefulness of the sinusoidal techniques to general cases.

The special methods which we shall develop for handling the solution of network problems in which the excitation is a steady-state sine wave are closely related to the methods we developed in earlier chapters for handling purely resistive networks. These "frequency-domain" methods make use of complex algebra to reduce steady-state sinusoidal problems to forms which are identical to those for resistive networks. An ability to use this technique is perhaps the most important characteristic of the electrical engineer. It is his "bread and butter" tool.

We have already encountered sinusoidal waves and damped sinusoidal waves in connection with the transient response functions of *L–C* and *R–L–C* networks. Indeed there is a close relationship between the response of a network to a steady-state sinusoidal excitation and its transient response. The relationship involves the fascinating phenomenon of resonance. Everyone is acquainted with resonance. A physical system which has a sinusoidal type of natural transient response reacts with great enthusiasm to an excitation function of the same frequency. The production of musical sounds and the vibrations of certain components of an automobile are familiar examples. By examining the phenomenon of resonance in some detail we shall be able to relate the transient response functions of previous chapters to the sinusoidal steady-state response functions of networks.

Among all the cogent arguments we have just given for devoting our time to the steady-state sinusoid, the best is still the simplicity of the resulting analysis! Other waveforms might have been used as building blocks, but none of these has the basic mathematical simplicity of the sine wave. The sums, differences, integrals, and derivatives of sine waves of a given frequency are themselves sine waves of the same frequency. No other waveform is preserved in this fashion.*

In summary, our reasons for devoting a good deal of attention to the responses of networks to sinusoidal steady-state excitation are:

1. Sinusoids occur in many practical generators.
2. Fourier's theorem allows sinusoidal analysis to be extended to the general case.
3. Sinusoids are easily handled mathematically.
4. The sinusoidal response is closely related to the natural transient behavior of the network.

In this chapter we devote our discussion to sinusoidal waves themselves. Some basic properties are analyzed, and a number of ways of representing them are presented. Of these the methods of phasor (vector) geometry and phasor (complex) algebra are singled out for special attention. The point of these methods is to replace the random formulas of trigonometry by the simple rules of geometry and algebra. Trigonometric quantities are easily manipulated in terms of geometric and algebraic vectors. The vectors are called *phasors,* and the manipulations of them are called *frequency-domain* methods.

15–2 Some properties of sinusoidal waves. We shall use the term "sinusoidal wave" to represent any waveform that can be expressed as a sine or a cosine. For our analytical work it will be convenient to use the cosine as the basic form. The cosine is perfectly general, since the addition of a phase angle of $-90°$ converts the cosine function into the sine. Thus

$$A \cos (\omega t - 90°) = A \sin \omega t. \tag{15–1}$$

Our basic waveform will have the mathematical form

$$f(t) = A \cos (\omega t + \theta). \tag{15–2}$$

---

* Actually, the most general case is the exponentially damped sine wave. The constant-amplitude sine wave and the exponential curve are special cases.

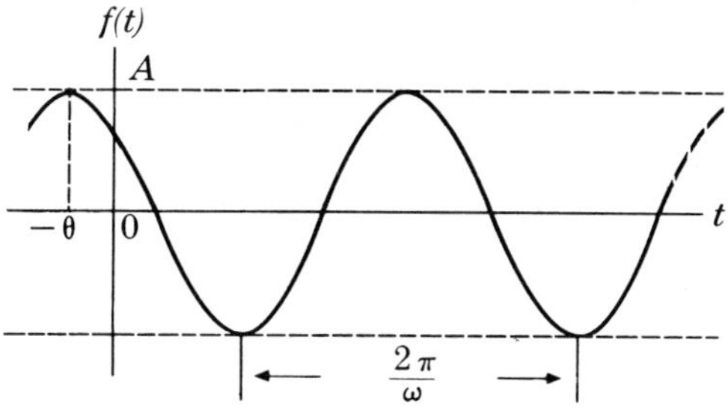

FIG. 15-1. The sinusoidal waveform $f(t) = A \cos(\omega t + \theta)$.

A sketch of this waveform is shown in Fig. 15-1. The operations required in linear circuits are addition, subtraction, integration, and differentiation. Let us see how these operations effect the waveform of Eq. (15-2).

Addition. The sum of two cosine waves can be obtained by the formulas of trigonometry. For example, let us find the sum

$$f(t) = A \cos \omega t + B \sin \omega t.$$

The easiest way to accomplish this addition is to define two new quantities by the equations

$$\begin{aligned} A &= C \cos \theta, \\ B &= C \sin \theta. \end{aligned} \tag{15-3}$$

From Eq. (15-3), the new quantities, C and θ, are determined by

$$\begin{aligned} C &= \sqrt{A^2 + B^2}, \\ \theta &= \tan^{-1} \frac{B}{A}. \end{aligned} \tag{15-4}$$

In terms of these quantities the sum of the two sinusoids is

$$\begin{aligned} f(t) &= C \cos \omega t \cos \theta + C \sin \omega t \sin \theta \\ &= C \cos(\omega t - \theta). \end{aligned} \tag{15-5}$$

The frequency of the resultant wave of Eq. (15-5) is the same as the frequency of the two waves added together to produce it, but the amplitude and the phase angle are different. Parenthetically, the awkwardness of the trigonometric solution should be noted. The vector methods we shall soon come to are much superior.

Differentiation. The derivative of a cosine wave is easy to obtain. It is

$$\begin{aligned} f(t) &= \frac{d}{dt}[A \cos (\omega t + \theta)] \\ &= -A\omega \sin (\omega t + \theta) \\ &= +A\omega \cos (\omega t + \theta + 90°). \end{aligned} \tag{15–6}$$

The resulting wave of Eq. (15–6) is again a sinusoid of the same frequency as the original wave, but the amplitude and the phase angle are different. The amplitude is multiplied by ω and the phase is advanced by 90°.

Integration. The integration of a cosine wave is more difficult than its differentiation because of the range of integration. Mathematically speaking, the integral from $-\infty$ to t is indeterminate. We evaluate it as follows:

$$\begin{aligned} f(t) &= \int_{(T)\to-\infty}^{t} A \cos (\omega t + \theta)\, dt \\ &= \left[\frac{A}{\omega} \sin (\omega t + \theta)\right]_{(T)\to-\infty}^{t} \\ &= \frac{A}{\omega}[\cos (\omega t + \theta - 90°) + \cos (\omega T + \theta - 90°)]_{T\to\infty}. \end{aligned} \tag{15–7}$$

In Eq. (15–7) the second term has a value which oscillates between +1 and −1 as $T \to \infty$. For steady-state sinusoidal circuits we are not interested in this term, *and we simply forget about it!* It represents a constant value. So far as the steady-state is concerned the net effect of integration on a cosine wave is to divide the amplitude by ω and to retard the phase by 90°.*

Amplitude and phase response. Each of the linear operations we have considered affects the amplitude and phase of a sinusoidal excitation function, but the frequency remains unchanged. A linear network performs a sequence of such operations, and we can therefore deduce that its steady-state response function is a sinusoid of the same frequency as the excitation function, but differing from it in amplitude and phase. The net effect of the circuit is thus to alter the amplitude and the phase of the excitation

* The constant dies out in practical circuits. For idealized circuits we must assume that the steady-state sinusoidal excitation is actually increasing slowly in amplitude from an original value of zero, i.e., its abscissa of absolute convergence is greater than zero. (See Chapter 20 for a discussion of convergence problems.)

function. The amount of amplitude change and the amount of phase shift which the network produces vary with the frequency of the excitation. Indeed, if we plot curves of the amplitude change and of the phase shift as functions of frequency, we can completely characterize the network.

The stimulus-response relationships of resistive networks are constants, independent of the input excitation. *The stimulus-response relationships of R–L–C networks are the amplitude and phase response curves plotted as functions of frequency.* From these curves the output response for any input frequency can be obtained. If the input is not a single frequency wave it can be resolved into frequency components, and the response for each component can be obtained from the amplitude and phase response curves. The over-all response function is the sum of all the component responses.

Since the amplitude and phase response curves are functions of frequency they are said to belong to the "frequency domain." The methods which we have just outlined for solving problems by means of a resolution into frequency components are called *frequency-domain methods.* The remainder of the textbook will be devoted to the exploration of frequency-domain methods. The relationship between frequency-domain methods and the time-domain methods we have been using up to this point is shown in Fig. 15–2. Sinusoidal analysis is represented at the center of the illustration. When inductance and capacitance are not present the purely resistive networks at the upper left are obtained. The methods of frequency-domain analysis which we shall develop reduce sinusoidal networks to the same form as resistive networks. These so-called frequency-domain networks are represented at the lower left of the illustration. If we wish to handle complex waves of excitation we must resort to the Fourier series

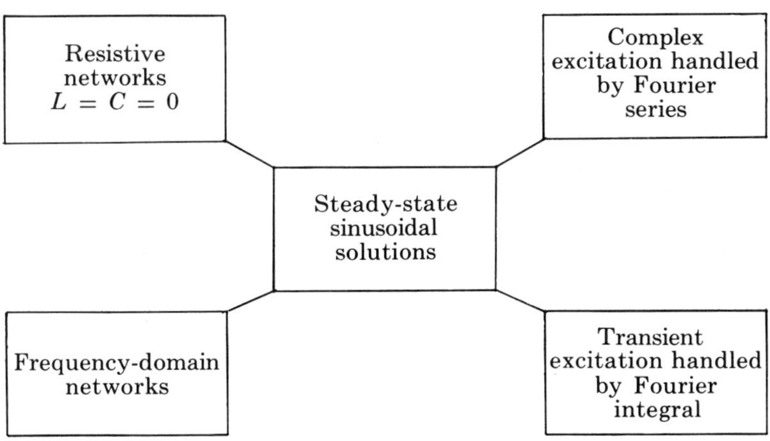

FIG. 15–2. The extent of the frequency-domain methods.

represented at the upper right of the illustration. For transient signals the Fourier series become the Fourier integrals represented at the lower right of the illustration. The power and generality of the frequency-domain methods are apparent at once from Fig. 15–2.*

15–3 Methods of representing sine waves. We have already represented the general sinusoid by *a pictorial plot* in Fig. 15–1. In some ways this drawing is the clearest of all representations, since it shows the real significance of the amplitude, the frequency, and the phase. To avoid making the drawing we can write the *trigonometric form* of the sinusoid:

$$f(t) = |F| \cos(\omega t + \theta). \tag{15–8}$$

The lower-case letter f is used to represent the time function, and the same letter in capital form is used to represent the maximum amplitude of the sinusoid. Magnitude bars are used to enclose F because later we shall use a complex number F to represent the amplitude and phase of the sinusoid.† If F is a real, positive number, the magnitude bars can be removed. The phase angle which we will use is the phase angle of a *cosine* wave. This is an arbitrary choice, and some authors use the sine wave as the basis of analysis. Once we make the choice we must stick to it, however, and the cosine has some analytical advantages. It is also worth recalling that the period of the wave of Eq. (15–8) is

$$T = \frac{2\pi}{\omega} \text{ sec}, \tag{15–9}$$

and its frequency is

$$f = \frac{1}{T} = \frac{\omega}{2\pi} \text{ cycles/sec}, \tag{15–10}$$

from which

$$\omega = 2\pi f \text{ radians/sec}.$$

For the 60-cycle frequency used in power systems, the value of ω is $2\pi \times 60 = 377$ radians per second.

---

* Some authors are so impressed that they cover nothing but frequency-domain methods! Resistive circuits, however, are the basis of frequency-domain methods, and the transient response is the basis of resonance phenomena. In fact, time-domain methods are equally as general

† In textbooks, vectors are frequently denoted by boldface letters. For the sake of simplicity in our notation, the vectors we shall study will be denoted by the customary capital italic letters.

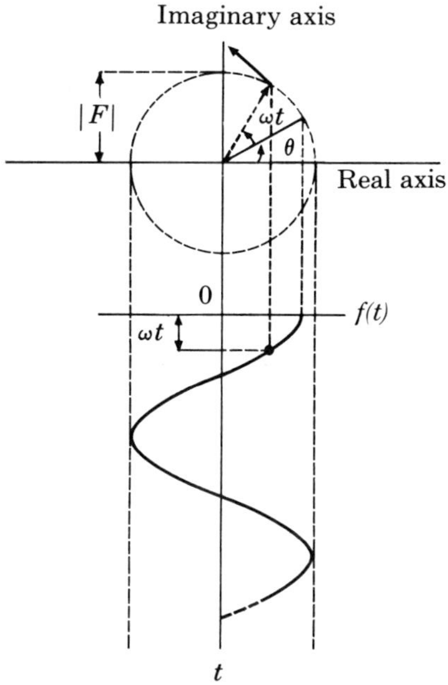

FIG. 15–3. A rotating phasor and its sinusoidal projection on the real axis.

The vector representation of sine waves. The projection of a vector on one of the coordinate axes is a cosine function of the angle which the vector makes with the axis. If the angle changes with time, the projection on the axis also changes, and its length continues to be equal to the cosine of the angle. If the projection on the axis is to represent the function of Eq. (15–8), the angle must be equal to θ at $t = 0$ and must increase linearly thereafter. The net result is a vector which rotates about the origin, as a spoke rotates about the hub of a wheel. To distinguish it from a vector in space coordinates, it is called a *rotating phasor*. The term *phasor*, alone, is used for the stationary vector at $t = 0$. The phasor which represents the sinusoid of Eq. (15–8) is shown in Fig. 15–3. It is a vector of length $|F|$, standing at an angle of θ from the real axis. If the phasor is made to rotate at an angular frequency of ω, its projection on the real axis is the cosine function. The cosine graph is sketched below the phasor, with time t increasing vertically downward. When the quantity $(\omega t + \theta)$ is 90° the projection is zero; when the quantity $(\omega t + \theta)$ is 180° the projection is $-|F|$; when the quantity $(\omega t + \theta)$ is 270° the projection is zero again, and so on.

In problems with only a single excitation function, all the sinusoidal functions will be of the same frequency, and thus all the rotating vectors will have the same angular velocity, or frequency. A sort of "merry-go-round" exists on which all the vectors are riding. The merry-go-round is the *frequency domain*. If one is in the frequency domain revolving with the vectors, they appear to stand still; hence currents and voltages can be represented by stationary vectors, or phasors. The phasor for a given sinusoid is a vector with a length equal to the magnitude of the sinusoid, and an angle equal to its phase angle. The frequency-domain phasor is essentially the initial position of the rotating phasor of Fig. 15–3. Since all the phasors in any given problem are on the same merry-go-round, they are stationary with respect to each other, and they can be manipulated by the ordinary rules of vector geometry or vector algebra. When the resultant vector, or phasor, has been obtained its projection on the real axis gives the time-function solution of the problem. Stepping on the merry-go-round, where the vectors are stationary, is called transforming the problem to the frequency domain. The resultant vector in the problem is called the frequency-domain solution, and the process of finding its projection on the real axis is called transforming the resultant to the time domain.

The reason for transforming trigonometric expressions into phasors is to economize our computational effort. To manipulate trigonometric expressions requires a host of complicated and unrelated formulas, while phasors are manipulated by the simple rules of vector algebra or vector geometry. The transformations linking the frequency domain and the time domain are relatively simple, and the balance of effort is much in favor of a frequency-domain solution. The method is akin to the use of logarithms, in which we transform real numbers to the domain of logarithms in order to perform multiplication by simply adding the logarithms.

The complex-algebra representation of sine waves. The full advantage of phasors is not realized until they are expressed in algebraic form. The vector form is useful for visualizing relationships, but the algebraic form is preferable for computations. There is, of course, a one-to-one correspondence between a two-dimensional vector, or phasor, and a complex number. The basis of the relationship is the equation (Euler's formula)

$$e^{\pm j\theta} = \cos\theta \pm j\sin\theta. \tag{15–11}$$

In Eq. (15–11) the quantity $e^{j\theta}$ is a vector with a real part of $\cos\theta$ and an imaginary part of $\sin\theta$.* The length of the vector is

$$\sqrt{\cos^2\theta + \sin^2\theta} = 1.$$

* The idea of considering the real part of a complex number as a value along the real axis and the imaginary part as a value along an axis at right angles to it is an arbitrary, but useful, concept.

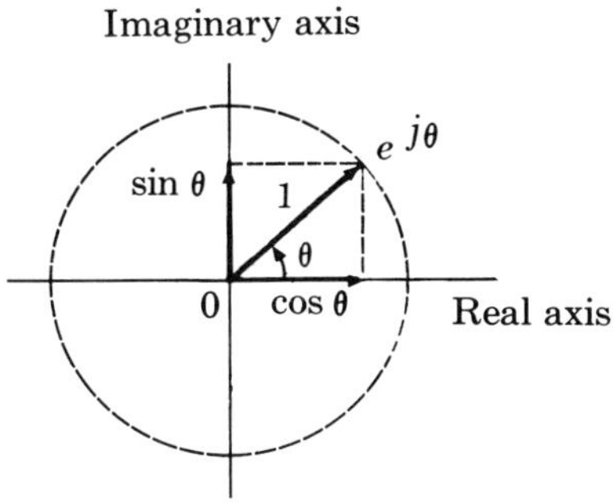

FIG. 15-4. The unit vector or phasor $e^{j\theta} = \underline{|\theta}$.

It is therefore a unit vector standing at an angle of θ from the real axis. The quantity $e^{-j\theta}$ is a similar unit vector, but stands at an angle of $-\theta$ from the real axis.

A unit vector is shown in Fig. 15-4. If this unit vector is caused to rotate from its initial position at $t = 0$, with an angular velocity of ω radians per second, it will have a projection on the real axis of

$$f(t) = \cos(\omega t + \theta). \tag{15-12}$$

The algebraic expression for a unit vector standing at an angle of θ is $e^{j\theta}$. If the vector has a length of $|F|$ its algebraic expression is

$$F = |F|e^{j\theta} = |F|\cos\theta + j|F|\sin\theta. \tag{15-13}$$

In Eq. (15-13) the complex number F represents the phasor. To obtain the corresponding time function we must rotate the vector and take its projection on the real axis. Rotating the vector changes θ to $\theta + \omega t$, and the projection on the real axis is the real part of the complex number. [We shall use the notation $\text{Re}(F)$ to denote the real part of a complex number.] The time function therefore is

$$f(t) = |F|\cos(\omega t + \theta). \tag{15-14}$$

In summary, the time-domain and frequency-domain expressions for a sinusoid are as follows:

Time-domain expression	*Frequency-domain expression*
$f(t) = \|F\|\cos(\omega t + \theta)$	$F = \|F\|e^{j\theta}$
	$= \|F\|\cos\theta + j\|F\|\sin\theta$

The transformation from the time domain to the frequency domain can also be discussed algebraically. We start with

$$f(t) = |F|\cos(\omega t + \theta), \tag{15-15}$$

and using Euler's formula, we replace the cosine function to obtain

$$f(t) = |F| \operatorname{Re}[e^{j(\omega t+\theta)}].$$

Since F is real it can be moved to within the real-part sign, and thus we can write

$$f(t) = \operatorname{Re}[|F|e^{j\theta}e^{j\omega t}]. \tag{15–16}$$

Equations (15–15) and (15–16) are equivalent expressions. To obtain the frequency-domain vector we drop the real-part sign and the $e^{j\omega t}$. The $e^{j\omega t}$ term causes the vector to rotate with time, and the real-part sign directs us to take the projection on the real axis. The stationary vector by itself is the frequency-domain phasor

$$F = |F|e^{j\theta}. \tag{15–17}$$

To transform from the frequency-domain vector, or phasor, to the time function we reverse this process. We multiply the phasor by $e^{j\omega t}$ and take the real part. Thus

$$\begin{aligned} f(t) &= \operatorname{Re}[|F|e^{j\theta}e^{j\omega t}] \\ &= |F| \cos(\omega t + \theta). \end{aligned} \tag{15–18}$$

When we multiply by $e^{j\omega t}$ we cause the vector to rotate with an angular velocity of ω radians per second, and when we take the real part we are taking the projection on the real axis.

The transformations back and forth between the time and the frequency domains are used so frequently that it is customary simply to memorize the results. *The magnitude of the frequency-domain phasor is the same as the amplitude of the sinusoid, and the angle of the phasor with respect to the real axis is the phase angle of the sinusoid written in cosine form.* These two rules allow rapid transformations in either direction. The summarized procedure for solving a sinusoidal steady-state problem is:

1. All the time functions in the problem are transformed to phasors in the frequency domain.
2. The problem is solved by complex algebra (or vector geometry) in the frequency domain.
3. The resulting vector is transformed back to the time domain.

In the remainder of this chapter we shall review vector geometry and complex algebra, postponing transformation of circuits to the frequency domain until Chapter 16.

The complex-frequency representation of sine waves. Before leaving this section we shall discuss yet another method of representing sinusoidal waves. This method makes use of Euler's formula in the form given by Eq. (15–11). If the sinusoid is

$$f(t) = |F| \cos (\omega t + \theta),$$

it can be written in the form

$$f(t) = |F| \frac{e^{j(\omega t+\theta)} + e^{-j(\omega t+\theta)}}{2}$$

$$= \frac{|F|}{2} e^{j\theta} e^{j\omega t} + \frac{|F|}{2} e^{-j\theta} e^{-j\omega t}. \tag{15–19}$$

In Eq. (15–19), the function $f(t)$ is expressed as the sum of two vectors, which are rotating in opposite directions. The first term is a vector of length $|F|/2$ which has an angle of θ at $t = 0$ and which rotates in the counterclockwise or positive direction at an angular velocity of ω radians per second. The second term is a vector of the same length which has an angle of $-\theta$ at $t = 0$ and which rotates in the clockwise or negative direction at an equal angular velocity. The two vectors are shown in Fig. 15–5(a). They form an instantaneous sum of

$$f(t) = |F| \cos (\omega t + \theta).$$

As the vectors rotate, the imaginary components cancel each other, and thus the sum is a purely real vector equal to the time function. In this representation of the sinusoid two "complex frequencies" are involved. These frequencies are the same as the complex transient frequencies we obtained for a pure L–C circuit. We represented them previously as two points on the imaginary ($j\omega$) axis of a "complex-frequency plane," as shown in Fig. 15–5(b).

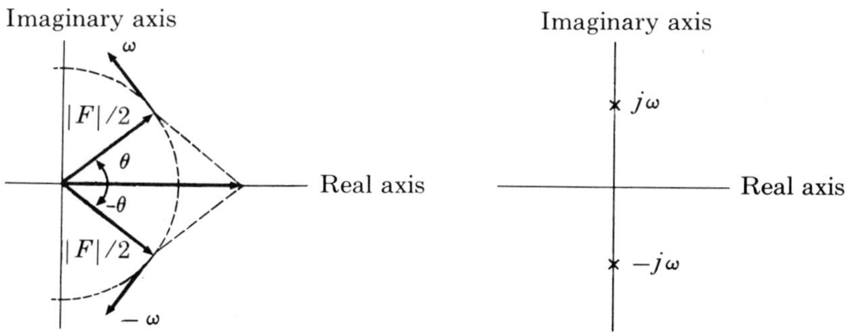

(a) A sinusoid produced by counter-rotating phasors

(b) A complex-frequency-plane representation of the two frequencies in (a)

FIG. 15–5. The representation of a sinusoid by two counter-rotating phasors.

It might appear that the representation of a sinusoid in terms of two counter-rotating phasors was preferable to the representation as the real part of a single rotating phasor. Such is not the case, however, since it is easier to take the real part of a vector than to deal simultaneously with two vectors rotating in opposite directions. Steady-state sinusoidal problems are usually solved by means of single rotating phasors. Fourier integrals and Laplace transforms, on the other hand, make use of both positive and negative frequencies of rotation. At least in their derivation we shall require the concept of counter-rotating phasors.

15–4 Vector (phasor) geometry.

Addition. Mathematical operations on sinusoidal functions can be accomplished by means of parallel operations on the corresponding phasors in the frequency domain. Since the phasors are vectors they can be added and subtracted by the usual methods for vectors. Consider, for example, the two time functions

$$i_1(t) = |I_1| \cos(\omega t + \theta),$$

$$i_2(t) = |I_2| \cos(\omega t + \phi).$$

The phasor for $i_1(t)$ has a length of $|I_1|$ and stands at an angle of θ from the real axis, and the phasor for $i_2(t)$ has a length of $|I_2|$ and stands at an angle of ϕ. The two phasors are shown in Fig. 15–6, along with their

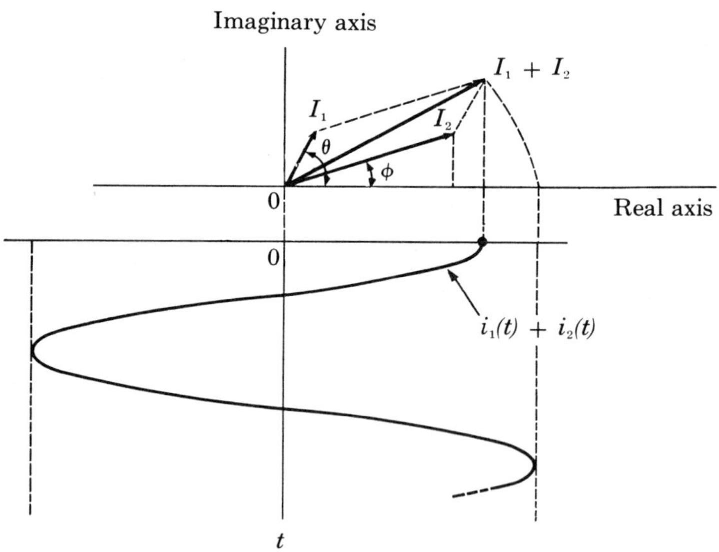

FIG. 15–6. The addition of sinusoids by means of the addition of their phasors.

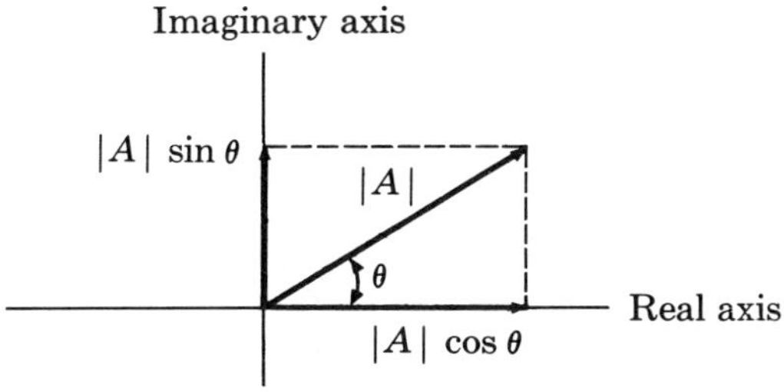

FIG. 15-7. The resolution of a phasor into component phasors.

vector sum. The projection of the resultant vector on the real axis traces out the resulting time function. The proof follows from the fact that the projection of the resultant vector on the real axis is the sum of the projections of the two component vectors.

Resolution. The operation inverse to summation is the resolution of a vector into components. Any two component vectors which provide the required sum are satisfactory, but the vectors are usually taken along the real and imaginary axes. For example, in Fig. 15-7 we have the phasor of length $|A|$ standing at an angle of θ from the real axis. Its components along the axes are

$$A_{\text{real}} = |A| \cos \theta,$$
$$A_{\text{imag}} = |A| \sin \theta.$$

The sum of these two components is certainly equal to the original phasor. Each of the components is a phasor in its own right, and each rotates at an angular velocity of ω radians per second, along with the original phasor. The corresponding time function is traced out by the projection of the original vector on the real axis.

Differentiation and integration. Phasors are implied functions of time, since they are rotating at an angular velocity of ω radians per second. The operations of differentiation and integration on the time functions have the effect of modifying the lengths of the phasors and advancing or retarding them by 90°. If the original time function is

$$f(t) = |F| \cos (\omega t + \theta), \tag{15-20}$$

its derivative, by Eq. (15-6), is

$$f't = |F|\omega \cos (\omega t + \theta + 90°).$$

Note here that the length of the phasor has been multiplied by ω and the phase has been advanced by 90°. The two phasors are shown as F and F' in Fig. 15-8.

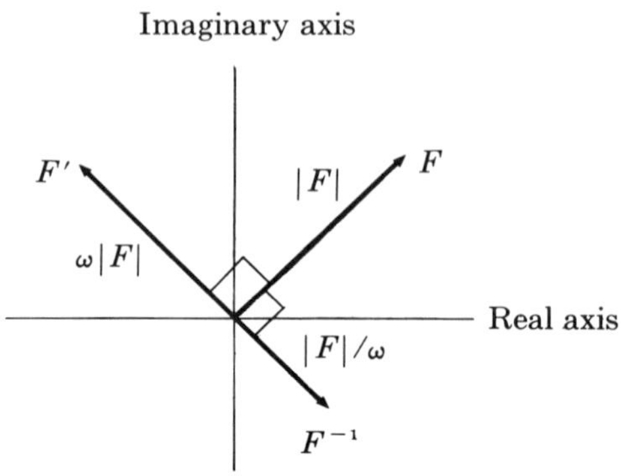

FIG. 15–8. Phasors for the derivatives and integrals of sinusoids.

By Eq. (15–7), and ignoring the second term, the integral of Eq. (15–20) is

$$f^{-1}(t) = \frac{|F|}{\omega} \cos(\omega t + \theta - 90^\circ).$$

Here we note that integration divides the length of the phasor by ω and retards the phase by 90°. The phasor for the integral is shown as F^{-1} in Fig. 15–8.

Phasor diagrams are useful in presenting a pictorial form of network relationships. There are certain problems in analysis and synthesis where such diagrams are invaluable. For the vast majority of routine analysis problems, however, phasor algebra is preferable, since it does not require the construction of a diagram and since the phases and amplitudes are automatically taken care of by the rules for complex algebra.

15–5 Complex (phasor) algebra. Complex algebra is more flexible than vector geometry because it does not require the construction or visualization of complicated geometrical figures. It has the further advantage that it reduces steady-state sinusoidal circuits to the form of pure resistance circuits. Once the problem has been converted to the frequency domain it is essentially a resistive problem. The only difference is that the real-number algebra used for resistive circuits becomes complex-number algebra for frequency-domain circuits. Thus before we undertake the study of frequency-domain circuits, it behooves us to review the rules of vector algebra.

There are certain standard conventions followed for frequency-domain circuits. Time functions are represented by lower-case letters, and the corresponding phasors are represented by the same letters in capital form. Furthermore, the same letter, in capital form and with magnitude bars

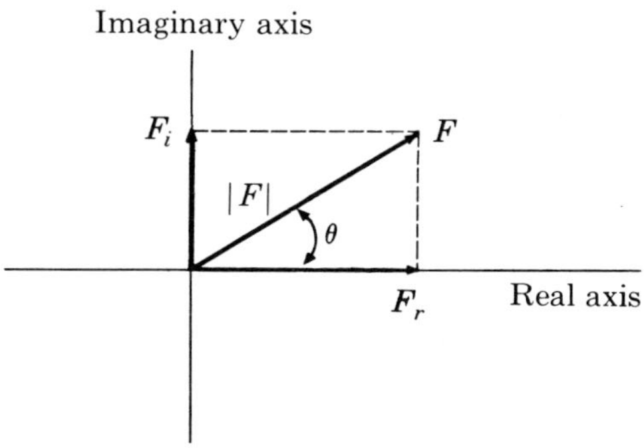

FIG. 15-9. A phasor in rectangular and polar forms.

around it, is used to represent the amplitude of the sinusoid. The length of the phasor, of course, is identical to the amplitude of the sinusoid. For example, suppose the time function is

$$f(t) = |F| \cos(\omega t + \theta). \tag{15-21}$$

The corresponding phasor is

$$F = |F|e^{j\theta} = |F| \underline{/\theta}. \tag{15-22}$$

The second form of the phasor in Eq. (15-22) is read "F magnitude at an angle of θ." It is in every way equivalent to the $e^{j\theta}$ notation and is actually the preferred form when the angle is in degrees rather than radians. In the following sections we shall review some of the rules of complex algebra as they apply to phasors.

Polar and rectangular forms. A complex number can be expressed in terms of its magnitude and angle, or in terms of its projection on the two axes. A typical example of a phasor is shown in Fig. 15-9. It has a length of $|F|$, and it stands at an angle of θ with respect to the real axis. The *polar* form of the phasor is

$$F = |F|e^{j\theta} = |F| \underline{/\theta}. \tag{15-23}$$

The *rectangular* form of the same phasor is

$$F = F_r + jF_i, \tag{15-24}$$

where F_r and F_i are the projections on the real and imaginary axes, respectively.

Polar to rectangular form. The conversion of a given phasor from polar to rectangular form can be accomplished by a formal manipulation of Eq. (15-23) or by resolving the vector geometrically along the two axes.

Thus

$$F = |F|e^{j\theta} = |F| \cos \theta + j|F| \sin \theta,$$

and the real and imaginary components are

$$F_r = |F| \cos \theta, \qquad F_i = |F| \sin \theta. \tag{15–25}$$

Rectangular to polar form. The conversion of a given phasor from rectangular to polar form is accomplished by applying the theorem of Pythagoras to the right triangle of Fig. 15–9. The length of the phasor is

$$|F| = \sqrt{F_r^2 + F_i^2}, \tag{15–26}$$

and the angle at which it stands is

$$\theta = \tan^{-1} \frac{F_i}{F_r}. \tag{15–27}$$

An alternative technique for obtaining the magnitude and angle is:

1. Take the ratio of F_i/F_r.
2. Obtain $\theta = \tan^{-1} F_i/F_r$.
3. Divide F_i by $\sin \theta$ to obtain $|F|$.

This technique should be used for numerical work on a slide rule. Since the tangent scale runs only up to 45°, the complementary triangle should be solved whenever F_i is greater than F_r. The value of $|F|$ is unchanged, and the phasor angle is: 90° minus the angle obtained.

The sum or difference of two phasors. In order to add two phasors it is convenient to write them in rectangular form. Suppose the two phasors are

$$F_1 = F_{r1} + jF_{i1},$$
$$F_2 = F_{r2} + jF_{i2}.$$

The sum is obtained by adding separately the real and imaginary parts. Thus

$$F_1 + F_2 = (F_{r1} + F_{r2}) + j(F_{i1} + F_{i2}). \tag{15–28}$$

If the original phasors are given in polar form it is necessary to convert them to rectangular form before performing the addition.

The product of two phasors. In order to multiply two phasors it is convenient to write them in polar form. Suppose the two phasors are

$$F_1 = |F_1|e^{j\theta_1},$$
$$F_2 = |F_2|e^{j\theta_2}.$$

The product is obtained by multiplying the magnitudes and adding the angles. Thus

$$F_1 \times F_2 = |F_1|\,|F_2|e^{j(\theta_1+\theta_2)}. \tag{15–29}$$

Since the product of two magnitudes is the magnitude of the product, Eq. (15–29) can also be written in the form

$$F_1 \times F_2 = |F_1F_2|\,\underline{|\theta_1 + \theta_2}\,. \tag{15–30}$$

Although it is convenient to have the phasors in polar form for multiplication, it is not necessary. In rectangular form they are multiplied by the well-known rules of algebra. If the phasors are

$$F_1 = F_{r1} + jF_{i1},$$
$$F_2 = F_{r2} + jF_{i2},$$

the product is

$$\begin{aligned} F_1 \times F_2 &= (F_{r1} + jF_{i1})(F_{r2} + jF_{i2}) \\ &= (F_{r1}F_{r2} - F_{i1}F_{i2}) + j(F_{r1}F_{i2} + F_{i1}F_{r2}). \end{aligned} \tag{15–31}$$

The quotient of two phasors. Let us suppose that we are dealing with the same two phasors which we considered for multiplication. If they are in polar form the quotient of F_1 divided by F_2 is

$$\frac{F_1}{F_2} = \frac{|F_1|e^{j\theta_1}}{|F_2|e^{j\theta_2}} = \frac{|F_1|}{|F_2|}\,e^{j(\theta_1-\theta_2)}. \tag{15–32}$$

Since the quotient of the magnitudes is the magnitude of the quotient, Eq. (15–32) can also be written in the form

$$\frac{F_1}{F_2} = \left|\frac{F_1}{F_2}\right|\,\underline{|\theta_1 - \theta_2}\,. \tag{15–33}$$

We note here that the magnitude of the quotient is the quotient of the magnitudes, and the angle of the quotient is the difference between the angles of the two phasors.*

If the two phasors are given in rectangular form the quotient can be obtained by the process called rationalization. That is, we multiply

---

* The magnitude of a complex number is the length of the vector representing the number. The magnitude of the sum of two vectors is less than the sum of the magnitudes. It is equal to the sum only if the two vectors have the same direction. The magnitude of a product, however, is always equal to the product of the magnitudes.

numerator and denominator by the conjugate of the denominator. (Conjugate phasors are defined in the following subsection.) Thus

$$\frac{F_1}{F_2} = \frac{F_{r1} + jF_{i1}}{F_{r2} + jF_{i2}} \times \frac{F_{r2} - jF_{i2}}{F_{r2} - jF_{i2}}$$

$$= \frac{F_{r1}F_{r2} + F_{i1}F_{i2} + j(-F_{r1}F_{i2} + F_{i1}F_{r2})}{F_{r2}^2 + F_{i2}^2}. \tag{15–34}$$

Alternatively, of course, the phasors can be changed to polar form before the division is attempted. In general, the polar form is preferred for multiplication and division, and the rectangular form for addition and subtraction.

Conjugate phasors. The definition of a pair of conjugate phasors is that they have the same magnitude and equal angles, each angle being the negative of the other. We shall represent the conjugate phasor by the symbol F^*. Thus, if

$$F = |F|e^{j\theta} = F_r + jF_i,$$

the conjugate phasor is

$$F^* = |F|e^{-j\theta} = F_r - jF_i. \tag{15–35}$$

In order to extend Eq. (15–35) to the general case we note that the real part of the vector remains unchanged when the conjugate is taken, but that the imaginary part is reversed in sign. No matter how complicated the expression for a vector or phasor, the conjugate can be obtained merely by changing the sign of all the j's which occur in the expression.

Conjugate phasors are useful for several reasons. For example, the sum of a phasor and its conjugate is

$$F + F^* = F_r + jF_i + F_r - jF_i = 2F_r, \tag{15–36}$$

and the difference is

$$F - F^* = F_r + jF_i - F_r + jF_i = 2jF_i. \tag{15–37}$$

The product of a phasor and its conjugate is

$$F \times F^* = |F|e^{j\theta} \times |F|e^{-j\theta} = |F|^2, \tag{15–38}$$

and the quotient is

$$\frac{F}{F^*} = \frac{|F|e^{j\theta}}{|F|e^{-j\theta}} = e^{j2\theta} = \underline{|2\theta}\,. \tag{15–39}$$

Equations (15–36), (15–37), (15–38), and (15–39) give explicit expressions for the real part, the imaginary part, the magnitude, and the phase of a phasor in terms of the phasor and its conjugate.

Numerical examples. Because of the importance of the phasor operations we have just discussed, we shall illustrate them by means of numerical examples involving the two phasors

$$F_1 = 3 + j4 = 5\,\underline{|53^\circ}\,,$$

$$F_2 = 1 + j1.732 = 2\,\underline{|60^\circ}\,.$$

(a) Polar to rectangular form.

$$\begin{aligned} F_1 &= 5\,\underline{|53^\circ} \\ F_r + jF_i &= 5\cos 53^\circ + j5\sin 53^\circ \\ &= 5 \times 0.6 + j5 \times 0.8 \\ &= 3 + j4 \end{aligned}$$

(b) Rectangular to polar form.

$$\begin{aligned} F_2 &= 1 + j1.732 \\ |F_2|\,\underline{|\theta_2} &= \sqrt{1^2 + (1.732)^2}\,\underline{|\tan^{-1} 1.732} \\ &= 2\,\underline{|60^\circ} \end{aligned}$$

(c) Sum and difference.

$$F_1 + F_2 = 3 + j4 + 1 + j1.732 = 4 + j5.73$$

$$F_1 - F_2 = 3 + j4 - 1 - j1.732 = 2 + j2.27$$

(d) Product in polar form.

$$F_1 \times F_2 = 5\,\underline{|53^\circ} \times 2\,\underline{|60^\circ} = 10\underline{|113^\circ}$$

(e) Product in rectangular form.

$$\begin{aligned} F_1 \times F_2 = (3 + j4)(1 + j1.732) &= (3 - 6.92) + j(4 + 5.19) \\ &= -3.92 + j9.19 \\ &= 10\,\underline{|113^\circ} \end{aligned}$$

(f) Quotient in polar form.

$$\frac{F_1}{F_2} = \frac{5\,\underline{|53^\circ}}{2\,\underline{|60^\circ}} = 2.5\,\underline{|-7^\circ}$$

(g) Quotient in rectangular form.

$$\begin{aligned}\frac{F_1}{F_2} &= \frac{3 + j4}{1 + j1.732} \times \frac{1 - j1.732}{1 - j1.732} \\ &= \frac{9.92 - j1.20}{4} \\ &= 2.48 - j0.30 \\ &= 2.5 \underline{/-7^\circ}\end{aligned}$$

(h) Conjugates.

$$\begin{aligned}F_1^* &= 3 - j4 = 5 \underline{/-53^\circ} \\ F_2^* &= 1 - j1.732 = 2 \underline{/-60^\circ} \\ F_1 + F_1^* &= 3 + j4 + 3 - j4 = 6 \\ F_1 - F_1^* &= 3 + j4 - 3 + j4 = j8 \\ F_1 \times F_1^* &= 5 \underline{/53^\circ} \times 5 \underline{/-53^\circ} = 25 \\ \frac{F_1}{F_1^*} &= \frac{5 \underline{/53^\circ}}{5 \underline{/-53^\circ}} = \underline{/106^\circ} = e^{j106^\circ}\end{aligned}$$

15–6 The manipulation of sine waves of a single frequency by means of phasors. We shall not consider the application of sinusoidal excitation to a network until the next chapter, but at this point we are prepared to perform a number of operations on the sinusoidal wave itself. We shall use the frequency-domain method of phasors, rather than the trigonometric method used in Section 15–2, and the superiority of the phasor method will be evident. There are three distinct steps in the method:

1. The sine waves are transformed to frequency-domain phasors.
2. The phasors are added or subtracted in the frequency domain.*
3. The resulting phasor is transformed back to the time domain.

Sine waves are represented in the frequency domain by stationary vectors or phasors which have magnitudes equal to the amplitudes of the sinusoids and phase angles equal to the phase angles of equivalent cosine waves at $t = 0$.

---

* Only the operations of addition and subtraction of two sine waves are linear and obey the phasor laws. Multiplication of two sine waves results in a double frequency term. The laws for multiplying phasors apply to linear "operations" which we shall discuss in Chapter 16.

The addition or subtraction of a pair of phasors in the frequency domain performs the equivalent operations on the corresponding sinusoidal waves in the time domain. Vector addition is required because both the angle and the magnitude are significant. For example, for the case of two sine waves of equal amplitude and a phase difference of 180°, the sum is zero.

All the phasors in the frequency domain have their direct counterparts in the time domain. In order to obtain the time functions we rotate the phasors and observe their projections on the real axis as functions of time.

As an example of the method let us consider the addition of the two sinusoids

$$i_1(t) = 10 \cos (377t + 60°),$$

$$i_2(t) = 10 \sin 377t = 10 \cos (377t - 90°).$$

Step 1. *Translation to the frequency domain.* The frequency-domain phasors which represent our time functions are

$$I_1 = 10 \underline{|60°},$$

$$I_2 = 10 \underline{|-90°}.$$

Step 2. *Addition of the phasors.* The sum of the two frequency-domain phasors is

$$\begin{aligned} I_1 + I_2 &= 10 \underline{|60°} + 10 \underline{|-90°} \\ &= 5 + j8.66 - j10 \\ &= 5 - 1.34j \\ &= 5.19 \underline{|-15°}. \end{aligned}$$

Step 3. *Translation back to the time domain.* The resulting wave in the time domain has a magnitude (amplitude) of 5.19 and a phase of −15°. Its analytical expression is

$$i_1(t) + i_2(t) = 5.19 \cos (377t - 15°).$$

Note that this result has been obtained with a great deal less effort than that required by a trigonometric solution.

Original expressions in the time domain will sometimes be written in terms of cosine functions and sometimes in terms of sine functions. The phase angles for the phasor method must be obtained from cosine expressions. It is important therefore to be able to transform rapidly and accurately from sine expressions to cosine expressions. In this regard, four particularly useful phasors to remember are those for $\cos \omega t$, $\sin \omega t$; $-\cos \omega t$ and $-\sin \omega t$. The phasor for $\cos \omega t$ is a unit vector at angle zero; the phasor for $\sin \omega t$ is a unit vector at −90°; the phasor for $-\cos \omega t$ is a unit

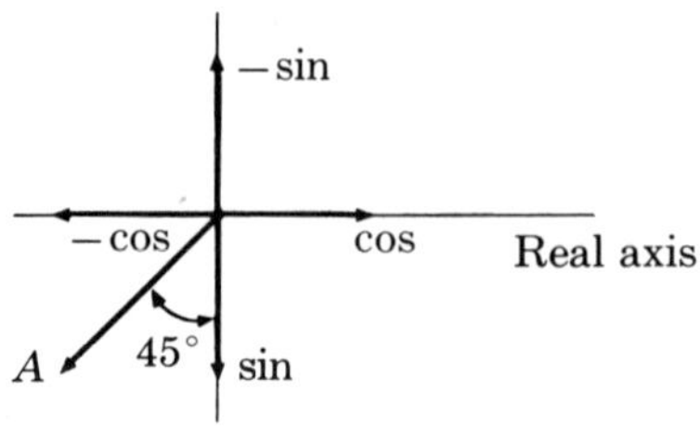

FIG. 15–10. The phasors for sines and cosines.

vector at ±180°; and the phasor for $-\sin \omega t$ is a unit vector at 90°. A sketch of these phasors is shown in Fig. 15–10.

Figure 15–10 can be used to transform from sine functions to cosine functions, and to obtain expressions for the time-domain equivalents of phasors. For example, phasor A in the illustration is 135° behind the cosine (the positive direction is the counterclockwise direction). Its time-domain expression is therefore

$$a(t) = |A| \cos (\omega t - 135°).$$

Note that the phasor is 45° behind the positive sine function. Thus another equivalent time-domain expression is

$$a(t) = |A| \sin (\omega t - 45°).$$

The phasor is 45° ahead of the negative cosine. Hence

$$a(t) = -|A| \cos (\omega t + 45°).$$

Since the phasor is 135° ahead of the negative sine, we also obtain

$$a(t) = -|A| \sin (\omega t + 135°).$$

All four of these expressions are, of course, equivalent. The sine-cosine phasor diagram permits us to pass easily from one form to another. The use of this simple diagram is highly recommended, since it avoids the necessity of remembering formulas. The whole point of complex algebra and the frequency domain is to avoid the use of hundreds of arbitrary trigonometric formulas by replacing them with a few simple rules of algebra. The diagram is easy to remember, too. The cosine phasor is at zero by definition. The sine phasor must be at −90° in order that when it starts rotating in the positive direction, its projection on the real axis will start to increase in the positive direction.

In this chapter we have introduced the notion of the frequency domain and have learned how to transform sinusoidal excitation functions into the frequency domain. In the next chapter we must investigate the effect which a network has on the frequency-domain phasors.

15–7 Summary.

The frequency domain

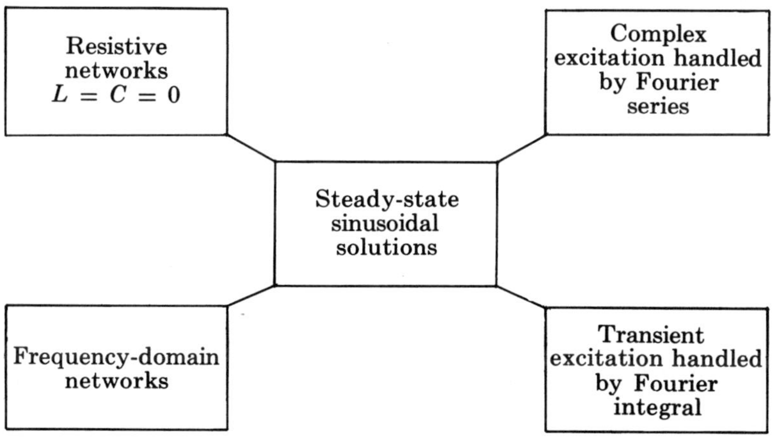

Methods of representing sinusoids

Time domain

(a) Graph

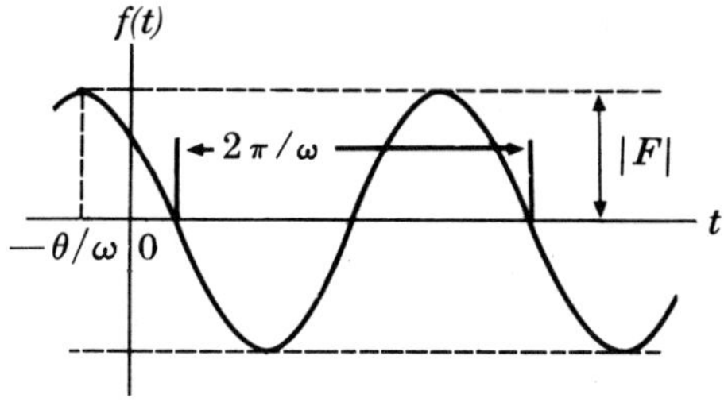

(b) Trigonometry $f(t) = |F| \cos (\omega t + \theta)$

Frequency domain

(c) Phasor

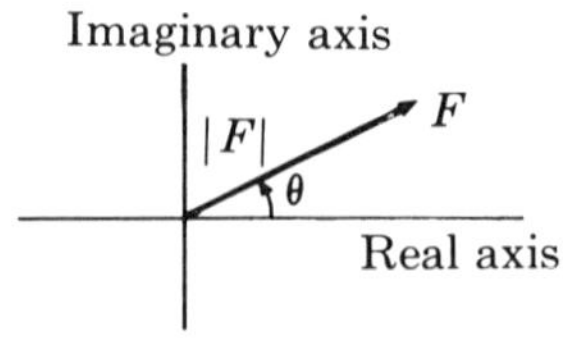

(d) Complex algebra $F = |F|e^{j\theta} = |F| \cos\theta + j|F| \sin\theta$

(e) Complex frequency

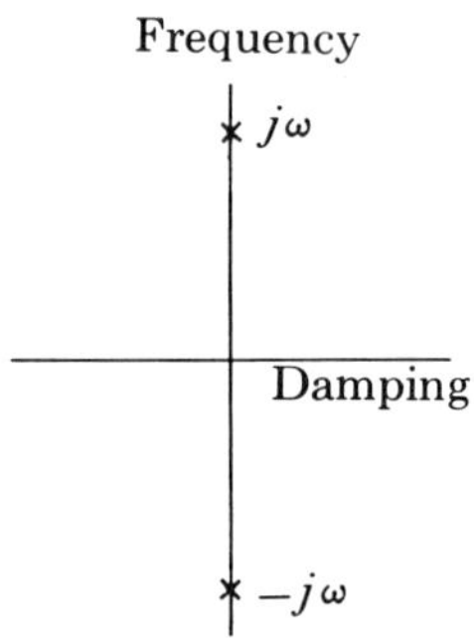

The properties of phasors

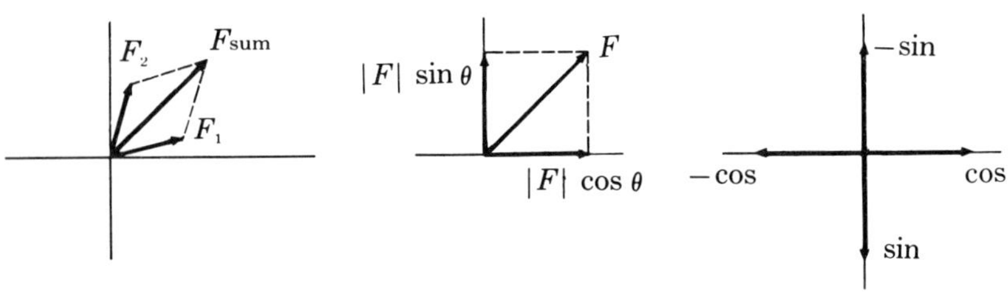

Addition Resolution Sine-cosine phasors

The rules of complex algebra

1. Polar form

$$F = |F|e^{j\theta} = |F| \underline{\lfloor\theta}$$

2. Rectangular form

$$F = F_r + jF_i$$

3. Polar to rectangular

$$F_r = |F| \cos \theta$$

$$F_i = |F| \sin \theta$$

4. Rectangular to polar

$$|F| = \sqrt{F_r^2 + F_i^2}$$

$$\theta = \tan^{-1} (F_i/F_r)$$

5. Addition and subtraction

$$F_1 \pm F_2 = (F_{r1} \pm F_{r2}) + j(F_{i1} \pm F_{i2})$$

6. Multiplication

$$F_1 \times F_2 = |F_1|\,|F_2|\,\underline{/\theta_1 + \theta_2}$$

7. Division

$$F_1/F_2 = |F_1|/|F_2|\,\underline{/\theta_1 - \theta_2}$$

8. Conjugate

$$F^* = |F|e^{-j\theta} = F_r - jF_i$$

Phasor addition and subtraction of sinuosids

1. Transform to the frequency domain.
2. Add or subtract the phasors.
3. Transform the resultant back to the time domain.

REFERENCES

Alternating-current circuit theory is the "bread and butter" tool of electrical engineering. Phasor methods were first used by Lord Rayleigh in his *Theory of Sound.* They were introduced to electrical engineers by Oliver Heaviside and popularized in America by Kennelly and Steinmetz. Most standard electrical-engineering circuit books cover phasors in great detail. In this text the phasor methods are treated in two parts. In this chapter we represent sinusoids by phasors, and in the next chapter we solve steady-state circuit problems by phasor methods. Of the references given here, the first four cite original sources; the remainder give a few selected present-day texts.

1. A. E. Kennelly, "Impedance," *AIEE Trans.*, 1893 (175–216).

2. C. P. Steinmetz, "Reactance," *AIEE Trans.*, 1894 (640–648).

3. Lord Rayleigh, *The Theory of Sound,* Vol. 1, Ch. 4 and 5, Macmillan and Co., London, Eng. 1894.

4. O. Heaviside, "On Resistance and Conductance Operators," *Phil. Mag.*, **XXVI,** 1878 (479–502).

5. MIT EE Staff, *Electric Circuits*, John Wiley & Sons, New York, 1944. Chapter 4 gives an excellent introduction to the use of frequency-domain vectors to solve sinusoidal circuit problems.

6. E. A. GUILLEMIN, *Introductory Circuit Theory*, John Wiley & Sons, New York, 1953. Chapter 6 gives an introduction to sinusoidal analysis and explains the importance of the sine wave to electrical engineers. The time domain and the frequency domain are clearly distinguished.

7. KERCHNER and CORCORAN, *Alternating Current Circuits*, John Wiley & Sons, New York, 3rd ed., 1951. This book begins Chapter 1 with alternating-current waveforms and by Chapter 4 has developed the phasor method of solution for sinusoidal problems.

8. LEPAGE and SEELY, *General Network Analysis*, McGraw-Hill, New York, 1951. Chapter 1 discusses the use of complex numbers to represent sine waves. The phasors are related to the RMS (root-mean-square) values of the sinusoids, rather than to the peak values as in this text.

9. H. H. SKILLING, Electrical Engineering Circuits, John Wiley & Sons, New York, 1957. Skilling discusses sinusoids and the use of phasors very early in the text (Chapter 3). Here also, the phasors are related to RMS values.

10. M. E. VAN VALKENBURG, *Network Analysis*, Prentice-Hall, New Jersey, 1955. The approach used by Van Valkenburg is completely different from the ones discussed above. The general Laplace transform theory is developed first, and the sinusoidal steady-state analysis is then treated as a special case.

11. G. B. THOMAS, JR., *Calculus and Analytical Geometry*, Addison-Wesley, Reading, Mass., 1954. The background material on complex numbers and complex algebra is well covered in this introductory mathematics text.

EXERCISES

Section 15–2

1. Plot several complete cycles of the waveform

$$f(t) = 10 \cos (20\pi t + 30°).$$

Use 30° increments of the argument. Give the following quantities

(a) Amplitude (b) Phase (relative to the cosine)
(c) Angular frequency (d) Cyclic frequency
(e) Period

2. Using 30° increments of the argument sketch the following two waveforms.

$$f_1(t) = 10 \cos 2\pi t, \qquad f_2(t) = 10 \sin 2\pi t.$$

Obtain their sum by adding the graphical waveforms, and obtain an analytical expression for the sum by trigonometry. Give the (a) amplitude, (b) phase, (c) angular frequency, (d) cyclic frequency, and (e) period of the resultant.

3. Resolve the waveform given below into a sine term and a cosine term. Sketch the original waveform and the two component waveforms.

$$f(t) = 10 \cos (2\pi t + 60°).$$

4. Find the derivative of the following waveform and sketch it. For the derivative give the (a) amplitude, (b) phase, (c) angular frequency, (d) frequency, and (e) period.

$$f(t) = 100 \cos (1000t + 30°).$$

5. Find and sketch the integral from zero to t of the following waveform. For the sinusoidal component of the integral give the (a) amplitude, (b) phase, (c) angular frequency, (d) cyclic frequency, and (e) period.

$$f(t) = 100 \cos (1000t + 30°).$$

6. (a) Find the integral from zero to t of the function

$$f(t) = 100 \cos (10t + \theta°).$$

(b) What is the maximum value of the constant term as θ varies?
(c) Give the amplitude of the sinusoidal component.
(d) Give the phase relative to the original waveform.
(e) Give the frequency and the period.
(f) How do these quantities change as θ is varied?

7. Evaluate the sinusoidal component of

$$f(t) = \int_0^t \left[30 \sin 30t + 30 \cos 30t + \frac{d}{dt}[1.41 \cos (30t + 45°)]\right] dt.$$

8. A certain network has the property of introducing delay into the output function. The output waveform at time t is identical to the input waveform at time $(t - 1)$ sec. The input waveform is given by

$$f(t) = \cos t + \cos 2t + \cos 3t + \cos 4t + \cos 5t.$$

Give the output waveform and the phase shift of each component relative to the same component at the input.

Section 15–3

9. Plot several cycles of the sinusoidal function given below by means of the graphical technique in which the projection of the rotating vector on the real axis represents the sinusoid. Give the amplitude, phase, frequency and period of the waveform.

$$f(t) = 100 \cos (377t + 30°).$$

10. Draw the phasors (rotating vectors at $t = 0$) which represent the two functions given below and obtain the resultant phasor by vector addition.

$$f_1(t) = 10 \cos 2\pi t, \qquad f_2(t) = 10 \sin 2\pi t.$$

Obtain the time function for the sum of the two functions by causing the resultant vector to rotate and by finding the projection on the real axis. Verify your results by trigonometry.

11. Sketch the phasors which represent the following sinusoids.

(a) $10 \cos t$ (b) $5 \cos (377t + 45°)$
(c) $10 \sin t$ (d) $-5 \cos (377t - 45°)$
(e) $10 \cos (t + 135°)$

12. Give the complex numbers in rectangular form which represent each of the phasors given below.

(a) $10 \underline{|0°}$ (b) $5 \underline{|45°}$
(c) $10 \underline{|-90°}$ (d) $5 \underline{|135°}$
(e) $10 \underline{|135°}$

13. Write the time functions which correspond to the following phasors.

(a) $10 \underline{|0°}$ $\omega = 1$
(b) $10 \underline{|90°}$ $\omega = 377$
(c) $5 \underline{|-60°}$ $\omega = 1000$
(d) $5 \underline{|53°}$ $\omega = 1$
(e) $2.82 \underline{|135°}$ $\omega = 1/10$

14. Write the time functions which correspond to the frequency-domain complex numbers given below. Use the general angular frequency ω in each case.

(a) $1 + j$ (b) $3 - j4$
(c) $-2 - j2$ (d) $-3 + j4$
(e) j

15. Write the time functions which correspond to the following phasors. Use the general frequency ω and phase angles with magnitudes of less than 90°.

(a) $3\,\underline{|135^\circ}$ (b) $-2 + j2$
(c) $-3 - j4$ (d) $3\,\underline{|300^\circ}$
(e) $4\,\underline{|-225^\circ}$

16. Sketch the phasor for $f(t)$ and for its derivative and its integral. Express each phasor as a magnitude and angle, as a polar complex number, and as rectangular complex number.

$$f(t) = 10 \cos (10t + 45^\circ).$$

17. Sinusoids can be represented as the instantaneous sum of two vectors rotating in opposite directions. Give the pairs of vectors at $t = 0$ which represent the following sinusoids.

(a) $10 \cos t$ (b) $5 \cos (t + 45^\circ)$
(c) $2 \sin t$ (d) $-10 \sin (t + 45^\circ)$
(e) $2 \cos (t + 225^\circ)$

18. The pairs of vectors given below represent the zero positions of counter-rotating vectors. In each case the first of the two vectors rotates in the positive direction. Give the time-domain sinusoidal functions in each case. Use the general angular frequency ω.

(a) $10\,\underline{|45^\circ}$ and $10\,\underline{|-45^\circ}$
(b) $10\,\underline{|-45^\circ}$ and $10\,\underline{|45^\circ}$
(c) j and $-j$
(d) $2\,\underline{|-225^\circ}$ and $2\,\underline{|225}$
(e) $-2 - j2$ and $-2 + j2$

Section 15–4

19. Find the sum of the following phasors by vector geometry.

(a) $1\,\underline{|0^\circ} + 1\,\underline{|90^\circ}$
(b) $1\,\underline{|\theta^\circ} + 1.732\,\underline{|90^\circ + \theta^\circ}$
(c) $2\,\underline{|-225^\circ} + 2\,\underline{|45^\circ}$
(d) $-3\,\underline{|215^\circ} + 6\,\underline{|35^\circ}$
(e) $3\,\underline{|70^\circ} + 4\,\underline{|160^\circ}$

20. Add the following phasors by means of geometrical diagrams.

(a) $5\,\underline{|0^\circ} + 5\,\underline{|180^\circ}$
(b) $10\,\underline{|0^\circ} + 10\,\underline{|-120^\circ} + 10\,\underline{|-240^\circ}$
(c) $5\,\underline{|30^\circ} + 5\,\underline{|-30^\circ} + 8.66\,\underline{|120^\circ} + 8.66\,\underline{|240^\circ}$
(d) $5\,\underline{|0^\circ} + 7.07\,\underline{|45^\circ} + 10\,\underline{|60^\circ} - 15\,\underline{|0^\circ} - 13.66\,\underline{|90^\circ}$
(e) $5\,\underline{|16^\circ} + 5\,\underline{|32^\circ} + 5\,\underline{|106^\circ} + 5\,\underline{|152^\circ} + 5\,\underline{|196^\circ} + 5\,\underline{|272^\circ} + 5\,\underline{|286^\circ}$

21. By geometrical methods resolve the following phasors into components along the real and the imaginary axes.

(a) $100\,\underline{/0°}$ (b) $100\,\underline{/35°}$

(c) $100\,\underline{/120°}$ (d) $100\,\underline{/210°}$

(e) $100\,\underline{/315°}$

22. Obtain the sum of the following phasors by resolving each one into components along the real and the imaginary axes and by adding the components.

$$100\,\underline{/0°} + 50\,\underline{/90°} + 70.7\,\underline{/225°} + 100\,\underline{/-60°} + 50\,\underline{/180°}.$$

23. Plot the phasors which represent the function

$$f(t) = 100 \cos (2t + 30°)$$

and its integral and derivative. Give the single phasor which represents

$$f(t) + f'(t) + 4f^{-1}(t).$$

24. Three phasors have a sum of zero. Their magnitudes are 3, 4, and 5 respectively. If the vector of length 3 is taken as the reference at zero phase angle, what possible phase angles can the other two phasors have?

25. Three phasors have a sum of zero. The first two are $100\,\underline{/60°}$ and $50\,\underline{/-30°}$. What is the third?

26. (a) Find the square roots of unity by solving the equation

$$x^2 = 1 = e^{jn\pi}, \qquad \text{where } n = 0, \pm 2, \pm 4, \ldots .$$

(b) Express the roots in phasor form and show that each one when multiplied by itself yields unity.

(c) Find the phasor sum of the two square roots of unity.

27. (a) Find the six sixth-roots of unity by solving the equation

$$x^6 = 1 = e^{jn\pi}, \qquad \text{where } n = 0, \pm 2, \pm 4, \ldots .$$

(b) Express the roots as phasors and show that each one multiplied by itself six times yields unity.

(c) Find the phasor sum of the six sixth-roots of unity.

Section 15–5

28. Change the following complex numbers from polar to rectangular form.

(a) $5\,\underline{/0°}$ (b) $10\,\underline{/45°}$ (c) $1.732\,\underline{/-120°}$

(d) $10\,\underline{/1°}$ (e) $10\,\underline{/-89°}$

29. Change the following complex numbers from polar to rectangular form.

(a) $10e^{j\pi}$ (b) $5e^{j\pi/3}$ (c) $10e^{-j0.6\pi}$

(d) $0.4e^{j16\pi/3}$ (e) $-10e^{-j\pi/4}$

30. Change the following complex numbers from rectangular to polar form.
(a) $j6$ (b) $1 - j2$ (c) $-1 - j2$
(d) $3 + j4$ (e) $-3 + j4$

31. Change the following complex numbers from rectangular to polar form.
(a) $0.345 + j0.216$ (b) $0.123 - j0.794$
(c) $-0.0134 + j0.637$ (d) $343 + j$
(e) $-1.73 \times 10^6 - j3.12 \times 10^6$

32. Find the magnitudes of the following complex numbers.

(a) $2 + j3$ (b) $2 - j3$ (c) $\dfrac{1}{2 + j3}$

(d) $(2 - j3)(2 + j3)$ (e) $e^{(1+j\pi/4)}$

33. Find the real part of each of the following complex numbers.
(a) $j17$ (b) $e^{j\pi/4}$ (c) $(3 + j4)(2 - j)$

(d) $\dfrac{1}{3 + j4}$ (e) $10\,\underline{|115^\circ}$

34. Write the polar form of the complex numbers which are complex-conjugate to the ones given below.

(a) $7 - j13$ (b) $(1 + j)(1 - j)$ (c) $\dfrac{1}{1 + j}$

(d) $4\,\underline{|60^\circ} + 2\,\underline{|-83^\circ}$ (e) $e^{(2+j\pi/2)}$

35. Evaluate the following expressions and write the answers in polar and rectangular form.

(a) $(3 + j4)(3 - j4)$ (b) $\dfrac{1}{3 + j4}$

(c) $3 + \dfrac{10}{1 + j}$ (d) $\dfrac{2 + j3}{3 + j2}$

(e) $\dfrac{3 - j2}{-2 + j3}$

36. Evaluate the following expressions and write the answers in polar and rectangular form.

(a) $3 + \dfrac{4 + j5}{6 + j7}$ (b) $\dfrac{1}{-1 - j} + \dfrac{1}{-4 + j5}$

(c) $3\,\underline{|-45^\circ} + \dfrac{1 + j2}{4 - j3}$ (d) $\dfrac{(1 + j2)(4 + j3)}{(5 + j4)}$

(e) $\dfrac{10\,\underline{|53^\circ} + 7\,\underline{|-13^\circ}}{4\,\underline{|3^\circ} - 10\,\underline{|120^\circ}}$.

37. By using the basic definitions

$$\cos\theta = \frac{e^{j\theta} + e^{-j\theta}}{2} \quad \text{and} \quad \sin\theta = \frac{e^{j\theta} - e^{-j\theta}}{2j},$$

find the meaning of (a) $\cos(x + jy)$ and (b) $\sin(x + jy)$.

Section 15–6

38. Translate the following two sinusoids to the frequency domain, add the phasors, and then translate the resulting phasor back to the time domain. Check your result by trigonometry.

$$e_1(t) = 100 \cos(\omega t - 30°),$$
$$e_2(t) = 100 \cos(\omega t + 60°).$$

39. Add the following three sinusoids by means of frequency-domain phasors.

$$e_1(t) = 100 \cos(377t),$$
$$e_2(t) = 100 \cos(377t + 120°),$$
$$e_3(t) = 100 \cos(377t + 240°).$$

40. Perform the following trigonometric operations using phasors in the frequency domain. Express the results as time functions.

(a) $4 \cos(t + 45°) - 3 \cos(t - 45°)$
(b) $5 \cos(t + 30°) + 5 \cos(t - 30°)$
(c) $8.66 \cos(t + 120°) - 8.66 \cos(t + 240°)$
(d) $7 \cos 13t + 4 \cos(13t + 70°)$
(e) $-\sin(3t + 45°) + \cos(3t + 45°) + \sin 3t$.

41. Express the following waveforms as positive cosine functions with a phase angle. In the process make use of phasor diagrams.

(a) $100 \sin t$ (b) $-100 \cos t$ (c) $-100 \sin t$
(d) $100 \sin(t + \pi)$ (e) $-100 \cos(t + \pi/2)$.

42. Give equivalent positive cosine functions for the following expressions.

(a) $4 \sin(377t + 45°)$ (b) $-6 \sin(t + 30°)$
(c) $-7 \cos(t - 135°)$ (d) $\sin(1000t + 7\pi/6)$
(e) $10 \sin 3(t - \pi/2)$.

43. An alternating voltage is given by

$$e(t) = 100 \cos 377t.$$

By phasor methods find the sum

$$e(t) + \frac{e'(t)}{100}.$$

44. An alternating current is given by

$$i(t) = 100 \cos(1000t + 45°).$$

By phasor methods find the sum

$$\frac{i(t)}{10} + 100 \int_{-\infty}^{t} i(t)\, dt.$$

Assume that the value of the integral at $t = -\infty$ is zero.

45. An alternating voltage is given by

$$e(t) = 100 \cos (2t + 45°).$$

By phasor methods find the sum

$$e(t) + \frac{de}{dt} + 4\int_{-\infty}^{t} e\, dt,$$

assuming that the value of the integral at $t = -\infty$ is zero.

46. Two voltages $e_1(t)$ and $e_2(t)$ are given below. By phasor methods find their sum, $e_1(t) + e_2(t)$, and also find their difference, $e_1(t) - e_2(t)$.

$$e_1(t) = 10 + 5 \sin \omega t + 3 \sin (3\omega t + 45),$$
$$e_2(t) = 5 + 5 \cos \omega t + 2 \cos (3\omega t - 45).$$

47. The current in the circuit of Fig. 15–11 is $i(t) = 100 \cos t$. What is the voltage $e(t)$? Neglect the value of the integral at $t = -\infty$.

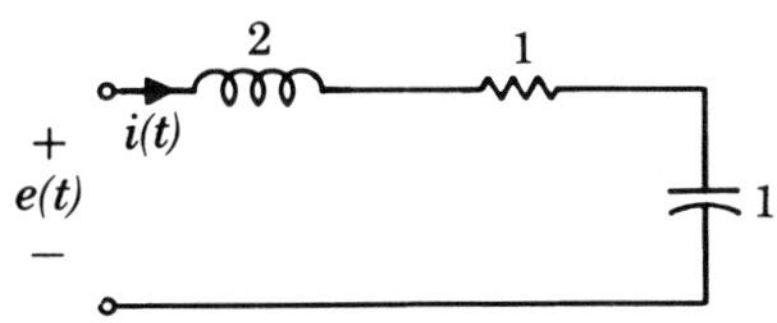

FIGURE 15–11

CHAPTER 16

FREQUENCY-DOMAIN NETWORKS

"... for thou shalt learn
The wisdom early to discern
True beauty in utility."

Longfellow

16–1 Introduction. Frequency-domain networks are certainly useful. They are the work horse of electrical engineering. And yet they are not without their beauty. Perhaps it lies in their utility, as Longfellow suggests, or perhaps in the internal consistency which allows us to treat steady-state sinusoidal networks, resistive networks, transient excitation, and arbitrary excitation by the same general techniques. Indeed, sinusoidal methods are so appealing that they are used in many fields remote from electrical engineering. Sinusoidal response functions have been obtained for human operators and for the economic system.

In Chapter 15 we defined the time and frequency domains and observed the usefulness of the frequency domain for the addition and subtraction of sinusoidal functions. In the present chapter we shall extend these concepts to the solution of network problems. In addition to the excitation function, we must transform the circuit itself to the frequency domain. In the frequency domain the circuit becomes a sort of operator which changes the magnitude and phase of the input phasor and transforms it into an output phasor. The operations required are addition, subtraction, and the sort of phasor multiplication which we discussed in Chapter 15.

In the frequency domain the effect of the circuit on the input phasor is most easily obtained by the use of phasor algebra. The procedures are identical with those which we used for resistive networks, except that the algebra is complex instead of real. There is a one-to-one correspondence, of course, between the phasor operations and the corresponding operations by means of phasor geometry. The geometrical approach is usually more difficult, but it may be more illuminating. The geometrical diagram provides a picture from which the operation of a circuit may often be visualized. We will discuss both the algebraic and the geometric method of solving a circuit problem, with perhaps slightly more emphasis on the algebraic method because of its generality.

When the exciting function is a single-frequency sinusoid, the output response function is also a sinusoid of the same frequency. The input and output differ only in magnitude and phase. The "response function" for the network at this frequency thus consists of two numbers, one of which gives the effect of the network on the amplitude and the other on the phase. In order to describe the effect of the network for all possible frequencies which might be applied to it, we would need curves of the amplitude and phase factors as functions of frequency. In effect these curves are the *system functions* which describe the behavior of the network for any input function.* A one-terminal-pair network is completely described by a single pair of curves. A two-terminal-pair network requires three pairs of curves for a general description, but each of these is a stimulus-response relationship in its own right which can be considered a separate problem. In this chapter we will consider both graphical and analytical methods of obtaining these response functions. The concepts involved are a simple extension of those which we already considered for resistive networks.

16–2 Transformation of networks to the frequency domain. Our general approach to the solution of steady-state sine wave excitation in a network will be similar to the frequency-domain approach which we developed in Chapter 15. That is,

1. The problem is translated to the frequency domain.
2. The solution is obtained for the phasor relationships in the frequency domain.
3. The resulting phasor is transformed back to the time domain.

Before studying the frequency-domain methods, the student should review the solution of steady-state sinusoidal problems by the methods of classical differential equations.† To appreciate the simplicity of the frequency-domain methods, the student must obtain the sinusoidal solution to at least one differential equation. The classical solution is straightforward, but it is laborious and it involves the use of many trigonometric formulas. The frequency-domain method we will now consider reduces the differential equation to an algebraic equation, and the trigonometry to complex algebra. It is a great deal easier!

The behavior of any network is completely determined by Kirchhoff's laws and the volt-ampere equations of the elements. We shall show how all of these equations are transformed to the frequency domain and, by inference therefore, how networks are transformed to the frequency domain.

---

* Some authors reserve the term "system function" for the stimulus-response relationship when the frequency is complex (see Chapter 21). This chapter represents a special case of the more general theory.

† See Section 9–4.

Kirchhoff's laws in the frequency domain. In the time domain the two Kirchhoff laws are

$$\sum e_n = 0, \qquad \sum i_n = 0. \tag{16–1}$$

In a steady-state sinusoidal problem, all the currents and all the voltages are sinusoids of a single frequency. If we write each sinusoid as a cosine with an arbitrary amplitude and an arbitrary phase, Kirchhoff's laws therefore become

$$\begin{aligned} \sum|E_n| \cos(\omega t + \theta_n) &= 0, \\ \sum|I_n| \cos(\omega t + \phi_n) &= 0. \end{aligned} \tag{16–2}$$

Each of the sinusoids in Eq. (16–2) can be represented in the frequency domain by a phasor with a length equal to the magnitude of the sinusoid, and an angle equal to the phase of the sinusoid. Equation (16–2) states that the sum of the projections of the phasors on the real axis must be zero as the phasors are rotated with an angular velocity of ω radians per second. The only phasor whose projection on the real axis would always be zero is that of zero length. Thus the resultant phasor in the frequency domain must be one of zero length. Kirchhoff's laws in the frequency domain therefore state that the sum of the voltage phasors around any closed path must be zero, and the sum of the current phasors at any junction point must be zero. Analytically

$$\sum E_n = 0, \qquad \sum I_n = 0, \tag{16–3}$$

where $E_n = |E_n| \underline{|\theta_n}, \qquad I_n = |I_n| \underline{|\phi_n}.$

It is perhaps worth emphasizing that the summations indicated in Eqs. (16–3) are vector summations, and that the quantities to be added are vectors or complex numbers. The relevant equations have the same form as Eqs. (16–1), but their meaning is quite different. The voltages and currents in Eqs. (16–1) are instantaneous functions of time.

An example which illustrates this difference is shown in Fig. 16–1. The three phasors in part (a) have equal lengths and are 120° apart. When added together they form an equilateral triangle with a zero sum. The three equivalent time functions are shown in part (b). They are the projections of the phasors on the real axis as the whole phasor diagram is rotated with an angular velocity of ω radians per second. The instantaneous sum of the time functions at any point along these waveforms will also sum to zero. This example is a rather specialized one, but it is a practical one which occurs in the three-phase power transmission system. In general, the voltages around a closed path will have arbitrary magnitudes and phases, but the vector sum must still be zero.

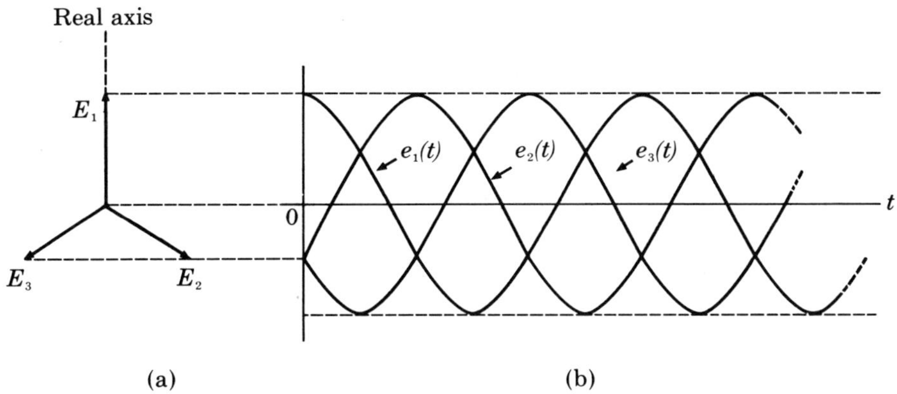

FIG. 16–1. Three phasors whose projections sum to zero at all times.

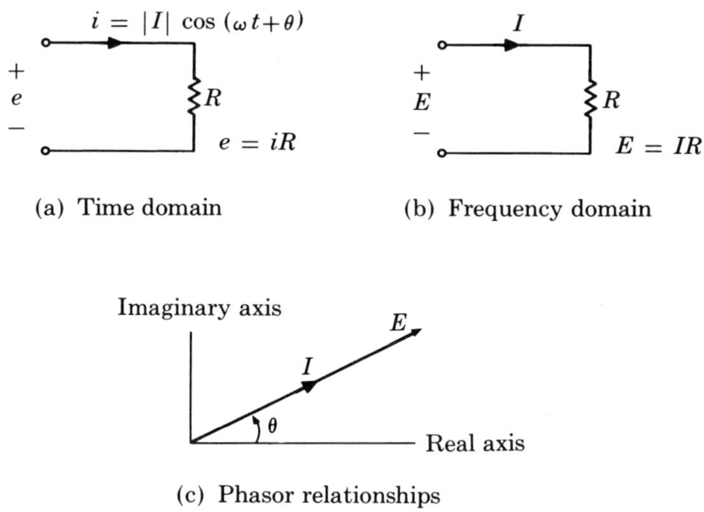

FIG. 16–2. The resistance element with sinusoidal excitation.

We have shown that Kirchhoff's laws can be transformed to the frequency domain, and that they have essentially the same form in the frequency domain as in the time domain. We must now transform the volt-ampere equations to the frequency domain.

The volt-ampere equations in the frequency domain.

Resistance. A resistance in the time domain is shown in Fig. 16–2(a). The current is a sinusoid of the form

$$i = |I| \cos(\omega t + \theta). \tag{16–4}$$

The volt-ampere equation for the resistance is

$$e = iR, \tag{16–5}$$

and hence the voltage is

$$e = |I|R \cos (\omega t + \theta). \tag{16–6}$$

The phasor representing the current of Eq. (16–4) is

$$I = |I| \underline{/\theta},$$

and the phasor representing the voltage in Eq. (16–6) is

$$E = |I|R \underline{/\theta}.$$

The relationship between the two phasors is simply

$$E = IR. \tag{16–7}$$

Equation (16–7) has the same form as Ohm's law, but it refers to two vectors, or two complex numbers, and not to a voltage and a current. The current and the voltage phasors are in the same straight line, and the length of the voltage phasor is R times the length of the current phasor. The frequency-domain diagram for the resistance element is shown in Fig. 16–2(b), and the relationship between the current and voltage phasors is shown in Fig. 16–2(c). The resistance satisfies the same equation in the frequency domain as it did in the time domain. The frequency-domain circuit for a resistance is still a constant value of R ohms. The current vector is multiplied by R to get the voltage vector, and the voltage vector is multiplied by $1/R$ to obtain the current vector.

Inductance. An inductance in the time domain is shown in Fig. 16–3(a). The excitation is a current

$$i(t) = |I| \cos (\omega t + \theta). \tag{16–8}$$

The volt-ampere equation for the inductance is

$$e = L \frac{di}{dt}, \tag{16–9}$$

and hence the voltage is

$$\begin{aligned} e(t) &= L|I| \frac{d}{dt} [\cos (\omega t + \theta)] \\ &= \omega L|I| \cos (\omega t + \theta + 90°). \end{aligned} \tag{16–10}$$

The phasor for the current of Eq. (16–8) is

$$I = |I| \underline{/\theta},$$

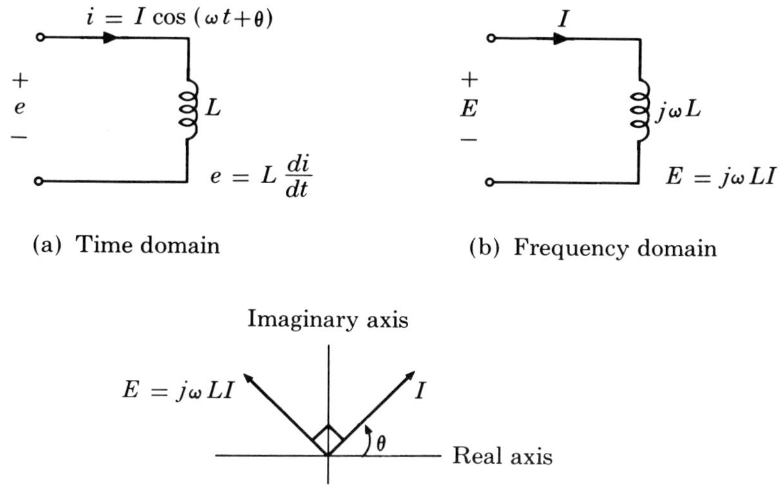

(a) Time domain

(b) Frequency domain

(c) Phasor relationships

FIG. 16–3. The inductance element with sinusoidal excitation.

and the phasor for the resulting voltage of Eq. (16–10) is

$$E = \omega L|I| \underline{/\theta + 90^\circ}. \tag{16–11}$$

Equation (16–11) shows that the voltage phasor is equal to the current phasor multiplied by ωL in magnitude and is shifted 90° ahead in phase. Thus

$$\begin{aligned} E &= (\omega L \times I) \underline{/90^\circ} \\ &= (\omega L \times I)(\cos 90^\circ + j \sin 90^\circ) \\ &= j\omega L \times I. \end{aligned} \tag{16–12}$$

The effect of the j factor in the last form of Eq. (16–12) is to advance the phase by 90° without changing the magnitude of the phasor I. The effect of the ωL is to multiply the magnitude of the phasor by a constant without changing the phase. The complete phasor equation is the volt-ampere relationship of the inductance in the frequency domain. If we reverse the roles of the stimulus and the response, we obtain

$$I = \frac{E}{j\omega L}. \tag{16–13}$$

The quantity $j\omega L$ in Eqs. (16–12) and (16–13) is the complex "resistance" of the inductance in the frequency domain. It is given the name *impedance*, and the symbol Z. An inductance is transformed to the frequency domain by replacing it by a complex resistance, or impedance, of value $Z = j\omega L$. In the frequency domain, this impedance has the formal properties of a resistance. It satisfies an equation which is identical in form to Ohm's law. That is,

$$E = IZ \qquad \text{with } Z \text{ in ohms.} \tag{16–14}$$

In Eq. (16–14) both E and I are phasors, and Z is a complex number or operator which changes the magnitude and phase of one of the phasors until it is identical to the other. In the frequency-domain diagram the inductance is replaced by its impedance, $Z = j\omega L$, as shown in Fig. 16–3(b). The corresponding current and voltage phasors are shown in Fig. 16–3(c). If the current phasor is given, the voltage phasor will lead it by 90° and will have a magnitude which is ωL times the magnitude of the current phasor. If the voltage phasor is given, the current phasor will lag it by 90° and will have a magnitude which is $1/\omega L$ of that of the voltage phasor. The operations can be performed algebraically by means of Eq. (16–14), and except for the complex nature of the algebra, they are identical to the operations for resistance.

An heuristic method of keeping track of the phase relationships in an inductance is provided by recalling that the voltage must exist across an inductance first, in order to force a current through it. The current phasor therefore lags the voltage phasor.

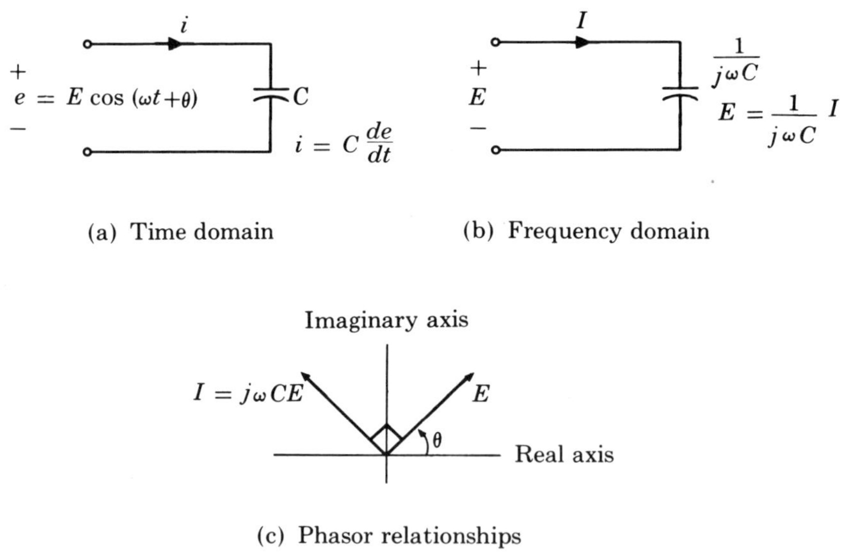

FIG. 16–4. The capacitance element with sinusoidal excitation.

Capacitance. A capacitance in the time domain is shown in Fig. 16–4(a). The excitation is a voltage

$$e(t) = |E| \cos (\omega t + \theta). \tag{16–15}$$

The volt-ampere equation for the capacitance is

$$i = C \frac{de}{dt}, \tag{16–16}$$

and hence the current is

$$\begin{aligned} i(t) &= C \frac{d}{dt} [|E| \cos (\omega t + \theta)] \\ &= \omega C |E| \cos (\omega t + \theta + 90°). \end{aligned} \tag{16–17}$$

The voltage phasor, from Eq. (16–15), is

$$E = |E| \underline{/\theta}, \tag{16–18}$$

and the resulting current phasor, from Eq. (16–17), is

$$I = \omega C |E| \underline{/\theta + 90°}. \tag{16–19}$$

From Eqs. (16–18) and (16–19) it is apparent that the current and voltage phasors are related by the following equation:

$$\begin{aligned} I &= (\omega C \times E) \underline{/90°} \\ &= j\omega C E. \end{aligned} \tag{16–20}$$

Equation (16–20) is the volt-ampere equation for the capacitance in the frequency domain. If we had applied a current and computed the resulting sinusoidal voltage, we would have obtained

$$E = \frac{I}{j\omega C}. \tag{16–21}$$

Equations (16–20) and (16–21) have the general Ohm's-law form

$$E = IZ,$$

if the impedance of the capacitance in the frequency domain is

$$Z = \frac{1}{j\omega C}. \tag{16–22}$$

Thus, it is seen that in the frequency domain the capacitance is replaced by a complex number, or phasor operator, $Z = 1/j\omega C$. This operator relates the phasor voltage and the phasor current for the capacitance, as shown in Eqs. (16–20) and (16–21). The frequency-domain diagram is shown in Fig. 16–4(b). The corresponding phasor relationships are shown in Fig. 16–4(c). If the voltage phasor is given, the current phasor leads it by 90° and has a magnitude which is ωC times as large as the voltage phasor. If the current phasor is given, the voltage phasor lags it by 90° and has a magnitude which is $1/\omega C$ times the magnitude of the current phasor. These phase relationships can be easily recalled if the student will remember that current must flow and charge a capacitance before a voltage appears. The voltage therefore lags the current.

We have now completed our transformation of the elements of a problem to the frequency domain. We have shown how to represent sinusoids by phasors in the frequency domain, and we have shown that the two Kirchhoff laws are valid for the phasors in the frequency domain, as well as for the sinusoidal time functions. Finally we have shown how to transform the elements themselves to the frequency domain as impedance operators which change the lengths and angles of the phasors applied to them. We have done all this because the networks in the frequency domain are particularly easy to handle. Essentially, they are resistive networks and all the methods which we have developed for resistive networks can be applied to them. The fact that the "resistances" are really complex impedances does not change the methods of solution. It merely requires that complex algebra, rather than real algebra, be used to obtain the numerical results. In the next section we will solve a few examples to illustrate the method.

16–3 Steady-state solutions by phasor algebra. The solution of a steady-state sinusoidal problem by means of phasor algebra involves the following steps:

1. The problem is transformed to the frequency domain.
2. The phasor solution is obtained by the methods of resistive circuits.
3. The resulting phasor is transformed back to a time function.

By transforming from the time domain to the frequency domain we change from differential equations to algebraic equations, and all the methods which we have learned for resistive circuits can be applied. The general methods of mesh currents and node voltages can be used, and furthermore all the special methods and network theorems are equally valid.

Example 1. As a first example we will solve the circuit of Fig. 16–5(a).

Step 1. *Transformation to the frequency domain.* The time-domain circuit is shown in Fig. 16–5(a) and the frequency-domain circuit in part (b). The phasor representing the voltage source has an angle of zero

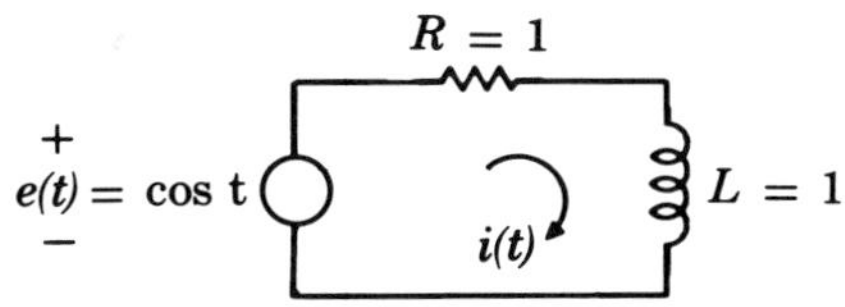

(a) A circuit excited by a sinusoidal time function

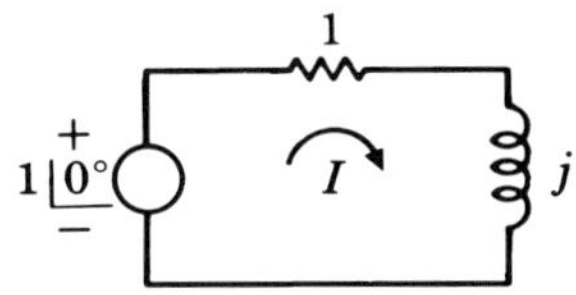

(b) Frequency-domain diagram

FIG. 16–5. An example solved in the frequency domain.

and a magnitude of unity. The 1-ohm resistance transforms to the frequency domain as an impedance of 1 real ohm; the angular frequency is $\omega = 1$; and the inductance transforms to the frequency domain as an impedance of $Z = j1$, or one "imaginary" ohm.

Step 2. Solution in the frequency domain. The solution in the frequency domain is obtained by the same methods we used for resistive circuits. Thus the total impedance of the circuit is the sum of the two series impedances:

$$Z = 1 + j.$$

The current is the voltage divided by the impedance:

$$I = \frac{E}{Z} = \frac{1\underline{|0}}{1 + j} = 0.707\underline{|-45^\circ}.$$

This current is a complex number or phasor, and the actual current as a function of time must be obtained by a transformation back to the time domain.

Step 3. Transformation to the time domain. The time function is a cosine wave with a frequency of $\omega = 1$, an amplitude equal to the amplitude of the current phasor, and a phase angle equal to the angle of the phasor. Thus

$$i(t) = 0.707 \cos (t - 45^\circ). \tag{16–23}$$

Example 2. A problem which can be solved by mesh currents is shown in Fig. 16–6(a), in the time domain. As the first step the circuit is trans-

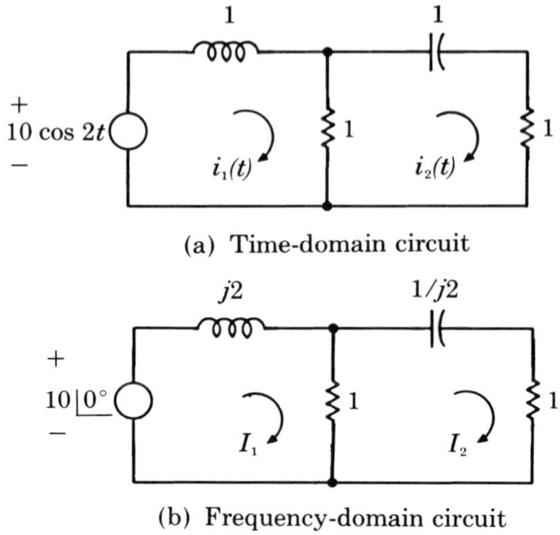

FIG. 16–6. A mesh-current problem in the frequency domain.

formed to the frequency domain, as shown in Fig. 16–6(b). The two resistances remain 1 ohm each. The source frequency is $\omega = 2$, and so the impedance of the inductance is $j2$ ohms, and the impedance of the capacitance is $1/j2$ ohm. The mesh currents i_1 and i_2 in the time domain transform to mesh-current phasors I_1 and I_2 in the frequency domain. The source voltage is a phasor with a magnitude of 10 and an angle of 0°.

The mesh-current equations in the frequency domain are

$$\begin{aligned} I_1(1 + 2j) - I_2 &= 10 \underline{|0^\circ}, \\ -I_1 + I_2\left(2 + \frac{1}{2j}\right) &= 0. \end{aligned} \tag{16–24}$$

The solutions for the currents are

$$\begin{aligned} I_1 &= \frac{\begin{vmatrix} 10 & -1 \\ 0 & 2 - \frac{1}{2}j \end{vmatrix}}{\begin{vmatrix} 1 + 2j & -1 \\ -1 & 2 - \frac{1}{2}j \end{vmatrix}} = -5.13 \underline{|105.8^\circ}, \\ I_2 &= \frac{\begin{vmatrix} 1 + 2j & 10 \\ -1 & 0 \end{vmatrix}}{\begin{vmatrix} 1 + 2j & -1 \\ -1 & 2 - \frac{1}{2}j \end{vmatrix}} = 2.48 \underline{|-60.2^\circ}. \end{aligned} \tag{16–25}$$

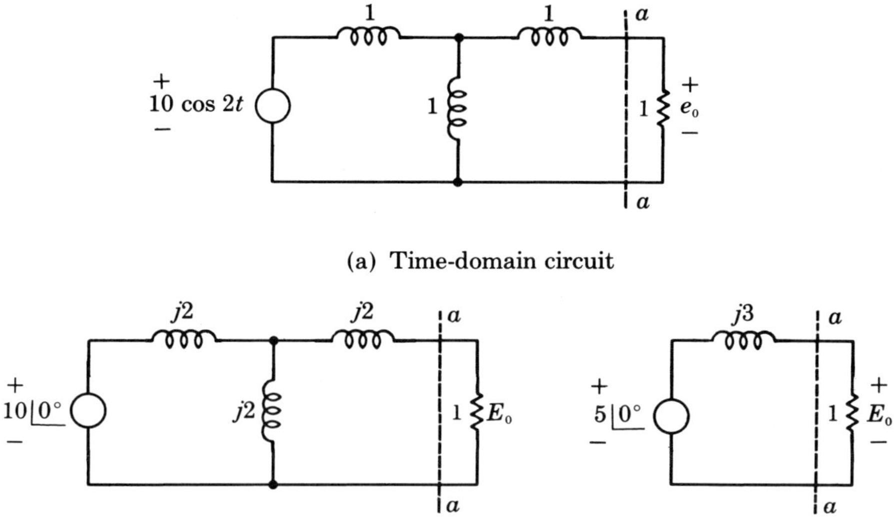

(a) Time-domain circuit

(b) Frequency-domain circuit

(c) Thévenin equivalent

FIG. 16–7. A sinusoidal problem solved by Thévenin's theorem.

As the final step the values of the current phasors in the frequency domain, as given by Eqs. (16–25), are transformed back to the time domain. The time expressions are

$$\begin{aligned} i_1(t) &= -5.13 \cos (2t + 105.8^\circ), \\ i_2(t) &= 2.48 \cos (2t - 60.2^\circ). \end{aligned} \tag{16–26}$$

Example 3. An example which can be conveniently treated by means of Thévenin's theorem is shown in Fig. 16–7(a), in the time domain. The first step in the solution is the transformation of the network to the frequency domain, as shown in Fig. 16–7(b). Since the source frequency is $\omega = 2$, the impedance of each of the inductances is $j2$ ohms. The source has a magnitude of 10 and angle of zero.

The frequency-domain circuit is most easily treated by means of Thévenin's theorem. The open-circuit voltage at terminals a–a is

$$E_{\text{oc}} = \frac{j2}{j2 + j2} \times 10 \angle 0^\circ = 5 \angle 0^\circ.$$

The "dead" network consists of an impedance $j2$ in series with a parallel combination of two $j2$ impedances. The net impedance is therefore

$$Z = 2j + \frac{2j \times 2j}{2j + 2j} = 3j.$$

The equivalent Thévenin circuit is shown in Fig. 16–7(c). The output voltage, E_0, of this circuit, is

$$E_0 = \frac{1}{1 + 3j} \times 5 \underline{|0°}$$
$$= 1.57 \underline{|-71.6°}.$$

Translated back to the time domain, this voltage becomes

$$e_0(t) = 1.57 \cos (2t - 71.6°).$$

All of the other methods and theorems which we developed for resistive circuits are directly applicable to the solution of frequency-domain circuits. From time to time, we will illustrate the other methods in the remainder of the book.

16–4 Phasor diagrams. All the operations which we discussed in the preceding section in terms of complex algebra can also be interpreted in terms of phasor diagrams. Often phasor diagrams provide insight to the operation of a circuit, and it is well worth while constructing the geometrical pictures. Occasionally, it is possible to solve a problem completely by means of geometrical diagrams, but more often it is easier to perform most of the computations algebraically, and to use phasor diagrams only for an interpretation of the results. For the simple series and parallel circuits which will be solved in this section, it is possible to perform the complete calculations geometrically.

Series resonant circuit. A series resonant circuit is shown in Fig. 16–8(a), in the time domain. A frequency-domain version of the same circuit is shown in Fig. 16–8(b). We will suppose that the input current is

$$i(t) = |I| \cos (\omega t + \theta).$$

The corresponding frequency-domain phasor is a vector of length $|I|$ standing at an angle θ, as shown in the phasor diagram of Fig. 16–8(c). The inductance and the capacitance have impedances of $j\omega L$ and $1/j\omega C$, respectively, and the resistance has a value of R ohms in the frequency domain. The voltage across the resistance, in phasor form, is

$$E_R = IR. \tag{16–27}$$

The voltage phasor of Eq. (16–27) is in phase with the current phasor, and it has a length of $|I|R$. The voltage across the inductance is

$$E_L = j\omega LI = \omega LI \underline{|90°}. \tag{16–28}$$

The phasor representing the voltage across the inductance, as given by Eq. (16–28), is 90° ahead of the current vector, and it has a magnitude

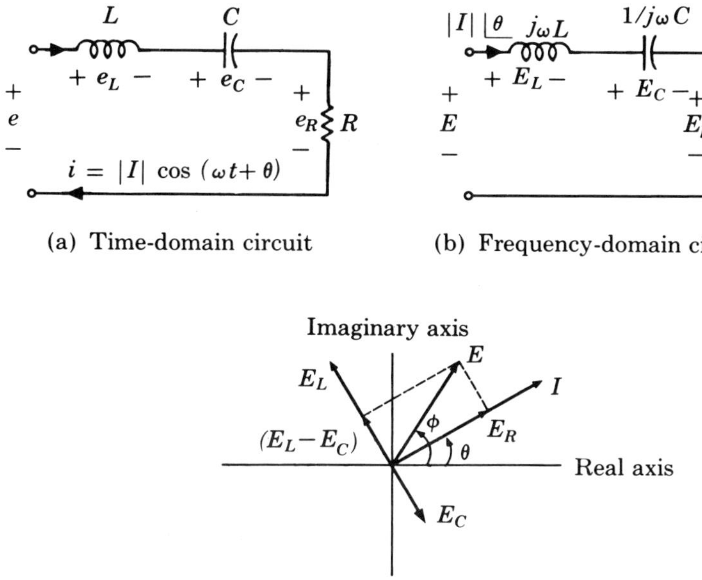

(a) Time-domain circuit (b) Frequency-domain circuit

(c) Phasor diagram

FIG. 16–8. The series resonant circuit.

which is ωL times the magnitude of the current phasor. The voltage across the capacitance is

$$E_C = \frac{1}{j\omega C} I = \frac{1}{\omega C} I \underline{/-90^\circ}. \tag{16–29}$$

The phasor for the voltage across the capacitance, as given by Eq. (16–29), is 90° behind the current phasor and has a magnitude which is $1/\omega C$ times the length of the current phasor. All these phasors are shown in the diagram of Fig. 16–8(c). The total voltage is the vector sum of the voltages across the inductance, the resistance, and the capacitance. It is shown as $|E| \underline{/\phi}$ in Fig. 16–8(c). As a final step this voltage is transformed to the time domain and becomes

$$e(t) = |E| \cos(\omega t + \phi). \tag{16–30}$$

When the voltage is the known stimulus and the current is the desired response, the problem solution must still be started with the current. One method is to start with an assumed value of current of $1 \underline{/0^\circ}$. The corresponding voltage is then computed, as in Fig. 16–8(c). The current which would be produced by the given excitation can then be found by linear scaling. It is

$$I\ (\text{actual}) = \frac{E\ (\text{actual})}{E\ (\text{computed})} \times 1 \underline{/0^\circ}. \tag{16–31}$$

Both of the voltages in Eq. (16–31) are complex numbers, and the division must be carried out by complex algebra.

The phasor diagram shows graphically the effect of changing the frequency in the circuit. When the inductive and capacitive voltages are equal in magnitude, they cancel completely and do not appear in the input voltage. The impedance of the circuit at this frequency is equal to the resistance alone, and a maximum value of current flows, which is

$$I_{\text{max}} = \frac{E}{R}. \tag{16–32}$$

This condition is called resonance in the circuit. Since the circuit is a series circuit, the phenomenon is called *series resonance.* The frequency at which resonance occurs is

$$\omega = \frac{1}{\sqrt{LC}}. \tag{16–33}$$

Parallel resonant circuit. A phasor diagram is also useful in the description of a parallel resonant circuit of the type shown in Fig. 16–9(a). Since the voltage is common to all the elements, we will assume initially that we have a voltage excitation given by

$$e(t) = |E| \cos(\omega t + \theta).$$

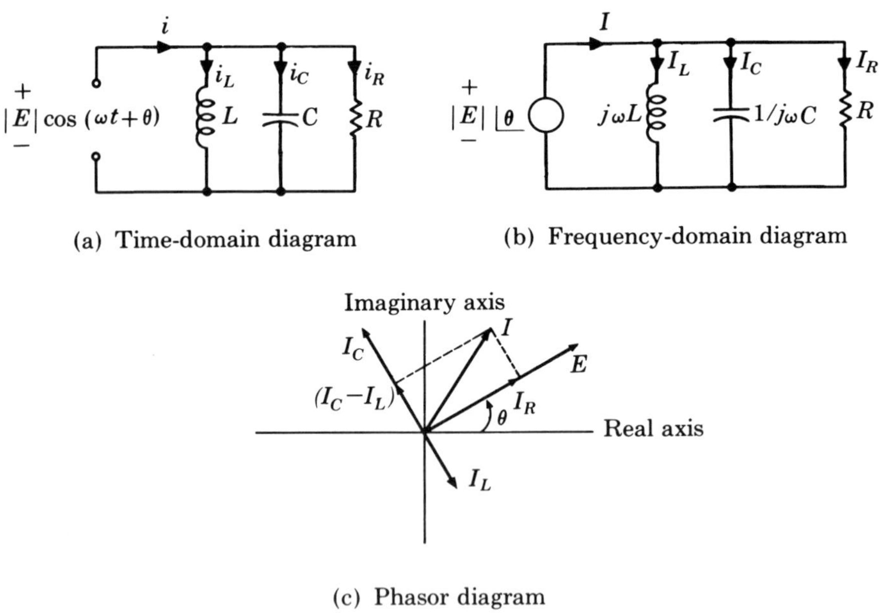

FIG. 16–9. The parallel resonant circuit.

The corresponding frequency-domain phasor is

$$E = |E| \underline{|\theta},$$

and it drives a frequency-domain circuit consisting of the parallel combination of three impedances R, $j\omega L$, and $1/j\omega C$, as shown in Fig. 16–9(b). The voltage phasor is a vector of length $|E|$ standing at an angle of θ, as shown in Fig. 16–9(c). The phasor representing the current through the resistance is

$$I_R = \frac{E}{R}, \tag{16–34}$$

It is a phasor of length $|E|/R$, and it has the same angle as the voltage phasor. The current through the inductance is, in phasor form,

$$I_L = \frac{E}{j\omega L} = \frac{E}{\omega L} \underline{|-90^\circ}, \tag{16–35}$$

which represents a phasor of length $|E|/\omega L$ and with an angle lagging the voltage angle by 90°. The current through the capacitance has the phasor expression

$$I_C = \frac{E}{1/j\omega C} = j\omega CE = \omega CE \underline{|90^\circ}, \tag{16–36}$$

and it is represented by a phasor of length $\omega C|E|$ which leads the voltage phasor by 90°. All these current phasors are shown on the diagram of Fig. 16–9(c). The total input current is the sum of the three component currents. It is shown as the phasor I on the diagram.

If the current is the known stimulus and the voltage is the desired response, the problem solution is started by assuming a voltage of $E = 1 \underline{|0^\circ}$. The corresponding current is obtained, as shown in Fig. 16–9(c). This computed current produces a voltage of $1 \underline{|0^\circ}$. Any other current produces a voltage

$$E\text{ (actual)} = \frac{I\text{ (actual)}}{I\text{ (computed)}} \times 1 \underline{|0^\circ}. \tag{16–37}$$

The interesting condition of *parallel resonance* occurs when the phasor I_L is exactly equal in magnitude to the phasor I_C. These two phasors then cancel, and the circuit behavior is determined entirely by the resistance. At this point the impedance of the parallel combination is a maximum, and the voltage produced by a given current is a maximum, with the value

$$E_{\max} = IR. \tag{16–38}$$

The frequency at which resonance occurs is

$$\omega = \frac{1}{\sqrt{LC}}. \tag{16-39}$$

We will examine series and parallel resonance in more detail in a later chapter. In particular, we shall be interested in the relationship between the response of a circuit to an input stimulus of a given frequency and the frequency of the natural transient which the circuit produces by itself. In general it can be said that the closer the stimulus to the natural behavior of the system, the greater the response.

Simple series resonance and simple parallel resonance are excellent examples of the use of phasor diagrams. The nature of the diagram almost forces us to consider the phenomenon of resonance. We might miss it entirely if we dealt only with complex algebra. It is possible to extend phasor methods to more complex problems, but the process rapidly becomes difficult. In the algebraic approach, it doesn't make any difference whether the current or the voltage is the stimulus. In the phasor approach we must start with a phasor which is common to several elements. Thus, in the series circuit it is necessary to start with the current phasor, and in the parallel circuit with the voltage phasor. Such restrictions make any general extension of the phasor method very difficult. Instead, we usually solve the problem by algebra and draw the phasor diagrams for the answers.

16–5 Driving-point impedances and transfer impedances. A one-terminal-pair resistive network can be reduced to a single equivalent resistance. This equivalent resistance is the response function of the network in the sense that it determines the response which will be obtained from any possible stimulus function. A one-terminal-pair network in the sinusoidal steady-state can be reduced to one equivalent impedance, called a *driving-point impedance*, in the frequency domain.

If the input excitation is a single-frequency sine wave, the driving-point impedance reduces to two terms, the real and imaginary parts of a complex number. *An equivalent network* can always be found which will represent this impedance in the form of a resistance and either an inductance or a capacitance, depending on the sign of the imaginary part of the impedance. This equivalent network, however, is not the system function of the original network. It will give the response only to a sinusoid of one particular frequency, and the response at any other frequency will be very different.

In order to have a driving-point impedance which is *a system function,* we must have one which is applicable for all possible input frequencies. Such an impedance is easy enough to obtain mathematically. The frequency ω is merely left as an unknown variable. The real part of the impedance is then a function of frequency and so is the imaginary part. *There is no simple equivalent network which will represent the original network at all frequencies.* We are forced to represent the network by its analytical impedance function. The best that we can do in the way of visualizing the impedance function is to plot its real and imaginary parts, or its magnitude and phase, as functions of frequency. In a later chapter we will discuss some short methods for making these frequency plots. In this chapter we will simply give an example of a driving-point impedance function, both at a single frequency and as a function of frequency.

A similar situation exists for two-terminal-pair networks. The general two-terminal-pair resistive network is described by three resistances, and the general two-terminal-pair network in the sinusoidal steady-state is described by three impedances. At any single frequency the three impedances can be separated into real and imaginary parts and an equivalent network in the form of either a T or a Π can be obtained. Three impedances which are functions of frequency will describe the network completely, but there is no simple equivalent circuit in the general case.

The driving-point impedance. The driving-point impedance of a one-terminal-pair network is the ratio of the voltage phasor applied to it divided by the resulting current phasor. Thus

$$Z = \frac{E}{I} \qquad \text{with } Z \text{ measured in ohms.} \tag{16–40}$$

Equation (16–40) is of the same form as Ohm's law, but the resulting impedance is a complex number rather than a constant. If we break the impedance up into its real and imaginary parts, we obtain

$$Z = R + jX. \tag{16–41}$$

In Eq. (16–41) the real part of the impedance is called *resistance*, and the coefficient of the imaginary part is called *reactance*. Both resistance and reactance are measured in ohms. The resistance is positive, but the reactance can have either sign.

The reciprocal of a resistance is a conductance, and it is measured in mhos. The reciprocal of an impedance also has a name. It is called *admittance*, and it has the symbol Y. Thus

$$Y = \frac{1}{Z} \qquad \text{with } Y \text{ measured in mhos.} \tag{16–42}$$

The admittance, like the impedance, is a complex number and it can be separated into a real and an imaginary part:

$$Y = G + jB. \tag{16–43}$$

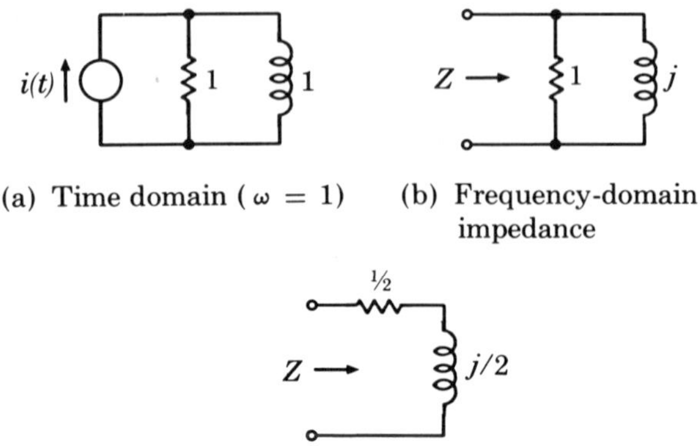

(a) Time domain ($\omega = 1$) (b) Frequency-domain impedance

(c) Equivalent impedance at $\omega = 1$

FIG. 16–10. Equivalent impedances at a single frequency.

In Eq. (16–43) the real part of the admittance is called *conductance*, and the coefficient of the imaginary part is called *susceptance*. Both conductance and susceptance are measured in mhos. Conductance, like resistance, is positive, but susceptance can have either sign.

The equivalent impedance of a network represents a resistance in series with a reactance. If the impedance refers to a single frequency, the circuit can be reduced to a resistance in series with an inductance when the reactance is positive, and to a resistance in series with a capacitance when the reactance is negative. The equivalent admittance represents the parallel combination of a resistive and a reactive element. At any single frequency, the circuit can be reduced to a single resistance in parallel with an inductance if the susceptance is negative, and to a single resistance in parallel with a capacitance if the susceptance is positive.

In order to enable us to handle inputs of arbitrary frequency, both Z and Y are written as functions of frequency. In this case the real and imaginary parts are also functions of frequency, and no simple equivalent circuit can be obtained.

Example of driving-point impedance. A simple circuit consisting of a resistance in parallel with an inductance is shown in Fig. 16–10(a). The driving-source frequency is $\omega = 1$, and the frequency-domain diagram is shown in Fig. 16–10(b). The impedance is obtained by treating the branch impedances as if they were resistances. Thus

$$Z = \frac{1 \times j}{1 + j} = \tfrac{1}{2} + j\tfrac{1}{2}. \tag{16–44}$$

At this particular frequency the impedance has a resistive part of $\frac{1}{2}$ ohm and a reactive part of $\frac{1}{2}$ ohm. An equivalent circuit which gives the same

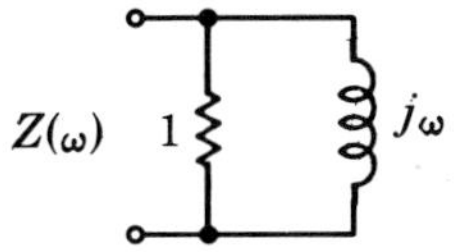

FIG. 16–11. The impedance of Fig. 16–10 for any value of ω.

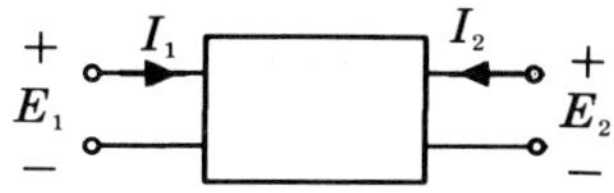

FIG. 16–12. General two-terminal-pair network in the frequency domain.

impedance is a $\frac{1}{2}$-ohm resistance in series with a $\frac{1}{2}$-henry inductance, as shown in Fig. 16–10(c). The equivalent circuit holds only at $\omega = 1$.

In order to obtain a driving-point impedance which is characteristic of the circuit for all frequencies we must obtain Z with the frequency ω as a variable. To accomplish this result we use the frequency-domain diagram of Fig. 16–11. The impedance of the 1-henry inductance is written as $j\omega$. For this diagram, the input impedance is

$$\begin{aligned} Z &= \frac{1 \times j\omega}{1 + j\omega} = \frac{j\omega(1 - j\omega)}{1 + \omega^2} \\ &= \frac{\omega^2}{1 + \omega^2} + j\frac{\omega}{1 + \omega^2}. \end{aligned} \tag{16–45}$$

In Eq. (16–45) both the resistance and the reactance are functions of frequency. There is no simple equivalent circuit. We can plot the two terms as functions of frequency, but more often the magnitude and phase of Z are plotted. The polar form of Z is

$$Z = \frac{\omega}{\sqrt{1 + \omega^2}} \underline{/\tan^{-1} 1/\omega}. \tag{16–46}$$

The polar form is preferred to the rectangular form principally because it is easier to measure the magnitude and the phase in the laboratory.

We have used impedance as our basic quantity in this example, and it is convenient always to do so. Thus the values given in frequency-domain diagrams should always be in ohms. For some problems it is convenient to use admittances for calculations; in these cases the admittances are obtained from the impedances of the elements.*

Impedances for two-terminal-pair networks. The impedances which describe a two-terminal-pair frequency-domain network are similar to the driving-point and transfer resistances which we used previously to describe a two-terminal-pair resistive network. For example, the phasor relationships for the general two-terminal-pair network of Fig. 16–12 are

$$E_1 = z_{11}I_1 + z_{12}I_2, \qquad E_2 = z_{12}I_1 + z_{22}I_2. \tag{16–47}$$

---

* We followed the same practice for resistive circuits. It avoids confusion when all labelled values are in ohms.

For these equations the definitions of the z's are similar to the definitions of the r's in the resistive case. That is, z_{11} is the input impedance with the output terminals open-circuited, z_{22} is the impedance at the output terminals with the input terminals open-circuited, and z_{12} is the output voltage which is produced by an input current of $1 \underline{|0^\circ}$. In writing Eqs. (16–47), the current phasors are considered to be the stimuli, and the voltage phasors the responses. It is also possible to consider the voltage phasors to be the stimuli and the current phasors the responses. In this case the resulting equations are

$$\begin{aligned} I_1 &= y_{11}E_1 + y_{12}E_2, \\ I_2 &= y_{12}E_1 + y_{22}E_2. \end{aligned} \tag{16–48}$$

This set of equations is analogous to the set for resistive networks in terms of the g parameters. Thus, y_{11} is the input admittance with a short circuit on the output terminals, y_{22} is the admittance at the output terminals with a short circuit on the input terminals, and y_{12} is the current which flows in a short circuit at the output terminals when a voltage of $1 \underline{|0^\circ}$ is applied to the input terminals.

If the z parameters of Eqs. (16–47) and the y parameters of Eqs. (16–48) are computed *at a single frequency*, it is possible to find an *equivalent* T *or* π *circuit* in terms of three resistances and three reactances. If we wish the z parameters or the y parameters to describe the behavior of a network for any input frequency, however, we must compute them as functions of

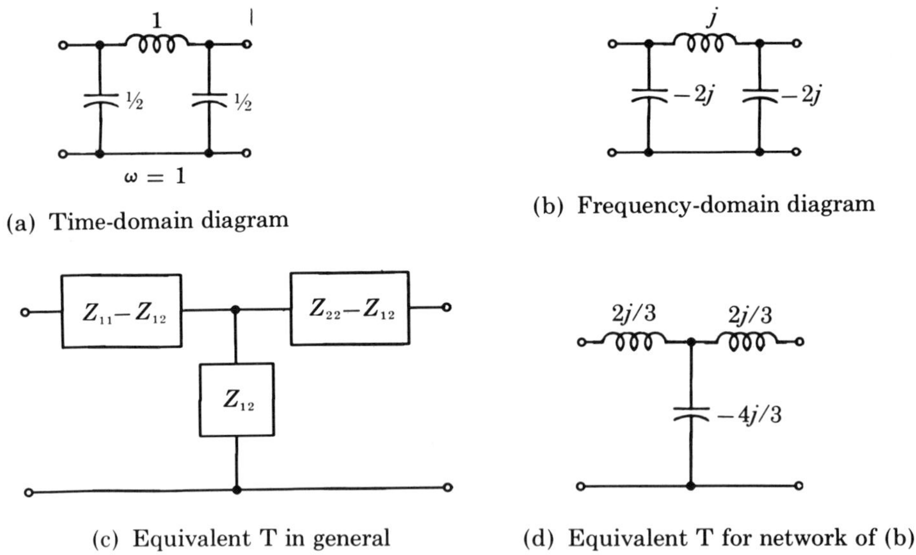

FIG. 16–13. Equivalent single-frequency, two-terminal-pair networks.

frequency. They are then true system functions which represent the network for a general input function. Since no general equivalent circuit exists, *the system functions must be used in algebraic form.*

As an example of the 3-impedance equivalent circuit which can be obtained at a single frequency, consider the network of Fig. 16–13(a). The input frequency is $\omega = 1$; the frequency-domain network is shown in Fig. 16–13(b). For this network

$$z_{11} = \frac{(-2j)(-j)}{-3j} = -\tfrac{2}{3}j = z_{22},$$

$$z_{12} = \frac{(-2j)(-2j)}{-3j} = -\tfrac{4}{3}j.$$

The equivalent T circuit has the general form shown in Fig. 16–13(c). The mutual element is z_{12} and the two arms are $(z_{11} - z_{12})$ and $(z_{22} - z_{12})$.* For this particular case the reactances of the arms have the values shown in Fig. 16–13(d). The mutual arm is a negative reactance, and hence a capacitance; the series arms are positive reactances, and hence inductances.

The frequency-domain diagram for the same circuit driven at an arbitrary frequency ω is shown in Fig. 16–14. The three impedances obtained from this circuit are the three response functions for this network. They have the following algebraic forms:

$$z_{22} = z_{11} = \frac{2/j\omega \times (j\omega + 2/j\omega)}{j\omega + 4/j\omega} = \frac{2}{j\omega}\,\frac{(-\omega^2 + 2)}{(-\omega^2 + 4)},$$

$$z_{12} = \frac{2/j\omega \times 2/j\omega}{j\omega + 4/j\omega} = \frac{2}{j\omega}\,\frac{2}{(-\omega^2 + 4)}. \tag{16–49}$$

The three functions represented in Eqs. (16–49) completely describe the given network for all possible frequencies of excitation. They can be used in algebraic form to represent the network. There is no equivalent T network, however, in this case. *The system functions must be used algebraically.* Since the excitation point and the response point are usually specified

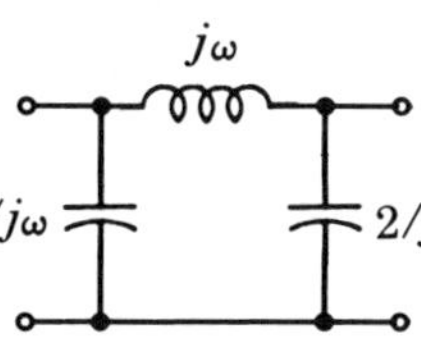

FIG. 16–14. The network of Fig. 16–13 driven by an arbitrary frequency.

---

* If there are resistances present, z_{12} may have a negative resistance. As in the resistive case, this can be removed by reversing the output polarity.

in any given problem only one of the functions is required at any one time. The three functions take care of all possible excitations and all possible responses, but only a single stimulus-response relationship is required for the solution of any one problem. For this reason, we will consider only a single function in many of the examples and problems in the remainder of the book. When the excitation is a current we will be dealing most naturally with a z-parameter, as in this example, but when the excitation is a voltage it will be desirable to switch over to the y-parameters. In fact, for the single stimulus and the single response, we do not need to pay much attention to whether we are using z-parameters or y-parameters. We simply solve the given problem for the desired response, and in the frequency domain we can do so by the same methods as we used in resistive circuits!

The principal point we have sought to emphasize in this section is that simple equivalent circuits for one-terminal-pair networks and for two-terminal-pair networks can be obtained only at a single frequency. If we wish to describe a network at all frequencies, we must use the algebraic system functions.

16–6 Frequency scaling. For resistive circuits we found that by the use of linear scaling we are able to reduce the values of the resistances in our circuits to the neighborhood of 1 ohm. A similar sort of normalization can be applied to the frequency variable to reduce the values of the inductances and capacitances in our circuits to the vicinity of 1 henry and 1 farad. The process of normalization is very simple. The frequency variable can appear in a circuit associated only with an inductance or a capacitance in an impedance expression of the form $j\omega L$ or $1/j\omega C$. The impedance in ohms will remain unchanged if the products ωL and ωC remain unchanged. Thus, if we wish to increase the frequency by some factor we can decrease inductances and capacitances by the same factor, and the branch impedances in ohms will be unchanged. Indeed, the whole frequency-domain diagram remains unchanged. The phasor currents and voltages are the same as they were, and the time-domain currents and voltages have the same magnitudes and phases but, of course, a new frequency.

The sort of frequency normalization we have just discussed is important from several points of view. In the first place, it lets us consider circuits with normalized values of inductance and capacitance in the vicinity of 1 henry and 1 farad. Such values are most convenient for calculations, although they are not convenient sizes to obtain physically. In the second place, normalized circuits provide a common ground for comparing the behavior of similar classes of circuits, such as series resonant circuits, parallel resonant circuits, and double-tuned circuits. One frequency-domain model covers a whole class of time-domain circuits. The notable

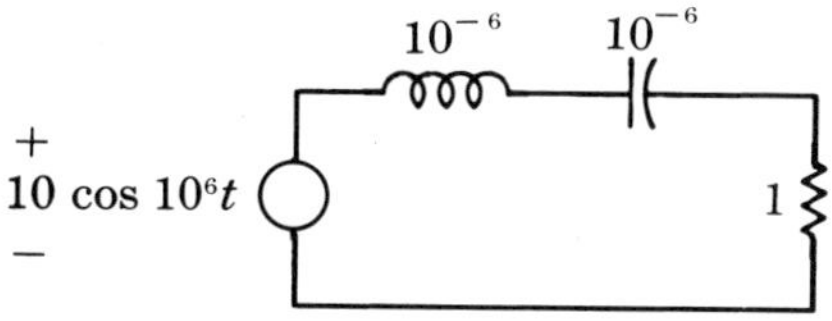

(a) Original circuit in the time domain

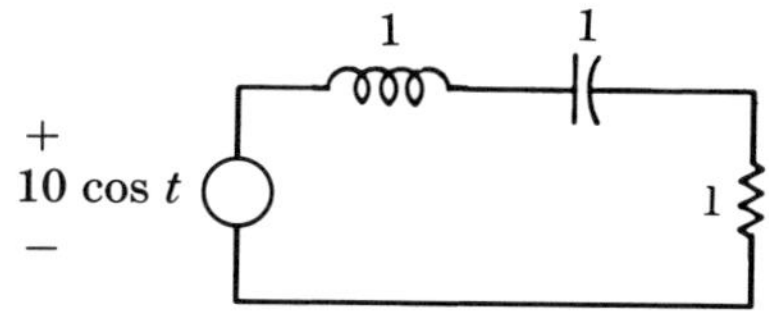

(b) Frequency-scaled circuit in the time domain

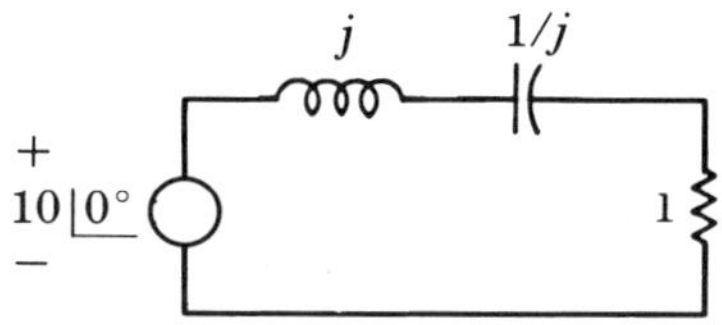

(c) Frequency-domain circuit

FIG. 16–15. Frequency scaling and the frequency domain.

importance of normalization is in network synthesis. No general method of network synthesis could be developed without it. For synthesis it is a necessity; for analysis it is merely a convenience.

As an example of frequency normalization, consider the circuit of Fig. 16–15(a). In the time domain the original circuit is driven by a frequency of $\omega = 10^6$ radians per second, and has physically realizable sizes of inductance and capacitance. A normalized time-domain circuit in which the frequency has been reduced by 10^6 and the inductance and capacitance have been increased by 10^6 is shown in Fig. 16–15(b). Both these circuits are represented in the frequency domain by the same network. The voltage source is a phasor of $10\,\underline{|0^\circ}$, the inductance is an impedance of j ohms, and the capacitance is an impedance of $1/j$ ohms, as shown in Fig. 16–15(c). The problem can be solved in the frequency domain, and the answer can be transformed to either the time domain or the normalized time domain. For example, the current in the frequency domain is

$$I = \frac{10}{1 + j + 1/j} = 10\,\underline{|0^\circ}. \tag{16–50}$$

The corresponding current in the time domain is

$$i(t) = 10 \cos 10^6 t, \tag{16–51}$$

while in the normalized time domain, it is

$$i(t) = 10 \cos t. \tag{16–52}$$

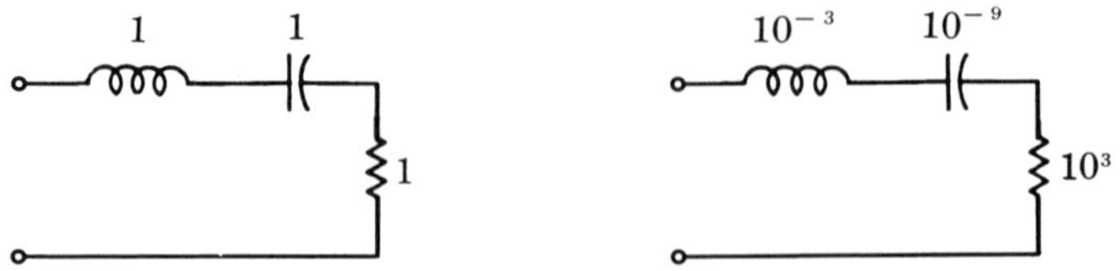

(a) Original circuit (b) Scaled version: frequency by 10^6 and impedance by 10^3

FIG. 16–16. Combined frequency and magnitude scaling.

Apart from the frequency, the normalized current of Eq. (16–52) is the same as the original current of Eq. (16–51). We will make use of normalization principally to justify the use of circuits with values of inductance, capacitance, and resistance in the vicinity of unity. Such circuits are scaled versions, or linear models, of the circuits encountered in practical problems.

Magnitude normalization in the frequency domain is also possible. It follows the same rules as magnitude scaling for resistive circuits. If all of the impedances are multiplied by a constant, and if the voltage is multiplied by the same constant, the currents will be unchanged. If all of the impedances are multiplied by a constant and an applied current is divided by this constant, the voltages everywhere will remain unchanged. The name "impedance levelling" is given to this process. In summary, to multiply the impedance level of a circuit by K:

1. All the R's are multiplied by K.
2. All the L's are multiplied by K.
3. All the C's are divided by K.

The capacitance element is treated differently from the inductance and resistance elements because it occurs in the denominator of an impedance expression.

It is also possible to combine amplitude and frequency scaling. A normalized time-domain circuit is shown in Fig. 16–16(a). It is desired to increase the frequency of this circuit by the factor 10^6, and to increase the impedance level by the factor 10^3. In effect, the resonant frequency is to be moved from $\omega = 1$ to $\omega = 10^6$ radians per second, and the impedance at resonance is to be increased from 1 to 10^3 ohms. The new circuit values will be

$$
\begin{aligned}
R &= 1 \times 10^3 &&= 10^3 \text{ ohms},\\
L &= 1 \times 10^3 \times 10^{-6} &&= 10^{-3} \text{ henry},\\
C &= 1 \times 10^{-3} \times 10^{-6} &&= 10^{-9} \text{ farad}.
\end{aligned}
$$

The scaled circuit is shown in Fig. 16–16(b). It is almost magical that the normalized 1-ohm, 1-henry, 1-farad, and 1-radian-per-second values result in practical element values when scaled both in frequency and impedance level!

16–7 Summary.

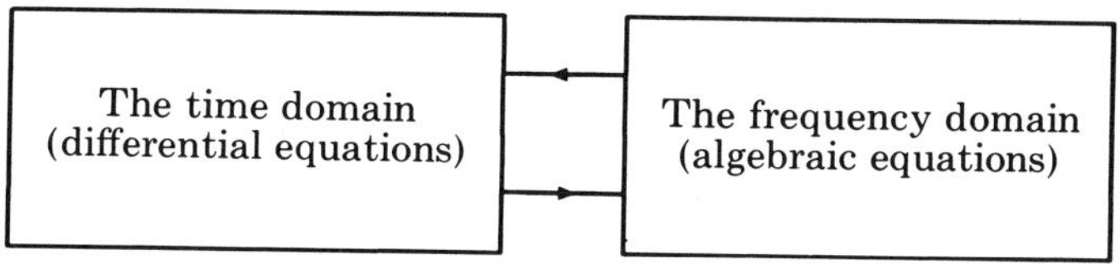

Transformations

	Time domain	*Frequency domain*
Variable	$f = \|F\| \cos(\omega t + \theta)$	$F = \|F\| \underline{/\theta°} = \|F\|e^{j\theta}$
Resistance	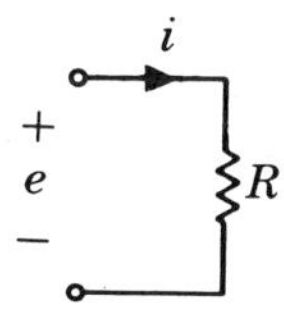 $e = iR$	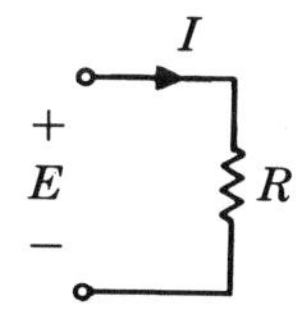 $E = IR$
Inductance	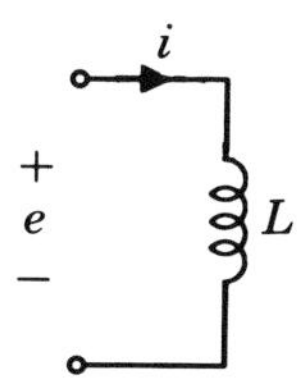 $e = L\dfrac{di}{dt}$	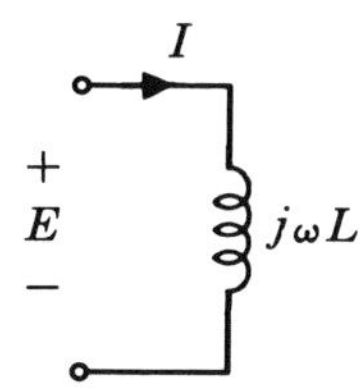 $E = j\omega LI$

	Time domain	*Frequency domain*
Capacitance	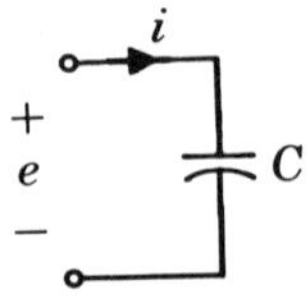	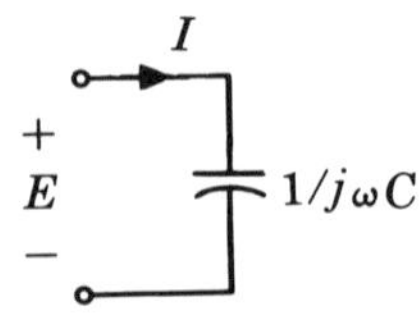
	$e = \frac{1}{C}\int_{-\infty}^{t} i\,dt$	$E = \frac{1}{j\omega C} I$
Kirchhoff's laws	$\begin{cases}\sum i = 0\\ \sum e = 0\end{cases}$	$\sum I = 0$ $\sum E = 0$

Phasor diagrams

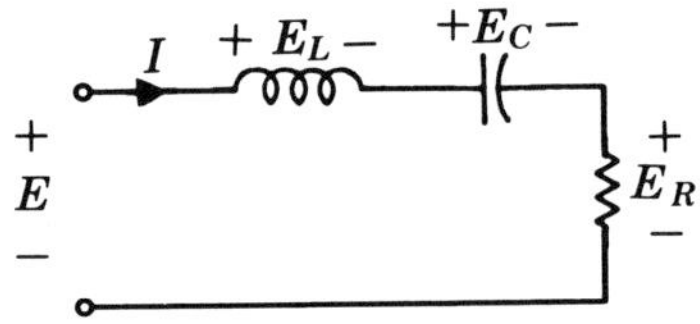

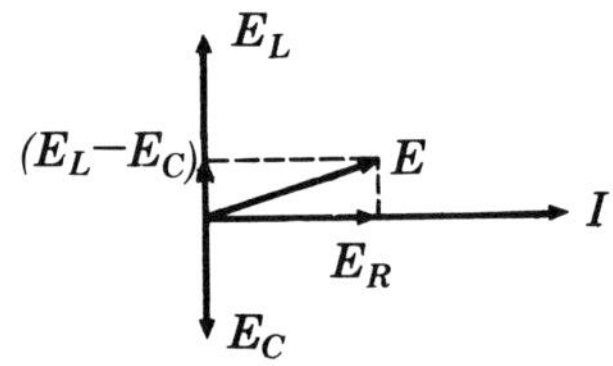

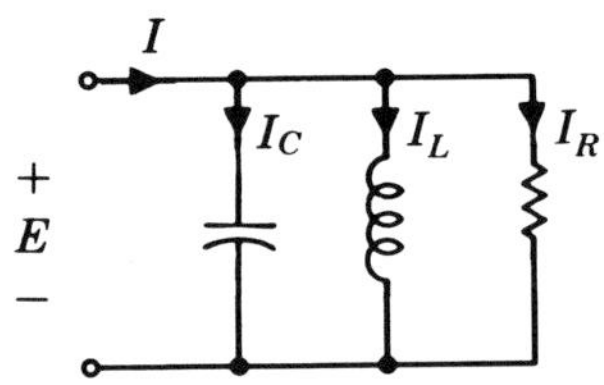

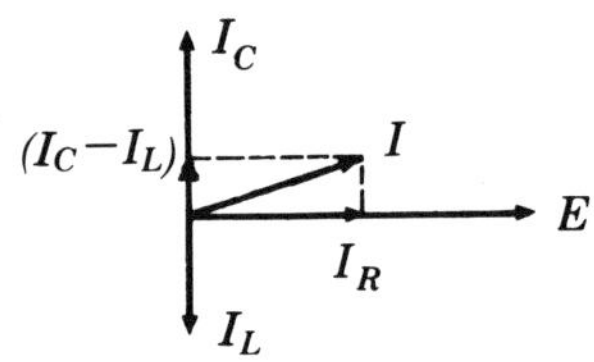

Driving-point impedance

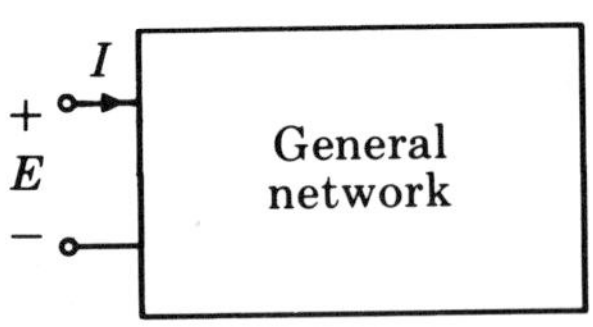

$$Z(\omega) = \frac{E}{I} = Z\,\underline{|\theta}$$
$$= R(\omega) + jX(\omega)$$

$$Y(\omega) = \frac{1}{Z(\omega)} = G(\omega) + jB(\omega)$$

$Z =$ impedance (ohms)
$R =$ resistance (ohms)
$X =$ reactance (ohms)
$Y =$ admittance (mhos)
$G =$ conductance (mhos)
$B =$ susceptance (mhos)

Single-frequency equivalent circuits

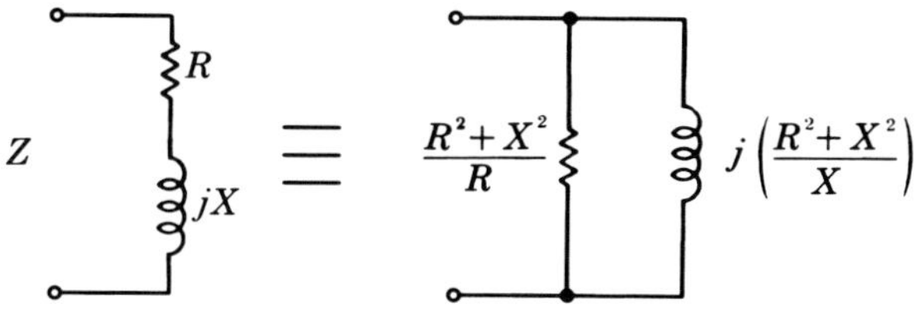

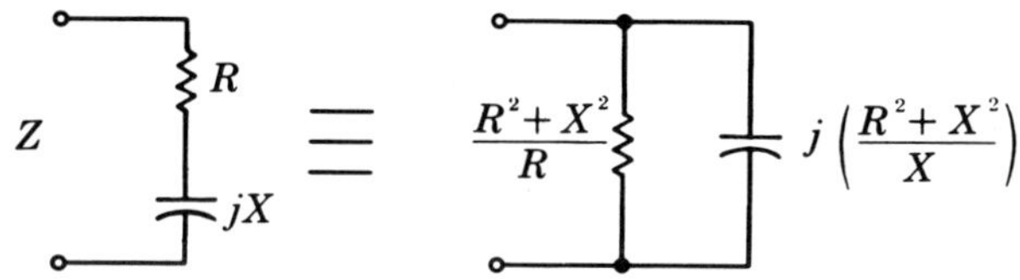

Two-terminal-pair networks

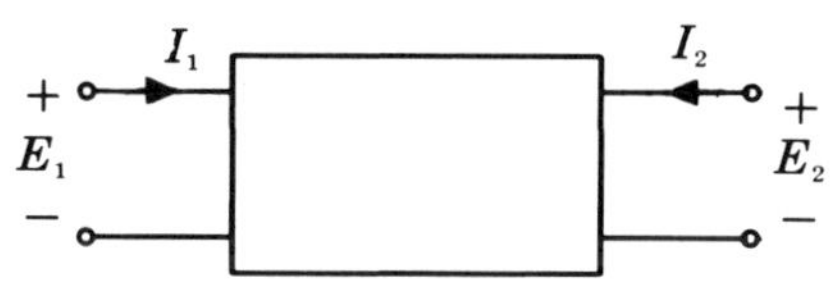

$$E_1 = z_{11}I_1 + z_{12}I_2$$
$$E_2 = z_{12}I_1 + z_{22}I_2$$

$$z_{11} = \left.\frac{E_1}{I_1}\right|_{I_2=0}, \qquad z_{12} = \left.\frac{E_2}{I_1}\right|_{I_2=0}, \qquad z_{22} = \left.\frac{E_2}{I_2}\right|_{I_1=0}$$

$$I_1 = y_{11}I_1 + y_{12}I_2$$
$$I_2 = y_{12}I_1 + y_{22}I_2$$

$$y_{11} = \left.\frac{I_1}{E_1}\right|_{E_2=0}, \qquad y_{12} = \left.\frac{I_2}{E_1}\right|_{E_2=0}, \qquad y_{22} = \left.\frac{I_2}{E_2}\right|_{E_1=0}$$

Frequency scaling

Multiply L and C by any constant.
Divide ω by the same constant.

Magnitude scaling

Multiply R by a constant.
Multiply L by the same constant.
Divide C by the same constant.
All impedances are multiplied by the constant.

REFERENCES

The vector, or phasor, methods for handling sine waves were discussed in Chapter 15. In Chapter 16 these methods are extended to the solution of linear circuit problems. The historical references for this material will be found in Chapter 15. The sharp distinction which we have made between time-domain networks and frequency-domain networks is not followed in some of the standard texts. A few selected references are given below.

1. MIT EE Staff, *Electric Circuits*, John Wiley & Sons, New York, 1943. Chapter 4 introduces steady-state circuits. The discussion of resonant circuits is particularly good.

2. E. A. GUILLEMIN, *Introductory Circuit Theory*, John Wiley & Sons, New York, 1953. In Chapter 6 Guillemin gives an excellent discussion of sinusoids, time- and frequency-domain networks, phasor diagrams and resonance. The complex frequency "s" is introduced at the outset.

3. H. H. SKILLING, *Electrical Engineering Circuits*, John Wiley & Sons, New York, 1957. Skilling's book is primarily concerned with frequency-domain networks, which he calls transform networks. They are introduced in Chapter 2 and used throughout the book.

4. BRENNER and JAVID, *Analysis of Electric Circuits*, McGraw-Hill, New York, 1959. Brenner and Javid introduce the steady-state sinusoidal response as a special case of the general operational response to exponentials.

5. M. E. VAN VALKENBURG, *Network Analysis*, Prentice-Hall, New Jersey, 1955. In this text the steady-state response is introduced as a special case of the general Laplace transform theory. This approach has the advantage of clearly separating the time and frequency domains.

6. M. B. REED, *Alternating-Current Circuit Theory*, Harper & Brothers, New York, 1956. Reed distinguishes the time and the frequency domains by using the R–L–C elements in the time domain and square boxes for impedances in the frequency domain.

7. KERCHNER and CORCORAN, *Alternating Current Circuits*, John Wiley & Sons, New York, 3rd ed., 1951. This book is devoted to frequency-domain methods. The last two chapters return to the time domain, with a discussion of some simple differential equations.

EXERCISES*

Section 16–2

1. In a certain network the voltage drops around a mesh are given by

$$e_1(t) = 100 \cos t,$$
$$e_2(t) = -100 \sin t,$$
$$e_3(t) = -141 \cos (t + 45°).$$

(a) Show that the sum of the voltages is zero in the time domain. (b) Show that the sum of the voltage phasors is zero in the frequency domain.

2. In a certain network the currents leaving a node are given by

$$i_1(t) = 100 \cos 377t,$$
$$i_2(t) = 173 \sin 377t,$$
$$i_3(t) = -200 \cos (377t - 60°).$$

(a) Show that the sum of the currents is zero at all times in the time domain. (b) Show that the sum of the current phasors is zero in the frequency domain.

3. A circuit consists of two elements in series with a source of voltage. The voltage drops across the two elements in the direction of an assumed current are

$$e_1(t) = 100 \cos (1000t + 60°),$$
$$e_2(t) = 100 \cos (1000t - 30°).$$

The voltage across the source is given by $e_1(t) + e_2(t)$. Find its value as a function of time.

4. A current source supplies current to two branches. The branch currents are

$$i_1(t) = 30 \cos (377t - 53°),$$
$$i_2(t) = 40 \cos (377t + 37°).$$

The source current is given by $i_1(t) + i_2(t)$. Find its value as a function of time.

5. In the circuit of Fig. 16–17 find the current in the inductance (a) by using the time-domain volt-ampere relationship and (b) by using the frequency-domain phasors (neglect the value of the integral at $t = -\infty$).

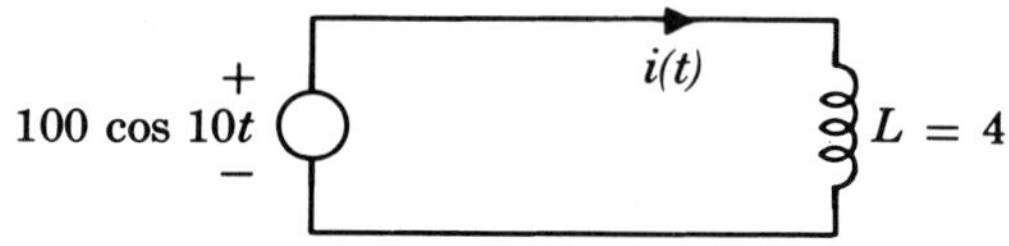

FIGURE 16–17

* Numerical values are in volts, amperes, ohms, henrys, and farads. Complex numbers for impedances are in ohms.

6. In the circuit of Fig. 16–18 find the current in the capacitance (a) by using the time-domain volt-ampere relationships and (b) by using the frequency-domain phasors.

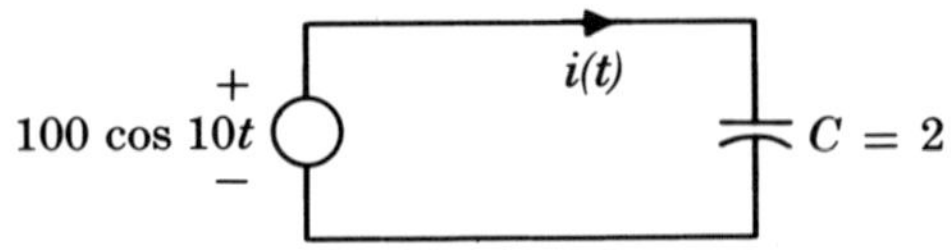

FIGURE 16–18

7. In the circuit of Fig. 16–19 find the voltage $e(t)$ (a) by means of the time-domain volt-ampere equations and (b) by means of phasors in the frequency domain.

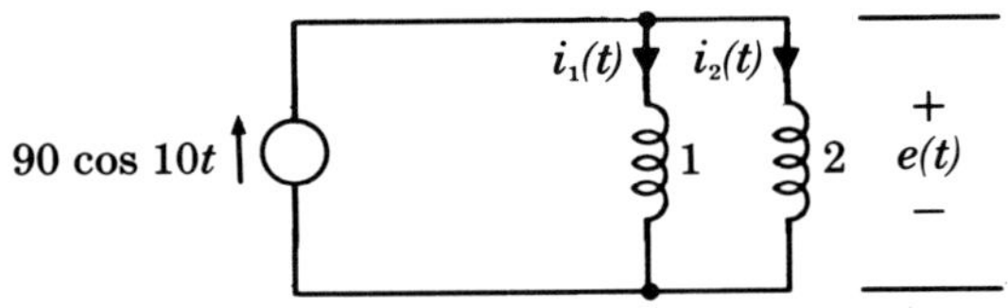

FIGURE 16–19

8. Find the current $i(t)$ in the circuit of Fig. 16–20 (a) by means of the time-domain volt-ampere equations and (b) by means of phasors in the frequency domain.

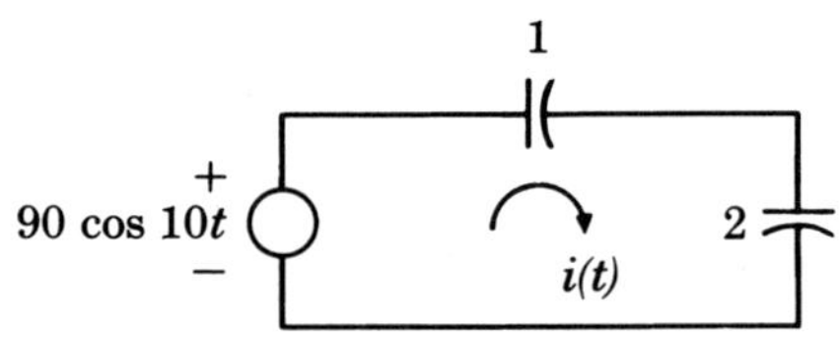

FIGURE 16–20

Section 16–3

9. Solve for the current in the circuit of Fig. 16–21 by the following steps:
   (a) make a frequency-domain diagram,
   (b) find the current phasor,
   (c) transform this phasor to the time domain.

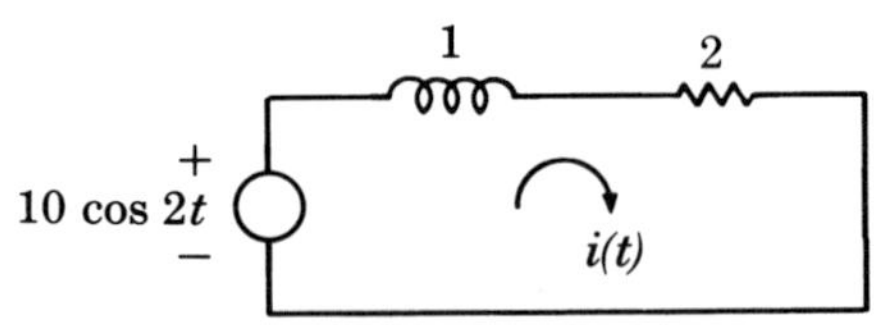

FIGURE 16–21

10. Find the voltage $e(t)$ in the circuit of Fig. 16–22 using the following steps:
   (a) make a frequency-domain diagram,
   (b) find the voltage phasor,
   (c) transform this phasor to the time domain.

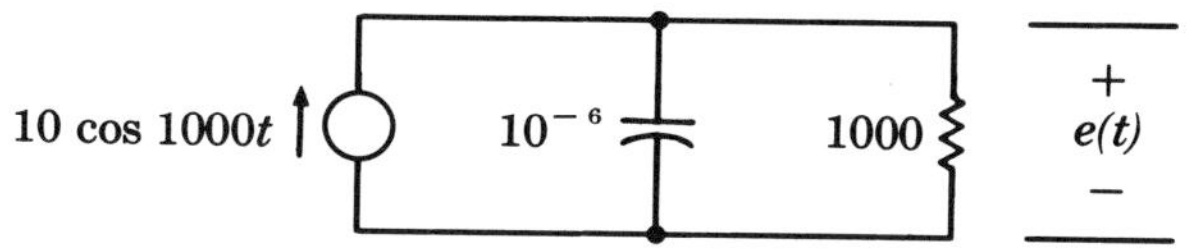

FIGURE 16–22

11. In the circuit of Fig. 16–23 find the current by means of frequency-domain methods.

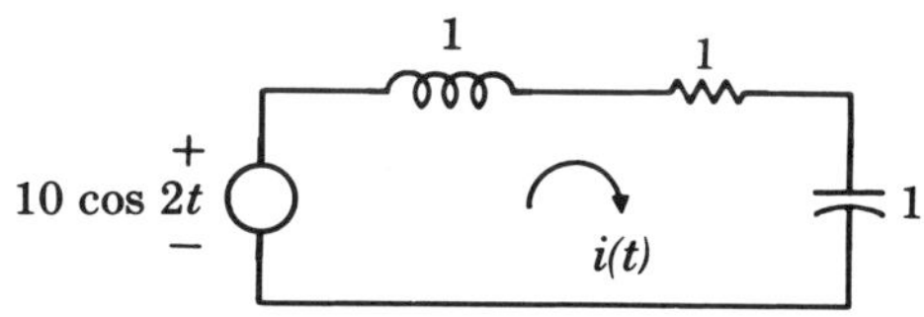

FIGURE 16–23

12. In the circuit of Fig. 16–24 find the voltage $e(t)$ by phasor methods.

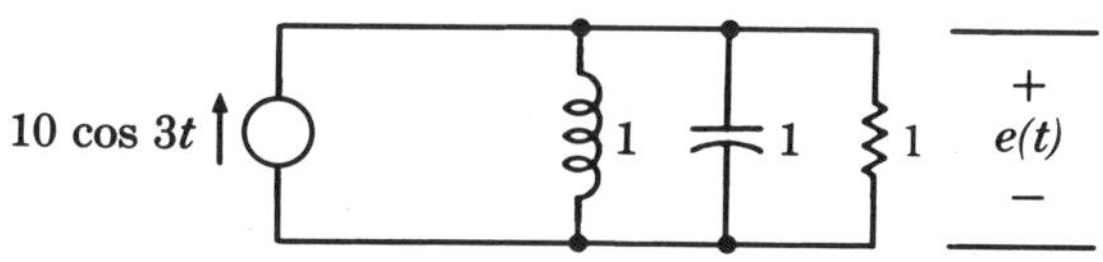

FIGURE 16–24

13. Transform the circuit of Fig. 16–25 to the frequency domain, write the mesh equations, and solve for the current phasor I_2. Transform this phasor back to the time domain in order to obtain $i_2(t)$.

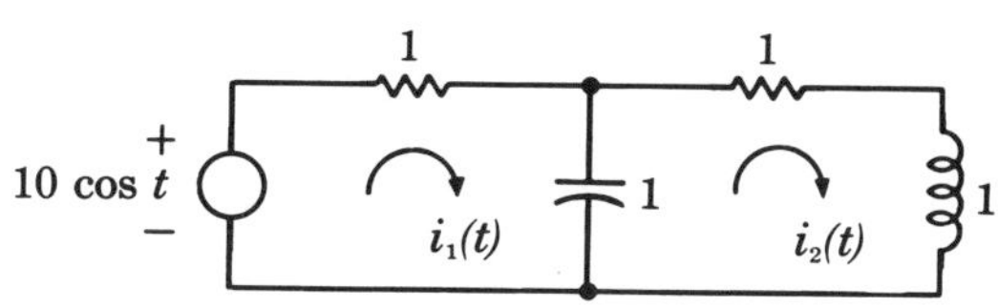

FIGURE 16–25

14. Transform the circuit of Fig. 16–26 to the frequency domain, write the node equations, and solve for the voltage phasor E_2. Transform this phasor to the time domain in order to obtain $e_2(t)$.

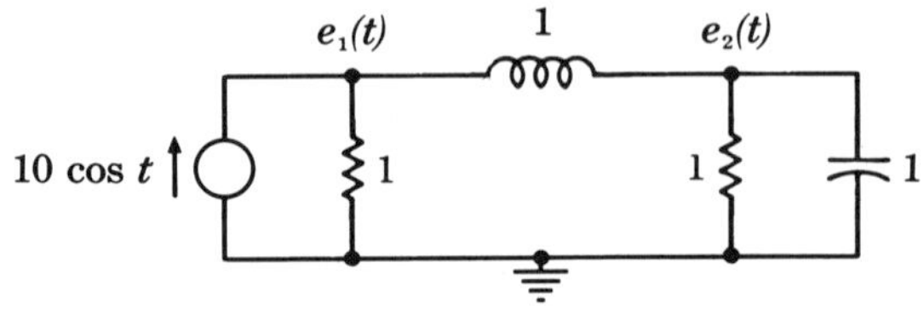

FIGURE 16–26

15. (a) Transform the circuit of Fig. 16–27 to the frequency domain and obtain the Thévenin equivalent circuit at terminal-pair a-b, in the form of a single voltage in series with a single impedance. (b) If a short circuit is placed across terminal-pair a-b, what current flows in it?

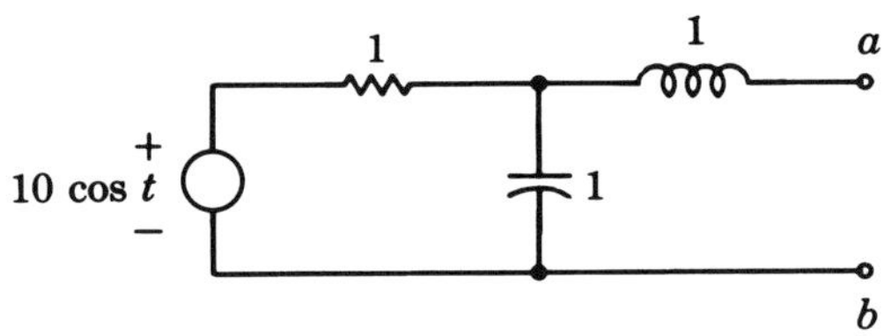

FIGURE 16–27

16. (a) Transform the circuit of Fig. 16–28 to the frequency domain and obtain the Norton equivalent circuit at terminal-pair a-b, in the form of a single current source in parallel with a single impedance. (b) If the terminals are left open-circuited, what voltage appears across them?

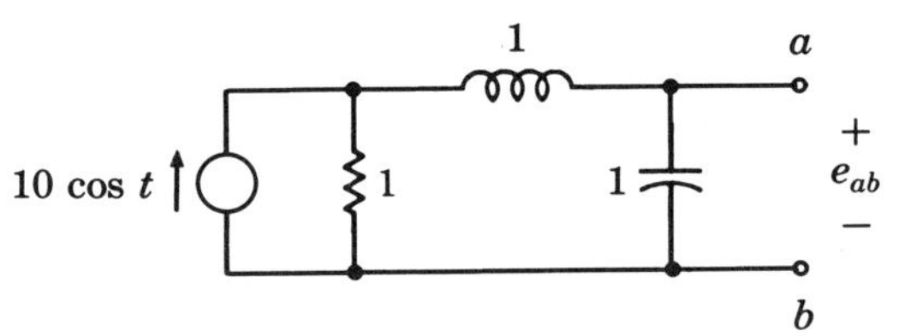

FIGURE 16–28

Section 16–4

17. A voltage source $e(t) = 100 \cos t$ is applied to a series circuit consisting of $R = 1$ ohm and $L = 1$ henry. Solve for the current by means of a phasor diagram. Follow the steps below:

(a) Make a frequency-domain diagram.

(b) Assume a current phasor of $1 \underline{|0°}$.

(c) Compute the associated voltage phasor from a phasor diagram.
(d) Scale the results to the original source value.
(e) Transform the current phasor to the time domain.

18. A current $i(t) = 100 \cos t$ is applied to a parallel circuit consisting of $R = \frac{1}{3}$ ohm, and $C = 4$ farads. Find the voltage phasor by means of a phasor diagram. Start with an assumed voltage phasor, compute the current phasor, and then scale.

19. A series R–L–C circuit is excited by the voltage source $e(t) = 100 \cos t$. The element values are $R = 1$ ohm, $L = 1$ henry, and $C = 1$ farad. Obtain the current by means of a phasor diagram. Start with an assumed value of current, compute the voltage, and then scale.

20. A parallel R–L–C circuit is excited by the current source $i(t) = 100 \cos t$. The element values are $R = 1$ ohm, $L = 1$ henry, and $C = 1$ farad. Find the voltage by means of a phasor diagram. Start with an assumed value of the voltage, compute the current, and then scale.

21. A series R–L–C circuit is excited by the voltage source $e(t) = 100 \cos \omega t$. The element values are $R = 1$ ohm, $L = 10^{-2}$ henry, and $C = 10^{-6}$ farad. The frequency of the source is a variable. Assume a current and draw a phasor diagram. Find the frequency at which a maximum current will flow (the resonant frequency). At this frequency find the ratio of the voltage phasor across the inductance to the voltage phasor across the resistance. (The magnitude of this ratio is the Q of the circuit.)

22. A parallel R–L–C circuit is excited by the current source

$$i(t) = 100 \cos 10^6 t.$$

The value of R is 100 ohms; the value of L is 10^{-4} henry. What value of C will result in a maximum magnitude of the voltage? At this value of C what is the ratio of the magnitude of the current phasor in the capacitance to the magnitude of the current phasor in the resistance? (The circuit is in parallel resonance, and this ratio is the Q.)

23. A series R–L circuit has a fixed reactance of 1 ohm and a fixed source voltage of $10 \underline{/0°}$. The resistance value is varied from zero to infinity. From a phasor diagram sketch the locus of the tip of the phasor representing the voltage across the resistance.

24. A parallel R–L circuit has a fixed resistance of 1 ohm and a fixed source current of $10 \underline{/0°}$. The reactance is varied from zero to infinity. Find the locus of the tip of the phasor representing the current through the inductance.

25. A series R–C circuit is driven by an a-c voltage source of magnitude 200 volts. The magnitude of the voltage across the resistance is 100 volts, and the magnitude of the voltage across the capacitance is 173 volts. Draw a phasor diagram for the circuit and convert the voltages to the time domain, if the frequency is 60 cps, and the source is taken to have zero phase angle.

26. In the circuit of Fig. 16–29 all three voltages have magnitudes of 100 volts, and the source is taken at the reference angle of zero degrees. The circuit in box B is known to be of an inductive nature. Draw the phasor diagram and write time expressions for $e_1(t)$ and $e_2(t)$, including their phase angles.

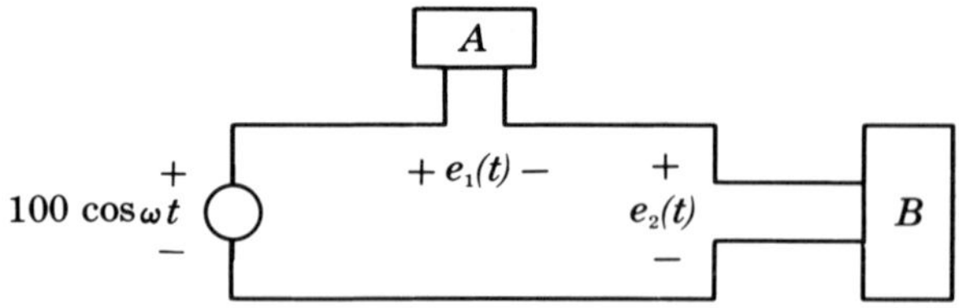

FIGURE 16–29

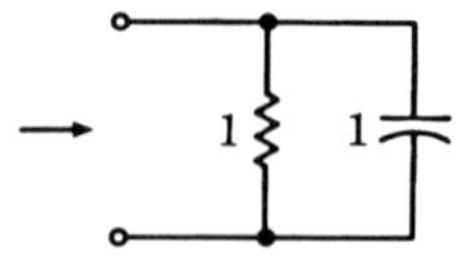

FIGURE 16–30

Section 16–5

27. For the circuit of Fig. 16–30:
   (a) Draw a frequency-domain diagram if the excitation frequency is $\omega = 1$ radian/sec.
   (b) Find the input impedance of the circuit at $\omega = 1$.
   (c) What equivalent series circuit will have the same impedance at $\omega = 1$.
   (d) Find the input impedance of the original circuit as a function of a general ω.
   (e) Give the real and imaginary parts of the input impedance as functions of frequency, and sketch graphs for them.
   (f) Why is there no equivalent circuit in the general case?
28. For the circuit of Fig. 16–31:
   (a) Make a frequency-domain circuit if $\omega = 1$.
   (b) Find the input admittance.
   (c) Give an equivalent parallel circuit with the same admittance at $\omega = 1$.
   (d) Find the input admittance as a function of ω.
   (e) Separate the admittance into its real and imaginary parts and sketch them.
   (f) Why is there no parallel equivalent circuit in the general case?

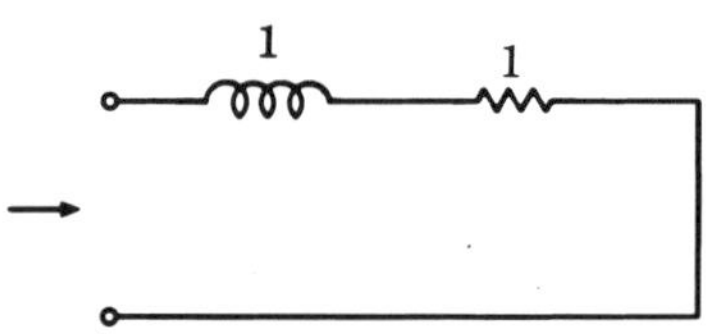

FIGURE 16–31

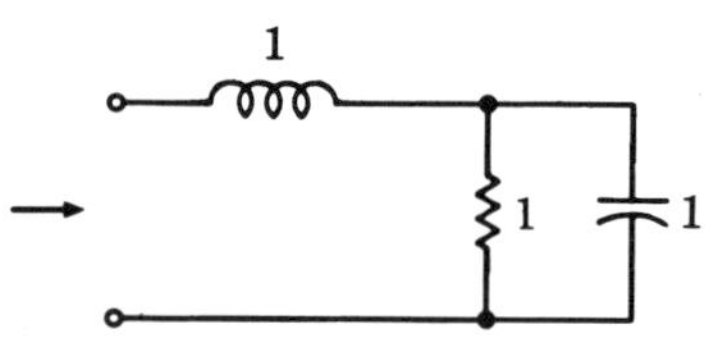

FIGURE 16–32

29. For the circuit of Fig. 16–32:
   (a) Find the input impedance at $\omega = 1$.
   (b) Give the equivalent two-element series circuit at $\omega = 1$.
   (c) Give the equivalent two-element parallel circuit at $\omega = 1$.
   (d) Find the impedance as a function of frequency and separate it into its real and imaginary parts.

30. For the circuit of Fig. 16–33:
   (a) Find the input impedance at $\omega = 1$.
   (b) Give the equivalent two-element series circuit at $\omega = 1$.
   (c) Give the equivalent two-element parallel circuit at $\omega = 1$.
   (d) Find the input impedance as a function of ω and separate it into its real and imaginary parts.

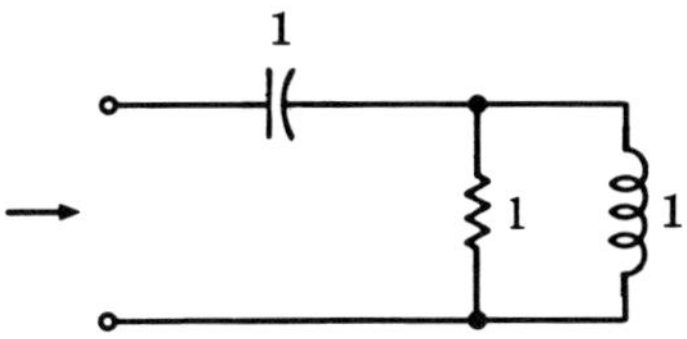

FIGURE 16–33

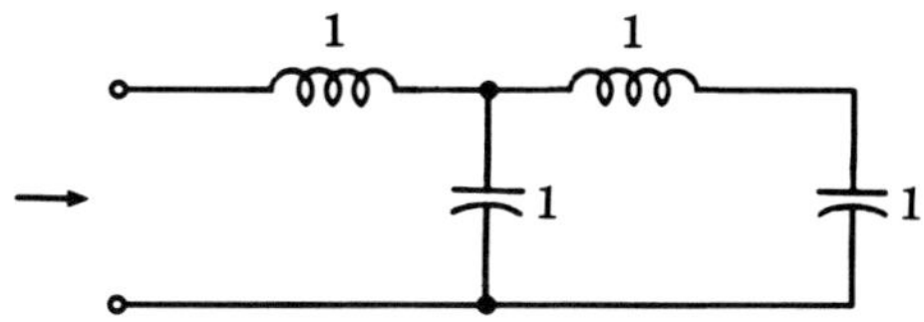

FIGURE 16–34

31. (a) Give the input impedance at $\omega = 1$ of the circuit in Fig. 16–34.
   (b) Give the input impedance of the circuit as a function of ω.

32. (a) Find the open-circuit impedance parameters z_{11}, z_{12}, and z_{22} for the circuit of Fig. 16–35 when the excitation frequency is $\omega = 1$.
   (b) Give the equivalent T network at $\omega = 1$.
   (c) Find z_{11}, z_{12}, and z_{22} as functions of ω.
   (d) Discuss the T-network reduction in the general case.

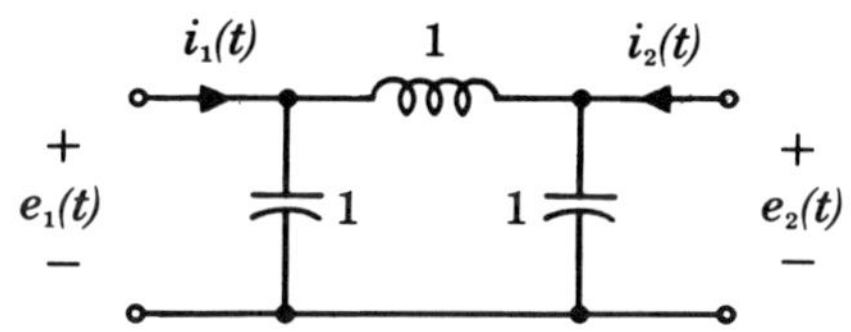

FIGURE 16–35

33. (a) Find the short-circuit admittance parameters y_{11}, y_{12}, and y_{22} for the circuit of Fig. 16–36 when the excitation frequency is $\omega = 1$.
   (b) Give the equivalent π circuit for $\omega = 1$.
   (c) Find y_{11}, y_{12} and y_{22} as functions of ω.
   (d) Discuss the π network in the general case.

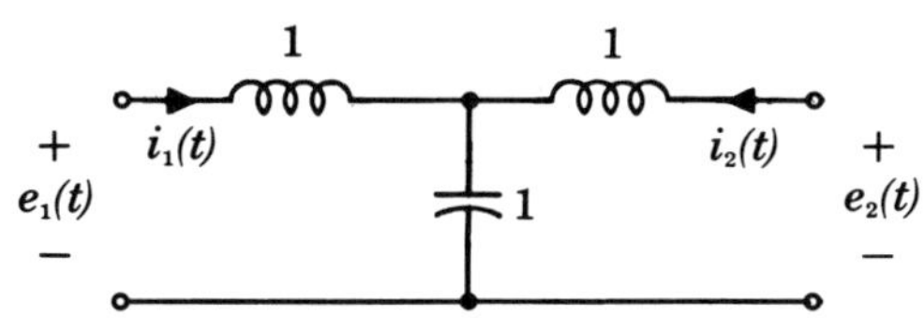

FIGURE 16–36

34. Repeat Exercise 33 if the shunt capacitance has a value of 2 farads, instead of the 1 farad originally used.

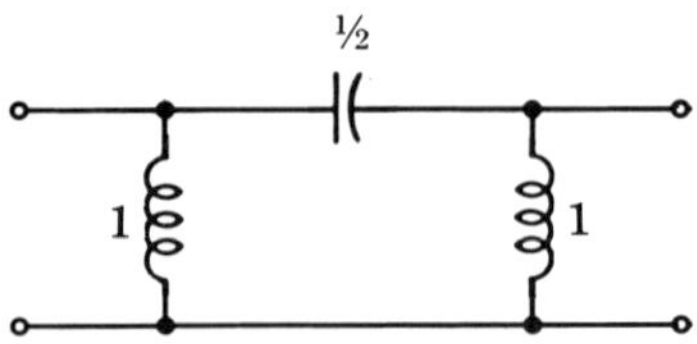

FIGURE 16–37

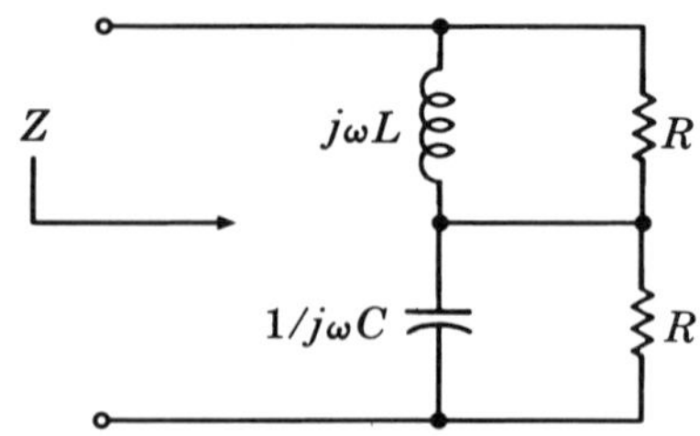

FIGURE 16–38

35. Make an equivalent T circuit for that of Fig. 16–37 at (a) $\omega = 0$, (b) $\omega = 1$, (c) $\omega = 1.414$, and (d) $\omega = \infty$.

36. Find the impedance of the circuit of Fig. 16–38 as a function of ω. Find the condition under which it will be a constant value independent of the frequency.

Section 16–6

37. The frequency-domain diagram of Fig. 16–39 has the units of ohms.
   (a) Give the corresponding time-domain diagram if the frequency is 1 radian/sec.
   (b) Give the time-domain diagram if the frequency is $\omega = 10^3$ rps.
   (c) Give the time-domain diagram if the frequency is $\omega = 10^6$ rps.

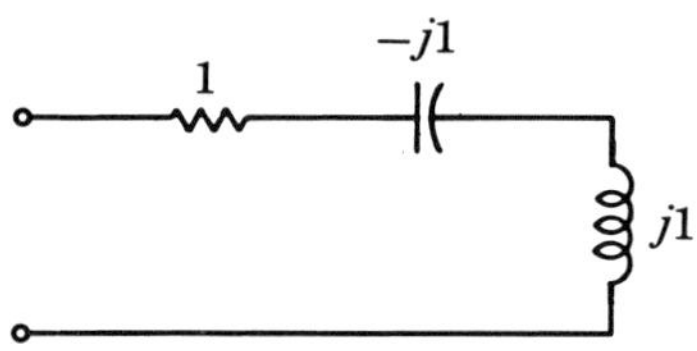

FIGURE 16–39

38. In the frequency-domain circuit of Fig. 16–40 the units are ohms.
   (a) Find the voltage phasor E.
   (b) If the frequency is $\omega = 1$, give the time-domain diagram and the voltage $e(t)$.
   (c) If the frequency is $\omega = 10^6$, give the time-domain diagram and the voltage $e(t)$.

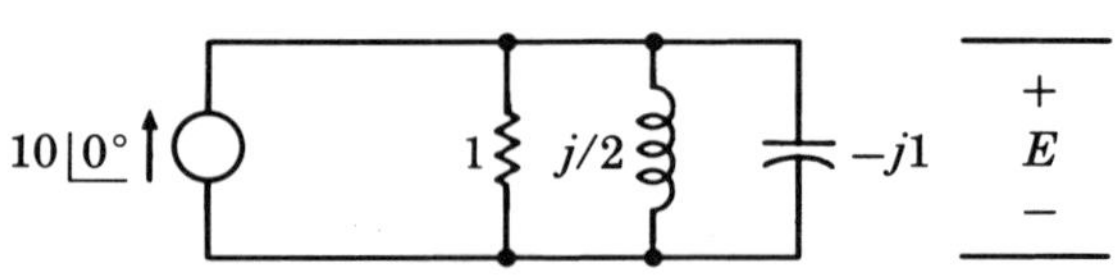

FIGURE 16–40

39. (a) Translate the circuit of Fig. 16–41 to the frequency domain and scale to an impedance level which gives a 1-ohm resistance.
(b) Solve the normalized circuit for the phasor E_0.
(c) Write the time function for the original voltage $e_0(t)$.

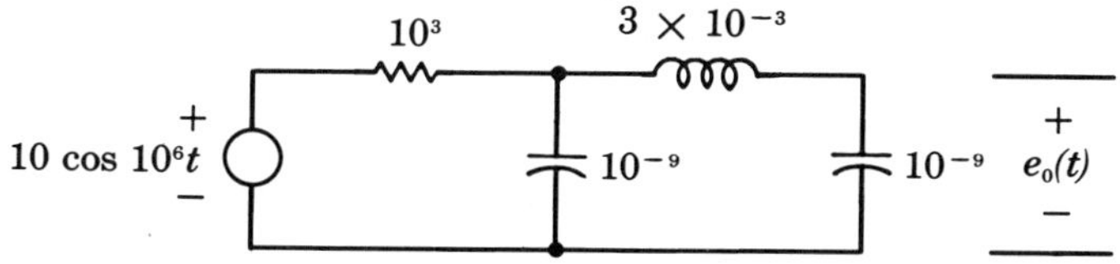

FIGURE 16–41

40. (a) Translate the circuit of Fig. 16–42 to the frequency domain and normalize the impedance level so that the resistance assumes a value of 1 ohm.
(b) Solve for the normalized voltage phasor E_2.
(c) Write the original time function $e_2(t)$.

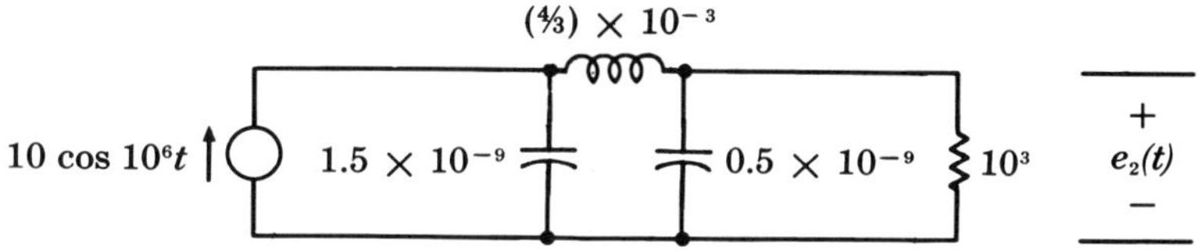

FIGURE 16–42

41. If the voltage is the excitation and the current is the response, the circuit of Fig. 16–43 is a low-pass filter.
(a) Write the impedance Z as a function of frequency.
(b) Scale the circuit so that the impedance is multiplied by 10^3.
(c) Make time-domain diagrams for the original circuit and for a circuit in which the frequency at which any given response occurs is increased by a factor of 10^6.

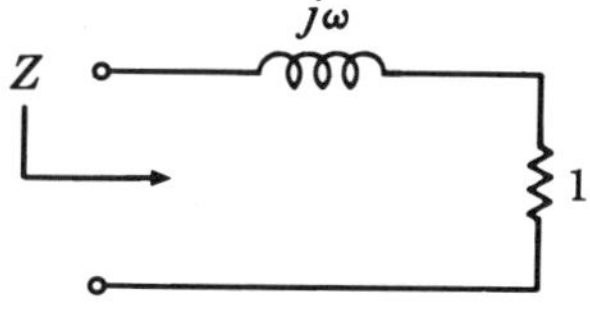

FIGURE 16–43

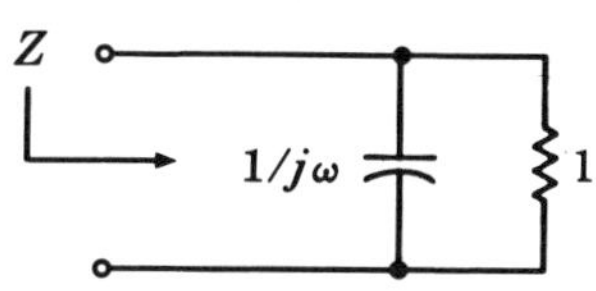

FIGURE 16–44

42. (a) Write the impedance of the circuit of Fig. 16–44 as a function of frequency.
(b) Scale the circuit in the time domain so that the impedance will be multiplied by 1000.

(c) Scale the circuit of part (b) so that the frequency of any given response is increased by a factor of 10^6.
(d) Sketch the magnitudes of the impedances of the circuits of parts (a) and (c) as functions of frequency.

43. (a) Write the impedance of the circuit of Fig. 16–45 as a function of ω.
(b) Scale the circuit so that the impedance magnitude will be increased by a factor of 1000, and write the new impedance.
(c) Scale the circuit so that the resonant frequency will be increased from $\omega = 1$ to $\omega = 10^6$, and again write the impedance. Include also the magnitude scaling of part (b).
(d) Sketch the magnitude of the impedances in parts (a) and (c) as functions of frequency.

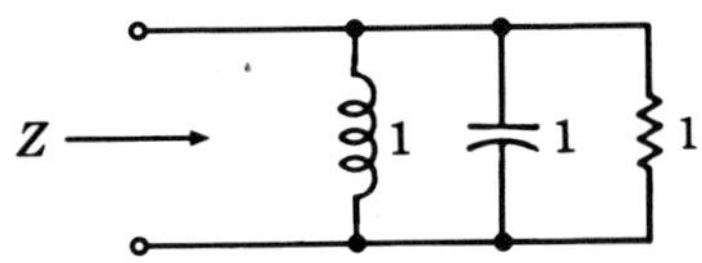

FIGURE 16–45

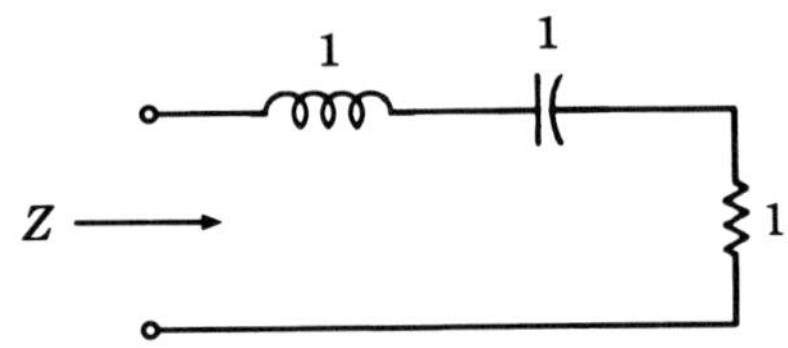

FIGURE 16–46

44. (a) Write the impedance of the circuit of Fig. 16–46 as a function of ω.
(b) Scale the circuit so that it will have an impedance 1000 times as great as the original circuit, and write the impedance.
(c) Scale the network of part (b) so that its resonant frequency will be increased from $\omega = 1$ to $\omega = 10^6$, and write the impedance.
(d) Sketch the magnitudes of the impedances in parts (a) and (c) as functions of frequency.

CHAPTER 17

MUTUAL INDUCTANCE, UNITY-COUPLED COILS, AND TRANSFORMERS

> "... he went down
> As when a lordly cedar, green with boughs,
> Goes down with a great shout upon the hills,
> And leaves a lonesome place against the sky."
>
> *Lincoln—by Edwin Markham*

17–1 Introduction. The ability to transform the landscape to a form which is useful to man is a measure of the advancement of civilization. So too, the ability to transform currents and voltages from one level to another is a measure of their usefulness to mankind. We did not encounter transformers in direct-current circuits, because no one has yet invented a satisfactory transformer for direct currents and voltages. Mutual inductance did occur in our transient problems, but the effect of the mutual inductance died out with time, and we were not lead to consider it as a "transformer." In a steady-state sinusoidal circuit, however, the currents and voltages continue to change in a cyclic fashion, and the mutual inductance continues to play a role. From a practical point of view, the role is so important that we will now devote a whole chapter to it.

Previously, we found it convenient to view a pair of inductances linked by mutual inductance as a primary inductance, a secondary inductance, and a mutual inductance. In the present chapter we will develop an alternative point of view in which two coupled coils with mutual inductance between them are considered to be a transformer which passes energy from one coil to the other and changes the energy level or voltage in the process. The concept of the transformer is similar to the lever in mechanics. The lever transforms energy from one place to another and at the same time it changes the nature of the energy. A small motion and a large force change into a large motion and a small force. A gear train provides the same function for rotary motions. As the shaft speed is increased by means of gearing, the torque which is available is decreased. The utility of gear trains is well known. They occur in many practical devices, from ancient clocks to modern automobiles. It should not be surprising that their electrical counterparts, transformers, should be useful in almost as many ways.

In mechanics we find it convenient to postulate ideal gears which have no losses of energy and which have no inertia and can therefore respond instantly to changes in the torque levels. Similarly, in electric circuits we will find it convenient to postulate *ideal transformers* which change the voltage and the current by fixed ratios, and which do not introduce either resistance or inductance in the circuit. Ideal transformers are not physically realizable, but within limits many practical transformers can be treated as ideal transformers. A *unity-coupled* coil is a practical transformer in which all the magnetic-field flux produced by one coil links the other coil. To obtain this effect many transformers are wound on iron cores which offer a very easy path for the magnetic flux and channel it from one core to the other. At high frequencies iron loses its effectiveness, and unity-coupled coils must be constructed by interleaving the turns of the two coils in the same region of space.

Any real transformer, composed of two coupled coils, can be represented as an ideal transformer in conjunction with resistances and inductances. Although an iron core is itself a nonlinear device, the resistances and inductances associated with its equivalent circuit are linear. The resistances are largely the resistances of the windings, and the inductances arise from "leakage" flux which exists in the air around the coils and does not link them.* The transformer equivalent circuit is therefore linear, although the self and mutual inductances of the coils are very nonlinear. Because of the importance of iron-core transformers in practical applications, we are justified in spending some time developing equivalent circuits for them. There are literally millions of iron-core transformers in our power distribution systems, and they are among the most efficient and reliable components used anywhere in the modern world.

17–2 Self and mutual inductance in the time and frequency domains. In the interest of completeness we will repeat the definitions of the self and mutual inductances of two coils which were given in Chapter 7. Two mutually coupled coils have two external terminal-pairs, and the volt-ampere equations which describe them must relate the input current and voltage to the output current and voltage. An additional unique feature of mutual coupling between coils is the arbitrariness of the sign of the mutual inductance. By changing the direction of the winding of the coils relative to each other, the manufacturer can change the polarity of the mutually induced voltages in the windings. Since the actual windings are usually hidden from view, the relative polarities are marked with dots on the case. The dots are placed so that if current enters at a dot in one winding it will induce a voltage in the other winding which is positive at

---

* This point is examined in great detail in the text by the MIT EE Staff, *Magnetic Circuits and Transformers*, John Wiley & Sons, New York, 1943. See Chapter 12, pp. 313–330.

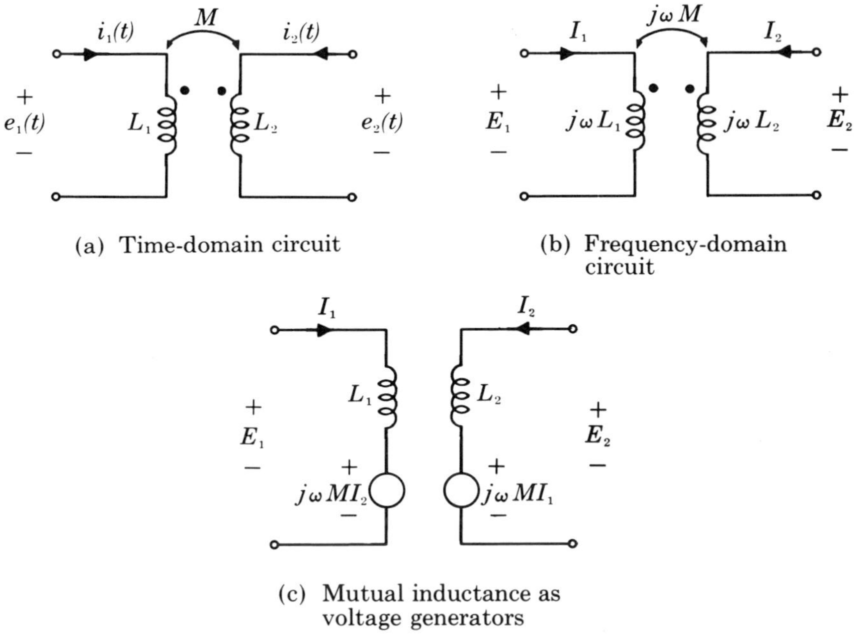

FIG. 17–1. The forms of mutual inductance.

the dot. The symbols used for two coupled coils in the time domain are shown in Fig. 17–1(a), and the equations which define them are

$$
\begin{aligned}
e_1 &= L_1 \frac{di_1}{dt} + M \frac{di_2}{dt}, \\
e_2 &= M \frac{di_1}{dt} + L_2 \frac{di_2}{dt}.
\end{aligned}
\tag{17–1}
$$

The signs of the mutually induced voltages of Eqs. (17–1) are obtained from the marked dots and the assumed positive directions of the currents and voltages in the two windings. When the current i_1 enters the dotted end of the L_1 inductance, it induces a voltage in the L_2 inductance which has the value $M\, di_1/dt$ and a polarity which is positive at the dotted end of L_2. This voltage adds a positive component to e_2 of Eq. (17–1). Similarly, the current i_2 enters L_2 at the dot, and produces a voltage $M\, di_2/dt$ which is positive at the dotted end of L_1 and therefore appears as a positive voltage in Eq. (17–1). Reversing the dots, or reversing the assumed directions of the currents or voltages in either winding, will change the sign of the mutual terms in Eq. (17–1).*

* It is also possible to define the currents in terms of the voltages by means of "reciprocal" inductances, but this is seldom done.

Mutual inductance transforms to the frequency domain in much the same fashion as self-inductance. The induced voltage, like the voltage across an inductance, leads the current producing it by 90°. In the frequency domain, Eqs. (17–1) become

$$\begin{aligned} E_1 &= j\omega L_1 I_1 + j\omega M I_2, \\ E_2 &= j\omega M I_1 + j\omega L_2 I_2. \end{aligned} \tag{17–2}$$

Equations (17–2) relate the phasor currents and phasor voltages of the coupled coils. The effect of the mutual inductance is to introduce a mutual reactance X_m, or a mutual impedance Z_m, expressed as

$$Z_m = jX_m = j\omega M. \tag{17–3}$$

When the current phasor in one winding is multiplied by the mutual impedance, the voltage phasor induced in the other winding is obtained. A frequency-domain diagram is shown in Fig. 17–1(b). Note that the sign convention and meaning of the dot notation are not changed by the transformation to the frequency domain.

Each of Eqs. (17–2) consists of two terms, one of which is the voltage drop produced in the winding by the current which flows in the winding, and the other of which is the voltage generated or induced in the winding by the current in the other winding. It is often advisable to draw the induced voltages as separate voltage generators, with the correct polarities assigned to them, as shown in Fig. 17–1(c). In this case both of the generators are positive at the top because both of the dots are at the top of the windings. Since both these generators are activated by currents, it is better to solve problems involving mutual inductance by means of current variables.

As an example, consider the two series inductances shown in Fig. 17–2(a). The equivalent frequency-domain diagram is shown in Fig. 17–2(b). The variable assumed is the mesh current $i(t)$, which has a frequency-domain value of I. This current flows through both L_1 and L_2 and produces voltage drops in them. In addition, there is a voltage generator in each coil produced by the flow of current in the other coil. The voltage generators are shown in Fig. 17–2(c). Because of the polarity of the dots, they both appear as voltage drops in the circuit. The voltage-law equation for the loop is

$$I(j\omega L_1 + j\omega L_2) + 2j\omega M I = E$$

or

$$I = \frac{E}{j\omega(L_1 + L_2 + 2M)}. \tag{17–4}$$

If the dot on one coil were reversed, the sign of the mutual term in Eq.

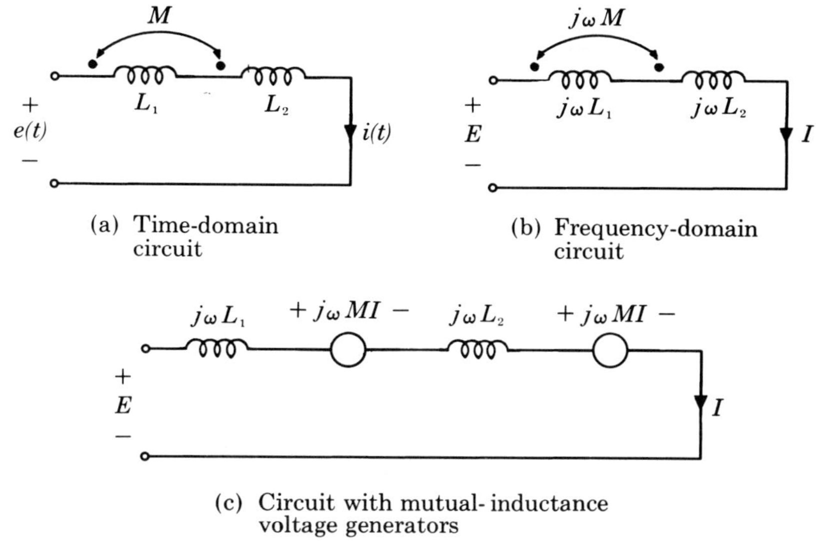

FIG. 17-2. Two inductors in series.

(17-4) would be minus. From this equation it is easy to verify that the equivalent inductance of two mutually coupled coils connected in series is

$$L_{eq} = L_1 + L_2 \pm 2M. \tag{17-5}$$

Since the net inductance must be positive*

$$L_1 + L_2 \geq 2M. \tag{17-6}$$

Two inductances in parallel are shown in Fig. 17-3(a) in the time domain, and in Fig. 17-3(b) in the frequency domain. The net current entering at the dotted end of L_1 is $I_1 - I_2$, and the net current entering at the dotted end of L_2 is I_2. These currents drive the generators in the two-generator equivalent circuit of Fig. 17-3(c). The mesh equations for this circuit are

$$\begin{aligned} I_1 j\omega L_1 - I_2 j\omega L_1 + j\omega M I_2 &= E, \\ -I_1 j\omega L_1 + j\omega M I_1 + I_2 j\omega L_1 + I_2 j\omega L_2 - 2I_2 j\omega M &= 0. \end{aligned} \tag{17-7}$$

Equations (17-7) are somewhat more complex than the usual mesh equations because of the mutual inductance terms. Unlike an ordinary impedance, the term for the mutual inductance can have either sign at any

* A negative inductance would supply infinite amounts of energy to a source of positively increasing current. See Chapter 7.

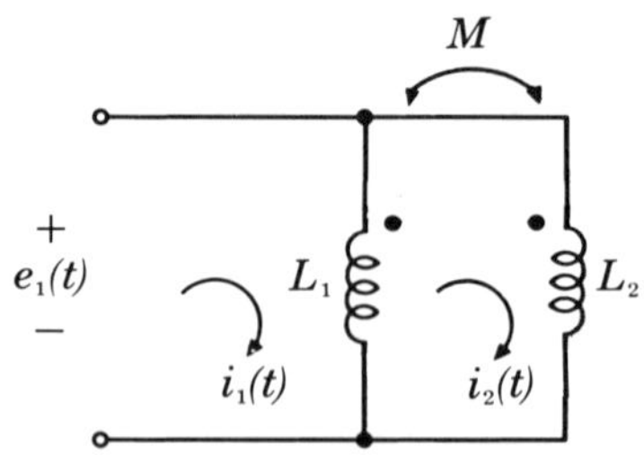

(a) Time-domain circuit

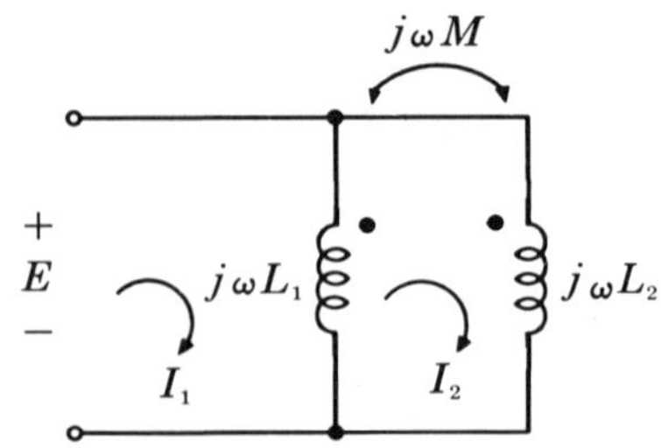

(b) Frequency-domain circuit

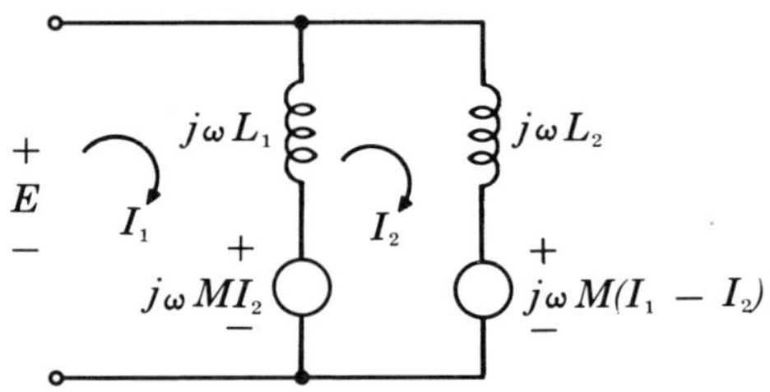

(c) Circuit with mutual-inductance voltage generators

FIG. 17–3. Two inductors in parallel.

point in the equation. The mutual coefficients in the two equations, must, however, be identical. A solution for I_1 is

$$I_1 = \frac{\begin{vmatrix} E & -j\omega L_1 + j\omega M \\ 0 & j\omega L_1 + j\omega L_2 - 2j\omega M \end{vmatrix}}{\begin{vmatrix} j\omega L_1 & -j\omega L_1 + j\omega M \\ -j\omega L_1 + j\omega M & j\omega L_1 + j\omega L_2 - 2j\omega M \end{vmatrix}}$$

$$= \frac{E(L_1 + L_2 - 2M)}{j\omega(L_1L_2 - M^2)}. \tag{17–8}$$

The equivalent inductance of the parallel combination is apparent from Eq. (17–8). It is

$$L_{\text{eq}} = \frac{L_1L_2 - M^2}{L_1 + L_2 - 2M}. \tag{17–9}$$

The sign of M in the denominator of Eq. (17–9) changes with the dot polarity, but the numerator does not change since M is squared. From

Eq. (17–6) we have established that the expression in the denominator is positive. Since the over-all inductance must be positive

$$L_1L_2 - M^2 \geq 0. \tag{17–10}$$

Equation (17–10) is more restrictive than Eq. (17–6). Essentially, Eq. (17–6) states that the mutual inductance must be less than the arithmetic mean of the two self-inductances, while Eq. (17–10) states that the mutual inductance must be less than the geometric mean of the two self-inductances. The geometric mean is less than the arithmetic mean except when the two inductances are equal, in which case there is no difference between the two means. The maximum value of the mutual inductance, from Eq. (17–10), is

$$M_{\text{max}} = \sqrt{L_1L_2}. \tag{17–11}$$

The ratio of the actual mutual inductance which is present in any given case to the maximum value which it can have, according to Eq. (17–11), is called the *coefficient of coupling*, K, and is given by

$$K = \frac{M}{\sqrt{L_1L_2}}\cdot \tag{17–12}$$

The coefficient of coupling K ranges from a minimum value of zero when there is no mutual inductance between the two coils to a value of unity when the maximum mutual inductance exists. The physical meaning of a unity coefficient of coupling is that all the flux produced by the current in one of the coils links the other. Nothing can be done to increase the mutual inductance beyond the value which it possesses at this point. The coefficients of coupling of the iron-core transformers used in power distribution systems are very nearly unity. Much lower values occur in the air-core coils used in radio and television receivers. In general, we can say that when the coefficient of coupling is low it is convenient to view coupled coils in terms of self and mutual inductances, and when it is high it is convenient to view the coupled coils as a transformer.

17–3 Reflected impedance. Mesh currents provide a perfectly general method for solving problems which involve mutual inductance. They do not shed much light on the special nature of mutual inductance, however, and for many problems the method of "reflected impedance," which we shall now discuss, is preferable. The presence of mutual inductance in an inductor causes a voltage to be induced in it which will tend to limit the current that flows in the inductor. The net effect of the induced voltage is thus identical to an impedance which is "reflected" into the circuit. We can investigate the nature of this impedance from the frequency-domain

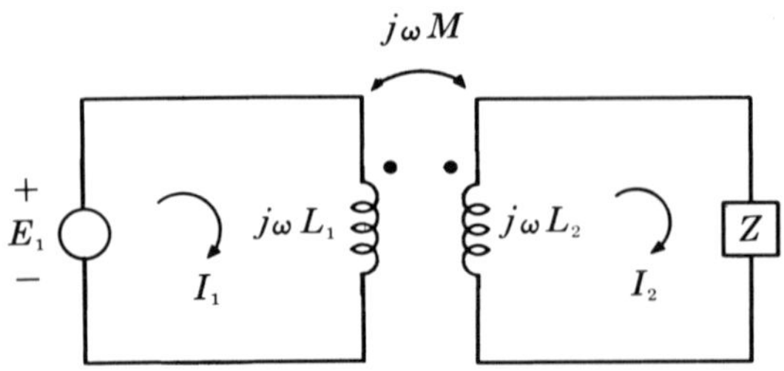

FIG. 17–4. A circuit for reflected-impedance computation.

circuit of Fig. 17–4. The mesh equations for this circuit are

$$\begin{aligned} I_1(j\omega L_1) - I_2(j\omega M) &= E_1, \\ -I_1(j\omega M) + I_2(j\omega L_2 + Z) &= 0, \end{aligned} \tag{17–13}$$

from which

$$I_1 = E_1 \frac{(j\omega L_2 + Z)}{(j\omega L_2 + Z)j\omega L_1 + \omega^2 M^2} \cdot \tag{17–14}$$

The total effective impedance in the primary, or L_1, circuit is the ratio of the input voltage to the input current:

$$Z_1 = \frac{E_1}{I_1} = j\omega L_1 + \frac{\omega^2 M^2}{j\omega L_2 + Z} \cdot \tag{17–15}$$

The first term in Eq. (17–15) is the impedance which would be present in the primary circuit if there were no mutual inductance. The second term represents the effect of the mutual inductance and is called the *reflected impedance* Z_f. According to Eq. (17–15)

$$Z_f = \frac{\omega^2 M^2}{j\omega L_2 + Z} \cdot \tag{17–16}$$

Equation (17–16) is relatively easy to remember. The reflected impedance is $\omega^2 M^2$ divided by the total impedance of the secondary, or L_2, circuit. The term $j\omega L_2$ represents the impedance of the transformer winding and the term Z represents the remainder of the impedances in the circuit. If the impedance Z equals $R + jX$, the reflected impedance is

$$\begin{aligned} Z_f &= \frac{\omega^2 M^2}{R + j(\omega L_2 + X)} \times \frac{R - j(\omega L_2 + X)}{R - j(\omega L_2 + X)} \\ &= \frac{\omega^2 M^2 R}{R^2 + (\omega L_2 + X)^2} - j\,\frac{\omega^2 M^2(\omega L_2 + X)}{R^2 + (\omega L_2 + X)^2} \cdot \end{aligned} \tag{17–17}$$

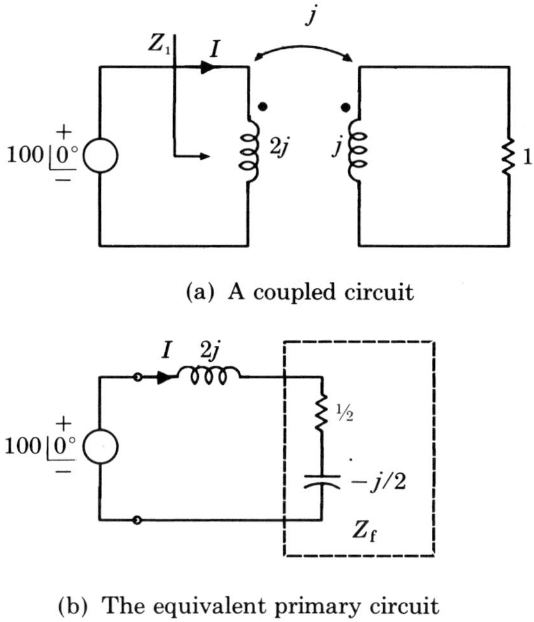

FIG. 17–5. A problem solved by reflected-impedance methods.

The first term in Eq. (17–17) is the reflected resistance. It is always positive since resistance R is positive. (The load is assumed to be passive and must have a positive R in order to absorb energy.) The second term is the reflected reactance. It has a sign which is opposite to that of the total reactance of the secondary circuit. Thus an inductance in the secondary reflects into the primary circuit as a negative reactance which cancels some of the positive reactance of the primary and allows more current to flow. The negative reactance is not actually capacitive since its magnitude increases with frequency, but at any one frequency it has much the same effect as a capacitance introduced into the primary circuit.

As an example of the use of reflected impedance, we will find the primary current in the circuit of Fig. 17–5(a). The total primary reactance is $2j$, the mutual reactance is j, the secondary reactance is j, and the secondary load resistance is 1. The total primary impedance is

$$Z_1 = j\omega L_1 + \frac{\omega^2 M^2}{j\omega L_2 + Z}$$

$$= 2j + \frac{1}{j+1} = 2j + \tfrac{1}{2} - j\tfrac{1}{2}$$

$$= \tfrac{1}{2} + j\tfrac{3}{2}.$$

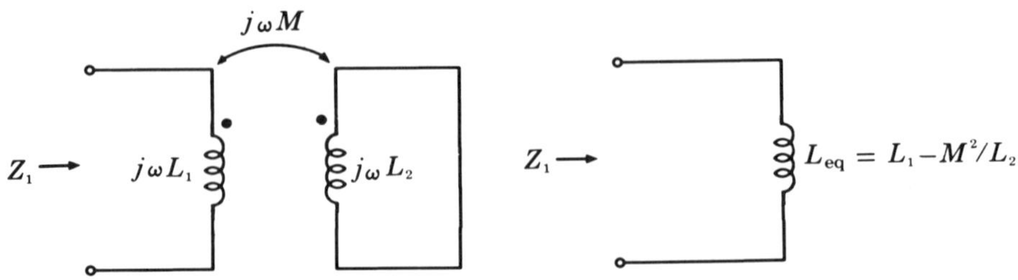

(a) Coupled coils with shorted secondary

(b) Equivalent primary circuit

FIG. 17–6. Coupled coils with a short circuit on the secondary.

The current in the primary circuit can be computed from the equivalent primary circuit of Fig. 17–5(b). It is

$$I = \frac{100\,\underline{|0^\circ}}{\frac{1}{2} + j\frac{3}{2}} = 20 - j60 = 63.4\,\underline{|-71.6^\circ}.$$

If there were no secondary circuit and no mutual inductance, the current would be

$$I = \frac{100\,\underline{|0^\circ}}{j2} = 50\,\underline{|-90^\circ}.$$

The effect of the mutual inductance is to reflect a negative reactance into the primary circuit which causes the net reactance to decrease, permitting more current to flow. The phase angle is also changed to draw some current in phase with the source, and hence some power from the source, which is eventually used up in the resistance of the load. The negative reactance is thus the mechanism which causes power to flow through the transformer.

An interesting special case occurs when the load on the secondary of a transformer is a short circuit, as shown in Fig. 17–6(a). Here the input impedance is

$$\begin{aligned} Z_1 &= j\omega L_1 + \frac{\omega^2 M^2}{j\omega L_2} \\ &= j\omega\left(L_1 - \frac{M^2}{L_2}\right). \end{aligned} \tag{17–18}$$

The input inductance of the shorted transformer cannot be negative, since a negative inductance is a source of energy. Hence, from Fig. 17–6(b)

$$\frac{M^2}{L_2} \le L_1 \qquad \text{or} \qquad M^2 \le L_1 L_2. \tag{17–19}$$

Equation (17–19) is a second derivation of the important formula for the maximum value of the mutual inductance, which was first obtained in Eq. (17–11). When M has its maximum value, the coefficient of coupling, $K = M/\sqrt{L_1L_2}$, is unity. An attempt is made in the design of real transformers to obtain a coefficient of coupling of unity. Unity-coupled coils, however, are not ideal transformers, the properties of which we will consider in the next section.

17–4 Ideal transformers. In mechanics a lever can transform a force applied at one place to a much larger force applied at another place. There is no net gain in energy, however, since the larger force acts through a proportionately smaller distance. Similarly, a gear train can transform a high-speed rotation to a slower-speed rotation and at the same time increase the torque which is available. In electric circuits the analog of the lever and the gear is the ideal transformer. The ideal transformer has two pairs of terminals. The input pair is called the primary and the output pair, the secondary. The voltage across the primary and the voltage across the secondary are related by a constant. In step-up transformers the secondary voltage is higher than the primary voltage; in step-down transformers the secondary voltage is less than the primary voltage. Whenever the voltage is stepped up, the current is stepped down by the same ratio, in order to have the power, or energy, remain unchanged.

The symbol for an ideal transformer is shown in Fig. 17–7(a). It is similar to a pair of coupled coils except for the bars between the coils and the ratio $1{:}N$ which is marked on the coils to show whether the transformer is a step-up or a step-down type. The following equations define an ideal transformer:

$$E_2 = NE_1, \qquad I_2 = \frac{1}{N} I_1. \tag{17–20}$$

where N is called the "turns-ratio" of the transformer.

In Eqs. (17–20) the voltages are both defined to be positive at the dots; the primary current is defined to enter at the dot in the primary winding; and the secondary current is defined to leave the dot in the secondary winding. These assumed directions agree with those which would occur in coupled coils if there were a source in the primary and a passive circuit in the se^condary. It is interesting to note that the input power and the output power are the same, that is

$$E_1I_1 = E_2I_2. \tag{17–21}$$

There is no power lost in the transformer itself, and there is no energy stored. The ideal transformer corresponds to an ideal lever (no mass and no friction) or to an ideal gear train (no inertia and no frictional losses). Although an ideal transformer can be used at any frequency, including direct current, the approximately ideal transformers available in practice have a shunt inductance which prevents them from operating on direct currents. In view of this we have discussed the ideal transformer in connection with alternating currents, and the diagram which we have shown for it is a frequency-domain diagram.

The effective impedance in the primary circuit is modified by the presence of the transformer. In the circuit of Fig. 17–7(b), a load impedance Z is connected in the secondary circuit, and a voltage is applied to the primary circuit. To obtain the effective primary impedance we note that

$$E_2 = I_2 Z. \tag{17–22}$$

If the definitions of E_1 and I_1, from Eqs. (17–20), are substituted in Eq. (17–22) the result is

$$NE_1 = \frac{I_1}{N} Z$$

or

$$\frac{E_1}{I_1} = \frac{1}{N^2} Z. \tag{17–23}$$

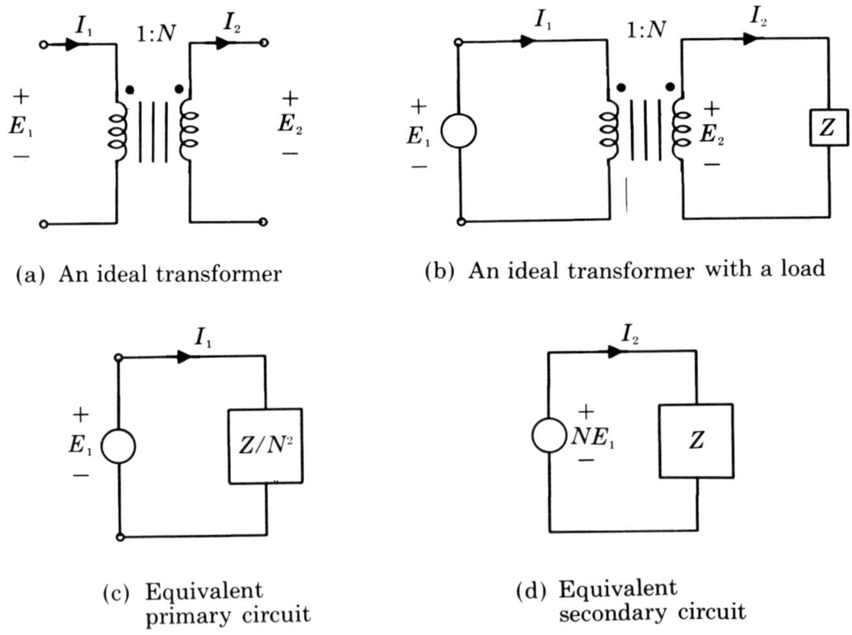

(a) An ideal transformer

(b) An ideal transformer with a load

(c) Equivalent primary circuit

(d) Equivalent secondary circuit

FIG. 17–7. The properties of an ideal transformer.

The ratio of the primary voltage to the primary current is the effective primary impedance, Z_1. Thus

$$Z_1 = \frac{1}{N^2} Z. \tag{17-24}$$

The equivalent primary circuit is shown in Fig. 17-7(c). The secondary impedance has been divided by the square of the turns-ratio of the transformer. If the transformer had been labelled N on the primary and 1 on the secondary, the impedance would have been divided by $1/N^2$ or multiplied by N^2. It is convenient for the student to note that *the impedance is always modified by the square of the turns-ratio*, and *that the high impedance occurs on the side with the larger number of turns.*

It is also possible to transform current and voltage sources from one side of a transformer to another. A voltage source is multiplied by the turns-ratio in the direction in which it is moved, and a current source by the reciprocal of this number. For example, in Fig. 17-7(a) if the source is moved to the right, through the transformer, it must be multiplied by the turns-ratio N. The equivalent secondary circuit is shown in Fig. 17-7(d). Both impedances and sources can be moved through ideal transformers at will. The sources are altered by the turns-ratio to the first power, and the impedances by the square of the turns-ratio.

A numerical example of a problem involving an ideal transformer is shown in Fig. 17-8(a). We will first solve the problem by means of mesh

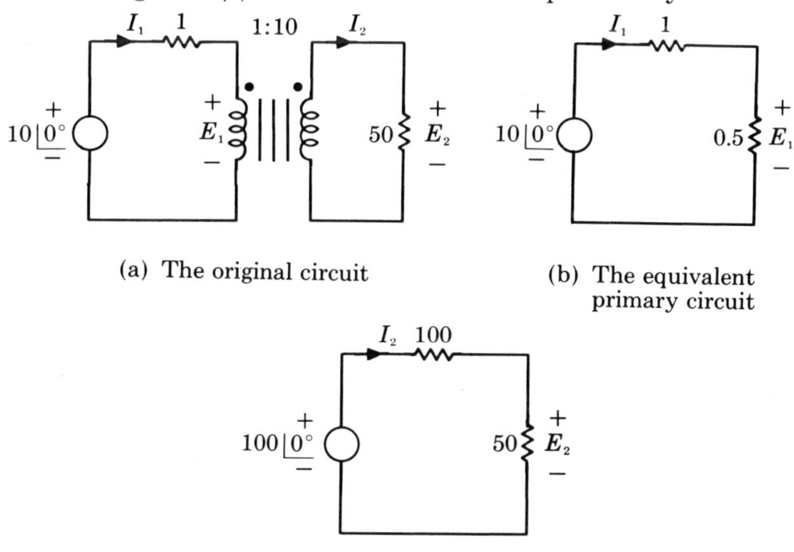

(a) The original circuit

(b) The equivalent primary circuit

(c) The equivalent secondary circuit

FIG. 17-8. An example involving an ideal transformer.

currents. In addition to the usual mesh equations, we must write the two constraint equations which define the transformer. The mesh equations are

$$I_1(1) = E - E_1,$$

$$I_2(50) = E_2.$$

The constraint equations are

$$E_2 = 10E_1,$$

$$I_2 = \frac{I_1}{10}.$$

From these equations we obtain

$$\begin{aligned} E_2 = 10E_1 &= 10E - 10I_1 \\ &= 100 - 100I_2 \\ &= 100 - \tfrac{100}{50}E_2 \end{aligned}$$

or

$$E_2 = 33.3 \text{ volts.}$$

Our second solution for the problem of Fig. 17–8(a) will be in terms of the equivalent primary circuit shown in Fig. 17–8(b). From this diagram the primary voltage E_1 can be easily obtained:

$$E_1 = \frac{0.5}{1 + 0.5} \times 10 = 3.33 \text{ volts.}$$

The secondary voltage is the turns-ratio multiplied by this value:

$$E_2 = 10 \times 3.33 = 33.3 \text{ volts.}$$

A third solution will be obtained for the equivalent secondary circuit of Fig. 17–8(c). The primary resistance is multiplied by 10^2 as it is moved through the transformer, and the primary voltage is multiplied by 10. The voltage E_2 is obtained directly from the secondary equivalent circuit, and is

$$E_2 = \frac{50}{50 + 100} \times 100 = 33.3 \text{ volts.}$$

In Chapter 6 we observed that a real source with internal resistance delivered maximum power to a load resistance connected to it when the value of the load resistance was equal to the internal resistance of the source. Sometimes the source resistance and the load resistance have

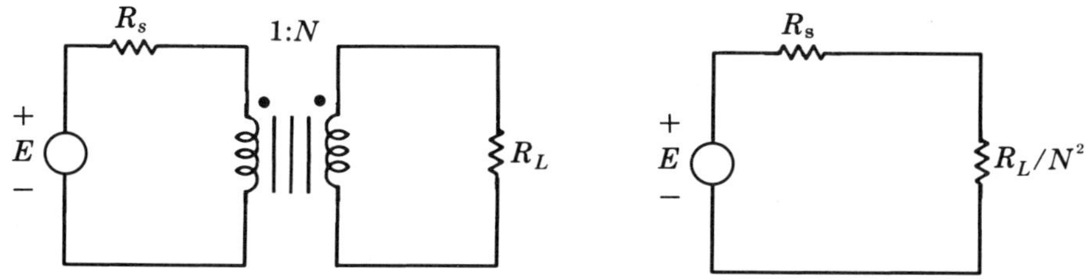

(a) The complete circuit

(b) The equivalent primary circuit

FIG. 17-9. Matching a source impedance and a load impedance with a transformer.

fixed values and cannot be adjusted to the optimum, or matched, condition. For such cases a transformer interposed between the source and the load can be used to achieve the desired result. An example is shown in Fig. 17-9(a). The turns-ratio of the transformer is selected to make the resistance reflected into the primary equal, or matched, to the source resistance. The required value of N is

$$N = \sqrt{\frac{R_L}{R_S}}. \tag{17-25}$$

The equivalent primary circuit is shown in Fig. 17-9(b).

The conditions for matching when the source and the load have reactive components will be discussed in Chapter 19. At this point we merely observe that some sort of matching can always be obtained by the use of a transformer.

17-5 Unity-coupled coils. The ideal transformer which we postulated in the previous section can be realized approximately by means of two coupled coils with a coefficient of coupling which is close to unity, and with inductance values which are as large as possible. In order to investigate this relationship, we will consider the nature of unity-coupled coils in the present section. A unity-coupled coil circuit is shown in Fig. 17-10(a) with an impedance Z connected to the secondary. We will compute the input impedance and obtain an equivalent primary circuit. The general formula for the input impedance, from Eq. (17-15), is

$$Z_1 = j\omega L_1 + \frac{\omega^2 M^2}{j\omega L_2 + Z}$$

$$= \frac{j\omega L_1 j\omega L_2 - (j\omega M)^2 + j\omega L_1 Z}{j\omega L_2 + Z}. \tag{17-26}$$

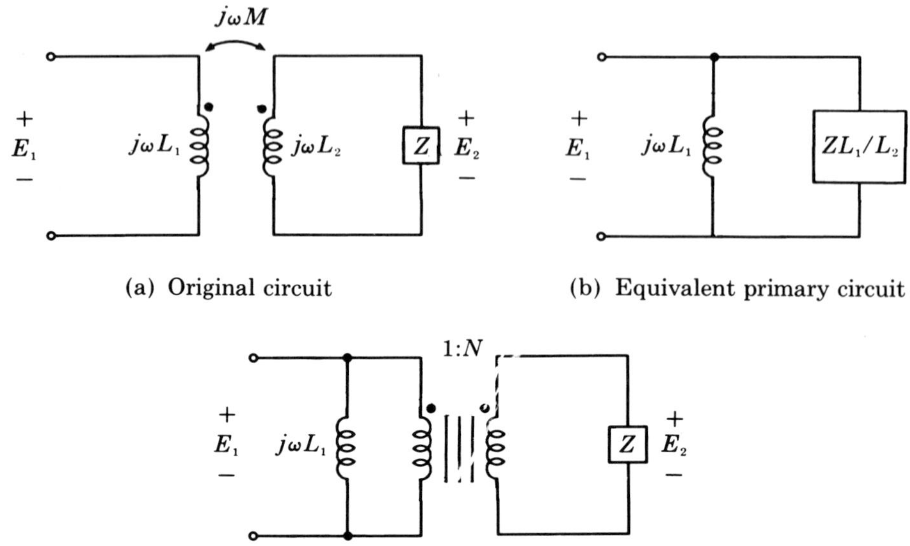

(c) Equivalent circuit with ideal transformer $N = \sqrt{L_2/L_1}$

FIG. 17–10. Unity-coupled coils ($L_1L_2 = M^2$).

For the particular case of unity-coupled coils, $L_1L_2 = M^2$, and Eq. (17–26) reduces to

$$Z_1 = \frac{j\omega L_1 Z}{j\omega L_2 + Z}. \tag{17–27}$$

From the standpoint of finding an equivalent circuit, we will consider input admittance, instead of the input impedance. The input impedance implies a series circuit, and the input admittance a parallel circuit. In this case the parallel circuit is simpler. Thus

$$Y_1 = \frac{1}{j\omega L_1} + \frac{1}{Z/(L_2/L_1)}. \tag{17–28}$$

Equation (17–28) shows that the equivalent primary circuit is the parallel combination of an inductance with impedance $j\omega L_1$ and an impedance ZL_1/L_2. As far as the primary circuit is concerned we can view the impedance ZL_1/L_2 as an impedance which has been reflected through the transformer *in parallel* with the impedance of the primary coil. [In Section 17–3 and in Eq. (17–26) the impedance reflected through the transformer was considered to be *in series* with the primary impedance.] An equivalent primary circuit is shown in Fig. 17–10(b).

With the background of ideal transformers which we have obtained in the previous section, we are lead to postulate a circuit of the form shown in Fig. 17–10(c) for the unity-coupled coil. In order to obtain the correct primary equivalent circuit it is only necessary that the turns-ratio of the ideal transformer be

$$N = \sqrt{\frac{L_2}{L_1}}. \tag{17–29}$$

The impedance reflected through the transformer is $Z/N^2 = ZL_1/L_2$.

We will now show that the turns-ratio N given by Eq. (17–29) will also give the correct secondary voltage in the circuit of Fig. 17–10(c), which therefore represents the unity-coupled coils from both the primary and the secondary terminals. We will start with the general equations for coupled coils:

$$\begin{aligned} E_1 &= j\omega L_1 I_1 + j\omega M I_2, \\ E_2 &= j\omega M I_1 + j\omega L_2 I_2, \end{aligned} \tag{17–30}$$

from which the ratio of E_2 to E_1 must be

$$\begin{aligned} \frac{E_2}{E_1} &= \frac{MI_1 + L_2 I_2}{L_1 I_1 + MI_2} \\ &= \frac{M}{L_1} \times \frac{I_1 + (L_2/M) I_2}{I_1 + (M/L_1) I_2}. \end{aligned} \tag{17–31}$$

Equation (17–31) holds in general. For the particular case of unity-coupled coils

$$L_1 L_2 - M^2 = 0$$

or

$$\frac{M}{L_1} = \frac{L_2}{M} = N \tag{17–32}$$

where $N = \sqrt{L_2/L_1}$ as before. Because of the particular conditions of Eq. (17–32), the ratio of the input and output voltages of Eq. (17–31) becomes

$$\frac{E_2}{E_1} = \frac{M}{L_1} = N.$$

Thus for unity-coupled coils the output voltage is N times the input voltage, where $N = \sqrt{L_2/L_1}$. This value of N is identical to the one assumed for the ideal transformer of Fig. 17–10(c), and we can therefore conclude that this equivalent circuit is correct for the secondary as well as for the primary terminals. In summary, a pair of unity-coupled coils is equivalent to an ideal transformer with a turns-ratio equal to the square

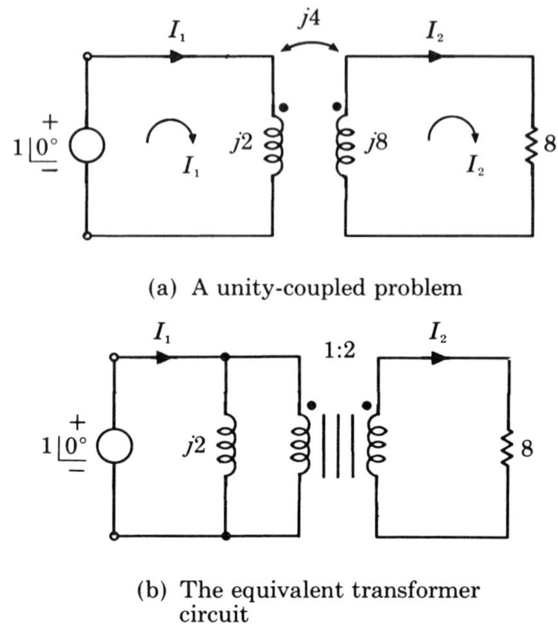

FIG. 17–11. The solution of a problem with unity coupling.

root of the ratio of the secondary and primary inductances, and a shunt inductance in the primary circuit of the same value as the primary inductance itself. An ideal transformer would not have this shunt inductance. Therefore, we can say that an *ideal transformer is a pair of unity-coupled coils in which the inductance values themselves are infinite.* In practice, unity-coupled coils behave like ideal transformers provided that the primary inductance is large enough to make a negligible path in parallel with the reflected load impedance.

An example of a unity-coupled problem is shown in Fig. 17–11(a). The primary reactance is 2 ohms, the secondary reactance is 8 ohms, and the mutual reactance is 4 ohms. The coefficient of coupling is

$$K = \frac{M}{\sqrt{L_1 L_2}} = \frac{4}{\sqrt{16}} = 1.$$

The equivalent transformer-type circuit is shown in Fig. 17–11(b). The primary shunt reactance is $j2$ ohms. The turns-ratio of the ideal transformer is

$$N = \sqrt{\frac{j\omega L_2}{j\omega L_1}} = \sqrt{\frac{j8}{j2}} = 2.$$

The impedance reflected into the primary through the ideal transformer is

$$Z_f = \frac{Z}{N^2} = \frac{8}{4} = 2 \text{ ohms.}$$

The current drawn from the source is thus $1/j2$ amp. through the shunt inductance and $\frac{1}{2}$ ampere through the reflected load impedance. The total current is therefore

$$I_1 = \tfrac{1}{2} - j\tfrac{1}{2} \quad \text{amp.}$$

The output voltage in the secondary is the turns-ratio times the primary voltage:

$$E_2 = 1 \times 2 = 2 \text{ volts.}$$

As a check on the results which we have obtained from the equivalent circuit of Fig. 17–11(b), we will solve the mesh equations for the original circuit of Fig. 17–11(a). The equations are

$$I_1(j2) - I_2(j4) = 1,$$

$$-I_1(j4) + I_2(8 + j8) = 0,$$

from which

$$I_1 = \frac{\begin{vmatrix} 1 & -j4 \\ 0 & 8 + j8 \end{vmatrix}}{\begin{vmatrix} j2 & -j4 \\ -j4 & 8 + j8 \end{vmatrix}} = \tfrac{1}{2} - j\tfrac{1}{2} \quad \text{amp,}$$

$$I_2 = \frac{\begin{vmatrix} j2 & 1 \\ -j4 & 0 \end{vmatrix}}{\begin{vmatrix} j2 & -j4 \\ -j4 & 8 + j8 \end{vmatrix}} = \tfrac{1}{4} \text{ amp.}$$

The output voltage is

$$E_2 = 8 \times I_2 = 2 \text{ volts.}$$

The answers which we have obtained by the mesh-current method are the same as the answers which we obtained from the transformer-type of equivalent circuit. The transformer solution was a good deal easier than the mesh-current solution, and it is preferred for unity-coupled coils, and for coupled coils in general when the coefficient of coupling is high. In the next section we will develop the transformer-type of equivalent circuit for coils which are not unity-coupled, and in the process we will discover the physical meaning of the turns-ratio N.

17–6 Transformers. Since unity-coupled coils lead to simpler computations than self and mutual inductances, it is often worth while replacing self and mutual inductances by an equivalent transformer-type of circuit. The more nearly the actual coils approach an ideal transformer, the more attractive is the substitution. Most iron-core transformers are very close to ideal transformers, and they are best treated as ideal transformers with finite shunt inductance, and with leakage reactances in the primary and

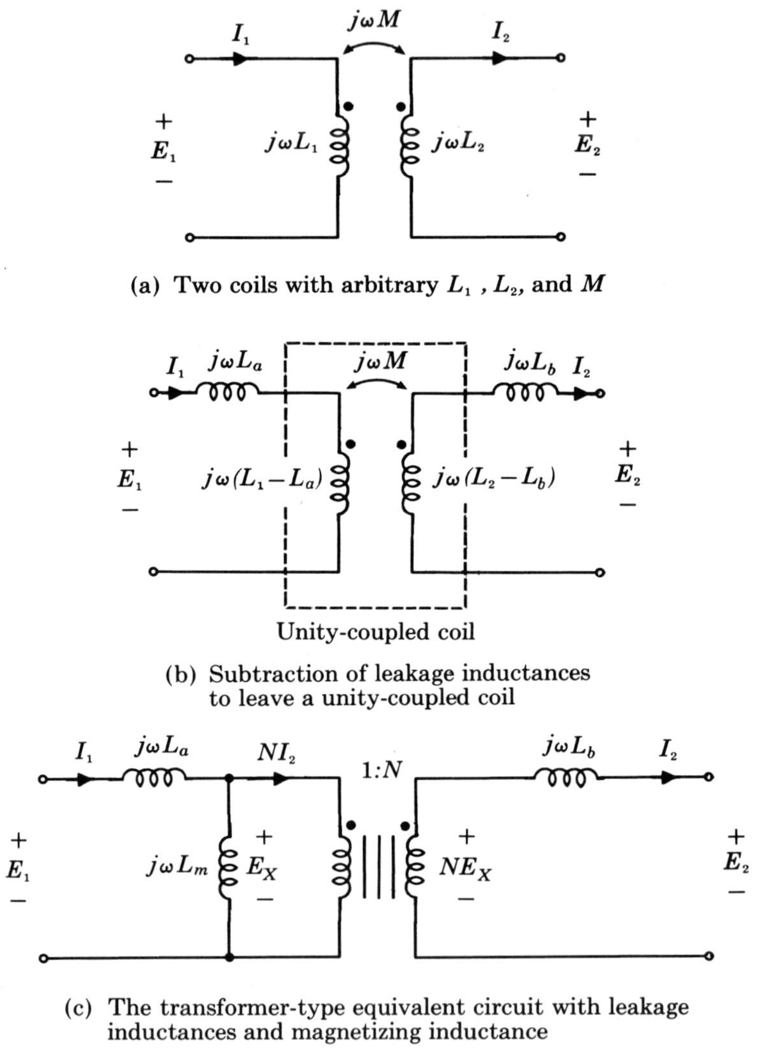

FIG. 17–12. The transformer-type equivalent circuit for coils with arbitrary coupling.

secondary windings. Indeed, it is almost a necessity to treat them this way since the primary and secondary inductances themselves are very nonlinear and could not be handled by ordinary linear circuit theory.

When the coefficient of coupling is less than unity, the mutual inductance is smaller than the square root of the product of the two self inductances. Effectively, the self-inductances are too large to form a unity-coupled coil. We can remedy this situation by subtracting series inductances from each of the windings until the remaining inductances form a unity-coupled coil with the given mutual inductance. The inductances which must be subtracted are called the *leakage inductances* of the windings. These leakage inductances represent the magnetic effects which belong to the individual coils and which are not associated with the mutual inductance, or the mutual flux which links them.

For an iron-core inductance the leakage flux is mostly in the air surrounding the inductor, and is therefore essentially linear. After the leakage inductances are removed, the remainder of the transformer consists of a unity-coupled coil, and the unity-coupled coil can be replaced by an ideal transformer with a shunt inductance. For an iron-core transformer the shunt inductance is nonlinear because of the nonlinear properties of the iron, but it is a large inductance which draws very little current and has very little effect on the operation of the transformer. The iron-core transformer is therefore almost linear in its operation, and is usually treated by linear circuit theory. The fact that the theory which we are about to develop can be applied to iron-core transformers is a major justification for its introduction at this point.

Two coupled coils with arbitrary values of L_1, L_2, and M are shown in Fig. 17–12(a). Leakage inductances of L_a and L_b are subtracted from the primary and secondary inductances respectively in order to leave a unity-coupled coil, as shown in Fig. 17–12(b). We will find the values which we must give to L_a and L_b later. At the moment we will suppose that we have a unity-coupled coil and that we can replace it by an ideal transformer with a turns-ratio of N and a shunt reactance $j\omega L_m$, as shown in Fig. 17–12(c). If the two circuits of Figs. 17–12(a) and 17–12(c) are to be equivalent, they must satisfy the same equations relating the terminal currents and voltages. The equations for the circuit of Fig. 17–12(a) are

$$\begin{aligned} E_1 &= j\omega L_1 I_1 - j\omega M I_2, \\ E_2 &= -j\omega M I_1 + j\omega L_2 I_2, \end{aligned} \tag{17–33}$$

and the equations for the circuit of Fig. 17–12(c) are

$$\begin{aligned} E_1 &= j\omega L_a I_1 + E_x = (j\omega L_a + j\omega L_m) I_1 - j\omega L_m N I_2, \\ E_2 &= N E_x + j\omega L_b I_2 = -j\omega L_m N I_1 + (j\omega L_m N^2 + j\omega L_b) I_2. \end{aligned} \tag{17–34}$$

If Eqs. (17–33) and (17–34) are to be identical, the corresponding coefficients must be identical. Thus

$$\begin{aligned} L_1 &= L_a + L_m, \\ NL_m &= M, \\ L_2 &= N^2 L_m + L_b. \end{aligned} \tag{17–35}$$

The quantities L_1, L_2, and M in Eqs. (17–35) can be considered to be known, with the quantities L_a, L_b, L_m, and N to be determined from them. Since there are four unknowns and only three equations, one of the unknowns can be picked arbitrarily. If we select a value for N and solve for the other three constants we obtain

$$\begin{aligned} L_a &= L_1 - \frac{M}{N}, \\ L_m &= \frac{M}{N}, \\ L_b &= L_2 - MN. \end{aligned} \tag{17–36}$$

Algebraically, Eqs. (17–36) are true for all values of N. However, if we wish to have the leakage inductances positive we must make N greater than M/L_1 and smaller than L_2/M. Any value of N in this range will be satisfactory. Thus

$$\frac{M}{L_1} \leq N \leq \frac{L_2}{M}.$$

For a number of reasons the geometric mean of the two extreme values is usually selected for N. Thus we choose to use

$$N = \sqrt{\frac{L_2}{L_1}}. \tag{17–37}$$

One advantage of this choice is that it agrees with the value of N which we used for the unity-coupled coils. Another advantage is that the N defined in Eq. (17–37) can be interpreted as the geometrical turns-ratio. Assuming that the same magnetic flux links all the turns of a coil, the total inductance of the coil is proportional to the square of the number of turns. Thus the ratio of the square root of the inductance of the secondary to the square root of the inductance of the primary is equal to the ratio of the number of turns on the secondary to the number of turns on the primary.

The name "turns-ratio" comes from the consideration which we have just discussed, although strictly speaking it applies only to unity-coupled coils in which the same flux links every turn. In a practical transformer

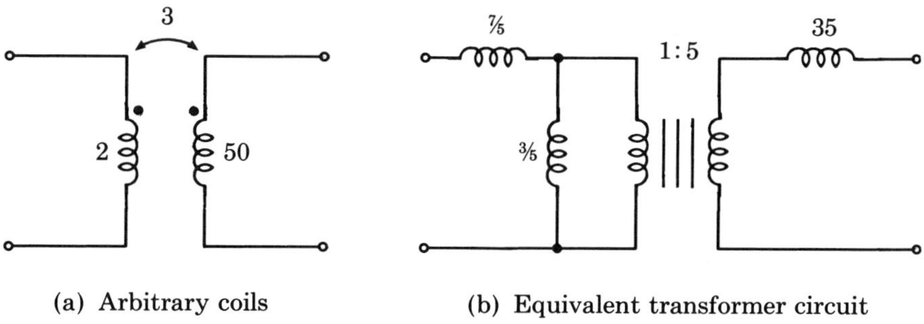

FIG. 17–13. An example of a general transformer-type equivalent circuit.

the turns-ratio is usually obtained by measuring the inductances, rather than by counting the turns, although the ratio of the turns will be approximately correct. An interesting corollary of the definition of N given by Eq. (17–37) is that the leakage inductances referred to the same side of the transformer are equal, since

$$L_b = N^2 L_a. \tag{17–38}$$

Some texts treat Eq. (17–38) as an experimentally verified postulate. Actually, it is a consequence of the definition of N. If a different definition of N were used, a different ratio of inductances would be obtained.

Let us now consider a numerical example in which the transformer-type of circuit is developed from self and mutual inductances. Two coupled coils are shown in Fig. 17–13(a). They have the parameters

$$L_1 = 2 \text{ henrys},$$

$$L_2 = 50 \text{ henrys},$$

$$M = 3 \text{ henrys}.$$

The equivalent turns-ratio is

$$N = \sqrt{\frac{L_2}{L_1}} = \sqrt{25} = 5.$$

The primary leakage inductance is

$$L_a = L_1 - \frac{M}{N} = 2 - \tfrac{3}{5} = \tfrac{7}{5} \text{ henrys}.$$

The secondary leakage inductance is

$$L_b = L_2 - MN = 50 - 3 \times 5 = 35 \text{ henrys}.$$

The shunt inductance (also called the magnetizing inductance) is

$$L_m = \frac{M}{N} = \tfrac{3}{5} \text{ henrys.}$$

The equivalent transformer circuit is shown in Fig. 17–13(b). The coefficient of coupling of the coils is low (0.3), and therefore the leakage inductances are large. This example is given to emphasize the fact the transformer-type of equivalent circuit is not restricted to the case of approximately unity coupling, but can be found for any pair of coupled coils. For iron-core transformers, the leakage inductances are much smaller and the shunt inductance is much larger. For approximate calculations it is often possible to neglect both the leakage inductances and the shunt inductances and to treat the iron-core transformer as an ideal transformer.

The subject of iron-core transformers is beyond the scope of this text, since it becomes involved with the nonlinear properties of iron cores and the practical problems of size, weight, materials, cost, and cooling. In this chapter we have attempted merely to provide a bridge between the coupled-circuit theory which is used for air-core coils, and the transformer theory which is used for iron-core coils. The student will have to be familiar with both theories in order to apply the more suitable one in any given problem.

17–7 Summary.

Mutual inductance

Time domain

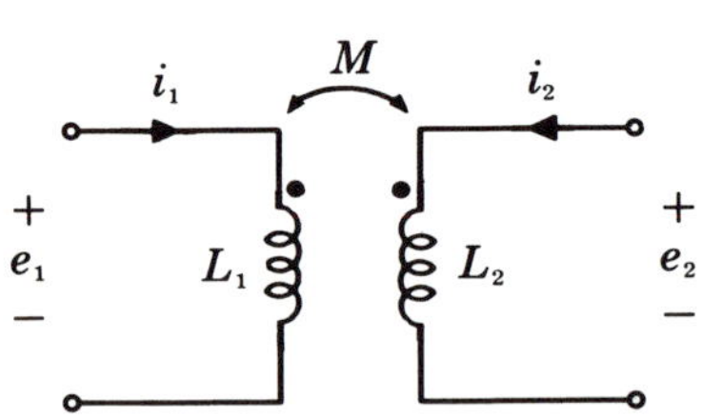

$$e_1 = L_1 \frac{di_1}{dt} + M \frac{di_2}{dt}$$

$$e_2 = M \frac{di_1}{dt} + L_2 \frac{di_2}{dt}$$

Frequency domain

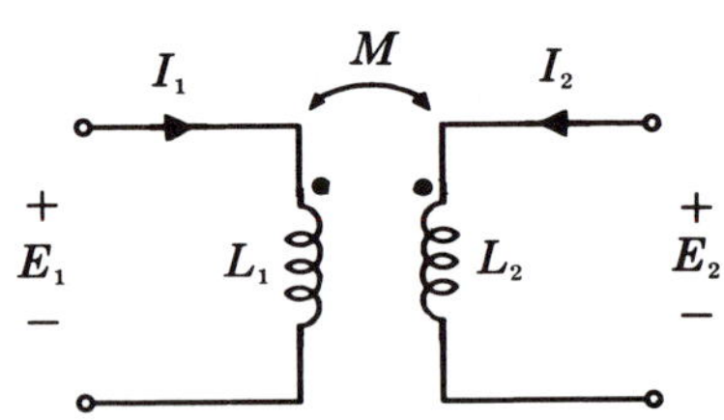

$$E_1 = j\omega L_1 I_1 + j\omega M I_2$$

$$E_2 = j\omega M I_1 + j\omega L_2 I_2$$

Series inductances

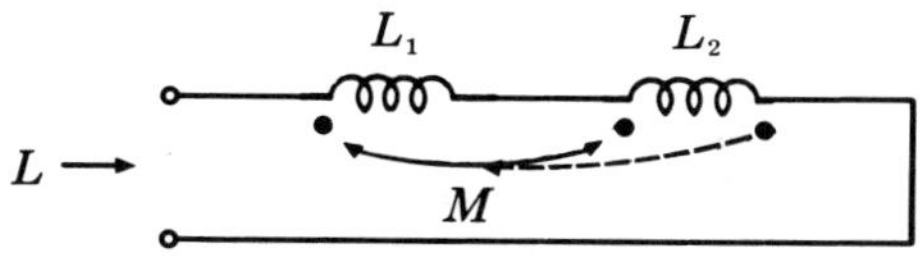

$$L = L_1 + L_2 \pm 2M$$
(+ sign for solid line)

Parallel inductances

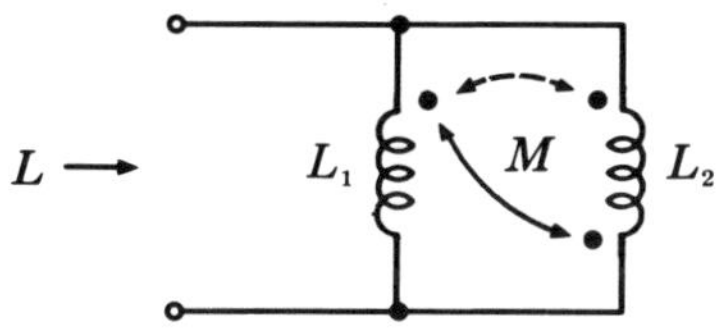

$$L = \frac{L_1 L_2 - M^2}{L_1 + L_2 \pm 2M}$$
(+ sign for solid line)

Coefficient of coupling

$$K = \frac{M}{\sqrt{L_1 L_2}} \qquad 0 \leq K \leq 1$$

Reflected impedance

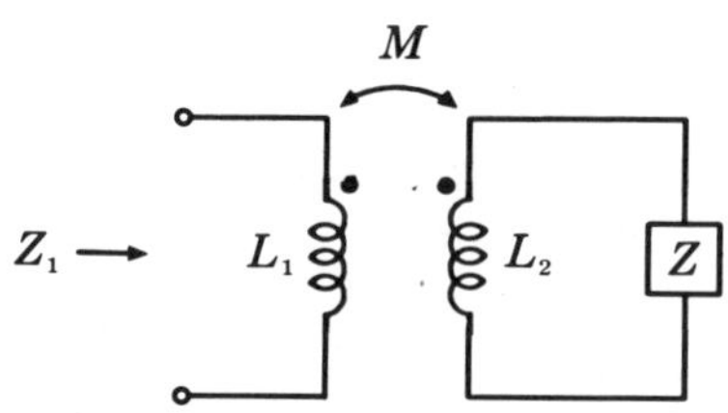

$$Z_1 = j\omega L_1 + \frac{\omega^2 M^2}{j\omega L_2 + Z}$$

Ideal transformer

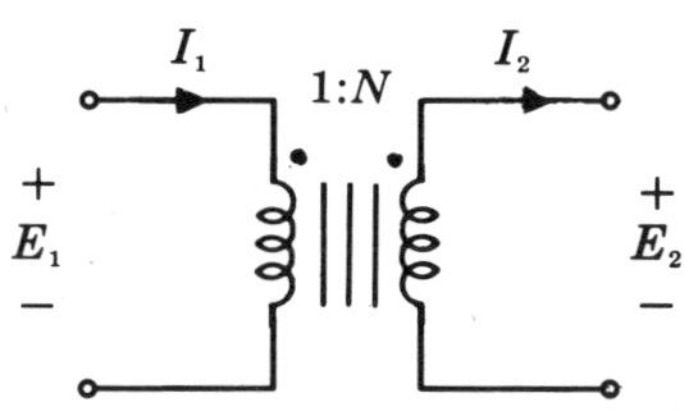

$$E_2 = NE_1$$
$$I_2 = I_1/N$$

Unity-coupled coils ($L_1L_2 - M^2 = 0$)

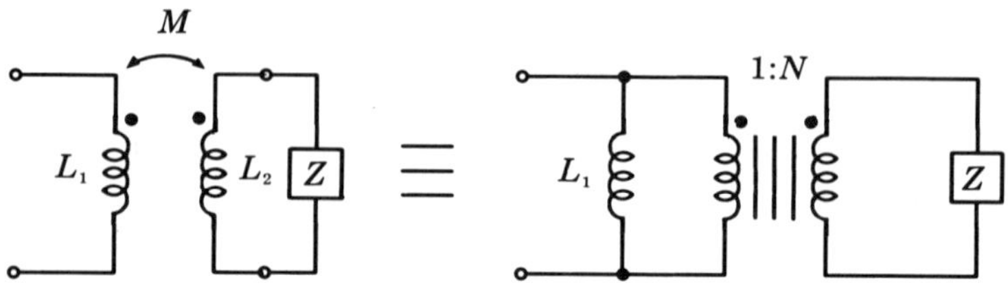

$$Y_1 = \frac{1}{j\omega L_1} + \frac{1}{Z/N^2} \quad \text{with} \quad N = \sqrt{\frac{L_2}{L_1}} = \frac{M}{L_1} = \frac{L_2}{M}$$

Real transformer

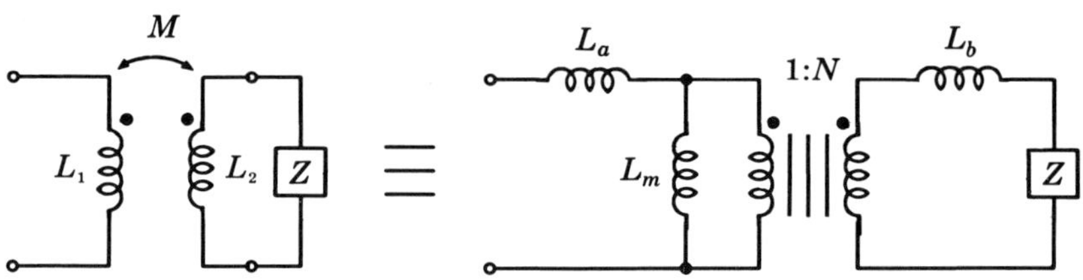

$$L_a = L_1 - M/N$$
$$L_b = L_2 - MN \qquad \text{where} \quad N = \sqrt{\frac{L_2}{L_1}}$$
$$L_b = N^2 L_a$$
$$L_m = M/N$$

REFERENCES

Two distinct view-points will be found in the literature concerning transformers. Some authors treat only self and mutual inductances. Other authors treat only leakage reactances and ideal transformers. In general the self and mutual inductances are found in communications theory texts, and the leakage reactances and ideal transformers in power and machinery texts. An attempt has been made in this chapter to present both points of view and to show the relationship between them. References to both points of view are given below.

1. E. A. GUILLEMIN, *Introductory Circuit Theory,* John Wiley & Sons, New York, 1953. Guillemin presents the general theory of coupled coils. He uses a matrix method to handle the polarities of the mutual inductances rather than the more common dot notation.

2. MIT EE Staff, *Magnetic Circuits and Transformers*, John Wiley & Sons, New York, 1943. This text is a very complete reference for the presentation of leakage reactances and ideal transformers. The equivalent circuits are derived from component fluxes rather than from self and mutual inductances. Practical problems of materials, heating, size, and cost are also included.

3. KERCHNER and CORCORAN, *Alternating Current Circuits,* John Wiley & Sons, New York, 3rd ed., 1951. Kerchner and Corcoran give an excellent treatment of the self- and mutual-inductance point of view. In addition to a general treatment of the dot convention they discuss a variety of practical tuned-transformer problems.

4. BRENNER and JAVID, *Analysis of Electrical Circuits,* McGraw-Hill, New York, 1959. Brenner and Javid give both points of view and explain the arbitrary choice of the turns-ratio, in the general case.

5. CLEMENT and JOHNSON, *Electrical Engineering Science,* McGraw-Hill, New York, 1960. Clement and Johnson consider component fluxes in a manner similar to that of Reference 2.

6. H. H. SKILLING, *Electrical Engineering Circuits,* John Wiley & Sons, New York, 1957. Skilling presents both of the points of view for transformers and coupled circuits and also exposes the arbitrariness of the "turns-ratio" in the general case.

7. FITZGERALD and HIGGINBOTHAM, *Basic Electrical Engineering,* McGraw-Hill, New York, 2nd ed., 1957. Chapters 8 and 9 give an excellent and somewhat simplified summary of the material of Reference 2.

EXERCISES*

Section 17–2

1. Two coils have self-inductances of $L_1 = 6$ henrys and $L_2 = 4$ henrys. The mutual inductance between them is 3 henrys.
   (a) Give the two values of inductance which can be obtained by connecting them in series.
   (b) Give the two values of inductance which can be obtained by connecting them in parallel.

2. Two coils have self-inductances of $L_1 = 6$ henrys and $L_2 = 4$ henrys. The mutual inductance between them is 3 henrys.
   (a) Find the inductance which is obtained at the terminals of L_1 if L_2 is shorted.
   (b) Find the inductance which is obtained at the terminals of L_2 if L_1 is shorted.

3. Two coils have self-inductances of $L_1 = 8$ henrys and $L_2 = 6$ henrys. The mutual inductance between them is 2 henrys. Find all the possible values of inductance which can be obtained by connecting them in various ways.

4. Two coils have a primary inductance of 1 henry and a secondary inductance of $\frac{1}{4}$ henry. The coefficient of coupling between them is $\frac{1}{2}$. A sinusoidal current $i(t) = 10 \cos t$ is flowing in the primary circuit and the secondary is open-circuited.
   (a) Give the voltage across the primary coil.
   (b) Give the voltage across the secondary coil.

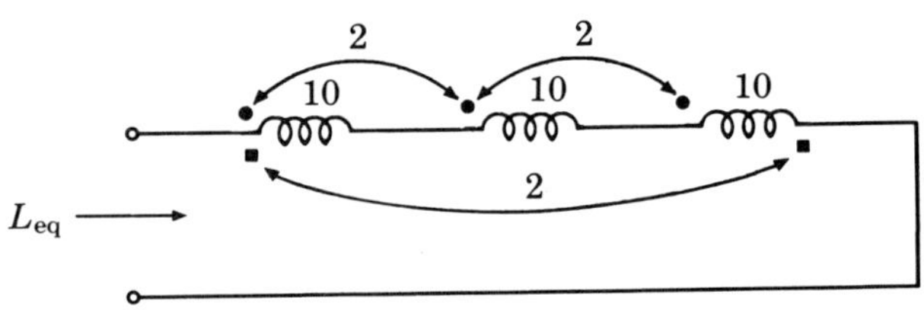

FIGURE 17–14

5. Find the net inductance of the three series-connected coils shown in Fig. 17–14. Note that the polarity dots on the three coils are not consistent and therefore must be represented by separate symbols.

6. Write the mesh-current equations for the circuit of Fig. 17–15 and find the magnitude and phase of the output voltage E_2.

7. In the circuit of Fig. 17–16 the coefficient of coupling between the coils is $K = \frac{1}{2}$. Find the output voltage E_2 in magnitude and phase.

---

* The units on the diagrams are ohms, henrys, farads, volts, and amperes. Complex numbers are in ohms.

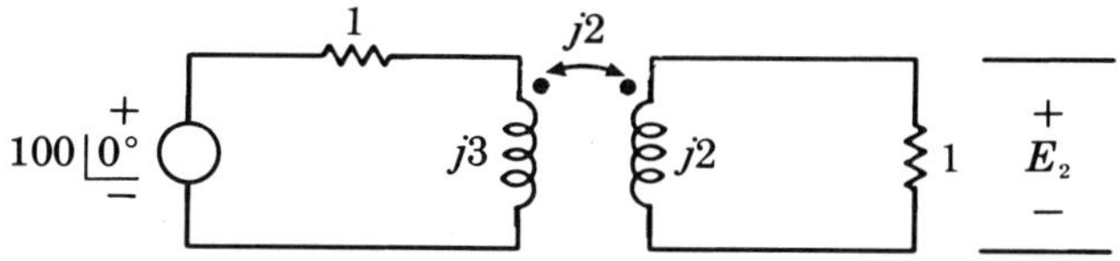

FIGURE 17–15

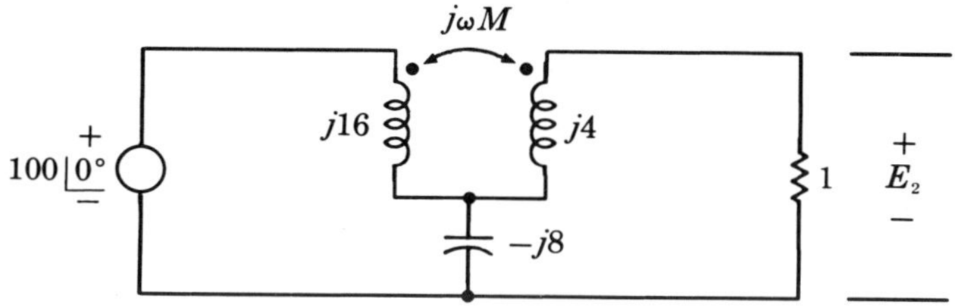

FIGURE 17–16

8. Find the input impedance Z of the circuit in Fig. 17–17.

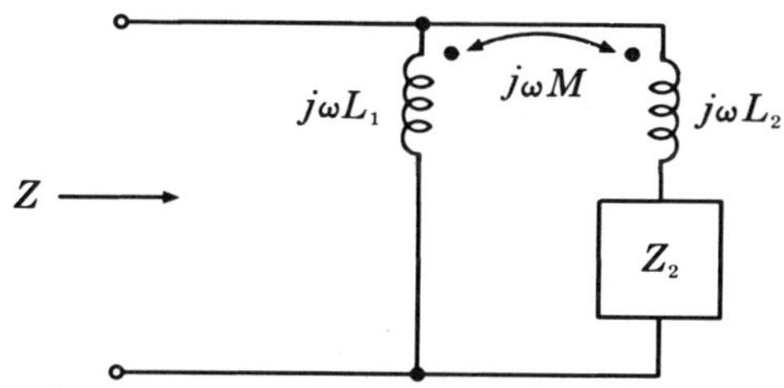

FIGURE 17–17

9. Find the input impedance Z of the circuit in Fig. 17–18.

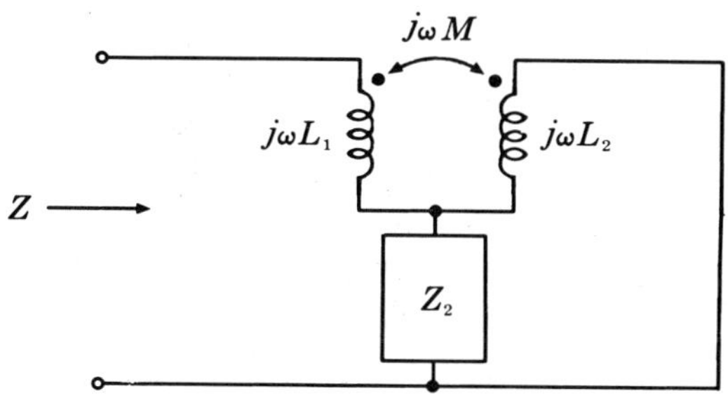

FIGURE 17–18

10. By means of mesh-current equations find the current I_R in the circuit of Fig. 17–19, in the form of a magnitude and a phase angle.

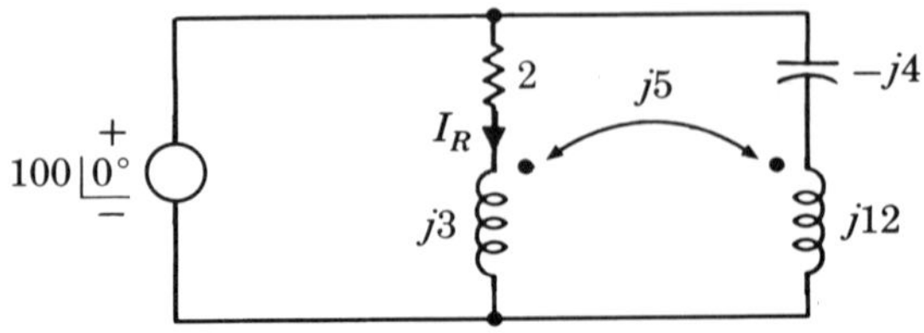

FIGURE 17–19

Section 17–3

11. Two coupled coils have a primary reactance of 20 ohms, a secondary reactance of 2000 ohms, and a coefficient of coupling of $\frac{1}{2}$.

(a) What is the impedance at the primary terminals with the secondary open-circuited?

(b) Give the impedance at the secondary terminals with the primary open-circuited.

(c) What is the primary input impedance with the secondary short-circuited?

(d) What is the secondary input impedance with the primary short-circuited?

(e) Give the input impedance at the primary terminals if there is a 2000-ohm resistance across the secondary terminals.

12. Find the net inductance at terminal-pair *a-b* in the circuit of Fig. 17–20. [*Hint:* apply a voltage source of frequency $\omega = 1$ and compute the input impedance.]

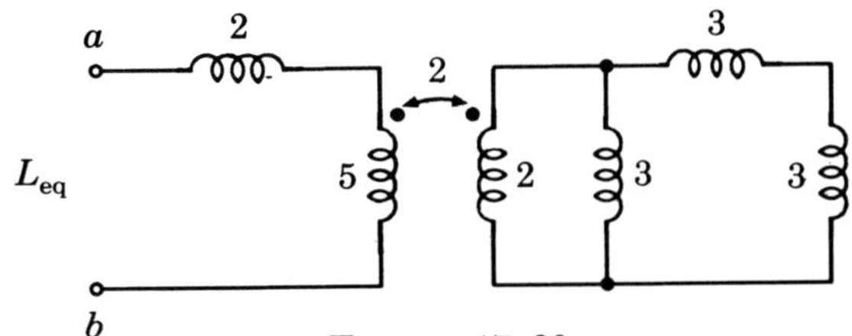

FIGURE 17–20

13. Find the input impedance at terminal-pair *a-b* of the circuit of Fig. 17–21.

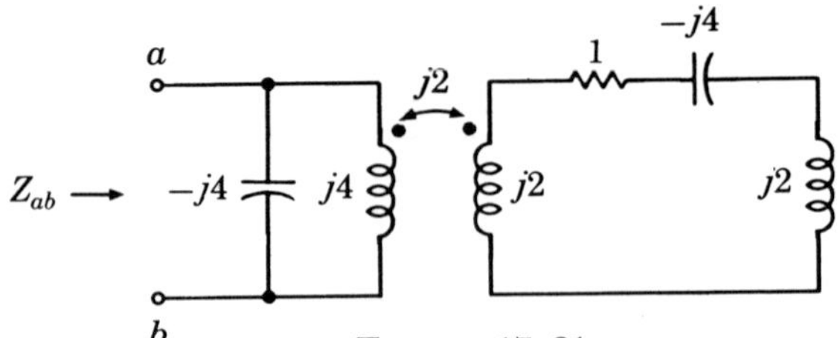

FIGURE 17–21

14. For the circuit of Fig. 17–22:
   (a) Find the Thévenin equivalent circuit at terminals *a-b*.
   (b) Find the current which would flow in a short circuit placed across terminals *a-b*.

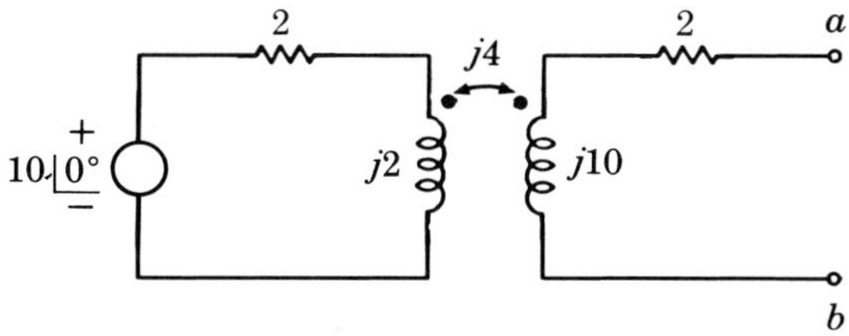

FIGURE 17–22

15. Find the Norton equivalent of the circuit in Fig. 17–23, in the form of a short-circuit current source and an equivalent impedance in parallel with it.

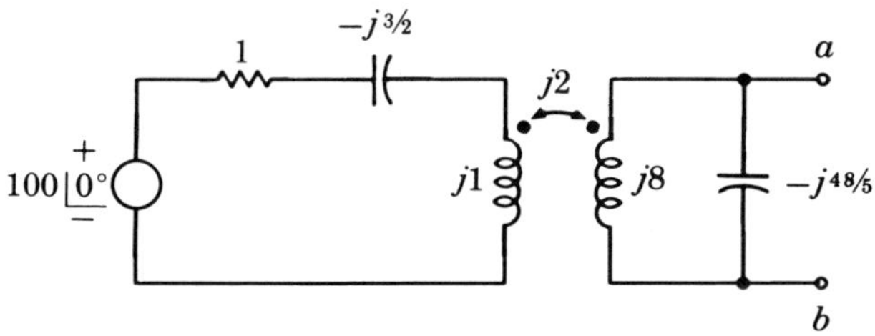

FIGURE 17–23

16. (a) In the circuit of Fig. 17–24 find the value of the capacitance C which will cause the net reactance in the primary circuit to become zero. (This condition is series resonance.)
   (b) Find the voltage E_2 for the value of C found in part (a) and convert it to a time function $e_2(t)$.
   (c) Show that the voltage obtained in part (b) is the maximum that can be attained by varying C.

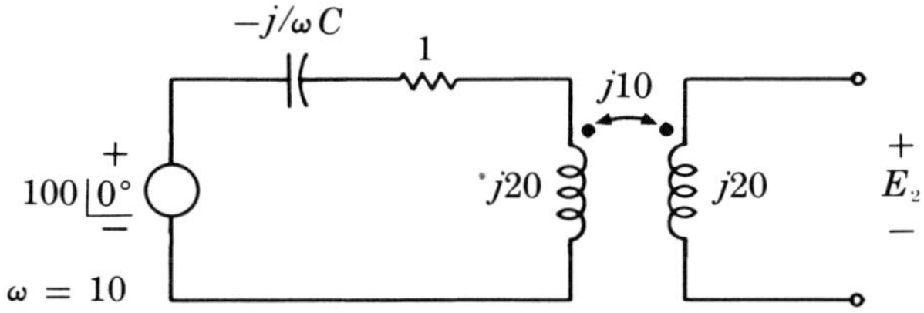

FIGURE 17–24

17. In the circuit of Fig. 17–25 find the value of the reactance X_C which will give a maximum magnitude to the voltage E_2. Give E_2 in magnitude and phase for this condition.

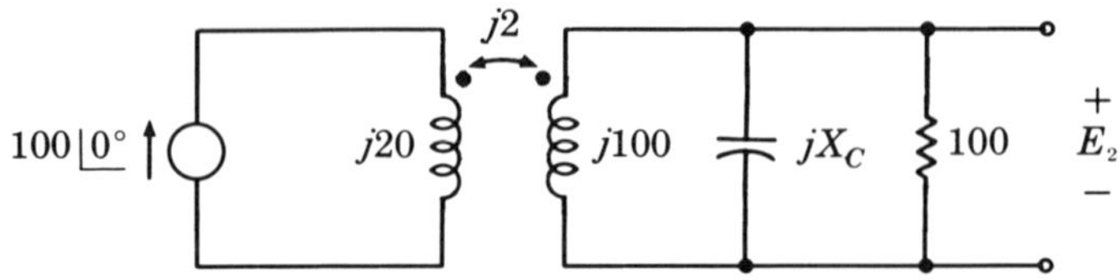

FIGURE 17–25

18. It is possible to represent a pair of coupled coils, like those of Fig. 17–26, by a set of inverse inductance parameters. These parameters are defined by the following equations:

$$I_1 = \frac{\Gamma_1}{j\omega} E_1 + \frac{\Gamma_m}{j\omega} E_2,$$

$$I_2 = \frac{\Gamma_m}{j\omega} E_1 + \frac{\Gamma_2}{j\omega} E_2.$$

Find the values of the inverse inductances (gamma's) in terms of the self- and mutual inductances of the coils.

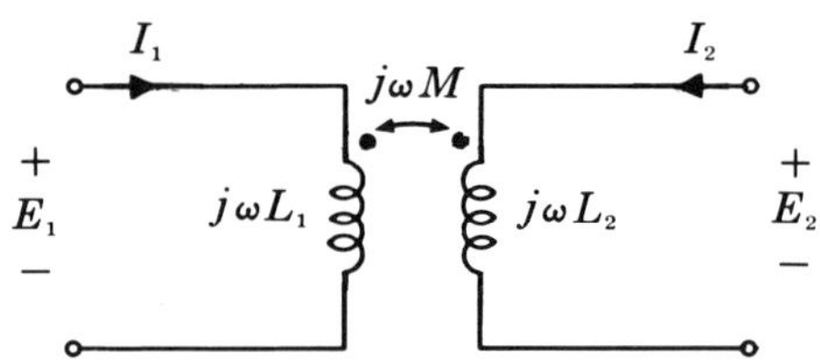

FIGURE 17–26

19. (a) For the circuit of Fig. 17–27 find the inverse inductances and write the general equations.
(b) Find the input impedance with the output voltage equal to zero.
(c) Find the impedance at the output terminals with the input voltage equal to zero.
(d) Check the answers to parts (b) and (c) by the reflected-impedance method.

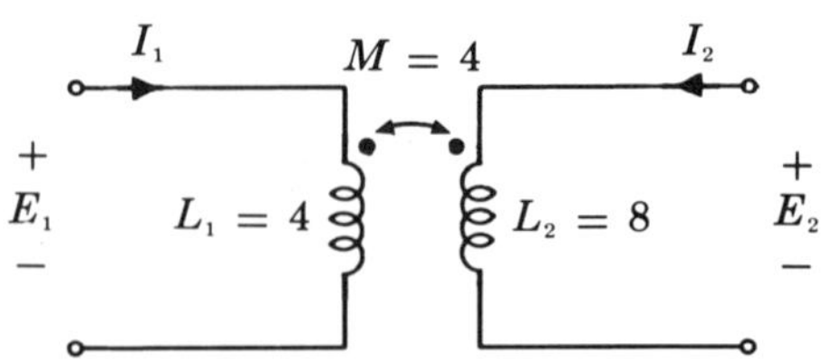

FIGURE 17–27

Section 17–4

20. The transformer in the circuit of Fig. 17–28 is "ideal." Solve for the voltage E_3 (a) by writing two mesh equations and two constraint equations and (b) by using an equivalent circuit referred to the output side of the transformer.

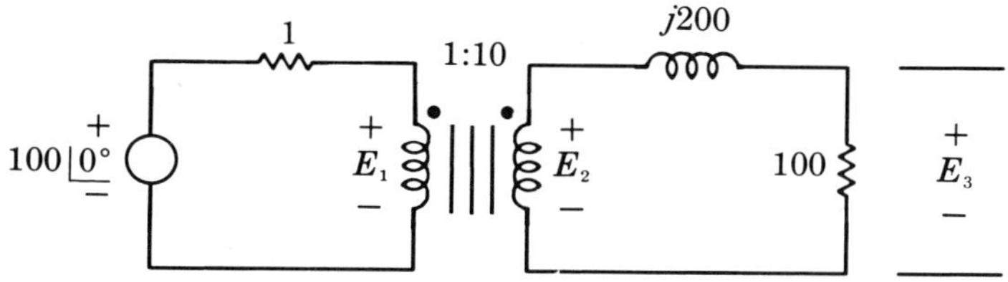

FIGURE 17–28

21. Find the input voltage E_1 in the circuit of Fig. 17–29 (a) by writing two node equations and two constraint equations and (b) by using an equivalent circuit referred to the input side of the ideal transformer.

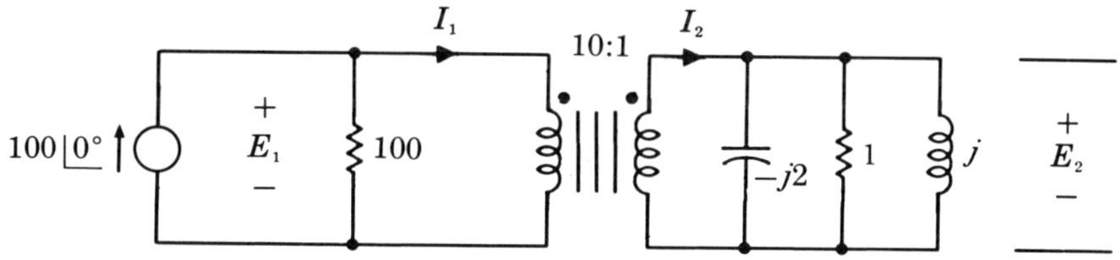

FIGURE 17–29

22. In the circuit of Fig. 17–30 the transformer is ideal. Find the current I_1 drawn from the source.

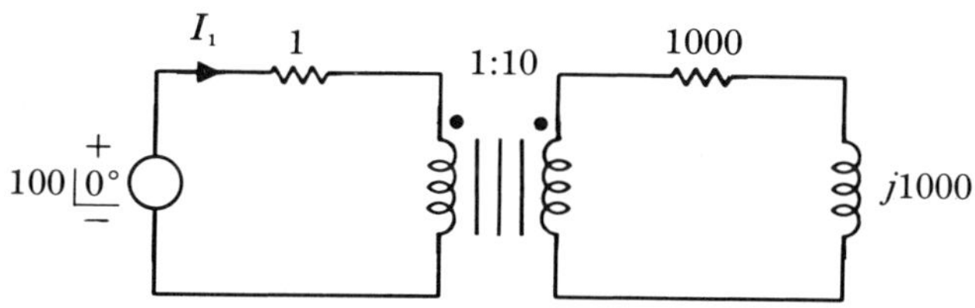

FIGURE 17–30

23. In the circuit of Fig. 17–31 an ideal transformer is driven by a current source. Find the output voltage E_2 in magnitude and phase.

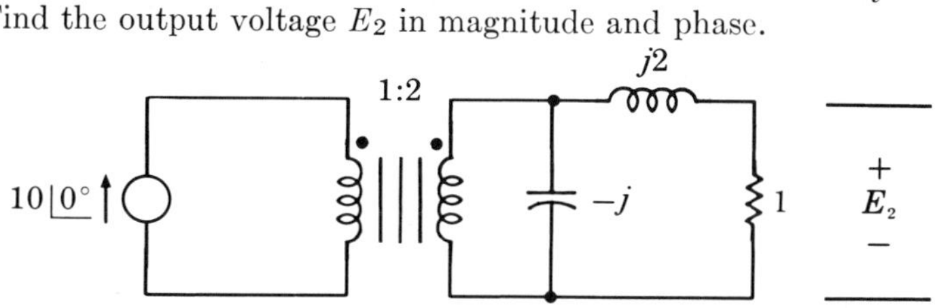

FIGURE 17–31

24. The transformer in the circuit of Fig. 17–32 is ideal. Find the voltage E_1 which is across its primary side.

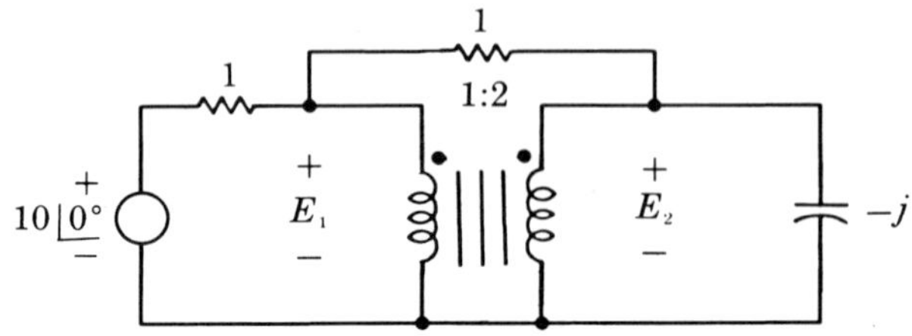

FIGURE 17–32

25. The transformer in the circuit of Fig. 17–33 is ideal. Find the current I_1 from the source.

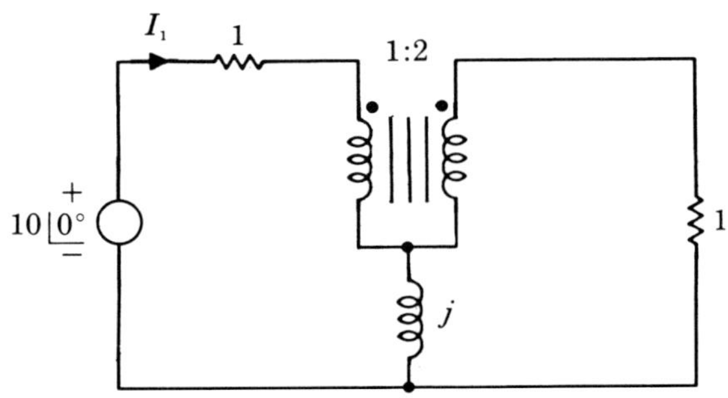

FIGURE 17–33

26. In the circuit of Fig. 17–34 the transformer is ideal. (a) Select the turns-ratio which will result in the maximum power being delivered to the load resistance R. (b) Give the value of the maximum power in R.

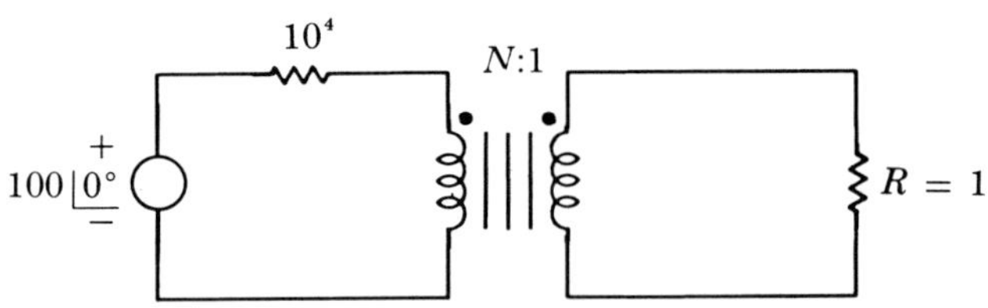

FIGURE 17–34

27. (a) Find the turns-ratio of the ideal transformer in the circuit of Fig. 17–35 which will result in maximum power being delivered to the resistance R. (b) Give the value of this maximum power.

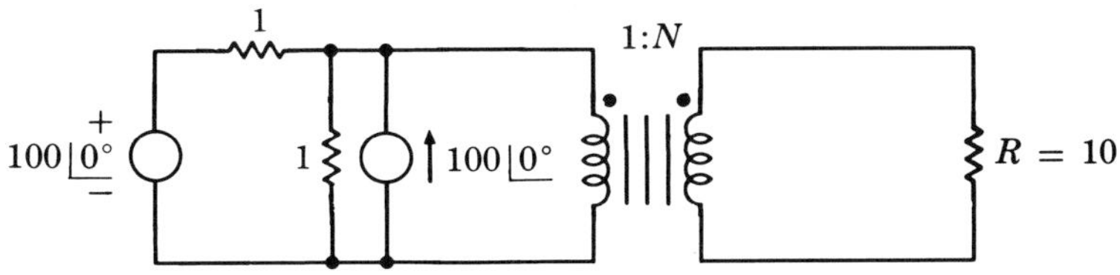

FIGURE 17–35

Section 17–5

28. Two coupled coils have a primary inductance of 10 henrys, a secondary inductance of 1 henry, and unity coupling.

(a) Give the two possible values of the inductance which can be obtained by connecting them in series.

(b) If the coils are connected in parallel what inductance values are obtained?

29. Two coupled coils have a primary inductance of 10 henrys, a secondary inductance of 1 henry, and unity coupling.

(a) A short circuit is placed across the secondary. What is the inductance at the primary terminals?

(b) A short circuit is placed across the primary terminals. What is the inductance at the secondary terminals?

(c) The secondary is open-circuited. What is the inductance at the primary terminals?

(d) The primary terminals are open-circuited. What is the inductance at the secondary terminals?

30. The coils in the circuit of Fig. 17–36 are unity-coupled.

(a) Solve for the voltage E_2 by making an equivalent primary circuit.

(b) Solve for the voltage E_2 by making an equivalent secondary circuit.

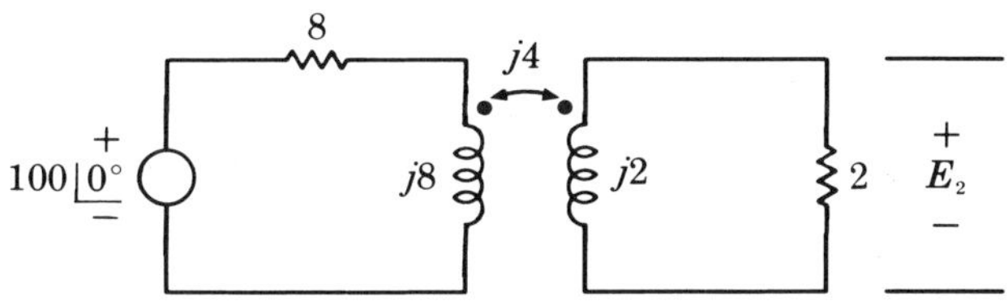

FIGURE 17–36

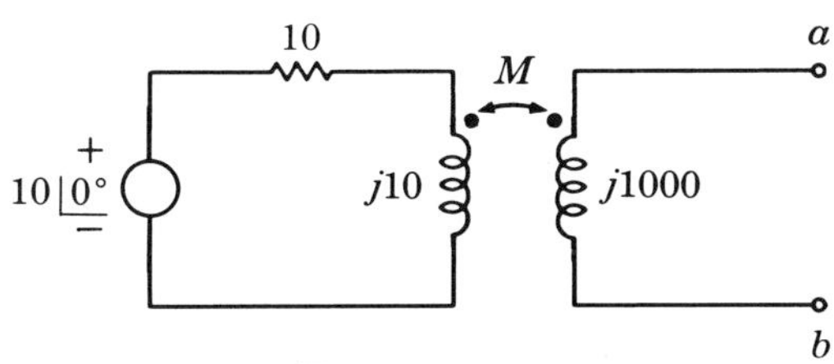

FIGURE 17–37

31. The coils in the circuit of Fig. 17–37 are unity-coupled.
   (a) Make an equivalent Thévenin circuit at terminals *a-b*.
   (b) Find the current which would flow if a short circuit were placed across terminals *a-b*.

32. In the circuit of Fig. 17–38 find the input current I_1 and the output voltage E_2.

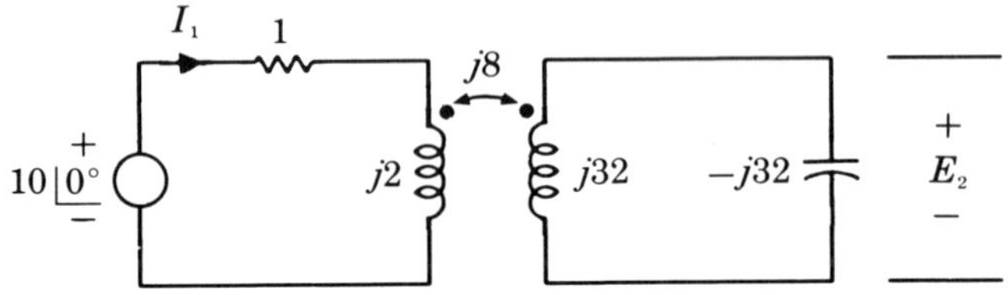

FIGURE 17–38

33. (a) In the circuit of Fig. 17–39 find the reactance X_C which will result in parallel resonance. (b) Compute the voltage E_2 with and without this reactance in place.

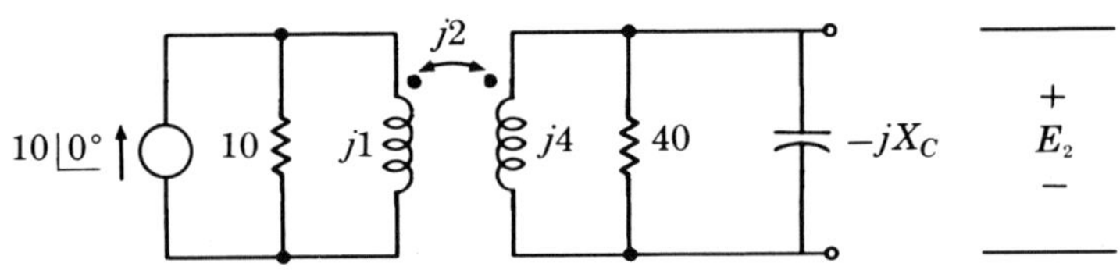

FIGURE 17–39

34. In the circuit of Fig. 17–40 it is desired to tune out the shunting effect of the finite inductance of the unity-coupled coils.
   (a) Find the value X_1 which will accomplish this result and compute the corresponding E_2.
   (b) Find the value of X_2 which will accomplish the same result and compute the corresponding E_2.
   (c) Comment on the need for tuning both primary and secondary of unity-coupled coils.

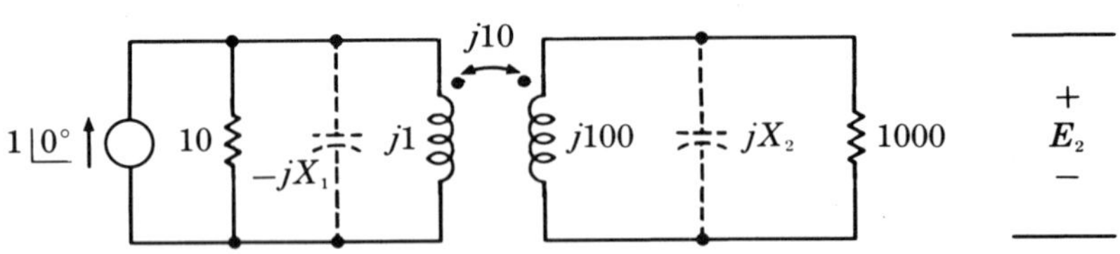

FIGURE 17–40

35. Find the input impedance of the unity-coupled-coil circuit of Fig. 17–41.

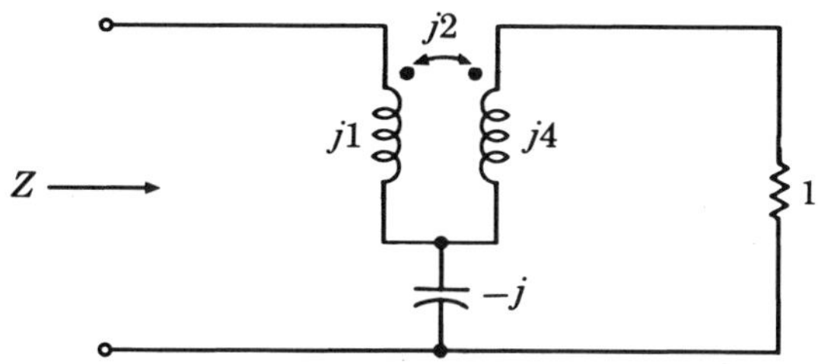

FIGURE 17–41

36. The circuit of Fig. 17–42 is called an "auto-transformer."
   (a) Find the input admittance Y of the circuit, considering that the two coils are unity-coupled.
   (b) Find the voltage ratio E_2/E_1.
   (c) Make an equivalent circuit for the auto-transformer, which includes an ideal transformer.

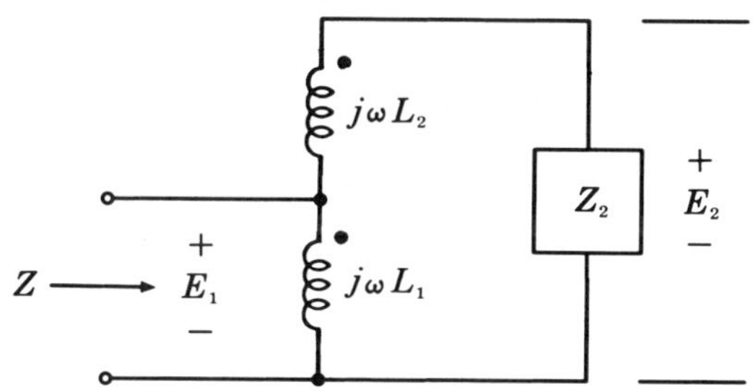

FIGURE 17–42

Section 17–6

37. The parameters of a pair of coupled coils are $L_1 = 1$ henry, $L_2 = 100$ henrys, and $M = 8$ henrys.
   (a) Find a transformer-type equivalent circuit in which the shunt inductance is on the L_1 side.
   (b) Find a transformer-type equivalent circuit in which the shunt inductance is on the L_2 side.

38. The primary winding of a transformer has a reactance X_1 of 100 ohms and a resistance R_1 of 10 ohms. The secondary winding has a reactance X_2 of 25 ohms and a resistance R_2 of 2 ohms. The coefficient of coupling is $\frac{1}{2}$.
   (a) Find an equivalent circuit involving an ideal transformer and having all the elements referred to the primary side.
   (b) Find an equivalent circuit involving an ideal transformer and having all the elements referred to the secondary side.

39. The circuit of Fig. 17–43 represents the equivalent circuit for a transformer. It includes the winding resistances and an extra resistance R_m to account for the energy lost in the iron core. The values of X_m and R_m are much larger than the other elements in the circuit.

(a) Find the input impedance when the output is short-circuited. Obtain an approximate value which utilizes the conditions stated above.

(b) Find the input impedance with the output open-circuited. Again use reasonable approximations.

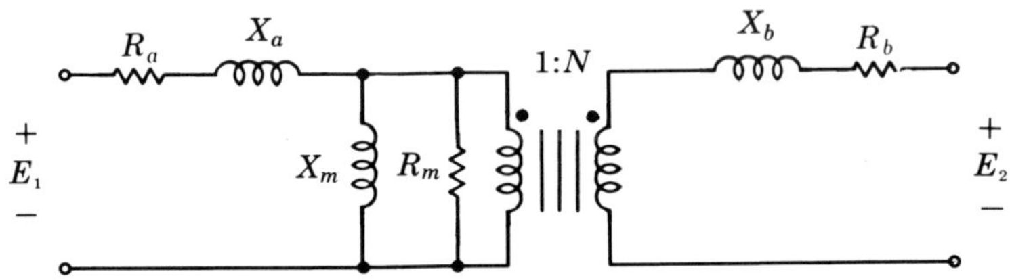

FIGURE 17–43

40. A 10-KVA, 60-cps, 2300/230-volt distribution transformer is tested with the secondary open and with the secondary shorted. The data are as follows:

Open-circuit data
Input voltage — 2300 volts (rated voltage)
Input current — $\frac{1}{10}$ amp
Phase angle — 45° lagging current.

Short-circuit data
Input voltage — 100 volts
Input current — 100 amp (rated current)
Phase angle — 45° lagging current.

Assume that the leakage reactances and the winding resistances are equal when referred to the same side of the transformer. Obtain an equivalent circuit for the transformer referred to the high-voltage winding.

41. An audio-frequency transformer is used to match a 10-ohm speaker coil to a 4000-ohm vacuum-tube source impedance. Referred to the high-voltage winding the transformer has leakage inductances and winding resistances given by

$$L_1 = L_2 = \tfrac{1}{20}\text{ henry},$$
$$R_1 = R_2 = 1\text{ ohm}.$$

The shunt reactance referred to the high-voltage winding is $L_m = 10$ henrys.

(a) Draw an equivalent circuit for the transformer.

(b) At what frequency is the shunt reactance equal in magnitude to the reflected load impedance?

(c) At what frequency is the total leakage reactance equal in magnitude to the reflected load impedance?

(d) Comment on the significance of the frequencies in parts (b) and (c).

42. Convert the circuit of Fig. 17–44 to one involving self- and mutual reactances.

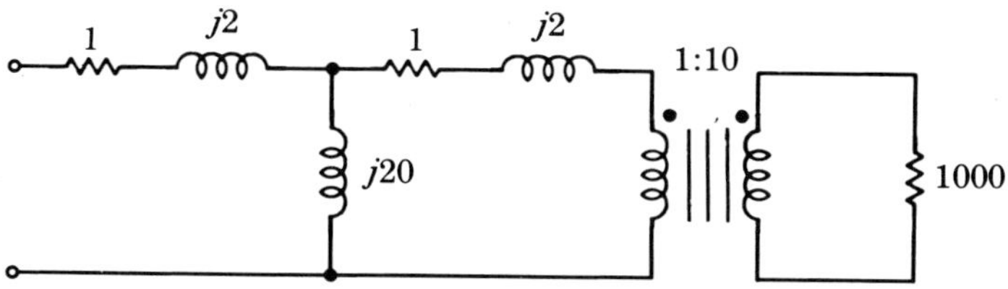

FIGURE 17–44

43. (a) Find the voltage E_2 in the circuit of Fig. 17–45 by writing mesh equations. (b) Convert the circuit to the transformer-type equivalent and again solve for E_2.

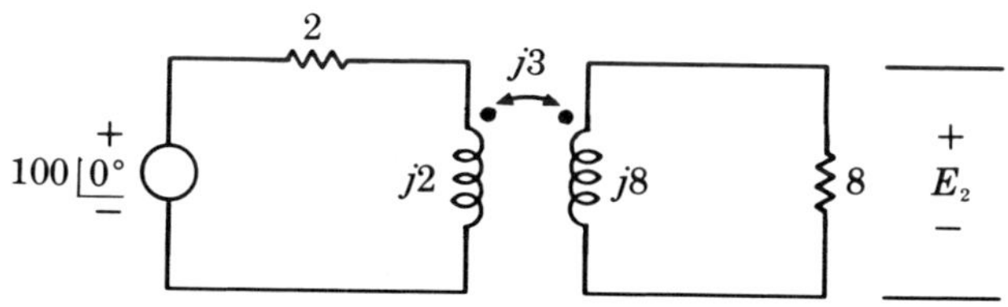

FIGURE 17–45

44. (a) Solve for the voltage E_2 in the circuit of Fig. 17–46 by using mesh equations. (b) Derive the transformer-type equivalent circuit and again find E_2.

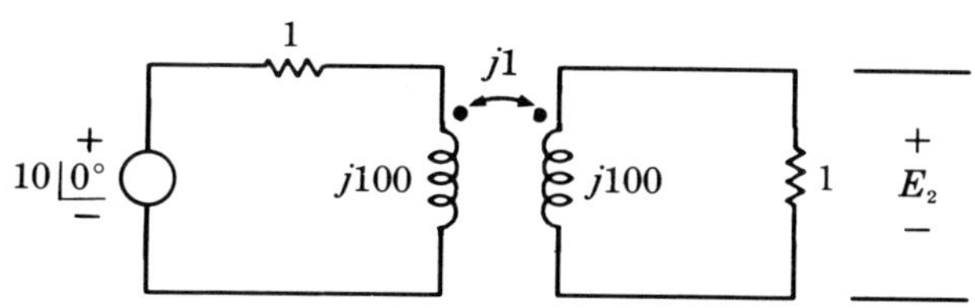

FIGURE 17–46

CHAPTER 18

MAGNITUDE AND PHASE RESPONSE CURVES

"I often say that when you can measure what you are speaking about, and express it in numbers, you know something about it; but when you cannot express it in numbers, your knowledge is of a meager and unsatisfactory kind; it may be the beginning of knowledge, but you have scarcely, in your thoughts, advanced to the stage of a science, whatever the matter may be."

Lord Kelvin, 1883

18–1 Introduction. The reduction of a concept to numerical form gives it a universal meaning which is the same for all men. Furthermore it provides for the possibility of experimental verification by an independent observer. By grounding itself firmly in this principle, science has avoided the semantic pitfalls which beset other areas of human thought and experience. It is not essential that all scientific results be verified experimentally and numerically, but it is essential that they can be verified if anyone desires to do so.

Network theory, as a branch of science, falls in this category. Not all of the stimulus-response relationships we have discussed in the previous chapters need to be verified experimentally, but all of them must be capable of verification if we so desire. Indeed, the form which the theory takes is strongly influenced by the experimental techniques which are available for checking it. In a steady-state a-c problem the output response function has the same frequency as the input stimulus, and differs from it only in the amplitude and the phase of the waveform. In the laboratory it is relatively easy to measure the amplitude ratio and the phase shift, and we conveniently represent them by a complex number of the same magnitude and angle. The components of the complex number in rectangular form are difficult to measure in the laboratory, and are therefore seldom used in theory.*

In order to describe a network completely we need to know its amplitude and phase response at every possible frequency. Two curves of the amplitude and the phase as functions of frequency are sufficient. Such curves can be obtained experimentally, but in this chapter we will be interested in obtaining them analytically. We have already obtained such functions by solving for the stimulus-response ratio with a stimulus of arbitrary

* Several writers have pointed out that real and imaginary parts would be better than magnitude and phase for theoretical work, but no one uses them.

frequency. These functions can be plotted by computing the magnitude and the phase at each frequency. Such a process is very laborious, however, and in this chapter we will develop some simpler methods.

The simple methods for obtaining amplitude and phase curves depend on the concept of "complex frequency." Physically, a complex frequency corresponds to a damped sine wave. Mathematically, it removes each $j\omega$ term from our response functions and replaces it by the simpler variable s. If we excite a network with a damped sinusoid, the response will have a transient term and a steady-state term which is a damped sine wave. The steady-state term can be computed from the sinusoidal response function with $j\omega$ replaced by s. The measurement of the steady-state component in the laboratory might be difficult because of the attendant transient term.* However, we are interested in it only as an analytical tool, and *the computation of a network response for a damped sinusoid is easier than the computation for an undamped sinusoid.* After the result is obtained the damping constant is allowed to become zero and the desired undamped result is obtained.

Two special techniques for utilizing the complex frequency will be discussed. The first of these is the log-db method, so-called because it results in a plot of the magnitude of the function in decibels against the logarithm of the frequency. The second method is called an s-plane vector plot, since the results are obtained from the angles and lengths of a set of vectors in the plane of the complex-frequency variable s. The s-plane method is based on the idea that the steady-state response is a function of the proximity of the applied frequency to the natural resonant frequency of the network. The vectors involved are a measure of the distance of the applied frequency from the natural transient frequency of the network. The log-db method and the s-plane method are complementary in the sense that the log-db method works best for networks which are primarily resistive, and the s-plane method works best for networks which are primarily energy-storage networks and which have pronounced resonance peaks.

18–2 Complex frequency. Our purpose in introducing the concept of complex frequency is to reduce the complexity of frequency-domain algebra. In effect we will replace the frequency variable $j\omega$ by a new complex-frequency variable, s. Although s is a complex variable, its com-

* Mathematically the transient term can be avoided by limiting the excitation to waves which do not decay as rapidly as the natural transients in the circuit. After a while the transients will be negligibly small. (That is, if $s = \sigma + j\omega$, σ is made greater than the abscissa of absolute convergence of the network function. See Chapter 20.) There is nothing to prevent us, however, from examining the *steady-state term* for other values of s.

plex nature is suppressed, and so far as frequency-domain algebra is concerned, it appears to be a real variable. The net result is that the frequency-domain algebra is changed from complex algebra to ordinary algebra, and the *frequency-domain procedures are identical with direct-current solutions.*

Although complex frequency can be viewed as a strictly algebraic substitution, it can also be given a physical interpretation. A complex frequency represents a sinusoidal wave multiplied by an exponential damping term. Such a wave is difficult to generate in the laboratory and is not actually used in practice, but from the standpoint of *interpretation* we may suppose that we have a generator of damped sine waves and that we can apply the damped sine waves to a network and obtain the response at some point. The steady-state response will be a damped sine wave of the same frequency and damping constant, but with a different amplitude and phase.* The response functions can be computed in the frequency domain in much the same way as the response functions for steady-state sinusoidal excitation were computed. There are some minor differences, however, and we will now investigate them.

A damped sinusoidal wave is expressed trigonometrically as

$$f(t) = |F|e^{\sigma t} \cos (\omega t + \theta). \tag{18–1}$$

By Euler's formula we can write Eq (18–1) in algebraic form and obtain

$$\begin{aligned} f(t) &= \text{Re}\,[|F|e^{\sigma t}e^{j(\omega t+\theta)}] \\ &= \text{Re}\,[|F|e^{j\theta}e^{(\sigma t+j\omega t)}] \\ &= \text{Re}\,[Fe^{st}]. \end{aligned} \tag{18–2}$$

In Eq. (18–2), F is a frequency-domain phasor with a magnitude of $|F|$ and an angle of θ. Thus

$$F = |F| \underline{/\theta}. \tag{18–3}$$

The phasor F is identical to the phasor which would have been obtained from an undamped sine wave. The frequency variable $j\omega$, however, has been replaced by a new variable s, defined as

$$s = \sigma + j\omega. \tag{18–4}$$

The real part of the complex frequency is a damping term, while the imaginary part is the usual angular frequency. There is nothing mysterious, therefore, about a complex frequency. It is simply a convenient method

---

* Note, however, that unless the natural transients in the circuit die out at a faster rate than the steady-state response, they would obscure it. We can examine the steady-state response analytically for any damping constant, however, and we will find it convenient to do so.

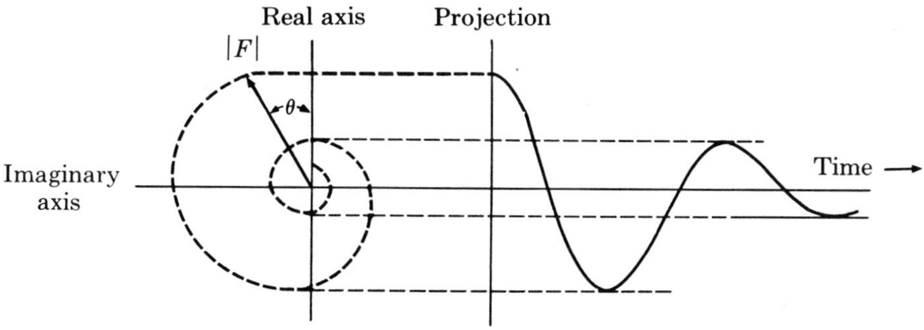

FIG. 18–1. A phasor which rotates and decays in length.

of keeping track of the damping constant and the frequency in a single complex number.

Equation (18–2) represents a phasor of length $|F|$ which starts at an angle θ and rotates with an angular velocity of ω, and at the same time decays in length (for a negative value of σ). The actual damped sine wave is the projection of this rotating and decaying phasor on the real axis, as shown in Fig. 18–1. All the phasors in a given problem rotate and decay at the same rate. The "complex-frequency domain" is a combination of a merry-go-round which is rotating and an Alice-in-wonderland domain which is shrinking. Since the whole domain rotates and shrinks, the phasors maintain their relative positions and lengths, and we can ignore the rotation and contraction until we wish to obtain the answer as a function of time. *The frequency-domain calculations*, as in the case of undamped sine waves, *are performed with the phasors as they appear at* $t = 0$. The variables are transformed to the frequency domain in terms of the phasor at $t = 0$, that is, with a magnitude equal to the undamped amplitude of the sine wave and an angle equal to the phase angle at $t = 0$, relative to a cosine function.

The frequency-domain networks are transformed to the complex-frequency domain by changing the frequency variable $j\omega$ to $s = \sigma + j\omega$. The frequency-domain forms of resistance, inductance, and capacitance are shown in Fig. 18–2. The complex impedances are now functions of s. Thus

$$Z_R = R, \qquad Z_L = Ls, \qquad Z_C = 1/Cs. \tag{18–5}$$

It is, of course, a simple matter to change these impedances back to the steady-state sinusoidal form, by replacing s by $j\omega$. The form with s is simpler, however, as far as computations in the frequency domain are concerned. In general, the variable s is used only for frequency-domain calculations, but in order to focus attention on the physical meaning of s, we will work an example in which a steady-state damped sinusoid is

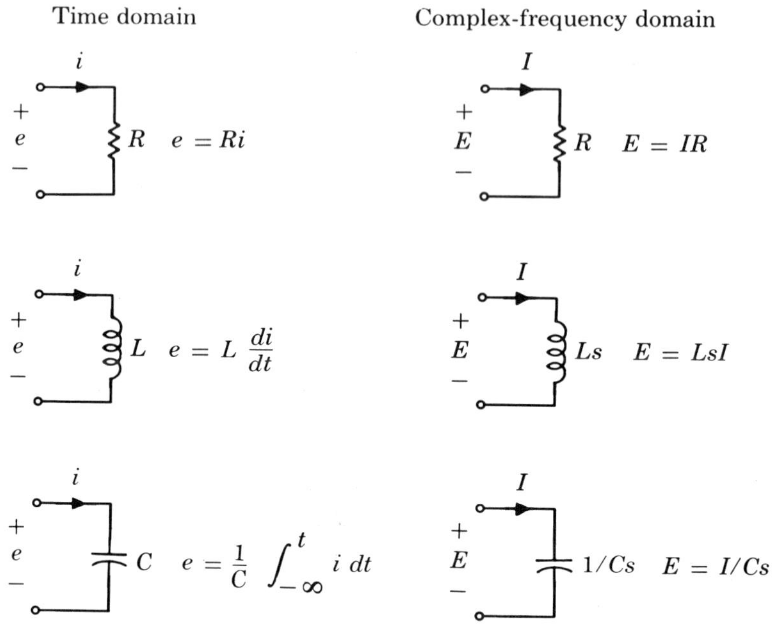

FIG. 18–2. Resistance, inductance, and capacitance in the complex-frequency domain.

applied to a network and the response is a steady-state damped sinusoidal function. The procedure is:

1. The exciting function and the network are transformed to the complex-frequency domain.
2. The response is obtained by simple algebra in the frequency domain.
3. The result is transformed back to the time domain.

Our example is shown in the time domain in Fig. 18–3(a). The corresponding frequency-domain diagram is shown in Fig. 18–3(b). In part (b) the input is a voltage phasor with a magnitude of 10 and an angle of zero; the inductance becomes a frequency-domain impedance of s complex ohms, and the resistance becomes an impedance of 1 complex ohm. The steady-state current in the frequency domain is

$$I_{ss} = \frac{10 \angle 0^\circ}{s+1}.$$

The value of s for this particular problem, with a damping constant $\sigma = -1$ and an angular frequency $\omega = 1$, is

$$s = -1 + j,$$

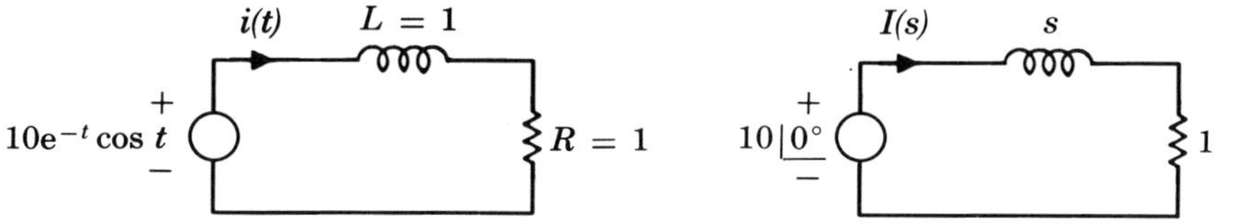

(a) A time-domain circuit (b) The same circuit in the complex-frequency domain

FIG. 18–3. A frequency-domain solution for a circuit driven by a damped sinusoid.

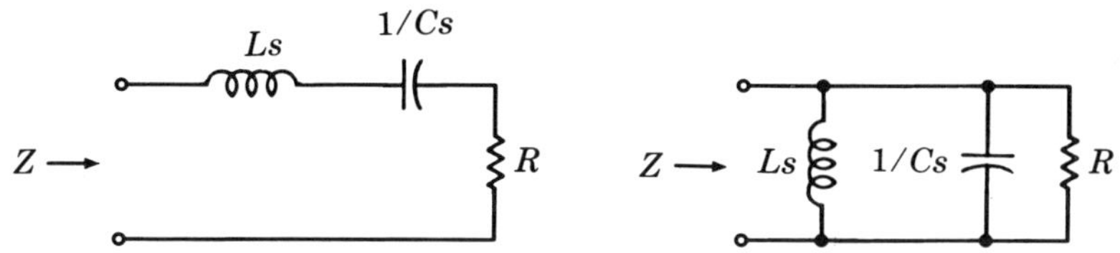

(a) Series circuit (b) Parallel circuit

FIG. 18–4. Impedances in the complex-frequency domain.

and thus I_{ss} becomes

$$I_{ss} = \frac{10\,\underline{|0^\circ}}{-1 + j + 1} = 10\,\underline{|-90^\circ}.$$

The frequency-domain value of I gives the initial magnitude and the initial phase of the function. The translation back to the time domain must insert the proper damping function and frequency, from the value of s. The steady-state component of the current is*

$$i_{ss}(t) = 10e^{-t}\cos(t - 90^\circ).$$

More often the complex frequency is used only for the computation of the response in the frequency domain, and the translation to the time domain is effected only for $s = j\omega$. As examples of the computation of response functions in terms of a complex-frequency variable, we will obtain the input impedances of the two networks in Fig. 18–4. The impedance of the series circuit of part (a) is

$$Z(s) = Ls + R + \frac{1}{Cs} = \frac{L[s^2 + (R/L)s + 1/LC]}{s}, \qquad (18\text{–}6)$$

* The transient term in this particular problem is zero. In general, a transient is excited when the damped sinusoid is applied. The transient can be found by the methods of the earlier chapters or by the methods of Chapter 20.

and the impedance of the parallel circuit of part (b), is

$$Z(s) = \frac{1}{Cs + 1/R + 1/Ls} = \frac{s}{C[s^2 + (1/RC)s + 1/LC]}. \tag{18–7}$$

Equations (18–6) and (18–7) are interesting as examples of the response functions of networks in the complex-frequency domain. These expressions represent the behavior of the networks of Fig. 18–4 for all frequencies (and for all damping constants as well), and therefore they are properly called system functions of the networks. For these two examples, the system functions turn out to be the ratio of two polynomials in the variable s. Such will always be the case, since we are dealing with linear equations with constant coefficients. Furthermore, all the system functions will be the ratios of simple determinants, if the problem is set up with mesh currents or node voltages. We can therefore infer that the general response function $\mathrm{T}(s)$ will have the form

$$\mathrm{T}(s) = \frac{P(s)}{Q(s)} = \frac{A(s - s_1)(s - s_3)\ldots}{(s - s_2)(s - s_4)\ldots}. \tag{18–8}$$

The functions $P(s)$ and $Q(s)$ represent polynomials in s. In the second form of the expression, the polynomials have been factored. The roots of the numerator, or the values of s which cause the numerator to become zero, are called the *zeros* of the function, and the values of s which cause the denominator to become zero and the function to become infinite are called the infinities or *poles* of the function. Thus

$$s_1, s_3, \ldots \text{ are zeros of } \mathrm{T}(s),$$

$$s_2, s_4, \ldots \text{ are poles of } \mathrm{T}(s).$$

In general, the poles and zeros will be complex numbers, and they must be complex conjugates since the polynomials $P(s)$ and $Q(s)$ have real coefficients. Apart from the constant A, the poles and the zeros of the function completely specify it. Throughout the remainder of the book we will make a good deal of use of poles and zeros.

It is now possible to explain in more detail the reason why we introduced the complex frequency variable s. The basic problem which we are faced with is the computation of the magnitude and phase curves of a response function such as $\mathrm{T}(s)$ of Eq. (18–8). *The introduction of the complex frequency s allows us to factor the expression for* $\mathrm{T}(s)$ *with some ease.* We can obtain the effect of each individual factor on the over-all response function and then we can combine the results. Both the log-db method and the s-plane method which we will discuss in later sections make use of this idea. The log-db method separates the factors by taking the logarithm of the response function, and the s-plane method separates the factors by viewing each of them as a separate vector.

18–3 Direct plots of magnitude and phase functions. In order to provide perspective for the log-db method and the s-plane method we will plot magnitude and phase response curves for two simple examples by the direct method. The first circuit to be considered is that shown in Fig. 18–5(a). The voltage is the stimulus and the current is the response, so that the response function is the admittance, Y, of the circuit. Thus

$$I = YE, \tag{18–9}$$

where

$$\begin{aligned} Y &= \frac{1}{R + j\omega L} \\ &= \frac{1}{\sqrt{R^2 + \omega^2 L^2}} \underline{\left| -\tan^{-1} \omega L/R \right.}. \end{aligned} \tag{18–10}$$

The magnitude of the admittance is

$$|Y| = \frac{1}{\sqrt{R^2 + \omega^2 L^2}}, \tag{18–11}$$

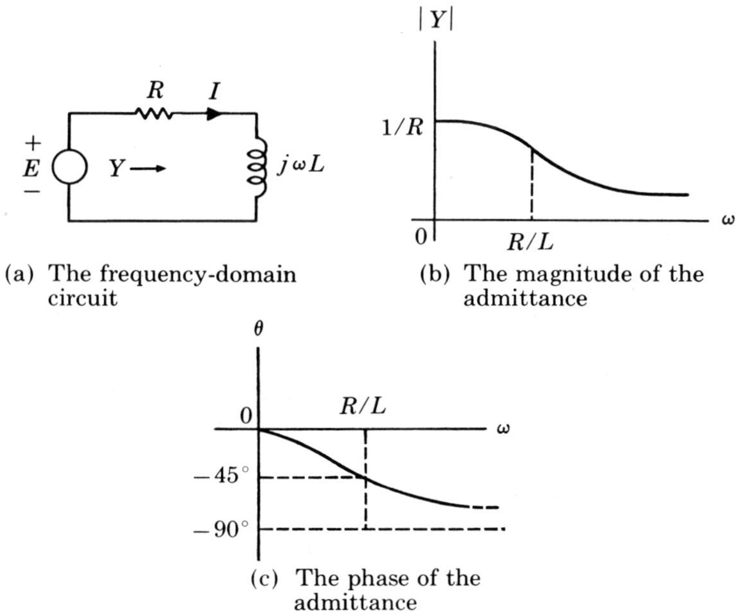

(a) The frequency-domain circuit

(b) The magnitude of the admittance

(c) The phase of the admittance

FIG. 18–5. The direct plotting of magnitude and phase curves.

and its phase is

$$\theta = -\tan^{-1} \omega L/R. \tag{18–12}$$

At $\omega = 0$ the magnitude is $1/R$, and at $\omega = R/L$ the magnitude is down to $0.707/R$. Since the voltage is 0.707 of its maximum value, the power, which is proportional to the square of the voltage, is down to half of its maximum value. Accordingly, the frequency $\omega = R/L$ is called the *half-power* frequency. For large values of ω the magnitude decays as $1/\omega$. A plot of the function is shown in Fig. 18–5(b). The corresponding phase curve, shown in Fig. 18–5(c), starts at zero when $\omega = 0$, drops to $-45°$ when $\omega = R/L$, and continues toward $-90°$ as ω becomes very large. The general forms of the magnitude and phase curves can be sketched in from the values at $\omega = 0$, $\omega = R/L$, and $\omega = \infty$. If more accuracy is required, additional points can be computed by substituting other values of ω into Eqs. (18–11) and (18–12).

Our second example is shown in Fig. 18–6(a). Again, we will suppose that the voltage is known and that the current is the desired response. The effect of the network is therefore best summarized in terms of the admittance function, which is

$$\begin{aligned} Y &= \frac{1}{j\omega L + 1/j\omega C + R} \\ &= \frac{j\omega C}{1 - \omega^2 LC + j\omega RC} \\ &= \frac{\omega C}{\sqrt{(1 - \omega^2 LC)^2 + (\omega RC)^2}} \underline{\left| 90° - \tan^{-1} \frac{\omega RC}{1 - \omega^2 LC} \right.}. \end{aligned} \tag{18–13}$$

The magnitude of Y is zero at $\omega = 0$ and at $\omega = \infty$. It reaches a maximum value, $1/R$, at $\omega = 1/\sqrt{LC}$, where the inductive and capacitive reactances cancel each other. A sketch of the magnitude as a function of frequency is shown in Fig. 18–6(b). The phase is 90° when $\omega = 0$, zero at $\omega = 1/\sqrt{LC}$, and $-90°$ when ω is very large. The phase curve is shown in Fig. 18–6(c). Additional points on the magnitude and phase curves can be obtained by substituting additional values of ω into the expressions for the magnitude and the phase in terms of ω. Even in this simple example, however, the process is laborious, and after he has computed a very few points, it is likely that the student will be willing to learn the more elegant methods of the following two sections.

In the process of producing amplitude and phase plots, however, the student should not lose sight of their purpose. They record the amplitude change and the phase shift which the network adds to the input stimulus in order to produce the output response. If the amplitude and phase of the stimulus are given, the amplitude and phase of the response function

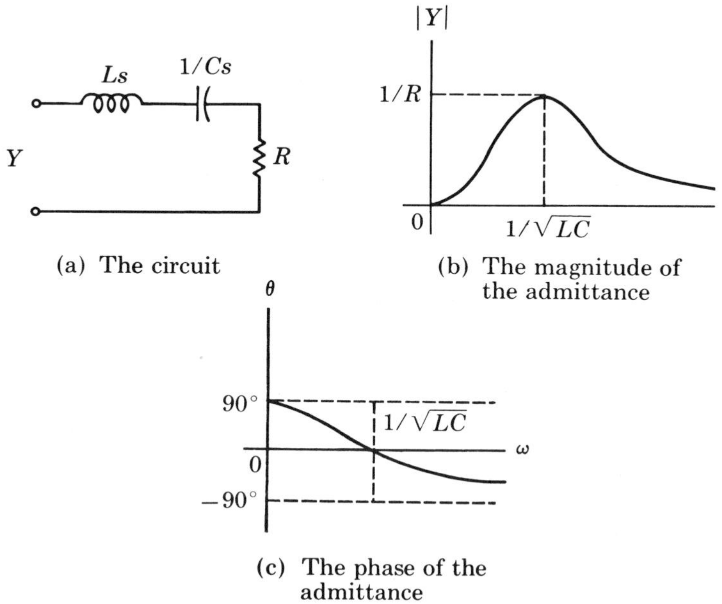

FIG. 18–6. Magnitude and phase curves for a series resonant circuit.

can be computed from the amplitude and phase curves. In a mathematical sense, these curves are a complete representation of the network. They contain all the information necessary to relate a given stimulus to its response.* They represent the system function of the network.

18–4 Logarithmic plots of magnitude and phase functions. In many cases it is more convenient to plot the frequency behavior of a response function in logarithmic form rather than in the original form. A logarithmic plot is just another way of presenting the data, but it does have the advantage of compressing the dynamic range of the data, thereby making it fit better on a sheet of graph paper. If we take the natural logarithm of the response function of Eq. (18–8) we obtain

$$\begin{aligned} \log_e \mathrm{T}(s) &= \log_e |\mathrm{T}(s)| + j \text{ angle } [\mathrm{T}(s)] \\ &= G + j\phi, \end{aligned} \tag{18–14}$$

since the logarithm of a complex quantity is the complex number with a real part equal to the logarithm of the magnitude of the complex quantity, and an imaginary part equal to the angle of the complex quantity. In

---

* Later, we will show that an arbitrary input function can be resolved into a summation of sine waves, all of which can be treated by the amplitude and phase curves.

circuit theory G is called the *gain function* (or sometimes the attenuation function) and it is measured in nepers. The quantity ϕ is called *the phase* and it is measured in radians. More often, in practice, the gain function is converted to decibels and the phase function to degrees, because these quantities are easier to find in tables. The number of decibels (abbreviated db) is obtained by taking the logarithm to the base 10 instead of the base e, and multiplying it by 20. Thus

$$\begin{aligned}\text{Number of db} &= 20 \log_{10} |\mathrm{T}(s)| \\ &= 8.68 \log_e |\mathrm{T}(s)| \\ &= 8.68 \times \text{number of nepers.}\end{aligned} \tag{18–15}$$

Similarly, the number of degrees is obtained from the number of radians:

$$\text{Number of degrees} = 57.3 \times \text{number of radians.} \tag{18–16}$$

The switch from nepers to decibels and from radians to degrees is simply a change of scale, analogous to plotting distance in feet instead of inches.

In addition to providing scale compression, the logarithmic transformation separates the product terms of the response function into a summation of terms and allows us to consider each one separately. Let us examine this feature with the general response function of Eq. (18–8):

$$\mathrm{T}(s) = \frac{A(s - s_1)(s - s_3)\ldots}{(s - s_2)(s - s_4)\ldots}. \tag{18–17}$$

For convenience we will first consider the case in which all the poles and zeros are negative real numbers, and we will write

$$s_1 = -\frac{1}{T_1}, \qquad s_2 = -\frac{1}{T_2}, \qquad s_3 = -\frac{1}{T_3}, \qquad \ldots.$$

In terms of the T's, Eq. (18–17) can be written

$$\mathrm{T}(s) = \frac{AT_2T_4\ldots}{T_1T_3\ldots} \times \frac{(1 + T_1s)(1 + T_3s)\ldots}{(1 + T_2s)(1 + T_4s)\ldots}. \tag{18–18}$$

The gain function G in db for Eq. (18–18) is

$$\begin{aligned}G = 20 \log_{10} \frac{AT_2T_4\cdots}{T_1T_3\cdots} &+ 20 \log_{10} |1 + T_1s| + \cdots \\ &- 20 \log_{10} |1 + T_2s| - \cdots,\end{aligned} \tag{18–19A}$$

and the phase function in degrees is,

$$\begin{aligned}\phi &= 57.3 \text{ angle } (1 + T_1s) + \cdots \\ &- 57.3 \text{ angle } (1 + T_2s) - \cdots.\end{aligned} \tag{18–19B}$$

The first term in the gain function is simply a constant. The remaining terms, whether added or subtracted, are of the following form:

$$20 \log_{10} |1 + Ts|.$$

Similarly, the phase is composed of positive and negative terms of the following form: angle $(1 + Ts)$. We will now discuss a short method for obtaining the amplitude and phase curves for such terms.

Linear factors. We will first obtain the gain function for one of the linear terms in Eq. (18–19A). It is easier to perform the factoring in terms of the complex frequency s, but now that we are ready to plot the function we must switch back to the undamped frequency $j\omega$ which we can generate in the laboratory. Thus we are interested in a term of the form

$$\begin{aligned} G \text{ (in db)} &= 20 \log_{10} |1 + j\omega T|, \\ &= 10 \log_{10} (1 + \omega^2 T^2). \end{aligned} \tag{18–20}$$

Low-frequency asymptote. For small values of ω, the quantity $\omega^2 T^2$ is negligible in comparison to unity and the gain function is

$$G = 10 \log_{10} 1 = 0 \text{ db}. \tag{18–21}$$

The low-frequency asymptote to which the curve tends as ω becomes small is therefore the zero db line shown in Fig. 18–7.

High-frequency asymptote. For large values of ω, the quantity $\omega^2 T^2$ is large in comparison to unity and the gain function is

$$\begin{aligned} G &= 10 \log_{10} \omega^2 T^2 \\ &= 20 \log_{10} \omega T \\ &= 20 \log_{10} T + 20 \log_{10} \omega. \end{aligned} \tag{18–22}$$

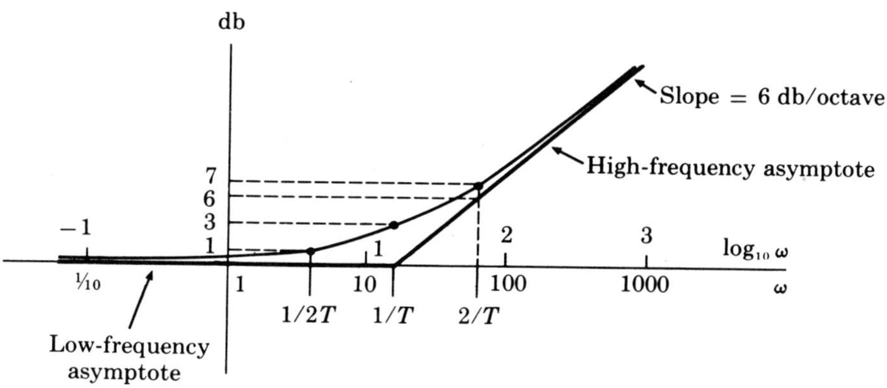

FIG. 18–7. The gain factor $10 \log_{10} (1 + \omega^2 T^2)$ plotted against log-frequency.

In Eq. (18–22) if we consider $\log_{10} \omega$ as the variable, instead of just ω, the equation is a straight line of the form

$$y = mx + b. \tag{18–23}$$

The slope m has a value of 20 db per unit of $\log_{10} \omega$, which is 20 db every time ω increases by a factor of 10, and which is sometimes stated as 20 db per decade of ω. Every time ω is doubled the number of decibels is also increased by a constant amount, which is given by

$$\begin{aligned} 20 \log_{10} 2\omega - 20 \log_{10} \omega \\ &= 20 \log_{10} 2 \\ &= 6 \text{ db}. \end{aligned}$$

Thus the high-frequency asymptote on the db versus $\log_{10} \omega$ scale is a straight line with a slope of 6 db per octave (doubling) of ω.

The break point. The intersection of the high- and low-frequency asymptotes is called the *break point* of the curves. From Eq. (18–22) it can be seen that the high-frequency asymptote is zero db when $\omega T = 1$, or $\omega = 1/T$. Hence the point $\omega = 1/T$ is the break point of the asymptotes, as shown in Fig. 18–7.

The true curve. The true curve lies above the asymptotes. For rough plots the asymptotes themselves are often sufficient to show the behavior of the function, but for more accuracy a few points on the true curve can be sketched in. At the break point the true curve is given by

$$\begin{aligned} G &= 20 \log_{10} |1 + j| \\ &= 10 \log_{10} 2 = 3 \text{ db}. \end{aligned} \tag{18–24}$$

The actual curve is thus above the intersection of the asymptotes by 3 db. At an octave below $\omega = 1/T$, that is, at $\omega = 1/2T$, the value of the true curve is

$$\begin{aligned} G &= 20 \log_{10} \left| 1 + \frac{j}{2} \right| \\ &= 1 \text{ db}. \end{aligned} \tag{18–25}$$

Similarly, at one octave above $\omega = 1/T$, the true curve is again 1 db above the asymptote, as shown in Fig. 18–8.

The Phase. The phase which corresponds to the amplitude function with which we have been working can also be found easily. It is given by

$$\phi = 57.3 \text{ angle } (1 + j\omega T). \tag{18–26}$$

For small values of ω the phase is zero. For large values of ω the phase is 90°. At the break point, $\omega = 1/T$, the phase is 45°. Plotted against

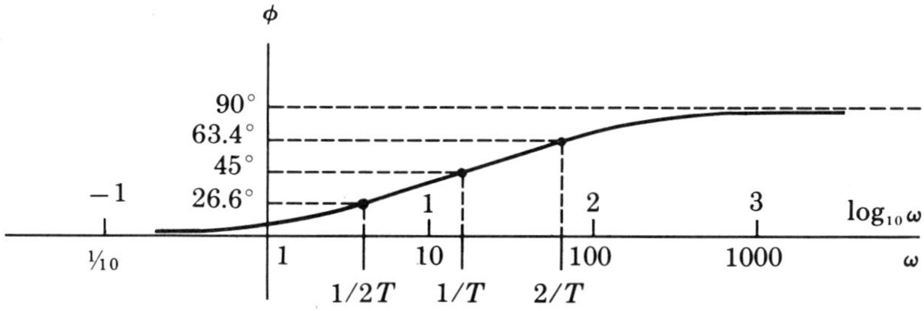

FIG. 18–8. The phase factor $57.3 \tan^{-1} \omega T$ plotted against log-frequency.

$\log_{10} \omega$, the phase is symmetrical about the break point. At $\omega = \frac{1}{2}T$ it is 26.6°, and at $\omega = 2/T$ it is 63.4°. The complete phase curve is shown in Fig. 18–8.

With the exception of the first term, which is merely a constant, all the terms in the gain and phase functions of Eqs. (18–19A) and (18–19B) are of the form $(1 + Ts)$. A somewhat special case occurs if the constant is missing in this expression and the s term is present by itself. The gain function for this case is

$$G = 20 \log_{10} \omega, \tag{18–27}$$

and the phase function is

$$\phi = 90°. \tag{18–28}$$

The gain function is a 6 db per octave sloping line which passes through the point $\omega = 1$ at zero db. The phase function is a constant, +90°, for all frequencies. (The T is absorbed in the constant term.)

Example. As an example of the log-db method we will plot the gain and phase curves for the response function

$$\mathrm{T}(s) = \frac{s}{(s+1)(s+2)} \cdot$$

The first step is to write the expression $\mathrm{T}(s)$ in the standard form in which the constant in each term is unity. Thus

$$\mathrm{T}(s) = \frac{1}{2} \frac{s}{(1+s)(1+s/2)} \cdot$$

The gain function is

$$G = 20 \log_{10} \tfrac{1}{2} + 20 \log_{10} \omega - 20 \log_{10} \sqrt{1+\omega^2} - 20 \log_{10} \sqrt{1+\omega^2/4}.$$

The first term is a constant of −6 db. The second term is a straight line with a slope of 6 db per octave, extending from minus infinity to plus infinity. The third term is a standard one with a break point at $\omega = 1$.

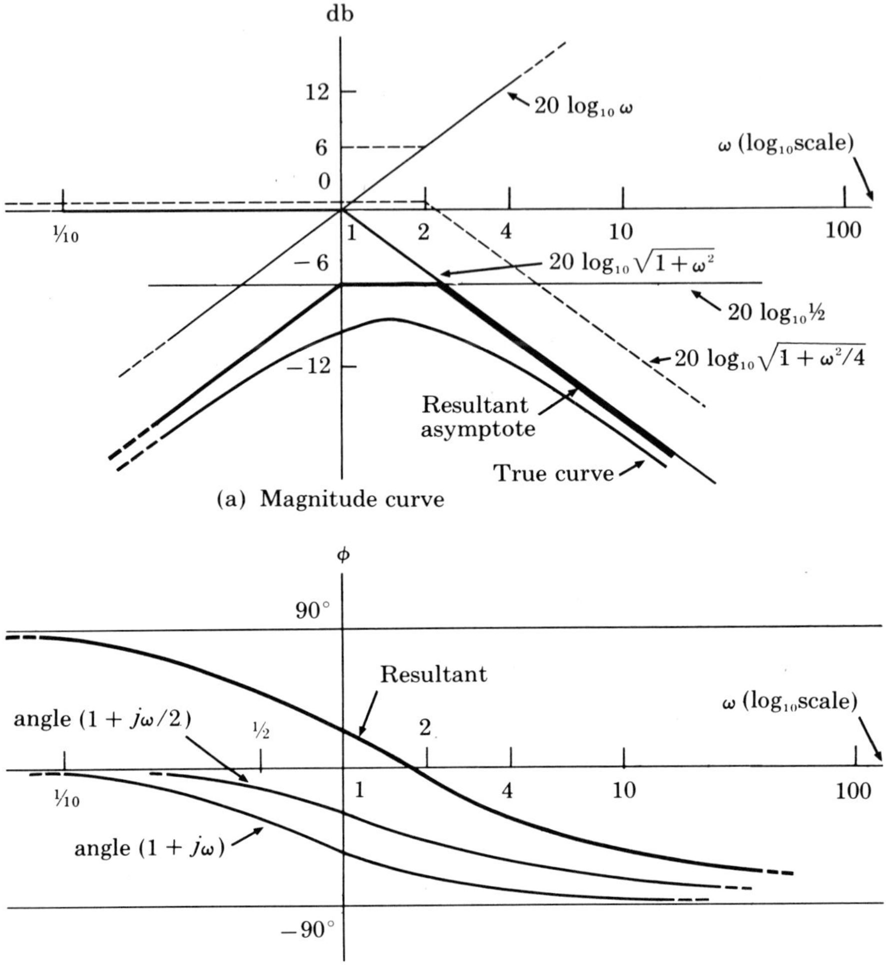

(a) Magnitude curve

(b) Phase curve

FIG. 18–9. Magnitude and phase curves for $T(s) = s/[(s+1)(s+2)]$.

The low-frequency asymptote is the zero db line, and the high-frequency asymptote is a straight line with a slope of −6 db per octave. The two asymptotes intersect at $\omega = 1$. The fourth term is similar to the third except that the break point occurs at $\omega = 2$. The four separate terms and their sum are shown in Fig. 18–9(a). For many applications, the asymptotes themselves provide a sufficiently accurate picture of the response function. If desired, the true curves can be sketched in, and the resultant found with greater accuracy. The principal effect is a rounding of the sharp corners of the asymptotic curve. In our example the true curve lies below the asymptotes by about 4 db at the break points.

The phase curve for the example response is given by

$$\phi = 90° - \text{angle}\,(1 + j\omega) - \text{angle}\,(1 + j\omega/2).$$

The first term is a constant, 90°. The second term is a standard phase curve with a break point, or 45° point, at $\omega = 1$. Because of the minus sign in the equation the phase is negative. The third term is similar, with a 45° point at $\omega = 2$. These three curves and their sum are shown in Fig. 18–9(b). The resultant phase curve drops from +90° at low frequencies to −90° at high frequencies.

The real utility of log-db plots lies in the approximate response curves which they yield from the asymptotes. In addition, of course, the log-db plots show that there is a definite relationship between the gain and phase functions. For example, an attenuation curve that is dropping at the rate of $6n$ db per octave has associated with it a negative phase of about $n \times 90°$.*

Quadratic terms. In the previous discussion we limited ourselves to gain and phase factors in which the poles or zeros lie on the negative real axis. It is possible to have complex roots as well, but the complex roots must lie in complex conjugate pairs and it is therefore convenient to consider them in pairs. If the two complex conjugate factors are combined, a quadratic factor results which we express in the standard form

$$T^2s^2 + 2\zeta Ts + 1. \tag{18–29}$$

We have made the constant term in Eq. (18–29) unity in order to have zero db when the frequency is low. The multiplying constant, as in the linear case, can be absorbed in a single constant added to the zero-frequency level in decibels. As with the linear factor, the frequency variable is Ts, but the quadratic factor is more complicated because of the range of values which the damping constant ζ can have. The low-frequency and the high-frequency asymptotes are no more difficult to obtain than in the case of the linear factor, but the departure of the true curve from the asymptotes now depends on the value of ζ. A family of curves can be plotted to include all values of ζ, but for ζ small the true curve departs a great deal from the asymptotes and the log-db method loses some of its simplicity.

---

* This condition holds for minimum phase-shift networks in which the poles and the zeros are in the left half of the s-plane. It is possible to increase the phase lag with terms of the form $(1 - j\omega T)/(1 + j\omega T)$ which do not change the gain function. These terms are called all-pass functions.

Gain for a quadratic factor. The gain function for a quadratic factor is

$$G = 20 \log_{10} |1 - \omega^2 T^2 + 2\zeta\omega T j|. \tag{18–30}$$

For very low frequencies the ω terms are negligible and the gain is

$$G = 0 \text{ db}.$$

The low-frequency asymptote is therefore the zero db line, just as for the linear term. For high frequencies the ω^2 term is the important one, and the gain is

$$G = 20 \log_{10} \omega^2 T^2 = 40 \log_{10} \omega T. \tag{18–31}$$

The graph of Eq. (18–31) is a straight line if it is plotted against $\log_{10} \omega$. The slope of the line is 12 db per octave, just twice the value for the linear factor, and the point at which it cuts the zero db axis is $\omega = 1/T$. The high-frequency asymptote is therefore a straight line with a 12 db per octave slope, and the break point is at $\omega = 1/T$.

The behavior of the true curve in the vicinity of the break point depends on ζ. The minimum value of ζ is zero, and the maximum value which we need to consider is unity, since above this value the roots are real and we can use two linear factors. At the break point, $\omega = 1/T$ and the gain is

$$G = 20 \log_{10} 2\zeta. \tag{18–32}$$

For $\zeta = 0$ the true curve drops sharply to $-\infty$ at the break point. For $\zeta = \frac{1}{2}$ the true curve passes through the break point, and for $\zeta = 1$ the true curve lies 6 db above the break point. When $\zeta = 1$ we have two linear factors and each of them contributes 3 db at the break point. A set of magnitude curves for three values of ζ is sketched in Fig. 18–10. They are most useful when the asymptotes can be used to approximate the function fairly closely. When $\zeta = 0$ the error is infinite at the break point!

Phase for a quadratic factor. The phase for a quadratic factor is given by

$$\phi = \tan^{-1} \frac{2\omega\zeta T}{1 - \omega^2 T^2}. \tag{18–33}$$

For small value of ω the phase is zero, and for large values it is 180°; at the break point, it is +90°. The curves for various values of ζ are symmetrical about the break point, and the change is more abrupt as ζ is made smaller. Indeed for $\zeta = 0$ the change from 0° to +180° occurs in one sudden jump at the break point. For $\zeta = 1$ the curve is simply the sum of two curves of the type shown in Fig. 18–9(b). Phase curves for three values of ζ are shown in Fig. 18–10.

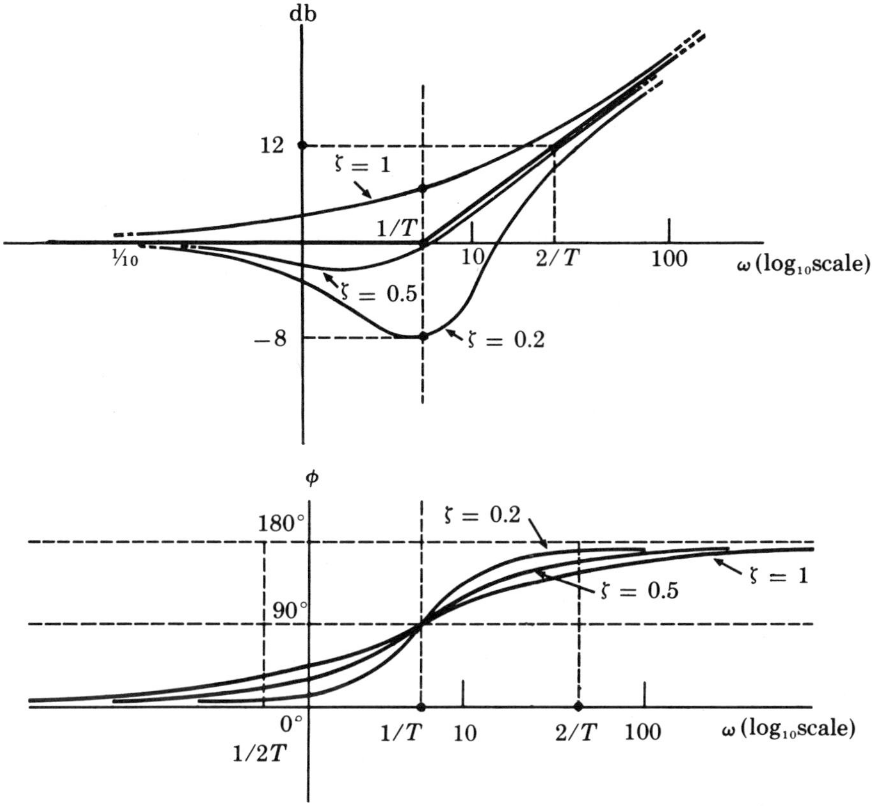

FIG. 18–10. Log magnitude and phase curves for the quadratic factor $T(s) = 1 + 2\zeta Ts + T^2s^2$.

The quadratic factors are used in the same fashion as the linear factors. In many problems the asymptotes are sufficiently accurate to describe the gain and phase functions. Whenever this accuracy is sufficient the log-db methods are superior to all others for obtaining the gain and phase functions. In addition, they show the relationship between the gain and the phase, and they can be used to obtain the phase curves from the gain curves and vice versa. One further application of the log-db curves exists, which we shall only mention briefly in this book. This application is the determination of an analytical expression for a given stimulus-response function obtained experimentally. The data is plotted in logarithmic form, and then approximated by 6 db and 12 db asymptotes. From the break points of the asymptotes and the desired overshoots at the break points the approximate analytical expression can be written.

When the damping factor is small, the true curve departs a great deal from the asymptotes, and the log-db method becomes rather unsatisfactory. Fortunately, for this case the s-plane vector method described in the next section is superb. The two methods therefore complement each other.

18–5 Magnitude and phase plots from s-plane vectors. The s-plane vector method for plotting magnitude and phase curves is similar to the log-db method in the sense that they both treat each individual factor of the response function by itself. The log-db method separated the factors by taking the logarithm of the product expression. The s-plane vector method separates the factors by considering each of them to be a separate vector whose length and angle can be found in the s-plane. The s-plane method is not particularly useful for making accurate plots, since the data is obtained geometrically, but it is an excellent method for sketching the approximate response, especially when one or more of the factors has a small damping constant.

Again, the over-all response function, from Eq. (18–8), will have the form

$$\mathrm{T}(s) = \frac{A(s - s_1)(s - s_3)\ldots}{(s - s_2)(s - s_4)\ldots}. \tag{18–34}$$

As a function of $s = j\omega$, for steady-state sinusoidal excitation, the response is

$$\mathrm{T}(j\omega) = \frac{A(j\omega - s_1)(j\omega - s_3)\ldots}{(j\omega - s_2)(j\omega - s_4)\ldots}. \tag{18–35}$$

The values $s_1, s_3, \ldots$ are complex numbers representing the zeros of the response function, and the values of $s_2, s_4, \ldots$ are complex numbers representing the poles of the response function. If we view the term $j\omega$ as a complex number, then each of the factors in Eq. (18–35) is simply the difference of two complex numbers. Alternatively, since complex numbers are in a one-to-one correspondence with two-dimensional vectors, each term is the difference of two vectors, which is the vector joining the tips of the two original vectors. The length of this difference vector is the magnitude of the corresponding term in Eq. (18–35), and the angle of this vector is the angle of the term in Eq. (18–35). The magnitude of the over-all response is the product of the lengths of the vectors in the numerator divided by the product of the lengths of the vectors in the denominator, and the phase of the response is the sum of the angles of the vectors in the numerator, less the sum of the angles of the vectors in the denominator. Of course, the whole calculation must be repeated for each value of ω, but a sketch of the vectors in the s-plane makes the process rather easy.

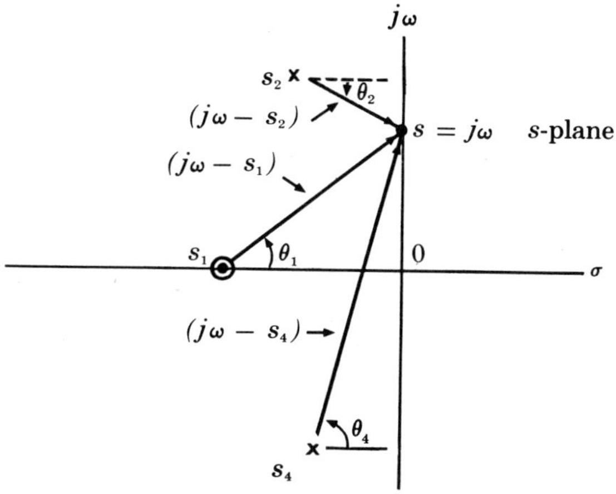

FIG. 18–11. The s-plane vectors for $T(j\omega) = (j\omega - s_1)/[(j\omega - s_2)(j\omega - s_4)]$.

The poles and zeros are plotted in the s-plane. The poles are marked with crosses and the zeros with circles. The point $s = j\omega$ which is common to all the vectors is shown with a dot at a representative point on the $j\omega$ axis. A plot is shown in Fig. 18–11 for a typical function

$$T(j\omega) = \frac{(j\omega - s_1)}{(j\omega - s_2)(j\omega - s_4)}. \tag{18–36}$$

The magnitude of the numerator in Eq. (18–36) is the length of the vector from s_1 to the point $j\omega$. The angle of the numerator is the angle θ_1 which this vector makes with the positive axis. The magnitudes of the two terms in the denominator of Eq. (18–36) are the lengths of the two vectors from s_2 and s_4 to the point $j\omega$, and the corresponding angles are the angles θ_2 and θ_4 which the vectors make with the real axis. The over-all magnitude is the length of the numerator vector divided by the product of the lengths of the two denominator vectors, and the over-all phase angle is the numerator-vector angle minus the two angles of the two denominator vectors.

The s-plane vectors do not contain any information about the constant multiplier, A, in Eq. (18–34). If the constant is known it is a simple matter to multiply by it. Sometimes the poles and zeros are known and the multiplying constant is not known explicitly, but is given instead by a known value of the complete response function at some frequency.

As an example of an s-plane plot we will obtain the magnitude and phase response curves for the current in the R–L–C circuit of Fig. 18–12. As a function of the complex-frequency variable s, the impedance is

$$Z(s) = Ls + R + \frac{1}{Cs} = L\,\frac{s^2 + (R/L)s + 1/LC}{s}. \tag{18–37}$$

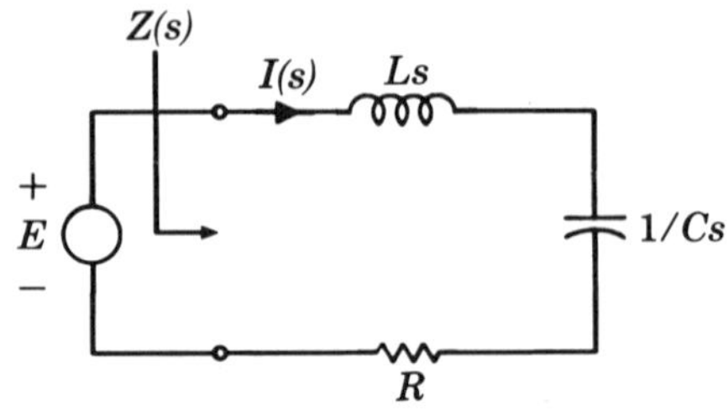

FIG. 18–12. A series R–L–C circuit driven by a voltage with a complex frequency.

We can put Eq. (18–37) in the standard form by factoring the numerator. Thus

$$Z(s) = L\frac{(s - s_1)(s - s_3)}{(s - 0)} \tag{18–38}$$

where

$$s_1 = -\alpha + j\omega_d,$$

$$s_3 = -\alpha - j\omega_d,$$

$$\alpha = \frac{R}{2L},$$

$$\omega_0^2 = \frac{1}{LC},$$

$$\omega_d = \sqrt{\omega_0^2 - \alpha^2}.$$

The s-plane vectors for Eq. (18–38) are plotted in Fig. 18–13. The point $s = j\omega$, which represents the frequency of the driving source, is taken to be in the vicinity of the resonant frequency s_1, and it is supposed that the circuit has a value of α which is much less than the value of ω_0. Under these assumptions, the length of the vector $(j\omega - s_1)$ is very much less than the length of the vector $(j\omega - s_3)$ or $(j\omega - 0)$. Indeed, with $j\omega$ in the vicinity of s_1, the two vectors $(j\omega - s_3)$ and $(j\omega - 0)$ are practically in the same direction, and the length of $(j\omega - s_3)$ is practically equal to twice the length of $(j\omega - 0)$. If we write

$$(j\omega - s_3) \cong 2(j\omega - 0), \tag{18–39}$$

an approximate expression for the impedance is

$$Z(s) = 2L(j\omega - s_1). \tag{18–40}$$

As long as the frequency is near the resonant frequency s_1, the current

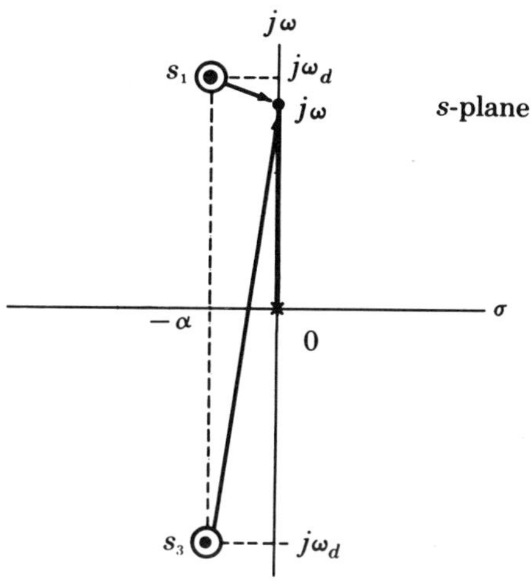

FIG. 18–13. The s-plane vectors for $Z(s) = L[(s+\alpha - j\omega_d)(s+\alpha+j\omega_d)/(s-0)]$.

will be given by

$$I = \frac{E}{2L(j\omega - s_1)}. \tag{18–41}$$

The maximum value of current occurs when the point $s = j\omega$ is directly opposite the position of s_1. This condition is called resonance in the circuit. It occurs at a frequency of ω_d, but since α is much less than ω_0 there is very little difference between ω_0 and ω_d. The minimum length of the vector $(j\omega - s_1)$ is α, and the maximum current is

$$I_{\max} = \frac{E}{2L\alpha} = \frac{E}{R}. \tag{18–42}$$

A sketch of the response current as a function of frequency is shown in Fig. 18–14. The current is down to 0.707 of its maximum value when the impedance is 1.414 of its minimum value. The impedance increases by 1.414 when the vector $(j\omega - s_1)$ increases by 1.414. From the geometry of Fig. 18–13 it is obvious that the vector which joins the point s_1 to the point on the $j\omega$ axis will have a value of 1.414α when it stands at an angle of $\pm 45°$. One of the frequencies on the $j\omega$ axis will be below the center frequency by α, and the other will be above the center frequency by α. These two frequencies are called the lower and upper half-power frequencies, since the power, which varies as the square of the voltage, is down to half of its maximum value when the voltage is down to 0.707 of its maximum value. At the lower half-power frequency

$$(j\omega - s_1) = \sqrt{2}\,\alpha \,\underline{|-45°},$$

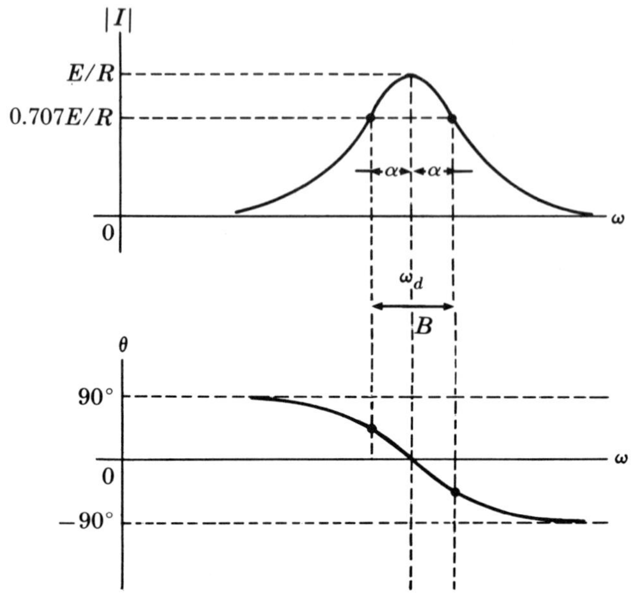

FIG. 18–14. Magnitude and phase responses obtained from the s-plane vector for $I = E/[2L(s + \alpha - j\omega_d)]$.

and

$$I = \frac{E}{2L\sqrt{2}\,\alpha\,\underline{/-45^\circ}} = \frac{E}{\sqrt{2}\,R}\,\underline{/45^\circ}. \tag{18–43}$$

At the upper half-power frequency

$$(j\omega - s_1) = \sqrt{2}\,\alpha\,\underline{/45^\circ},$$

and

$$I = \frac{E}{2L\sqrt{2}\,\alpha\,\underline{/45^\circ}} = \frac{E}{\sqrt{2}\,R}\,\underline{/-45^\circ}. \tag{18–44}$$

The *bandwidth* is usually taken to be the distance between the two half-power frequencies. It can be seen from the geometry of Fig. 18–13 that the bandwidth is

$$B = 2\alpha = \frac{R}{L}\cdot \tag{18–45}$$

The ratio of the center frequency to the bandwidth is a measure of the *quality* or *sharpness* of the resonance, and it is called the Q of the circuit. For the series resonant circuit which we have been discussing the Q is

$$Q\ (\text{series}) = \frac{\omega_d}{B} = \frac{\omega_d L}{R}\cdot \tag{18–46}$$

The concept of the quality factor Q is somewhat more general than the definition given in Eq. (18–46). Whenever the resonance is produced by a single pole or zero in the s-plane the Q will be twice the radian angle subtended by the pole or the zero and the $j\omega$ axis at the origin. Thus a pole or a zero has a Q which is given by*

$$Q = \frac{\omega_d}{2\alpha}. \tag{18–47}$$

The series resonant circuit which we have discussed here is important in its own right, as well as for providing an excellent example of the utility of the s-plane vectors. At the resonant frequency the response of the circuit is a maximum. If the resistance had been zero, in fact, the response would have been infinite. In this case we would describe the circuit as being in *perfect series resonance.* Physically, the perfect series resonance occurs when we drive the circuit at its own natural resonance frequency. The circuit is almost human. When we ask it to produce an output of the sort which it likes to produce it responds with great enthusiasm!

A parallel resonant circuit is the dual of the series resonant circuit. The Q of the parallel R–L–C circuit is

$$Q \text{ (parallel)} = \omega_d CR. \tag{18–48}$$

At resonance the admittance is $1/R$, which becomes zero if the parallel R is infinite. *A perfect parallel resonance is therefore characterized by an infinite impedance.* Again, the circuit is responding with great enthusiasm, because the frequency of the stimulus is the natural resonant frequency of the circuit, but in this case the impedance becomes infinite instead of zero. In the next section we will relate the poles and zeros of the complex impedance function to the natural resonant frequencies of the circuit. In doing so we will bring to light the interesting relationship between the steady-state response of a network and its transient response.

18–6 Poles and zeros. In the two preceding sections we used the poles and zeros of a response function as purely algebraic factors which enabled us to break up the over-all function into smaller components which we could consider separately. To facilitate the factoring we replaced the frequency variable $j\omega$ by the complex frequency variable $s = \sigma + j\omega$. We observed that this change means, in physical terms, that we were exciting the network with a steady-state damped sinusoidal function instead of a

---

* The idea of Q belongs to a single singularity, or an equivalent single singularity with the same bandwidth and center band value.

constant amplitude sinusoid. Up to now we have made no use of the physical meaning of the variable s, but in this section we will use it to examine the nature of the poles and zeros and to relate the steady-state sinusoidal behavior of our networks to their natural transient behavior.

The bridge which relates the steady-state sinusoidal behavior to the natural transient behavior is the concept of resonance. Everyone has some experience of resonance. A physical system produces its maximum response when it is excited by a source whose frequency is close to the natural frequency of the system. We have all observed the sympathetic vibrations of automobile components and accessories in response to the engine vibrations at different speeds. We have read about great bridges collapsing when excited at their natural resonant frequency by unusual wind conditions. We have seen clock pendulums swinging over large arcs and yet requiring infinitesimal amounts of energy from the escapement mechanism. And in the area of electrical devices, every one has operated the tuning controls of a radio or a television set. In the tuning process the natural resonant frequency of the receiving set is matched to the frequency of the radio waves from the desired transmitting station, and a maximum response is obtained.

In the preceding section we observed that the series resonant circuit with zero resistance has a zero of impedance at the resonant frequency. When resistance is added the impedance is no longer zero, but it is a minimum. We should like to view the change as resulting from the fact that we are no longer exciting the circuit at its natural resonant frequency. The natural resonant frequency with resistance present is a damped sinusoid, and if we are to excite the circuit at its natural resonant frequency we must excite it with a damped sinusoid. If we do so we will again obtain a zero of impedance when the exciting frequency is equal to the natural resonant frequency of the circuit. At this frequency an infinite current response would occur for a voltage excitation. When the circuit is required to produce the exact waveform which it likes to generate, it responds with infinite enthusiasm!

Similarly, a parallel circuit with no shunt resistance produces a perfect parallel resonance for an undamped exciting frequency. When resistance is added the impedance drops from infinity to a finite value. Again, we would like to view the change as the result of no longer exciting the circuit at its natural resonant frequency. If we excite the circuit with a damped sinusoidal current the impedance will be infinite when the damping constant and the frequency agree with the natural damping constant and frequency of the circuit. *The condition for resonance is thus the coincidence of the driving complex frequency with the complex frequency which represents the natural behavior of the circuit.*

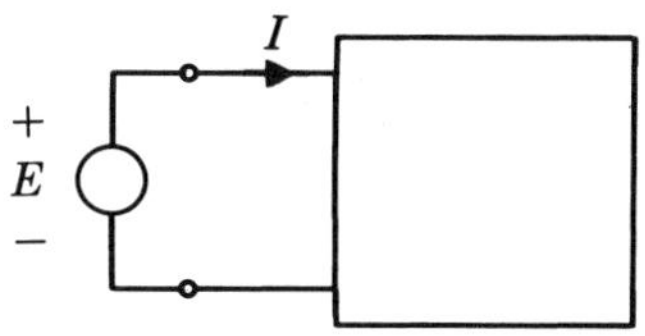

(a) Voltage source drive

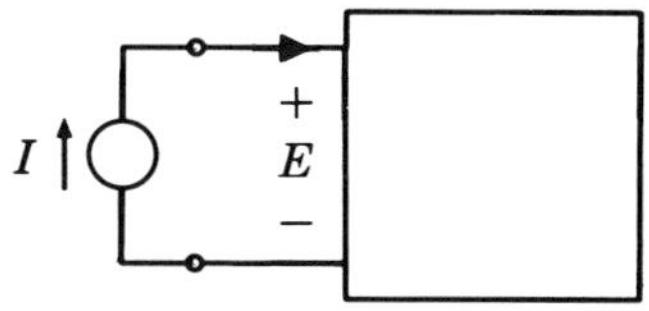

(b) Current source drive

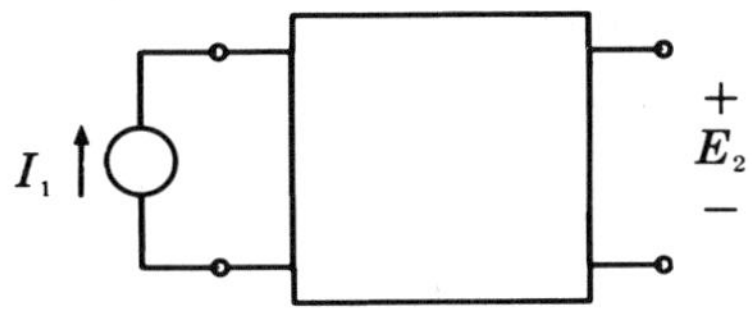

(c) Transfer impedance

FIG. 18–15. Complex impedance and transient natural frequencies.

To be more specific, let us consider a driving-point impedance function given by the following function of the complex frequency s:

$$Z(s) = \frac{A(s - s_1)(s - s_3)\ldots}{(s - s_2)(s - s_4)\ldots},$$

where $s_1, s_2, \ldots$ are the zeros,
$s_2, s_4, \ldots$ are the poles.

We can drive this impedance with a voltage source, as shown in Fig. 18–15(a), or with a current source, as shown in Fig. 18–15(b). If we drive it with the voltage source the current response will be

$$I = \frac{E(s - s_2)(s - s_4)\ldots}{A(s - s_1)(s - s_3)\ldots}. \tag{18–49}$$

A resonance will occur when Eq. (18–49) is infinite,* or at the values of exciting frequency s given by the zeros of the impedance function. So far as the natural transient frequencies of the circuit are concerned the voltage source represents a short circuit across the terminals of the impedance. We have therefore shown that

the zeros of the impedance function are the natural transient frequencies of the circuit with a short circuit across the terminals where the impedance is measured.

* Alternatively, when $E = 0$ a transient I can exist if $s = s_1, s_3 \cdots$.

Similarly, if we drive the impedance with the current source shown in Fig. 18–15(b) we will obtain a voltage response which is given by

$$E = \frac{IA(s - s_1)(s - s_3)\dots}{(s - s_2)(s - s_4)\dots}. \tag{18–50}$$

An infinite response,* or a resonance, will occur in Eq. (18–50) when the denominator becomes zero. This occurs at any one of the pole frequencies. The current source has an infinite internal impedance and is an open circuit so far as the natural transients, produced by internal energy stored in the circuit, are concerned. Hence we have shown that

the poles of the impedance function are the natural transient frequencies of the circuit with an open circuit across the terminals where the impedance is measured.

It obviously does not make any difference whether we consider an impedance function or an admittance function, since the admittance function is simply the reciprocal of the impedance function. A transfer function, however, poses a slightly different problem. For example, a transfer impedance of the form

$$Z_{12} = \frac{E_2}{I_1} = \frac{A(s - s_1)(s - s_3)\dots}{(s - s_2)(s - s_4)\dots} \tag{18–51}$$

has poles and zeros just as a driving-point impedance does. If a current is applied at the input terminals, as shown in Fig. 18–15(c), the response E_2 is measured at the output terminals. The poles again represent a resonant condition in the circuit since at the frequency of a pole, the transfer function Z_{12} is infinite, and the output response is infinitely greater than the input stimulus. Indeed, the situation is essentially no different from the driving-point impedance so far as the poles are concerned. The natural resonant frequencies of the circuit are excited by the current source at I_1 for both cases. For the driving-point impedance the response is observed at the same pair of terminals, while for the transfer impedance it is measured at a different pair of terminals. However, any frequencies which are present in the network will be observable at any terminal-pair in the network.† Thus

the poles of a transfer impedance are the same as the poles of a driving-point impedance, and represent the natural frequencies of the network with the input terminals open-circuited.

---

* Alternatively, when $I = 0$ a transient E can exist if $s = s_2, s_4 \cdots$.

† A small, but finite, possibility exists of excluding some natural frequency from a part of the network. A pole and a zero of the transfer function then cancel and one of the natural frequencies does not appear in the transfer function.

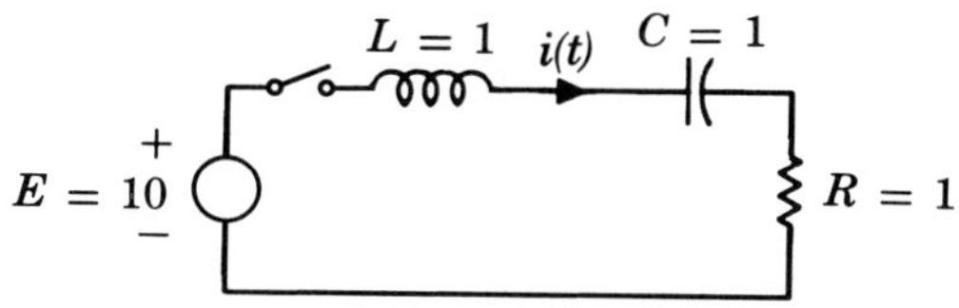

FIG. 18–16. A transient problem solved by means of complex impedance.

The zeros of the transfer function, however, have no simple interpretation in terms of the natural transient frequencies of the network. They simply represent frequencies at which no output is obtained even when an input is present. The zeros of transfer functions are related to the topology of the network. A series impedance which goes to infinity at the complex frequency s will produce a zero at this frequency in the transfer function. Similarly, a shunt element which goes to zero at frequency s will produce a zero in the output. A bridge which is balanced at the given frequency will likewise produce a zero of transmission. Hence

the zeros of a transfer impedance are frequencies at which zero output is obtained, and are not related to the natural transients of the circuit.

A similar argument would show that the poles of an admittance transfer function are the short-circuit transient natural frequencies of a network, while the zeros are simply zeros of transmission.

The interpretation which we have given to the poles and zeros of network steady-state response functions provides a relationship between the steady-state response and the transient response. The natural transient frequencies are easier to obtain from the steady-state impedance functions than they are by classical methods. The following example illustrates the solution of a transient problem by the use of poles and zeros. For the circuit shown in Fig. 18–16, the driving-point impedance function is

$$\begin{aligned} Z(s) &= s + 1 + \frac{1}{s} \\ &= \frac{s^2 + s + 1}{s} \\ &= \frac{A(s - s_1)(s - s_3)}{(s - s_2)}. \end{aligned}$$

The zeros are $s_1 = -0.5 + j0.866, \qquad s_3 = -0.5 - j0.866.$

The pole is $s_2 = 0.$

Since the circuit is driven by a voltage source, the transient terms which will be excited are the short-circuit transients, or the zeros. The only possible steady-state response is a constant. Hence the response must be of the form

$$i(t) = A_1e^{s_1t} + A_2e^{s_3t} + B,$$

where A_1, A_2, and B are constants to be determined from the initial conditions and from the value at infinity.

The initial conditions are

$$i(0) = 0,$$

$$\frac{di(0)}{dt} = 10,$$

and the final value is

$$i(\infty) = B = 0,$$

from which

$$i(0) = 0 = A_1 + A_2,$$

and

$$\frac{di(0)}{dt} = 10 = A_1s_1 + A_2s_3,$$

so that

$$A_1 = -A_2 = \frac{10}{s_1 - s_3} = \frac{10}{2j0.866} = 5.77\,\underline{|-90^\circ}.$$

Thus

$$\begin{aligned} i(t) &= 5.77e^{j90^\circ}e^{-0.5t}e^{j0.866t} + 5.77e^{-j90^\circ}e^{-0.5t}e^{-j0.866t} \\ &= 11.54e^{-0.5t}\cos(0.866t + 90^\circ) \\ &= 11.54e^{-0.5t}\sin(0.866t). \end{aligned}$$

The solution which has been obtained here is similar to the transient solutions which were given in Chapter 12. The difference lies only in that we have obtained the transient natural frequencies from the steady-state solution. Each transient natural frequency corresponds to a term of the form Ae^{pt}.

At this point we have gone half way in the process of converting transient problems to steady-state problems. In the remainder of the book we will go even further. We will show that transient excitations can be expanded into summations of sinusoidal and damped sinusoidal functions, and that transient problems can therefore be solved as a superposition of a number of steady-state problems. It is important, however, that the student should understand the nature of the transient solutions before he learns the steady-state procedures for arriving at them.

It will be instructive to make one further comment on the relationship of the steady-state and transient responses. This comment concerns the method by which we computed the steady-state response functions in the preceding section, in terms of vectors which joined the poles and the zeros to the steady-state applied frequency. These vectors represent the distances separating the applied frequency from the natural transient frequencies of the circuit. The method is therefore one which predicts the steady-state behavior of the network from a knowledge of the natural transient frequencies of the circuit and of the distances by which the steady-state frequency deviates from the resonance points. The distances are vector distances, and it is possible to predict not only the magnitude of the steady-state response, but also its phase. In this process the relative responses at various frequencies are given directly, but the multiplying constant which includes the level of the excitation must be obtained at some particular frequency, usually at zero or infinity.

18–7 Summary.

Complex frequency $(s = \sigma + j\omega)$

Time domain	*Complex-frequency domain*
$f(t) = \|F\|e^{\sigma t}\cos(\omega t + \theta)$	$F = \|F\|e^{j\theta} = F\angle\theta$

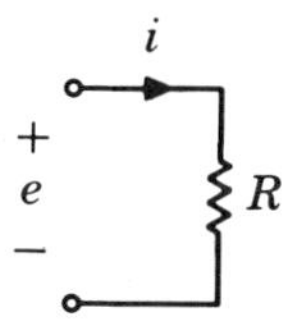

$e = iR$

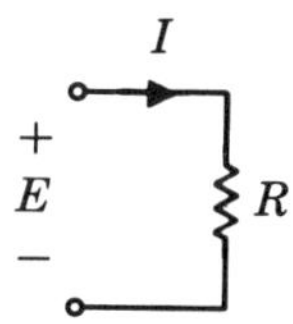

$E = IR$

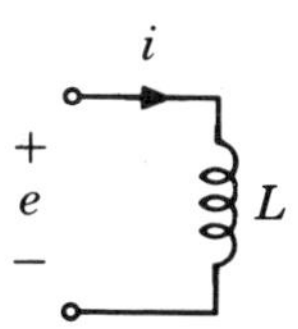

$e = L\dfrac{di}{dt}$

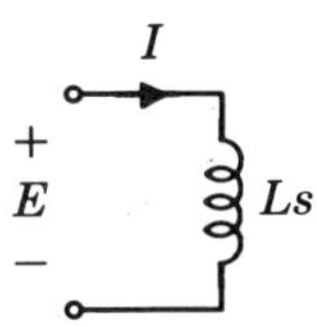

$E = ILs$

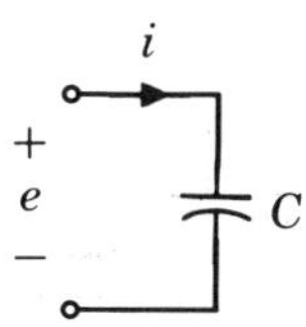

$e = \dfrac{1}{C}\displaystyle\int_{-\infty}^{t} i\,dt$

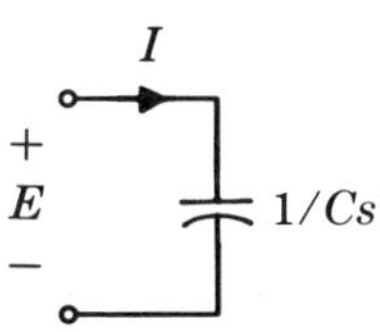

$E = \dfrac{I}{Cs}$

Response functions of the complex frequency s

$$\mathrm{T}(s) = \frac{\text{Output}}{\text{Input}} = \frac{P(s)}{Q(s)} = \frac{A(s-s_1)(s-s_3)\ldots}{(s-s_2)(s-s_4)\ldots}$$

$s_1, s_3 \ldots$ are zeros and $s_2, s_4, \ldots$ are poles

Direct plots of response functions

$$\mathrm{T}(j\omega) = \frac{1}{1+j\omega T} \qquad\qquad \mathrm{T}(j\omega) = \frac{1}{1+2\zeta j\omega T + (j\omega T)^2}$$

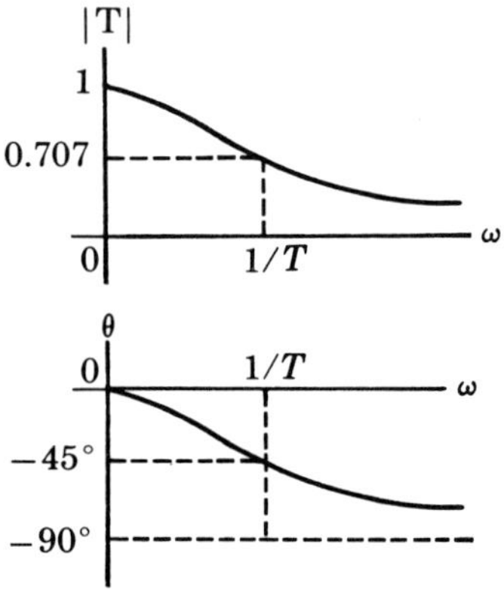

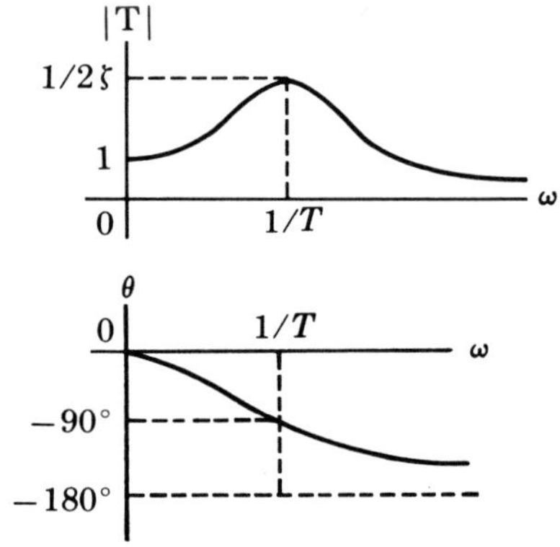

Logarithmic plots of response functions

$$\mathrm{T}(j\omega) = \frac{1}{1+j\omega T} \qquad\qquad \mathrm{T}(j\omega) = \frac{1}{1+2\zeta j\omega T + (j\omega T)^2}$$

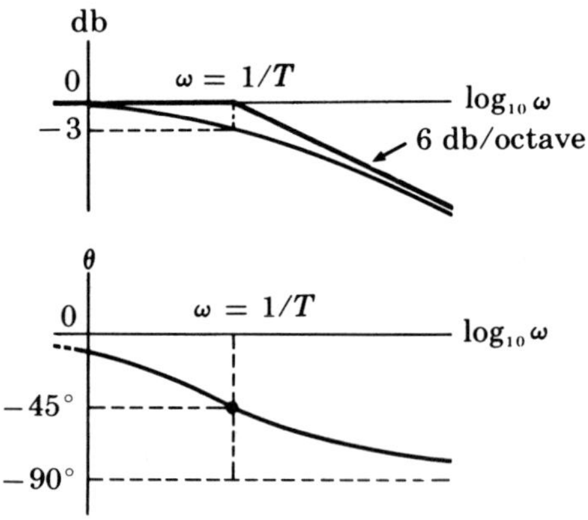

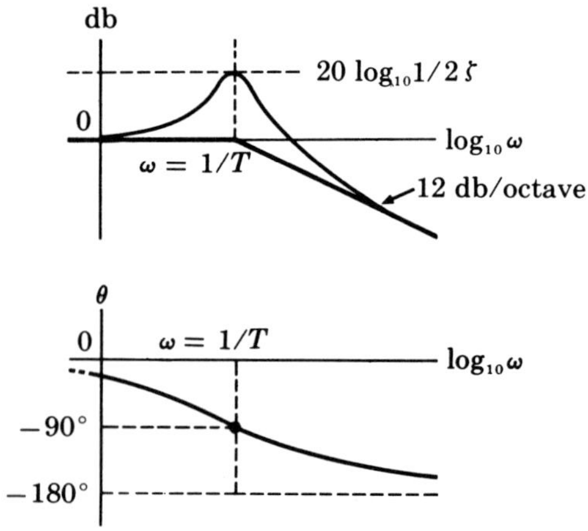

***s*-plane plot of response function**

$$\mathrm{T}(j\omega) = \frac{(j\omega - s_1)}{(j\omega - s_2)(j\omega - s_4)}$$

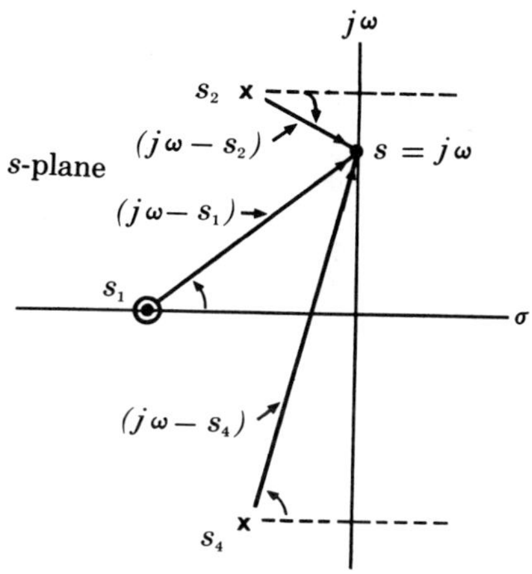

Poles and zeros of response functions

Driving-point impedance

Poles: open-circuit natural resonant frequencies
Zeros: short-circuit natural resonant frequencies

Driving-point admittance

Poles: short-circuit natural resonant frequencies
Zeros: open-circuit natural resonant frequencies

Transfer impedance

Poles: open-circuit natural resonant frequencies
Zeros: zeros of transmission

Transfer admittance

Poles: short-circuit natural resonant frequencies
Zeros: zeros of transmission

REFERENCES

In this chapter we have discussed two simple techniques for characterizing networks with magnitude and phase curves plotted against frequency. The log-db method is useful for low-Q circuits, and the s-plane vector method is useful for high-Q circuits. Many authors fail to appreciate the complementary nature of these methods and treat only one of them. References to both methods are given below.

1. H. W. BODE, *Network Analysis and Feedback Amplifier Design,* D. Van Nostrand, New York, 1945. Bode was the first author to use log-db plots, or semi-infinite slopes, as he called them. In Chapter 15 they are introduced as a technique for obtaining the phase curve from the amplitude curve.

2. JAMES, NICHOLS and PHILIPS, *Theory of Servomechanisms,* McGraw-Hill, New York, 1947. In this text the result of much war-time work at MIT and the Bell Telephone Laboratories is summarized, and the log-db method is presented in detail.

3. BROWN and CAMPBELL, *Principles of Servomechanisms,* John Wiley & Sons, New York, 1948. Chapter 8 gives an excellent summary of the log-db method.

4. CHESTNUT and MAYER, *Servomechanisms and Regulating Systems,* John Wiley & Sons, New York, 1951. As in most modern servomechanisms texts, there is here a good discussion of the log-db method, with many examples.

5. E. A. GUILLEMIN, *Introductory Circuit Theory,* John Wiley & Sons, New York, 1953. Guillemin has taught the s-plane vector method for many years and it is presented in detail in this elementary text.

6. E. J. ANGELO, *Electronic Circuits,* McGraw-Hill, New York, 1958. Angelo's text is one of the few that give an adequate treatment of both the s-plane vector method and the log-db method.

7. BRENNER and JAVID, *Analysis of Electric Circuits,* McGraw-Hill, New York, 1959. The elements of both methods are presented under the heading "Adjustable Circuits."

8. CLEMENT and JOHNSON, *Electrical Engineering Science,* McGraw-Hill, New York, 1960. Specific examples of both log-db plots and s-plane vectors are given in this text.

EXERCISES*

Section 18–2

1. (a) For the circuit of Fig. 18–17 make a complex-frequency-domain circuit.
   (b) Find the steady-state current which is of the same nature as the exciting voltage.
   (c) Find the transient term which would be excited by the sudden application of the voltage at $t = 0$. (The transient term adds to the steady-state term to produce the actual current in the circuit at $t = 0^+$.)

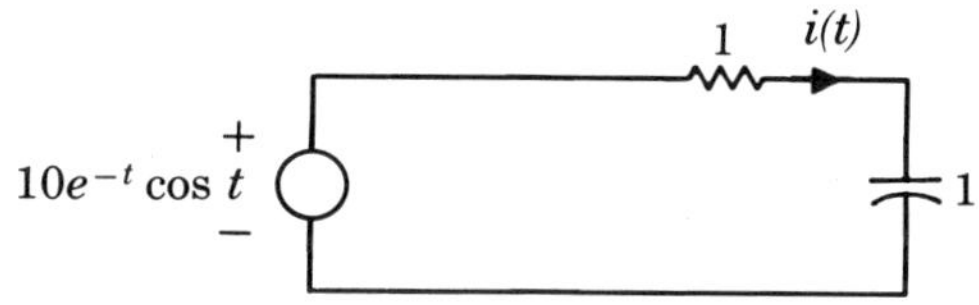

FIGURE 18–17

2. (a) Make a complex-frequency-domain circuit for Fig. 18–18.
   (b) Find the steady-state component of the current.
   (c) The voltage is applied at $t = 0^+$. Find the transient current which, when added to the steady-state current, produces the current actually in the circuit at $t = 0^+$.

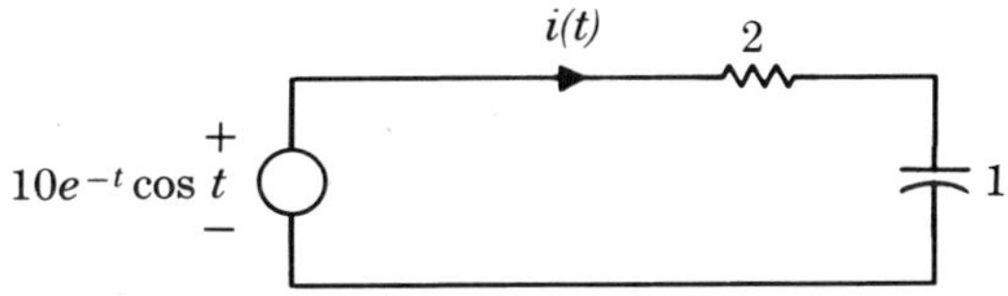

FIGURE 18–18

3. (a) Make a complex-frequency-domain circuit for Fig. 18–19.
   (b) Find the steady-state component of the current.
   (c) Find the transient current which must be added to the steady-state value, if the voltage is applied at $t = 0^+$.
   (d) In this case would it be possible to measure the steady-state component of the current?

---

* The units on the diagrams are ohms, henrys, farads, volts, and amperes.

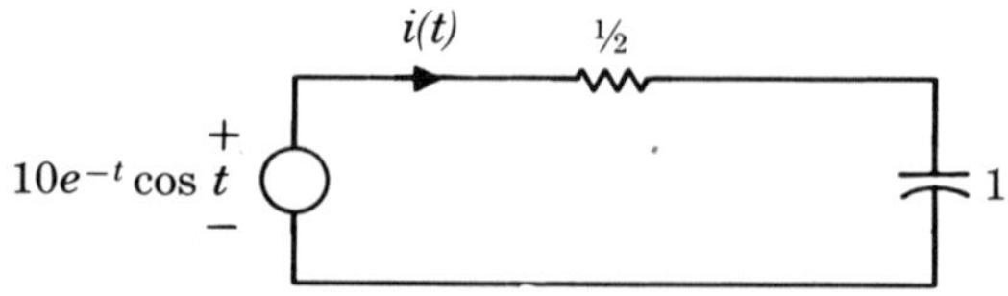

FIGURE 18–19

4. Show that if the driving function for a network is a sinusoid with a rising exponential envelope, the steady-state response function can always be observed. What is the general criterion for the availability of the steady-state term?

5. Give the impedance of each of the networks in Fig. 18–20 as a function of complex frequency, and list the poles and zeros.

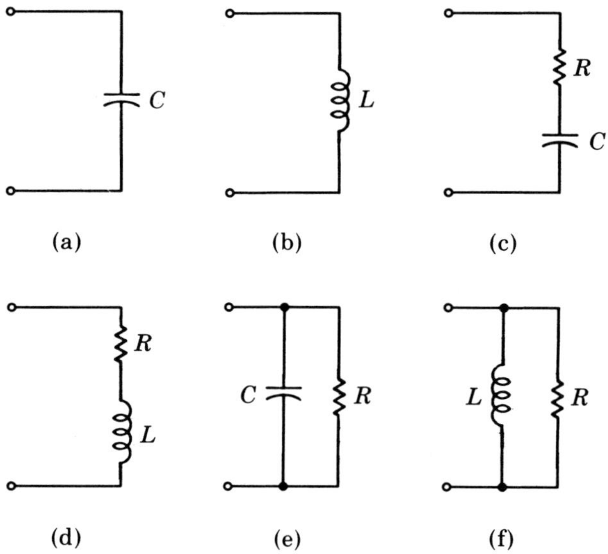

FIGURE 18–20

6. (a) Find the impedance of the network in Fig. 18–21 as a function of the complex frequency s. (b) Give its poles and zeros.

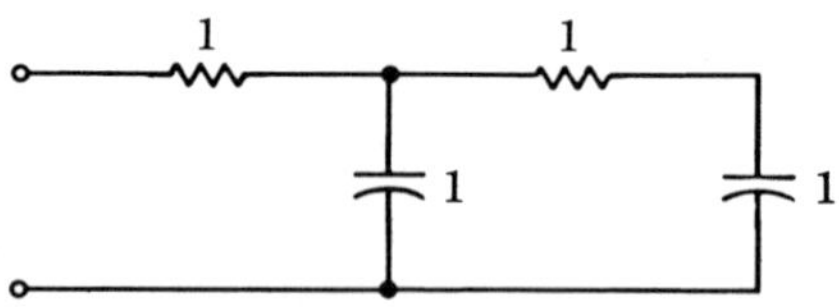

FIGURE 18–21

7. (a) Find the impedance of the network in Fig. 18–22 as a function of the complex frequency s. (b) Give the poles and zeros.

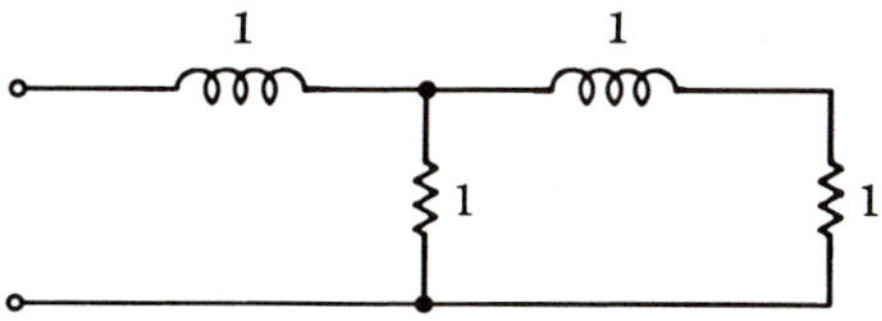

FIGURE 18–22

8. (a) Find the driving-point impedance of the network in Fig. 18–23 as a function of the complex frequency s. (b) Give the poles and zeros.

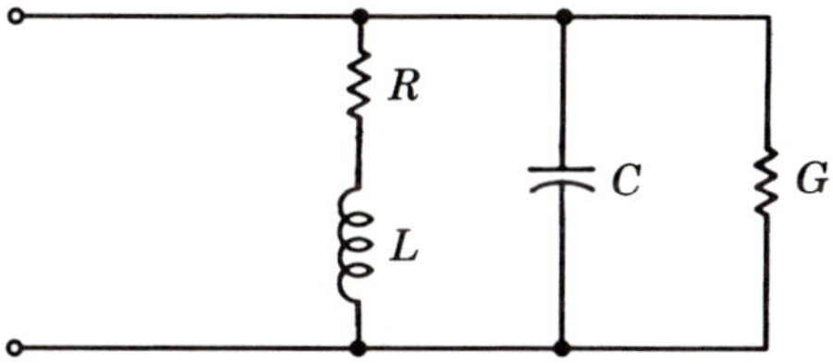

FIGURE 18–23

9. (a) Find the driving-point impedance of the network in Fig. 18–24 as a function of the complex frequency s. (b) Give the poles and zeros.

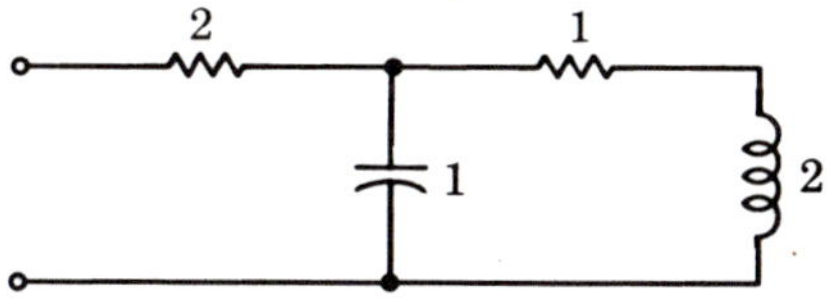

FIGURE 18–24

10. (a) Find the driving-point admittance of the circuit in Fig. 18–25. (b) Find the poles and zeros of the admittance.

11. (a) Find the z parameters of the two-terminal-pair network in Fig. 18–26. (b) Find the y parameters of the network.

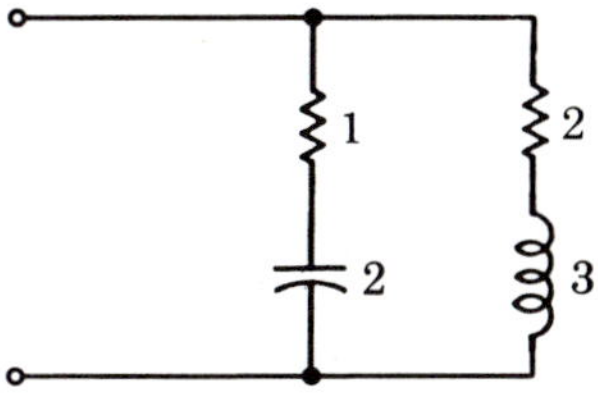

FIGURE 18–25

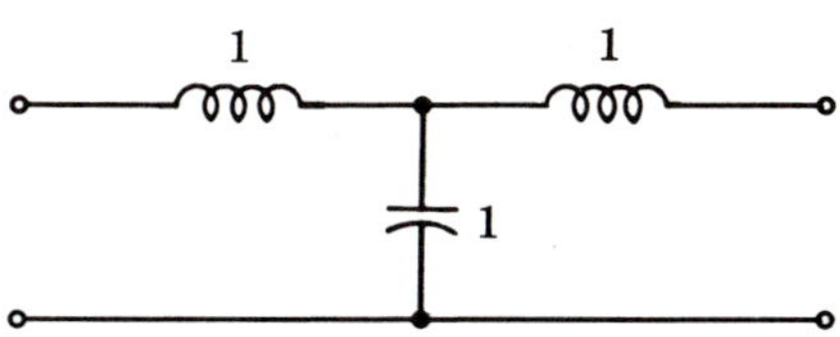

FIGURE 18–26

Section 18–3

12. (a) For the network of Fig. 18–27 find the ratio E_2/E_1 as a function of ω.
    (b) Sketch the magnitude of the ratio in part (a) as a function of frequency.
    (c) Sketch the phase of the ratio in part (a) as a function of frequency.
    (d) Scale the network values to give a half-power frequency of 10 kc and a resistance of 100K ohms.

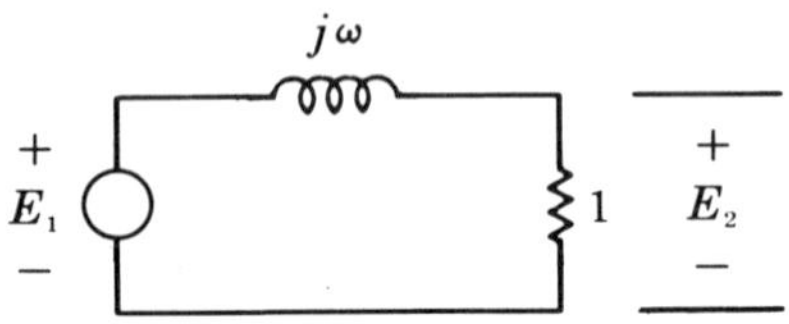

FIGURE 18–27

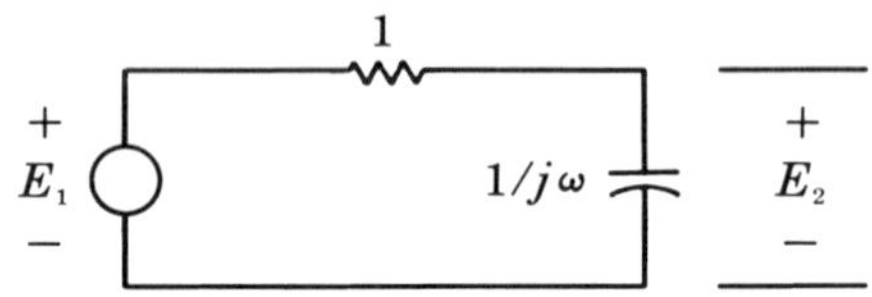

FIGURE 18–28

13. (a) Find the ratio E_2/E_1 for the network of Fig. 18–28.
    (b) Sketch magnitude and phase curves as functions of frequency.
    (c) Scale the network values to give a half-power frequency of $10^6/2\pi$ cps and a resistance of 10K ohms.
14. (a) A low-pass filter with two energy storage elements is shown in Fig. 18–29. Find the ratio E_2/I_1.
    (b) Sketch magnitude and phase curves for this ratio.
    (c) Scale the network to give a half-power frequency of $100/2\pi$ cps and a resistance of 1K ohms.

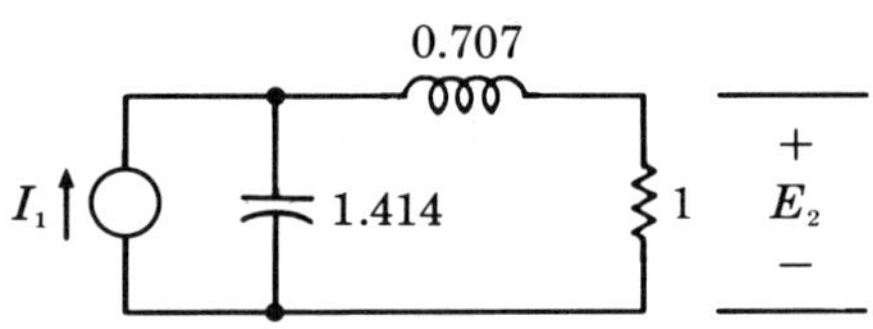

FIGURE 18–29

15. (a) A low-pass filter with three energy storages is shown in Fig. 18–30. Find the ratio E_2/I_1.
    (b) Sketch magnitude and phase curves for this ratio.
    (c) Scale the element values to give a half-power frequency of $1000/2\pi$ cps and an impedance of 1000 ohms at dc.

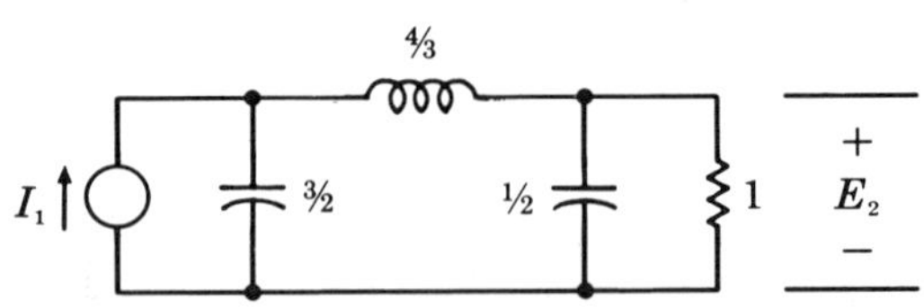

FIGURE 18–30

16. A series R–L–C circuit has $R = 10$ ohms, $L = 0.01$ henry, and $C = 10^{-6}$ farad.

(a) Find the impedance as a function of ω.
(b) Give the resonant frequency.
(c) What is the bandwidth of the resonance curve between the half-power frequencies?
(d) Give the Q of the circuit.
(e) Sketch the magnitude and phase of the impedance as functions of frequency.

17. A series R–L–C circuit is to have a resonant frequency of $1000/2\pi$ cps and a bandwidth of $100/2\pi$ cps. The impedance at resonance is to be 100 ohms. Find the values of R, L, and C.

18. A parallel R–L–C circuit has $R = 10^5$ ohms, $L = 0.001$ henry, and $C = 10^{-7}$ farad.

(a) Find the impedance as a function of ω.
(b) Give the resonant frequency.
(c) What is the bandwidth of the resonance curve between the half-power frequencies?
(d) Give the Q of the circuit.
(e) Sketch the magnitude and phase of the impedance as functions of frequency.

19. A parallel R–L–C circuit is to have a resonant frequency of $1000/2\pi$ cps. The impedance at resonance is to be 10^5 ohms, and the bandwidth between the half-power frequencies is to be $100/2\pi$ cps. Find the values of R, L, and C.

Section 18–4

20. A transfer function for a network (output response divided by input stimulus) is given by

$$T(s) = \frac{1}{s + 10}.$$

(a) Sketch a curve of log-magnitude versus log-frequency.
(b) Sketch the phase versus log-frequency curve.
(c) Give the values of the curve of (a) at $\omega = 5$, 10, and 20 radians.

21. A transfer function is given by

$$T(s) = \frac{s}{s + 10}.$$

(a) Sketch the log-magnitude versus log-frequency curve
(b) Sketch the phase versus log-frequency curve.
(c) Give the values of the curve in db at $\omega = 5$, 10, and 20 radians.

22. A transfer function is given by

$$T(s) = \frac{1}{s^2 + 5s + 4}.$$

(a) Sketch the log-magnitude versus log-frequency curve.
(b) Sketch the phase versus log-frequency curve.
(c) Give the values of the curve in db at $\omega = 1$ and $\omega = 4$ radians.

23. A transfer function is given by

$$T(s) = \frac{s+1}{s+10}.$$

(a) Sketch the log-magnitude versus log-frequency curve.
(b) Sketch the phase versus log-frequency curve.
(c) Give the values of the curve in db at $\omega = 1$ and $\omega = 10$ radians.

24. A transfer function is given by

$$T(s) = \frac{s-10}{s+10}.$$

(a) Sketch the log-magnitude versus log-frequency curve.
(b) Sketch the phase versus log-frequency curve.

25. The transfer function of a resistance-capacitance coupled vacuum-tube amplifier is given by

$$\frac{E_2(s)}{E_1(s)} = \frac{100}{(1+100/s)(1+s/10^5)}.$$

(a) Sketch the log-magnitude versus log-frequency curve.
(b) Give the lower and the upper half-power frequencies.
(c) What is the mid-band gain?
(d) Sketch the phase versus log-frequency curve.

26. A transfer function is given by

$$T(s) = \frac{s}{s^2+100}.$$

(a) Sketch the log-magnitude versus log-frequency curve.
(b) Sketch the phase versus log-frequency curve.

27. A transfer function is given by

$$T(s) = \frac{s}{s^2+s+1}.$$

(a) Sketch the log-magnitude versus log-frequency curve.
(b) Sketch the phase versus log-frequency curve.

28. A transfer function is given by

$$T(s) = \frac{1}{s^3+2s^2+2s+1}.$$

(a) Sketch the log-magnitude versus log-frequency curve.
(b) Sketch the phase versus log-frequency curve.

Section 18–5

29. An impedance has a zero at zero and a pole at $s = -1$. At very high frequencies its value is 2. (a) Write the impedance. (b) Give the phase at $\omega = \infty$ and at $\omega = 0$. (c) Find the magnitude and phase at $s = j1$, from the frequency-domain vectors.

30. An impedance has a zero at zero and a zero at infinity, and a pair of complex conjugate poles at $s = -1 \pm j$. It is further known that $sZ(s) \rightarrow 1$ as $s \rightarrow \infty$.

(a) Give $Z(s)$.
(b) Give the phase at $\omega = \infty$, and at $\omega = 0$.
(c) Give the magnitude and phase at $s = j$.

31. (a) Find the impedance represented by the s-plane plot of Fig. 18–31, given that the value at $s = 0$ is 1.
(b) From the s-plane vectors evaluate the impedance at $s = j$.
(c) If a current $i(t) = 10 \cos t$ is applied to the impedance, what voltage will occur across it.

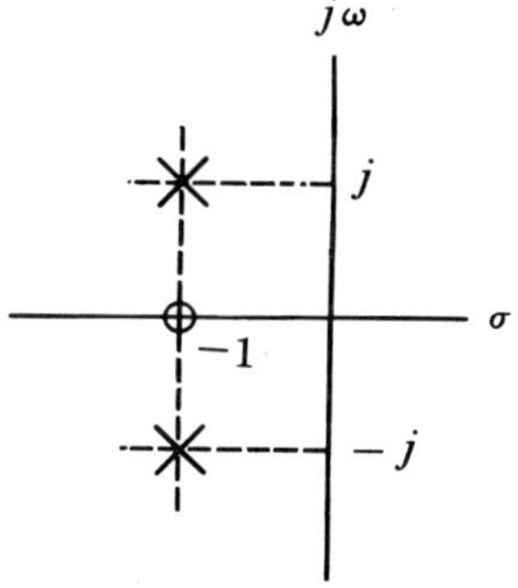

FIGURE 18–31

32. The impedance of a series R–L–C circuit is

$$Z(s) = \frac{10(s + 1 + j10)(s + 1 - j10)}{s}.$$

(a) Make use of the s-plane vectors to obtain an approximate expression for the impedance in the vicinity of the resonance.
(b) Give the frequency of the resonance.
(c) What is the bandwidth?
(d) What is the Q.
(e) Give the impedance at resonance.
(f) Sketch the resonance curve in magnitude and phase.

33. A series R–L–C impedance is given by

$$Z(s) = \frac{5(s + 1 + j100)(s + 1 - j100)}{s}.$$

(a) Find the values of R, L, and C.
(b) Find the element values which will increase the resonant frequency by a factor of 1000 by frequency scaling.
(c) Sketch the poles and zeros in parts (a) and (b) and in each case give the Q.

34. A parallel resonant circuit has the impedance

$$Z(s) = \frac{1}{10} \frac{s}{(s+1+j10)(s+1-j10)}.$$

(a) Find an approximate form of $Z(s)$ which holds in the vicinity of resonance.
(b) Give the resonant frequency.
(c) What is the impedance at the resonant frequency?
(d) Give the bandwidth.
(e) Give the Q.

35. The series parallel circuit of Fig. 18–32 has the impedance

$$Z(s) = 10^6 \frac{(s+1)}{(s+1+j100)(s+1-j100)}.$$

(a) Find the values of R, L, C, and G.
(b) Find the element values if the resonant frequency is scaled up by a factor of 1000.
(c) Sketch the poles and zeros in parts (a) and (b) and give the Q in each case.

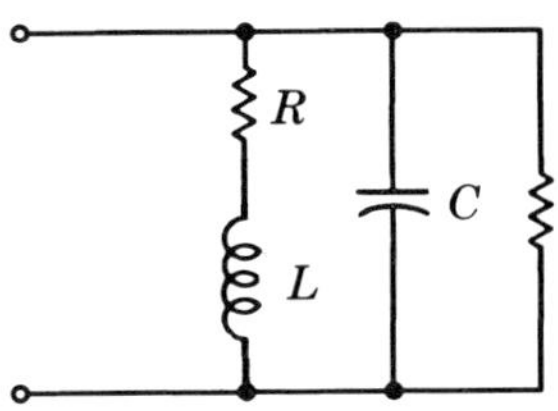

FIGURE 18–32

36. An impedance is given by

$$Z(s) = \frac{10(s+1+j100)(s+1-j100)}{(s+1)}.$$

A voltage source

$$e(t) = 1.0 + \cos 50t + \cos 100t$$

is applied to the impedance. Compute the resulting current at each frequency.

Section 18–6

37. A network impedance has a pole at $s = -2$ and a zero at $s = -1$. Write the general form of the transients which will occur in the circuit (a) if the circuit is driven by a voltage source and (b) if the circuit is driven by a current source.

38. The impedance of a network is given by

$$Z(s) = \frac{s+1}{s^2+2s+100}.$$

(a) Write the general form of the transient which will occur if the circuit is driven by a voltage source. (b) Write the general form of the transient which will occur if the circuit is driven by a current source.

39. An impedance is given by

$$Z = \frac{s^2+2s+100}{(s+1)(s+2)}.$$

(a) Write the impedance of the dual circuit.
(b) Give the general nature of the short-circuit transients in the dual circuit.
(c) Give the general nature of the open-circuit transients in the dual circuit.

40. (a) Find the driving-point impedance of the network in Fig. 18–33.
(b) Write the general form of its open-circuit transient.
(c) Write the general form of its short-circuit transient.

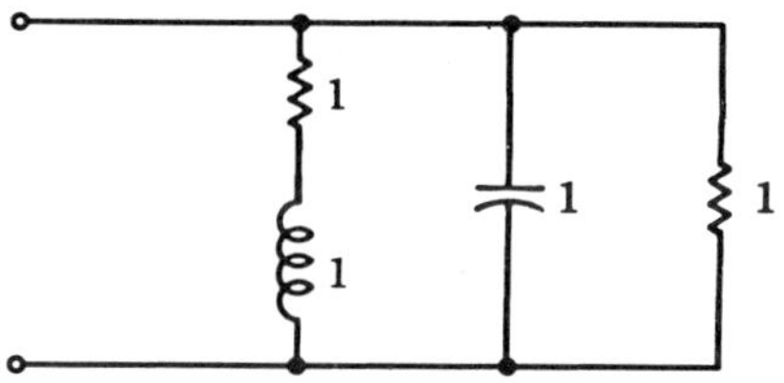

FIGURE 18–33

41. (a) Find the driving-point impedance of the circuit in Fig. 18–34.
(b) Give the poles and zeros and the short- and open-circuit transients.
(c) If the switch is opened at $t = 0$, after having been closed for a long time, find $e_1(t)$, $e_2(t)$, and $e_1(t) + e_2(t)$.

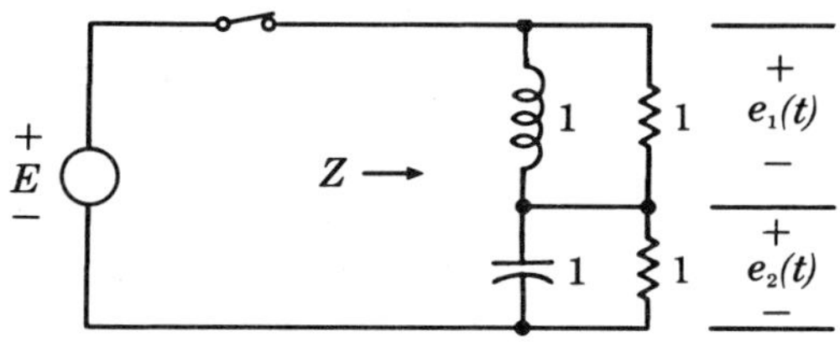

FIGURE 18–34

42. (a) Find the driving-point impedance of the circuit in Fig. 18–35.
(b) Give the poles and zeros and the short- and open-circuit transients.
(c) If the switch is closed at $t = 0$, find $i_1(t)$, $i_2(t)$, and $i_1(t) + i_2(t)$.

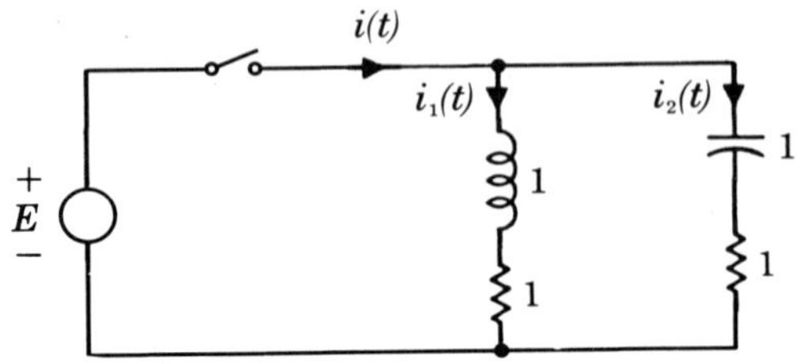

FIGURE 18–35

43. (a) Compute the z parameters for the two-terminal-pair network of Fig. 18–36. Use the complex-frequency variable s.
(b) If a 1-ohm resistance is added to the output of the network compute the ratio $E_2/I_1 = Z_{12}$.
(c) Comment on the poles and zeros of z_{11}, z_{12}, z_{22}, and Z_{12}.

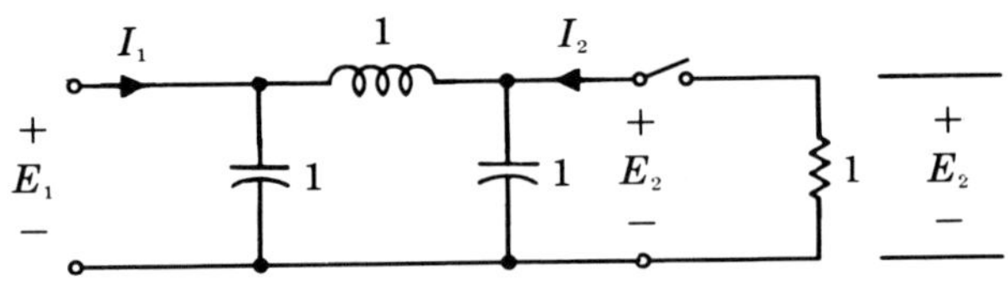

FIGURE 18–36

44. (a) Compute the y parameters for the two-terminal-pair circuit of Fig. 18–37 as functions of the complex frequency variable s.
(b) If a 1-ohm resistance is added to the output terminals, what is the transfer ratio $Y_{12} = I_2/E_1$.
(c) Comment on the poles and zeros of y_{11}, y_{12}, y_{22}, and Y_{12}.

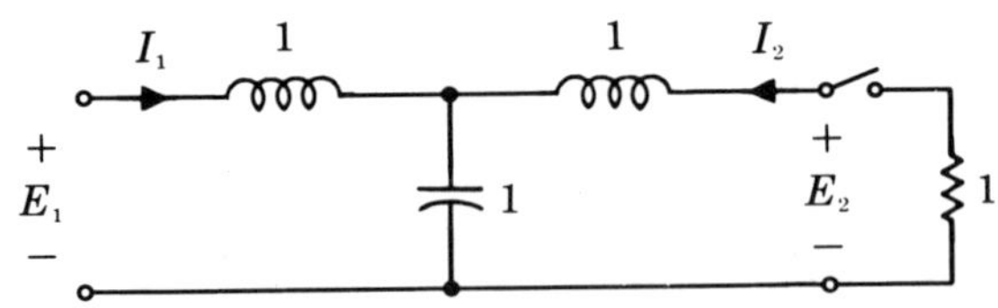

FIGURE 18–37

CHAPTER 19

POWER AND ENERGY IN ALTERNATING-CURRENT CIRCUITS

"Take warning! Alternating currents are dangerous! They are fit only for powering the electric chair. The only similarity between an a-c and a d-c lighting system is that they both start from the same coal pile."

Thomas Edison, Pamphlet of 1887

19–1 Introduction. Edison's first lighting systems employed direct currents, and he remained a firm advocate of direct current all his life. It was his rival George Westinghouse who, with the technical help of Nicola Tesla, promoted and developed the alternating-current system. The superiority of a-c systems lies in the higher voltages which they can use for transmission purposes and the ease with which the voltage can be transformed to a low value for the ultimate user. At the present time perhaps 25% of the total energy used in the United States is converted to electricity for transportation purposes. The reason that the percentage is not higher is that it is easier and more efficient to transport energy in the form of oil and coal. A large oil pipe line can transport about 20 times the energy-per-hour that a high voltage transmission line can handle, and more efficiently too. The losses in the pipe line run about 1% of the energy transported, while the losses in the transmission line may be as large as 20% of the energy transmitted. However, we are rapidly running out of oil and coal, and it has been estimated that in 50 years essentially all the energy used in the United States will be electrical and that the prime generators will be atomic in nature.*

One of the reasons that Edison mistrusted alternating currents is that he didn't really understand them. They are more difficult to predict than direct currents, and it wasn't until the work of Steinmetz at the General Electric Company that a good method for solving a-c circuits became available. In 1897, complex algebra and phasors were applied to electric circuits in a classical paper by Steinmetz, and since that time no one has had any excuse for not understanding a-c circuits!

The power and energy relations in a-c circuits are more complex than in d-c circuits, but they can be obtained readily enough when the currents and voltages in the circuit are known. Part of the difficulty is simply a

---

* See Putnam, *Energy for the Future*, D. Van Nostrand, New York, 1953.

matter of notation. The introduction of RMS, or root-mean-square, values falls in this category. The instantaneous power dissipated in a resistance by alternating current falls to zero twice every cycle, and reaches a maximum value between the zeros. (Since the current is squared, the reversal of current does not change the sign of the instantaneous power.) Thus it is convenient to define the average power dissipated in the resistance and the equivalent constant current which would dissipate the same amount of power. This equivalent current value is the RMS value of the alternating current.

A second example of new notation concerns the energy stored in the inductances and capacitances of the circuit. Even though the current changes sign there is an average amount of energy stored in an inductance, and even though the voltage changes sign there is an average amount of energy stored in a capacitance. It is not customary to compute these average energies directly, but rather to define a new quantity called imaginary power which is equal to the frequency multiplied by a function of the stored energy.

The expressions for the instantaneous power and energy in an a-c circuit are complex functions of time. Usually, we are not interested in instantaneous power and energy, but only in the average values which we must supply or which we are consuming. A real simplification occurs here, for the average power and the average energy stored can be computed from frequency-domain phasors, while the instantaneous values must be computed in the time domain. There is no a priori reason why phasors should contain information about the power and energy in the circuit. Phasors were defined for linear operations, and power is a nonlinear quantity. It is only a happy *coincidence* that the product of a current phasor and a voltage phasor is related to the power and energy in a circuit. In this chapter we shall be concerned with exploiting this coincidence as fully as possible.

19–2 Root-mean-square (RMS) values.

When a sinusoidal current is passed through a resistance, the instantaneous power dissipated is a complex function of the time. It has a value of zero twice each cycle and reaches a peak between the two zeros. Although the current reverses on alternate half-cycles, the power does not since it is a function of the square of the current. For most applications we are interested only in the average power, and the average power is a constant so long as the amplitude of the current is a constant. As an example of instantaneous and average power, we will compute the power in the resistance of Fig. 19–1. The current is

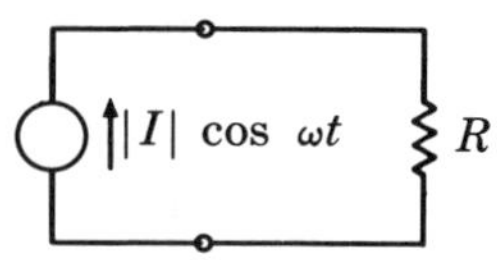

FIG. 19–1. A resistance excited by an alternating current.

$$i(t) = |I| \cos \omega t,$$

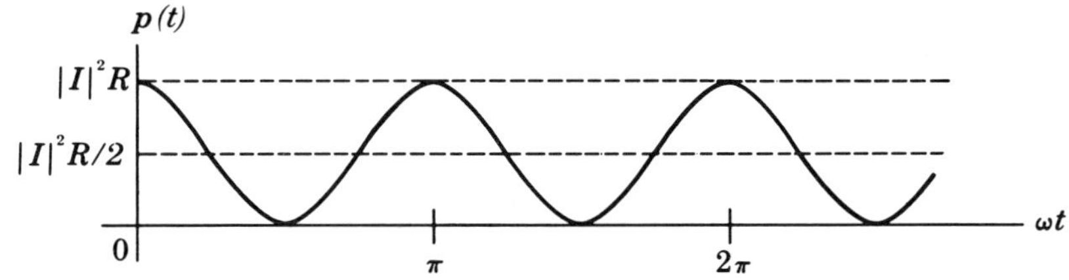

FIG. 19–2. The instantaneous power in the resistance of Fig. 19–1.

and the power as a function of time is

$$p = e \cdot i = i^2 R = |I|^2 R \cos^2 \omega t$$

$$= \frac{|I|^2 R}{2} (1 + \cos 2\omega t). \tag{19–1}$$

The instantaneous power of Eq. (19–1) is plotted in Fig. 19–2.* It varies from a peak of $|I|^2 R$ to a minimum of zero. The average value, denoted by P, is

$$P = \frac{|I|^2 R}{2}. \tag{19–2}$$

In order to make Eq. (19–2) agree in form with the expression for power in a d-c circuit, it is customary to write

$$P = I_{\text{rms}}^2 R. \tag{19–3}$$

The meaning of I_{rms} is that it is the equivalent direct current which would produce the same amount of heating. For the sinusoidal example which we have just computed, the RMS value is

$$I_{\text{rms}} = 0.707\,|I|. \tag{19–4}$$

The equivalent heating value, or RMS value, is a function of the wave-shape. Let us observe the process by which it was obtained. The first step was to square the waveform, as in Eq. (19–1). The second step was to take the average value of the squared function, as in Eq. (19–2). The third step was to take the root of this squared value, as in Eq. (19–4). What we have done is to take the Root of the Mean of the Squared function. The abbreviation RMS is thus a description of the process by which

* The student should observe for future reference that the average value of the square of any sinusoidal wave is one-half the peak value of the squared wave.

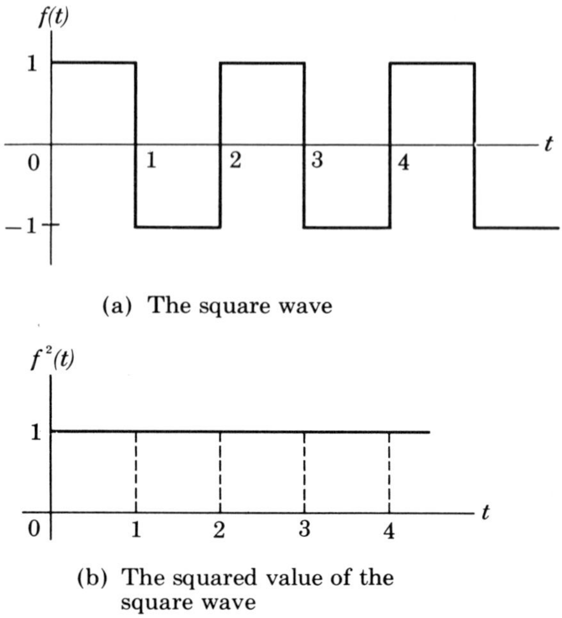

FIG. 19–3. Computation of the RMS value of a square wave.

the equivalent heating value is obtained. It is also clear that the RMS property is a function of the waveshape and can be found for voltages as well as for currents.

As a second example of RMS calculations, consider the square wave shown in Fig. 19–3(a). The first step in the process is to square the waveform, as shown in Fig. 19–3(b). The squared waveform has a constant value of unity. Its average is therefore unity. Finally, the root of the average is obtained and in this case the value is unity. Thus *the RMS value of the square wave is unity.* It is just as effective for heating as a constant wave of the same peak value.

The RMS value can always be obtained graphically by the process which we have used in the above example. If the waveform is known analytically the RMS value can also be obtained analytically. The waveform is squared and integrated over a period in order to obtain the area under the squared curve. This area is then divided by the period, T, in order to obtain the average value, and the square root of the average value is taken. The formula which expresses these operations is

$$F_{\text{rms}} = \sqrt{\frac{1}{T}\int_0^T f^2(t)\,dt}. \tag{19–5}$$

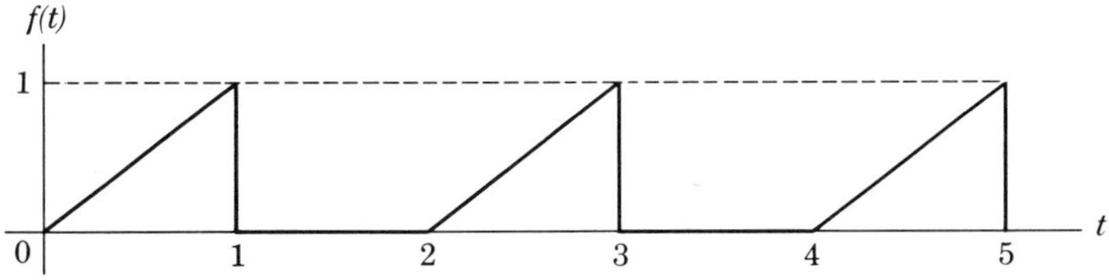

FIG. 19–4. A function whose RMS value is obtained analytically.

As an example of the use of Eq. (19–5) we will obtain the RMS value of the waveform shown in Fig. 19–4. The function is defined by

$$f(t) = \begin{cases} t & 0 < t < 1, \\ 0 & 1 < t < 2. \end{cases} \tag{19–6}$$

The period is $T = 2$, and the RMS value is

$$\begin{aligned} F_{\text{rms}} &= \sqrt{\tfrac{1}{2}\int_0^1 t^2\,dt} \\ &= \sqrt{\tfrac{1}{2} \times \tfrac{1}{3}} \\ &= 0.408. \end{aligned}$$

RMS values are commonly used for the computation of power and energy in the sinusoidal steady-state. Most meters for use on 60-cps systems are calibrated to read RMS values. Indeed, if a 60-cps voltage or current is stated without any qualifications, it is safe to assume that it is an RMS value. The 110-volt a-c service for household lighting is 110 volts RMS. The peak value of the voltage is about 155 volts.

A second and less obvious use for RMS values is for the addition of two waves of different frequencies. The time expression for the sum of two sine waves is complex in form, and we are often glad to settle for the equivalent heating effect of the combination. Suppose the two sine waves are

$$\begin{aligned} i_1(t) &= |I_1| \cos(\omega_1 t + \theta_1), \\ i_2(t) &= |I_2| \cos(\omega_2 t + \theta_2). \end{aligned} \tag{19–7}$$

The sum of the waves is

$$i(t) = |I_1| \cos(\omega_1 t + \theta_1) + |I_2| \cos(\omega_2 t + \theta_2). \tag{19–8}$$

The heating value of the sum is expressed by the RMS value. To obtain the RMS value we square the wave, average the squared waveform, and take the square root. The square of Eq. (19–8) is

$$i^2(t) = |I_1|^2 \cos^2 (\omega_1 t + \theta_1) + |I_2|^2 \cos^2 (\omega_2 t + \theta_2) + 2|I_1|\,|I_2| \cos (\omega_1 t + \theta_1) \cos (\omega_2 t + \theta_2). \tag{19–9}$$

The average value of each of the cosine-squared terms in Eq. (19–9) is $\frac{1}{2}$. The average value of the product term is zero.* If we denote the average value by a bar, we can write

$$\overline{i^2(t)} = \frac{|I_1|^2}{2} + \frac{|I_2|^2}{2},$$

and

$$I_{\text{rms}} = \sqrt{I_{1_{\text{rms}}}^2 + I_{2_{\text{rms}}}^2}. \tag{19–10}$$

Equation (19–10) shows that the heating effect of the sum of two sinusoids is the root of the sum of the squares of the separate RMS values. A summation of this sort is called *RMS addition.* It is important that the student understand when it is possible to perform RMS addition and when it is not. Essentially, RMS addition is used for sine waves with different frequencies.† Two direct-current values, or two identical waveforms of any kind, are added by *arithmetic addition.* Two sinusoids of the same frequency are added by *vector addition.* The "order of preference" for addition is

1. Arithmetic addition.
2. Vector addition.
3. RMS addition.

Arithmetic addition gives us the complete resultant waveform; vector addition gives us the resultant phasor; and RMS addition gives us only the equivalent heating effect. Not only is RMS addition less informative than the other two, but also it is *wrong* for waveforms which can be treated by the other methods. For example, two equal and opposite waveforms with a resultant of zero would have an apparent RMS sum of 1.414 times the value of either one of them!

---

* The sum of the two cosine waves exhibits the phenomenon of beats. The envelope of the sum has a maximum when the waves are in phase and a minimum when they are out of phase. The average must extend over a beat period (1/difference frequency), and therefore the case of two equal frequencies is ruled out.

† Mathematically, the waveforms must be orthogonal. They must have a commensurable period and the integral of their product over this period must be zero. See Chapter 20.

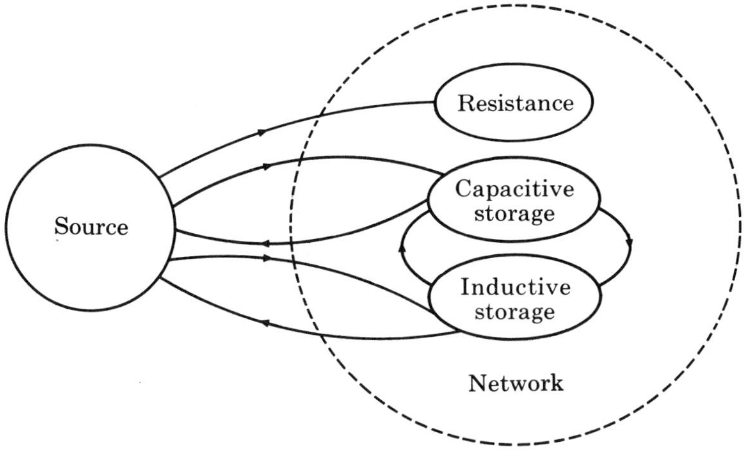

FIG. 19–5. An energy flow chart for a general a-c network.

19–3 Active and reactive power in the time domain. In d-c circuits we are not concerned to any great extent with the direction of power flow. Usually, power originates in a source and flows into a load resistance. When two sources are present, there is the possibility that power might flow from one source to the other, but such examples are relatively rare. In alternating-current circuits, however, the direction of power flow is more important, since on one half-cycle a source may be delivering power to a circuit, and on the other half-cycle it may be receiving power from the circuit.

A resistance always absorbs energy, and the direction of power flow is always from the source to the resistance element. An energy-storage element, however, cannot continue to absorb power indefinitely, and in the sinusoidal steady-state the energy stored must drop to zero twice each cycle, when the pertinent voltage or current drops to zero. Thus energy must flow into and out of energy-storage elements at a rate of twice the supply frequency. On one half-cycle the source supplies energy to the energy-storage element, and on the next half-cycle the energy-storage element returns the energy to the source. A typical situation is diagrammed in Fig. 19–5. The direction of energy flow is always from the source to the resistance. The amount of energy flowing is not a constant, however, but pulsates from zero to a maximum value twice during each cycle of the source frequency. The direction of energy flow to the energy-storage elements reverses twice during each cycle. First, the energy flows into the energy-storage element, and then it flows back to the source again. Energy is also interchanged between capacitive and inductive energy-storage elements within the network.

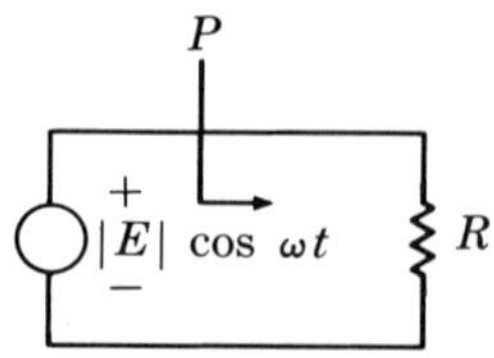

(a) A resistance excited by an a-c voltage

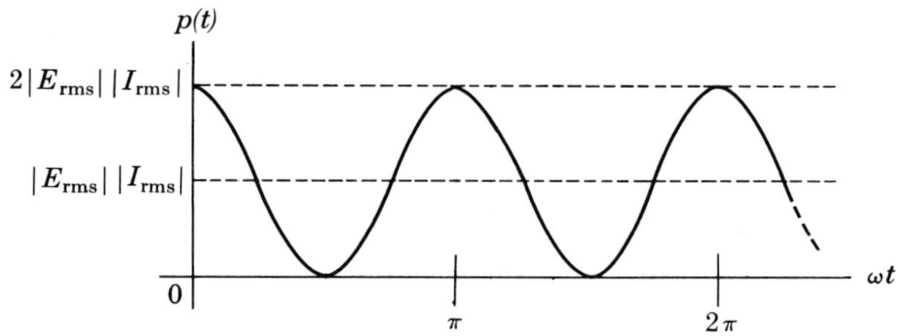

(b) The instantaneous power dissipated in the resistance

FIG. 19–6. Power and energy in the resistance element.

The student may be inclined to ignore the energy fluctuating back and forth between the source and the energy-storage elements on the grounds that it represents no net energy supplied to the network. Such a conclusion is hasty, however, since the power company must supply generating capacity and transmission facilities for the power loaned to the network, as well as for the power absorbed by the network. Furthermore, the fluctuating currents required to supply the stored energy produce losses in the generating and transmitting system, and somebody must pay for them! Thus the power company is forced to keep track not only of the *real power* which it is supplying, but also of the energy which is "swapped" back and forth. The name *reactive-power*, or *imaginary power*, is given to the power associated with this energy interchange.

We shall begin our consideration of the power and energy relationships of a-c circuits by considering the purely resistive case shown in Fig. 19–6(a). We shall suppose that the voltage and current are given by

$$e(t) = |E| \cos \omega t,$$

$$i(t) = |I| \cos \omega t.$$

The instantaneous power, the product of the instantaneous current and voltage, is

$$\begin{aligned} p = e \cdot i &= |E|\,|I| \cos^2 \omega t \\ &= \frac{|E|\,|I|}{2}(1 + \cos 2\omega t) \\ &= |E_{\rm rms}|\,|I_{\rm rms}| + |E_{\rm rms}|\,|I_{\rm rms}| \cos 2\omega t. \end{aligned} \tag{19–11}$$

Equation (19–11) is plotted in Fig. 19–6(b). The instantaneous power fluctuates from zero to a maximum value of $2E_{\rm rms}I_{\rm rms}$ twice during each cycle of the input waveform. The average value of the power is

$$\begin{aligned} P &= |E_{\rm rms}|\,|I_{\rm rms}| \\ &= \frac{|E_{\rm rms}|^2}{R} \\ &= |I_{\rm rms}|^2 R. \end{aligned} \tag{19–12}$$

Except for the change to RMS values, the formulas for the a-c power in a resistance are identical to the formulas for the d-c case. The direction of power flow is always from the source to the resistance.

An inductance driven by a sinusoidal voltage is shown in Fig. 19–7(a). If the voltage is

$$e(t) = |E| \cos \omega t,$$

the alternating component of the current is*

$$i(t) = \frac{|E|}{\omega L} \sin \omega t = |I| \sin \omega t, \tag{19–13}$$

and the instantaneous power is

$$\begin{aligned} p = e \cdot i &= |E|\,|I| \cos \omega t \sin \omega t \\ &= |E_{\rm rms}|\,|I_{\rm rms}| \sin 2\omega t. \end{aligned} \tag{19–14}$$

A plot of the instantaneous power of Eq. (19–14) is given in Fig. 19–7(b). For the first half-cycle the power is positive, representing a flow from the source to the inductance. For the second half-cycle the power is negative, representing a flow from the inductance back to the source. The peak value of this fluctuating component is called *the reactive power*, Q.† In

---

* As in all alternating-current circuits, the transients are ignored.

† The real power P of Eq. (19–11) can also be defined as the peak value of the fluctuating component of the power absorbed.

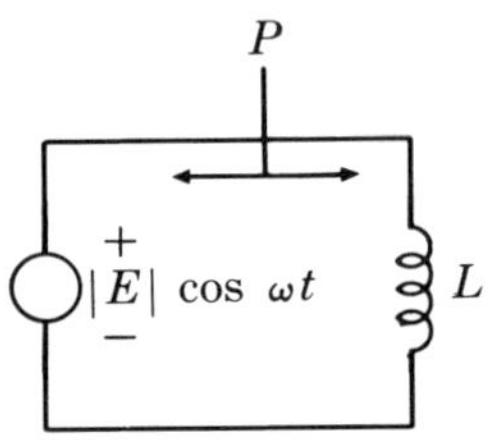

(a) An inductance excited by an a-c voltage

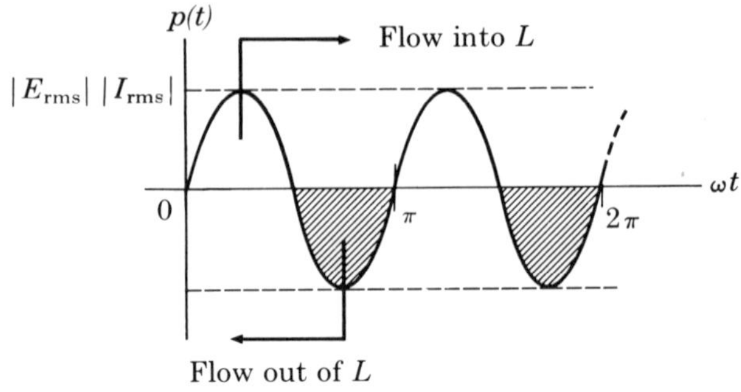

(b) The instantaneous power flow into and out of the inductance

FIG. 19–7. Power and energy in the inductance element.

the case of the pure inductance

$$Q = |E_{\text{rms}}|\,|I_{\text{rms}}| = \frac{|E_{\text{rms}}|^2}{\omega L} = |I_{\text{rms}}|^2 \omega L. \tag{19–15}$$

The reactive power Q, which is flowing back and forth to charge the inductance, is also related to the average amount of energy which is stored in the inductance. The instantaneous energy in the inductance is

$$w_L = \frac{Li^2}{2} = \frac{L|I|^2}{2}\sin^2\omega t = \frac{L|I_{\text{rms}}|^2}{2}(1 - \cos 2\omega t). \tag{19–16}$$

The average value of the energy stored is*

$$W_L = \tfrac{1}{2}L|I_{\text{rms}}|^2. \tag{19–17}$$

A comparison of Eqs. (19–17) and (19–15) reveals that

$$Q = 2\omega(W_L). \tag{19–18}$$

Stated in words, Eq. (19–18) shows that the *reactive power Q drawn by an inductance is 2ω times the average value of the magnetic energy stored in the inductance.* The importance of this conclusion for future arguments lies in the fact that average energy is a scalar quantity which can be combined

* The symbol T is sometimes used for the average inductive stored energy.

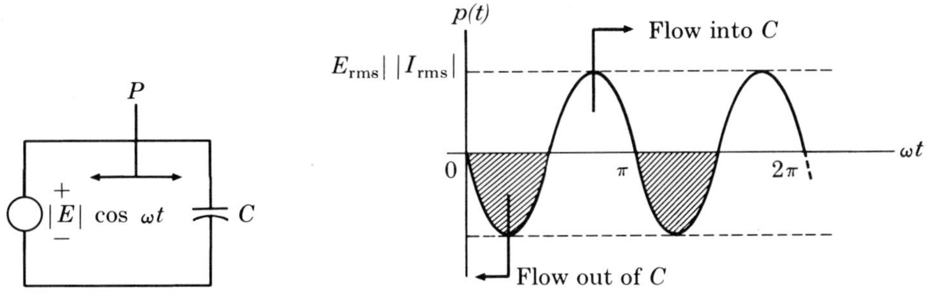

(a) A capacitance excited by an a-c voltage

(b) The instantaneous power flow into and out of the capacitance

FIG. 19–8. Power and energy in the capacitance element.

with other scalar quantities without any regard for the relative phases of the time functions which are producing the energy.

The energy and power relationships of the capacitance of Fig. 19–8(a) are similar to the relationships for the inductance. In order to have comparable energy expressions, we will also excite the capacitance with the voltage

$$e(t) = |E| \cos \omega t.$$

The current in the capacitance is

$$i(t) = C\frac{de}{dt} = -\omega C|E| \sin \omega t = -|I| \sin \omega t, \tag{19–19}$$

and the instantaneous power is

$$\begin{aligned} p = e \cdot i &= -|E|\,|I| \cos \omega t \sin \omega t \\ &= -|E_{\text{rms}}|\,|I_{\text{rms}}| \sin 2\omega t. \end{aligned} \tag{19–20}$$

The instantaneous power given by Eq. (19–20) is plotted in Fig. 19–8(b). On the first half-cycle the energy is negative, representing a flow of energy from the capacitance to the source, and on the second half-cycle it is positive, representing a flow of energy from the source to the capacitance. If the student is wondering how the capacitance can give up energy to the source before it gets energy from the source, he should recall that the system is in the steady-state condition, and that the energy has been swapping back and forth for some time before $t = 0$. The original energy got into the capacitance during the original transient state of the network, while the steady-state was being established. The energy in the capacitance, however, is different from the energy in the inductance.

The two energy flows are 180° out of phase. While energy is flowing into the inductance, it is flowing out of the capacitance, and while energy is flowing into the capacitance, it is flowing out of the inductance. Furthermore, if the two energy flows occurred at the same time, they would tend to cancel each other. Thus, if we take the reactive power for the inductance to be positive, we should take the reactive power for the capacitance to be negative.* For a capacitance

$$\begin{aligned} Q &= -|E_{\text{rms}}|\,|I_{\text{rms}}| \\ &= -|E_{\text{rms}}|^2\omega C \\ &= -\frac{|I_{\text{rms}}|^2}{\omega C}. \end{aligned} \tag{19–21}$$

Again, there is a relationship between the reactive power and the average energy stored in the capacitance. The instantaneous energy is

$$\begin{aligned} w_C &= \frac{Ce^2}{2} = \frac{C|E|^2}{2}\cos^2\omega t \\ &= \frac{C|E_{\text{rms}}|^2}{2}(1+\cos 2\omega t), \end{aligned} \tag{19–22}$$

and the average value of the energy stored is†

$$W_C = \frac{C|E_{\text{rms}}|^2}{2}. \tag{19–23}$$

From Eqs. (19–21) and (19–23) it is apparent that for a capacitance

$$Q = -2\omega(W_C). \tag{19–24}$$

According to Eq. (19–24), the reactive power in a capacitance is negative and has a value which is 2ω times the average energy stored in the capacitance.

We are now in a position to consider a more general branch in which resistance and reactance are combined, as shown in Fig. 19–9(a). The voltage is

$$e(t) = |E|\cos(\omega t),$$

and the current is

$$i(t) = |I|\cos(\omega t - \theta).$$

---

* The choice of the sign of reactive power is arbitrary. Common usage in the USA assigns the positive sign to inductive reactive power.

† The symbol V is sometimes used for the average capacitive stored energy.

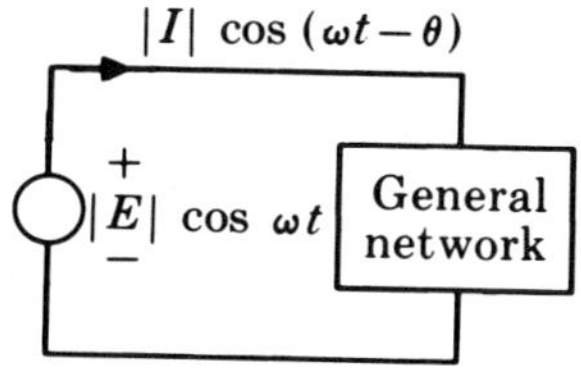

(a) A general network branch excited by an a-c voltage

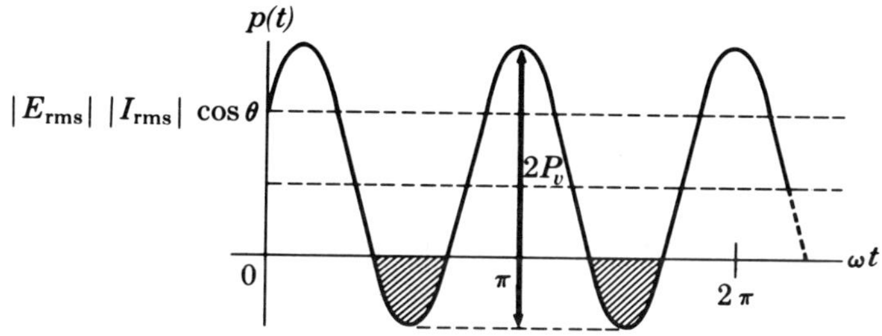

(b) The instantaneous power $p(t)$ and the vector power P_v

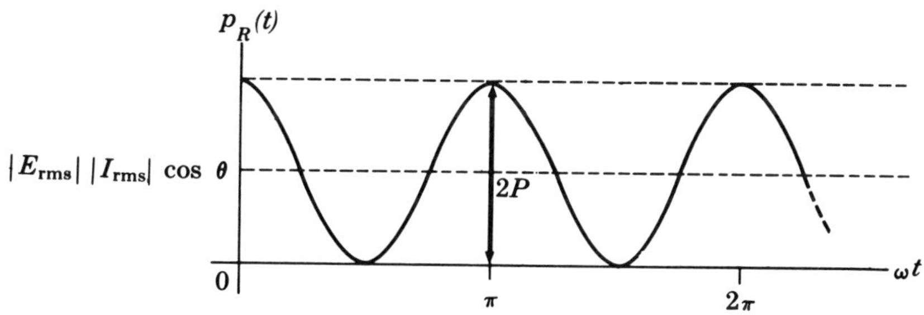

(c) The real power $p_R(t)$ and its average value P

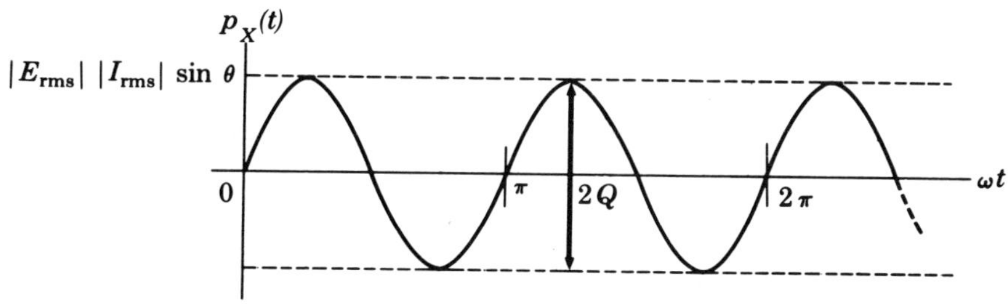

(d) The reactive power $p_X(t)$ and its peak value Q

FIG. 19-9. Definition of real and reactive power in the time domain.

Since the current lags the voltage for an inductive load, a positive value of θ corresponds to an inductive load. The instantaneous power is

$$\begin{aligned} p = e \cdot i &= |E|\,|I| \cos \omega t \cos (\omega t - \theta) \\ &= \frac{|E|\,|I|}{2} \cos \theta + \frac{|E|\,|I|}{2} \cos (2\omega t - \theta) \\ &= \underbrace{|E_{\text{rms}}|\,|I_{\text{rms}}| \cos \theta + |E_{\text{rms}}|\,|I_{\text{rms}}| \cos \theta \cos 2\omega t}_{\text{Real power}} \\ &\quad + \underbrace{|E_{\text{rms}}|\,|I_{\text{rms}}| \sin \theta \sin 2\omega t}_{\text{Reactive power}} \end{aligned} \tag{19–25}$$

Equation (19–25) is plotted in Fig. 19–9(b). The power flow has both positive and negative components, but there is a net average positive value which represents the power absorbed in the resistance. The first two terms in Eq. (19–25), plotted in Fig. 19–9(c), represent the real power which fluctuates from zero to a maximum of $2E_{\text{rms}}I_{\text{rms}} \cos \theta$. The third term represents the reactive power which is surging back and forth, and which has no net average value. The peak value of the reactive power in the circuit is $E_{\text{rms}}I_{\text{rms}} \sin \theta$, as shown in Fig. 19–9(d). From this resolution we see that

$$\begin{aligned} P &= |E_{\text{rms}}|\,|I_{\text{rms}}| \cos \theta, \\ Q &= |E_{\text{rms}}|\,|I_{\text{rms}}| \sin \theta, \end{aligned} \tag{19–26}$$

where P represents the real power absorbed by the circuit and Q represents the peak value of the reactive component, which is surging back and forth. The sign of Q is automatically taken care of by the sign of $\sin \theta$, with the assumption that θ is positive for an inductive load, and negative for a capacitive load. The factor $\cos \theta$ is called the *power factor*, and the factor $\sin \theta$ is called the *reactive factor*, since these factors modify the volt-amperes in each case.

One of the principal advantages of working with power and energy in a circuit is that these quantities are scalars and can be added together arithmetically, no matter where they arise in the circuit. The real power P is always a positive quantity, and the average real power absorbed by a circuit is simply the sum of the average real powers absorbed in each branch. This conclusion is based on the conservation of energy. Thus

$$P_{\text{total}} = \text{Sum of the branch } P\text{'s}. \tag{19–27}$$

There is no similar conservation law for reactive power, and although we can reach a similar conclusion regarding the reactive power, the argument is more complex. First, let us suppose that the energy-storage elements are all inductive. Then the reactive power drawn by each branch is 2ω times the average inductive stored energy. The sum of all the average inductive stored energies must be supplied by the source (by the conservation of energy), and hence the total reactive power supplied to the network must be the sum of the branch reactive powers absorbed.

Now let us consider a single branch which has both inductive and capacitive elements. For a common voltage across them (or a common current through them), the two reactive powers will have opposite signs, as shown in Eqs. (19–15) and (19–21). The net average energy which must be supplied to the branch is therefore the difference between the average inductive energy and the average capacitive energy. When a number of these branches are combined only the net average energy is supplied to each branch. Since the reactive power is 2ω times the average stored energy, the total reactive power which must be supplied is the sum of the total inductive reactive power less the sum of the total capacitive reactive power, or more simply the sum of the net reactive powers of the branches. Thus

$$Q_{\text{total}} = \text{Sum of the branch } Q\text{'s}. \tag{19–28}$$

Implicit in Eq. (19–28) is the sign convention which makes inductive reactive power positive and capacitive reactive power negative.

Equation (19–28) is a rather remarkable result, and it is anything but obvious. It states in effect that a capacitance added at any point in a network can be used to cancel the effects of an inductance added at any other point, and that the cancellation will be complete provided that they both draw the same amounts of reactive power. The phenomenon is one of resonance, but it is a good deal more general than the series or parallel resonance which occurs within a single branch. As a practical matter, of course, if the problem is to supply negative reactive power to cancel some undesirable positive reactive power, the capacitance should be as close to the inductance as possible in order to minimize circuit losses produced in the intervening resistances by the additional currents introduced when the capacitance is added.

The actual computation of active and reactive power is much easier in the frequency domain than in the time domain, and hence we will defer the solution of a practical example until we have covered the computation of power in the frequency domain.

19–4 Active and reactive power in the frequency domain. In the previous section we developed expressions for the average power dissipated in a network and for the peak value of the power which fluctuates back and forth between the source and the energy-storage elements in the network. Since the computation of voltage and current in the network is easier in the frequency domain than in the time domain, it is natural to inquire about the computation of the power in the frequency domain. There is no a priori reason to suspect that power can be computed from the frequency-domain phasors, however. They were defined for the linear operations of circuit theory, and there is no reason to suppose that they will also hold for the nonlinear operation of obtaining power. We will be agreeably surprised to find that they do, but we must "doctor" them in a rather arbitrary way before we can obtain power from them.

The expressions which we obtained for the real, or active, power P, and the imaginary, or reactive, power Q are

$$P = |E_{\text{rms}}|\,|I_{\text{rms}}| \cos\theta,$$

$$Q = |E_{\text{rms}}|\,|I_{\text{rms}}| \sin\theta,$$

where $|E_{\text{rms}}|$ and $|I_{\text{rms}}|$ are the RMS values of the magnitudes of the voltage and the current waveforms, and θ is the angle between the current and the voltage. It is an easy step to change our frequency-domain vectors to RMS values by dividing their lengths by 1.414. The angle θ poses a slightly greater problem. If we simply multiply the two phasors, we obtain an angle which is the sum of the angles of E and I and not the angle between them. Suppose

$$E_{\text{rms}} = |E_{\text{rms}}|\,\underline{|\phi_1},$$

$$I_{\text{rms}} = |I_{\text{rms}}|\,\underline{|\phi_2},$$

as shown in Fig. 19–10. Now the conjugate vector for I_{rms} is

$$I^*_{\text{rms}} = |I_{\text{rms}}|\,\underline{|-\phi_2},$$

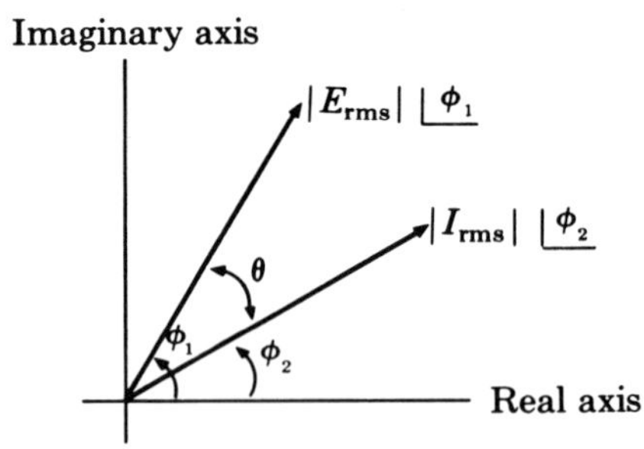

FIG. 19–10. RMS vectors.

and the product of the RMS voltage phasor and the conjugate of the RMS current phasor is

$$E_{\text{rms}}I_{\text{rms}}^* = |E_{\text{rms}}|\,|I_{\text{rms}}|\,\underline{/\phi_1 - \phi_2} = |E_{\text{rms}}|\,|I_{\text{rms}}|\,\underline{/\theta}, \tag{19–29}$$

in which the angle θ is the difference angle between the voltage and the current phasors. The expansion of Eq. (19–29) is

$$\begin{aligned} E_{\text{rms}}I_{\text{rms}}^* &= |E_{\text{rms}}|\,|I_{\text{rms}}| \cos\theta + j|E_{\text{rms}}|\,|I_{\text{rms}}| \sin\theta \\ &= P + jQ. \end{aligned} \tag{19–30}$$

The phasor defined by Eq. (19–30) is called the *vector (or phasor) power* and is designated by P_v. Thus

$$P_v = E_{\text{rms}}I_{\text{rms}}^* = P + jQ. \tag{19–31}$$

The phasor product of the RMS voltage and the conjugate of the RMS current gives with one inclusive sweep the real power and the reactive power. The meaning of Eq. (19–31) is best grasped by referring to Eq. (19–25), which is repeated here for convenience:

$$\begin{aligned} p = |E_{\text{rms}}|\,|I_{\text{rms}}| \cos\theta &+ |E_{\text{rms}}|\,|I_{\text{rms}}| \cos\theta \cos 2\omega t \\ &+ |E_{\text{rms}}|\,|I_{\text{rms}}| \sin\theta \sin 2\omega t. \end{aligned} \tag{19–32}$$

The first term is the real average power. The second term has an amplitude which is the real power and the third term has an amplitude which is the reactive power. Both of the latter terms have a frequency of $2\omega t$. The phase angle of the real power is zero and the phase angle of the reactive power is 90°. The vector sum of $P + jQ$ is the phasor which represents the sum of the two fluctuating terms in Eq. (19–32). The symbol P is the phasor for the time function shown in Fig. 19–9(c), Q is the phasor for the time function shown in Fig. 19–9(d), and P_v, the resultant phasor, represents the time function shown in Fig. 19–9(b), apart from its constant term.

Equation (19–31) is the basic formula for computing power in the frequency domain. The three quantities P, Q, and P_v are all products of voltage and current and have the dimensions of volt-amperes. In order to distinguish them, the real power P is measured in *watts*, the reactive power Q is measured in *reactive volt-amperes, or vars,* and the vector power is measured in *volt-amperes*, or *va*.

A common problem which occurs is the computation of the input power to a network from a source. At the frequency of the source, the network

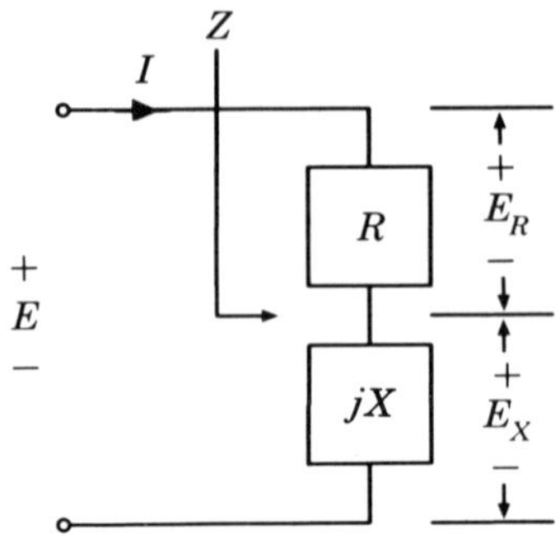

FIG. 19-11. A general impedance branch.

has an equivalent impedance

$$Z = R + jX. \tag{19-33}$$

The power absorbed by this impedance is the same as the power absorbed by the network, since both draw the same power from the source. A special set of formulas is often developed for this case, based on the "magic triangle." A typical network is shown in Fig. 19-11, and the corresponding impedance triangle is plotted in Fig. 19-12(a). If the impedance is multiplied by I, Eq. (19-33) becomes

$$IZ = IR + jIX$$

or

$$E = E_R + jE_X. \tag{19-34}$$

The vector diagram for the voltage of Eq. (19-34) is identical to the vector diagram for the impedance of Eq. (19-33), except for a change in scale and a rotation to line up the E_R vector with the current vector.

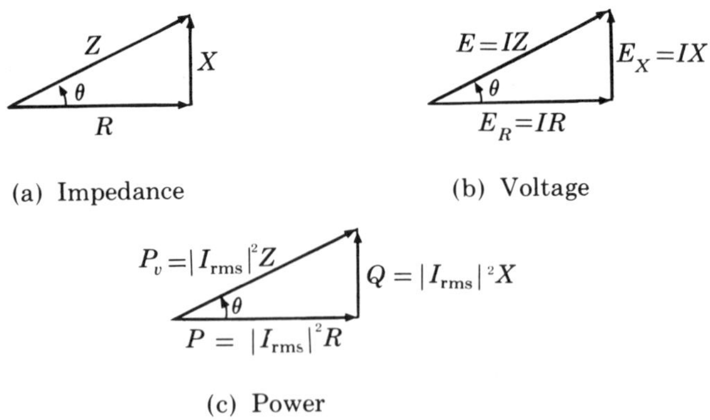

FIG. 19-12. The impedance-voltage-power triangle.

As a second step, let us multiply Eq. (19–33) by $|I_{\text{rms}}|^2$. We then get

$$\begin{aligned} |I_{\text{rms}}|^2 Z &= |I_{\text{rms}}|^2 R + j|I_{\text{rms}}|^2 X \\ &= |I_{\text{rms}}|^2 |Z| \cos\theta + j|I_{\text{rms}}|^2 |Z| \sin\theta \\ &= |I_{\text{rms}}|\,|E_{\text{rms}}| \cos\theta + j|I_{\text{rms}}|\,|E_{\text{rms}}| \sin\theta, \end{aligned}$$

or

$$P_v = P + jQ. \tag{19–35}$$

The vector diagram for the power of Eq. (19–35) is identical to the vector diagram for the impedance, except for a change in scale. The voltage vector diagram and the power vector diagram are shown in Figs. 19–12(b) and 19–12(c). The magic impedance diagram becomes voltage or power just by a change in scale!

As a numerical example of the computation of vector power, real power, and reactive power, consider the circuit of Fig. 19–13. For this circuit, the impedance is

$$Z = 1 + j \text{ ohms.}$$

The current is

$$I_{\text{rms}} = \frac{1\,\underline{|0^\circ}}{1+j} = 0.707\,\underline{|-45^\circ} \text{ amp,}$$

for an input voltage of

$$E_{\text{rms}} = 1\,\underline{|0^\circ} \text{ volts.}$$

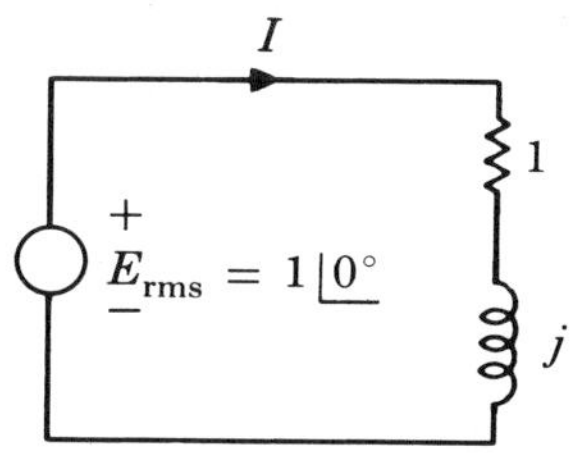

FIG. 19–13. A simple example for the calculation of power quantities.

Vector power. The basic formula for the vector power is

$$P_v = E_{\text{rms}} I^*_{\text{rms}} = (1\,\underline{|0^\circ})(0.707\,\underline{|45^\circ}) = 0.5 + j0.5 \text{ va.}$$

An alternative method from the power triangle is

$$P_v = |I_{\text{rms}}|^2 (R + jX) = 0.5(1 + j) = 0.5 + j0.5 \text{ va.}$$

Real power. The real power can be obtained from the vector power. Thus

$$P = \text{Re}\,(E_{\text{rms}} I^*_{\text{rms}}) = 0.5 \text{ watt.}$$

It can also be obtained from the power triangle, in the form

$$P = |I_{\text{rms}}|^2 R = 0.5(1) = 0.5 \text{ watt.}$$

Another formula, which is essentially the same as the real part of the vector power, is

$$P = |E_{\text{rms}}|\,|I_{\text{rms}}| \cos\theta = (1)(0.707)\cos 45° = 0.5 \text{ watt}.$$

Reactive power. The reactive power can be obtained from the vector power:

$$Q = \text{Im}\,(E_{\text{rms}} I^*_{\text{rms}}) = 0.5 \text{ var}.$$

It can also be obtained from the power triangle, in the form

$$Q = |I_{\text{rms}}|^2 X = 0.5(1) = 0.5 \text{ var}.$$

Another formula, which is essentially the same as the imaginary part of the vector power, is

$$Q = |E_{\text{rms}}|\,|I_{\text{rms}}| \sin\theta = (1)(0.707)\sin 45° = 0.5 \text{ var}.$$

Power factor. Sometimes an explicit statement of the power factor is desired. The basic definition is

$$\text{pf} = \cos\theta = \cos 45° = 0.707.$$

It can also be obtained from the power triangle as

$$\text{pf} = \frac{R}{|Z|} = \frac{1}{1.414} = 0.707.$$

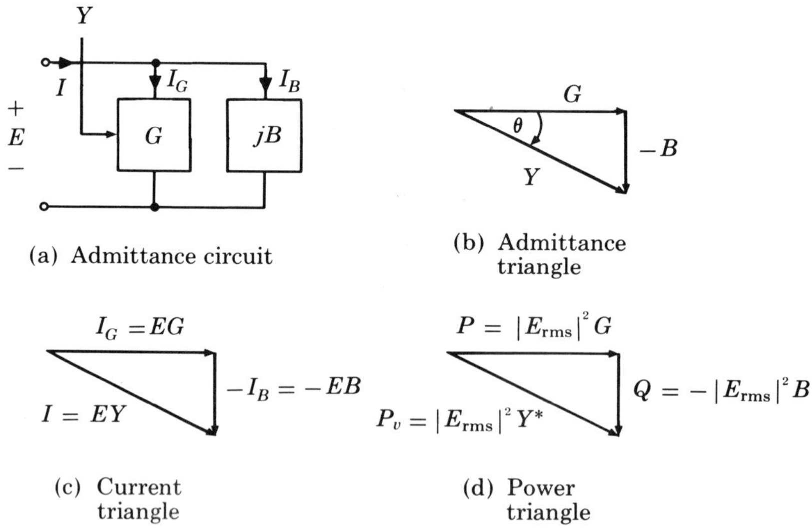

FIG. 19–14. The admittance-current-power triangle.

Another form of the power triangle expression is

$$\text{pf} = \frac{P}{|P_v|} = \frac{0.5}{0.707} = 0.707.$$

In the preceding discussion we have assumed that a network is replaced by an equivalent impedance at its input terminals. It is also possible to replace the network by its equivalent admittance, and when this is done the impedance-power triangle becomes an admittance-power triangle. The form of the equivalent circuit is shown in Fig. 19-14(a), and the input admittance is

$$Y = G + jB. \tag{19-36}$$

The admittance triangle is plotted in Fig. 19-14(b). The susceptance is plotted in the negative direction, because it is negative for an inductive circuit. If we multiply each side of the admittance triangle by E, we obtain the current triangle of Fig. 19-14(c), since

$$EY = EG + jEB$$

is the same as

$$I = I_G + jI_B. \tag{19-37}$$

If we multiply each side of the admittance triangle by $|E_{\text{rms}}|^2$, we obtain the power triangle of Fig. 19-14(d), since

$$|E_{\text{rms}}|^2 Y = |E_{\text{rms}}|^2 G + j|E_{\text{rms}}|^2 B$$

is the same as

$$P_v^* = P - jQ. \tag{19-38}$$

Note that when the sides of the admittance triangle are multiplied by $|E_{\text{rms}}|^2$, it becomes a triangle for the conjugate of the vector power. The only difference here is in the sign of the reactive power. Since this sign is arbitrary, the student can best keep track of it by always making sure that the reactive power is positive for an inductive load.

To work a numerical example, we will solve for the power in the circuit of Fig. 19-15. The input admittance is

$$Y = 1 - j \text{ mhos}.$$

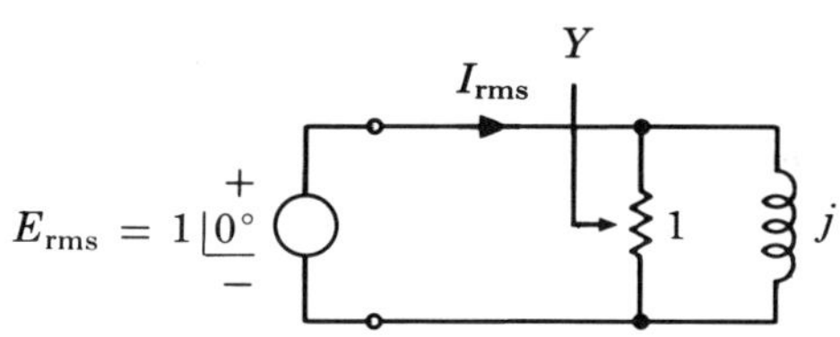

FIG. 19-15. An example for admittance-power calculations.

For an RMS input voltage of $1 \underline{|0^\circ}$ volts, the current is

$$I_{\text{rms}} = E_{\text{rms}} Y = (1 - j) \text{ amp.}$$

Vector power. The basic formula for the vector power is

$$P_v = E_{\text{rms}} I^*_{\text{rms}} = (1 \underline{|0^\circ})(1 + j) = (1 + j) \text{ va.}$$

An alternative formula, from the admittance-power triangle, is

$$P_v = |E_{\text{rms}}|^2 (G - jB) = (1 \underline{|0^\circ})(1 + j) \text{ va.}$$

Real power. The real power can be obtained from the vector power in the form

$$P = \text{Re}\,(E_{\text{rms}} I^*_{\text{rms}}) = |E_{\text{rms}}|\,|I_{\text{rms}}| \cos \theta = 1 \text{ watt.}$$

It can also be obtained from the power triangle in the form

$$P = |E_{\text{rms}}|^2 G = \frac{|E_{\text{rms}}|^2}{R} = 1 \text{ watt,}$$

where $R = 1/G$.

Reactive power. The reactive power can be obtained from the vector power in the form

$$Q = \text{Im}\,(E_{\text{rms}} I^*_{\text{rms}}) = |E_{\text{rms}}|\,|I_{\text{rms}}| \sin \theta = 1 \text{ var.}$$

It can also be obtained from the power triangle in the form

$$Q = -|E_{\text{rms}}|^2 B = -1 \times (-1) = 1 \text{ var.}$$

Power factor. The power factor is

$$\text{pf} = \cos \theta = \cos 45^\circ = 0.707.$$

It can also be obtained from the admittance triangle:

$$\text{pf} = \frac{G}{|Y|} = \frac{1}{1.414} = 0.707.$$

The power factor does not contain any information about the nature of the reactive part of the load. The reactive factor, $\sin \theta$, changes sign with θ, and therefore does contain information about the sign of the reactance. It is common practice, however, to compress the sign information into the power factor expression and not to compute the reactive factor. To accomplish this result, it is necessary to know only whether the current is leading or lagging the voltage. If it is leading, the circuit is capacitive and the reactive power is negative. If it is lagging, the circuit is inductive and the reactive power is positive. In power systems, the voltage is the stimulus and the current is the response. Hence a statement that the power factor is 0.5 lagging means that the magnitude of the power factor is 0.5 and that *the current* is lagging the voltage, or the circuit is inductive.

The student is cautioned against memorizing all the possible formulas for power. There are too many of them! He should memorize the basic formula for vector power, should know the convention for the sign of reactive power, and should understand the reasoning behind the impedance-power and admittance-power triangles. Beyond this, he should tackle each problem on its own merits.

19–5 Power in terms of branch currents and voltages. A power company is interested in real and reactive power, because it is paid for the real power which it delivers and often also for the reactive power which it delivers. The communications engineer is interested in real and reactive power, because they are scalar quantities and can be added by ordinary addition no matter where they arise in a circuit. In fact, the input impedance of a network (or the input admittance) can be found in terms of the total real power and total reactive power dissipated or used in the network. This approach is much used at high frequencies and in field theory problems. In this section we will derive the basic formulas and illustrate them by application to simple networks.

The total real power in a network is

$$P = \sum_{1}^{n} P_i, \qquad (19\text{–}39)$$

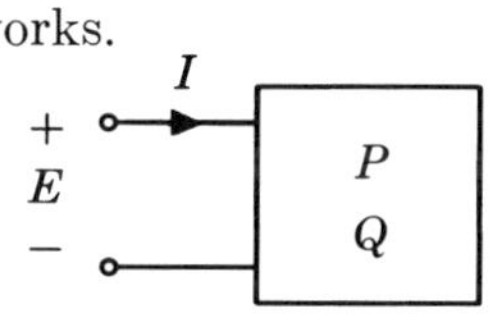

FIG. 19–16. Impedance and admittance in terms of P and Q.

where P_i is the real power absorbed in the ith branch of a network. Similarly, the total reactive power in a network is

$$Q = \sum_{1}^{n} Q_i, \qquad (19\text{–}40)$$

where Q_i is the reactive power absorbed in the ith branch of a network. The proof of Eqs. (19–39) and (19–40) is given in Section 19–3. The real-power formula was easy to prove, but the reactive-power formula was rather difficult. It depended on the identification of reactive power with the average energy stored in the energy-storage elements of the circuit. Now let us suppose that we have a circuit, as shown in Fig. 19–16, in which all the branch powers have been computed and the total real power and the total reactive power have been obtained. (All of the real powers will be positive, but the reactive powers will be positive for inductive branches, and negative for capacitive branches.) The total vector power drawn from the source is therefore

$$P_v = P + jQ. \tag{19–41}$$

However, from the impedance-power triangle the total vector power must also be

$$P_v = |I_{\text{rms}}|^2 Z. \tag{19–42}$$

According to Eqs. (19–41) and (19–42) the input impedance to the network must be

$$Z = \frac{P + jQ}{|I_{\text{rms}}|^2} \tag{19–43}$$

As an alternative, from the admittance-power triangle the total vector power is

$$P_v = |E_{\text{rms}}|^2 Y, \tag{19–44}$$

and the admittance is given by

$$Y = \frac{P - jQ}{|E_{\text{rms}}|^2}. \tag{19–45}$$

As an example of computation of the input power and input impedance from the branch quantities, consider the network of Fig. 19–17. The entire computation will be performed for an assumed output voltage of 1 volt. The input impedance does not depend on the power level of the network. The computations (in RMS values) are as follows:

1. $I_1 = 1$ amp, $\quad P_1 = |I_1|^2 R = 1^2 \times 1 = 1$ watt.
2. $I_2 = 1$ amp, $\quad Q_2 = |I_2|^2 X = 1^2 \times 1 = 1$ var.
3. $I_3 = (1 + j)(+j) = (j - 1)$ amp, $\quad Q_3 = |I_3|^2 X = 2(-1) = -2$ vars.
4. $I_4 = j$ amp, $\quad Q_4 = |I_4|^2 X = 1 \times 2 = 2$ vars.
5. Total $P = 1$ watt.
6. Total $Q = 1 + 2 - 2 = 1$ var.
7. Total $P_v = 1 + j$.
8. Input $Z = \dfrac{P_v}{|I_4|^2} = \dfrac{1 + j}{1} = (1 + j)$ ohms.

As a second example of the branch method of computing power, we shall consider the "power-factor-correction" problem. Most industrial loads

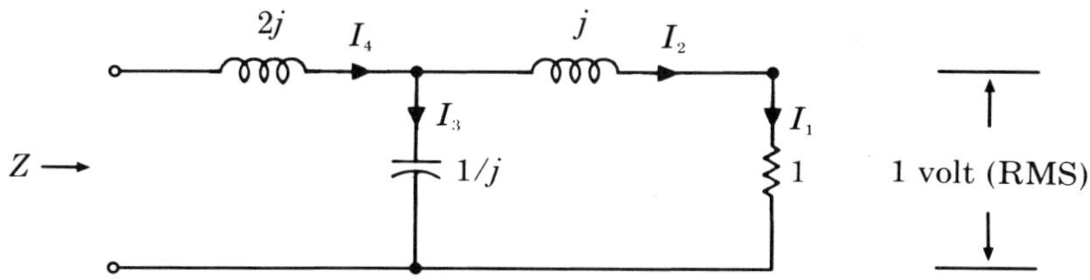

FIG. 19–17. Impedance in terms of branch power.

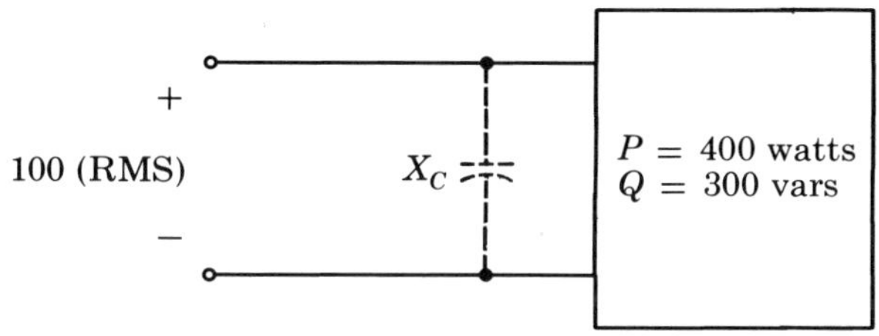

FIG. 19–18. The power-factor-correction problem.

are inductive, and since the power company charges for the reactive power consumed, it may be desirable for the user to buy a large capacitance and to connect it in parallel with the inductive load to supply some of the reactive power. The sole purpose of the capacitance is to improve the power factor of the combined load. Consider the circuit shown in Fig. 19–18. The inductive load draws 400 watts of real power and 300 vars of reactive power (typical industrial loads may be many times as great). The problem is to find the values of capacitive reactance which will bring the power factor of the combined load to unity or 0.9.

Solution 1. For unity power factor, the branch reactive power of the capacitance must be equal to the reactive power of the load. Thus

$$Q_C = -300 \text{ vars}.$$

From the admittance-power triangle

$$Q_C = -E_{\text{rms}}^2 B_C = -100^2 B_C.$$

Thus

$$B_C = \frac{300}{10{,}000}\cdot$$

The reactance of the pure capacitance is the reciprocal of its susceptance and therefore

$$X_C = -33.3 \text{ ohms}.$$

Solution 2. For a power factor of 0.9, the power-factor angle is

$$\theta = \cos^{-1} 0.9 = 26°.$$

From the power triangle, the allowable reactive power after correction is

$$Q = P \tan \theta = 400 \tan 26° = 195 \text{ vars}.$$

The capacitance must supply

$$300 - 195 = 105 \text{ vars}.$$

Thus the necessary capacitive reactance is

$$X_C = \frac{-100^2}{105} = -95 \text{ ohms}.$$

The capacitor required in the second solution is about $\frac{1}{3}$ the size, and has therefore about $\frac{1}{3}$ the cost, of the capacitor required to bring the power factor up to a value of unity. The saving in power cost might not warrant the use of the larger capacitance.

19–6 Power matching in a-c circuits. For purely resistive circuits we were able to derive the conditions under which the maximum amount of power would be drawn from a source of constant voltage and with an internal series resistance. We found that maximum power was drawn from a source when the load resistance was equal to the source resistance. Under these circumstances, the load is said to be "matched" to the source. When the load resistance is matched to the source half of the generated power is lost in the internal resistance of the source. The efficiency of the energy transfer is thus only 50%. Such a system would not be desirable for a power system where large amounts of power are involved. Matching is important only in communications systems where the power levels are small, where the efficiency is not important, and where the maximum value of an output signal is the sole criterion of usefulness. In this section we will investigate the matching conditions for maximum power when the source impedance and the load are not resistive but complex impedances. The production of maximum power in the load resistance is still the governing condition.

The general situation is shown in Fig. 19–19. A source of constant voltage has an internal impedance of $R_s + jX_s$, and the load connected to it has a value of $Z_L = R_L + jX_L$. The condition for maximum power in the load resistance depends on what is fixed and what is variable in the circuit. We will consider two cases. In the first, the source and its internal impedance are given, and both the load resistance and load reactance can

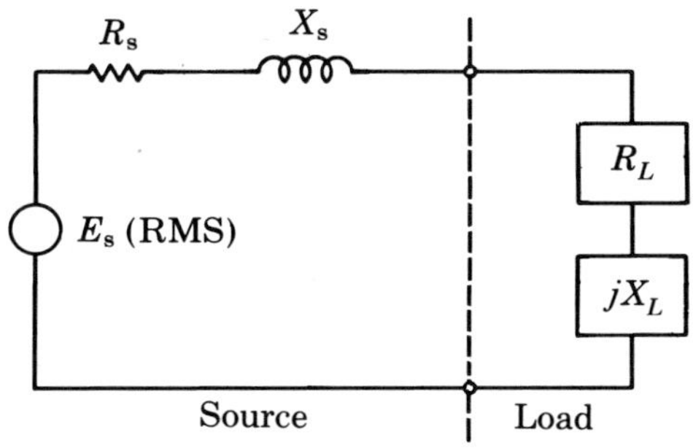

FIG. 19–19. Matching for maximum power transfer.

be varied independently. In the second case, we will suppose that the angle of the load impedance is fixed and that it is possible only to vary its magnitude.

Case 1: both R_L and X_L variable. The current in the circuit is

$$I = \frac{E}{(R_s + R_L) + j(X_s + X_L)}. \tag{19–46}$$

The power in the load resistance is given by

$$P_L = |I_{\rm rms}|^2 R_L = \frac{E^2 R_L}{(R_s + R_L)^2 + (X_s + X_L)^2}.$$

Let us first find the maximum which can be obtained by varying X_L. When we take the derivative with respect to X_L, we obtain

$$\frac{dP_L}{dX_L} = E^2 R_L \left\{\left[\frac{-2(X_s + X_L)}{(R_s + R_L)^2 + (X_s + X_L)^2}\right]^2\right\} \tag{19–47}$$

For a maximum value Eq. (19–47) must be zero, which results in

$$X_L = -X_s. \tag{19–48}$$

This condition could have been obtained by simple reasoning. A maximum current occurs in the circuit at resonance, and hence maximum power is produced in the load. With $X_s = -X_L$, we will proceed to find the value of the load resistance R_L which results in maximum power in the load. The power is now given by

$$P_L = \frac{E^2 R_L}{(R_s + R_L)^2},$$

from which

$$\frac{dP_L}{dR_L} = E^2 \left[\frac{(R_s + R_L)^2 - 2R_L(R_s + R_L)}{(R_s + R_L)^4} \right]. \tag{19–49}$$

For maximum power the numerator of Eq. (19–49) is zero, and we obtain

$$R_L = R_s. \tag{19–50}$$

We see therefore that the condition for maximum power transfer when the resistance and reactance of the load are separately variable is that the reactances in the circuit should resonate and the remaining resistances should be equal.

Case 2: only Z_L variable. If we are producing the matching with a transformer (see Chapter 17), we have control of the magnitude of the reflected load impedance, but not of the separate resistive and reactive components. Under these circumstances, the conditions for matching are somewhat different from the ideal conditions discussed in case 1. Let us denote the reflected load impedance as

$$Z_L = A \underline{/\theta^\circ}. \tag{19–51}$$

Then the resistance is $A \cos \theta$ and the reactance is $A \sin \theta$. The current in the circuit is

$$I = \frac{E}{(R_s + A \cos \theta) + j(X_s + A \sin \theta)}. \tag{19–52}$$

The power in the resistive component of the load is

$$P = |I_{\text{rms}}|^2 R = \frac{E^2 A \cos \theta}{(R_s + A \cos \theta)^2 + (X_s + A \sin \theta)^2}. \tag{19–53}$$

The variable in Eq. (19–53) is the magnitude of the reflected load impedance, A. We will differentiate the expression with respect to A in order to find the maximum value of P as A is varied.

$$\frac{dP}{dA} = E^2 \left[\frac{\begin{array}{l}[(R_s + A \cos \theta)^2 + (X_s + A \sin \theta)^2] \cos \theta \\ \quad - 2A \cos \theta [(R_s + A \cos \theta) \cos \theta + (X_s + A \sin \theta) \sin \theta]\end{array}}{[(R_s + A \cos \theta)^2 + (X_s + A \sin \theta)^2]^2} \right] \tag{19–54}$$

The condition for a maximum is that the numerator of Eq. (19–54) be zero. Thus, with cancellation of E^2 and the common $\cos \theta$ factor,

$$\begin{aligned}(R_s + A \cos \theta)^2 + (X_s + A \sin \theta)^2 &- 2A \cos \theta (R_s + A \cos \theta) \\ &- 2A \sin \theta (X_s + A \sin \theta) = 0,\end{aligned}$$

from which

$$A^2 = R_s^2 + X_s^2,$$

and

$$A = \sqrt{R_s^2 + X_s^2}. \tag{19–55}$$

Equation (19–55) shows that the condition for matching when only the magnitude of the load impedance is variable is for the magnitude of the load impedance to be equal to the magnitude of the source impedance. The power obtained in this case is, of course, less than the power obtained for the ideal matching considered in case 1.

Additional matching problems can arise under different constraints. It is possible that the source-voltage value may vary, and also that the source impedance may vary. If the nature of the variation is known, it is possible to find the conditions for maximum power subject to the given constraints. The general methods used here for deriving the matching conditions will always be applicable. The student should not have the impression that matching always occurs for $R_s = R_L$!

In this chapter on power in a-c circuits, we have attempted to present the power considerations in a fashion general enough to apply to both 60-cps power systems and high-frequency communications systems. The general principles are the same in both cases, but some of the practical details are different. In power systems, efficiency is a prime consideration, while in communications systems, maximum power transfer is more important.

19–7 Summary.

RMS values

$$F_{\text{rms}} = \sqrt{\frac{1}{T}\int_0^T f^2(t)\,dt} = \sqrt{\overline{f^2(t)}}$$

where the bar denotes average value.

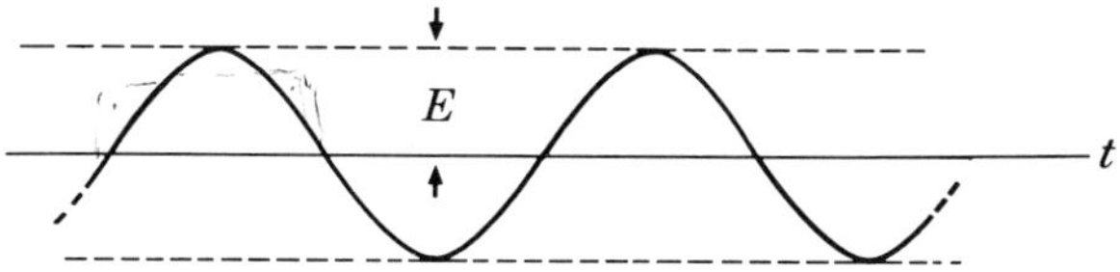

RMS value $= 0.707E$

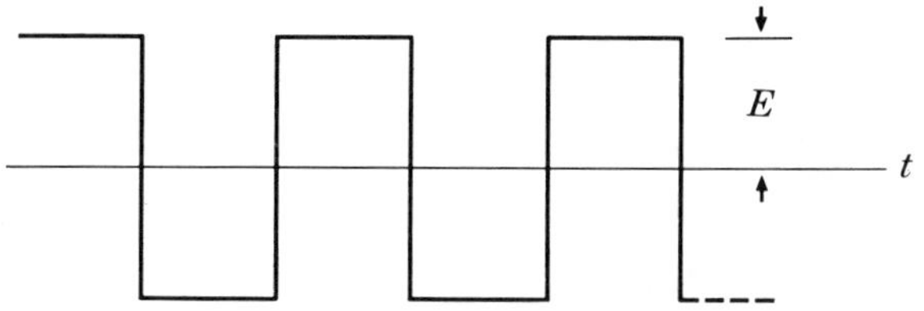

RMS value $= E$

RMS addition

$$F_{\mathrm{rms}} = \sqrt{f_{1_{\mathrm{rms}}}^2 + f_{2_{\mathrm{rms}}}^2 + \cdots}$$

Power in the time domain

$$p = \overbrace{\frac{|E|\,|I|}{2}\cos\theta + \frac{|E|\,|I|}{2}\cos\theta\cos 2\omega t}^{p_r} + \overbrace{\frac{|E|\,|I|}{2}\sin\theta\sin 2\omega t}^{p_X}$$

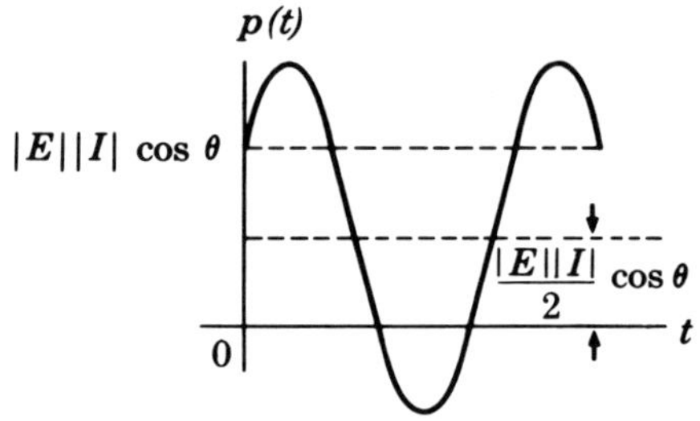

Instantaneous power

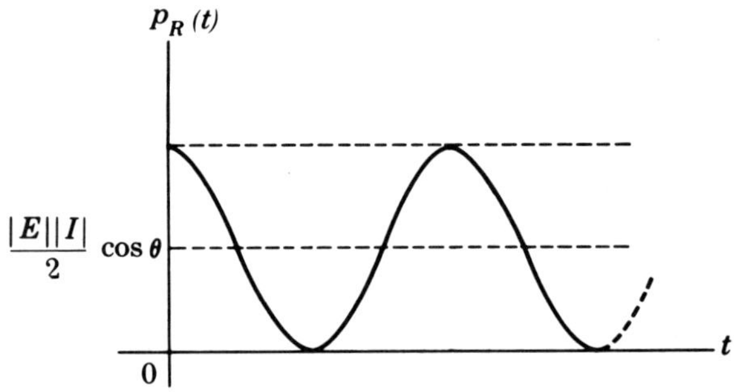

Real power

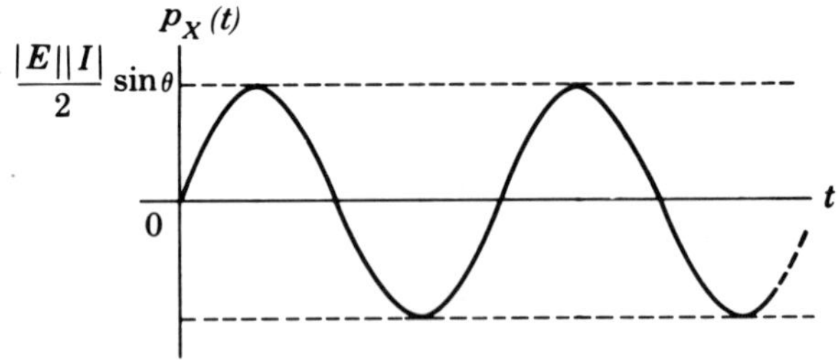

Reactive power

Power in the frequency domain

Vector power

$$P_v = E_{\text{rms}} I^*_{\text{rms}} = |I_{\text{rms}}|^2 Z = |E_{\text{rms}}|^2 Y = |E_{\text{rms}}|\,|I_{\text{rms}}|(\cos\theta + j\sin\theta)$$

Real power

$$P = \text{Re}\,(E_{\text{rms}} I^*_{\text{rms}}) = |I_{\text{rms}}|^2 R = |E_{\text{rms}}|^2 G = |E_{\text{rms}}|\,|I_{\text{rms}}|\cos\theta$$

Reactive power

$$Q = \text{Im}\,(E_{\text{rms}} I^*_{\text{rms}}) = |I_{\text{rms}}|^2 X = -|E_{\text{rms}}|^2 B = |E_{\text{rms}}|\,|I_{\text{rms}}|\sin\theta$$

Power factor

$$\text{pf} = \cos\theta = \frac{P}{|P_v|} = \frac{R}{|Z|} = \frac{G}{|Y|}$$

Reactive factor

$$\text{rf} = \sin\theta = \frac{Q}{|P_v|} = \frac{X}{|Z|} = \frac{-B}{|Y|}$$

Lagging power factor: Current lags voltage; Load is inductive; Reactive power is positive

Leading power factor: Current leads voltage; Load is capacitive; Reactive power is negative

Impedance-power triangle

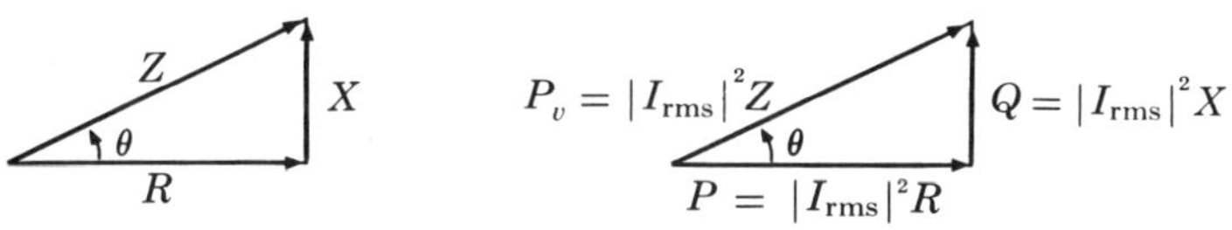

Impedance Power

Admittance-power triangle

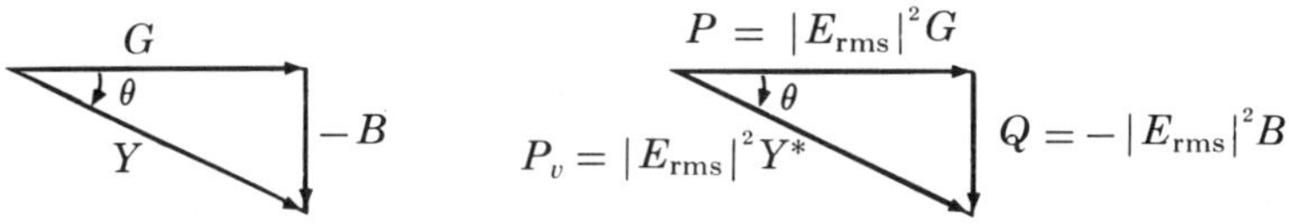

Admittance Power

Branch power

$P = \sum P_i$ where P_i equals the real power in branch i

$Q = \sum Q_i$ where Q_i equals the reactive power in branch i

$$\text{Input impedance: } Z = \frac{P + jQ}{|I_{\text{rms}}|^2}$$

$$\text{Input admittance: } Y = \frac{P - jQ}{|E_{\text{rms}}|^2}$$

$$P_v = P + jQ = P + j(W_L - W_C)$$

where W_L is the average energy stored in inductances
W_C is the average energy stored in capacitances

Power matching

Case 1: both R_L and X_L variable

$$X_L = X_s \qquad \text{and} \qquad R_L = R_s$$

Case 2: magnitude of load impedance variable

$$|Z_L| = |Z_s|$$

REFERENCES

Two extreme points of view will be found in the literature on power and energy in alternating-current circuits. Some authors are concerned solely with 60-cps systems, and all of their examples and exercises concern 60-cps systems. Other authors strive to be very general and to treat all frequencies, including damped sinusoids. Mesh and node equations and quadratic forms are the tools of the general analysis. In this chapter we have tried to steer a course midway between these two extremes. References to other points of view are given below.

1. Kerchner and Corcoran, *Alternating Current Circuits,* John Wiley & Sons, New York, 3rd ed., 1951. This text is concerned almost completely with alternating currents and RMS values. It contains a particularly good chapter on RMS values and Fourier series.

2. Fitzgerald and Higginbotham, *Basic Electrical Engineering,* McGraw-Hill, New York, 2nd ed., 1957. A concise treatment of the usual 60-cps power problems, including RMS values, real and reactive power, and power factor correction.

3. E. A. Guillemin, *Introductory Circuit Theory,* John Wiley & Sons, New York, 1953. Chapter 7 of this text discusses real and reactive power in the time and frequency domains. Chapter 10 gives an introduction to quadratic forms and general energy functions.

4. E. A. Guillemin, *Network Synthesis,* John Wiley & Sons, New York, 1959. In this text Guillemin gives a very complete discussion of the general energy relationships of R–L–C networks under steady-state excitation. Excitation by damped sine waves is also included. Reference is made to original work by Otto Brune.

5. Seshu and Balabanian, *Network Analysis,* John Wiley & Sons, New York, 1959. This text presents a more general form of the energy relationships in the general case.

6. Clement and Johnson, *Electrical Engineering Science,* McGraw-Hill, New York, 1960. A presentation of a-c power from a middle-of-the-road point of view.

7. Brenner and Javid, *Analysis of Electrical Circuits,* McGraw-Hill, New York, 1959. Typical of many transient-circuit-theory texts Brenner and Javid de-emphasize power. Branch power is entirely left out.

EXERCISES*

Section 19–2

1. Find the RMS value of the waveform shown in Fig. 19–20.

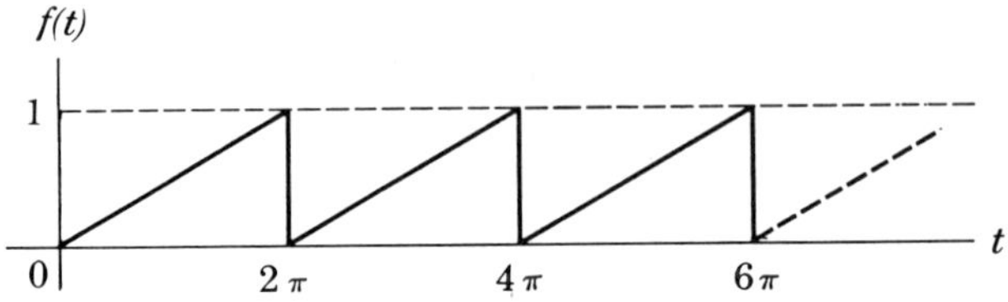

FIGURE 19–20

2. Find the RMS value of the half-wave rectified sine wave shown in Fig. 19–21.

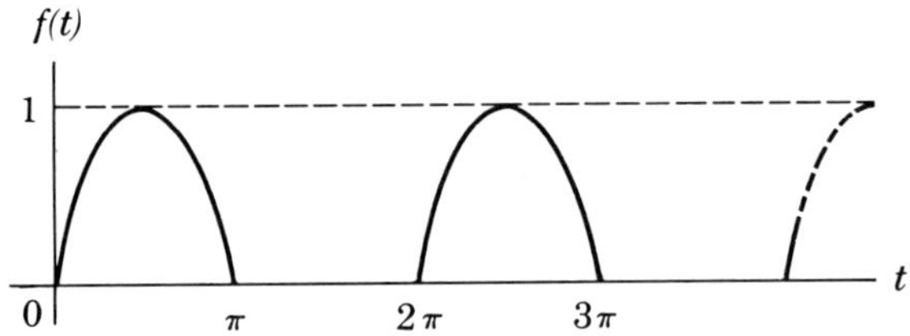

FIGURE 19–21

3. The waveform of Fig. 19–22 repeats with a period of 5 sec. Find its RMS value.

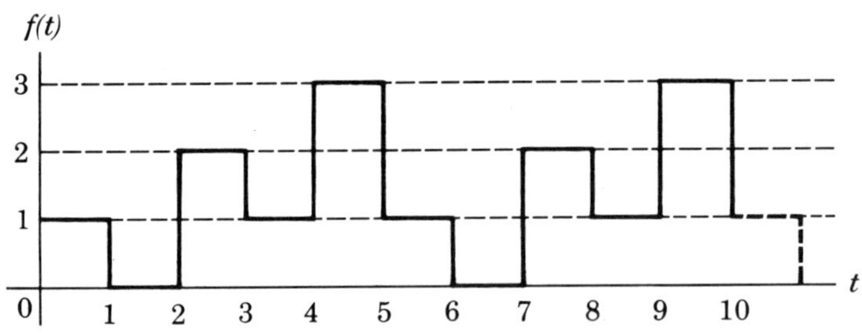

FIGURE 19–22

* Numerical values are in volts, amperes, ohms, henrys, and farads. Complex quantities are in ohms.

4. Find the RMS value of the waveform shown in Fig. 19–23.

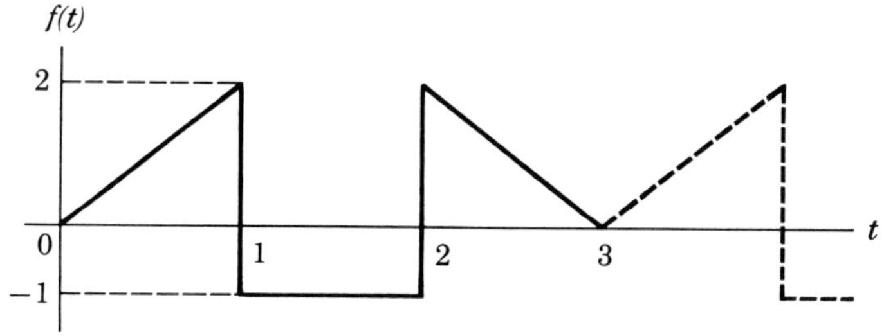

FIGURE 19–23

5. Three currents are given by

$$\begin{aligned} i_1(t) &= 10 \cos (t + 45°), \\ i_2(t) &= 5 \sin (2t - 13°), \\ i_3(t) &= 10 \sin (13t + 65°). \end{aligned}$$

Find the RMS value of the sum of the three currents.

6. Find the RMS value of the sum of the following three currents:

$$\begin{aligned} i_1(t) &= 10, \\ i_2(t) &= 6 \cos t + 5 \cos (3t + 45°), \\ i_3(t) &= 5 + 6 \sin t + 5 \cos (3t - 45°). \end{aligned}$$

7. The voltage across a resistance of 2 ohms is

$$e(t) = 1 + 2 \cos t + 1.414 \cos (3t + 45°).$$

(a) Find the RMS value of the voltage. (b) What is the average power dissipated in the resistance?

8. The voltage and the current across a terminal-pair are given by

$$\begin{aligned} e(t) &= 1 + 1.414 \cos t, \\ i(t) &= 3 + 5.64 \cos t. \end{aligned}$$

(a) Find the RMS value of the voltage and of the current. (b) Find the average power dissipated in the device. [The instantaneous value is $e(t) \times i(t)$.]

9. If the voltage and the current at a terminal-pair are given by

$$\begin{aligned} e(t) &= E_0 + E_1 \cos t + E_2 \cos 2t + \cdots, \\ i(t) &= I_0 + I_1 \cos t + I_2 \cos 2t + \cdots. \end{aligned}$$

(a) Find the RMS voltage and the RMS current.
(b) Find the average value of the power if the instantaneous power is $e(t) \times i(t)$.
(c) Show that the average power is *not* $E_{\text{rms}} \times I_{\text{rms}}$.

10. The voltage and current at a terminal-pair are given by

$$e(t) = 100\,[\cos(1+\delta)t + \cos(1-\delta)t],$$
$$i(t) = 50\,[\cos(1+\delta)t + \cos(1-\delta)t].$$

(a) If the instantaneous power is $e(t) \times i(t)$ find the average power entering the terminal-pair.
(b) Find the average power when $\delta = 0$.
(c) Reconcile the answers of parts (a) and (b).

Section 19–3

11. A voltage $e(t) = 100 \cos 377t$ is applied to a resistance of 10 ohms.
(a) Find the instantaneous power absorbed.
(b) Find the average power absorbed.
(c) Sketch $p(t)$.
(d) Discuss the direction of power flow.

12. A voltage $e(t) = 100 \cos 10t$ is applied to an inductance with a value of 10 henrys.
(a) Find $p(t)$, the instantaneous power absorbed.
(b) Find $w(t)$, the instantaneous energy stored.
(c) Find W_L, the average energy stored.
(d) Sketch $p(t)$.
(e) Discuss the direction of power flow.

13. A voltage $e(t) = 100 \cos 10t$ is applied to a capacitance with a value of 0.001 farad.
(a) Find $p(t)$.
(b) Find $w(t)$.
(c) Find W_C, the average energy stored.
(d) Sketch $p(t)$.
(e) Discuss the direction of power flow.

14. A voltage $e(t) = 10 \cos t$ is applied to the series combination of $R = 1$ ohm and $L = 1$ henry.
(a) Find the instantaneous power in the form of a constant term, a cosine term of frequency $\omega = 2$, and a sine term of frequency $\omega = 2$.
(b) Sketch $p(t)$.
(c) Give P, the average power absorbed.
(d) Give Q, the reactive power absorbed.

15. A voltage $e(t) = 10 \cos t$ is applied to the series combination of $R = 1$ ohm and $C = 1$ farad.
(a) Find the instantaneous power in the form of a constant term, a cosine term of double frequency, and a sine term of double frequency.
(b) Sketch $p(t)$.
(c) Give P, the average power absorbed.
(d) Give Q, the reactive power absorbed.

16. For the circuits of Exercises 14 and 15 verify the formula for the reactive power in terms of the average energy stored in the elements, that is

$$Q = 2\omega(W_L - W_C).$$

17. For the circuit of Fig. 19–24.
   (a) Find $p(t)$ the instantaneous power absorbed.
   (b) Sketch $p(t)$.
   (c) Give P, the average power absorbed.
   (d) Give Q, the average reactive power absorbed.

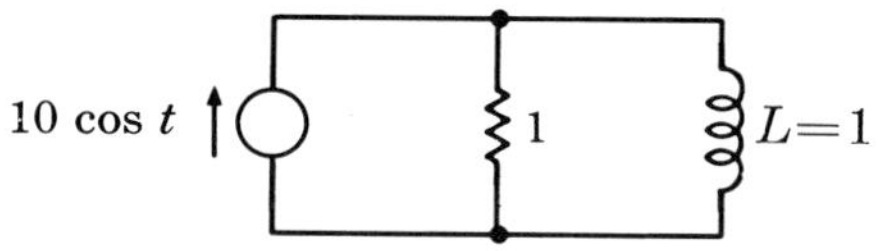

FIGURE 19–24

18. For the circuit of Fig. 19–25 find the average power delivered by each source and the direction of power flow for each of the following cases.

(a) $Z = 1$ ohm (b) $Z = j1$ ohm (c) $Z = -j1$ ohm

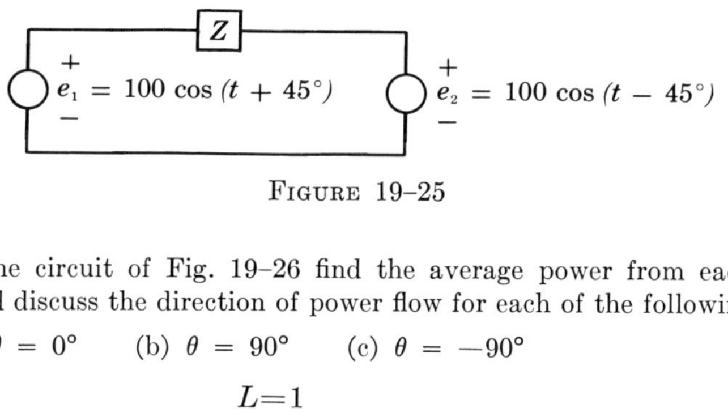

FIGURE 19–25

19. In the circuit of Fig. 19–26 find the average power from each of the sources and discuss the direction of power flow for each of the following cases.

(a) $\theta = 0°$ (b) $\theta = 90°$ (c) $\theta = -90°$

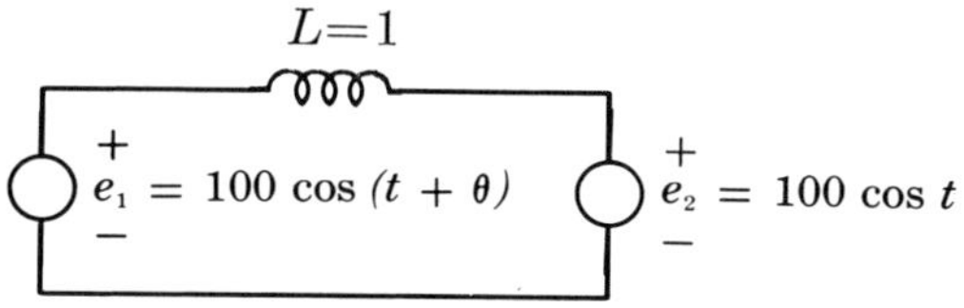

FIGURE 19–26

Section 19–4

20. An RMS voltage of 100 volts is applied to a circuit with an effective resistance of 10 ohms and an effective reactance of $j10$ ohms.
   (a) Find the vector power, real power, reactive power, and power factor.
   (b) Draw the power vector diagram.

21. Repeat Exercise 20 but convert the impedance into an equivalent admittance and find the power in each of the parallel branches.

22. An RMS voltage of 100 volts is applied to circuit represented by a resistance of 10 ohms in parallel with a reactance of $j10$ ohms.

(a) Find the vector power, real power, reactive power, and the power factor.

(b) Draw the power vector diagram.

23. Solve Exercise 22 by converting the admittance into an equivalent series impedance and finding the power in each of the two series elements.

24. A circuit is connected to a voltage source of 100 volts, RMS. The total volt-amperes drawn from the source is 2000, and the power factor is 0.866 with the current lagging the voltage. Find the equivalent series resistance and the reactance of the circuit.

25. A circuit is connected to a voltage source of value 100 volts, RMS. The power factor of the circuit is 0.866 capacitive and the component of the load current in phase with the source is 10 amp. Find P and Q, and P_v for the circuit.

26. In the circuit of Fig. 19–27 the voltage across the generator and the voltage across the load are both 100 volts, RMS. The total vector power from the source is 1000 volt-amperes. Find the real and reactive powers absorbed in the load.

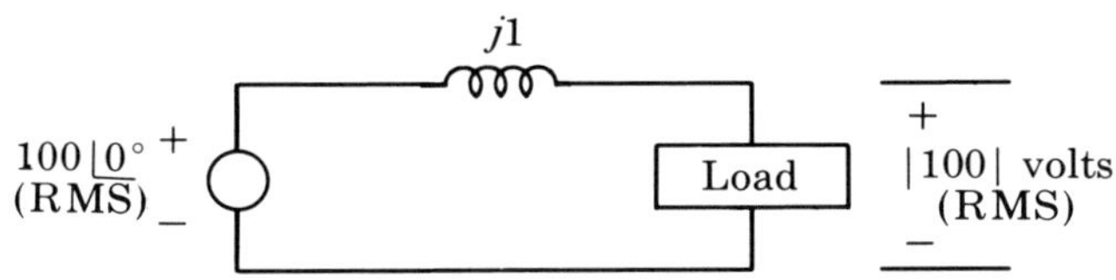

FIGURE 19–27

27. In Fig. 19–28 a generator with a voltage of 120 volts, RMS, is feeding a load through a transmission line represented by the impedance $1 + j$. The power factor of the load is known to be 0.8 and the voltage across the load is 100 volts, RMS. Find the current and the power factor at the generator.

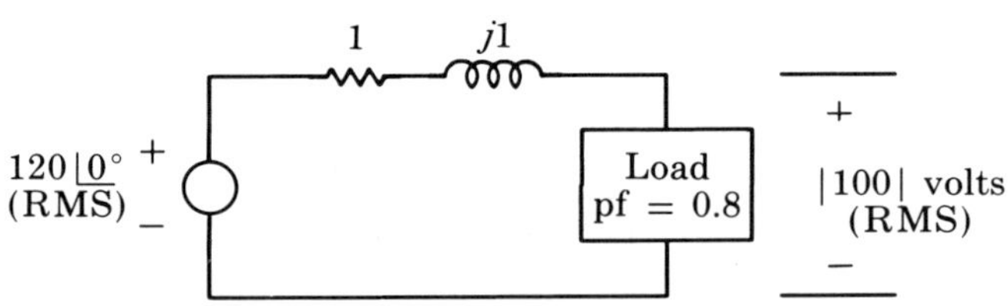

FIGURE 19–28

Section 19–5

28. In the circuit of Fig. 19–29 find the currents and voltages which would be compatible with an output voltage of $E = 1\,\underline{|0°}$.

(a) Compute the net P and the net Q for the circuit from the branch quantities, for the conditions described.
(b) Obtain the input Z from P_v. Check your answer by current and voltage computations.

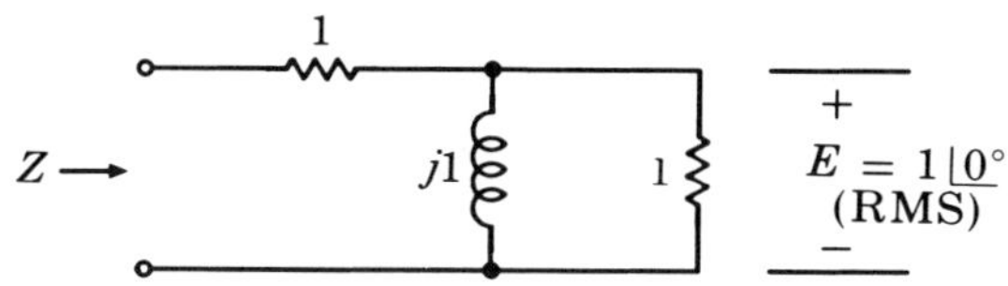

FIGURE 19–29

29. In the circuit of Fig. 19–30 find the currents and voltages which would result in an output voltage $E = 1\underline{|0°}$.
(a) Compute the net P and the net Q for the circuit from the branch quantities, for the conditions described.
(b) Obtain the input Z from P_v. Check your answer from the current and voltage relationships.

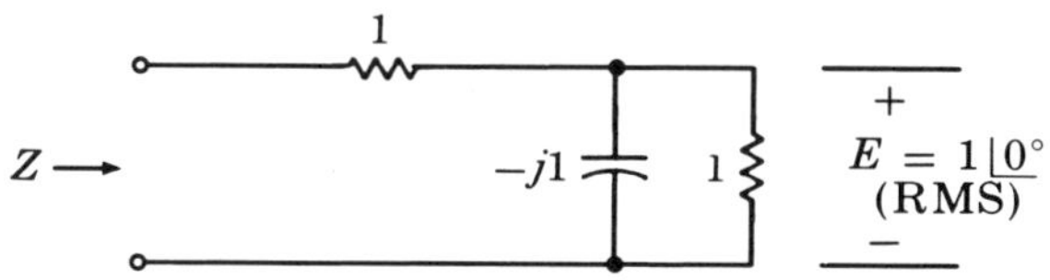

FIGURE 19–30

30. (a) Find the total P and the total Q in the circuit of Fig. 19–31.
(b) Find the input impedance Z from P_v.

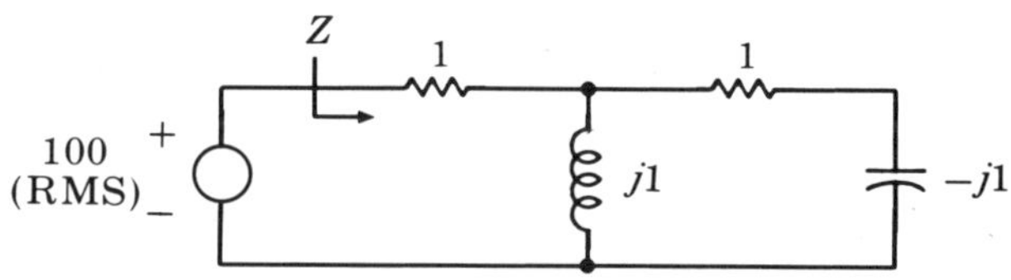

FIGURE 19–31

31. Find the input impedance of the circuit of Fig. 19–32 by computing the branch powers for an assumed output voltage of $E = 1\underline{|0°}$.

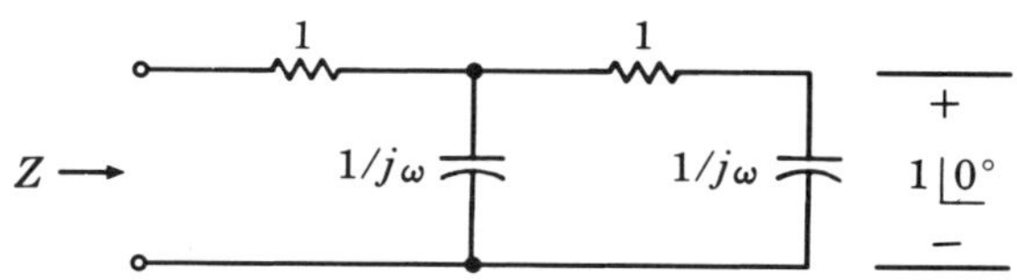

FIGURE 19–32

32. A load which draws 100 watts and 100 vars (inductive) is to be connected in parallel with a load which draws 100 watts and 50 vars (capacitive). The voltage source remains constant in each case.

(a) Give the power factor of the combined load.
(b) What is the reactive power of an element added in parallel with the other loads which will bring the power factor to unity?

33. In the circuit of Fig. 19–33 compute the value of C which will bring the power factor to 0.95.

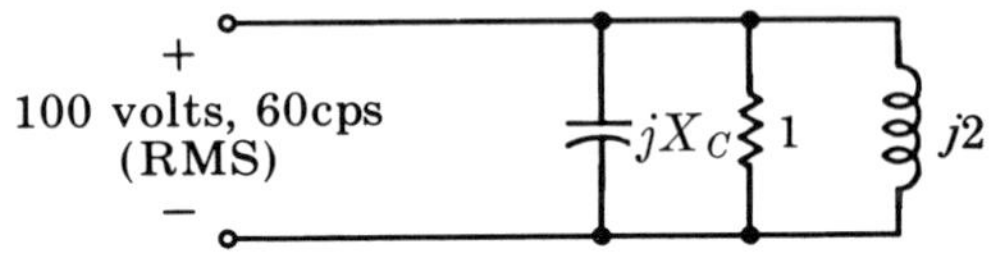

FIGURE 19–33

34. The circuit of Fig. 19–34 represents a generator feeding a load through a transmission line. The power factor of the load is 0.8 and the power factor at the generator is 0.5, without the capacitance. Find the capacitive reactance which when added across the generator will bring the power factor at the generator to unity.

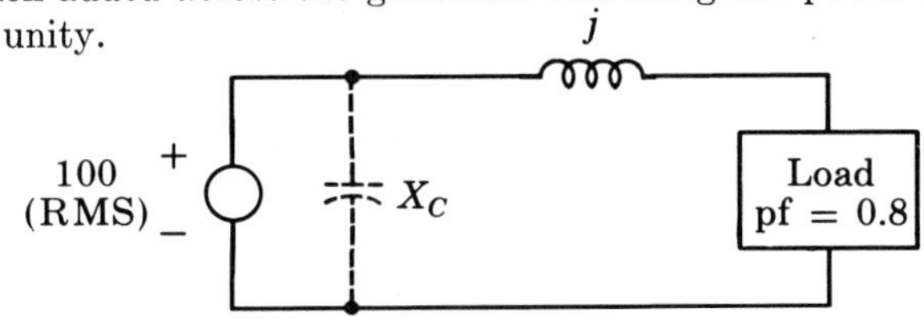

FIGURE 19–34

35. In the circuit of Fig. 19–35 find the value of the capacitive reactance which will give unity power factor at the terminals of the generator, although connected directly across the load.

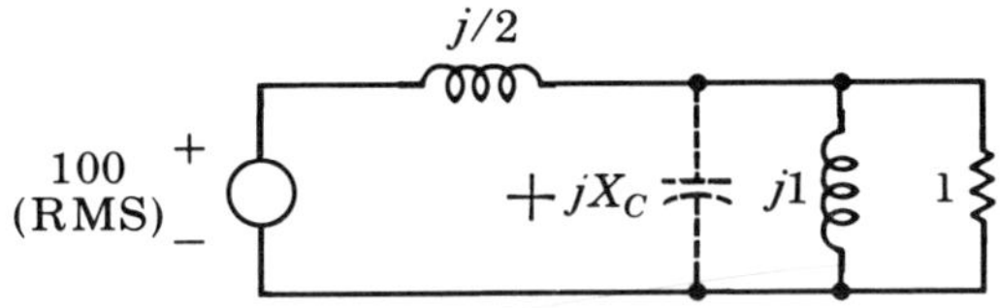

FIGURE 19–35

Section 19–6

36. Both the resistive and reactive parts of the impedance in the circuit of Fig. 19–36 can be varied.

(a) Find the value of Z which will absorb maximum power.
(b) Give the value of the maximum power.

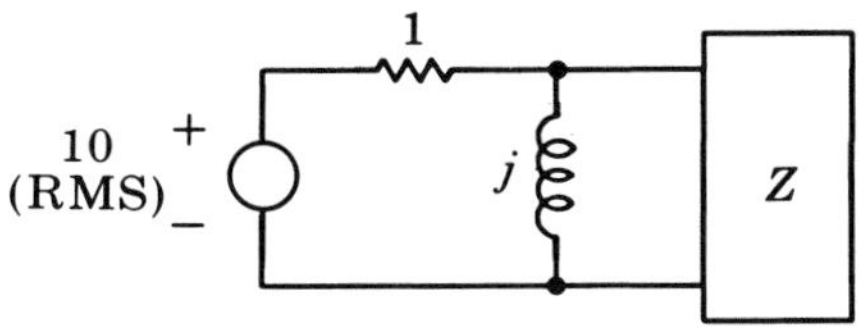

FIGURE 19–36

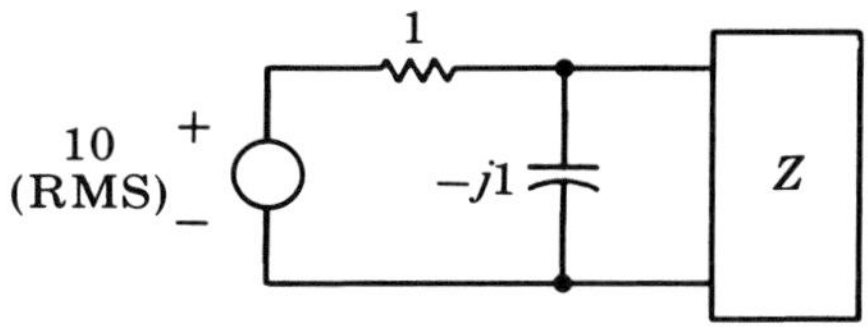

FIGURE 19–37

37. Both the real and imaginary parts of the impedance in the circuit of Fig. 19–37 can be varied.
   (a) Find the value of Z which absorbs maximum power.
   (b) Give the value of the maximum power.

38. In the circuit of Fig. 19–38 both the real and imaginary parts of Z_L can be varied.
   (a) Find the conditions for maximum power in Z_L.
   (b) Give the value of the maximum power.

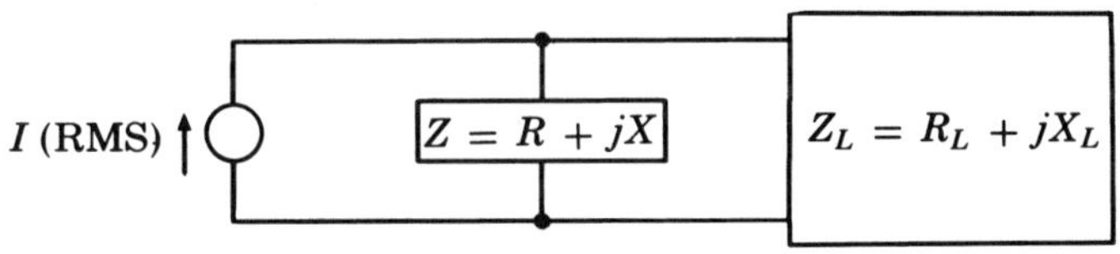

FIGURE 19–38

39. In the circuit of Fig. 19–39 only the turns-ratio of the ideal transformer can be varied.
   (a) Find the turns-ratio which gives maximum power in the load resistance on the right side of the transformer.
   (b) Give the value of this maximum power.

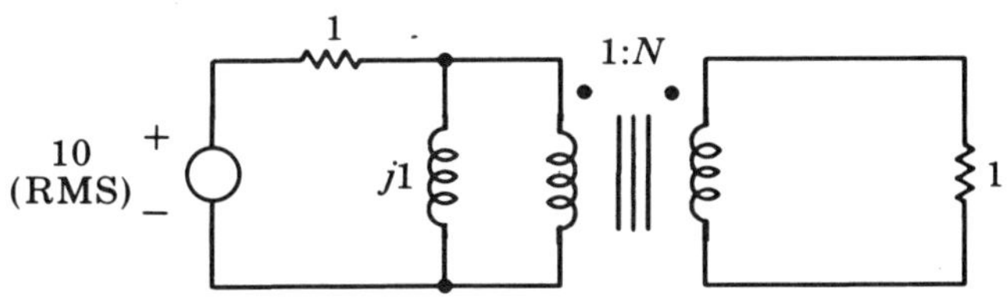

FIGURE 19–39

40. (a) Find the value of N in the circuit of Fig. 19–40 which results in the maximum power in the resistance R.
   (b) Give the value of this maximum power.

41. In the circuit of Fig. 19–41 only the turns-ratio N is variable.
   (a) Find the value of N which gives maximum power in the load Z_L.
   (b) Give the value of this maximum power.

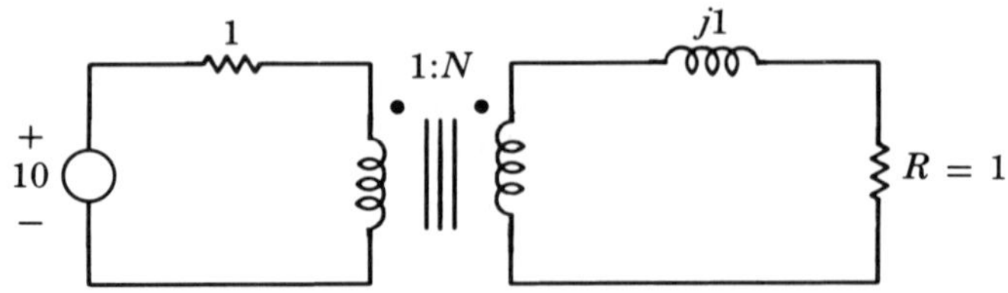

FIGURE 19–40

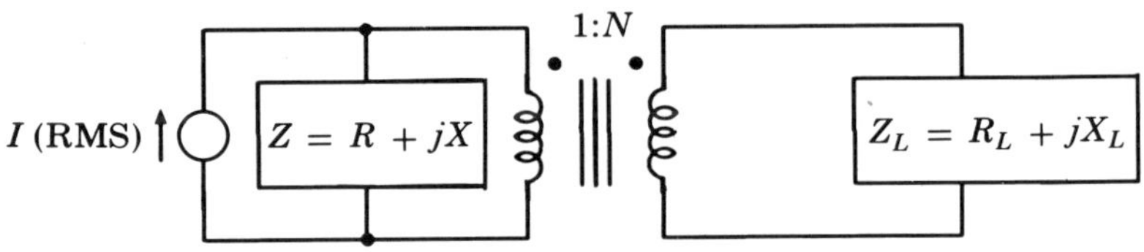

FIGURE 19–41

42. Matching in the circuit of Fig. 19–42 is done with a unity-coupled coil instead of an ideal transformer. At very low frequencies the power is shunted by the transformer and is not delivered to the load. The matching is adjusted to give a maximum power in the load for high frequencies, where the shunting effect of the transformer can be neglected.

(a) Find the turns-ratio. The primary inductance remains at 1 henry as the secondary turns are varied. Give also L_2 and M.

(b) What is the half-power frequency at which the power delivered to the load is one-half of its high-frequency value?

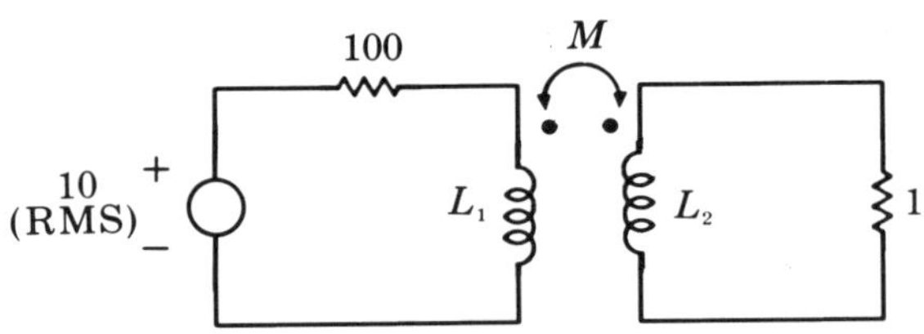

FIGURE 19–42

CHAPTER 20

SIGNAL ANALYSIS IN THE FREQUENCY DOMAIN

"As great Pythagoras of yore
Standing beside the blacksmith's door,
And hearing the hammers, as they smote
The anvils with a different note,
Stole from the varying tones, that hung
Vibrant on every iron tongue,
The secret of the sounding wire,
And formed the seven-chorded lyre."

Longfellow

20–1 Introduction. The concept of frequency and the components of frequency arose in connection with music and musical instruments. Musical scales are built around the harmonics, or octaves, of a given tone, and musical chords depend on the resolution of complex sounds into individual components. An extension of these concepts to the analysis of more general signals first occurred in the unlikely field of heat transfer. Working in this field, Fourier announced his famous theorem which stated that any signal could be resolved into sinusoidal components. Fourier's theorem is at the base of most network and communications theory. In general, a signal is resolved into its sinusoidal components, networks are designed to pass these components, and the desired output is obtained. Because of the simplicity which it lends to network theory, signal analysis is a vital part of the process.

In this chapter we will consider the basic Fourier theorem and apply it to periodic waves. The resulting summation of harmonic sinusoids is called a *Fourier Series.* Certain simplifications occur when the waves are symmetrical, and these are considered in some detail. In addition, numerical procedures are discussed for finding the Fourier series of a given periodic wave when it is given in graphical rather than analytic form. As a by-product of the discussion of numerical procedures for finding Fourier series, we will verify a theorem given by C. E. Shannon which states that a complex waveform with no frequencies at f_0 or above can be completely determined if it is sampled at a rate of $2f_0$.

We will also consider, on a heuristic basis, the extension of Fourier's theorem to nonperiodic waveforms. The period of repetition of a given pulse is allowed to increase without limit. The fundamental frequency of the Fourier series approaches zero, and the spacing between the harmonics approaches zero as well. The amplitudes of the harmonics also approach zero, but their distribution as a function of frequency does not. The Fourier series becomes *the Fourier integral* of the pulse. The frequency distribution of the harmonics in a Fourier series is a *line spectrum;* the frequency distribution in a Fourier integral is a *continuous, or band, spectrum.*

Certain convergence problems exist in Fourier integrals for step functions and other similar nonperiodic waveforms which exist over infinite periods of time. These problems can be avoided if the Fourier expansions are performed in terms of rising sine waves instead of constant-amplitude sine waves. When this step is taken, the Fourier integral becomes the *Laplace integral,* or *Laplace transform.*

20–2 The Fourier theorem. In the year 1822 Fourier published his famous *Théorie Analytique de la Chaleur.* He was concerned primarily with the flow of heat, but in the course of his analysis he was lead to expand arbitrary functions in terms of an infinite series of sinusoidal components. The expansion of a function into an infinite series of algebraic functions was well known. For example,

$$\frac{1}{1-x} = 1 + x + x^2 + x^3 + \cdots x^n + \cdots, \qquad |x| < 1.$$

The expansion of even a periodic function into an infinite series of sine waves seemed strange to the mathematicians of the day. And when Fourier went further and claimed to expand nonperiodic functions in the same fashion, some of the most famous mathematicians stated categorically that it could not be done. The mathematical storm that Fourier aroused did not subside for a hundred years. The exact conditions under which the expansions were valid were given by Norbert Wiener in 1933.* We will consider first the case of periodic functions.

All periodic, physically realizable† functions with a period T can be expanded into a Fourier series of the form

$$\begin{aligned} f(t) = a_0 &+ a_1 \cos \omega_0 t + a_2 \cos 2\omega_0 t + \cdots a_n \cos n\omega_0 t + \cdots \\ &+ b_1 \sin \omega_0 t + b_2 \sin 2\omega_0 t + \cdots b_n \sin n\omega_0 t + \cdots, \end{aligned} \tag{20–1}$$

where $\omega_0 = 2\pi/T$ is the fundamental frequency and the a's and b's are constants which must be found. The process of finding the constants $a_0 \ldots a_n$ and $b_1 \ldots b_n$ is called *Fourier analysis.* In practice the utility of Fourier series derives principally from the fact that most series con-

---

* N. Wiener, *The Fourier Integral and Certain of its Applications,* Cambridge University Press, New York, 1933. See also the Dover Publications edition.

† Later in the chapter, the realizability conditions will be extended to include idealized input functions such as steps, impulses, and doublets.

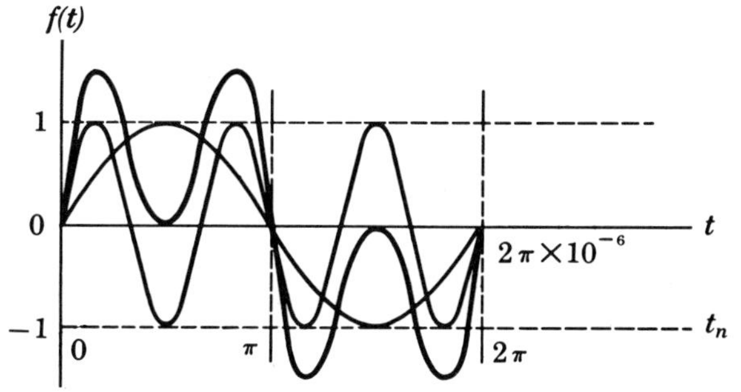

FIG. 20–1. A normalized time scale.

verge rather rapidly and can be represented adequately by a very few terms called *partial sums.*

The shape of a waveform and its resolution into sinusoidal components are independent of the units which are used. It is convenient, therefore, to work always in a normalized time scale in which the fundamental period of the wave is 2π units, and the corresponding normalized angular frequency is $\omega_0 = 1$. In order to make the result applicable in the original time scale it is necessary only to insert the correct value of ω_0 in the normalized Fourier series. It is important to understand and use the normalized time scale. Consider the example shown in Fig. 20–1. The actual wave consists of two sinusoidal components of frequencies $\omega_0 = 10^6$ and $3\omega_0 = 3 \times 10^6$. It can be written

$$f(t) = \sin 10^6 t + \sin 3 \times 10^6 t.$$

In the normalized time scale, in which the period is always 2π and the frequency of the fundamental component is $\omega_0 = 1$, the relative frequencies of the two sine waves remain 1 to 3, and the relative amplitudes remain the same. The normalized time function is

$$f_n(t) = \sin t + \sin 3t.$$

Note that *the actual time function can be obtained from the normalized one simply by inserting the correct fundamental frequency* in place of $\omega_0 = 1$ in the terms of the normalized series. Thus we are perfectly general in considering Fourier series always to have fundamental periods of 2π. Thus the basic series is

$$\begin{aligned} f(t) = a_0 + a_1 \cos t + a_2 \cos 2t + \cdots a_n \cos nt + \cdots \\ + b_1 \sin t + b_2 \sin 2t + \cdots b_n \sin nt + \cdots. \end{aligned} \qquad (20\text{–}2)$$

The coefficients of Eq. (20–2) are found by making use of the so-called "orthogonality"* properties of sinusoidal waves. The orthogonality conditions, which can be verified by direct integration, are

$$\int_0^{2\pi} \cos nt\, dt = \int_0^{2\pi} \sin nt\, dt = \int_0^{2\pi} \sin nt \cos nt\, dt = 0,$$

$$\int_0^{2\pi} \cos mt \sin nt\, dt = \int_0^{2\pi} \cos mt \cos nt\, dt = \int_0^{2\pi} \sin mt \sin nt\, dt = 0 \quad \text{for } n \neq m,$$

$$\int_0^{2\pi} \cos^2 nt\, dt = \int_0^{2\pi} \sin^2 nt\, dt = \pi \qquad \text{for } n = m, \tag{20–3}$$

where the integrals can be taken over any complete period.

In Eqs. (20–3) the constants m and n can have any integral value from 1 to ∞. If we divide all the integrals by the period, 2π, we obtain the average value over the period. Another way to state the orthogonality conditions is that the average value over a period of any sinusoidal wave, or of the product of any two sinusoidal waves, is zero unless the waves are identical, under which circumstance the average value is one-half. The conditions are also true for nonintegral values of m and n if the average value is taken over all time instead of just one period. (Note that the average values of the sine-squared and cosine-squared functions were used previously to define RMS values.)

To find the value of the constant a_0, we integrate both sides of Eq. (20–2) over a complete period and use the results of Eqs. (20–3) to simplify the expression. We obtain

$$\int_0^{2\pi} f(t)\, dt = \int_0^{2\pi} a_0\, dt + \int_0^{2\pi} a_1 \cos t\, dt + \cdots + \int_0^{2\pi} b_1 \sin t\, dt + \cdots.$$

All the integrals on the right side of the equation, except the first, are zero. Thus

$$\int_0^{2\pi} f(t)\, dt = a_0 2\pi,$$

or

$$a_0 = \frac{1}{2\pi} \int_0^{2\pi} f(t)\, dt. \tag{20–4}$$

Equation (20–4) gives the constant a_0 in terms of the original function $f(t)$. In fact, a_0 is simply the average or d-c value of the function.

---

* The name originates from a similar set of conditions which hold for the direction cosines of sets of orthogonal vectors.

To find a_1 we multiply both sides of Eq. (20-2) by $\cos t$ and then integrate all the terms over one period. Again, the results of Eqs. (20-3) allow us to simplify the expression. The integration gives

$$\int_0^{2\pi} f(t)\cos t\,dt = \int_0^{2\pi} a_0 \cos t\,dt + \int_0^{2\pi} a_1 \cos^2 t\,dt + \int_0^{2\pi} a_2 \cos t \cos 2t\,dt + \cdots + \int_0^{2\pi} b_1 \cos t \sin t\,dt + \cdots.$$

All the integrals on the right side of this expression are zero, with the exception of that involving a_1. Thus

$$\int_0^{2\pi} f(t)\cos t\,dt = a_1\pi,$$

or

$$a_1 = \frac{1}{\pi}\int_0^{2\pi} f(t)\cos t\,dt. \tag{20-5}$$

Equation (20-5) shows that a_1 is twice the average value of the function $f(t)$ weighted by the multiplying factor $\cos t$. By a similar process, we can obtain the other a's and b's. In summary

$$a_n = \frac{1}{\pi}\int_0^{2\pi} f(t)\cos nt\,dt, \tag{20-6}$$

$$b_n = \frac{1}{\pi}\int_0^{2\pi} f(t)\sin nt\,dt. \tag{20-7}$$

It should be emphasized that the d-c term a_0 is not the value of Eq. (20-6) for $n = 0$, but is instead

$$a_0 = \frac{1}{2\pi}\int_0^{2\pi} f(t)\,dt. \tag{20-8}$$

As an example of the direct computation of the Fourier coefficients, con-

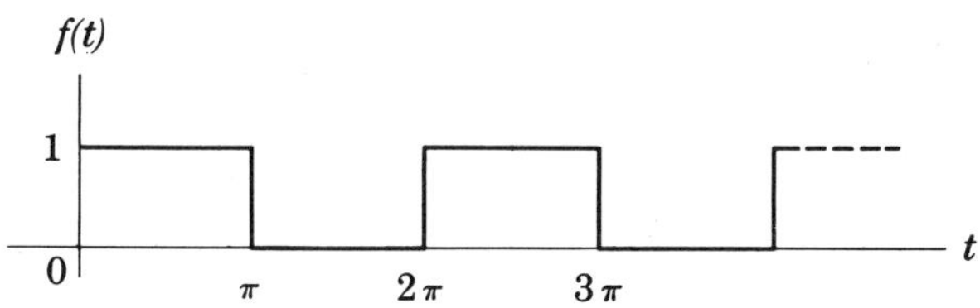

FIG. 20-2. Fourier analysis of a square wave.

sider the square wave of Fig. 20–2. The function $f(t)$ is defined by

$$f(t) = \begin{cases} 1 & 0 < t < \pi, \\ 0 & \pi < t < 2\pi. \end{cases}$$

The computations are as follows:

$$\begin{aligned}
a_0 &= \frac{1}{2\pi}\int_0^{2\pi} f(t)\,dt = \frac{1}{2\pi}\int_0^{\pi} (1)\,dt + \frac{1}{2\pi}\int_{\pi}^{2\pi} (0)\,dt \\
&= \frac{1}{2\pi}\,\pi + 0 \\
&= \tfrac{1}{2}, \\
a_1 &= \frac{1}{\pi}\int_0^{2\pi} f(t)\cos t\,dt = \frac{1}{\pi}\int_0^{\pi} \cos t\,dt + \frac{1}{\pi}\int_{\pi}^{2\pi} (0)\,dt \\
&= \frac{1}{\pi}\,[\sin t]_0^{\pi} \\
&= 0, \\
a_n &= \frac{1}{\pi}\int_0^{2\pi} f(t)\cos nt\,dt = \frac{1}{\pi}\int_0^{\pi} \cos nt\,dt + \frac{1}{\pi}\int_{\pi}^{2\pi} (0)\,dt \\
&= \frac{1}{\pi}\left[\frac{\sin nt}{n}\right]_0^{\pi} \\
&= 0, \\
b_1 &= \frac{1}{\pi}\int_0^{2\pi} f(t)\sin t\,dt = \frac{1}{\pi}\int_0^{\pi} \sin t\,dt + \frac{1}{\pi}\int_{\pi}^{2\pi} (0)\,dt \\
&= \frac{1}{\pi}\,[-\cos t]_0^{\pi} \\
&= \frac{2}{\pi}, \\
b_n &= \frac{1}{\pi}\int_0^{2\pi} f(t)\sin nt\,dt = \frac{1}{\pi}\int_0^{\pi} \sin nt\,dt + \frac{1}{\pi}\int_{\pi}^{2\pi} (0)\,dt \\
&= \frac{1}{\pi}\left[\frac{-\cos nt}{n}\right]_0^{\pi} \\
&= \begin{cases} 0 & \text{for } n \text{ even}, \\ \dfrac{2}{n\pi} & \text{for } n \text{ odd}. \end{cases}
\end{aligned}$$

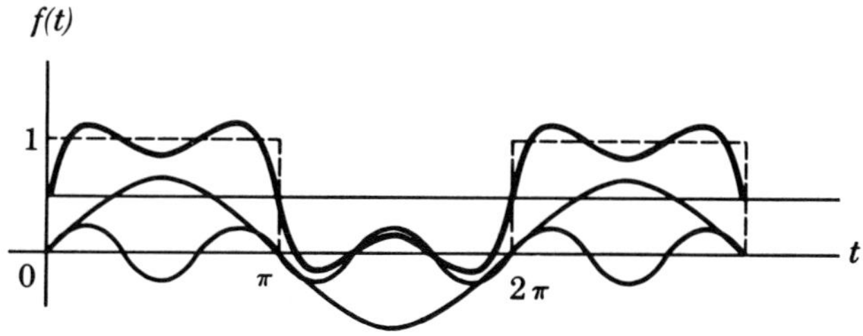

FIG. 20–3. The partial sum of the Fourier series for a square wave.

The complete series is

$$f(t) = \tfrac{1}{2} + \frac{2}{\pi}\left(\sin t + \frac{\sin 3t}{3} + \frac{\sin 5t}{5} + \cdots\right). \tag{20–9}$$

A sketch of the d-c value, the fundamental, the third harmonic and of their partial sum is given in Fig. 20–3. Even with these few terms, it is apparent that the square waveform is being approached. The addition of more terms to the partial sum would cause the resultant wave to approach closer and closer to a square wave, except at the points of discontinuity. The square wave of Fig. 20–3 is not a physically realizable waveform, and the Fourier series for it does not converge to the expected form in the neighborhood of the discontinuity. There are many times when we wish to use idealized waveforms of this type, and we must therefore consider the problem of convergence in more detail.

20–3 Convergence of Fourier series. There are really three separate problems involved in the convergence of Fourier series. They are

1. Can we find the coefficients a_n and b_n?
2. Can we sum the resulting series for $f(t)$?
3. Can we approximate $f(t)$ with a small number of terms of the series?

The conditions for being able to find a_n and b_n are often called the *weak Dirichlet conditions*. They can be seen from the definitions of the constants. For example,

$$a_n = \frac{1}{\pi}\int_0^{2\pi} f(t)\cos nt\,dt.$$

At the very worst, the cosine term could cause the integrand to be always positive. Hence

$$a_n \lesseqgtr \frac{1}{\pi}\int_0^{2\pi} |f(t)|\,dt < \infty. \tag{20–10}$$

The existence of this integral guarantees the existence of a_n, and similarly also of b_n. Another integral, very similar to that of Eq. (20–10) but easier to evaluate, is often used as the test. It is

$$\int_0^{2\pi} [f(t)]^2 \, dt < \infty. \tag{20–11}$$

If $f(t)$ is a current or a voltage, it is apparent that Eq. (20–11) represents the energy which would be supplied by the source in one cycle. The condition of Eq. (20–11) thus reduces to the condition that the energy in the waveform for each cycle must be finite. All physical waveforms would, of course, satisfy this criterion. The weak Dirichlet conditions do not restrict the function $f(t)$ to be finite. In particular, they allow impulses to be present which have infinite values but finite areas under them. Whenever $f(t)$ is infinite, the Fourier series for it will not converge at that point. If we want a series which converges everywhere, we must apply some additional constraints. These are known as the *strong Dirichlet conditions.*

For a uniformly convergent Fourier series, the function $f(t)$ must remain finite and must have a finite number of maxima and minima. It can have a finite number of finite discontinuities. The reason for the first additional condition is obvious. If the function is infinite at some point, the series representing it will also be infinite and nonconvergent. The reason for the second restriction is fairly obvious. If there were an infinite number of maxima and minima in a given region, there would be a Fourier term of infinite frequency in the series, containing an appreciable amount of energy. The series would therefore require the use of an infinite number of terms and would not converge at infinity.

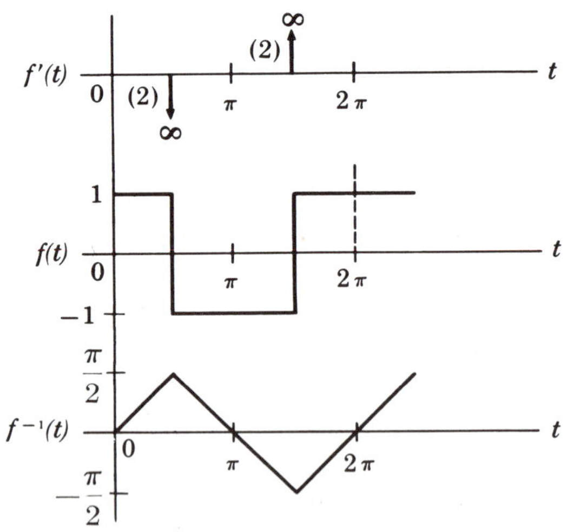

FIG. 20–4. Fourier series for the square-wave family.

If a function satisfies the weak Dirichlet conditions, it is possible to find the coefficients in the Fourier series. It may not be possible to sum the series, however. If a function satisfies the strong Dirichlet conditions it is possible both to find the series and to find its sum. All real physical waveforms satisfy the strong Dirichlet conditions, but many of the idealized waveforms which we use satisfy only the weak Dirichlet conditions. Even for the convergent series the question concerning the number of terms which must be used in order to obtain a good approximation must still be discussed. Consider the square wave $f(t)$ shown in Fig. 20–4. The derivative of this waveform is shown above it, and its integral below. The Fourier series for the square wave is

$$f(t) = \frac{4}{\pi}\left(\cos t - \frac{\cos 3t}{3} + \frac{\cos 5t}{5} - \frac{\cos 7t}{7} + \cdots\right). \tag{20–12}$$

The Fourier series for the waveform of the indefinite integral $f^{-1}(t)$ can be obtained by integrating the Fourier series for $f(t)$. The result is

$$f^{-1}(t) = \frac{4}{\pi}\left(\sin t - \frac{\sin 3t}{3^2} + \frac{\sin 5t}{5^2} - \cdots\right). \tag{20–13}$$

If we take the derivative of both sides of Eq. (20–12), we obtain

$$f'(t) = \frac{4}{\pi}\,(-\sin t + \sin 3t - \sin 5t + \cdots)\cdot \tag{20–14}$$

The functions $f(t)$ and $f^{-1}(t)$ satisfy the strong *Dirichlet* conditions and the series for them are uniformly convergent. The function $f'(t)$ satisfies only the weak Dirichlet conditions, and the Fourier series for it is not, strictly speaking, convergent. The series is infinite at the points where the function $f'(t)$ is infinite, and it is zero everywhere else.

When we wish to use a finite number of terms in a Fourier series, we are interested in the rate of convergence of the series, or essentially in the ratio of the coefficients of successive terms. It is apparent from this example that integration improves the rate of convergence, and differentiation impairs the rate of convergence. We can say more. If there are infinite discontinuities in a function, its Fourier series will not converge. If there are infinite discontinuities in the derivative of a function, its Fourier series will converge as $1/n$. If there are infinite discontinuities in the second derivative of a function, its Fourier series will converge as $1/n^2$, and so on. The more derivatives of a function which exist, the more rapidly will its Fourier series converge. In the limit, a single sine wave has an infinite number of derivatives and requires only one term in a Fourier series for an exact representation. *Thus the smoother a function is, the fewer the terms required in its Fourier series.*

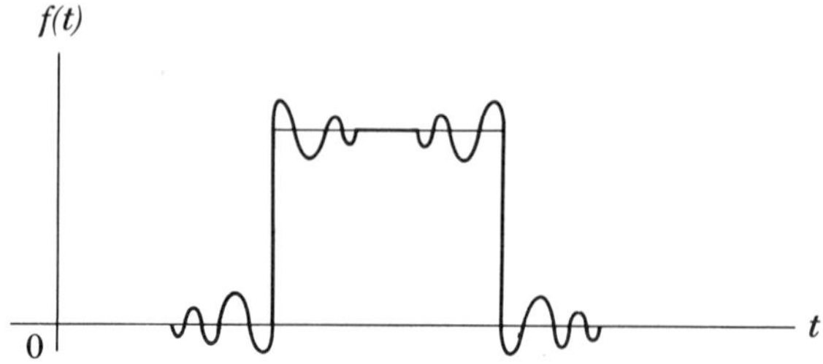

FIG. 20–5. The Gibbs phenomenon.

A sudden discontinuity in a function requires an infinite number of terms in the corresponding Fourier series. But even this is not enough! At the discontinuity, the Fourier series produces a high-frequency oscillation which overshoots the desired peak value by about 18 percent. With an infinite number of terms this overshoot takes place right at the discontinuity, but with a finite number of terms the overshoot is spread over a considerable section of the wave. The nonuniform convergence of the Fourier series was first investigated by Sir Willard Gibbs, and the characteristic overshoot is referred to as the Gibbs phenomenon. An example showing the overshoot in the vicinity of a discontinuity is shown in Fig. 20–5. In some examples this effect is important and trigonometric polynomials other than Fourier partial sums must be used.*

20–4 Symmetry in Fourier series. The existence of symmetry in a waveform can be exploited to reduce the amount of labor required to find the Fourier coefficients. Usually, the symmetry is obvious by inspection. Mathematical tests are available for testing one's intuition. In general, symmetry exists either about a vertical line at some point in the waveform or about a horizontal line. If the symmetry is about a vertical line the origin should be taken at that point. If the symmetry is about a horizontal line, the value of this line must be the direct component of the waveform. The direct component should be removed, leaving the symmetry about the horizontal axis of the waveform. Before we discuss symmetry in general, it is well to note that a cosine wave is identical for positive and negative values of time, for it satisfies the condition

$$f(t) = f(-t). \tag{20–15}$$

Any waveform which satisfies this condition is called an *even function.* On the other hand, a sine wave is identical in shape for positive and nega-

* For Fejer polynomials and Cesaro summations, see E. A. Guillemin, *The Mathematics of Circuit Analysis,* John Wiley & Sons, New York, 1949.

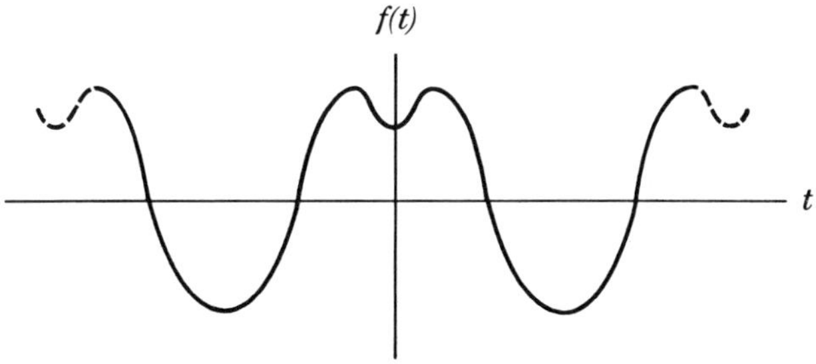

FIG. 20–6. Even-function symmetry.

tive values of time, but there is a reversal of sign between the values for positive t and the values for negative t. A sine wave satisfies the condition

$$f(t) = -f(-t). \tag{20–16}$$

Any waveform which satisfies this condition is called an *odd function.* Symmetry about the origin will be of either the even or odd type. We can classify the types of symmetry possessed by waveforms as follows:

1. Even-function symmetry.
2. Odd-function symmetry.
3. Rotation symmetry.
4. Quarter-wave symmetry.

Even-function symmetry. Even functions satisfy the condition $f(t) = f(-t)$. If the function for positive time is folded about the origin, it falls exactly on the function for negative time. A sketch of an even function is shown in Fig. 20–6. All cosine waves are even functions, and therefore a series made up of cosine terms is an even function. Since all sine waves are odd functions, the presence of just one of them spoils the even-function symmetry. We are thus led to conclude that *the Fourier series for an even*

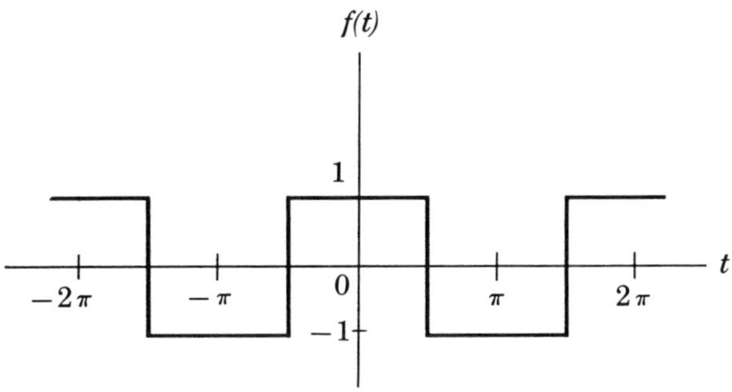

FIG. 20–7. A square wave with even-function symmetry.

function has only cosine terms. As an example of the exploitation of even-function symmetry, consider the problem of finding the Fourier series of the waveform shown in Fig. 20–7. The function is defined by

$$f(t) = \begin{cases} 1 & -\pi/2 < t < \pi/2, \\ -1 & \pi/2 < t < 3\pi/2. \end{cases}$$

It possesses no d-c term and no sine terms because of the even-function symmetry. For the cosine terms

$$\begin{aligned} \pi \cdot a_n &= \int_{-\pi/2}^{3\pi/2} f(t) \cos nt\, dt = \int_{-\pi/2}^{\pi/2} \cos nt\, dt - \int_{\pi/2}^{3\pi/2} \cos nt\, dt \\ &= \left[\frac{\sin nt}{n}\right]_{-\pi/2}^{\pi/2} - \left[\frac{\sin nt}{n}\right]_{\pi/2}^{3\pi/2} \\ &= \begin{cases} (-1)^{(n+3)/2}\left[\dfrac{2}{n} - \left(-\dfrac{2}{n}\right)\right] & \text{for } n \text{ odd}, \\ 0 - 0 \quad \text{for } n \text{ even}. \end{cases} \end{aligned}$$

Thus

$$a_n = \begin{cases} \dfrac{4(-1)^{(n+3)/2}}{\pi n} & \text{for } n \text{ odd}, \\ 0 \quad \text{for } n \text{ even}. \end{cases}$$

Note that the symmetry makes it possible to integrate over one-half the wave and to simply double the answer obtained. The complete series for the function is

$$f(t) = \frac{4}{\pi}\left(\cos t - \frac{\cos 3t}{3} + \frac{\cos 5t}{5} - \cdots\right). \tag{20–17}$$

Odd-function symmetry. Odd functions satisfy the condition $f(t) = -f(-t)$. If the function for positive time is folded about the origin, and then folded about the zero axis, it will fall exactly on the function for negative time. A typical odd function is shown in Fig. 20–8. All sine

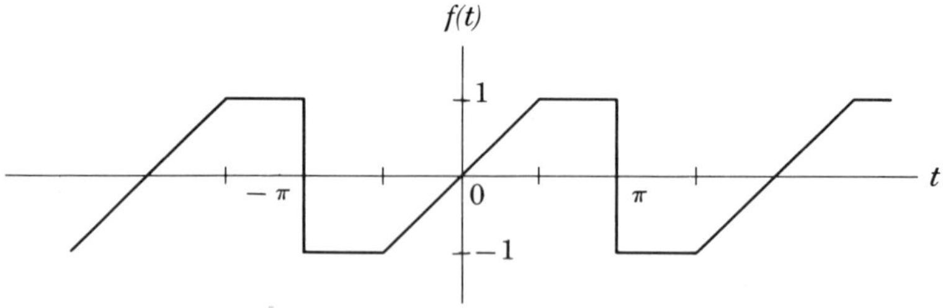

FIG. 20–8. Odd-function symmetry.

waves are odd functions and the sum of a number of sine waves is still an odd function. All cosine functions are even and the presence of just one in a series would spoil odd-function symmetry. We thus conclude that *the Fourier series for an odd function has only sine terms.*

If the origin of the wave in Fig. 20–7 is shifted to the left by $\pi/2$, the wave will possess odd-function symmetry. It can then be defined as

$$f(t) = \begin{cases} 1 & 0 < t < \pi, \\ -1 & \pi < t < 2\pi. \end{cases}$$

The Fourier series for this wave has only sine terms. The coefficients are given by

$$\begin{aligned} \pi \cdot b_n &= \int_0^{2\pi} f(t) \sin nt\, dt = \int_0^{\pi} \sin nt\, dt - \int_{\pi}^{2\pi} \sin nt\, dt \\ &= \left[\frac{-\cos nt}{n}\right]_0^{\pi} - \left[\frac{-\cos nt}{n}\right]_{\pi}^{2\pi} \\ &= \begin{cases} \dfrac{2}{n} - \dfrac{-2}{n} & \text{for } n \text{ odd}, \\ 0 - 0 & \text{for } n \text{ even}. \end{cases} \end{aligned}$$

Thus

$$b_n = \begin{cases} \dfrac{4}{\pi n} & \text{for } n \text{ odd}, \\ 0 & \text{for } n \text{ even}. \end{cases}$$

Note again that symmetry makes it possible to integrate over one-half of the period and to simply double the answer obtained. The complete series for the function is

$$f(t) = \frac{4}{\pi}\left(\sin t + \frac{\sin 3t}{3} + \frac{\sin 5t}{5} + \cdots\right). \tag{20–18}$$

The waves represented by Eqs. (20–17) and (20–18) are identical except for a shift of the origin. The shift is one-quarter of the fundamental period. The shift represents $\pi/2$ radians for the fundamental term in the Fourier series but *$n\pi/2$ radians for the nth harmonic.* To change from Eq. (20–17) to (20–18), therefore, we must subtract $\pi/2$ from the fundamental, $3\pi/2$ from the third harmonic, $5\pi/2$ from the fifth harmonic, and so on.

Rotation symmetry. Odd-function symmetry is a combination of symmetry about the origin and symmetry about the zero axis. A second form of symmetry about the zero axis can exist when there is no odd-function symmetry. It is defined by the condition that $f(t) = -f(t + \pi)$. For this type of symmetry, the waveshapes on alternate half-cycles are identical but reversed in sign. Waveforms of this type are produced in rotating electrical machinery when a north pole and a south pole alternately pass a given point. The term rotation symmetry is used to describe any waveform of this general type. For a wave with rotation symmetry, there is no d-c term. The coefficients of the cosine terms are given by

$$\begin{aligned} a_n &= \frac{1}{\pi}\int_0^{2\pi} f(t)\cos nt\,dt = \frac{1}{\pi}\int_0^{\pi} f(t)\cos nt\,dt + \frac{1}{\pi}\int_{\pi}^{2\pi} f(t)\cos nt\,dt \\ &= \frac{1}{\pi}\int_0^{\pi} f(t)\cos nt\,dt + \frac{1}{\pi}\int_0^{\pi} f(t+\pi)\cos n(t+\pi)\,dt \\ &= \frac{1}{\pi}\int_0^{\pi} [f(t)\cos nt + f(t+\pi)(\cos nt\cos n\pi - \sin nt\sin n\pi)]\,dt \\ &= \frac{1}{\pi}\int_0^{\pi} [f(t)\cos nt - f(t)(\cos nt\cos n\pi)]\,dt \\ &= \begin{cases} 0 & \text{for } n \text{ even,} \\ \dfrac{2}{\pi}\displaystyle\int_0^{\pi} f(t)\cos nt\,dt & \text{for } n \text{ odd.} \end{cases} \end{aligned} \tag{20–19A}$$

In a similar fashion, we obtain

$$b_n = \begin{cases} 0 & \text{for } n \text{ even,} \\ \dfrac{2}{\pi}\displaystyle\int_0^{\pi} f(t)\sin nt\,dt & \text{for } n \text{ odd.} \end{cases} \tag{20–19B}$$

Thus we can conclude that *the Fourier series for a wave with rotation symmetry will have only odd harmonics.* An example of a typical waveform with rotation symmetry is given in Fig. 20–9.

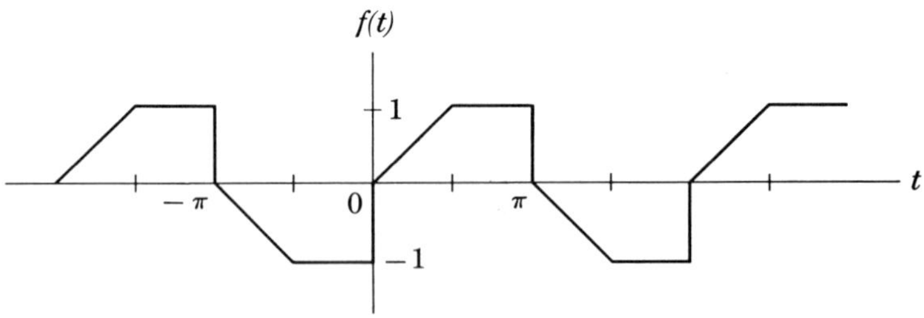

FIG. 20–9. Rotation symmetry.

Both of the square waves represented by Eqs. (20–17) and (20–18) have rotation symmetry. We know immediately, without any calculation, that they cannot contain any even harmonic terms. The problem reduces in the case of Eq. (20–17) to finding the odd cosine terms, and in the case of Eq. (20–18) to finding the odd sine terms.

Rotation symmetry is also called *half-wave symmetry,* since the integrals for the coefficients in Eqs. (20–19A) and (20–19B) need be taken over only half a period. No error will be committed if the general integrals are used over a whole period, but the value obtained for the integral over the second half of the period will be identical to the value over the first half of the period; it is simpler to multiply by two.

Quarter-wave symmetry. A combination of rotation symmetry with either even-function or odd-function symmetry results in a wave with quarter-wave symmetry. If a template were cut out for the first quarter of a period, it could be used to trace the other sections of the wave as well. Analytically, the test for quarter-wave symmetry is

$$f(t) = -f(\pi + t) \qquad \text{and} \qquad f(t) = \pm f(-t).$$

It is easier to detect, however, as a combination of the other types of symmetry. The waveform in Fig. 20–7 possesses quarter-wave symmetry. It thus has even-function symmetry and rotation symmetry. The Fourier series can therefore have only odd cosine terms. Furthermore, each quarter-cycle is the same, and the integration need only be taken over a quarter-cycle and the result multiplied by 4. Thus we can find the coefficients as

$$a_n = \frac{4}{\pi}\int_0^{\pi/2} f(t)\cos nt\,dt = \frac{4}{\pi}\int_0^{\pi/2} \cos nt\,dt$$

$$= \frac{4}{\pi}\left[\frac{+\sin nt}{n}\right]_0^{\pi/2}.$$

The value at the lower limit of this integral is always zero. The value at the upper limit is plus one when $n = 1$, minus one for $n = 3$, plus one for $n = 5$, and so on. We can therefore write a_n as

$$a_n = \frac{4}{\pi n}(-1)^{(n+3)/2},$$

and the over-all series as

$$f(t) = \frac{4}{\pi}\left(\cos t - \frac{\cos 3t}{3} + \frac{\cos 5t}{5} - \cdots\right),$$

which is identical to Eq. (20–17). Whenever quarter-wave symmetry exists, it pays to take advantage of it!

Half-range series. Rotation symmetry is odd-harmonic symmetry, since only odd harmonics are present in the Fourier series. It seems reasonable to inquire whether an even-harmonic type of symmetry exists in which the Fourier series has only even harmonics. (The student should not confuse even and odd functions of time with even and odd orders of harmonics.) The condition under which even-harmonic symmetry exists is

$$f(t) = f(t + \pi). \tag{20–20}$$

Stated in words, Eq. (20–20) describes a function which has the same form in the interval from zero to π as it has in the interval π to 2π. Clearly, we have taken a period which is twice too long for the function. The corresponding fundamental frequency is twice too low, and we naturally obtain a factor of 2 in all the harmonics.

The Fourier phase-angle and magnitude series. We previously wrote the general Fourier series in the form

$$\begin{aligned} f(t) = a_0 &+ a_1 \cos t + a_2 \cos 2t + \cdots + a_n \cos nt + \cdots \\ &+ b_1 \sin t + b_2 \sin 2t + \cdots + b_n \sin nt + \cdots. \end{aligned}$$

It is sometimes convenient to combine the sine and cosine terms in each harmonic and to write the series in the form

$$\begin{aligned} f(t) = a_0 &+ A_1 \cos (t + \theta_1) + A_2 \cos (2t + \theta_2) + \cdots \\ &+ A_n \cos (nt + \theta_n) + \cdots, \end{aligned} \tag{20–21}$$

where A_n = amplitude of nth harmonic,
θ_n = phase of nth harmonic.

The relationship between the amplitude and phase constants for the nth harmonic and the amplitude of the corresponding cosine and sine terms in the original series is obtained by expanding the nth cosine term of Eq. (20–21). Thus

$$A_n \cos (nt + \theta_n) = A_n \cos nt \cos \theta_n - A_n \sin nt \sin \theta_n,$$

which must be equal to

$$a_n \cos nt + b_n \sin nt.$$

From this equality, we obtain

$$\begin{aligned} A_n \cos \theta_n &= a_n, \\ A_n \sin \theta_n &= -b_n. \end{aligned} \tag{20–22}$$

To solve for A_n we square both equations and add. To obtain θ_n, we take

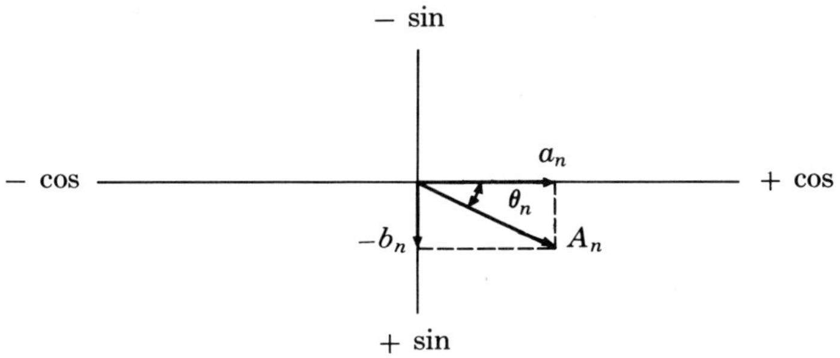

FIG. 20–10. Harmonic vector diagram.

the ratio of the two equations. Thus

$$A_n^2 = a_n^2 + b_n^2, \qquad \theta_n = \tan^{-1}\frac{-b_n}{a_n}. \tag{20–23}$$

The vector relationships are shown in Fig. 20–10.

There are no convenient direct integral formulas for the magnitude and phase of a harmonic in Eq. (20–21). They are usually obtained analytically from Eqs. (20–23). On the other hand, if the coefficients are being obtained graphically, or if the Fourier series is to be plotted, it is more convenient to work with the magnitude and phase.

20–5 Numerical methods for Fourier analysis. Students often regard numerical methods as tedious and boring. They are confusing the practice of numerical methods (which is best left to machines) with the principles of numerical methods, which can be fascinating and instructive. Such is the case with Fourier analysis. By a simple extension of numerical Fourier analysis, we will be able to prove the Shannon sampling theorem, on which the whole theoretical structure of modern information theory and communications theory rests. We will discuss two methods of analysis. The first is simply the numerical or graphical integration of the basic formulas for the Fourier coefficients. The second method is less general, but often more useful in practical problems. The Fourier series is assumed to consist of a finite number of terms with a finite number of unknown coefficients. By sampling the function at various instants of time, a set of equations can be written from which the unknown coefficients can be found. The latter method is usually used in problems involving electrical machinery where only the fundamental, third, fifth and seventh harmonics occur, and in vacuum tube problems where only the second and third harmonics have significant values.

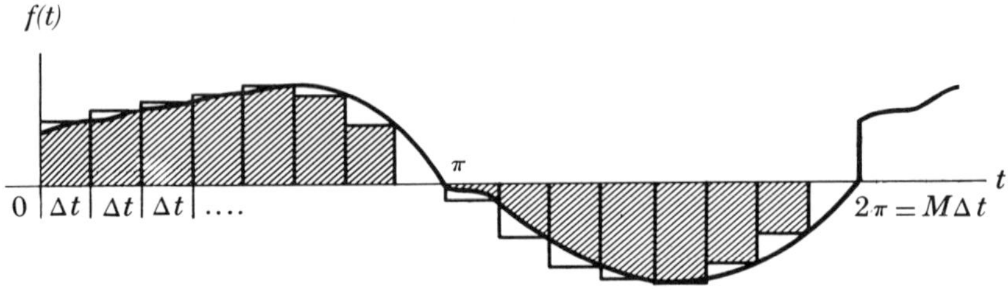

FIG. 20–11. A sampled waveform.

Method 1. Numerical integration. In order to proceed with numerical integration, the basic waveform is divided into a series of uniformly spaced sections, as shown in Fig. 20–11. This process is also called *sampling* the wave. We will suppose that the function is sampled at the points Δt, $2\,\Delta t$, $3\,\Delta t, \ldots m\,\Delta t, \ldots$ up to $M\,\Delta t = 2\pi$. We thus have M points and the spacing between adjacent points is uniform and equal to $2\pi/M$.

The first coefficient in the Fourier series is given by

$$a_0 = \frac{1}{2\pi}\int_0^{2\pi} f(t)\,dt.$$

We will replace the area under the integral by the area of the rectangles obtained by multiplying the sampled values of the function by Δt. This area is shown shaded in Fig. 20–11. The integral sign becomes a summation, and we have

$$a_0 \cong \frac{1}{2\pi}\sum_{m=1}^{M} f(m\,\Delta t)\,\Delta t. \tag{20–24}$$

A similar process can be carried out for the other Fourier coefficients. Consider the general coefficient

$$a_n = \frac{1}{\pi}\int_0^{2\pi} f(t)\cos nt\,dt.$$

We imagine that the product $f(t)\cos nt$ is first plotted and then sampled at the same points as before. A similar trapezoidal wave is obtained which can be integrated by summation. Thus

$$a_n \cong \frac{1}{\pi}\sum_{m=1}^{M} f(m\,\Delta t)\cos(nm\,\Delta t)\,\Delta t. \tag{20–25}$$

Similarly

$$b_n \cong \frac{1}{\pi}\sum_{m=1}^{M} f(m\,\Delta t)\sin(nm\,\Delta t)\,\Delta t. \tag{20–26}$$

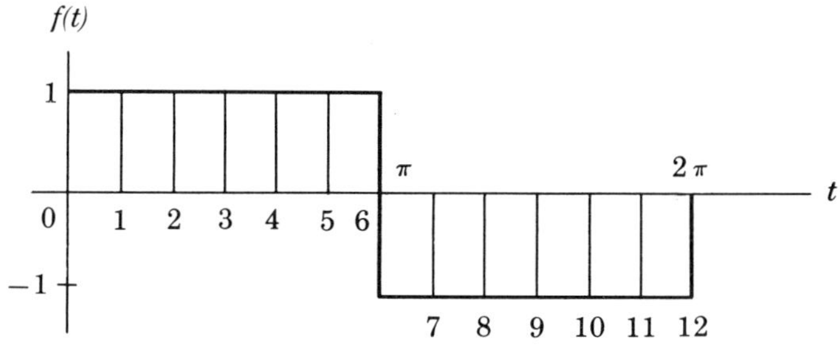

FIG. 20–12. Numerical Fourier series.

The numerical formulas for the Fourier coefficients require only the sampled values of the original function and certain discrete values of the cosines and sines. To evaluate such expressions, it is almost essential to use an organized tabular form. As an example, the first two nonzero coefficients of the Fourier series for the square wave in Fig. 20–12 will be obtained. Twelve sample points will be used, with an angle of 30° between sample points. More sample points must be used if higher harmonics are required, as will be discussed below.

Both the function in Fig. 20–12, and its sampled approximation are symmetrical so far as the two half-waves are concerned. The sampled approximation is not symmetrical about the given origin in the quarter-wave sense, and hence the quarter-wave equations cannot be used. (It would be possible to shift the sampling points to obtain quarter-wave symmetry.) The wave has rotation symmetry and possesses only odd sine terms. The half-wave formulas for the fundamental and the third harmonic are

$$b_1 \cong \frac{2}{\pi} \sum_{m=1}^{M/2} f(m\,\Delta t) \sin (m\,\Delta t)\,\Delta t,$$

$$b_3 \cong \frac{2}{\pi} \sum_{m=1}^{M/2} f(m\,\Delta t) \sin (3\,m\,\Delta t)\,\Delta t.$$

A tabular solution for b_1 is given in Table 20–1, and a tabular solution for b_3 in Table 20–2. Both values agree within sliderule accuracy with the values obtained analytically by Eq. (20–18).

Note that at sample points where discontinuity occurs, as for point 6 in Fig. 20–12, the value at the sample is taken to be the value just to the left of the discontinuity. A similar situation occurs at point 12. The reason for this rule is that we are using the sampled value to represent the area under the function to the left of each sample point. A value of zero at sample point 6, for example, would give a totally erroneous area for the trapezoid between points 5 and 6. The value of the fundamental is prac-

TABLE 20–1

NUMERICAL EVALUATION OF FUNDAMENTAL

m	$m\,\Delta t$	$\sin m\,\Delta t$	$f(m\,\Delta t)$	$f(m\,\Delta t)\sin m\,\Delta t$
1	30	0.500	1	0.500
2	60	0.866	1	0.866
3	90	1.000	1	1.000
4	120	0.866	1	0.866
5	150	0.500	1	0.500
6	180	0.000	1	0.000
				Sum 3.732

$$b_1 = \frac{2}{\pi} \times 3.732 \times \frac{\pi}{6} = 1.24$$

TABLE 20–2

NUMERICAL EVALUATION OF THIRD HARMONIC

m	$3m\,\Delta t$	$\sin 3m\,\Delta t$	$f(m\,\Delta t)$	$f(m\,\Delta t)\sin 3m\,\Delta t$
1	90	1.00	1	1.00
2	180	0.00	1	0.00
3	270	−1.00	1	−1.00
4	360	0.00	1	0.00
5	450	1.00	1	1.00
6	540	0.00	1	0.00
				Sum 1.00

$$b_3 = \frac{2}{\pi} \times 1.00 \times \frac{\pi}{6} = 0.33$$

tically perfect. Because the sample rate is too small, the value of the third harmonic is very approximate. We will investigate this question further in the next section.

Method 2. Algebraic equations. In many physical waveforms it is known that there is an insignificant amount of energy in the higher harmonics. It seems reasonable therefore to neglect them entirely and to write a Fourier partial sum, instead of an infinite series, as an approximation to the function. If we decide to include n harmonics in the partial sum, there will be n constants to find for the sine terms and n constants to find for the cosine terms, and one additional constant for the d-c value of the waveform. The d-c value is trivial and can always be obtained by a direct integration. Let us suppose that it has been removed. There are then $2n$ constants in our partial sum, and we must obtain $2n$ equations to solve for them. Roughly speaking, we can do this by sampling the function $f(t)$ at $2n$ places and equating the values to the assumed series. The only question which arises is the independence of the equations which we obtain, and the consistency of the equations. For example, we might sample at $2n$ points and find that two of our equations turn out to be identical. Or, if we have neglected higher harmonics which should have been included, we might obtain two different equations for the same constant.

The question of the consistency of the equations is easy to resolve theoretically. We just assume that there are no harmonics present higher than the ones we are considering! In practice we may have to try several times to arrive at this happy state of affairs. The question of the independence of the equations is harder to resolve. If we sample uniformly, with $2n$ samples and n harmonics we will have sampled the highest harmonic twice in its period. Clearly, two samples of the nth harmonic cannot determine the wave. They might both occur at the points where the wave passes through zero. Considering the nth harmonic alone, we might write

$$f(t) = a_n \cos nt + b_n \sin nt.$$

If we sample at $\theta = nt$ and $\theta + \pi = nt$ we will have the required two uniformly spaced samples per period. The two equations obtained will be

$$\begin{aligned} f(t_1) &= a_n \cos \theta + b_n \sin \theta, \\ f(t_2) &= a_n \cos (\theta + \pi) + b_n \sin (\theta + \pi) \\ &= -a_n \cos \theta - b_n \sin \theta. \end{aligned}$$

Thus the two equations are not independent. There are two ways out of the dilemma. The first is to use nonuniform sampling. The second is to sample more often. Shannon* has shown in a general way that any rate of sampling greater than $2n$ per period will be satisfactory. An older solution discovered many years ago by Fischer and Hinnen† sampled at twice this rate. If one set of sample points missed the nth harmonic, the other set was sure to get it!

We will apply the method to the harmonic analysis of the distorted waveform typically found in a vacuum-tube amplifier. The basic wave-

---

* C. E. Shannon, *Proc. IRE*, Vol. 36, p. 1234, (Nov. 1948). Note particularly that the usual statement of the Shannon theorem, "the highest frequency sine wave must be sampled twice," is not correct.

† See the Fischer-Hinnen method given in many standard texts, e.g. Frazier, *Elementary Electric Circuit Theory*, McGraw-Hill, New York, 1945.

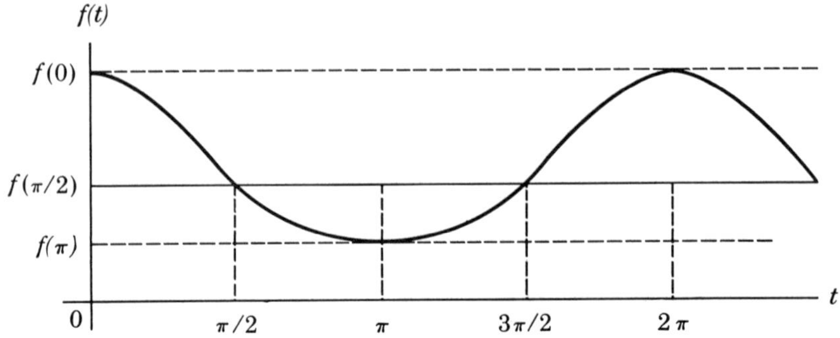

FIG. 20–13. Partial sum harmonic analysis.

form is shown in Fig. 20–13. If the origin is taken at the point shown in the illustration, the waveform possesses even-function symmetry and the Fourier series will have only cosine terms. If we neglect all harmonics above the second, we can write

$$f(t) = a_0 + a_1 \cos t + a_2 \cos 2t.$$

Since there are three constants to be obtained, we require three equations. Any three instants of time that give independent equations are satisfactory. We shall select $t = 0$, $t = \pi/2$, and $t = \pi$. The equations are

$$f(0) = a_0 + a_1 + a_2,$$
$$f(\pi/2) = a_0 - a_2,$$
$$f(\pi) = a_0 - a_1 + a_2.$$

These equations can be solved to give

$$a_1 = \frac{f(0) - f(\pi)}{2},$$
$$a_0 = \frac{[f(0) + f(\pi)]/2 + f(\pi/2)}{2},$$
$$a_2 = \frac{[f(0) + f(\pi)]/2 - f(\pi/2)}{2}.$$

Even when a larger number of harmonics is assumed, this method gives a rapid solution for the coefficients. Various special formulas have been worked out in the literature to cover particular choices of the sampling points. We are interested here primarily in the principle underlying the method.

20–6 The Fourier integral. In this section we will extend the concept of the Fourier series for a periodic function to a Fourier integral for a single pulse of arbitrary shape. We will adopt a slightly different point of view toward a periodic wave. We will now think of it as a "repeated transient." For example, the square wave in Fig. 20–14 can be described as a square wave of period 2π, or it can be described as a pulse of width π repeated with a period of 2π. It is natural to inquire what happens to the Fourier series if we leave the pulse alone and increase the repetition period. All the harmonics will have smaller amplitudes because the integrals which describe them are averaged over the full period. The distribution of the harmonics relative to each other, however, can be found, since the actual pulse integrated will not change.

The first step in deriving the Fourier integral involves the expression of the general Fourier series in the form of complex exponentials instead of trigonometric functions. The basic substitutions are of the form

$$\sin\theta = \frac{e^{j\theta} - e^{-j\theta}}{2j} \qquad \text{and} \qquad \cos\theta = \frac{e^{j\theta} + e^{-j\theta}}{2}.$$

We first write the Fourier series in the following form, for an arbitrary period T:

$$f(t) = a_0 + \sum_{n=1}^{\infty} (a_n \cos n\omega_0 t + b_n \sin n\omega_0 t),$$

where

$$a_0 = \frac{1}{T}\int_0^T f(t)\,dt,$$

$$a_n = \frac{2}{T}\int_0^T f(t)\cos n\omega_0 t\,dt,$$

$$b_n = \frac{2}{T}\int_0^T f(t)\sin n\omega_0 t\,dt,$$

$$\omega_0 = \frac{2\pi}{T}.$$

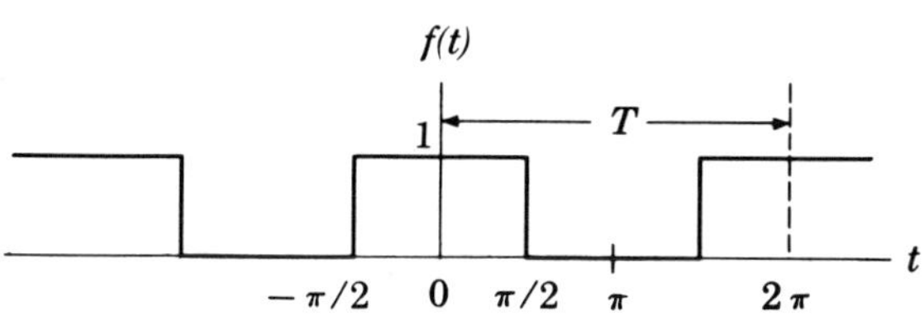

FIG. 20–14. A repeated transient.

Substitution of the exponential expressions gives

$$f(t) = a_0 + \sum_{n=1}^{\infty} \left[a_n \frac{(e^{jn\omega_0 t} + e^{-jn\omega_0 t})}{2} + b_n \frac{(e^{jn\omega_0 t} - e^{-jn\omega_0 t})}{2j} \right]$$

$$= a_0 + \sum_{n=1}^{\infty} \left[\frac{(a_n - jb_n)}{2} e^{jn\omega_0 t} + \frac{(a_n + jb_n)}{2} e^{-jn\omega_0 t} \right]$$

$$= a_0 + \sum_{n=1}^{\infty} [c_n e^{jn\omega_0 t} + c_n^* e^{-jn\omega_0 t}],$$

where c_n^* is the complex conjugate of c_n and

$$c_n = \frac{a_n - jb_n}{2}$$

$$= \frac{2}{T} \int_0^T f(t) \left[\frac{(e^{jn\omega_0 t} + e^{-jn\omega_0 t})}{4} - j \frac{(e^{jn\omega_0 t} - e^{-jn\omega_0 t})}{4j} \right] dt$$

$$= \frac{1}{T} \int_0^T f(t) e^{-jn\omega_0 t}\, dt. \tag{20–27}$$

The expression for the Fourier series can be simplified if we note that the value of c_n given by Eq. (20–27) for $n = 0$ is identical to the value of a_0, and that the terms in the series of the form $c_n^* e^{-jn\omega_0 t}$ are identical to the terms which would have been obtained if we had taken the terms involving c_n and made n negative. Accordingly, instead of a constant term and two infinite sums from plus one to infinity, we can use a single sum from minus infinity to plus infinity. The result is the complex Fourier series

$$f(t) = \sum_{n=-\infty}^{\infty} c_n e^{jn\omega_0 t}, \tag{20–28}$$

where

$$c_n = \frac{1}{T} \int_0^T f(t) e^{-jn\omega_0 t}\, dt. \tag{20–29}$$

The complex Fourier series of Eq. (20–28) expresses a periodic function of time as a sum of terms involving positive and negative complex exponentials. The actual sinusoidal harmonics in the series are composed of pairs of positive and negative terms at each frequency. The amplitude of an actual harmonic is twice the amplitude of either of the corresponding exponential terms alone. It is also possible to view the positive exponential terms as corresponding to a "positive" frequency and the negative exponential terms as corresponding to a "negative" frequency. The amplitude spectrum for the positive and negative frequencies is just a

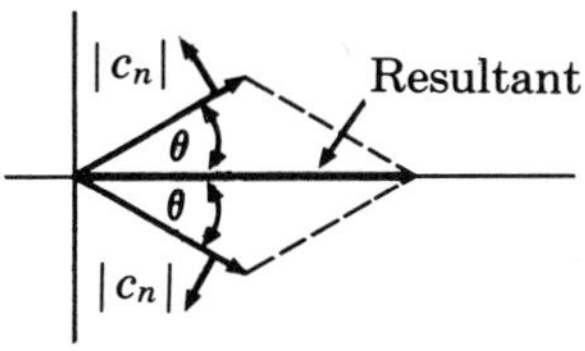

FIG. 20–15. Counter-rotating vectors.

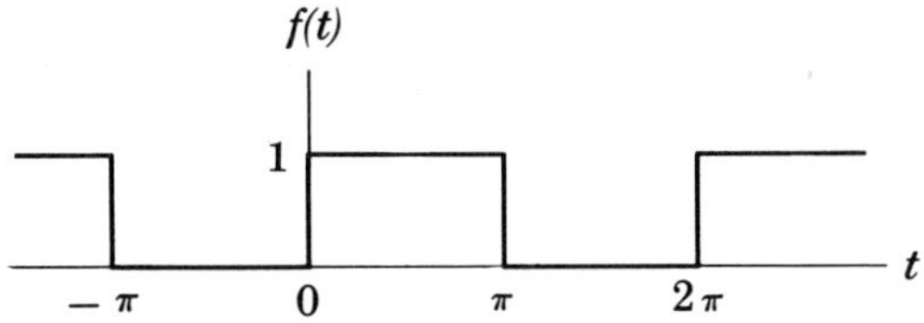

FIG. 20–16. An example for the complex Fourier series.

plot of the values of the magnitudes of c_n. The phase of the actual harmonic is the angle of the complex number c_n. To arrive at this fact, let us add together two harmonic terms in the series. We will suppose that the coefficients c_n are complex (in general they are) and are given by

$$c_n = |c_n|e^{j\theta},$$
$$c_n^* = |c_n|e^{-j\theta}.$$

The sum of the two terms for the nth harmonic in the series is thus expressed by

$$|c_n|e^{j\theta}e^{jn\omega_0 t} + |c_n|e^{-j\theta}e^{-jn\omega_0 t}.$$

From the vector diagram of Fig. 20–15, the sum of these two vectors is

$$2|c_n| \cos (n\omega_0 t + \theta). \tag{20–30}$$

Equation (20–30) shows that if we plot a graph of the phase of the complex number c_n for each harmonic, we will be plotting the phase angles of the actual harmonics.

Example. As an example of the complex Fourier series, consider the waveform of Fig. 20–16. The coefficients c_n are given by

$$\begin{aligned} c_n &= \frac{1}{2\pi}\int_0^{2\pi} f(t)e^{-jnt}\,dt = \frac{1}{2\pi}\int_0^{\pi} e^{-jnt}\,dt \\ &= \frac{1}{2\pi}\left[\frac{e^{-jnt}}{-jn}\right]_0^{\pi} \\ &= \begin{cases} \dfrac{1}{2\pi}\cdot\dfrac{2}{jn} & \text{for } n \text{ odd,} \\ 0 & \text{for } n \text{ even.} \end{cases} \end{aligned} \tag{20–31}$$

The expression is indeterminate for $n = 0$, and the d-c value is more easily determined by inspection. Here it is $a_0 = \frac{1}{2}$. The amplitude and

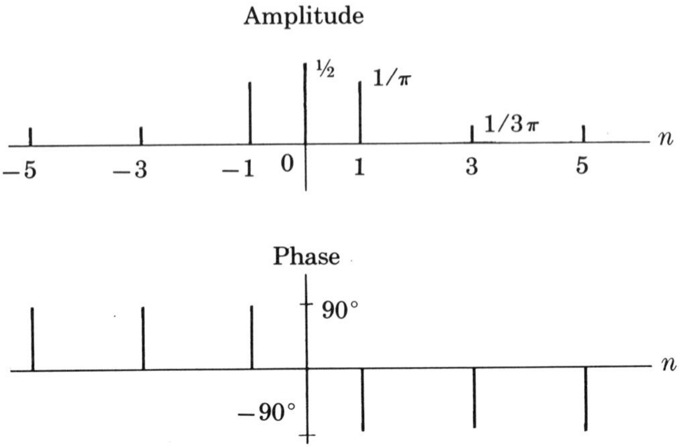

FIG. 20–17. Amplitude and phase spectra.

phase spectra of the c_n's are given in Fig. 20–17. The phase is $-90°$ for all positive-frequency harmonics. Since the cosine wave represents zero phase, all the harmonics will be sine waves of zero phase angle. The actual Fourier series can be recovered from the complex coefficients of Eq. (20–30). It is given by

$$f(t) = \sum_{n=-\infty}^{+\infty} c_n e^{jn\omega_0 t}$$
$$= \cdots - \frac{1}{3\pi j} e^{-j3t} - \frac{1}{\pi j} e^{-jt} + \frac{1}{2} + \frac{1}{\pi j} e^{jt} + \frac{1}{3\pi j} e^{3jt} + \cdots.$$

Combining the positive and negative exponentials in pairs, we obtain

$$f(t) = \frac{1}{2} + \frac{2}{\pi}\left(\sin t + \frac{\sin 3t}{3} + \frac{\sin 5t}{5} + \cdots\right),$$

as in Eq. (20–9).

The mathematical expressions for the complex Fourier series are beautifully simple and symmetrical. The simplicity is, however, deceptive. If the problem is merely to find the terms in a Fourier series, it is usually easier to use the formulas which have been given before. We have developed the complex form of the series, because it is a logical stepping stone toward the Fourier integral and it adds a great deal to the understanding of the integral.

The Fourier integral. The formal extension from the complex Fourier series to the Fourier integral is quite simple. The limiting processes cannot be rigorously justified, however, and mathematicians prefer to proceed in the other direction. The Fourier integral is postulated and its properties are derived. The straightforward process, however, lends more insight into Fourier methods and we will follow it. We will view the complex Fourier series as a representation of a set of repeated transients and then we will let the period of the repetition tend to infinity. The fundamental frequency of the wave approaches zero, and so do the values of the harmonics. In order to focus attention on the distribution of the harmonics relative to each other, we will multiply the value of the harmonics in the series by the period T. The original complex series is

$$f(t) = \sum_{n=-\infty}^{\infty} c_n e^{jn\omega_0 t}, \tag{20–32}$$

with

$$c_n = \frac{1}{T}\int_{-T/2}^{T/2} f(t)e^{-jn\omega_0 t}\,dt. \tag{20–33}$$

The new distribution function we will call $F(\omega)$. It is defined by

$$F(\omega) = c_n T = \int_{-T/2}^{T/2} f(t)e^{-jn\omega_0 t}\,dt, \tag{20–34}$$

and in terms of it the Fourier series is

$$f(t) = \sum_{n=-\infty}^{\infty} \frac{F(\omega)}{T}\, e^{jn\omega_0 t} \tag{20–35}$$

Now we are in position to let the period approach infinity. The frequency, ω_0, of the fundamental will approach zero, but we will never let it quite reach zero. Instead, it will become $\Delta\omega$. We do not let it become zero, because we will want to think of some high harmonic, given by $n\omega_0$ as occurring at a finite frequency ω. The period of this harmonic is approaching zero, but we can let the number of harmonics we use approach infinity. The product is indeterminate, and we must suppose that the limiting process is such that the product remains finite and is equal to some frequency ω. A further consequence of the limiting process is that the summation in Eq. (20–35) becomes an integral. In summary

$$T \to \infty$$

$$\omega_0 = 2\pi/T \to \Delta\omega \to d\omega$$

$$n\,\Delta\omega \to \omega,$$

and Eqs. (20–34) and (20–35) become

$$F(\omega) = \int_{-\infty}^{\infty} f(t)e^{-j\omega t}\, dt, \tag{20–36}$$

$$f(t) = \int_{-\infty}^{\infty} \frac{F(\omega)}{2\pi}\, e^{j\omega t}\, \omega_0$$

$$= \frac{1}{2\pi}\int_{-\infty}^{\infty} F(\omega)e^{j\omega t}\, d\omega. \tag{20–37}$$

Equations (20–36) and (20–37) are the *Fourier transforms.** Equation (20–36) tells us how to find the harmonic components or the frequency components of a single transient pulse. The basic thought involved is that if we can find the frequency components of a pulse and design our networks to pass all these frequencies with no distortion, then we will also pass the pulse with no distortion.

Equation (20–37) expresses a time function as an infinite sum of an infinite number of infinitesimal harmonics. The actual amplitude of any frequency component is $F(\omega)\, d\omega/2\pi$. Since the sinusoidal wave is made up of two of the exponential terms, the sine-wave terms have amplitudes of twice this value. Furthermore, the phase of the harmonic term is given by the angle of $F(\omega)$, considered as a complex number at that frequency.

The transform which gives the frequency components in terms of the time integral is called the *direct transform.* In practice, it is somewhat more important than the other transform, which is called the *inverse transform.* The inverse transform involves the integration of complex numbers, and is for this reason alone somewhat more complicated than the direct transform.

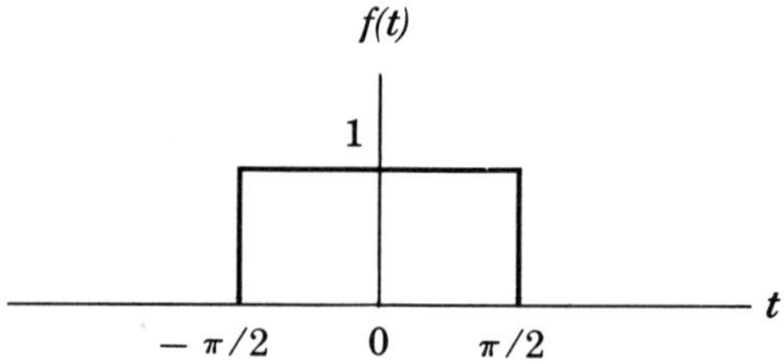

FIG. 20–18. A single pulse.

* Some authors call $F(\omega)$ *the* Fourier transform. We prefer to call $F(\omega)$ the direct transform and $f(t)$, in Eq. (20–37), the inverse transform. The factor 2π is treated differently by some authors. Equations (20–36) and (20–37) follow Laplace transform practice.

As an example of the resolution of a pulse into its Fourier components, consider the square wave of Fig. 20–18. The direct transform, or frequency spectrum, of the pulse is given by

$$\begin{aligned} F(\omega) &= \int_{-\infty}^{\infty} f(t)e^{-j\omega t}\,dt = \int_{-\pi/2}^{\pi/2} e^{-j\omega t}\,dt \\ &= \left[\frac{e^{-j\omega t}}{-j\omega}\right]_{-\pi/2}^{\pi/2} = \frac{e^{j\omega\pi/2} - e^{-j\omega\pi/2}}{j\omega} \\ &= 2\,\frac{\sin \omega\pi/2}{\omega}. \end{aligned}$$

The amplitude versus frequency plot for this function gives the spectral distribution of the Fourier components in the square pulse. The function is a pure real and therefore the components are all cosine functions, in phase at $t = 0$. From $-\pi/2$ to $\pi/2$ they are sufficiently in phase to add up to the pulse. Beyond these points they have random phases and cancel out completely to zero!

The "Fourier series" for the pulse of Fig. 20–18 is now a Fourier integral:

$$f(t) = \int_{-\infty}^{\infty} 2\,\frac{\sin \omega\pi/2}{2\pi\omega}\,d\omega e^{j\omega t}.$$

The exponential terms combine in pairs to give real sinusoids with zero phase angle and amplitudes of

$$2\,\frac{\sin \omega\pi/2}{\pi\omega}\,d\omega.$$

We will further investigate the nature of this distribution of harmonics in the next section.

20–7 The Fourier integral and the Fourier series. The idea that all types of waveforms, periodic, transient, and random, can be resolved into summations of steady-state sine waves lasting from minus infinity to plus infinity is basic to all circuit theory. If we design our circuits to pass all the components which are present in the input wave, we can be reasonably sure that the output will be a faithful reproduction of the input. In the transient case the individual components at any given frequency will have vanishingly small amplitudes, but they are real nevertheless, and we must cater to them in the design of our networks! An interesting relationship exists between the Fourier spectrum for a periodic wave and the spectral distribution function for a transient pulse equal to one period of the periodic wave. The integrals for the Fourier

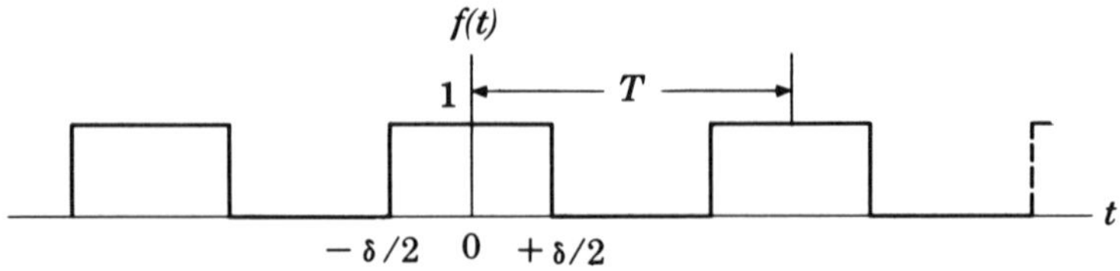

FIG. 20–19. Repeated pulses.

integral, Eqs. (20–36) and (20–37), are practically identical to the integrals for the complex Fourier series, Eqs. (20–33) and (20–32). In particular, the spectrum, as given by Eq. (20–36), and the value of the harmonic amplitude times the period, as given by Eq. (20–34), are identical, with the exception that the variable in the former case is continuous, while in the latter case it takes on discrete values ω_0, $2\omega_0$, $\ldots n\omega_0$. *The Fourier integral spectrum is the envelope of the corresponding Fourier series spectrum.* The continuous spectrum is called a *band spectrum*, while the discrete spectrum of the Fourier series is called a *line spectrum*. The way in which the line spectrum for a series of pulses coalesces into a band spectrum as the repetition period is increased is both interesting and instructive. We will observe the effect for the pulse of Fig. 20–19.

The complex Fourier series for the waveform of Fig. 20–19 has coefficients given by

$$c_n = \frac{1}{T}\int_{-\delta/2}^{+\delta/2} f(t)e^{-jn\omega_0 t}\,dt.$$

The harmonic distribution for the Fourier integral, which holds for the waveform as the period is allowed to increase without limit, is

$$F(\omega) = \int_{-\delta/2}^{\delta/2} f(t)e^{-j\omega t}\,dt.$$

The similarity of these two equations is apparent. We will obtain the values for the Fourier series.

$$\begin{aligned} c_n &= \frac{1}{T}\int_{-\delta/2}^{\delta/2} e^{-jn\omega_0 t}\,dt = \frac{1}{T}\left[\frac{e^{-jn\omega_0 t}}{-jn\omega_0}\right]_{-\delta/2}^{\delta/2} \\ &= \frac{1}{T}\,\frac{\delta}{2}\,\frac{e^{jn\omega_0\delta/2} - e^{-jn\omega_0\delta/2}}{jn\omega_0\delta/2} \\ &= \frac{\delta}{T}\,\frac{\sin n\omega_0\delta/2}{n\omega_0\delta/2}, \qquad \text{where } T = \frac{2\pi}{\omega_0}, \end{aligned}$$

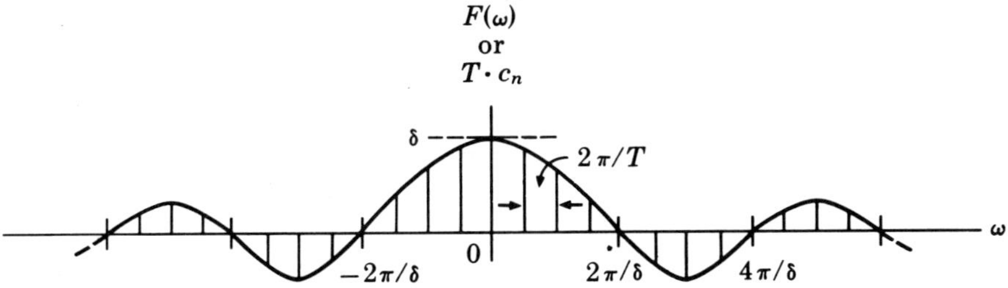

FIG. 20–20. The spectrum of a single pulse.

and similarly

$$F(\omega) = \delta \frac{\sin \omega\delta/2}{\omega\delta/2}.$$

A plot of the continuous spectrum for $F(\omega)$ is given in Fig. 20–20. The limiting value of $(\sin x)/x$ as x approaches zero is unity. The zeros of the function occur when

$$\frac{\omega\delta}{2} = k\pi \qquad \text{for} \quad k = 1, 2, 3, \ldots.$$

The first zero occurs at $\omega = 2\pi/\delta$. The function oscillates about the zero line with a period of $4\pi/\delta$ and with a gradually decreasing amplitude.

If we consider $T \cdot c_n$ we will obtain an expression which is identical to the Fourier integral at the points where $\omega = n\omega_0$. For a symmetrical wave of period $T = 2\delta$, the fundamental frequency is $\omega_0 = 2\pi/T = 2\pi/2\delta = \pi/\delta$. The line spectrum for this wave has a component at zero, one at π/δ, nothing at $2\pi/\delta$ because the envelope is zero, a negative component at $3\pi/\delta$, and so on. The components for the Fourier series are obtained by sampling the waveform for the Fourier integral and dividing by the period T. As T is increased, the fundamental frequency, $\omega_0 = 2\pi/T$, decreases and the lines in the band spectrum come closer together. The harmonic amplitudes decrease to zero, but their relative sizes still fall on the same envelope curve so long as the pulse width remains the same. If we plot $T \cdot c_n$ the amplitude does not decrease to zero, but remains constant. It is this distribution function which is used in Fourier integral analysis. As the period approaches infinity, the sampling points come closer and closer together and the spectrum becomes more and more nearly continuous. The spectrum probably never becomes "really" continuous, but as soon as the harmonics are so close that we cannot separate them with our measuring apparatus, we may as well consider that they are. Furthermore, the "almost continuous" viewpoint is helpful in resolving any mathematical problems which may arise.

20–8 The Laplace transform. All real physical periodic functions can be expanded in Fourier series, and similarly all real physical pulses can be expanded in Fourier integrals. In some cases, however, we like to use idealized pulses which we pretend will last forever. The unit step function is one such pulse. The Fourier integral does not converge for any such pulse which is finite at $t = \infty$. Because of their utility, we would like to be able to include these pulses in our Fourier analysis. We will never be able to expand them in steady-state sine waves, but we may be able to expand them in a similar type of series using rising sine waves. Mathematically, it is very easy to obtain rising sine waves from undamped sine waves. We simply replace the usual exponential frequency variable $e^{j\omega t}$ by e^{st}, where $s = \sigma + j\omega$. We previously found it convenient to make a similar substitution in our steady-state networks in order to observe resonances occurring at the damped natural frequencies of the network. When we make this substitution in the integral series we obtain the complex Fourier transform, which is often known as the *Laplace transform*. Naturally enough we pay some price for the convergence which we obtain. If we select a value of σ to make our integrals converge for positive values of time, they will, of course, grow infinitely large for negative values of time. Fortunately, many problems start at $t = 0$, and we are not interested in investigating the functions for negative time. For such problems the convergence factor in the Laplace transform is very useful.

An example of a pulse which cannot be expanded directly in a Fourier series is the unit step $f(t) = u_{-1}(t)$. The formal transform for the frequency components is

$$\begin{aligned} F(\omega) &= \int_{-\infty}^{\infty} f(t)e^{-j\omega t}\,dt = \int_0^{\infty} e^{-j\omega t}\,dt \\ &= \int_0^{\infty} (\cos\omega t - j\sin\omega t)\,dt. \end{aligned} \tag{20–38}$$

Neither the real nor the imaginary integral in Eq. (20–38) converges at $t = \infty$. The integral under a sinusoidal wave will vary from zero to $2/\omega$, depending on where the integration is stopped. Infinity gives us no information on where to stop. Similar troubles occur with most functions which have values over an infinite range.

If we decide to settle for a resolution into rising sine waves instead of undamped sine waves, we need merely to change the $j\omega$ in Eq. (20–38) to $\sigma + j\omega = s$. Formally, then,

$$\begin{aligned} F(s) &= \int_0^{\infty} f(t)e^{-st}\,dt = \int_0^{\infty} e^{-st}\,dt \\ &= \left[\frac{e^{-st}}{-s}\right]_0^{\infty} = \left[\frac{e^{-\sigma t}e^{-j\omega t}}{-(\sigma + j\omega)}\right]_0^{\infty}. \end{aligned} \tag{20–39}$$

The expression of Eq. (20–39) converges at infinity for a value of σ greater than zero. The meaning of this statement is that we can find the coefficients for a summation of exponentially rising sine waves only. At infinity the integral becomes zero and thus we are left with

$$F(s) = \frac{1}{s}\cdot \tag{20–40}$$

Equation (20–40) gives us the value of the coefficients in the series of rising sine waves into which we can resolve the unit step function. Its generality is perhaps a little frightening! We can pick any damping constant we like, provided it is greater than zero, and Eq. (20–40) will give us the corresponding resolution into rising sine waves. We have not just one Fourier series, but rather an infinite number of them available. This problem is not serious, however, since all the information to generate any one of these Fourier summations is contained in the single expression for the general term in Eq. (20–40).

The boundary on the value of σ above which we must go in order to obtain a convergent series is called *the abscissa of absolute convergence* of the function. For the unit step function, the abscissa of absolute convergence is zero. A list of the values which would be obtained if we integrated other functions is given in Table 20–3. For the function e^{-at} the

TABLE 20–3

ABSCISSAE OF ABSOLUTE CONVERGENCE

$f(t)$	Abscissa of absolute convergence
$u_{-1}(t)$	0
$(\sin \omega t)u_{-1}(t)$	0
e^{-at}	$-a$
e^{at}	$+a$
e^{t^2}	None

abscissa of absolute convergence is negative. This function decays to zero by itself and presents no convergence problems for a Fourier series. Not only can it be expanded in a summation of undamped sine waves, but even in a series of damped sine waves, up to a damping constant of a. On the other hand the function e^{at} does present convergence problems for the Fourier summation. It cannot be expanded in a summation of undamped sine waves. The waves must be positively damped, or increasing, with a damping constant of at least $+a$. As the last function in Table 20–3 shows, it may not be possible to expand a function in damped

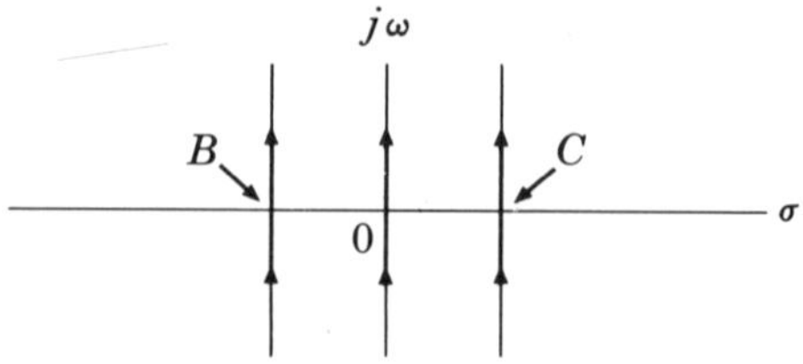

FIG. 20–21. Paths for integration in the complex s-plane.

sine waves under any conditions. Such functions are rather rare and never occur in physical problems.

We will summarize the foregoing discussion by writing down the transforms for complex frequency:

$$F(s) = \int_0^\infty f(t)e^{-st}\,dt, \tag{20–41}$$

$$f(t) = \frac{1}{2\pi}\int_C F(\sigma + j\omega)e^{(\sigma+j\omega)t}\,d\omega. \tag{20–42}$$

Equation (20–41) has limits which run only from zero to plus infinity. It is assumed that the function $f(t)$ is either zero, or at least of no interest, for negative time. It is necessary to cut off the function at zero, in this way, in order to avoid convergence problems at $t = -\infty$. A convergence factor which causes the integral to go to zero at $+\infty$ will make it diverge wildly at $t = -\infty$. Equation (20–42) represents a summation of damped sine waves with a damping constant σ and a radian frequency ω. The contour C along which the integration should be carried out can be described as having a constant value of σ and a value of frequency running from $-\infty$ to $+\infty$. Since σ is a constant, the derivative of $s = \sigma + j\omega$ is $jd\omega$ and the integral of Eq. (20–42) can be written

$$f(t) = \frac{1}{2\pi j}\int_{-j\infty+\sigma_C}^{+j\infty+\sigma_C} F(s)e^{st}\,ds. \tag{20–43}$$

The path of integration of Eq. (20–43) is along a line parallel to the $j\omega$ axis and displaced from it by the abscissa of absolute convergence σ_C. This path for various possible values of σ_C is shown in Fig. 20–21.

An integration or summation along the $j\omega$ axis in Fig. 20–21 would result in the usual Fourier components, or a series in terms of undamped sine waves. An integration along a path to the left of the $j\omega$ axis, such as path B, represents a summation of damped sine waves. In general, a function must be zero, or must converge rapidly, at infinity to be representable by such a sum. An integration along a path C to the right of the

$j\omega$ axis represents a summation of rising sine waves. Such a summation can represent functions which are constant at infinity, or even infinite. There are some functions which cannot be represented by any series of rising or falling sinusoids, however.

The integrals of Eqs. (20–41) and (20–42) are called the *Laplace transforms*. Equation (20–41) is the direct transform. It resolves the function $f(t)$ in terms of a summation of damped sinusoids. It is a real integration and relatively easy to perform. Equation (20–43) is the inverse transform. It expresses the function as the summation of its frequency components. It is a complex integral and relatively difficult to evaluate. Usually, the direct transform is tabulated, and since there is a one-to-one correspondence between the function and its frequency components, tables are used instead of the integral for the inverse transformation.

The function $f(t)$ is said to be *in the time domain*. The function $F(s)$, which represents the frequency components of $f(t)$ resolved into some sort of Fourier summation, is said to be *in the frequency domain*. The nomenclature here is an extension of that we used for a single steady-state sine wave, where the function of time is said to be in the time domain, and the complex number representing the magnitude and phase of the sine wave is said to be in the frequency domain. The function $F(s)$ is likewise a complex number and at any given frequency it gives the magnitude and phase of the damped sine-wave component of $f(t)$.

We have considered the Laplace transform as a method of resolving an input waveform into an infinite summation of sinusoidal components with exponential damping factors to make sure that the series converges for large values of time. The damping factor which is necessary to insure convergence in the series varies with the nature of the input waveform. The value which we must exceed is called the abscissa of absolute convergence. If the damping factor is kept constant and the frequency is allowed to change, we can form a convergent infinite summation which converges to $f(t)$. In the frequency domain we are actually using values of s which fall on a line parallel to the $j\omega$ axis and displaced from it by the damping exponent. The meaning of the abscissa of absolute convergence in the frequency domain is easy to determine. Consider for example the function $f(t) = e^{at}$. The integral for $F(s)$ is

$$F(s) = \int_0^\infty e^{at}e^{-st}\,dt$$

$$= \left[\frac{e^{(-\sigma+a)t}e^{-j\omega t}}{-(s-a)}\right]_0^\infty. \qquad (20\text{–}44)$$

Obviously, Eq. (20–44) will not converge unless σ is greater than a. Thus $\sigma = a$ is the abscissa of absolute convergence. The integral then vanishes at infinity and

$$F(s) = \frac{1}{s - a}.$$

It is also clear that in the frequency domain, if we were to use values of damping equal to a in our summation of damped exponential terms the function would become infinite at $s = a$. Clearly, we must avoid this point! One way of doing so is always to use a damping constant which is greater than a. On the other hand, $F(s)$ is perfectly well defined for any value except $s = a$, and we can bend our contour in various ways so long as we do not pass through $s = a$, and so long as we go to the right of it eventually.* The point to be made at this moment is that $F(s)$ is a perfectly well defined function for every value of s except $s = a$. Any point where $F(s)$ goes to infinity, as at the point $s = a$ here, is called a *pole of the function.*

The meaning of a pole, or infinity, in the spectrum of an excitation function is not too difficult to discover. We must remember that in a continuous (or band) spectrum, which is the one we are using in the Laplace transform, the amplitudes of the individual frequency components are vanishingly small. Essentially, we are adding up an infinite number of infinitesimal harmonics, with practically no spacing between them, to form the function $f(t)$. In order to focus our attention on the distribution of the harmonics, we multiplied their amplitudes by the value of the period which we then allowed to approach infinity. If there had been any finite components in the spectrum, corresponding to lines in a line spectrum, they would likewise have been multiplied by T, and they would appear as infinities in the distribution function for the band spectrum. The meaning of the poles, or infinities, in the band spectrum is thus clear. They are line-spectrum components in the input wave. We might be encouraged to look for all these line spectral components, and to try to expand the input function in terms of a finite number of these instead of an infinite number of components in some arbitrary band spectrum.† We will not actually carry out this process, but it should be apparent that the poles are important properties of any waveform. The student should distinguish the poles of a waveform (the line spectra in it) from the previously encountered poles of a network function (the transient natural frequencies).

We will close this chapter on the analysis of waveforms by calculating one more Laplace transform, and by giving a short table of transforms for

* The selection of the best contour for integration in the complex plane is a mathematical subject which is still occupying some of the finest mathematicians of our time.

† The mathematical technique of expanding an excitation function in terms of its inherent line spectra is called the calculus of residues.

Table 20-4

A Short Table of Laplace Transforms

$f(t)$	$F(s)$
$u_{-1}(t)$	$\dfrac{1}{s}$
e^{at}	$\dfrac{1}{s-a}$
e^{-at}	$\dfrac{1}{s+a}$
$\cos \beta t$	$\dfrac{s}{s^2+\beta^2}$
$\sin \beta t$	$\dfrac{\beta}{s^2+\beta^2}$
t or $u_{-2}(t)$	$\dfrac{1}{s^2}$
$u_0(t)$	1
t^n	$\dfrac{n!}{s^{n+1}}$
$e^{-at} \sin \beta t$	$\dfrac{\beta}{(s+a)^2+\beta^2}$
$e^{-at} \cos \beta t$	$\dfrac{s+a}{(s+a)^2+\beta^2}$

some of the most frequently occurring functions. Consider the function $f(t) = (\cos at)u_{-1}(t)$. Because it is cut off for $t < 0$, the Fourier spectrum is not just a simple cosine wave. In order to expand this function in a series of sinusoids which run over all time, we must use the transform method, and the result is a band spectrum given by

$$F(s) = \int_0^\infty \cos at \cdot e^{-st}\,dt = \int_0^\infty \frac{e^{jat} + e^{-jat}}{2} e^{-st}\,dt$$

$$= \frac{1}{2}\left[\frac{e^{(ja-s)t}}{ja - s} + \frac{e^{(-ja-s)t}}{-ja - s}\right]_0^\infty .$$

The abscissa of absolute convergence is chosen to make the integrals vanish for $t = \infty$. The values at the lower limit give

$$F(s) = \frac{1/2}{s + ja} + \frac{1/2}{s - ja} = \frac{s}{s^2 + a^2} \cdot \tag{20-45}$$

So long as we stay away from the points in Eq. (20–45) which make $F(s) = \infty$, the function is well defined. It is the Laplace transform, or frequency resolution, of the input function $f(t)$. Note that we cannot obtain a Fourier integral for this function because the poles are on the $j\omega$ axis. If we attempt an expansion with $s = j\omega$ we would encounter these discontinuities. Table 20–4 lists a number of common transforms. We will use the table hereafter, rather than computing inverse transforms.

Now that we have been able to resolve arbitrary input waveforms into sinusoidal and damped sinusoidal components we can apply them to networks using the same theory which we used for steady-state analysis. Each component of the input waveform is multiplied by the corresponding steady-state transfer function of the network, and the output component at that frequency is obtained. The summation of the output components represents the output waveform. The easiest way to find the output waveform is to look in a table of transforms for the function of time which would have those frequency components. The direct integration is usually more difficult and, as we shall see in the subsequent chapters, unnecessary.

20–9 Summary.

Fourier's theorem (normalized to base period of 2π)

$$f(t) = a_0 + a_1 \cos t + a_2 \cos 2t + \cdots$$
$$+ b_1 \sin t + b_2 \sin 2t + \cdots$$

where

$$a_0 = \frac{1}{2\pi} \int_0^{2\pi} f(t)\, dt$$

$$a_n = \frac{1}{\pi} \int_0^{2\pi} f(t) \cos nt\, dt$$

$$b_n = \frac{1}{\pi} \int_0^{2\pi} f(t) \sin nt\, dt$$

provided

$$\int_0^{2\pi} |f(t)|\, dt \qquad \text{is finite}$$

Examples of Fourier series

Square wave

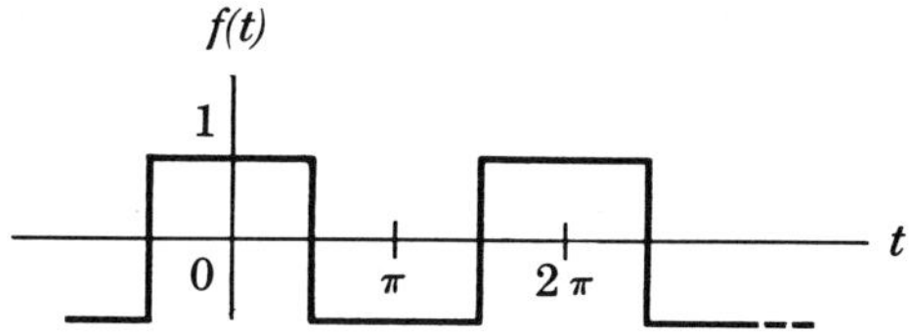

$$f(t) = \frac{4}{\pi}\left(\cos t - \tfrac{1}{3}\cos 3t + \tfrac{1}{5}\cos 5t - \cdots\right)$$

Triangular wave

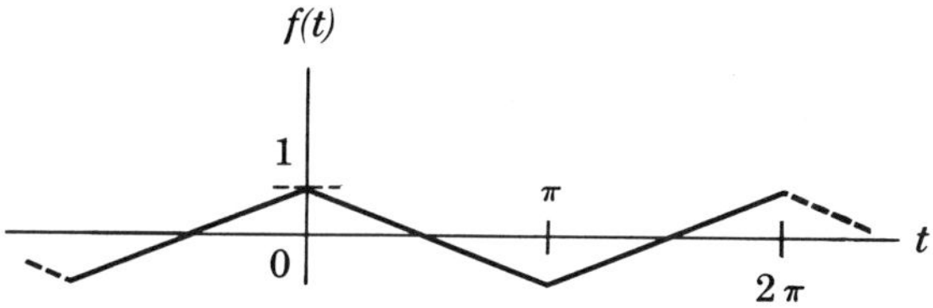

$$f(t) = \frac{8}{\pi^2}\left(\cos t + \tfrac{1}{9}\cos 3t + \tfrac{1}{25}\cos 5t + \cdots\right)$$

Impulse train

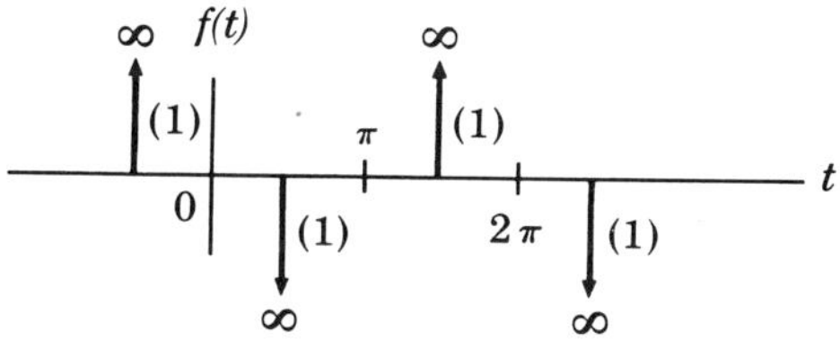

$$f(t) = \frac{-2}{\pi}\left(\sin t - \sin 3t + \sin 5t - \cdots\right)$$

Saw-tooth wave

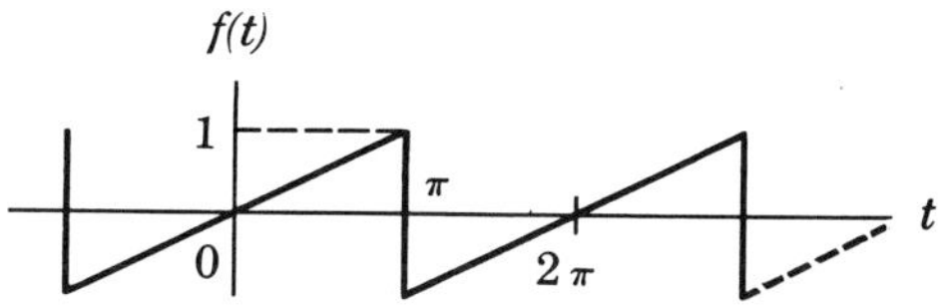

$$f(t) = \frac{2}{\pi}\left(\sin t - \tfrac{1}{2}\sin 2t + \tfrac{1}{3}\sin 3t - \cdots\right)$$

Half-wave rectified sine wave

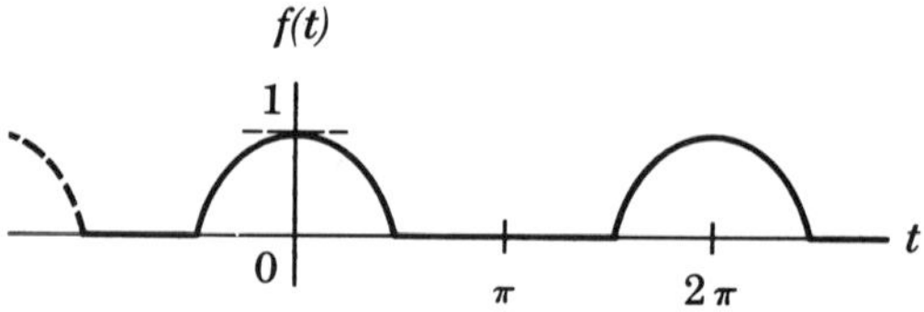

$$f(t) = \frac{1}{\pi}\left(1 + \frac{\pi}{2}\cos t + \tfrac{2}{3}\cos 2t - \tfrac{2}{15}\cos 4t \ldots\right)$$

Full-wave rectified sine wave

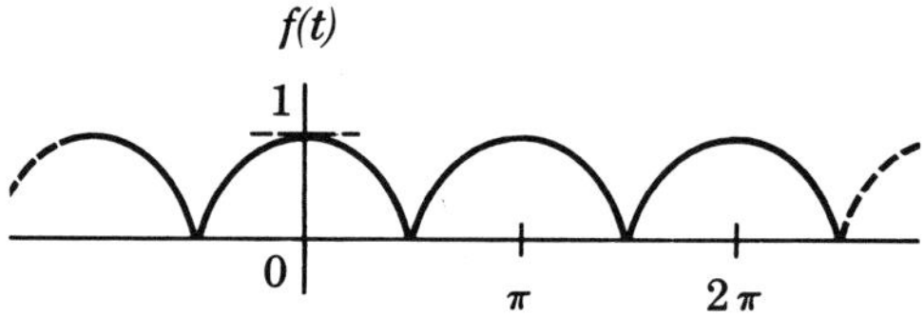

$$f(t) = \frac{2}{\pi}\,(1 + \tfrac{2}{3}\cos 2t - \tfrac{2}{15}\cos 4t + \ldots)$$

Symmetry in Fourier series

1. Even function: $f(t) = f(-t)$
   (Only cosine terms)

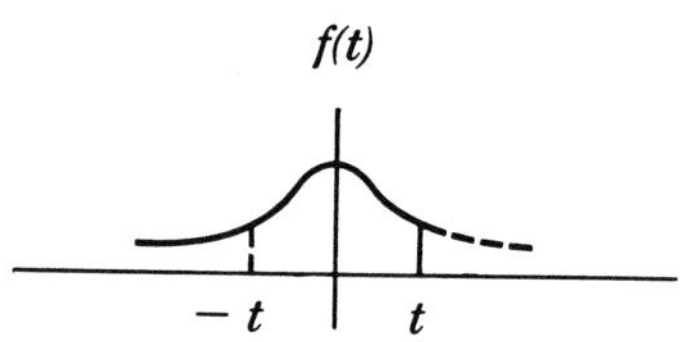

2. Odd function: $f(t) = -f(-t)$
   (Only sine terms)

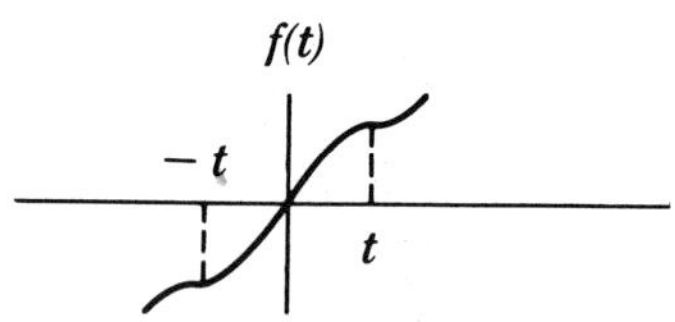

3. Rotation: $f(t) = -f(t + \pi)$
   (Only odd harmonics)

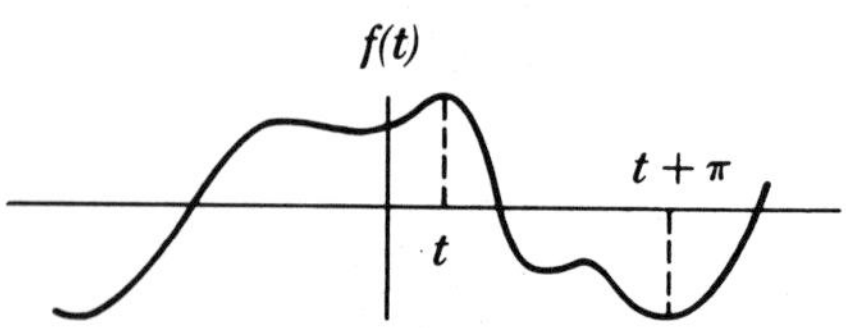

4. Quarter wave
   (a) Rotation + even function
   (b) Rotation + odd function

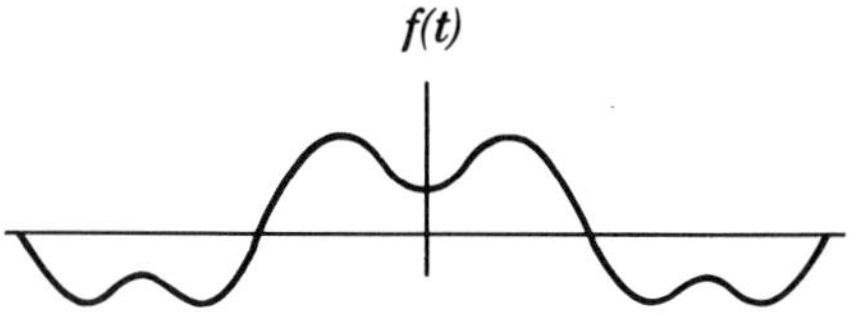

Numerical Fourier series

$$a_0 = \frac{1}{2\pi} \sum_{m=1} f(m\,\Delta t)\,\Delta t$$

$$a_n = \frac{1}{\pi} \sum_{m=1} f(m\,\Delta t)\cos(nm\,\Delta t)\,\Delta t$$

$$b_n = \frac{1}{\pi} \sum_{m=1} f(m\,\Delta t)\sin(nm\,\Delta t)\,\Delta t$$

Sampling theorem

To expose the nth harmonic of the fundamental frequency, the waveform must be sampled more than $2n$ times per period of the fundamental.

Complex Fourier series

$$f(t) = \sum_{-\infty}^{\infty} c_n e^{jn\omega t}$$

$$c_n = \frac{1}{T}\int_{-T/2}^{T/2} f(t)e^{-jn\omega t}\,dt$$

Example

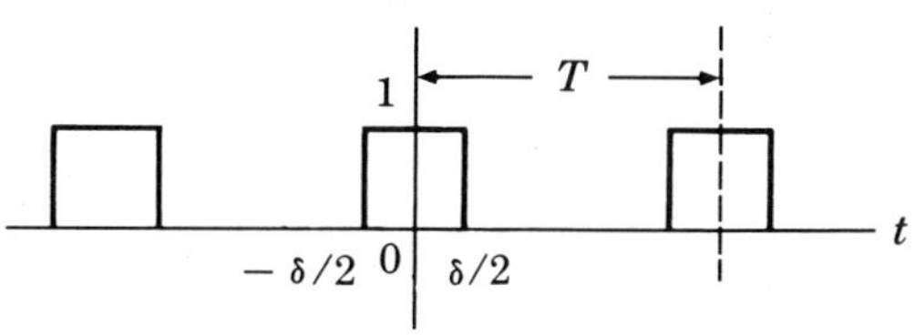

$$T \cdot c_n = \delta\,\frac{\sin n\omega_0\,\delta/2}{n\omega_0\,\delta/2}$$

Fourier integrals (Fourier transforms)

$$f(t) = \frac{1}{2\pi}\int_{-\infty}^{\infty} F(\omega)e^{j\omega t}\,d\omega$$

$$F(\omega) = \int_{-\infty}^{\infty} f(t)e^{-j\omega t}\,dt$$

Example

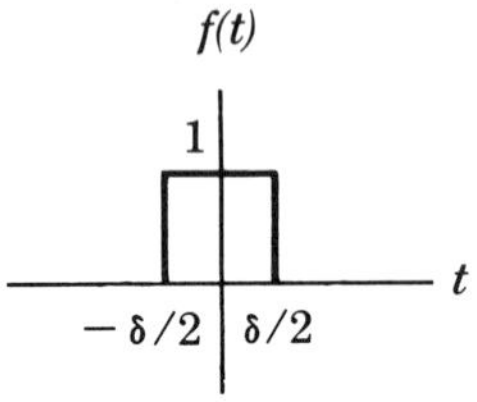

$$F(\omega) = \delta \frac{\sin \omega\delta/2}{\omega\delta/2}$$

Laplace transforms

$$f(t) = \frac{1}{2\pi j} \int_{-j\infty+\sigma_C}^{+j\infty+\sigma_C} F(s)e^{st}\, ds$$

$$F(s) = \int_0^{\infty} f(t)e^{-st}\, dt$$

$f(t)$	$F(s)$
$u_0(t)$	1
$u_{-1}(t)$	$\frac{1}{s}$
$u_{-2}(t)$ or t	$\frac{1}{s^2}$
$u_{-(n+1)}t$ or t^n	$\frac{n!}{s^{n+1}}$
$e^{-\alpha t}$	$\frac{1}{s+\alpha}$
$\cos \beta t$	$\frac{s}{s^2+\beta^2}$
$\sin \beta t$	$\frac{\beta}{s^2+\beta^2}$
$e^{-\alpha t} \cos \beta t$	$\frac{s+\alpha}{(s+\alpha)^2+\beta^2}$
$e^{-\alpha t} \sin \beta t$	$\frac{\beta}{(s+\alpha)^2+\beta^2}$

REFERENCES

In the literature there are three quite different approaches to the Fourier integral and the Laplace transform:

1. The operational calculus introduced by Heaviside.
2. The integral transform introduced by Laplace.
3. Signal analysis introduced by Fourier.

In this text we are following the signal-analysis point of view. Complex signals are resolved into their frequency components, the output response for each component is obtained, and then the total output is obtained by summing the separate components. References for the other approaches are given below.

1. VALLEY and WALLMAN, "Vacuum-Tube Amplifiers," *Radiation Laboratory Series,* Vol. 18, McGraw-Hill, New York, 1948. The first chapter of this volume presents an excellent discussion of the Laplace transform as an integral transform.

2. E. A. GUILLEMIN, *The Mathematics of Circuit Analysis,* John Wiley & Sons, New York, 1948. This volume is the standard reference for the signal-analysis point of view. It covers Fourier series and Fourier integrals, but does not cover Laplace transforms.

3. V. BUSH, *Operational Circuit Analysis,* John Wiley & Sons, New York, 1929. Bush presents the Heaviside operational methods in a framework of transients in electric circuits.

4. GARDNER and BARNES, *Transients in Linear Systems,* John Wiley & Sons, New York, 1942. This popular text presents the Laplace transform from a point of view midway between that of an integral transform and that of Fourier signal analysis.

5. KERCHNER and CORCORAN, *Alternating Current Circuits,* John Wiley & Sons, New York, 3rd ed., 1951. Kerchner and Corcoran give an excellent discussion of Fourier series, complex waveforms, and symmetry conditions.

6. S. FICH, *Transient Analysis in Electrical Engineering,* Prentice-Hall, New York, 1951. Fich gives a good discussion of the Fourier series and the Fourier integral spectra, and of the relationship between them.

7. CAMPBELL and FOSTER, "Fourier Integrals for Practical Applications," *Bell System Monograph Series,* No. B-584, 1931. In addition to the theory of the Fourier integral, Campbell and Foster give an extensive table of transforms.* A reprint (D. Van Nostrand, New York, 1950) is also available.

8. M. SCHWARTZ, *Information Transmission, Modulation, and Noise,* McGraw-Hill, New York, 1959. In order to treat noise and random signals Schwartz develops the Fourier integral at some length. Since power and energy are in-

---

* These tables and many others in the literature include a factor of 2π which does not appear in our table of Fourier transforms, which we have normalized to agree with the more common Laplace transforms.

volved, we have reserved our discussion of these topics until Chapter 23, in which further references of this type are given.

9. S. GOLDMAN, *Frequency Analysis, Modulation, and Noise,* McGraw-Hill, New York, 1948. Goldman considers signal analysis and its application to communications problems, rather than to network problems.

10. C. CHERRY, *Pulses and Transients in Communications Circuits,* Chapman and Hall, Ltd., London, 1949. See also the Dover publication, New York, 1950. Cherry uses the Fourier series and the Fourier integral to solve a number of problems in pulse transmission circuits.

EXERCISES

Section 20–2

1. Find the Fourier series for the waveform of Fig. 20–22.

2. If the period of the waveform in Fig. 20–22 is 1 microsec, instead of 2π sec, what is the Fourier series?

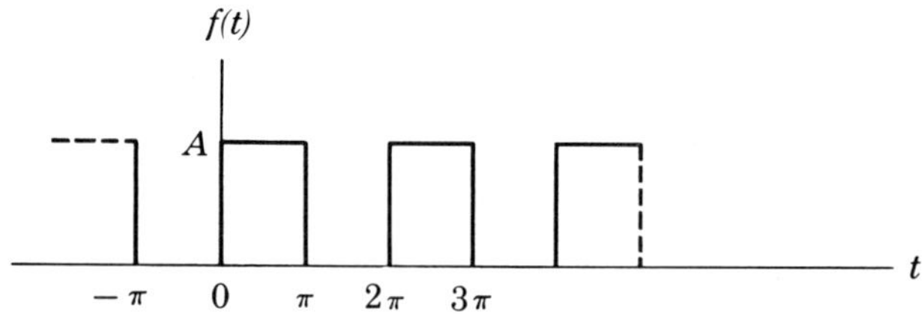

FIGURE 20–22

3. Find the Fourier series for the waveform of Fig. 20–23.

4. If the period of the waveform in Fig. 20–23 is 100 sec instead of 2π sec, what is the Fourier series.

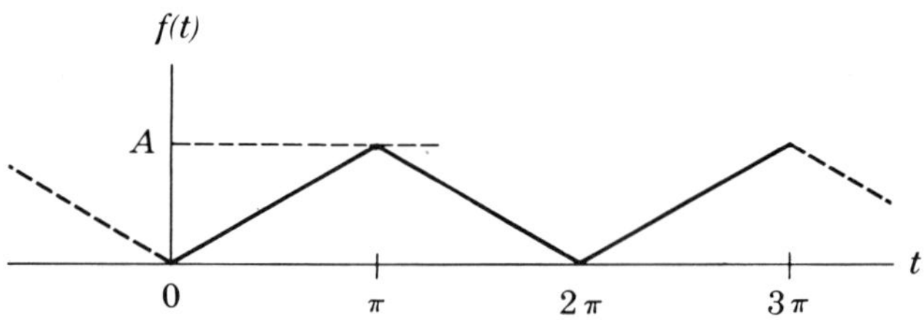

FIGURE 20–23

5. The impulse train shown in Fig. 20–24 is important in the theory of Class C amplifiers and in many modulation problems. Find its Fourier series, and show that each frequency component is present with the same strength.

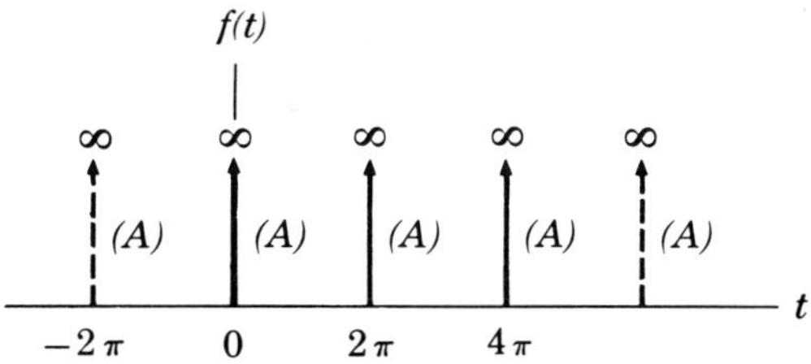

FIGURE 20–24

6. Find the Fourier series for the alternating impulse train shown in Fig. 20–25.

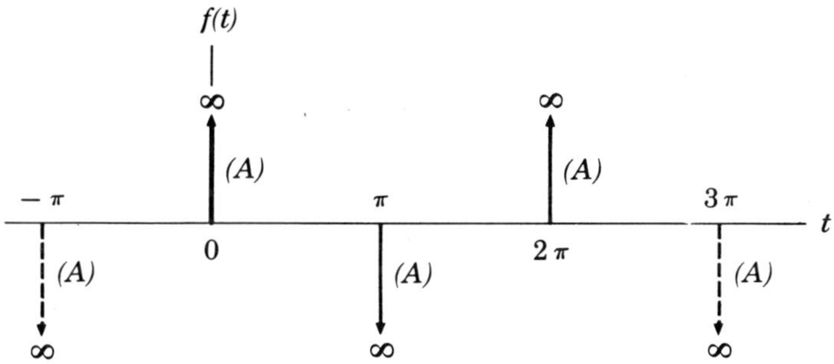

FIGURE 20–25

7. The waveform of Fig. 20–26 is used as a "sweep" in radar and television circuits. Find its Fourier series.

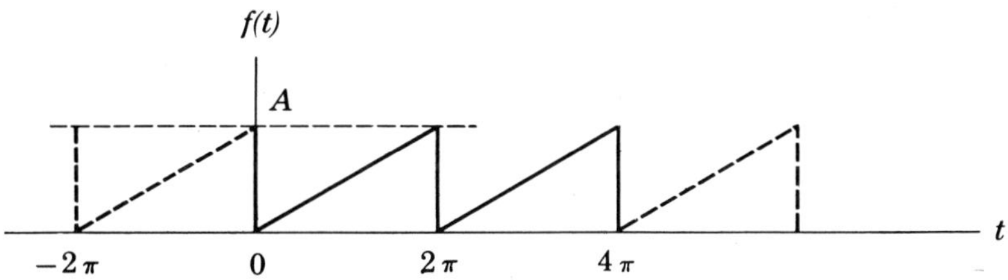

FIGURE 20–26

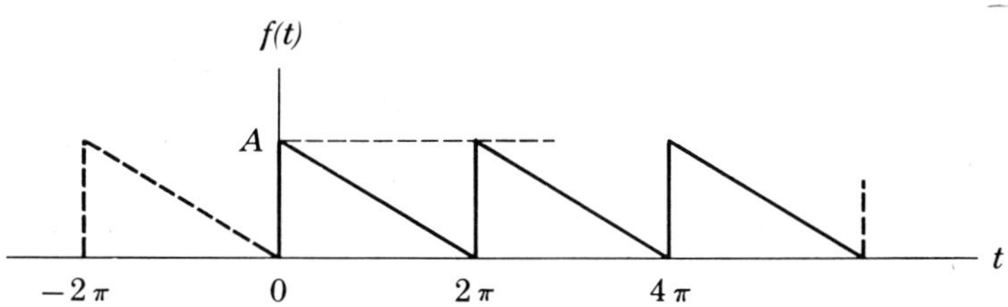

FIGURE 20–27

8. The waveform of Fig. 20–27 is the mirror image of the one in Fig. 20–26. Find its Fourier series and compare it with the Fourier series for Exercise 7.

9. Find the Fourier series for the full-wave rectified sine wave shown in Fig. 20–28.

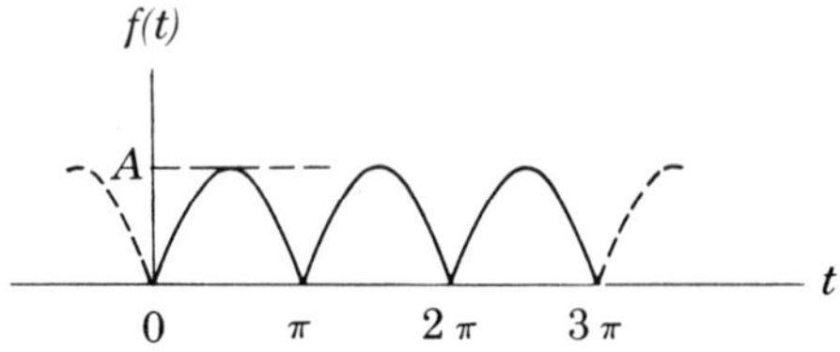

FIGURE 20–28

10. Find the Fourier series of the half-wave rectified sine wave shown in Fig. 20–29.

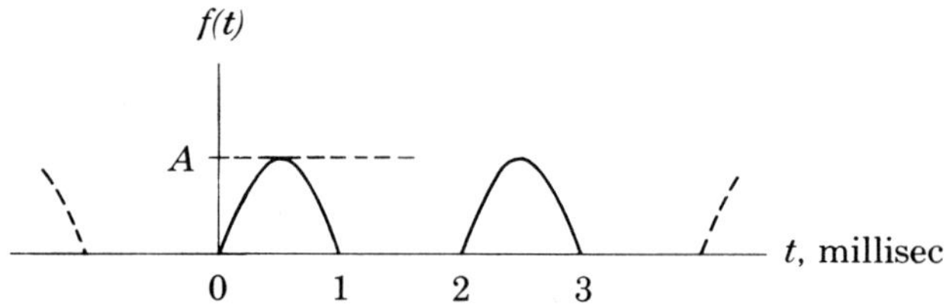

FIGURE 20–29

Section 20–3

11. For the square wave of Fig. 20–22 sketch the following partial sums:
   (a) The d-c term plus the fundamental.
   (b) The d-c term plus the fundamental and the third harmonic.
   (c) The d-c term, the fundamental, the third harmonic, and the fifth harmonic.

12. For the triangular wave of Fig. 20–23 sketch the following partial sums:
   (a) The d-c term and the fundamental.
   (b) The d-c term, the fundamental, and the third harmonic.
   (c) The d-c term, the fundamental, the third harmonic, and the fifth harmonic.

13. For the impulse train of Fig. 20–24 sketch the partial sum consisting of the d-c term, the fundamental, the second harmonic, and the third harmonic.

14. For the sweep waveform shown in Fig. 20–26 sketch the partial sum consisting of the d-c term, the fundamental, the second harmonic, and the third harmonic.

15. Sketch the integral from 0 to t of the waveform in Fig. 20–30 and also sketch the derivative of the waveform. Find the Fourier series for each of the three waveforms. Show that the original function has terms which decrease in amplitude as $1/n$, where n is the order of the harmonic; that the integral

waveform has terms which decrease in amplitude as $1/n^2$; and that the derivative has terms which remain of the same order of magnitude out to the nth harmonic.

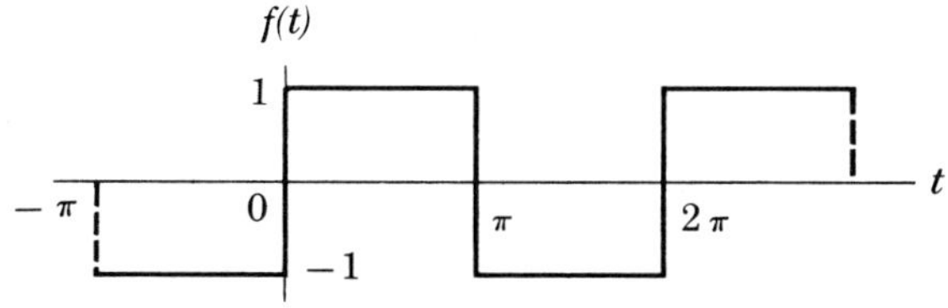

FIGURE 20–30

16. Consider a function $f(t)$ which is smooth and has no infinities in itself or in its first two derivatives. The third derivative, however, does go to infinity.

(a) In the Fourier series for $f(t)$, how will the harmonics decrease in amplitude?
(b) How many times can the Fourier series for $f(t)$ be integrated without spoiling the convergence?
(c) How many times can the Fourier series be differentiated before the resultant will contain infinite values, that is, will be divergent.
(d) Show that in general the harmonics will decrease as $1/n^m$, where n is the order of the harmonic and m is the lowest derivative of the function which has infinite values.

17. The Fourier series for a triangular wave with no d-c component can be written as

$$f(t) = \sum_{n=1,3,5\ldots}^{\infty} \frac{\cos nt}{n^2}.$$

(a) Sketch the form of the function $f(t)$. (b) Give an analytic expression for it. (c) Show that the series is less than the convergent series

$$\sum_{n=1,2,3\ldots}^{\infty} \frac{1}{n^2},$$

and is therefore convergent.

18. The Fourier series for a square wave with no d-c component can be written as

$$f(t) = \sum_{n=1,3,5\ldots}^{\infty} \frac{\sin nt}{n}.$$

(a) Sketch the waveform of $f(t)$. (b) Give an analytic expression for it. (c) Show that for any value of t except $t = 0, 2\pi, 4\pi, \ldots$, the series is less than the known convergent series

$$\sum_{n=1,2,3\ldots}^{\infty} (-1)^{n+1} \frac{1}{n},$$

and is therefore convergent.

19. The Fourier series for an impulse train with no d-c value can be written as

$$f(t) = \sum_{n=1,2,3\ldots}^{\infty} \cos nt.$$

(a) Sketch the waveform of $f(t)$. (b) Give an analytic expression for it. (c) This series is not convergent in the usual sense. If the terms are grouped properly, however, the series does sum to zero except at the points $t = 0, 2\pi, 4\pi, \ldots$. To verify this conclusion pick a value of t and draw vectors to represent each term in the series.

20. The convergence problems in Exercise 19 occur because of the idealized nature of the impulse function. The pulse train shown in Fig. 20–31 is more representative of the waveform which could be generated in a physical problem. Show that the Fourier series for this waveform is

$$f(t) = \frac{\delta E}{2\pi} + \frac{\delta E}{\pi} \sum_{n=1,2,3\ldots}^{\infty} \left(\frac{\sin n\delta/2}{n\delta/2}\right) \cos nt.$$

Show that for large n this series converges more rapidly than the known convergent series $(1 - \frac{1}{2} + \frac{1}{3} - \frac{1}{4} + \frac{1}{5} + \ldots)$.*

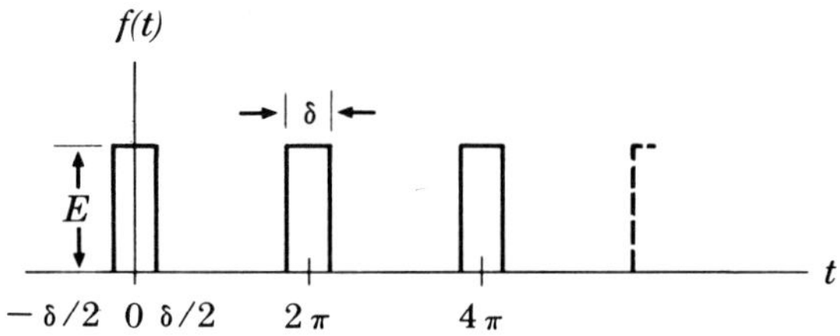

FIGURE 20–31

Section 20–4

21. For the waveform shown in Fig. 20–32:
   (a) Give the types of symmetry present.
   (b) Obtain the Fourier series by integrating over a period.
   (c) Obtain the Fourier series by integrating over a half-period.
   (d) Obtain the Fourier series by integrating over a quarter-period.

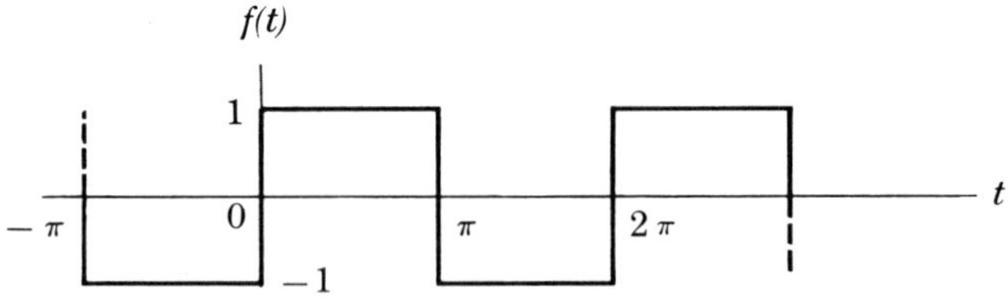

FIGURE 20–32

---

* This series is conditionally convergent since the series $(1 + \frac{1}{2} + \frac{1}{3} + \frac{1}{4} + \cdots)$ is divergent.

22. When the d-c term is removed from the waveform of Fig. 20–33 it possesses quarter-wave symmetry. Take advantage of the symmetry and obtain the Fourier series by integrating over a quarter of the period.

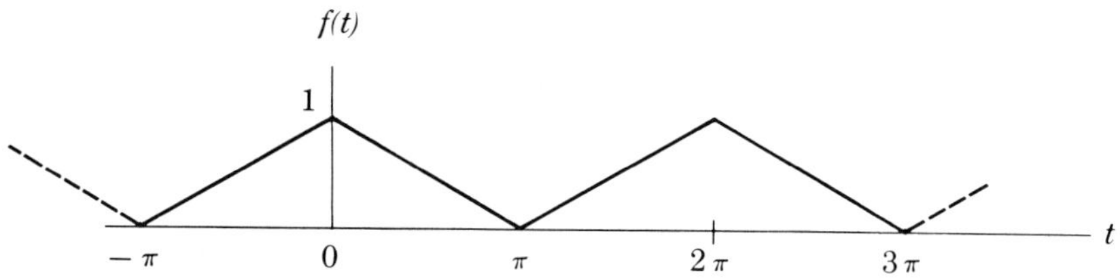

FIGURE 20–33

23. What type of symmetry exists in the waveform of Fig. 20–34? Obtain the Fourier series by integrating over the shortest possible range.

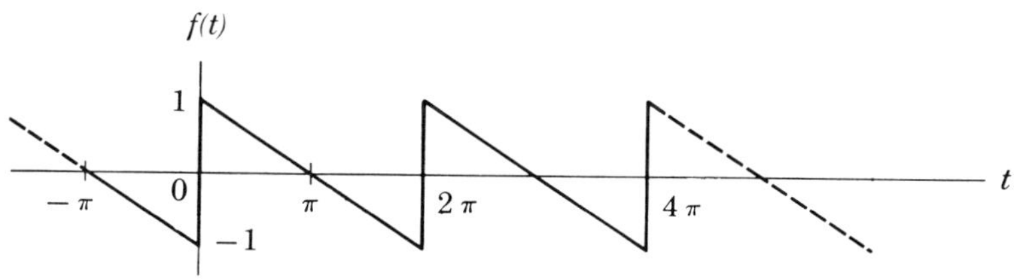

FIGURE 20–34

24. Sketch periodic functions over a range of one period when the Fourier series of the functions have the following properties:

(a) Only sine terms, but all harmonics present.
(b) Only cosine terms, but all harmonics present.
(c) Only odd sine terms present.
(d) Only odd cosine terms present.
(e) Only odd harmonics, but both sine and cosine terms present.
(f) Only even harmonics, but both sine and cosine terms present.
(g) The odd harmonics are sines, and the even are cosines.
(h) The even harmonics are sines, and the odd are cosines.

25. Discuss the symmetry of the following waveforms with respect to the origins given:

(a) $f(t) = 2 + \cos t + \cos 3t + \cos 5t + \cdots$
(b) $f(t) = \cos t + \cos 2t + \cos 3t + \cos 4t + \cdots$
(c) $f(t) = \sin t + \sin 2t + \sin 3t + \sin 4t + \cdots$
(d) $f(t) = \cos (t + 45°) + \sin (3t + 15°) + \cos (5t + 45°) + \cdots$
(e) $f(t) = 2 + \cos (2t + 15°) + \cos (4t + 35°) + \cdots$
(f) $f(t) = 2 + \cos (t + 45°) + \sin (2t + 30°) + \cos (3t + 15°) + \cdots$

26. The waveforms given below are not symmetrical in the form shown. Removing the d-c terms and shifting the origin will expose some hidden symmetries. Discuss each case.

(a) $f(t) = 2 + \cos(t + 45°) + \sin(3t + 15°) + \cos 5t + \cdots$
(b) $f(t) = 2 + \cos(t + 45°) + \cos(3t + 135°) + \cos(5t + 225°) + \cdots$
(c) $f(t) = 2 + \cos(t + 45°) - \sin(3t + 45°) + \sin(5t - 45°) + \cdots$
(d) $f(t) = \cos(t + 15°) - \cos(2t - 150°) - \sin(3t - 45°)$
$+ \sin(4t + 150°) + \cdots$
(e) $f(t) = \cos t + \sin t + \sin 2t - \cos 3t + \sin 3t - \cos 4t + \cdots$

27. The waveform of Fig. 20–35 exists in the region from $t = 0$ to $t = \pi$. Complete the waveform over a period of 2π and write the Fourier series if (a) $f(t)$ is an even function and (b) $f(t)$ is an odd function.

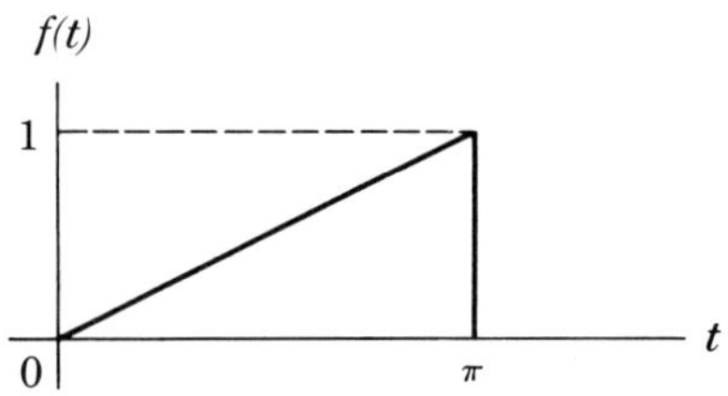

FIGURE 20–35

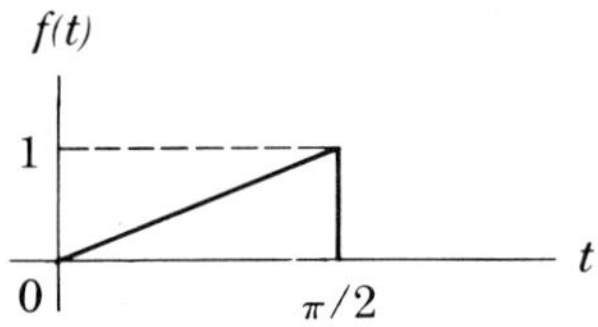

FIGURE 20–36

28. The waveform shown in Fig. 20–36 is known over only a quarter of a period. It can be completed in a variety of ways in order to obtain a symmetrical Fourier series. Sketch the waveform for the following cases:

(a) $f(t)$ is an even function with rotation symmetry.
(b) $f(t)$ is an even function without rotation symmetry.
(c) $f(t)$ is an odd function with rotation symmetry.
(d) $f(t)$ is an odd function without rotation symmetry.
(e) $f(t)$ is an even function and the Fourier series for it has only even harmonics.
(f) $f(t)$ is an odd function and the Fourier series for it has only even harmonics.

29. Discuss the symmetry represented by each of the following functions and give the terms left in the Fourier series in each case.

(a) $f(t) = f(-t)$
(b) $f(t) = -f(-t)$
(c) $f(t) = -f(t + \pi)$
(d) $f(t) = f(t + \pi)$
(e) $f(t) = f(\pi - t)$ and $f(t) = -f(-t)$
(f) $f(t) = -f(\pi - t)$ and $f(t) = f(-t)$
(g) $f(t) = f(\pi - t)$
(h) $f(t) = -f(\pi - t)$

Section 20–5

30. The waveform of Fig. 20–37 has quarter-wave symmetry and hence the Fourier series is of the form

$$f(t) = a_1 \cos t + a_3 \cos 3t + a_5 \cos 5t + \cdots.$$

The wave is sampled every 30° ($\Delta t = \pi/6$) starting at $t = \pi/6$. The sampled wave has half-wave symmetry, but not quarter-wave symmetry. Hence

$$a_1 = \frac{2}{\pi} \sum_{m=1}^{6} f(m\,\Delta t) \cos(m\,\Delta t)\,\Delta t,$$

and so on. Find a_1 and a_3 by approximate numerical integration.

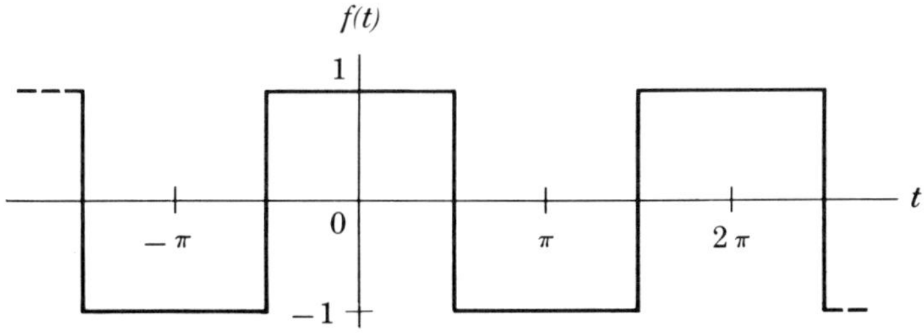

FIGURE 20–37

31. To take advantage of the quarter-wave symmetry in the waveform of Fig. 20–37, the sampling points should be at the center of the regions which they represent. Thus, for $\Delta t = \pi/6$, the first sample should be at $t = \pi/12$, and the next one at $t = 3\pi/12$, and so on. Show that the formulas for the harmonics are

$$a_1 = \frac{4}{\pi} \sum_{m=1,3,5\ldots} f\left(\frac{m\,\Delta t}{2}\right) \cos\frac{m\,\Delta t}{2}\,\Delta t,$$

$$a_3 = \frac{4}{\pi} \sum_{m=1,3,5\ldots} f\left(\frac{m\,\Delta t}{2}\right) \cos\frac{3m\,\Delta t}{2}\,\Delta t.$$

Find the values of a_1 and a_3 by this method.

32. Repeat Exercise 30 using twice as many sampling points per period.

33. Repeat Exercise 31 using twice as many sampling points per period.

34. Use 24 sampling points per period and take advantage of the symmetry to find the first three harmonics of the waveform in Fig. 20–38.

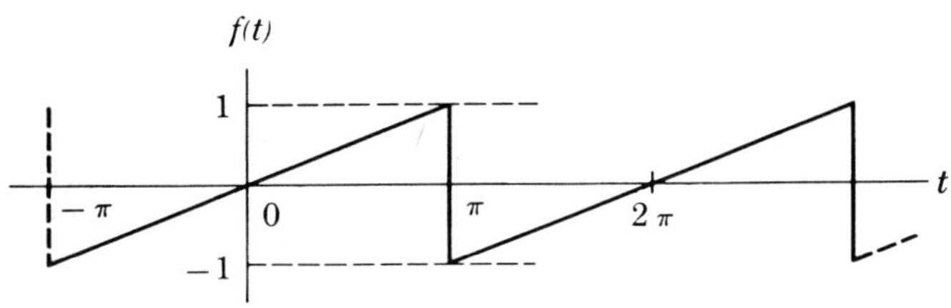

FIGURE 20–38

35. A periodic wave is represented over one-half period by the following tabular data:

ωt (degrees)	$f(t)$
0	0
30	1
60	2
90	2
120	3
150	3
180	0

The wave has odd-function symmetry. Find the first three harmonics.

36. When the d-c term is removed the waveform of Fig. 20–39 has even function symmetry. Assume that it can be represented by

$$f(t) = a_0 + a_1 \cos t + a_2 \cos 2t.$$

Find a_0, a_1 and a_2 by sampling the wave at $t = 0$, $t = \pi/2$, and $t = \pi$. (The neglected higher harmonics introduce some error into these results.)

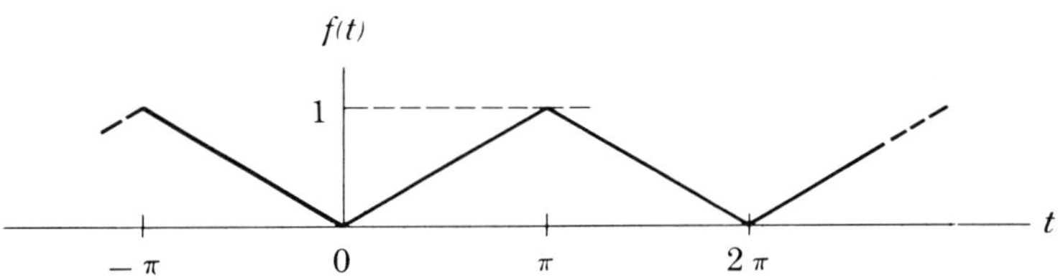

FIGURE 20–39

37. Repeat Exercise 36, but assume a series of the form

$$f(t) = a_0 + a_1 \cos t + a_2 \cos 2t + a_3 \cos 3t.$$

Find the values of a_0, a_1, a_2, and a_3 by sampling the wave at the points $t = 0$, $\pi/3$, $2\pi/3$, and π.

38. Assume that the waveform of Fig. 20–38 can be represented by

$$f(t) = b_1 \sin t + b_2 \sin 2t + b_3 \sin 3t.$$

Find b_1, b_2, and b_3 by means of sampled values at $t = \pi/4$, $\pi/2$, and $3\pi/4$.

Section 20–6

39. Discuss the effect of even, odd, rotation, and quarter-wave symmetry on the coefficients of the complex Fourier series. For each case give the minimum range over which the function must be integrated.

40. Find the complex Fourier series for the waveform of Fig. 20–40. Plot the frequency spectrum as a function of positive and negative frequency. Convert the complex series to the usual sine-cosine series. (Show both amplitude and phase in the spectral plot.)

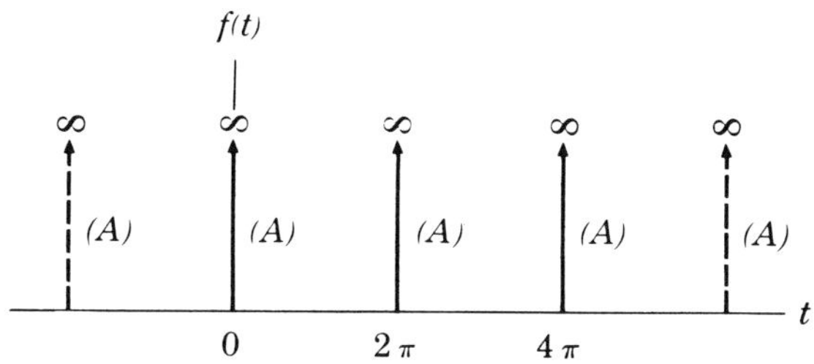

FIGURE 20–40

41. Find the complex Fourier series for the waveform of Fig. 20–41. Sketch the frequency spectrum and convert the series to the sine-cosine form. (In the spectral plot show both amplitude and phase.)

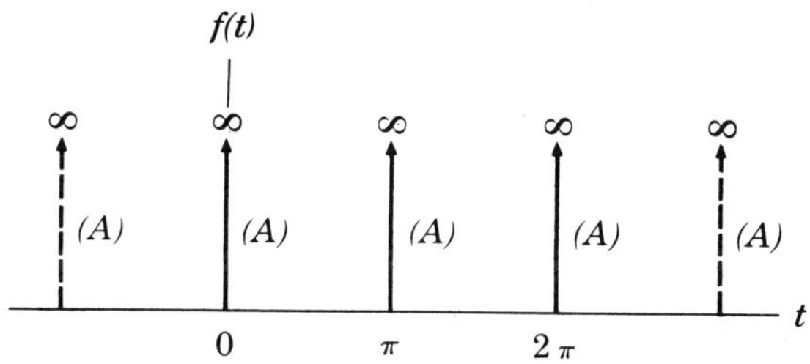

FIGURE 20–41

42. Find the complex Fourier series for the waveform of Fig. 20–42. Sketch the amplitude and phase spectra as functions of positive and negative frequency. Reduce the waveform to the sine-cosine expression.

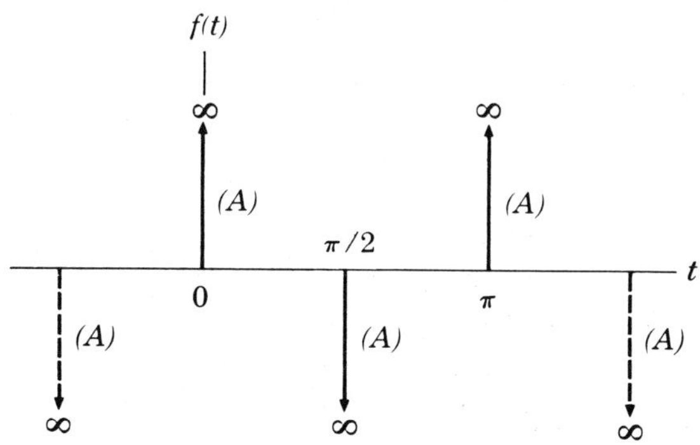

FIGURE 20–42

43. Find the complex Fourier series for the waveform of Fig. 20–43. Sketch the amplitude and phase spectra as functions of positive and negative frequency. Reduce the series to the sine-cosine form.

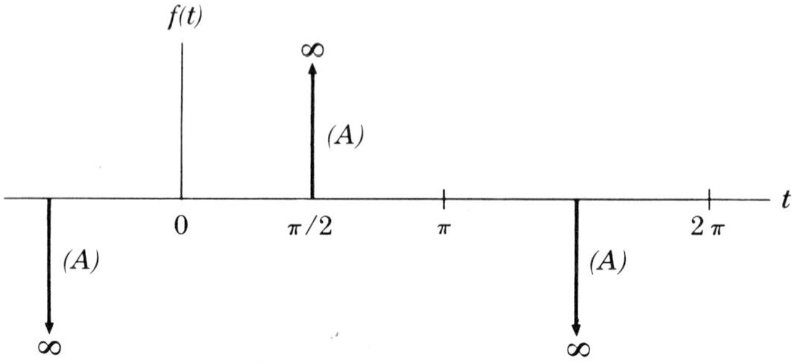

FIGURE 20–43

44. Find the complex Fourier series for the waveform of Fig. 20–44. Plot the amplitude and phase spectra as functions of positive and negative frequency. Reduce the series to a sine-cosine series.

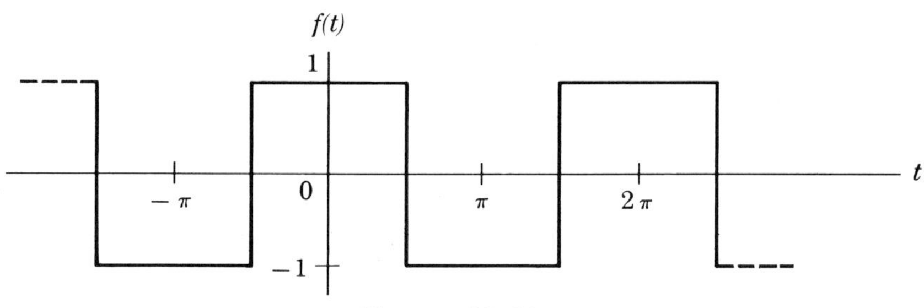

FIGURE 20–44

45. Find the complex Fourier series for the waveform of Fig. 20–45. Plot the amplitude and phase spectra as functions of positive and negative frequency. Reduce the series to a sine-cosine form.

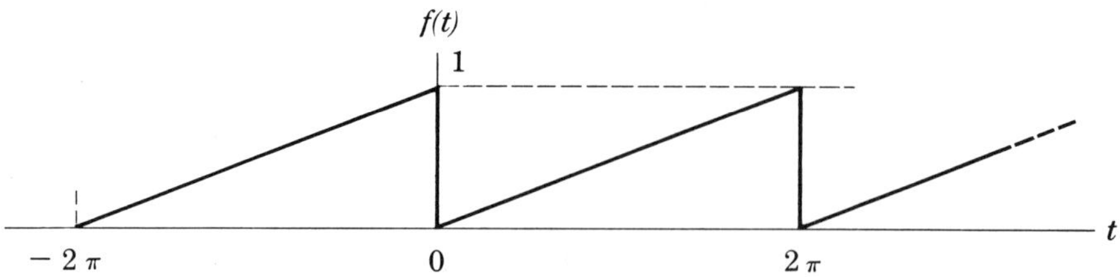

FIGURE 20–45

46. For the repeated pulse shown in Fig. 20–46 find the coefficients of the complex Fourier series, and sketch the amplitude and phase spectra.

47. Find the Fourier integral of a single pulse of the sort shown in Fig. 20–46. Sketch the amplitude and phase spectra.

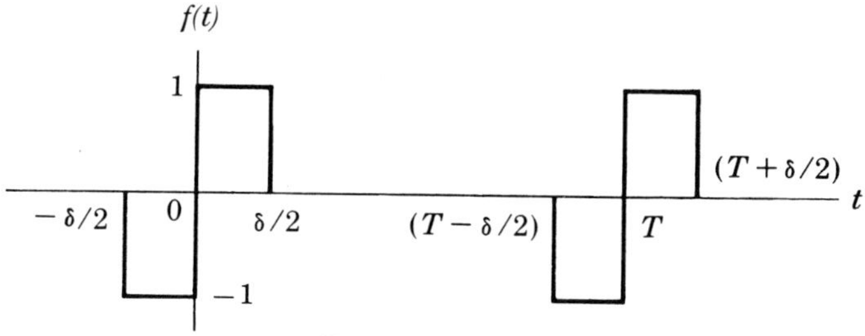

FIGURE 20–46

48. For the repeated pulse shown in Fig. 20–47 find the coefficients of the complex Fourier series. Write the general term and separate it into real and imaginary parts.

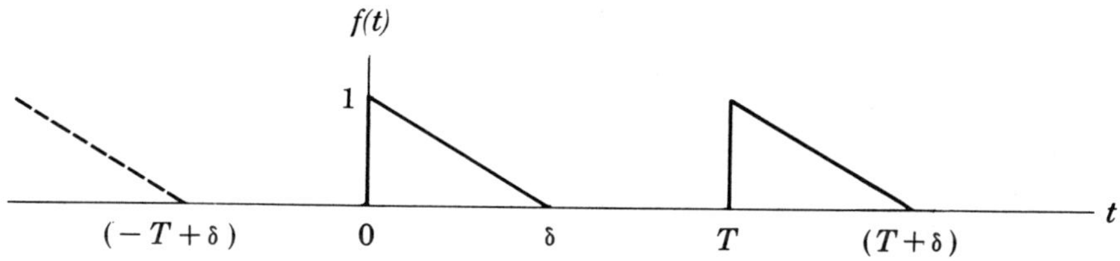

FIGURE 20–47

Section 20–7

49. Find the coefficient c_n in the complex Fourier series for the waveform of Fig. 20–48. Sketch the line spectra representing the amplitude and the phase of c_n as a function of $\omega = n\omega_0$, which here equals n since $\omega_0 = 1$. At what frequency could the first harmonic with zero amplitude occur?

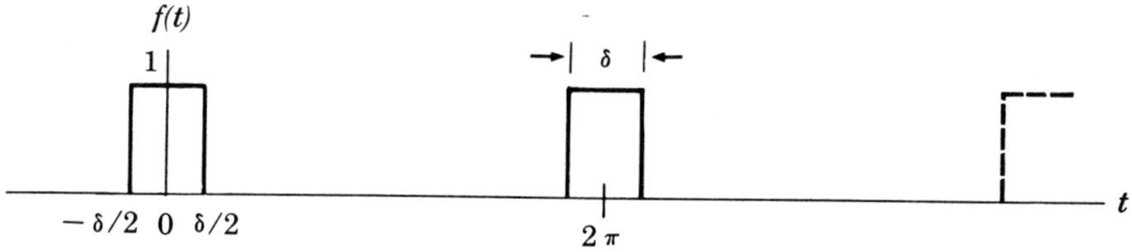

FIGURE 20–48

50. The waveform of Fig. 20–48 is normalized to give a period of $T = 2\pi$ sec. Let the period become T and the pulse retain the same percentage of the period

as in the normalized diagram. Find the complex Fourier series for the new waveform and plot the amplitude and phase of c_n as a function of $n\omega_0$. At what frequency could the first harmonic with zero amplitude occur?

51. In the waveform of Fig. 20–49 the period T has been increased from its normalized value of 2π, but the width of the pulse has been left at the original value of δ. Find the complex Fourier series for this waveform. Sketch the amplitude and phase spectra of c_n for $T = 2\delta$, 5δ, and 10δ. For each of the above cases sketch also the spectra of $c_n \times T$.

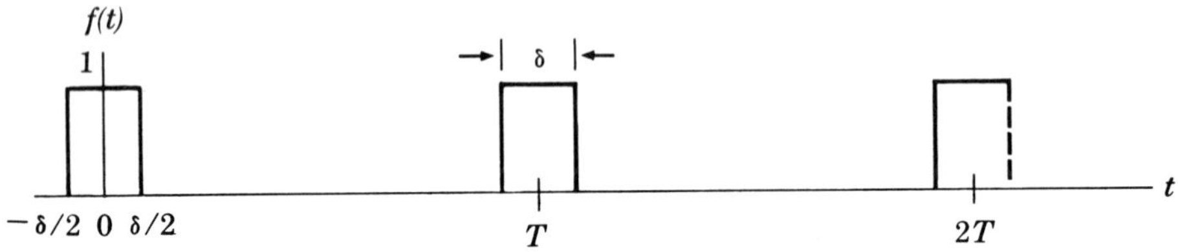

FIGURE 20–49

52. (a) Find the Fourier integral of a single pulse of the waveform of Fig. 20–49. Sketch the magnitude and the phase of the frequency distribution function $F(\omega)$ for $\delta = 1$ sec and for $\delta = 1$ microsec. (b) What bandwidth is required in a circuit to pass all the frequency components in the waveforms of part (a) out to the first zero in the spectrum, for $\delta = 1$ sec and for $\delta = 1$ microsec. (c) A criterion often used for the bandwidth required to pass a pulse is $f_0 = 2/\delta$, where f_0 is the bandwidth (in cycles) required and δ is the width of the pulse in seconds. How much of the spectrum is passed by this criterion?

53. Find the complex Fourier series of the impulse train shown in Fig. 20–50. Sketch the line spectra of the amplitude and the phase of c_n as functions of $\omega = n\omega_0$ for $T = 1$ sec, $1/10$ sec, and $1/100$ sec. Sketch also the spectra for $c_n \times T$. Finally, obtain the Fourier integral of a single impulse and compare its spectra with those of $c_n \times T$.

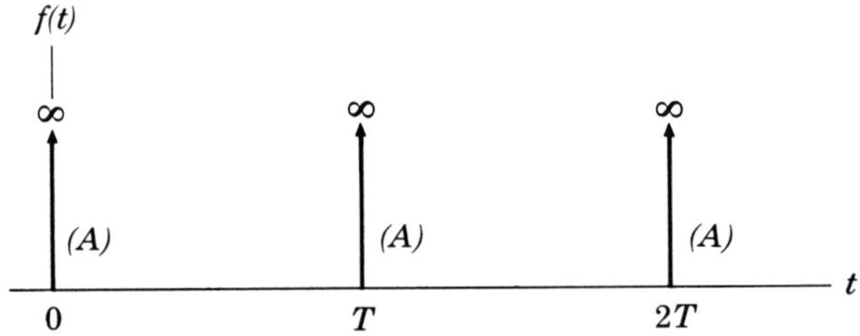

FIGURE 20–50

54. (a) Find the complex Fourier series for the waveform shown in Fig. 20–51 and sketch the amplitude and phase spectra as functions of $\omega = n$. (b) Scale the waveform of part (a) to a new period of T, instead of 2π, leaving the pulse width equal to the same percentage of the period as in part (a). Find the same amplitude and phase spectra for the new waveform.

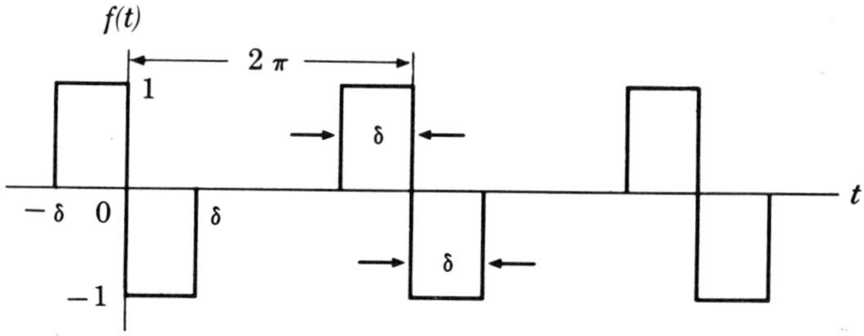

FIGURE 20–51

55. Find the complex Fourier series for the waveform of Fig. 20–52. Plot the spectra of c_n as functions of $n\omega_0$ and also plot $c_n \times T$.

56. Find the Fourier integral for a single pulse of the waveform of Fig. 20–52. Sketch the amplitude and phase spectra.

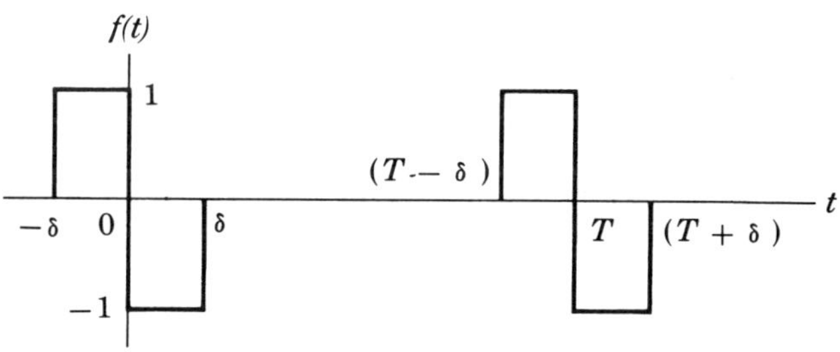

FIGURE 20–52

57. Find the Fourier integral of the pulse shown in Fig. 20–53. Show that if δ is allowed to approach zero, the pulse becomes a unit doublet. By letting δ approach zero in the Fourier transform, find the transform of a unit doublet function.

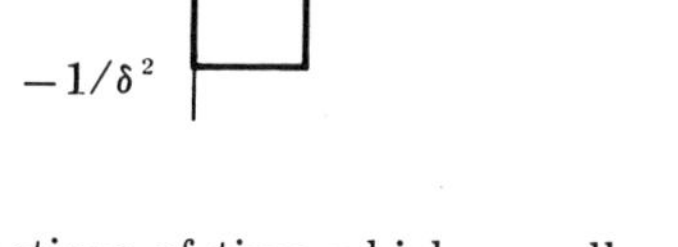

FIGURE 20–53

Section 20–8

58. For the following functions of time which are all zero for $t < 0$ (a) give the abscissa of absolute convergence, (b) find the Laplace transform, and (c) give the poles of the transform in the frequency domain.

(1) $f(t) = \sin \beta t$

(2) $f(t) = e^{-\alpha t} \sin \beta t$

(3) $f(t) = te^{-\alpha t} \sin \beta t$

(4) $f(t) = te^{\alpha t}$

(5) $f(t) = (1 - \cos \beta t)$

59. The following functions of time are zero for negative values of t. For each function find (a) the abscissa of absolute convergence, (b) the Laplace transform, and (c) the poles of the transform.

(1) $(1 - e^{-\alpha t})$ (4) $\sin(\beta t + \theta)$

(2) $\sinh \beta t$ (5) $\cos(\beta t + \theta)$

(3) $\cosh \beta t$

60. Find the abscissa of absolute convergence and the Laplace transform of each of the following functions ($\alpha < \beta$).

(1) a^t (3) $\sin \beta t / t$

(2) $\dfrac{e^{-\alpha t} - e^{-\beta t}}{t}$ (4) e^{t^2}

61. Show that the following formulas hold if $F(s) = \mathcal{L}[f(t)]$, where $\mathcal{L}$ is read "the Laplace transform of".

(1) $\mathcal{L}[f'(t)] = sF(s) - f(0^+)$

(2) $\mathcal{L}[f^{-1}(t)] = \dfrac{F(s) + f^{-1}(0^+)}{s}$

(3) $\mathcal{L}[e^{-\alpha t} f(t)] = F(s + \alpha)$

(4) $\mathcal{L}[f(t - a)u_{-1}(t - a)] = e^{-as}F(s)$

(5) $f(t)\ (t \to 0) = sF(s)\ (s \to \infty) \quad (\sigma \leq 0)$

(6) $f(t)\ (t \to \infty) = sF(s)\ (s \to 0)$

62. Find the Laplace transforms of the following functions.

(1) $e^{-at} \cdot u_{-1}(t)$ (5) $e^{-at} \cdot u_{-1}(t + 1/a)$

(2) $e^{-at} \cdot u_{-1}(t - 1/a)$ (6) $e^{-a(t+1/a)} \cdot u_{-1}(t + 1/a)$

(3) $e^{-a(t-1/a)} \cdot u_{-1}(t - 1/a)$ (7) $e^{-a(t+1/a)} \cdot u_{-1}(t)$

(4) $e^{-a(t-1/a)} \cdot u_{-1}(t)$

63. Sketch the following functions and find their Laplace transforms.*

(1) $u_{-1}(t) - u_{-1}(t - 1)$

(2) $u_{-1}(t) - u_{-1}(t - 1) + u_{-1}(t - 2) + \cdots$

(3) $2 \sum_{n=0}^{n=\infty} u_{-1}(t - 2n - 1)$

(4) $2 \sum_{n=0}^{n=\infty} (-1)^n u_{-1}(t - 2n - 1)$

(5) $u_{-2}(t) - \sum_{n=1}^{n=\infty} u_{-1}(t - n)$

* Note that $1/(1 - x) = 1 + x + x^2 + x^3 + \cdots$.

64. For the function sketched in Fig. 20–54:
  (a) Find the Laplace transform.
  (b) Find the abscissa of absolute convergence.
  (c) Find the poles in the complex frequency domain.

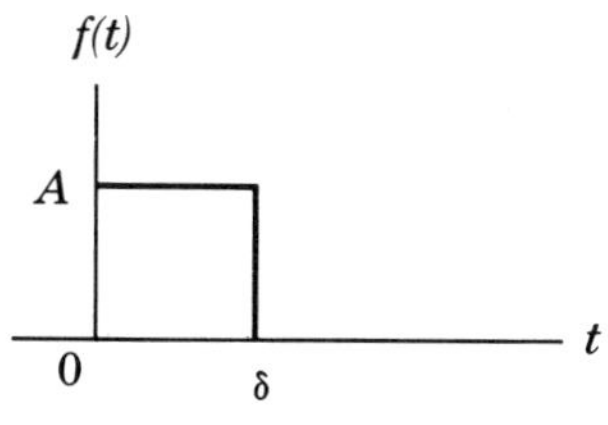

FIGURE 20–54

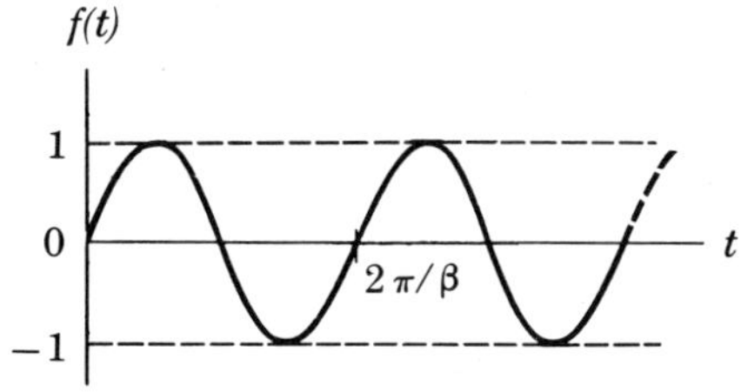

FIGURE 20–55

65. The transform of a sine wave sectioned at $t = 0$, shown in Fig. 20–55, is

$$F(s) = \frac{\beta}{s^2 + \beta^2}.$$

Show that it is permissible to replace s by $j\omega$ in this expression, and sketch the amplitude and phase spectrums for $s = j\omega$. What interpretation do you give to the infinities at $s = \pm j\beta$?

66. Repeat Exercise 65 for a cosine wave sectioned in the same fashion at $t = 0$.

CHAPTER 21

NETWORK RESPONSE TO COMPLEX WAVES OF EXCITATION

"Words are the daughters of earth,
Things are the sons of heaven."

Samuel Johnson

21–1 Introduction. Our interest in Fourier methods lies in their application to real physical devices. In Chapter 20 we resolved arbitrary waveforms into summations of sinusoidal and damped sinusoidal components. If the original waveform is periodic the components are discrete harmonic frequencies of the repetition rate of the waveform; if the waveform is a single pulse the components form a band spectrum consisting of an infinite number of harmonics of infinitesimal amplitude and very small separation. We are interested in the sinusoidal resolutions of arbitrary waveforms, because we have very efficient methods for solving network problems for sinusoidal steady-state inputs.

If the input function to a network is a periodic waveform which can be expanded in a Fourier series, we first find the response of the network to each component in the series and then obtain the over-all response function by the superposition of each of the component responses. This process may sound long and complicated to the student. In practice, however, it may involve much less work than any alternative solution, and in addition it is part of a general procedure which can be used with any input waveform.

If the input to a network is a single pulse, we can resolve the waveform into Fourier-integral components, and find the response of the network to each of these components. The output is again obtained by superposition, although it is now an integral rather than a summation.

When we use the Laplace integral instead of the Fourier integral, the input is resolved into damped sinusoids rather than constant-amplitude sinusoids. The network response to damped sinusoidal functions is obtained by replacing $j\omega$ by s in the network functions. The output is again obtained by superposition, and appears as an integral involving the complex frequency variable s. The evaluation of this integral is usually a difficult process which is circumvented by the use of *tables of transforms.* Transform tables are used in much the same fashion as tables of integrals, and special methods (such as partial fraction expansions) can be employed to increase the scope of the tables.

21–2 Networks excited by Fourier series. When the input waveform is periodic we can resolve it into a Fourier series. Two cases of practical interest occur. The first is when the waveform can be expressed mathematically, and the general term can be found in the series. In this case a general response function can also be found, and a complete expression for the output can be written. In the second case the general mathematical form for the input wave may not be known, but it is known that the principal components of energy in the spectrum occur in the first few harmonics. We can then truncate the series after a few terms and find the response for each of the terms in the partial sum. The addition of these response functions will give a good approximation to the response from the original waveform.

Case 1. Complete series known. As an example of the case in which the complete series is known, consider the application of a square wave of voltage to a capacitance, as shown in Fig. 21–1. The Fourier series for the waveform has been obtained before. [See Eq. (20–12).] It has the form

$$e(t) = \frac{4}{\pi}\left(\cos t - \frac{\cos 3t}{3} + \frac{\cos 5t}{5} - \cdots\right). \tag{21–1}$$

The amplitude and the phase of the nth harmonic can be written directly in the frequency domain.

$$E_n = \frac{4}{n\pi}\,\angle\frac{n+3}{2}\pi \qquad \text{for } n \text{ odd.}$$

The nth harmonic current is obtained by dividing the nth harmonic voltage by the impedance to the nth harmonic. This impedance is

$$Z_n = \frac{1}{j\omega_0 nC} = \frac{1}{jn},$$

since $\omega_0 = C = 1$. The nth harmonic current is

$$I_n = \frac{E_n}{Z_n} = \frac{4n}{n\pi}\,\angle\frac{n+3}{2}\pi + \frac{\pi}{2} = \frac{4}{\pi}\,\angle n\pi/2 \qquad \text{for } n \text{ odd.} \tag{21–2}$$

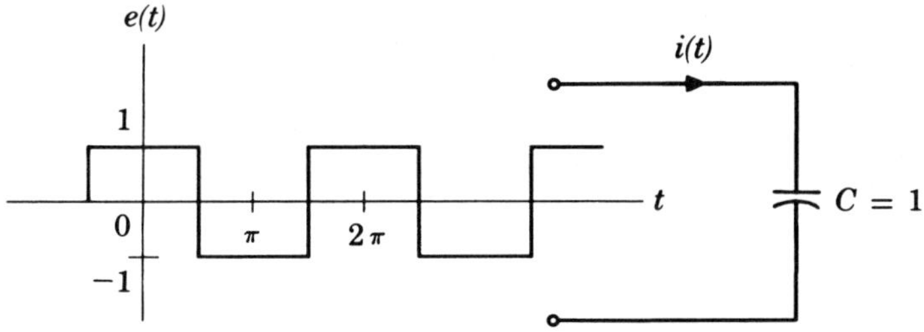

FIG. 21–1. A square wave applied to a capacitance.

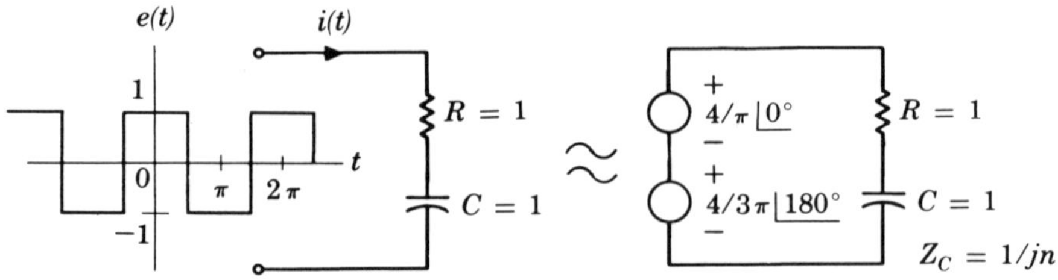

FIG. 21–2. A square wave applied to an R–C circuit.

Expanded, Eq. (21–2) becomes

$$i(t) = \frac{4}{\pi}(-\sin t + \sin 3t - \sin 5t + \cdots). \tag{21–3}$$

In this particular case we might recognize the sum in Eq. (21–3) as a series of impulses occurring at the discontinuities in the wave of Fig. 21–1. It is identical to the series for the impulse chain which we obtained previously in Eq. (20–14). If the resulting series is not recognizable as a function whose Fourier series has been previously obtained, it is very difficult to obtain an explicit answer for the resulting time function. As a practical matter, we can plot a few terms and get an idea of the form of the output waveform. Machines have been constructed for the Fourier synthesis problem represented by the addition of a number of terms like those of Eq. (21–3). The first of these was constructed by Lord Kelvin from cranks and rods to sum the Fourier series he had obtained for the motion of the tides.

Case 2. Only a few harmonics known. As an example of the procedure used in a case where only a finite number of harmonics of the input wave are of interest, consider the R–C circuit shown in Fig. 21–2. The circuit is excited by the same square wave used in the previous problem. We will suppose that we wish only an approximate idea of the nature of the output waveform and that for this purpose it will be necessary only to consider the first two harmonics of the input waveform. The first two terms in the input function are

$$e(t) = \frac{4}{\pi}\cos t - \frac{4}{3\pi}\cos 3t.$$

We will obtain the responses for the two component voltages and add them to obtain an approximation to the response of the square wave.

The fundamental. The fundamental voltage is

$$E_1 = \frac{4}{\pi}\underline{|0^\circ}.$$

The fundamental impedance is

$$Z_1 = 1 - j,$$

and thus the fundamental current is

$$\frac{E_1}{Z_1} = \frac{4}{\pi(1 - j)} \underline{|0^\circ},$$

or

$$I_1 = \frac{4}{\pi\sqrt{2}} \underline{|45^\circ} \cdot$$

The third harmonic. The third harmonic voltage is

$$E_3 = \frac{4}{3\pi} \underline{|180^\circ}.$$

The third harmonic impedance is

$$Z_3 = 1 - \frac{j}{3} = 1.05 \underline{|-18.4^\circ},$$

and hence the third harmonic current is

$$I_3 = \frac{E_3}{Z_3} = \frac{4}{3\pi 1.05} \underline{|180^\circ + 18.^\circ 4}.$$

The output current. Insofar as it can be represented by the first two harmonic terms, the output current is

$$i(t) = \frac{4}{\pi}[0.707 \cos(t + 45^\circ) - 0.318 \cos(3t + 18.4^\circ)]. \quad (21\text{–}4)$$

It is difficult to infer the form of $i(t)$ from Eq. (21–4). To aid the process, it is sketched in Fig. 21–3. The graphical addition of sine waves is labor-

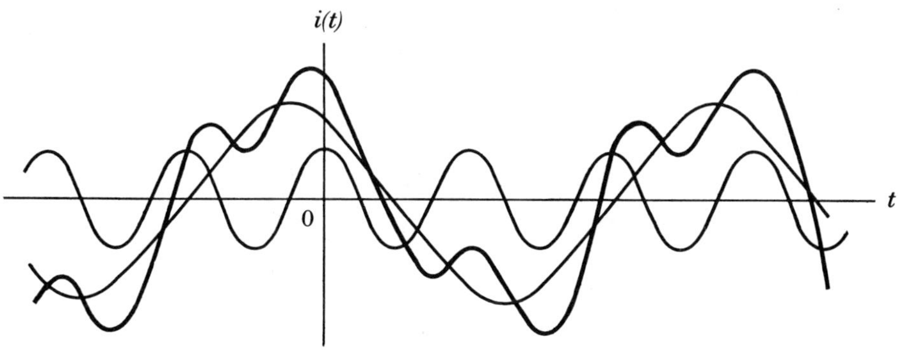

FIG. 21–3. The output as the sum of two harmonics.

ious but straightforward. Often it presents the only practical method for obtaining a solution to a problem. The mere statement of the solution as a sum of sine waves is not very useful, since the maxima, minima, and general behavior of the wave are not apparent in that form.

If the input to a network is resolved into a Fourier integral instead of a Fourier series, there are an infinite number of infinitesimal harmonics and it is not practical to treat them individually. We must then deal with the general term and trust a table of integrals to sum the resulting output function for us.

21–3 Networks excited by a Fourier integral. When a network is driven by a single transient pulse, the Fourier method can still be used. The input is resolved into an infinite sum of infinitesimal harmonics. Each of these is multiplied by the network response function for a steady-state sine wave at the appropriate frequency, and an infinite number of infinitesimal responses are obtained. The sum of these responses represents the output waveform. Because of the infinite number of terms involved, the whole process must be carried out in analytic form. In effect, the general term of the infinite Fourier summation must be multiplied by a general term which represents the network response, and the product is then summed in a new Fourier expression to give the output time function. The situation is portrayed schematically in Fig. 21–4.

The input waveform to the circuit is $f_1(t)$. The corresponding summation of the frequency components in $f_1(t)$ is called $F_1(\omega)$. The response function of the network at frequency ω is given by $H(\omega)$. The function $H(\omega)$ is the complex number representing the amplitude and phase of the output function $f_2(t)$ when the input is driven with a steady-state sine wave of amplitude unity and phase zero. $H(\omega)$ may have the dimensions of impedance if the input is a current and the output a voltage, or of admittance if the input is a voltage and the output a current, or of a constant if the input and the output have the same dimensions. The output spectral function $F_2(\omega)$ is obtained by multiplying the response function at any frequency by the amount of the input at that frequency. Thus

$$F_2(\omega) = F_1(\omega) \times H(\omega). \tag{21–5}$$

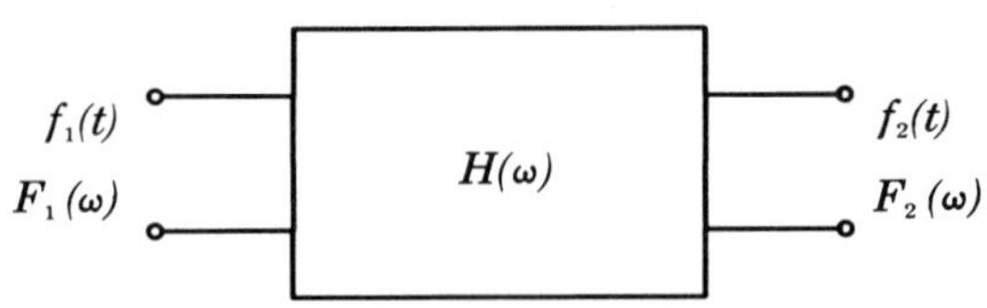

FIG. 21–4. The response of a general network to a general input.

If $F_1(\omega)$ and $F_2(\omega)$, in Eq. (21-5), are each multiplied by $d\omega/\pi$ they represent the actual values of the harmonics present in the input and output waveforms. It is convenient, however, to drop the $d\omega/\pi$ on each side and to deal directly with the distribution functions. The time function for the output is the sum of all the sinusoidal terms with frequencies from zero to infinity and with amplitudes and phases given by $F_2(\omega)$. To merely write this summation is not too useful. It is necessary to perform the integration and to obtain an explicit time function for $f_2(t)$. As an alternative it is possible to tabulate the functions of ω which belong with the usual functions of t which are obtained, and to use a table to find the desired function of time when the function of frequency is known. In summary, there are four steps in the Fourier method:

Step 1. The input function $f_1(t)$ is resolved into a frequency function $F_1(\omega)$.

$$F_1(\omega) = \int_{-\infty}^{\infty} f(t)e^{-j\omega t}\,dt. \tag{21-6}$$

Step 2. The network response function $H(\omega)$ is obtained.

$$\left.\begin{aligned}\text{Inductors} &\rightarrow j\omega L\\ \text{Capacitors} &\rightarrow 1/j\omega C\\ \text{Resistors} &\rightarrow R\end{aligned}\right\}\text{ohms}$$

$$H(\omega) = \frac{F_2(\omega)}{F_1(\omega)}\ [= F_2(\omega) \text{ for } F_1(\omega) = 1\,\underline{|0^\circ}]. \tag{21-7}$$

Step 3. The output frequency spectrum is obtained.

$$F_2(\omega) = H(\omega) \times F_1(\omega). \tag{21-8}$$

Step 4. The output time function is written as an integral.

$$f_2(t) = \frac{1}{2\pi}\int_{-\infty}^{\infty} F_2(\omega)e^{j\omega t}\,d\omega. \tag{21-9}$$

The last step in this process is usually the most difficult. Extensive tables*† are available which make it unnecessary to evaluate the integrals usually encountered. The existence of these tables (and of the methods of using them which we will discuss more fully later in the chapter) was essential for any widespread use of transform methods. The general

---

* Campbell and Foster, "Fourier Integrals for Practical Applications," Bell System Monograph, No. B-584 (1931).

† Gardner and Barnes, *Transients in Linear Systems*, John Wiley & Sons, New York, 1942.

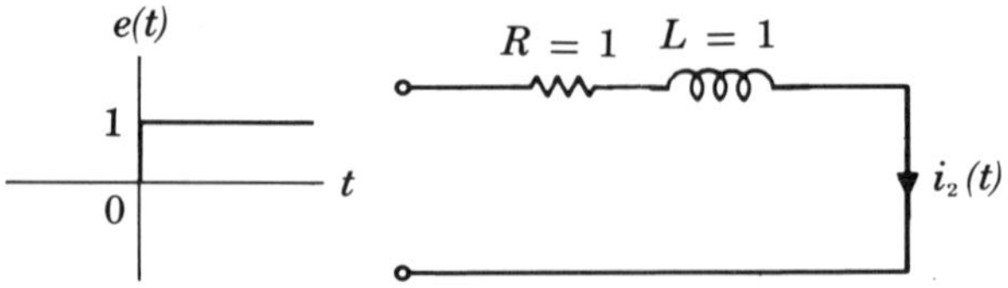

FIG. 21–5. Circuit excited by a Fourier integral.

methods were worked out by Laplace in 1779.* They did not receive wide use, however, until the work of Oliver Heaviside† in the field of electrical engineering, and until the publishing of Campbell and Foster's tables in 1931.

As an example of the Fourier integral method of solving network problems, we will consider the circuit of Fig. 21–5, which is excited by a unit step function of voltage. Strictly speaking, the Fourier integral for the unit step does not converge, but so long as we treat it as the limiting case of a Laplace transform in which the abscissa of absolute convergence has been allowed to approach arbitrarily close to zero, the value of the integral at infinity will be zero.‡ The solution to the problem will be obtained in four steps.

Step 1. Obtain $E_1(\omega)$. To obtain $E_1(\omega)$ we integrate as follows:

$$E_1(\omega) = \int_{-\infty}^{\infty} f(t)e^{-j\omega t}\,dt = \int_0^{\infty} e^{-j\omega t}\,dt$$

$$= \left[\frac{e^{-j\omega t}}{-j\omega}\right]_0^{\infty}.$$

If we consider that the value at infinity is zero, we obtain

$$E_1(\omega) = \frac{1}{j\omega}.$$

Step 2. Obtain $H(\omega)$. To obtain $H(\omega)$ we excite the network with the sinusoid $E_1(\omega) = 1\,\underline{|0^\circ}$ and compute the output current. Thus

$$I_2 = \frac{E_1}{Z_1} = \frac{1\,\underline{|0^\circ}}{j\omega + 1}.$$

---

* P. S. Laplace, *Complete Works*, pp. 1–89, Academy of Science, Paris, 1779.

† O. Heaviside, *Electrical Papers*, Macmillan, New York, 1892. See also the Dover Publications edition.

‡ An alternative point of view is that the step function is really the limiting case of a damped exponential as the damping becomes very small.

Hence

$$H(\omega) = \frac{1}{1 + j\omega}.$$

Step 3. Obtain $I_2(\omega)$. The output spectral function is

$$I_2(\omega) = H(\omega) \times E_1(\omega) = \frac{1}{1 + j\omega} \times \frac{1}{j\omega}.$$

Step 4. Obtain $i_2(t)$. The output current as an integral is

$$i_2(t) = \frac{1}{2\pi} \int_{-\infty}^{\infty} I_2(\omega) e^{j\omega t}\, d\omega,$$

or

$$i_2(t) = \frac{1}{2\pi} \int_{-\infty}^{\infty} \frac{1}{(1 + j\omega)(j\omega)} e^{j\omega t}\, d\omega.$$

The simplest way to evaluate this integral is to find a function $i_2(t)$, in a table, which has a transform in the frequency domain of $I_2(\omega) = 1/(1 + j\omega)(j\omega)$. Such a function is $i_2(t) = 1 - e^{-t}$. In the next section we will consider the evaluation of integrals of the type given by Eq. (21–9).

21–4 Cosine and sine transforms. The last step in the process of solving a network problem by Fourier methods is the summation of the harmonic terms in the output to find a function of time which can be recognized and utilized. If this summation is available from a table of transforms the problem is trivial. If it is not, a direct integration of the inverse Fourier transform can be performed and the output obtained in this way. For experimental data, and for the computation of inverse transforms by digital and analog computers, it is necessary to solve the inverse integral.

First, let us take another look at the direct transform,

$$F(\omega) = \int_{-\infty}^{\infty} f(t) e^{-j\omega t}\, dt.$$

If the function $f(t)$ is zero for negative t it seems unnecessary to integrate for negative values of t. We might like to perform the integration over the whole range of t, however, in order to take advantage of the even-, and odd-function symmetry of the cosine and sine functions. To accomplish this, we will think of $f(t)$ as being made up of two functions. For t greater than zero, each of them will be exactly equal to one-half of $f(t)$. Thus together they add up to give $f(t)$. For t negative, however, they must add up to zero. Hence one of the functions is taken to be even, and the other odd. A resolution of a time function into even and odd

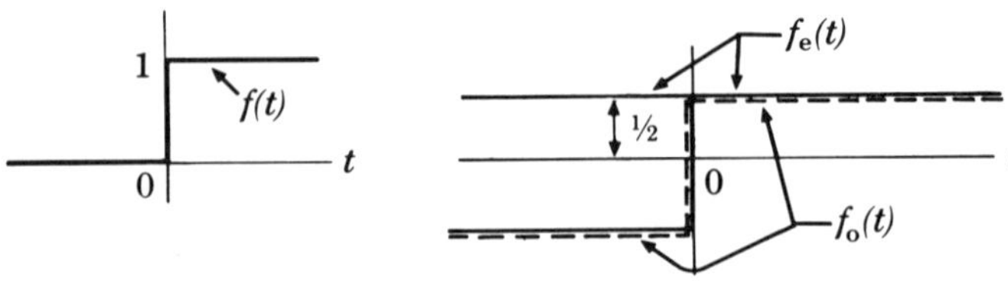

FIG. 21–6. Even and odd functions of time.

components is shown in Fig. 21–6. Analytically, we write

$$f(t) = f_e(t) + f_o(t)$$

where $f_e(t)$ is an even function of t equal to $f(t)/2$ for $t > 0$,

$f_o(t)$ is an odd function of t equal to $f(t)/2$ for $t > 0$.

In terms of the even and odd functions of time, we can write

$$\begin{aligned} F(\omega) &= \int_{-\infty}^{\infty} \{[f_e(t) + f_o(t)](\cos \omega t - j \sin \omega t)\}\, dt \\ &= \int_{-\infty}^{\infty} [f_e(t) \cos \omega t - jf_o(t) \sin \omega t + f_o(t) \cos \omega t - jf_e(t) \sin \omega t]\, dt. \end{aligned} \tag{21–10}$$

The product of an even function by an even function is even; the product of an odd function by an odd function is even; and the product of an odd and an even function is odd. Thus the first two integrals in Eq. (21–10) are even and the second two are odd. The integral of an odd function over all time is zero and the integral of an even function over all time is twice the value of the integral from zero to infinity. Accordingly, we obtain

$$F(\omega) = 2 \int_0^{\infty} [f_e(t) \cos \omega t - jf_o(t) \sin \omega t]\, dt.$$

For the particular case of $f(t) = 0$ for t less than zero,

$$f_e(t) = f_o(t) = \frac{f(t)}{2} \text{ for } t > 0,$$

and

$$F(\omega) = \int_0^{\infty} f(t) \cos \omega t\, dt - j \int_0^{\infty} f(t) \sin \omega t\, dt. \tag{21–11}$$

The real part of $F(\omega)$ is the cosine transform and the imaginary part is the sine transform. As an example of the use of the formula of Eq. (21–11), we will compute the Fourier transform of a step function. We let

$$f(t) = u_{-1}(t).$$

Then

$$F(\omega) = \int_0^\infty \cos \omega t \, dt - j \int_0^\infty \sin \omega t \, dt$$

$$= \left[\frac{\sin \omega t}{\omega}\right]_0^\infty - j \left[\frac{-\cos \omega t}{\omega}\right]_0^\infty. \tag{21–12}$$

Strictly speaking Eq. (21–12) does not converge. If we view it as a limiting case of a Laplace transform in which the abscissa of absolute convergence approaches zero, we can neglect the values at infinity. The first integral is zero and the second gives

$$F(\omega) = \frac{-j}{\omega} = \frac{1}{j\omega}. \tag{21–13}$$

The inverse transform. A similar set of simplifications can be effected in the inverse transformation. Again, they are useful for numerical or machine computation, and they provide a certain amount of insight into the nature of the transforms. The general expression for the inverse transform is

$$f(t) = \frac{1}{2\pi} \int_{-\infty}^{\infty} F(\omega) e^{j\omega t} \, d\omega. \tag{21–14}$$

The function $F(\omega)$, as can be seen from Eq. (21–11), has a real and an imaginary part. The real part is an even function of ω, since the ω term enters only in the cosine. The imaginary part is an odd function of ω, since the ω term enters only in the sine term. Another way to view the preceding argument is to note that ω always occurs with j, and that even powers of j are real, while odd powers of j are imaginary. Thus

$$F(\omega) = R(\omega) + jI(\omega),$$

where $R(\omega)$ is an even function of ω,
$I(\omega)$ is an odd function of ω.

Equation (21–14) can now be written in the form

$$f(t) = \frac{1}{2\pi} \int_{-\infty}^{\infty} \{[R(\omega) + jI(\omega)](\cos \omega t + j \sin \omega t)\} \, d\omega$$

$$= \frac{1}{2\pi} \int_{-\infty}^{\infty} [R(\omega) \cos \omega t - I(\omega) \sin \omega t] \, d\omega$$

$$+ j \frac{1}{2\pi} \int_{-\infty}^{\infty} [R(\omega) \sin \omega t + I(\omega) \cos \omega t] \, d\omega.$$

Since $f(t)$ is a real function, the second integral must vanish and since both terms in the first integral are even functions of ω, we can write

$$f(t) = \frac{1}{\pi}\int_0^\infty [R(\omega)\cos\omega t - I(\omega)\sin\omega t]\,d\omega. \tag{21–15}$$

One further simplification is possible if the function $f(t)$ is zero for t less than zero. The first integral in Eq. (21–15) is an even function of t, and the second integral is an odd function of t. If the function of t is to vanish for t negative, the even and the odd portions must be equal for t greater than zero. Hence either the cosine integral or the sine integral may be used to obtain $f(t)$:*

$$f(t) = \frac{2}{\pi}\int_0^\infty R(\omega)\cos\omega t\,d\omega = \frac{-2}{\pi}\int_0^\infty I(\omega)\sin\omega t\,d\omega. \tag{21–16}$$

As an example of the use of Eq. (21–16) we will obtain the inverse transformation of the function $F(\omega)$ for the unit step, as obtained in Eq. (21–13). We let

$$F(\omega) = \frac{1}{j\omega} = \frac{-j}{\omega}.$$

Then

$$f(t) = \frac{-2}{\pi}\int_0^\infty I(\omega)\sin\omega t\,d\omega$$

$$= \frac{-2}{\pi}\int_0^\infty \frac{-\sin\omega t}{\omega}\,d\omega.$$

This integral is tabulated. It has a value of $-\pi/2$ for $t > 0$. Hence

$$f(t) = \left(-\frac{2}{\pi}\right)\times\left(-\frac{\pi}{2}\right) = 1 \qquad \text{for } t > 0.$$

As a second example we will evaluate the inverse transform of the function

$$F(\omega) = \frac{1}{1 + j\omega}.$$

The real part of the function is

$$F(\omega) = \text{Re}\left(\frac{1}{1 + j\omega}\times\frac{1 - j\omega}{1 - j\omega}\right)$$

$$= \frac{1}{1 + \omega^2}.$$

---

* For the singularity functions $R(\omega)$ is apparently zero, but it can be shown to contain an impulse, so that both integrals still hold.

Using the first integral of Eq. (21–16), we write

$$f(t) = \frac{2}{\pi}\int_0^\infty R(\omega)\cos\omega t\,d\omega$$

$$= \frac{2}{\pi}\int_0^\infty \frac{\cos\omega t}{\omega^2+1}\,d\omega.$$

This integral is tabulated. It has a value of $\pi\, e^{-t}/2$ for $t > 0$. Thus

$$f(t) = \frac{2}{\pi}\times\frac{\pi}{2}e^{-t} = e^{-t} \qquad \text{for } t > 0.$$

From the results of the two examples just completed, we can evaluate the inverse transform of the function previously obtained from a table, that is, the function

$$F(\omega) = \frac{1}{j\omega(1+j\omega)}\cdot \tag{21–17}$$

It is convenient to break this function into two terms by a process which is the inverse of bringing two terms to a common denominator. The student can easily verify that

$$F(\omega) = \frac{1}{j\omega} - \frac{1}{1+j\omega}\cdot \tag{21–18}$$

This process is called expanding in partial fractions, and we will consider it in more detail in the next section. The inverse transform of $F(\omega)$ is the sum of the inverse transforms of its two components, and is therefore

$$f(t) = 1 - e^{-t} \qquad \text{for } t > 0.$$

This answer can also be obtained by a direct integration of the imaginary part of Eq. (21–17). The integral for the real part is not applicable unless the impulse at $\omega = 0$ is included.

21–5 Circuits excited by Laplace transforms. The Laplace transform resolves an input function into a series of infinitesimal sinusoidal harmonics with arbitrary damping factors. We have already discussed the response of networks to damped sine waves in connection with resonance phenomena. We now have occasion to use this type of response in conjunction with the Laplace transform of the input function in order to obtain an output spectrum of damped sine waves. As in the case of Fourier integrals, the output spectrum itself is of little use until it is summed to obtain an explicit function of time. First, let us review briefly the nature of damped sine waves in the time domain and the frequency domain. In the time domain an undamped sine wave is given by

$$f(t) = |F| \cos(\omega t + \theta) = \text{Re}\,[|F|e^{j\theta}e^{j\omega t}].$$

In the frequency domain we drop the real-part sign and the $e^{j\omega t}$ factor and observe the vector $|F|e^{j\theta}$. The magnitude of this vector is the amplitude of the sinusoid and the angle of the vector is the phase of the sinusoid. The frequency does not appear explicitly in the vector F, but it affects it indirectly since the network must also be transformed to the frequency domain and the inductance becomes an impedance of $j\omega L$, and the capacitance an impedance of $1/j\omega C$.

In the time domain a damped sine wave is given by

$$f(t) = |F|e^{\sigma t} \cos(\omega t + \theta) = \text{Re}\,[|F|e^{j\theta}e^{(\sigma + j\omega)t}].$$

In the frequency domain we drop the real-part sign and the rotating and expanding vector $e^{(\sigma+j\omega)t}$ and again concentrate on the vector $|F|e^{j\theta}$. Here too, the frequency does not affect F directly, but it does so indirectly since the inductance now transforms to the frequency domain as Ls and the capacitance as $1/Cs$, where $s = \sigma + j\omega$. In summary, the change from steady-state sine waves to damped sine waves entails the following changes in the input function and the network:

Steady-state sine waves		*Damped sine waves*
$j\omega$	$\rightarrow$	$\sigma + j\omega = s$
$j\omega L$	$\rightarrow$	Ls
$1/j\omega C$	$\rightarrow$	$1/Cs$

Contrary to expectations, the networks for complex frequencies are actually simpler than the networks for undamped frequencies. The volt-ampere relationships are essentially those for direct-current circuits. Kirchhoff's laws hold for the frequency-domain vectors in the complex-frequency domain, and the response functions are obtained exactly as they would be for resistive circuits.

We return now to our main topic, the Laplace transform solution. A block diagram of the over-all situation is shown in Fig. 21–7. The input

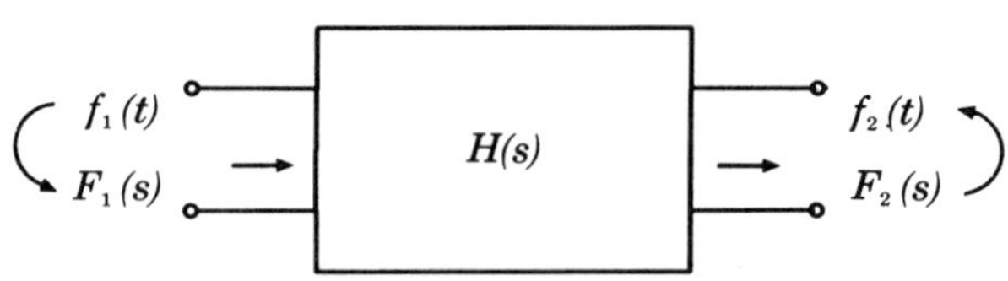

FIG. 21–7. The general Laplace transform solution.

function $f_1(t)$ is resolved into a general sum of damped sine waves whose frequency spectrum is given by $F_1(s)$. It is not necessary to specify the value of σ in the spectrum any more than it is necessary to specify the value of ω. The function $F_1(s)$ will give the spectrum for any σ greater than the abscissa of absolute convergence, that is, for any σ for which a convergent series is obtained.

The network response function $H(s)$ is the ratio of the output vector to the input vector for an arbitrary frequency ω and an arbitrary damping constant σ. It can be obtained by applying an input of $1\,\underline{|0°}$ and computing the output. The output spectrum $F_2(s)$ which results from the application of all the components of the input is obtained by multiplying each component of $F_1(s)$ by the response for that particular frequency, $H(s)$. The time function $f_2(t)$ is the summation of the damped sinusoids represented by $F_2(s)$. It should be noted that while $f_1(t)$ and $f_2(t)$ are in the time domain, the input $F_1(s)$, the network itself, and the output $F_2(s)$ are in the frequency domain, in the usual steady-state sense. In summary, the method of using Laplace transforms has four steps.

Step 1. The input time function is resolved into a sum of damped sine waves.

$$F_1(s) = \int_0^\infty f_1(t)e^{-st}\,dt.$$

It is essential that the integration be performed from zero to infinity. The damping constant in the exponential makes the integral converge at plus infinity, but would cause it to diverge wildly at minus infinity. No time functions with values before $t = 0$ can be handled.

Step 2. The network transfer function $H(s)$ is obtained.

$$\left.\begin{array}{r} \text{Inductors} \to Ls \\ \text{Capacitors} \to 1/Cs \\ \text{Resistors} \to R \end{array}\right\}\text{ohms}$$

$$H(s) = \frac{F_2(s)}{F_1(s)}\ \left[= F_2(s)\ \text{for}\ F_1(s) = 1\,\underline{|0°}\right].$$

Step 3. The output frequency spectrum of the damped sine waves is obtained.

$$F_2(s) = H(s) \times F_1(s).$$

Step 4. The output time function is written as an integral and evaluated.

$$f_2(t) = \frac{1}{2\pi j}\int_{-\infty+\sigma_c}^{\infty+\sigma_c} F_2(s)e^{st}\,ds.$$

This integration is inherently with respect to a complex variable. It is

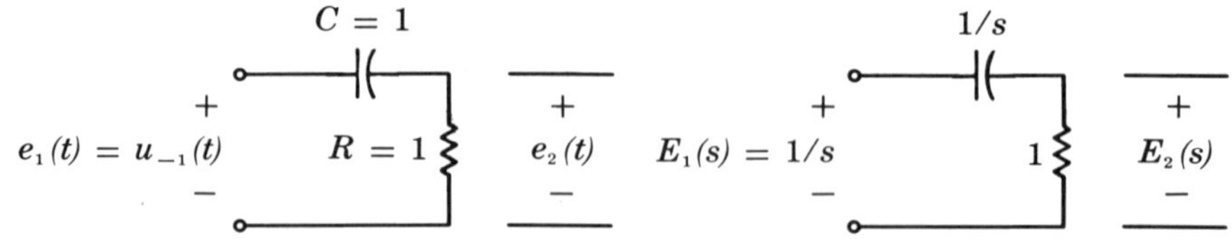

(a) Time-domain circuit (b) Frequency-domain circuit

FIG. 21–8. An example of a problem solved by the Laplace transform.

more difficult than real integration. We will avoid the use of this formula by using tables of transforms. When a given $F_2(s)$ is obtained, we will turn to tables to find an $f_2(t)$ which has the required form for $F_2(s)$.

In concept, the difference between Fourier and Laplace transforms lies in the fact that the Fourier transform resolves a time function into a summation of constant-amplitude sine waves, while the Laplace transform resolves a time function into a summation of damped sine waves. The Laplace transform is more general, since it includes the Fourier transform as the special case of damping equal to zero. Formally, a Laplace transform is changed to a Fourier transform by replacing s by $j\omega$, and vice versa. For real physical waveforms there is never any trouble with the convergence of the Fourier transforms. Tables of Fourier transforms and tables of Laplace transforms can thus be used interchangeably. The Laplace transforms are more convenient for many network problems, but the Fourier transforms are preferred for theoretical work because of their symmetry, and also for numerical integration because of simplicity of the sine and cosine transforms. We will now consider an example of the Laplace method for solving a problem, utilizing the circuit of Fig. 21–8(a).

Step 1. The frequency spectrum of the input function is

$$F_1(s) = \int_0^\infty f_1(t)e^{-st}\,dt = \frac{1}{s}\cdot \qquad \text{(See Table 20–4.)}$$

Step 2. The network response function is obtained by applying an input of $1\,\underline{|0^\circ}$ and computing the output. It is

$$H(s) = \frac{1\,\underline{|0^\circ}}{1 + 1/s} \times 1 = \frac{s}{s+1}\cdot$$

Step 3. The output for an input of $F_1(s)$ is

$$F_2(s) = F_1(s) \times H(s) = \frac{1}{s+1}\cdot$$

Step 4. The time function corresponding to $F_2(s)$ is obtained from Table 20–4. It is

$$f_2(t) = e^{-t}.$$

It is worth noting that Steps 2 and 3 can be combined. Rather than find the response for unit input and then multiply it by the actual input, it is possible to find the output for the actual input directly. Thus

$$F_2(s) = \frac{1}{1 + 1/s} \times \frac{1}{s} = \frac{1}{s+1}\cdot$$

The transform, or frequency-domain, circuit from which this solution is obtained is shown in Fig. 21–8(b).

21–6 Laplace transforms, initial energy storages, and impulse functions. In the preceding section, the Laplace transform is presented as a method of resolving an arbitrary input wave into a sum of damped sine waves. The network response to damped sine waves is easily obtained by replacing $j\omega$ by s in the steady-state frequency-domain network. The frequency spectrum of the output wave is the product of the input frequency spectrum and the network response function, and the output time function is the summation of the output harmonic terms. Nowhere in this process have we allowed for the presence of initial energy storages in inductors or capacitors. The way in which we will handle initial energy storage is similar to the method used in our previous transient analysis. We will convert the initial energy storages to equivalent sources and include the effect of these sources in our results by means of superposition.

The equivalent sources for initial energy storages are called into being at $t = 0$. They represent the accumulated energy from the past which is present in the storage element at $t = 0$, and their presence in the circuit expresses the effect of this energy storage for $t > 0$. Since these functions are zero for negative time, they can be resolved into damped spectra by means of the Laplace transform. They are thus treated in all respects identically as are "real" sources. The exact form of the source depends on the nature of the energy-storage element and whether we wish to represent the energy by a current source or a voltage source. From our previous work with transients we recall the equivalent networks shown in Fig. 21–9.

These networks show that the initial energy storage on a capacitance can be considered as a step source of voltage in series with the capacitance, or an impulse current source in parallel with the capacitance. Similarly, the initial energy stored in an inductance can be considered as a step current source in parallel with the inductance, or as an impulse voltage source in series with the inductance. We previously worked out the transformation of a unit step to the frequency domain by means of the Laplace trans-

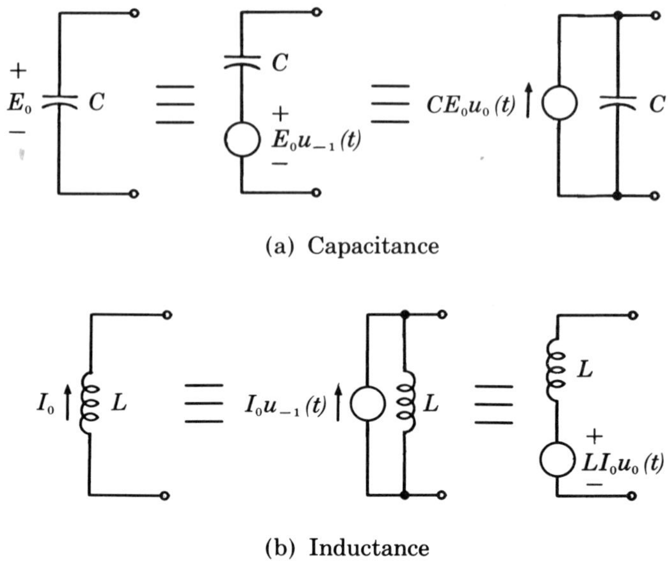

FIG. 21–9. Equivalent networks for initial energy storage.

form, but we have not yet considered the unit impulse function. Accordingly, before we transform the circuits of Fig. 21–9 to the frequency domain, it will be necessary to obtain the resolution of an impulse function into damped frequency components by means of the Laplace transform.

The impulse. A formal expression for the transform of the impulse $u_0(t)$ is

$$F(s) = \int_0^\infty u_0(t)e^{-st}\,dt. \tag{21–19}$$

The impulse is a pulse of area unity, occurring between $t = 0^-$ and $t = 0^+$. If the integration is carried out from $t = 0^+$, we would lose the impulse altogether and obtain only zero for the transform.* In order to include the impulse we will therefore integrate from $t = 0^-$ to $t = \infty$. The value of the impulse is zero except at $t = 0$. Hence, in performing the integration, the factor e^{-st} can be replaced by its value at $t = 0$, namely unity. Thus the integral becomes

$$F(s) = \int_{0-}^\infty u_0(t)\,dt. \tag{21–20}$$

Equation (21–20) tells us to find the area under a curve whose area is unity!

---

* Some authors integrate from $t = 0^+$, and then are forced to use very qualitative reasoning to bring the impulse back!

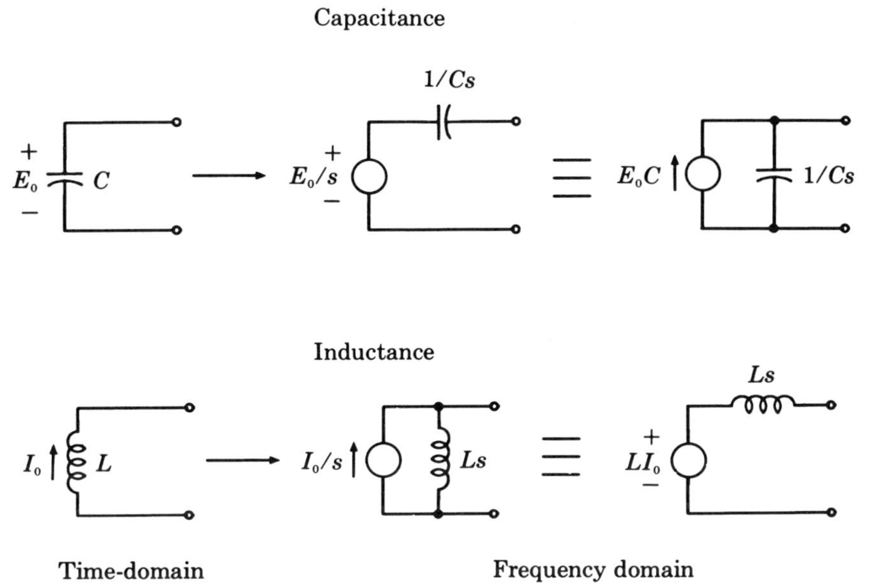

FIG. 21–10. Frequency-domain circuits for initial energy storage.

The obvious answer is unity, and as a result

$$F(s) = 1. \tag{21–21}$$

The meaning of Eq. (21–21) is that the frequency spectrum of a unit impulse is a constant value of unity. All frequencies are represented equally. A spectrum of this sort is called "white" from the analogous situation in the area of light where white light is a combination of all the other colors or frequencies. The importance of the impulse is perhaps better appreciated in terms of its spectrum than in terms of its time representation. It is the only function which contains all frequencies from zero to infinity in equal amounts. For this reason, it has been chosen as the basic singularity function and labelled $u_0(t)$.

Since the sources of Fig. 21–9 can be treated in the same manner as regular sources, we can transform them to the frequency domain along with the associated circuits. The frequency-domain circuits, including both sources and networks, are shown in Fig. 21–10. It is interesting to note that the two frequency-domain circuits in each case are related by Thévenin's theorem, or by Norton's theorem. For example, let us apply Norton's theorem to the first frequency-domain circuit for the charged capacitor. We must compute the short-circuit current and the impedance

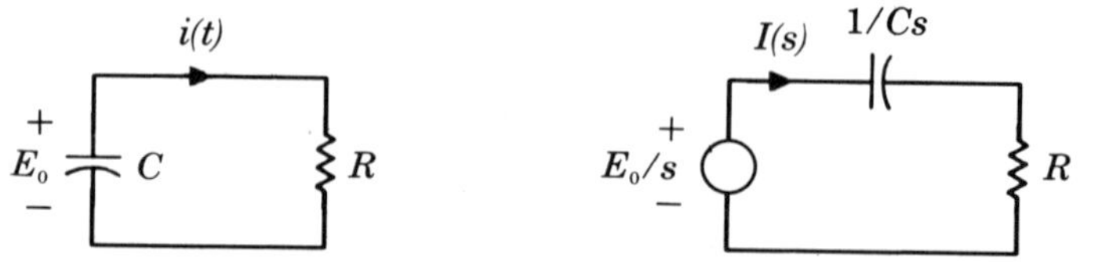

(a) Time-domain circuit (b) Frequency-domain circuit

FIG. 21–11. An R–C problem solved with the Laplace transform.

looking back into the network. The short-circuit current is

$$I_{sc} = \frac{E(s)}{Z(s)} = \frac{E_0}{s} \times \frac{1}{1/Cs} = E_0 C,$$

and the impedance looking back, with the voltage source shorted, is

$$Z_{eq} = \frac{1}{Cs} \cdot$$

The short-circuit current in parallel with the network impedance is identical to the network which we would have obtained if we had transformed the impulse form of the equivalent source. We will now solve two simple problems involving initial energy storage by means of the Laplace transform.

Example 1. A series R–C circuit has an initial voltage of E_0 on the capacitance. Find the time expression for the current for $t > 0$. The time-domain circuit is shown in Fig. 21–11(a) and the frequency-domain circuit in part (b). The current in the frequency-domain circuit, produced by the frequency-domain voltage source, is

$$\begin{aligned} I(s) &= \frac{E_0}{s} \times \frac{1}{R + 1/Cs} = \frac{E_0 C}{1 + RCs} \\ &= \frac{E_0}{R} \times \frac{1}{s + 1/RC} \cdot \end{aligned} \tag{21–22}$$

The current as a function of time is obtained as the inverse transformation of Eq. (21–22). The basic transform for $1/(s + a)$ is e^{-at}. Thus the answer to our problem is

$$i(t) = \frac{E_0}{R} e^{-t/RC} \qquad \text{for} \qquad t > 0.$$

Example 2. A series R–L circuit has an initial current of I_0 amperes in the inductance at $t = 0$. Find the current as a function of time for $t > 0$. The time-domain circuit is shown in Fig. 21–12(a) and the frequency-

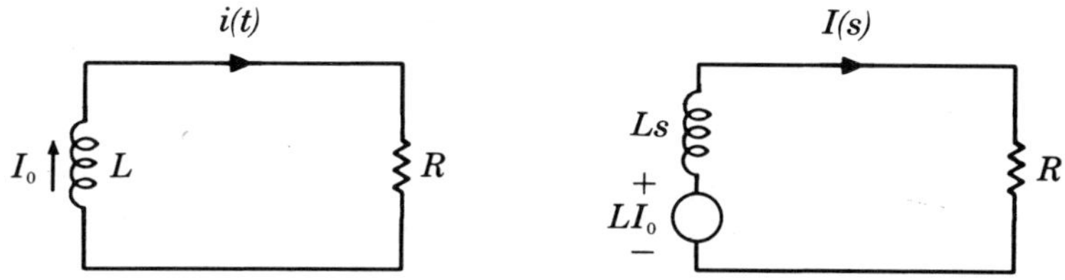

(a) Time-domain circuit (b) Frequency-domain circuit

FIG. 21–12. An R–L problem solved with the Laplace transform.

domain circuit in part (b). The current in the frequency-domain circuit is

$$I(s) = \frac{LI_0}{Ls + R} = \frac{I_0}{s + R/L}\cdot$$

Since the basic transform is the same as in the previous problem, we have for the current in the time domain

$$i(t) = I_0 e^{-Rt/L} \qquad \text{for} \qquad t > 0.$$

21–7 The method of partial fractions. Inverse Laplace transforms are almost always obtained from tables and not by direct integration. Since it is not practical to print tables containing all possible transforms, we must inquire into methods for extending the use of a small number of basic transforms. The principal tool which we will use is called the partial fraction expansion. The method of partial fractions is essentially the inverse operation of bringing terms over a common denominator. Like most inverse operations it is more difficult than the direct operation, but fortunately it can be carried out under rather general conditions. We will restrict our discussion to functions which can be expressed as the ratio of two polynomials in s. Usually, in circuit problems the polynomials can be factored into a finite number of factors in the numerator and the denominator. In a few special cases the number of terms may be infinite. The roots of both numerator and denominator can be repeated any number of times, although if a finite answer is expected they must repeat a finite number of times.

We will make the assumption that the function of s which we wish to transform into the time domain can be written as follows:

$$\begin{aligned} F(s) &= \frac{P(s)}{Q(s)} \\ &= \cdots + b_2 s^2 + b_1 s + b_0 + \frac{a_1}{s - s_1} + \frac{a_2}{s - s_2} + \cdots. \end{aligned} \tag{21–23}$$

This equation is assumed to be an identity, and therefore true for all values of s and for all similar operations performed on either side of the equation.* We will not prove this equation in general, but it is easy enough to verify in any given case. The terms involving the b's occurred because the numerator $P(s)$ was of higher order than the denominator. They can be removed by ordinary division until a remainder function is obtained with a numerator of lower order than the denominator. The partial fraction expansion involves the a's. If all the terms involving a are brought to a common denominator, the order of the numerator will be less than the denominator. Thus the partial fraction expansion is performed only for a function whose numerator is one degree less than the denominator, at most. In the discussion below we will assume that there are a finite number of roots in $Q(s)$. The corresponding factors are $(s - s_1)$, $(s - s_2)$, and so forth. The roots can be real or conjugate complex, and they can be separate and distinct or they can be repeated. As an example of the removal of higher-order powers from the numerator, consider the reduction

$$F(s) = \frac{s^3 + 2s^2 + 2s + 2}{s^2 + s + 1} = s + 1 + \frac{1}{s^2 + s + 1},$$

which results from the following division:

$$\begin{array}{r|l} & \underline{s + 1 \qquad\qquad\qquad} \\ s^2 + s + 1 & s^3 + 2s^2 + 2s + 2 \\ & \underline{s^3 + s^2 + s \qquad} \\ & \qquad s^2 + s + 2 \\ & \qquad \underline{s + s + 1} \\ & \qquad\qquad\qquad 1\ . \end{array}$$

The terms involving the b's in Eq. (21–23) are obtained by such division. The remainder function is used to obtain the a's. We will consider first the case of no repeated roots. No restriction is made as to whether the roots are real or complex.

Case 1. No repeated roots. If there are no repeated roots and if the numerator is of lower order than the denominator we can write $F(s)$ as

$$F(s) = \frac{P(s)}{Q(s)} = \frac{a_1}{s - s_1} + \frac{a_2}{s - s_2} + \cdots \frac{a_n}{s - s_n}. \tag{21–24}$$

Since Eq. (21–24) is an identity, it is true for all values of s. We could select n values at random and by substituting them into the equation obtain n independent simultaneous equations for the a's. A much simpler way, however, is to evaluate each term by itself in the neighborhood of the point

* A rigorous justification of Eq. (21–23) requires some knowledge of functions of a complex variable and the calculus of residues.

where the function becomes infinite. Near $s = s_1$ for example, both sides of Eq. (21–24) become infinite. The only term which contributes much on the right-hand side is the first term. Thus in the neighborhood of $s = s_1$, $F(s)$ is approximately equal to $a_1/(s - s_1)$. To exploit this fact we multiply both sides by $(s - s_1)$, and obtain

$$\frac{P(s)(s - s_1)}{Q(s)} = a_1 + (s - s_1)\left(\frac{a_2}{s - s_2} + \cdots \frac{a_n}{s - s_n}\right). \tag{21–25}$$

It is clear that if we then let $s = s_1$ in Eq. (21–25), all the terms on the right side will be zero with the exception of a_1, which can then be written in the form

$$a_1 = \left[\frac{P(s)(s - s_1)}{Q(s)}\right]_{s=s_1}.$$

Similarly,

$$a_n = \left[\frac{P(s)(s - s_n)}{Q(s)}\right]_{s=s_n}. \tag{21–26}$$

Example. As an example of the use of Eq. (21–26) we will expand the following function $F(s)$ into partial fractions:

$$F(s) = \frac{s}{(s+1)(s+2)} = \frac{a_1}{s+1} + \frac{a_2}{s+2},$$

$$a_1 = \left[\frac{s}{s+2}\right]_{s=-1} = \frac{-1}{1} = -1,$$

$$a_2 = \left[\frac{s}{s+1}\right]_{s=-2} = \frac{-2}{-1} = 2.$$

Hence

$$F(s) = \frac{-1}{s+1} + \frac{2}{s+2}.$$

As a check

$$F(s) = \frac{-1(s+2) + 2(s+1)}{(s+1)(s+2)} = \frac{s}{(s+1)(s+2)}.$$

The real proof of the partial fraction expansion is the fact that the terms can be recombined to give the original function. In fact, any way of obtaining the terms is legitimate, so long as they recombine to give the original function. In simple cases, guessing the answer is often the fastest approach!

Case 2. Repeated roots. Repeated roots are never possible in a real physical problem. No two things in nature are ever quite identical. In our idealized networks, however, we may find it convenient to postulate identical roots. They can arise if we drive a circuit at exactly its resonant frequency, or if we cascade two identical networks with an ideal vacuum tube or transistor between them. We will first consider repeated roots with no other roots. Later we will extend the argument to include both repeated and nonrepeated roots. For a root $s = s_1$ which occurs k times, we write

$$F(s) = \frac{P(s)}{(s - s_1)^k} = \frac{K_1}{(s - s_1)^k} + \frac{K_2}{(s - s_1)^{k-1}} + \cdots \frac{K_k}{(s - s_1)}. \quad (21\text{–}27)$$

There are other forms in which Eq. (21–27) can be written, but the form we have selected is convenient for most Laplace transforms. When the right-hand side of Eq. (21–27) is brought to a common denominator, the numerator will be a polynomial containing the k constants $K_1, K_2, \ldots K_k$. These terms can be used to match any polynomial $P(s)$. An alternative procedure for finding the values of the K's follows the general method of Eq. (21–26). The highest-order term is easy to find. Both sides of the equation are multiplied by $(s - s_1)^k$, giving

$$P(s) = K_1 + K_2(s - s_1) + \cdots K_k(s - s_1)^{k-1}. \quad (21\text{–}28)$$

The substitution of $s = s_1$ in Eq. (21–28) gives

$$K_1 = \left[\frac{P(s)}{(s - s_1)^k}(s - s_1)^k\right]_{s=s_1}.$$

Obviously some other trick must be used to obtain the other terms. We might subtract the term involving K_1, and repeat the process with the remainder. Another procedure is to differentiate both sides of Eq. (21–28), giving

$$\frac{d}{ds}P(s) = 0 + K_2 + 2(s - s_1)K_3 + \cdots K_k(k - 1)(s - s_1)^{k-2}.$$

Now the substitution of $s = s_1$ gives

$$K_2 = \left[\frac{d}{ds}\left(\frac{P(s)(s - s_1)^k}{(s - s_1)^k}\right)\right]_{s=s_1}.$$

Since the process can be repeated, the next term is

$$K_3 = \frac{1}{2!}\left[\frac{d^2}{ds^2}\left(\frac{P(s)(s - s_1)^k}{(s - s_1)^k}\right)\right]_{s=s_1},$$

and

$$K_r = \frac{1}{(r - 1)!}\left[\frac{d^{r-1}}{ds^{r-1}}\left(\frac{P(s)(s - s_1)^k}{(s - s_1)^k}\right)\right]_{s=s_1}.$$

The general term is useful in theoretical work, but in a numerical example it is easier to follow the process by which we arrived at the general term.

Example 1. As an example of repeated roots, consider the function

$$F(s) = \frac{s^2}{(s+1)^3} = \frac{K_1}{(s+1)^3} + \frac{K_2}{(s+1)^2} + \frac{K_3}{(s+1)}.$$

We first multiply both sides of the equation by $(s+1)^3$, obtaining

$$s^2 = K_1 + (s+1)K_2 + (s+1)^2K_3.$$

At $s = -1$

$$K_1 = [s^2]_{s=-1} = (-1^2) = 1.$$

By differentiation

$$2s = 0 + K_2 + 2(s+1)K_3.$$

At $s = -1$

$$K_2 = [2s]_{s=-1} = -2.$$

A second differentiation gives

$$2 = 0 + 0 + 2K_3,$$

from which

$$K_3 = 1.$$

In summary

$$\frac{s^2}{(s+1)^3} = \frac{1}{(s+1)^3} + \frac{-2}{(s+1)^2} + \frac{1}{(s+1)},$$

as can be verified easily by recombining the terms.

Example 2. The same process can be used when there are both repeated and nonrepeated roots. Thus

$$F(s) = \frac{s}{(s+2)(s+1)^2} = \frac{a_1}{(s+2)} + \frac{K_1}{(s+1)^2} + \frac{K_2}{(s+1)},$$

where

$$a_1 = \left[\frac{s}{(s+1)^2}\right]_{s=-2} = \frac{-2}{(-1)^2} = -2.$$

Multiplication by the highest power of the repeated root gives

$$\frac{s}{s+2} = \frac{a_1}{s+2}(s+1)^2 + K_1 + (s+1)K_2,$$

from which

$$K_1 = \left[\frac{s}{s+2}\right]_{s=-1} = \frac{-1}{+1} = -1.$$

By differentiation

$$\frac{(s+2)-s}{(s+2)^2} = (s+1)(\text{function of } s) + 0 + K_2,$$

from which

$$K_2 = \left[\frac{2}{(s+2)^2}\right]_{s=-1} = \frac{2}{1^2} = 2.$$

Thus

$$\frac{s}{(s+2)(s+1)^2} = \frac{-2}{s+2} - \frac{1}{(s+1)^2} + \frac{2}{s+1}.$$

Note that all the terms involving the nonrepeated roots will be zero after the substitutions, and it is not necessary actually to perform the differentiation of these terms.

21–8 Inverse transforms by the method of partial fractions. By the method of partial fractions, we can reduce a complicated function of s to the sum of a number of simple terms. These terms are of a relatively few simple types, and it is possible to catalog them in a rather short table of transforms. In fact the table is so short that it is easily committed to memory, with the result that we are then independent of a table for the usual types of functions found in network problems. The basic transforms which occur in partial fraction expansions are given in Table 21–1. The positive powers of s in the table are used to transform the b terms in Eq. (21–23). They correspond to the singularity functions in the time domain. The constant term in the frequency domain is analogous to the unit impulse in the time domain. The power of s corresponds to the order of the singularity function. Each time we multiply by s in the frequency domain, we differentiate in the time domain. Similarly, the negative powers of s correspond to the negative-exponent singularity functions. Each time we divide by s in the frequency domain we integrate the function in the time domain.

The basic transform of the partial fraction part of Eq. 21–23 is represented by $1/(s+a)$ when there are no repeated roots, and its value in the time domain is e^{-at}. This transform occurs most frequently, and can be used to transform the majority of the terms which occur in practical problems. When there are repeated roots in the transform, the time function is multiplied by a power of t which is one less than the number of repeated roots, divided by the factorial of the same power.

TABLE 21–1

LAPLACE TRANSFORMS USED WITH PARTIAL FRACTION EXPANSIONS

$F(s)$	$f(t)$
$\frac{1}{s}$	$u_{-1}(t)$
1	$u_0(t)$
s	$u_1(t)$
s^n	$u_n(t)$
$\frac{1}{s^2}$	$u_{-2}(t) = t$
s^{-n}	$u_{-n}(t) = \frac{t^{n-1}}{(n-1)!}$
$\frac{1}{s+a}$	e^{-at}
$\frac{1}{(s+a)^2}$	$e^{-at}(t)$
$\frac{1}{(s+a)^n}$	$e^{-at}\left[\frac{t^{n-1}}{(n-1)!}\right]$

Example 1. Two simple poles on the negative real axis. The first example we will consider of the inverse transformation of a function by means of partial fractions will be the simple case of two roots on the negative real axis. The function is, from the previous section,

$$F(s) = \frac{s}{(s+1)(s+2)} = \frac{-1}{s+1} + \frac{2}{s+2}.$$

The corresponding time function is

$$f(t) = -e^{-t} + 2e^{-2t}.$$

Example 2. The special case when one of the roots is at the origin. For this example we will consider the function

$$F(s) = \frac{1}{s(s+2)} = \frac{a_1}{s} + \frac{a_2}{s+2},$$

where

$$a_1 = \left[\frac{1}{s+2}\right]_{s=0} = \tfrac{1}{2},$$

$$a_2 = \left[\frac{1}{s}\right]_{s=-2} = -\tfrac{1}{2}.$$

Thus
$$F(s) = \frac{\frac{1}{2}}{s} + \frac{-\frac{1}{2}}{s+2},$$

and the corresponding time function is

$$f(t) = \tfrac{1}{2}(1 - e^{-2t}) \qquad \text{for} \qquad t > 0.$$

Since Laplace transforms are defined only for $t > 0$, it is understood that the time functions are zero for negative t without explicitly stating the fact in each equation.

Example 3. Imaginary roots. A simple case involving imaginary roots is

$$F(s) = \frac{1}{s^2+1} = \frac{a_1}{s-j} + \frac{a_2}{s+j},$$

where

$$a_1 = \left[\frac{1}{s+j}\right]_{s=j} = \frac{1}{2j},$$

$$a_2 = \left[\frac{1}{s-j}\right]_{s=-j} = \frac{-1}{2j}.$$

We can write

$$F(s) = \frac{1/2j}{s-j} + \frac{-1/2j}{s+j}.$$

The corresponding time function is

$$f(t) = \frac{e^{jt} - e^{-jt}}{2j} = \sin t.$$

An alternative way to handle the case of complex roots and imaginary roots is to notice that the coefficients of the complex conjugate root must be complex conjugates. The expressions are identical except for a change in the sign of all the j's. Hence it is not necessary to find the second term; it will always be the complex conjugate of the first term. Furthermore, the corresponding time functions are also complex conjugates. If we now recall that the sum of a number and its complex conjugate is two times the real part of the number, we can obtain $f(t)$ from the time function for the first factor (the one with the negative imaginary part, or the positive time exponent). Thus in the above problem

$$f(t) = 2\,\mathrm{Re}\left[\frac{e^{jt}}{2j}\right] = \sin t.$$

This second method is preferable when there is a phase angle in the coefficient a_1, because it saves a good deal of complex algebra.

Example 4. Complex conjugate roots. The easiest method of handling complex conjugate roots is an extension of the method just discussed. For example,

$$F(s) = \frac{1}{s^2+s+1} = \frac{1}{(s+0.5+j0.866)(s+0.5-j0.866)}$$
$$= \frac{a_1}{s+0.5-j0.866} + \frac{a_2}{s+0.5+j0.866}.$$

The constants a_1 and a_2 are complex conjugates, as we can see by writing

$$a_1 = \left[\frac{1}{(s+0.5+j0.866)}\right]_{s=-0.5+j0.866} = \frac{1}{2j0.866},$$

$$a_2 = \left[\frac{1}{(s+0.5-j0.866)}\right]_{s=-0.5-j0.866} = \frac{-1}{2j0.866}.$$

The time functions are also complex conjugates, having the form

$$f(t) = \frac{1}{2j0.866}e^{-0.5t}e^{j0.866t} + \frac{-1}{2j0.866}e^{-0.5t}e^{-j0.866t}.$$

Thus the function $f(t)$ can be obtained as twice the real part of the first term. That is, we may write

$$f(t) = 2\,\mathrm{Re}\left[\frac{1}{2j0.866}e^{-0.5t}e^{j0.866t}\right] = 1.15e^{-0.5t}\sin 0.866t.$$

It can be seen from the example that it is not even necessary to find a_2.

21–9 Summary.

Fourier series and networks

$$f_1(t) = a_0 + a_1\cos(\omega t+\theta_1) + a_2\cos(2\omega t+\theta_2) + \cdots$$

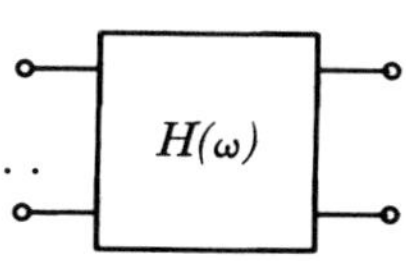

$$f_2(t) = b_0 + b_1\cos(\omega t+\phi_1) + b_2\cos(2\omega t+\phi_2) + \cdots$$

Fourier integral excitation

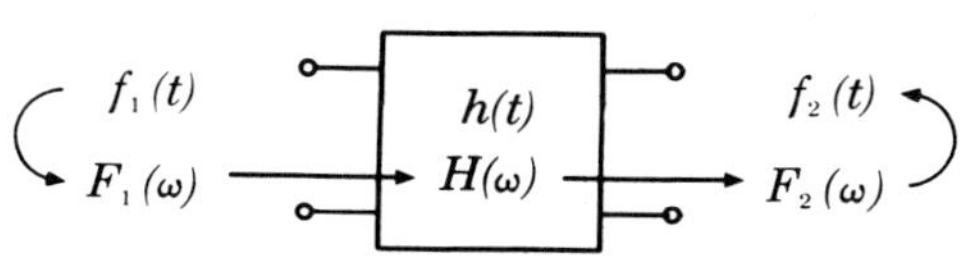

Fourier sine and cosine transforms

$$\text{When } f(t) = 0 \text{ for } t < 0$$

$$F(\omega) = \int_0^\infty f(t) \cos \omega t \, dt - j \int_0^\infty f(t) \sin \omega t \, dt$$

$$= R(\omega) + jI(\omega)$$

$$f(t) = \frac{2}{\pi} \int_0^\infty R(\omega) \cos \omega t \, d\omega$$

$$= \frac{-2}{\pi} \int_0^\infty I(\omega) \sin \omega t \, d\omega$$

Laplace transform excitation

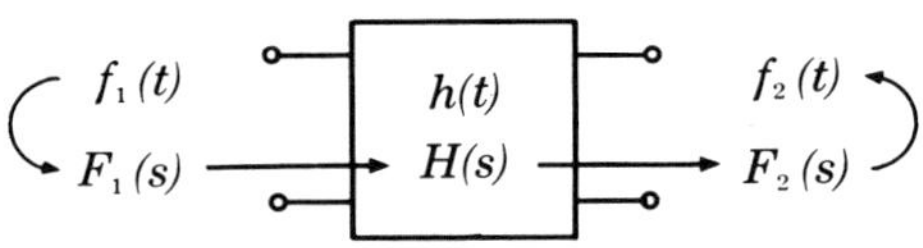

Initial energy storage

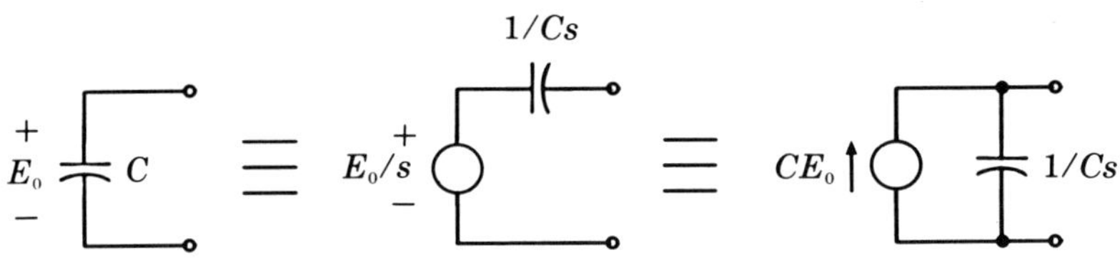

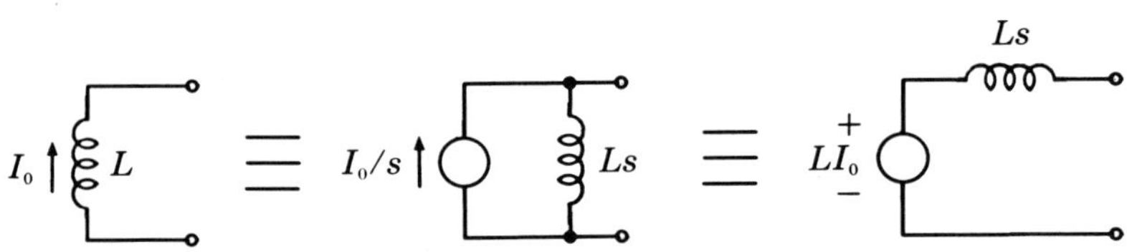

Partial fractions

(a) *No repeated roots*

$$\frac{P(s)}{Q(s)} = \frac{a_1}{s - s_1} + \frac{s_2}{s - s_2} + \cdots$$

$$a_1 = \left[\frac{P(s)}{Q(s)} (s - s_1)\right]_{s=s_1} \text{ etc.}$$

(b) *Repeated roots*

$$\frac{P(s)}{(s-s_1)^k} = \frac{K_1}{(s-s_1)^k} + \frac{K_2}{(s-s_1)^{k-1}} + \cdots + \frac{K_k}{(s-s_1)}$$

$$K_1 = \left[\frac{P(s)}{(s-s_1)^k}(s-s_1)^k\right]_{s=s_1}$$

$$K_r = \frac{1}{(r-1)!}\left[\frac{d^{r-1}}{ds^{r-1}}\left(\frac{P(s)(s-s_1)^k}{(s-s_1)^k}\right)\right]_{s=s_1}$$

Repeated root transforms

$F(s)$	$f(t)$
s^n	$u_n(t)$
$1/s^n$	$u_{-n}(t) = \dfrac{t^{n-1}}{(n-1)!}$
$1/(s+\alpha)^n$	$e^{-\alpha t}\left[\dfrac{t^{n-1}}{(n-1)!}\right]$
$\dfrac{1}{(s+\alpha \pm j\beta)^n}$	$e^{-\alpha t}e^{\mp j\beta t}\left[\dfrac{t^{n-1}}{(n-1)!}\right]$

REFERENCES

In Chapter 20 complex signals are resolved into sinusoidal components. In Chapter 21 network responses to such signals are obtained. In the end we solve transient problems by frequency-domain methods. A number of excellent texts are devoted entirely to the solution of transient problems by Fourier or Laplace methods.

1. O. Heaviside, *Electromagnetic Theory,* Dover Publications, New York, 1950. The original Heaviside work is available in this excellent reprint. In addition to lumped-element problems, Heaviside applied his methods to distributed-parameter systems, such as cables and transmission lines.

2. Gardner and Barnes, *Transients in Linear Systems,* John Wiley & Sons, New York, 1942. Gardner and Barnes apply the Laplace transform to the solution of transient problems in electrical and mechanical systems.

3. M. E. Van Valkenburg, *Network Analysis,* Prentice-Hall, New Jersey, 1955. Van Valkenburg presents essentially the same material as Gardner and Barnes at a level more suitable for undergraduate students.

4. D. K. Cheng, *Analysis of Linear Systems,* Addison-Wesley, Reading, Mass., 1959. Cheng gives a slightly more modern treatment of the ground covered by Gardner and Barnes. The Laplace transform is the basic tool used, but a heuristic development of the Laplace transform from Fourier series and Fourier integrals is included.

5. R. V. Churchill, *Modern Operational Methods in Engineering,* McGraw-Hill, New York, 1951. Churchill covers transform methods in detail, from the point of view of an applied mathematician.

6. S. Goldman, *Transformation Calculus and Electrical Transients,* Prentice-Hall, New Jersey, 1949. The methods of Laplace transforms are applied to lumped-parameter circuits and distributed-parameter circuits. The relationship of the methods to the calculus of residues is covered in some detail.

7. E. A. Guillemin, *The Mathematics of Circuit Analysis,* John Wiley & Sons, New York, 1948. This text represents a good summary of the Fourier methods, as opposed to the Laplace methods. In particular, pulses which exist for both positive and negative time are considered as excitations for a network.

8. J. Truxal, *Control System Synthesis,* McGraw-Hill, New York, 1955. Truxal discusses the two-sided Laplace transform which can be used for both positive and negative time.

9. Van der Pol and Bremmer, *Operational Calculus Based on the Two-sided Laplace Integral,* Cambridge University Press, London, 1950. This text is the original presentation of the two-sided Laplace transform, which applies to pulses defined for both positive and negative time. It is very readable and contains many examples.

10. Brenner and Javid, *Analysis of Electric Circuits,* McGraw-Hill, New York, 1959. The Brenner and Javid book is an example of a modern text which presents the Heaviside operational approach to the solution of transient problems.

EXERCISES*

Section 21–2

1. In the circuit of Fig. 21–13 the steady-state square wave has been applied for some time. Write the Fourier series for the input voltage and solve for the nth-harmonic current phasor. Write the Fourier series for the current.

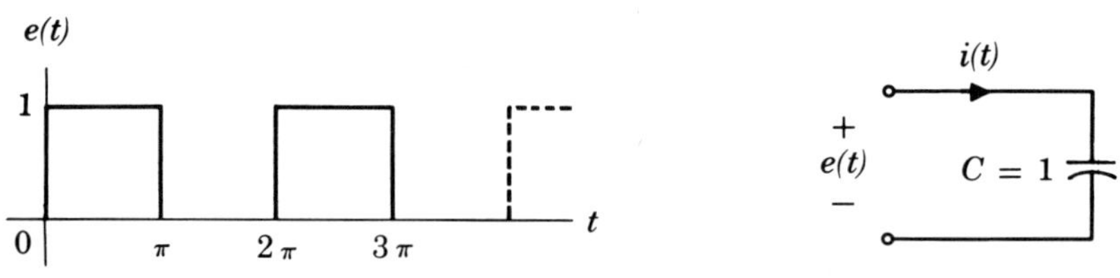

Figure 21–13

2. Write the Fourier series for the input-voltage waveform of Fig. 21–14. Obtain the current phasor for the nth harmonic and write the Fourier series for the current.

---

* Unless otherwise specified, numerical values are in ohms, henrys, and farads.

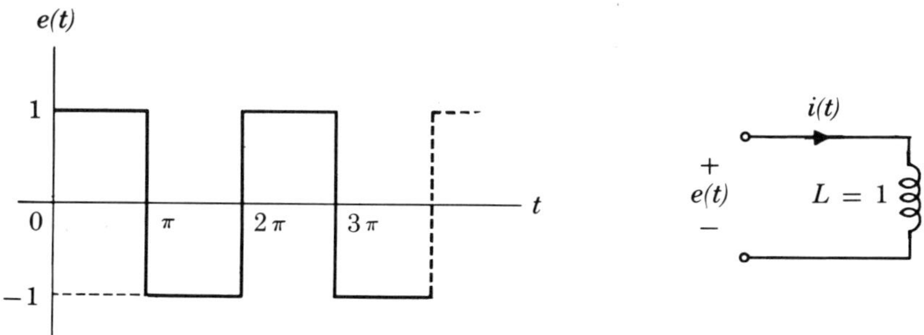

FIGURE 21–14

3. For the circuit of Fig. 21–15 give the nth-harmonic current phasor, the nth-harmonic voltage phasor, and the Fourier series for the voltage.

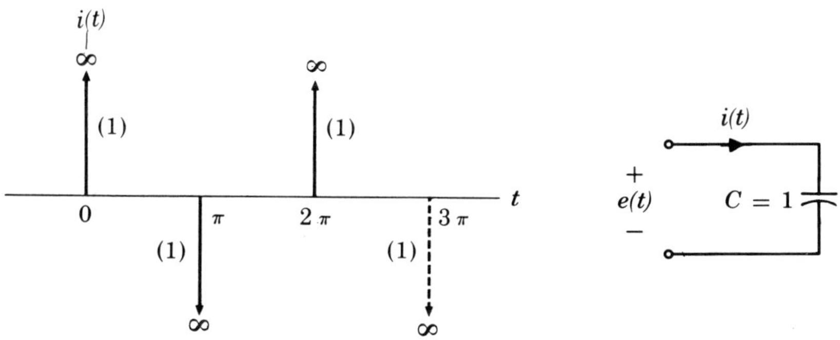

FIGURE 21–15

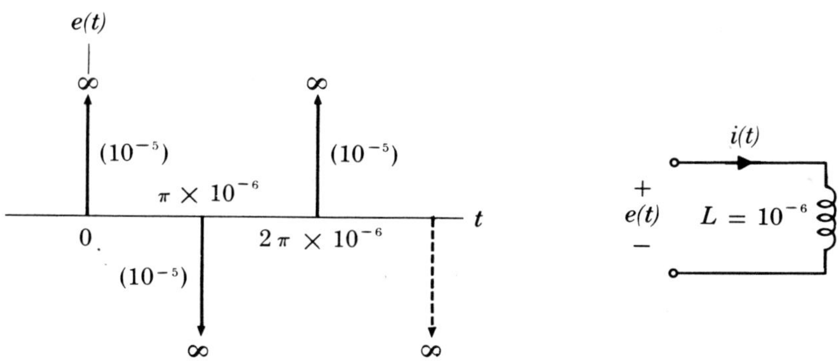

FIGURE 21–16

4. Normalize the problem in Fig. 21–16 so that the period of the waveform is $T = 2\pi$. (Do not forget to multiply the area of the impulse by the appropriate

normalizing factor.) Use Fourier methods and obtain the current which flows in the circuit in the form of a Fourier series with the correct frequency reinserted.

5. In the circuit of Fig. 21–17 resolve the input into Fourier components and find the output Fourier components by multiplying by the appropriate impedance. Write the Fourier series for the output $e(t)$.

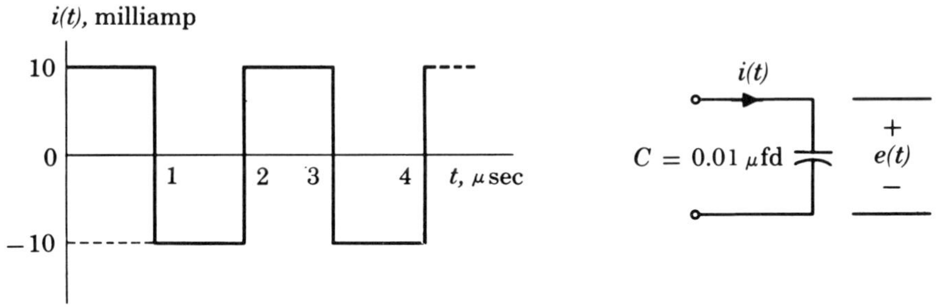

FIGURE 21–17

6. In the circuit of Fig. 21–18 resolve the input voltage into its harmonics. Give the input voltage phasor and the output voltage phasor for the nth harmonic. Write the first few terms for the Fourier series representing the output voltage.

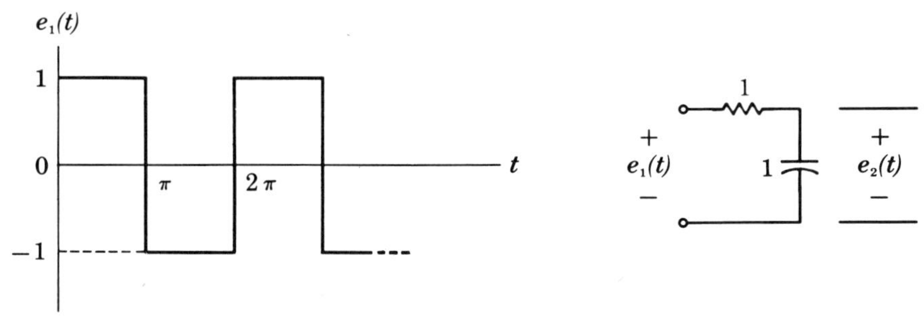

FIGURE 21–18

7. In order to produce direct-voltage from a sinusoid, the sinusoid is rectified and the resulting waveform is filtered. The circuit in Fig. 21–19 is a possible filter for the half-wave rectified sine wave. Compute the d-c component of the voltage $e_2(t)$ and the amplitude of the fundamental component of the a-c voltage. From these obtain the percentage ripple in the output d-c voltage.

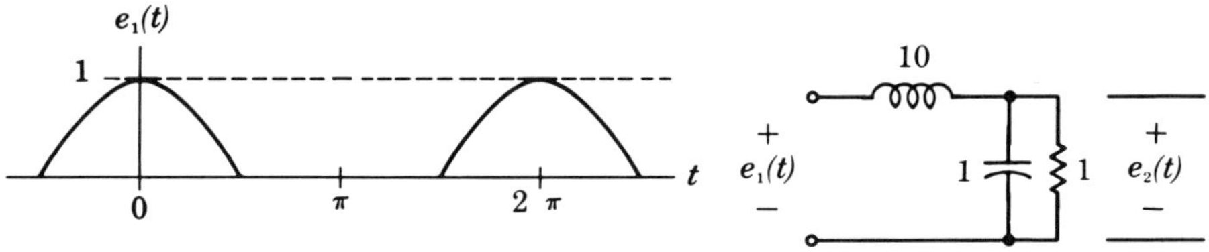

FIGURE 21–19

8. The circuit of Fig. 21–20 is designed to filter the alternating components out of the input rectified sine wave and to allow only the direct voltage to appear at the output, $e_2(t)$. Compute the ratio of the first a-c component of $e_2(t)$ to the d-c component of $e_2(t)$.

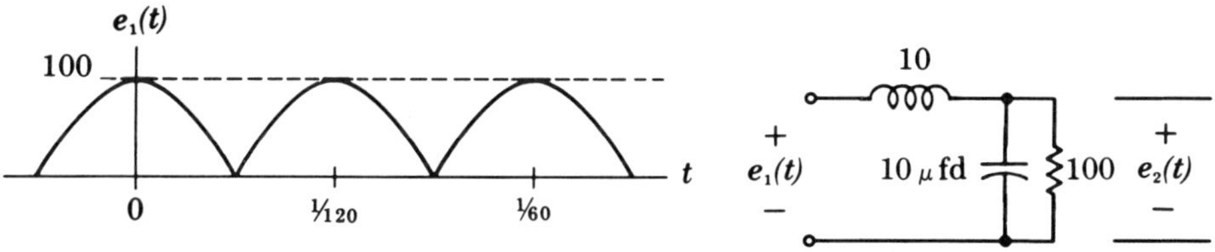

FIGURE 21–20

9. Repeat Exercise 8 if alternate half-cycles of the input rectified sine wave are missing.

Section 21–3

10. In the circuit of Fig. 21–21 the stimulus is a step-function voltage and the response is the current.

(a) Obtain the Fourier transform of the input, $E(\omega)$, by finding the transform of $e(t) = e^{-at}$, and then letting a approach zero.
(b) Find $H(\omega)$, the network transfer function in the frequency domain.
(c) Find the Fourier transform for the current, $I(\omega)$.
(d) Use Table 20–4, with $s = j\omega$, to find the $i(t)$ corresponding to $I(\omega)$.

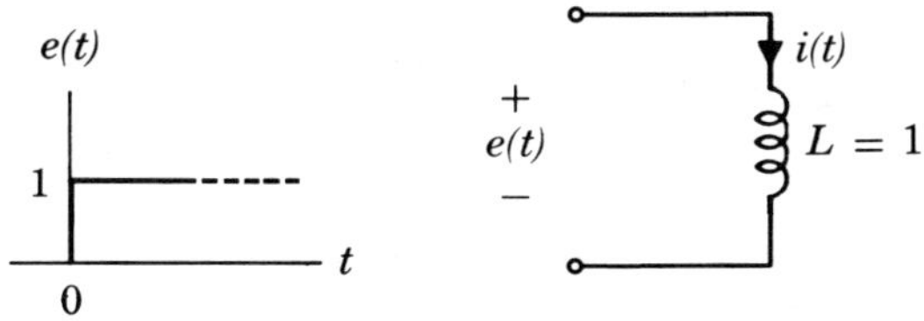

FIGURE 21–21

11. (a) Find the Fourier transform of a pulse of height $1/\delta$ and width δ, centered at the time origin. (b) Let δ approach zero and thus find the transform of an impulse.

12. The circuit of Fig. 21–22 is excited by an impulse voltage. The desired response is the current.

(a) Find the Fourier transform $E(\omega)$.
(b) Find the network transfer function $H(\omega)$.
(c) Find the transform of the current, $I(\omega)$.
(d) Evaluate $i(t)$ by means of Table 20–4, with $s = j\omega$.

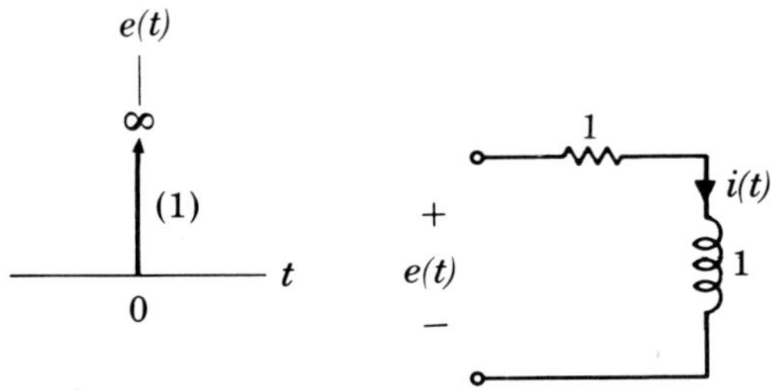

FIGURE 21–22

13. The input stimulus to the circuit of Fig. 21–23 is an exponentially decaying current sectioned at $t = 0$. The response is the voltage. (a) Find $I(\omega)$. (b) Find $H(\omega)$. (c) Find $E(\omega)$. (d) Find $e(t)$ by means of the table in Section 20–8.

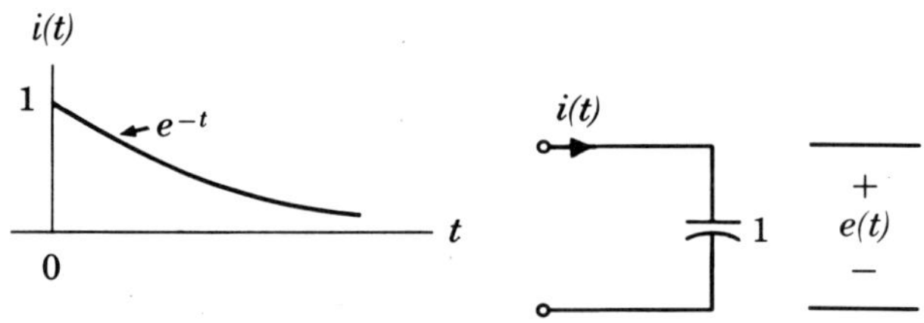

FIGURE 21–23

14. The input stimulus to the circuit of Fig. 21–24 is a step-function current. (a) Find $I(\omega)$. (b) Find $H(\omega)$ for the voltage response. (c) Find $E(\omega)$. (d) Find $e(t)$ by means of Table 20–4.

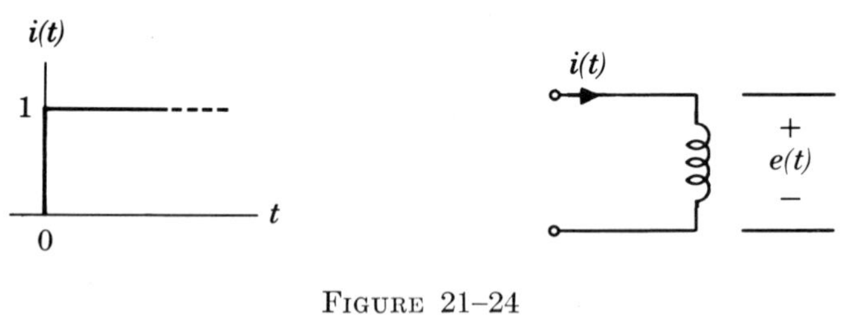

FIGURE 21–24

15. In the circuit of Fig. 21–25 the current is the excitation and the voltage is the desired response. The current is given by

$$i(t) = (e^{-t} - e^{-10t})u_{-1}(t).$$

(a) Find the Fourier transform $I(\omega)$. (b) Give $H(\omega)$. (c) Find $E(\omega)$. (d) Evaluate $e(t)$ by means of Table 20–4, with $s = j\omega$.

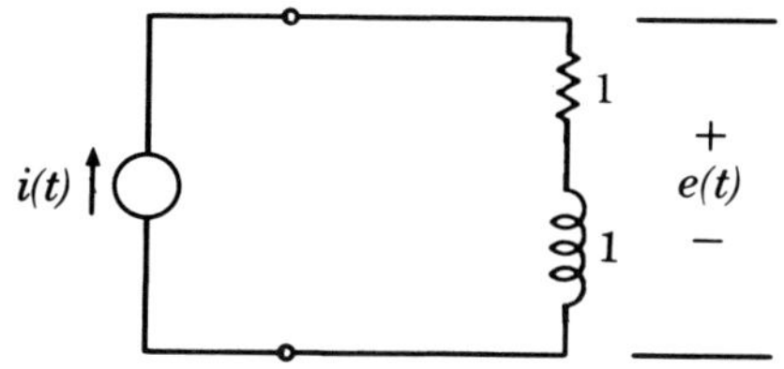

FIGURE 21–25

16. In the circuit of Fig. 21–26 the input stimulus is the voltage and the desired response is the current. (a) Give $E(\omega)$. (b) Give $H(\omega)$. (c) Give $I(\omega)$. (d) Find $i(t)$ by means of Table 20–4.

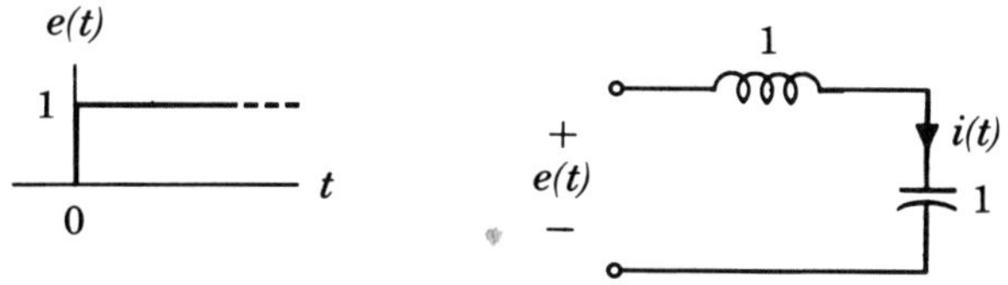

FIGURE 21–26

17. Fourier transforms can represent functions which exist over all time. Laplace transforms represent functions only for $t > 0$. The input voltage in the circuit of Fig. 21–27 is defined by

$$e(t) = \begin{cases} e^{t} & \text{for } t < 0, \\ e^{-t} & \text{for } t > 0. \end{cases}$$

(a) Find the Fourier transform $E(\omega)$ of $e(t)$. (b) Find the network transfer function $H(\omega)$. (c) Find the response function $I(\omega)$. (d) Obtain $i(t)$ from the Laplace transform table, with $s = j\omega$.

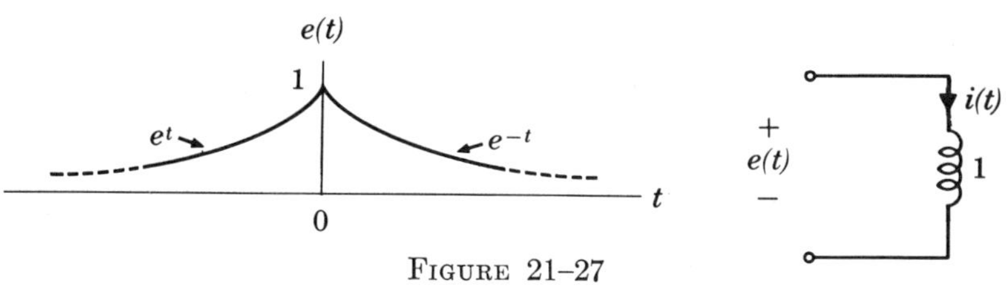

FIGURE 21–27

18. For the circuit of Fig. 21–28 sketch the magnitude and phase curves for the input voltage spectrum and the output voltage spectrum.

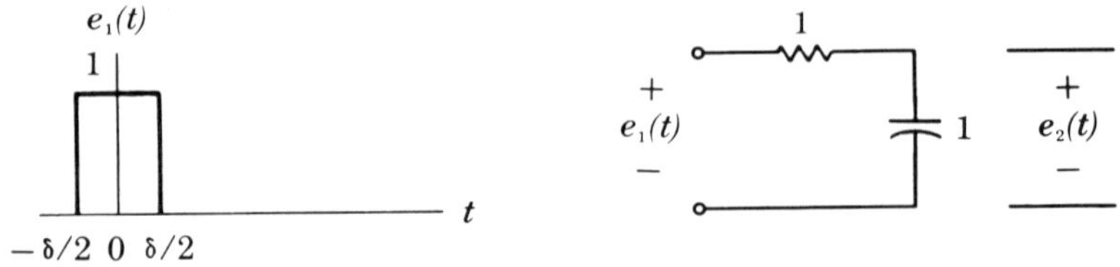

FIGURE 21–28

Section 21–4

19. For functions which are zero for $t < 0$ the Fourier transforms can be written as

$$F(\omega) = \int_0^\infty f(t) \cos \omega t \, dt - j \int_0^\infty f(t) \sin \omega t \, dt,$$

$$f(t) = \frac{2}{\pi} \int_0^\infty R(\omega) \cos \omega t \, d\omega = \frac{-2}{\pi} \int_0^\infty I(\omega) \sin \omega t \, d\omega.$$

Show that the same formulas hold for the Laplace transforms evaluated at $s = j\omega$.

20. For the function $f(t) = e^{-t} \cdot u_{-1}(t)$, find the Fourier transform $F(\omega)$ by using the formula of Exercise 19. Check the result against the Laplace transform of the function as listed in Table 20–4.

21. Using the formula of Exercise 19, find the Fourier transform $F(\omega)$ of the time function

$$f(t) = \sin t \cdot u_{-1}(t).$$

Check the result by putting $s = j\omega$ in the appropriate transform in Table 20–4.

22. Find the Fourier transforms for each of the functions shown in Fig. 21–29. Show that the time functions in parts (b) and (c) are the even and odd parts of the time function in part (a), and that the sum of the Fourier transforms for parts (b) and (c) gives the transform of part (a).

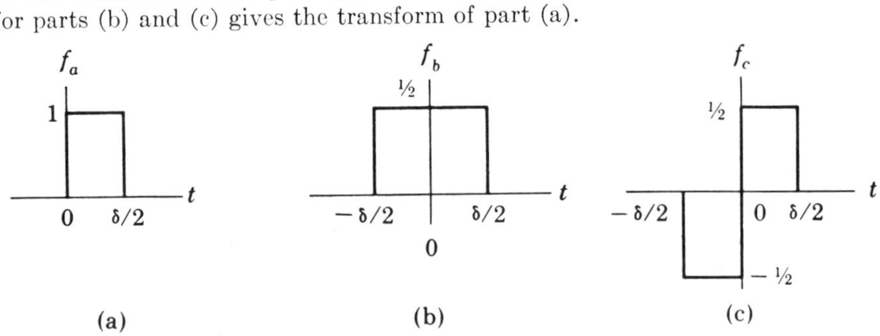

FIGURE 21–29

23. Show that for an even function the Fourier transform will be a pure real number. Evaluate the transform of the function shown in Fig. 21–30.

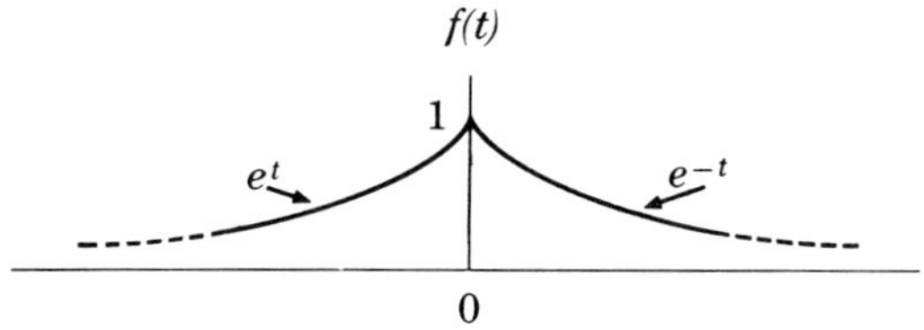

FIGURE 21–30

24. Show that for an odd function of time the Fourier transform is a pure imaginary quantity. Find the Fourier transform for the function shown in Fig. 21–31.

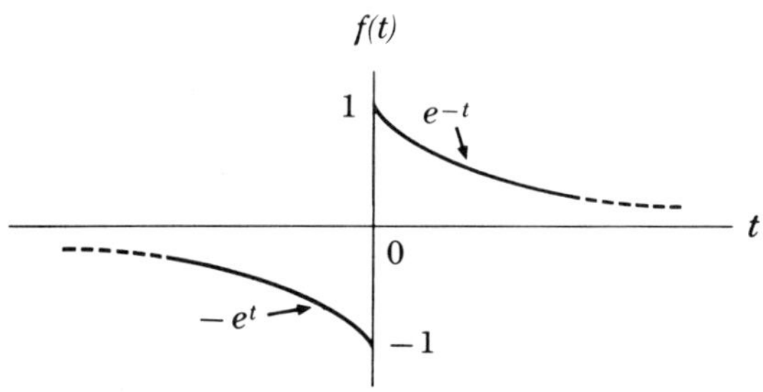

FIGURE 21–31

25. What time function is the sum of the even and odd functions in Exercises 23 and 24? What is its Fourier transform?

26. A Fourier transform is given by $F(\omega) = 1/(1 + j\omega)$. Find the corresponding $f(t)$ by (a) integrating the real part of $F(\omega)$ and (b) by integrating the imaginary part of $F(\omega)$.

27. A Fourier transform is given by $F(\omega) = 1/(1 + \omega^2)$. Show that the corresponding time function is an even function of t, and find an analytic expression for it. Sketch the function.

28. A Fourier transform is given by $F(\omega) = -j/(1 + \omega^2)$. Show that the corresponding time function is an odd function of t, and find an analytic expression for it. Sketch the function.

29. The singularity functions pose some problems if we wish to deal with both the real function of ω and the imaginary function of ω. The function $F(\omega) = 1/j\omega$ has no apparent real part. This function is really, however,

$$F(\omega) = \left[\frac{1}{a + j\omega}\right]_{a \to 0}$$

in order to satisfy the convergence requirements of the Fourier integral. Find the real part of this latter expression and show that the time function can be obtained from this real part. (Do not let $a \to 0$ until the final time function has been obtained.)

30. Show that the functions

$$f(t) = e^{-t^2/2} \qquad \text{and} \qquad F(\omega) = \sqrt{2\pi}\, e^{-\omega^2/2}$$

are a Fourier-transform pair if both functions extend from $-\infty$ to $+\infty$.

Section 21–5

31. Translate the circuit of Fig. 21–32 to the complex-frequency domain and solve for the current $i(t)$ by means of the Laplace transform.

32. Translate the circuit of Fig. 21–33 to the complex-frequency domain and solve for the current $i(t)$ by means of the Laplace transform.

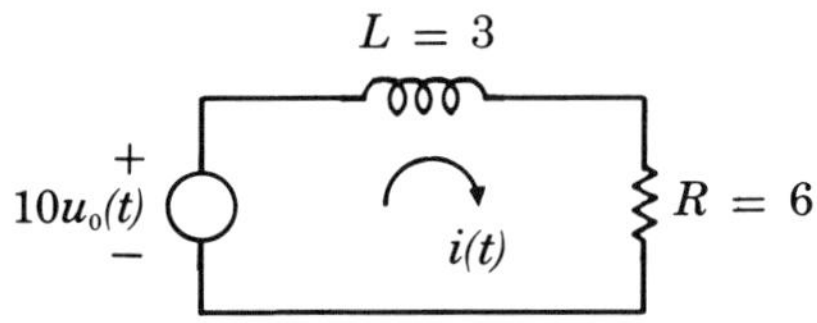

FIGURE 21–32

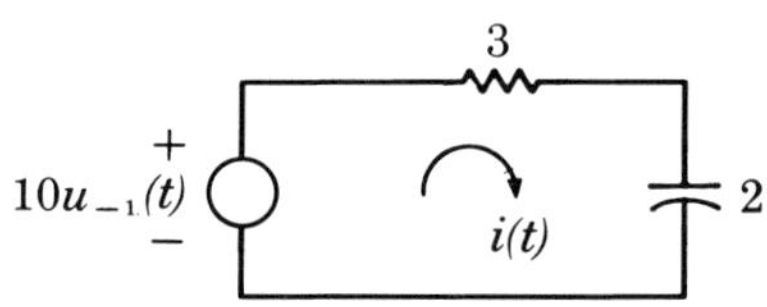

FIGURE 21–33

33. Translate the circuit of Fig. 21–34 to the complex-frequency domain and solve for the voltage $e(t)$ by means of the Laplace transform.

34. Translate the circuit of Fig. 21–35 to the complex-frequency domain and solve for the current $i(t)$ by means of the Laplace transform.

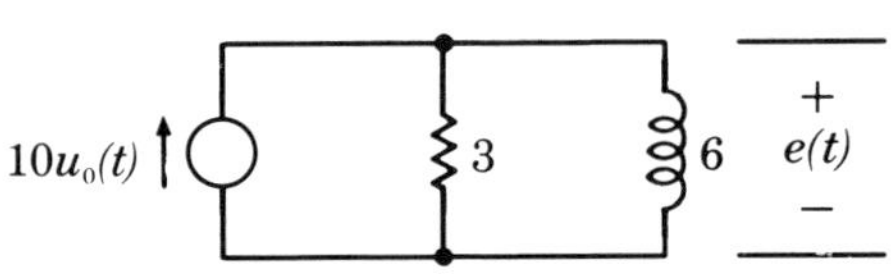

FIGURE 21–34

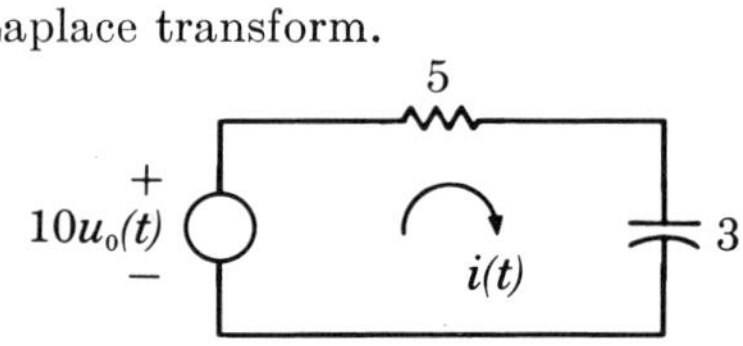

FIGURE 21–35

35. Translate the circuit of Fig. 21–36 to the complex-frequency domain and solve for the current $i(t)$ by means of the Laplace transform.

36. Translate the circuit of Fig. 21–37 to the complex-frequency domain and solve for the voltage $e(t)$ by means of the Laplace transform.

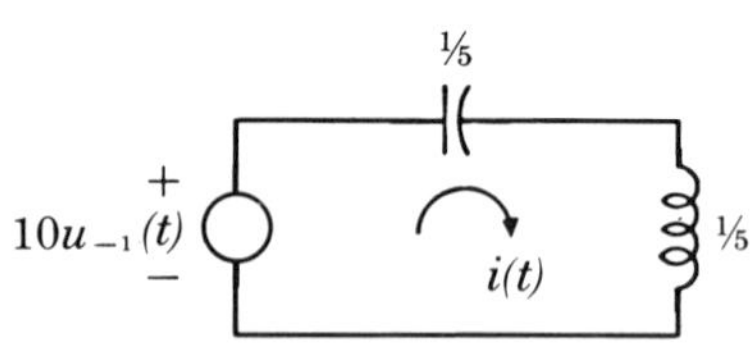

FIGURE 21–36

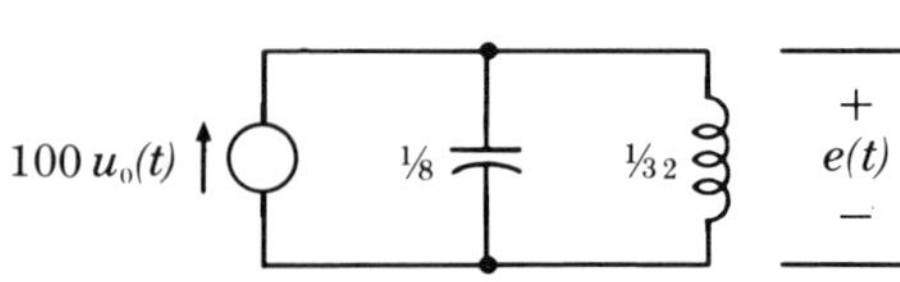

FIGURE 21–37

37. Solve the R–L–C circuit of Fig. 21–38 for the current $i(t)$ by means of the Laplace transform.

38. Use the Laplace transform method to obtain the voltage $e(t)$ in the circuit of Fig. 21–39.

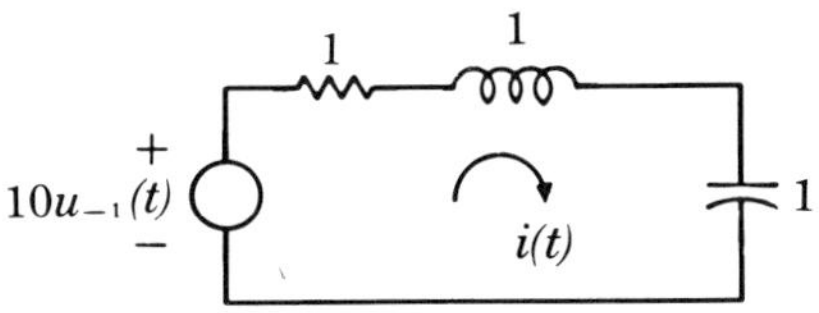

FIGURE 21–38

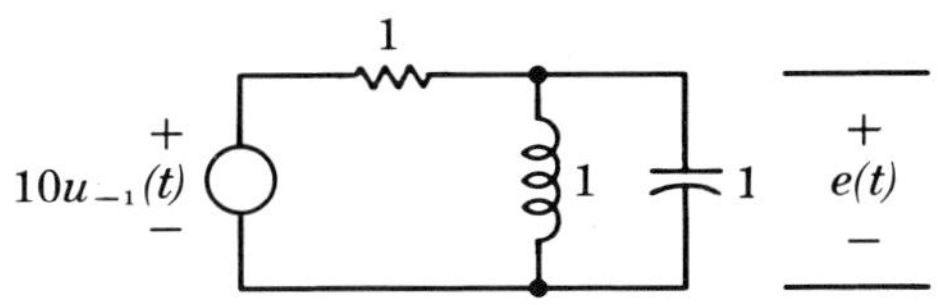

FIGURE 21–39

Section 21–6

39. (a) In the circuit of Fig. 21–40 convert the initial energy storage to a step source in series with the capacitance and solve for the current by means of the Laplace transform method. (b) Repeat part (a) but convert the initial energy storage into an impulse source in parallel with the capacitance. (c) Show that the circuit for part (b) is the Thévenin circuit for part (a).

40. (a) In the circuit of Fig. 21–41 convert the initial energy stored in the inductance to a step current in parallel with the inductance and solve for the voltage $e(t)$ by means of the Laplace transform method. (b) Repeat part (a) but represent the initial energy storage by an impulse voltage source in series with the inductance. (c) Show that the circuit for part (b) is the Norton circuit for part (a).

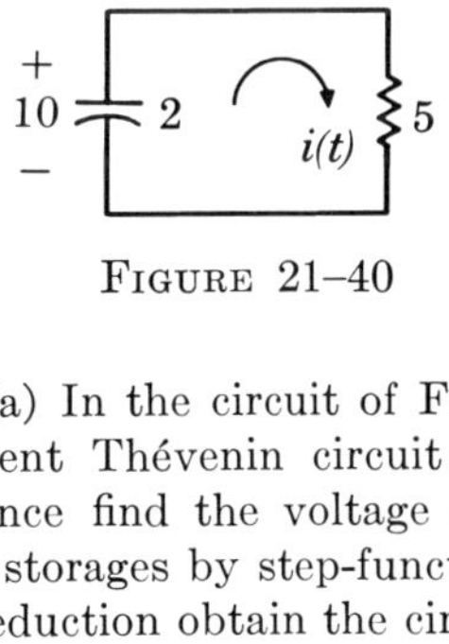

FIGURE 21–40

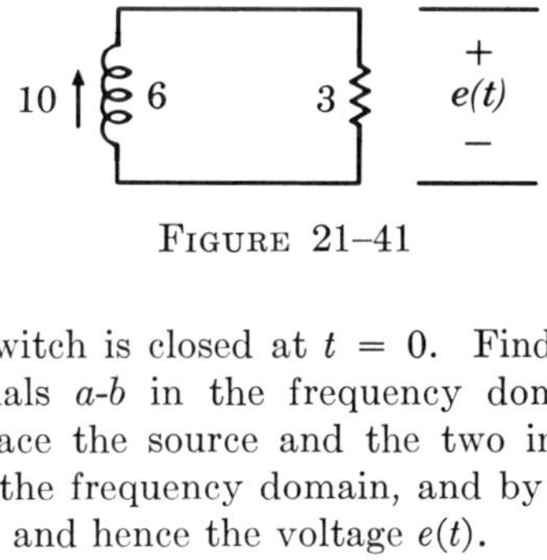

FIGURE 21–41

41. (a) In the circuit of Fig. 21–42 the switch is closed at $t = 0$. Find the equivalent Thévenin circuit at the terminals a-b in the frequency domain, and hence find the voltage $e(t)$. (b) Replace the source and the two initial energy storages by step-function sources in the frequency domain, and by network reduction obtain the circuit of part (a) and hence the voltage $e(t)$.

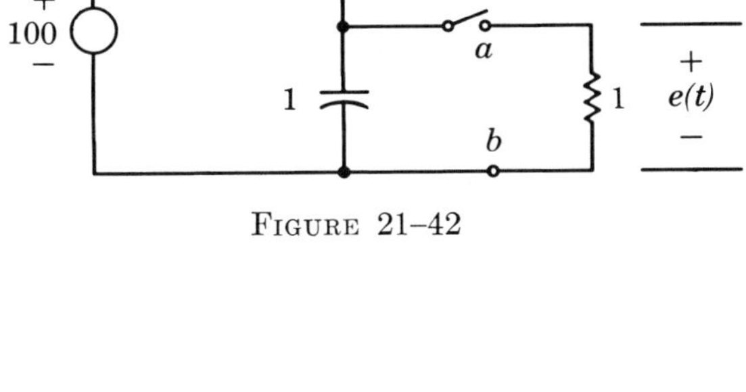

FIGURE 21–42

42. The switch in the circuit of Fig. 21–43 is closed at $t = 0$. Use the Laplace transform method to find the current which flows through the switch for $t > 0$. Previous to $t = 0$ a steady-state has been reached.

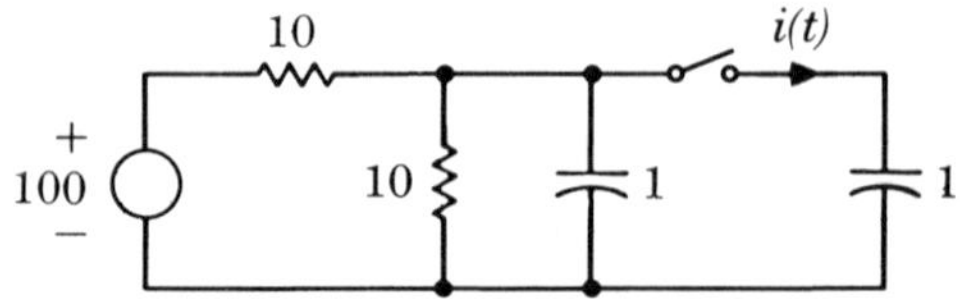

FIGURE 21–43

43. The switch in the circuit of Fig. 21–44 is closed and a steady-state is reached in the network. Then at $t = 0$ the switch is opened. Find the voltage across the switch for $t > 0$.

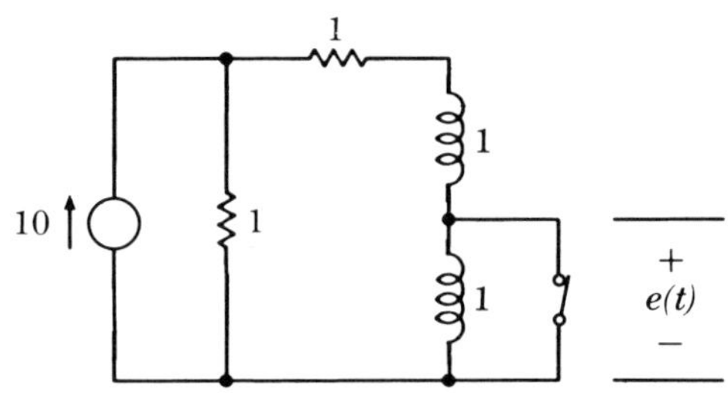

FIGURE 21–44

44. At $t = 0$ in the circuit of Fig. 21–45 there is a current of 10 amp flowing and a voltage of 10 volts across the capacitance. Using the Laplace transform method and a frequency-domain diagram, find the current which flows for $t > 0$.

45. For the circuit of Exercise 44 find the voltage $e(t)$ across the capacitance (a) by converting the energy storages to current sources in parallel with the elements and (b) by integrating the current obtained in Exercise 44.

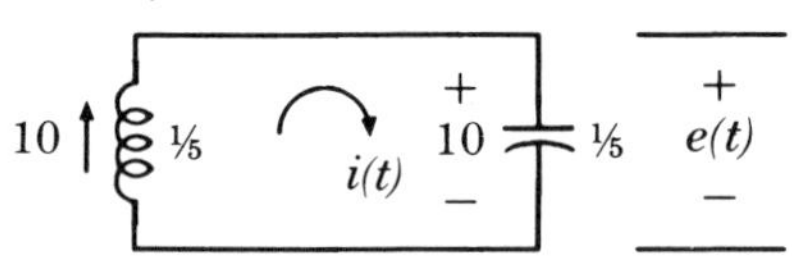

FIGURE 21–45

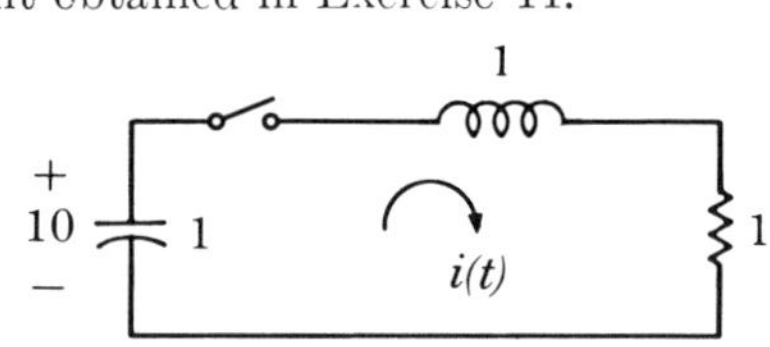

FIGURE 21–46

46. The switch in the circuit of Fig. 21–46 is closed at $t = 0$. Use the Laplace transform method to find the current which flows in the circuit for $t > 0$.

47. Prior to $t = 0$ in the circuit of Fig. 21–47 there is a current of 10 amp in the inductance and the switch, and no energy stored in the capacitance. At

$t = 0$ the switch is opened. Find the current $i(t)$ by means of the Laplace transform.

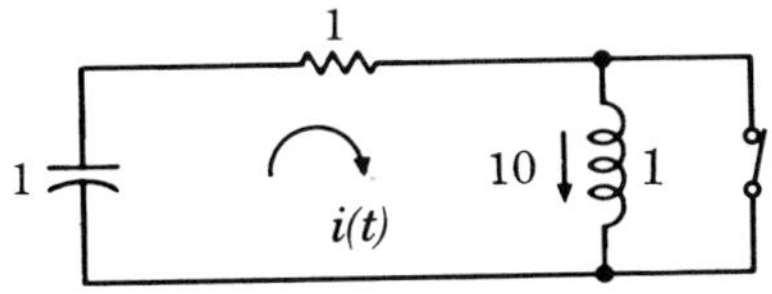

FIGURE 21–47

48. The switch is closed in the circuit of Fig. 21–48 at $t = 0$. At this time a steady-state has been reached in the section of the network to the left of the switch. Find the voltage $e(t)$ for $t > 0$.

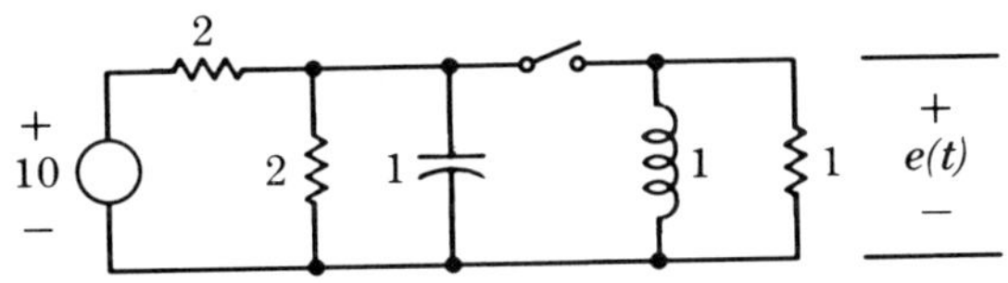

FIGURE 21–48

Section 21–7

49. Expand the following function into partial fraction form and verify the result by recombination of the terms:

$$F(s) = \frac{1}{s(s+1)(s+2)}.$$

50. Find the partial fraction expansion for

$$F(s) = \frac{s^3 + 3s^2 + 2s + 1}{s^3 + 3s^2 + 2s}.$$

51. Find the partial fraction expansion of

$$F(s) = \frac{s^3}{s^3 + 6s^2 + 11s + 6}.$$

52. Give the partial fraction expansion for the following function, and express the complex numbers in polar form:

$$F(s) = \frac{s}{s^2 + s + 1}.$$

53. Give the partial fraction expansion of the following function, with the complex constants in polar form:

$$F(s) = \frac{s^4 + s^3 + s^2 + s + 1}{s^2 + s + 1}.$$

54. Find a partial fraction for the following function, in which the numerator of each term is a constant:

$$F(s) = \frac{s}{(s+1)^3}.$$

55. Find the partial fraction expansion of

$$F(s) = \frac{s^3}{(s+1)^3}.$$

56. Find a partial fraction expansion for

$$F(s) = \frac{s^2+2}{(s+1)(s+2)^2}.$$

57. For the following function find a partial fraction expansion in the form of four terms with complex constants in their numerators; express the complex constants in polar form:

$$F(s) = \frac{s}{(s^2+2s+2)^2}.$$

58. Find a partial fraction expansion for

$$F(s) = \frac{s}{(s+1)(s^2+2s+2)^2}.$$

Section 21–8

59. Find the inverse transform of

$$F(s) = \frac{1}{s(s+1)(s+2)}.$$

60. Expand the following function in partial fractions and thus obtain its inverse transform:

$$F(s) = \frac{s}{s^2+1}.$$

61. Using the method of partial fractions find the inverse Laplace transform of

$$F(s) = \frac{s}{s^2+s+1}.$$

62. Using the method of partial fractions find the inverse Laplace transform of

$$F(s) = \frac{s}{(s+1)^3}.$$

63. Using the method of partial fractions verify the Laplace-transform pair

$$\frac{1}{(s+a)(s+b)(s+c)} \sim \frac{e^{-at}}{(b-a)(c-a)} + \frac{e^{-bt}}{(a-b)(c-b)} + \frac{e^{-ct}}{(a-c)(b-c)}.$$

64. Using the method of partial fractions verify the Laplace-transform pair

$$\frac{s}{(s^2+\beta^2)(s^2+\mu^2)} \sim \frac{\cos \beta t - \cos \mu t}{\mu^2 - \beta^2}.$$

65. Using the method of partial fractions verify the Laplace transform pair

$$\frac{1}{s(s^2+2as+\omega^2)} \sim \frac{1}{\omega^2} + \frac{1}{\beta\omega} e^{-at} \sin(\beta t - \theta),$$

where

$$\beta^2 = \omega^2 - a^2 \qquad \text{and} \qquad \theta = \tan^{-1} \beta/-a.$$

66. By means of a partial fraction expansion verify the transform pair

$$\frac{1}{s^2(s+a)} \sim \frac{e^{-at} + at - 1}{a^2}.$$

67. By means of a partial fraction expansion verify the transform pair

$$\frac{1}{s^2(s^2+\beta^2)} \sim \frac{t}{\beta^2} - \frac{\sin \beta t}{\beta^3}.$$

68. By means of a partial fraction expansion verify the transform pair

$$\frac{s^2}{(s^2+\beta^2)^2} \sim \frac{1}{2\beta}(\sin \beta t + \beta t \cos \beta t).$$

69. By means of a partial fraction expansion verify the transform pair

$$\frac{1}{s(s+a)^2} \sim \frac{1}{a^2} - \frac{e^{-at}}{a^2}(1+at).$$

70. By means of a partial fraction expansion verify the transform pair

$$\frac{(s+a)^2 - \beta^2}{[(s+a)^2+\beta^2]^2} \sim te^{-at} \cdot \cos \beta t.$$

CHAPTER 22

PROBLEMS SOLVED BY LAPLACE TRANSFORMS

"War talk by men who have been to war is always interesting; whereas moon talk by a poet who has not been to the moon is likely to be quite dull."

Mark Twain

22–1 Introduction. There is no substitute for practice and direct personal experience. So far as the theory of the Laplace transform is concerned, Chapters 20 and 21 suffice. The addition of this chapter is dictated by the student's need to apply the Laplace transform methods to a variety of problems. Many of the problems in this chapter we have solved before by other methods. They are included in order to give the student the opportunity to compare the transform methods with those used earlier, and also to demonstrate the tremendous generality of the transform approach.

In this chapter we will begin by solving some transient problems which we previously solved by classical methods. For the first time we will combine transient and steady-state solutions in a single problem. Repeated transients will be handled by means of delay factors. Both the initial transient and the ultimate periodic behavior will be related to each other.

We will look at driving-point functions and transfer functions and use some of the transform concepts, such as the frequency domain, zeros, and poles, to examine the general properties of these functions, including the criteria for physical realizability and stability. Most of the more advanced methods for the analysis and synthesis of networks, and linear systems in general, start from the driving-point and transfer functions expressed in terms of the complex frequency variable s.

Finally, we will relate the general frequency-domain methods to the general time-domain methods which were covered in the earlier chapters. In particular, we will discuss the relationship between the transform methods and convolution theory, and the relationship between the transform methods and differential equations.

Chapter 22 represents a culmination and a summary. Of all the methods presented in this text, those described here are the most advanced. Much of the previous material was included in order that the student might properly appreciate the meaning and usefulness of the transform methods. Now, in applying them to problems treated earlier by other methods we summarize a good deal of the ground we already have covered.

(a) Time-domain circuit (b) Frequency-domain circuit

FIG. 22–1. A single energy-storage circuit.

22–2 Laplace transform solutions of transient problems previously solved. We will not attempt to solve all the transient problems which we solved previously, but will give representative examples of single and double energy-storage circuits excited by initial energy storages, steps, and impulses.

Example 1. Let us consider a single-energy-storage circuit excited by initial energy storage. The basic time-domain circuit is shown in Fig. 22–1(a).

The first step is to transform the input function and the network to the frequency domain. From the circuits of Fig. 21–10 the initial energy storage transforms to the frequency domain as a voltage of E/s in series with an uncharged capacitor. The remainder of the circuit transforms by the usual rules of Section 21–5. The complete frequency-domain circuit is given in Fig. 22–1(b).

The second step is to obtain the response function of the system in the frequency domain. If the input voltage is $1\,\underline{|0^\circ}$ the current produced is

$$H(s) = \frac{1}{R + 1/Cs}. \tag{22–1}$$

The third step is to obtain the output for the actual input present in the problem. It is

$$I(s) = \frac{E}{s}\,\frac{1}{R + 1/Cs} = \frac{E}{R}\,\frac{1}{s + 1/RC}. \tag{22–2}$$

It should be noted that steps two and three are easily combined. The voltage E/s is treated as the source in the network and the output produced by it can be obtained directly. In the following examples we will use the shorter method and refrain from computing $H(s)$ separately.

In the fourth or last step the time function for the output is obtained from tables or by means of partial fraction expansions. Here it is obvious by inspection. Thus

$$i(t) = \frac{E}{R}\,e^{-t/RC} \tag{22–3}$$

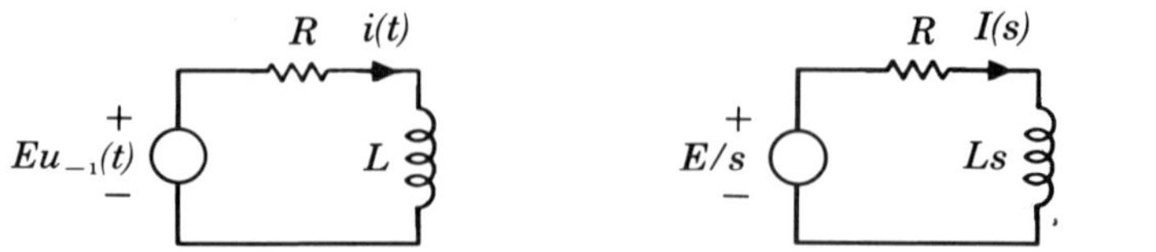

(a) Time-domain circuit (b) Frequency-domain circuit

FIG. 22–2. Single-time-constant circuit driven by a step function.

Example 2. For our second example we will take a single-time-constant circuit driven by a step function of voltage, as shown in Fig. 22–2(a). The input function is easily transformed to the frequency domain as E/s. The remainder of the circuit transforms by the usual rules. The complete transformed circuit is shown in Fig. 22–2(b). The current in the frequency domain is given by

$$I(s) = \frac{E}{s}\frac{1}{R + Ls} = \frac{E}{L}\frac{1}{s(s + R/L)}.$$

This function is not found in Table 20–4. We will therefore use partial fractions, and write

$$I(s) = \frac{E}{L}\frac{1}{s(s + R/L)} = \frac{a_1}{s} + \frac{a_2}{s + R/L},$$

where

$$a_1 = \left[\frac{E}{L}\frac{1}{s + R/L}\right]_{s=0} = \frac{E}{R},$$

$$a_2 = \left[\frac{E}{L}\frac{1}{s}\right]_{s=-R/L} = \frac{-E}{R}.$$

Thus

$$I(s) = \frac{E/R}{s} - \frac{E/R}{s + R/L}.$$

The current as a function of time is

$$i(t) = \frac{E}{R}(1 - e^{-Rt/L}). \tag{22–4}$$

Equations (22–3) and (22–4) represent the two general types of behavior which can be found in single-time-constant circuits. As in all of our previous work, the time constant of the R–C circuit is RC, and of the R–L circuit, L/R. The direct solution of problems of this sort by noting the initial value, final value, and time constant is so easy that it does not pay to use the Laplace transform. If we use the transform method for such

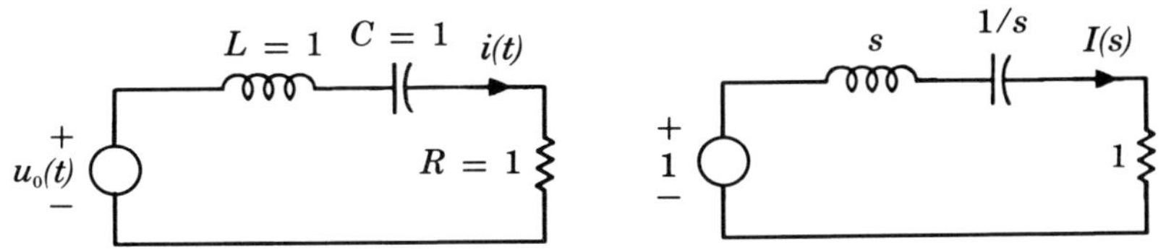

(a) Time-domain circuit (b) Frequency-domain circuit

FIG. 22–3. Two-energy-storage circuit driven by an impulse.

simple problems we are driving a tack with a sledge hammer! The tack will go in all right, but some of the immediate environs may be obscured.

Example 3. The natural behavior of a two-energy-storage circuit is so much more complicated than that of a single-energy-storage circuit that it is often more convenient to solve it by the method of Laplace transforms than by the method of matching initial conditions which we discussed earlier. The transform solution will automatically tell us whether the circuit is overdamped, underdamped, or critically damped. As an example we will consider the series circuit of Fig. 22–3(a), which is driven by a unit impulse of voltage. The impulse source transforms to the frequency domain as a constant. The remainder of the transformed circuit is shown in Fig. 22–3(b). The current in the frequency domain is given by

$$I(s) = \frac{1}{s + 1 + 1/s} = \frac{s}{s^2 + s + 1}.$$

This function is not found in the usual short tables of transforms, and so we will use the method of partial fractions. The roots of the denominator are

$$s_1 = -0.5 + j0.866, \qquad s_2 = -0.5 - j0.866.$$

Thus

$$\begin{aligned} I(s) &= \frac{s}{(s + 0.5 - j0.866)(s + 0.5 + j0.866)} \\ &= \frac{a_1}{s + 0.5 - j0.866} + \text{conjugate}, \end{aligned}$$

where

$$\begin{aligned} a_1 &= \left[\frac{s}{s + 0.5 + j0.866}\right]_{s=-0.5+j0.866} \\ &= \frac{-0.5 + j0.866}{2j0.866} \\ &= \frac{1\,\angle 120^\circ - 90^\circ}{2(0.866)}. \end{aligned}$$

We can now write

$$I(s) = \frac{1/2(0.866)\,\underline{|30^\circ}}{s + 0.5 - j0.866} + \text{conjugate},$$

and

$$i(t) = 2\,\mathrm{Re}\left[\frac{1}{2(0.866)}\,e^{j30^\circ}e^{-0.5t}e^{j0.866t}\right]$$

$$= \frac{1}{0.866}\,e^{-0.5t}\cos(0.866t + 30^\circ).$$

In this example the student should note particularly the advantage of working with the polar form of the constant a_1, and of treating a_2 as the conjugate of a_1.

22–3 Combined steady-state and transient. The application of a driving source at some point in a network results in a transient disturbance which occurs over a short period of time, and eventually in a steady-state behavior of the network in which all the currents and voltages are of the same general type as the input source. The constant current which flows from a d-c battery into a resistive circuit is an example of a steady-state condition. Likewise, all the sinusoidal components which we computed by means of vector algebra or vector geometry for sinusoidal problems also represent steady-state conditions. The transient behavior of the network is determined by the nature of the network and to a lesser extent by the magnitude of the excitation. The steady-state behavior is determined in nature by the type of the source, and only in magnitude by the network. In the present section we will solve a problem in which a steady-state sinusoid is suddenly applied to a network at $t = 0$. It might be supposed that the steady-state behavior could be set up instantly, but such is not the case. It takes some time for the energy to redistribute itself and for the final steady-state condition to occur. The first thing we must observe is that a sinusoid which starts at $t = 0$ has a whole band of frequencies in it, and not just a single frequency such as is in a sine wave which occurs over all time. It is these components which excite the transient in the circuit. A steady-state sine wave and a sine wave sectioned at $t = 0$ are shown in Fig. 22–4.

Since the method of Laplace transforms does not consider a function for t negative, it cannot distinguish between the two types of sine waves shown in Fig. 22–4. The steady-state sine wave, from a Fourier point of view, represents a line spectrum and the output which it produces can be obtained by the usual methods of steady-state sinusoidal analysis. The sectioned sine wave is the only one treated by the Laplace transform, and it has a band spectrum consisting of an infinite number of infinitesimal

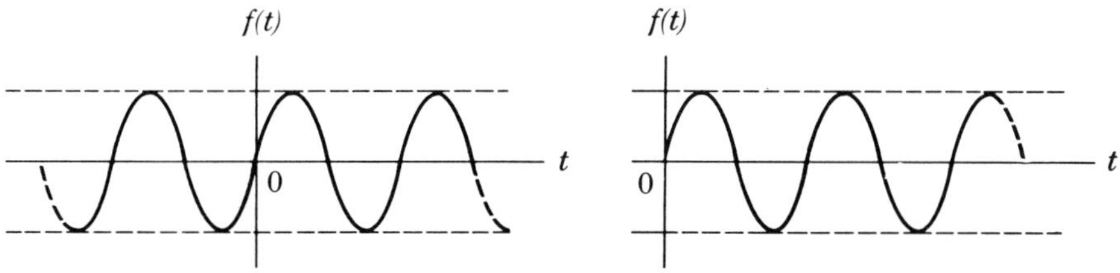

(a) Steady-state sine wave (b) Sectioned sine wave

FIG. 22-4. Steady-state and sectioned sine waves.

harmonics at all frequencies. The spectrum, however, becomes infinite at the frequency of the sine wave, and after some time this component takes over and the output in the network is proportional to it alone. This final state in the network is the sinusoidal steady-state. We will begin by taking the transform of the cosine wave:

$$
\begin{aligned}
F(s) &= \int_0^\infty \cos te^{-st}\,dt \\
&= \int_0^\infty \left[\frac{e^{(j-s)t} + e^{-(j+s)t}}{2}\right] dt \\
&= \left[\frac{e^{(j-s)t}}{2(j-s)}\right]_0^\infty - \left[\frac{e^{-(j+s)t}}{2(j+s)}\right]_0^\infty \\
&= \frac{1}{2(s-j)} + \frac{1}{2(s+j)} = \frac{s}{s^2+1}.
\end{aligned}
$$

For $s = j\omega$, the function $F(s)$ represents the amplitudes of the harmonic terms necessary to represent a sectioned cosine wave in terms of sinusoidal waves which run from $-\infty$ to $+\infty$. The function $F(s)$ becomes in this case

$$
F(j\omega) = \frac{j\omega}{1-\omega^2}. \tag{22-5}
$$

The amplitude function is

$$
|F(\omega)| = \frac{\omega}{1-\omega^2},
$$

and the phase of each harmonic is 90°. Plots of the amplitude and phase functions are given in Fig. 22-5.

The amplitude spectrum goes to infinity at $\omega = 1$. This value corresponds to the frequency of the wave, and a line spectrum at this point would be obtained if the original sinusoid ran from $-\infty$ to $+\infty$. The infinite value in the band spectrum represents a line-spectrum component, and after the transient dies out this component produces the principal

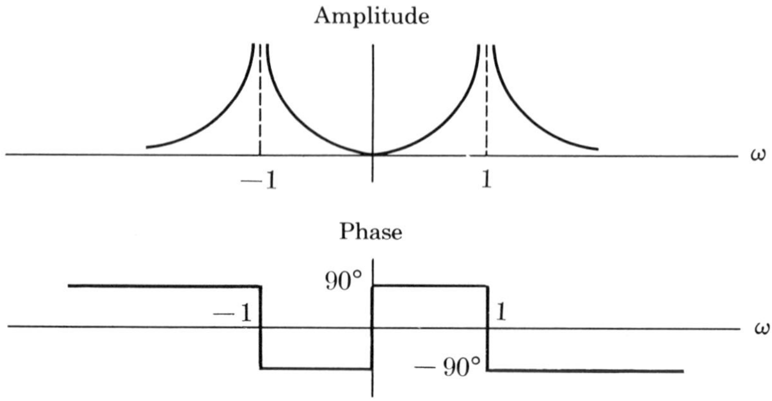

FIG. 22–5. Amplitude and phase spectra for a sectioned cosine wave.

response. The transient response is produced by the band spectrum, and the steady-state response by the line spectrum.

Let us now apply a sectioned cosine wave to a network and obtain the steady-state and transient components of the response. The network is shown in the time domain in Fig. 22–6(a), and in the frequency domain in part (b). The current in the frequency domain is given by

$$I(s) = \frac{E(s)}{Z(s)} = \frac{s}{s^2+1}\,\frac{1}{s+1}\,.$$

Resolved into partial fractions, this expression becomes

$$I(s) = \frac{s}{(s+1)(s-j)(s+j)} = \frac{a_1}{s+1} + \frac{a_2}{s-j} + \text{conjugate},$$

where

$$a_1 = \left[\frac{s}{s^2+1}\right]_{s=-1} = -\tfrac{1}{2},$$

$$a_2 = \left[\frac{s}{(s+1)(s+j)}\right]_{s=j} = \frac{j}{(1+j)2j} = \frac{1}{2(1.414)}\underline{|-45^\circ}\,.$$

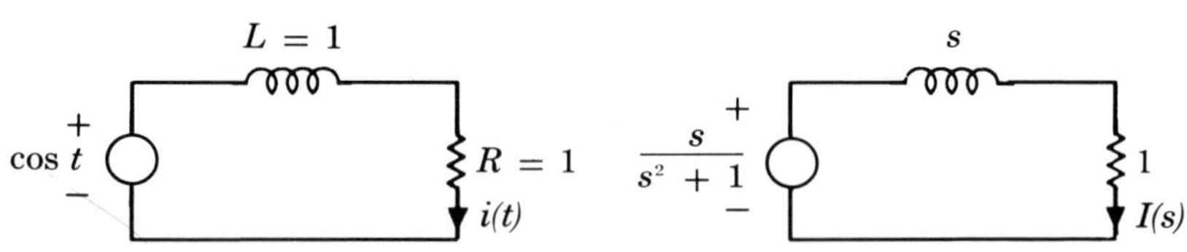

(a) Time-domain circuit (b) Frequency-domain circuit

FIG. 22–6. Circuit for steady-state and transient responses.

Thus

$$I(s) = \frac{-\frac{1}{2}}{s+1} + \frac{1/2(1.414)\,\underline{|-45^\circ}}{s-j} + \text{conjugate},$$

and

$$\begin{aligned} i(t) &= -\tfrac{1}{2}e^{-t} + 2\,\mathrm{Re}\left[\frac{1\,\underline{|-45^\circ}}{2(1.414)}\,e^{jt}\right] \\ &= -\tfrac{1}{2}e^{-t} + 0.707\cos(t - 45^\circ). \end{aligned} \tag{22–6}$$

The first term in Eq. (22–6) is the transient term. It dies out with a time constant of unity, and after about five time constants it is essentially zero. The second term is the steady-state response. It is of the same nature as the input function and it does not die out with time. After the transient dies out this term is left. It is the usual steady-state response which we would calculate by using complex algebra. For the steady-state computation, the input is $1\,\underline{|0^\circ}$, and the current in the frequency domain is given by

$$I(s) = \frac{1\,\underline{|0^\circ}}{s+1} = 0.707\,\underline{|-45^\circ}\cdot$$

Transformed back to the time domain the line-spectrum response is

$$i(t) = 0.707\cos(t - 45^\circ). \tag{22–7}$$

The amplitude of the transient term in Eq. (22–6) is not obvious at first glance. At $t = 0$, a step of unity is suddenly applied to the circuit, but the transient term is not the total response to this step. Some of the total response goes to make up the steady-state response which should occur at $t = 0$. A little thought along this line will convince us that the transient component in the circuit must supply the difference between what the circuit is actually doing at $t = 0$ and what the steady-state response would like it to be doing. The transient acts as a matching device between the source and the circuit at the initial instant, and it gradually retires as the circuit begins to respond in the way in which the source would like it to. The value of the steady-state expression in Eq. (22–7) at $t = 0$ is

$$i(0) = 0.707\cos(-45^\circ) = \tfrac{1}{2}. \tag{22–8}$$

Equation (22–8) tells us that the steady-state response of the network at $t = 0$ requires a current of value $\frac{1}{2}$ in the circuit. The actual current in the circuit at $t = 0$ must be zero since the response to a step (the initial value of the cosine wave) in a circuit with an inductance and a resistance is zero. The transient must match between the desired value of $\frac{1}{2}$ and the actual value of zero. It must have a value of $-\frac{1}{2}$ at $t = 0$, and it will of course be retired at the time constant of the circuit. In circuits with more than one time constant, the transient supplies the difference between the actual behavior and the desired behavior for as many derivatives at $t = 0$ as is possible. Each additional energy storage accounts for an additional derivative.

In some high-power circuits the existence of a switching transient is not desirable. It is interesting therefore to inquire if there is some instant at which the switch can be closed on an a-c source with the immediate establishment of the steady-state current and with no transient. To investigate this question let us suppose that the input voltage in Fig. 22–6 has a phase angle of θ. Then the steady-state output term would be

$$i(t) = 0.707 \cos (t + \theta - 45°). \tag{22–9}$$

Since the transient must balance this value at $t = 0$, there will be no transient if the value for Eq. (22–9) at $t = 0$ is zero. This occurs if the angle θ is $-45°$. Thus if we close the switch 45 degrees before the input cosine wave reaches its peak, there will be no transient in the circuit. As a check we will compute the output for this input wave. The time expression for the input is

$$\begin{aligned} e(t) &= \cos (t - 45°) = \cos t \cos 45° + \sin t \sin 45° \\ &= 0.707 (\cos t + \sin t), \end{aligned}$$

and its transform is

$$\begin{aligned} E(s) &= 0.707 \left(\frac{s}{s^2 + 1} + \frac{1}{s^2 + 1}\right) \\ &= 0.707 \frac{s + 1}{s^2 + 1}. \end{aligned}$$

The impedance in the frequency domain, as before, is $Z = s + 1$, and the current is

$$\begin{aligned} I(s) = \frac{E(s)}{Z(s)} &= \frac{0.707(s + 1)}{s^2 + 1} \frac{1}{s + 1} \\ &= 0.707 \frac{1}{s^2 + 1}. \end{aligned}$$

The corresponding inverse transform is

$$i(t) = 0.707 \sin t,$$

which is the steady-state response without a transient term.

The transform of the more general cosine wave

$$e(t) = \cos (\omega t + \theta)$$

is

$$E(s) = \frac{s \cos \theta - \omega \sin \theta}{s^2 + \omega^2}. \tag{22-10}$$

A judicious choice of the angle θ can always make the numerator of Eq. (22-10) cancel with a first-order factor from a network with one energy storage. If the network is a quadratic, however, it is no longer possible to obtain an exact cancellation. Fortunately, the networks in most power systems are primarily resistive and inductive.

22-4 Problems involving delayed functions. All the transforms which we considered previously started at $t = 0$. We cannot shift the time functions to the left of the origin, because the transform ignores anything which occurs before $t = 0$. It is possible, however, to shift time functions to the right, or to delay them, and still have them exist in the region considered by the Laplace transform. Examples of time functions which have been shifted are shown in Fig. 22-7. The time expression for the delayed step function in Fig. 22-7 is

$$f(t) = u_{-1}(t - a). \tag{22-11}$$

The time expression for the delayed sine wave is

$$f(t) = \sin (t - a)u_{-1}(t - a). \tag{22-12}$$

Consider a function which is zero for $t < 0$ and which can be written as

$$\phi(t) = f(t)u_{-1}(t).$$

If this function is shifted to the right by an amount a, it becomes

$$\phi(t - a) = f(t - a)u_{-1}(t - a). \tag{22-13}$$

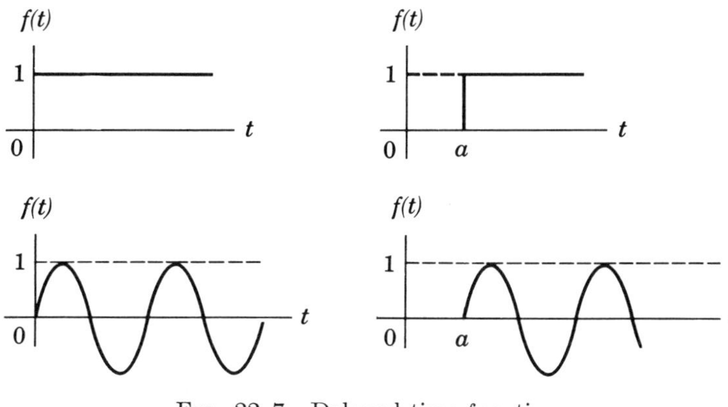

FIG. 22-7. Delayed time functions.

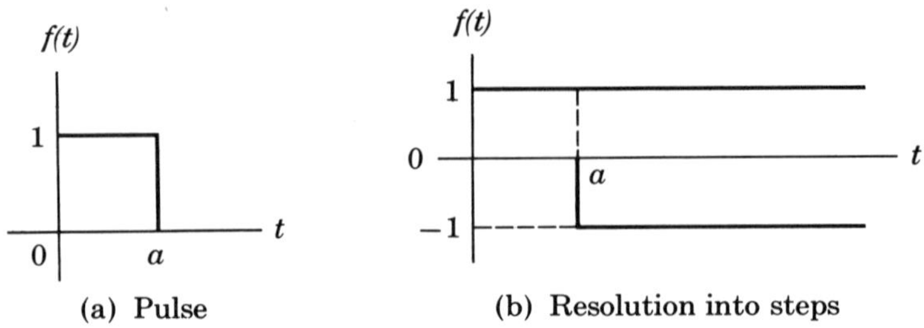

FIG. 22–8. A finite-width pulse.

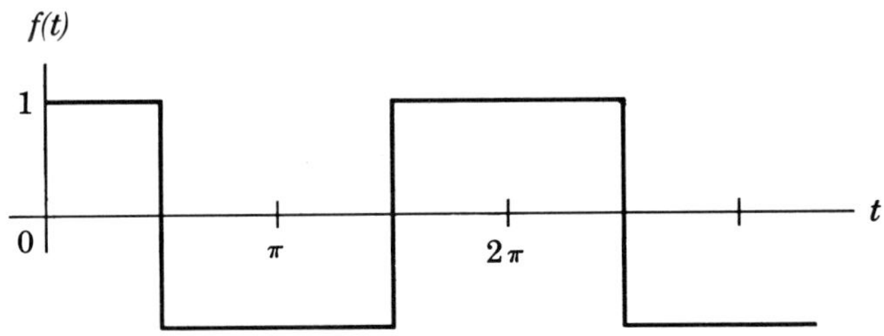

FIG. 22–9. A square wave.

It is possible to express a number of useful functions in terms of delayed components in the form of Eq. (22–13). One example is the finite-width pulse shown in Fig. 22–8. The finite pulse of part (a) can be considered as the superposition of the two steps shown in part (b). Until $t = a$ only the positive step exists. After this instant both steps exist and cancel each other to give a resultant of zero. An expression for the pulse is

$$f(t) = u_{-1}(t) - u_{-1}(t - a). \tag{22–14}$$

A square wave consisting of alternate positive and negative lobes can likewise be expressed in terms of unit step functions. The wave shown in Fig. 22–9 is that which we applied previously as an example of Fourier series. The square wave can be resolved into a series of step functions expressed by

$$f(t) = u_{-1}(t) - 2u_{-1}(t - \pi/2) + 2u_{-1}(t - 3\pi/2) - \cdots. \tag{22–15}$$

Similarly, any pulse $p(t)$ which is zero except for a region between $t = 0$ and $t = a$ and which is then repeated with period a, as shown in Fig. 22–10, can be expressed in terms of delayed functions. The expression for the repeated pulse can be written in the series

$$f(t) = p(t)u_{-1}(t) + p(t - a)u_{-1}(t - a) + p(t - 2a)u_{-1}(t - 2a) + \cdots.$$

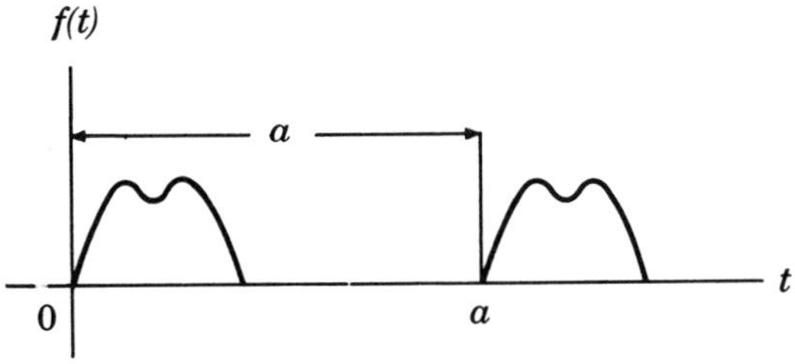

FIG. 22–10. A repeated pulse.

Because there are a number of common waveforms which can be expressed in terms of delayed time expressions, we will find it desirable to be able to obtain the Laplace transform of functions of the sort

$$g(t) = f(t - a)u_{-1}(t - a).$$

A formal expression for the transform is

$$G(s) = \int_0^\infty f(t - a)u_{-1}(t - a)e^{-st}\,dt.$$

The unit step function in the integral insures that the integrand is zero for $t < a$, and therefore the integration need be performed only from $t = a$ to $t = \infty$. Thus

$$G(s) = \int_a^\infty f(t - a)e^{-st}\,dt.$$

In this integral we will now make the substitutions

$$x = (t - a), \qquad dx = dt,$$

and since $t = a$ when $x = 0$ we then obtain

$$G(s) = \int_0^\infty f(x)e^{-(a+x)s}\,dx = e^{-as}\int_0^\infty f(x)e^{-xs}\,dx.$$

The second integral is the Laplace transform of $f(t)$. The variable x is a dummy and can be changed to t without changing the value of the integral. Thus

$$G(s) = e^{-as}F(s). \tag{22–16}$$

The meaning of Eq. (22–16) is that a shift to the right of a seconds in the time domain corresponds to multiplying by the factor e^{-as} in the frequency domain. We have essentially deduced this rule previously in connection with the Fourier series. When $s = j\omega$, Eq. (22–16) becomes

$$G(j\omega) = F(j\omega)e^{-j\omega a} = F(j\omega)\,\underline{\left|-(\omega a)^\circ\right.}. \tag{22–17}$$

In shifting the origin of the function, we shift only the phases of the harmonics and do not change their amplitudes. Furthermore, the phase shift is proportional to frequency. If the fundamental is shifted one degree, the third harmonic will be in shifted three degrees, and so on. For a continuous spectrum the phase shift will still be proportional to the frequency. The proportionality constant is the delay in seconds.

We now make use of Eq. (22–16) to obtain the Laplace transforms of the time functions in Figs. 22–8, 22–9, and 22–10. For the pulse of Fig. 22–8 the time function is

$$f(t) = u_{-1}(t) - u_{-1}(t - a).$$

The corresponding transform is

$$F(s) = \frac{1}{s} - \frac{e^{-as}}{s}. \tag{22–18}$$

For the square-wave train of Fig. 22–9 the time function is

$$f(t) = u_{-1}(t) - 2u_{-1}(t - \pi/2) + 2u_{-1}(t - 3\pi/2) - \cdots.$$

The corresponding transform is

$$\begin{aligned} F(s) &= \frac{1}{s} - 2\,\frac{e^{-\pi s/2}}{s} + \frac{2e^{-3\pi s/2}}{s} - \cdots \\ &= \frac{1}{s} - \frac{2}{s}e^{-\pi s/2}\,(1 - e^{-\pi s} + e^{-2\pi s} - \cdots). \end{aligned}$$

The term within parenthesis can be summed as a geometric progression. The formulas for infinite geometrical progressions can be obtained by division; that is

$$\frac{1}{1 - x} = 1 + x + x^2 + x^3 + \cdots, \tag{22–19}$$

$$\frac{1}{1 + x} = 1 - x + x^2 - x^3 + \cdots. \tag{22–20}$$

Using the identity of Eq. (22–20) we can write the transform of Fig. 22–9 as

$$F(s) = \frac{1}{s} - \frac{2e^{-\pi s/2}}{s(1 + e^{-\pi s})}. \tag{22–21}$$

The general case of a repeated pulse, as shown in Fig. 22–10, can be treated similarly. The time function is

$$f(t) = p(t)u_{-1}(t) + p(t - a)u_{-1}(t - a) + p(t - 2a)u_{-1}(t - 2a) + \cdots,$$

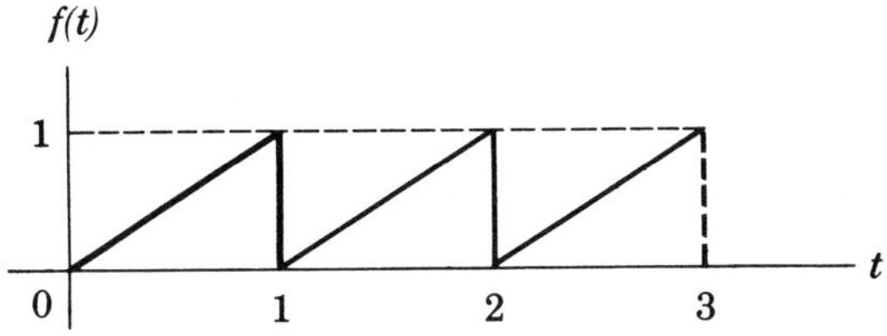

FIG. 22-11. A repeated pulse.

and the transform is

$$F(s) = P(s) + P(s)e^{-as} + P(s)e^{-2as} + \cdots$$
$$= P(s)(1 + e^{-as} + e^{-2as} + \cdots)$$
$$= \frac{P(s)}{1 - e^{-as}}. \tag{22-22}$$

Equation (22-22) is a rather important formula, since it allows us to convert the band spectrum of a single pulse into the line spectrum of a Fourier series, in which the pulse repeats with a period of a, simply by dividing by $(1 - e^{-as})$. The transform of Eq. (22-22) is more general than the spectrum for a Fourier series. It includes the transient part which occurs, because the series does not start until $t = 0$. As we showed in the previous section, however, the steady-state components are the same in the Laplace transform as in the usual steady-state analysis. As an example of the use of the formula, consider the wave of Fig. 22-11. The time function of the pulse is

$$p(t) = tu_{-1}(t) - (t - 1)u_{-1}(t - 1) - u_{-1}(t - 1).$$

The transform of the pulse is

$$P(s) = \frac{1}{s^2} - \frac{e^{-s}}{s^2} - \frac{e^{-s}}{s},$$

and the transform of the repeated pulse is

$$F(s) = \frac{1 - e^{-s} - se^{-s}}{s^2} \frac{1}{1 - e^{-s}}.$$

22-5 The application of delayed functions to networks. In the previous section we observed that many waveforms can be expressed in terms of delayed time functions in the time domain and in terms of e^{-as} factors in the frequency domain. We will proceed to apply these waveforms to simple networks and to compute the response functions by means of Laplace transforms. As a first example we will consider a pulse applied to a series

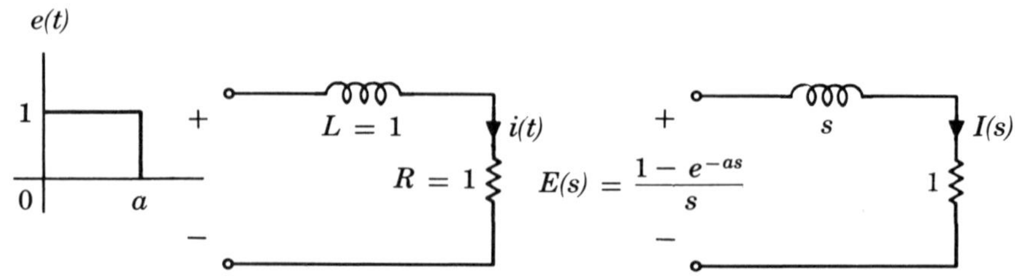

(a) Time-domain circuit (b) Frequency-domain circuit

FIG. 22–12. A pulse applied to a series R–L circuit.

R–L circuit, as shown in Fig. 22–12(a). The current is to be found. The circuit in the frequency domain has the usual form. The exciting function is considered to be a positive step at $t = 0$ followed by a negative step at $t = a$. The frequency-domain function is

$$E(s) = \frac{1 - e^{-as}}{s}.$$

The current in the frequency domain is

$$I(s) = \frac{E(s)}{Z(s)} = \frac{1 - e^{-as}}{s(s + 1)}. \tag{22–23}$$

It is not possible to expand Eq. (22–23) directly into partial fractions, because the numerator is of higher order than the denominator. Actually, the factor e^{-as} contains all powers of s from zero to infinity. It is not possible to divide these terms out. We must seek another interpretation of the transform. For this purpose we write

$$I(s) = \frac{1}{s(s + 1)}(1 - e^{-as}). \tag{22–24}$$

Written in the form of Eq. (22–24), $I(s)$ can be interpreted as the sum of the inverse transform of the function

$$\frac{1}{s(s + 1)}$$

and the same function delayed by a seconds in the time domain. Factors of the type e^{-as} occurring in the numerator of transforms can always be interpreted in this way. Thus in our example the time function is

$$i(t) = (1 - e^{-t})u_{-1}(t) - [1 - e^{-(t-a)}]u_{-1}(t - a). \tag{22–25}$$

A sketch of Eq. (22–25) is shown in Fig. 22–13.

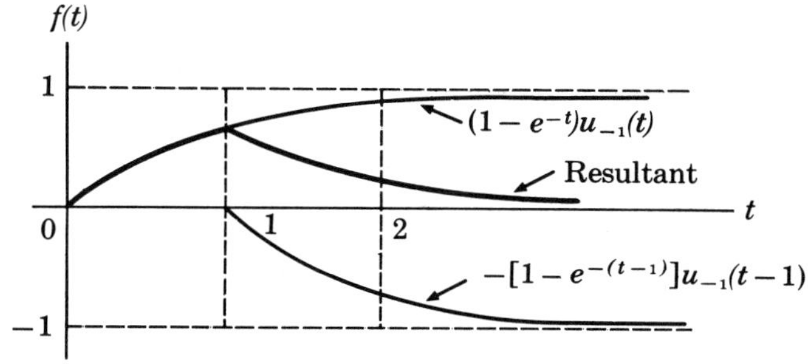

FIG. 22–13. The response to a delayed step plus a step.

When the response function involves only a few terms, as in Eq. (22–25), it is often convenient to plot the resultant graphically. When it contains a large number of terms, however, it may no longer be convenient to carry out this process. A particular case of the latter situation occurs when we are dealing with an infinite number of terms, for example with the series for a repeating square wave. In this case, however, we know that the output should contain a transient term plus a series of steady-state terms representing the Fourier series responses of the network. An interesting result occurs in this case. *The partial fraction expansion of the inverse transform gives the Fourier series response term* for the network. As a result, in many problems the e^{-as} factors can be interpreted as delays in the time domain, or they can be resolved into harmonic terms of the form of the Fourier series. An example of this dual interpretation will be given for the network of Fig. 22–14(a). The input wave can be written in the time domain as

$$e(t) = u_{-1}(t) - u_{-1}(t-1) + u_{-1}(t-2) - u_{-1}(t-3) + \cdots.$$

The corresponding frequency-domain form is

$$E(s) = \frac{1}{s}(1 - e^{-s} + e^{-2s} - e^{-3s} + \cdots) = \frac{1}{s(1+e^{-s})}\cdot$$

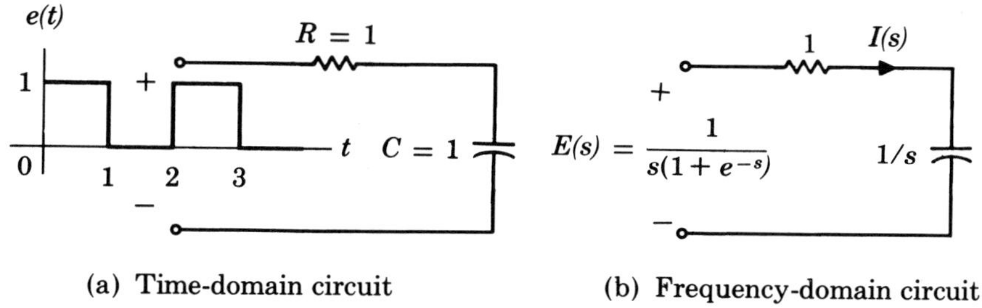

FIG. 22–14. The square-wave response of a network.

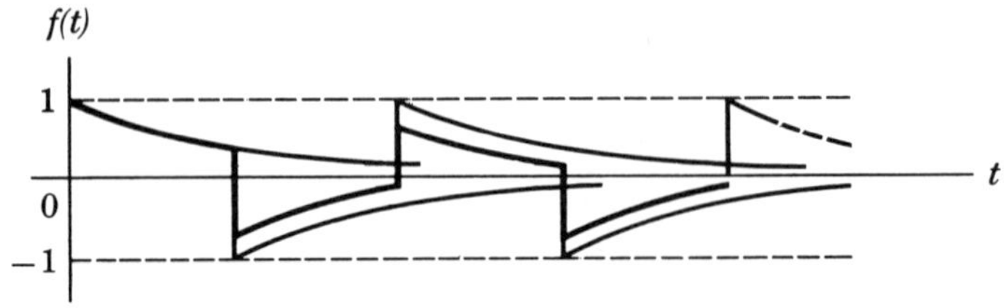

FIG. 22–15. The response to a series of square waves.

The frequency-domain current is

$$I(s) = \frac{1}{s(1 + e^{-s})}\,\frac{s}{s + 1} = \frac{1}{(s + 1)(1 + e^{-s})}\cdot \tag{22–26}$$

Interpretation as delay factors. We will first interpret the current in terms of frequency-domain delay factors of the form e^{-as}, which produce an equivalent delay in the time domain of a seconds. To accomplish this result we write

$$I(s) = \frac{1}{s + 1}(1 - e^{-s} + e^{-2s} - e^{-3s} + \cdots)\cdot$$

The corresponding time function is

$$i(t) = e^{-t}u_{-1}(t) - e^{-(t-1)}u_{-1}(t - 1) + e^{-(t-2)}u_{-1}(t - 2)\cdots. \tag{22–27}$$

A sketch of Eq. (22–27) is shown in Fig. 22–15.

Whenever each individual time function in a series like Eq. (22–27) dies out before the next one occurs, it is relatively easy to sketch the function. However, when the individual responses overlap over large regions of time, it may be entirely impractical to use this method. The alternative is to interpret the inverse transform in terms of partial fraction terms, which are the Fourier series components of the output function.

Interpretation as Fourier series components. The interpretation of Eq. (22–26) in terms of Fourier series components is accomplished by expanding the function in partial fraction form. The first step in this process is the determination of the roots of the denominator of $I(s)$. The first of these occurs at $s = -1$. The others occur at the roots of

$$1 + e^{-s} = 0$$

or

$$e^{-s} = -1 = e^{\pm jn\pi} = (\cos n\pi \pm j \sin n\pi) \quad \text{for } n \text{ odd.}$$

The s values which give zeros in the denominator are

$$s = \pm jn\pi \quad \text{for } n \text{ odd.}$$

There are an infinite number of roots and therefore an infinite number of terms in the partial fraction expansion. The roots are pure imaginary, and occur in complex conjugate pairs. Each pair of roots corresponds to an undamped sine wave term in the time domain, and therefore to a term in the Fourier series of the output response. The fundamental frequency is $\omega = \pi$, and the other frequencies are harmonically related to the fundamental. Since only odd terms exist in the roots, only odd harmonics exist in the Fourier series. The partial fraction expansion is of the form

$$I(s) = \frac{1}{(s+1)(1+e^{-s})} = \frac{a_0}{s+1} + \frac{a_1}{s-j\pi} + \text{conjugate} + \frac{a_3}{s-j3\pi} + \text{conjugate} + \cdots, \tag{22–28}$$

where

$$a_0 = \left[\frac{1}{1+e^{-s}}\right]_{s=-1} = \frac{1}{e+1}, \tag{22–29}$$

$$a_1 = \left[\frac{s-j\pi}{(s+1)(1+e^{-s})}\right]_{s=j\pi}. \tag{22–30}$$

Equation (22–30) is indeterminate, since both numerator and denominator go to zero at $s = j\pi$. The expression can be evaluated by differentiating the numerator and the denominator before making the substitution. Thus

$$a_1 = \left[\frac{1}{(s+1)(-e^{-s}) + (1+e^{-s})}\right]_{s=j\pi} = \frac{1}{1+j\pi} = (1+\pi^2)^{-1/2} \underline{/-\tan^{-1}\pi}.$$

Similarly

$$a_n = [1 + (n\pi)^2]^{-1/2} \underline{/-\tan^{-1} n\pi}. \tag{22–31}$$

The time function corresponding to Eq. (22–28) is

$$\begin{aligned} i(t) &= \frac{1}{e+1} e^{-t} + 2\,\mathrm{Re}\,[(1+\pi^2)^{-1/2} e^{j\pi t} e^{j(-\tan^{-1}\pi)}] + \cdots \\ &= \frac{1}{e+1} e^{-t} + \frac{2}{(1+\pi^2)^{1/2}} \cos(\pi t - \tan^{-1}\pi) + \cdots \\ &\quad + \frac{2}{[1+(n\pi)^2]^{1/2}} \cos(n\pi t - \tan^{-1} n\pi) + \cdots. \end{aligned} \tag{22–32}$$

Equation (22–32) represents the sum of the transient and the steady-state responses. The first term is the transient. The remaining terms are the Fourier series terms in the output. As a check we will compute the response for the fundamental term in the Fourier series of the square wave which produced Eq. (22–32). The input voltage at the fundamental frequency has the value

$$e_1(t) = \frac{2}{\pi}\sin \pi t \qquad \text{or} \qquad E_1(s) = \frac{2}{\pi}\underline{|-90^\circ}.$$

The impedance at frequency $\omega = \pi$ is $1 - j/\pi$. The output current at this frequency is

$$I_1(s) = \frac{2/\pi\ \underline{|-90^\circ}}{1 - j/\pi} = \frac{2}{1 + j\pi}.$$

The corresponding time-domain expression is

$$i_1(t) = \frac{2}{(1 + \pi^2)^{1/2}} \cos(\pi t - \tan^{-1}\pi). \tag{22–33}$$

Equation (22–33) agrees with the first term in the steady-state response of Eq. (22–32). We have thus verified the fact that the partial fraction expansion for delayed functions gives the Fourier series for the corresponding time functions.

22–6 Driving-point and transfer functions. It is interesting to note that everytime one learns a new method for solving an old problem some new light is shed on the process. For the analysis of circuits it is necessary to know only one method of solution. For the synthesis of circuits, or the design of new devices, it is desirable to know many ways of formulating and attacking circuit problems. The Laplace transform does more than provide us with a method for solving a wide variety of circuit problems. It gives us a new way to formulate circuit problems, and in the process introduces concepts which unify the subject and which clearly classify many types of design problems.

One such concept is the "s-plane". Networks can be characterized by their response functions in terms of the damped frequency variable s. In the process of analysis we mentioned the "system (or response) function" of the network which related the output frequency-domain function to the input frequency-domain function. Thus we wrote

$$F_2(s) = H(s)F_1(s), \tag{22–34}$$

in which $H(s)$ is the system function of the network. In the analysis problem we did not find it necessary to compute $H(s)$ directly. Rather, it was

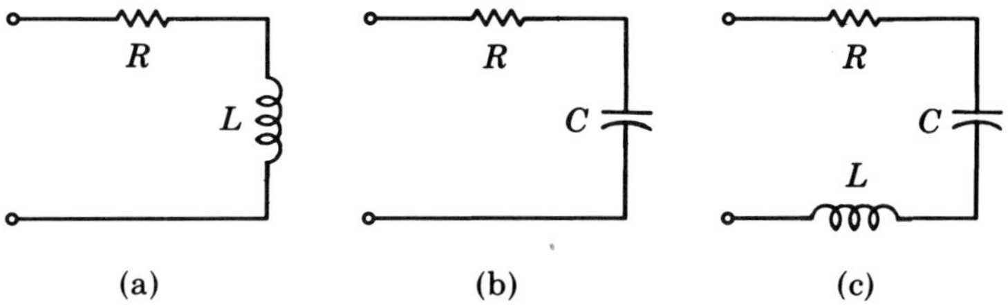

FIG. 22–16. Typical driving-point impedances.

usually easier to compute the output for the given input, $F_1(s)$, and to bypass the calculation of $H(s)$. Now, however, we would like to examine the $H(s)$ function itself to see what information it contains about the nature of the network. If the input and the output quantities are measured at the same pair of terminals, $H(s)$ is a driving-point function. It can be either an admittance or an impedance. If the input is applied at one pair of terminals, and the output is measured at another, $H(s)$ is a transfer function. Transfer functions can have the dimensions of admittance or impedance, or they can be dimensionless when the input and the output quantities have the same dimensions. We will consider driving-point functions first.

Driving-point functions. Driving-point functions can be either impedances or admittances. If the input is a current and the output is a voltage, $H(s)$ will have the dimensions of impedance. Conversely, if the input is a voltage and the output a current, $H(s)$ will have the dimensions of admittance. For the same pair of terminals, the impedance and admittance functions are reciprocals. Some typical driving-point impedances, for the networks of Fig. 22–16, are as follows: For the network of Fig. 22–16(a)

$$Z(s) = Ls + R = L\left(s + \frac{R}{L}\right).$$

For the network of Fig. 22–16(b)

$$Z(s) = R + \frac{1}{Cs} = \frac{R(s + 1/RC)}{s}.$$

For the network of Fig. 22–16(c)

$$Z(s) = R + Ls + \frac{1}{Cs} = \frac{L[s^2 + (R/L)s + 1/LC]}{s}.$$

All driving-point impedance functions (and admittance functions) can be written in the form

$$Z(s) = \frac{K(s - s_1)(s - s_3)\cdots}{(s - s_2)(s - s_4)\cdots}, \tag{22–35}$$

where K is a constant,

$s_1, s_3, \ldots$ are the zeros of the network impedance,

$s_2, s_4, \ldots$ are the poles of the network impedance.

We can investigate the meaning of the poles and zeros by exciting the network with an impulse function and observing the transient response which is obtained. With impulse excitation, there is no steady-state, and the entire behavior can be classified as "transient." First, let us excite the network represented by Eq. (22–35) with an impulse current. The output voltage is

$$E(s) = \frac{K(s - s_1)(s - s_3)\cdots}{(s - s_2)(s - s_4)\cdots} = \frac{a_2}{(s - s_2)} + \frac{a_4}{(s - s_4)} + \cdots,$$

and the corresponding time function is

$$e(t) = a_2 e^{s_2 t} + a_4 e^{s_4 t} + \cdots. \tag{22–36}$$

When the values of the poles in Eq. (22–35) are real, the transient response will contain exponential terms. When the roots are pure imaginaries, the transient response will consist of pure sine waves, and when the roots are complex the transient response will contain damped sine waves. The complex numbers representing the poles give the damping constant and the natural frequency of the transient. From this argument we conclude that the poles of a network impedance function are the open-circuit, or current source, transient natural frequencies of the network.

Alternatively, suppose we have an admittance function of the form

$$Y(s) = \frac{(s - s_2)(s - s_4)\cdots}{K(s - s_1)(s - s_3)\cdots} = \frac{1}{Z(s)},$$

and suppose we apply an impulse of voltage to the network. The resulting current would be

$$I(s) = \frac{(s - s_2)(s - s_4)\cdots}{K(s - s_1)(s - s_3)\cdots} = \frac{a_1}{s - s_1} + \frac{a_3}{s - s_3} + \cdots,$$

and the corresponding time function is

$$i(t) = a_1 e^{s_1 t} + a_3 e^{s_3 t} + \cdots. \tag{22–37}$$

If the zeros of the original impedance are real, the transient terms in Eq. (22–37) will be damped exponentials; if they are pure imaginaries, the transients will be pure sine waves; and if the zeros are complex, the transients will be damped sine waves. Thus the zeros of an impedance function are the voltage source, or short-circuit, transient natural frequencies. In summary

$$\text{Poles of impedance} = \text{open-circuit transients,}$$
$$\text{Zeros of impedance} = \text{short-circuit transients.}$$

We can make use of the results summarized here to deduce some general properties of the impedance functions of linear passive networks. In the first place, if we excite the network with an impulse we will impart to it a finite amount of energy, and if the network is a real physical one this energy will gradually be dissipated and the transient will die out. The transients of Eqs. (22–36) and (22–37) will die out only if the exponential terms have negative real parts. *Thus the poles and zeros of the impedance functions of passive networks must lie in the left half of the s-plane.*

We can also deduce something about the impedance when it is in the form

$$Z(s) = \frac{a_n s^n + a_{n-1}s^{n-1} + \cdots + a_0}{b_m s^m + b_{m-1}s^{m-1} + \cdots + b_0}. \tag{22–38}$$

If the lowest-order terms are both constants, there will be no a/s term in the transient response. If b_0 is zero there will be an a/s term and a step function in the transient response. The term b_1 cannot also be zero, since there would be an a/s^2 term in the partial fraction expansion and a growing $u_{-2}(t)$ term in the transient response. Such a term would represent an unrealizable behavior in a passive circuit, since a source of energy would be required to produce a growing transient. *From this argument, we can conclude that the lowest-order term in the denominator can differ from the lowest-order term in the numerator by only one power of s.* A similar argument holds for the reciprocal function $Y(s) = 1/Z(s)$, showing that the higher order could occur in either the numerator or denominator, but reinforcing the conclusion that the lowest-order terms can differ at most by one power of s.

If the highest-order terms in numerator and denominator are of the same order, we divide out and obtain a constant. The time response for this constant is an impulse. If the numerator were of higher order we would divide out the terms and obtain higher-order singularity functions. To see whether or not these would correspond to a realizable network, we replace s by $1/s$ in Eq. (22–38). This substitution involves replacing all L's by C's and vice versa. The network is still a physical one, and by the argument we have used before its lowest-order terms can differ by only plus or minus one. The lowest-order terms of this function are the highest-order terms of the original function. *Thus we conclude that the highest- and lowest-order terms of an impedance function can differ by only plus or minus one from numerator to denominator.*

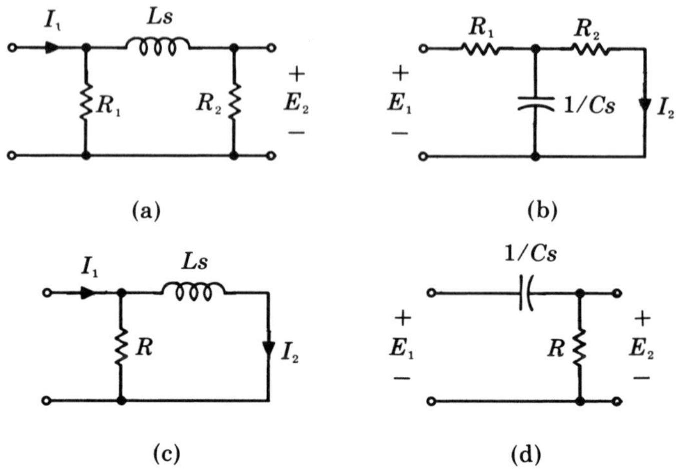

FIG. 22–17. Examples of transfer functions.

One further general property of networks will be discussed. The natural transient frequencies under open-circuit conditions must be the same at every pair of terminals in a network. *Thus the poles of the impedance functions obtained at every pair of terminals in a network must be the same.** Similarly, if the impedance is measured by cutting a wire and "looking in," the natural transients for a short circuit at this point of entry will be the same for all such points of entry. These natural transients would be the zeros of the impedance functions obtained by opening up the network in this fashion. *The zeros of all of these impedances in a network will be the same, wherever the impedance is measured by opening up the network.*

Special networks have special properties. For example, purely *reactive networks can produce only undamped transients and therefore have poles and zeros on the jw axis. R–C or R–L networks can produce only damped exponential transients and therefore have their poles and zeros on the negative real axis.*

Transfer functions. When the input, or stimulus, is applied at one terminal-pair and the output or response is measured at another terminal-pair, the system function which relates the two is called a *transfer function.* There are four different types of transfer functions. If the input is a current and the output a voltage, the transfer function is an impedance. If the input is a voltage and the output a current, the transfer function is an admittance. If the input is a voltage and the output a voltage, or if the

* This statement is not completely general. A possibility exists of excluding or filtering a given frequency from some part of a network.

input is a current and the output a current, the ratio is dimensionless. The general expression $H(s)$ is used to represent all four types, and also as special degenerate cases, the driving-point functions. Examples of the four types of transfer functions are shown in Fig. 22–17. The network of part (a) is driven by a current source and the output is a voltage. The transfer function is an impedance given by

$$H(s) = \frac{E_2}{I_1} = \frac{R_1R_2}{R_1 + R_2 + Ls} = \frac{R_1R_2}{L} \frac{1}{s + \dfrac{R_1 + R_2}{L}}. \tag{22–39}$$

Equation (22–39) has the dimension of ohms. The frequency factor has the dimensions of frequency. The Ls term has the dimensions of ohms. The denominator has the dimensions of ohms, but the numerator has the dimensions of ohms-squared. Hence the over-all expression has the dimensions of ohms.

The network of Fig. 22–17(b) is driven by a voltage source and the output is a current. The transfer function is an admittance given by

$$H(s) = \frac{I_2}{E_1} = \frac{1}{R_2} \frac{\dfrac{R_2}{1 + R_2Cs}}{R_1 + \dfrac{R_2}{1 + R_2Cs}} = \frac{1}{R_1R_2C} \frac{1}{s + \dfrac{R_1 + R_2}{R_1R_2C}}. \tag{22–40}$$

The linear factor in the denominator of Eq. (22–40) has the dimensions of frequency. The Cs term has the dimensions of admittance, or 1/ohms. Hence the over-all expression has the dimensions 1/ohms, as expected.

The network of Fig. 22–17(c) is driven by a current and the output is a current. The transfer function is a dimensionless ratio given by

$$H(s) = \frac{I_2}{I_1} = \frac{R}{Ls + R} = \frac{R}{L} \frac{1}{s + R/L}. \tag{22–41}$$

In Eq. (22–41) both numerator and denominator have the dimensions of ohms. Thus the quotient is dimensionless.

The network of Fig. 22–17(d) is driven by a voltage source and the output is a voltage. The transfer function is a dimensionless ratio given by

$$H(s) = \frac{E_2}{E_1} = \frac{R}{R + 1/Cs} = \frac{s}{s + 1/RC}. \tag{22–42}$$

Both numerator and denominator of Eq. (22–42) have the dimensions of frequency and the ratio is dimensionless, as required.

One further comment will be made on dimensional checking of transfer functions. If a polynomial in s occurs in either the numerator or denominator of the transfer function, every term will have the same dimensions. As the power of s changes, so must the dimensions of the corresponding coefficient. For example, consider the polynomial

$$s^n + a_{n-1}s^{n-1} + a_{n-2}s^{n-2} + \cdots + a_0. \tag{22–43}$$

Every term in Eq. (22–43) must have the dimensions of frequency to the nth power. The coefficient a_{n-1} must have the dimensions of frequency, a_{n-2} of frequency squared, and so on. Finally, the coefficient a_0 also must have the dimensions of frequency to the nth power.

Transfer functions, like driving-point functions, can be factored into expressions of the form

$$H(s) = \frac{K(s - s_1)(s - s_3)\ldots}{(s - s_2)(s - s_4)\ldots}, \tag{22–44}$$

where K is a constant,

$s_1, s_3 \ldots$ are zeros of the function,

$s_2, s_4 \ldots$ are poles of the function.

If an impulse is applied at the input of a network, $H(s)$ represents the output in the frequency domain. If it is expanded into partial fractions, the poles will represent the separate terms in the inverse transform. Since the transient cannot grow indefinitely, these poles must lie in the left half of the s-plane.

No such similar statement can be made about the zeros. They represent frequencies for which no output would be obtained even if an input were applied. They are zeros of signal transmission. They could occur, for example, at the frequency for which a wheatstone bridge is balanced. In general, there are no restrictions on the locations of the zeros of a transfer function. They can occur anywhere. The total number of zeros, however, must not exceed the total number of poles by more than one, in the finite portion of the s-plane. It is customary, in fact, to say that the total number of poles and zeros is the same. If they do not balance in the finite part of the plane, the remaining zeros are at infinity. For example,

$$H(s) = \frac{1}{s^2 + s + 1} \tag{22–45}$$

is a perfectly valid function. It has two complex poles, and apparently no zeros. But note that as s approaches infinity, $H(s)$ approaches zero as $1/s^2$. Thus, in a sense, it has two zeros at infinity.

22–7 Convolution and Laplace transforms. The solution of a problem by means of Laplace transforms involves three or four steps. First the

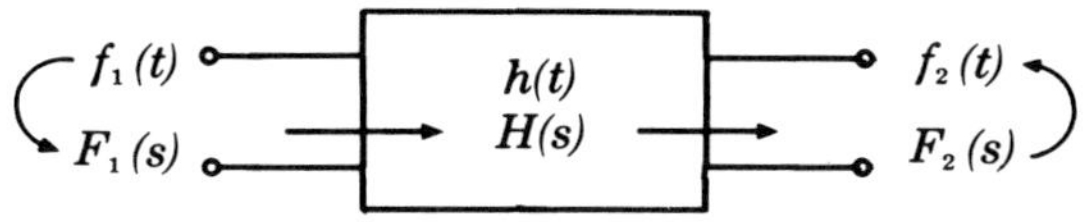

FIG. 22–18. The solution of a general network problem.

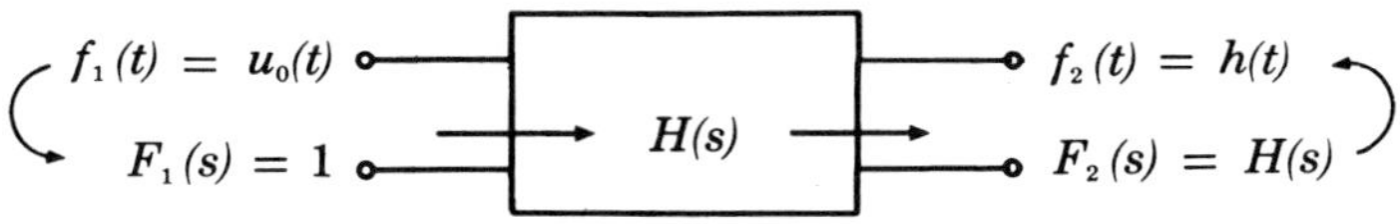

FIG. 22–19. The impulse response of a network.

input function is resolved into the frequency domain. Then the frequency response function of the network is obtained. The output as a function of frequency is obtained by multiplying the input frequency spectrum by the frequency response functions of the network. Finally, the output as a function of time is obtained by finding the inverse transform of the output frequency function. For convenience, a diagram of the process is repeated in Fig. 22–18.

It is natural to ask if there is a way to express the output time function directly in terms of the input time function without going through the transformations to the frequency domain. Such an expression does exist. It is formulated in terms of the inverse transform of the frequency response function $H(s)$. To discover the meaning of this function, $h(t)$, we perform the following experiment: We drive the network of Fig. 22–18 with an impulse function, as shown in Fig. 22–19. The spectrum of the input function is a constant. Thus

$$F_1(s) = 1.$$

The output spectrum, obtained by multiplying the input spectrum by the network function, is therefore

$$F_2(s) = H(s).$$

The corresponding output time function is the inverse transform of $H(s)$. Thus

$$f_2(t) = h(t). \tag{22–46}$$

Equation (22–46) expresses this important result:

The impulse response equals the inverse transform of the system function.

Another way of stating this result is to observe that if we were to record the impulse response function measured in a network, and then to take the

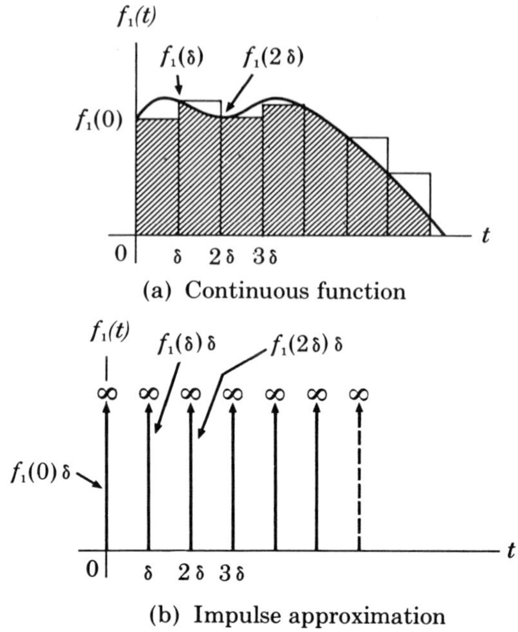

FIG. 22–20. The approximation of a continuous function by impulses.

Laplace transform of this time function, we would obtain the system function $H(s)$ of the network. At times, when the interior of a network is not accessible, this process is very useful. For example, transfer functions of human operators are obtained in this fashion.

In order to exploit this new information for our network functions, we will review convolution theory. Let us divide the input time function into a large number of impulses. We have already learned that there is very little difference between the response to a finite pulse and the response to an impulse if the period of the pulse is short relative to the significant time constants of the system to which it is applied. Our original justification for finding impulse responses was that they would be approximately the same as the response to large, short pulses. Thus we will first replace a continuous function, as shown in Fig. 22–20(a), by a series of contiguous rectangles, and then replace each of the rectangles by an impulse which has the same area. The impulse type of approximation is very good for most physical systems, since these inevitably contain energy-storage elements which cause some averaging of the results. The area of the first rectangle is $f_1(0)\,\delta$, of the second $f_1(\delta)\,\delta$, of the third $f_1(2\delta)\,\delta$, and so on. For simplicity we will take the approximate impulses as occurring at the beginning of the pulses which they represent. Thus the first impulse will be $f_1(0)\,\delta u_0(t)$, the second $f_1(\delta)\,\delta u_0(t - \delta)$, and so on. The expression for the approximate time function is

$$f_1(t) \cong f_1(0)\,\delta u_0(t) + f_1(\delta)\,\delta u_0(t - \delta) + \cdots f_1(n\delta)\,\delta u_0(t - n\delta) + \cdots. \tag{22–47}$$

Equation (22–47) expresses the input, in an approximate way, as a series of delayed impulses. The output for each impulse will be the impulse response of the network multiplied by the appropriate value of the impulse, and delayed by the amount of the delay of the particular input impulse. We can write the output, by inspection, as

$$\begin{aligned} f_2(t) &= f_1(0)\delta h(t) + f_1(\delta)\delta h(t - \delta) + \cdots f_1(n\delta)\delta h(t - n\delta) + \cdots \\ &= \sum_{n=0}^{N} f_1(n\delta)h(t - n\delta)\delta. \end{aligned} \tag{22–48}$$

We now let the spacing between the samples approach zero and the number of samples approach infinity, so that we will still cover the time function. This implies

$$n\delta \to T, \qquad \delta \to dT.$$

It is necessary to use a different variable here for time, since we have two times to consider. The time t is the time at which we view the output, while the time T is the time at which we sample the input function. As the number of samples approaches infinity, the total number of them, N, also approaches infinity, but the new variable T runs only over the range which is being sampled. Suppose the end of the time function $f_1(T)$ occurs at $T = T_N$. Then, in the limit, the summation of Eq. (22–48) becomes the integral

$$f_2(t) = \int_0^{T_N} f_1(T)h(t - T)\,dT. \tag{22–49}$$

Since the impulse response $h(t - T)$ is zero when its argument is negative, it will cause the whole function to be zero for $T > t$. Hence, it is necessary only to perform the integration out to $T = t$. We can write therefore

$$f_2(t) = \int_0^{t} f_1(T)h(t - T)\,dT. \tag{22–50}$$

On the other hand, even though nothing is contributed to the integral for $T > t$, it would not hurt to integrate beyond this point, and we can write

$$f_2(t) = \int_0^{\infty} f_1(T)h(t - T)\,dT. \tag{22–51}$$

Equations (22–50) and (22–51) are the convolution integrals. They are the expressions which give the output time function of the network in terms of the input time function of the network and the impulse response of the network. They are much more complicated than the equivalent relation in the frequency domain, which is simply

$$F_2(s) = F_1(s)H(s). \tag{22–52}$$

It is the simplicity of Eq. (22–52) which makes the frequency-domain solution so desirable! There are relatively few examples in which a direct integration of the convolution integral is at all practical.

The convolution integral* also provides a way of finding the time function which corresponds to the product of two frequency-domain functions whose separate time functions are known. For example, suppose we wish the inverse transform of

$$F(s) = \frac{1}{(s+1)^2}. \tag{22–53}$$

We make the identification

$$F_1(s) = H(s) = \frac{1}{s+1},$$

from which

$$f_1(t) = h(t) = e^{-t}u_{-1}(t).$$

The convolution integral for $f_2(t)$ is

$$f_2(t) = \int_0^\infty e^{-T}u_{-1}(T)e^{-(t-T)}u_{-1}(t-T)\,dT.$$

Since the effect of the unit step functions is to reduce the interval of integration to the range from zero to t, we can write

$$f_2(t) = \int_0^t e^{-t}\,dT = e^{-t}\int_0^t dT = te^{-t}. \tag{22–54}$$

Thus the inverse transform of the original $F(s)$ is te^{-t}.

22–8 The direct transformation of differential equations. We have presented Fourier and Laplace transforms as methods of resolving input time functions into summations of sinusoidal and damped sinusoidal waves. Since we previously obtained network solutions for steady-state sinusoidal and damped sinusoidal inputs, it was therefore a simple matter to obtain the output function for the Fourier or Laplace sum. The transformation of the output back into a time function completed the process. This ap-

* The uses of convolution in the time domain are covered in Chapter 13.

proach is the traditional Fourier transform method. The Laplace transform, however, originated in a slightly different fashion, which we will briefly discuss. Network solutions can be formulated in terms of sets of differential equations in the time domain. The solution of these equations gives the solution to the problem. The Laplace transform originated as an integrating factor used to multiply differential equations and to convert them into algebraic equations more easily solved. After many years the Laplace and Fourier transforms were recognized to be identical. Some books still present Laplace transforms as integrating factors for differential equations. This approach suppresses the physical insight which an engineer must have for a problem, an insight that enables him to synthesize, or to design, as well as merely to analyze. For this reason we have preferred the approach in terms of Fourier components. However, we will develop the formulas for applying Laplace transforms directly to differential equations to round out our treatment of the subject. In the following development we will use the symbol $\mathcal{L}$ to mean "Laplace transform of."

Theorem 1. The Laplace transform of the sum of two functions is the sum of the Laplace transforms of the functions:

$$\mathcal{L}[af_1(t) + bf_2(t)] = aF_1(s) + bF_2(s).$$

Proof:

$$\mathcal{L}[af_1(t) + bf_2(t)] = \int_0^\infty [af_1(t) + bf_2(t)]e^{-st}\,dt$$

$$= a\int_0^\infty f_1(t)e^{-st}\,dt + b\int_0^\infty f_2(t)e^{-st}\,dt$$

$$= aF_1(s) + bF_2(s).$$

Theorem 2. The Laplace transform of the derivative of a time function is s times the Laplace transform of the function minus the value of the function at $t = 0^+$:

$$\mathcal{L}\left[\frac{df(t)}{dt}\right] = sF(s) - f(0^+).$$

Proof:

$$\mathcal{L}\left[\frac{df(t)}{dt}\right] = \int_0^\infty \frac{df(t)}{dt}\,e^{-st}\,dt$$

$$= \int_0^\infty e^{-st}\,d[f(t)].$$

We can integrate this expression by parts, using the formula

$$\int u\,dv = uv - \int v\,du.$$

The result is

$$[e^{-st}f(t)]_0^\infty + s\int_0^\infty f(t)e^{-st}\,dt.$$

The first term is zero at $t = \infty$ because of the convergence factor in the s. At zero, the first term has the value $f(0)$. The value at $t = 0^+$ should be used in order to include the effect of any jumps or impulses which occur at $t = 0$. Thus

$$\mathcal{L}\left[\frac{df(t)}{dt}\right] = sF(s) - f(0^+).$$

Repeated applications of this formula give the transforms of derivatives of order higher than one. For example,

$$\mathcal{L}\left[\frac{d^2f(t)}{dt^2}\right] = s^2F(s) - sf(0^+) - \frac{df(0^+)}{dt}.$$

Theorem 3. The Laplace transform of the integral of a function is the Laplace transform of the function divided by s, plus the value of the integral at zero divided by s:

$$\mathcal{L}\left[\int_{-\infty}^t f(t)\,dt\right] = \frac{F(s)}{s} + \frac{f^{-1}(0^+)}{s}.$$

Proof:

$$\mathcal{L}\left[\int_{-\infty}^t f(t)\,dt\right] = \int_0^\infty \left[\int_{-\infty}^t f(t)\,dt\right] e^{-st}\,dt.$$

$$= \int_0^\infty \left[\int_{-\infty}^t f(t)\,dt\right] \frac{d(e^{-st})}{-s}.$$

Integration by parts gives

$$\left[\frac{\int_{-\infty}^t f(t)\,dt\,e^{-st}}{-s}\right]_0^\infty + \frac{1}{s}\int_0^\infty f(t)e^{-st}\,dt.$$

The value of the first term at $t = \infty$ is zero. We will write the value at $t = 0$ in terms of $f^{-1}(t)$ to represent the integral. Thus

$$\mathcal{L}\left[\int_{-\infty}^t f(t)\,dt\right] = \frac{F(s)}{s} + \frac{f^{-1}(0^+)}{s}.$$

Further applications of this formula give the transforms for higher integrals. For example,

$$\mathcal{L}\left[\int_{-\infty}^{t}\int_{-\infty}^{t} f(t)\,dt\,dt\right] = \frac{F(s)}{s^2} + \frac{f^{-1}(0^+)}{s^2} + \frac{f^{-2}(0^+)}{s}.$$

We will illustrate the application of Laplace transforms to the solution of differential equations by two examples.

Example 1. Given

$$L\frac{di}{dt} + Ri = u_{-1}(t),$$

$$i(0^+) = 1 \text{ amp},$$

find $i(t)$. We will write the transform of the unknown current as $I(s)$. In terms of this transform we can write the differential equation as

$$L[sI(s) - i(0^+)] + RI(s) = \frac{1}{s}.$$

The solution for $I(s)$ is

$$I(s)(Ls + R) = \frac{1}{s} + L. \tag{22-55}$$

From a circuit point of view, Eq. (22–55) is the frequency-domain equation for an R–L circuit driven by a unit step of voltage, and with an initial current in the coil of 1 amp which has been automatically transformed into an impulse source of $Li(0^+)$ in series with the coil. The solution for $I(s)$ is

$$I(s) = \frac{1/L}{s(s + R/L)} + \frac{1}{s + R/L},$$

and the solution in the time domain is

$$i(t) = \frac{1}{R}(1 - e^{-Rt/L}) + e^{-Rt/L}.$$

Example 2. Given

$$\frac{d^2i}{dt^2} + 2\frac{di}{dt} + i = 0,$$

$$i(0^+) = 0, \qquad \frac{di(0^+)}{dt} = 1,$$

find $i(t)$. The transform of this equation and its derivatives gives

$$I(s)(s^2 + 2s + 1) = 1, \qquad I(s) = \frac{1}{(s + 1)^2}.$$

Since $I(s)$ is identical to Eq. (22–53) the corresponding $i(t)$, from Eq. (22–54), is

$$i(t) = te^{-t}.$$

This second example illustrates one advantage of Laplace transforms for solving linear equations. The special case of equal roots, which is a singular solution in classical differential equations, is carried through by the same process as is any other solution. Although it would be theoretically possible to write the differential equations for all our network problems and then to transform them by the methods of this section, it usually is much easier to work with the circuits in the frequency domain, and to forget about the differential equations entirely.

There are two special cases, however, in which the differential equations yield simple and useful formulas. These two cases relate to the initial and final values of a time function. We shall consider the initial value first.

Initial value theorem. From Theorem 3

$$sF(s) - f(0^+) = \int_0^\infty f'(t)e^{-st}\,dt. \tag{22–56}$$

If we let s approach infinity in Eq. (22–56), the convergence factor, e^{-st}, becomes zero everywhere except at $t = 0$. Hence the integral becomes zero, unless there is an impulse at the origin, and we obtain the following formula for the initial value of the time function:

$$f(0^+) = [sF(s)]_{s\to\infty} \tag{22–57}$$

An example of a function which can be treated by means of Eq. (22–57) is the unit step function, $u_{-1}(t)$, whose Laplace transform is $1/s$. The initial value is given by

$$\begin{aligned} f(0^+) &= \left[s \times \frac{1}{s}\right]_{s\to\infty} \\ &= 1. \end{aligned}$$

An example which cannot be treated by means of Eq. (22–57) is the unit impulse function, $u_0(t)$, whose Laplace transform is 1. According to Eq. (22–57), the initial value of this function at $t = 0^+$ should be

$$f(0^+) = [s \times 1]_{s\to\infty} = \infty.$$

This is not the correct value since the impulse has returned to zero by $t = 0^+$. Since most functions do not have impulses at the origin, this restriction is not a serious one for practical problems.

Final value theorem. The initial value theorem gives the initial value of a time function in terms of the final value of the frequency function.

The final value theorem gives the final, or constant, value of the time function in terms of the zero frequency behavior of the frequency function. Suppose that the function $F(s)$ has been expanded in partial fraction form:

$$F(s) = \frac{A_0}{s} + \frac{A_1}{s - s_1} + \cdots. \tag{22-58}$$

The constant value of the corresponding time function, $f(t)$, is given by A_0, and if the other terms die out at $t = \infty$, then

$$f(\infty) = A_0 = [sF(s)]_{s\to 0}. \tag{22-59}$$

An example to which Eq. (22-59) is applicable is the unit step function, $u_{-1}(t)$, with the transform $1/s$. The value at infinity is given by

$$f(\infty) = \left[s \times \frac{1}{s}\right]_{s\to 0} = 1.$$

An example which cannot be treated by means of Eq. (22-59) is the function $\sin t$, with the Laplace transform $1/(s^2 + 1)$. According to the formula the final value should be

$$f(\infty) = \left[\frac{s}{s^2 + 1}\right]_{s\to 0} = 0.$$

This value is not correct since the sine function is indeterminate at infinity. The formula gives the correct answer only if the transient terms die out at infinity.

The initial value theorem and the final value theorem are often useful in practical problems. They show that the high frequencies in the frequency domain are associated with a sudden jump at $t = 0$, and that the low frequencies in the frequency domain are associated with the behavior over long periods of time. Although the two ends of the time and frequency scales are thus related there is unfortunately no simple relationship covering other points in the two domains.

22–9 Summary.

The transform method of solution

Time domain *Frequency domain*

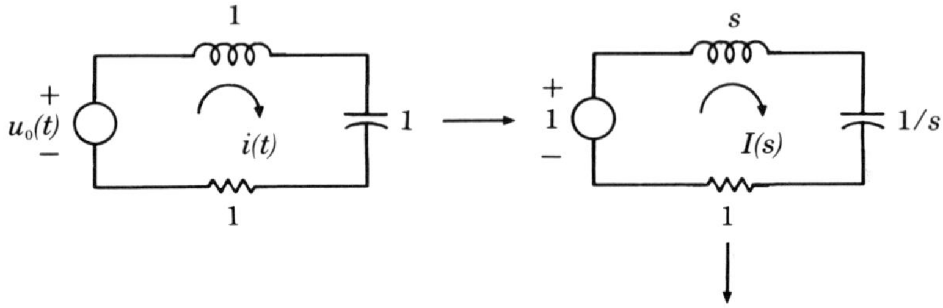

$$i(t) = \frac{1}{0.866} e^{-0.5t} \cos(0.866t + 30°) \quad \leftarrow \quad I(s) = \frac{s}{s^2 + s + 1}$$

Time functions *Frequency functions*

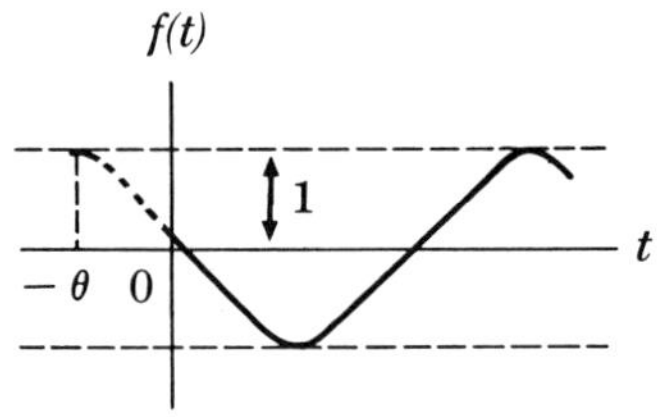

$$f(t) = \cos(\omega_0 t + \theta)u_{-1}(t) \qquad F(s) = \frac{s\cos\theta - \omega_0 \sin\theta}{s^2 + \omega_0^2}$$

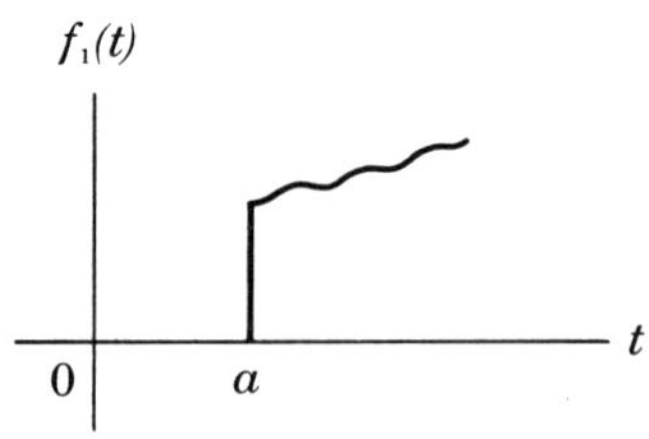

$$f_1(t) = f(t - a)u_{-1}(t - a) \qquad F_1(s) = e^{-as}F(s)$$

Time functions *Frequency functions*

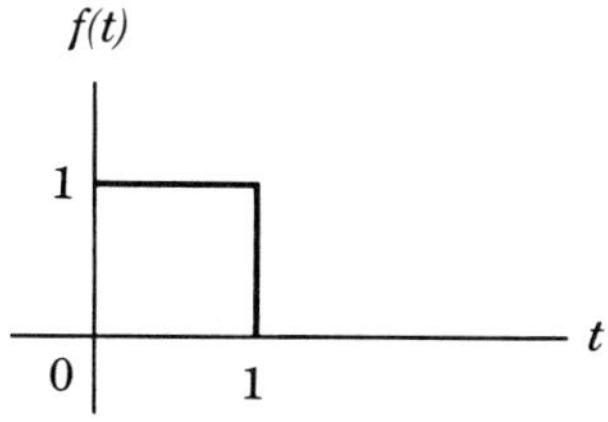

$$f(t) = u_{-1}(t) - u_{-1}(t-1) \qquad F(s) = \frac{1 - e^{-as}}{s}$$

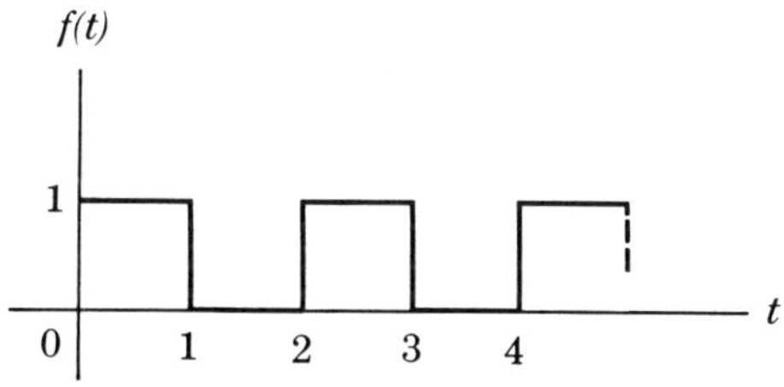

$$f(t) = u_{-1}(t) - u_{-1}(t-1) + u_{-1}(t-2) + \ldots \qquad F(s) = \frac{1}{s(1 - e^{-s})}$$

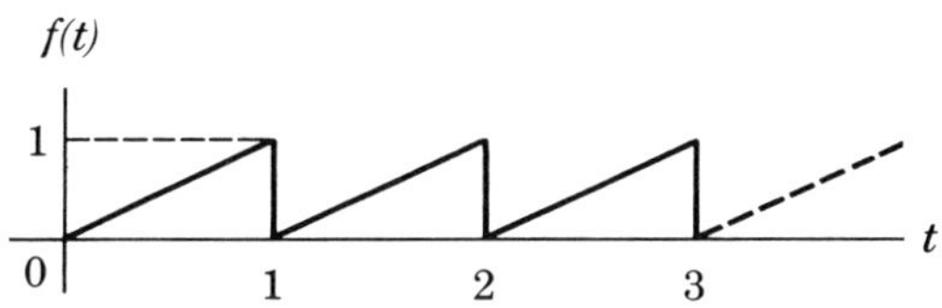

$$f(t) = tu_{-1}(t) - u_{-1}(t-1) - u_{-1}(t-2) + \cdots \qquad F(s) = \frac{1 - e^{-s} - se^{-s}}{s^2(1 - e^{-s})}$$

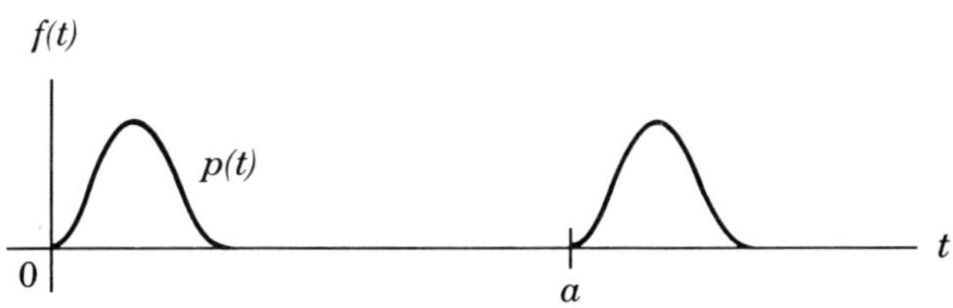

$$f(t) = p(t) + p(t-a)u_{-1}(t-a) + p(t-2a)\,u_{-1}(t-2a) + \cdots \qquad F(s) = \frac{P(s)}{1 - e^{-as}}$$

Driving-point impedance

$$Z(s) = \frac{K(s - s_1)(s - s_3)\ldots}{(s - s_2)(s - s_4)\ldots}$$

$s_1, s_3, \ldots$ are zeros
$s_2, s_4, \ldots$ are poles

Zeros = short-circuit natural resonant frequencies
Poles = open-circuit natural resonant frequencies

Passive network: poles and zeros are in left half of plane
L–C network: poles and zeros are on $j\omega$ axis
R–C, *R–L* networks: poles and zeros are on negative real axis

Transfer functions

$$H(s) = \frac{K(s - s_1)(s - s_3)\ldots}{(s - s_2)(s - s_4)\ldots}$$

$s_1, s_3, \ldots$ are zeros
$s_2, s_4, \ldots$ are poles

Zeros = zeros of transmission and *not* transient resonant frequencies
Poles = open-circuit natural resonant frequencies

Passive network: poles are in left half of plane
zeros occur anywhere

Convolution

Time domain *Frequency domain*

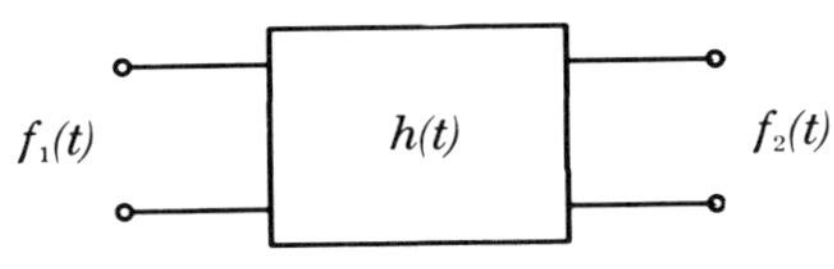

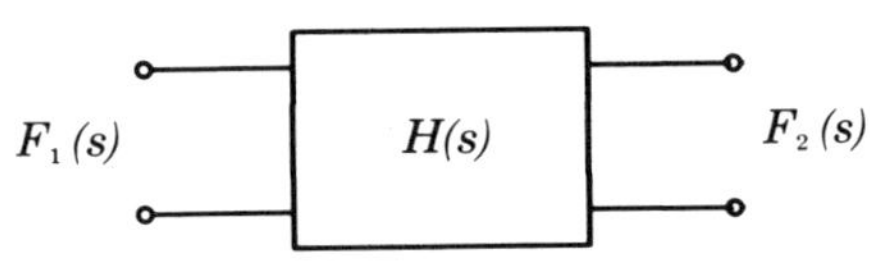

$$f_2(t) = \int_0^\infty f_1(T)h(t - T)\,dT \qquad F_2(s) = F_1(s)H(s)$$

Transformation of differential equations

$$\mathcal{L}[af_1(t) + bf_2(t)] = aF_1(s) + bF_2(s)$$

$$\mathcal{L}\left[\frac{df(t)}{dt}\right] = sF(s) - f(0^+)$$

$$\mathcal{L}\left[\frac{d^2f(t)}{dt^2}\right] = s^2F(s) - sf(0^+) - f'(0^+)$$

$$\mathcal{L}\left[\int_{-\infty}^{t} f(t)\,dt\right] = \frac{F(s)}{s} + \frac{f^{-1}}{s}$$

$$\mathcal{L}\left[\int_{-\infty}^{t}\int_{-\infty}^{t} f(t)\,dt\,dt\right] = \frac{F(s)}{s^2} + \frac{f^{-1}(0^+)}{s^2} + \frac{f^{-2}(0^+)}{s}$$

Initial value theorem

$$f(0^+) = [sF(s)]_{s\to\infty} \qquad \text{if there are no impulses at } t = 0$$

Final value theorem

$$f(\infty) = [sF(s)]_{s\to 0} \qquad \text{if the other terms die out}$$

REFERENCES

Chapter 22 is devoted to the very useful techniques and routine procedures of the Laplace-transform method. The frequency-domain response functions are the basis for modern network synthesis, and some references are given below. Laplace transforms can also be applied directly to differential equations. Some references to this technique are also given.

1. E. A. Guillemin, *Synthesis of Passive Networks*, John Wiley & Sons, New York, 1957. A complete presentation of the mathematical background for network synthesis and filter design in terms of the Laplace transform response functions of the networks.

2. D. F. Tuttle, *Network Synthesis*, Vol. 1, John Wiley & Sons, New York, 1958. Tuttle devotes 1163 pages just to the consideration of two-terminal networks. The book is beautifully written and monumentally complete.

3. M. E. Van Valkenburg, *Introduction to Modern Network Synthesis*, John Wiley & Sons, New York, 1960. Van Valkenburg presents the material of network synthesis at a level which can be appreciated by good undergraduates.

4. N. Balabanian, *Network Synthesis*, Prentice-Hall, New Jersey, 1958. Another book presenting network synthesis in terms of the Laplace-transform response functions of the systems.

5. Gardner and Barnes, *Transients in Linear Systems*, John Wiley & Sons, New York, 1942. Gardner and Barnes make the Laplace transformation of differential equations the central theme of their presentation.

6. R. V. Churchill, *Modern Operational Methods in Engineering*, McGraw-Hill, New York, 1951.

7. C. R. WYLIE, *Advanced Engineering Mathematics*, McGraw-Hill, New York, 1951.

8. W. T. THOMPSON, *Laplace Transformation*, Prentice-Hall, New Jersey, 1950 (2nd edition 1960). This and the two preceding texts apply Laplace transforms directly to differential equations.

EXERCISES

Section 22–2

1. In the circuit of Fig. 22–21 the stimulus is a step current of amplitude 10 amp and the desired response is the voltage $e(t)$. (a) Obtain $E(s)$ from a frequency-domain diagram. (b) Obtain $e(t)$ from $E(s)$.

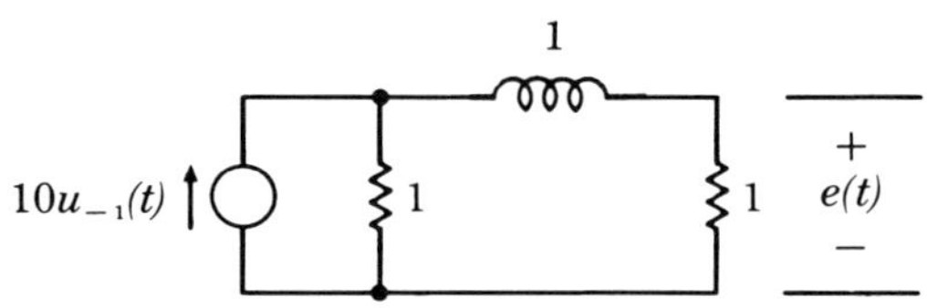

FIGURE 22–21

2. Find the voltage $e(t)$ in the circuit of Fig. 22–22 by the Laplace transform method.

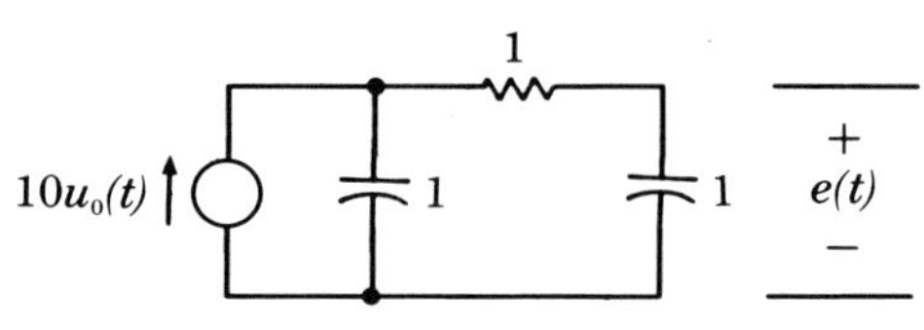

FIGURE 22–22

3. For the circuit of Fig. 22–23 draw the frequency-domain diagram and obtain the equivalent Thévenin circuit to the left of terminal-pair a-b. From the Thévenin circuit find the current $i(t)$.

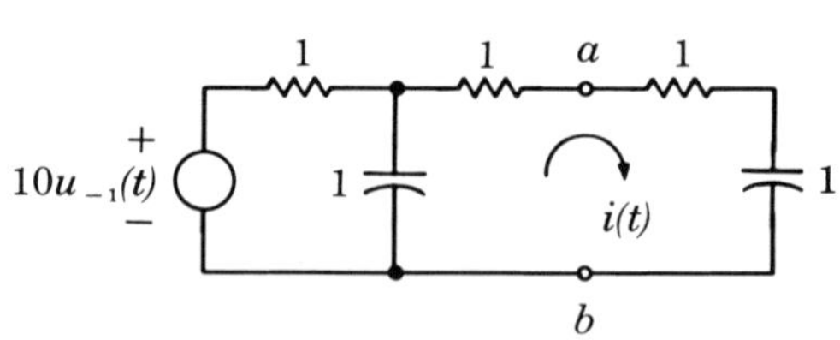

FIGURE 22–23

4. For the circuit of Fig. 22–24 find the Thévenin equivalent at terminal-pair a-b in the frequency domain. Use this circuit to solve for the current $i(t)$.

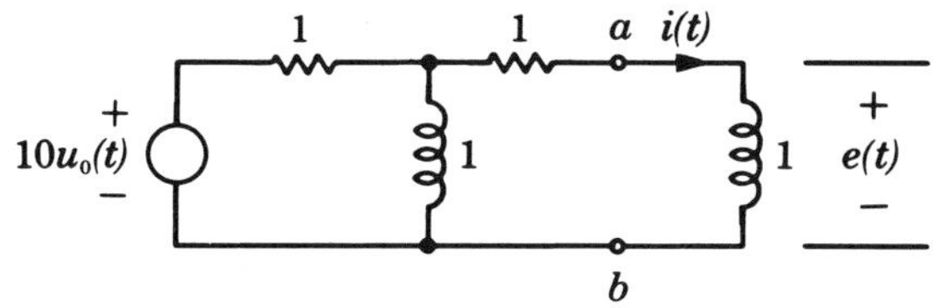

FIGURE 22–24

5. Use the Laplace transform method to find the voltage $e(t)$ in the circuit of Fig. 22–25.

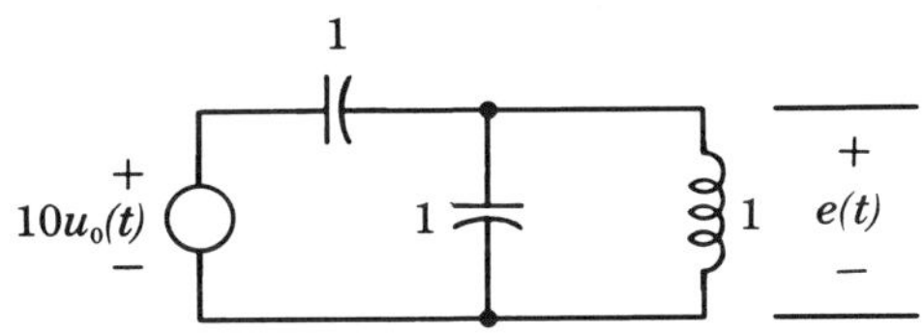

FIGURE 22–25

6. Find the current $i(t)$ in the circuit of Fig. 22–26 by means of the Laplace transform.

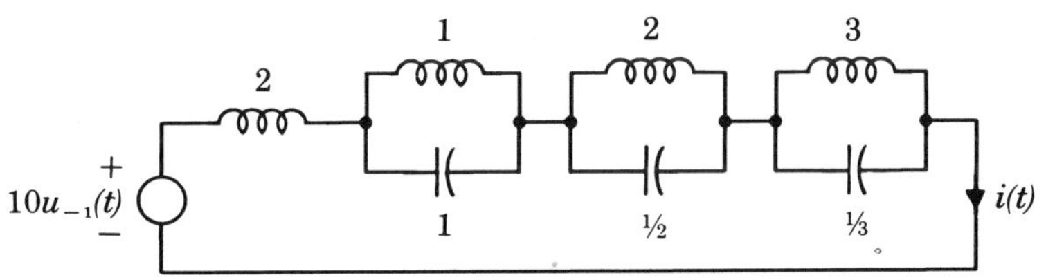

FIGURE 22–26

7. The circuit of Fig. 22–27 is a "low-pass" filter. If all frequencies are present at the input, only the low frequencies will appear at the output. The transition region between those frequencies which are passed and those which are rejected occurs around 1 radian per second. Find the impulse response $e(t)$ of the filter.

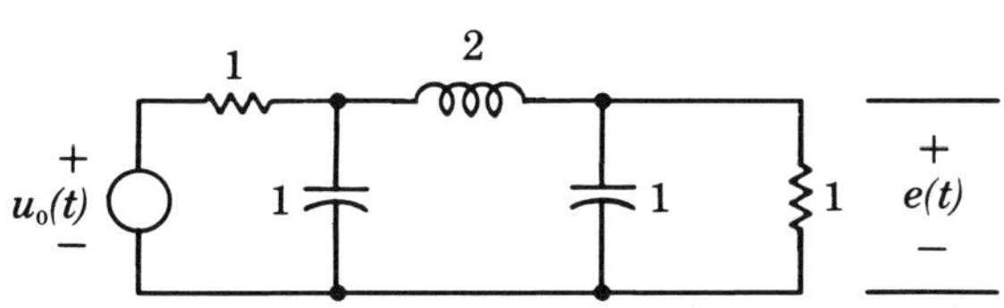

FIGURE 22–27

8. The switch in the circuit of Fig. 22–28 is closed at $t = 0$. (a) Make an equivalent Thévenin circuit and thus find the current $i(t)$ through the switch. (b) Replace the source and the initial energy storages by step or impulse sources and solve the complete problem for $i(t)$. (Previous to $t = 0$ a steady-state has been reached in the circuit.)

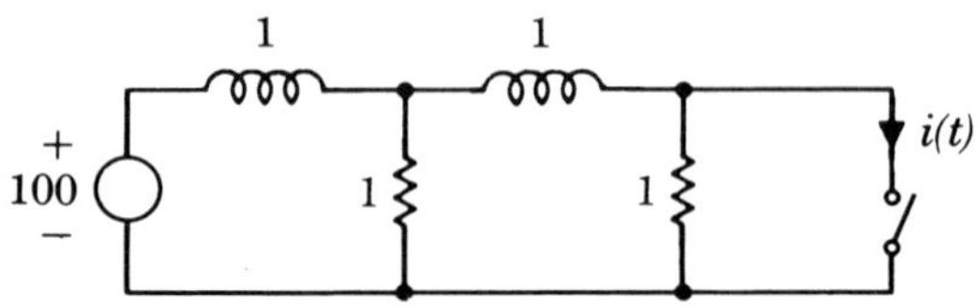

FIGURE 22–28

9. The switch in the circuit of Fig. 22–29 is opened at $t = 0$. (a) Make an equivalent Norton circuit and thus find the voltage $e(t)$ across the open switch. (b) Replace the source and the initial energy storages by step or impulse sources and solve for $e(t)$ from the original circuit. (Previous to $t = 0$ a steady-state has been reached in the circuit.)

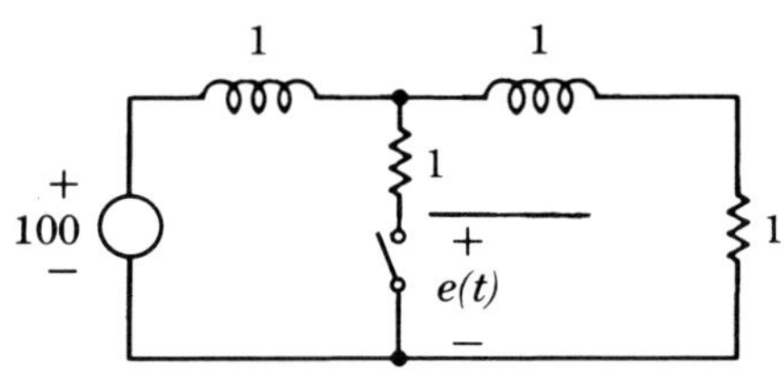

FIGURE 22–29

Section 22–3

10. The voltage source in the circuit of Fig. 22–30 is suddenly applied at $t = 0$. Use the Laplace transform method to find the current which flows for $t > 0$ and separate it into its steady-state and transient components. Check the steady-state value by phasor algebra.

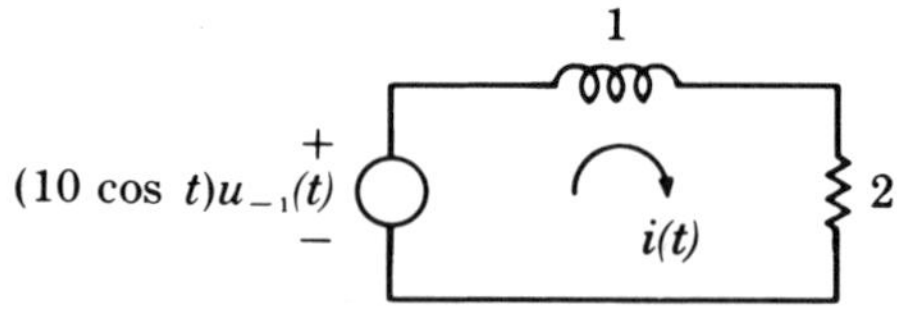

FIGURE 22–30

11. The current source in the circuit of Fig. 22–31 is applied at $t = 0$. Use the Laplace-transform method to find the voltage which is produced for $t > 0$ and separate it into its steady-state and transient components. Check the steady-state value by phasor algebra.

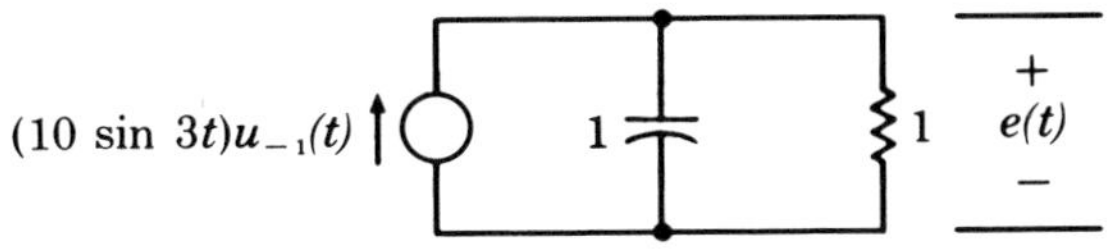

FIGURE 22–31

12. The voltage source in the circuit of Fig. 22–32 is applied at $t = 0$. Use the Laplace transform method to find the current which flows and separate it into its steady-state and transient components. Check the steady-state value by phasor algebra.

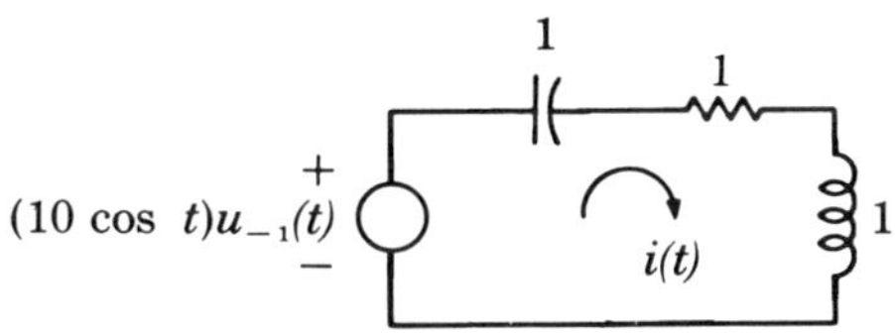

FIGURE 22–32

13. The sinusoidal current applied to the parallel L–C circuit of Fig. 22–33 is at the resonant frequency of the circuit, and hence there is no steady-state component in the output voltage. Find the output voltage by the method of Laplace transforms. What term replaces the steady-state in the response?

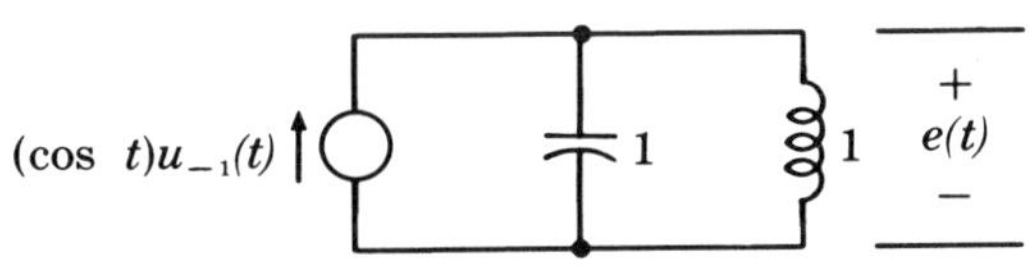

FIGURE 22–33

14. In the circuit of Fig. 22–34 it is possible to vary the phase angle of the input sinusoidal voltage which is connected to the circuit at $t = 0$. Find the value of the phase angle θ which will result in no transient term in the solution for the current in the circuit.

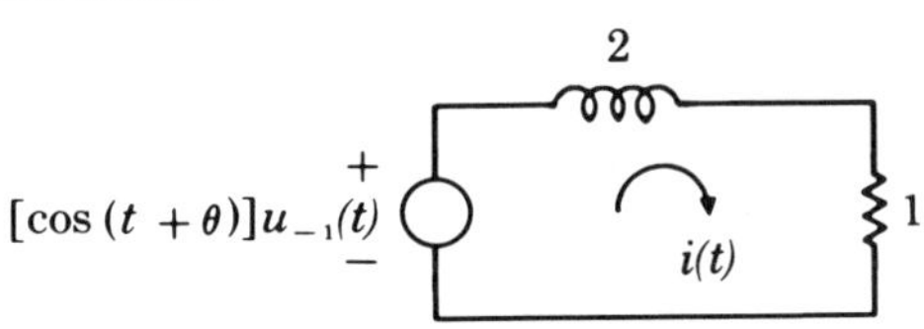

FIGURE 22–34

15. An alternative way of reducing the transient to zero in the circuit of Exercise 14 is to apply an impulse along with the sinusoid at $t = 0$. If the input voltage to the circuit is

$$e(t) = \cos t \cdot u_{-1}(t) + Au_0(t),$$

what value of A will result in no transient term in the current solution.

16. In a two-energy-storage circuit it is not possible to reduce the transient term to zero by adjusting the phase angle of the input stimulus. For the circuit of Fig. 22–35 find the value of the phase angle θ which results in the maximum value of the transient term, and the value of θ which results in the minimum value of the transient term.

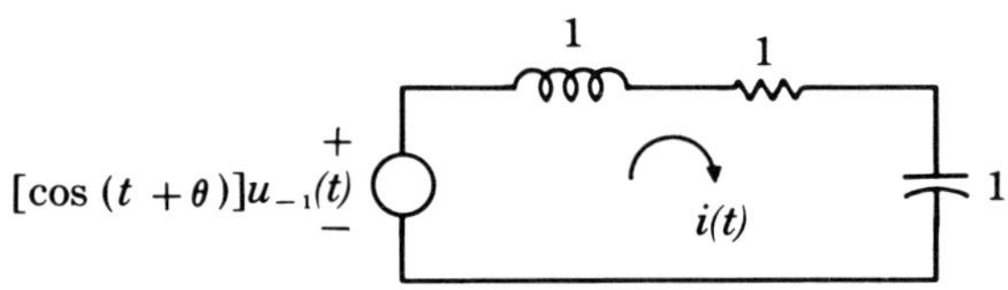

FIGURE 22–35

17. The addition of an impulse voltage and a step voltage to a sinusoid at $t = 0$ can reduce the transient term which is produced to zero. For the circuit of Fig. 22–36 write the general equations which must be satisfied by A and E if there is to be no transient term. Show that there is a solution with real values of A and E.

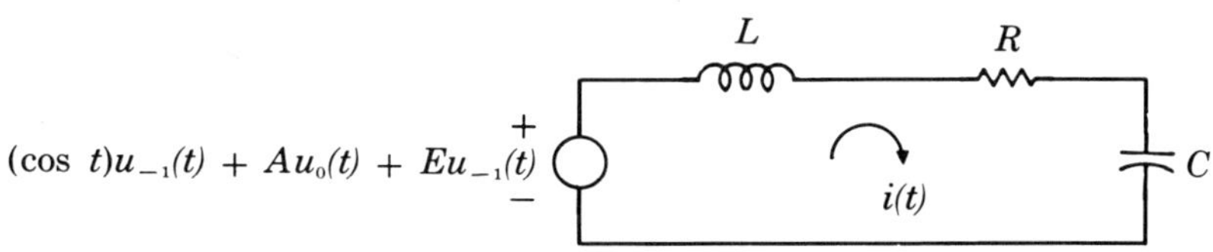

FIGURE 22–36

Section 22–4

18. Sketch the following functions and write their Laplace transforms.

(a) $\sin t$ (b) $\sin t \cdot u_{-1}(t)$
(c) $\sin t \cdot u_{-1}(t - \pi)$ (d) $\sin (t - \pi) \cdot u_{-1}(t - \pi)$
(e) $\sin (t - \pi) \cdot u_{-1}(t)$.

19. The waveform of Fig. 22–37 is one-half cycle of a sine wave. Find its Laplace transform.

20. Find the Laplace transform of the waveform shown in Fig. 22–38.

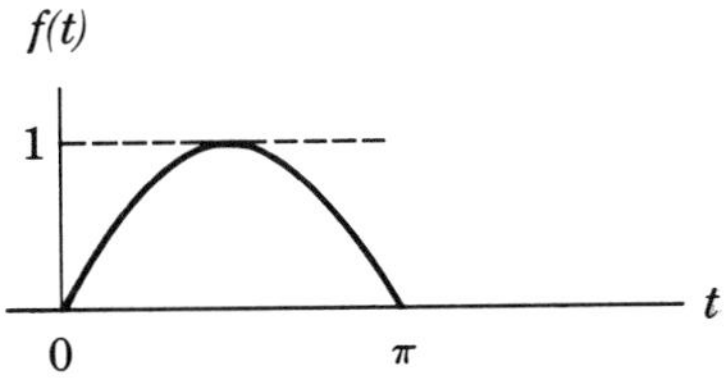

FIGURE 22–37

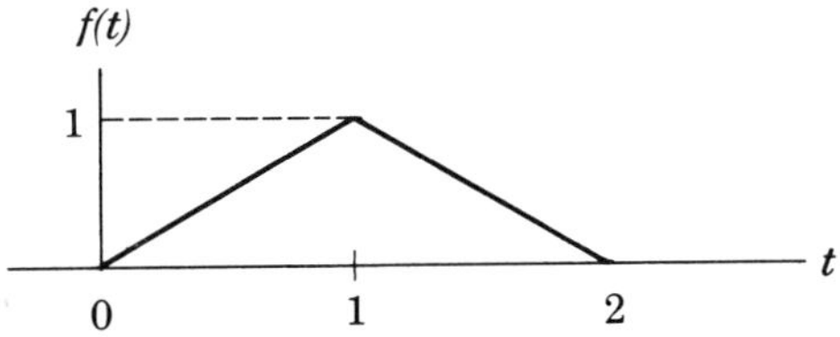

FIGURE 22–38

21. (a) Write the expression for the waveform of Fig. 22–39 in terms of delayed singularity functions. (b) Obtain its Laplace transform.

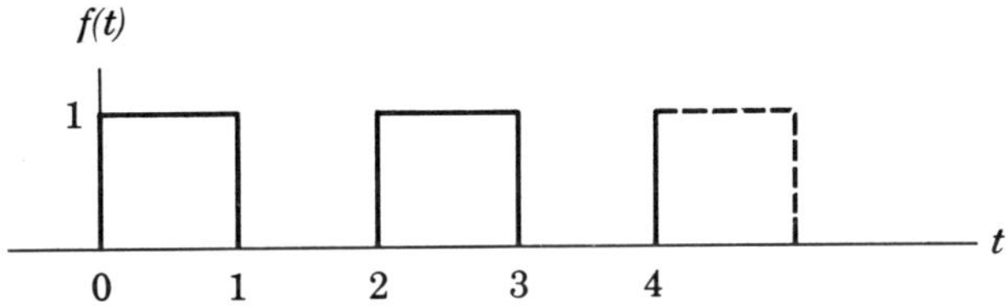

FIGURE 22–39

22. (a) Write the expression for the waveform of Fig. 22–40 in terms of delayed singularity functions. (b) Obtain its Laplace transform.

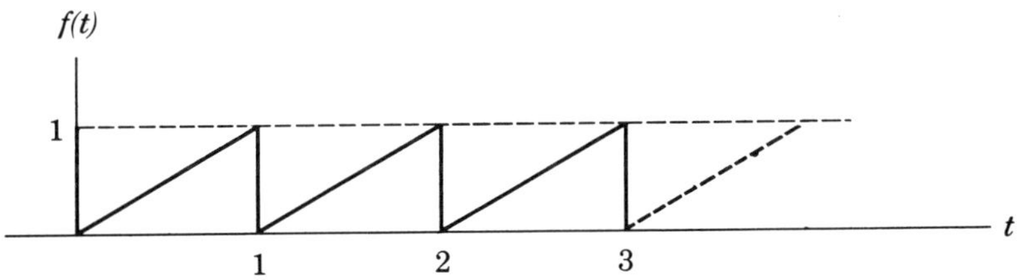

FIGURE 22–40

23. (a) Write the time function for the first pulse of the waveform in Fig. 22–41. (b) Give the Laplace transform for the pulse. (c) Give the Laplace transform of the periodic waveform.

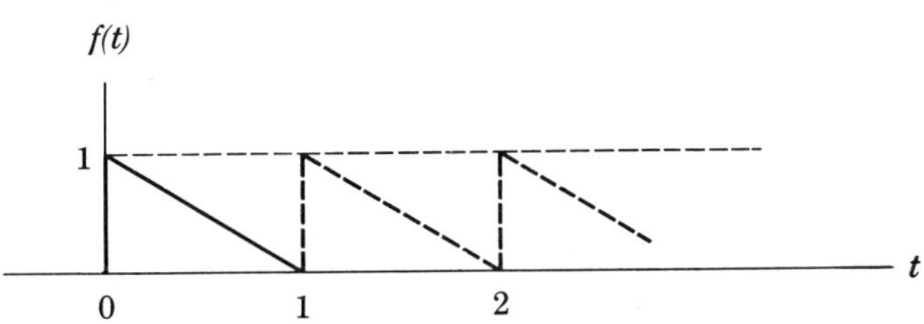

FIGURE 22–41

24. (a) Write the time expression for a single pulse of the waveform in Fig. 22–42. (b) Give the Laplace transform of the single pulse. (c) Give the Laplace transform of the periodic waveform.

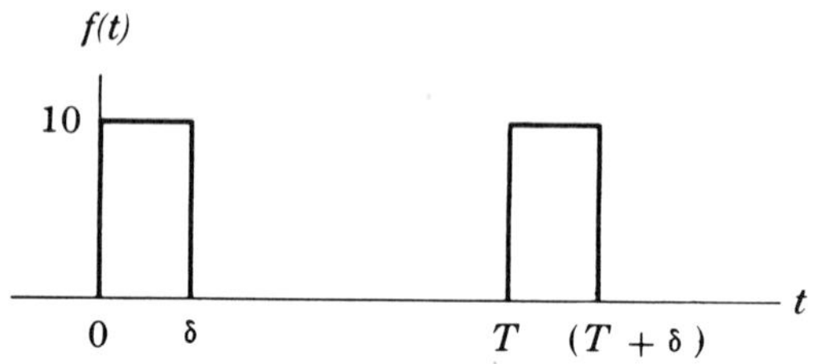

FIGURE 22–42

25. Give and sketch the function $f(t)$ whose Laplace transform is

$$F(s) = \left(\frac{1 - e^{-s}}{s}\right)^3.$$

26. Give and sketch the time function whose Laplace transform is

$$F(s) = \frac{1}{s \sinh s}.$$

27. Give and sketch the time function whose Laplace transform is

$$F(s) = \frac{1}{s \cosh s}.$$

Section 22–5

28. A complex-frequency-domain function is given by

$$F(s) = \frac{1}{1 - e^{-s}}.$$

(a) Find and sketch the corresponding time function in terms of delayed singularity functions. (b) Resolve $F(s)$ into partial fractions and thus find the Fourier series for the time function.

29. Repeat Exercise 28 for the frequency-domain function

$$F(s) = \frac{1}{1 + e^{-s}}.$$

30. (a) Write the Laplace transform for the waveform of Fig. 22–43. (b) Expand the transform in partial fractions and thus obtain the Fourier series for the original waveform.

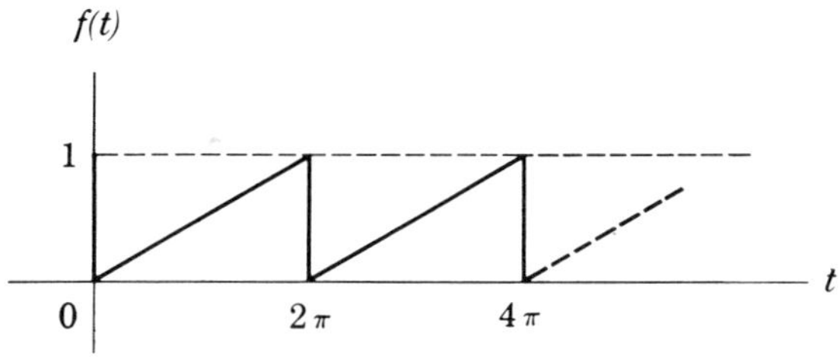

FIGURE 22–43

31. (a) Using delayed factors write the Laplace transform for the half-wave rectified sine wave of Fig. 22–44. (b) By resolving the transform into partial fractions obtain the Fourier series for the original waveform.

32. Repeat Exercise 31 with the waveform a full-wave rectified sine wave instead of the half-wave rectified sine wave which is shown.

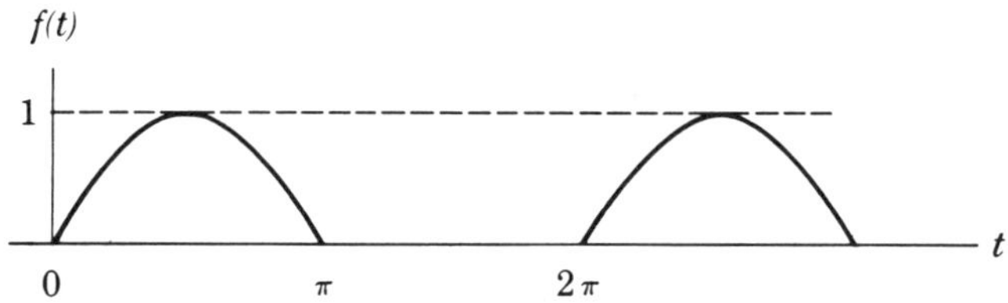

FIGURE 22–44

33. The current in the circuit of Fig. 22–45 is an impulse chain. Find the voltage across the capacitance (a) as a series of singularity functions and (b) as a Fourier series.

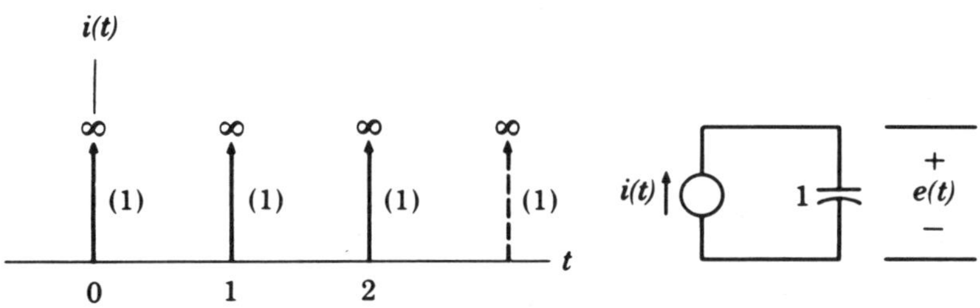

FIGURE 22–45

34. The voltage applied to the circuit of Fig. 22–46 is a symmetrical square wave as shown. (a) Find the component in the current which occurs at the natural frequency of the circuit. (b) Find the component of the current which occurs at the fundamental frequency of the applied square wave.

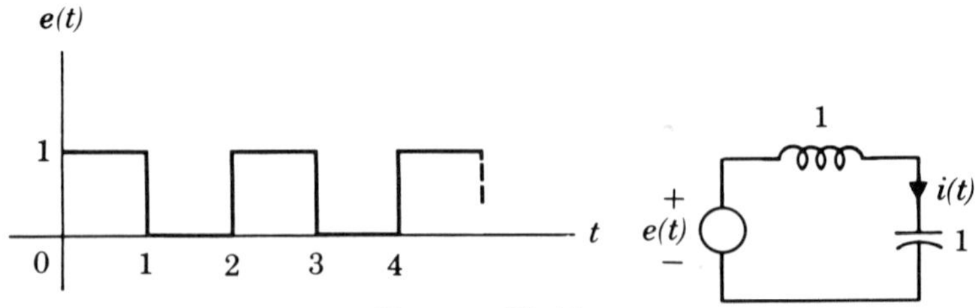

FIGURE 22–46

35. In the circuit of Fig. 22–47:

(a) Find the Laplace transform of the current.
(b) Write the time function for the current in terms of delayed functions, and sketch it.
(c) By expanding the transform in partial fractions obtain the transient and steady-state components of the current.

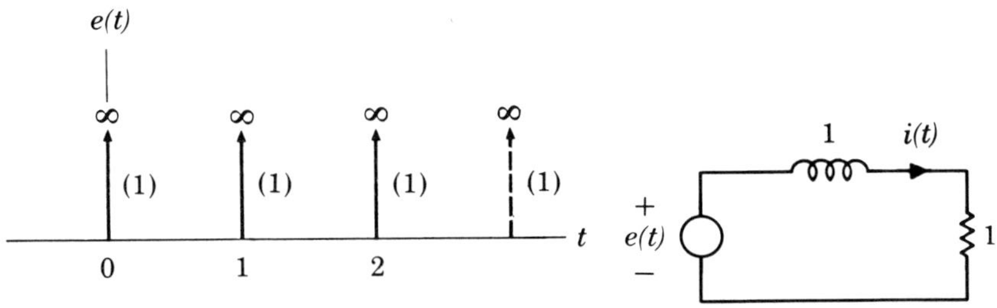

FIGURE 22–47

Section 22–6

36. (a) Find the input impedance for the network of Fig. 22–48. (b) Give the poles and zeros. (c) Give the meaning of the poles and zeros in terms of the natural transients in the circuit.

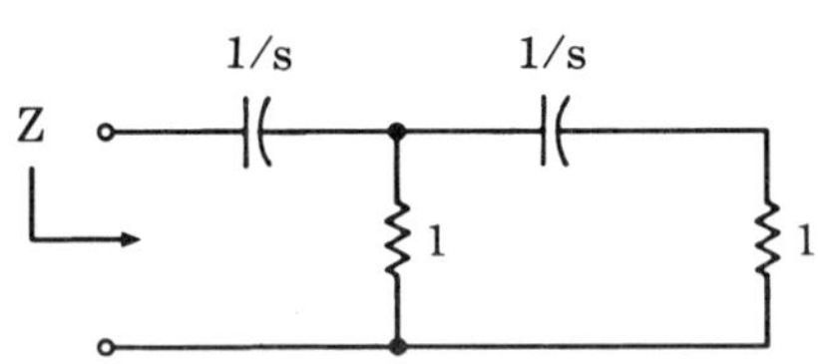

FIGURE 22–48

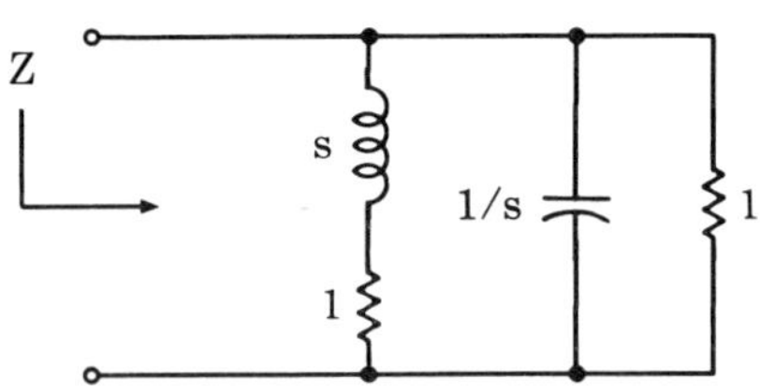

FIGURE 22–49

37. (a) Find the driving-point impedance of the network of Fig. 22–49. (b) Give the poles and zeros. (c) Explain the meaning of the poles and zeros in terms of the natural transient frequencies of the circuit.

38. (a) Find the driving-point impedance of the circuit in Fig. 22–50, subject to the condition that $R^2 = L/C$. (b) A constant voltage of 100 volts is applied to the circuit and allowed to remain until a steady-state is reached. It is then removed at $t = 0$. Find $e_2(t)$, $e_3(t)$, and $e_1(t)$ as functions of time.

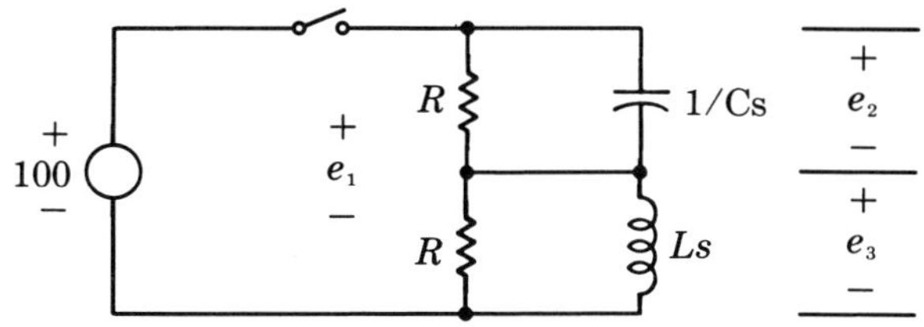

FIGURE 22–50

39. (a) Find the input impedance of the symmetrical lattice shown in Fig. 22–51. (b) Under what conditions is the input impedance equal to 1 ohm?

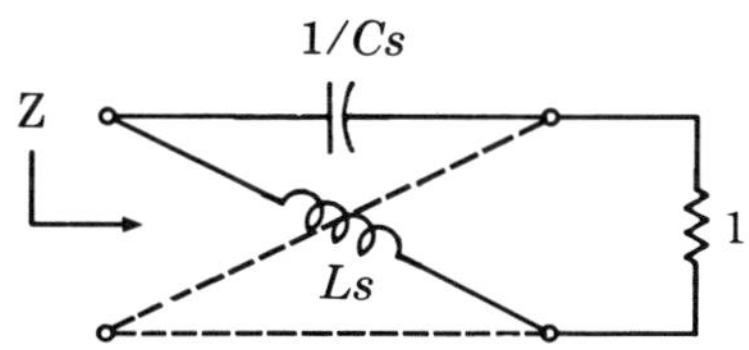

FIGURE 22–51

40. Find the input impedance of each of the networks of Fig. 22–52.

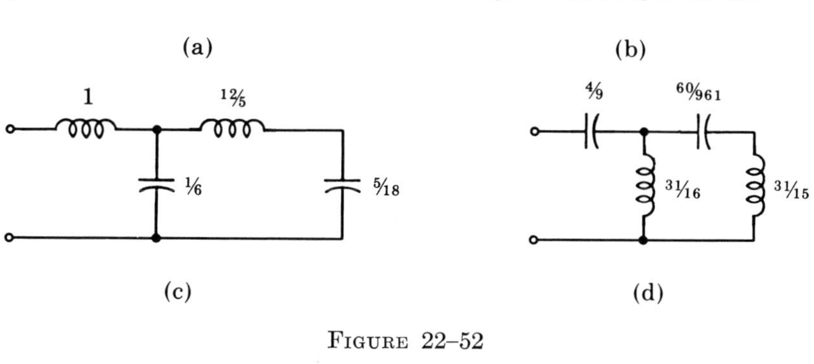

FIGURE 22–52

41. (a) Find the ratio of E_2/I_1, that is, Z_{12}, for the network of Fig. 22–53 (b) Give the zeros and the poles of Z_{12}.

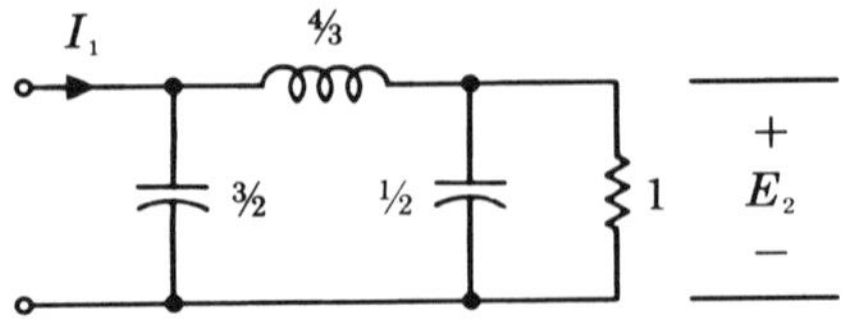

FIGURE 22–53

42. The general two-terminal-pair network of Fig. 22–54 is described by the z parameters according to the equations

$$E_1 = z_{11}I_1 + z_{12}I_2,$$
$$E_2 = z_{12}I_1 + z_{22}I_2.$$

A 1-ohm resistance is then added across the output terminals. Find (a) the input impedance E_1/I_1 and (b) the transfer impedance E_2/I_1.

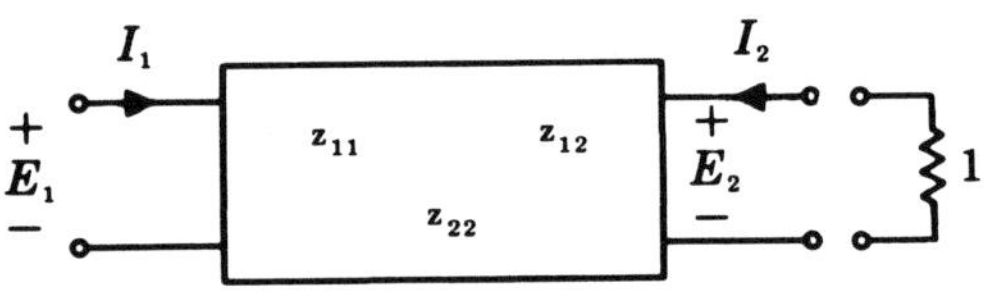

FIGURE 22–54

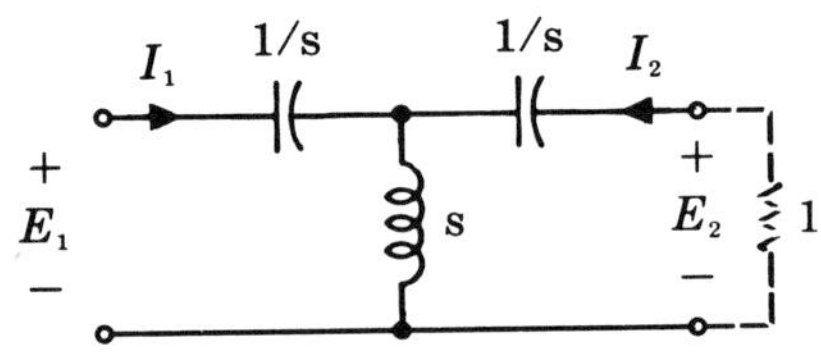

FIGURE 22–55

43. (a) Find the z parameters, z_{11}, z_{12}, and z_{22}, for the circuit of Fig. 22–55. (b) If the circuit is loaded with a 1-ohm resistance at the output, find Z_{12}.

Section 22–7

44. By means of the convolution integral evaluate the inverse Laplace transforms of the functions given below.

(a) $\dfrac{1}{s^2}$ (b) $\dfrac{1}{(s+a)^2}$ (c) $\dfrac{1}{s(s+a)}$

(d) $\dfrac{1}{s(s+a)^2}$ (e) $\dfrac{s}{(s+a)^2}$

45. Find the inverse transforms of the following Laplace transforms by making use of Table 20–4 and the convolution integral.

(a) $\dfrac{1}{(s+a)(s+b)}$ (b) $\dfrac{1}{(s+a)(s+b)(s+c)}$ (c) $\dfrac{1}{(s^2+1)^2}$

46. (a) Show that the triple convolution product is the product of the three transforms:

$$f_1(t)*f_2(t)*f_3(t) = F_1(s)\cdot F_2(s)\cdot F_3(s).$$

(b) Use the result of part (a) to evaluate the inverse transform of

$$F(s) = \frac{1}{(s+a)^3}.$$

47. (a) Show that the convolution of a function with a unit doublet is the derivative of the function. (b) Use the result of part (a) to evaluate the following transforms.

(1) $\dfrac{s}{(s+a)}$ (2) $\dfrac{s}{s^2+1}$ (3) $\dfrac{s}{(s+a)^2+\beta^2}$

48. (a) Show that the convolution of a function with a unit step is the integral of the function. (b) Use the result of part (a) to evaluate the inverse transform of

$$F(s) = \frac{1}{s^2(s+a)}.$$

49. (a) Show that if $f(t)$ is the inverse transform of $1/F(s)$, then the inverse transform of

$$\frac{as^2+bs+c}{F(s)}$$

is

$$f_1(t) = af''(t) + bf'(t) + cf(t).$$

(b) Use the result of part (a) to evaluate the inverse transform of

$$F(s) = \frac{s^2+s+1}{s^2+1}.$$

50. (a) Find $H(s)$ for the circuit of Fig. 22–56 if the input is $e_1(t)$ and the output is $e_2(t)$. (b) Find the corresponding $h(t)$. (c) Convolve the derivative of $e_1(t)$ with the integral of $h(t)$ to obtain $e_2(t)$. (d) Check the answer by the Laplace transform partial fraction expansion.

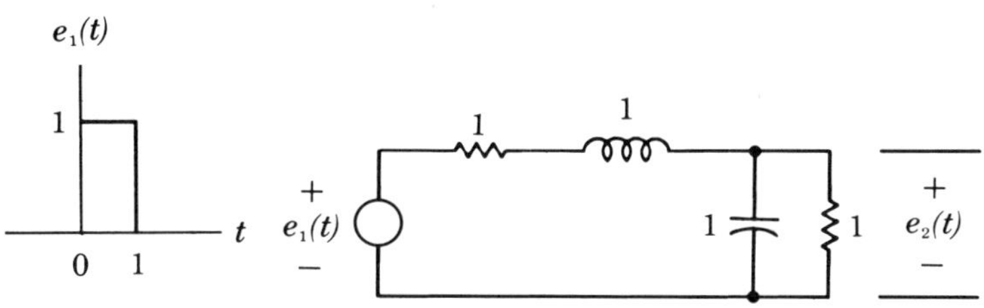

FIGURE 22–56

51. (a) Find $H(s)$ for the circuit of Fig. 22–57 if the input is $e_1(t)$ and the output is $e_2(t)$. (b) Find $h(t)$. (c) Take derivatives of the input until impulses appear and convolve these impulses with the appropriate integrals of the function $h(t)$ to obtain $e_2(t)$. (d) Check your results by the Laplace method.

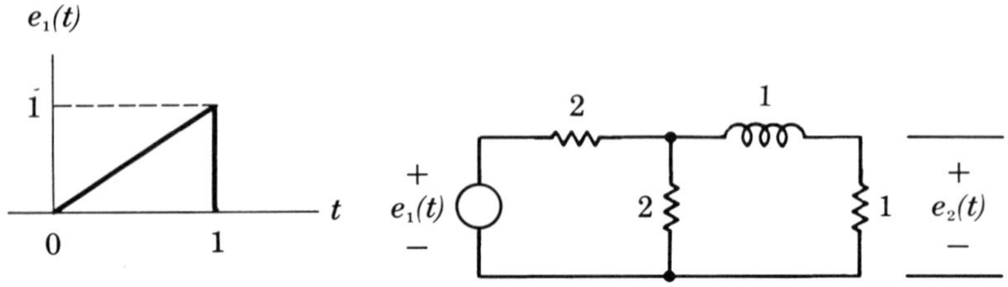

FIGURE 22–57

52. The input function to a given circuit is a unit step function and the output is

$$f_2(t) = \sin 3t \cdot u_{-1}(t).$$

(a) Find the system function $H(s)$. (b) Find the output if the input is

$$f_1(t) = \sin 3t \cdot u_{-1}(t).$$

53. The input function for a certain circuit is

$$f_1(t) = (2e^{-t} - e^{-2t})u_{-1}(t).$$

The corresponding output is

$$f_2(t) = e^{-3t} \cdot u_{-1}(t).$$

(a) Find the system function $H(s)$ for the network. (b) Find the output time function if the input is

$$f_1(t) = e^{-3t} \cdot u_{-1}(t).$$

Section 22–8

54. Whenever a constant final value of a function exists it can be obtained from the Laplace transform by the formula

$$f(\infty) = [sF(s)]_{s\to 0}.$$

To which of the following functions does the formula apply? What are the final values in these cases?

(a) $\dfrac{1}{s^2+s+1}$ (b) $\dfrac{s^2+1}{(s^2+s+1)s}$ (c) $\dfrac{s^2}{s^2+s+1}$

(d) $\dfrac{1}{s^2+1}$ (e) $\dfrac{s^2}{s^2+1}$

55. If there are no impulses at the origin the initial value of a function at $t = 0^+$ can be obtained from the Laplace transform according to the formula

$$f(0^+) = [sF(s)]_{s\to\infty}$$

Tell which of the functions in Exercise 54 can be treated by this formula, and for these give the initial values.

56. In the circuit of Fig. 22–58 the input is a step current. Find the transform $E_2(s)$ and from it obtain the initial value of $e_2(t)$ and its first three derivatives.

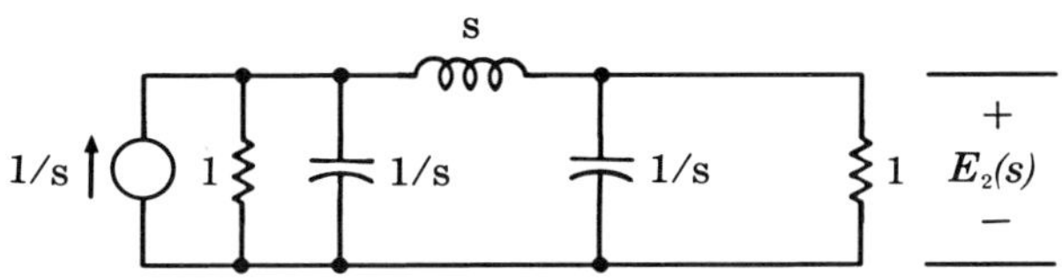

FIGURE 22–58

57. From the Laplace transform $E_2(s)$ in Exercise 56 find the final value of $e_2(t)$ and of its first three derivatives.

58. In the differential equation given below the current and its derivative and its first integral are zero at $t = 0^-$ just before the step is applied. Transform the equation directly, and find $i(t)$ by means of the Laplace transform.

$$2\frac{di}{dt} + 12i + 68\int_0^t i\,dt = 10u_{-1}(t)$$

59. For the differential equation shown below the following initial conditions hold: $i(0^+) = 10$, $i^{-1}(0^+) = 10$. Transform the equation and thus find $i(t)$.

$$2\frac{di}{dt} + 8\int_{-\infty}^t i\,dt = 0$$

60. The driving function for the differential equation shown below is an impulse of value 5. Before the impulse strikes the current is zero.

(a) Solve for the current by transforming the differential equation directly.

(b) Find the initial value of the current which the impulse causes in the circuit and transform the equation for $t = 0^+$ and thus obtain $i(t)$.

$$2\frac{di}{dt} + 4i = 5u_0(t)$$

61. In the differential equation

$$\frac{di}{dt} + 2i + 6\int_0^t i\,dt = 10u_0(t),$$

the current and its derivatives are zero at $t = 0^-$ and the impulse is applied at $t = 0$. (a) Find the value of the current considering the impulse as a source. (b) Compute the initial value of the current produced by the impulse and write the differential equations which holds for $t > 0$. Again solve for $i(t)$.

62. (a) For the circuit of Fig. 22–59 write the differential equation which holds for $t > 0$. Transform the equation and solve for the current. (b) Transform the initial conditions in the circuit to energy sources and write the differential equation in terms of these singularity functions. Again solve for $i(t)$.

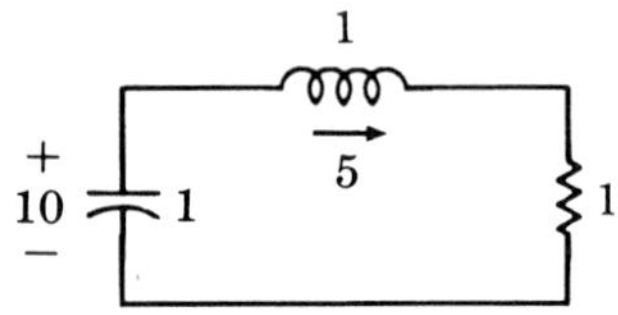

FIGURE 22–59

63. Reconstruct differential equations for the transforms given below.

$$\text{(a)}\ I(s) = \frac{s+1}{s^2+3s+1} \qquad \text{(b)}\ I(s) = \frac{s^2+s+1}{2s^3+2s^2+s+1}$$

64. The singularity functions representing the derivatives of $u_0(t)$ are equivalent to initial energy storages in a problem. From the singularity functions in the equation below compute the initial value of the current, and of its first three derivatives. Show that the same differential equation would be obtained for $t > 0$ if these initial values were used when the equation was transformed, and the singularity functions were ignored.

$$\frac{d^3i}{dt^3} + 2\frac{d^2i}{dt^2} + 2\frac{di}{dt} + 2i = u_0(t) + 2u_1(t) + 3u_2(t)$$

CHAPTER 23

POWER AND ENERGY RELATIONSHIPS IN NETWORKS EXCITED BY ARBITRARY FUNCTIONS

"Knowledge is a ladder; its rungs are never meant to rest upon, but only to hold a man's foot long enough to enable him to put the other somewhat higher."

Thomas Huxley

23–1 Introduction. In this last chapter we shall discuss the power and energy relationships that hold when more general signals are applied to linear networks. In one sense this material is a conclusion, rounding out our earlier discussions of power in direct-current circuits, and of power and energy in steady-state sinusoidal networks. In another sense, the material of this chapter is an introduction and not a conclusion. The methods presented here are the basis of the statistical theory of networks which was formally introduced by Norbert Wiener in 1948, with the publication of his book *Cybernetics*. Section 23–5 in this chapter may serve as an introduction to this field.

For simplicity, we will restrict ourselves in this chapter to the consideration of a network with an input function, an output function, and a transfer function of frequency which relates the input and output. In the first section we will consider the case in which the input is a single sinusoid. Previously we used an RMS value to characterize such a sine wave. In this chapter we will more often use the square of the RMS value or *the mean-square value* of the wave. If we plot the power in the input signal as a function of frequency, we will obtain a line spectrum with a value at the frequency of the input. A more convenient plot is that of power-per-cycle versus frequency. Such a plot contains an impulse at the frequency of the input, and is called *a power-density spectrum.* We will see that the power-density spectrum of the function applied to the network is related to the power-density spectrum of the output by the *squared magnitude* of the transfer function of the network. This relationship is the power response function of the network.

In the second section we will consider an input which is a periodic function, given by a Fourier series. The power-density spectrum of the input signal consists of a series of impulses at the frequencies of the harmonics in the signal. The power-density spectrum of the output from the network is again obtained by multiplying the input power-density spectrum by the squared magnitude of the transfer function of the network.

In the third section we will consider an excitation which is a single pulse whose frequency components are given by the Fourier integral. A band spectrum results in which the power per cycle approaches zero, but in which the energy per cycle is finite and follows the distribution of the spectrum. For this case the input is described by an *energy-density spectrum.* The output energy-density spectrum is obtained by multiplying the input energy-density spectrum by the squared magnitude of the network transfer function.

Finally, we will consider the case in which the input function to the network is known only in a statistical fashion. For example, in designing a system to receive speech or music, we do not know the exact waveforms which will occur. We know only that the frequency components will lie in a given band, from 20 to 15,000 cps. If we assume that all frequencies will occur with equal probability, we have completed a statistical description of the input function. More often the distribution is not flat, as in our example, and we describe the input by a power-density spectrum. This spectrum is very much like that of a Fourier series, except that the harmonics coalesce and form a continuous band spectrum. The output power-density spectrum is again obtained by multiplying by the squared magnitude of the transfer function. Except in a few special cases, the analytical computation of power-density spectra lies beyond the scope of this text.

23–2 Power response function for a single sine wave. We will begin our discussion of power response functions with a consideration of a single sine wave. In Chapter 19 we covered the power and energy relationships of circuits in the sinusoidal steady-state, but the emphasis there was on voltage and current, and power and energy were computed from these variables. In this chapter we propose to deal with power and energy directly and to obtain network response functions which can be used with them. Actually, it is not the power itself in which we will be interested, but the RMS, or equivalent heating, values of the stimulus and response functions. Furthermore, we usually will not take the square root, but will deal directly with the mean-square values of the functions. In addition to its interpretation as the equivalent heating value of a function, the mean-square value can be interpreted as the power which the function produces in a 1-ohm resistance, whether the function be a current or voltage. The mean-square value can therefore be viewed as a normalized power expression valid for both current and voltage.

Our purpose in dealing with mean-square values is to enable us to add together sinusoids of arbitrary frequencies. The resulting time functions are too complex to handle, and we must settle for an RMS type of addition which preserves only the equivalent heating value. In this proc-

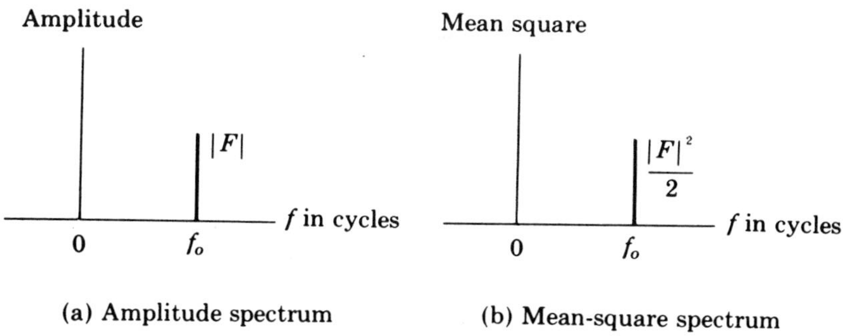

FIG. 23-1. The one-sided spectra of the sine wave $f(t) = |F| \cos \omega_0 t$.

ess a good deal of information is lost, including all the phase information. The RMS value can always be found from the waveform, but the waveform cannot be reconstructed from the RMS value.

Let us consider now the single-frequency sinusoid

$$f(t) = |F| \cos \omega_0 t. \tag{23-1}$$

The first concept which we will develop is the power-density spectrum. The amplitude spectrum for the single-frequency sine wave of Eq. (23-1) is plotted in Fig. 23-1(a). It is a single harmonic of value $|F|$ occurring at a frequency of f_0. The equivalent RMS vector length is

$$\text{RMS vector length} = 0.707|F|. \tag{23-2}$$

The mean-square value is the square of Eq. (23-2). Thus

$$\text{Mean-square vector length} = \frac{|F|^2}{2}. \tag{23-3}$$

The mean-square spectrum is plotted in Fig. 23-1(b).

Instead of plotting the one-sided spectra shown in Fig. 23-1 we can plot spectra for positive and negative frequencies. (Negative frequencies represent vectors rotating in the clockwise, or negative, direction.) For such two-sided spectra the amplitudes are half those of the single-sided spectra, according to the equation

$$|F| \cos \omega_0 t = \frac{|F|}{2} (e^{j\omega_0 t} + e^{-j\omega_0 t}).$$

The two-sided spectrum for the amplitude of Eq. (23-1) is shown in Fig. 23-2(a). The two-sided mean-square spectrum is shown in part (b). The student should note that the mean-square values are obtained by taking half of the true mean-square value, as shown in Fig. 23-1(b), and

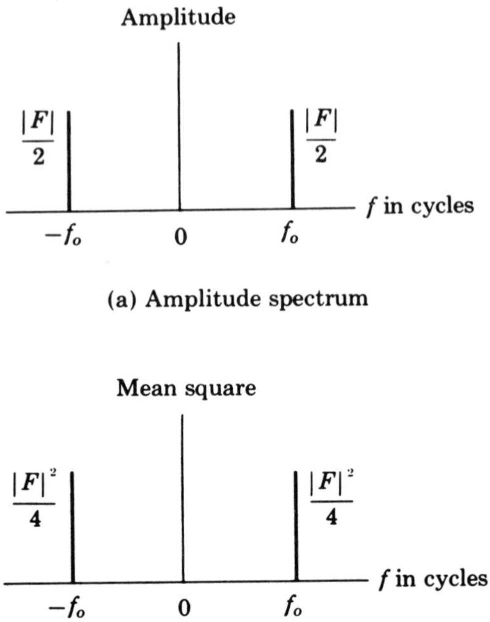

(a) Amplitude spectrum

(b) Mean-square spectrum

FIG. 23–2. The two-sided spectra of the sine wave $f(t) = |F| \cos \omega_0 t$.

not by squaring the separate terms in the double-sided amplitude spectrum of Fig. 23–2(a).

One problem with the double-sided spectrum of Fig. 23-2(b) remains to be discussed. The ordinate which we have used is power, and the value of a line in the spectrum represents the power associated with a given frequency. These lines have essentially zero width and look like impulse functions. However, a true impulse function is represented by the value of the area under it and not by its amplitude. We can treat the spectrum as a series of impulses if we change the ordinate scale to power per cycle instead of power.* The power in the sine wave is finite, but since the frequency range over which it occurs is essentially zero, the power density in watts per cycle approaches infinity and a true impulse is obtained. The area under the impulse is the power in the sine wave at that frequency, and it has the dimensions of watts per cycle times cycles, or watts. A spectrum plotted as in Fig. 23–3, with true impulses, is called a power-density spectrum. Here the ordinate is the power density G, in watts per

* We can use power per radian just as well, but power per cycle is the more basic quantity and will be used throughout this chapter.

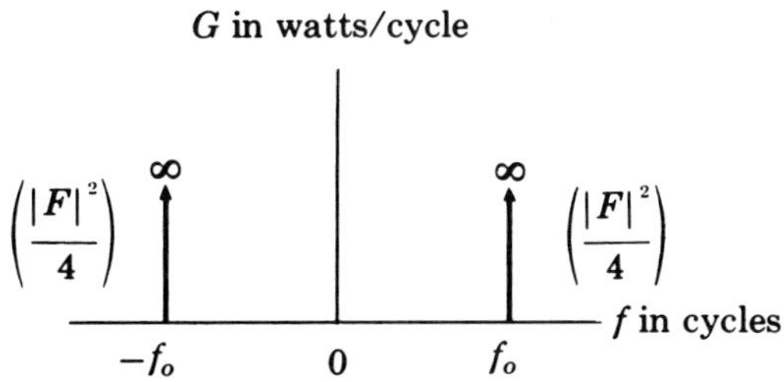

FIG. 23-3. The power-density spectrum for the sine wave $f(t) = |F| \cos \omega_0 t$.

cycle, and there are two impulses: one at a frequency of $+f_0$ and one at $-f_0$. The area under each impulse is the power associated with that frequency component, or one-half the power associated with the actual sine wave, that is, $|F|^2/4$ watts. An analytical expression for the power-density spectrum of a single sine wave is

$$G(f) = \frac{|F|^2}{4} u_0(f + f_0) + \frac{|F|^2}{4} u_0(f - f_0)$$

or

$$G(\omega) = \frac{|F|^2}{4} u_0(\omega + \omega_0) + \frac{|F|^2}{4} u_0(\omega - \omega_0). \tag{23-4}$$

Although denoted by $G(\omega)$, the power density is still measured *in watts per cycle*. Changing the scale from f to ω doesn't change the ordinate value.

The actual power between any two frequencies is the integral of the power density between the frequencies. Thus

$$P = \int_{f_1}^{f_2} G(\omega)\, df = \frac{1}{2\pi} \int_{\omega_1}^{\omega_2} G(\omega)\, d\omega. \tag{23-5}$$

Since the total average power in the sine wave must be the same whether it is computed in the time domain or the frequency domain, there is an interesting relationship between the integral of Eq. (23-5) and the time expression for the mean-square value, or the normalized power of the wave. The relationship is

$$\frac{1}{T} \int_{-T/2}^{T/2} f^2(t)\, dt = \int_{-\infty}^{\infty} G(\omega)\, df. \tag{23-6}$$

Equation (23-6) is a special case of *Parseval's theorem.**

* This theorem can also be proved from the mathematical properties of the Fourier transform.

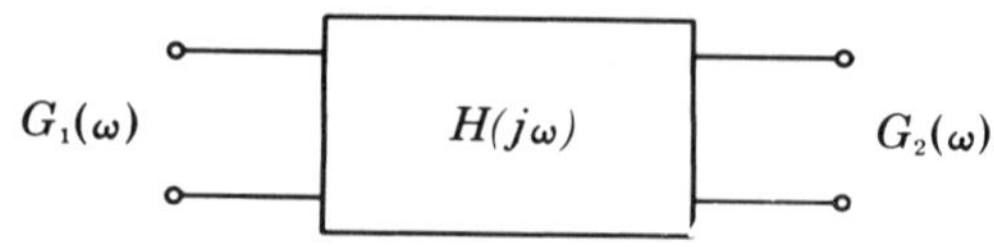

FIG. 23–4. A network excited by a power-density spectrum.

We will proceed now to investigate the power response function of a network of the form shown in Fig. 23–4. The input is described by its power-density spectrum, and the network is described by a system function $H(j\omega)$ which is the customary frequency-domain transfer function giving the amplitude and phase of the output response at any frequency in terms of the input amplitude and phase at that frequency. Since we are describing the input stimulus by its power-density spectrum, we are not interested in the phase of the output response. In fact, we must settle for the output power-density spectrum which gives the distribution of the output heating values as a function of frequency. In order to find the relationship of the output power-density spectrum $G_2(\omega)$ to the input power-density spectrum $G_1(\omega)$, we will consider a single-frequency input to the network of the form

$$f_1(t) = |F_1| \cos \omega_0 t, \tag{23–7}$$

and with the frequency-domain phasor,

$$F_1 = |F_1| \underline{|0^\circ}.$$

If the network response function is $H(j\omega)$, the output in the frequency domain is

$$F_2 = F_1 \underline{|0^\circ} \times H(j\omega). \tag{23–8}$$

Equation (23–8) represents a single-frequency sinusoid. According to Eq. (23–4) it has an output power-density spectrum given by

$$\begin{aligned} G_2(\omega) &= \frac{|F_2|^2}{4} [u_0(\omega - \omega_0) + u_0(\omega + \omega_0)] \\ &= \frac{|F_1|^2 |H(\omega)|^2}{4} [u_0(\omega - \omega_0) + u_0(\omega + \omega_0)]. \end{aligned} \tag{23–9}$$

Since the input power-density spectrum is

$$G_1(\omega) = \frac{|F_1|^2}{4} [u_0(\omega - \omega_0) + u_0(\omega + \omega_0)], \tag{23–10}$$

it is apparent that the output power-density spectrum $G_2(\omega)$ is related to

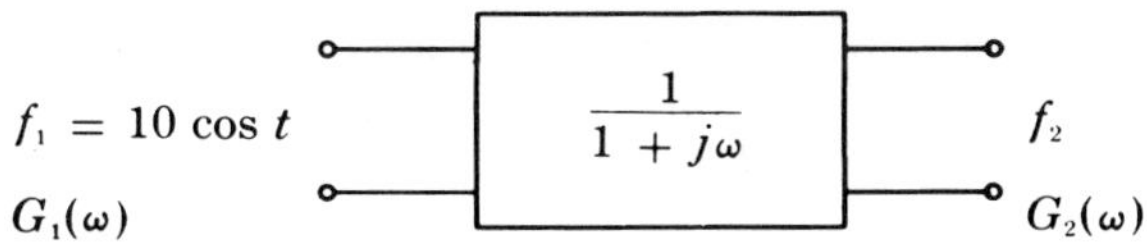

FIG. 23–5. An example of power-density calculations.

the input power-density spectrum by

$$G_2(\omega) = G_1(\omega)|H(\omega)|^2. \tag{23–11}$$

We have derived Eq. (23–11) specifically for the case of a single-frequency input, or a single pair of lines in the spectrum. Later we will show it to be true for all frequencies, and in fact for a band spectrum as well as a line spectrum.

As a numerical example of the results obtained thus far, we will consider the power relationships in the circuit of Fig. 23–5. The input power-density spectrum is

$$G_1(\omega) = \frac{10^2}{4}[u_0(\omega - 1) + u_0(\omega + 1)].$$

The transfer function is

$$H(j\omega) = \frac{1}{1 + j\omega},$$

and the squared magnitude of the transfer function is

$$|H(\omega)|^2 = \frac{1}{1 + \omega^2}.$$

The output power-density function is

$$\begin{aligned} G_2(\omega) &= G_1(\omega)|H(\omega)|^2 \\ &= 25[u_0(\omega - 1) + u_0(\omega + 1)]\left(\frac{1}{1 + \omega^2}\right). \end{aligned}$$

The impulses are zero except at $\omega = 1$ and $\omega = -1$, and therefore

$$G_2(\omega) = \tfrac{25}{2}[u_0(\omega - 1) + u_0(\omega + 1)].$$

As a check on the value of the output power density, we will integrate it to obtain the output power, or mean-square value. Thus

$$P_2 = \int_{-\infty}^{\infty} [12.5u_0(\omega - 1) + 12.5u_0(\omega + 1)]\, df.$$

The area under each impulse function is 12.5 watts, and hence

$$P_2 = 25 \text{ watts.}$$

We can compute the output power in the time domain and check Parseval's theorem [Eq. (23–6)]. The output time function is

$$f_2(t) = 7.07 \cos (t - 45°).$$

The average power in this function is

$$\begin{aligned} P_2 &= \frac{1}{T}\int_{-T/2}^{T/2} (7.07)^2 \cos^2 (t - 45°)\, dt \\ &= 25 \text{ watts.} \end{aligned}$$

So far in this chapter we have considered nothing more difficult than a single sine wave. The student may have the feeling that we are making things unnecessarily complicated by introducing power-density functions and computing the network responses as power functions. True, for the single sine wave the results are more easily obtained by vector algebra. Vector algebra cannot be extended to random functions, however, as can power-density spectra. We have not presented power-density spectra as a method of solving the single-sine-wave problem, but rather as an introduction to the more complicated random-variable problem.

23–3 Power response function for Fourier series. As a second step in our development of the theory of power response functions we will consider the case of an excitation which is a periodic wave in the time domain and which can be represented as a summation of sinusoids by means of Fourier's theorem. We will use the Fourier series in amplitude and phase form. If the function $f(t)$ is periodic with a period $T = 2\pi/\omega_0$, it can be represented by the series

$$f(t) = A_0 + A_1 \cos (\omega_0 t + \theta_1) + A_2 \cos (2\omega_0 t + \theta_2) + \cdots, \qquad (23\text{–}12)$$

where the A's are positive real numbers representing the amplitudes of the various harmonics and the constant term. The mean-square value of $f(t)$ is

$$\overline{f(t)^2} = \frac{1}{T}\int_{-T/2}^{T/2} [A_0 + A_1 \cos (\omega_0 t + \theta_1) + A_2 \cos (2\omega_0 t + \theta_2) + \cdots]^2\, dt. \qquad (23\text{–}13)$$

All the cross-product terms in Eq. (23–13) average to zero; only the cosine-

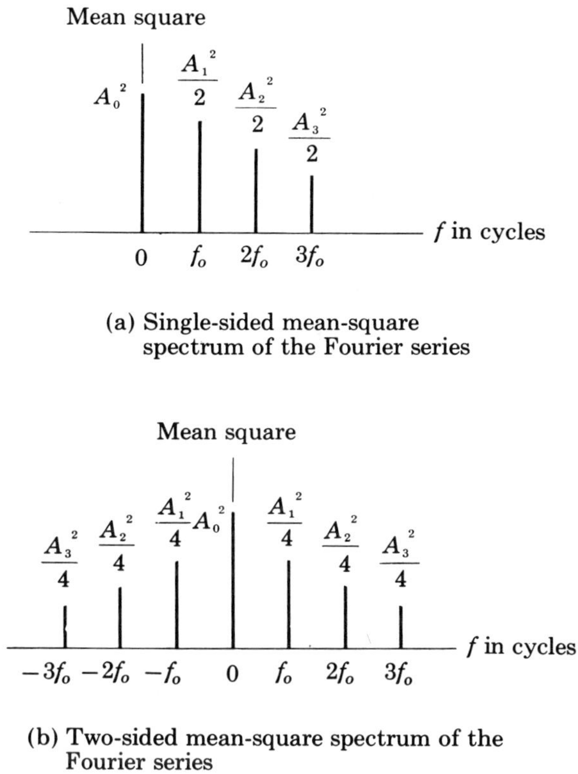

FIG. 23–6. The mean-square (power) spectra of a Fourier series.

squared terms and A_0^2 have an average value. Since each cosine-squared term has an average value of $\frac{1}{2}$

$$\overline{f(t)^2} = A_0^2 + \frac{A_1^2}{2} + \frac{A_2^2}{2} + \cdots. \tag{23–14}$$

The various components of the average power, on the right side of Eq. (23–14), occur at different frequencies and can be plotted as shown in Fig. 23–6(a). However, the equivalent double-sided spectrum shown in Fig. 23–6(b) is more convenient mathematically. Here both positive and negative components are shown for each frequency, and half of the power is assigned to each of them.

If we wish to obtain an analytical expression for a spectrum like that shown in Fig. 23–6(b), we must plot the power density rather than the power. Then, the contribution of each term can be represented by an impulse. The power-density spectrum for Eq. (23–14) is plotted in

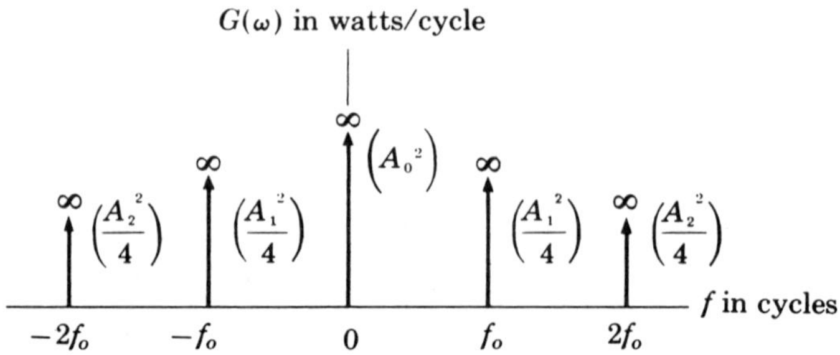

FIG. 23–7. The power-density spectrum for a Fourier series. Series of impulses.

Fig. 23–7. The analytic expression for the power density in watts/cycle is

$$
\begin{aligned}
G(\omega) = A_0^2 u_0(\omega) &+ \frac{A_1^2}{4}[u_0(\omega + \omega_0) + u_0(\omega - \omega_0)] \\
&+ \frac{A_2^2}{4}[u_0(\omega + 2\omega_0) + u_0(\omega - 2\omega_0)] \\
&+ \cdots
\end{aligned}
\tag{23–15}
$$

The power density of Eq. (23–15) is a series of impulses, occurring at $\omega = 0$ and at each of the harmonic frequencies of the wave. The total power value of the wave, or the mean-square value, is obtained by integrating the power density $G(\omega)$ from minus infinity to plus infinity. Thus

$$
P = \int_{-\infty}^{\infty} G(\omega)\, df. \tag{23–16}
$$

Note that the units of $G(\omega)$ are *watts per cycle*, and the integration of Eq. (23–16) is performed with respect to the cyclic frequency f.

The average power in the Fourier series can be computed in the time domain, as in Eq. (23–13), or in the frequency domain, as in Eq. (23–16). The equality of the two results,

$$
\frac{1}{T}\int_{-T/2}^{T/2} f^2(t)\, dt = \int_{-\infty}^{\infty} G(\omega)\, df, \tag{23–17}
$$

is again an example of Parseval's theorem.

If a complex waveform given by a Fourier series is applied to a network as the exciting function, the response function at each frequency can be obtained separately, and the over-all result can be obtained by the superposition of the partial results. The output power density at each frequency

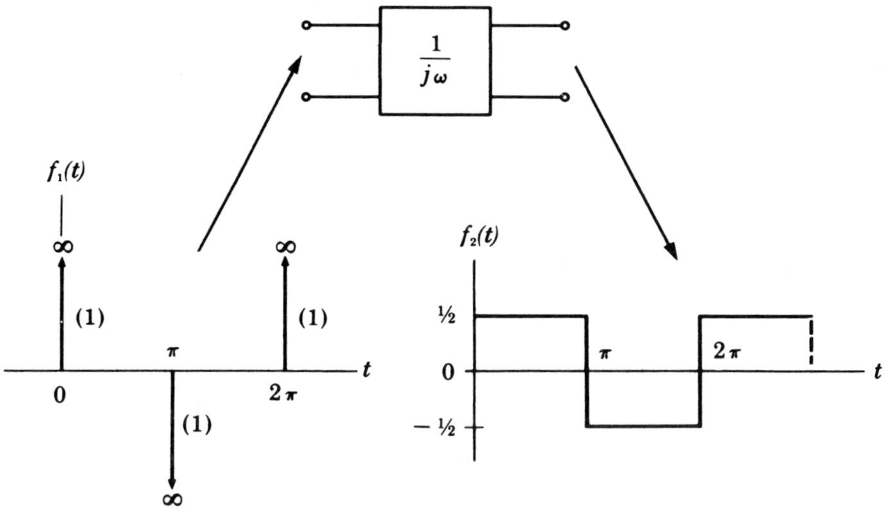

FIG. 23–8. An example of power-density computations for a Fourier series.

is obtained by multiplying the input power density by the square of the absolute magnitude of the system transfer function, as in Eq. (23–11). The squared absolute magnitude appears in each term and can be factored out, giving the output power-density function

$$G_2(\omega) = G_1(\omega)|H(\omega)|^2, \tag{23–18}$$

as for the single sine wave. The difference lies in the more complex nature of the input power-density function $G_1(\omega)$.

As an example of the computation of an output power-density function when a network is excited by a periodic wave, consider the situation shown in Fig. 23–8. The input waveform is a series of impulses of unity area and alternating sign. The system function of the network is $1/j\omega$. The Fourier series for the input waveform is

$$f_1(t) = \frac{2}{\pi}(\cos t + \cos 3t + \cos 5t + \cdots).$$

The input power-density spectrum, from Eq. (23–4), is

$$G_1(\omega) = \frac{4}{4\pi^2}[u_0(\omega + 1) + u_0(\omega - 1) + u_0(\omega + 3) + u_0(\omega - 3) + \cdots] \quad \text{watts/cycle.}$$

The output power-density spectrum is obtained by multiplying $G_1(\omega)$ by the squared magnitude of the transfer function, that is, $1/\omega^2$. Thus

$$G_2(\omega) = \frac{1}{\pi^2\omega^2}[u_0(\omega+1) + u_0(\omega-1) + u_0(\omega+3) + u_0(\omega-3) + \cdots] \quad \text{watts/cycle.}$$

As a check on the output power-density function, we obtain the average power, or mean-square value, of the output function. It is

$$\begin{aligned} P_{\text{av}} &= \int_{-\infty}^{\infty} G(\omega)\,df \\ &= \int_{-\infty}^{\infty} \frac{1}{\pi^2\omega^2}[u_0(\omega+1) + u_0(\omega-1) + u_0(\omega+3) + \cdots]\,d\omega. \end{aligned}$$

The impulses are zero except when their arguments are zero, and at this point they contribute an area equal to the constant by which they are multiplied. Hence

$$\begin{aligned} P_{\text{av}} &= \frac{1}{\pi^2}\left(1 + 1 + \frac{1}{3^2} + \frac{1}{3^2} + \cdots\right) \\ &= \frac{2}{\pi^2}\left(1 + \frac{1}{3^2} + \frac{1}{5^2} + \cdots\right)^* \\ &= \frac{2}{\pi^2}\left(\frac{\pi^2}{8}\right) \\ &= \tfrac{1}{4}\ \text{watt.} \end{aligned}$$

The average power can be checked by Parseval's theorem. The network is effectively an integrator and the output is a square wave which is alternately plus and minus $\frac{1}{2}$. (We assume that the input has been applied for a very long time and that the constant value, which depends on whether a positive or a negative impulse occurred first, has decayed to zero.) The mean-square value of this waveform is

$$\frac{1}{T}\int_{-T/2}^{T/2} f^2(t)\,dt = \tfrac{1}{4}\ \text{watt.}$$

In this particular case it is easier to compute the output power from the time function, but in general the power approach in the frequency domain will be easier, since the form of the output time function will not be known explicitly.

---

* This series, one of Bernoulli's, will be found in any good book of tables.

23–4 Energy response functions for Fourier integrals. In Chapter 20 we showed that a single transient pulse can be resolved into an infinite sum of sinusoidal harmonics with infinitesimal amplitudes and infinitesimal spacing between them. The sinusoids exist over all time, but they sum to zero at all instants except for the duration of the pulse. Since there can be no finite amount of net power in the harmonics, the power-density spectrum approaches zero everywhere, and we must turn to an energy-density spectrum. In order to show the derivation of the energy-density spectrum, we repeat here the definitions of the Fourier transforms:

$$f(t) = \frac{1}{2\pi}\int_{-\infty}^{\infty} F(\omega)e^{j\omega t}\,d\omega,$$

$$F(\omega) = \int_{-\infty}^{\infty} f(t)e^{-j\omega t}\,dt. \tag{23–19}$$

The first of Eqs. (23–19) expresses the time function $f(t)$ as a summation of an infinite number of infinitesimal sinusoids. In order to make the sinusoids more obvious we can rewrite the equation as

$$f(t) = \int_0^{\infty}\left(\frac{F(\omega)\,d\omega}{\pi}\right)\left(\frac{e^{j\omega t} + e^{-j\omega t}}{2}\right), \tag{23–20}$$

since $F(\omega)$ is an even function of ω and $F(\omega) = F(-\omega)$. The sum of a pair of positive and negative frequency terms thus represents the actual sinusoid. The amplitude and phase information is contained in the factor $F(\omega)\,d\omega/\pi$. In the derivation of the Fourier integral we assumed that the pulse repeated with a period of T. By letting T approach infinity we achieved the effect of a single pulse. With a period of T, as shown in Fig. 23–9(a), the fundamental frequency is

$$\omega_0 = \frac{2\pi}{T},$$

or

$$f_0 = \frac{1}{T}\cdot \tag{23–21}$$

Thus, in the spectral plot the spacing between harmonics is $\Delta\omega = 2\pi/T$ and the corresponding length Δf is $1/T$.

The function $F(\omega)$, plotted as a function of cyclic frequency, is shown in Fig. 23–9(b). Small sections, Δf in width, are crosshatched for $\pm f_1$. The width Δf is taken to be just wide enough to contain one harmonic of the pulse of part (a), namely,

$$\Delta f = \frac{1}{T}\cdot \tag{23–22}$$

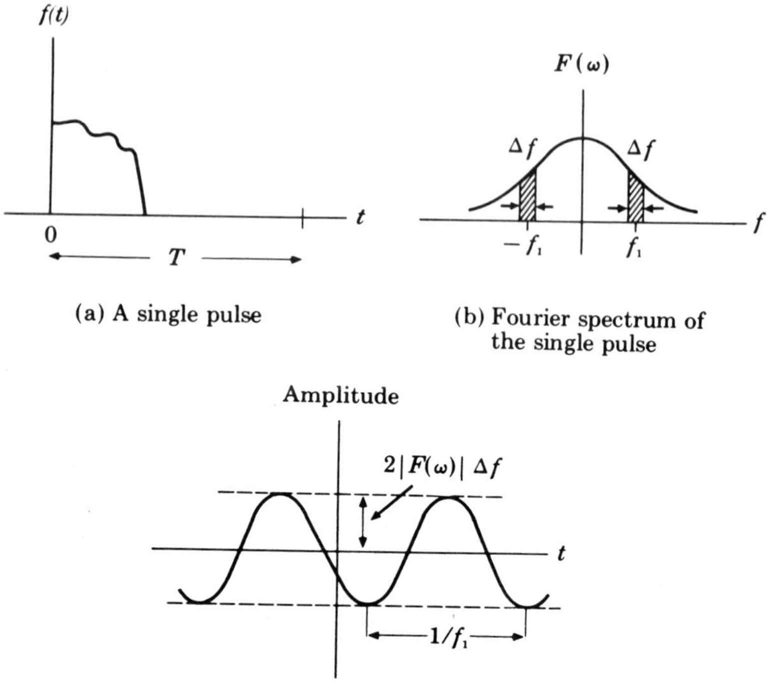

(a) A single pulse

(b) Fourier spectrum of the single pulse

(c) Component corresponding to the section of width Δf

FIG. 23–9. The spectral components in the Fourier integral of a pulse.

The actual amplitude of the sinusoid in this frequency band, from Eq. (23–20), is

$$\text{Harmonic amplitude} = \frac{|F(\omega)|\,\Delta\omega}{\pi}, \tag{23–23}$$

or

$$\text{Harmonic amplitude} = 2|F(\omega)|\,\Delta f. \tag{23–24}$$

The equivalent sinusoid is shown in Fig. 23–9(c). The phase of the sinusoid is determined by the phase angle of $F(\omega)$, but it has no significance in a power and energy analysis.

We will now see why we cannot define a power-density spectrum for Fourier integrals. The power in one of the harmonics given by Eq. (23–24), or the mean-square value, is

$$P = \text{Mean-square value} = \frac{4|F(\omega)|^2\,\Delta f^2}{2}\ \text{watts}. \tag{23–25}$$

As $\Delta\omega$ approaches zero the power in the relevant frequency region approaches zero. This result is expected since we have here a band spectrum, and the power goes to zero as the width of the spectrum goes to zero. However, the power density also goes to zero. If we divide the power of Eq. (23–25) by the frequency interval in which it occurs, we obtain the power density. That is,

$$G = \frac{P}{\Delta f} = 2|F(\omega)|^2\,\Delta f \text{ watts/cycle.} \tag{23–26}$$

From Eq. (23–26) we can see that not only the power but also the power density goes to zero as Δf approaches zero.

The remedy for this difficulty is to plot the energy-density spectrum rather than the power-density spectrum. We can select some time T over which we wish the analysis to be valid, and we can find the energy in each of the sinusoids in this interval. The time T is the period of the pulse repetition of Fig. 23–9(a). It is related to the space occupied in our spectrum by one harmonic according to Eq. (23–22). The energy in a single harmonic, with an amplitude given by Eq. (23–24) and considered over an interval T, is the power times the time interval. Analytically

$$\begin{aligned} W = P \times T &= 2|F(\omega)|^2\,\Delta f^2 T \\ &= 2|F(\omega)|^2\,\Delta f \qquad \text{watt·sec or joules.} \end{aligned} \tag{23–27}$$

It can be seen here that the energy in the harmonic considered for time T still goes to zero. However, the energy density, which is the value of Eq. (23–27) divided by the frequency interval over which it occurs, does not go to zero, for the energy density E_s is

$$E_s = 2|F(\omega)|^2 \text{ joules/cycle.} \tag{23–28}$$

Equation (23–28) expresses the energy per cycle which occurs in the real sinusoid covering this frequency region. [It should be noted that Eq. (23–28) expresses energy per cycle of the frequency spectrum, and not energy per cycle of oscillation.] If we plot a double-sided spectrum with lines at positive and negative frequencies to represent a real sinusoid, we obtain the rather remarkable result that the energy density as a function of positive and negative frequency is

$$E(\omega) = |F(\omega)|^2 \text{ joules/cycle.} \tag{23–29}$$

The final result is independent of the period T which we assumed for the original transient pulse, and is valid as T approaches infinity. The student should note that even though $E(\omega)$ is a function of ω, it has the units of *joules per cycle*. The natural occurrence of this unit in Eq. (23–29) is the governing reason for our earlier choice of watts per cycle as the unit for power density.

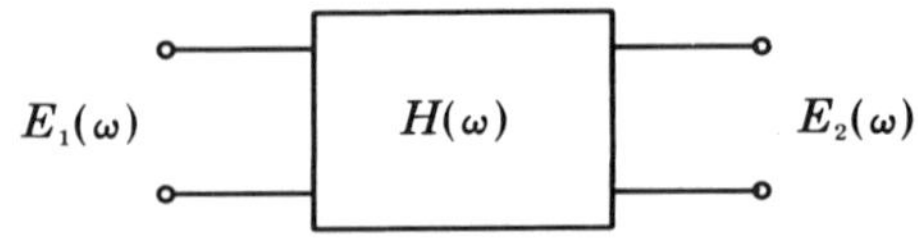

FIG. 23–10. An example for energy-density spectral computations.

The total energy of the pulse computed in the time domain must agree with the total energy of the pulse computed in the frequency domain. Hence

$$\int_{-\infty}^{\infty} f^2(t)\,dt = \int_{-\infty}^{\infty} |F(\omega)|^2\,df. \tag{23–30}$$

The energy in the time domain is the integral of the squared time function, and the energy in the frequency domain is the integral of the energy-density function of Eq. (23–29). A formal statement of Eq. (23–30) is called *Plancherel's theorem.* It is an extension of Parseval's theorem to the case of band spectra.

We can trace energy-density spectra through networks in much the same way that we previously traced power-density spectra. Consider the network of Fig. 23–10. We assume that the input energy-density spectrum $E_1(\omega)$ is given, that the network is described by the system function $H(\omega)$, and that the output energy-density spectrum $E_2(\omega)$ is to be found. Let the input function in the frequency domain be $F_1(\omega)$ and the output function be $F_2(\omega)$. Then

$$F_2(\omega) = F_1(\omega)H(j\omega),$$
$$F_2(-\omega) = F_1(-\omega)H(-j\omega).$$

Since j and ω always occur together, changing the sign of ω in a function $F_2(\omega)$ is the same thing as taking the conjugate function. The output energy-density spectrum is

$$\begin{aligned} E_2(\omega) = |F_2(\omega)|^2 &= F_2(\omega)F_2(-\omega) \\ &= F_1(\omega)F_1(-\omega)H(j\omega)H(-j\omega) \\ &= |F_1(\omega)|^2|H(\omega)|^2 \\ &= E_1(\omega)|H(\omega)|^2. \end{aligned} \tag{23–31}$$

The output energy-density function is thus the input energy-density function multiplied by the squared magnitude of the system transfer function. We note that energy-density functions are treated in the same manner as power-density functions.

For a numerical example of energy-density computations, consider the network shown in Fig. 23–11. We take the input time function to be the

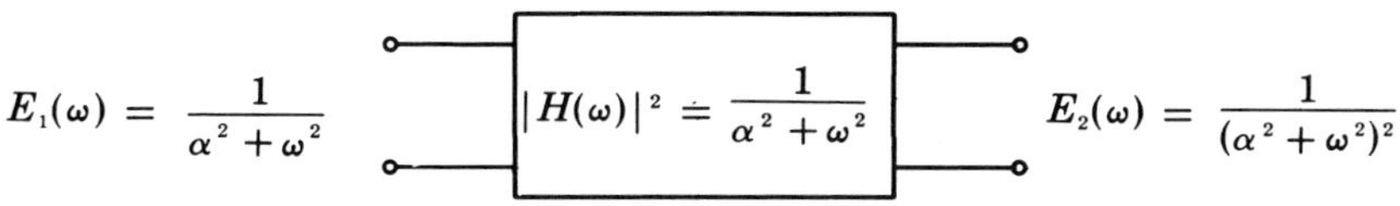

FIG. 23–11. A numerical example of energy-density computations.

transient pulse

$$f_1(t) = e^{-\alpha t} \cdot u_{-1}(t),$$

with the Fourier transform

$$F_1(\omega) = \frac{1}{\alpha + j\omega} \cdot$$

The network is assumed to have a system function which, like the transform, is

$$H(\omega) = \frac{1}{\alpha + j\omega} \cdot$$

The input energy-density function is

$$E_1(\omega) = |F_1(\omega)|^2 = \frac{1}{\alpha^2 + \omega^2} \cdot$$

The squared magnitude of the system function is

$$|H(\omega)|^2 = \frac{1}{\alpha^2 + \omega^2},$$

and the output energy-density function is

$$E_2(\omega) = |F_2(\omega)|^2 = E_1(\omega)|H(\omega)|^2 = \frac{1}{(\alpha^2 + \omega^2)^2} \cdot$$

Since the time functions in this example are easy to obtain, we can verify Plancherel's theorem. The output time function is

$$f_2(t) = te^{-\alpha t} \cdot u_{-1}(t).$$

The total energy, computed from the time function, is

$$W = \int_{-\infty}^{\infty} f^2(t)\, dt = \int_0^{\infty} t^2 e^{-2\alpha t}\, dt = \frac{1}{8\alpha^3} \int_0^{\infty} (\alpha t)^2 e^{-2\alpha t}\, d(\alpha t)^*$$

$$= \frac{1}{8\alpha^3} \times 2 = \frac{1}{4\alpha^3} \cdot$$

---

* The integral for the third-order gamma function.

In the frequency domain the total energy for the output function is

$$W = \int_{-\infty}^{\infty} E_2(\omega)\,df = \frac{1}{2\pi}\int_{-\infty}^{\infty} E_2(\omega)\,d\omega$$

$$= \frac{1}{2\pi}\int_{-\infty}^{\infty} \frac{1}{(\alpha^2+\omega^2)^2}\,d\omega$$

$$= \frac{1}{\pi}\int_{0}^{\infty} \frac{1}{(\alpha^2+\omega^2)^2}\,d\omega \qquad \text{(since the integrand is even)}$$

$$= \frac{1}{\pi}\times\frac{\pi}{4\alpha^3} = \frac{1}{4\alpha^3}.$$

Thus the total energy computed in the frequency domain is the same as the total energy computed in the time domain. At times one of the integrals in such a problem is a great deal easier to evaluate than the other, and in that case is to be preferred.

23–5 Power response functions for random excitations. A communications engineer must deal with many signals whose exact form he cannot know. The waveforms of speech and music are in this category. Any given speech waveform or any given musical waveform is more or less random in nature. There are certain similarities between the various waveforms of speech and music, however, which make it possible to design circuits which will operate for all signals of a given class. These similarities are called the statistical parameters which describe the class of signals. A number of such parameters are possible, but we will consider only the simplest of them, which is the distribution of the energy, on the average, as a function of frequency. In fact, we can describe a random signal by a *power-density spectrum* of the same sort we used for a periodic signal, except that no definite lines exist in the spectrum and it is spread over all frequencies as a *continuous band spectrum.* (We have already encountered

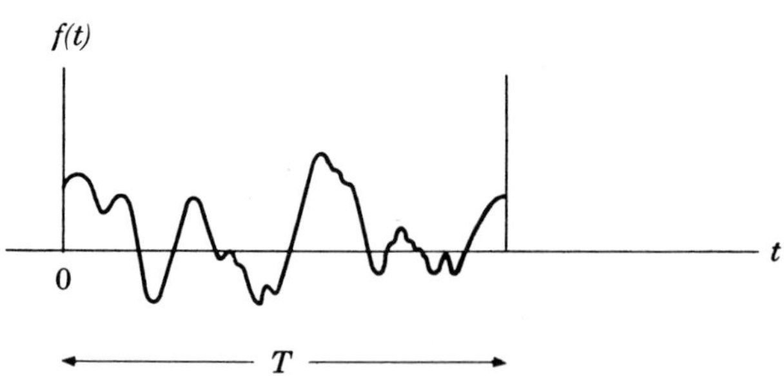

FIG. 23–12. A section T seconds long of a random time function.

band spectra for Fourier integrals, but in that case they are energy-density spectra, rather than power-density spectra.)

In defining the power-density spectrum for random signals, we shall take the experimental rather than the mathematical point of view. If we have a random signal which is recorded over a time T, as shown in Fig. 23–12, we can repeat the same signal over and over again and thus obtain an approximation to a completely random signal existing over all time. Intuitively, we feel that if the sample is sufficiently long the results we obtain will be very little different from a completely random signal.* We shall express the repeated wave as a Fourier series, written in complex form to facilitate the limiting process as T is allowed to increase without limit. The complex Fourier series is

$$f(t) = \sum_{-\infty}^{\infty} \frac{F(n\omega_0)}{T} e^{jn\omega_0 t}, \tag{23–32A}$$

where

$$F(n\omega_0) = \int_{-T/2}^{+T/2} f(t)e^{-jn\omega_0 t}\, dt, \qquad \omega_0 = \frac{2\pi}{T}. \tag{23–32B}$$

The actual sinusoids in the complex Fourier series consist of a positive and a negative frequency term, and therefore have amplitudes of

$$A_n = \frac{2F(n\omega_0)}{T}. \tag{23–33}$$

The power-density spectrum for the Fourier series is a set of impulses occurring at the harmonic frequencies, and given by

$$G(\omega) = \sum_{-\infty}^{\infty} \left[\frac{2F(n\omega_0)}{T}\right]^2 \frac{u_0(\omega + n\omega_0)}{4} = \sum_{-\infty}^{\infty} \left[\frac{F(n\omega_0)}{T}\right]^2 u_0(\omega + n\omega_0). \tag{23–34}$$

The impulses in Eq. (23–34) have areas of $[F(n\omega_0)/T]^2$ and occur at f_0 and its harmonic frequencies. As an approximation when the period T is very large and the spacing of the harmonics correspondingly small, we can replace the impulses by pulses of width f_0. The height of the pulse at

* The frequency spectrum obtained by such a process does not converge for random Gaussian signals which are much used in theoretical work as idealizations of real physical random signals.

the nth harmonic is

$$h(n\omega_0) = \left[\frac{F(n\omega_0)}{T}\right]^2 \times \frac{1}{f_0} = \frac{|F(n\omega_0)|^2}{T}.$$

Since the spacing between the pulses is f_0 and the width of a pulse is f_0, these pulses run together as a continuous curve of the frequency variable $\omega = n\omega_0$. As the period T is allowed to approach infinity the approximation becomes very good, and we obtain a continuous spectrum $G(\omega)$ given by

$$G(\omega) = h(\omega) = \frac{|F(\omega)|^2}{T} \qquad \text{watts/cycle,} \tag{23–35}$$

where

$$F(\omega) = \int_{-T/2}^{T/2} f(t)e^{-j\omega t}\, dt \qquad \text{with } T \to \infty. \tag{23–36}$$

We have derived the power-density spectrum from the Fourier series, but the frequency distribution function $F(\omega)$ defined by Eq. (23–36) is the Fourier transform of a single period of the wave. The student should recall that a relationship always exists between the spectrum of a Fourier series and the spectrum of a pulse equal to one period of the series. The convergence of the process is assured if the mathematical expression truly represents a physical waveform.*

The total power in the waveform can be found by integrating the power-density function with respect to frequency, or by integrating and averaging the square of the time function. Thus

$$P(\omega) = \int_{-\infty}^{\infty} G(\omega)\, df = \lim_{T\to\infty} \frac{1}{T}\int_{-T/2}^{T/2} f^2(t)\, dt \qquad \text{watts.} \tag{23–37}$$

Equation (23–37) is an expression of Parseval's theorem for the random case. The evaluation of the time integral is not usually attempted directly, since the time function is random and not known explicitly as a function of time. The integral can be evaluated experimentally, however, by its heating effect averaged over a time long enough for it to settle down to a constant value.

If a random function is applied to a network, as shown in Fig. 23–13, each frequency component in the input power-density spectrum is multi-

* The process does not converge for Gaussian random functions which are not band-limited. For these functions, which are used to approximate random noise, the definition of the power-density spectrum in terms of autocorrelation functions is preferred. The autocorrelation function is defined later in the chapter.

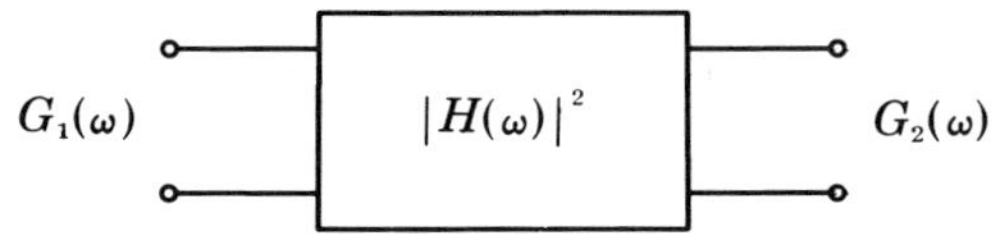

FIG. 23–13. The power-density transfer function for random signals.

plied by the square of the transfer function in order to give the corresponding frequency component in the output power-density spectrum. Thus

$$G_2(\omega) = |H(\omega)|^2 G_1(\omega). \tag{23–38}$$

Equation (23–38) has the same form as the like equation for the Fourier series. The only difference is the added complexity of the input power-density spectrum. We will not actually evaluate such a spectrum from the statistical description of the signal. Instead, we will deal with the rather common case of *white noise.*

For white noise the power-density spectrum as a function of frequency is flat.* Every frequency increment of the same value from zero frequency to infinite frequency has the same power associated with it. Of course, such a definition implies a total power in the signal which is infinite. Hence, true white noise is not physically realizable. It is convenient, however, to treat a signal as white noise if it has a flat spectrum over the frequency band of interest in a given problem. (The name "white noise" originates from the optical example of white light which has a band spectrum of constant power density over the visible spectrum.)

A physical example of white noise is the Johnson noise generated in a resistance by the thermal agitation of the electron cloud. The mean-square voltage generated in a resistance of value R ohms is

$$|E_{\text{rms}}|^2 = 4kTBR, \tag{23–39}$$

where k = Boltzmann's constant (1.37×10^{-23} watt·sec/deg),
T = absolute temperature in degrees Kelvin,
B = bandwidth in cycles,
R = resistance in ohms.

* A flat frequency spectrum can belong to a wide variety of time-domain waveforms. Additional phase information is required to specify the waveform, except in the case of Gaussian random noise, which is a series of random impulses occurring at random times. Such a function is so completely random that it has no phase information and is completely specified by its flat amplitude spectrum.

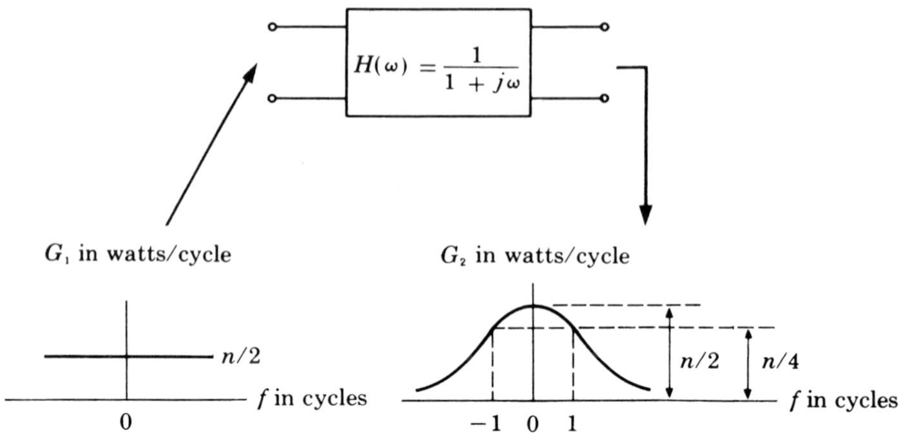

FIG. 23–14. The application of white noise to a low-pass circuit.

The noise power is directly proportional to the bandwidth, and the power density in watts per cycle is

$$G(\omega) = \frac{4kTR}{2} = \frac{n}{2} \text{ watts/cycle.} \tag{23–40}$$

The factor of 2 in the denominator of Eq. (23–40) converts the spectrum to positive and negative frequencies. As an example, in Fig. 23–14 random noise is applied to a network with a transfer function $H(\omega) = 1/(1 + j\omega)$. The power-density spectrum at the output of the network is

$$G_2(\omega) = |H(\omega)|^2 G_1(\omega).$$

The input spectrum is

$$G_1(\omega) = \frac{n}{2} \text{ watts/cycle,}$$

and the output power-density spectrum is

$$G_2(\omega) = \frac{n/2}{1 + \omega^2} \text{ watts/cycle.*} \tag{23–41}$$

The output power-density spectrum is now a low-pass type which decays

* This result gives the correct amplitude spectrum of the output signal whether the input is Gaussian or not. If the input is Gaussian, the output is Gaussian and both are completely described by their power-density spectra. If the input is not Gaussian, the output will tend to be Gaussian by the central limit theorem.

to zero at infinity. We can compute the total power in the signal by means of Eq. (23-37). Thus

$$P = \int_{-\infty}^{\infty} \frac{n/2}{1+\omega^2} \frac{d\omega}{2\pi} = \frac{n/2}{2\pi} [\tan^{-1} \omega]_{-\infty}^{\infty} = \frac{n/2}{2\pi} \times \pi = \frac{n}{4} \text{ watts.}$$

A spectrum such as that of Eq. (23-41) is called band-limited noise. It has the shape of the squared magnitude of the transfer function of the network. A large variety of power-density spectra can be generated by passing white noise through appropriate networks.

A more sophisticated discussion of random signals would start with the time functions and would define correlation functions of the signals as convenient statistical measures of their properties.* The *autocorrelation function* is the Fourier transform of the power-density spectrum according to a theorem proved independently by Weiner and Kintchine. We will introduce the autocorrelation function from this point of view, and thus obtain its definition, but we will not delve deeply enough into statistics to compute autocorrelation functions from random time functions.

Let us imagine that we have a sample of a random function which exists over a time T. Then, according to Eq. (23-35) the power density is

$$G(\omega) = \frac{|F(\omega)|^2}{T} \text{ watts/cycle,} \tag{23-42}$$

and the function $F(\omega)$ is the Fourier transform of this single section of the waveform. According to Eq. (23-36) it is given by

$$F(\omega) = \int_{-T/2}^{T/2} f(t)e^{-j\omega t}\,dt \qquad \text{with } T \to \infty. \tag{23-43}$$

An alternative way of writing Eq. (23-42) is

$$G(\omega) = \frac{F(j\omega)F(-j\omega)}{T}, \tag{23-44}$$

since the product of a function and its conjugate gives the squared magnitude of the function. Moreover, Eq. (23-43) establishes that $F(\omega)$ is a Fourier transform and the convolution theorem holds for it. Thus apart from the factor $1/T$, the inverse transform of $G(\omega)$, which we will call $g(t)$, is the convolution of the time function corresponding to $f(t)$ and the time function corresponding to the inverse transform of $F(-j\omega)$. Changing the sign of ω in an equation of the sort given by Eq. (23-43) is the same thing as changing the sign of t. Hence the inverse transform of $F(-j\omega)$

* When the signal is Gaussian it is completely described by its autocorrelation function.

is $f(-t)$. Thus

$$
\begin{aligned}
g(t) &= \frac{1}{T}\,[f(t)*f(-t)] \\
&= \frac{1}{T}\int_{-T/2}^{T/2} f(x)f(x - t)\,dx \qquad \text{with } T \to \infty. \qquad (23\text{–}45)
\end{aligned}
$$

The function $g(t)$ of Eq. (23–45) is defined as the autocorrelation function of the time function $f(t)$. Usually it will not matter which way $f(t)$ is shifted before it is multiplied and averaged,* and hence the autocorrelation function can be written

$$
g(t) = \frac{1}{T}\int_{-T/2}^{T/2} f(x)f(x \pm t)\,dx \qquad \text{with } T \to \infty. \qquad (23\text{–}46)
$$

The autocorrelation function of Eqs. (23–45) and (23–46) is the Fourier transform of the power-density spectrum. The converse relationship is also true. The power-density spectrum is the Fourier transform of the autocorrelation function. Whenever the function is known in an analytical and statistical way, it is easier to compute its autocorrelation function than its power-density spectrum. The power-density spectrum is obtained by taking the transform of the correlation function.

We will not compute any autocorrelation functions from the statistical properties of a waveform, but we will compute the autocorrelation function of the band-limited random noise which we obtained in Eq. (23–41). For this power-density spectrum the autocorrelation function is

$$
\begin{aligned}
g(t) &= \frac{1}{2\pi}\int_{-\infty}^{\infty} \frac{n/2}{1+\omega^2}\, e^{j\omega t}\,d\omega \\
&= \frac{n}{4}\int_{-\infty}^{\infty} \frac{\cos \omega t}{1+\omega^2}\,d\omega \\
&= \frac{n}{4}\, e^{-|t|}.
\end{aligned}
$$

In statistical analysis this autocorrelation function and the power-density spectrum from which it is derived play much the same roles as the time- and frequency-domain functions in the usual transform theory. The student who wishes to pursue this fascinating subject further will find it adequately treated in a number of modern texts.

---

* Functions whose statistical properties do not shift with time are called stationary random functions. Many physical waveforms, such as thermal noise, fall in this category, and elementary discussions of statistical methods usually restrict themselves to stationary functions.

23–6 Summary

Power-density spectra [$G(\omega)$ in watts/cycle]

$$f(t) = |F| \cos \omega_0 t \qquad G(\omega) = \frac{|F|^2}{4} u_0(\omega \pm \omega_0)$$

$$f(t) = A_0 + \sum A_n \cos (n\omega_0 t + \theta_n) \qquad G(\omega) = A_0^2 + \sum \frac{A_n^2}{4} u_0(\omega \pm n\omega_0)$$

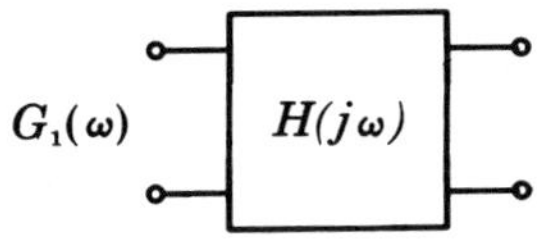

$$G_2(\omega) = G_1(\omega)|H(\omega)|^2$$

Parseval's theorem

$$\frac{1}{T}\int_{-T/2}^{T/2} f^2(t)\, dt = \int_{-\infty}^{\infty} G(\omega)\, df$$

Energy-density spectra [$E(\omega)$ in joules/cycle]

$$E(\omega) = |F(\omega)|^2 \quad \text{where } F(\omega) \text{ is the Fourier transform of } f(t)$$

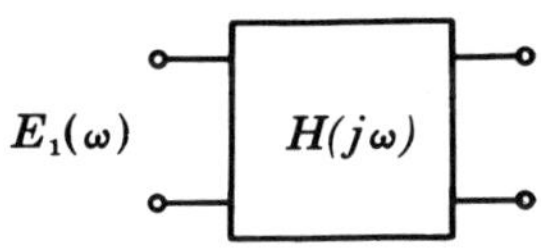

$$E_2(\omega) = E_1(\omega)|H(\omega)|^2$$

Plancherel's theorem

$$\int_{-\infty}^{\infty} f^2(t)\, dt = \int_{-\infty}^{\infty} |F(\omega)|^2\, df$$

Random signals

$$\left.\begin{aligned} G(\omega) &= \frac{|F(\omega)|^2}{T} \\ \text{where}\quad F(\omega) &= \int_{-T/2}^{T/2} f(t)e^{-j\omega t}\, dt \end{aligned}\right\} \begin{array}{c}\text{as } T \to \infty \\ \text{(for most signals)}\end{array}$$

$$\begin{aligned} \text{and}\qquad g(t) &= \text{the inverse Fourier transform of } G(\omega) \\ &= \frac{1}{T}\int_{-T/2}^{T/2} f(x)f(x \pm t)\, dx \\ &= \text{the autocorrelation function of } f(t) \end{aligned}$$

Noise voltages

$$|E_{\text{rms}}|^2 = 4kTBR \quad \text{and} \quad G(\omega) = 2kTR \text{ watts/cycle}$$

where k = Boltzmann's constant (1.37×10^{-23} watt·sec/degree)
T = absolute temperature in degrees Kelvin
B = Bandwidth in cycles
R = Resistance in ohms

REFERENCES

Chapter 23 extends the previous discussions of power and energy to waveforms described by Fourier series and Fourier integrals. A brief mention is made of random waveforms, since it is in this area that the power and energy functions have found their chief use. References to this work are given below.

1. NORBERT WIENER, *Extrapolation, Interpolation, and Smoothing of Stationary Time Series*, John Wiley & Sons, New York, 1949. In this book Wiener records the results of a life-time of work in the area of Fourier methods and their applications to statistically random time functions.

2. NELSON WAX (edit.), *Selected Papers on Noise, and Stochastic Processes*, Dover Publications, New York, 1954. This collection of original papers is perhaps the best introduction which there is to the general subject of statistical communications theory.

3. M. SCHWARTZ, *Communications Theory*, McGraw-Hill, New York, 1959. Schwartz presents the elements of statistical communications theory at the level of undergraduate students in engineering.

4. DAVENPORT and ROOT, *An Introduction to the Theory of Random Series and Noise*, McGraw-Hill, New York, 1958. Davenport and Root give an excellent summary of statistical communication theory at the level of a graduate engineering student.

5. J. L. DOOB, *Stochastic Processes*, John Wiley & Sons, New York, 1953. Doob's book is a definitive work on the subject of statistical communications theory, but it is highly mathematical and cannot be read by the average graduate engineering student.

6. JAMES, NICHOLS, and PHILLIPS, *Theory of Servomechanisms*, MIT Rad. Lab. Series No. 25, McGraw-Hill, New York, 1947. One of the earliest practical applications of Wiener's work to the theory of servomechanisms is found in this text.

7. T. C. FRY, *Probability and Its Engineering Uses*, D. Van Nostrand, New Jersey, 1928.

8. W. FELLER, *Probability Theory and Its Applications*, John Wiley & Sons, New York, 1950. This text and that of Reference 7 are good introductions to probability for engineers.

9. Y. W. LEE, *Statistical Theory of Communication*, John Wiley & Sons, New York, 1960. An authoritative and readable account of the field of statistical communications.

EXERCISES

Section 23–2

1. A function is given by $f(t) = 100 \cos 100t$.
   (a) If $f(t)$ is a voltage give its mean-square value with units.
   (b) If $f(t)$ is the voltage across a 1-ohm resistance give the power in the resistance.
   (c) If $f(t)$ is a current give its mean-square value and the units.
   (d) If $f(t)$ is the current in a 1-ohm resistance give the power in the resistance.
   (e) Comment on the use of watts for the dimensions of mean-square values.
2. A voltage is given by $e(t) = 100 \cos 1000t$.
   (a) Obtain an analytic expression for its power-density spectrum and sketch it.
   (b) Verify Parseval's theorem by finding the mean-square value in the time domain and also from the power-density spectrum.
3. A current is given by $i(t) = 10^{-3} \cos 10^6 t$.
   (a) Find its power-density spectrum and sketch it.
   (b) Find the mean-square value from the time function and also from the power-density spectrum. Verify Parseval's theorem.
4. Find $|H(\omega)|^2$ for the network of Fig. 23–15 [$H(s) = E_2/E_1$].

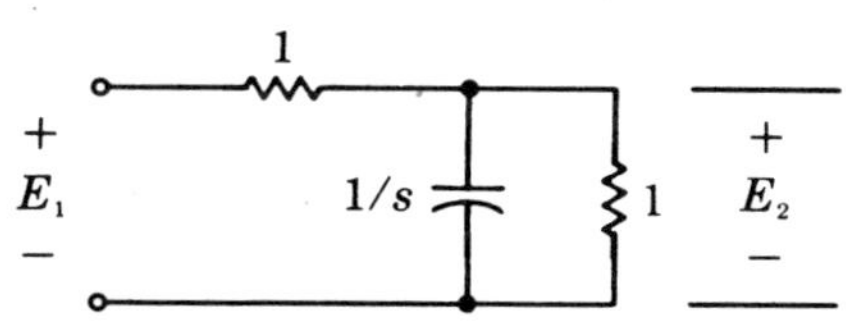

FIGURE 23–15

5. Find $|H(\omega)|^2$ for the network of Fig. 23–16 [$H(s) = E_2/E_1$].

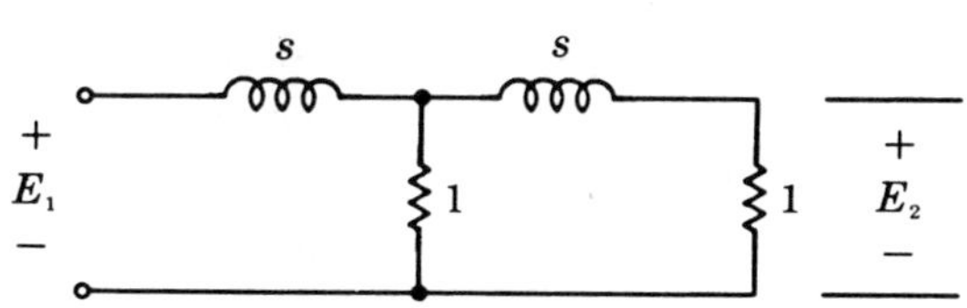

FIGURE 23–16

6. Find $|H(\omega)|^2$ for the network of Fig. 23–17 [$H(s) = E_2/E_1$].

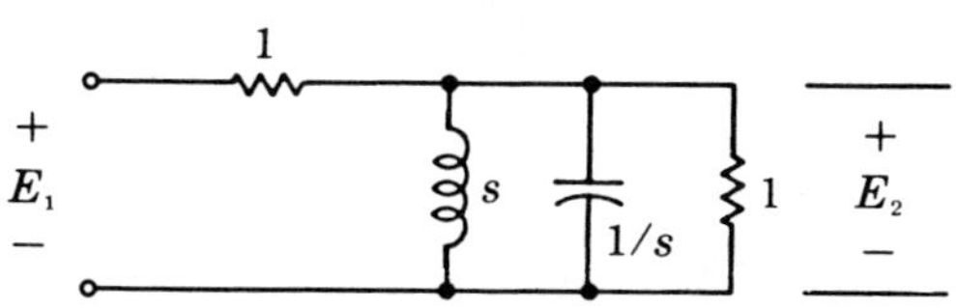

FIGURE 23–17

7. Find the output power-density function for the circuit of Fig. 23–18, in each of the following two ways:
  (a) Find $G_1(\omega)$, $|H(\omega)|^2$, and hence $G_2(\omega)$.
  (b) Find $F_1(\omega)$, $H(j\omega)$, $F_2(\omega)$, and hence $G_2(\omega)$.

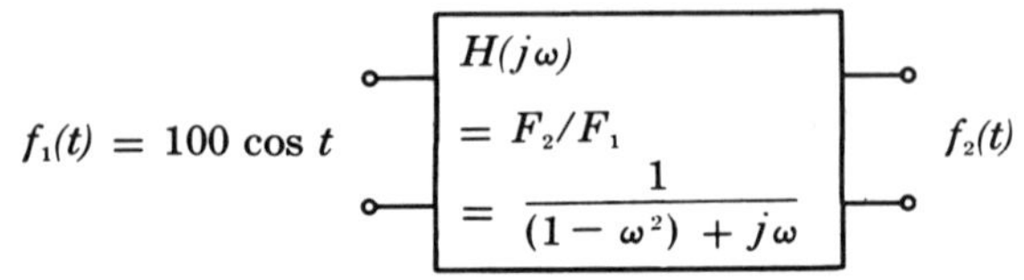

FIGURE 23–18

8. Find the power-density function for the voltage E_2 in the circuit of Fig. 23–19, in each of the following two ways:
  (a) Find $G_1(\omega)$, $|H(\omega)|^2$, and hence $G_2(\omega)$.
  (b) Find $I_1(\omega)$, $H(j\omega)$, $E_2(\omega)$, and hence $G_2(\omega)$.

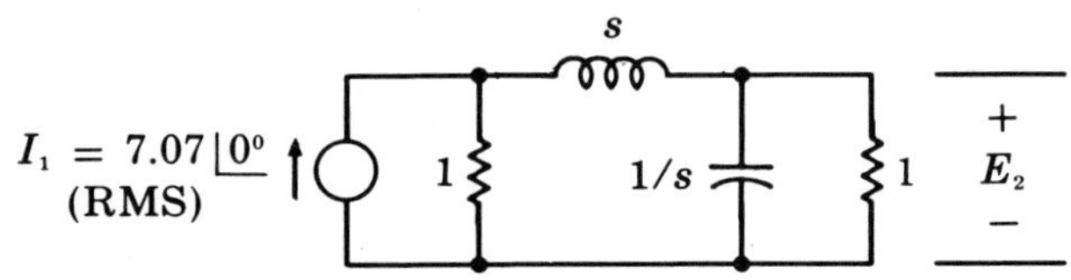

FIGURE 23–19

Section 23–3

9. The square wave of Fig. 23–20 has the Fourier series

$$f(t) = \frac{4}{\pi}\left(\sin t + \frac{\sin 3t}{3} + \frac{\sin 5t}{5} + \cdots\right).$$

  (a) Find the RMS value.
  (b) Find the mean-square value of the first 6 nonzero terms of $f(t)$.
  (c) What percentage error does the result of part (a) possess so far as the true mean-square value is concerned.

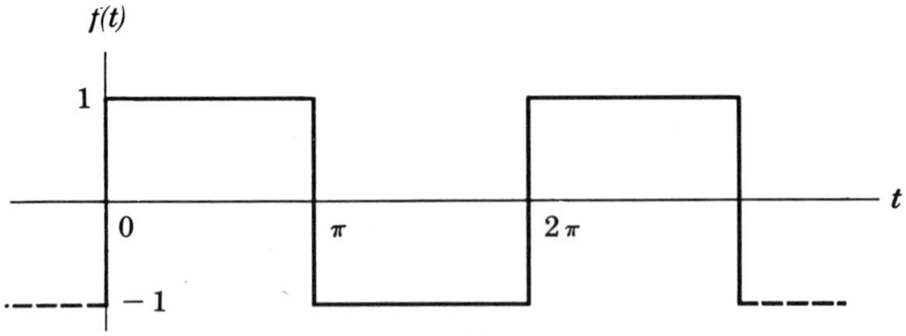

FIGURE 23–20

10. What is the power-density spectrum of a term which is constant over all time, such as $f(t) = A$? What are the units?

11. (a) Find the power-density spectrum of $f(t) = 2 + 2\cos t$.
(b) Verify Parseval's theorem by computing the mean-square value of the function from the time expression, and also from the power-density spectrum.

12. A function is given by the three terms

$$f(t) = \frac{4}{\pi}\left(\cos t - \frac{\cos 3t}{3} + \frac{\cos 5t}{5}\right).$$

(a) Give its power-density spectrum.
(b) Verify Parseval's theorem by computing the mean-square value from the time expression, and also from the power-density spectrum.

13. The waveform of Fig. 23–21 has the Fourier series

$$f(t) = \pi - 2\left(\sin t + \frac{\sin 2t}{2} + \frac{\sin 3t}{3} + \cdots\right).$$

(a) Find the mean-square value of the function from the exact waveform, and also from the constant term and the first three harmonics of the Fourier series.
(b) Find the power-density spectrum and then find the mean-square value of the function from the zero-frequency term and the first three harmonic terms of the power-density spectrum.

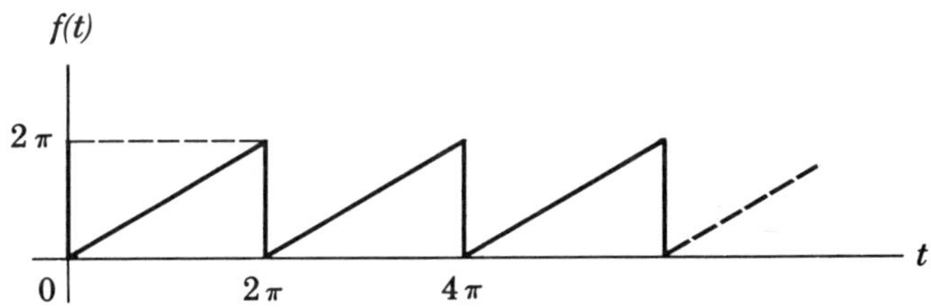

FIGURE 23–21

14. A square wave of voltage is applied to the circuit of Fig. 23–22. Find the power-density spectrum of the current. The wave extends over all time.

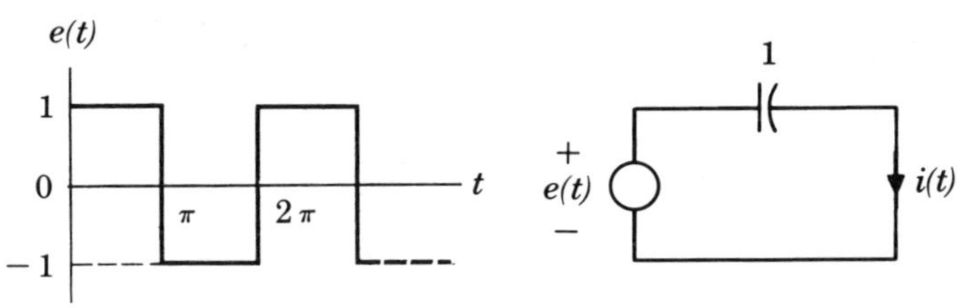

FIGURE 23–22

15. A square wave of voltage is applied to the inductance in the circuit of Fig. 23–23. The wave extends over all time.

(a) Find the power-density spectrum of the current.

(b) Find the time waveform of the current.

(c) Verify Parseval's theorem by computing the mean-square value from the time function and from the power-density spectrum. (See Bernoulli's numbers for the summation of the series.)

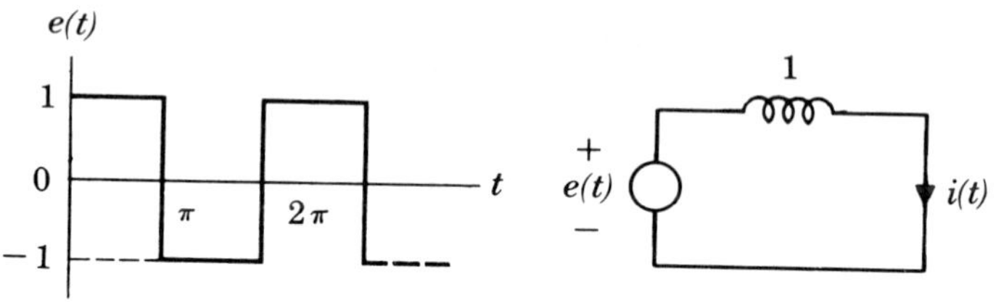

FIGURE 23–23

16. The complex form of a Fourier series is given by

$$f(t) = \sum_{-\infty}^{\infty} c_n e^{jn\omega_0 t},$$

with

$$c_n = \frac{1}{T}\int_{-T/2}^{T/2} f(t)e^{-jn\omega_0 t}\,dt.$$

Find the power-density spectrum in terms of the c_n's.

17. (a) Find the complex Fourier series for the waveform of Fig. 23–24.

(b) Give the power-density spectrum and sketch it.

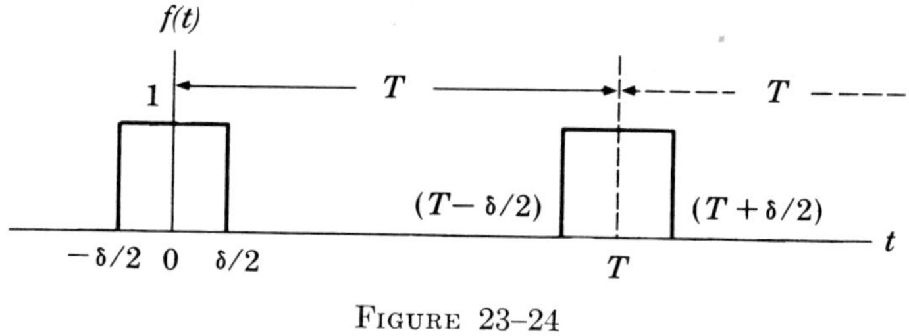

FIGURE 23–24

ction 23–4

18. A pulse is given by $f(t) = e^{-t} \cdot u_{-1}(t)$.

(a) Find its energy-density spectrum in joules per cycle and also in joules per radian.

(b) Verify Plancherel's theorem by finding the total energy which $f(t)$ would dissipate in a 1-ohm resistance if it were either a current or a voltage, and also by integrating the energy-density spectrum.

19. (a) Find the energy-density spectrum for the unit step function $f(t) = u_{-1}(t)$.
    (b) Verify Plancherel's theorem. It is necessary to treat the step function as the limit of $e^{-\alpha t}$ as α tends to zero.

20. Obtain and sketch the energy-density functions for
    (a) a unit impulse,
    (b) a unit doublet (the derivative of an impulse),
    (c) a unit triplet (the derivative of a doublet).

21. (a) Obtain the energy-density function for the pulse shown in Fig. 23–25.
    (b) At what frequency does the first zero occur in the energy-density spectrum.
    (c) At what frequency has the spectrum dropped 1% from its zero-frequency value. (The pulse is a good approximation for an impulse out to this frequency.)

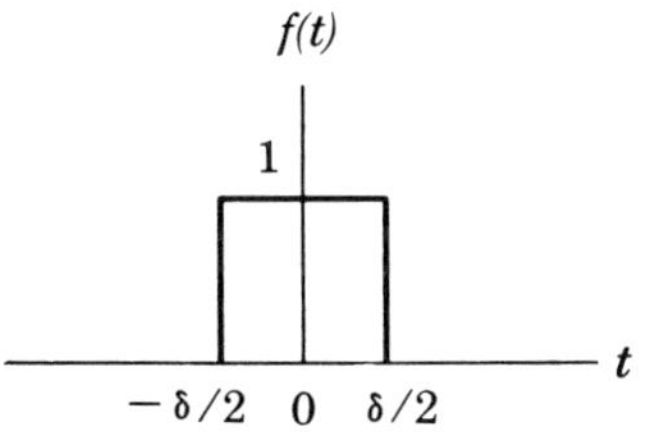

FIGURE 23–25

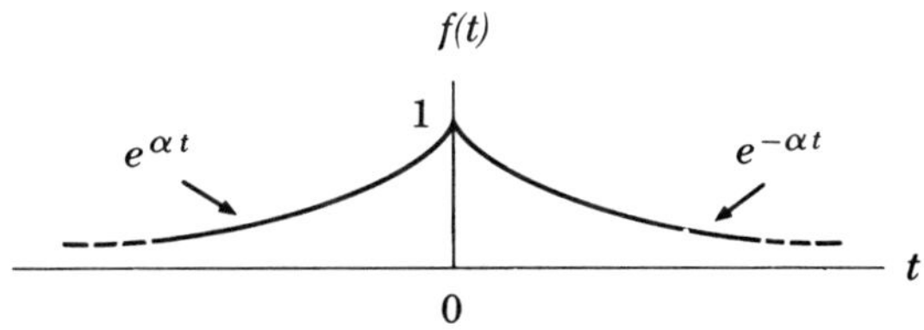

FIGURE 23–26

22. Verify Plancherel's theorem for the pulse of Exercise 21 by computing the energy which the pulse would dissipate in a 1-ohm resistance, and also by integrating the energy-density spectrum.

23. Compute the Fourier integral for the pulse shown in Fig. 23–26. Compute its energy-density spectrum. Verify Plancherel's theorem for the pulse.

24. Compute the energy-density spectrum for the output voltage $e_2(t)$ in the circuit of Fig. 23–27 by each of the following two methods:
    (a) Find the input energy-density spectrum, and the squared magnitude of the system function.
    (b) Find the output by the usual Laplace transform methods, and find the magnitude squared as a function of ω.

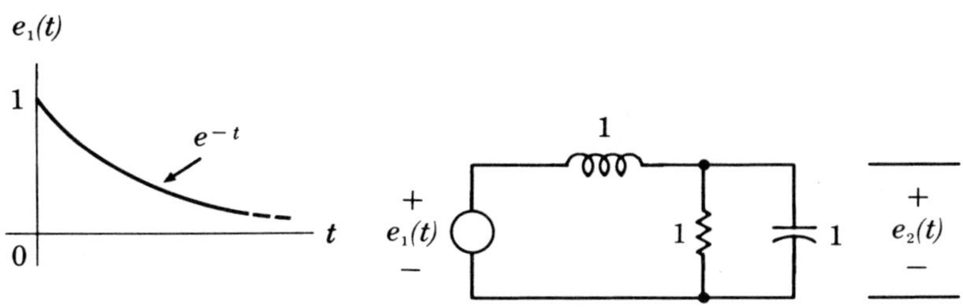

FIGURE 23–27

25. Compute the energy-density spectrum of the output function $f_2(t)$ in the circuit of Fig. 23–28. Verify Plancherel's theorem for the output function.

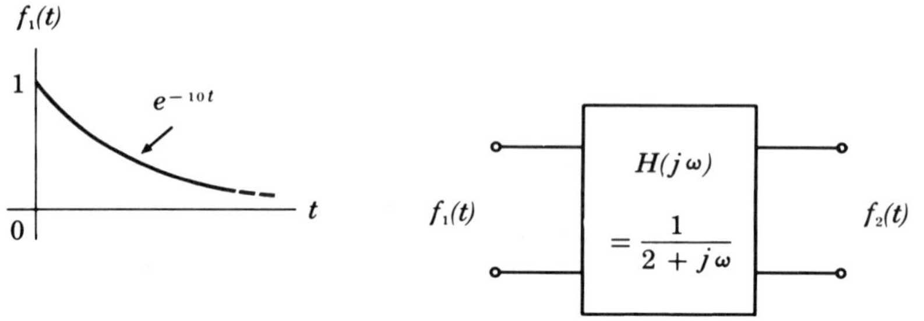

FIGURE 23–28

Section 23–5

26. A function $f(t)$ is given by a series of positive and negative impulses as shown in Fig. 23–29.

(a) Write the Fourier series for the function and give its power-density function. Give the units.

(b) Sketch the power-density spectrum as a function of f in cycles, and give the height of the curve if the impulses are replaced by pulses which run together to form a continuous curve.

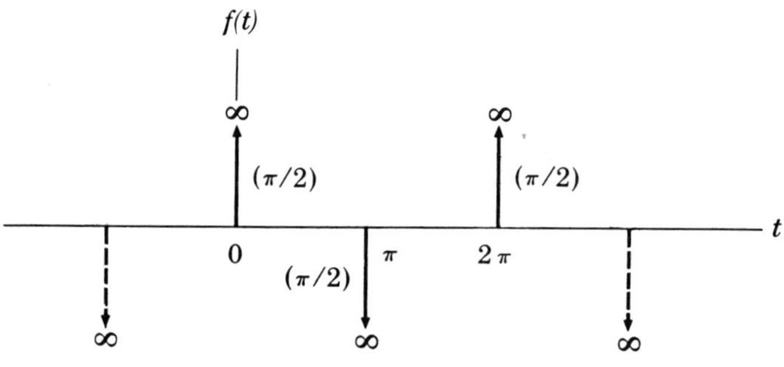

FIGURE 23–29

27. Figure 23–30 shows a train of positive impulses of unit area and of spacing $t = 2\pi$ seconds.

(a) Consider a single pulse at the origin and obtain its energy-density spectrum.

(b) Write the Fourier series and obtain the power-density spectrum for the impulse train.

(c) Show that the energy-density for the single pulse, divided by the period, gives the same value for a continuous power spectrum as is obtained by replacing the impulses by pulses in the power-density spectrum of the Fourier series.

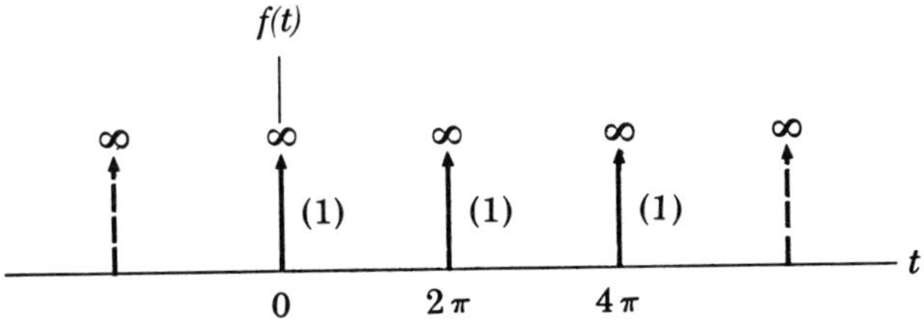

FIGURE 23–30

28. White noise with a power-density spectrum $G_1(\omega) = 0.001$ watt per cycle is applied to a filter with a transfer function

$$H(\omega) = \frac{1000}{1000 + j\omega}.$$

(a) Find the output power-density spectrum.
(b) Sketch the filter transfer function and find the percentage of the output power which is in the pass-band of the filter out to the half-power frequency.

29. White noise with a power-density spectrum $G_1(\omega) = 0.001$ watt per cycle is passed through a low-pass filter with a transfer function

$$H(\omega) = \frac{10^6}{(10^6 - \omega^2) + j\omega 1.414 \times 10^3}.$$

(a) Find the output power-density spectrum.
(b) Sketch the filter transfer function and find the percentage of the output power which is in the pass-band of the filter out to the half-power frequency.

30. The voltage source in the circuit of Fig. 23–31 is a noise voltage with a squared magnitude given by

$$|E_{1_{\text{rms}}}|^2 = 100B \quad \text{microwatts},$$

where B = bandwidth in cycles. Find the energy-density spectrum of the output voltage E_2, and the total power dissipated in the output resistance.

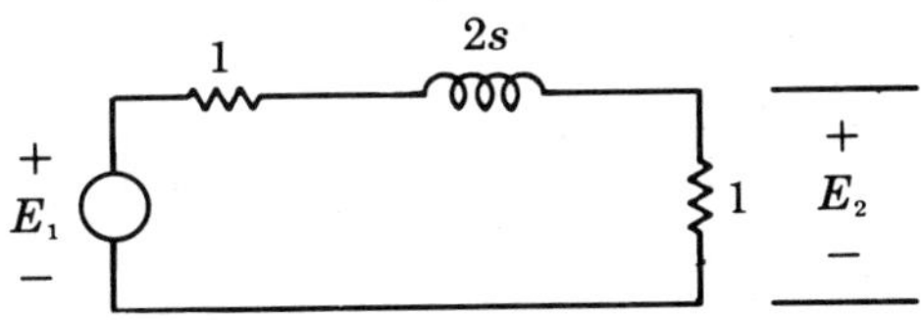

FIGURE 23–31

31. (a) Find $|E_{2_{\text{rms}}}|^2$ in the circuit of Fig. 23–32.
(b) Show that the value obtained in part (a) is the same as would be obtained from $|E_{2_{\text{rms}}}|^2 = 4kTBR_{\text{eq}}$, if R_{eq} is the real part of the equivalent output impedance of the circuit. (This general result is sometimes called Nyquist's theorem.)

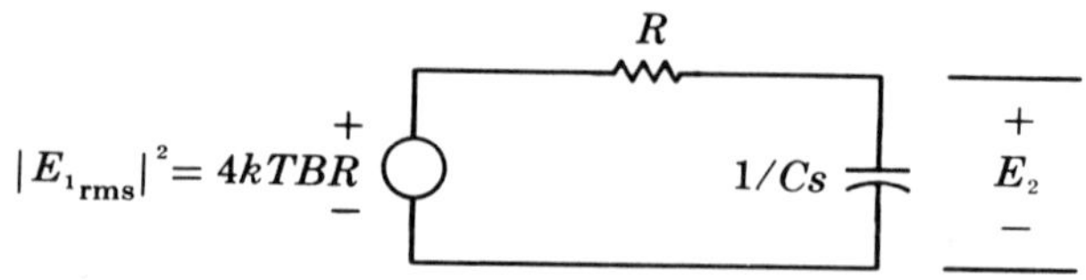

FIGURE 23–32

32. (a) Find the total mean-square voltage $|E_{rms}|^2$ in the circuit of Fig. 23–33.
(b) Show that the result in part (a) is the same as would be obtained from $|E_{rms}|^2 = 4kTBR_{eq}$, where R_{eq} is the parallel resistance of the two elements.

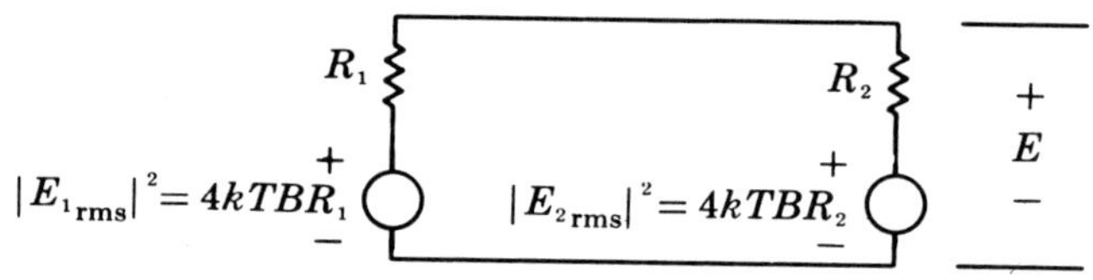

FIGURE 23–33

33. The noise figure of a system is the ratio of the total mean-square output noise to that component of it which is due to the source alone. Thus

$$F = \frac{(\text{RMS output noise voltage})^2}{(\text{RMS output noise voltage from source resistance alone})^2}.$$

If R_1 in Fig. 23–33 is considered to be the source and if the output is the voltage E, what is the noise figure of the system.

34. In the circuit of Fig. 23–34 the source resistance is R_1 and the output is E_2. Find the noise figure.

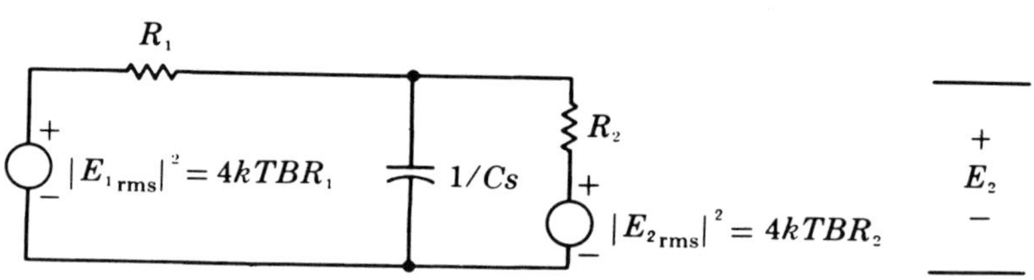

FIGURE 23–34

35. The autocorrelation function of a time function $f(t)$ is defined to be

$$g(t) = \frac{1}{T}\int_{-T/2}^{T/2} f(x)f(x-t)\,dx.$$

(a) If $f(t)$ is the voltage applied to a resistance R of 1 ohm, show that $g(0)$ is the average power dissipated in the resistance.

(b) If the expression for the time function is

$$f(t) = A_0 + \sum A_k \sin (\omega_k t + \theta_k),$$

show that

$$g(t) = A_0^2 + \sum \frac{A_k^2}{2} \cos \omega_k t.$$

APPENDIX

TRANSIENTS IN MORE COMPLEX CIRCUITS

A–1 Introduction. The discussion of transient response functions presented in Chapter 12 can be extended fairly easily, and we will indicate the extensions in this appendix. The actual mechanics of a transient solution in a complicated network are best left to the method of Laplace transforms, which is discussed in Chapter 22.

The first extension we can make relates to the natural transient frequencies of a network. When the network is simple, or when it can be reduced to a simple form, the natural frequencies, and hence the form of the natural behavior, are obvious by inspection. When the network is more complicated, the natural frequencies must be found from the differential equations which are written for it. The process involves an assumption of solutions in the form

$$f(t) = Ae^{pt},$$

and the determination of the allowable values of p.

In addition to the natural frequencies of the network, a number of arbitrary constants must be obtained, one for each of the allowable natural frequencies. These constants are determined by the initial energy stored in the network. There will be a separate p-value and an arbitrary constant for each independent energy storage in the network. The method we will use to evaluate the constants is called the method of Vandermonde equations. The natural response function and its first $(n - 1)$ derivatives are matched to the initial values of the variable and its first $(n - 1)$ derivative, which are computed from the resistive circuit which holds at $t = 0$ in the network. This method is straightforward, but it can be laborious if the system is one of high order and if the p-values are complex conjugates.

A general expression for the natural response of a complicated network is

$$f(t) = A_1e^{p_1t} + A_2e^{p_2t} + A_3e^{p_3t} + \cdots + A_2e^{p_nt}. \qquad \text{(A–1)}$$

The p-values in Eq. (A–1) are the natural frequencies of the network. If they are negative real numbers, the response is a sum of damped exponential terms. If they are pure imaginary numbers, the p-values will exist in conjugate pairs which can be combined to give an undamped sinusoidal response. If they are complex numbers, the p-values will exist in complex conjugate pairs which can be combined to give damped sinusoidal response functions.

The A-values in Eq. (A–1) are obtained from the initial value of $f(0^+)$ and from its first $(n - 1)$ derivatives at $t = 0^+$. The analytical expressions are obtained by differentiating the equation $(n - 1)$ times. The values which the derivatives must have at $t = 0^+$ can be obtained from the resistive circuit which holds at $t = 0$. The n equations for the n constants of Eq. (A–1) are called the *Vandermonde equations.* The solution is straightforward, but it may be laborious.

B–2 The transient natural frequencies. It is most natural to form the differential equations of a complex network in terms of the mesh equations or the node voltage equations. The result is a set of simultaneous differential equations involving the mesh currents or the node voltages, and their integrals and derivatives. If a natural frequency appears at any point in the network, it will usually appear everywhere in the network.* Thus all the variables can be taken to have the same natural frequencies, and they are distinguished only by the multiplying constants associated with them. Let us assume that

$$\begin{aligned} i_1(t) &= I_1 e^{pt}, \\ i_2(t) &= I_2 e^{pt}, \end{aligned} \tag{A–2}$$

and so on.

When the values of all the variables of Eq. (A–2) are substituted in the differential equations, it is seen that the assumed values can be valid only if the determinant of the coefficients is zero. This condition results in an equation in the variable p (of order n), and from it the n p-values are obtained. As soon as the p-values are known the equation for the desired variable can be written in the form of Eq. (A–1). From the resulting equation the A-values can be obtained.

As an example of the determination of the natural frequencies of a circuit, consider the circuit of Fig. A–1. The mesh currents i_1, i_2, and i_3

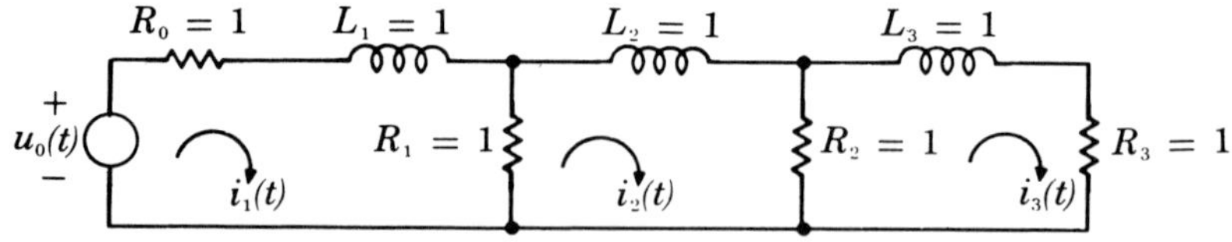

Fig. A–1. A circuit with three independent energy-storage elements.

* If a frequency is trapped from some part of the network, some of the constants will be zero in the general solution.

are assumed as the variables, and the differential equations are written. The circuit is excited by an initial current of 1 ampere in the inductance L_1, produced by the impulse source $u_0(t)$. For $t > 0$ the differential equations are

$$\begin{aligned}
\frac{di_1}{dt} + 2i_1 \qquad - i_2 \qquad\qquad &= 0, \\
- i_1 \qquad + \frac{di_2}{dt} + 2i_2 \qquad - i_3 &= 0, \\
- i_2 \qquad + \frac{di_3}{dt} + 2i_3 &= 0.
\end{aligned} \tag{A-3}$$

To find the p-values, we assume

$$\begin{aligned}
i_1 &= I_1 e^{pt}, \\
i_2 &= I_2 e^{pt}, \\
i_3 &= I_3 e^{pt},
\end{aligned}$$

where I_1, I_2, and I_3 are constants. When these values are substituted into Eqs. (A–3), the e^{pt} term is common and can be removed, leaving

$$\begin{aligned}
I_1(p+2) \qquad - I_2 \qquad\qquad &= 0, \\
-I_1 \qquad + I_2(p+2) \qquad - I_3 &= 0, \\
-I_2 \qquad + I_3(p+2) &= 0.
\end{aligned} \tag{A-4}$$

The only way that Eqs. (A–4) can be satisfied for finite values of I_1, I_2, and I_3 is for the determinant of the coefficients to be equal to zero. This determinant is

$$\begin{vmatrix} p+2 & -1 & 0 \\ -1 & p+2 & -1 \\ 0 & -1 & p+2 \end{vmatrix} = 0, \tag{A-5}$$

from which

$$(p+2)(p^2 + 4p + 2) = 0. \tag{A-6}$$

There are three p-roots for Eq. (A–6). There are three energy-storage elements in the circuit. There must be three exponential terms in the solution with three arbitrary constants to take care of the three arbitrary energy storages. The p-values are

$$\begin{aligned}
p_1 &= -2, \\
p_2 &= -2 + 1.414 = -0.59, \\
p_3 &= -2 - 1.414 = -3.41.
\end{aligned} \tag{A-7}$$

Each of the currents in the network will have three exponential terms involving the three values of p of Eqs. (A–7). The current i_3, for example, can be written as

$$i_3(t) = A_1e^{-2t} + A_2e^{-0.59t} + A_3e^{-3.41t}. \tag{A–8}$$

The three A's in Eq. (A–8) are evaluated from the initial conditions of the problem. One method of achieving this result is discussed in the next section.

A–3 Evaluating the arbitrary constants in the general transient solution. The arbitrary constants in the general transient solution are necessary to give the solution sufficient flexibility to meet the initial conditions imposed on it by the various energy-storage elements present in the network. If there are n energy-storage elements, there will be n arbitrary constants. Essentially, the values of the energy storages determine the constants, but we will find it convenient to accomplish the process in two steps. First, we will use the initial energy storages and the resistive circuit which holds at $t = 0$ to determine the initial value of the variable of interest, and the initial values of its first $(n - 1)$ derivatives. Since the circuit is resistive, this process is relatively easy.

The general expression for the transient response, in the form given by Eq. (A–1), is then differentiated $(n - 1)$ times to obtain a total of n equations which it must satisfy. The n values of the function and its derivatives at $t = 0$ are then matched to the n equations for the function and its derivatives which have been obtained from Eq. (A–1). The result is a set of n algebraic equations for the n arbitrary constants. These are the Vandermonde equations. They have special properties which can be exploited in the solution, but since we can achieve the same results by means of the Laplace transform method, we will not dwell on the Vandermonde equations, but only write them and solve a particular example.

For convenience, we repeat the general expression of Eq. (A–1):

$$f(t) = A_1e^{p_1t} + A_2e^{p_2t} + \cdots + A_ne^{p_nt}.$$

Its $(n - 1)$ derivatives are

$$\begin{aligned}
f'(t) &= A_1p_1e^{p_1t} + A_2p_2e^{p_2t} + \cdots + A_np_ne^{p_nt},\\
f''(t) &= A_1p_1^2e^{p_1t} + A_2p_2^2e^{p_2t} + \cdots + A_np_n^2e^{p_nt},\\
&\vdots\\
f^{n-1}(t) &= A_1p_1^{n-1}e^{p_1t} + A_2p_2^{n-1}e^{p_2t} + \cdots + A_np_n^{n-1}e^{p_nt}.
\end{aligned} \tag{A–9}$$

When Eqs. (A–9) are equated to the initial values of the function and its $(n-1)$ derivatives, the Vandermonde equations are obtained:

$$\begin{aligned}
A_1 \quad &+ A_2 \quad + \cdots + A_n \quad = f(0^+),\\
A_1 p_1 \quad &+ A_2 p_2 \quad + \cdots + A_n p_n \quad = f'(0^+),\\
A_1 p_1^2 \quad &+ A_2 p_2^2 \quad + \cdots + A_n p_n^2 \quad = f''(0^+),\\
&\vdots\\
A_1 p_1^{n-1} &+ A_2 p_2^{n-1} + \cdots + A_n p_n^{n-1} = f^{n-1}(0^+).
\end{aligned} \tag{A–10}$$

Equations (A–10) are a set of n algebraic equations for the n unknown constants $A_1, A_2, \ldots A_n$. A solution can always be obtained. They are certainly independent equations, because each one contains new information. As an example, we will complete the solution of Eq. (A–8) for the current $i_3(t)$ of Fig. A–1. In the preceding section we found the natural frequencies and the form of the current,

$$i_3(t) = A_1 e^{-2t} + A_2 e^{-0.59t} + A_3 e^{-3.41t}, \tag{A–11a}$$

from which

$$\begin{aligned}
i_3'(t) &= -2A_1 e^{-2t} - 0.59 A_2 e^{-0.59t} - 3.41 A_3 e^{-3.41t},\\
i_3''(t) &= 4A_1 e^{-2t} + 0.348 A_2 e^{-0.59t} + 11.6 A_3 e^{-3.41t}.
\end{aligned} \tag{A–11b}$$

Before we can proceed we must compute the initial values of i_3 and its first two derivatives from the circuit of Fig. A–1. At $t = 0$ the impulse source appears across L_1 and produces a current in it of 1 ampere. The voltage across R_1 is finite, and therefore there are no currents in the other inductances. Hence

$$\begin{aligned}
i_1(0) &= 1 \text{ amp},\\
i_2(0) &= 0,\\
i_3(0) &= 0.
\end{aligned} \tag{A–12}$$

At $t = 0^+$ there is a voltage of 1 volt in mesh 2 and zero volts in mesh 3. Hence

$$\begin{aligned}
i_2'(0^+) &= \frac{1}{L_2} = 1 \text{ amp/sec},\\
i_3'(0^+) &= 0.
\end{aligned} \tag{A–13}$$

A rate of change of 1 volt/sec appears across R_2 and also across L_3. The second derivative of the current i_3 is thus

$$i_3''(0^+) = \frac{e_{L3}'}{L_3} = 1 \text{ amp/sec/sec}. \tag{A–14}$$

If we substitute the values of i_3 and its derivatives at $t = 0^+$, from Eqs. (A–12), (A–13), and (A–14) in the general expressions of Eqs. (A–11a) and (A–11b), with $t = 0$, we obtain

$$A_1 + A_2 + A_3 = 0,$$
$$-2A_1 - 0.59A_2 - 3.41A_3 = 0,$$
$$4A_1 + 0.348A_2 + 11.6A_3 = 1,$$

from which

$$A_1 = -\tfrac{1}{2}, \qquad A_2 = \tfrac{1}{4}, \qquad A_3 = \tfrac{1}{4}.$$

The final solution for $i_3(t)$ is thus

$$i_3(t) = -\tfrac{1}{2}e^{-2t} + \tfrac{1}{4}e^{-0.59t} + \tfrac{1}{4}e^{-3.41t}. \tag{A–15}$$

The expression for $i_3(t)$ given by Eq. (A–15) is the sum of three damped exponential terms. The arithmetic in this example is elementary, because the roots are real (as they will always be in R–L and R–C circuits). When the root are complex, the algebra is more difficult. Furthermore, complex exponential terms must be grouped in pairs to form the real damped sinusoids in the final solution. The method is straightforward; the difficulties are purely computational. Since we have a better method for handling the general case, we will not pursue the matter any further.

It is interesting to note that the step response can be obtained by this method, as well as the natural behavior and the impulse response. The step response merely contributes a constant to the general expression of Eq. (A–1) which is the value of the variable at $t = \infty$. This constant must be added before the Vandermonde equations are solved. It changes the values of the A's, but it does not change the method of finding them.

The material of this appendix is included in the text primarily to round out the discussion of natural transient frequencies. Actual solutions are usually achieved by means of the Laplace transform. This is the reason why most authors rest their discussion of transients with the two-energy-storage system. For further reading, the interested student may refer to the references of Chapter 12, especially E. A. Guillemin, *Introductory Circuit Theory*, John Wiley & Sons, New York, 1953, Chapter 9.

ANSWERS TO EXERCISES: PART 1

Chapter 1

1. (a) Mechanical-acoustic (b) Mechanical-electrical-acoustic (c) Hydraulic-mechanical-electrical (d) Chemical-mechanical-electrical (e) Chemical-mechanical-electrical (f) Chemical-mechanical-electrical-mechanical (g) Solar-electrical (h) Electrical-thermal-mechanical-electrical (i) Chemical-mechanical-electrical-light

2. Atom, Electron, Electric circuit, Map, Economic system, Inertia, Particle, Line, Circle, Point, Energy, Equator.

3. $1 + 1 = 1$ (closed path); $1 + 0 = 0$ (open path); $0 + 0 = 0$ (closed path).

4. War–stock boom; Elections–less graft; Government spending–inflation; High birth rate–expanding economy; Labor government–higher wages; Business government–higher profits; Fear and uncertainty–recession.

5. Stimulus: demand for power; Response: additional power generated; Measure power demanded and adjust steam power to meet demand.

6. Store prices are constant but stock prices rise; shares bought are not proportional to money spent.

7. (a) 30 amp (b) —30 amp (c) signs reverse

8. (a) 10 volts (b) —10 volts (c) —10 volts (d) +10 volts

9. (a) 100 watts (b) —100 watts (c) —100 watts

10. (b) and (d)

11. 34 ohms

12. 0.0144×10^{-6} farad

13. 1 henry

14. 3 amp; 18 watts.

15. 18×10^{-6} joule; nothing.

16. 50 joules; nothing.

17. (a) none (b) 100 watts (c) infinity (d) —10 watts

18. (a) infinity (b) none (c) 100 watts (d) —10 watts

19.

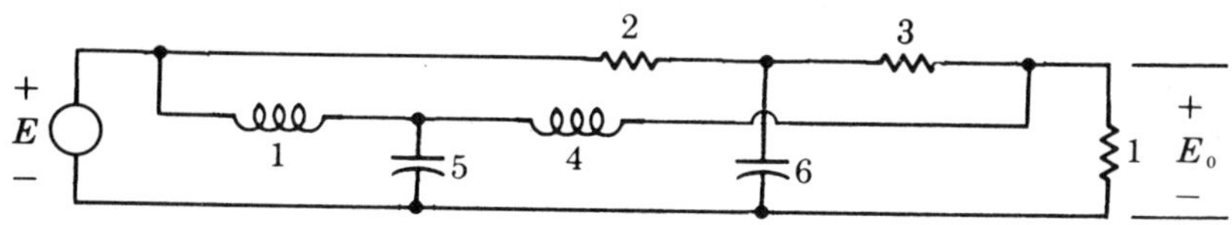

20. $V_{cd} = -3$ volts

21. $I_{do} = -11$ amp

22. $e_{ba} = 6$ volts; $e_{ca} = -7$ volts; $e_{da} = -4$ volts.

23. $i_{10} = i_a - i_b$; $i_{20} = i_b - i_c$; $i_{30} = i_c - i_d$; $i_{40} = i_d - i_a$.

Chapter 2

1. (a) 2 (b) 2×10^6 (c) 1 (d) ∞ (e) 9
2. (a) $\frac{1}{2}$ (b) $\frac{1}{2} \times 10^6$ (c) 0 (d) 1 (e) 2
3. $\dfrac{R_1R_2R_3}{R_1R_2 + R_2R_3 + R_3R_1}$; (a) 1 (b) $\frac{1}{3}$ (c) $\frac{1}{2}$ (d) 1 (e) $\frac{6}{11}$
4. $\frac{9}{5}$
5. $\frac{34}{21}$
6. $\frac{13}{21}$
7. $\dfrac{R_5[(R_1 + R_2)(R_3 + R_4) + R_3R_4]}{(R_1 + R_2)(R_3 + R_4 + R_5) + R_3(R_4 + R_5)}$
8. (a) $e = 5 + 10i$ (b) $e = 10 + 5i$ (c) $e = 10 + 5i$ (d) $e = -10 + 5i$.
9. 10 volts; 6 amp; 2 volts; +8 amp.
10. 5 amp, 2 ohms; —2 amp, 3 ohms; $\frac{5}{2}$ amp, 2 ohms; impossible.
11. 10 volts, 2 ohms; —15 volts, 3 ohms; —12 volts, 2 ohms; impossible.
12. 5 volts, $\frac{10}{3}$ ohms; 0 volts, 15 ohms; 66 volts, 10 ohms.
13. 1 amp, 7 ohms; 0 amp, 2 ohms; 8 amp, 5 ohms.
14. 10 volts in series with R_2
15. 10 amp in parallel with R_2
16. (a) 10 volts in series with R_1 and R_2 in parallel (b) the 10-volt source (c) a 2-amp source across R_5.
17. 25 volts
18. 5 amp
19. 20 volts
20. 10 amp
21. $\frac{20}{31}$ volts
22. $\frac{10}{3}$ volts
23. 5 volts
24. $\frac{25}{4}$ volts
25. $b = 8$, $n = 6$, $l = 3$, 5 current equations, 3 voltage equations, 8 v-a equations.
26. $b = 18$, $n = 12$, $l = 7$, 11 current equations, 7 voltage equations, 18 v-a equations.
27. $b = 12$, $n = 7$, $l = 6$, 6 current equations, 6 voltage equations, 12 v-a equations.
28. $b = 8$, $n = 7$, $l = 2$.
29. (a) and (b) are planar.
30. (a) $b = 0$, $n = 1$, $l = 0$ (b) $b = 10$, $n = 7$, $l = 4$
31. (a) $b = 6$, $n = 4$, $l = 3$ (b) $b = 7$, $n = 5$, $l = 3$
32. $b = 4$, $n = 4$, $l = 1$; dummy $b = 7$; dummy $n = 3$; dummy $l = 4$.
33. $b = 0$, $n = 0$, $l = 0$; dummy $b = 11$; dummy $n = 5$; dummy $l = 7$.
34. $e_1 = 4$, $e_2 = 2$, $e_3 = 2$; $i_1 = -2$, $i_2 = 2$, $i_3 = 2$.
35. $i_1 = -3$, $i_2 = 1.5$, $i_3 = 1.5$; $e_1 = 3$, $e_2 = 3$, $e_3 = 3$.

36. $i_1 = 1, i_2 = -1; e_1 = 12, e_2 = 12.$
37. $i_1 = -1, i_2 = -2, i_3 = 1.5, i_4 = 1.5, i_s = 3.$
38. Same as 37.
39. $e_1 = 6, e_2 = -12, e_3 = 12, e_4 = 6, e_s = 18.$
40. Same as 39.
41. $i_1 = 10, i_2 = 0, i_3 = \frac{15}{2}, i_4 = \frac{5}{2}.$
42. Same as 41.

Chapter 3

1. $i_1 = 1, i_2 = 1.$
2. $i_1 = 4, i_2 = 9.$
3. $i_1 = 2, i_2 = 1.$
4. $v_{ab} = -11$
5. 10 amp from right to left.
6. 16 watts in 1-ohm resistor, 18 watts in the 2-ohm resistors.
7. 20.8 watts supplied
8. $i_1 = \frac{1}{3}, i_2 = 1.$
13. $i_1 = 3, i_2 = \frac{3}{4}, i_3 = \frac{15}{4}.$
14. $i_1 = -5, i_2 = 2.5.$
15. $i_1 = 6, i_2 = 4, i_3 = 3.$
16. $i_1 = 0, i_2 = 2, i_3 = 0, i_4 = -1.$
17. $e_a = 1, e_b = 0.$
18. $e_a = 4, e_b = -2.$
19. $e_a = -1, e_b = 3.$
20. $e_a = -6, e_b = 1, e_c = 3.$
21. $e_a = 11, e_b = 23$; power from 10-volt source = -10 watts.
22. $e_2 = 150$
23. $e_1 = 3, e_2 = 2.$
24. $e_a = \frac{5}{3}, e_b = \frac{9}{7}.$
29. $e_a = 5, e_b = 10, e_c = 10.$
30. $e_a = 10, e_b = 5, e_c = -5.$
31. $e_a = -2, e_b = 10, e_c = 4.$
32. $e_a = 1, e_b = 6, e_c = 3, e_d = -20.$
33. Node 1: 3-amp source in, $\frac{1}{2}$ and 1 ohms to ground, $\frac{1}{4}$ ohm to node 2.
    Node 2: 5-amp source out, $\frac{1}{3}$ and 1 ohms to ground, $\frac{1}{4}$ ohm to node 1.
35. Node 1: dummy, value 5 volts, 1 ohm to node 2.
    Node 2: $\frac{1}{2}$ ohm to ground, 1 ohm to node 1, $\frac{1}{3}$ ohm to node 3.
    Node 3: $\frac{1}{4}$ ohm to ground, $\frac{1}{3}$ ohm to node 2, $\frac{1}{5}$ ohm to node 4.
    Node 4: dummy, value -10 volts, $\frac{1}{5}$ ohm to node 3.
37. Same as original network.
38. Node 1: 5-amp source in, $\frac{1}{3}$ ohm to ground, $\frac{1}{4}$ ohm to node 2.
    Node 2: 1 ohm to ground, $\frac{1}{4}$ ohm to node 1, 2-volt source to node 3 (plus end at 3).

Node 3: $\frac{1}{3}$ ohm to node 4, $\frac{1}{2}$ ohm to ground, plus end of 2-volt source.
Node 4: $\frac{1}{3}$ ohm to node 3, $\frac{1}{5}$ ohm and $\frac{1}{4}$ ohm to ground.

40. Original network has no independent nodes; dual has no independent meshes.
41. (a) $\frac{2}{11}$ amp (b) $\frac{200}{11}$ amp (c) $\frac{2}{11} \times 10^{-4}$ amp.
42. (a) $\frac{1}{2}$ volt (b) $\frac{1}{2} \times 10^{-3}$ volt (c) $\frac{1}{2} \times 10^{3}$ volts.
43. (a) $\frac{1}{6}$ amp (b) linear from zero to 167 microamp.
44. (a) volts (b) megohms (c) megohms (d) milliamperes.
45. $i_1 = 10$. No current from voltage sources.
46. (a) $e = \frac{2}{3}$ (b) $e = 0$
47. (a) $e = 0$ (b) $e = \frac{2}{3}$
48. (a) $i = 1$ (b) $P_R = 1$ watt, $P_e = 0$, $P_i = 1$ watt.
49. $i = 17$
50. 20 volts
51. —5 volts
52. 99.99 amp

Chapter 4

1. 3 ohms
2. 3 ohms
3. 3 ohms
4. 3 ohms
5. $\frac{13}{9}$ ohms
6. $\frac{13}{9}$ ohms
7. $\frac{13}{9}$ ohms
8. $\frac{13}{9}$ ohms
9. $\frac{16}{3}$ ohms
10. $\frac{16}{3}$ ohms
11. $\frac{41}{57}$ ohms
12. (a) 12, 12, 12 ohms (b) $\frac{19}{3}$, $\frac{19}{4}$, 19 ohms (c) $\frac{3}{2}$, $\frac{3}{4}$, $\frac{9}{8}$ ohms
13. (a) $\frac{7}{3}$, $\frac{7}{3}$, $\frac{7}{3}$ ohms (b) 1, $\frac{1}{3}$, $\frac{1}{2}$ ohm (c) $\frac{1}{11}$, $\frac{2}{11}$, $\frac{3}{11}$ ohm
14. All three resistances equal $\frac{5}{9}$ ohms.
15. 3 ohms
16. $\frac{13}{9}$ ohms
17. $\frac{16}{3}$ ohms
18. $\frac{41}{57}$ ohms
19. 8 ohms
20. $\frac{13}{4}$ ohms
21. $\frac{16}{3}$ ohms
22. 8 ohms
23. $\frac{13}{4}$ ohms
24. $\frac{180}{37}$ ohms
25. $\frac{7}{12}$ ohm
26. $\frac{5}{6}$ ohm
27. $\frac{3}{4}$ ohm
28. $E_1 = 3I/8$; $E_2 = I/8$; $R = \frac{1}{2}$.
29. $\frac{89}{55}$ ohms
30. $\frac{89}{55}$ ohms
31. $\frac{5}{11}$ ohms
32. $\frac{5}{11}$ ohms
33. $\frac{181}{21}$ ohms
34. $\frac{118}{93}$ ohms
35. $\frac{153}{56}$ ohms
36. 18 ohms
37. $E_{oc} = 5$ volts; $R_{eq} = 1$ ohm; $I_{sc} = 5$ amp.
38. $E_{oc} = 50 \cos (377t)$; $R_{eq} = 40$ ohms; $I_{sc} = 1.25 \cos (377t)$.
39. $E_{oc} = 3$ volts; $R_{eq} = \frac{5}{3}$ ohms; $I_{sc} = \frac{9}{5}$ amp.
40. $E_{oc} = -2$ volts; $R_{eq} = 2$ ohms; $I_{sc} = -1$ amp.
41. $E_{oc} = \frac{65}{8}$ volts; $R_{eq} = \frac{31}{8}$ ohms; $I_{sc} = \frac{65}{31}$ amp.
42. $E_{oc} = 2$ volts; $R_{eq} = \frac{7}{8}$ ohm; $I_{sc} = \frac{16}{7}$ amp.
43. $E_{oc} = 6$ volts; $R_{eq} = \frac{8}{5}$ ohms; $I_{sc} = \frac{15}{4}$ amp.
44. $E_{oc} = 0$; $R_{eq} = \frac{3}{4}$ ohm; $I_{sc} = 0$.
45. (a) $E_{oc} = \frac{10}{3}$ volts; $R_{eq} = \frac{4}{3}$ ohms (b) 2.5 amp (c) $\frac{10}{7}$ amp
46. $E_{oc} = 88$ volts; $R_{eq} = 36$ ohms; $I_{sc} = \frac{22}{9}$ amp.

47. $E_{oc} = \frac{40}{21}$ volts; $R_{eq} = \frac{22}{21}$ volts; $I_{sc} = \frac{20}{11}$ amp.
48. $E_{oc} = 17$ volts; $R_{eq} = 1$ ohm; $I_{sc} = 17$ amp.
49. (a) $R = 2$ ohms (b) $E_2 = 6$ volts (c) $E_0 = 1$ volt
50. (a) $R = \frac{3}{5}$ ohms (b) $E_2 = \frac{6}{5}$ volts (c) $I_0 = \frac{8}{5}$ amp.
51. $e_0 = 1$ volt
52. $e_0 = 3$ volts, $i_a = 24$ amp.
53. $E_2 = 1$ volt
54. Current decreases by $\frac{5}{13}$ amp.
55. Current increases by $\frac{5}{13}$ amp.

Chapter 5

1. 10 amp
2. —4 volts
3. $\frac{12}{17}$ amp
4. 10 volts
5. $\frac{1}{2}$ amp
6. 1 volt
7. $r_{11} = 2$, $r_{12} = \frac{2}{5}$, $r_{22} = \frac{27}{25}$, $I_1 = -\frac{100}{77}$.
8. $-\frac{50}{79}$ amp
9. π circuit; $\frac{5}{7}$, $\frac{15}{2}$, and $\frac{5}{7}$ ohms.
10. π circuit; 2, 2, and 2 ohms.
11. T values: $\frac{1}{3}$, $\frac{9}{2}$, and 1 ohms.
12. Six resistances: five of 2 ohms, one of $\frac{4}{3}$ ohms.
13. Four resistances of 6 ohms; one of 3 ohms; and one of 12 ohms.
14. T values: $\frac{33}{7}$, $\frac{9}{14}$, and $\frac{33}{7}$ ohms.
15. T values: 30, $\frac{5}{2}$, and 30 ohms.
16. Four resistances of 12 ohms; two of $\frac{12}{5}$ ohms.
17. Four resistances of 4 ohms; two of 2 ohms.
18. $r_{11} = r_{22} = \frac{11}{5}$; $r_{12} = \frac{9}{5}$. T values: $\frac{2}{5}$, $\frac{9}{5}$, and $\frac{2}{5}$ ohms.
19. $g_{11} = g_{22} = \frac{11}{8}$; $g_{12} = -\frac{9}{8}$. π values: 4, $\frac{8}{9}$, and 4 ohms.
20. $r_{11} = \frac{21}{10}$; $r_{22} = \frac{12}{5}$; $r_{12} = \frac{1}{5}$. T values: $\frac{19}{10}$, $\frac{1}{5}$, and $\frac{11}{5}$ ohms.
21. $g_{11} = \frac{12}{25}$; $g_{22} = \frac{21}{50}$; $g_{12} = -\frac{1}{25}$. π values: $\frac{25}{11}$, 25, and $\frac{50}{19}$ ohms.
22. $r_{11} = \frac{26}{21}$; $r_{22} = \frac{68}{21}$; $r_{12} = \frac{2}{21}$. T values: $\frac{8}{7}$, $\frac{2}{21}$, and $\frac{22}{7}$ ohms.
23. $g_{11} = \frac{34}{42}$; $g_{22} = \frac{13}{42}$; $g_{12} = -\frac{1}{42}$. π values: $\frac{14}{11}$, 42, and $\frac{7}{2}$ ohms.
26. T values: $\frac{8}{3}$, $\frac{2}{3}$, and $\frac{8}{3}$ ohms.
27. T values: $\frac{5}{2}$, $\frac{1}{4}$, and $\frac{5}{2}$ ohms.
28. π values: $\frac{6}{7}$, 6, and $\frac{6}{7}$ ohms.
29. $\frac{20}{7}$ volts.
30. zero volts
31. $\frac{25}{4}$ amp.
32. $\frac{290}{7}$ volts.
33. 12.5 amp.
34. $r_{11} = r_{22} = \frac{68}{55}$; $r_{12} = \frac{2}{55}$.
35. $g_{11} = g_{22} = \frac{17}{21}$; $g_{12} = -\frac{1}{42}$.
36. $r_{11} = r_{22} = 3$; $r_{12} = 1$. $g_{11} = g_{22} = \frac{3}{8}$; $g_{12} = -\frac{1}{8}$. T values: 2, 1, and 2 ohms. π values: 4, 8, and 4 ohms.
37. $r_{12} = -1$; $g_{12} = \frac{1}{8}$. T values: 4, —1, and 4 ohms. π values: 2, —8, and 2 ohms.
39. T values: 1, $\frac{1}{6}$, and 1 ohms.
40. π values: $\frac{4}{3}$, 8, and $\frac{4}{3}$ ohms.
41. $R_x = 2$, $R_y = 6$.
42. $R_x = \frac{2}{3}$, $R_y = 2$.
43. $R_x = 1$, $R_y = \frac{4}{3}$.
44. $R_x = \frac{1}{2}$, $R_y = \frac{3}{2}$.
45. T values: 2, 3, and 2 ohms.
46. T values: 4, 2, and 4 ohms.
47. π values: 8, $\frac{16}{3}$, and 8 ohms; 8, 16, and 8 ohms.

Chapter 6

1. $P_{R_1} = 0$; $P_{R_2} = 100$ watts; $P_V = 0$; $P_I = -100$ watts.
2. 416 watts
3. 100 watts
4. 200 watts; source powers indeterminate.
5. 171 watts
6. 1470 watts
7. Sink: $E_1 = 70$ watts; sources: $E_2 = 12$ watts, $I_1 = 50$ watts, $I_2 = 8$ watts.
8. $E_1 = 3$ watts; $E_2 = 2$ watts; $E_3 = 1$ watt; $E_4 = 0$; $I_1 = 3$ watts; $I_2 = -4$ watts; $I_3 = -3$ watts; $I_4 = -2$ watts.
9. 3 watts
10. Three 1-volt sources; each delivers 1 watt.
11. 2 watts
12. Two 1-amp sources at nodes e_1 and e_2. One watt each.
13. 50 watts
14. 54 watts
15. 20 watts
16. 20 watts
20. $P_{\text{load}} = 4$ watts; $P_{\text{int}} = 56$ watts; P_{int} (Thévenin) $= 6$ watts.
21. $P_{\text{load}} = 16$ watts; $P_{\text{int}} = 124$ watts; P_{int} (Thévenin) $= \frac{208}{15}$ watts.
22. Equal if $R = R_L$. Voltage-source losses greater if $R > R_L$; current-source losses greater if $R < R_L$.
25. 240 watts
26. 48 watts
27. $\frac{248}{3}$ watts
28. 56 watts
29. 56 watts.
30. $\frac{248}{3}$ watts.
31. 925 watts
32. Reverse polarity of output source for positive elements. T values: 2, 3, and 2 ohms.
33. 25 watts.
34. 12.5 watts
35. $\frac{25}{6}$ watts
36. $\frac{27}{32}$ watts
37. $\frac{27}{4}$ watts
38. $R = 2$; $P_{\max} = 12.5$ watts.
39. $R = \frac{3}{2}$; $P_{\max} = \frac{25}{6}$ watts.
40. $R = \frac{1}{2}$; $P_{\max} = 12.5$ watts.
41. $R = \frac{2}{3}$; $P_{\max} = \frac{25}{6}$ watts.
42. Correct power $= 200$ watts. $P_{E_1} = \frac{200}{3}$; $P_{E_2} = \frac{200}{3}$.
43. Correct power $= 150$ watts. $P_{I_1} = P_{I_2} = 50$ watts.
44. 120 watts.
45. Voltage source power $= 0$ at $I = 1$ amp. Supplies power below 1 amp and absorbs power above 1 amp.
46. $P_I = 0$ at $E = -100$. Supplies power for E greater than -100 volts.
47. When $R = 0$ no net power is supplied from the two sources.
48. When $R = \infty$ no net power is supplied by the two sources.

Chapter 7

1. (a) Step of height 5 (b) Ramp of slope 2
2. (a) Step of height 20 (b) Ramp of slope $\frac{5}{2}$
3. (a) $5 \sin 3t$ (b) $90 \cos 3t$ (c) $\frac{5}{6}(1 - \cos 3t)$
4. (a) $20 \cos 2t$ (b) $\frac{5}{3} \sin 2t$ (d) $-80 \sin 2t$

5. $Q = 30$; time $= 0$.
6. $\lambda = 10$; time $= 0$.
7. $W(10) = 400$; $W(100) = 40{,}000$.
8. $W(5) = 225$; $W(10) = 900$.
9. Four series inductances of value 3; four shunt capacitances of value 2.
10. Voltage $= 9$; current $= 10$; $R = \frac{1}{8}, \frac{1}{3}$; $C = 7, 6$; $L = 4, 5$.
11. Inductances: 6 volts; Capacitances: 6 volts.
12. $i = 0; i' = \frac{5}{4}; i(\infty) = 0.$
13. $e = 0; e' = \frac{5}{4}; e(\infty) = 0.$
14. $i' = -\frac{20}{3}; i'' = \frac{40}{9}.$
15. $i = 5; i' = -\frac{5}{6}; i'' = \frac{5}{36}.$
16. $e = 5; e' = \frac{5}{4}; e(\infty) = 10.$
17. $i = 5; i' = -\frac{5}{6}; i(\infty) = \frac{5}{2}.$
18. $e = 20; e' = \frac{5}{3}, e(\infty) = 40.$
19. $i = 2; i' = -\frac{1}{3}; i(\infty) = \frac{5}{4}.$
20. $e = 0, e' = 0; e'' = 0; e''' = 5.$
21. $e_1 = 2i_1 + 10; e_2 = 4i_2'; e_3 = \frac{1}{3}i_3^{-1}; i_1 + i_2 + i_3 = 0; e_1 = e_2 = e_3.$
22. $2i_a + 4i_a' - 4i_b' = 10; -4i_a' + 4i_b' + \frac{1}{3}i_b^{-1} = -5.$
23. $0.5e_a + 3e_a' - 0.5e_b = 10; -0.5e_a + 0.25e_b^{-1} + 0.5e_b = -3.$
24. $2i_2 + 3i_2' + 0.25i_2^{-1} - 0.25i_3^{-1} = 20;$
$-0.25i_2^{-1} + 0.25i_3^{-1} + 0.2i_3^{-1} = -10.$
25. $3e_2' + 0.5e_2 + 0.25e_2^{-1} - 0.5e_3 = 2.5t; -0.5e_2 + 0.5e_3 + 0.2e_3^{-1} = 0.$
26. $Li' + (1/C)i^{-1} = (-1/C)i_1^{-1}; Ce_a' + (1/L)e_a^{-1} = (1/L)e_1^{-1}.$
27. $3i_1' + 0.5i_1^{-1} - 0.5i_2^{-1} = -10; -0.5i_1^{-1} + 0.5i_2^{-1} + 2i_2' + 4i_2 = 10.$
28. $0.5e_a + 0.5e_a^{-1} + 0.33e_a - 0.33e_b = 5; -e_a/3 + e_b/3 + 4e_b' = 0.$
29. $i(0) = I_0; i'(0) = -I_0R/L - E_0/L.$
30. $e(0) = E_0; e'(0) = -E_0/RC - I_0/C.$
31. $e = 2 + t/3 +$ impulse across L
32. $4i' + 2i + \frac{1}{3}i^{-1} = 1$
33. $i = t/2 + \frac{1}{5} +$ impulse into C
34. $3e' + e/5 + 0.5e^{-1} = 1$
35. $e_{11} = e_{22} = 1 + t/2; e_{12} = t/2.$
36. $i_{11} = i_{22} = t/2 + \frac{1}{3}; i_{12} = -\frac{1}{3}.$
37. $3i_1' + 0.5i_1^{-1} + 0.5i_2^{-1} = 1; +3i_1' + 3i_2' + 0.5i_2^{-1} = 0.$
38. $4e_1' + 0.2e_1^{-1} - 0.2e_2^{-1} = 1; -0.2e_1^{-1} + 0.2e_2^{-1} + 4e_2' = 0.$
39. $L_1i_1' - Mi_2' = e_1(t); -Mi_1' + L_2i_2' + Ri_2 + C^{-1}i_2^{-1} = 0.$
40. $L_1i_1' + R_1i_1 + C^{-1}i_1^{-1} - i_2R_1 - L_1i_2' + Mi_2' = e_1;$
$-L_1i_1' - R_1i_1 + Mi_1' + L_1i_2' + L_2i_2' + (R_1 + R_2)i_2 - 2Mi_2^{-1} = 0.$
41. $L_1 + L_2 - 2M$
42. Change sign of all M's.
43. $(L_1L_2 - M^2)$ divided by $L_1 + L_2 - 2M$
44. $(L_1L_2 - M^2)$ divided by $L_1 + L_2 + 2M$
45. $(L_1 - M); M; (L_2 - M).$

46. $(L_1 + M)$; $-M$; $(L_2 + M)$.
47. $\Gamma_1 = L_2/\Delta$; $\Gamma_2 = L_1/\Delta$; $\Gamma_M = -M/\Delta$; where $\Delta = L_1L_2 - M^2$.
48. $\Gamma_1 = L_2/\Delta$; $\Gamma_2 = L_1/\Delta$; $\Gamma_M = M/\Delta$.

Chapter 8

1. In the five intervals, the derivative is 1, 0, —1, —1, 1.
2. Positive impulse at 0; negative impulse at 1; positive impulse at 2.
3. From 0 to 3, 1; $-\infty$ at 1 and at 3.
4. Derivative $= \cos t$.
5. Derivative $= e^{-t}$.
6. $f' = 2e^{2t}$; $f'' = 4e^{2t}$; $f''' = 8e^{2t}$.
7. Peak value is 2 at $t = 3$; final value is 1.
8. From 0 to 1, a linear rise from 0 to +1; from 1 to 2, a linear drop from +1 to 0.
9. From 0 to 1, a parabolic rise from 0 to $+\frac{1}{2}$; from 1 to 2, a parabolic fall from $+\frac{1}{2}$ to 0; from 2 to 3, a parabolic rise from 0 to $-\frac{1}{2}$.
10. Integral is $(1 - \cos t)$
11. Integral is $10(1 - e^{-t/10})$
12. f^{-1} = linear slope; $f^{-2} = t^2/2$; $f^{-3} = t^3/6$.
13. Derivative is 2 from 0 to 1 and —2 from 1 to 2.
14. Derivative is $\cos t$; integral of derivative is $\sin t$.
15. Integral is $(1 - \cos t)$; derivative of integral is $\sin t$.
16. Derivative is $-\sin t$; integral is $(-1 + \cos t)$; not inverse.
17. Derivative is e^t; integral is e^t. (Obtain original curve.)
18. Derivative is e^t; integral is $e^t - 1$; Error because curve starts at $t = 0$.
19. Impulse at $t = 0$ of area unity.
20. Area is unity; derivative becomes a unit impulse, $u_0(t)$.
21. $13u_{-1}(t)$ 22. $13u_0(t)$ 23. $13u_{-2}(t)$
24. Derivative $= u_0(t) - \sin t \cdot u_{-1}(t)$
25. Integral $= (1 - \cos t)u_{-1}(t)$
26. (a) Derivative $= u_1(t)$; integral $= u_{-1}(t)$ (b) $f(t) = 0$, $f'(t) = u_0(t)$, $f^{-1}(t) = 0$
27. $u_{-1}(t) - 2u_{-1}(t-1) + u_{-1}(t-2)$; derivative: replace u_{-1} by u_0; integral: replace u_{-1} by u_{-2}.
28. $u_{-2}(t) - 2u_{-1}(t-1) - u_{-1}(t-3) - u_{-2}(t-3)$; integral index down one; derivative index up one.
29. $u_{-2}(t) - 2u_{-2}(t-1) + u_{-2}(t-2)$
30. $u_{-2}(t) - 2u_{-2}(t-1) + 2u_{-2}(t-2) - 2u_{-2}(t-3) + \cdots$
31. $\sin t \cdot u_{-1}(t) - \sin t \cdot u_{-1}(t-\pi)$ or $\sin t \cdot u_{-1}(t) + \sin(t-\pi) \cdot u_{-1}(t-\pi)$
32. A staircase rising one unit for each unit of time.
33. A triangle rising from zero to 1 at $t = 1$ and then falling back to zero at $t = 2$.

34. Linear rise from zero to 1. At $t = 1$ jumps to 2 and decays linearly to zero at 2.
35. Exact expression: $u_{-2}(t) - u_{-2}(t-4) - 4u_{-1}(t-4)$.
Approximate expression: $u_{-1}(t) + u_{-1}(t-1) + u_{-1}(t-2) + u_{-1}(t-3) - 4u_{-1}(t-4)$.
36. The integral is a good approximation; the derivative a poor one.

Chapter 9

1. $10, 10/e, 10/e^2, 0; T = 10$ sec.
2. $-e^{-t/10}; T = 10$ sec; 10 sec.
3. $5e^{-2t/3}$
4. $-10e^{-t/2}$
5. $6e^{-3t/2}$
6. $2e^{-t/15}$
7. $i_1 = 3e^{-2t/5}; i_2 = 2e^{-2t/5}$.
8. $e = 2e^{-t/5}$
9. 0; 6.32; 8.65; 10; $T = 10$ sec.
10. $e^{-t/10}; T = 10$ sec; 10 sec.
11. $i = \frac{7}{3}(1 - e^{-3t/5})$
12. $e = \frac{10}{3}(1 - e^{-3t})$
13. $i = 2(1 - e^{-5t/3})$
14. $i = 5e^{-t/10}$
15. $e = 60(1 - e^{-t/30})$
16. $i = 4(1 - e^{-t})$
17. 5; 10; $T = \frac{1}{3}$ sec; $f = 3$ cps; $\omega = 6\pi$ rps; $\theta = -60°$.
18. $f'(t) = 60\pi \cos (6\pi t + 30°)$; f, ω, and T remain the same; θ and A are different.
19. Transient term $= 10e^{-t/10}$; steady-state term $= 5 \sin t$; 53 sec.
20. $i = 0.707 \cos (t - 135°) + \frac{1}{2}e^{-t}$
21. $i = 0.707 \cos (t - 45°) - \frac{1}{2}e^{-t}$
22. $i = 0.707 (\cos t - e^{-t})$
23. $i = 7.07 \cos (3t + 45°) + 5e^{-3t}$
24. $e = 6e^{-2t} - 12te^{-2t}$
25. 0; 0; Max $= 3.5$ at $t = 1.58$.
26. 0; 0; Max $= 4/e$ at $t = 2$.
27. 0.707; 0.428; 0.262; 0; $T = 2$ sec; period $= 1$ sec; $f = 1$ cps; $\theta = 45°$.
28. $i = 10 (e^{-2t} - e^{-3t})$
29. $e = 10te^{-2t}$
30. $i = 10e^{-4t}(\cos 3t + \sin 3t)$
31. $i = \frac{50}{3}e^{-5t} - \frac{20}{3}e^{-2t}$
32. $e = e^{-6t}(5 - 25t)$
33. $i = e^{-5t}(10 \cos 12t + \frac{25}{12} \sin 12t)$
34. 0; 10; steady-state term $= 10$; transient term $= -10e^{-t/2} \cos 2\pi t$; $T = 2$ sec; $f = 1$ cps.
35. 10; 10; T's $= 1$ and 10 sec.
36. Transient term $= 5e^{-t}(\cos 2t + \sin 2t)$;
steady-state term $= 10(\cos 6t + \sin 6t)$; $e(0) = 15$; $e'(0) = 65$;
amplitude of transient term $= 7.07$;
amplitude of steady-state term $= 14.1$.
37. $i = 2 + \frac{2}{3}e^{-4t} - \frac{8}{3}e^{-t}$
38. $i = 1 - e^{-3t}(1 + 3t)$
39. $i = 10(1 - e^{-0.707t} \cos 0.707t - e^{-0.707t} \sin 0.707t)$
40. $i = \frac{10}{3}(e^{-2t} - e^{-5t})$
41. $i = 10 - e^{-4t}(10 + 40t)$
42. $i = 10e^{-2t} \sin 4t$

Chapter 10

1. (a) $10u_{-1}(t)$ (b) $\frac{100}{3}u_{-2}(t)$ (c) $\frac{100}{6}u_0(t)$
2. (a) $300u_{-1}(t)$ (b) $700u_0(t)$ (c) $25u_{-2}(t)$
3. (a) $u_{-1}(t)$ (b) $\frac{10}{6}u_{-2}(t)$ (c) $50u_0(t)$
4. (a) $50u_{-1}(t)$ (b) $70u_0(t)$ (c) $\frac{10}{3}u_{-2}(t)$
5. $2u_{-1}(t) - 2u_{-1}(t-1) + 2u_{-2}(t) - 2u_{-2}(t-1) + 2u_0(t) - 2u_0(t-1)$
6. Same as Exercise 5.
7. (a) $\cos 2t \cdot u_{-1}(t)$ (b) $\sin 2t \cdot u_{-1}(t)$ (c) $5u_0(t) - 10 \sin 2t \cdot u_{-1}(t)$
8. (a) $25 \sin 3t \cdot u_{-1}(t)$ (b) $45 \cos 3t \cdot u_{-1}(t)$ (c) $-\frac{5}{6} \cos 3t \cdot u_{-1}(t)$
9. Parallel $R = \frac{1}{5}$, $L = \frac{1}{3}$, $C = 6$
10. Series $R = 3$, $L = \frac{1}{3}$, $C = \frac{8}{7}$
11. (a) $10u_0(t)$ (b) $\frac{100}{3}u_{-1}(t)$ (c) $500u_1(t)$
12. (a) $52u_0(t)$ (b) $65u_1(t)$ (c) $39u_{-1}(t)$
13. $5u_0(t) - 5u_0(t-1) + 2u_{-1}(t) - 2u_{-1}(t-1) + 2.5u_1(t) - 2.5u_1(t-1)$
14. Same as Exercise 13.
15. $5u_{-1}(t) - 5u_2(t)$
16. $5u_1(t) - 40u_{-2}(t)$
17. Parallel $R = 0.2$, $C = 3$, $L = \frac{1}{10}$
18. Series $R = \frac{1}{3}$, $L = 7$, $C = 15$
19. $i_{C_1} = 6u_0(t)$; $i_{C_2} = 4u_0(t)$; $e = 2u_{-1}(t)$.
20. $e_{L_1} = 6u_0(t)$; $e_{L_2} = 12u_0(t)$; $i = 2u_{-1}(t)$.
21. (a) $C = 3$; $e_{oc} = 10u_{-1}(t)$ (b) $C = 3$; $i_{sc} = 30u_0(t)$
22. (a) $L = 5$, $e_{oc} = 50u_0(t)$ (b) $L = 5$, $i_{sc} = 10u_{-1}(t)$
24. $i = 5$; $i' = -\frac{5}{6}$.
25. $e = 20$; $e' = -8$.
26. (a) 2 (b) 2
27. (a) 2 (b) 2
28. $e = -45$; $e' = 67.5$.
29. $i = -\frac{1}{2}$; $i' = \frac{1}{20}$.
30. $i = 2$; $i' = -\frac{8}{5}$.
31. $e = 2$; $e' = -\frac{1}{5}$.
32. $C = 2$
33. $L = 1$
34. $C = \frac{21}{13}$
35. $L = \frac{13}{21}$
36. $C = \frac{11}{24}$
37. $L = \frac{11}{24}$
38. $e = 6$; $i = 12u_0(t)$.
39. $i = 6$; $e = 12u_0(t)$.
40. $e = \frac{7}{6}$
41. $i = -\frac{13}{6}$

Chapter 11

1. (a) 10, $10/e$, $10/e^2$, 0 (c) $-\frac{5}{3}$ (d) 6
2. $T = \frac{4}{3}$; $i(0) = \frac{3}{2}$; $i(\infty) = 0$; $i(t) = \frac{3}{2}e^{-3t/4} \cdot u_{-1}(t)$.
3. $16u_0(t) - \frac{64}{5}e^{-4t/5} \cdot u_{-1}(t)$
4. $6e^{-5t/2} \cdot u_{-1}(t)$
5. $160e^{-t/3} \cdot u_{-1}(t)$ for $0 < t < 3$;
   $29.3e^{-(t-3)/6} \cdot u_{-1}(t-3)$ for $3 < t < \infty$
6. $6e^{-t} \cdot u_{-1}(t)$
7. $8e^{-2t/3} \cdot u_{-1}(t)$
8. (a) 736 volt·sec (b) 368 amp·sec

9. $e = 4e^{-t} \cdot u_{-1}(t)$; $i = -2(1 - e^{-t})u_{-1}(t)$
10. $e = 20e^{-t} \cdot u_{-1}(t)$; $i = 6.67(1 - e^{-t})u_{-1}(t)$
11. $10e^{-5t/3} \cdot u_{-1}(t)$
12. (a) 0, 6.3, 8.65, 10 (c) $\frac{5}{3}$ (d) 6 sec
13. $5(1 - e^{-2t/3})u_{-1}(t)$
14. $24e^{-2t} \cdot u_{-1}(t)$
15. (a) $2e^{-3t/4} \cdot u_{-1}(t)$ (b) Same
16. $i = [-6 + 26(1 - e^{-5t/3})]u_{-1}(t)$
17. $A = 10$
18. $(1 + e^{-6t/5})u_{-1}(t)$
19. $(10 - 6e^{-2t/3})u_{-1}(t)$
20. $6(1 - e^{-25t/6})u_{-1}(t)$
21. $(10 - 5e^{-2t})u_{-1}(t)$
22. $(-5 + 10e^{-5t/6})u_{-1}(t)$
23. $e = [10t + 20(e^{-t/2} - 1)]u_{-1}(t)$; $e_{ss} = 10(t - 2)$
24. 1 volt/sec
25. 10^{-9} volt; 2×10^{-10} sec.
26. $100(1 - e^{-t/5})u_{-1}(t)$
27. $(10t - 10 + 10e^{-t/3})u_{-1}(t)$
28. $\frac{5}{2}t \cdot u_{-1}(t)$
29. $50u_{-1}(t) + 30u_{-2}(t)$
30. $e(t) = 5e^{-t/6} \cdot u_{-1}(t)$
31. $i(t) = 5e^{-t/4} \cdot u_{-1}(t)$
32. $5u_0(t) - 5e^{-t} \cdot u_{-1}(t)$
33. $4e^{-t/25} \cdot u_{-1}(t)$
34. $5e^{-t/15} \cdot u_{-1}(t)$
35. 55 amp·sec; 110 volt·sec.
36. $e(t) = 20e^{-t/6} \cdot u_{-1}(t)$; $e_x = 20(1 - e^{-t/6})u_{-1}(t)$
37. $i = 2e^{-t/6} \cdot u_{-1}(t)$; $e = 6(1 - e^{-t/6})u_{-1}(t)$
38. $\frac{5}{3}e^{-5t/12} \cdot u_{-1}(t)$
39. $(1 - e^{-2t/15})u_{-1}(t)$
40. $i(t) = 3e^{-t/6} \cdot u_{-1}(t)$
41. $15(1 - e^{-t/6})u_{-1}(t)$
42. $3e^{-t/15} \cdot u_{-1}(t)$
43. $(30 - 36e^{-t/6})u_{-1}(t)$
44. $(10 - 5e^{-t/10})u_{-1}(t)$
45. $(5 + 25e^{-15t})u_{-1}(t)$
46. $(10 - 9e^{-t/15})u_{-1}(t)$
47. $0.4(1 - e^{-t/6})u_{-1}(t)$
48. $(30 + 15e^{-t/50})u_{-1}(t)$
49. $(75 + 25e^{-t/10})u_{-1}(t)$
50. $6R_1e^{-(R_1+R_2)t/L} \cdot u_{-1}(t)$
51. $[3C_2/R(C_2 + C_3)]e^{-t/R(C_2+C_3)} \cdot u_{-1}(t)$
52. $(3R/L_1)e^{-t/T} \cdot u_{-1}(t)$, where $T = L_1L_2/(L_1 + L_2)R$
53. $(4/C_1)e^{-t/T} \cdot u_{-1}(t)$, where $T = RC_1C_2/(C_1 + C_2)$
54. $e_{ab} = \frac{4}{5}e^{-9t/25} \cdot u_{-1}(t)$; $e = \frac{4}{25}e^{-9t/25} \cdot u_{-1}(t)$
55. $e_{ab} = \frac{6}{5}(1 - e^{-t})u_{-1}(t)$; $i = \frac{2}{5}(1 - e^{-t})u_{-1}(t)$
56. $e_a = \frac{5}{3}(1 - e^{-t})u_{-1}(t)$; $e_b = e_c = \frac{2}{5}e_a$; $e_c = e_d = \frac{1}{5}e_a$.
57. $e_a = (\frac{10}{3} + \frac{2}{3}e^{-9t/25})u_{-1}(t)$; $e_b = 2e^{-9t/25} \cdot u_{-1}(t)$; $e_c = 2e_b/5$; $e_d = e_b/5$
58. $e_a = \frac{1}{3}e^{-t}$; $e_b = e^{-t}$; $e_c = (e^{-t} - 1)$; $e_d = \frac{1}{3}(e^{-t} - 1)$; all for $t > 0$
59. $e_a = (\frac{20}{3} - \frac{2}{3}e^{-9t/25})u_{-1}(t)$; $e_b = (\frac{10}{3} - \frac{4}{3}e^{-9t/25})u_{-1}(t)$;
$e_c = [\frac{10}{3}(1 - e^{-9t/25})]u_{-1}(t)$; $e_d = e_c/3$.

Chapter 12

1. (b) 10 (c) 1 (d) 1 (e) 2π (f) $+45°$
2. 1.4×10^{-6}; 10^6; $-45°$.
3. $i = 1$; $i' = 0$; $\omega_0 = 8$; $i(t) = \cos 8t \cdot u_{-1}(t)$.
4. $i = 0$; $i' = 4$; $\omega_0 = 8$; $i(t) = \frac{1}{2} \sin 8t \cdot u_{-1}(t)$
5. $i(t) = 1.4 \cos (t/4 + 45°)u_{-1}(t)$

6. $i = 40; i' = 0; \omega_0 = 4; i(t) = 40 \cos 4t \cdot u_{-1}(t)$
7. $e = 40; e' = 0; \omega_0 = 4; e(t) = 40 \cos 4t \cdot u_{-1}(t)$
8. $e = 10 \sin t/2 \cdot u_{-1}(t); i = \frac{20}{3}(1 - \cos t/2)u_{-1}(t)$
9. $i = -2 \sin t/6 \cdot u_{-1}(t)$
10. (a) $i = 0; i' = \frac{10}{3}; \omega_0 = \frac{1}{6}$; (b) $i(\infty) = 0$ (c) $i = 20 \sin t/6 \cdot u_{-1}(t)$
11. $i = 0; i' = \frac{10}{6}; \omega_0 = \frac{1}{6}; i(\infty) = 0; i(t) = 10 \sin t/6 \cdot u_{-1}(t)$
12. $i = 5; i' = \frac{10}{3}; \omega_0 = \frac{1}{6}; i = 20.6 \cos (t/6 - 76°)u_{-1}(t)$
13. $e = 100(1 - \cos 4t)u_{-1}(t)$
14. $i = 10(1 - \cos 16t)u_{-1}(t)$
15. $i_{L_1} = (10 - 8 \cos t/5)u_{-1}(t); i_{L_2} = 8 \cos t/5 \cdot u_{-1}(t)$
16. $e = 10(1 - \cos t/5)u_{-1}(t)$
17. $e = (10 - 4 \cos t/5)u_{-1}(t)$
18. $e = (72 \cos 12t + 10 \sin 12t)u_{-1}(t)$
19. $i = (80 \cos 12t - 10 \sin 12t)u_{-1}(t)$
20. (b) $\omega = 2\pi, f = 1$, period $= 1$ (c) $T = 1$ (d) $+45°$
21. $i = 0; i' = 1; i(\infty) = 0; i(t) = 10e^{-t/10} \sin t/10 \cdot u_{-1}(t)$.
22. $i = 10; i' = -1; i(\infty) = 0; i(t) = 10e^{-t/10} \cos t/10 \cdot u_{-1}(t)$
23. Same as 22.
24. $i = (10e^{-t/2} \cos 0.866t - 5.77e^{-t/2} \sin 0.866t)u_{-1}(t)$
25. $i = 2; i' = 20; i(\infty) = 0; i(t) = (6e^{-2t} - 4e^{-8t})u_{-1}(t)$
26. $i = (13.3e^{-8t} - 3.3e^{-2t})u_{-1}(t)$
27. (a) $\alpha = 0.5, \omega_0 = 1, \omega_d = 0.866$ (b) $L = \frac{1}{4}$ (c) $C = 4$ (d) $R = 2$
(e) $i = 10e^{-t}(1 - t)u_{-1}(t)$
28. $i = 3te^{-t/6} \cdot u_{-1}(t)$
29. (a) $e = 0, e' = \frac{1}{2}, e(\infty) = 0$ (b) $e = 5e^{-t/10} \sin t/10 \cdot u_{-1}(t)$
30. $e = 7.07e^{-t/10} \cos (t/10 + 45°)u_{-1}(t)$
31. $e = 5e^{-t/10} \cos t/10 \cdot u_{-1}(t)$
32. $e = (6e^{-t/2} \cos 0.866t - 3.46e^{-t/2} \sin 0.866t)u_{-1}(t)$
33. $e = 11.5e^{-t/2} \sin 0.866t \cdot u_{-1}(t)$
34. $\alpha = 10; \omega_0 = 8; e(t) = (5e^{-16t} + 5e^{-4t})u_{-1}(t)$
35. $e = (-2.05e^{-0.41t} + 12.05e^{-2.41t})u_{-1}(t)$
36. $e = (40e^{-8t} - 320te^{-8t})u_{-1}(t)$
37. $C = \frac{1}{64}; i = [(12 - 96t)e^{-8t}]u_{-1}(t)$
38. $i = 0, i' = 10; i(\infty) = 0; i = 11.55e^{-0.5t} \sin 0.866t \cdot u_{-1}(t)$
39. $e = [10 - 11.55e^{-0.5t} \cos (0.866t - 30°)]u_{-1}(t)$
40. $e = [100 - 70.7e^{-t/10} \cos (t/10 - 45°)]u_{-1}(t)$
41. $e = (100 - 100e^{-t/10} \cos t/10)u_{-1}(t)$
42. $i = 0; i' = 240; i(\infty) = 0; i = (40e^{-2t} - 40e^{-8t})u_{-1}(t)$
43. $e = (60 - 80e^{-2t} + 20e^{-8t})u_{-1}(t)$
45. $C = 4; i(t) = 100te^{-t/2} \cdot u_{-1}(t)$
46. $e = [100 + e^{-t/6}(-50 - 50t/12)]u_{-1}(t)$
47. $e = 0; e' = 10; e(\infty) = 0; e = 11.55e^{-0.5t} \sin 0.866t \cdot u_{-1}(t)$
48. $i = [10 - 11.55e^{-0.5t} \cos (0.866t - 30°)]u_{-1}(t)$

49. $i = [50 - 212e^{-t/10} \cos (t/10 - 45°)]u_{-1}(t)$
50. $i = (100 + 100e^{-t/10} \sin t/10)u_{-1}(t)$
51. $e = 0; e' = 150; e(\infty) = 0; e = (25e^{-2t} - 25e^{-8t})u_{-1}(t)$
52. $i = (30 - 40e^{-2t} + 10e^{-8t})u_{-1}(t)$
53. $R = 5; e(t) = 10^7te^{-10^5t} \cdot u_{-1}(t)$
54. $i = [10 - 10e^{-10^5t}(1 + 10^5t)]u_{-1}(t)$
56. $p = -\frac{1}{2}; f = Ae^{-t/2}\ (t > 0)$.
57. $p = -\frac{1}{12}; f = Ae^{-t/12}\ (t > 0)$.
58. $p = \pm j\frac{1}{10}; f = A_1e^{jt/10} + A_2e^{-jt/10}\ (t > 0)$.
59. $p = -0.5 \pm j0.866; f = e^{-0.5t}(A_1e^{j0.866t} + A_2e^{-j0.866t})\ (t > 0)$.
60. $p = -\frac{1}{6}; f = Ae^{-t/6}\ (t > 0)$.
61. $p = -2; f = Ae^{-2t}\ (t > 0)$.
62. $p = \pm j8; f = A_1e^{j8t} + A_2e^{-j8t}\ (t>0)$.
63. $p = -0.5 \pm j0.866; f = e^{-0.5t}(A_1e^{j0.866t} + A_2e^{-j0.866t})\ (t > 0)$.
64. $6e^{-t/2}\ (t > 0)$
65. $5e^{-t} \sin 2t\ (t > 0)$
66. (a) $i = Ae^{-3t}\ (t > 0)$ (b) $e = Ae^{-2t}\ (t > 0)$ (c) Add a constant

Chapter 13

1. $i = 2$ in both cases.
2. e (pulse) $= (50/\delta)\, u_0(t) - (50/\delta)u_0(t - \delta)$; e (impulse) $= 50u_1(t)$
3. Pulse: $e = 2t/\delta$ for $t < \delta$; $e = 2$ for $t > \delta$. Impulse: $e = 2$.
4. Pulse: $i = (50/\delta)u_0(t) - (50/\delta)u_0(t - \delta)$. Impulse: $i = 50u_1(t)$.
5. Pulse: $e_2 = (1 - \delta/20)e^{-t/10}$. Impulse: $e_2 = e^{-t/10}$ for $\delta < \frac{1}{5}$.
6. (a) $\delta = \frac{1}{5}$ (b) $\delta = \frac{1}{5}$
7. Both pulse and impulse give $i(t) = (10/L) \cos t/\sqrt{LC}\ (t > 0)$.
8. $\delta = 0.28\sqrt{LC}$
9. $i = 6t$ for $t < 4$; $i = 24$ for $t > 4$. Approximately: $i = 0, 6, 12, 18,$ and 24.
10. $e = 0.33e^{-t/10}$; e (approx) $= 0.346e^{-t/10}$.
11. (a) $e = 0.313e^{-t/10}$ (b) $e = 0.33e^{-t/10}$
12. Max $i = 0.416$ (approx) (accurate value $= 0.37$)
13. Zero
14. (a) 1, 0 (b) 0, 1
15. (a) -1, 0 (b) 1, 0
16. $f_2 = t$ for $t < 1; f_2 = 1$ for $t > 1$.
17. $f_2 = 10^4(1 - e^{-t/100})$ for $t < 1; f_2 = 100e^{-t/100}$ for $t > 1$.
18. $f_2 = t$ for $t < 1; f_2 = 2 - t$ for $1 < t < 2; f_2 = 0$ for $t > 2$.
19. $f_2 = t + e^{-t} - 1$ for $t > 0$
20. $\frac{1}{5}(e^{-5t} - e^{-10t})$ for $t > 0$
21. $[1 - e^{-(t-1)}]u_{-1}(t - 1)$

22. $[e^{-1} - e^{-(t-1)}]u_{-1}(t-2)$
23. $(1 - e^{-t})u_{-1}(t) - [1 - e^{-(t-1)}]u_{-1}(t-1)$
24. $\frac{1}{2}(\sin t + t \cos t)$
25. $\frac{1}{2}(-t \cos t + \sin t)$
26. $\frac{1}{2}(\cos t + \sin t) - \frac{1}{2}e^{-t}$
27. $\frac{1}{2}(\sin t - \cos t) + \frac{1}{2}e^{-t}$
28. $\sin t$
29. (a) $(1 - e^{-t})u_{-1}(t) - [1 - e^{-(t-1)}]u_{-1}(t-1)$ (b) Same
30. $(1 - \cos t)u_{-1}(t) - 2[1 - \cos(t-\pi)]u_{-1}(t-\pi) +$
$[1 - \cos(t - 2\pi)]u_{-1}(t - 2\pi)$
31. $e^{-t+2} \cdot u_{-1}(t-2) + (1 - e^{-t})u_{-1}(t) - [1 - e^{-(t-1)}]u_{-1}(t-1)$
32. $(3e^{-t} + t - 1)u_{-1}(t) - (e^{-t+1} + t - 2)u_{-1}(t-1) +$
$(e^{-t+2} - 1)u_{-1}(t-2)$
33. $A = 0$
34. Parabolic arc from $t = 0$ to $t = 2$. Correct derivative for $t > 2$.
35. For $t < \delta$: $(t^2/2\delta) - t$. For $t > \delta$: $(\delta/2)e^{-t}$.
36. Same as Exercise 35.
37. Successive values: 0, 1, 1, —1, 1, —3, —3, 3, 1, 0, 0, . . .

Chapter 14

1. (a) 6, 4, 2, 2, (b) 36, 16, 4, 4 (c) 600 (d) 600 (e) $\frac{5}{3}$ ohms, 600 joules
2. (a) $I = -5$ (b) $I = \frac{10}{3}$ (c) 1.45 or —11.45
3. 360, 250, 10, 490, 320 joules
4. 18, 8, 2, 2 joules
5. $13.5t^2$, $3t^2$, $6t^2$ joules. $22.5t^2$ joules.
6. 6, $\frac{8}{3}$, $\frac{2}{3}$, $\frac{2}{3}$ joules; 10 joules.
7. $\frac{1800}{121}$, $\frac{900}{121}$, $\frac{600}{121}$ joules
8. $\frac{25}{18}t^2$, $\frac{50}{18}t^2$, $\frac{75}{18}t^2$ joules; $\frac{25}{3}t^2$ joules. $\frac{25}{9}t$, $\frac{50}{9}t$, $\frac{50}{6}t$ watts. $\frac{50}{3}t$ watts.
9. $100t$ watts; $50t^2$ joules; $50t$ watts; $25t^2$ joules; $33.3t$ watts; $16.7t^2$ joules. Source: $183.3t$ watts; $91.7t^2$ joules.
10. 200 joules
11. $9(1 - 2e^{-2t} + e^{-4t})$; $9(-3 + 4t + 4e^{-2t} - e^{-4t})$;
$(36t + 18e^{-2t} - 18)$ joules.
12. 600 joules; 300 joules; 300 joules; $\frac{1}{2}$.
13. $W_L(\max) = 8$ joules; $W(\text{source}) = 14.7$ joules.
14. 150 joules; 150 joules.
15. $W_C = 100$ joules; $W_R = 100$ joules; Equal.
16. $P_s = 200(1 - e^{-t/6})$ watts; $P_R = 200(1 - 2e^{-t/6} + e^{-t/3})$ watts;
$P_C = 200(e^{-t/6} - e^{-t/3})$ watts.
17. $W_{C_1} = 12.5$; $W_{C_2} = \frac{25}{8}$; $W_{C_3} = \frac{25}{8}$; $W_R = \infty$.
18. $W_C(\max) = 18$ joules $= W_L(\max)$; $i = 24 \sin 8t$.
19. $i(t) = 12 \cos 4t$; $e(t) = 24 \sin 4t$.

20. $i(t) = 8.49 \cos 6t$; $e(t) = 16.9 \sin 6t$.
21. $W_L = 600 \sin^2 t/6$; $W_C = 600(1 - 2 \cos t/6 + \cos^2 t/6)$;
 $W_s = 1200(1 - \cos t/6)$.
22. $P_L = 288 (\sin 4t - \sin 4t \cos 4t)$ watts; $P_C = 288 (\sin 4t \cos 4t)$ watts;
 $P_s = 288 \sin 4t$ watts.
23. $W_{L_1} = 12.5(t^2 + 2t \sin t + \sin^2 t)$; $W_{L_2} = 12.5(t^2 - 2t \sin t + \sin^2 t)$;
 $W_C = 25(1 - 2 \cos t + \cos^2 t)$.
24. 0.46 joule
25. W (total) $= 223$ joules; $i_{\max} = 12.2$; $\omega_0 = \frac{1}{6}$.
26. 50 joules; 50 joules.
27. 50 joules; $R = 2$ ohms.
28. 50 joules
29. 50 joules; $e(t) = 11.5e^{-t/2} \sin 0.866t$; 50 joules.
30. 50 joules; $e(t) = 10te^{-t}$; 50 joules.
31. 50 joules; $e(t) = 5(e^{-0.41t} - e^{-2.41t})$; 50 joules.
32. $i(t) = 16.6e^{-4t} \cos (3t + 53°)$;
 $P_R = 222e^{-8t} - 62.2e^{-8t} \cos 6t - 214e^{-8t} \sin 6t$; $W_R = 10$ joules.
33. $W_s = 4$ joules; $W_L = 2$ joules; $W_R = 2$ joules.
34. $W_s = 25$ joules; $W_L = 12.5$ joules; $W_R = 12.5$ joules.
35. $W_s = \frac{50}{8}$ joules; $W_R = \frac{25}{8}$ joules; $W_C = \frac{25}{8}$ joules.
36. (a) 2×10^{-2} joule (b) 4×10^{-2} joule (c) ∞
37. (a) 2.5×10^{-9} joule (b) 2.5×10^{-9} joule (c) 2.5×10^{-9} joule.
38. (a) 2.5×10^{-9} joule (b) 2.5×10^{-9} joule (c) 2.5×10^{-9} joule.
39. $W_s = 200$ joules; $W_C = 100$ joules; W(switch) $= 100$ joules.
40. $W_s = 500$ joules; $W_L = 250$ joules; W(switch) $= 250$ joules.
41. (a) $i = (100/R)e^{t/T} \cdot u_{-1}(t)$ (b) $W_s = 5000[C_1C_2/(C_1 + C_2)]$
 (c) $W_{C_1} = 5000[C_1^3/(C_1 + C_2)^2]$; $W_{C_2} = 5000[C_1^2C_2/(C_1 + C_2)^2]$
 (d) Same energy is lost in $R = 0$ $[T = RC_1C_2/(C_1 + C_2)]$
42. (a) $e = 100Re^{-t/T} \cdot u_{-1}(t)$ (b) $W_s = 5000[L_1L_2/(L_1 + L_2)]$
 (c) $W_{L_1} = 5000[L_1L_2^2/(L_1 + L_2)^2]$; $W_{L_2} = 5000[L_2^3/(L_1 + L_2)^2]$
 (d) Same energy is lost in $R = \infty$ $[T = L_1L_2/R(L_1 + L_2)]$
43. (a) $2e^{-t/5}$ (b) $W_s = 10$ joules (c) $W_L(0) = 10$ joules
 (d) $W_R = 10$ joules

ANSWERS TO EXERCISES: PART 2

Chapter 15

1. (a) 10 (b) 30° (c) 20π rps (d) 10 cps (e) $\frac{1}{10}$ sec
2. (a) 14.1 (b) $-45°$ (c) 2π rps (d) 1 cps (e) 1 sec
3. $5 \cos 2\pi t - 8.66 \sin 2\pi t$
4. (a) 10^5 (b) 120° (c) 1000 rps (d) $1000/2\pi$ cps (e) $2\pi/1000$ sec
5. (a) $\frac{1}{10}$ (b) $-60°$ (c) 1000 rps (d) $1000/2\pi$ cps (e) $2\pi/1000$ sec
6. (a) $f^{-1}(t) = 10 \cos (10t - 90° + \theta) - 10 \sin \theta$ (b) 10 (c) 10 (d) $-90°$ (e) $10/2\pi$ cps, $2\pi/10$ sec (f) Only the d-c term changes
7. Zero
8. $f_2(t) = \cos (t - 1) + \cos (2t - 2) + \cos (3t - 3) + \cos (4t - 4) + \cos (5t - 5)$, Phase shifts: -1, -2, -3, -4, and -5 radians
9. Amplitude = 100; phase = 30°; frequency = 60 cps; period = $\frac{1}{60}$ sec
10. $f(\text{sum}) = 14.1 \cos (2\pi t - 45°)$
11. (a) length 10, angle 0° (b) length 5, angle 45° (c) length 10, angle $-90°$ (d) length 5, angle 135° (e) length 10, angle 135°
12. (a) $10 + j0$ (b) $3.54 + j3.54$ (c) $-j$ (d) $-3.54 + j3.54$ (e) $-7.07 + j7.07$
13. (a) $10 \cos t$ (b) $-10 \sin 377t$ (c) $5 \cos (1000t - 60°)$ (d) $5 \cos (t + 53°)$ (e) $2.82 \cos (t/10 + 135°)$
14. (a) $1.41 \cos (\omega t + 45°)$ (b) $5 \cos (\omega t - 53°)$ (c) $2.82 \cos (\omega t + 225°)$ (d) $5 \cos (\omega t + 127°)$ (e) $-\sin \omega t$ or $\cos (\omega t + 90°)$
15. (a) $-3 \cos (\omega t - 45°)$ (b) $-2.82 \cos (\omega t - 45°)$ (c) $-5 \cos (\omega t + 53°)$ (d) $3 \cos (\omega t - 60°)$ (e) $-4 \cos (\omega t - 45°)$
16. $F = 10 \underline{|45°} = 10e^{j45°} = 7.07 + j7.07$
$F' = 100 \underline{|135°} = 100e^{j135°} = -70.7 + j70.7$
$F^{-1} = 1 \underline{|-45°} = e^{-j45°} = 0.707 - j0.707$
17. (a) $5 \underline{|0°} + 5 \underline{|0°}$ (b) $2.5 \underline{|45°} + 2.5 \underline{|-45°}$ (c) $1 \underline{|-90°} + 1 \underline{|90°}$ (d) $5 \underline{|135°} + 5 \underline{|-135°}$ (e) $1 \underline{|225°} + 1 \underline{|-225°}$
18. (a) $20 \cos (\omega t + 45°)$ (b) $20 \cos (\omega t - 45°)$ (c) $-2 \sin \omega t$ (d) $4 \cos (\omega t + 135°)$ (e) $5.64 \cos (\omega t - 135°)$
19. (a) $1.41 \underline{|45°}$ (b) $2 \underline{|60° + \theta°}$ (c) $2.82 \underline{|90°}$ (d) $9 \underline{|35°}$ (e) $5 \underline{|123°}$
20. All answers are zero.

21. (a) Re $=$ 100, Im $=$ 0 (b) Re $=$ 81.9, Im $=$ 57.3
(c) Re $=$ -50, Im $=$ 86.6 (d) Re $=$ -86.6, Im $=$ -50
(e) Re $=$ 70.7, Im $=$ -70.7

22. $50 - j86.6 = 100\,\underline{|-60^\circ}$

23. Sum $= 100\,\underline{|30^\circ}$

24. $3\,\underline{|0^\circ}$, $4\,\underline{|-90^\circ}$, $5\,\underline{|-233^\circ}$ or $3\,\underline{|0^\circ}$, $4\,\underline{|90^\circ}$, $5\,\underline{|233^\circ}$

25. $-112\,\underline{|33.5^\circ}$

26. (a) $x = \pm 1$ (b) $1\,\underline{|0^\circ}$, $1\,\underline{|180^\circ}$ (c) zero

27. (a) $x = 1\,\underline{|n\pi/6}$ (b) $1\,\underline{|0^\circ}$, $1\,\underline{|60^\circ}$, $1\,\underline{|120^\circ}$, $1\,\underline{|180^\circ}$, $1\,\underline{|240^\circ}$, $1\,\underline{|300^\circ}$
(c) Zero

28. (a) $5 + j0$ (b) $7.07 + j7.07$ (c) $-0.866 - j1.5$ (d) $10 + j0.174$
(e) $0.174 - j10$

29. (a) $-10 + j0$ (b) $2.4 + j4.33$ (c) $-3.1 - j9.45$ (d) $-0.2 - j0.336$
(e) $-7.07 + j7.07$

30. (a) $6\,\underline{|90^\circ}$ (b) $2.24\,\underline{|-63.5^\circ}$ (c) $-2.24\,\underline{|63.5^\circ}$ (d) $5\,\underline{|53^\circ}$ (e) $-5\,\underline{|-53^\circ}$

31. (a) $0.406\,\underline{|32^\circ}$ (b) $0.805\,\underline{|-81.2^\circ}$ (c) $-0.637\,\underline{|-88.78^\circ}$ (d) $343\,\underline{|0.167^\circ}$
(e) $-3.57 \times 10^6\,\underline{|61^\circ}$

32. (a) 3.61 (b) 3.61 (c) 0.277 (d) 13 (e) 2.72

33. (a) zero (b) 0.707 (c) 10 (d) 0.12 (e) -4.23

34. (a) $14.7\,\underline{|61.7^\circ}$ (b) $2\,\underline{|0^\circ}$ (c) $0.707\,\underline{|45^\circ}$ (d) $2.69\,\underline{|-33.4^\circ}$ (e) $7.4\,\underline{|-90^\circ}$

35. (a) 25 (b) $0.2\,\underline{|-53^\circ}$ (c) $9.43\,\underline{|-32^\circ}$ (d) $1\,\underline{|21.6^\circ}$ (e) $-1\underline{|21.6^\circ}$
or (a) 25 (b) $0.12 - j0.16$ (c) $8 - j5$ (d) $0.928 + j0.369$
(e) $-0.928 - j0.369$

36. (a) $3.7\,\underline{|-0.365^\circ}$ (b) $-0.71\,\underline{|-32.2^\circ}$ (c) $2.64\,\underline{|-39.5^\circ}$ (d) $1.74\,\underline{|62^\circ}$
(e) $1.15\,\underline{|-16.4^\circ}$ or (a) $3.7 - j0.0235$ (b) $-0.598 + j0.378$
(c) $2.04 - j1.68$ (d) $0.816 + j1.54$ (e) $1.15\,\underline{|69.8^\circ}$ or $0.395 + j1.08$

37. (a) $\cos x \cosh y - j \sin x \sinh y$ (b) $\sin x \cosh y + j \cos x \sinh y$

38. Sum $= e(t) = 141 \cos(\omega t + 15^\circ)$

39. Sum $=$ zero

40. (a) $5 \cos(t + 82^\circ)$ (b) $8.66 \cos t$ (c) $15 \cos(t + 90^\circ)$
(d) $9.1 \cos(13t + 24.4^\circ)$ (e) $0.41 \cos(3t + 90^\circ)$

41. (a) $100 \cos(t - 90^\circ)$ (b) $100 \cos(t + \pi)$ (c) $100 \cos(t + \pi/2)$
(d) $100 \cos(t + \pi/2)$ (e) $100 \cos(t - \pi/2)$

42. (a) $4 \cos(377t - 45^\circ)$ (b) $6 \cos(t + 120^\circ)$ (c) $7 \cos(t + 45^\circ)$
(d) $\cos(1000t + 120^\circ)$ (e) $10 \cos 3t$

43. Sum $= 389 \cos(377t + 75.1^\circ)$

44. Sum $= 14.1 \cos 1000t$

45. Sum $= 100 \cos (2t + 45°)$

46. Sum $= 15 + 7.07 \cos (\omega t - 45°) + 5 \cos (\omega t - 45°)$
Difference $= 5 - 7.07 \cos (\omega t + 45°) + \cos (\omega t - 45°)$

47. $e(t) = 141 \cos (t + 45°)$

Chapter 16

3. $e = 141 \cos (1000t + 15°)$

4. $i = 50 \cos 377t$

5. (a) $i = 2.5 \sin 10t$ (b) same

6. (a) $i = -2000 \sin 10t$ (b) same

7. (a) $e = -600 \sin 10t$ (b) same

8. (a) $i = -600 \sin 10t$ (b) same

9. (b) $I = 3.53 \underline{/-45°}$ (c) $i(t) = 3.5 \cos (2t - 45°)$

10. (b) $E = 7070 \underline{/-45°}$ (c) $e(t) = 7070 \cos (1000t - 45°)$

11. $i(t) = 5.53 \cos (2t - 56.4°)$

12. $e(t) = 3.53 \cos (3t - 69.5°)$

13. $I_2 = +4.48 \underline{/-63.5°}$, $i(t) = +4.48 \cos (t - 63.5°)$

14. $E = +4.48 \underline{/-63.5°}$, $e(t) = +4.48 \cos (t - 63.5°)$

15. (a) $E_{oc} = 7.07 \underline{/-45°}$, $Z_{eq} = 0.707 \underline{/45°}$ (b) $i_{sc} = 10 \cos (t - 90°)$

16. (a) $I_{sc} = 7.07 \underline{/-45°}$, $Z_{eq} = 1.4 \underline{/-45°}$ (b) $e_{ab} = 10 \cos (t - 90°)$

17. $i(t) = 70.7 \cos (t - 45°)$

18. $e(t) = 20 \cos (t - 53°)$, $E = 20 \underline{/-53°}$

19. $i(t) = 100 \cos t$

20. $e(t) = 100 \cos t$

21. $\omega = 10^4$ (resonance), $Q = 100$

22. $C = 10^{-8}$ farad, $Q = 10$

23. A semicircle of diameter 10 volts

24. A semicircle of diameter 10 amp

25. e (source) $= 200 \cos 377t$, $e_R = 100 \cos (377t + 60°)$,
$e_C = 173 \cos (377t - 30°)$

26. $e_1(t) = 100 \cos (\omega t - 60°)$, $e_2(t) = 100 \cos (\omega t + 60°)$

27. (b) $Z = \frac{1}{2} - j\frac{1}{2}$ (c) $R = \frac{1}{2}$, $C = 2$ (d) $Z = (1 - j\omega)/(1 + \omega^2)$
(e) $1/(1 + \omega^2)$, $\omega/(1 + \omega^2)$ (f) Elements are functions of frequency

28. (b) $Y = \frac{1}{2} - j\frac{1}{2}$ (c) 2 ohms, 2 henrys

(d) $Y(\omega) = 1/(1+\omega^2) - j[\omega/(1+\omega^2)]$ (e) $1/(1+\omega^2)$, $\omega/(1+\omega^2)$
(f) The elements are functions of frequency

29. (a) $Z = 0.5 + j0.5$ (b) $R = 0.5$, $L = 0.5$ (c) $R = 1$, $L = 1$
(d) $Z(\omega) = 1/(1+\omega^2) + j[\omega^3/(1+\omega^2)]$

30. (a) $Z = 0.5 - j0.5$ (b) $R = \frac{1}{2}$, $C = 2$ (c) $R = 1$, $C = 1$
(d) $Z(\omega) = \dfrac{\omega^2}{1+\omega^2} + j\,\dfrac{1}{\omega+\omega^3}$

31. (a) $Z = j$ (b) $Z(\omega) = j\,\dfrac{\omega^4 - 3\omega^2 + 1}{\omega^3 - 2\omega}$

32. (a) $z_{11} = z_{22} = 0$, $z_{12} = -j$ (b) Shunt C, $-j$ ohms; series L's, j ohms
(c) $z_{11} = z_{22} = \dfrac{(\omega^2 - 1)}{j\omega(\omega^2 - 2)}$, $z_{12} = \dfrac{1}{j\omega(2-\omega^2)}$

33. (a) $y_{11} = y_{22} = 0$, $y_{12} = +j$ (b) Shunt C's, $-j$ ohms; series L, j ohms
(c) $y_{11} = y_{22} = \dfrac{1-\omega^2}{j\omega(2-\omega^2)}$, $y_{12} = \dfrac{-1}{j\omega(2-\omega^2)}$

34. (a) $y_{11} = y_{22} = \infty$, $y_{12} = \infty$ (b) Three short circuits
(c) $y_{11} = y_{22} = \dfrac{j(2\omega^2 - 1)}{2\omega - 2\omega^3}$, $y_{12} = \dfrac{j}{2\omega - 2\omega^3}$

35. (a) Three short circuits (b) Three open circuits
(c) Series arm: $-j1.4$; shunt arm: $j1.4$ (d) Two series shorts; one shunt open circuit.

36. $Z = R$ for $L/R = RC$

37. (a) $R = 1$ ohm, $C = 1$ farad, $L = 1$ henry
(b) $R = 1$ ohm, $C = 10^{-3}$ farad, $L = 10^{-3}$ henry
(c) $R = 1$ ohm, $C = 10^{-6}$ farad, $L = 10^{-6}$ henry

38. (a) $E = 7.07\,\underline{|45^\circ}$
(b) $e(t) = 7.07\cos(t + 45^\circ)$, $R = 1$ ohm, $L = \frac{1}{2}$ henry, $C = 1$ farad
(c) $e(t) = 7.07\cos(10^6t + 45^\circ)$, $R = 1$ ohm, $C = 10^{-6}$ farad, $L = 0.5 \times 10^{-6}$ henry.

39. (a) Series $R = 1$ ohm, shunt $C = -j$ ohms, series $L = j3$ ohms, shunt $C = -j$ ohms
(b) $E_0 = 4.47\,\underline{|153.5^\circ}$ (c) $e_0(t) = 4.47\cos(10^6t + 153.5^\circ)$

40. (a) Shunt $C = -2j/3$ ohms, series $L = 4j/3$ ohms, shunt $C = -2j$ ohms, shunt $R = 1$ ohm
(b) E_2 (normalized) $= 7.07\,\underline{|225^\circ}$ (c) $e_2(t) = 7070\cos(10^6t + 225^\circ)$

41. (a) $Z = 1 + j\omega$ (b) $Z = 1000 + j1000$, $R = 10^3$ ohms, $L = 10^3$ henrys
(c) Original circuit: $R = 1$ ohm, $L = 1$ henry; Frequency-scaled circuit: $R = 1$ ohm, $L = 10^{-6}$ henry

42. (a) $Z = 1/(1 + j\omega)$ (b) Scaled values: $R = 10^3$ ohms, $C = 10^{-3}$ farad
(c) Scaled values, magnitude and frequency: $R = 10^3$ ohms, $C = 10^{-9}$ farad.

43. (a) $Z = \dfrac{1}{1 + j(\omega - 1/\omega)}$
(b) New values are: $R = 1000$ ohms, $L = 1000$ henrys, $C = 10^{-3}$ farad
(c) Values scaled in magnitude and frequency: $R = 1000$ ohms, $L = 10^{-3}$ henry, $C = 10^{-9}$ farad

44. (a) $Z = 1 + j(\omega - 1/\omega)$
(b) Impedance-scaled values: $R = 1000$ ohms, $L = 1000$ henrys, $C = 10^{-3}$ farad
(c) Values scaled in impedance and frequency: $R = 1000$ ohms, $L = 10^{-3}$ henry, $C = 10^{-9}$ farad

Chapter 17

1. (a) 16 henrys, 4 henrys (b) $\frac{15}{4}$ henrys, $\frac{15}{16}$ henry
2. (a) $\frac{15}{4}$ henrys (b) $\frac{5}{2}$ henrys
3. Alone: 8 and 6 henrys; Series: 18 and 10 henrys;
Parallel: $\frac{22}{9}$ and $\frac{22}{5}$ henrys; Shorted: $\frac{22}{3}$ and $\frac{11}{2}$ henrys
4. (a) $e_1(t) = 10 \cos (t + 90°)$ (b) $e_2(t) = 2.5 \cos (t + 90°)$
5. $L_{eq} = 34$ henrys
6. $E_2 = 3.9 \angle{-11°}$
7. $E_2 = 8.15 \angle{-99.4°}$
8. $Z = j\omega L_1 + \dfrac{\omega^2(L_1 - M)^2}{j\omega(L_1 + L_2 - 2M) + Z_2}$
9. $Z = j\omega L_1 + Z_2 - \dfrac{(j\omega M + Z_2)^2}{j\omega L_2 + Z_2}$
10. $I_R = 18.6 \angle{3.6°}$
11. (a) $j20$ ohms (b) $j2000$ ohms (c) $j15$ ohms (d) $j1500$ ohms
(e) $2.5 + j17.5$ ohms
12. $L_{eq} = 6$ henrys
13. $Z_{ab} = 4 - 4j$ ohms
14. (a) $E_{oc} = 14 \angle{45°}$, $Z_{eq} = 6 + j6$ (b) $I_{sc} = 1.65 \angle{0°}$
15. $I_{sc} = 17.6 \angle{45°}$, $Z_{eq} = \frac{144}{5} - j\frac{48}{5}$
16. (a) $C = \frac{1}{200}$ farad (b) $E_2 = 1000j$, $e_2(t) = 1000 \cos (10t + 90°)$
(c) Maximum primary current, therefore maximum secondary voltage
17. $X_C = -100$, $E_2 = 200 \angle{0°}$

18. $\Gamma_1 = \dfrac{L_2}{L_1L_2 - M^2}, \quad \Gamma_2 = \dfrac{L_1}{L_1L_2 - M^2}, \quad \Gamma_m = \dfrac{-M}{L_1L_2 - M^2}$

19. (a) $\Gamma_1 = \frac{1}{2}$, $\Gamma_2 = \frac{1}{4}$, $\Gamma_m = -\frac{1}{4}$ (b) $Z = 2j\omega$ ohms (c) $Z = 4j\omega$ ohms

20. (a) $E_3 = 353 \underline{|-45^\circ}$ (b) Same

21. (a) $E_1 = 4840 \underline{|14^\circ}$ (b) Same

22. $I_1 = 6.7 \underline{|-42.3^\circ}$

23. $E_2 = 3.53 \underline{|-135^\circ}$

24. $E_1 = 2.24 \underline{|-63.5^\circ}$

25. $I_1 = 7.84 \underline{|-11.3^\circ}$

26. (a) $N = 100$ (b) $P = \frac{1}{4}$ watt

27. (a) $N = 4.45$ (b) $P = 5000$ watts

28. (a) 17.32 henrys, 4.68 henrys (b) Zero for both connections

29. (a) Zero (b) Zero (c) 10 henrys (d) 1 henry

30. (a) $E_2 = 22.6 \underline{|26.5^\circ}$ (b) Same

31. (a) $E_{oc} = 70.7 \underline{|45^\circ}$, $Z_{eq} = 707 \underline{|45^\circ}$ (b) $I_{sc} = 0.1 \underline{|0^\circ}$

32. $I_1 = 0$, $E_2 = 40 \underline{|0^\circ}$

33. (a) $X_C = 4$ (b) E_2 (without C) $= 19.6 \underline{|78.7^\circ}$, E_2 (with C) $= 100 \underline{|0^\circ}$

34. (a) $X_1 = 1$, $E_2 = 50 \underline{|0^\circ}$ (b) $X_2 = 100$, $E_2 = 50 \underline{|0^\circ}$
(c) No advantage in tuning both sides of unity-coupled coils

35. $Z = \dfrac{1}{1 + j3} = 0.316 \underline{|-71.6^\circ}$

36. (a) $Y_1 = \dfrac{1}{Z} = \dfrac{1}{j\omega L_1} + \dfrac{(N+1)^2}{Z_2}$ where $N^2 = L_2/L_1$
(b) $E_2/E_1 = N + 1$
(c) A shunt inductance L_1 and an ideal transformer with a turns-ratio of 1 to $(N + 1)$.

37. (a) Series $L = \frac{1}{5}$, shunt $L = \frac{4}{5}$, ideal transformer of ratio 10, series inductance $L = 20$
(b) Shunt inductance moved through the transformer becomes 80 henrys.

38. (a) Series $10 + j50$, shunt $j50$, series $8 + j50$, ideal transformer with $N = \frac{1}{2}$
(b) Ideal transformer with $N = \frac{1}{2}$, series $2.5 + j12.5$, shunt $j12.5$, series $2 + j12.5$

39. (a) $Z_{sc} = R_a + \dfrac{R_b}{N^2} + j\left(X_a + \dfrac{X_b}{N^2}\right)$ (b) $Z_{oc} = \dfrac{1}{1/R_m + 1/jX_m}$

40. $R_a = 0.353$, $X_a = 0.353$, $R_m = 32{,}400$, $X_m = 32{,}400$, $X_b/N^2 = 0.353$, $R_b/N^2 = 0.353$, $N = \frac{1}{10}$

41. (a) $N = \frac{1}{20}$ (b) $f = 63.8$ cps (c) $f = 6380$ cps
(d) Transformer passes the frequency band between these two values.

42. Primary: $R = 1$ ohm, $X_L = j22$ ohms; Mutual: $j200$ ohms;
Secondary: $1100 + j2200$ ohms

43. (a) $E_2 = 72.2 \underline{|15.7°}$ (b) Same

44. (a) $E_2 = 0.001 \underline{|-90°}$ (b) Same

Chapter 18

1. (a) Source: $10 \underline{|0°}$; circuit: $Z = 1 + \dfrac{1}{s}$
(b) $i(t) = 14.1e^{-t} \cos (t + 45°)$ (c) No transient

2. (a) Source: $10 \underline{|0°}$; circuit: $Z = 2 + \dfrac{1}{s}$
(b) $i(t) = 6.32e^{-t} \cos (t + 18.5°)$ (c) Transient: $i(t) = -e^{-t}$

3. (a) Source: $10 \underline{|0°}$; circuit: $Z = \frac{1}{2} + \dfrac{1}{s}$
(b) $i(t) = 20e^{-t} \cos (t + 90°)$ (c) Transient: $i(t) = 20e^{-2t}$
(d) The transient would obscure the steady-state term

4. Steady-state grows; transients die out

5. (a) $Z = 1/Cs$, pole at 0, zero at ∞ (b) $Z = Ls$, zero at 0, pole at ∞
(c) $Z = R + 1/Cs$, zero at $-1/RC$, pole at 0
(d) $Z = R + Ls$, zero at $-R/L$, pole at ∞
(e) $Z = R/(1 + RCs)$, zero at ∞, pole at $s = -1/RC$
(f) $Z = RLs/(R + Ls)$, zero at $s = 0$, pole at $s = -R/L$

6. (a) $Z = \dfrac{s^2 + 3s + 1}{s^2 + 2s}$
(b) Zeros: $s = -2.62, -0.38$; poles: $s = 0, -2$

7. (a) $Z = \dfrac{s^2 + 3s + 1}{s + 2}$
(b) Zeros: $s = -2.62, -0.38$; poles: $s = -2, \infty$

8. (a) $Z = \dfrac{1}{C} \dfrac{(s + R/L)}{[s^2 + s(R/L + G/C) + (1 + RG)/LC]}$
(b) Zeros: $s = -R/L, \infty$

$$\text{Poles: } s = -\tfrac{1}{2}\left(\frac{R}{L} + \frac{G}{C}\right) \pm j\sqrt{\frac{1 + RG}{LC} - \tfrac{1}{4}\left(\frac{R}{L} + \frac{G}{C}\right)^2}$$

9. $Z(s) = \dfrac{4s^2 + 4s + 3}{2s^2 + s + 1}$
(b) Zeros: $s = -\frac{1}{2} \pm j0.707$ (c) Poles: $s = -0.25 \pm j0.66$

10. (a) $Y(s) = \dfrac{6s^2 + 6s + 1}{6s^2 + 5s + 1}$
(b) Zeros: $s = -0.783, -0.217$ (c) Poles: $s = -\frac{2}{3}, -\frac{1}{2}$

11. (a) $z_{11} = z_{22} = \dfrac{s^2 + 1}{s}$, $z_{12} = \dfrac{1}{s}$
(b) $y_{11} = y_{22} = \dfrac{s^2 + 1}{s(s^2 + 2)}$, $y_{12} = \dfrac{-1}{s(s^2 + 2)}$

12. (a) $1/(1 + j\omega)$ (d) $R = 10^5$, $L = 10/2\pi$

13. (a) $1/(1 + j\omega)$ (c) $R = 10^4$, $C = 10^{-10}$

14. (a) $1/(-\omega^2 + j1.414\omega + 1)$ (c) $R = 10^3$, $L = 7.07$, $C = 14.1 \times 10^{-6}$

15. (a) $1/(-j\omega^3 - 2\omega^2 + 2j\omega + 1)$
(c) $R = 10^3$, $C_1 = 1.5 \times 10^{-6}$, $C_2 = 0.5 \times 10^{-6}$, $L = 1.33$

16. (a) $10 + 0.01j(\omega - 10^8/\omega)$ (b) $\omega = 10^4$ (c) $B = 1000$ rps (d) $Q = 10$

17. $R = 100$, $L = 1$, $C = 10^{-6}$

18. (a) $10^7 j\omega/(-\omega^2 + 10^2 j\omega + 10^{10})$ (b) $\omega = 10^5$ (c) $B = 100$ rps
(d) $Q = 1000$

19. $R = 10^5$, $C = 10^{-7}$, $L = 10$

20. (a) Asymptote: -20 db to $\omega = 10$; -6 db/octave for $\omega > 10$
(b) Phase: zero for small ω, $-90°$ for large ω
(c) $\omega = 5$: -21 db; $\omega = 10$: -23 db; $\omega = 20$: -27 db

21. (a) Asymptote: $+6$ db/octave for $\omega < 10$; 0 db for $\omega > 10$
(b) Phase: $90°$ for small ω, zero for large ω
(c) $\omega = 5$: -7 db; $\omega = 10$: -3 db; $\omega = 20$: -1 db

22. (a) Asymptote: -12 db for small ω, -6 db/octave for $\omega > 1$, -12 db/octave for $\omega > 4$
(b) Phase: zero for small ω, $-180°$ for large ω
(c) $\omega = 1$: -15.26 db; $\omega = 4$: -27.26 db

23. (a) Asymptote: -20 db to $\omega = 1$; 6 db/octave to $\omega = 10$, 0 db to $\omega = \infty$
(b) Phase: small ω, zero phase; positive phase from $\omega = 1$ to $\omega = 10$, zero phase for $\omega > 10$.
(c) $\omega = 1$: -17 db; $\omega = 10$: -3 db

24. (a) Zero db everywhere
(b) Low frequencies: phase = zero; high frequencies: phase = $-180°$

25. (b) $\omega = 100$ radians = lower half-power point; $\omega = 10^5$ radians = upper half-power point
(c) Gain = 100 (d) $+90°$ to $0°$ to $-90°$

26. (a) Goes to ∞ at $\omega = 10$ (b) Phase jumps from $+90°$ to $-90°$ at $\omega = 10$

27. (a) +6 db/octave for ω less than 1, -6 db/octave for $\omega > 1$, 0 db at $\omega = 1$
(b) $+90°$ for small ω, $-90°$ for large ω

28. (a) Zero db to $\omega = 1$, -18 db/octave for $\omega > 1$; down 3 db at $\omega = 1$
(b) Zero for small ω, 135° at $\omega = 1$, $-270°$ for large ω

29. (a) $Z = 2s/(s + 1)$ (b) Phase ($\omega = \infty$) = 0°, phase ($\omega = 0$) = 90°
(c) $Z(j1) = 1.41 \underline{|45°}$

30. (a) $Z(s) = s/(s^2 + 2s + 2)$
(b) Phase ($\omega = \infty$) = $-90°$, phase ($\omega = 0$) = 90°
(c) $Z(j1) = 0.45 \underline{|26.5°}$

31. (a) $Z(s) = \dfrac{2(s + 1)}{s^2 + 2s + 2}$ (b) $Z(j) = 1.26 \underline{|-18.5°}$
(c) $e(t) = 12.6 \cos (t - 18.5°)$

32. (a) $Z(s) = 20(s + 1 - j10)$ (b) $s = j10$ (c) $B = 2$ rps (d) $Q = 5$
(e) Z (restive) $-$ 20 ohms

33. (a) $R = 10$, $L = 5$, $C = 2 \times 10^{-5}$
(b) $R = 10$, $L = 5 \times 10^{-3}$, $C = 2 \times 10^{-8}$ (c) $Q = 50$ in each case

34. (a) $Z = \dfrac{1}{20(s + 1 - j10)}$
(b) $s = j10$ (c) $Z = \frac{1}{20}$ (d) $B = 2$ (e) $Q = 5$

35. (a) $G = C = 10^{-6}$ (b) $R = L = 100$ (c) $Q = 50$ in each case

36. $i(0) = 10^{-5}$ amp, $I(50) = 6.6 \times 10^{-4} \underline{|90°}$, $i(100) = 5 \times 10^{-2}$ amp

37. (a) Ae^{-t} (b) Ae^{-2t}

38. (a) Ae^{-t} (b) $Ae^{-t} \cos (10t + \theta°)$

39. (a) $Z = \dfrac{(s + 1)(s + 2)}{s^2 + 2s + 100}$
(b) $Ae^{-t} + Be^{-2t}$ (c) $Ae^{-t} \cos (10t + \theta°)$

40. (a) $Z = \dfrac{s + 1}{s^2 + 2s + 2}$ (b) $Ae^{-t} \cos (t + \theta°)$ (c) Ae^{-t}

41. (a) $Z = 1$ (b) No poles or zeros
(c) $e_1(t) = -Ee^{-t}$, $e_2(t) = Ee^{-t}$, $e_1(t) + e_2(t) = 0$

42. (a) $Z = 1$ (b) No poles or zeros
(c) $i_1(t) = E(1 - e^{-t})$, $i_2(t) = Ee^{-t}$, $i_1(t) + i_2(t) = E$

43. (a) $z_{11} = z_{22} = \dfrac{s^2 + 1}{s(s^2 + 2)}$, $z_{12} = \dfrac{1}{s(s^2 + 2)}$
(b) $Z_{12} = \dfrac{1}{s^3 + s^2 + 2s + 1}$
(c) The z parameters have the same poles; z_{12} and Z_{12} have the same zeros

44. (a) $y_{11} = y_{22} = \dfrac{s^2+1}{s(s^2+2)}$, $y_{12} = \dfrac{-1}{s(s^2+2)}$

(b) $Y_{12} = \dfrac{-1}{s^3+s^2+2s+1}$

(c) The y parameters have the same poles; y_{12} and Y_{12} have the same zeros.

Chapter 19

1. RMS value $= 0.577$
2. RMS value $= 0.5$
3. RMS value $= 1.732$
4. RMS value $= 1.104$
5. $I_{\text{rms}} = 10.6$ amp
6. $I_{\text{rms}} = 16.9$ amp
7. (a) $E_{\text{rms}} = 2$ volts (b) $P = 2$ watts
8. (a) $E_{\text{rms}} = 1.41$ volts, $I_{\text{rms}} = 5.0$ amp, (b) $P = 7$ watts
9. (a) $E^2_{\text{rms}} = E_0^2 + E_1^2/2 + E_2^2/2 + \cdots$ (b) $I^2_{\text{rms}} = I_0^2 + I_1^2/2 + \cdots$
   (c) $P = E_0I_0 + E_1I_1/2 + E_2I_2/2 + \cdots$
10. (a) $P = 5000$ watts (b) $P = 10{,}000$ watts
    (c) The averaging must be done over an infinite period
11. (a) $p = 10^3 \cos^2 377t$ watts (b) $P = 500$ watts
    (d) Power fluctuates but flows always from source to resistance
12. (a) $p = 50 \sin 20t$ (b) $w = 2.5\,(1 - \cos 20t)$ (c) $W_L = 2.5$ joules
    (e) Power flows into and out of L at double the source frequency
13. (a) $p = -50 \sin 20t$ watts (b) $w = 2.5\,(\cos 20t + 1)$ joules
    (c) $W_C = 2.5$ joules
    (e) Power flows out of and into C at double the source frequency
14. (a) $p = 25 + 25 \cos 2t + 25 \sin 2t$ watts (c) $P = 25$ watts
    (d) $Q = +25$ vars
15. (a) $p = 25 + 25 \cos 2t - 25 \sin 2t$ watts (c) $P = 25$ watts
    (d) $Q = -25$ vars
16. (a) $W_L = 12.5$ joules, $Q = 25$ vars (b) $W_C = 12.5$ joules, $Q = -25$ vars
17. (a) $p = 25 + 25 \cos 2t - 25 \sin 2t$ watts (c) $P = 25$ watts
    (d) $Q = +25$ vars
18. (a) $P_{e_1} = 5000$ watts; $P_{e_2} = 5000$ watts; power flows from both sources to the resistance load

(b) $P_{e_1} = 5000$ watts; $P_{e_2} = -5000$ watts; power flows from left to right
(c) $P_{e_1} = -5000$ watts; $P_{e_2} = 5000$ watts; power flows from right to left.

19. (a) No power drawn
(b) $P_{e_1} = 10{,}000$ watts; $P_{e_2} = -10{,}000$ watts; power flows from left to right
(c) $P_{e_1} = -10{,}000$ watts; $P_{e_2} = 10{,}000$ watts; power flows from right to left

20. (a) $P_v = 500 + j500$ watts, pf $= 0.707$

21. (a) $P_v = 500 + j500$ watts, pf $= 0.707$

22. (a) $P_v = 1000 + j1000$ watts, pf $= 0.707$

23. Same as Exercise 22

24. $R = 4.33$ ohms, $X = 2.5$ ohms

25. $P_v = 1000 - j577$ watts, $|P_v| = 1154$ watts

26. $P \cong 1000$ watts, $Q = -50$ vars

27. $I = 14\,\underline{|-38^\circ}$, pf $= 0.78$

28. (a) $P_v = 3 + \text{j}1$ (b) $Z = 1.5 + j0.5$

29. (a) $P_v = 3 - j1$ (b) $Z = 1.5 - j0.5$

30. (a) $P_v = 4000 + j2000$ (b) $Z = 2 + j1$

31. $Z = \dfrac{\omega^3 + 5\omega - j(\omega^2 + 2)}{\omega^3 + 4\omega}$

32. (a) pf $= 0.974$ (b) Add reactance $= -50$ vars

33. $C = 461\ \mu$fd

34. $X_C = -2.36$ ohms

35. $X_C = \frac{1}{2}$ ohm

36. (a) $Z = 0.5 - j0.5$ (b) $P_{\max} = 25$ watts

37. (a) $Z = 0.5 + j0.5$ (b) 25 watts

38. (a) $R = R_L$, $X = -X_L$ (b) $P_{\max} = \dfrac{I^2(R^2 + X^2)}{4R}$

39. (a) $N = 1.18$ (b) $P_{\max} = 20.8$ watts

40. (a) $N = 1.18$ (b) $P_{\max} = 20.8$ watts

41. (a) $N^4 = \dfrac{X_L^2 + R_L^2}{X^2 + R^2}$ (b) $P_{\max} = \dfrac{(R^2 + X^2)R_LN^2}{(N^2R + R_L)^2 + (N^2X + X_L)^2}$

42. (a) $N = 10{:}1$, $L_1 = 1$ henry, $L_2 = 0.01$ henry, $M = 0.1$ henry
(b) Half-power frequency: $\omega = 50$ rps

Chapter 20

1. $f(t) = \frac{A}{2} + \frac{2A}{\pi}\left(\sin t + \frac{\sin 3t}{3} + \frac{\sin 5t}{5} + \cdots\right)$

2. $f(t) = \frac{A}{2} + \frac{2A}{\pi}\left(\sin 2\pi \times 10^6 t + \frac{\sin 6\pi \times 10^6 t}{3} + \cdots\right)$

3. $f(t) = \frac{A}{2} - \frac{4A}{\pi^2}\left(\cos t + \frac{\cos 3t}{9} + \frac{\cos 5t}{25} + \cdots\right)$

4. $f(t) = \frac{A}{2} - \frac{4A}{\pi^2}\left(\cos \frac{2\pi t}{100} + \frac{\cos 6\pi t/100}{9} + \cdots\right)$

5. $f(t) = \frac{A}{2\pi} + \frac{A}{\pi}(\cos t + \cos 2t + \cos 3t + \cdots)$

6. $f(t) = \frac{2A}{\pi}(\cos t + \cos 3t + \cos 5t + \cdots)$

7. $f(t) = \frac{A}{2} + \frac{A}{\pi}\left(-\sin t - \frac{\sin 2t}{2} - \frac{\sin 3t}{3} - \cdots\right)$

8. $f(t) = \frac{A}{2} + \frac{A}{\pi}\left(\sin t + \frac{\sin 2t}{2} + \frac{\sin 3t}{3} + \cdots\right)$

9. $f(t) = \frac{2A}{\pi}\left(1 - \frac{2\cos 2t}{3} - \frac{2\cos 4t}{15} - \frac{2\cos 6t}{35} - \cdots\right)$

10. $f(t) = \frac{A}{\pi}\left(1 + \frac{\pi}{2}\sin 1000\pi t - \frac{2\cos 2000\pi t}{3} - \frac{2\cos 4000\pi t}{15} - \cdots\right)$

15. $f(t) = \frac{4}{\pi}\left(\sin t + \frac{\sin 3t}{3} + \frac{\sin 5t}{5} + \cdots\right)$

$f^{-1}(t) = -\frac{4}{\pi}\left(\cos t + \frac{\cos 3t}{9} + \frac{\cos 5t}{25} + \cdots\right)$

$f'(t) = \frac{4}{\pi}(\cos t + \cos 3t + \cos 5t + \cdots)$

16. (a) $1/n^3$ (b) Any number (c) Twice

17. (b) $(\pi - 2t)(\pi/8)$ for $0 < t < \pi$ and $(t - 3\pi/2)(\pi/4)$ for $\pi < t < 2\pi$

18. (b) $\pi/4$ for $0 < t < \pi$ and $-\pi/4$ for $\pi < t < 2\pi$

19. (b) $-\frac{1}{2} + u_0(t) + u_0(t - 2\pi) + u_0(t - 4\pi) + \cdots$

21. (a) Odd function plus rotation

(b) $f(t) = (4/\pi)(\sin t + (\sin 3t)/3 + (\sin 5t)/5 + \cdots)$

22. $f(t) = \frac{1}{2} + (4/\pi^2)(\cos t + (\cos 3t)/9 + (\cos 5t)/25 + \cdots)$

23. (a) Odd function only (b) $f(t) = (2/\pi)(\sin t + (\sin 2t)/2 + (\sin 3t)/3 + \cdots)$

24. (a) Odd symmetry (b) Even (c) Quarter-wave odd (d) Quarter-wave even (e) Rotation (f) Half range (g) $f(t) = f(\pi - t)$ (h) $f(t) = -f(\pi - t)$

25. (a) D-c + even + rotation (b) Even (c) Odd (d) Rotation
(e) D-c + half range (f) None

26. (a) D-c + rotation (b) D-c + rotation + even function about $t = -45°$
(c) Same as (b) (d) Even function about $t = -15°$
(e) Even about $t = -45°$

27. (a) $f(t) = \frac{1}{2} - (4/\pi^2)(\cos t + \frac{1}{9}\cos 3t + \frac{1}{25}\cos 5t + \cdots)$
(b) $f(t) = (2/\pi)(\sin t - \frac{1}{2}\sin 2t + \frac{1}{3}\sin 3t - \cdots)$

29. (a) Even–cosines (b) Odd–sines (c) Rotation–odd harmonics
(d) Half range–even harmonics (e) Quarter wave–odd sines
(f) Quarter wave–odd cosines
(g) Symmetrical plus loops and different symmetrical minus loops–odd sines and even cosines
(h) Same as in (g) with odd cosines and even sines

30. $a_1 = 1.24$, $a_3 = -0.33$; exact values: 1.26, −0.42

31. $a_1 = 1.28$, $a_3 = -0.42$

32. $a_1 = 1.26$, $a_3 = -0.41$

33. $a_1 = 1.27$, $a_3 = -0.43$

34. $b_1 = 0.632$, $b_2 = -0.312$, $b_3 = 0.118$; exact values: 0.636, −0.318, 0.212

35. $b_1 = 2.77$, $b_2 = -0.866$, $b_3 = 0.666$

36. $a_0 = 0.5$, $a_1 = -0.5$, $a_2 = 0$; exact values: 0.5, 0.405, 0

37. $a_0 = 0.5$, $a_1 = -0.44$, $a_2 = 0$, $a_3 = -0.08$; exact values: 0.5, 0.405, 0, −0.13.

38. $b_1 = 0.78$, $b_2 = -0.5$, $b_3 = 0.28$; exact values: $b_1 = 0.636$, $b_2 = -0.318$, $b_3 = 0.212$

39. Even-terms, real; odd-terms, imaginary; rotation-terms only, odd harmonics; quarter-wave, odd real terms or even real terms. Use full period always.

40. $f(t) = (A/2\pi)(1 + e^{jt} + e^{-jt} + e^{2jt} + e^{-2jt} + \cdots)$
$= A/2\pi + (A/\pi)(\cos t + \cos 2t + \cdots)$

41. $f(t) = (A/\pi)(1 + e^{-2jt} + e^{2jt} + e^{-4jt} + e^{4jt} + \cdots)$
$= A/\pi + (2A/\pi)(\cos 2t + \cos 4t + \cdots)$

42. $f(t) = (2A/\pi)(e^{-2jt} + e^{2jt} + e^{-6jt} + e^{6jt} + \cdots)$
$= (4A/\pi)(\cos 2t + \cos 6t + \cdots)$

43. $f(t) = (A/\pi)(-je^{jt} + je^{-jt} + je^{3jt} - je^{-3jt} + \cdots)$
$= (2A/\pi)(\sin t - \sin 3t + \cdots)$

44. $f(t) = (2/\pi)[e^{jt} + e^{-jt} - (e^{j3t} + e^{-j3t})/3 + \cdots]$
$= (4/\pi)(\cos t - (\cos 3t)/3 + \cdots)$

45. $f(t) = \frac{1}{2} - (1/2\pi)[-je^{jt} + je^{-jt} + (-je^{j2t} + je^{-j2t})/2 + \cdots]$
$= \frac{1}{2} - (1/\pi)(\sin t + (\sin 2t)/2 + \cdots)$

46. $c_n = \dfrac{1}{Tjn\omega_0}\left(2 - 2\cos\dfrac{n\omega_0\delta}{2}\right), \ \omega_0 = \dfrac{2\pi}{T}$

47. $F(\omega) = \dfrac{1}{j\omega}\left(2 - 2\cos\dfrac{\omega\delta}{2}\right)$

48. $c_n = \dfrac{1}{T}\left(\dfrac{1 - \cos(n\omega_0\delta)}{n^2\omega_0^2\delta} + j\,\dfrac{\sin(n\omega_0\delta) - n\omega_0\delta}{n^2\omega_0^2\delta}\right)$

49. $c_n = \dfrac{\delta}{2\pi}\,\dfrac{\sin n\delta/2}{n\delta/2}$; Zero at $\dfrac{2\pi}{\delta}$

50. $c_n = \dfrac{\delta}{2\pi}\,\dfrac{\sin n\delta/2}{n\delta/2}$; Zero at $\dfrac{2\pi}{\delta}\,\omega_0$

51. $c_n = \dfrac{\delta}{T}\,\dfrac{\sin n\omega_0\delta/2}{n\omega_0\delta/2}$; Envelope of $c_n \times T$ is a constant form

52. (a) $F(\omega) \cdot \dfrac{\delta \sin \omega\delta/2}{\omega\delta/2}$
(b) 2π radians, $2\pi \times 10^6$ radians
(c) To $\omega = 4\pi/\delta$ (one positive and one negative loop)

53. $c_n = A/T = A\omega_0/2\pi$; $c_n \times T = A$; $F(\omega) = A$

54. (a) $c_n = (1/2\pi jn)(2\cos n\delta - 2)$ (b) $c_n = (1/2\pi jn)(2\cos n\delta - 2)$

55. $c_n = (1/Tjn\omega_0)(2\cos n\omega_0\delta - 2)$; $T \times c_n = (1/jn\omega_0)(2\cos n\omega_0\delta - 2)$

56. $F(\omega) = (2\cos\omega\delta - 2)/j\omega$

57. $F(\omega) = (1/\delta^2)[(2\cos\omega\delta - 2)/j\omega]$; $F(\omega)_{\delta=0} = j\omega$

58. (1) $\sigma_c = 0$; $F(s) = \beta/(s^2 + \beta^2)$; poles at $s = \pm j\beta$
(2) $\sigma_c = -\alpha$; $F(s) = \beta/[(s + \alpha)^2 + \beta^2]$; poles at $s = -\alpha \pm j\beta$
(3) $\sigma_c = -\alpha$; $F(s) = 2\beta(s + \alpha)/[(s + \alpha)^2 + \beta^2]^2$; poles at $s = -\alpha \pm j\beta$
(4) $\sigma_c = \alpha$; $F(s) = 1/(s - \alpha)^2$; poles at $s = \alpha$
(5) $\sigma_c = 0$; $F(s) = \beta^2/s(s^2 + \beta^2)$; poles at $s = 0$ and $s = \pm j\beta$

59. (1) $\sigma_c = 0$; $F(s) = \alpha/s(s + \alpha)$; poles $s = 0$ and $-\alpha$
(2) $\sigma_c = \beta$; $F(s) = \beta/(s^2 - \beta^2)$; poles at $s = \pm\beta$
(3) $\sigma_c = \beta$; $F(s) = s/(s^2 - \beta^2)$; poles at $s = \pm\beta$
(4) $\sigma_c = 0$; $F(s) = (s\sin\theta + \beta\cos\theta)/(s^2 + \beta^2)$; poles at $s = \pm j\beta$
(5) $\sigma_c = 0$; $F(s) = (s\cos\theta - \beta\sin\theta)/(s^2 + \beta^2)$; poles at $s = \pm j\beta$

60. (1) $\sigma_c = \log_e a$; $F(s) = 1/(s - \log_e a)$
(2) $\sigma_c = -\alpha$; $F(s) = \log_e[(s + \beta)/(s + \alpha)]$
(3) $\sigma_c = 0$; $F(s) = \tan^{-1}(\beta/s)$ (d) No σ_c; no $F(s)$

62. (1) $1/(s + a)$ (2) $e^{-1}e^{-s/a}/(s + a)$ (3) $e^{-s/a}/(s + a)$ (4) $e/(s + a)$
(5) $1/(s + a)$ (6) $e^{-1}/(s + a)$ (7) Same

63. (1) $(1 - e^{-s})/s$ (2) $1/s(1 + e^{-s})$ (3) $1/s \sinh s$ (4) $1/s \cosh s$
(5) $(e^s - s - 1)/s^2(e^s - 1)$

64. (a) $A(1 - e^{-\delta s})/s$ (b) $s = +\infty$ (c) None

65. $\sigma_c = 0$, all right except at $s = +j\beta$. Line spectrum at $s = \pm j\beta$. Band-spectrum harmonics are infinitesimal in comparison.

66. Same as Exercise 65 except for the plot.

Chapter 21

1. $I_n = \dfrac{2}{\pi}$, $i(t) = \dfrac{2}{\pi}(\cos t + \cos 3t + \cos 5t + \cdots)$

2. $I_n = \left(\dfrac{-4}{n^2\pi}\right)_{n \text{ odd}}$, $i(t) = \dfrac{-4}{\pi}\left(\cos t + \dfrac{\cos 3t}{3^2} + \dfrac{\cos 5t}{5^2} + \cdots\right)$

3. $I_n = \dfrac{2}{\pi}$ (n odd), $E_n = \dfrac{2}{\pi n}\underline{|-90^\circ}$, $e(t) = \dfrac{2}{\pi}\left(\sin t + \dfrac{\sin 3t}{3} + \cdots\right)$

4. $i(t) = \dfrac{20}{\pi}\left(\sin 10^6 t + \dfrac{\sin 3 \times 10^6 t}{3} + \dfrac{\sin 5 \times 10^6 t}{5} + \cdots\right)$

5. $e(t) = \dfrac{-4}{\pi^2}\left(\cos \pi \times 10^6 t + \dfrac{\cos 3\pi \times 10^6 t}{9} + \dfrac{\cos 5\pi \times 10^6 t}{25} + \cdots\right)$

6. $E_1 = \dfrac{4}{n\pi}\underline{|-90^\circ}$, $E_n = \dfrac{4}{\pi n\sqrt{1 + n^2}}\underline{|-90^\circ - \tan^{-1} n}$ (n odd),

$E_2 = \dfrac{4}{\pi n\sqrt{1 + n^2}}\underline{|-\tan^{-1} n}$;

$e_2(t) = (4/\pi)[0.707 \cos (t - 45^\circ) + 0.105 \cos (3t - 71.5^\circ) + \cdots]$

7. D-c $= 1/\pi$, Fundamental $= \frac{1}{27}$, Percentage ripple $= 11.6\%$

8. D-c $= 200/\pi$, Second harmonic $= 400/281\pi$, Percentage ripple $= 0.71\%$

9. D-c $= 100/\pi$, Fundamental $= 1/39.5$, Percentage ripple $= 3.98\%$

10. (a) $1/j\omega$ (b) $1/j\omega$ (c) $1/(j\omega)^2$ (d) $u_{-2}(t)$

11. (a) $\dfrac{\sin \omega\delta/2}{\omega\delta/2}$ (b) $F(\omega) = 1$

12. (a) 1 (b) $1/(1 + j\omega)$ (c) $1/(1 + j\omega)$ (d) $e^{-t} \cdot u_{-1}(t)$

13. (a) $1/(1 + j\omega)$ (b) $1/j\omega$ (c) $1/j\omega - 1/(j\omega + 1)$ (d) $(1 - e^{-t})u_{-1}(t)$

14. (a) $1/j\omega$ (b) $j\omega$ (c) 1 (d) $u_0(t)$

15. (a) $1/(1 + j\omega) - 1/(10 + j\omega)$ (b) $1 + j\omega$ (c) $9/(10 + j\omega)$
(d) $9e^{-10t} \cdot u_{-1}(t)$

16. (a) $1/j\omega$ (b) $j\omega/(1 - \omega^2)$ (c) $1/(1 - \omega^2)$ (c) $\sin t \cdot u_{-1}(t)$

17. (a) $-1/(-1 + j\omega) + 1/(1 + j\omega) = 2/(1 + \omega^2)$ (b) $1/j\omega$
(c) $2/j\omega - 1/(j\omega + 1) - 1/(-1 + j\omega)$
(d) $i(t) = 2u_{-1}(t) - e^{-t} \cdot u_{-1}(t) + e^t \cdot u_{-1}(-t)$ [for term 3 see part (a)]

18. $E_1(\omega) = \dfrac{\delta \sin \omega\delta/2}{\omega\delta/2}$, $E_2(\omega) = \dfrac{\delta \sin \omega\delta/2}{(\omega\delta/2)(1 + j\omega)}$,

$E_1(\omega) = \dfrac{2 \cos \omega\delta - 2}{j\omega}$, $E_2(\omega) = \dfrac{2 \cos \omega\delta - 2}{j\omega(1 + j\omega)}$

20. (a) $F(\omega) = 1/(1 + \omega^2) - j[\omega/(1 + \omega^2)]$ (b) $F(\omega) = 1/(1 + j\omega)$

21. (a) $F(\omega) = 1/(1 - \omega^2)$ (b) $F(\omega) = 1/(1 - \omega^2)$

22. (a) $(1/j\omega)(1 - \cos \omega\delta/2 + j \sin \omega\delta/2)$ (b) $(\delta/2)[\sin (\omega\delta/2)/(\omega\delta/2)]$ (c) $(1 - \cos \omega\delta/2)/j\omega$

23. $F(\omega) = 2/(1 + \omega^2)$

24. $F(\omega) = -j2\omega/(1 + \omega^2)$

25. (a) $f(t) = 2e^{-t} \cdot u_{-1}(t)$ (b) $F(\omega) = 2/(1 + j\omega)$

26. (a) $f(t) = e^{-t} \cdot u_{-1}(t)$ (b) $f(t) = e^{-t} \cdot u_{-1}(t)$

27. $f(t) = \begin{cases} \dfrac{e^{-t}}{2} & \text{for } t > 0 \\ \dfrac{e^{t}}{2} & \text{for } t < 0 \end{cases}$

28. $f(t) = \begin{cases} \dfrac{e^{-t}}{2} & \text{for } t > 0 \\ \dfrac{-e^{t}}{2} & \text{for } t < 0 \end{cases}$

29. $R(\omega) = a/(a^2 + \omega^2)$, $f(t) = u_{-1}(t)$

31. $i(t) = \frac{10}{3}e^{-2t} \cdot u_{-1}(t)$

32. $i(t) = \frac{10}{3}e^{-t/6} \cdot u_{-1}(t)$

33. $e(t) = 30u_0(t) - 15e^{-t/2} \cdot u_{-1}(t)$

34. $i(t) = 2u_0(t) - \frac{2}{15}e^{-t/15} \cdot u_{-1}(t)$

35. $i(t) = 10 \sin 5t \cdot u_{-1}(t)$

36. $e(t) = 800 \cos 16t \cdot u_{-1}(t)$

37. $i(t) = 11.5e^{-t/2} \sin 0.866t$

38. $e(t) = 11.5e^{-t/2} \sin 0.866t$

39. (a) $i(t) = 2e^{-t/10} \cdot u_{-1}(t)$ (b) Same

40. (a) $e(t) = 30e^{-t/2} \cdot u_{-1}(t)$ (b) Same

41. (a) $e(t) = 50e^{-t/2} \cdot u_{-1}(t)$ (b) Same

42. $i(t) = 25u_0(t) + 2.5e^{-t/10} \cdot u_{-1}(t)$

43. $e(t) = 2.5u_0(t) + 2.5e^{-t} \cdot u_{-1}(t)$

44. $i(t) = 10 \cos 5t - 10 \sin 5t \quad \text{for } t > 0$

45. (a) $10 \sin 5t + 10 \cos 5t$ (b) Same

46. $i(t) = 11.5e^{-t/2} \sin 0.866t \quad \text{for } t > 0$

47. $i(t) = 11.5e^{-t/2} \cos (0.866t + 30°) \quad \text{for } t > 0$

48. $e(t) = 5e^{-t} \cdot u_{-1}(t)$

49. $F(s) = \dfrac{\frac{1}{2}}{s} - \dfrac{1}{s+1} + \dfrac{\frac{1}{2}}{s+2}$

50. $F(s) = 1 + \dfrac{\frac{1}{2}}{s} - \dfrac{1}{s+1} + \dfrac{\frac{1}{2}}{s+2}$

51. $F(s) = 1 - \dfrac{\frac{1}{2}}{s+1} + \dfrac{8}{s+2} - \dfrac{\frac{27}{2}}{s+3}$

52. $F(s) = \dfrac{0.577 \underline{|30°}}{s + \frac{1}{2} - j0.866} + \dfrac{0.577 \underline{|-30°}}{s + \frac{1}{2} + j0.866}$

53. $F(s) = s^2 + \dfrac{0.577 \underline{|-30°}}{s + \frac{1}{2} - j0.866} + \dfrac{0.577 \underline{|30°}}{s + \frac{1}{2} + j0.866}$

54. $F(s) = \dfrac{-1}{(s+1)^3} + \dfrac{1}{(s+1)^2}$

55. $F(s) = 1 - \dfrac{1}{(s+1)^3} + \dfrac{3}{(s+1)^2} - \dfrac{3}{s+1}$

56. $F(s) = \dfrac{3}{s+1} - \dfrac{6}{(s+2)^2} - \dfrac{2}{s+2}$

57. $F(s) = \dfrac{0.353 \underline{|-45°}}{(s+1-j)^2} + \dfrac{0.353 \underline{|45°}}{(s+1+j)^2} + \dfrac{0.25 \underline{|90°}}{s+1-j} + \dfrac{0.25 \underline{|-90°}}{s+1+j}$

58. $F(s) = \dfrac{-1}{s+1} + \dfrac{-0.25 - j0.25}{(s+1-j)^2} + \dfrac{-0.25 + j0.25}{(s+1+j)^2} + \dfrac{0.5 - j0.25}{s+1-j}$
$\qquad + \dfrac{0.5 + j0.25}{s+1+j}$

59. $f(t) = \frac{1}{2} - e^{-t} + \frac{1}{2}e^{-2t} \quad [t > 0]$

60. $f(t) = \cos t \quad (t > 0)$

61. $f(t) = 1.15e^{-t/2} \cos (0.866t + 30°) \quad (t > 0)$

62. $f(t) = te^{-t}\left(1 - \dfrac{t}{2}\right) \quad (t > 0)$

Chapter 22

1. (a) $E(s) = 10/s(s+2)$ (b) $e(t) = 5(1 - e^{-2t})u_{-1}(t)$
2. $e(t) = 5(1 - e^{-2t})u_{-1}(t)$
3. $E_{oc} = 10/s(s+1)$, $Z_{eq} = (s+2)/(s+1)$,
   $i(t) = 3.54(e^{-0.3t} - e^{-1.7t}) \quad \text{for } t > 0$

4. $E_{oc} = 10s/(s+1)$, $Z_{eq} = (2s+1)/(s+1)$,
$i(t) = (11.7e^{-2.62t} - 1.7e^{-0.38t})u_{-1}(t)$

5. $e(t) = 5u_0(t) - 3.535 \sin 0.707t \cdot u_{-1}(t)$

6. $i(t) = \frac{5}{4}t + \frac{15}{8} \sin 2t \quad \text{for } t > 0$

7. $e_2(t) = \frac{1}{2}e^{-t} - 0.577e^{-t/2} \cos (0.866t + 30°) \quad \text{for } t > 0$

8. $i(t) = 50u_{-2}(t) + 125u_{-1}(t) - 25e^{-2t} \cdot u_{-1}(t)$

9. $e(t) = 50u_0(t) + 100u_{-1}(t) + 25e^{-t/2} \cdot u_{-1}(t)$

10. $i(t) = [4.48 \cos (t - 26.5°) - 4e^{-2t}]u_{-1}(t)$.
First term is steady-state; second term is transient.

11. $e(t) = [3e^{-t} - 3.16 \cos (3t + 18.5°)]u_{-1}(t)$.
First term is transient; second term is steady-state.

12. $i(t) = [10 \cos t - 11.5 \cos (0.866t + 30°)e^{-t/2}]u_{-1}(t)$.
First term is steady-state; second term is transient.

13. $e(t) = [(t \cos t + \sin t)/2]u_{-1}(t)$.
First term is infinite as $t \to \infty$, as the steady-state should be.

14. $\theta = -26.6°$

15. $A = \frac{2}{5}$

16. Maximum transient at $\theta = 45°$. Minimum transient at $\theta = 0°$ and $90°$.

18. (a) $1/(s^2+1)$ (b) $1/(s^2+1)$ (c) $-e^{-\pi s}/(s^2+1)$ (d) $e^{-\pi s}/(s^2+1)$
(e) $-1/(s^2+1)$

19. $(1 + e^{-\pi s})/(s^2+1)$

20. $[(1 - e^{-s})/s]^2$

21. (a) $f(t) = u_{-1}(t) - u_{-1}(t-1) + u_{-1}(t-2) - \cdots$
(b) $F(s) = 1/s(1 + e^{-s})$

22. (a) $f(t) = u_{-2}(t) - u_{-1}(t-1) - u_{-1}(t-2) - u_{-1}(t-3) - \cdots$

(b) $F(s) = \dfrac{1 - e^{-s} - se^{-s}}{s^2(1 - e^{-s})}$

23. (a) $p(t) = u_{-1}(t) - u_{-2}(t) + u_{-2}(t-1)$

(b) $P(s) = \dfrac{s - 1 + e^{-s}}{s^2}$

(c) $F(s) = \dfrac{s - 1 + e^{-s}}{s^2(1 - e^{-s})}$

24. (a) $p(t) = 10u_{-1}(t) - 10u_{-1}(t - \delta)$

(b) $P(s) = \dfrac{10(1 - e^{-\delta s})}{s}$

(c) $F(s) = \dfrac{10(1 - e^{-\delta s})}{s(1 - e^{-Ts})}$

25. $f(t) = \begin{cases} t^2/2 & \text{for } 0 < t < 1 \\ \frac{3}{4} - (t - \frac{3}{2})^2 & \text{for } 1 < t < 2 \\ \frac{1}{2}(t - 3)^2 & \text{for } 2 < t < 3 \\ 0 & \text{for } 3 < t < \infty \end{cases}$

26. $f(t) = 2u_{-1}(t - 1) + 2u_{-1}(t - 3) + 2u_{-1}(t - 5) + \cdots$

27. $f(t) = 2u_{-1}(t - 1) - 2u_{-1}(t - 3) + 2u_{-1}(t - 5) - \cdots$

28. (a) $f(t) = u_0(t) + u_0(t - 1) + u_0(t - 2) + \cdots$
(b) $f(t) = 1 + 2 \cos 2\pi t + 2 \cos 4\pi t + \cdots$

29. (a) $f(t) = u_0(t) - u_0(t - 1) + u_0(t - 2) + \cdots$
(b) $f(t) = 2 \cos \pi t + 2 \cos 3\pi t + 2 \cos 5\pi t + \cdots$

30. (a) $F(s) = \dfrac{1 - e^{-2\pi s} - se^{-2\pi s}}{s^2(1 - e^{-2\pi s})}$
(b) $f(t) = \frac{1}{2} - \dfrac{1}{\pi}\left(\sin t + \dfrac{\sin 2t}{2} + \dfrac{\sin 3t}{3} + \cdots\right)$

31. (a) $F(s) = \dfrac{1 + e^{-\pi s}}{(s^2 + 1)(1 - e^{-2\pi s})}$
(b) $f(t) = \dfrac{1}{\pi}\left(1 + \dfrac{\pi}{2} \sin t - \frac{2}{3} \cos 2t - \frac{2}{15} \cos 4t + \cdots\right)$

32. (a) $F(s) = \dfrac{1 + e^{-\pi s}}{(s^2 + 1)(1 - e^{-\pi s})}$
(b) $f(t) = \dfrac{2}{\pi}\left(1 - \frac{2}{3} \cos 2t - \frac{2}{15} \cos 4t - \frac{2}{35} \cos 6t - \cdots\right)$

33. (a) $e(t) = u_{-1}(t) + u_{-1}(t - 1) + u_{-1}(t - 2) + \cdots$
(b) $e(t) = t + \frac{1}{2} + (1/\pi) \sin 2\pi t + (1/2\pi) \sin 4\pi t + \cdots$

34. (a) i (transient) $= 0.565 \sin (t + 28.5°)$
(b) i (steady-state) $= [2/(1 - \pi^2)] \cos \pi t$

35. (a) $I(s) = \dfrac{1}{(s + 1)(1 - e^{-s})}$
(b) $i(t) = e^{-t}u_{-1}(t) + e^{-(t-1)}u_{-1}(t - 1) + e^{-(t-2)}u_{-1}(t - 2) + \cdots$
(c) $i(t) = 1 + \dfrac{e^{-t}}{1 - e} + \dfrac{1}{[1 + (2\pi)^2]^{1/2}} \cos (2\pi t - \tan^{-1} 2\pi)$
$+ \dfrac{1}{[1 + (4\pi)^2]^{1/2}} \cos (4\pi t - \tan^{-1} 4\pi) + \cdots$

36. (a) $\dfrac{s^2 + 3s + 1}{2s^2 + s}$
(b) Zeros: $s = -2.62, -0.38$; poles: $s = 0, -0.5$
(c) Zeros: short-circuit transients; poles: open-circuit transients

37. (a) $\dfrac{s + 1}{s^2 + 2s + 2}$

(b) Zeros: $s = -1, \infty$; poles: $s = -1 \pm j$

(c) Zeros: short-circuit transients; poles: open-circuit transients

38. (a) $Z = R$ (b) $e_2 = 100e^{-t/RC}$, $e_3 = -100e^{-Rt/L}$, $e_1 = 0$

39. (a) $\dfrac{LCs^2 + 2Ls + 1}{LCs^2 + 2Cs + 1}$ (b) $L = C$

40. $Z = \dfrac{(s^2 + 1)(s^2 + 9)}{s(s^2 + 4)}$ for each network

41. (a) $Z_{12} = \dfrac{1}{s^3 + 2s^2 + 2s + 1}$

(b) Zeros: three at ∞; poles: $s = -1$, $s = -0.5 \pm j0.866$

42. (a) $Z_1 = z_{11} - \dfrac{z_{12}^2}{1 + z_{22}}$ (b) $Z_{12} = \dfrac{z_{12}}{1 + z_{22}}$

43. (a) $z_{11} = z_{22} = s + \dfrac{1}{s}$; $z_{12} = s$ (b) $Z_{12} = \dfrac{s^2}{s^2 + s + 1}$

44. (a) $f(t) = t$ (b) $f(t) = te^{-at}$ (c) $f(t) = \dfrac{(1 - e^{-at})}{a}$

(d) $f(t) = \dfrac{1 - (1 + at)e^{-at}}{a^2}$ (e) $(1 - at)e^{-at}$

45. (a) $\dfrac{e^{-at} - e^{-bt}}{b - a}$ (b) $\dfrac{e^{-at}}{(b - a)(c - a)} + \dfrac{e^{-bt}}{(a - b)(c - b)} + \dfrac{e^{-ct}}{(a - c)(b - c)}$

(c) $\frac{1}{2}(\sin t - t \cos t)$

46. (b) $e^{-at}(t^2/2)$

47. (b) [1] $u_0(t) - ae^{-at} \cdot u_{-1}(t)$ [2] $\cos t \cdot u_{-1}(t)$

[3] $\dfrac{(a^2 + \beta^2)^{1/2}}{\beta} e^{-at} \sin(\beta t + \tan^{-1} \beta/-a)u_{-1}(t)$

48. (b) $\dfrac{e^{-at} + at - 1}{a^2} u_{-1}(t)$

49. (b) $u_0(t) + \cos t \cdot u_{-1}(t)$

50. (a) $H(s) = \dfrac{1}{s^2 + 2s + 2}$ (b) $h(t) = e^{-t} \sin t \cdot u_{-1}(t)$

(c) $\left[\dfrac{1 - e^{-t}(\sin t + \cos t)}{2}\right] u_{-1}(t)$

$$- \left[\frac{1 - e^{-(t-1)} \sin(t - 1) + \cos(t - 1)}{2}\right] u_{-1}(t - 1)$$

51. (a) $H(s) = \dfrac{1}{2(s + 2)}$ (b) $h(t) = \dfrac{e^{-2t}}{2} \cdot u_{-1}(t)$

(c) $\left[\frac{-1}{8}+\frac{t}{4}+\frac{e^{-2t}}{8}\right]u_{-1}(t) - \left[\frac{-1}{8}+\frac{t-1}{4}+\frac{e^{-2(t-1)}}{8}\right]u_{-1}(t-1)$

$-\left[\frac{1-e^{-2t+2}}{4}\right]u_{-1}(t-1)$

52. (a) $H(s) = \frac{3s}{s^2+9}$ (b) $e_2(t) = \frac{9t}{6}\sin 3t \cdot u_{-1}(t)$

53. (a) $H(s) = \frac{(s+1)(s+2)}{(s+3)^2}$ (b) $f_2(t) = e^{-3t}(2t^2 - 3t + 1)u_{-1}(t)$

54. (a) 0 (b) 1 (c) 0 (d) Does not apply (e) Does not apply

55. (a) 0 (b) 1 (c) Does not apply (d) 0 (e) Does not apply

56. $e_2(0) = 0,\ e_2'(0) = 0,\ e_2''(0) = 0,\ e_2'''(0) = 1$

57. $e_2(\infty) = \frac{1}{2},\ e_2'(\infty) = 0,\ e_2''(\infty) = 0,\ e_2'''(\infty) = 0$

58. $i(t) = e^{-3t}\sin 5t \cdot u_{-1}(t)$

59. $i(t) = (10\cos 2t - 40\sin 2t)u_{-1}(t)$

60. $i(t) = 2.5e^{-2t}\cdot u_{-1}(t)$

61. $i(t) = 11.2e^{-t}\cos(2t + 26°)u_{-1}(t)$

62. $i(t) = [5e^{-t/2}\cos 0.866t + 8.66e^{-t/2}\sin 0.866t]u_{-1}(t)$

63. (a) $\frac{d^2i}{dt^2} + 3\frac{di}{dt} + i = u_0(t) + u_1(t)$

(b) $2\frac{d^3i}{dt^3} + 2\frac{d^2i}{dt^2} + \frac{di}{dt} + i = u_0(t) + u_1(t) + u_2(t)$

64. $i(0^+) = 3,\ i'(0^+) = -4,\ i''(0^+) = 3,\ i'''(0^+) = -4$

Chapter 23

1. (a) 5000 $(\text{volts})^2$ (b) 5000 watts (c) 5000 $(\text{amps})^2$ (d) 5000 watts
   (e) It implies normalization to 1 ohm
2. (a) $G(\omega) = 2500\,[u_0(\omega + 1000) + u_0(\omega - 1000)]$ (b) $P = 5000$ watts
3. (a) $G(\omega) = 0.25 \times 10^{-6}\,[u_0(\omega + 10^6) + u_0(\omega - 10^6)]$
   (b) $P = 0.5 \times 10^{-6}$ watt
4. $1/(4 + \omega^2)$
5. $1/(\omega^4 + 7\omega^2 + 1)$
6. $\omega^2/(\omega^4 + 2\omega^2 + 1)$
7. $G_2(\omega) = 2500\,[u_0(\omega + 1) + u_0(\omega - 1)]$
8. $G_2(\omega) = 25\,[u_0(\omega + 1) + u_0(\omega - 1)]$
9. (a) $F_{\text{rms}} = 1$ (b) $F_{\text{rms}}^2 = 0.965$ (c) 3.5%
10. $G(\omega) = A^2u_0(\omega)$ watts/cycle

11. (a) $G(\omega) = 4u_0(\omega) + u_0(\omega + 1) + u_0(\omega - 1)$ (b) $P = 6$ watts

12. (a) $G(\omega) = \dfrac{4}{\pi^2}\left[u_0(\omega \pm 1) + \dfrac{u_0(\omega \mp 3)}{9} + \dfrac{u_0(\omega \mp 5)}{25}\right]$

(b) $F_{\text{rms}}^2 = 0.934$ watts

13. (a) F_{rms}^2 (exact) $= 13.1$ watts; F_{rms}^2 (approx) $= 12.5$ watts

(b) $G(\omega) = \pi^2 u_0(\omega) + u_0(\omega \pm 1) + \dfrac{u_0(\omega \mp 2)}{4} + \dfrac{u_0(\omega \mp 3)}{9} + \cdots;$

F_{rms}^2 (approx) $= 12.5$ watts

14. $G(\omega) = \dfrac{4}{\pi^2}\left[u_0(\omega \pm 1) + u_0(\omega \pm 3) + u_0(\omega \pm 5) + \cdots\right]$

15. (a) $G(\omega) = \dfrac{4}{\pi^2}\left[u_0(\omega \pm 1) + \dfrac{u_0(\omega \mp 3)}{81} + \dfrac{u_0(\omega \mp 5)}{625} + \cdots\right]$

(b) Saw-tooth wave; peak value $= \pm\pi/2$; period $= 2\pi$

(c) $P = \pi^2/12$ watts

16. $G(\omega) = \displaystyle\sum_{n=-\infty}^{\infty} |c_n|^2[u_0(\omega + n\omega_0]$

17. (a) $c_n = \dfrac{\delta}{T}\dfrac{\sin n\omega_0\delta/2}{n\omega_0\delta/2}$ (b) $G(\omega) = \displaystyle\sum_{n=-\infty}^{\infty} \frac{\delta^2}{T^2}\left(\frac{\sin n\omega_0\delta/2}{n\omega_0\delta/2}\right)^2 u_0(\omega + n\omega_0)$

18. (a) $1/(1 + \omega^2)$ joules/cycle; $\frac{1}{2}\pi/(1 + \omega^2)$ joules/radian

(b) $P = \frac{1}{2}$ joule

19. (a) $1/\omega^2$ (b) $W = 1/2\alpha$ joules

20. (a) 1 (b) ω^2 (c) ω^4 joules/cycle

21. (a) $|F(\omega)|^2 = \delta^2 \dfrac{\sin^2 \omega\delta/2}{(\omega\delta/2)^2}$ joules/cycle (b) $\omega = 2\pi/\delta$ (c) $\omega = 0.346/\delta$

22. $W = \delta$ joules

23. $F(\omega) = \dfrac{2\alpha}{\alpha^2 + \omega^2}$; $|F(\omega)|^2 = \dfrac{4\alpha^2}{(\alpha^2 + \omega^2)^2}$; $W = \dfrac{1}{\alpha}$ joules.

24. $|F_2(\omega)|^2 = 1/(1 + \omega^6)$

25. (a) $|F_2(\omega)|^2 = 1/(\omega^4 + 104\omega^2 + 400)$ (b) $W = 0.0021$ joule

26. (a) $f(t) = \cos t + \cos 3t + \cos 5t + \cdots$

$G(\omega) = \frac{1}{4}[u_0(\omega \pm 1) + u_0(\omega \pm 3) + \cdots]$ watts/cycle

(b) Height of continuous curve $= \pi/4$ watts/cycle

27. (a) $F(\omega) = 1$ (b) $G(\omega) = \dfrac{1}{4\pi^2}\displaystyle\sum_{n=-\infty}^{\infty} u_0(\omega + n)$

(c) $\dfrac{F(\omega)^2}{T} = \dfrac{1}{2\pi}$ watts/cycle;

average height $= \dfrac{1}{2\pi}$ watts/cycle $= \dfrac{1}{(2\pi)^2}$ watts/radian

28. (a) $G_2(\omega) = 10^3/(\omega^2 + 10^6)$ (b) 50% below 1000 rps

29. (a) $G_2(\omega) = 10^9/(\omega^4 + 10^{12})$ (b) 75% below 1000 rps

30. $|E_2(\omega)|^2 = \dfrac{100 \times 10^{-6}}{4 + 4\omega^2}$; $P = 12.5$ microwatts

31. $|E_2(\omega)|^2 = \dfrac{4kTBR}{1 + (\omega CR)^2}$

32. $|E(\omega)|^2 = \dfrac{4kTBR_1R_2}{(R_1 + R_2)}$

33. $F = 1 + \dfrac{R_1}{R_2}$

34. $F = 1 + \dfrac{R_1}{R_2}$

INDEX